The MOTHS of LANCASHIRE

Stephen Palmer and Ben Smart

Published by

The MOTHS of LANCASHIRE

Published 2024 for the Tanyptera Trust and National Museums Liverpool by Pisces Publications

Copyright © S. Palmer, B. Smart and National Museums Liverpool
Copyright © of the photographs and illustrations remains with the photographers and illustrators

All rights reserved. No part of this publication may be reproduced, stored in a retrieval system or transmitted, in any form or by any means electronic, mechanical, photocopying, recording or otherwise, without the prior permission of the publishers

First published 2024

British-Library-in-Publication Data
A catalogue record for this book is available from the British Library

ISBN 978-1-913994-13-6

Designed by Pisces Publications in consultation with S. Palmer, B. Smart, National Museums Liverpool and M. Eweda

Pisces Publications is the imprint of NatureBureau, 2C The Votec Centre, Hambridge Lane, Newbury, Berkshire RG14 5TN
Visit our bookshop
www.naturebureau.co.uk

Printed and bound in the UK by Gomer Press Ltd

Funded by the Tanyptera Trust through the Tanyptera Project based at World Museum, Liverpool

COVER PHOTOGRAPHS
Front cover:
Belted Beauty © Jack Morris

Back cover:
Leighton Moss © Jenny Mackness
Inset photos from top to bottom:
Anania funebris © Jack Morris
Puss Moth larva © Trevor Davenport
Emperor Moth © Trevor Davenport

Contents

- v   **Foreword**
- vii   **Authors**
- viii   **Acknowledgements**
- 1   **Introduction**
- 8   **Moth Habitats in Lancashire** by Steve Garland
  - 8   Climate
  - 9   Geology and history
  - 10   General habitat summary
  - 16   Natural Character Areas of Lancashire
- 30   **Species Accounts Introduction**
  - 30   Introduction
  - 31   Species text
  - 32   Distribution maps
  - 32   Flight graph
  - 32   Photographs
  - 32   Terminology and abbreviations
  - 32   Historic data – dates and foodplants
  - 32   Historic data – geographical coverage
  - 33   Data and coverage
  - 33   Excluded records
- 34   **Species Accounts**
- 572   **Addendum: New Lancashire records for 2023**
- 576   **Appendix 1** Checklist of the Moths of Lancashire
- 589   **Appendix 2** Moth record contributors
- 598   **Appendix 3** Photographic contributors and photograph details
- 621   **Appendix 4** Scientific names of foodplants mentioned in the text
- 625   **Appendix 5** Unaccepted published data
- 630   **Appendix 6** Aggregated macro-moth species
- 631   **Appendix 7** Recording moths
- 639   **Bibliography and References**
- 649   **Index (of species)**

Foreword

Lancashire is one of the largest, most densely populated and geologically varied parts of the UK. It is a land of great contrasts and contains many nationally important wildlife sites. From the rugged beauty of its upland fells and moors, to extensive dune systems and wide estuaries; from floristically-rich limestone pavements and hills clad in deciduous woodlands, to intensively-farmed lowland plains and great conurbations with parks, gardens and brownfield sites. Taken as a whole, the county (as defined in this book) is even more impressive than its varied parts, but these landscapes and their wildlife have collectively never been more threatened. The more we document, understand and interpret Lancashire's biodiversity, the more we can help to protect it.

There is a long, proud, deeply rooted tradition of studying natural history in Lancashire, dating back to 'artisan naturalists' who organised themselves into local societies, nearly 200 years ago. Since then, there have been many significant contributions to the recording of moths and these are referenced in this book. I don't think that any of these past lepidopterists could have envisaged a day when a book on Lancashire moths would be based on nearly two million records, submitted by over 2,400 recorders. Or, that it would cover 1,570 species and be accompanied by colour photographs, detailed biological and distributional information. This book represents the culmination of so much recording, by so many dedicated people, past and present. In a time of rapid change, it marks the beginning of an even more important period in the study and conservation of Lancashire moths. Incredibly, whilst the text was being completed in 2023, an additional eleven species were recorded in the county (listed in an Addendum).

There are so many reasons why moths matter. To many, they are mysterious creatures, inhabiting a nocturnal world. But the more attention they are given, the more captivating they become. They are infinitely varied in form, subtly or strikingly coloured, have fascinating life histories and are ecologically significant as pollinators and an important food source. Moths stimulate our curiosity, enrich our world and inform our environmental decision making.

This book contains a wealth of biological information about Lancashire moths. It presents a snapshot in time, against which future surveys and records will be compared. It will be of value to anyone interested in natural history. For specialist readers it contains much ecological information of relevance within and beyond the county borders. The data will also inform targeted species conservation management, regionally and nationally. When aggregated with larger databases, it will help scientists monitor environmental change. By caring for and protecting moths, we help to protect our natural places for future generations.

National Museums Liverpool and the Tanyptera Trust have been privileged to support this project. Through a combination of sheer hard work, immense personal knowledge, careful planning and the support of so many people, Steve and Ben have produced this book in just three years - what an astonishing achievement! It is a very significant faunal publication and they 'stand tall' in the history of natural history in this special county.

Steve Judd
World Museum, National Museums Liverpool
November 2023

Authors

Stephen Michael Palmer

Born in Hemel Hempstead, Hertfordshire in 1952 and a keen lepidopterist since his father made him a moth trap in c.1960. Now retired, he worked as an Air Traffic Controller, an occupation that took him to North-east Scotland in the late 1970s and early 1980s, Wiltshire from 1983 to 1993 and finally Lancashire where he still lives, in Preston. He and his wife Carolyn have three grown up children. Previous publications have involved editing a *Birds of Wiltshire* supplement, producing *The Microlepidoptera of Wiltshire* and co-editing, and authoring parts of, *A Field Guide to the Smaller Moths of Great Britain and Ireland*. In later life, his particular interest related to the British micro-moth family, the Gelechiidae, and he and Carolyn have run the national Gelechiid Recording Scheme since 2011. Steve acts as a national verifier on iRecord, covering the Gelechiidae.

Since arriving in Lancashire Steve took on various County Moth Recorder (CMR) roles and was the VC59 Micro-moth CMR until May 2024. He instigated the formation of the Lancashire Moth Group in 1995, producing its Newletters until 2013.

Ben Smart

Ben was born in Beckenham, Greater London in 1965 but did not remain a southerner for long. At eight months old, his parents moved the family to Rochdale. Apart from three years at the University of Hull studying Politics, Ben has remained in Greater Manchester, always taking an interest in natural history.

Ben has been a Nurse since 1987 and remains in the profession. He is married with three grown-up children, and they helped to rekindle his interest in the wildlife to be found locally, particularly the insect life. From 1999 to 2003 he was the editor of the Cheshire and Greater Manchester Butterfly Conservation newsletter, also illustrating the front covers.

His particular interest is in the micro-lepidoptera and their early stages. His first publication was in 2017 and reflected this interest. The book, *Micro-moth Field Tips: A Guide to Finding the Early Stages in Lancashire and Cheshire*, was well received and sold out the initial print run, requiring the publication of a second edition. A second volume followed in 2021, describing and photographing the life cycles of a further 200 species of micro-lepidoptera.

Acknowledgements

Three years after the initial proposal by the Tanyptera Trust to finance and provide logistical support for *The Moths of Lancashire* publication, the project has now been completed. Throughout this time, the authors have had the strategic advice of Steve Judd and full-time help, support and expertise of Gary Hedges, both of the Trust, to guide us. It is no exaggeration to say that this book would not have seen the light of day without their help and Gary's calm guidance, enthusiasm, technical knowledge and assistance in so many aspects of the project.

From the planning phase onwards, the work and support received from Carolyn Palmer has proved invaluable. The large volume of preparatory typing, cross-checking of material to ensure it was ready for use and so many other tasks has freed up a considerable amount of time for the authors. Carolyn, together with Trevor Davenport, also provided assistance with the acquisition, storage and documentation of the photographs through to the latter stages of the project.

The assistance of the Lancashire County Moth Recorder team (John Girdley, Justine Patton and Richard Walker) has allowed us to plan, develop and deliver the project to a particularly tight time schedule. Members of this team have also helped in other ways, including the photographing of specimens using World Museum's excellent camera facilities (with the support and kind permission of Museum staff). They were also part of our guest author team for some of the particularly significant moths receiving additional text coverage. In this task, they were joined by Richard Burkmar, Brian Hancock, Graham Jones, Pete Marsh and Brian Ridout. Full details of the content authored by this team are listed in the introduction to the species accounts.

The authors would also like to take this opportunity to express our particular thanks to the following: former and recent County Moth Recorders, Derek and Jeremy Steeden, Simon Hayhow, Chris Darbyshire, Graham Jones and Pete Marsh for their commitment, hard work and diligence over the years; Dr El-Moustafa Eweda for the creation and development of the distribution maps and flight graphs and to Steve Garland for designing, developing and writing the Habitat section. Steve's chapter benefitted from advice and assistance provided by John Lamb, Stuart Marsden, Pete Marsh, Kevin McCabe and Phil Smith.

The book has also benefitted from considerable assistance in many other areas. Mark Young for his much appreciated proof reading and for many helpful comments throughout the project; Brian Hancock for advice on, and proof reading of, the Pug content and Pete Marsh for similarly assessing and providing useful comments on upland macro-moth text content; Dave Bickerton for creating graphics to keep our recorders updated on areas that would benefit from extra recording and for keeping the Lancashire Moth website updated with material listing our progress; Kevin McCabe for his considerable support over many years, in so many ways, including the marathon task of digitising all of the early paper-based data; and Jack Morris for very kindly providing us with the perfect front cover photograph for this book.

The continuing list of those assisting the project is large, and to each of you we pass our thanks for the time and assistance you have provided. These include, Dave Earl BSBI Recorder for VC59 for information relating to several plant species distributions; the Trustees of the British Museum of Natural History, London for permissions to use various specimen photographs http://creativecommons.org/licenses/by/4.0/ and also Museum staff, Geoff Martin and Alberto Zilli, for assistance with confirmation of identifications of some historic material in the Museum collection; the staff of the three Lancashire area and one Cumbrian Biological Record Centres, for data provision (GMEU, LERN and Merseyside Biobank and the Cumbria Biodiversity Data Centre); World Museum and staff, including Tony Hunter, Guy Knight and Ian Wallace for facilitating access to World Liverpool Museum's collections and data over the years; Dmitri Loganov, Phil Rispin and Diana Arzuza Buelvas for similar arrangements at Manchester Museum; Bolton Museum staff, particularly Lauren Field, and Stuart Sharp of Lancaster University, both for assistance with specimen research; Mark Parsons and Phil Sterling for taxonomic and identification advice respectively; Les Evans-Hill of Butterfly Conservation for advice on graph design, data handling and other aspects of the project; Jeremy Steeden for Fylde region data provision, recent targeted fieldwork and advice on some aspects of the project; Reuben Neville of the Lancashire Wildlife Trust for advice on plants at Heysham NR; Mark Shaw for assistance with identification of parasitic wasps; Danny Fitzpatrick for assistance with map corrections and staff at Cliffe Castle Museum, Keighley for granting access to examine their Lancashire moth specimens.

To those families and estates that have allowed long term light trapping arrangements we extend our thanks and acknowledge their exceptional help. These include Val May at Mill Houses, Christopher Mason-Hornby and Keith Scott at Green Bank, the Gilchrist family at Sunderland Point, Lord Reay and 'Raggy' at Docker Moor.

And last, but by no means least, a huge thank you to each and every moth recorder who has sent in records (a few of you for three or four decades!) and to all photographic contributors (whether or not your photos were used). Details of all contributors of records and photos are listed elsewhere in the book. Recently, considerable extra effort was made by a few recorders to visit under-recorded areas, with Steve Hind, Dave Bickerton, Geoff Turner and Jeremy Steeden really pulling out the stops.

This book reflects all of your efforts which, combined with the unique opportunity afforded to us by the Tanyptera Trust, will hopefully fuel your enthusiasm to keep recording and studying this fascinating group of insects.

Stephen Palmer and Ben Smart

Personal dedications
On a personal note, I wish to thank all of my family, and particularly my wife Carolyn, for their support and understanding during the production of this book.
Stephen Palmer

I would like to thank Katherine for all her support and encouragement, particularly over the last few years during the preparation of this book, and her patience as it took up rather more of our time than originally envisaged. Here's to the post-book years!
Ben Smart

Introduction

The idea of producing a book detailing the moths of Lancashire had been informally discussed for a few years prior to commencement of the project. There was a recognition that, with the last detailed county list being produced in 1940, such an undertaking was long overdue. However, the amount of work anticipated to bring such a project to fruition was rather off-putting to say the least.

A formal approach by Steve Judd on behalf of the Tanyptera Trust and World Museum to Steve Palmer (the County Recorder for micro-moths in VC59 (south Lancashire)) led to an initial meeting in February 2021 with Gary Hedges from the Trust and Ben Smart, where the decision was made to push ahead with the creation of this examination of the Lancashire moth fauna. Steve Palmer and Ben Smart agreed to provide the text, whilst inviting input from other county lepidopterists. The decision was made that every single county species was deserving of a description of its history and current status in the county, a distribution map, a flight graph and, where possible, a photograph of the moth and of its early stages, usually a larva or a leaf-mine. A decision was made early in the project that only photographs taken of Lancashire specimens would be used in the book. This meant that we would struggle to provide images for some species (and for 100 or so we did not succeed), but it meant that every photograph used represented a real Lancashire record, the details of which are provided in Appendix 3, and they are preserved here for posterity. Also, the attempt to obtain such photographs resulted in dedicated fieldwork to this end, which may not necessarily have been the case had we given ourselves the freedom to include photographs irrespective of their origins. It also meant we were able to highlight the work, dedication and skill of a large army of Lancashire lepidopterists, without whom this Fauna would not have been possible. The book, detailing historical and current records, also allowed us to credit the many lepidopterists who went before us, detailing their fascinating finds, research and discoveries.

Where is Lancashire?
There are many possible answers to this question. Certainly, Lancashire is in north-west England, just below the mid-point of the mainland United Kingdom between Portland Bill (Dorset, England) and John O' Groats (Caithness, Scotland) (see **Map 1**). It is bounded by the Irish Sea on its western coast, Cumbria to the north, Yorkshire to the east and Cheshire to the south. However, the exact location of the latter three boundaries differs depending on the definition used and has varied over time.

Historically, Lancashire was a far more extensive county than the much-reduced administrative body that remains today. In the days of Gregson (1857), Ellis (1890) and Mansbridge (1940), the northern reaches of the county encompassed Barrow, Grange-over-Sands, Coniston and the banks of much of Windermere, now all parts of Cumbria. The southern border was defined by the River Mersey, providing a clear marker between Lancashire and Cheshire all the way from the River's origins in Stockport to its end point separating Liverpool and the Wirral as it flows into the Irish Sea. Similarly, the Pennines provided a border between Lancashire and Yorkshire in the east of the county. This has proved rather more malleable though, as certain towns, e.g., Todmorden, Saddleworth, and parts of Bowland, etc., have at different times found themselves on either side of this border.

The Local Government Act 1972 created Merseyside and Greater Manchester from districts formerly considered parts of Lancashire and Cheshire, and in the case of Greater Manchester, some from Yorkshire too. The same reorganisation saw Todmorden and Cornholme move from Lancashire into West Yorkshire. Further reforms in 1998 removed Blackpool, Blackburn and Darwen from the Lancashire administrative council, although they have remained part of the ceremonial county.

The Watsonian Vice-county recording system (Watson, 1852) offers a way to ensure stability of borders in the field of biological recording, rather than modern county boundaries, which remain subject to revision. This, therefore, provides the geographical basis for current moth recording in the UK. The system divides the whole of Great Britain along ancient county lines, with larger counties, such as Lancashire, being subdivided into numbered Vice-counties. Their use, and descriptions of their borders was outlined in a Lancashire and Cheshire Fauna Society journal (Sumner, 1996).

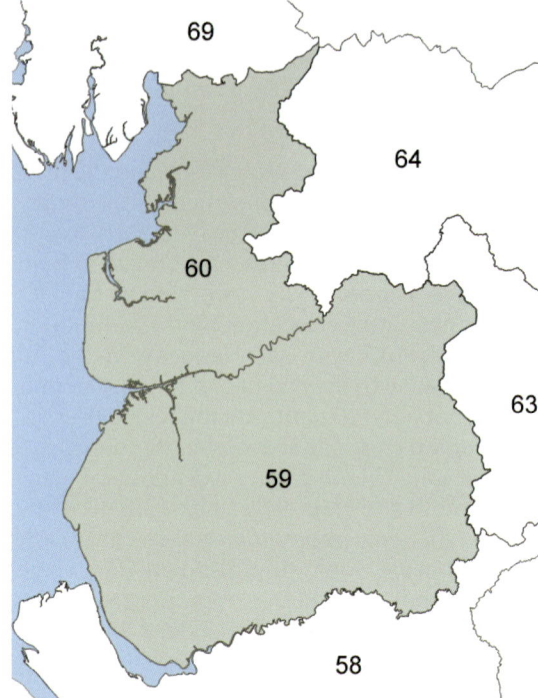

Map 1. The position of Lancashire within the United Kingdom **Map 2.** The boundaries of VC59 and VC60

Vice-county 59 (VC59) encompasses the southern part of the historic county, bordered by the Irish Sea, the River Ribble to the north, the River Mersey to the south, and south-west Yorkshire (VC63) to the east. It includes much of modern-day Greater Manchester and Merseyside.

Vice-county 60 (VC60) contains the northern portion of the county, north of the River Ribble, and is bordered by the Irish Sea, south Cumbria (VC69) to the north, and mid-west Yorkshire (VC64) to the east.

The boundaries of VC59 and VC60 are shown in **Map 2**. These are the borders used for moth recording nationally and locally, and the boundaries by which Lancashire is defined in this book. Therefore, any further mention of Lancashire should be taken to mean Watsonian Vice-counties 59 and 60 and not a past or present administrative district. C. W. Plant, in his study of *The Moths of Hertfordshire* (2008), made the point in relation to his use of Vice-county 20 to represent Hertfordshire, rather than an administrative area, that 'this will undoubtedly annoy some people, but unless the same boundary is used throughout history there would be little point in using the county at all.' Undoubtedly, the same applies to Lancashire.

The History of Moth Recording and Recorders in Lancashire

The recording of moths within the county goes back almost two centuries, based on documented data, and probably sometime before that. This has led to an extensive database of nearly two million records covering, to the end of 2022, 1,559 different species, comprised of 968 micro-moths and 591 macro-moths. This is a significant percentage of the 2,564 resident and migrant species on the British list (Agassiz *et al.*, 2022). Eleven further species have been added to the Lancashire list during 2023. (See Addendum, p. 572).

One of the earliest reports we have of a moth being recorded in Lancashire is of a Silver-striped Hawk-moth, noted at Newton Heath, Manchester in 1838 by **Robert S. Edleston** (1819–1872), a calico-maker, who resided at Fearn Acre, Cheetham Hill, Manchester. He was one of our earliest entomologists, and one of the first to have his records preserved for posterity by virtue of his frequent notes published in mid-19th century journals such as *The Entomologist*, *The Zoologist* and *The Entomologist's Weekly Intelligencer*. These records included county firsts such as *Antispila* *metallella*, *Ochsenheimeria taurella*, *Cydia coniferana* and *Dahlica inconspicuella*, all from the 1840s and 50s, species that have rarely, if at all, been recorded in the county since. One of his most notable records was said to relate to the Peppered Moth. It is widely reported that in 1848, he 'caught and pinned a rare, dark form of *Biston betularia*' near the centre of Manchester (Hooper, 2002). This seems to have been the earliest documented report of a melanic Peppered Moth, although Edleston did not document this until 1864, referring to melanic specimens noted 16 years earlier (Edleston, 1864), without exactly stating that his was one of them. He observed that 'some sixteen years ago,' the melanic 'aberration of this common species was almost unknown; more recently it has been had by several parties.'

In 1854, **Isaac Byerley** (1813–1897), a doctor from Seacombe, Cheshire, published *The Fauna of Liverpool*. As well as descriptions of visits to inspect fishermen's nets, looking for maritime fauna, the book also contained an extensive list of Lepidoptera. In compiling the list, Byerley credits 'the entomologists of the district, almost without exception, especially Messrs. Brockholes, Warrington, Diggles, Almond, and other industrious investigators, [who] have kindly supplied abundant facts with reference to Lepidoptera.' However, as the vast majority of sites mentioned throughout the book are in the Wirral (in VC58), even the records that simply state 'common', are considered unconfirmed for Lancashire, apart from on the very few occasions where a VC59 site is explicitly named.

A further list of Lepidoptera from the Liverpool district was completed by **Charles S. Gregson** (1817–1899) in 1857, having been published in sections since 1855 (Gregson, 1857). His list was based on his own experiences and those of other entomologists of the day. Unlike Byerley's work, very many of the records related to sites north of the Mersey, in the (pre-1972) Lancashire portion of what is now Merseyside. Gregson was born in Lancaster and worked for many years as a ship painter in Liverpool. His contribution to the local natural history societies, and prodigious contributions to the entomological journals of the time, added much to the understanding of distribution, particularly among the micro-moths. Described by Salmon (2000) as 'strongly parochial, forthright, energetic, testy and argumentative', he was dedicated to his collection, and to rarities in particular. Instances of this included purchasing birch trees containing Welsh Clearwing larvae and transporting them by rail to Liverpool, and on seeing a Clouded Yellow whilst out rabbiting without his net, shot it instead. An intriguing and dramatically written episode, which we noted whilst preparing this book, relates to a sighting of Kentish Glory at Withnell Birch Clough, near Chorley, on 20 and 21 April 1856. 'Saw a moth on the wing, which I was afraid to name (to his companions J.B. Hodgkinson and W. Ashworth); not being satisfied about it I went back from Preston next day. The sun being over-clouded for a short time, I had the pleasure of seeing *Endromis versicolor* (Kentish Glory) sitting at the end of a birch-bough – a fine male' (Gregson, 1856). The lack of any known specimen (from a prodigious collector), as well as the failure of Ellis and Mansbridge to subsequently include it in their county lists, and the lack of any further sightings, forced us to consider this record as unconfirmed. Nevertheless, his contribution to entomology in Lancashire was immense and he was responsible for recording many county firsts, including Common Swift, Small China-mark, Humming-bird Hawk-moth, Small Chocolate-tip, Gothic and the sole county record of *Povolnya leucapennella*.

James B. Hodgkinson (1823–1897) was born in Preston. He was an engineer, working in Manchester and Kent, and later in the cotton trade as a yarn agent. His interest in nature, encompassing birds and beetles, in addition to Lepidoptera, was sparked during childhood by a collection of pinned butterflies at the Carlisle home of T. C. Heysham, a Cumbrian naturalist and insect collector. Most of his Lancashire recording was done in the Preston area, often with his friend **J. H. Threlfall**, another esteemed Lancashire lepidopterist, with notable hunting grounds including a 'little wet sheltered corner on the moors' (Hodgkinson, 1880) at Dutton, near Ribchester, where he reported finding approximately 200 *Elachista alpinella* moths, and also a number of county rarities on Dyer's Greenweed. Hodgkinson was an extremely frequent contributor of novel sightings and discoveries to the entomological journals of the time. Nationally, it was reported, during his latter years, that he had added about 30 species to the British list (Robson, 1893). These included two species named after him, *Pterophorus hodgkinsoni* and *Nepticula hodgkinsoni*, although both turned out to be varieties of already named species, *Stenoptilia zophodactylus* and *Stigmella centifoliella*.

Henry Tibbats Stainton (1822–1892), an entomologist from Lewisham, documented many of the early Lancashire records, as the Manchester district was one of the localities used in his pioneering work, *A Manual of British Butterflies and Moths* (1859). However, whether Stainton even visited the county is unknown and the records were likely provided by local entomologists of the time such as J. Chappell.

John W. Ellis (1857–1916) was a Liverpool doctor, and a keen botanist and entomologist. In 1880, he began a list of the *Coleopterous Fauna of the Liverpool District*, as well as the first Lepidoptera list covering the whole of Lancashire (and Cheshire). This was initially published in parts in *The Naturalist* from 1885 to 1890 as *The Lepidopterous Fauna of Lancashire and Cheshire*, and later collected as a single volume, within which he 'endeavoured to present, as completely as possible, the facts known with reference to the occurrence in Lancashire and Cheshire of the British species of Lepidoptera' (Ellis, 1890). The reports within Ellis' list provide an extremely significant collection of records from the late 19th century, and are based on the records of important Victorian entomologists, such as J. Chappell, J. B. Hodgkinson, J. H. Threlfall, J. C. Melvill, J. Sidebotham, F. N. Pierce, etc.

T. Baxter, of Lytham St Annes, was an active lepidopterist from the late 1880s into the first two decades of the 20th century. Many of his records survive via specimens from museum and other collections, and others were noted in entomological journals of the time. An example of the latter was a specimen of Sandhill Rustic (*Luperina nickerlii*), the first for the county, taken at St Annes in 1889, although its identity was not fully recognised at the time (South, 1889). It was considered intermediate between what were thought to be two forms of Flounced Rustic (*L. testacea*), *gueneei* and *nickerlii*, and only later correctly understood as the *gueneei* form of Sandhill Rustic.

William Mansbridge (1870–1955) was a particularly important figure. In 1913 he began to revise Ellis's 1890 Lancashire and Cheshire list. The first instalment of the Mansbridge revision was published by the Lancashire and Cheshire Entomological Society (LCES) in 1913 and further instalments published up to 1940. The completed list was then made available as a separate publication (Mansbridge, 1940).

Mansbridge's early entomological work was carried out in the South of England, but in 1904, business brought him to Liverpool. In December 1905, he was elected one of the Honorary Recording Secretaries of the LCES, a post he held until 1950, and he was also President of the Society for eight years. His many papers presented to the Society discussed, amongst other topics, variation, melanism, his breeding experiments, the results of his active fieldwork around Liverpool and elsewhere. Examples included *Variation due to Chemical Feeding*, presented to the Society on 20 January 1931, *Collecting on Simonswood Moss*, presented later the same year on 17 November, and *Collecting Spring Larvae*, presented on 19 April 1932 (*LCES Annual Report and Proceedings*, 1931–33).

In 1914 William Mansbridge was also appointed as one of the two referees for Lepidoptera of the Lancashire and Cheshire Fauna Committee, the precursor to the Lancashire and Cheshire Fauna Society. His annual Lepidoptera reports featured in the latter Society's Annual Report from 1915 to 1950.

His obituary, in the LCES Journal, notes that 'to other entomologists, expert or tyro, he was always helpful and he would devote hours to the identification of smaller moths sent to him by less experienced collectors' (*LCES Annual Report and Proceedings*, 1953–55).

The lists produced by Ellis and Mansbridge included all of the species found in Cheshire, Lancashire and included southern parts of Cumbria. For each species, a north-west regional

William Mansbridge, 1954 (Photo taken from LCES report, 1955)

status was given, significant records were usually listed by location and recorder, and sometimes with a year, infrequently with a full date. Most species were summarised and generalised locations were often used. In most cases foodplants were not listed. Despite these limitations, these books listed large amounts of material that have not been relocated in any other publications or in Museum collections.

As is the case today, much of the above recording took place at favoured sites. This meant that large parts of the county remained unvisited from the 19th to the mid-20th century. However, one aspect of recording that rectified this to some extent, was the local societies or individuals who produced regional lists, some of whom, such as the **Nelson Naturalists**, remain active. Until the 1950s there were plenty of local area lists, usually in pamphlet style publications, such as **W. G. Clutten**'s *Moths and Butterflies of the Burnley District* (1918), that supplied more detailed local information, usually in list form but often with added notes of interest.

These became less frequent as the years rolled on and the last main local area list was *The Lepidoptera of Formby*, by **M. J. Leech** and **H. N. Michaelis**, published by The Raven Society in 1957.

G. de C. Fraser (1872–1952), was an important figure in moth recording in the Sefton Coast area, establishing the Raven Society, and producing the first Formby Lepidoptera list in 1946. In the foreword of the Leech and Michaelis update, published after Fraser's death, A. Brindle wrote this extremely evocative piece about an occasion at Warren Mount, the home of the Frasers, just prior to the onset of WW2. 'It was in August, 1939, that I first had the pleasure of meeting the Frasers. Together with a small party of entomologists from N.E. Lancashire invited to collect at light near the shore at Formby, I experienced the somewhat overwhelming enthusiasm and hospitality extended to all at "Warren Mount." Fed, fêted, and equipped, we were taken by cars to the shore, sheets were laid down in front of the headlights, and a memorable night ensued. The night was dark and warm, the moths arrived in great numbers and, true to the Fraser tradition, work did not cease until a few minutes before the last train left Formby, when a hurried dash to the station deposited a happy band of entomologists on the first stage of the journey home. The next meeting was long delayed. The following week-end saw the black-out in force, and it was not until 1946 that I found the Frasers, still enthusiastic, establishing the Raven Society.'

Hugh Nicholas Michaelis (1904–1995) was brought up in Manchester where his family was involved in the cotton trade. During the First World War, he spent much of his summer wandering over the dunes of Deganwy and Conway, learning the names of the wild flowers, and later the moths and the butterflies. He attended Manchester Grammar School where George Sidney Kloet (a fellow, future entomologist) was a fellow pupil, and together they would explore the countryside for specimens (Morgan, 1996). Later, Kloet acquired a motor bike and, with Michaelis on the parcel rack, they were able to search further afield. His membership of the Manchester Entomological Society brought him into contact with William Mansbridge, who encouraged Michaelis' study of micro-moths, a far more difficult undertaking then than now, when copious material exists for identification purposes.

Michaelis returned from his Second World War service, including over three and a half years in India, and became a bank manager in Manchester. His love of entomology remained, and his micro-moth recording was prodigious and responsible for many county firsts, including *Etainia decentella*, *Coleophora orbitella* and *Pseudococcyx turionella*, all taken during the 1950s, the first two from south Manchester, the third from Formby. His extensive field work on the Sefton Coast provided many of the most interesting records detailed in *The Lepidoptera of Formby*.

Morgan (1996) observed that Michaelis 'was a welcome companion at field meetings, always adding a good list of species to the records for the day. Children would cluster round him with their finds or enjoy an evening's light trapping in his company, and many have been inspired by his enthusiasm and encouraged to retain an interest for life in natural history.'

Michaelis was prominent in the **Manchester Entomological Society**, serving as President on two occasions, as did his old school-friend, Kloet. The Society, active between 1902 and 1991, was one of the successors to the many 19th century Manchester and district natural history societies, their profusion explained by Cook and Logunov (2017). 'The frequently horrific conditions of the 19th century industrial working class often encouraged a keen interest in natural history among its members. Many factory workers used what spare time they had visiting open spaces to study and collect minerals, wildflowers and insects'.

Two of the most prominent of the county societies, both also covering Cheshire, and both continuing to this day, are the **Lancashire and Cheshire Entomological Society** and the **Lancashire and Cheshire Fauna Society**.

The Lancashire and Cheshire Entomological Society was founded in 1877, in order to further the study of entomology by amateurs and professionals alike, its activities centred upon Lancashire, Cheshire and adjoining parts of North Wales. Between 1881 to 2010 the society produced a (mostly) annual journal with other occasional publications covering all aspects of regional entomology, with numerous articles and checklists. The Society owns an entomology library and a British Lepidoptera specimen collection, both of which are housed at World Museum, Liverpool. The Society continues to offer an annual programme of talks, and occasional field meetings.

The first president, S. J. Capper (1881), explained the Society's origins as lying in 'the meeting of a few ardent entomologists at the house of our vice-president (Mr. Nicholas Cooke), Gorsey Hey, Liscard. I did not know of the meeting till afterwards, and was surprised when I heard such a society was formed, and still more so when I was told that I had been nominated as president. The first meeting was held at my house, Huyton Park, on the 24th February, 1877. The society consisted of only eleven members, viz., Messrs. N. Cooke, Mountfield, Birchall, Carrington, N. Greening, T. J. Moore, Roxburgh, Whitby, Johnson, Cross and myself. We elected our officers, drew up our rules, and christened our society The Lancashire and Cheshire Entomological Society.'

The Lancashire and Cheshire Fauna Society (initially the Lancashire and Cheshire Fauna Committee) was formed in 1914. Initially based in the museums and universities of Manchester and Liverpool, it evolved to become an informative and accessible organisation recording and presenting data on the status and distribution of all faunal groups in the region, including annual reports from the Recording Officer for Lancashire Lepidoptera. The society's other publications have included annual bird reports for Lancashire, but also entomological publications including *Butterflies and Day-flying Moths of Lancashire* (Marsh & White, 2019), *Pug Moths of North-west England* (Hancock, 2018), two volumes of *Micro-moth Field Tips* (Smart, 2018; 2021), and studies of other orders, such as Hymenoptera and Odonata.

The Lancashire Moth Group is an informal grouping, with annual social events featuring presentations from members, and producing annual reports via the County Moth Recorders. The group was established c.1997. Its need was indicated by C. I. Rutherford (1994) in his *Checklist of the Macro-moths in Cheshire 1961 to 1993*, in which he explained that co-opting the whole of Warrington and Widnes into the Cheshire recording area was necessary as 'much interesting and valuable entomological data could have been lost if we had not included these results in our records; perhaps if there had been as much active moth recording in Lancashire as there is in Cheshire, these records would have been omitted from the Cheshire List.' The group has produced Newsletters, at least annually, since 1999, some of which are referenced within this book. A booklet was also produced by **Steve Palmer** (2002a) for use by members of the society, containing the first list of micro- and macro-lepidoptera based solely on Lancashire records, using the Vice-counties 59 and 60 as the recording area. Field meetings are occasional but always include an annual count at the Belted Beauty site at Potts Corner, where all are welcome to assist.

There is also a Lancashire Moths website (https://www.lancashiremoths.co.uk/home), established by Mark Palmer c.2006 and currently maintained by Dave Bickerton, and a very active Lancashire Lepidoptera Facebook group. The website contains all previous newsletters, detailed information on distribution, guidelines for recorders and details of field meetings.

The Tanyptera Project is an initiative established in 2017, initially intended to run until 2022 (now extended to 2029), and funded by the Tanyptera Trust to promote the study and conservation of insects and other invertebrates in Lancashire and Cheshire. The Project is run from the Entomology Department at World Museum, Liverpool. As part of its remit, the Project has organised field trips, surveys for specific species (e.g., Least Minor and *Anacampsis temerella*), site and assemblage surveys, loaned equipment, provided grants, organised indoor meetings, workshops and web-based seminars, and facilitated access to the museum's invertebrate collections. They also run a grant scheme, lend microscopes and field equipment, and invest in facilities for recorders (e.g., books, microscopes, camera set-ups). Their website, amongst much else, contains links to a monthly newsletter *Invertebrate Notes* as well as many of the historical documents used in the preparation of this book, and can be accessed at https://www.northwestinvertebrates.org.uk/. The existence of this book would not be possible without the very generous support of the Trust.

How to record moths

This can be done in a number of ways, from just making a note of any moths encountered, on walls, tree-trunks, in the house, often through an open window into a lit room, or flying in the day. Far larger quantities of moths can be recorded by setting up a light trap in the garden and keeping a note of the numbers of each species that visit the trap. It is quite possible that on a good night in midsummer, a garden trap may be occupied by a total in the region of 100 different species, including micro-moths. New species will occur throughout the year. Successive years will see a reduction in this number of new species. This can be counteracted by taking the trap into the field, powered by a generator or a portable lithium battery, into new habitats where a different range of species are likely to appear. Another type of trapping, increasingly used, is with pheromone lures. These are primarily used for pest tortricid species, such as the Plum Fruit Moth *Grapholita funebrana* and the Codling Moth *Cydia pomonella*, and for the Clearwings, a fascinating, but elusive, group of moths. The use of lures has resulted in county first records of Six-belted Clearwing and Red-belted Clearwing during the last few years. Add to this, new county records of *Pammene giganteana* and *Pammene suspectana*, and it is probable that further developments of this recording method will produce further exceptional records in the near future.

All the methods mentioned above involve recording the adult stage of the moth. However, fieldwork offers the opportunity to also record the early stages of moths, and to understand a little more about the biology and life history of the moth, as well as its preferred habitat. This allows recording to take place throughout the year. Certain species, such as many of the leaf-mining moths, rarely attend light traps, and so would be significantly under-recorded if solely reliant on light-trapping of the adults. Hopefully, the species accounts contained within this book, including known county foodplants, and in many cases images of the early stages, will facilitate such fieldwork. Further details on recording moths can be found in Appendix 7.

How to submit records

We strongly urge all moth recorders to submit their records. Nearly 1.9 million records have been collected, verified and collated since the formation of the Lancashire Moth Group, and it is these, that have allowed us to build up this picture of Lancashire's moth fauna. Continued submission of all records remains strongly recommended, to maximise understanding of our Lepidoptera and to highlight any changes in abundance and distribution. Details of how to submit can be found on the Lancashire Moths website pages (https://www.lancashiremoths.co.uk/guidelines-for-recorders). The preferred method is to use excel spreadsheets, which can be downloaded from the website. Alternative methods include entering records onto Mapmate and submitting sync files, or uploading records to iRecord. The latter picks up 'research grade' records from iNaturalist. Please avoid duplication of records (and unnecessary work for the County Moth Recorders (CMRs)) caused by uploading records to these sites AND submitting excel files – just one or the other please. If none of these methods are possible for you, then please speak to the CMRs about alternative methods of record submission.

Moth Habitats in Lancashire

by Steve Garland

Introduction

Lancashire (as defined in the introduction) stretches from the River Mersey in the south to the Silverdale area in the north. It is bounded by the Irish Sea, Cumbria, Yorkshire and Cheshire. It has a long coast comprising mostly sand and saltmarsh, with very few rocky stretches, and uplands comprising the West Pennines, the Bowland Fells and the Pennines around Leck in the north-east. It is

Map 3. Major cities and towns of Lancashire and key sites

divided by three large river valleys; the Mersey basin and the Ribble and Lune Valleys, and also by the River Wyre, which flows across the Fylde to Fleetwood. The larger conurbations, as well as many of the important recording sites covered in this section, and elsewhere, are detailed above in **Map 3**.

Climate
The climate is dominated by a prevailing wind from the south-west, a slightly above national average annual precipitation level and mean winter and summer temperatures about mid-point of the national range. Localised variations are affected by adjacency to the coast, particularly the nearby, shallow Morecambe Bay, with colder and wetter conditions further inland and upland.

Geology and history
The underlying geology consists mostly of Upper Carboniferous Millstone Grit and Coal Measures, Lower Carboniferous Limestone, and Permian and Triassic sandstones and mudstones. However, most of the lowland geology is buried beneath thick deposits of glacial drift from the Ice Ages, alluvium and silt from rivers, blown sand along the coast and peat accumulated in lakes and bogs.

The Lower Carboniferous Limestone outcrops in several areas around Clitheroe, such as at Cross Hill and Salthill Quarry Nature Reserves, but is most significant around Morecambe Bay where it forms hills and crags as well as limestone pavement at sites such as Warton Crag and Gait Barrows Nature Reserves. Other exposures just inside the Lancashire recording area occur at Dalton Crags as part of Cumbria's Hutton Roof area and on Leck Fell, as part of the extensive Yorkshire Dales limestone area. Limestone quarrying has impacted these areas, but several disused quarries have become valuable wildlife sites.

The Upper Carboniferous Millstone Grit forms the uplands of Bowland and the West Pennines, as well as outliers such as Pendle Hill. The Coal Measures are found in a large part of the area around Wigan, Leigh, Bolton and Salford. The coal-mining legacy of these areas has had a huge impact,

Dalton Crags, April 2022

Pearson's Flash, August 2023

partly through some open-cast mining, but especially around the Wigan Flashes, where mining subsidence and subsequent flooding has resulted in the creation of many important lakes and wetlands.

Other mineral extraction has included extensive peat excavation that destroyed or damaged many lowland mosses, and some sand or gravel extraction, although this created opportunities for nature reserves at Mere Sands Wood, Brockholes and Rixton Claypits.

Industrialisation had an enormous impact during the 19th and 20th centuries, especially around Manchester, Liverpool, Warrington, Preston and in the Ribble Valley from Blackburn to Colne. Smoke and sulphur dioxide levels resulted in an almost complete absence of lichens. Changes in agriculture, extensive sheep-grazing on the fells, and the establishment of coniferous plantations, have adversely affected areas of herb-rich grassland and deciduous woodland.

The population of Lancashire is over 5.6 million and urbanisation and development has had a huge impact with the growth of the Greater Manchester and Liverpool conurbations, as well as Blackpool, Preston, Southport, Blackburn and many others. The 1901 census showed a population below 1.6 million, demonstrating a remarkable increase in a very short historical period of time. These expanding communities, their infrastructure and amenities, have almost certainly been detrimental to biodiversity in the county, including Lepidoptera. Sea-rise mitigation measures to protect coastal communities also have the potential to damage saltmarshes, dunes and other important habitats. However, the fauna and flora that has developed at post-industrial sites, country parks, gardens, etc., show that the impact of humanity need not be wholly negative.

General habitat summary

Natural England's Natural Character Areas (NCAs) divide England into 159 distinct natural areas, 13 of which have a significant element in Lancashire as covered in this book, and are listed on p.16. Each NCA has unique characteristics related to its landscape and habitat range.

Ainsdale NNR pine plantation, June 2023

The NCAs, therefore, provide a convenient framework for a discussion of Lancashire's habitats in more detail, but first it is useful to briefly consider the general distribution of these biotopes across Lancashire.

Woodland: Ancient woodland is particularly scarce in Lancashire. It is mostly found around the Morecambe Bay area, in the valleys of the north, west and south slopes of Bowland and in many West Pennine cloughs. There is also a concentration around Chorley and northern Wigan. Old parkland with wood pasture habitat is uncommon across Lancashire, with elements present around halls and castles such as Speke, Stoneyhurst, Towneley, Rufford, Rivington, Scarisbrook, Thursland and Knowsley. Few of these are intact parkland and most are private, so have not been much surveyed for moths.

Non-ancient broadleaved woodland is more frequent in the countryside and urban fringes of all the cities and towns in the county. In the last 50 years there have been extensive schemes to plant new trees in Country Parks such as Croxteth (Liverpool), Moses Gate (Bolton/Bury) and on old spoil-heaps. More recently, Community Forest initiatives, and programmes of tree-planting to combat climate change have become more widespread. These young plantations, if planted in the right locations and properly managed, should become significant for moths as they mature and as woodland plants become established beneath the trees.

Conifer woodland is not native to Lancashire, and plantations are not common, although scattered conifers occur throughout and there are many small (<250 hectare) plantations. Conifer plantations of over 1 km^2 include the extensive, mature pine on the coastal dunes around Formby and plantations on Longridge Fell. Other significant plantations occur in the West Pennines at Charter's Moss and Entwistle, in west Bowland at Beacon Fell and in the north at Docker Moor, Thrushgill and Lord's Lot Wood. Conifer plantations tend to be monocultures with a poor woodland flora, but do support some moths, which feed specifically on them, and the wider access tracks/rides support flora which can be good for moths, such as the Small Argent and Sable on Heath Bedstraw.

Acid grassland at Longworth Clough in the West Pennines, September 2023

Grassland: Grassland is a common habitat across Lancashire, but the overwhelming majority (especially in the lowlands) is improved and intensively managed, often dominated by rye-grass cultivars. The fauna and flora of these areas is extremely impoverished in every way, and this applies to the moths too. The largest concentrations of the commoner grass moths are from upland sites with floristically rich grassland, such as Green Bank in Upper Hindburndale.

Unimproved grasslands have become extremely rare and are scattered and often small, mostly less than five hectares. Many parts of Lancashire have virtually none left, such as the Lancashire Plain, the Fylde, the Lune Valley floodplain, the Liverpool and Manchester conurbations, Mersey Valley and the Lancashire Coal Measures. In other areas, the last remnants are predominantly in upland areas, including the West Pennine Moors and the slopes of Bowland. There are also a number of calcareous grasslands of importance on the Morecambe Bay Limestone, many at low altitude.

Upland areas such as Bowland, Leck Fell and the West Pennines have large areas of acidic, upland grassland dominated by species such as Mat-grass and Purple Moor-grass with rushes in wetter areas. Many are heavily grazed and quite poor in plant species, but are extensive and can be enriched by wet flushes and where grazing is reduced or absent. They have a distinctive, but limited moth fauna, which may be enriched if plants such as bedstraw are frequent.

Many grassland moths are now most numerous on brownfield sites or in Country Parks and other urban fringe habitat, where grasslands are increasingly being managed for wildlife. There are also some rich roadside verges, which can provide valuable connections between sites.

Mosses and bogs: Historically, lowland raised bogs (mosses) dominated huge swathes of the Lancashire Plain and the Manchester area. Isolated remnants remain, scattered throughout the county. Upland bogs, referred to as blanket bogs, cover large areas of Bowland and the West Pennines, but have been affected by drainage, grazing and burning in many areas. However, there are wetter areas with active bog-moss, cottongrass and a more diverse flora, including those with historical names such as Blaze Moss.

Heaths: Lowland heaths are now found mainly on the Sefton Coast, those on the Fylde being tiny remnants. Upland heaths (moorland) are dominated by Heather and Bilberry and are mostly a habitat maintained by regimes of burning and/or grazing. Intensity of management varies, which impacts the vegetation and the diversity of moths present. These are particularly extensive across Bowland, in the Leck Valley and in the West Pennines. The Docker Moor area has an isolated example of unmanaged

Cockerham Moss, August 2010

View from Longridge Fell, March 2015

Leck Fell, April 2015

Bilberry/Heather and supports really good numbers of typical moorland species, such as Heath and Neglected Rustics.

Coastal dunes: The Lancashire coast contains a number of excellent dune systems. These contain several different habitats, from the embryo dunes forming at the foreshore, with mobile dunes, semi-fixed and fixed dunes developing away from the beach towards heath and woodland. Temporary pools known as dune slacks form in the dips between the dunes, and provide a vital habitat for many organisms, including a rich mixture of plants adapted to the saline and non-saline conditions.

The Sefton Coast Dune system stretches for around 20km and is the largest dune system in England, much of it designated as a Site of Special Scientific Interest. Closest to the foreshore is the 'Green Beach' at Birkdale. Similar phenomena have occurred previously along the Sefton Coast, but are prone to temporary status due to being washed away (Smith & Lockford, 2021).

Uplands: Lamb (2018) described the Lancashire uplands as 'the hills, fells, moors and mountains and typically support unenclosed areas of natural and semi-natural vegetation, but can also include fields and enclosures that support one or more principal upland habitats.' These habitats comprise acidic grassland, blanket bog, heathland and marshy grassland. However, the term 'upland' is defined in this book in relation to the specific ecology of the land, the climatic conditions, and the resultant moth fauna (Averis *et al.*, 2004). This characteristic upland fauna can be found in the Leck area, Bowland, Pendle Hill, Longridge Fell, the Pennine fringes of Rochdale and Oldham and the West Pennines. Whilst most of these sites will peak at well over 200m above sea level, categorisation of these areas as 'upland' is influenced mostly by geology, weather, land use, climate and terrain, rather than by altitude alone.

Saltmarsh: Internationally important tidal mud-flats around the estuaries of the Ribble, Lune, Wyre and Kent and in Morecambe Bay, are bounded by extensive areas of saltmarsh. Much is grazed, but the Wyre estuary has some of the best (and largest) ungrazed saltmarsh in Lancashire, with particularly important sites including Burrows Marsh and Barnaby's Sands. Inland edges of the saltmarshes are rich in reeds and rushes, and are home to specialist species such as *Elachista scirpi*.

Wetlands: Until the late 1700s, Martin Mere was an enormous lake of up to 1,500 hectares and was the largest wetland in Lancashire. It was then drained, and remaining wetlands were restricted to smaller areas around ponds, rivers and lakes. Although drainage work has destroyed further wetlands in the river valleys and on the Lancashire Plain, there have been some increases. Mineral extraction (Brockholes and Rixton Clay Pits), mining subsidence (Wigan Flashes), flood-prevention schemes (Lunt Meadows) and areas managed for wildlife (Leighton Moss) have all created new wetlands, to the benefit of wetland moths. Other wetland areas within the county include the streams and river valleys, with rivers of variable size and containment, such as the Irwell, Alt, Mersey, Ribble, Lune, Wyre and Kent, the many reservoirs, including Anglezarke, Upper and Lower Rivington and Watergrove, and an extensive canal network.

Farmland: Much of lowland Lancashire, outside the urban areas, consists of intensive farmland with mixed size, sometimes large, fields and few substantial hedgerows, as are the flood-plains of all the major river valleys. Much is intensively managed for grazing pasture, while in more southern parts of the county, root crops, cereal and arable farming predominates on the easily cultivated peat, light alluvial and sandy soils. Most pasture is heavily enriched with fertiliser/slurry and much is not grazed but cut regularly for silage. Surrounding the major urban centres are many less-intensively farmed areas, mostly comprising grazing for cows, sheep and horses. In upland areas, sheep-farming is the major agricultural land-use.

Burrows Marsh, August 2023

Sheep grazing in the valley between Turton Moor and Darwen Moor, September 2023

Sutton Manor, St Helens April 2021

Ash Hill, Flixton, November 2013

Urban: Lancashire has many large towns and cities, especially in the south. These developed around industries including coal-mining, engineering, textiles and the ports associated with these. The urban area stretches from Rochdale and Oldham in the east, across to Liverpool in the west, including Warrington, Wigan, Bolton, Salford and Manchester. Further industrial urban areas stretch from Preston along the northern edge of the West Pennines through Blackburn to Burnley. The south Lancashire coast has large centres at Southport and Formby, while that of the Fylde is dominated by Fleetwood, Blackpool and Lytham St Annes. North Lancashire is more rural with Lancaster and Morecambe bordering Morecambe Bay.

Brownfield sites: These are most extensive around the major urban areas, and are predominantly ex-industrial sites, such as factories, mines, quarries and tips. Despite being important and diverse sites for moths, many are threatened by redevelopment.

Further information about Lancashire habitats is provided in Greenwood (2012) for modern Lancashire north of the River Ribble, and in Edmunds, Mitcham & Morries (2004) and in Lamb (2018) for modern Lancashire.

Natural Character Areas of Lancashire

Each of Natural England's Natural Character Areas (NCAs) represent an area of distinct and recognisable character; each defined by a unique combination of landscape, biodiversity, geodiversity, history, and cultural and economic activity. Their boundaries follow natural lines in the landscape, not county or district boundaries. This makes them a good framework for decision-making and planning for future change at a national level.

Full details of these Natural Character Areas can be found online (Natural England, 2014) and on a site under development (Natural England, 2023). These resources include boundaries, descriptions, key data, changes, trends and conservation priorities in greater detail than can be considered here. References to the relevant NCA numbers are in parentheses in section headings.

Map 4. National Character Areas

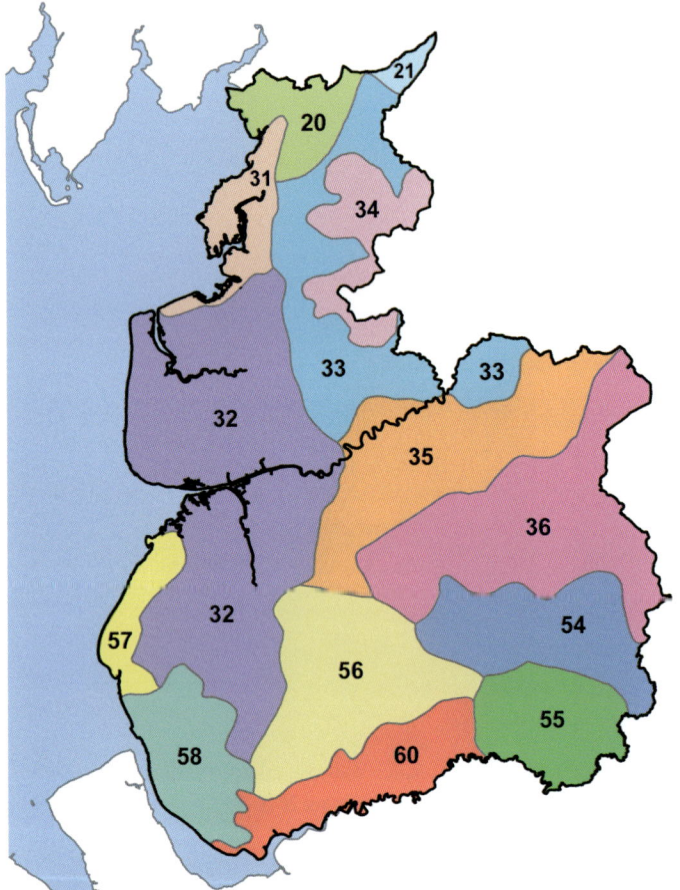

National Character Areas

20. Morecambe Bay Limestones
31. Morecambe Coast and Lune Estuary
32. Lancashire Plain
33. Bowland Fringe & Pendle Hill
34. Bowland Fells
35. Lancashire Valleys
36. Southern Pennines
54. Manchester Pennine Fringe
55. Manchester Conurbation
56. Lancashire Coal Measures
57. Sefton Coast
58. Merseyside Conurbation
60. Mersey Valley

There is also a small spike from NCA 21: Yorkshire Dales, based on the Leck Fell area in the north-eastern tip of VC60. This is discussed in the Uplands section (above), and on page 23.

Lancashire Plain and Morecambe Coast (NCAs 32 & 31)

The Lancashire Plain is a large area of flat countryside stretching from the northern edge of Liverpool, north to Preston and then across the Fylde peninsula. Around the Ribble and Wyre estuaries there are also extensive areas of saltmarsh. The Morecambe Coast borders the Lancashire side of Morecambe Bay and stretches from the Wyre estuary near Fleetwood up to the Cumbria border near Silverdale. The latter is a narrow ribbon of flat coastal land with a small mossland (Heysham) and extensive areas of saltmarsh. Most of this area is farmland, but with a number of towns and cities, such as Blackpool, Lancaster, and Preston. Most of Lancashire's saltmarsh lies in this area with extensive areas around the estuaries of the Ribble, Wyre, Lune and edges of Morecambe Bay.

Woodland is sparse across this area, mostly comprising small plantations and is especially lacking on the Fylde. There are no designated Ancient Woodlands larger than 2 hectares on the Fylde and very few in other areas of the Plain and Morecambe Coast. Some woodland species survive in a few older, less-intensively managed hedgerows and around older, well-established ponds, such as *Coleophora siccifolia*, found on apple in an old hedgerow.

The Lancashire Plain includes the majority of Lancashire's best agricultural land, and much of the farmed area is used for rearing sheep, cattle and pigs, and for growing cereals and root crops, although unimproved patches survive. Some of the moth species typical of farmland habitats are noted in **Table 1**. As with all tables 1–10, typical plants of the habitat are also listed, many of which will provide food for the local moth fauna.

Large areas of this part of the county were originally covered by extensive lowland bogs, lakes and wetlands. The vast expanses of historic lowland bog can be traced by the extensive areas of peat soils in this region and also on John Speed's 1610 map of Lancashire (Speed & Hondius, 1610). One stretched from Hightown and Maghull, northwards to the Ribble estuary, another extended

Carr House Green Common, Fylde, September 2023

Table 1. Farmland

MOTHS	PLANTS
Calybites phasianipennella	Goosefoot
Diamond-back Moth	Common Sorrel
Bedellia somnulentella	Peas and vetches
Coleophora spinella	Wild Mustard
Plum Fruit Moth	St John's-wort
Pea Moth	Burdocks
Barred Rivulet	Red Bartsia
Sloe Pug	Apple
Treble-bar	Bindweeds
Brimstone Moth	Hawthorn
Frosted Orange	Blackthorn
Heart and Dart	

Hedgerows near Inskip, Fylde, September 2023

The Moths of Lancashire

from Knowsley to Skelmersdale, and on the Fylde there were areas south-east of Blackpool and an extensive area stretching from Pilling to Garstang. These were largely drained, especially during the 1800s, and now only small remnants survive.

Lamb (2018) lists 17 mossland fragments covering 239 hectares. Along with the mosslands in the Salford and Manchester areas, these would have been very important habitats for moths and other wildlife, but many of these 17 areas are now much drier and have minimal active peat-formation. Nearly half of this hectarage is accounted for by Winmarleigh and Cockerham Moss near Garstang, an SSSI and a Lancashire Wildlife Trust Nature Reserve. Another protected remnant is Heysham Moss west of Lancaster, also a LWT Reserve and SSSI.

On Winmarleigh, there are extensive areas of Purple Moor-grass, with cottongrass, Bog-myrtle, Cross-leaved Heath, Heather, bog-moss, Round-leaved Sundew, Bog-rosemary and Cranberry. Major re-wetting works have been carried out by Wildlife Trusts on Winmarleigh and other mosslands to renew active peat deposition by bog-mosses and cottongrass, changing the nature of the habitat, having a mixed effect on moth species, depending upon their habitat requirements. Typical

Winmarleigh Moss, September 2022

Table 2. Wetland & Reedbeds

MOTHS	TYPICAL PLANTS
Orthotelia sparganella	Crack-willow
Depressaria daucella	Alder
Limnaecia phragmitella	Common Reed
Elachista maculicerusella	Yellow Iris
Chilo phragmitella	Bulrushes
Beautiful China-mark	Reed Canary-grass
Donacaula forficella	Reed Sweet-grass
Silky Wainscot	Common Sedge
Crescent	Pondweeds
Bulrush Wainscot	
Southern Wainscot	
Obscure Wainscot	

Mere Sands Wood, July 2023

18 *The Moths of Lancashire*

mossland moth species survive here, such as Grass Wave, Purple-bordered Gold, *Amphisbatis incongruella* and *Phiaris schulziana*. In addition, many upland moorland species are also found on these sites, such as Beautiful Yellow Underwing, Smoky Wave, Emperor Moth, Clouded Buff, Oak Eggar and Fox Moth. Other typical wetland species are listed in **Table 2**.

Although most of the original wetlands were drained, a few areas survived. Common Reed is often a dominant plant in lowland wetlands. Major reedbeds in this area include Marton Mere near Blackpool, and a major new area is developing at Lunt Meadows Nature Reserve. Smaller areas of Common Reed and other marsh vegetation survive in wet areas and along drains and ditches across the plain, as well as on nature reserves such as Mere Sands Wood and Martin Mere. The largest reedbed in Lancashire is just north of this area, at Leighton Moss RSPB Nature Reserve. Other large reedbeds are developing at Brockholes in the Ribble Valley, east of Preston, and on several sites in the Flashes of Wigan and Leigh National Nature Reserve. Apart from reedbeds, other wetland types are scarce and one of the few moths still widespread across the plain is *Acrolepia autumnitella* on Bittersweet, a wetland plant which grows in ditches across the area.

Barnaby Sands NR, August 2023

Table 3. Saltmarshes

MOTHS	TYPICAL PLANTS
Bucculatrix maritima	Oraches and Goosefoot
Monochroa tetragonella	Sea Aster
Scrobipalpa instabilella	Sea Club-rush
Scrobipalpa nitentella	Sea Arrowgrass
Coleophora adjunctella	Saltmarsh-grasses
Elachista scirpi	Sea-milkwort
Gynnidomorpha vectisana	Sea Plantain
Bactra robustana	Saltmarsh Rush
Eucosma tripoliana	Sea-rush
Pediasia aridella	Glasswort
Belted Beauty	
Crescent Striped	

Leighton Moss, May 2012

Lancashire's coast has no major rocky cliffs, but boasts extensive saltmarsh. The largest areas are around the mouth of the Ribble, including Warton Bank, Hesketh Out Marsh and Banks/Crossens Marshes. The Wyre has the best ungrazed saltmarsh in the county at Burrows Marsh Nature Reserve and Barnaby Sands Nature Reserve. There are also good areas around the edge of Morecambe Bay, especially around Carnforth, Bolton-le-Sands, Potts Corner and the mouth of the River Lune. Important foodplants and their dependant species, based on nationally documented usage, are noted in **Table 3**.

Lancashire coastal dunes (NCAs 57 and 32: south-west part)

The Sefton Coast NCA includes most of the remaining sand-dune habitat in Lancashire, extending from Crosby northwards to the Ribble Estuary at Southport. Covering about 2000 hectares, this is the largest coastal dune system in England and arguably the most diverse. Remnants of the once equally extensive Fylde dune system also survive at Lytham St. Anne's but around 84% was destroyed by development, mainly in the 19th century. Despite this, a number of typical dune moth species do survive on the Fylde. Similarly, about half of the Sefton duneland was lost, while large areas were modified by the creation of golf courses and pine plantations. Most of the surviving dunes north and south of the Ribble are now protected by SSSI designation, while a sizeable proportion is managed in National or Local Nature Reserves.

These dunelands are extremely important, as the only Lancashire localities for many specialist dune and coastal moths. The Sefton dunes, in particular, have a recorded total of around 1,000 species, at least 43 being considered 'Notable'.

Nearest the sea, the strandline, with vegetation such as Sea Rocket, Prickly Saltwort and orache, provides the habitat of Sand Dart.

Ainsdale LNR frontal dunes, August 2019

Moving inland, the broader zone of embryo and mobile (or yellow) dunes is dominated by Sand Couch, Lyme-grass and Marram, these being the foodplants for Lyme Grass, Sandhill Rustic, etc. Between the dunes are seasonally-flooded hollows or 'slacks'. These often contain Creeping Willow, which supports the Red-tipped Clearwing, Portland Moth, *Anacampsis temerella*, and a number of other important species, some of which are listed in **Table 4**.

Further inland, the oldest fixed-dunes have been around long enough for all the calcium in the soil to have been washed out by rainfall. The result is acid grassland and dune heath, the latter being a rare habitat favouring Heather, Gorse, Sheep's Sorrel and other plants characteristic of acid, well-drained soils. Moths that are found in these conditions include Forester and Archer's Dart along with some lowland populations of otherwise upland heath/moor species, such as Emperor Moth, Oak Eggar, Beautiful Yellow Underwing and Fox Moth.

Corsican Pine plantations on the Sefton dunes are well known for their Red Squirrels but also support many pine-feeding moths (see **Table 5**). Although most frequently recorded here, many of these also occur in pinewoods in other parts of Lancashire.

Morecambe Bay Limestone (NCA 20: western part)

The lowland Carboniferous Limestone areas near Morecambe Bay contain some of the most biologically diverse habitat in Lancashire. Sites such as Warton Crag, Gait Barrows and Dalton Crags comprise mosaics of woodland, scrub and grassland, with extensive areas of limestone pavement in several areas. Many moth species have their Lancashire headquarters here, also occurring in the southern limestone areas of Cumbria. The area is well wooded, with Ash as the

Birkdale Dunes strandline, July 2023

Devil's Hole, Formby after a record wet winter, June 2021

Table 4. Coastal dunes

MOTHS	TYPICAL PLANTS
Phyllonorycter quinqueguttella	Sea Rocket
Anacampsis temerella	Prickly Saltwort
Caryocolum marmorea	Orache
Willow Tortrix	Sand Couch
Red-tipped Clearwing	Lyme-grass
Forester	Marram
Anerastia lotella	Creeping Willow
Grass Eggar	Gorse
Sandhill Rustic	Sheep's Sorrel
Lyme Grass	Yellow-rattle
Portland Moth	
Shore Wainscot	
Archer's Dart	
Sand Dart	

Table 5. Coniferous woodland

MOTHS	TYPICAL PLANTS
Ocnerostoma piniariella	Corsican Pine
Piniphila bifasciana	Scots Pine
Pine Leaf-mining Moth	Spruces
Pine Shoot Moth	European Larch
Pine Hawk-moth	
Grey Pine Carpet	
Ochreous Pug	
Tawny-barred Angle	
Satin Beauty	
Bordered White	
Barred Red	
Pine Beauty	

dominant tree, but also Wych Elm, Yew, birch, oak, Small-leaved Lime, Sycamore and willow. Hazel coppice dominates large parts, and the scrub layer includes important foodplant and nectar species such as Buckthorn, Blackthorn, Hawthorn, Guelder-rose, rose and occasional Spindle, Wild Privet and Traveller's-joy. In a few places there are good patches of Juniper, though not as extensive as on south Cumbrian sites. Many of these sites were more open in the past, but woodland growth has encroached onto previously open areas. Most protected sites are actively managed by people and cattle to preserve open areas and rides, which are vital for many species of insect. In these areas, important foodplants include Common Rock-rose, Marjoram, thyme, Mouse-ear-hawkweed, Goldenrod, Burnet-saxifrage, bedstraw and the dominant grass is Blue Moor-grass.

Limestone woodland, scrub and grassland holds populations of species such as Barred Carpet, *Elachista cingillella*, *Endothenia ustulana*, *Semioscopis avellanella* and a few surviving populations of Four-dotted Footman, Lesser Cream Wave and Reddish Light Arches. Other important moths found in the Lancashire portion of the AONB limestone are noted in **Table 6**.

Gait Barrows, August 2015

Table 6. Limestone grassland, scrub and woodland

MOTHS	TYPICAL PLANTS
Ectoedemia arcuatella	Small-leaved Lime
Sorhagenia janiszewskae	Hazel
Elachista gangabella	Common Juniper
Neotelphusa sequax	Common Rock-rose
Mompha miscella	Marjoram
Scythris fallacella	Thyme
Cistus Forester	Great Knapweed
Pyrausta ostrinalis	Buckthorn
Anania funebris	Traveller's-joy
Speckled Yellow	Goldenrod
Bleached Pug	Blue Moor-grass
Barred Tooth-striped	Wood Sage
Least Minor	
Blossom Underwing	

Warton Crag, May 2010

Lune Valley (NCAs 20: eastern part & 33: northern part)

The source of the River Lune is in eastern Cumbria. In Lancashire it is fed by the Leck Beck, River Wenning and River Greta. It is a relatively undeveloped river valley surrounded by pasture, although most of it is heavily improved, with very little natural or semi-natural grassland, and the valley floor floods most years. There are numbers of hedgerows and field trees, and willows grow along exposed shingle and by ox-bow lakes and river-banks.

The lower Lune Valley is a broad river valley with an extensive floodplain, comprising improved grassland, much of which is still grazed, but also intensively enriched with slurry from livestock. There are no lowland meadows remaining on the floodplain and only tiny, isolated patches of semi-natural or unimproved grassland on the northern slopes of Bowland, and northwards, between the Lune and Hutton Roof.

The Valley holds a number of important ancient woodlands including Aughton Wood/Burton Wood SSSI and LWT Nature Reserve, Lord's Lot Wood, and a number of wooded valleys on the north slopes of Bowland, including Roeburndale, Mill Houses, Artle Dale and Calf Hill and Cragg Woods SAC. These woods support many important Lancashire species, such as Oak Lutestring, some of which are listed in **Table 7**.

North of the Lune, the River Keer runs from Docker Moor down to Carnforth. Docker Moor has several specialities, including Satyr Pug and Striped Wainscot.

The Leck Beck begins in Ease Gill, which reaches the highest point in Lancashire at 628m on Green Hill. The area is mostly moorland and upland grassland, although significant areas are on Carboniferous limestone rocks. It has a moth fauna similar to that of the Bowland moors, although also supporting a few species found nowhere else in the county, such as Pale Eggar and *Eudonia murana*.

Docker Moor, October 2023

Table 7. Deciduous woodland

MOTHS	TYPICAL PLANTS
Heliozela sericiella	Pedunculate Oak
Green Oak Tortrix	Wych Elm
Pammene fasciana	Rowan
Lunar Hornet Moth	Whitebeam
Scoparia ancipitella	Hawthorn
Oak Lutestring	Goat Willow
Northern Winter Moth	Birch
Clouded Magpie	Beech
Small Brindled Beauty	Hazel
Oak Beauty	
Great Prominent	
Merveille du Jour	

Aughton Wood, August 2023

Roeburndale Woods, August 2023

Bowland Fells, Pendle Hill, Upper Ribble Valley, West Pennines and Manchester Pennine Fringe (NCAs 34, 33, 35, 36 & 54)

The Forest of Bowland is a misleading name in that it comprises mostly moorland and much is managed for grouse-shooting. Although most of it is now in modern Lancashire, the central and south-eastern parts are in the Mid-west Yorkshire VC64, so are outside the scope of this book. The West Pennines has similar habitats, but is not managed for grouse, much being grazed by sheep, and whilst under-recorded, appears to have a less specialist moth fauna than northern Lancashire moorland sites. Both areas, along with Pendle Hill and parts of Longridge have extensive acidic grassland dominated by Mat-grass or Purple Moor-grass. There are areas of wetter, rush-dominated grassland and some upland peat-bog with cottongrass and *Sphagnum* moss. The Forest of Bowland, Pendle Hill and Longridge Fell all fall within the boundaries of the extensive Bowland AONB.

The moth fauna is typical of many northern England uplands with species such as Smoky Wave, Emperor Moth, Oak Eggar, *Ancylis myrtillana* and others listed in **Table 8**; many of which are associated with Bilberry and heather. Where the grassland has plentiful bedstraw, species such as Striped Twin-spot Carpet and Small Argent and Sable occur. Welsh Wave is also found where Rowan is common.

Rushy pasture at Roeburndale, August 2023

Table 8. Moorland and upland grassland

MOTHS	TYPICAL PLANTS
Bryotropha politella	Heather
Philedonides lunana	Purple Moor-grass
Bilberry Tortrix	Cranberry
Acleris hyemana	Cowberry
Small Argent and Sable	Bilberry
Striped Twin-spot Carpet	Gorse
Small Autumnal Moth	Ash
Manchester Treble-bar	Honeysuckle
Scarce Silver Y	Yew
Light Knot Grass	Birch
Haworth's Minor	Bracken
Glaucous Shears	

Loftshaw Moss, August 2023

24   *The Moths of Lancashire*

Turton Moor, September 2023

Cheesden Valley, Heywood, August 2023

    In wetter areas, where its favoured foodplant Cranberry grows, there are good colonies of Manchester Treble-bar. The species is named from its original discovery on the lowland mosses of Manchester, where it is now extinct.

    The edges of the Bowland Fells include many small rivers and valleys, which hold a small number of hay meadows and many woodlands, such as Calder Vale and Brock Bottoms in the south-west, the Wyre Valley, Abbeystead and Grizedale in the west. This area also contains some of Lancashire's larger conifer plantations.

    The Manchester Pennine Fringe contains pockets of clough woodland in the narrow, steep-sided, heather-lined valleys, with areas of acid grassland and reminders of the region's industrial heritage. Examples include Ashworth Valley, Cheesden Valley, Tame Valley and Lower Red Lees Pasture.

Lancashire Coal Measures and mosses (NCAs 56 & 60)

Much of south Lancashire is on the Coal Measures (a geological area known for its rich deposits of coal), and coal was mined across large areas, including the Ribble Valley and West Pennines around Accrington, Burnley, Nelson and Bacup, Rochdale and Oldham and eastern Manchester. The most active area, with the most numerous and largest mines, covered an area from St Helens and Newton-le-Willows in the south-west, north to Chorley and east to southern Bolton and Bury. The last active mines were at Agecroft in Salford (closed 1991) and Parkgate in Newton-le-Willows (closed 1993).

While the collieries were operating, they were huge industrial sites and many areas were dominated by colliery spoil tips such as the Three Sisters in Wigan, Sutton Manor in St Helens (now Country Parks) and what was the largest colliery spoil heap in Europe at Cutacre (now in part a Wildlife Trust Nature Reserve). The legacy of this is a number of brownfield sites, with a moth fauna similar to that mentioned under the Manchester/Liverpool section. Species such as *Coleophora artemisicolella* on Mugwort are especially widespread in the area.

However, the biggest effect on the area south of Wigan was caused by mining subsidence in the early 1900s after coal was extracted and mine tunnels collapsed. Water bodies called Flashes formed including Pennington, Scotsman's and Pearson's Flashes and Ince Moss. This complex of lakes, wetlands and wet woodland became Leigh and Wigan Flashes National Nature Reserve in 2022.

The Flashes now hold significant marshes and reedbeds with a moth fauna similar to those on the Lunt Meadows, Marton Mere, Brockholes and Leighton Moss. Some of the species that are most numerous around The Flashes, include those associated with Alder, birch, willows and oak.

In the early 1800s the area from Irlam and Warrington to Wigan, was one of the most important, accessible and well-known mosslands in England, often referred to as Chat Moss, although actually comprising many named areas. Gradually it was destroyed through drainage, dumping of human

Holcroft Moss, June 2003

Little Woolden Moss, June 2022

Scotsman's Flash, Wigan, August 2023

Table 9. Mosses

MOTHS	TYPICAL PLANTS
Lampronia fuscatella	Bog-myrtle
Glyphipterix haworthana	Cross-leaved Heath
Phyllonorycter anderidae	Heather
Amphisbatis incongruella	Bog-moss
Monochroa suffusella	Round-leaved Sundew
Elachista albidella	Bog-rosemary
Crambus hamella	Cottongrasses
Emperor Moth	*Sphagnum* moss
Purple-bordered Gold	Cranberry
Common Heath	Grey Willow
Clouded Buff	Downy Birch
Beautiful Yellow Underwing	

The Moths of Lancashire

Astley Moss, March 2008

waste (night soil), peat-cutting, conversion to agricultural land, encroachment by development and air pollution. A few fragments survived at Astley, Holiday, Highfield, Risley and Holcroft Mosses in varying states. Sadly, the eponymous Manchester Treble-bar has been extinct in the area for around 90 years with the Purple-bordered Gold last seen on Chat Moss in 1964, (although it does remain resident in the Winmarleigh and Cockerham Mosses in VC60). However, a few mossland species have survived, some of which are noted in **Table 9**.

In recent years, several significant areas have been acquired for nature conservation, and Lancashire Wildlife Trust has been managing Little Woolden Moss since peat-cutting ended. It is hoped that, when in favourable condition, some species may recolonise these new areas.

Manchester and Liverpool city regions (NCAs 55 and 58)

Greater Manchester (north of the Mersey) and north Merseyside are two huge conurbations, homes to millions of people and heavily industrialised. The growth of cities destroys huge swathes of countryside and can also affect the environment through air pollution and the pressure of public access in wild areas. However, the two city regions have also generated new habitats, introduced new species and created new environmental conditions. Liverpool and Manchester are also at the southern edge of Lancashire, so many species undergoing a northwards range expansion are first recorded in these areas. Most of these observations also apply to the other cities and towns across Lancashire, such as Blackburn, Blackpool, Lancaster, Burnley and Preston.

Manchester has several valleys that penetrate the urban area and provide varied habitats for moths within an otherwise built-up environment. Many of these are managed for recreation and, to a limited extent, wildlife, such as the Croal/Irwell, Medlock and Roch Valleys. Across both conurbations there are also numerous Country Parks, including Sankey Valley, Rock Hall, Wellacre, Daisy Nook, Tandle Hill, Chorlton Water Park, Rimrose Valley and Croxteth.

Two special habitats in urban areas are brownfield sites and gardens/parks. Brownfield sites can result from abandoned industrial sites, contaminated land, or as areas of marginal land along railways, roads and other developments. Parks and gardens are unusual in creating areas of managed land which often have large numbers of plants and habitats that would not naturally occur in the area and are often extremely diverse in their structure. Inner-city sites also have their own microclimates, being usually warmer than the surrounding countryside.

Brownfield sites can be extremely diverse, with many plant species, soil-types and micro-habitats in relatively small areas. However, they are also very dynamic with many transitory features. A widespread decline of UK industry in the 1980s and 90s led to the creation of large areas of brownfield sites. These were colonised by ruderal species, such as Rosebay Willowherb, Mugwort and Wormwood, and some moths, such as the Wormwood, became frequent. Eventually some sites were developed, or ruderal plants were displaced by perennials or dense scrub, and the Wormwood moth became extinct. The areas still support many local species of interest, including those listed in **Table 10**.

Some areas develop into dry grassland with calcareous soils, due to the presence of mortar and concrete or chemical waste/ash, such as Ash Hill/Green Hill, Flixton. Typical species include trefoil feeders such as Burnet Companion, mouse-ear feeders such as Small Yellow Underwing and, on Common Toadflax, Toadflax Pug.

Outside Liverpool and Manchester, there are many other important brownfield areas, especially around the south Wigan area, and a particularly well-recorded site at Middleton Nature Reserve, near Lancaster, which has an amazingly diverse mosaic of habitats on an old factory site.

Urban and suburban gardens cover over 30,000 hectares in Lancashire (estimated from Office of National Statistics, 2020 figures), but this will include a lot of paved areas, and other unsuitable habitat. Garden areas contain many foodplants that would not naturally occur in the area, sometimes aliens from other countries. The horticultural and agricultural trade is international and regular shipments of plants and seeds have been responsible for the introduction of many

Chorlton Water Park, Manchester, November 2015

Table 10. Brownfield sites

MOTHS	TYPICAL PLANTS
Trifurcula cryptella	Mugwort
Coleophora artemisicolella	Tansy
Dichrorampha simpliciana	Colt's-foot
Tansy Plume	Clover
Narrow-bordered Five-spot Burnet	Vetch
Shaded Broad-bar	Bird's-foot-trefoil
Marsh Pug	Mouse-ear
Slender Pug	Marjoram
Blackneck	Red Campion
Mother Shipton	Goat Willow
Blair's Shoulder-knot	Alder
Small Ranunculus	Prickly Lettuce

Table 11. Lancashire garden colonists

Species	Foodplant	1st UK	1st Lancs
Golden Plusia	Delphinium/Monk's-hood	1890	1917
Light Brown Apple Moth	Polyphagous	1930s	1986
Blair's Shoulder-knot	Cypresses	1951	1980
Acleris abietana	Spruce/Fir	1965	2001
Argyresthia trifasciata	Cypress/Juniper	1982	2000
Azalea Leaf Miner	Azalea	1984	1999
Firethorn Leaf Miner	Firethorn (*Pyracantha*)	1989	1999
Cypress Tip Moth	Cypress/Juniper	1997	2007
Horse Chestnut Leaf-miner	Horse Chestnut	2002	2007
Box-tree Moth	Box	2007	2014

Flixton garden, 2023

Trap in action, Flixton garden, 2009

new insect species to the UK, including moths, such as the Light Brown Apple Moth. It is also the probable source of pathogens, such as Ash Dieback, the likely effect of which on the many Ash-feeding moths is still to be assessed.

Many garden plants support specialised moths. Some previously rare moths that fed on local wild plants, such as Juniper, have happily adopted garden varieties and so have become increasingly widespread and common, such as Juniper Carpet and Juniper Pug. Other native species that have been able to expand their range by using garden plants include *Pyrausta aurata* (on mint and Marjoram) and *Lobesia littoralis* (on Thrift).

Other species could not have colonised Lancashire without garden plants. Abundant planted Cypresses in gardens have enabled species such as Blair's Shoulder-knot, Freyer's Pug and *Argyresthia trifasciata* to colonise the county. Golden Plusia would not exist in Lancashire without garden Delphinium and Monk's-hood.

Given that their foodplants can be abundant and that in recent decades a huge commercial trade in garden plants has developed, their initial speed of spread can be astonishing! A few examples of UK and Lancashire colonists are listed in **Table 11**.

One special garden within the county is that of Kevin McCabe in Flixton, Manchester. Kevin has trapped almost every night since 1994, and during that time has managed to record 858 different species in his garden alone, with approximately 640,000 individual moths, all of which have made their way onto the county database. The garden is approximately 35x25m. Highlights have included Lilac Beauty, Stout Dart, Cream-bordered Green Pea, Waved Black, *Lampronia fuscatella*, *Gelechia cuneatella* and *Crambus hamella*.

Garden records of this nature can provide extremely valuable data sets, allowing comparisons to be made over time regarding changes in the status of particular species.

The most frequent moths in gardens are those with larvae that can feed on a broad range of foodplants, such as Angle Shades, Large Yellow Underwing, Lesser Yellow Underwing, Heart and Dart, Hebrew Character, Bright-line Brown-eye and Light-brown Apple Moth. Despite being common, these caterpillars are rarely numerous enough to be considered severe pests, although Angle Shades seems to have a particular liking for geraniums and the Bright-line Brown-eye loves tomato plants.

Other potential pests of crops include Garden Pebble, Currant Clearwing, Leek Moth, Codling Moth, Cabbage Moth and Diamond-back Moth.

Urban areas are also where a number of new wildlife-friendly management innovations are taking place, including pesticide-free zones, urban wildflower meadows and other pro-active projects to encourage wildlife. These are a relatively recent trend, but are becoming significant and may allow these areas to develop their own characteristic moth fauna in the future. In urban settings, wildlife sites can easily become isolated, so linear features such as road verges, cycle routes, canals and railways can provide important links between sites. New developments can address this, through innovative planning and design followed by sympathetic management.

SPECIES ACCOUNTS

Introduction

Each species has a detailed summary of its presence in Lancashire, including the first and last known year of occurrence. These are based on all verified records received and accepted up to the end of November 2022. Any species discovered new for the county during 2023 have been included in an Addendum (page 572). Additionally, where space permitted, particularly notable records of existing species in 2023 have been added to the main species text, the record mapped and the date range extended to 2023. It should be noted, however, that it did not prove possible to add 2023 records to the flight graphs. All species, where possible, have been given a distribution map, flight graph and photographs of Lancashire-found adults and larvae, or their feeding signs.

The area covered is based on the Watsonian Vice-counties VC59 (South Lancaster) and VC60 (West Lancaster) as described in Dandy (1969). Any reference to Lancashire or 'the county' in the text refers to these two Vice-counties, combined. The rationale behind the use of Vice-county boundaries for biological recording and a map of the two areas can be found in the main Introduction chapter (page 2).

The space allocated in this section for most species is one-third of a page. However, some have been allotted a larger amount of space. Three species, with complex or uncertain taxonomic status, have been allocated two-thirds of a page each, and 17 species of conservation significance have been given a full page (**see Box 1**). To facilitate the inclusion of these variable page sizes and to maintain the correct taxonomic order, 36 species have been allocated half-page coverage.

Box 1. Species with full page coverage and their respective authors

Species	Authors	Species	Authors
9.006 *Lampronia fuscatella*	B. Smart	66.006 Grass Eggar	G. Jones
34.003 *Euclemensia woodiella*	B. V. Ridout	70.088 Netted Carpet	B. Hancock
35.013 *Anacampsis temerella*	B. Smart	70.201 Barred Tooth-striped	J. Patton
35.068 *Monochroa tetragonella*	S. Palmer	70.215 V-Moth	P. Marsh
43.001 *Scythris fallacella*	S. Palmer	70.250 Belted Beauty	S. Palmer
54.002 Forester	R. Walker	73.132 Sandhill Rustic	R. Burkmar
54.003 Cistus Forester	B. Hancock	73.146 Least Minor	J. Patton
63.024 *Anania funebris*	J. Girdley	73.308 Portland Moth	R. Walker
65.014 Oak Lutestring	P. Marsh		

Species text

The Code, scientific and vernacular names used are as published in Agassiz *et al.* (2013), and as subsequently amended in the *Entomologist's Record and Journal of Variation* to February 2024. The species authority has been omitted here to save space but can be found in Appendix 1 - Lancashire Moth Checklist. Recent synonyms can be found in Agassiz *et al.* (loc. cit.). In late 2023, a new set of vernacular names were published in the *Pyralidae and Crambidae* (Parsons & Clancy, 2023) and the *Field Guide to the Micro-moths of Great Britain and Ireland* (second edition), by P. Sterling *et al.* in December 2023. These came too late for inclusion in this book.

Although the text is based on data received up to the end of November 2022, some notable additions found during 2023 have been added to the species text, where space permitted. In these cases, the date range has been extended to 2023.

The Lancashire Status for each species has been specifically developed for this publication and refers to the frequency, distribution and residency of each species at the end of 2022

Box 2. Lancs Status criteria

Frequency	Distribution	Residency
Abundant	Ubiquitous	Resident
Frequent	Widespread	Migrant
Occasional	Local	Adventive
Scarce	Very local	Vagrant
Rare		Unknown

The Moths of Lancashire

Box 3. Layout of the species account

[ABH no.] [Vernacular name] [Scientific name] [Flight graph] [Family name]

NOLIDAE

74.008 Green Silver-lines *Pseudoips prasinana*

■ **LANCS STATUS** Frequent / widespread resident   c.1857–2022
First in Hale (Gregson, 1857), followed by singles in Bolton and Silverdale (1886), and Sabden (1939). Light trapping has helped to determine its distribution, but it also appears to have expanded its range in recent years, avoiding treeless upland areas. A second brood was first suspected in 1982 at Salwick and on eleven occasions since then. Generally found in low single numbers, with eight at Docker Moor the largest single count. An early individual was noted in Flixton on 24 March 2020 (K. McCabe).
■ **HABITAT** Deciduous woods, parks, gardens and tree-lined edges of mosses and moors.
■ **FOODPLANTS** Beech, birch, oak. Larva found in August and September.
Single brooded, with an occasional second brood. Readily attracted to light.

[Photographer's name]

Detailed view of the map legend

● 2000+
■ 1970–1999   □ Pre-1970
▲ 1970–1999 10km²   △ Pre-1970 10km²

(**Box 2**). Changes of status, compared to those previously published in historical documents are, where appropriate, discussed within the text. For the species' national status, readers are referred to those published in Randle, *et al.* (2019) for the macro-moths and Davis (2012) for the micro-moths.

The Lancashire Status assessment process considered a wide range of differing factors. These included abundance, distribution, the nature of its presence in the county (e.g. as a resident, migrant etc.), the extent of suitable habitats, the range and distribution of larval foodplants known, or probably utilised, in the county, the number of records, identification issues, how readily attracted to light a species is, larval and adult phenology, ease/difficulty of access to potentially occupied areas for surveying purposes, ease or otherwise of larval and adult detectability etc.. The authors' extensive experience and knowledge of the county fauna has then been applied to the three main criteria elements (Frequency, Distribution and Residency) and the listed subdivisions within each element. Specific definitions of the criteria have not been applied, although Rare is used generally when five or less records are known, and other factors suggest it is not overlooked. Ubiquitous signifies that it is considered likely that such species will be present in nearly all tetrads.

A broad summary of known habitat associations in Lancashire is listed for each species. The larval foodplants listed are those confirmed as utilised in the county. In the text, only vernacular plant names are utilised, but a full list of all plants documented in the species text, with their scientific names is listed in **Appendix 4** and follows Stace (2019). Where a foodplant such as oak or birch is noted, but was not identified to species level, the plant is listed with a lower-case first letter. Those where the plant has been identified to species level commence with an upper-case letter. If the reader wishes to investigate wider foodplant associations in the British Isles, we recommend using Langmaid *et al.* (2018) for micro-moths and Henwood *et al.* (2020) for macro-moths.

A list of all recorders who have submitted moth records can be found in **Appendix 2**.

Locations listed are those provided by recorders or as documented in publications. Where an Ordnance Survey (OS) grid reference was supplied but the site names were missing or vague (e.g., 'My Garden') the nearest location named on an OS map has been utilised. Where site names submitted included additional information, the main site or location name was utilised.

Distribution maps
The Tetrad (2 x 2km) scale distribution map, with symbol legend (**Box 3**), includes dots for all verified records up to the end of November 2022 whenever possible. Records of any species group submitted at aggregate level (e.g., Common Rustic / Lesser Common Rustic) are not included.

Flight graph
The flight (phenology) graph is based on all imago records with full date information up to the end of November 2022 but excludes bred dates. Readers interested in exploring phenology changes in the county over the years for the macro-moths are referred to the flight graphs in Randle *et al.* (2019). Where obvious changes in the number of broods (each being a single cycle from egg to imago) have occurred over the recording period, these are detailed. Those species without flight graphs lack any imago dates, usually as they constitute old records without full dates or are made up of larval stage records only.

Photographs
Photographs of Lancashire adults and larvae (or their feeding signs) have been added wherever possible. If photos of live specimens were not available, photographs of a few set specimens have been included, if available. Details of each photograph and all photographers who have contributed material are listed in **Appendix 3**. Habitat photographs have been added for full-page coverage species and, in a few other cases, elsewhere.

Terminology and abbreviations
The definition of terms and abbreviations used in this section are as follows:

AONB	Area of Outstanding Natural Beauty	**NR**	Nature Reserve
BRC	Biological Record Centre	**R**	River
MV	Mercury Vapour light	**SSSI**	Site of special scientific interest
NHM	Natural History Museum (London)	**Upland**	See definition in Habitat chapter, p.13
NNR	National Nature Reserve		

Historic data - dates and foodplants
Each species is given an earliest and latest recorded year wherever possible. Where the first Lancashire report originated in Byerley (1854), Gregson (1857), Ellis (1890) and Mansbridge (1940) and lacked a specific date, the year listed here refers to that in which the relevant book was published and is preceded by c. (circa).

Larval foodplants are rarely mentioned in any of the historic texts. This disappointing omission prompted the decision to only list foodplants known to be utilised by larvae in Lancashire, thereby providing a starting point for future research.

Historic data - geographical coverage
Locations where species have been noted in historic texts are well known to lack any precision. In fact, many species, common at that time (but maybe no longer) have no sites listed at all. To cater for this in the maps, a centralised, and often approximated, 10km symbol is used. Examples occur in Gregson (1857), where his list of species covers a 12-mile radius of the centre of Liverpool (which includes parts of Cheshire) and J. B. Hodgkinson whose 'Preston area' may have extended from Pilling Moss to the western edge of Longridge Fell. In Ellis (1890) and Mansbridge (1940), coverage includes Cheshire, Lancashire and some southern parts of south Cumbria (then listed as North Lancashire). Stainton (1859) used M (= Manchester) for moths found in that region, an area that includes some parts of Cheshire, Yorkshire and possibly Derbyshire. This area was undefined and

for a very few critical species it has not been possible to locate the exact area (or modern county) in which a moth was found.

Where there is doubt about a record originating from within a Lancashire VC, we have followed the decisions made by Ellis and Mansbridge in their works, where possible. If uncertainty remains, this is mentioned in the species text.

Data and coverage

Despite the variety of methods employed by recorders to find moths, most of the modern county database originates from light trapping activities in gardens. This has benefits, such as long-term monitoring at specific sites, and picks up species expanding their ranges, particularly if associated with garden plants. The downside is that many areas remain unvisited and unrecorded. Similarly, species that do not readily come to light can be significantly under-recorded. Searches for larval leaf-mines has considerably bolstered the knowledge of this otherwise poorly recorded group of very small and often difficult to identify insects. Attempts have been made during the two years of fieldwork for this project to address these deficiencies by targeting under-recorded areas and some individual species.

To assist with locating under-recorded areas in the county, **Map 1** depicts the number of records per tetrad and **Map 2** the number of species per tetrad, on the county dataset up to the end of 2022.

Excluded records

County Moth Recorders over the years have assessed all submitted moth records and this process has included published historic records as well. Any historic records found to lack enough detail to enable confirmation of a record have been removed from the county dataset. As these have been published elsewhere, it was thought advisable to document them individually outside the species text blocks. Details can be found in **Appendix 5**.

Map 1. Records per tetrad

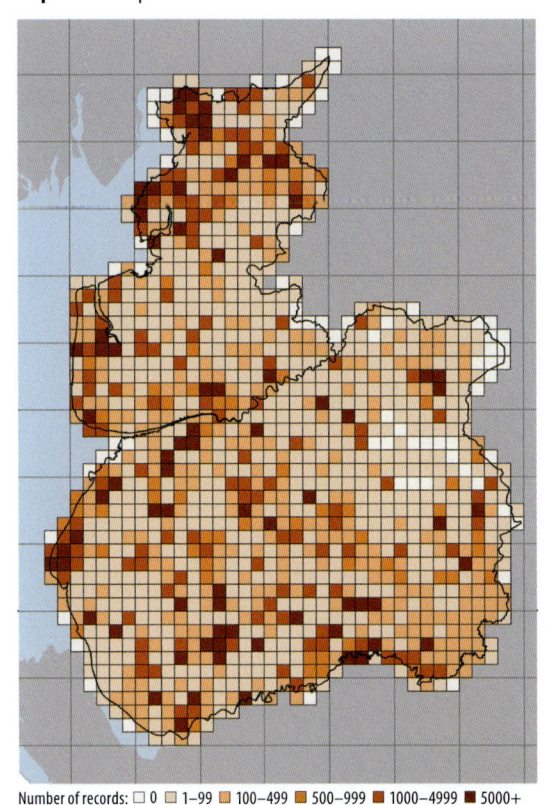

Map 2. Species per tetrad

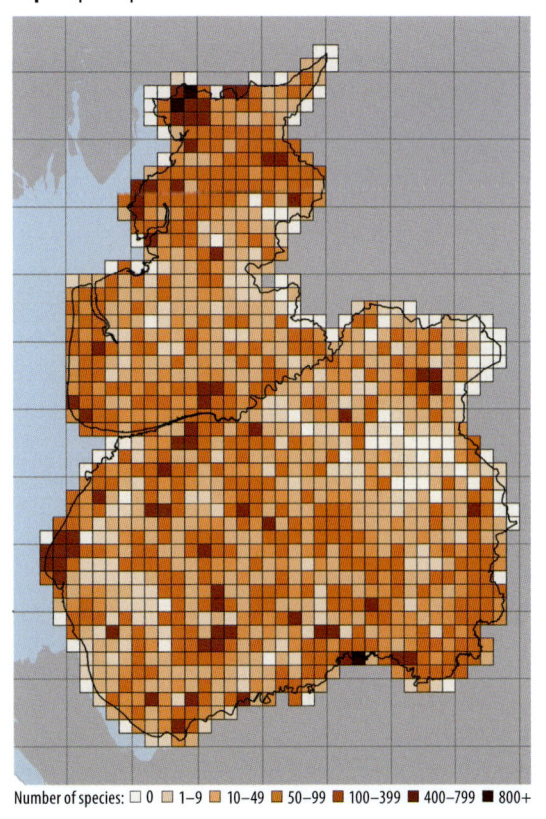

Number of records: ☐ 0 ■ 1–99 ■ 100–499 ■ 500–999 ■ 1000–4999 ■ 5000+

Number of species: ☐ 0 ■ 1–9 ■ 10–49 ■ 50–99 ■ 100–399 ■ 400–799 ■ 800+

MICROPTERIGIDAE

1.001 *Micropterix tunbergella*

■ **LANCS STATUS** Scarce / very local resident c.1857–2022

Early reports were received from Croxteth (Gregson, 1857), Manchester district c.1859 and Silverdale in 1914. Next seen at Martin Mere (M. Harrop) and Gait Barrows (M. Shaw), in 1989. A singleton was seen at Red Scar Wood, Preston on 24 May 1998, but all subsequent records were from the limestone areas of Dalton Crag, Gait Barrows (where 100 were seen on 21 May 2008), Silverdale and Warton Crag.
■ **HABITAT** Woods and scrub, predominantly on limestone, both lowland and upland.
■ **FOODPLANTS** Not recorded in Lancashire.
Single brooded. All records have been made during the day, except that three were netted close to a sheet and light run at Gait Barrows on 14 May 2007.

● 2000+ □ 1970–1999 △ 1970–1999 10km² □ Pre-1970 △ Pre-1970 10km²

B. SMART

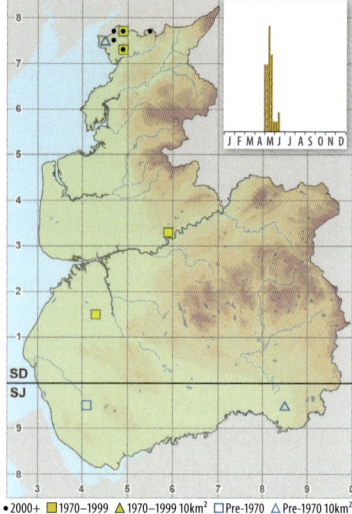

1.002 *Micropterix mansuetella*

■ **LANCS STATUS** Considered Extinct pre-1859

The validity of this as a Lancashire species has proved impossible to confirm with any certainty. It is referred to in Stainton (1859), where, on the basis of information supplied to him by contact/s unknown, he adds 'M' (= Manchester district), to the localities where the moth has been noted. Although most of the Manchester district is within VC59, parts do extend into adjacent portions of Cheshire (VC58) and/or Yorkshire (VC63). It is worthwhile noting that neither of these counties include the record in their respective county datasets (Fletcher, C. & Hind, S.H., pers. comm.).
■ **HABITAT** Unknown in the county.
■ **FOODPLANTS** Not recorded in Lancashire.
The species was recorded once in adjacent south-west Yorkshire (VC63) in 1997 (Beaumont, 2002). There is also a dot in VC59 (assumed to be the Manchester area record) on the distribution map in Heath (1976). Nationally, it is reported to be associated with very wet situations in woodland, carrs and fens and it is recommended that searches are best for females visiting the pollen of sedges in such situations (Heath, loc. cit.). Possibly overlooked.

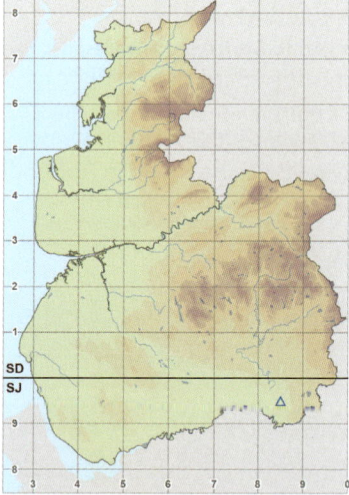

● 2000+ □ 1970–1999 △ 1970–1999 10km² □ Pre-1970 △ Pre-1970 10km²

1.003 *Micropterix aureatella*

■ **LANCS STATUS** Occasional / local resident c.1859–2022

First noted at Chat Moss c.1859 by J. Chappell, and then Burnley by A.E. Wright (Mansbridge, 1940). Next seen in 1991 on Winmarleigh Moss, with subsequent sightings at Astley Moss, Beacon Fell, Birk Bank, Cockerham Moss, Colne, Grize Dale, Longridge Fell and Lord's Lot Wood over the following decades. One was found at Rixton Claypits in 2003 (G. Jones) and may be a wanderer from the nearby mosses. Usually seen in single figures, although 15 were on Cockerham Moss on 2 June 2010.
■ **HABITAT** Mosses and moorland.
■ **FOODPLANTS** Not recorded in Lancashire.
Single brooded. All records relate to daytime observations, including net sweeping.

B. SMART    M. MEMORY

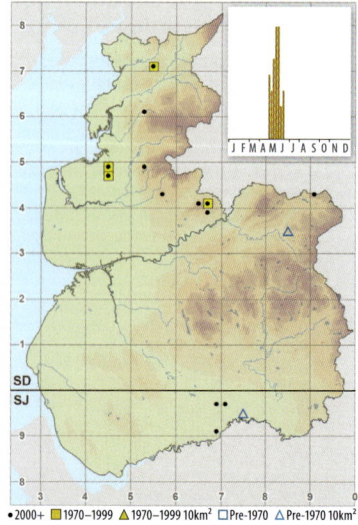

● 2000+ □ 1970–1999 △ 1970–1999 10km² □ Pre-1970 △ Pre-1970 10km²

34   The Moths of Lancashire

1.004 *Micropterix aruncella*

- **LANCS STATUS** Frequent / widespread resident   c.1859–2022

Recorded as common in the Manchester district (Stainton, 1859). Between then and 1980, only reported from Preston in 1889 (Ellis, 1890), Silverdale in 1914 and Formby pre-1940 (Mansbridge, 1940). Since then, noted widely across lowland parts and more locally on higher ground, such as Jackhouse N.R., Oswaldtwistle (M. Memory).
- **HABITAT** Dry, open grassland across most habitats.
- **FOODPLANTS** Not recorded in Lancashire.

Single brooded. Regularly visits flower-heads to feed, where it can occur in good numbers. It is likely that confusion between the female of this species and both sexes of *M. calthella* occurred historically, as it has on occasions more recently.

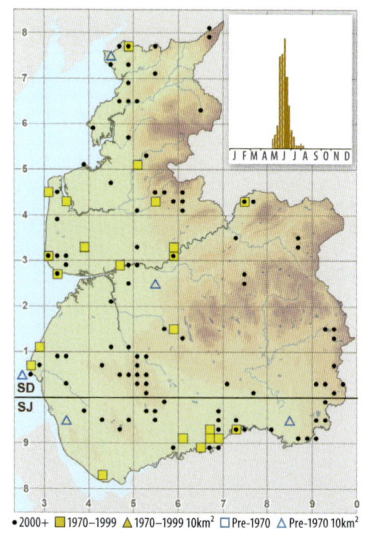

• 2000+   ■ 1970–1999   ▲ 1970–1999 10km^2 □ Pre-1970   △ Pre-1970 10km^2

1.005 *Micropterix calthella*

- **LANCS STATUS** Frequent / widespread resident   c.1857–2022

Recorded from Childwall, Liverpool (Gregson, 1857), Withington in 1859 and Haskayne in 1973. Reported more widely, but not annually, in recent years. Sites with more than a few reports include Freshfield, Gait Barrows and Woolston Eyes, with Burnley, in particular, producing good counts almost annually since 2015 (G. Turner).
- **HABITAT** Hedgerows, woodland edge, verges, brownfield sites and other grassy areas.
- **FOODPLANTS** Not recorded in Lancashire.

Single brooded. Visits flowering plants by day to feed, such as sedges and buttercups, sometimes in numbers. 520 were recorded east of Burnley on 24 May 2017. Clear, close-up views with a hand lens are needed to separate this and female *M. aruncella*.

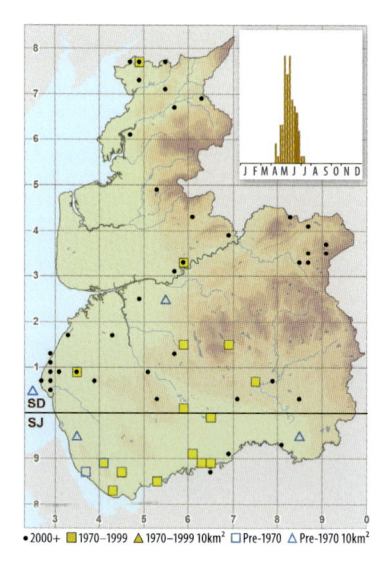

• 2000+   ■ 1970–1999   ▲ 1970–1999 10km^2 □ Pre-1970   △ Pre-1970 10km^2

2.001 *Dyseriocrania subpurpurella*

- **LANCS STATUS** Abundant / widespread resident   c.1889–2022

This leaf-mining species was first noted in the late 19th century, with records from Chat Moss and Prestwich by J. Chappell, and from Preston by J. Threlfall (Ellis, 1890). Mansbridge (1940) noted its abundance in oak woods throughout. However, no other sites noted until the mid-1980s. Over 950 records since, primarily in lowland areas.
- **HABITAT** Woodland, parks, gardens and other habitats where oak is present.
- **FOODPLANTS** Pedunculate Oak and potentially other oaks. Tenanted mines noted May and June, often in abundance, including 100 at Skelmersdale 2004 (C.A. Darbyshire). Single brooded. Adults can be found flying around oak in early spring sunshine in numbers. Also recorded at light, with 73 in a Billinge light trap on 24 April 2010.

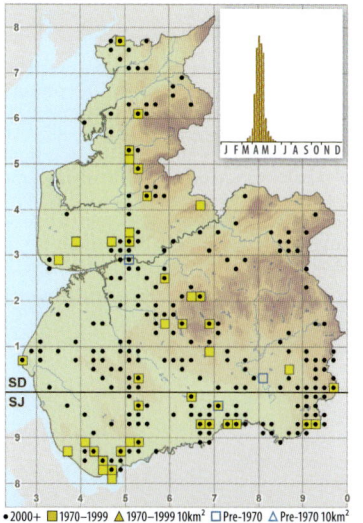

• 2000+   ■ 1970–1999   ▲ 1970–1999 10km^2 □ Pre-1970   △ Pre-1970 10km^2

MICROPTERIGIDAE

ERIOCRANIIDAE

ERIOCRANIIDAE

2.002 *Paracrania chrysolepidella*

■ **LANCS STATUS** Scarce / very local resident   1994–2023

Recorded in 1994 by K. Bland and M. Shaw, as a leaf-mine on Hazel at Gait Barrows. This was the first of ten county records, all of leaf-mines from the Morecambe Bay area. Tenanted mines noted during May on a couple of occasions, at Gait Barrows and Warton Crag, with one moth reared from the latter site in 2010. Unrecorded from 2016 to 2 June 2023, when a few vacated mines were found at Warton Crag.
■ **HABITAT** Hazel woodland on lowland limestone.
■ **FOODPLANTS** Hazel. There are no records in the county of leaf-mines on Hornbeam, the other recognised foodplant of this species.
Single brooded.

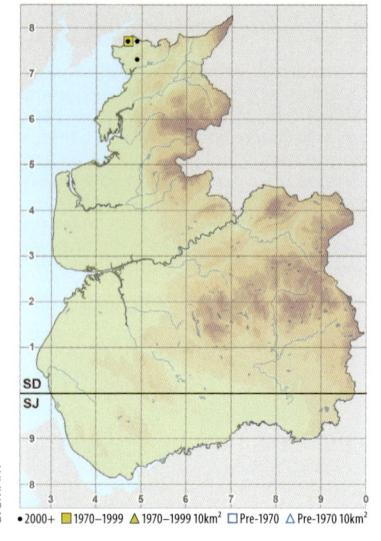

• 2000+   ☐ 1970–1999   ▲ 1970–1999 10km²   ☐ Pre-1970   △ Pre-1970 10km²

2.003 *Eriocrania unimaculella*

■ **LANCS STATUS** Occasional / widespread resident   1995–2022

One of six birch-mining Eriocraniidae, all found in Lancs. Reported from young birches on the mosses by Gregson (1857). However, due to lack of information on methods used to separate members of the genus, early records of the *Eriocrania* spp. cannot be validated. First confirmed record at Gait Barrows 1995 by A.M. Emmet. Over 60 subsequent records, including one of 60 mines at Winmarleigh Moss in May 2003.
■ **HABITAT** Woodland edges, dunes, heaths, mosses. Found in lowland and upland areas.
■ **FOODPLANTS** Birch, including Silver Birch, with a preference for seedlings and saplings. Single brooded. Most records are of the early stages. Adults noted at light, flying around birch, and swept from Heather growing amongst birch seedlings.

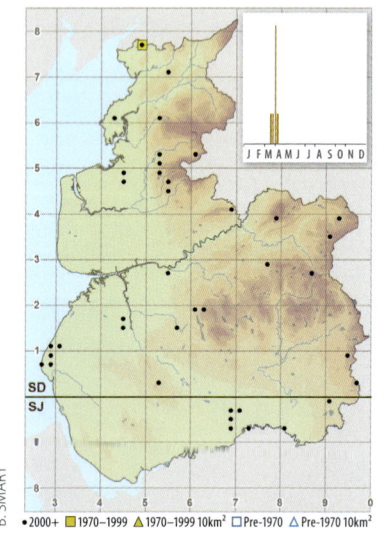

• 2000+   ☐ 1970–1999   ▲ 1970–1999 10km²   ☐ Pre-1970   △ Pre-1970 10km²

2.004 *Eriocrania sparrmannella*

■ **LANCS STATUS** Occasional / widespread resident   2000–2022

As with the previous species, 19th century reports are considered unconfirmed. The first confirmed records, therefore, were in 2000, at Holcroft Moss by S. McWilliam, and at Gait Barrows by J.R. Langmaid. All but six of 67 records since are from south Lancashire (VC59), although this may be due to limited searches in VC60.
■ **HABITAT** Woodland edges, heaths and mosses. Primarily in lowland habitats.
■ **FOODPLANTS** Birch. Highest count of 14 mines at Ash Hill, Flixton on 23 June 2012. Single brooded. The reddish mines, starting away from the leaf edge, have been noted from mid-June to August, later in the year than those of its close relatives. Adults netted during the day and attracted to light; recorded in April and May.

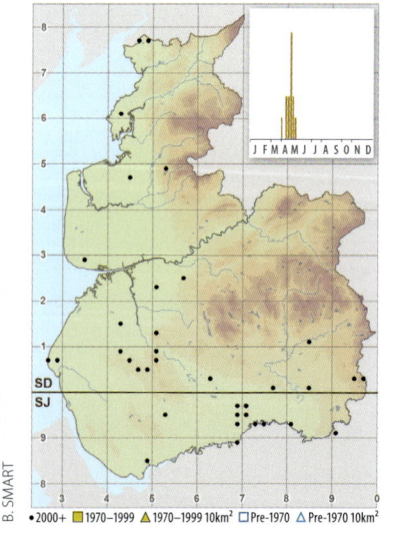

• 2000+   ☐ 1970–1999   ▲ 1970–1999 10km²   ☐ Pre-1970   △ Pre-1970 10km²

36   *The Moths of Lancashire*

2.005 *Eriocrania salopiella*

■ **LANCS STATUS** Occasional / widespread resident   2003–2022

This species was first noted on 5 May 2003 as a tenanted leaf-mine at Holcroft Moss (K. McCabe). The absence of even unconfirmed historical records may indicate recent arrival in the county. Subsequent distribution is concentrated in the south, with approximately 30 known sites in VC59. Gait Barrows is the only known site in VC60.
■ **HABITAT** Lowland woodland edges, scrub, dunes, heaths, mosses.
■ **FOODPLANTS** Silver Birch, Downy Birch.
Single brooded. All records are of leaf-mines, with the only adults being those reared from the larval stage. The mines, which start away from the leaf edge, have been recorded during May and June, with a maximum count of 30 at Astley Moss in 2006.

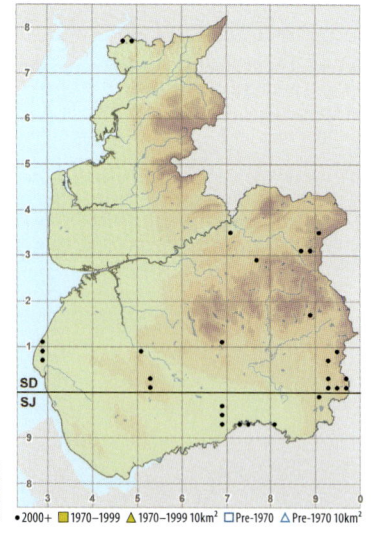

● 2000+ ■ 1970–1999 ▲ 1970–1999 10km² □ Pre-1970 △ Pre-1970 10km²

2.006 *Eriocrania cicatricella*

■ **LANCS STATUS** Occasional / widespread resident   1995–2022

First noted as a leaf-mine at Warton Crag and Gait Barrows, both on 23 May 1995, by A.M. Emmet. Subsequently, recorded regularly throughout; more often in the south of the county. Present in lowland areas, such as the Mersey Valley, Astley Moss and Winmarleigh Moss, and in upland areas, such as Longridge Fell and Birk Bank.
■ **HABITAT** Birch woodland in limestone areas and elsewhere, mosses and scrub.
■ **FOODPLANTS** Birch, including Silver Birch.
Single brooded. Most records are of mines, found in May and June; easy to separate from other Eriocraniidae as contain multiple white larvae. Adults recorded flying and resting around young birch trees, and rarely at light, during late March and April.

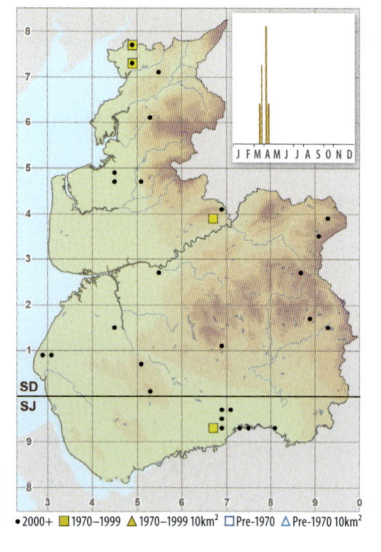

● 2000+ ■ 1970–1999 ▲ 1970–1999 10km² □ Pre-1970 △ Pre-1970 10km²

2.007 *Eriocrania semipurpurella*

■ **LANCS STATUS** Frequent / widespread resident   1995–2022

The validity of 19th century reports from Chat Moss cannot be ascertained. First confirmed from Gait Barrows 1995 by A.M. Emmet. Subsequent observations have shown the moth to be well distributed at lowland and upland sites. Again, most records are in south Lancs, possibly due to greater study of leaf-mines in these areas.
■ **HABITAT** Woodland, gardens, dunes, heaths, mosses.
■ **FOODPLANTS** Birch, including Silver Birch.
Single brooded. As with the other birch-feeding Eriocraniidae, most records are of leaf-mines, found in April and May. The moth has also been recorded at light on at least nine occasions, with dissection required to separate from the following species.

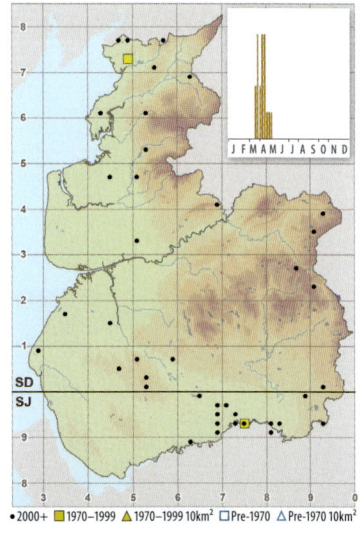

● 2000+ ■ 1970–1999 ▲ 1970–1999 10km² □ Pre-1970 △ Pre-1970 10km²

ERIOCRANIIDAE

2.008 *Eriocrania sangii*

■ **LANCS STATUS** Frequent / widespread resident   1985–2022

First recorded at Risley Moss in 1985, by L.W. Hardwick, followed by one in 1994 at Longridge Fell (S. Palmer), then in 2003. From this point, records were far more numerous with 5–10 reports per year, likely related to increased awareness of the species. Found in lowland and upland sites, such as Grizedale Fell, Birk Bank etc.
■ **HABITAT** Woodland edges, heaths, mosses, moorland.
■ **FOODPLANTS** Birch, including Silver Birch. Seedlings and saplings are preferred.
Single brooded. Tenanted mines in late April and May are easy to record due to dark grey larva; larvae of other Eriocraniidae are white. Whilst most records are of mines, adults also recorded in the field during the day and at light in late March and April.

• 2000+   ☐ 1970–1999   △ 1970–1999 10km²   ☐ Pre-1970   △ Pre-1970 10km²

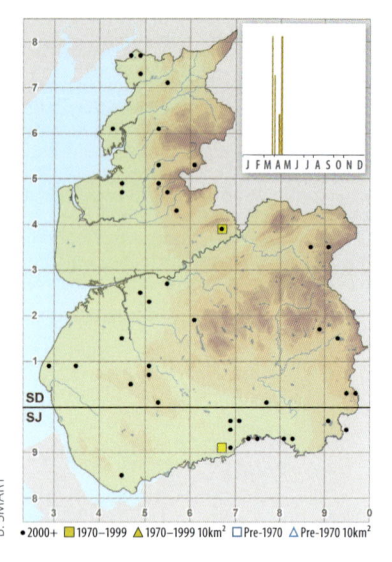

HEPIALIDAE

3.001 Orange Swift *Triodia sylvina*

■ **LANCS STATUS** Frequent / local resident   1880–2022

First reported in 1880 at Turton Toppings by J.W. Baldwin and thereafter, widely noted from around 18 lowland and upland sites, up to the mid-20th century. Warton Crag was found to be a particularly good location by W. Mansbridge c.1920. Since then, recorded most frequently in lowland areas such as Bispham, Hale, Heysham (26 on one August night in 2002), Leighton Moss, Morecambe and Sunderland Point.
■ **HABITAT** Rough grassland, near the coast, on limestone, commons, brownfield sites and, to a lesser extent, in gardens and moors.
■ **FOODPLANTS** Not recorded in Lancashire.
Single brooded. Comes to light. Sometimes confused with Common Swift.

• 2000+   ☐ 1970–1999   △ 1970–1999 10km²   ☐ Pre-1970   △ Pre-1970 10km²

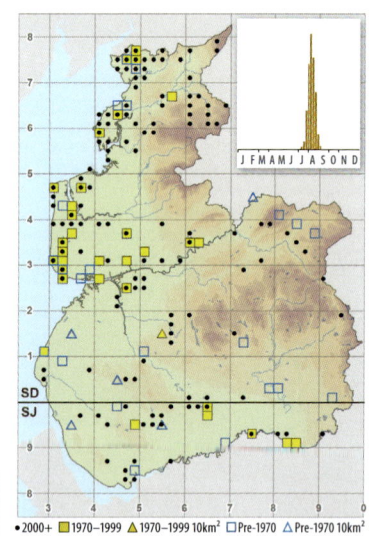

3.002 Common Swift *Korscheltellus lupulina*

■ **LANCS STATUS** Frequent / ubiquitous resident   c.1857–2022

Abundant in old lanes in the Liverpool district (Gregson, 1857) and later, noted as widespread and plentiful across the region by Mansbridge (1940). Increased light trapping from the 1990s led to many more records, with regular reports from both lowland and upland sites, including Billinge, Briercliffe, Claughton, Flixton, Fulwood, Hale, Lancaster, Longton, Pennington and Yealand Conyers. Usually encountered in single figures, but 163 came to light in Ainsdale on 8 June 2016 (R. Moyes, C. Daly).
■ **HABITAT** Rough, open grassland across the county in most habitats.
■ **FOODPLANTS** Mugwort, on the roots. A wider range of plants are used elsewhere.
Single brooded. Comes to light. Readily found flying at dusk. Larva found in February.

• 2000+   ☐ 1970–1999   △ 1970–1999 10km²   ☐ Pre-1970   △ Pre-1970 10km²

The Moths of Lancashire

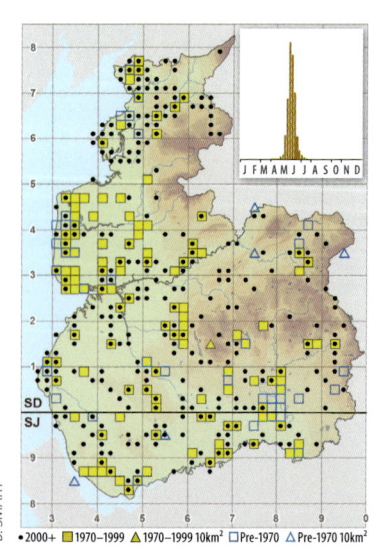

3.003 Map-winged Swift *Korscheltellus fusconebulosa*

■ **LANCS STATUS** Frequent / widespread resident   c.1857–2022
First noted on Simonswood Moss (Gregson, 1857). Later, as very numerous in the Oldham area (J.T. Rogers) in 1883 and, more generally, as abundant on moors and mosses (Mansbridge, 1940). During 1961, R. Leverton found it to be very common at several sites in the Clifton area of Manchester and it was noted as common at Leighton Moss during 1977 and Heysham 1989. Since then, lowland records have increased, probably as a result of more extensive light trapping, from sites such as Ainsdale, Billinge, Gait Barrows, Heysham, Martin Mere, Mere Sands Wood and Swinton. Upland reports have continued from a wide range of sites, including Botton Mill, Briercliffe, Burnley, Claughton, Docker Moor, Green Bank, Littleborough, Rishton and Royton. The absence of this species from much of the Fylde appears genuine and not due to under-recording. Most reports of numbers to light have been in single figures, but larger counts are not uncommon including 32 at Ainsdale in 2018, 35 at Leck Fell (2005) and Thrushgill (2009) and 40 at Trough Summit in 2001. The largest count came during the day, on 21 June 2011, when over 500 were seen at Belmont (S. Martin), many being predated by a large flock of Black-headed Gulls.
■ **HABITAT** Mosses, moors, often where bracken grows, including limestone grassland.
■ **FOODPLANTS** Not recorded in Lancashire.
Single brooded. Comes to light. The plain f. *gallica* is noted occasionally in small numbers throughout the region (see photo below left).

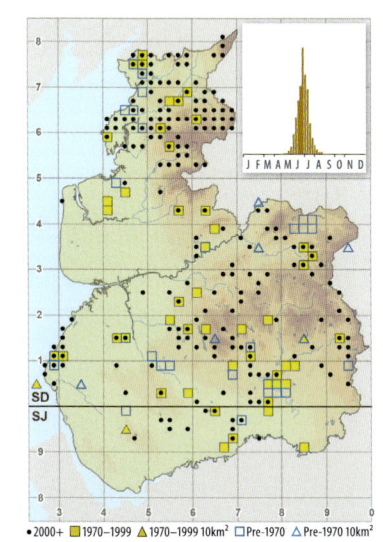

● 2000+   ◻ 1970–1999   △ 1970–1999 10km²   ▢ Pre-1970   △ Pre-1970 10km²

3.004 Gold Swift *Phymatopus hecta*

■ **LANCS STATUS** Occasional / local resident   c.1857–2022
Reported from Croxteth and Hale (Gregson, 1857) and later, from several sites such as Bolton, Burnley, Huyton, Prestwich, Sabden, Silverdale and St Helens, all prior to 1950. In the 1960s, noted as local, but common to abundant where it occurred, in the Clifton area, near Manchester (R. Leverton). More recently, recorded widely but at only a few sites with any regularity, such as Astley Moss, Docker Moor, Herring Head Wood, Leighton Moss and Warton Crag. Found mostly in small numbers when noted at light. Searches from late afternoon to dusk in areas of deciduous woodland with extensive bracken can be more productive. In such circumstances, 25 were seen at Billinge plantations on 12 June 2002 (C.A. Darbyshire), 40 were noted at Leighton Moss at dusk on 21 June 2003 (S.H. Hind & K. McCabe) and around 50 were recorded flying in Red Scar Wood, Preston, on a cloudy late afternoon, on 21 June 1996 (S. Palmer). Specimens with obscure markings are rare, with single records from sites such as Herring Head Wood and Leighton Moss.
■ **HABITAT** Damp woodland and scrub often associated with steep river banks, mosses, moors and limestone.
■ **FOODPLANTS** Not recorded in Lancashire. A larva was found under a log at Longworth Clough on 16 May 2002 (I.F. Smith) adjacent to an area with extensive Bracken.
Single brooded. Comes to light. Flies at dusk.

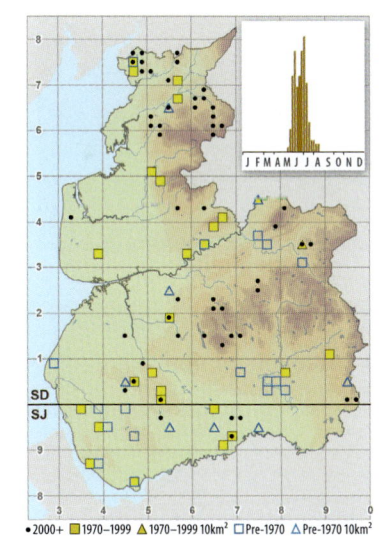

● 2000+   ◻ 1970–1999   △ 1970–1999 10km²   ▢ Pre-1970   △ Pre-1970 10km²

The Moths of Lancashire

HEPIALIDAE

3.005 Ghost Moth *Hepialus humuli*

■ **LANCS STATUS** Frequent / widespread resident   c.1857–2022

First noted in the Liverpool district (Gregson, 1857), it was subsequently reported as abundant throughout the region by all early authors, including Mansbridge (1940). From the 1990s, recorded regularly at many sites including Hale, Hoghton, Leigh, Orrell and nr. Longton. Larger counts have become increasingly unusual, with none in double figures since 2016. The last time exceptional numbers were noted, was when 100 came to light at Pickering's Pasture on 11 July 2000 (P. Hillyer).

■ **HABITAT** Rough grassland, including meadows and verges.
■ **FOODPLANTS** Not recorded in Lancashire.

Single brooded. Comes to light. Occasionally reported 'dancing' at dusk over grass.

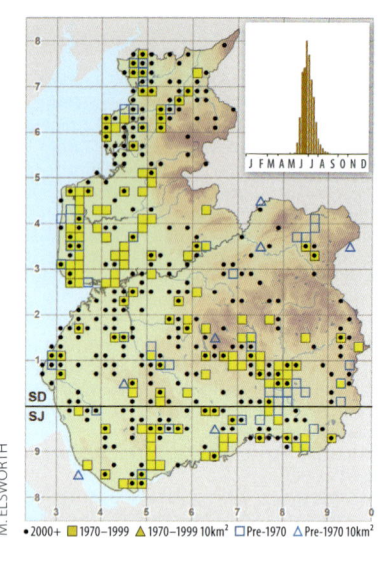

• 2000+   ▢ 1970–1999   △ 1970–1999 10km²   ▢ Pre-1970   △ Pre-1970 10km²

NEPTICULIDAE

4.002 *Stigmella lapponica*

■ **LANCS STATUS** Frequent / widespread resident   1995–2022

Leaf-mines first noted on Silver Birch at Winmarleigh Moss by S. Palmer, and at Gait Barrows by A.M. Emmet, both in 1995. Records since confirm the moth to be widely distributed throughout the county in most habitats where the foodplants occur, such as Ash Hill (Flixton), Cadishead Moss, Wigan Flashes, Dalton, Hollingworth Lake, Mere Sands Wood, and a few upland sites such as Beacon Fell and Cragg Wood.

■ **HABITAT** Mosses, woodland edges, scrub, heaths, dunes etc.
■ **FOODPLANTS** Silver Birch, Downy Birch.

Single brooded. All records are of leaf-mines, tenanted in June. The feeding gallery contains a narrow frass line with the initial section filled with cloudy green frass.

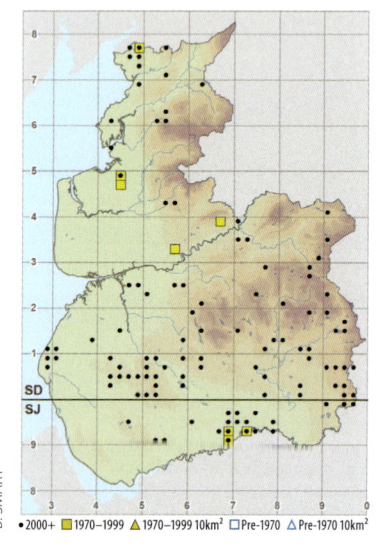

• 2000+   ▢ 1970–1999   △ 1970–1999 10km²   ▢ Pre-1970   △ Pre-1970 10km²

4.003 *Stigmella confusella*

■ **LANCS STATUS** Occasional / widespread resident   1996–2022

S. Palmer noted vacated mines on Silver Birch at Red Scar Wood L.W.T. in 1996. Subsequently, recorded throughout Lancashire, but a little less frequently than *S. lapponica*, and less commonly in the north. As with the previous species, it is difficult to be certain if the lack of earlier records means the moth was absent or missed.

■ **HABITAT** Mosses, woodland edges, scrub, dunes, moorland.
■ **FOODPLANTS** Silver Birch, Downy Birch.

Single brooded. All records have been of leaf-mines, usually in ones or twos, with tenanted mines recorded from July to September. These differ from the previous species in that the narrow frass line is black throughout.

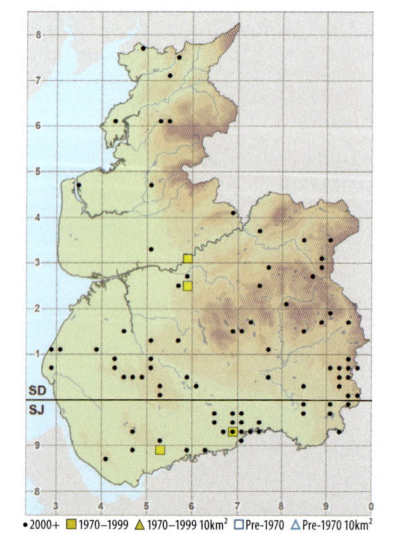

• 2000+   ▢ 1970–1999   △ 1970–1999 10km²   ▢ Pre-1970   △ Pre-1970 10km²

The Moths of Lancashire

4.004 *Stigmella tiliae*

■ **LANCS STATUS** Scarce / local resident   2004–2022

The first Lancashire record was at Eaves Wood, Silverdale in 2004, when leaf-mines were noted on Small-leaved Lime by K. McCabe and S. Palmer. There have been only nine further records, all of leaf-mines. The highest count was of approximately 20 mines on lime at Cringlebarrow, nr. Yealand Redmayne in 2009. All records have been in the northernmost areas of the county around Silverdale and Warton.
■ **HABITAT** Woodland on limestone.
■ **FOODPLANTS** Small-leaved Lime, and possibly other lime species.
Some evidence of the species being double brooded, as is the case nationally. A single mine was found in July; all others recorded from September to October.

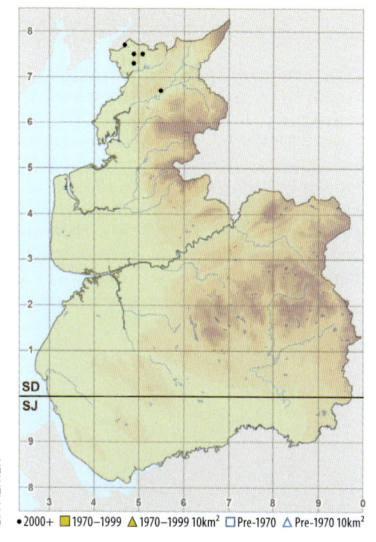

• 2000+   □ 1970–1999   ▲ 1970–1999 10km²   □ Pre-1970   △ Pre-1970 10km²

4.005 *Stigmella betulicola*

■ **LANCS STATUS** Occasional / local resident   1995–2022

First reported in 1995 by L.W. Hardwick at Risley Moss, where mines were frequent. Next, recorded on the birch seedlings and saplings at many of the Lancashire Mosses, with a maximum count of 50 mines at Astley Moss on 12 October 2009 (K. McCabe). Further noted at Gait Barrows, Winmarleigh Moss, Rixton Claypits, and in 2015, at a burial woodland site in Rainford on Downy Birch. Unrecorded from upland areas.
■ **HABITAT** Lowland mosses, woodland edges.
■ **FOODPLANTS** Downy Birch, Silver Birch.
Double brooded. All records are of leaf-mines, often up to four or five on a single leaf. Tenanted mines have been recorded from mid-June to late October.

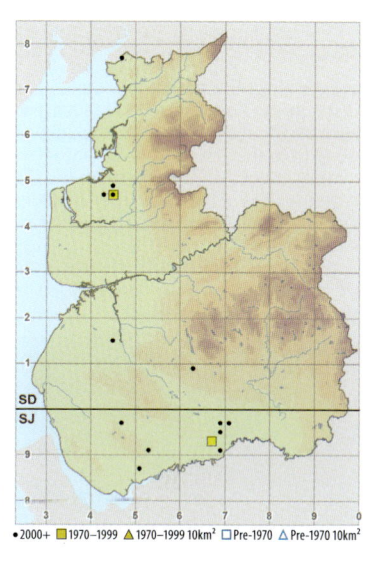

• 2000+   □ 1970–1999   ▲ 1970–1999 10km²   □ Pre-1970   △ Pre-1970 10km²

4.006 *Stigmella sakhalinella*

■ **LANCS STATUS** Occasional / local resident   1995–2022

This lowland species was first recorded at Winmarleigh Moss by S. Palmer in 1995, still the only known VC60 site. Over 75 subsequent records from VC59, almost all from Greater Manchester, including 30 from the east of the conurbation noted by S.H. Hind 2021–2022. Also, noted at Warrington, St Helens and Mere Sands Wood.
■ **HABITAT** Woodland edges, scrub, former industrial sites.
■ **FOODPLANTS** Silver Birch, Downy Birch.
All records are of leaf-mines, characterised by a long gallery with coiled frass. The highest count is of 14, with 6 tenanted, at Chorlton, 13 September 2022. There appears to be one extended generation, with tenanted mines August to October.

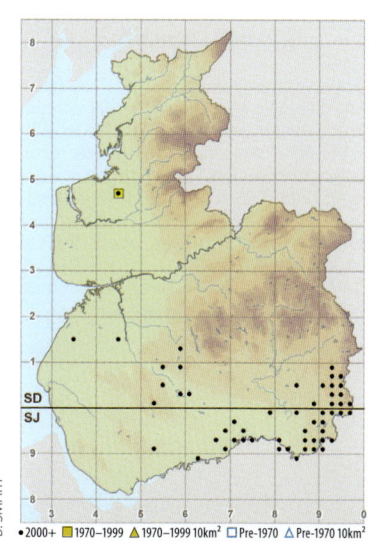

• 2000+   □ 1970–1999   ▲ 1970–1999 10km²   □ Pre-1970   △ Pre-1970 10km²

The Moths of Lancashire   **41**

4.007 Stigmella luteella

■ **LANCS STATUS** Frequent / widespread resident   1993–2022
As with other birch *Stigmella*, there are no confirmed pre-1990 records. As three of these species, including *S. luteella*, were recorded from Cumbrian sites in late 1800s, it seems likely they have been here all along, with the change due to recorder effort and increased understanding of mine morphology. First noted by A.M. Emmet at Gait Barrows in 1993. Subsequently, recorded over 200 times from a variety of habitats.
■ **HABITAT** Likely to be found wherever foodplants occur in lowland and upland areas.
■ **FOODPLANTS** Silver Birch, Downy Birch with tenanted mines from July to October. One extended generation. All records are of the scallop-edged gallery mines, other than a pair found *in-cop* on a birch trunk at Flixton on 6 July 2002 by K. McCabe.

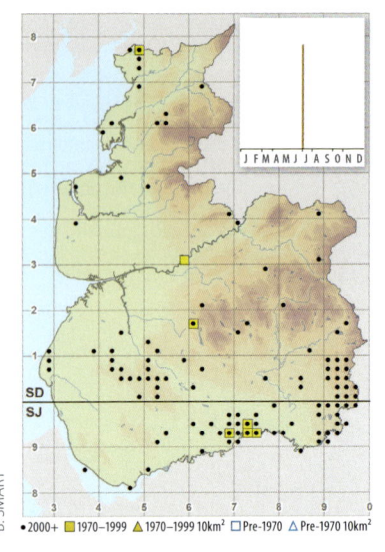

4.008 Stigmella glutinosae

■ **LANCS STATUS** Occasional / widespread resident   c.1889–2022
First noted in Preston by J.H. Threlfall (Ellis, 1890), although the method used to separate this from *S. alnetella* is unknown. The 25 widely distributed records since are all from the 21st century. Tenanted mines can be identified by noting the dark central spot of the ventral prothoracic plate of the larva; pale in *S. alnetella*. As vacated mines of the two cannot be differentiated, both are likely under recorded.
■ **HABITAT** Damp woodland, banks of lakes, ponds, rivers. Primarily in lowland areas.
■ **FOODPLANTS** Alder.
Single brooded with tenanted mines noted August to early November. All records are of leaf-mines, with a count of seven in a single leaf at Middleton N.R. in 2019.

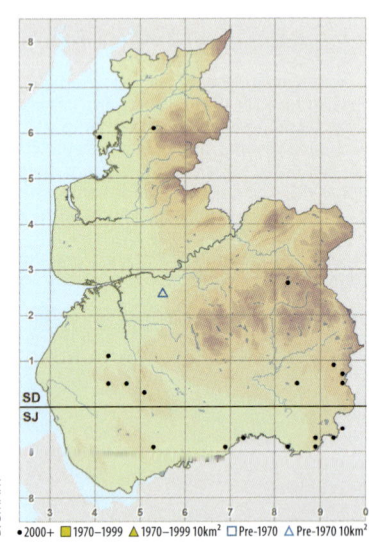

4.009 Stigmella alnetella

■ **LANCS STATUS** Occasional / local resident   c.1889–2022
Another alder species, with larval examination required to differentiate the mine from *S. glutinosae*. In this case, the larva has a pale, ventral prothoracic plate, visible in the mine with a hand-lens. Fewer records (16) than the previous species, but with similar distribution. First recorded at Preston by J.H. Threlfall (Ellis, 1890), but not again until 1995 at Scorton. Mostly from lowland areas such as the Mersey Valley and Middleton N.R., although with a single record from Longridge Fell.
■ **HABITAT** Damp woodland, ponds, lakes and riverbanks.
■ **FOODPLANTS** Alder.
Single brooded with tenanted mines recorded in October and early November.

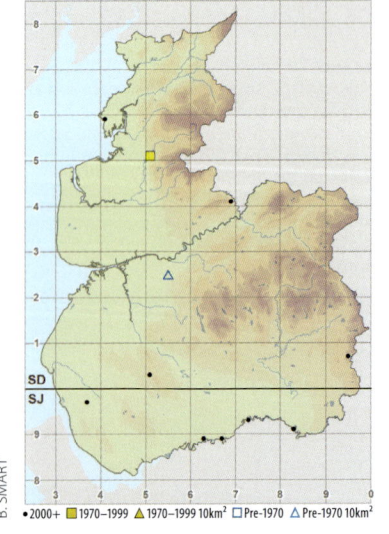

42   *The Moths of Lancashire*

4.010 *Stigmella microtheriella*

■ **LANCS STATUS** Abundant / widespread resident 1918–2022
First noted by W.A. Tyerman at Bowring Park, Knowsley in 1918 (Mansbridge, 1940). Not recorded again until 1995, with over 550 subsequent records throughout the county, possibly as a result of increased interest in leaf-mining species rather than a dramatic change in abundance. Present in numbers in lowland and upland areas, with a maximum count of 100 mines on Hazel at Flixton, October 2014 (K. McCabe).
■ **HABITAT** Likely to be found wherever the foodplant occurs.
■ **FOODPLANTS** Hazel, Hornbeam.
Double brooded. Mines noted from July onwards, with tenanted mines from September to November. The adult moth remains unrecorded, other than when bred.

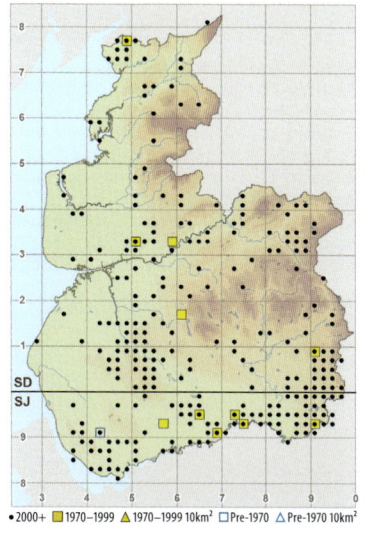

• 2000+   ■ 1970–1999   ▲ 1970–1999 10km²   □ Pre-1970   △ Pre-1970 10km²

4.011 *Stigmella prunetorum*

■ **LANCS STATUS** Scarce / very local resident 2018–2022
A long-vacated mine was found by B. Smart at Gait Barrows on 18 August 2018. A further visit six weeks later resulted in a find of 20 much fresher, although vacated, mines, with a single mine containing an early instar larva. This fed for a further seven days but did not successfully pupate. Eight mines noted at Gait Barrows from 25 July 2022 to 5 August 2022. A moth emerged from the latter batch on 21 May 2023.
■ **HABITAT** Woodland path edges on limestone. Gait Barrows is the only known site.
■ **FOODPLANTS** Blackthorn. There appears a preference for stunted, small-leaved plants. Nationally, considered double brooded. It is unclear if this is also the case locally, or whether there is a single, extended generation. All records relate to leaf-mines.

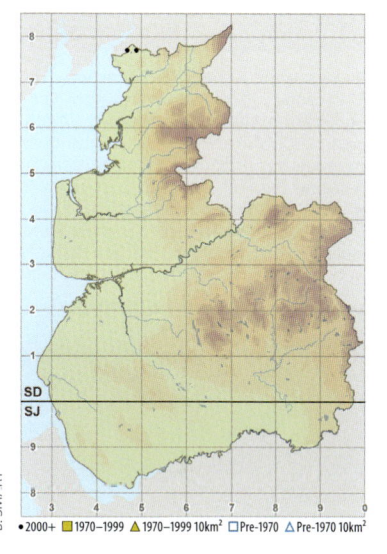

• 2000+   ■ 1970–1999   ▲ 1970–1999 10km²   □ Pre-1970   △ Pre-1970 10km²

4.013 **Apple Pygmy** *Stigmella malella*

■ **LANCS STATUS** Rare / very local resident 1977–2011
This leaf-mining species was first recorded at Dalton, nr. Parbold by E. Pelham-Clinton on 2 November 1977, when five vacated mines were noted on Apple. Further mines were found by C.A. Darbyshire at White Moss tip, Skelmersdale on 30 May 2002 and 1 October 2003. The same recorder noted mines on Crab Apple at Douglas Valley, nr. Appley Bridge on 14 November 2008 and at King's Moss, St Helens on 11 July 2011. Confirmed records have therefore been limited to this area in the south-west of the county. However, VC60 was also circled as a location for this species on the Micro-moth Distribution Maps kept by A.M. Emmet. Unfortunately, we have been unable to locate any further details of this possible record. It is noted that in the distribution map in *Moths and Butterflies of Great Britain and Ireland Vol. 1* (Heath, 1976), VC60 is not marked for this species. The moth appears to be local in the neighbouring counties of Cheshire and Yorkshire.
■ **HABITAT** Woodland edges, hedgerows.
■ **FOODPLANTS** Apple, Crab Apple.
Double brooded. All records are of vacated mines, with the earliest of the year those from White Moss in May 2002.

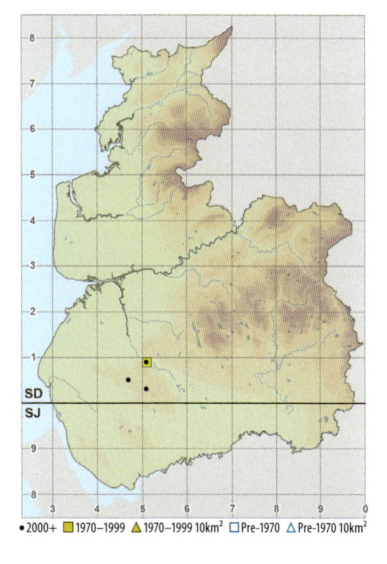

• 2000+   ■ 1970–1999   ▲ 1970–1999 10km²   □ Pre-1970   △ Pre-1970 10km²

4.014 Stigmella catharticella

■ **LANCS STATUS** Scarce / very local   1998–2020

Restricted to sites in the north around Morecambe Bay. First recorded by S. Palmer at Gait Barrows on 14 September 1998. Other sites since noted include Hawes Water, Dalton Crags, Yealand Hall Allotment and Heysham N.R. Never common, with highest counts of just six mines from Eaves Wood 2004 and at Gait Barrows 2008 and 2015.
■ **HABITAT** Limestone woodland edges where the foodplant can be found.
■ **FOODPLANTS** Buckthorn.
Double brooded. All records have been of leaf-mines, with tenanted mines found in July and September to early October. A single adult emerged in June 2019 from a mine collected at Gait Barrows on 30 September 2018.

● 2000+   ■ 1970–1999   ▲ 1970–1999 10km²   ☐ Pre-1970   △ Pre-1970 10km²

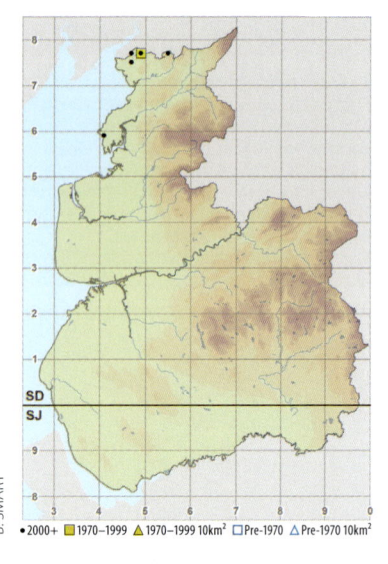

4.015 Rose Leaf Miner *Stigmella anomalella*

■ **LANCS STATUS** Abundant / widespread resident   c.1857–2022

19th century records from Liverpool (Gregson, 1857), Manchester (Stainton, 1859), and Preston. Stainton described the moth as abundant. However, it was not recorded again until 1984 at Risley Moss. Subsequent records show the moth to be found in numbers throughout, with mines easily found in gardens and wilder habitats.
■ **HABITAT** Gardens, hedgerows, scrub, woodland edges. In lowland and upland areas.
■ **FOODPLANTS** Roses, including Dog, Field and Burnet Rose. Garden roses.
Double brooded. All records, where the stage is documented, relate to leaf-mines, including a count of over 100 in a Rochdale back-yard in 2021. Adults reared on a few occasions, confirming that the mines relate to this species rather than the following.

● 2000+   ■ 1970–1999   ▲ 1970–1999 10km²   ☐ Pre-1970   △ Pre-1970 10km²

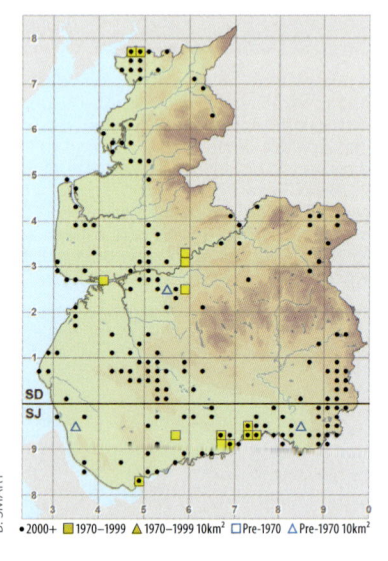

4.017 Stigmella centifoliella

■ **LANCS STATUS** Considered Extinct   1883

The only record is from the Leyland area, where three adults emerged in 1884, bred from leaf-mines on rose, collected in autumn 1883 by J.B. Hodgkinson (Stainton, 1884). Stainton reported these to be 'three brilliant *Nepticula* new to science', noting that 'Mr Hodgkinson's first impression was that he had simply detected a new locality for *N. centifoliella* (*S. centifoliella*), which Mr W.C. Boyd has repeatedly met with at Cheshunt; but the very first glance I had… satisfied me (as it did also Mr Boyd) that it was a very different species from *centifoliella*', and named the species as *N. hodgkinsoni*. However, despite this initial assessment, the moths were later confirmed as black-headed specimens of *S. centifoliella*. Given the lack of any subsequent records, it is likely the moth is extinct in the county. However, due to the difficulty in separating mines of this species from those of *Stigmella anomalella* – rearing through of the tenanted mines is required – it is possible that the species may still be present but overlooked.
■ **HABITAT** Unknown in the county.
■ **FOODPLANTS** Rose.

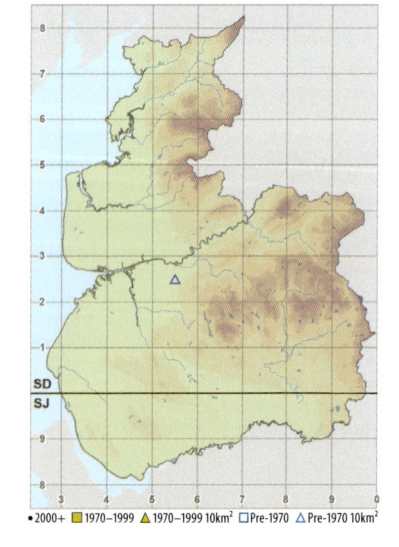

● 2000+   ■ 1970–1999   ▲ 1970–1999 10km²   ☐ Pre-1970   △ Pre-1970 10km²

4.020 *Stigmella paradoxa*

■ **LANCS STATUS** Rare / local resident   2000–2013

This species was first recorded on 19 July 2000 by S.H. Hind and K. McCabe at Warton Crag. There have been only three further records of this species; at Great Moss nr. Rainford in June 2002, at Billinge Hill in September 2013 (both recorded by C.A. Darbyshire), and at Yealand Storrs. The latter is the only one known to have been tenanted. The very small mine, collected by S. Palmer on 5 August 2001, contained a dehydrated larva fitting written descriptions. The lack of any other records, particularly in view of the distinctive blotch mine and larva, seems to confirm this as a rarity in the county. Recorded in all the neighbouring counties with the most recent records from Cheshire in 2020 and in Yorkshire from 2022 (cheshire-chart-maps.co.uk; yorkshiremoths.co.uk, both accessed 1 September 2023).
■ **HABITAT** Woodland.
■ **FOODPLANTS** Hawthorn.
Single brooded. All records are of leaf-mines.

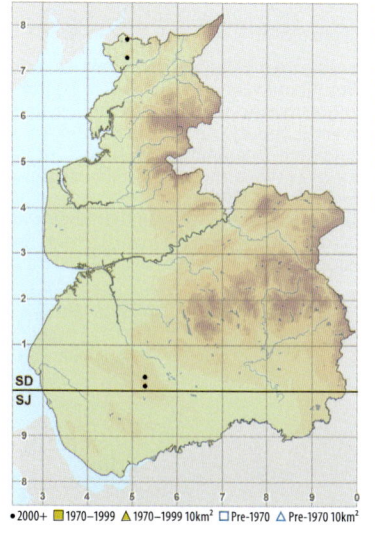

• 2000+   □ 1970–1999   △ 1970–1999 10km²   □ Pre-1970   △ Pre-1970 10km²

4.023 *Stigmella crataegella*

■ **LANCS STATUS** Frequent / widespread resident   1924–2022

Leaf-mines first recorded in Wavertree, Liverpool by W. Mansbridge 1924, with the emerged adult residing in Manchester Museum. Next, recorded at Parbold by E.C. Pelham-Clinton in 1973, but not again until 1995, since when records have increased markedly with the species noted on over 160 occasions. High counts of 20 mines at Skelmersdale, Upholland and Billinge Hill in 2022 by C.A. Darbyshire.
■ **HABITAT** Hedgerows, woodland edges; primarily from lowland sites.
■ **FOODPLANTS** Hawthorn.
Single brooded; tenanted mines in July and early August. Identification aided by presence of green larva, with the similar *S. oxyacanthella* appearing later in the year.

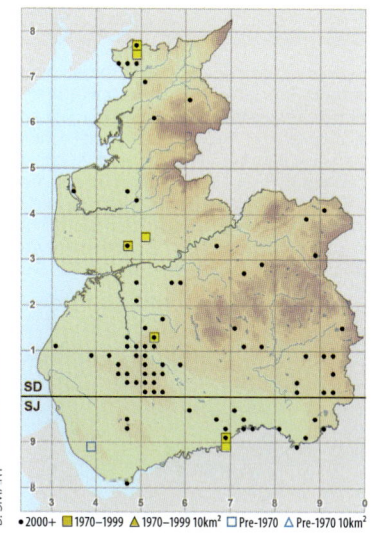

• 2000+   □ 1970–1999   △ 1970–1999 10km²   □ Pre-1970   △ Pre-1970 10km²

4.024 *Stigmella magdalenae*

■ **LANCS STATUS** Occasional / local resident   2010–2022

Mines on Rowan were recorded at Hurstwood, nr. Burnley by S.H. Hind and K. McCabe on 1 September 2010. Less widespread than the following species, with only eleven records. Largely restricted to upland areas, although with a few lowland limestone records, such as at Gait Barrows. Absent from the far south of the county, with the Burnley mines remaining the southernmost records.
■ **HABITAT** Moorland woods and hedgerows. Occasional in limestone woodland.
■ **FOODPLANTS** Rowan.
Single brooded. All records have been of leaf-mines, with the only tenanted mines those located by T. Ward at Lord's Lot Wood 12 August 2022.

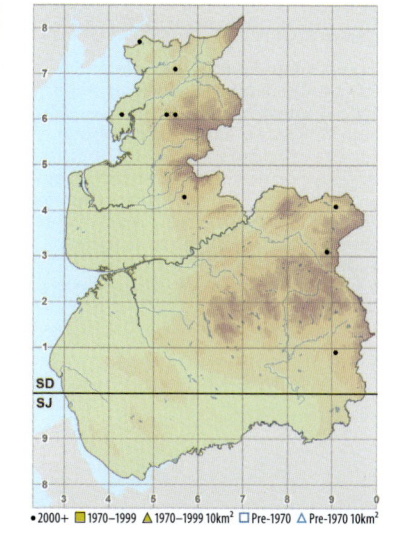

• 2000+   □ 1970–1999   △ 1970–1999 10km²   □ Pre-1970   △ Pre-1970 10km²

The Moths of Lancashire

NEPTICULIDAE

4.025 Stigmella nylandriella

■ **LANCS STATUS** Frequent / widespread resident   1995–2022

First recorded at Lightfoot Green, Preston on 28 June 1995 by S. Palmer. Far more frequently encountered than the previous species, with over 200 records spread over the county, but with a southern bias. Largest counts are of 15 vacated mines at Rigg Lane, Quernmore on 1 August 2018, and of 12 mines at Ash Hill, Flixton in 2011.
■ **HABITAT** Woodland edges and hedgerows in lowland and upland areas.
■ **FOODPLANTS** Rowan.
Single brooded. Most tenanted mines from June and July, occasionally September to October, suggesting extended larval period or possibly a small second generation. All records are of leaf-mines; once reared from mine collected Chorlton 27 June 2009.

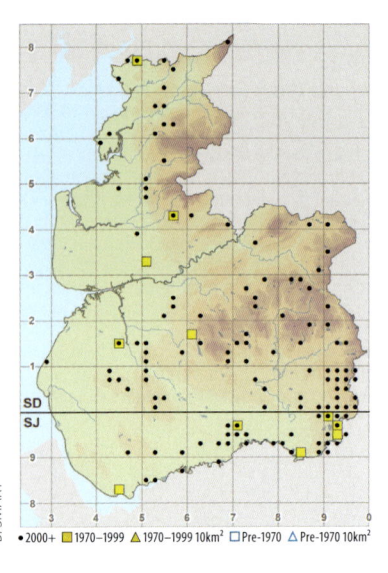

● 2000+   ■ 1970–1999   ▲ 1970–1999 10km²   □ Pre-1970   △ Pre-1970 10km²

4.026 Stigmella oxyacanthella

■ **LANCS STATUS** Frequent / widespread resident   1973–2022

Recorded by J. Threlfall at Preston in 1889 (Ellis, 1890). However, previous confusion with *S. crataegella* means this record is considered uncertain. The first confirmed record was by E. Pelham-Clinton at Parbold in 1973. Next, from Risley Moss 1984, then Gait Barrows 1994, and on over 360 subsequent occasions. Most records from lowland sites, but also noted in upland areas such as Worsaw Hill and Longridge Fell.
■ **HABITAT** Woodland edges, orchards, hedgerows.
■ **FOODPLANTS** Hawthorn, Apple, Pear, Whitebeam. Also, a few records on Rowan which could potentially relate to misidentified late *S. nylandriella* mines.
Single brooded. All Lancashire records are of leaf-mines.

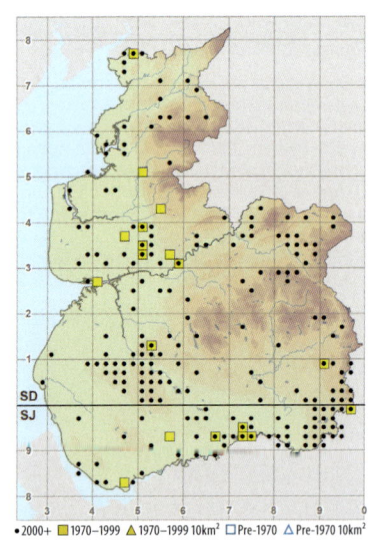

● 2000+   ■ 1970–1999   ▲ 1970–1999 10km²   □ Pre-1970   △ Pre-1970 10km²

4.028 Stigmella minusculella

■ **LANCS STATUS** Considered Extinct   c.1887

This pear-mining species was recorded in the late 19th century by J.B. Hodgkinson and J.H. Threlfall in the Preston area (Ellis, 1890). As there are dots present for both VC59 and VC60 on the distribution maps in *The Moths and Butterflies of Great Britain and Ireland Vol. 1* (Heath, 1976), it appears that records of this species in the two vice-counties were accepted by A.M. Emmet. The text describes national distribution as 'local and rather uncommon in England as far north as Lancashire.' We have been unable to track down details of any further records. The moth is unrecorded in Cheshire and Yorkshire.
■ **HABITAT** Unknown in the county.
■ **FOODPLANTS** Not recorded in Lancashire.

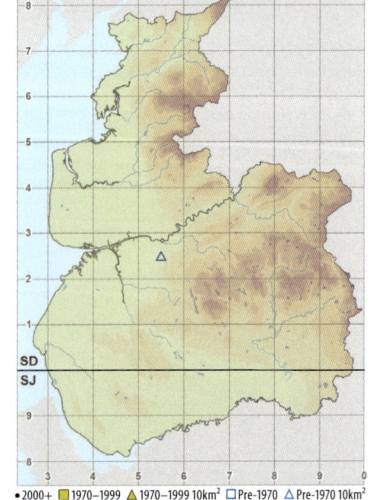

● 2000+   ■ 1970–1999   ▲ 1970–1999 10km²   □ Pre-1970   △ Pre-1970 10km²

The Moths of Lancashire

4.030 *Stigmella hybnerella*

■ **LANCS STATUS** Frequent / widespread resident   c.1889–2022

First recorded in Preston during the late 19th century by J.B. Hodgkinson and J.H. Threlfall (Ellis, 1890), but not again until noted at Parbold in 1973 by E.C. Pelham-Clinton. Since then, there have been over 720 records, mostly from the last 20 years, and widely distributed throughout lowland and upland areas. Recent records from Appley Bridge, Heysham Moss, Briercliffe, Whitworth, Towneley, Worsaw Hill, etc.
■ **HABITAT** Woodland, hedgerows, parks, brownfield sites, etc.
■ **FOODPLANTS** Hawthorn.

Double brooded. All records are of leaf-mines, tenanted from April to June and August to October, other than a female trapped at light on 23 July 2012 in Preston.

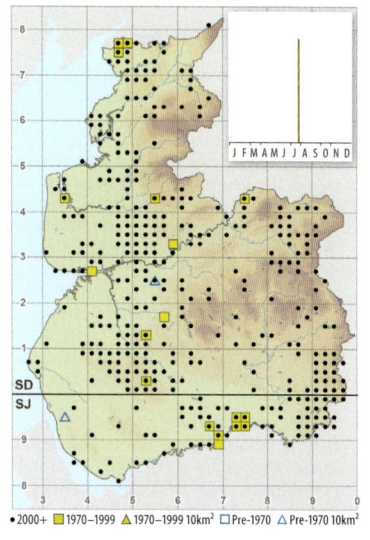

● 2000+ ▫ 1970–1999 △ 1970–1999 10km² ▫ Pre-1970 △ Pre-1970 10km²

4.032 *Stigmella floslactella*

■ **LANCS STATUS** Frequent / widespread resident   c.1889–2022

Noted by Ellis (1890), and later Mansbridge (1940), as common everywhere, without specifying any locations. However, not recorded again until 1995, surely a reflection on the paucity of Nepticulidae recording throughout much of the 20th century, rather than absence of the species. Since 1995, the leaf-mines have been recorded on over 250 occasions throughout most of the county in upland and lowland areas.
■ **HABITAT** Woodland edges, parks, gardens.
■ **FOODPLANTS** Hazel, Hornbeam.

Double brooded. All records are of leaf-mines, tenanted in June and September to October; more numerous in the second generation. Reared from both foodplants.

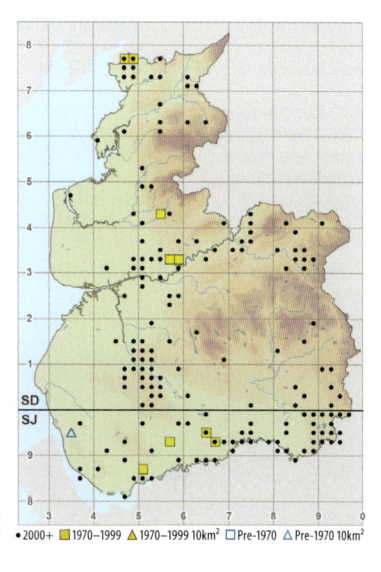

● 2000+ ▫ 1970–1999 △ 1970–1999 10km² ▫ Pre-1970 △ Pre-1970 10km²

4.034 *Stigmella tityrella*

■ **LANCS STATUS** Abundant / widespread resident   1995–2022

In Ellis (1890) and Mansbridge (1940), the range of synonyms and their different usage means it has not been possible to separate historic records of this species and *S. hemargyrella*. Therefore, the first confirmed record is of a vacated mine found by S. Palmer at Lancaster in 1995. There are over 450 subsequent records of the leaf-mines. Noted wherever foodplant occurs, although higher counts in lowland areas.
■ **HABITAT** Hedgerows, woodland, parks.
■ **FOODPLANTS** Beech.

Double brooded. All records are of leaf-mines, with one moth bred from a mine collected two weeks earlier on 28 June 2009 in Chorlton by B. Smart.

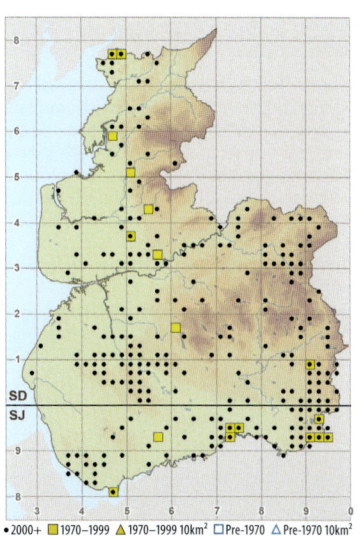

● 2000+ ▫ 1970–1999 △ 1970–1999 10km² ▫ Pre-1970 △ Pre-1970 10km²

The Moths of Lancashire

NEPTICULIDAE

4.035 *Stigmella salicis*

■ **LANCS STATUS** Frequent / widespread resident   1995–2022

Described as common in the Manchester district (Stainton, 1859) and also noted from Chat Moss and Preston (Ellis, 1890). However, historical records are considered likely to include records of *S. obliquella*. The first confirmed record of *S. salicis* was therefore of leaf-mines on Grey Willow at Broughton by S. Palmer in 1995. More recently, the separation of this former single species into clusters, representing six not yet formally named species (Nieukerken *et al.*, 2012; Nieukerken & Hartman, 2019), has also cast doubt on many modern records.

The map used here includes all *S. salicis* records, irrespective of the cluster to which they belong. In order to produce maps for the different clusters found within the county, it would be helpful if all recorders of this group could note the species of tree, the position of the egg on the leaf and the larval appearance.

■ **LANCS STATUS** The frequency, distribution and phenology of the two clusters known to occur in the county are uncertain.

4.035 *Stigmella salicis* (sensu stricto) (cluster 1)

Cluster 1 specimens are associated with the egg being on the underside of a leaf of one of the rough-leaved willows. A mine fitting this description was noted at Pennington Flash by B. Smart on 12 October 2013. Rarely recorded at light, although a moth trapped in Preston on 31 May 2021 was dissected and assigned to this cluster (S. Palmer). Seems likely to be the most abundant of the group within the county. The leaf-mines and resultant moths shown here appear to belong to this cluster, as in each case the egg was noted on the underside.

■ **FOODPLANTS** Grey Willow, Goat Willow.

4.035x *Stigmella "salicis"* (cluster 6)

Leaf-mines of this cluster are found on rough-leaved willows with the egg on the upperside, usually near the midrib. Leaf-mines typical of this cluster were collected from Grey Willow at Heysham Moss on 15 August 2017 by J. Patton, S. & C. Palmer, and on Goat Willow at Cringle Park, Levenshulme by S.H. Hind on 13 October 2022.

■ **FOODPLANTS** Grey Willow, Goat Willow.

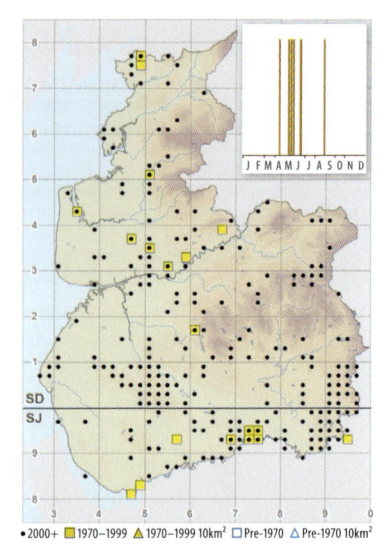

• 2000+   ☐ 1970–1999   ▲ 1970–1999 10km²   ☐ Pre-1970   △ Pre-1970 10km²

4.036 *Stigmella myrtillella*

■ **LANCS STATUS** Occasional / local resident   2001–2022

First recorded by I. Kimber at Littleborough on 12 October 2001. All of the 36 records come from upland areas in the east of the county. The highest count is of nine leaf-mines, four of which were tenanted, at Cragg Wood, Littledale on 4 September 2020. The mines were found on the same shoots as those of *Incurvaria oehlmanniella*, with adults of both species subsequently reared (Smart, 2021).

■ **HABITAT** Moorland and damp woodland.

■ **FOODPLANTS** Bilberry.

Double brooded. All records are of leaf-mines, other than a male to light at Bay Horse, nr. Dolphinholme on 28 May 2014, dissected and recorded by N.A.J. Rogers.

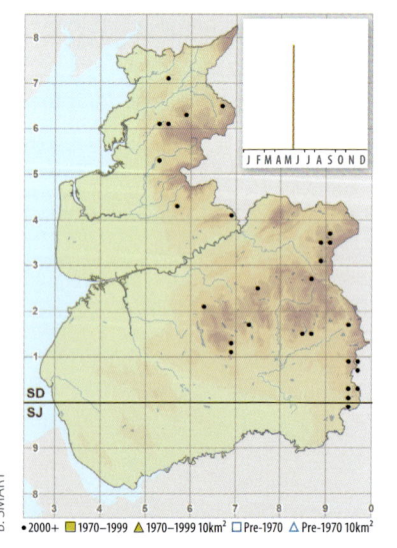

• 2000+   ☐ 1970–1999   ▲ 1970–1999 10km²   ☐ Pre-1970   △ Pre-1970 10km²

48   *The Moths of Lancashire*

4.038 *Stigmella obliquella*

■ **LANCS STATUS** Occasional / local resident   1999–2022

First recorded in 1999 at Burtonwood by S.H. Hind, and at Flixton by K. McCabe. Next at Cuerden, Skelmersdale, Billinge and Littleborough in 2001. Since, recorded on over 100 occasions. Likely to be found where foodplants occur, particularly in south of the county, with VC59 sites providing 90% of all records. Possibly under-recorded in VC60, where sites include Lightfoot Green, Fulwood, Scorton and Leighton Moss.
■ **HABITAT** Woodland edges, parks, rivers, ponds and other wetland habitats.
■ **FOODPLANTS** Osier, Crack-willow, White Willow and Babylon (Weeping) Willow. Vacated mines noted in July, with tenanted mines September to November, confirming species as double brooded. All Lancashire records are of leaf-mines.

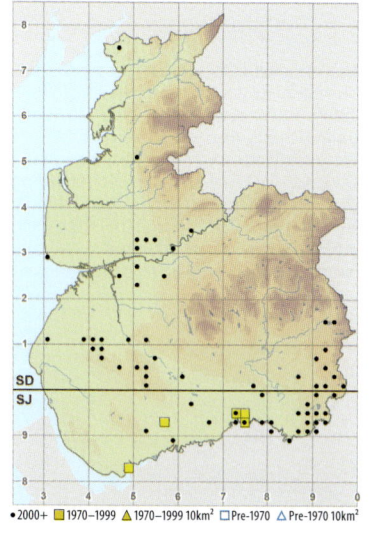

4.039 *Stigmella trimaculella*

■ **LANCS STATUS** Frequent / widespread resident   c.1889–2022

First recorded in Lancashire at Leyland by J.H. Threlfall (Ellis, 1890), but not again until 16 August 1997, when a moth was trapped at light at Lytham St Annes by D. & J. Steeden. Noted on over 150 occasions since, with records far more numerous in the southern half of the county and only a handful north of Preston. Mainly in lowland areas, although also present in some of the more elevated areas east of Oldham.
■ **HABITAT** Woodland edges, country parks, hedgerows.
■ **FOODPLANTS** Black-poplar, Canadian Poplar, Lombardy Poplar, Balsam Poplar. Double brooded, with tenanted mines from June to July, and again from September to October. The distinctive adults have been trapped at light from June to August.

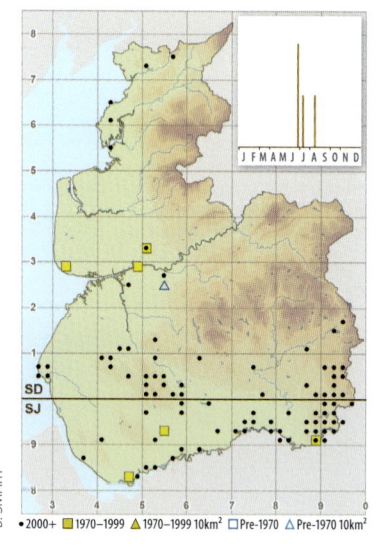

4.040 *Stigmella assimilella*

■ **LANCS STATUS** Occasional / local resident   1999–2022

A relatively recent arrival in the county; first recorded on White Poplar by I.F. Smith on 11 September 1999 at Freshfield. The species has since been noted on 50 occasions. All have been of leaf-mines in the south of the county; at Halewood, Flixton, Chorlton, St Helens, Oldham Edge, etc., with the most northerly Lancashire records being from Mere Sands Wood in 2019. No obvious evidence of range expansion subsequently, presumably limited by availability of the foodplants.
■ **HABITAT** Woodland edges, hedgerows, scrub, dunes, etc.
■ **FOODPLANTS** Aspen, White Poplar. Moth bred from Aspen mine at Chorlton 2018. Single brooded. Tenanted mines have been recorded from August to November.

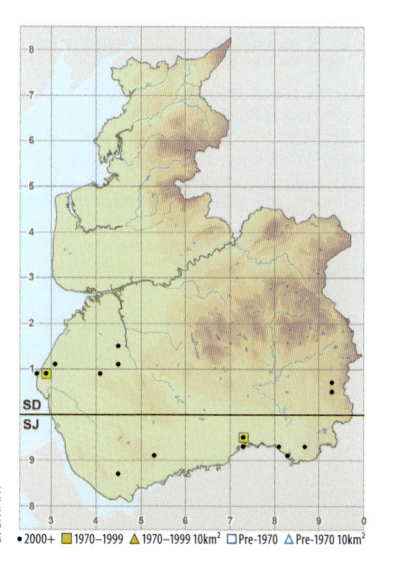

4.041 *Stigmella sorbi*

■ **LANCS STATUS** Occasional / local resident   c.1889–2022

The earliest Lancs records were from Preston and Longridge by J.H. Threlfall (Ellis, 1890), and at Silverdale in 1905. Modern records, post-1995, show preference for upland areas. Also, recorded on some of the mosses, but not for 20 years. This, and the overall reduced counts from the last 15 years, compared to the decade prior, is of concern. However, increased recording effort over the last couple of years has shown mines still occurring at Oldham, Ashton, Rochdale and Mossley.
■ **HABITAT** Moorland, woodland edges, scrub, mosses.
■ **FOODPLANTS** Rowan. Tenanted mines have been recorded from June to early August. Single brooded, with all records being of leaf-mines.

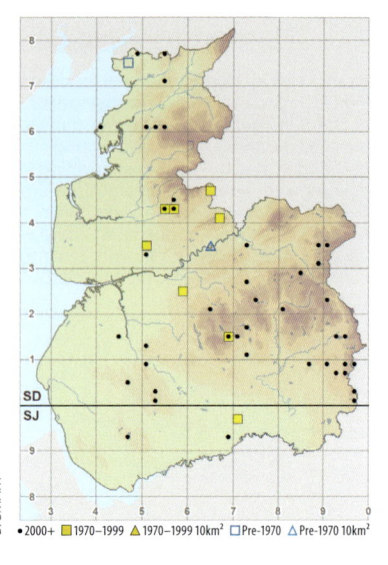

• 2000+   ☐ 1970–1999   △ 1970–1999 10km²   ☐ Pre-1970   △ Pre-1970 10km²

4.042 *Stigmella plagicolella*

■ **LANCS STATUS** Frequent / widespread resident   c.1859–2022

First noted in the Manchester area and considered abundant (Stainton, 1859). Remains common in southern parts. Records are fewer in the north, with the mines largely absent from the Fylde and much of the uplands. Found on the same stunted blackthorns at Gait Barrows where *S. prunetorum* noted. A high count of 20 leaf-mines, surprisingly already vacated, was noted at Chorlton Ees 20 June 2017.
■ **HABITAT** Woodland edges, scrub, hedgerows, amenity planted sites.
■ **FOODPLANTS** Blackthorn, Plum, Cherry. Tenanted mines recorded June to October. Double brooded. All records are of leaf-mines, primarily on Blackthorn.

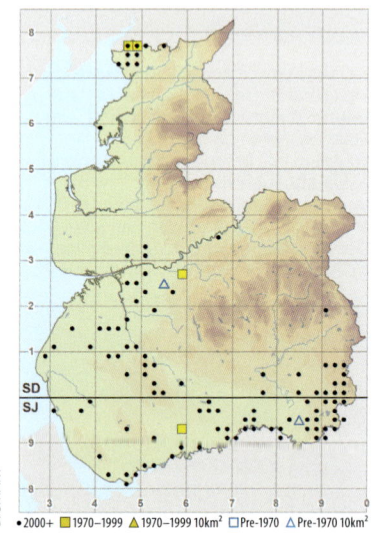

• 2000+   ☐ 1970–1999   △ 1970–1999 10km²   ☐ Pre-1970   △ Pre-1970 10km²

4.043 *Stigmella lemniscella*

■ **LANCS STATUS** Frequent / widespread resident   c.1889–2022

Recorded at Preston in the late 19th century by J.B. Hodgkinson and J.H. Threlfall (Ellis, 1890). Unrecorded again until 1995, but with 230 well-distributed records since. This is more likely to be a result of increasing leaf-mine recording than increased abundance of the moth. It is possible that some increase in elm numbers following the ravages of Dutch elm disease has played a part too, as mines are often found on the 'suckers' that are relatively unaffected by the disease.
■ **HABITAT** Hedgerows, woodland edges.
■ **FOODPLANTS** Wych Elm, Small-leaved Elm.
Double brooded. All records are of leaf-mines, tenanted from July to October.

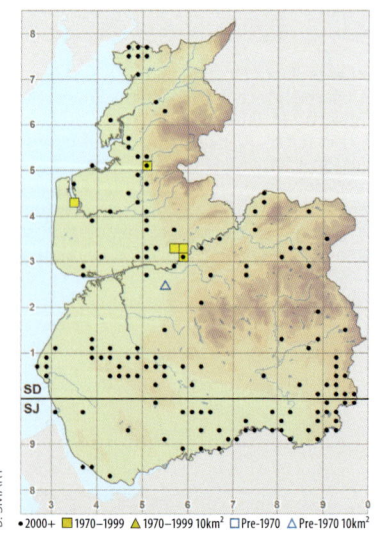

• 2000+   ☐ 1970–1999   △ 1970–1999 10km²   ☐ Pre-1970   △ Pre-1970 10km²

4.044 *Stigmella continuella*

■ **LANCS STATUS** Occasional / local resident 2003–2022
This birch miner was first found by S. McWilliam at Culcheth in 2003. Thereafter, recorded at around ten lowland sites, including Rixton, Cadishead Moss, Harpurhey Ponds, Prescot Reservoirs and Ash Hill, Flixton. Distribution appears to have a southern bias within the county, with Gait Barrows the only site known in VC60.
■ **HABITAT** Mosses, woodland edges, scrub.
■ **FOODPLANTS** Birch, with Silver Birch the only confirmed species in the county. Double brooded. All records are of leaf-mines, with tenanted mines found from July to October. Bred from mines collected on 10 September 2017 at Little Woolden Moss, with emergence on 27 June 2018 (B. Smart).

● 2000+ ■ 1970–1999 ▲ 1970–1999 10km² □ Pre-1970 △ Pre-1970 10km²

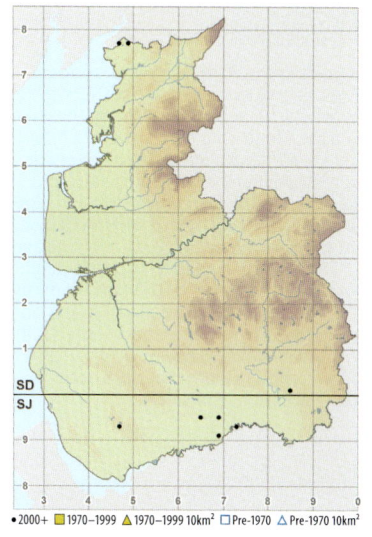

4.045 *Stigmella aurella*

■ **LANCS STATUS** Abundant / widespread resident c.1857–2022
Recorded at Sales Wood, Prescot in 1919 (Mansbridge, 1940). Gregson (1857) earlier noted the species as 'plentiful where brambles grow'. With over 2000 records, there remains an expectation of finding the mines wherever Bramble occurs, although harder to find in the uplands. The high counts of 100 mines or so, at Roby Mill in 2002 by C.A. Darbyshire, at Urmston and at Flixton in 2003 by K. McCabe, have not been replicated since, suggesting a possible reduction in abundance.
■ **HABITAT** Wherever Bramble occurs.
■ **FOODPLANTS** Bramble, Raspberry, Agrimony, Wood Avens, Water Avens.
Likely at least triple brooded, with tenanted mines in all months except May and July.

● 2000+ ■ 1970–1999 ▲ 1970–1999 10km² □ Pre-1970 △ Pre-1970 10km²

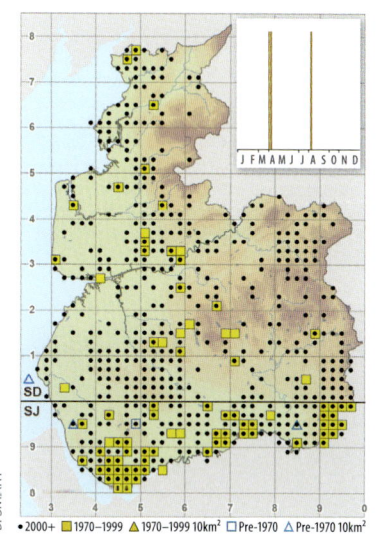

4.049 *Stigmella aeneofasciella*

■ **LANCS STATUS** Scarce / very local resident c.1889–2022
First noted at Lytham St Annes by J.H. Threlfall (Ellis, 1890). Next, reported from Liverpool by W. Mansbridge (1940). Not noted again until 1995 at Gait Barrows by A.M. Emmet, and on six further occasions, all from the north of the county. These include a record from Red Scar Wood, Preston and four from Warton Crag.
■ **HABITAT** Scrub, meadows, path edges, limestone grassland.
■ **FOODPLANTS** Agrimony, Creeping Cinquefoil.
Double brooded, with tenanted mines in June and August to October. All records are of leaf-mines, mainly on Agrimony; the only known exception being the Preston record which was on Creeping Cinquefoil, noted on 12 August 1996 (S. Palmer).

● 2000+ ■ 1970–1999 ▲ 1970–1999 10km² □ Pre-1970 △ Pre-1970 10km²

NEPTICULIDAE

The Moths of Lancashire **51**

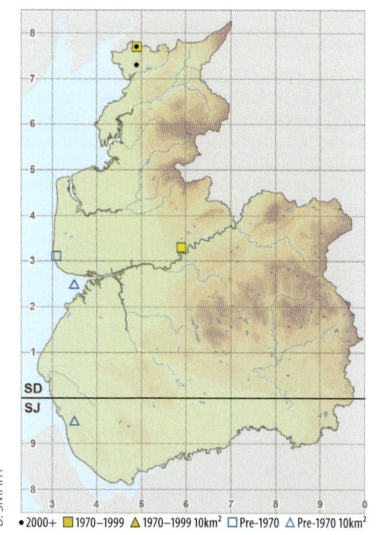

4.051 Stigmella poterii

■ **LANCS STATUS** Rare / very local resident **pre-1976**

The only evidence we have of the moth's historical presence in the county is the hand-written record for VC60 on the Micro-moth Distribution Maps kept by A.M. Emmet, identifying the Lancashire record as f. *serella*, and the circling of VC60 in the map in *Moths and Butterflies of Great Britain and Ireland Vol. 1* (Heath, 1976). The text accompanying the latter map states that the moth is 'widespread but local in the south-east of England to Oxfordshire and in Lancashire and Westmorland.' The icon (a star) is centrally placed within VC60 on the map, although details of the specific location are unknown, as are the date and recorder.

The species appears very rare in neighbouring counties, with the last Yorkshire record coming from 1982 (yorkshiremoths.co.uk, accessed 1 September 2023), and a single Cheshire record of leaf-mines on Tormentil at Lyme Park in 2014 (cheshire-chart-maps.co.uk, accessed 1 September 2023). Elsewhere, tenanted mines were located in east Cumbria in October 2018 on the leaves of Great Burnet (S. Palmer, pers. comm.), a plant that does occur locally within the Lune, Ribble and Hodder valleys (Greenwood, 2012).

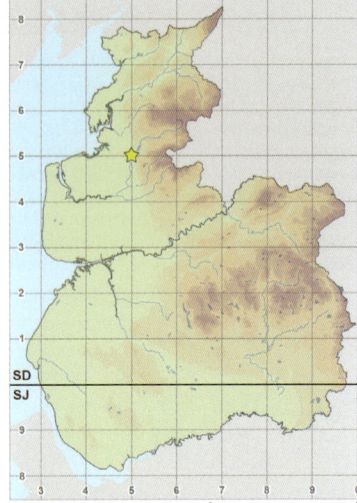

● 2000+ ■ 1970–1999 ▲ 1970–1999 10km² □ Pre-1970 △ Pre-1970 10km²

4.053 Stigmella incognitella

■ **LANCS STATUS** Scarce / very local resident **c.1889–2022**

First recorded in the county on apple by J.B. Hodgkinson at Preston (Ellis, 1890), but not at all during the 20th century. Mines collected in the early 2000s around Skelmersdale and Appley Bridge were initially thought to be this species, although recent re-examination has proved inconclusive. However, on 10 October 2022, five small, vacated mines were found on a single apple tree near Appley Bridge Station by B. Smart and confirmed as this species by E. van Nieurkerken.

■ **HABITAT** Orchards and woodland edges.
■ **FOODPLANTS** Apple.
Double brooded.

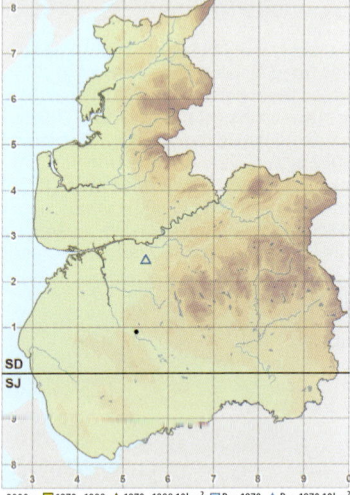

● 2000+ ■ 1970–1999 ▲ 1970–1999 10km² □ Pre-1970 △ Pre-1970 10km²

4.054 Stigmella perpygmaeella

■ **LANCS STATUS** Occasional / local resident **1973–2022**

Victorian records exist from Liverpool and Preston. However, it seems the structural differences separating hawthorn mines were not fully understood at the time, and so the first confirmed record is of mines at Parbold by E.C. Pelham-Clinton on 25 October 1973. Of the 75 21st century records, a stronghold appears to be the east of Greater Manchester, where S.H. Hind recorded mines at 16 locations in 2021–2022.

■ **HABITAT** Woodland edges and hedgerows in lowland and upland areas.
■ **FOODPLANTS** Hawthorn.

Nationally, considered double brooded. Lancs records, with tenanted mines from mid-August to October, could also be consistent with an extended single generation.

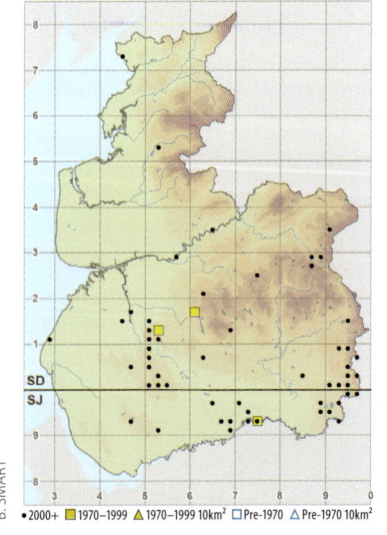

● 2000+ ■ 1970–1999 ▲ 1970–1999 10km² □ Pre-1970 △ Pre-1970 10km²

4.055 *Stigmella hemargyrella*

■ **LANCS STATUS** Occasional / local resident 1919–2022

As noted with *S. tityrella*, early records of the two could not be separated. However, there is a 1919 example of this species from Burnley in Manchester Museum, finder unknown. Next recorded, as a mine, at Claughton by S. Palmer in 1995. Noted on 324 occasions since. High counts of 20 mines from Billinge and Dalton by C.A. Darbyshire in 2002. Maximum counts of just two mines since 2010 suggest declining abundance.

■ **HABITAT** Beech woodland and hedges; in lowland and upland areas.

■ **FOODPLANTS** Beech.

Double brooded. Most records are of leaf-mines, the only exception being the 1919 specimen.

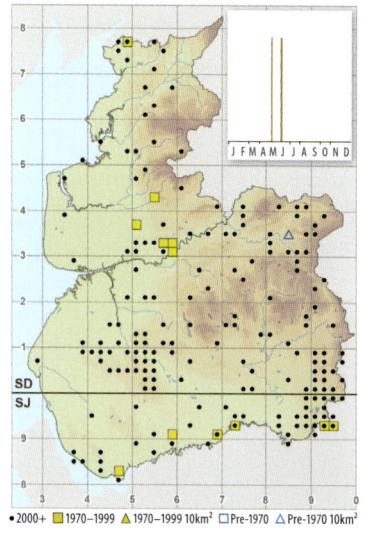

4.056 *Stigmella speciosa*

■ **LANCS STATUS** Occasional / local resident 1999–2022

Range has expanded nationally. First noted locally at Flixton by K. McCabe on 28 September 1999. Over 250 subsequent records, with a high count of 23 at Ainsdale in September 2022. Scattered throughout with a southern bias in distribution, although spreading north. Mainly lowland, but with a few records from the Pennine valleys.

■ **HABITAT** Woodland edges, hedgerows, parks, riverbanks.

■ **FOODPLANTS** Sycamore, Montpelier Maple.

Double brooded. All records are of leaf-mines. Tenanted mines sometimes elusive but have been recorded in July and from September to October. An interesting 2022 find was of a mine on Montpelier Maple at Upholland, recorded by C.A. Darbyshire.

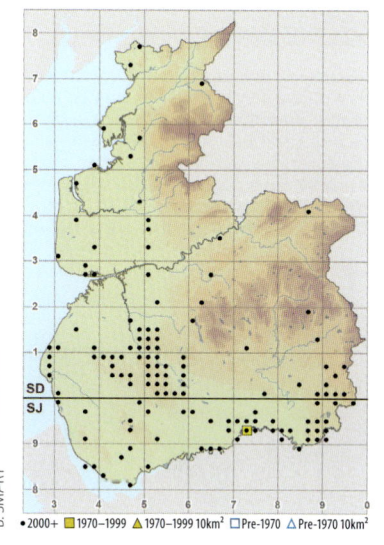

4.059 *Stigmella svenssoni*

■ **LANCS STATUS** Rare / very local resident 2001

A male emerged in May 2002 from leaf-mines on oak, collected by I. Kimber at Littleborough during September 2001. Dissection confirmed identity. A slide of the genitalia was made and was re-examined as part of the preparation for this book. Previous records of this species related to mines with no adult bred through and are considered unreliable. This is therefore the sole accepted record of this moth.

■ **HABITAT** Oak woodland.

■ **FOODPLANTS** Oak. Leaf-mines of the oak-feeding *Stigmella* species are often impossible to identify. Larval appearance and egg position can help, but in many cases, the only way to safely identify the species is to breed through and subsequently dissect.

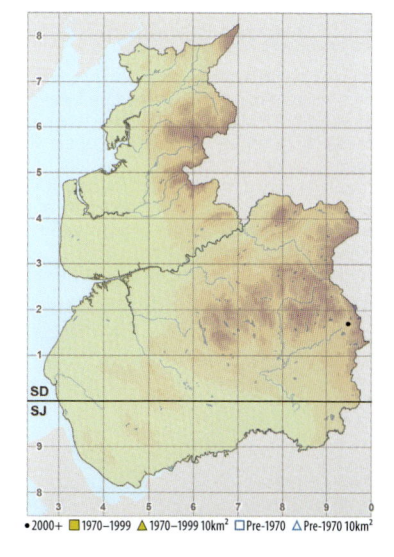

4.060 *Stigmella ruficapitella*

■ **LANCS STATUS** Frequent / widespread resident   1995–2022

Recorded in Preston by J.H. Threlfall (Ellis, 1890). However, due to taxonomic changes it is uncertain which species the historic records of oak-feeding *Stigmella* refer to, and so they have been omitted from the database. Our first confirmed record is from Gait Barrows in 1995 by A.M. Emmet. Identification difficulties likely mean that the 55 subsequent records significantly underestimate actual frequency.
■ **HABITAT** Oak woodland, parks, scrub; most habitats where oak is present.
■ **FOODPLANTS** Pedunculate Oak; likely other deciduous oak species too.
Double brooded. Almost all records are of the leaf-mine, with the egg noted on the upperside of the leaf. Noted at light on one occasion; at Flixton 30 July 2000.

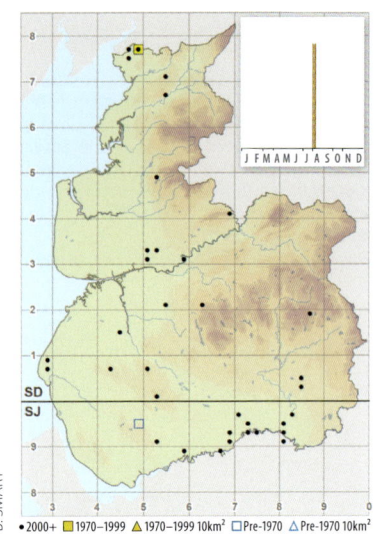

● 2000+   ☐ 1970–1999   △ 1970–1999 10km²   ☐ Pre-1970   △ Pre-1970 10km²

4.061 *Stigmella atricapitella*

■ **LANCS STATUS** Frequent / widespread resident   1995–2022

As with the previous species, historic records cannot be reliably assigned to this species. Recorded at Gait Barrows in 1995 by A.M. Emmet. There are relatively few records since (24) for what is probably a common leaf-miner, presumably because of the difficulty in verifying vacated mines.
■ **HABITAT** Oak woodland, parks, scrub; most habitats where oak is present.
■ **FOODPLANTS** Pedunculate Oak; likely other deciduous oak species too.
Double brooded. All records are of leaf-mines. Tenanted mines contain a yellow larva with blackish prothoracic sclerites. This, combined with the egg on the underside, confirms identity. Adults bred on a few occasions; once dissected (in 2007).

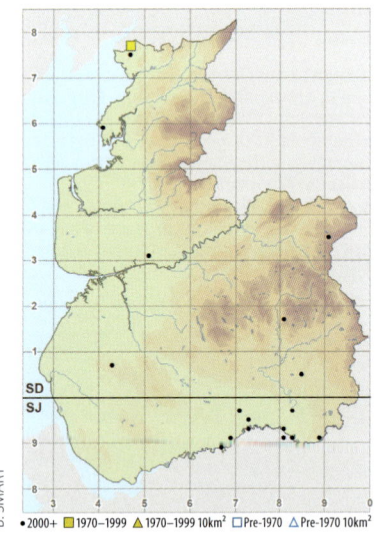

● 2000+   ☐ 1970–1999   △ 1970–1999 10km²   ☐ Pre-1970   △ Pre-1970 10km²

4.062 *Stigmella samiatella*

■ **LANCS STATUS** Scarce / very local resident   2021–2022

Five vacated mines were noted on Sweet Chestnut at Chorlton Water Park by B. Smart on 12 November 2021. Checks of the same trees in mid-September 2022 revealed 24 mines, three of which were still tenanted. Other oak-feeding *Stigmella* may occasionally use Sweet Chestnut, but *S. samiatella* is the only one to do so regularly and in the numbers found at this site. Four vacated mines were also found on Sweet Chestnut by C.A. Darbyshire at Carr Mill, St Helens on 18 October 2022.
■ **HABITAT** Woodland edges of ponds and lakes.
■ **FOODPLANTS** Sweet Chestnut.
Nationally, double brooded. All records are of leaf-mines.

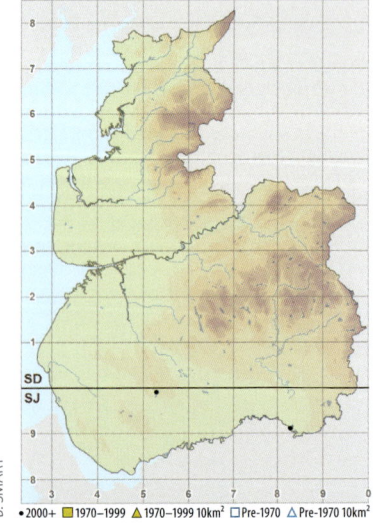

● 2000+   ☐ 1970–1999   △ 1970–1999 10km²   ☐ Pre-1970   △ Pre-1970 10km²

4.063 *Stigmella roborella*

■ **LANCS STATUS** Occasional / local resident   1973–2016
Leaf-mine recorded 27 October 1973 by E.C. Pelham-Clinton at Parbold, with moth emerging 1974. Records since are few, but likely to reflect difficulty in confirming identity rather than true scarcity of the moth. Moths bred on a few occasions since, once from an oak sapling. All records are from the southern half of the county.
■ **HABITAT** Oak woodland.
■ **FOODPLANTS** Oak.
Double brooded. All records are of leaf-mines. Egg on underside of leaf. Larvae lack the dark prothoracic sclerites of *S. atricapitella* but are difficult to identify. Continues to mine until late autumn, with tenanted mines recorded as late as 10 November.

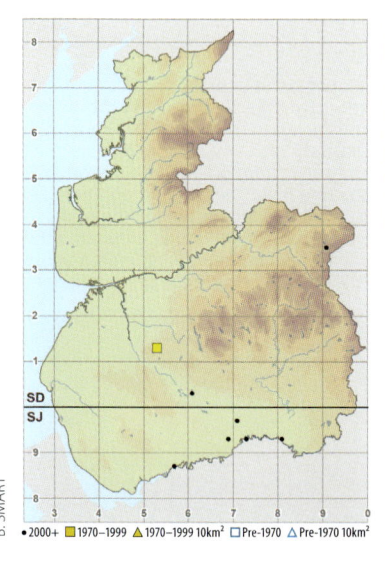

4.065 *Trifurcula cryptella*

■ **LANCS STATUS** Scarce / very local resident   1922–2021
A moth was netted at Formby on 20 May 1922 by W. Mansbridge. Doubts remain over identification of mines alone, partly due to potential foodplant misidentification, and European records suggesting that *T. cryptella* and *T. eurema* are known to mine 'each other's' foodplant as well as their 'own'. It was with some relief then that three adults emerged in June 2022 from Greater Bird's-foot-trefoil mines found at Rixton, 16 July 2021 (B. Smart). Dissection and DNA analysis confirmed these as *T. cryptella*.
■ **HABITAT** Damp meadows and woodland edges where foodplant present.
■ **FOODPLANTS** Greater Bird's-foot-trefoil.
Single brooded. Mines also noted at Chorlton and Dobcroft N.R., although not bred.

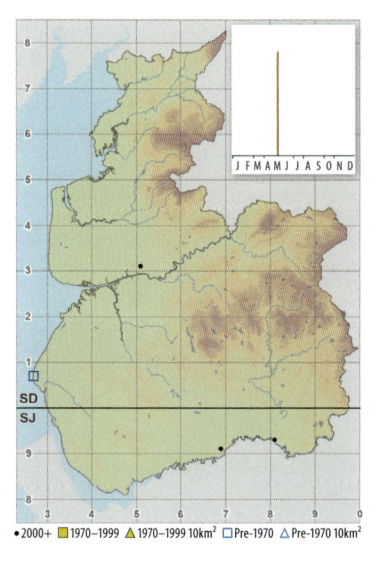

4.066 *Trifurcula eurema*

■ **LANCS STATUS** Scarce / very local resident   2006–2021
Mines noted on Common Bird's-foot-trefoil at Dalton, nr. Parbold by C.A. Darbyshire in 2006. Two tenanted mines found on the same plant at Middleton N.R. on 14 June 2018 by J.R. Langmaid, S. Palmer and B. Smart. One cocoon formed within the leaf, the other on the side of the pot. Tenanted mines found again at the latter site in 2021. Unfortunately, rearing unsuccessful in all cases. The reservations mentioned in the previous species account, regarding recording from mines alone, apply here too.
■ **HABITAT** Short sward open grassland with foodplant present.
■ **FOODPLANTS** Common Bird's-foot-trefoil.
Tenanted mines recorded from mid-June to mid-July.

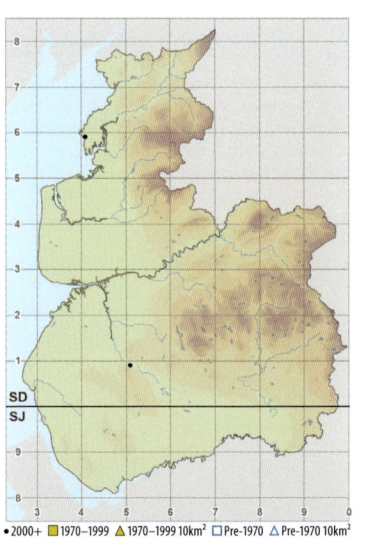

The Moths of Lancashire

NEPTICULIDAE

4.068 *Trifurcula immundella*

■ **LANCS STATUS** Occasional / local resident c.1889–2020

Recorded from Dutton, nr. Ribchester by J.B. Hodgkinson, and at Lytham St Annes and Rivington by J.H. Threlfall (Ellis, 1890). Also, recorded at Formby 1920 and Didsbury 1953. 21st century records include mines on Broom from the Formby area, and adults from Billinge, Littleborough, Southport and Briercliffe.

■ **HABITAT** Scrub, heathland. Potentially wherever foodplant occurs.
■ **FOODPLANTS** Broom, mining the twigs.

Single brooded. Adults recorded from July to early September at light and during day. Mined twigs can be separated from those of *Leucoptera spartifoliella* and *Phyllonorycter scopariella* by presence of a shiny, black egg at the start of the mine.

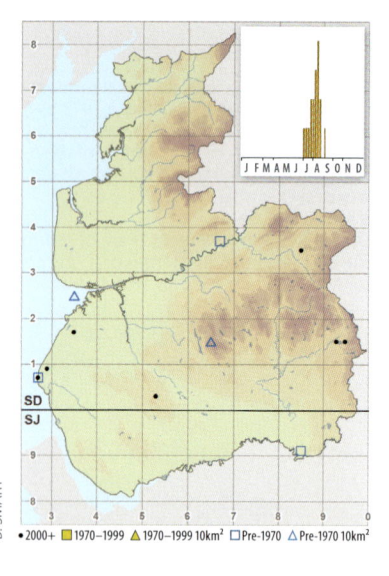

4.069 *Trifurcula beirnei*

■ **LANCS STATUS** Considered Extinct 1879

There is a single record of this moth, under *T. pallidella*, from Dutton, nr. Ribchester, swept by J.B Hodgkinson from a patch of 'rough open ground in a young plantation' in the fourth week of August 1879 (Hodgkinson, 1880a). Dyer's Greenweed, the known foodplant nationally, was formerly known from this location (as was *Agonopterix atomella*, *Mirificarma lentiginosella* and *Grapholita lathyrana*, other species using this host). The plant has since declined considerably in the county (E. Greenwood, pers. comm.). Because of this, and the lack of subsequent records of the moth, it appears probable that this species is now extinct in Lancashire. The moth is unrecorded in Yorkshire and Cheshire.

■ **HABITAT** Rough open ground.
■ **FOODPLANTS** Not recorded in Lancashire.

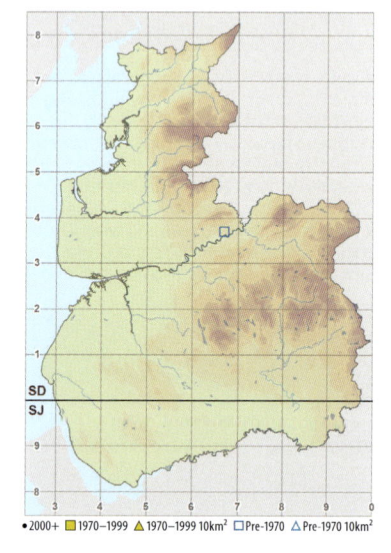

4.071 *Bohemannia pulverosella*

■ **LANCS STATUS** Occasional / local resident c.1889–2022

First recorded by J.B. Hodgkinson at Dutton, nr. Ribchester (Ellis, 1890). Not recorded again until leaf-mines found at Flixton in 1997. There are over 70 21st century records, showing the moth to be fairly well-distributed over the south of the county, but rarer north of the Ribble with only nine records. Mainly in lowland areas, but also found on the edges of the Pennines at Shaw, Oldham Edge, Littleborough, etc.

■ **HABITAT** Woodland edges, orchards, hedgerows.
■ **FOODPLANTS** Apple, Crab Apple.

Single brooded. All records are of leaf-mines, tenanted from June to early July. The highest count is of ten mines, including four tenanted, from Chorlton 28 June 2020.

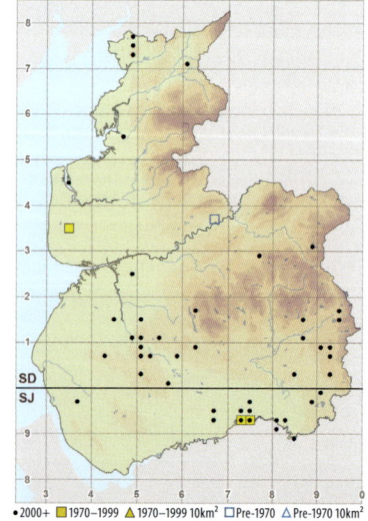

56 *The Moths of Lancashire*

4.072 *Bohemannia quadrimaculella*

■ **LANCS STATUS** Occasional / local resident　　　　c.1889–2015

First reported by J.H. Threlfall at Preston (Ellis, 1890). Moth not recorded again until noted on Grey Alder at Flixton by K. McCabe in 2000. Subsequently, recorded at Lathom, Billinge, Freshfield, Chorlton and Preston. The Preston records represent the northern edge of the known range of the moth within the county.
■ **HABITAT** Alder woodland.
Single brooded. Almost certainly under-recorded; all records are of adults. An interesting species, whose early stages are still not understood, although there seems to be a link with Alder. Adults have been beaten or netted from Alder and Grey Alder and are also attracted to light. Recorded in July and August.

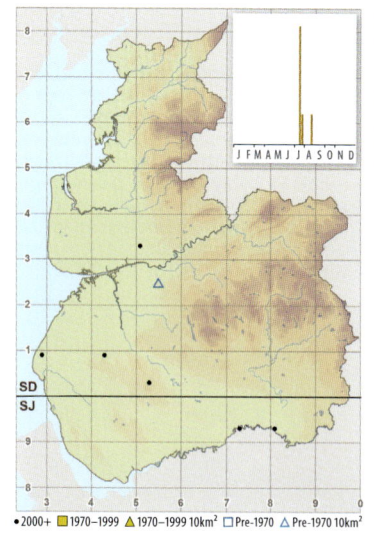

4.074 *Etainia sericopeza*

■ **LANCS STATUS** Occasional / local resident　　　　2006–2022

The first of 29 records was of a moth to light, recorded by D. Owen at St Helens on 21 September 2006. The first mines were found the following year in June at Flixton by K. McCabe, with the adult emerging three weeks later. Also recorded at Fazakerley, Chorlton, Gorton, Denton, Oldham and Heyrod, nr. Stalybridge. There have been nine records from two garden traps in Preston, the only sites known in VC60.
■ **HABITAT** Woodland edges, hedgerows, amenity planted areas.
■ **FOODPLANTS** Norway Maple, mining the samaras rather than the leaves.
Double brooded. Mines from late May onwards. Cocoon found on samara 26 August 2018; moth emerged six days later. Moths at light from late June to early October.

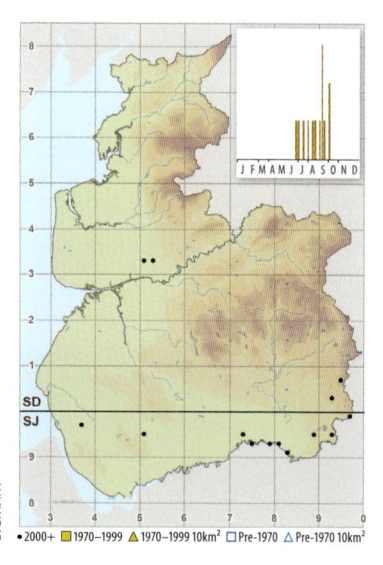

4.075 *Etainia louisella*

■ **LANCS STATUS** Occasional / local resident　　　　2015–2022

A recent arrival in the county, with finds so far limited to sites around south Manchester. Mined samaras noted in Mersey Valley from Flixton and Chorlton, through to Stockport and the Tame Valley. Mines first noted on 17 July 2015 at Chorlton Ees, Manchester by B. Smart, with emergence on 7 August 2015. Recorded on 13 further occasions, with highest count of 12 in 2018, also at Chorlton Ees.
■ **HABITAT** Woodland edges, hedgerows, amenity planted areas.
■ **FOODPLANTS** Field Maple, mining the samaras rather than the leaves.
Double brooded. All records are of the early stages with mines noted from July to October. The cocoon was recorded on a mined samara on 29 July 2022.

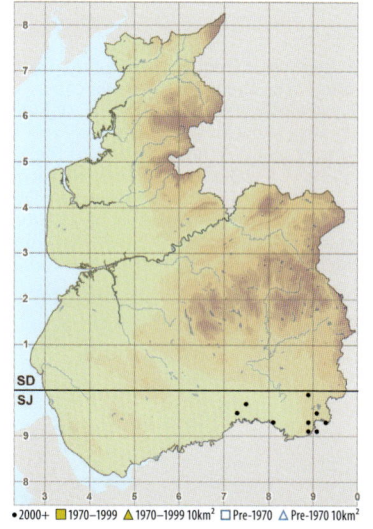

The Moths of Lancashire

4.076 *Etainia decentella*

■ **LANCS STATUS** Frequent / widespread resident   1953–2022
An unusual nepticulid in that all 83 records are of adults. First noted by H.N. Michaelis at Heaton Mersey in 1953, and at Didsbury in 1957, with a 40-year gap before next recorded. The 21st century records suggest this is a predominantly lowland moth, with 80% of records coming from VC59, and most from garden traps.
■ **HABITAT** Probably wherever Sycamore grows.
■ **FOODPLANTS** Unrecorded within the county.
Double brooded. Numbers highest in first generation. Recorded more often at light than other nepticulids, although the moth's distinctiveness means the record will likely be submitted, not always the case for others where dissection usually required.

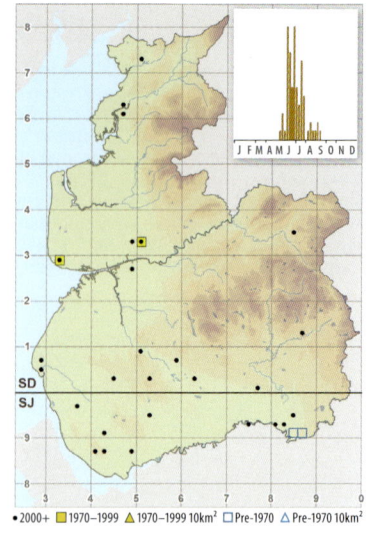

4.077 *Fomoria weaveri*

■ **LANCS STATUS** Scarce / very local resident   1856–2022
This upland species was first recorded on 21 April 1856 by C.S. Gregson near Chorley (Gregson, 1856a). Next found at Green Thorn Fell, Ribble Valley on 28 June 1879 (Hodgkinson, 1879c). Not recorded again until three mines noted at Clougha Pike on 5 May 2002 (S. Palmer). Over 20 mines recorded on 28 March 2022 by S. Palmer & B. Smart at Windy Clough, Littledale, with one adult emerging on 4 July 2022.
■ **HABITAT** Moorland, where the foodplant grows.
■ **FOODPLANTS** Cowberry. Mines are likely to occur wherever the foodplant is noted. Single brooded. Cowberry is found on the highest parts of the damp moorlands and is more frequent on the western slopes of the Bowland Fells than elsewhere.

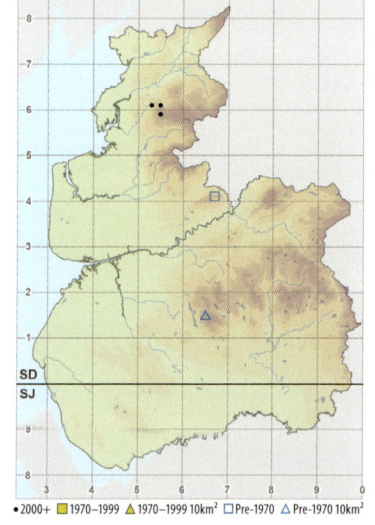

4.078 *Fomoria septembrella*

■ **LANCS STATUS** Frequent / widespread resident   c.1859–2022
Noted to be abundant in the Manchester district (Stainton, 1859). Also recorded at Silverdale by J.H. Threlfall (Ellis, 1890). Subsequent records from 1997 onwards, all of mines, show the moth to be fairly well-distributed, although unrecorded from Sefton Coast and thinly spread over the north of the county. Mostly lowland records, but also a few Oldham and Rochdale records from the western edge of the Pennines.
■ **HABITAT** Wherever foodplants present, including gardens and amenity planted areas, such as supermarket car parks, sport centres and cemeteries.
■ **FOODPLANTS** Perforate St John's-wort, Tutsan, Rose of Sharon, *Hypericum* shrubs. Double brooded. Tenanted mines recorded in March, April and August to December.

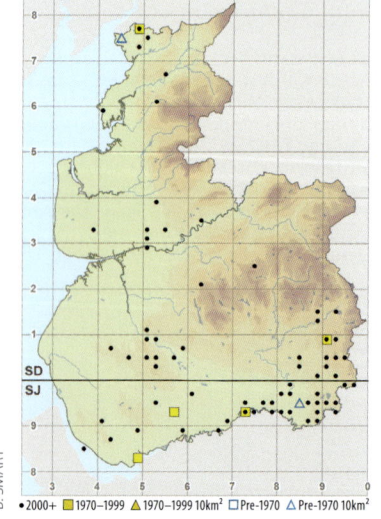

4.082 *Ectoedemia intimella*

■ **LANCS STATUS** Occasional / local resident 2001–2022
A tenanted mine of this previously unrecorded species was noted at Risley Moss on 17 November 2001 by K. McCabe, S.H. Hind *et al*. There have been 86 subsequent records, all of mines, from Formby, St Helens, Skelmersdale and from southern parts of Greater Manchester. The only VC60 records are from Preston and Dalton Crags.
■ **HABITAT** Woodland edges, damp meadows, hedges.
■ **FOODPLANTS** Goat Willow, Grey Willow.
Single brooded. Mined leaves may be found on the trees in late autumn, but also on the ground where 'green islands' on the leaf-blades are conspicuous. Larvae mine from the midrib into the blade. Adults reared on a few occasions.

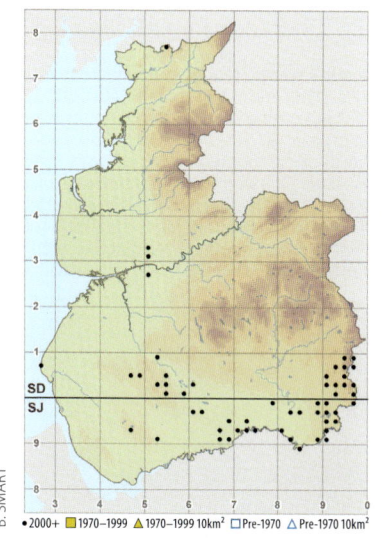

4.085 *Ectoedemia argyropeza*

■ **LANCS STATUS** Occasional / local resident 2003–2022
The first record was of leaf-mines at Flixton, noted by K. McCabe on 8 November 2003, with moths emerging in 2004. Most records are from south of the county, at St Helens, Orrell, Manchester, Stockport, Oldham, Mere Sands Wood, etc. Only twice noted in VC60, at Herring Head Wood, nr. Tatham 2017 and at Halton Green 2019.
■ **HABITAT** Woodland edges where foodplant grows.
■ **FOODPLANTS** Aspen.
Single brooded. All records are of leaf-mines. Tenanted mines mid-October to 1 December. Most detected following leaf-fall by looking for conspicuous green islands surrounding the mine at base of the leaf. Highest count of 50 at Billinge Hill in 2016.

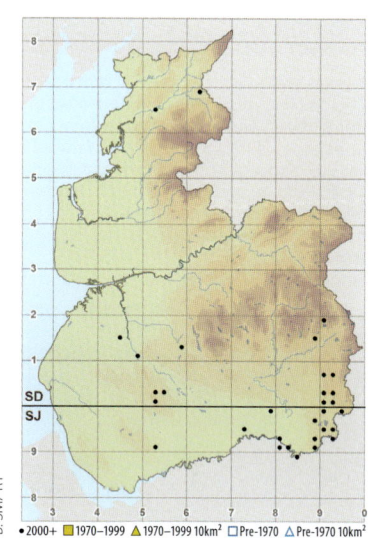

4.089 *Ectoedemia albifasciella*

■ **LANCS STATUS** Frequent / widespread resident 1995–2022
Surprisingly absent from early texts, possibly due to confusion with the following two species. First recorded 31 August 1995 by S. Palmer at Brock Bottom. With over 360 subsequent records, and presence in upland, lowland and coastal areas, this species can be expected to be encountered wherever the host tree occurs.
■ **HABITAT** Oak woodland; also on isolated trees.
■ **FOODPLANTS** Pedunculate Oak; likely other deciduous oak species too.
Single brooded with tenanted leaf-mines August to October. Any from November will need to be closely examined and the following two species excluded. Most records are of leaf-mines. Adults trapped or netted by day on six occasions, all during June.

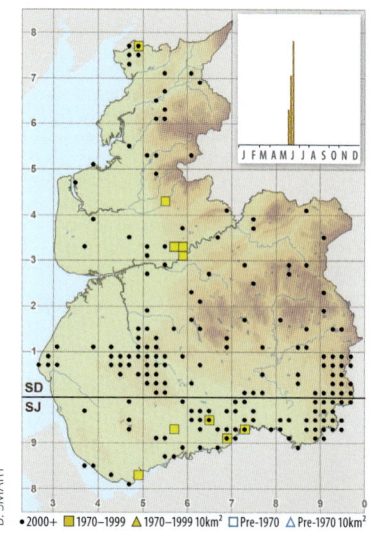

The Moths of Lancashire

NEPTICULIDAE

4.090 *Ectoedemia subbimaculella*

■ **LANCS STATUS** Frequent / widespread resident   c.1889–2022

Ellis (1890) and Mansbridge (1940) both describe this species as common everywhere without noting specific sites. Next recorded at Dalton, nr. Parbold by E.C. Pelham-Clinton in 1977. Recorded on 170 occasions. Appears more thinly distributed in the north of the county, and is, perhaps surprisingly, unrecorded from the Sefton Coast.
■ **HABITAT** Oak woodland, hedgerows, isolated trees.
■ **FOODPLANTS** Pedunculate Oak; likely other deciduous oak species too.
Single brooded with tenanted leaf-mines from October to November. Most records are of mines, although the adult has been recorded at light on four occasions, all from June. The blotch mines are easy to identify, having a slit on the underside.

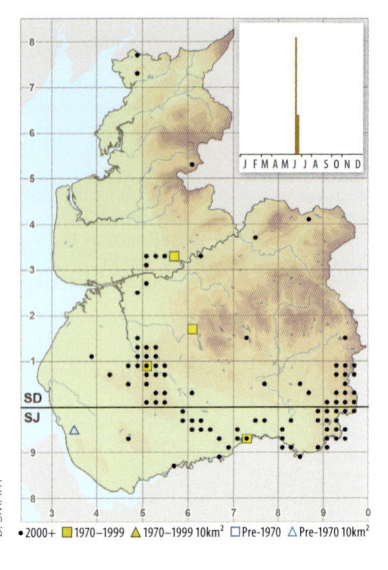

4.091 *Ectoedemia heringi*

■ **LANCS STATUS** Scarce / very local resident   1995–2022

This oak leaf-miner has been recorded on very few occasions within the county, but is always worth considering if encountering tenanted blotch mines without an underside slit in November. A.M. Emmet first recorded this species at Gait Barrows in 1995. Subsequently, recorded at Chorlton in 2015 (still feeding as late as 15 November) and Billinge on 27 October 2020. A female emerged from the Billinge mines on 6 May 2021 and was dissected to confirm identity.
■ **HABITAT** Oak woodland
■ **FOODPLANTS** Oak.
Single brooded. All records are of leaf-mines.

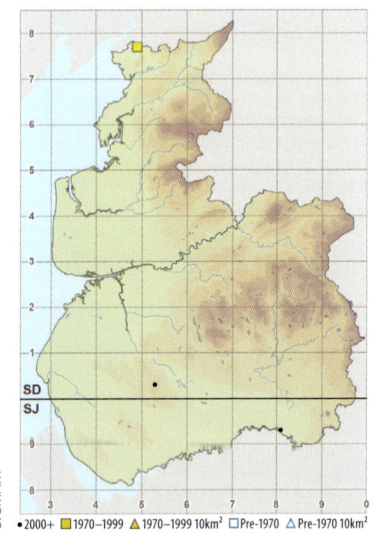

4.094 *Ectoedemia angulifasciella*

■ **LANCS STATUS** Occasional / local resident   c.1859–2022

Considered abundant in the Manchester district by H.T. Stainton (1859). Reported at Preston by J.B. Hodgkinson and J.H. Threlfall (Ellis, 1890), but not noted again until 1995 when recorded at Gait Barrows by A.M. Emmet. Has been found in central Lancs e.g., Dalton, Colne, Chorley, etc., but since 2006 only in the north, on limestone at Gait Barrows, Trowbarrow Quarry and Jack Scout, suggesting range retraction. Noted at Risley Moss, 2006. Otherwise, appears absent from Mersey Valley.
■ **HABITAT** Scrub, hedgerows, on limestone.
■ **FOODPLANTS** Rose, including Dog Rose.
Single brooded. All 40 records have been of leaf-mines.

60   *The Moths of Lancashire*

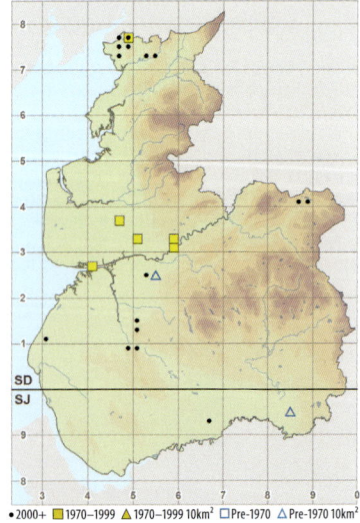

4.095 *Ectoedemia atricollis*

■ **LANCS STATUS** Frequent / widespread resident   c.1859–2022
Described as abundant in Manchester district by H.T. Stainton (1859), but with a long absence until the next record, from 1973 at Parbold by E.C. Pelham-Clinton. As with many other Nepticulidae, it is probable that the lack of records during this period is due to under-recording rather than true absence. Subsequently found throughout, primarily in lowland areas; most frequently in VC59 with 207 of 226 Lancs records.
■ **HABITAT** Woodland edges, hedgerows, parks, scrub.
■ **FOODPLANTS** Hawthorn, Pear, Apple, Crab Apple. Once on Blackthorn.
All records of this single brooded species have been of the leaf-mining stage. Tenanted mines found from September to October.

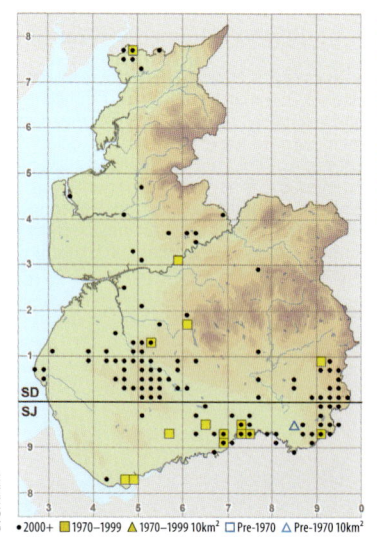

4.096 *Ectoedemia arcuatella*

■ **LANCS STATUS** Scarce / very local resident   2017–2022
A recent discovery in the county, but likely an overlooked, long-term resident. The first record was of leaf-mines at Warton Crag on 15 October 2017, noted by J. Patton. Three weeks later, vacated mines were found at Gait Barrows by the same recorder. Several tenanted mines also found at Dalton Crags on 15 October 2022, with moths bred. So far, these are the only sites where the species has been recorded.
■ **HABITAT** Shaded locations in limestone woodland.
■ **FOODPLANTS** Wild Strawberry.
Single brooded. All records are of the leaf-mines, tenanted from the end of August to late October. The highest count is of 20 mines at Warton Crag on 31 October 2017.

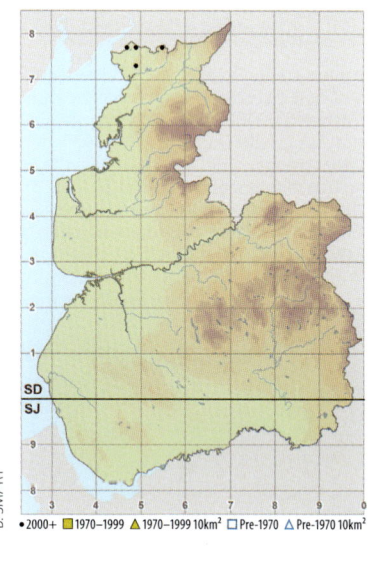

4.099 *Ectoedemia occultella*

■ **LANCS STATUS** Frequent / widespread resident   1918–2022
First recorded at Formby in 1918 by W.A. Tyerman, then at Silverdale (Mansbridge, 1940). The next record was by S. Palmer at Haighton in 1995. Subsequent fieldwork has shown the moth to be found throughout most of the county, but more thinly so in the north, with less than one-tenth of the 236 records coming from VC60.
■ **HABITAT** Woodland, mosses, parks, etc., in lowland and upland areas.
■ **FOODPLANTS** Silver Birch, Downy Birch.
Single brooded. Tenanted mines from September to November, with occasional early record, e.g., 30 July 2017 at Chorlton. Adult found once, at Claughton during day on 27 June 2006. High count of 50 mines found at Astley Moss 2009 by K. McCabe.

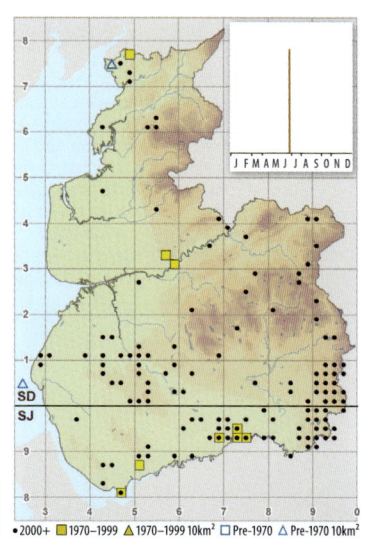

The Moths of Lancashire **61**

NEPITICULIDAE

4.100 *Ectoedemia minimella*

■ **LANCS STATUS** Occasional / local resident   1999–2022

Whilst VC59 is circled in the distribution map in *Moths and Butterflies of Great Britain and Ireland Vol. 1* (Heath, 1976), further details are unknown. The first confirmed record we are aware of was of leaf-mines found on birch at Flixton by K. McCabe on 22 August 1999. Found in lowland and upland areas. Only once on Sefton Coast. Most numerous in the south of the county with only 17 of 140 records coming from VC60.
■ **HABITAT** Woodland edges, mosses, country parks.
■ **FOODPLANTS** Silver Birch, Downy Birch and Hazel.
Single brooded. Records on birch outnumber those on Hazel by a ratio of around 4:1. Adult recorded seven times, more than other *Ectoedemia*, during day and at light.

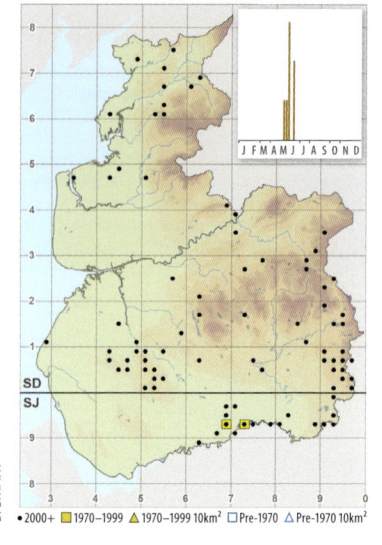

OPOSTEGIDAE

5.001 *Opostega salaciella*

■ **LANCS STATUS** Occasional / local resident   c.1859–2022

First noted at Barton Moss c.1859 by J. Chappell, then in the Preston district c.1889, Freshfield (1940), Tyldesley (1946) and Leighton Moss on 23 August 1952. With the increase in light trapping in the 1980s, it has been recorded more widely in both lowland and upland areas. Some of the upland sites have produced larger catches, such as nine at Green Bank on 28 June 2021 and 20 at Tarnbrook Fell on 10 June 2016 (G. Jones). A more general increase in reports has occurred since about 2009.
■ **HABITAT** Moorland edge, mosses, dune heathland, gardens and brownfield sites.
■ **FOODPLANTS** Not recorded in Lancashire.
Single brooded. Comes to light mostly in small numbers.

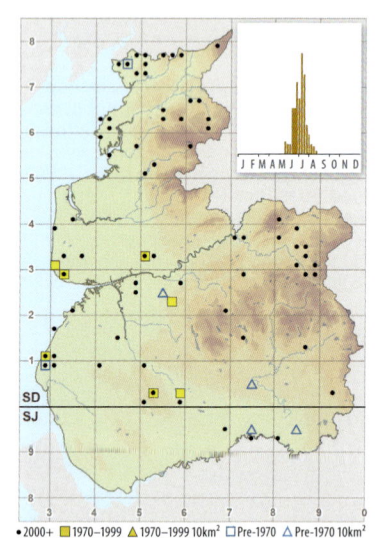

5.004 *Pseudopostega crepusculella*

■ **LANCS STATUS** Scarce / very local resident   c.1859–2022

First noted from the Manchester area (Stainton, 1859), and then Preston c.1889, by J.B. Hodgkinson. In 1925, W. Mansbridge reported two in the Formby area, 'among *Mentha*'; a specimen (same recorder and location) is in Manchester Museum, dated 10 July 1929. Not seen again until one came to light at Leighton Moss in 2006. Since then, found at four sites from 2014 to 2022, these being Altcar (twice), Flixton, Birkdale (twice) and Freshfield, the last site on 30 June 2022 (R. Walker).
■ **HABITAT** Wetland and marshy habitats.
■ **FOODPLANTS** The link with *Mentha* is reported as a possibility nationally.
Single brooded. Comes to light in small numbers.

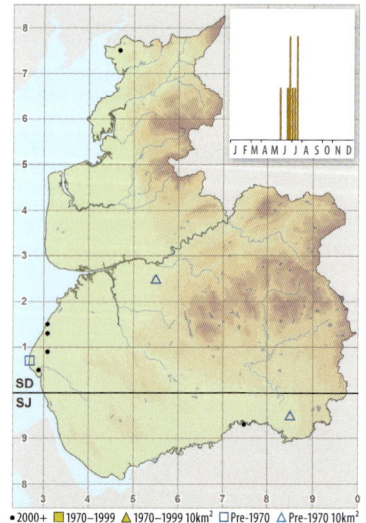

The Moths of Lancashire

HELIOZELIDAE

6.001 *Antispila metallella*

■ **LANCS STATUS** Uncertain   1845

A single record under *Microsetia pfeifferella* was included in a list of moths found in the Pendlebury area during July 1845 by R.S. Edleston (Edleston, 1846). Despite the stage not being mentioned by the recorder, Ellis (1890) and Mansbridge (1940) accepted the record as *A. pfeifferella* Hübn., now known as *A. metallella*, but don't question the month, a little late for an adult. A VC59 map dot for *A. pfeifferella* appears in Heath (1976) which is assumed to refer to this record, but the species is not listed in Stainton (1859) from Manchester, which would have been expected for such a significant record for the region. As a native dogwood feeder, its origins are also in some doubt, as Preston *et al.* (2002), consider this plant an alien species in the Manchester area. No supporting specimen has been located.

■ **HABITAT** Unknown.

The species has been reported from VC63, most recently in 2020, and in VC64, date unknown, vice counties that abut Lancashire. The localities however are nowhere near the Yorkshire/Lancashire border (yorkshiremoths.co.uk – accessed 29.8.2023).

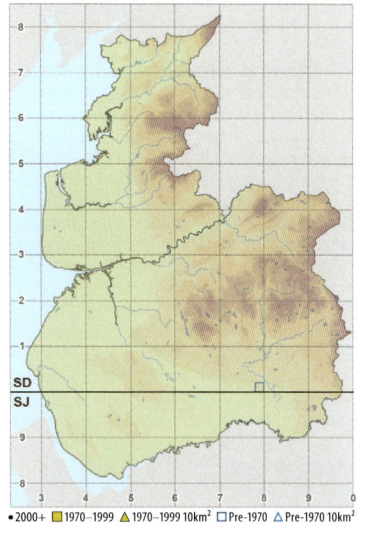

6.003 *Heliozela sericiella*

■ **LANCS STATUS** Occasional / local resident   c.1857–2022

First recorded in 'old lanes among flowers' in the Liverpool district (Gregson, 1857) and then in Prestwich by J. Chappell c.1859. Ellis (1890) reported it as local while Mansbridge (1940) listed it as locally abundant. Next noted at Parbold in 1973 (P. Summers) and Preston in 1998. Since then, found fairly widely where mature oaks occur, such as Billinge, Dalton (nr. Wigan), Flixton, Littledale and Oldham.

■ **HABITAT** Woodland, parkland and hedgerows with oak standards.

■ **FOODPLANTS** Oak species, including Pedunculate Oak.

Single brooded. Comes to light. Most frequently noted by the presence of larval cut-outs on oak leaves. Flies around oaks, sometimes in numbers, in midday sunshine.

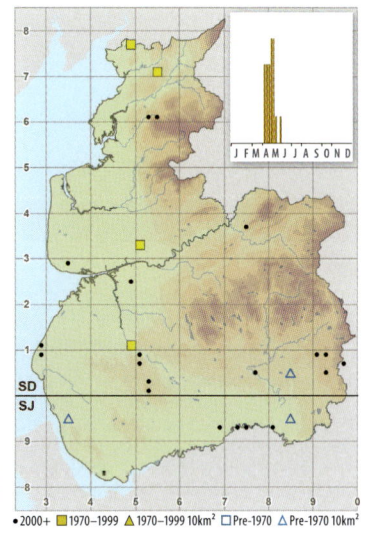

6.004 *Heliozela resplendella*

■ **LANCS STATUS** Occasional / local resident   c.1859–2022

First found in the Manchester district (Stainton, 1859) and then in Preston (Ellis, 1890). Since 2001, leaf mine searches have contributed most of the records, such as from Dalton, nr. Wigan, Walton summit, Flixton and Chorlton. In 2017 it was found in northern parts of the county and extensive searches in south-east Lancs during 2021 and 2022 (S.H. Hind) suggest it is under-recorded in much of the county.

■ **HABITAT** Riversides, damp woodland and other areas where Alder grows.

■ **FOODPLANTS** Alder.

Single brooded. Netted once, as an adult, from Alder. All other records relate to larval feeding signs, particularly the distinctive small larval cut-out in the leaf blade.

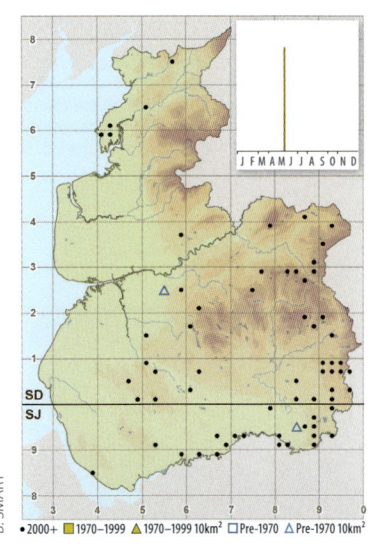

The Moths of Lancashire

HELIOZELIDAE

6.005 *Heliozela hammoniella*

■ **LANCS STATUS** Occasional / local resident   1995–2022

Not identified as a British species until 1890 (Heath, 1976), it is considered a previously overlooked resident in Lancashire, having been noted in south Cumbria in the past. It seems possible that, due to the similarity with the other two resident *Heliozela* species, it may have been misidentified among any caught adults. None of the early authors make reference to birch as a larval foodplant for any members of this genus. It was first noted at Gait Barrows (A.M. Emmet) in 1995, but the national vice county maps, annotated by Emmet, don't mention this record. In the same year, it was found at Winmarleigh Moss as a vacated mine on birch on 20 October 1995 (S. Palmer). Most sightings, since then, have been on lowland raised bogs (mosses), such as Astley, Cadishead, Cockerham, Heysham and Little Woolden, but it is also fairly regular among birch scrub at Gait Barrows. Most records relate to mines and cut outs, often observed on birch seedlings, saplings and scrubby regrowth. Searches in south-east Lancashire during 2022, at sites such as Heyrod, Park Bridge, Royton and Medlock Vale demonstrated it was more widespread than previously suspected.

■ **HABITAT** Mosses, moorland edge, woods and scrub on acidic and limestone soils.
■ **FOODPLANTS** Downy and Silver Birch.

Single brooded. Recorded as a leaf miner or day-flying adult; not seen at light.

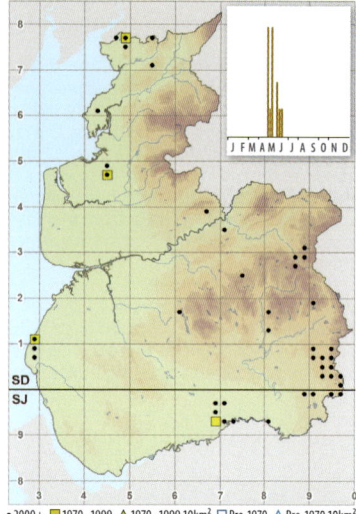

ADELIDAE

7.001 *Nemophora degeerella*

■ **LANCS STATUS** Frequent / widespread resident   c.1857–2022

First recorded on Rainford Moss (Gregson, 1857), with other early records from Manchester, Glazebrook, Worsley, Prescot, Nelson and Whalley. Records were few and far between until the 1990s, since when it has been noted at scattered locations across much of the county, but not with any great regularity. Sites with more than a few records include Ainsdale, Astley Moss, Billinge, Flixton, Gait Barrows and Oswaldtwistle. Its absence from much of the Fylde and some other near-coastal areas appears genuine, rather than due to under-recording.

■ **HABITAT** Woodland rides, hedgerows, edges of mosses and scrub.
■ **FOODPLANTS** A larval case was found amongst oak leaf litter in Chorlton on 31 March 2018, emerging on 27 May of the same year (B. Smart).

Single brooded, with the vast majority of records by day and during June. The latest, rather isolated record, was of one seen at Worsthorne on 8 August 2006 (G. Gavaghan). On occasions, it can be found 'dancing' in larger numbers in sheltered, sunny locations. In these circumstances, counts of up to 100 are not unknown and, exceptionally, larger congregations have been noted. These included 200 at Rivington on 7 June 2018 (C.A. Darbyshire) and 350 at Irlam Moss on 15 June 2018 (D. Steel).

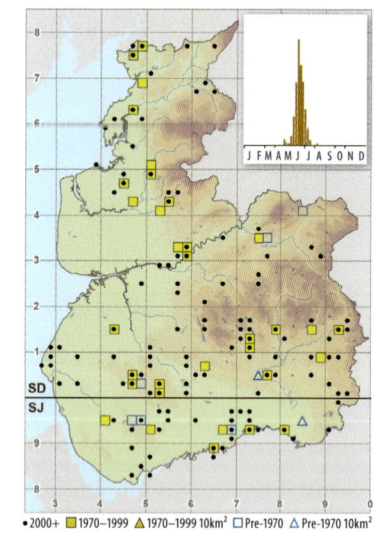

The Moths of Lancashire

7.003 *Nemophora cupriacella*

- **LANCS STATUS** Rare / very local resident   **c.1859–2019**

First reported in the Manchester area (Stainton, 1859), which may relate to a locality outside VC59. All other records are of daytime sightings, at Gait Barrows, on 27 July 2000 (J.R. Langmaid, R.M. & S. Palmer) and 11 August 2012 (M. & J. Clerk). During 2023, details were received of an adult found resting on a flower of common knapweed at Clitheroe by A. Holmes on 29 July 2019.
- **HABITAT** Flower-rich limestone grassland.
- **FOODPLANTS** Not recorded in Lancashire.

Single brooded. A day-flier in sunny conditions. The adults visit flowers of the larval foodplants, which might include Field Scabious or Devil's-bit Scabious in Lancashire.

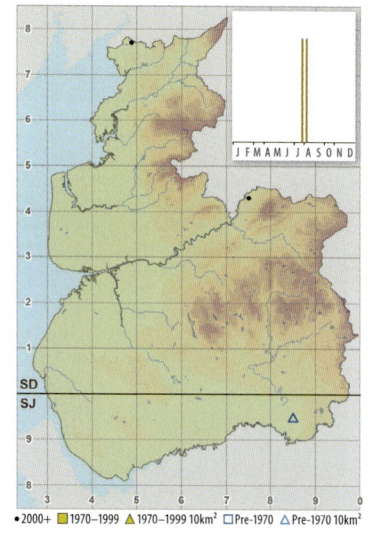

7.005 *Nemophora minimella*

- **LANCS STATUS** Rare / very local resident   **1879–2023**

Netted in 1879 (Hodgkinson, 1880a) from the 'grandest little patch in the neighbourhood....50 yards of rough open ground in a young plantation'. Ellis (1890) mentioned two sites relating to this find, Longridge and nr. Ribchester, referring to Hodgkinson, *loc. cit.* and *'in litt.'* If more than one site was involved, no further details are known. On 19 July 2023, D. Nelson found a male and female at Barley, Pendle, nectaring on Sneezewort, the first seen in Lancashire for 144 years.
- **HABITAT** Rough grassland on moorland edge.
- **FOODPLANTS** Not recorded in Lancashire.

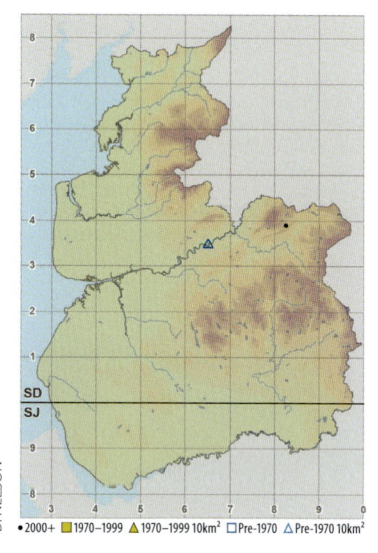

7.006 *Adela reaumurella*

- **LANCS STATUS** Frequent, occasionally abundant / widespread resident   **c.1859–2022**

First recorded in the Manchester district (Stainton, 1859) and later, noted as locally common by Ellis (1890), while Mansbridge (1940) reports it as common and generally distributed. From the late 1930s, it was reported widely, including at sites such as Ainsdale, Blackpool, Burnley, Formby, Nelson and Sabden. It has since been regularly encountered, often in reasonable numbers, in lowland and sheltered upland sites.
- **HABITAT** Hedgerows, scrub and edges of woodland, including brownfield areas.
- **FOODPLANTS** Not recorded in Lancashire.

Single brooded. A daytime flier, in sunny conditions, often seen flitting around trees. Can occur in large numbers; around 1,000 seen in Skelmersdale on 9 May 2002.

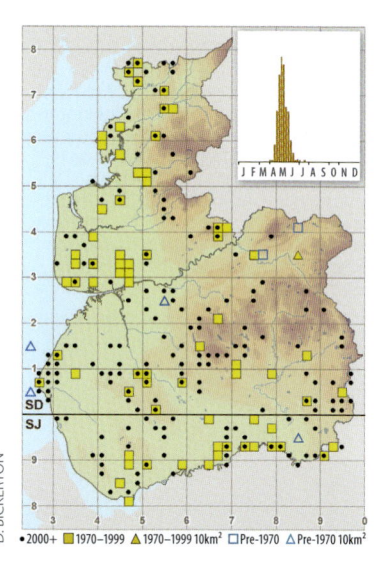

ADELIDAE

7.007 *Adela cuprella*

■ **LANCS STATUS** Rare / very local resident 1999–2015

All records relate to singletons from three scattered localities. A dead moth was found at Gait Barrows by R. Petley-Jones in 1999. A decade later, one was observed at Middleton Wood, Heysham by D. Taylor on 19 April 2009. Finally, one was noted at Ainsdale NNR on the 13 May 2015 by R. Moyes. Considered an overlooked species.

■ **HABITAT** Scrub and copses where mature, sometimes isolated sallows occur.
■ **FOODPLANTS** Not recorded in Lancashire.

Single brooded. Sometimes seen flying around the tops of tall sallow trees in early spring sunshine elsewhere. This could mean the moth is overlooked in Lancashire and searches during calm, sunny spells in April could well produce more records.

• 2000+ ☐ 1970–1999 △ 1970–1999 10km² ☐ Pre-1970 △ Pre-1970 10km²

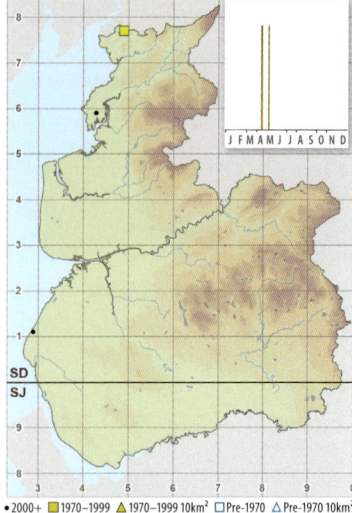

7.008 *Adela croesella*

■ **LANCS STATUS** Rare / very local resident c.1890–2020

As with several scarce moths noted in the historic literature, no year is given for this sighting at Brockholes Wood, Preston by J.B. Hodgkinson, when documented in Ellis (1890), and repeated in Mansbridge (1940). The only other record was when the moth was seen during the day at Gait Barrows on 23 June 2020 (S. Garland).

■ **HABITAT** Open limestone grassland with scrub.
■ **FOODPLANTS** Not recorded in Lancashire.

Single brooded. A day-flyer, associated with Wild Privet bushes, and sometimes Ash, in other parts of the British Isles. It may be under-recorded in Lancashire and searches in suitable habitat on sunny days in May would be worthwhile.

• 2000+ ☐ 1970–1999 △ 1970–1999 10km² ☐ Pre-1970 △ Pre-1970 10km²

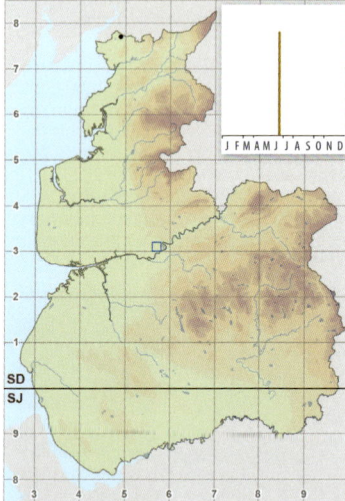

7.009 *Cauchas fibulella*

■ **LANCS STATUS** Occasional / local resident c.1859–2022

Reported in Glazebrook and Withington c.1859 (J. Chappell) and Preston district by J.B. Hodgkinson (Ellis, 1890). The only other early record was from Gait Barrows on 20 June 1914 (W. Mansbridge). Since 1995, almost all records have been from the northern limestone areas, including Dalton Crags, Jack Scout and Warton Crag. The exceptions were in Standish (1995), Clitheroe (1999) and Mere Sands Wood in 2010.

■ **HABITAT** Sheltered rides and sunny banks, particularly in areas of limestone grassland.
■ **FOODPLANTS** Not recorded in Lancashire.

Single brooded. A daytime flier in sunny weather. Usually found close to, or on, Germander Speedwell flowers, on sunny sheltered banks.

• 2000+ ☐ 1970–1999 △ 1970–1999 10km² ☐ Pre-1970 △ Pre-1970 10km²

The Moths of Lancashire

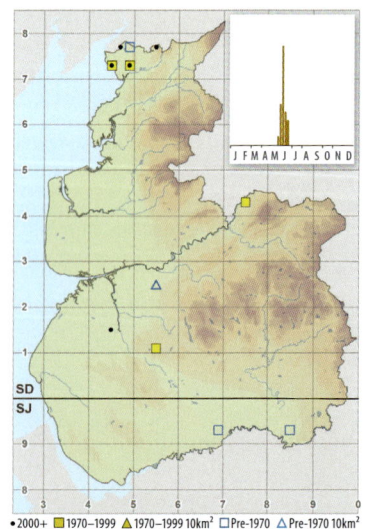

7.010 *Cauchas rufimitrella*

- **LANCS STATUS** Frequent / local resident　　　　　c.1859–2022

Formerly considered uncommon with few early records, the first at Glazebrook in 1859 (J. Chappell). Since 1992, recorded widely across lowland and upland areas, including Broughton, Chorlton, Pennington Flash, Stanah, Rochdale and Towneley. About 200 were seen flying around scurvy-grass sp. at Marshside on 16 May 2014 (K. McCabe & S. Palmer) and, in smaller numbers, at Sunderland Point marsh in 2016.
- **HABITAT** Damp meadows, hedgerows, verges, brownfield sites and saltmarsh.
- **FOODPLANTS** Cuckooflower, Garlic Mustard and, possibly, scurvy-grass sp.

Single brooded. A daytime flier, often seen on flowerheads, particularly Cuckooflower. Larvae and their cases found in mid-June and July (B. Smart).

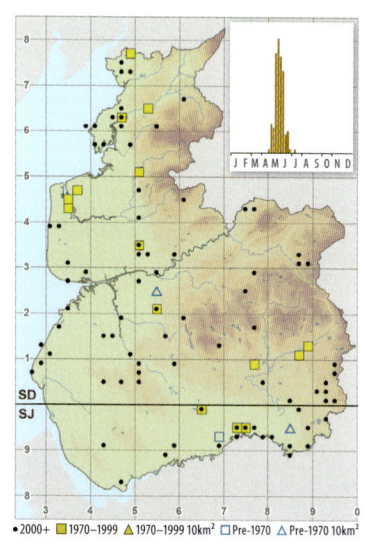

7.011 *Nematopogon pilella*

- **LANCS STATUS** Scarce / very local resident　　　　　1880–2021

Several seen amongst Bilberry above Stonyhurst College in May 1880 (Hodgkinson, 1880b) and same site, same recorder, in 1882, 1889 and 1896 (K. Bland pers. comm.). Next, 31 May 1919 (A.E. Wright) at Sabden. More recently, few seen Jeffrey Hill 30 May 2012 (S. Palmer). Four found 30 May 2021 at Bishop Park, Grains Bar, Oldham (S.H. Hind), the sites being on the edges of, and within, hollows containing a Bilberry and heather mix, with a south to west aspect (S.H. Hind pers. comm.)
- **HABITAT** Moorland.
- **FOODPLANTS** Life history unknown nationally, but probably associated with Bilberry.

Single brooded. A day-flyer, found in small numbers. Probably overlooked.

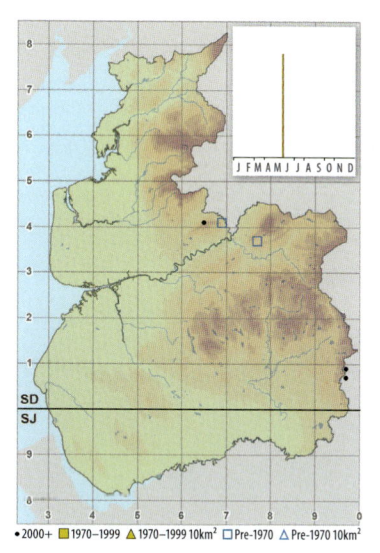

7.012 *Nematopogon schwarziellus*

- **LANCS STATUS** Occasional / widespread resident　　　　　c.1857–2022

Recorded commonly in the Liverpool area (Gregson, 1857), with later records from Prestwich, Formby and Rufford. Infrequently noted in the 1980s and '90s, but with records then increasing to 2009. During that time, it was regular at several sites, including Billinge, Gait Barrows, Haydock and Stanley Bank Meadow. Records have declined slightly since then, but it remains regular on the limestone.
- **HABITAT** Woods, mosses, scrubland, brownfield sites, gardens and moorland edge.
- **FOODPLANTS** Not recorded in Lancashire.

Single brooded. Comes to light and flies during day. Usually encountered in small numbers, but 50 were observed 'dancing' at Cragg Wood on 17 May 2021 (R. Foster).

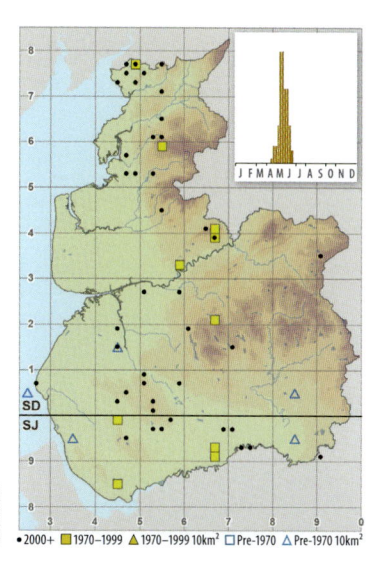

The Moths of Lancashire

ADELIDAE

7.014 *Nematopogon metaxella*

■ **LANCS STATUS** Occasional / local resident   c.1887–2022

The least common of the pale members of this genus, it also flies a little later in the year than the others. The first were reported in Salwick by J.H. Threlfall and by J.B. Hodgkinson nr. Longridge, both prior to 1888. The next was not until 27 June 1986 nr. Churchtown (M. Evans) and then in Ormskirk, in 2005 (C.A. Darbyshire). Subsequent records have been in the north, at Leighton Moss, Silverdale Moss, White Moss (nr. Yealand) and Trowbarrow Quarry, between 2008 and 2022.
■ **HABITAT** Damp woodland and scrub.
■ **FOODPLANTS** Not recorded in Lancashire.
Single brooded. Most recent records relate to light trapped moths, in small numbers.

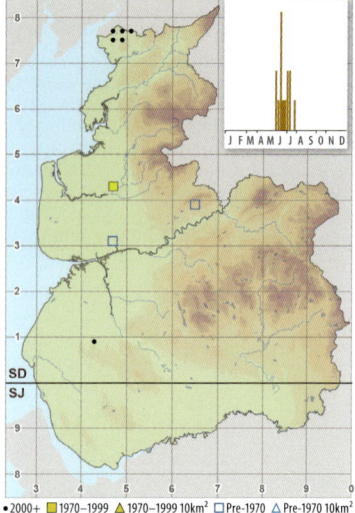

● 2000+   ◻ 1970–1999   △ 1970–1999 10km²   ◻ Pre-1970   △ Pre-1970 10km²

7.015 *Nematopogon swammerdamella*

■ **LANCS STATUS** Frequent / widespread resident   c.1857–2022

The largest member of this genus, it is also the most regularly encountered. The first was found in Liverpool (Gregson, 1857) and it was reported to be common, to locally abundant, by early authors. Since the late 1980s, it has been noted from many lowland locations, with regular reports from Astley Moss, Flixton, Gait Barrows and Leighton Moss. Records from upland areas remain fairly few and far between.
■ **HABITAT** Woods, hedges, scrub and mosses, occasionally in gardens.
■ **FOODPLANTS** Not recorded in Lancashire.
Single brooded. Readily disturbed by day and attracted to light in small numbers. Up to 25 were noted at Holcroft Moss in 2005 and Yealand Storrs in 2013.

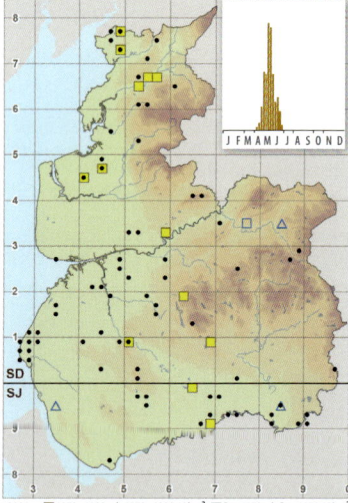

● 2000+   ◻ 1970–1999   △ 1970–1999 10km²   ◻ Pre-1970   △ Pre-1970 10km²

INCURVARIIDAE

8.001 *Incurvaria pectinea*

■ **LANCS STATUS** Scarce / local resident   c.1857–2022

Described as plentiful amongst birches in the Liverpool district (Gregson, 1857). However, not recorded again until 1984, at Gait Barrows. The subsequent eleven 21st century records are from lowland sites and the Pennine fringes. Widnes, Newhey and Crown Point, nr. Burnley are the only known sites in VC59, with Winmarleigh Moss and Gait Barrows the only known northern sites. The population seems healthy at the latter site with feeding signs found in 2022 on all three foodplants noted below.
■ **HABITAT** Woodland edges, mosses.
■ **FOODPLANTS** Birch, Hazel, Hornbeam. Tenanted mines on birch late May to early June. Single brooded. All records are of occupied mines or larval cut-outs, up to 50 per leaf.

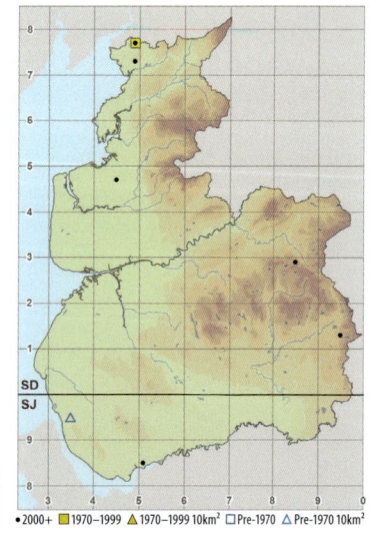

● 2000+   ◻ 1970–1999   △ 1970–1999 10km²   ◻ Pre-1970   △ Pre-1970 10km²

68   The Moths of Lancashire

8.002 *Incurvaria masculella*

■ **LANCS STATUS** Occasional / widespread resident   c.1857–2022

Documented as 'common' by Byerley in his book *The Fauna of Liverpool* (1854). However, as the vast majority of sites mentioned throughout the book are in Wirral (in VC58), this is considered an unconfirmed record for Lancashire. The moth was later recorded at Bootle, where it was 'noted to be plentiful in hedges in May, especially where birch grows' (Gregson, 1857). Also, recorded in the Manchester district by Stainton (1859). This is the most numerous and widely distributed member of the genus, with 145 records, all but 17 of these coming from the present century. Most of these are from lowland areas such as Rindle Wood, Flixton, Lightfoot Green, Hoghton, Winmarleigh Moss, and on limestone at Warton Crag, Gait Barrows and Yeisland Hall Allotments. There is also the occasional record from more elevated sites such as Longridge Fell, Grains Bar and Oswaldtwistle.

■ **HABITAT** Woodland edges, hedgerows, mosses.

■ **FOODPLANTS** Not recorded in Lancashire. The moth was bred from a larval case found by A. Barker at Higher Penwortham in January 2008, foodplant unknown. Single brooded. Moths trapped at light and netted during day.

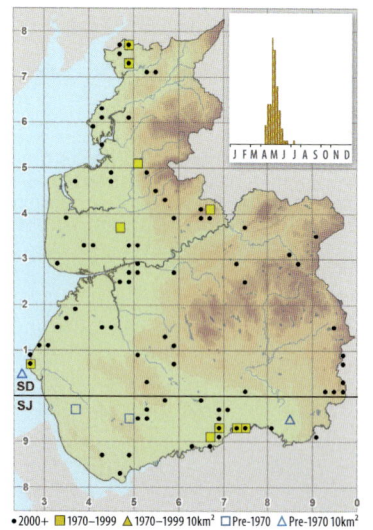

● 2000+   ☐ 1970–1999   ▲ 1970–1999 10km²   ☐ Pre-1970   △ Pre-1970 10km²

8.003 *Incurvaria oehlmanniella*

■ **LANCS STATUS** Occasional / local resident   c.1845–2022

First recorded in mid-19th century at Manchester by R.S. Edleston and at Woolton (Gregson, 1857). Thereafter, from Preston, Burnley, Simonswood Moss and Sales Wood, Prescot (Mansbridge, 1940). More recent records from Little Woolden Moss, Docker Moor, Heysham Moss, Rainford, etc. Quite a local species with scattered distribution in lowland and upland areas. Absent from the Sefton and Fylde coasts.

■ **HABITAT** Woodland, mosses.

■ **FOODPLANTS** Bilberry. Over 40 mines noted in shade beneath oaks at Cragg Wood, Littledale on 4 September 2020, with up to 14 on a single leaf. Larvae from these mines were then observed feeding from within cases cut out from leaves, later pupating within (Smart, 2021). Adults emerged from 8 May 2021 to 9 May 2022. Single brooded. Moths recorded flying during the day. Also noted at light.

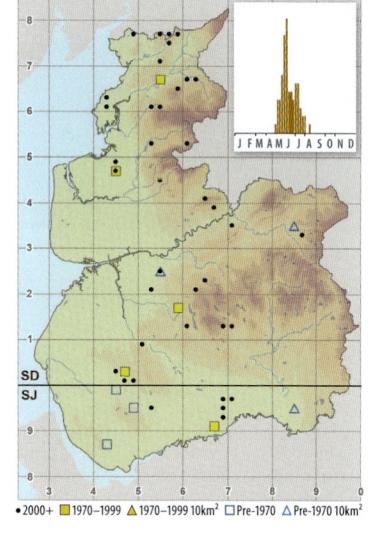

● 2000+   ☐ 1970–1999   ▲ 1970–1999 10km²   ☐ Pre-1970   △ Pre-1970 10km²

The Moths of Lancashire

INCURVARIIDAE

8.004 *Incurvaria praelatella*

■ **LANCS STATUS** Occasional / local resident   c.1859–2022

First noted in Manchester c.1859 by J. Chappell and by H.T. Stainton. Next, seen at Brockholes Wood, Preston by J.B. Hodgkinson (Ellis, 1890) and at Silverdale in 1909. Described as local and scarce (Mansbridge, 1940), with no further records until Gait Barrows 1987. Other than a few VC59 records from Bolton, Billinge, Skelmersdale and Worsley from 2000 to 2005, all records this century from Morecambe Bay limestone, e.g., Leighton Moss, Warton Crag, Yealand Redmayne, Challan Hall Allotment, etc.
■ **HABITAT** Understorey of limestone woodland, mosses.
■ **FOODPLANTS** Fiddle-shaped larval cases on underside of Wild Strawberry leaves. Single brooded. Moths netted during day; also at light. Cases noted in October.

● 2000+   ▫ 1970–1999   △ 1970–1999 10km²   ▫ Pre-1970   △ Pre-1970 10km²

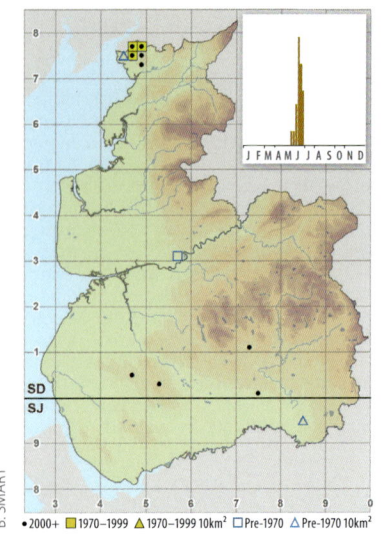

8.005 *Phylloporia bistrigella*

■ **LANCS STATUS** Occasional / local resident   c.1857–2022

First reported at Woolton, Liverpool (Gregson, 1857). Mansbridge (1940) found it scarce and infrequent. Noted at Formby by H.N. Michaelis and R. Pritchard in 1953, but not again until 1982 at Winmarleigh Moss. Leaf-mines, mainly vacated with oval cut-outs, first noted 2001 at Heywood and Middleton, with others at Astley, Holcroft and Heysham Mosses shortly after. Absent from Fylde, Ribble Valley, Bowland Fells.
■ **HABITAT** Mosses, woodland edges, dunes. A few upland records from West Pennines.
■ **FOODPLANTS** Birch. Leaf-mines noted from July onwards.
Single brooded, with occasional small second generation, likely represented by Astley Moss record, 4 August 2009. 31 of 77 records are of moths at light or netted in day.

● 2000+   ▫ 1970–1999   △ 1970–1999 10km²   ▫ Pre-1970   △ Pre-1970 10km²

PRODOXIDAE

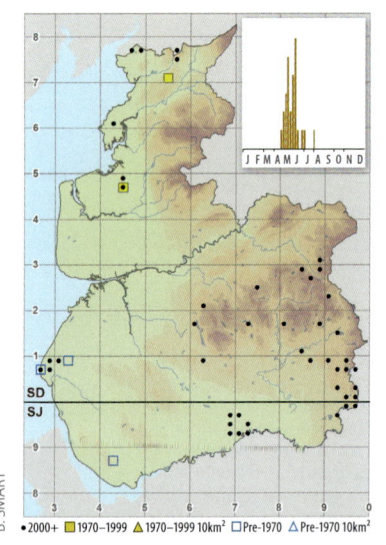

9.001 Currant Shoot Borer *Lampronia capitella*

■ **LANCS STATUS** Considered Extinct   c.1859

This species was reported by Stainton (1859) to have been found in the Manchester district during the mid-19th century. As no further details are given as to location, it is possible this could have originated from a part of Manchester outside VC59. Ellis (1890) and Mansbridge (1940) both refer to this record and make no mention of any subsequent sightings.
Neighbouring county records show the moth to be unrecorded from Cheshire and Cumbria, and recorded just twice in Yorkshire, although not since 1883. Given the lack of further records, it seems likely that the species is now extinct in Lancashire.
■ **HABITAT** Unknown in the county.
■ **FOODPLANTS** Not recorded in Lancashire.

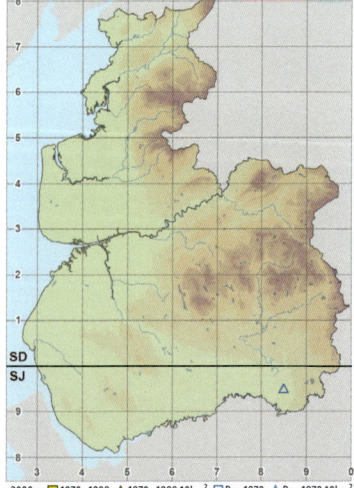

● 2000+   ▫ 1970–1999   △ 1970–1999 10km²   ▫ Pre-1970   △ Pre-1970 10km²

9.002 Lampronia luzella

■ **LANCS STATUS** Scarce / very local resident   c.1859–2010

The first Lancashire record was from Withington, Manchester by J. Chappell in the mid-19th century. It was next noted at Burnley by W.G. Clutten in 1916 and in Liverpool by W.A. Tyerman in 1919 (Mansbridge, 1921). Status was described as 'local and usually scarce' (Mansbridge, 1940). The moth was subsequently recorded at Newfield Clough, Horwich on 1 June 1982 by E.G. Hancock, at Winmarleigh Moss on 5 June 1988 by M. Evans and at Cockerham Moss on 7 June 2010 by S. Palmer, suggesting the moth may still be present in the county, albeit in very low numbers. The picture in neighbouring counties is not dissimilar. Last recorded in Cheshire in 1982 (cheshire-chart-maps.co.uk, accessed 1 September 2023). Noted at Rievaulx, Yorkshire in 2022 (yorkshiremoths.co.uk, accessed 1 September 2023) which was the first record in the county for 18 years.

■ **HABITAT** Woodland edges, mosses.
■ **FOODPLANTS** Unrecorded within the county.

Single brooded. All Lancashire sightings are of moths recorded during daytime fieldwork.

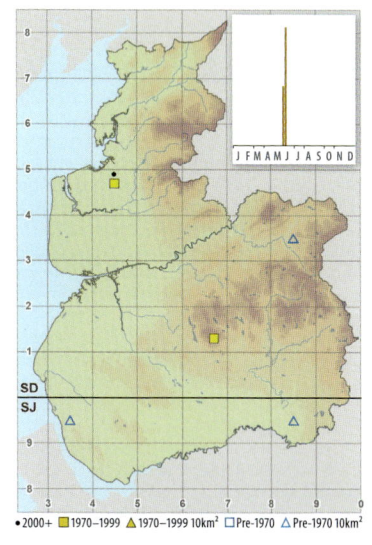

● 2000+   ■ 1970–1999   ▲ 1970–1999 10km²   □ Pre-1970   △ Pre-1970 10km²

9.003 Raspberry Moth Lampronia corticella

■ **LANCS STATUS** Scarce / very local resident   c.1857–1993

First recorded by C.S. Gregson (1857) in the Liverpool district. He described the moth as occurring in profusion in old gardens among the raspberry trees, and on the mosses where raspberries grow. Later, noted from Brockholes Wood, Formby and Burnley (Mansbridge, 1940). The most recent records are of moths from Claughton in 1992 and 1993, once indoors and once at light. Worryingly, this very distinctive moth has now remained unrecorded in the county for over 30 years.

■ **HABITAT** Woodland edges, mosses.
■ **FOODPLANTS** Raspberry.

Single brooded. Larvae recorded as common on Raspberry in 1950 in Formby area.

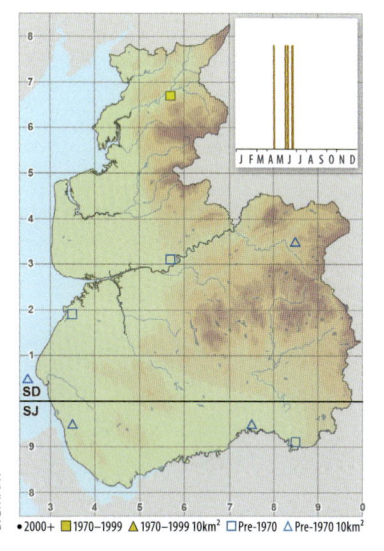

● 2000+   ■ 1970–1999   ▲ 1970–1999 10km²   □ Pre-1970   △ Pre-1970 10km²

9.004 Lampronia morosa

■ **LANCS STATUS** Scarce / very local resident   c.1857–2019

Early records note the moth to have been plentiful in the Liverpool district around wild roses (Gregson, 1857) and abundant in Manchester (Stainton, 1859). It was later described as common and generally distributed (Ellis, 1890). Three moths were noted at Formby in 1953. However, it now appears absent from these areas, with the only post-1953 records, an adult in 2011 and a larva in 2019, both from Sunderland Point. The moth was beaten from the foodplant, with the larva noted feeding in a shoot.

■ **HABITAT** Exposed scrub at edge of saltmarsh.
■ **FOODPLANTS** Burnet Rose.

Single brooded. Larva noted in late April.

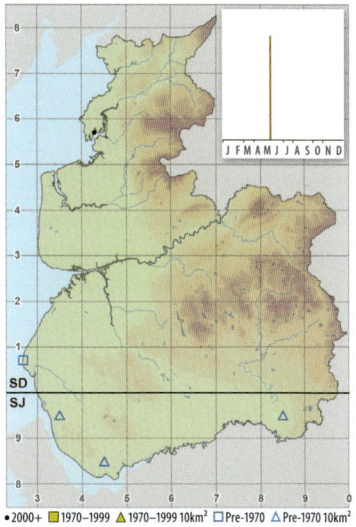

● 2000+   ■ 1970–1999   ▲ 1970–1999 10km²   □ Pre-1970   △ Pre-1970 10km²

9.006 *Lampronia fuscatella*

■ LANCS STATUS Scarce / very local resident   c.1859–2022

First recorded in the county from the Manchester district (Stainton, 1859). The exact location is uncertain, but the likelihood is that this was on part of the Chat Moss complex, encompassing parts of Salford, Wigan and Warrington.

This formerly huge area of lowland raised peat bog dates back 10,000 years or so, but by 1859 was on the verge of being reclaimed by the use of drainage ditches, with conversion of much of the land to agriculture and peat extraction. Despite the destruction of much of the Moss, some relatively good areas of mossland habitat remain; at Astley and Bedford Mosses, Holcroft Moss, Risley Moss and Cadishead Moss for example. The wetter areas host *Sphagnum* mosses and cottongrasses. The vegetation is dominated by Purple Moor-grass, Heather and saplings of Downy Birch.

It is in these birches that the early stages of *Lampronia fuscatella* can be found at Astley Moss and presumably elsewhere. The larva feeds over winter within a node of a young birch twig and a gall is formed as a spherical swelling. In early spring the larva creates an exit-hole in the gall capped over with red-brown frass. Pupation later occurs in the gall, with the adult emerging from May onwards.

These galls, tenanted by overwintering larvae, were first noted at Astley Moss by K. McCabe in late March 2007, following his daytime capture of a moth flying around birch at the same site on 17 May 2005. These feeding signs have not, as yet, been located elsewhere in the county.

Other Lancashire records of this moth include an 1889 record from the Preston district by J.B. Hodgkinson, and more recently at Cockerham Moss in the north of the county on 2 June 2010 by S. Palmer, when an adult was netted during the daytime. A surprise in May 2022 was a visit by this species to a Flixton garden light trap. It is considered likelier for the moth to have wandered the few miles south-east from the mosses, rather than being of local origin.

Nationally, this Nationally Scarce pRDB3 moth (Davis, 2012) is known from four distinct geographical areas: eastern Scotland, southern England, south Wales and north-west England. As well as the Lancashire records, the moth has been recorded in the surrounding counties of Yorkshire, Cheshire and Cumbria.

Recent work intending to restore areas of the mosses to former glories may affect, to an unknown extent, the presence of this species. The removal of much of the invasive scrub at Little Woolden Moss, etc., to promote rewetting and *Sphagnum* growth, may involve removal of some of the larval foodplant. However, this should also prevent the longer-term conversion of mossland to woodland, which may potentially aid the longer-term survival of this species.

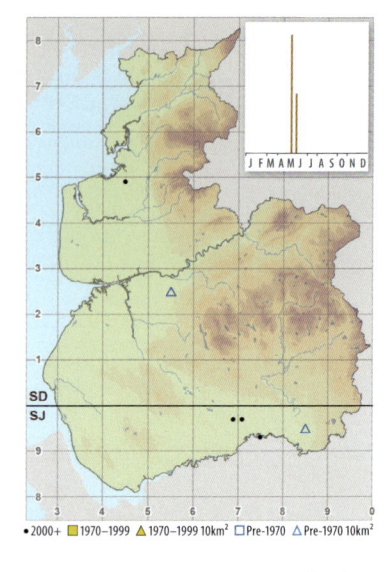

• 2000+   ▢ 1970–1999   △ 1970–1999 10km²   ▢ Pre-1970   △ Pre-1970 10km²

10.001 *Tischeria ekebladella*

■ **LANCS STATUS** Frequent / widespread resident   c.1857–2022

First noted in the Liverpool area (Gregson, 1857) and later, from Burnley in 1920 (W.G. Clutten) and Dalton, nr. Wigan in 1977 (E.C. Pelham-Clinton). During 1995, an increase in leaf mine recording produced records from Anglezarke, Brock Bottom, Kirkby and Risley Moss. Most modern records are from VC59, with mines recorded commonly in much of the Mersey Valley. It appears to be more local in VC60, but may be under-recorded. It is, however, regular at Gait Barrows.

■ **HABITAT** Woodland, hedgerows with standards, parks and gardens with oaks.
■ **FOODPLANTS** Pedunculate and Turkey Oak and Sweet Chestnut.

Single brooded. Comes to light. Tenanted mines found late August to November.

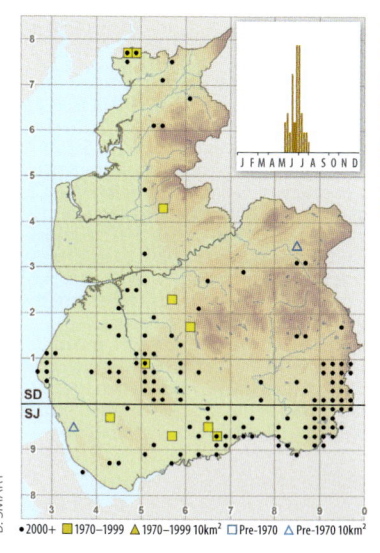

10.002 *Tischeria dodonaea*

■ **LANCS STATUS** Rare / very local resident   c.1976–2022

A single mine was found on an oak tree leaf at Barlow Eye tip, Chorlton on 30 October 2022 by B. Smart, the first documented record for the county. The site is a landscaped former tip, the mine being the only one found on a line of mature oaks. The site has been examined regularly for leaf mines over the last 20 years.

■ **HABITAT** Mature oaks on the edge of a brownfield site.
■ **FOODPLANTS** Oak.

Heath (1976) includes a VC59 dot on the distribution map for this species. The original map, produced by Maitland Emmet for this volume, has been examined, but does not give any details relating to this record.

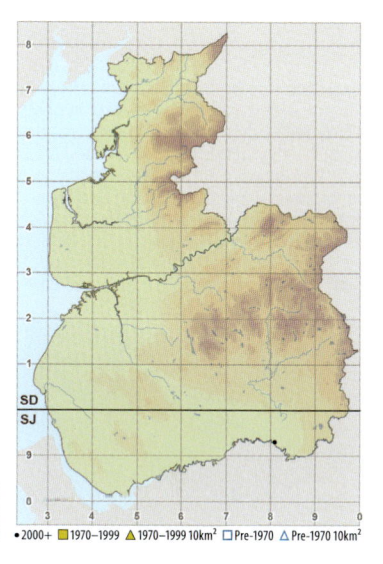

10.003 *Coptotriche marginea*

■ **LANCS STATUS** Frequent / widespread resident   c.1857–2022

Recorded in the Liverpool area by Gregson (1857), it remained a moth of the Sefton Coast and the Mersey Valley for many years. Despite a Preston record c.1889 and Parbold in 1976, it was not until after 1995 that it was found to be moving further inland and northward. Noted at St Annes in 1996, where it is now easily found in the dunes mining dewberry and in the Silverdale area, from 1997, where it is regularly found in scrubby areas. Has spread widely across southern VC59 since the late 1990s.

■ **HABITAT** VC59, wide range, including mosses, hillsides. VC60, dunes, limestone scrub.
■ **FOODPLANTS** Bramble spp. and Dewberry.

Double brooded. Comes to light. Tenanted mines mainly Sept.-April; less so, July.

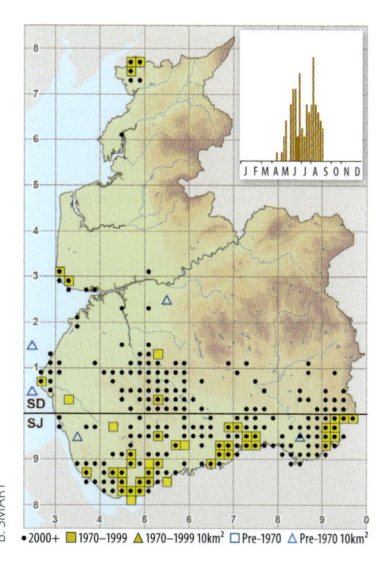

The Moths of Lancashire   73

TISCHERIIDAE

10.006 *Coptotriche angusticollella*

■ **LANCS STATUS** Unknown   c.1859

The validity, or otherwise, of this as a Lancashire species has not been possible to assess with certainty due to the lack of any specific record details. The only feasible record was that listed in Stainton (1859) as occurring in the Manchester area. Although most of the city region is within VC59, some parts do extend into adjacent vice counties, including VC63 (South-west Yorkshire). Neither Cheshire or Yorkshire list the Manchester record in their databases. The closest confirmed national records are a few pre-1983 ones from well east of the Lancashire border, (Sutton & Beaumont, 1989). The distinctive larval feeding signs on rose suggest that the species has not been overlooked in recent years as considerable leaf-mine fieldwork has been taking place since the early 1990s.

■ **HABITAT** Unknown in the county.
■ **FOODPLANTS** Not recorded in Lancashire.

Margaret Dempsey's list (c.1990) includes a VC60 record which appears on the original national VC map for this species. It was accepted by A.M. Emmet, as submitted at the time, but is now considered unconfirmed and no specimen or photo was available for examination.

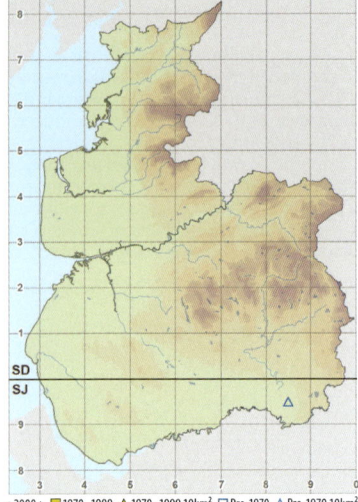

• 2000+   ■ 1970–1999   ▲ 1970–1999 10km²   □ Pre-1970   △ Pre-1970 10km²

PSYCHIDAE

11.001 *Diplodoma laichartingella*

■ **LANCS STATUS** Occasional / local resident   c.1857–2022

First recorded as single cases on old posts at Simonswood Moss, Kirkby (Gregson, 1857), with other 19th century records from Manchester (Stainton, 1859), Preston and Cleveleys (Ellis, 1890). Not noted again until 1996 when a case was found at Red Scar Wood by S. Palmer. Cases since noted at Astley Moss, Broughton, Rixton. Adults recorded at Lancaster, Preston and Silverdale; surprisingly, twice noted indoors.

■ **HABITAT** On tree-trunks and wooden posts in well-established woodland and mosses.
■ **FOODPLANTS** Larvae found on bracket fungi, birch, Alder and rotten Beech trunks, although feeding not directly observed.

Single brooded. Largest count is of 5 cases on Alder at Broughton in 1998 (S. Palmer).

• 2000+   ■ 1970–1999   ▲ 1970–1999 10km²   □ Pre-1970   △ Pre-1970 10km²

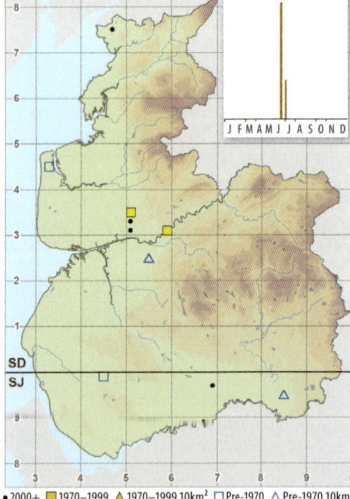

11.002 *Narycia duplicella*

■ **LANCS STATUS** Occasional / local resident   c.1859–2022

Reported as common in the Manchester district by Stainton (1859), although the moth was not listed by Ellis (1890) or Mansbridge (1940). There were no further records until 1996, when cases were found on a Sycamore trunk at Red Scar Wood by S. Palmer. Almost all the 47 subsequent records have been from the south of the county; from the Manchester district, St Helens, Rixton, Astley Moss, Billinge, Longton, Euxton and Halewood. Unrecorded from upland sites.

■ **HABITAT** *Lepraria*-coated surfaces, e.g., tree-trunks, gravestones, walls, etc.
■ **FOODPLANTS** *Lepraria* sp. Noted on oak, Hawthorn, Sycamore, Ash, poplar, birch.

Single brooded. Most records are of cases, with adults infrequently recorded at light.

• 2000+   ■ 1970–1999   ▲ 1970–1999 10km²   □ Pre-1970   △ Pre-1970 10km²

74   *The Moths of Lancashire*

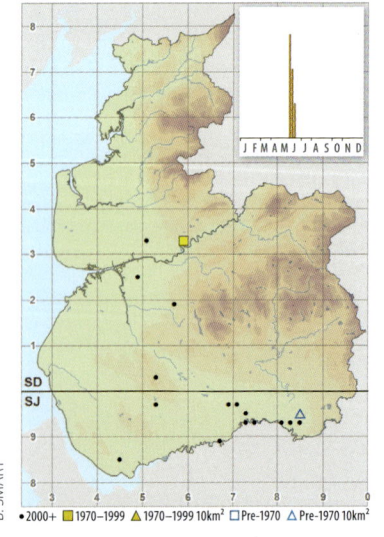

11.004 Lesser Lichen Case-bearer *Dahlica inconspicuella*

■ **LANCS STATUS** Considered Extinct c.1858–1889

The moth was first reported by R.S. Edleston (1859). He reported that 'the cases of *S. inconspicuella* are found here on beech trees in Prestwich Wood, and the moths appear early in April, and are most sluggish creatures.' Edleston also noted that these larvae 'regularly produce both sexes every year.' The species was later noted at Pendlebury by J. Chappell and at Preston by J.B. Hodgkinson (Ellis, 1890). It was described as common by Stainton (1859), and as local, sometimes common, by Mansbridge (1940), although no further records were added by the latter. Whilst the method of determining the species involved in the Victorian records is unclear, this is the only one of the three *Dahlica* species with a male as well as a female, particularly relevant when considering Edleston's observations. There have been no confirmed records of this species during the 20th and 21st centuries. Twice recorded in Cumbria during 2017 (S. Colgate, pers. comm.), but with no recent sightings in Yorkshire or Cheshire.
■ **HABITAT** Woodland.
■ **FOODPLANTS** Unrecorded in the county.
Single brooded.

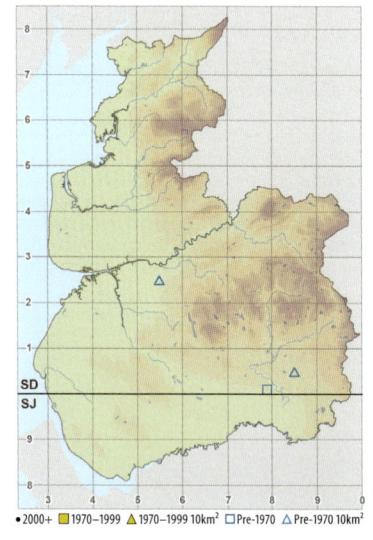

11.005 Lichen Case-bearer *Dahlica lichenella*

■ **LANCS STATUS** Rare / very local resident 2018–2023

Ten larval cases were noted on a metal fence at the base of a railway embankment near Morecambe Golf Course, Bare on 18 October 2018, with a further case found at the same site in November 2019. In April 2022, further cases were noted. Attempted rearing was unsuccessful. However, DNA barcoding confirmed the deceased larva as this species. A further case was found at the same site on 28 January 2023. All cases were found by J. Patton on a 250m stretch of the embankment.
■ **HABITAT** The only known site is described above.
■ **FOODPLANTS** Lichens and algae.
Single brooded.

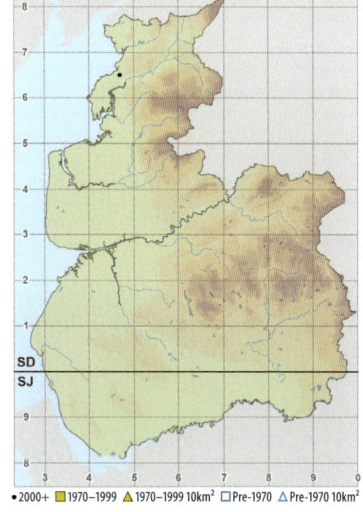

11.006 *Taleporia tubulosa*

■ **LANCS STATUS** Unknown c.1859

The species was documented as abundant in the Manchester district by Stainton (1859), although no specific sites were mentioned, and it has not been recorded since. Ellis (1890) later noted that the species was known only from Cheshire, with Mansbridge (1940) stating that there were no Lancashire records. Possibly as a result of Stainton's comments, VC59 was circled on the Micro-moth Distribution Map for this species by A.M. Emmet, with the source of the record given as H.N. Michaelis, and also on the distribution map printed in *Moths and Butterflies of Great Britain and Ireland Vol. 2* (Heath & Emmet, 1985). As a result, its former presence in the county is considered uncertain, but worthy of note to assist with future research. The species is reasonably well-recorded in Cheshire with 21 records this century, including 295 cases found at Withington, nr. Jodrell Bank in 2014, with 103 on a single oak tree (cheshire-chart-maps.co.uk, accessed 1 September 2023). Local in Yorkshire with all records in the south and east (yorkshiremoths.co.uk, accessed 1 September 2023).
■ **HABITAT** Unknown in the county.
■ **FOODPLANTS** Not recorded in Lancashire.

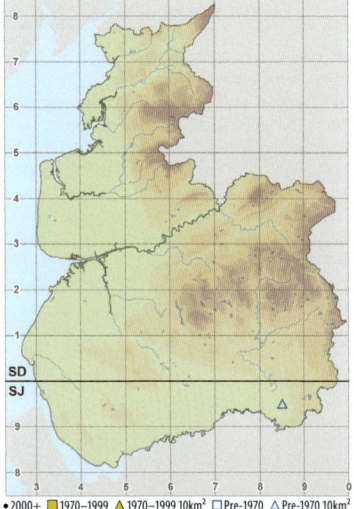

PSYCHIDAE

11.007 *Bankesia conspurcatella*

■ **LANCS STATUS** Scarce / very local resident   2016–2023

Males of this species (females are wingless) noted in 2016 by T. Hutchison on 27 January and 9 March on a wall at a VC60 train station. Subsequent investigation revealed cases with counts of up to 30 on one single wall, close to an area of scrub. The roughly triangular cases, up to 8mm long, are coated with soil, sand, lichen, detritus and pieces of dead insects. Cases and moths from this site are the only confirmed records. Potentially vulnerable to repainting or resurfacing of the wall.
■ **HABITAT** Cases on a painted brick wall; partially sheltered and south-east facing.
■ **FOODPLANTS** Probably feeding on dead insects from vacated spider webs.
Single brooded. Males likely attracted to external light. 18 cases noted, May 2023.

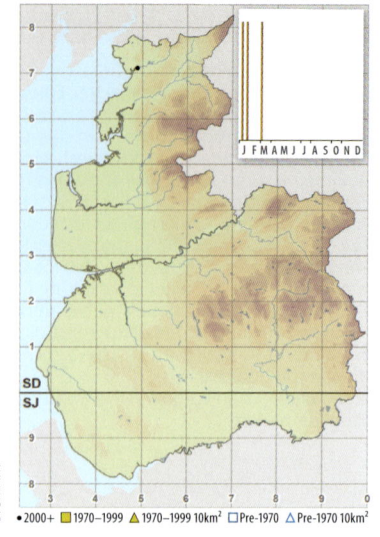

11.009 *Luffia lapidella*

■ **LANCS STATUS** Occasional / local resident   2002–2023

A recent discovery, but possibly an over-looked, long-term resident. Cases were found on a mature Horse Chestnut trunk in Chorlton, Manchester, 10 May 2002, by B Smart. Numbers were huge, estimated at 100 on the single tree, increasing to approx. 1000 the following year. Subsequently recorded at other south Manchester sites (West Didsbury, Whalley Range, Flixton), and at Leyland and Halewood. Cases beaten from pine at Ainsdale in 2022. First VC60 record at Preston in 2023 (S. Palmer).
■ **HABITAT** Parks, street trees, cemeteries, woodland, etc.
■ **FOODPLANTS** *Lepraria* sp. on gravestones, tree-trunks.
Single brooded. The parthenogenetic female was reared in 2016 (Smart, 2017).

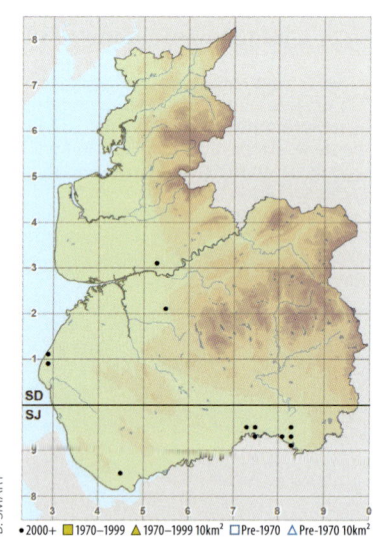

11.012 *Psyche casta*

■ **LANCS STATUS** Occasional / local resident   c.1859–2022

First recorded at Chat Moss by J. Chappell around 1859. Also noted from Preston, Pilling Moss and Formby (Mansbridge, 1940). No further records until found at Risley Moss in 1983 by L.W. Hardwick and M. Hull. Subsequent lowland and upland records from Astley Moss, Freshfield, Longridge Fell, Rivington Pike, Clougha, Birk Bank, etc., with high count of 16 cases from the latter site in May 2003 (S. Palmer).
■ **HABITAT** Woodland, grassland, bogs and mosses. Cases on posts, rocks, tree-trunks.
■ **FOODPLANTS** Likely feeding on lichens on tree trunks and other surfaces.
Single brooded. Most records are of cases. Males occasionally at light. Apterous female found attached to its case on a fence post at Gait Barrows in June 2022.

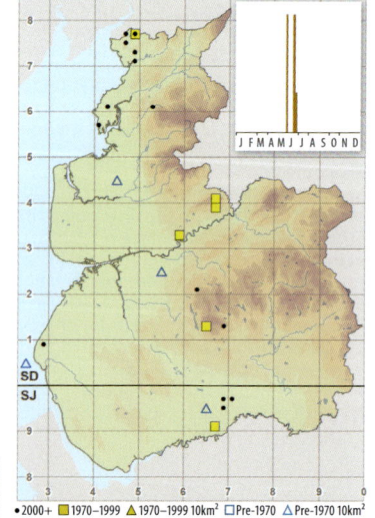

76   *The Moths of Lancashire*

11.016 *Acanthopsyche atra*

■ **LANCS STATUS** Unknown   c.1963

Not mentioned by Gregson, Ellis or Mansbridge in their lists of Lancashire Lepidoptera. However, H.L. Burrows wrote in the Manchester Entomological Society journal (1964) that *A. atra* (then *Pachythelia opacella*) 'has a more extensive distribution than the meagre records suggest. I have seen it on the Witherslack mosses and on the Pennine moors of the West Riding of Yorkshire, Lancashire, Cheshire, Derbyshire and Staffordshire…' Neither VC59 or VC60 are circled on the Micro-moth Distribution Map for this species by A.M. Emmet, or on the map in *Moths and Butterflies of Great Britain and Ireland Vol. 2* (Heath, Emmet, 1985). The blue square on the map indicates an approximated area, with no precise location known. There are a number of Cheshire records from the mid-20th century, all from the Goyt Valley, close to the western Derbyshire border (cheshire-chart-maps.co.uk, accessed 1 September 2023). From Yorkshire, there is a 1900 record from Dunford Bridge, north-west of Sheffield close to the north-eastern Derbyshire border. More recently, a larval case was found at Thirsk, North Yorkshire in 2017 (yorkshiremoths.co.uk, accessed 1 September 2023).
■ **HABITAT** Moorland.
■ **FOODPLANTS** Not recorded in Lancashire.

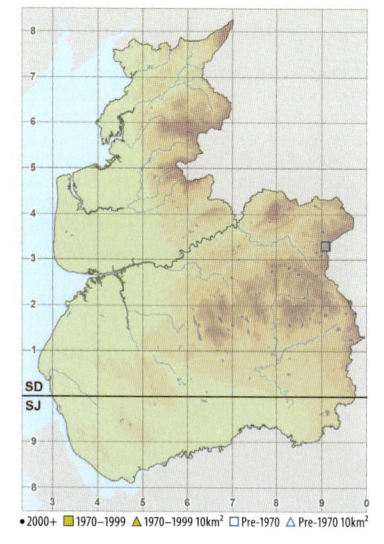

• 2000+   ▢ 1970–1999   ▲ 1970–1999 10km²   ☐ Pre-1970   △ Pre-1970 10km²

12.006 *Infurcitinea argentimaculella*

■ **LANCS STATUS** Scarce / very local resident   c.1859–2022

First noted in Stainton's Manual (1859) as occurring in the Manchester area but not recorded again until 30 July 2001, when a male came to light in St Annes (D. & J. Steeden). On 23 March 2021, seven larval tubes were found amongst lichen on a stone wall in Stretford, Manchester (B. Smart), with eight tubes there a week later. A further visit to the same site on 26 March 2022 produced a count of ten larval tubes. Probably under-recorded.
■ **HABITAT** Old stone walls with suitable lichen growth.
■ **FOODPLANTS** *Lepraria* sp. lichens.
Comes to light on occasions; best searched for by examining stone walls.

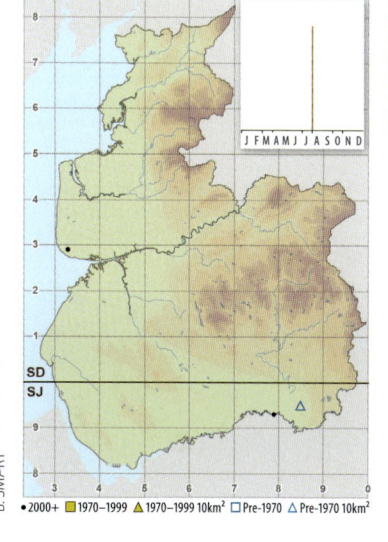

• 2000+   ▢ 1970–1999   ▲ 1970–1999 10km²   ☐ Pre-1970   △ Pre-1970 10km²

12.010 *Morophaga choragella*

■ **LANCS STATUS** Occasional / local resident   2007–2022

A recent arrival in the county as part of a national expansion of range, it had been first noted in Cheshire in 1986. In Lancashire, it was first observed on 21 August 2007 at Flixton (K. McCabe), this site producing over half of the county's subsequent 22 records. A larva was found on 24 March 2008 in bracket fungus on an old birch stump at Astley Moss (B. Smart). Since then, it has been seen at Hough Green in 2010, Swinton in 2011, Ainsdale in 2012, Chorlton in 2013 and Croxteth Park in 2016.
■ **HABITAT** Mosses and other sites where bracket fungi grow on birch.
■ **FOODPLANTS** On an unidentified bracket fungus.
Comes to light.

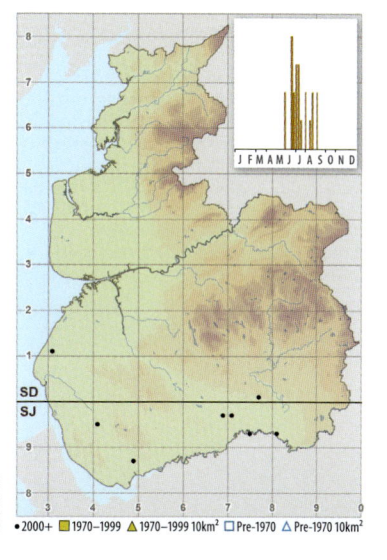

• 2000+   ▢ 1970–1999   ▲ 1970–1999 10km²   ☐ Pre-1970   △ Pre-1970 10km²

The Moths of Lancashire

12.011 *Triaxomera fulvimitrella*

■ **LANCS STATUS** Occasional / very local resident   c.1857–2018

First reported from Pigue Lane, Wavertree, Liverpool by P.H. Newnham on an old decaying oak tree (Gregson, 1857), subsequently becoming plentiful at this site. Later records came from Agecroft and Prestwich c.1859, by J. Chappell. Next noted in Fallowfield on 3 July 1912 and near Burnley in 1918. Only five recent records are known, these from Stanley Bank Meadow in 2005, Dalton (VC59) in 2006, Astley Moss in 2011, Lytham Hall in 2012 and Holmeswood on 3 June 2018 (S. Priestley).
■ **HABITAT** Where older trees are left standing to rot, formerly in lanes and parks.
■ **FOODPLANTS** Not recorded in Lancashire.
Comes to light. Observed in the day, active on decorticated bark of an old beech tree.

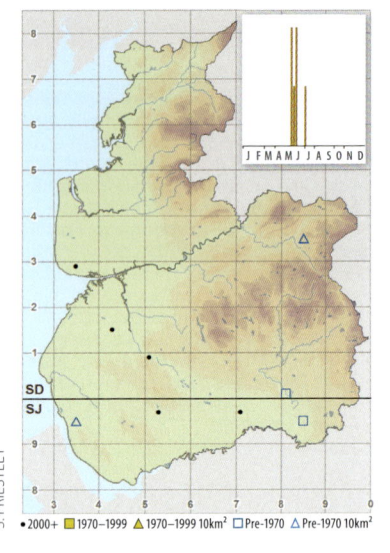

12.012 *Triaxomera parasitella*

■ **LANCS STATUS** Scarce / very local resident   2009–2019

Although found in Cheshire in the 19th century, the first Lancashire record was not until 14 June 2009, in Fulwood, Preston (A. Powell), when two came to light. The following year it was found at the same site on four occasions, and these were followed by single records in 2013, 2014, 2016, 2017 and 2019. It has only been recorded at three other sites, in Hoghton on two occasions in June 2015 (G. Dixon), Formby in 2015 (R. Walker *et al*.) and Heysham Moss on 1 June 2017 (J. Patton).
■ **HABITAT** In Fulwood, the light trap was run alongside a log pile, under mature Beech.
■ **FOODPLANTS** Not recorded in Lancashire.
Comes to light. Netted at dusk on one occasion.

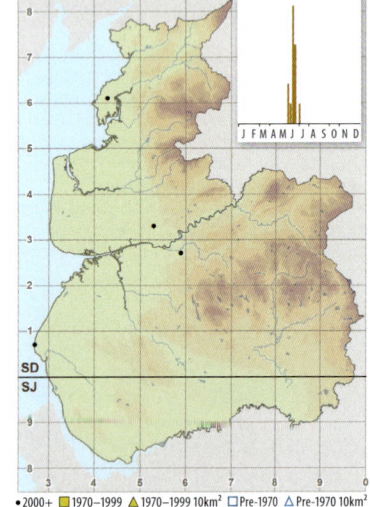

12.014 *Nemaxera betulinella*

■ **LANCS STATUS** Scarce / local resident   c.1857–2022

First reported from the Liverpool district (Gregson, 1857), followed by sites such as Penwortham, Warrington and Preston (Ellis, 1890). During the early 20th century, it was noted as common and widely distributed in woods (Mansbridge, 1940). Since then, however, records have declined, but include one in Preston (1997), in Billinge from 2000–2004, Birkbank (2001), Gait Barrows (2015) and Chorlton (2022). Seven were seen at Anglezarke on 23 June 2001 (C.A. Darbyshire) on an oak trunk.
■ **HABITAT** Woodland and gardens.
■ **FOODPLANTS** Not recorded in Lancashire.
Comes to light. Seen on an oak trunk and once at a Clearwing pheromone lure.

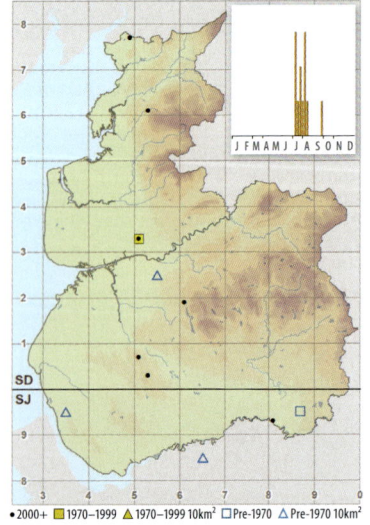

12.015 Corn Moth *Nemapogon granella*

■ **LANCS STATUS** Adventive   c.1853–1922

This was formerly reported as being found in profusion within grain stores around Liverpool Docks area and was first noted as such by Byerley (1854), followed by Gregson (1857). After 1890, only a single reference to this species has been found. It appears in Mansbridge (1940), where he notes it as abundant round a flour mill on the Dock Road, Liverpool in 1922. Specimens kept by Mansbridge are present in Manchester Museum, dated July 1922.
■ **HABITAT** Indoors, amongst stored grain.
■ **FOODPLANTS** Grain or possibly other stored products.
Changes in food hygiene processes and transport methods will have led to its demise.

● 2000+   ■ 1970–1999   ▲ 1970–1999 10km²   □ Pre-1970   △ Pre-1970 10km²

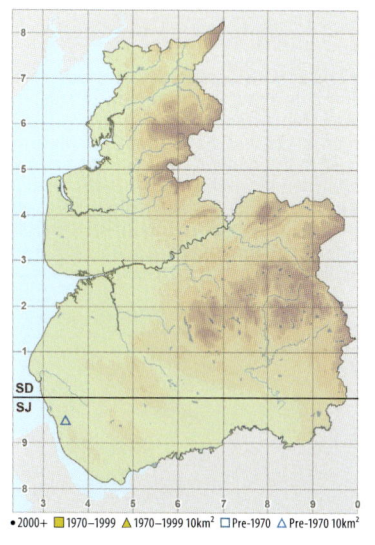

12.016 Cork Moth *Nemapogon cloacella*

■ **LANCS STATUS** Frequent / widespread resident   c.1857–2022

Early authors note it as very common throughout the region, the first being in Kirkby (Gregson, 1857). Since then, reported widely in lowland areas, but less frequently encountered on higher ground. Has been noted almost annually in the past three decades at a few sites, such as Flixton and Preston. Light trapped numbers are low, but up to 30 have been seen by day, such as Middleton N.R. in 2016 (J. Patton).
■ **HABITAT** Hedgerows, gardens and woodland, where dead or dying wood occur.
■ **FOODPLANTS** Bracket fungus on Silver Birch.
Single brooded; a possible small second brood in some years. Comes to light; rests on tree trunks; flies in sunshine and has a particularly strong evening flight.

● 2000+   ■ 1970–1999   ▲ 1970–1999 10km²   □ Pre-1970   △ Pre-1970 10km²

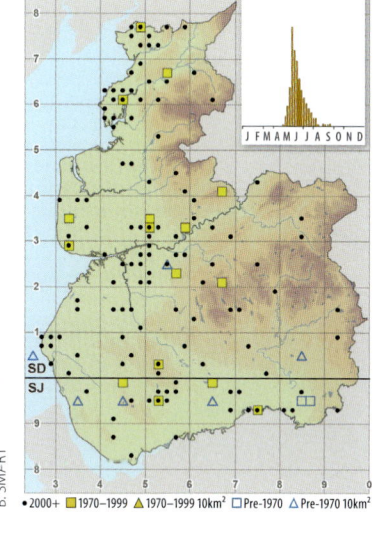

12.017 *Nemapogon koenigi*

■ **LANCS STATUS** Scarce / very local resident   c.1890–2020

Noted as very local and infrequent, it was first found in Brockholes Wood (Preston) and at Cleveleys by J.H. Threlfall (Ellis, 1890). The next sighting was not until 21 June 1996, at Red Scar Wood (east of Brockholes Wood) by S. Palmer. A series of five reports came from Gait Barrows between 1999 and 2015, with an indoor sighting in Silverdale on 26 June 2011 (T. Riden). One was observed in Warton, near Silverdale, on 12 August 2020 (P. Stevens).
■ **HABITAT** Deciduous woodland, particularly but not exclusively on the limestone.
■ **FOODPLANTS** Not recorded in Lancashire.
Single brooded. Disturbed during the day as singletons and once found indoors.

● 2000+   ■ 1970–1999   ▲ 1970–1999 10km²   □ Pre-1970   △ Pre-1970 10km²

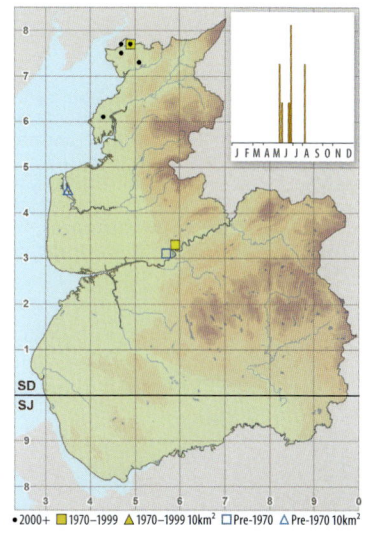

TINEIDAE

12.021 *Nemapogon clematella*

■ **LANCS STATUS** Occasional / very local resident   c.1857–2022

Formerly noted as uncommon, the first record was in Kirkby (Gregson, 1857), and then Manchester, Longridge and Preston, all prior to 1890 (Ellis, 1890). It was next seen in 1986 at Gait Barrows (L.W. Hardwick) and most subsequent records have been in this general area, including at Leighton Moss and Challan Hall Allotment, often as a larva. The moth was also seen in Torrisholme in 2009 (C.A. Darbyshire).

■ **HABITAT** Deciduous woodland, copses and old hedgerows.
■ **FOODPLANTS** *Diatrype* fungus on Hazel, and once on fungus on Birch.
Single brooded. Comes sparingly to light. Larvae feed on crust fungi, mainly on dead or dying hazel, with frass covered tubes between the fruiting bodies often visible.

• 2000+   □ 1970–1999   ▲ 1970–1999 10km²   □ Pre-1970   △ Pre-1970 10km²

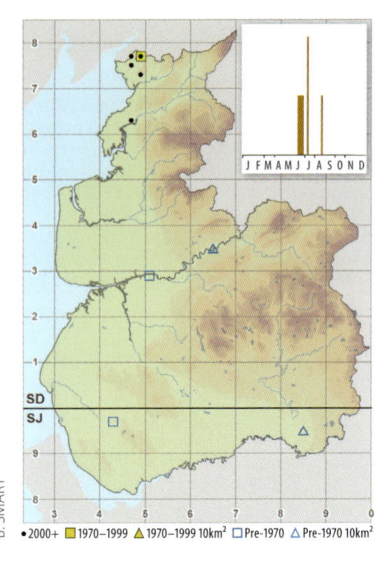

12.022 *Nemapogon picarella*

■ **LANCS STATUS** Unknown   1857

The only reference to the possible presence of this species in Lancashire comes from Stainton (1859), where it is reported from the Manchester district, without date, recorder or more precise location. In Ellis (1890), it is noted that the year in question was 1857, also listing Stainton as the source, but not giving a specific reference. The use of 1857 by Ellis suggests more information was available to him but, if so, this has not been located. The area covered by Stainton's Manchester correspondents (and who they were) are not detailed, so it is possible that this could have originated from a part of the Manchester district outside VC59. This reference is also noted in the Cheshire database without any additional information and the same reference. Additionally, it is noted from another Cheshire locality, possibly between 1960 and 1985, which has proved impossible to trace (Cheshire-moth-charts.co.uk, website accessed 29.8.2023).

■ **HABITAT AND FOODPLANTS** Unknown in the county. Elsewhere, associated with the fungus Alder Bracket on dead Alder trunks. May be present and overlooked, although its main areas nationally are in North-east England and Central Highlands of Scotland.

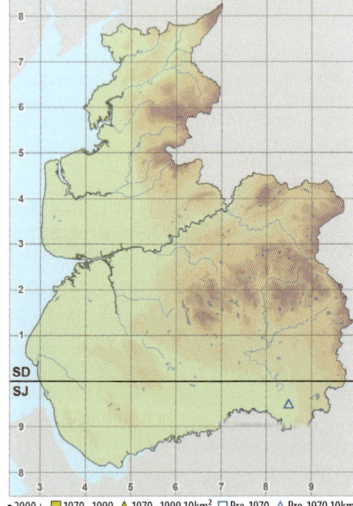

• 2000+   □ 1970–1999   ▲ 1970–1999 10km²   □ Pre-1970   △ Pre-1970 10km²

12.025 Tapestry Moth *Trichophaga tapetzella*

■ **LANCS STATUS** Adventive   1856–1922

First noted on 24 June 1856, as 'abundant in the hold of the cargo ship Ferris', containing wool and hides from Montevideo (Cooke, 1856). Reported as a destructive insect (Gregson, 1857) and abundant everywhere in woollen materials (Ellis, 1890), with nearly all records from Liverpool and its Docks. Also reported from the Manchester area (Stainton, 1859). By the early 20th century, it was described as not often recorded (Mansbridge, 1940) and the last record appears to be in 1922, by Mansbridge, who bred two specimens, now in Manchester Museum.

■ **HABITAT** Formerly, a temporary inhabitant of wool warehouses.
■ **FOODPLANTS** Wool products.

• 2000+   □ 1970–1999   ▲ 1970–1999 10km²   □ Pre-1970   △ Pre-1970 10km²

80   *The Moths of Lancashire*

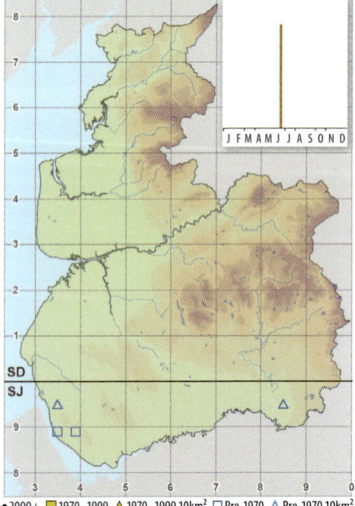

12.026 Common Clothes Moth *Tineola bisselliella*

■ **LANCS STATUS** Occasional / local resident c.1857–2022

First noted in the Liverpool district (Gregson, 1857) and Manchester area (Stainton, 1859); thereafter, as a common pest throughout, including Didsbury in 1953 (H.N. Michaelis). The next record was not until 2003, from St Annes, followed by many reports from a relatively small number of sites, including Chorlton, Fallowfield, Fulwood (Preston) and Victoria Park (Manchester). Almost certainly under-recorded.

■ **HABITAT** In houses.

■ **FOODPLANTS** Woollen fabrics, including carpets, jackets and jumpers.

Continuously brooded depending on the conditions in the house. Comes to light. Is sometimes found in plague proportions but much less frequently than in the past.

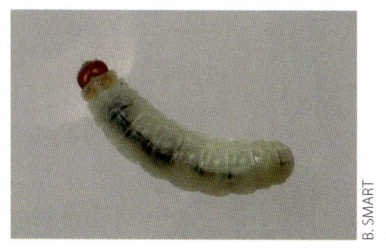

• 2000+ ☐ 1970–1999 ▲ 1970–1999 10km² ☐ Pre-1970 △ Pre-1970 10km²

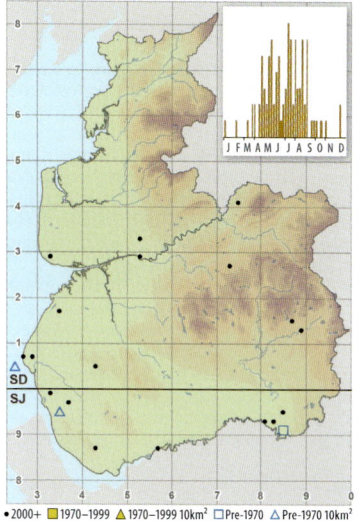

12.027 Case-bearing Clothes Moth *Tinea pellionella*

■ **LANCS STATUS** Occasional / widespread resident c.1857–2022

First recorded in the Liverpool district (Gregson, 1857), it was later noted as a very common pest by other early authors. One from Liverpool on 3 July 1918 (W.A. Tyerman) was, much later, dissected to confirm the identity, a practise continued with modern records. Regularly recorded in moth-recorder's houses, such as in Flixton, Preston and St Helens but is almost certainly under-recorded elsewhere.

■ **HABITAT** In houses and out-buildings.

■ **FOODPLANTS** Case found indoors, larva not feeding. Historic accounts give wool & fur.

Indoor records suggest it has multiple broods but with a peak during the summer months. Occurs mostly indoors in small numbers and does come to light.

• 2000+ ☐ 1970–1999 ▲ 1970–1999 10km² ☐ Pre-1970 △ Pre-1970 10km²

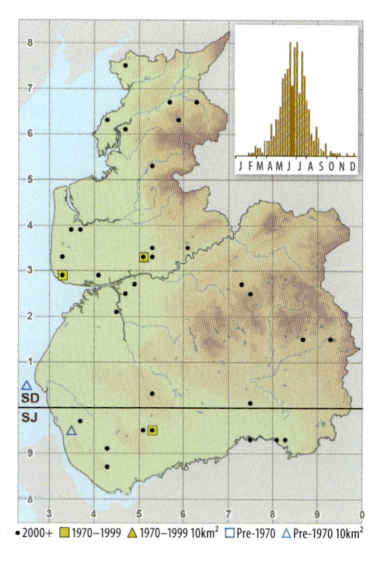

12.0272 *Tinea translucens*

■ **LANCS STATUS** Adventive 1856

First noted under *T. merdella* Zell. by N. Cooke (Cooke, 1856a), who took several specimens in a warehouse containing wool from Naples, Trieste, Peru and Australia on 11 July and 14 July 1856. Gregson (1857) and Ellis (1890) repeat this, under *T. merdella*. As part of a study on the *Tinea* species, Robinson (1979) critically examined Cooke's 1856 *T. merdella* specimens from Liverpool which, at this stage, had been renamed *T. metonella* by Pierce & Metcalfe (1934). Robinson's work finally proved them to be *T. translucens*, and one was designated as a species lectotype.

■ **HABITAT** In wool warehouses, following importation.

■ **FOODPLANTS** Believed to feed on wool.

This is very similar externally to the common and widespread *Tinea pellionella*. As with other similar looking, but apparently rare, members of this genus, it will only be possible to establish their true status by retention and microscopic examination of more potential *T. pellionella*.

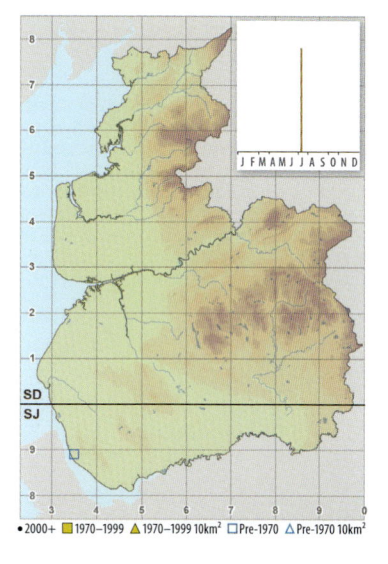

• 2000+ ☐ 1970–1999 ▲ 1970–1999 10km² ☐ Pre-1970 △ Pre-1970 10km²

The Moths of Lancashire

TINEIDAE

12.028 Tinea dubiella

■ **LANCS STATUS** Rare / very local resident   1859–1999

This species was described, new to science, by Stainton (1859a), based on specimens caught in the Liverpool area in 1859 by C.S. Gregson. Later, six specimens (two males and four females) from the same area and recorder were critically examined by Robinson (1979), and one female, caught on 18 August 1859, was designated a species lectotype. Since then, there has been only one other record, this being of a male found alive, indoors at a window in RSPB Leighton Moss's staff rest room and kitchen on 12 August 1999 (S. Palmer). The subsequent permanent genitalia slide was sent to G.S. Robinson, who confirmed it as *T. dubiella*.

■ **HABITAT** Found indoors.

■ **FOODPLANTS** Not recorded in Lancashire.

The genitalia prep. slide for the 1999 record was returned to the recorder following examination by G.S. Robinson. However, several years later, the set specimen, retained in the recorders collection, was destroyed following a problem with an insect infestation. The genitalia slide, number 0130999, has been permanently lodged with Liverpool World Museum. Microscopic examination of more potential *Tinea pellionella* specimens, which are externally identical, may prove this species to be more widespread than is currently thought.

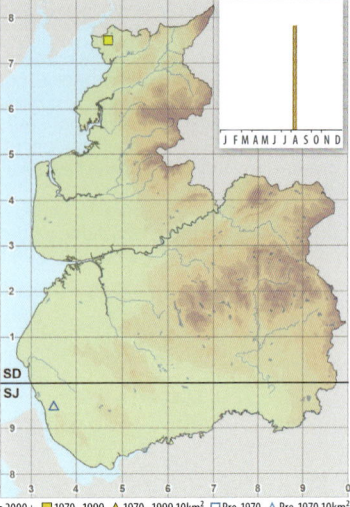

• 2000+   □ 1970–1999   △ 1970–1999 10km^2 □ Pre-1970   △ Pre-1970 10km^2

12.0291 Tinea lanella

■ **LANCS STATUS** Adventive   1922

Bred from larvae on Alpaca wool stored in a wool warehouse in Liverpool collected in, or emerging during, June to July 1922 by W. Mansbridge. He originally identified the specimens as *T. merdella* Zell., but by the time he published his updated regional list in 1940, the moths had been reidentified as *T. lanella*, a species new to science (Pierce & Metcalfe, 1934). Specimens were later critically examined by G.S. Robinson in 1979 and a male was designated lectotype. The type-series are also comprised of the Liverpool specimens (Robinson, 1979).

■ **HABITAT** Found indoors.

■ **FOODPLANTS** Imported woollen materials, although only Alpaca has been documented.

Heath and Emmet (1985) report that, in the British Isles, it is only known from the moths collected in a wool warehouse in Liverpool and that it has subsequently been found in Spain and Romania. They comment further that its true home is most probably elsewhere. A continental publication, *Microlepidoptera of Europe, volume 9, Tineidae II*, by R. Gaedike (2019) repeats the above and adds, 'the large gaps in the hitherto known distribution are zoogeographically not explainable'.

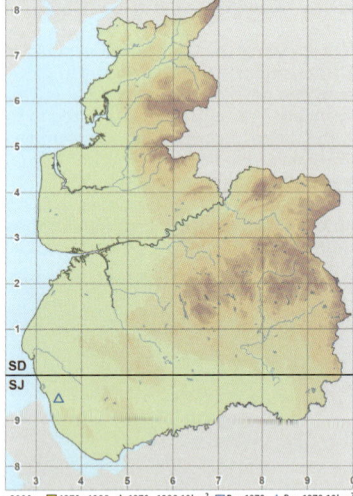

• 2000+   □ 1970–1999   △ 1970–1999 10km^2 □ Pre-1970   △ Pre-1970 10km^2

12.030 Large Pale Clothes Moth *Tinea pallescentella*

■ **LANCS STATUS** Occasional / local resident   c.1857–2021

First noted in the Liverpool district (Gregson, 1857) and then in the Manchester area (Stainton, 1859). During 1917, it was reported from Liverpool Dock warehouses, where it was abundant, although considered only common by 1940 (Mansbridge, 1940). Thereafter, not recorded until 1989 at Martin Mere, and on 27 January 1999 in Flixton, both in light traps. Since then, it has been noted at eight widely separated localities, most frequently in Billinge, Chorlton and Morecambe.

■ **HABITAT** Formerly indoors in wool and skin warehouses; recently in gardens.

■ **FOODPLANTS** Not recorded in Lancashire.

Capable of occurring at any time of year, indoors or outdoors; comes to light.

• 2000+   □ 1970–1999   △ 1970–1999 10km^2 □ Pre-1970   △ Pre-1970 10km^2

82   *The Moths of Lancashire*

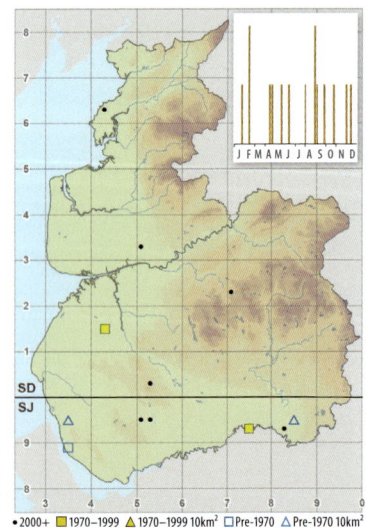

12.032 *Tinea semifulvella*

■ **LANCS STATUS** Occasional / widespread resident   c.1857–2022
Recorded from the Liverpool district (Gregson, 1857) and later, in Knowsley in 1905. Up to 1990, it was noted at several sites, including Formby and Leighton Moss. From 1995 onwards, widespread light trapping produced records from many lowland areas, particularly in VC59, with Flixton (K. McCabe) producing over half the counties 460 records. Other regular sites included Haydock, Parr and Preston. Most reports are of low numbers, but 167 came to a light trap in Hale in August 2019 (C. Cockbain).
■ **HABITAT** Hedges and gardens.
■ **FOODPLANTS** Bred from a bird's nest.
Appears to be double brooded, the second smaller than the first. Comes to light.

• 2000+   ■ 1970–1999   ▲ 1970–1999 10km²   □ Pre-1970   △ Pre-1970 10km²

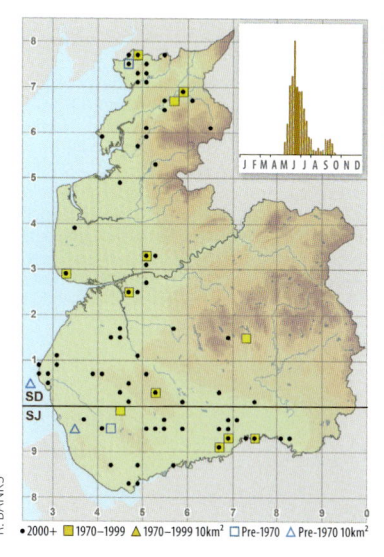

12.033 *Tinea trinotella*

■ **LANCS STATUS** Frequent / widespread resident   1844–2022
First noted in the Manchester area on 2 June 1844 (Edleston, 1844) and later in Liverpool, Pendleton, Eccles and Preston. Few other reports were received until 1995, when the expanding use of light traps across the county led to a large increase in records. These were predominantly from lowland areas, particularly in VC59, and mostly involved low single figures. Sites with more regular reports include Billinge, Fazakerley, Flixton, Parr, St Annes, St Helens, Swinton and Woolton.
■ **HABITAT** Hedges and gardens.
■ **FOODPLANTS** Bred in numbers from a Long-tailed Tit's nest, by D. Miller in 2010.
Appears to be double brooded. Comes readily to light.

• 2000+   ■ 1970–1999   ▲ 1970–1999 10km²   □ Pre-1970   △ Pre-1970 10km²

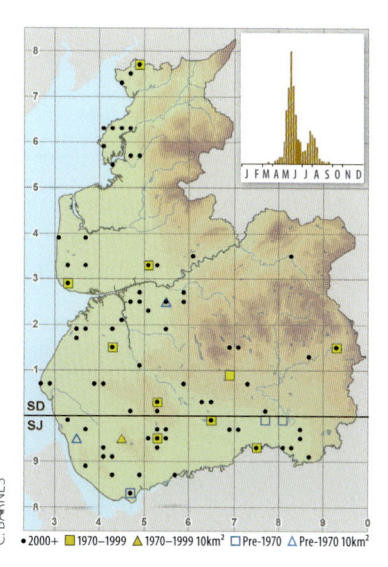

12.034 Brown-dotted Clothes Moth *Niditinea fuscella*

■ **LANCS STATUS** Scarce / local resident   c.1857–2018
Formerly abundant in Liverpool and Manchester areas, as noted by Gregson (1857) and Ellis (1890). It appeared to decline dramatically during the early 20th century, with single records from Liverpool in 1905, and Burnley in 1910. It was next recorded in Lytham (1976) and since then has been erratic in appearance. It was occasional in Billinge between 2000 and 2007 and, in Dolphinholme, was regular from 2005 to 2009 (N. Rogers). Only noted at two sites since, the last in Woolton (S. McWilliam).
■ **HABITAT** In houses, outbuildings and gardens.
■ **FOODPLANTS** Not recorded in Lancashire.
Brood situation unknown. Comes to light and found, slightly less frequently, indoors.

• 2000+   ■ 1970–1999   ▲ 1970–1999 10km²   □ Pre-1970   △ Pre-1970 10km²

The Moths of Lancashire

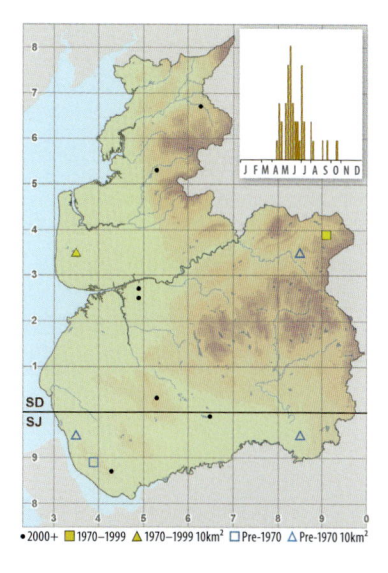

TINEIDAE

12.036 Skin Moth *Monopis laevigella*

■ **LANCS STATUS** Occasional / widespread resident   1950–2022

Not widely known to be separated from *M. weaverella* until 1910 (Heath & Emmet, 1985) and, as a result, early records from Liverpool are considered unverified. The first confirmed record is one from Old Trafford on 18 May 1950 (H.L. Burrows), followed by one in Hardhorn in 1985. Since 1995 it has been noted in light traps mainly, but not exclusively, operated in lowland areas. Sites with more regular records include, Billinge, Flixton, Haydock, Morecambe, Parr, Preston and St Annes.
■ **HABITAT** Hedgerows, copses and gardens.
■ **FOODPLANTS** Bird's nests, including Great Tit and Long-tailed Tit.
Double brooded. Comes to light. Ten reared from single nests on three occasions.

● 2000+ ☐ 1970–1999 ▲ 1970–1999 10km² ☐ Pre-1970 △ Pre-1970 10km²

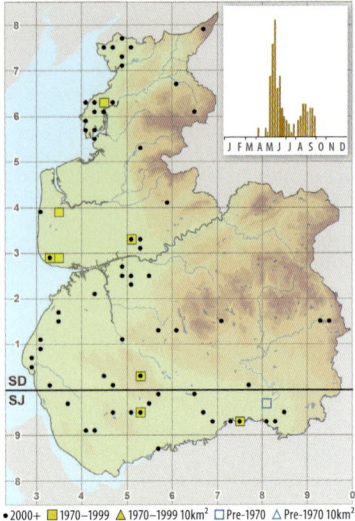

12.038 *Monopis obviella*

■ **LANCS STATUS** Scarce / very local resident   1999–2006

This and *M. crocicapitella* have been confused for a long time, including in recent years, and the records listed involved critical examination of at least one specimen from each site. The first was in Flixton on 17 July 1999, followed by one on 16 June 2000 (K. McCabe). On the 4 July 2000, one came to light in Parr, and five more singletons appeared here over the next few years, the last on 4 July 2006 (all R. Banks). The only other record was from Bold Moss on 9 July 2005 (R. Banks).
■ **HABITAT** Gardens and brownfield sites
■ **FOODPLANTS** Not recorded in Lancashire.
Single brooded. An occasional visitor to light.

● 2000+ ☐ 1970–1999 ▲ 1970–1999 10km² ☐ Pre-1970 △ Pre-1970 10km²

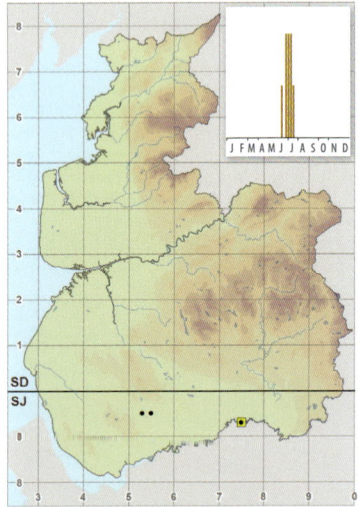

12.039 *Monopis crocicapitella*

■ **LANCS STATUS** Scarce / very local resident   2013–2021

This and *M. obviella* have been confused for a considerable time, including in recent years. The records listed have therefore involved critical examination of at least one specimen from each site. The first came to light in West Derby, Liverpool on 6 October 2013 (C. Daly). It was not seen again until 29 May 2020 in Fazakerley and then subsequently at the same site on three occasions during 2021. The last of these was on 7 July 2021 (L. Ward) and all were singletons.
■ **HABITAT** Gardens.
■ **FOODPLANTS** Not recorded in Lancashire.
Possibly double brooded and comes occasionally to light.

● 2000+ ☐ 1970–1999 ▲ 1970–1999 10km² ☐ Pre-1970 △ Pre-1970 10km²

84 *The Moths of Lancashire*

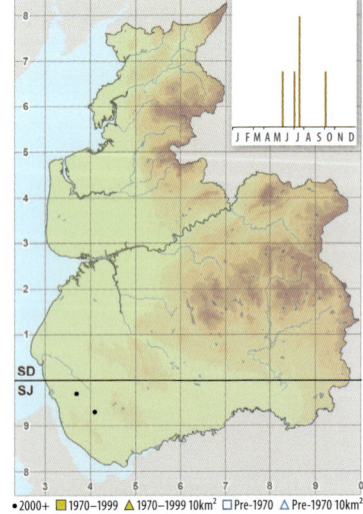

12.040 *Monopis imella*

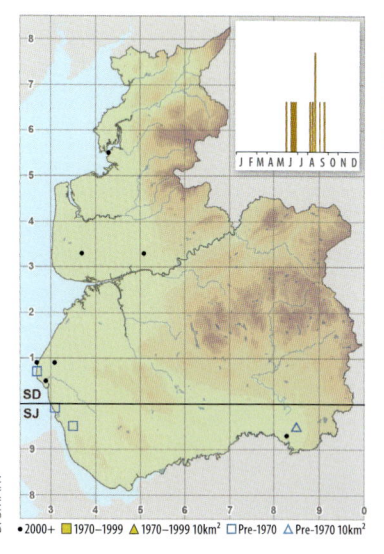

■ **LANCS STATUS** Scarce / local resident   c.1857–2022

First found at Linacre Marsh on old posts (Gregson, 1857), then in the Manchester district (1859) and on Crosby dunes (1912). A larva was found (and bred) in Formby in 1921 (W. Mansbridge) and an adult, also seen here, on 19 June 2000. Thereafter, sites included Chorlton in 2002, Altcar in 2003, Freshfield in 2004, Preston in 2005 and Sunderland Point in 2008. The only other record came when one was attracted to a light trap run in a nursery, near Westby on 11 June 2022 (J. Steeden).
■ **HABITAT** Mainly coastal, on sand dunes or estuaries; also gardens and a plant nursery.
■ **FOODPLANTS** Although bred in 1921, no note was published detailing the pabulum. Double brooded. Comes to light, almost always as a singleton.

● 2000+   ☐ 1970–1999   ▲ 1970–1999 10km²   ☐ Pre-1970   △ Pre-1970 10km²

12.044 *Haplotinea insectella*

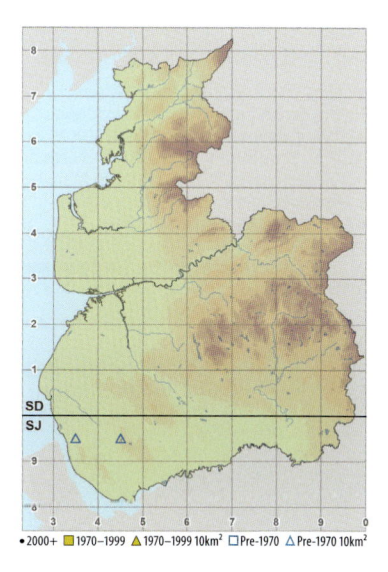

■ **LANCS STATUS** Considered Extinct   1880 and/or 1917

The only direct clue to the presence of *H. insectella* in Lancashire is a dot on a map of 'confirmed' records (but with no supporting details) for VC59 in Heath & Emmet (1985). Prior to this, two records of *Tinea misella* Zell. are listed in Mansbridge (1940), these being from Huyton in 1880 (J.H. Threlfall) and the Liverpool district in 1917 (W.A. Tyerman). In 1938, it was discovered that *T. misella* consisted of two superficially indistinguishable species, *H. insectella* and *H. ditella* (Pierce, Metcalfe & Diakonoff, 1938). It is therefore assumed, in the absence of any other known records, that the VC59 dot refers to one or both of the above records but at present it has not been possible to establish any more details relating to this.

● 2000+   ☐ 1970–1999   ▲ 1970–1999 10km²   ☐ Pre-1970   △ Pre-1970 10km²

12.0442 *Lindera tessellatella*

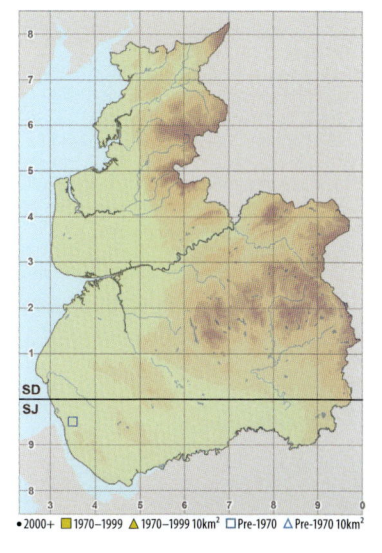

■ **LANCS STATUS** Adventive   1943

First reported, as new to the British Isles, breeding in floor sweepings in a mill at Bootle, Lancashire in January 1943. It is noted that the size and head scale arrangement give it a superficial appearance of an Oecophorid. The diet of this species is said to include remains of mites and lepidopterous larvae, and its geographical range encompasses the subtropical regions of both the Old and New World (E.C. Pelham Clinton in Heath & Emmet, 1985a). The only other British record of this pantropical species is believed to be from a warehouse in London in 1943 (Agassiz *et al.*, 2013).
■ **HABITAT** Indoors, in a mill.
■ **FOODPLANTS** Floor sweepings.
Access to the original paper covering this (Stringer, 1943), has not proved possible.

● 2000+   ☐ 1970–1999   ▲ 1970–1999 10km²   ☐ Pre-1970   △ Pre-1970 10km²

TINEIDAE

12.046 Yellow V Moth *Oinophila v-flava*

■ **LANCS STATUS** Considered Extinct   c.1890–1917

The first report came from Ashton-on-Ribble, prior to 1890, by J.B. Hodgkinson, but no mention is made of the date in Ellis (1890). The only other documented reports are of two in Liverpool by F.N. Pierce, during 1917, inside his place of work in Castle Street, Liverpool, one of which was retained. The Lancashire and Cheshire Fauna Society, Annual Lepidoptera Report of 1917 by W. Mansbridge, pages 13–16, detail the sightings.

■ **HABITAT** Indoors.
■ **FOODPLANTS** Not recorded in Lancashire.

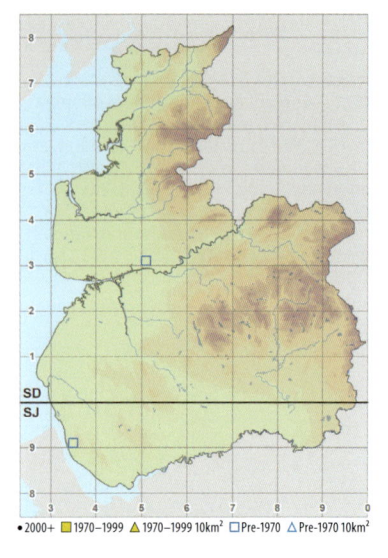

• 2000+   □ 1970–1999   △ 1970–1999 10km^2 □ Pre-1970   △ Pre-1970 10km^2

12.047 *Psychoides verhuella*

■ **LANCS STATUS** Scarce / local resident   c.1890–2022

First noted in Cleveleys by J.B. Hodgkinson (Ellis, 1890). In April 2004 it was found, as a larva, in Eaves Wood and next, also as a larva, in the Douglas Valley on 26 February 2008 (C. Darbyshire). Most other records are from lowland limestone woods in VC60, but has also been found at Martin Mere between 2013 and 2016.

■ **HABITAT** Where the foodplant occurs, in shady, damp woodland edges and lanes.
■ **FOODPLANTS** Hart's-tongue Fern.

Not recorded at light; has been attracted to a TIP pheromone lure. Most records are of larval cases in the Spring. The foodplant is widespread, but this species appears to be restricted in distribution by other factors.

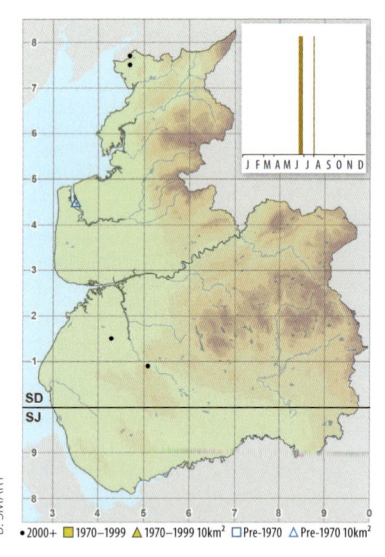

• 2000+   □ 1970–1999   △ 1970–1999 10km^2 □ Pre-1970   △ Pre-1970 10km^2

12.048 *Psychoides filicivora*

■ **LANCS STATUS** Frequent / widespread resident   2003–2022

It was first noted in Ireland in 1909, arrived in England by 1940 and reached Cheshire in 2000. The first Lancashire record was in St Annes, at light, on 8 August 2003 (D. & J. Steeden). It has since established itself in scattered localities across the county, but wasn't recorded in VC59, in Flixton, until February 2007. Often numerous, up to 100 larvae have been seen, such as in Bolton-le-Sands in August 2015 (S. & B. Garland).

■ **HABITAT** Woodland, shady hedgerows and gardens, wherever ferns occur.
■ **FOODPLANTS** Hart's-tongue, Male and Soft Shield Ferns and Common Polypody.

Has a succession of generations. It has been noted in numbers as an adult, including at light, and as a larva, in almost every month of the year. Will fly into houses.

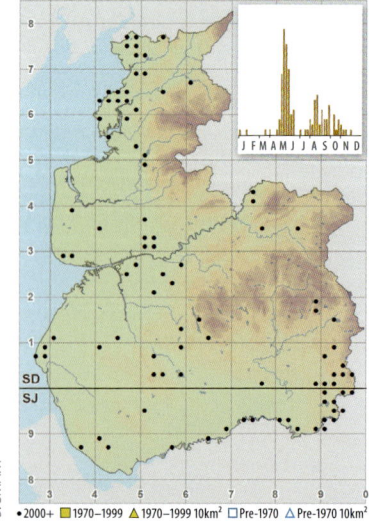

• 2000+   □ 1970–1999   △ 1970–1999 10km^2 □ Pre-1970   △ Pre-1970 10km^2

The Moths of Lancashire

13.002 *Roeslerstammia erxlebella*

■ **LANCS STATUS** Occasional / local resident 2003–2022

First noted by C.A. Darbyshire at Billinge Hill on 22 April 2003. Most records since have been from VC59, e.g., Wigan, Flixton, Chorlton, Swinton, with more northerly records from Preston, Warton, Gait Barrows, etc. Noted 76 times, roughly half as adults netted or at light, and half as the mines created by early instar larvae in leaf-tips, including 20+ from Oldham and Tameside by S.H. Hind in 2021–2022.
■ **HABITAT** Woodland edges where the foodplants occur. In upland and lowland areas.
■ **FOODPLANTS** Lime, including Small-leaved Lime, and birch.
Single brooded. A tenanted birch mine found by K. McCabe and B. Smart at Clock Face C.P., St Helens on 10 September 2020 was successfully bred through.

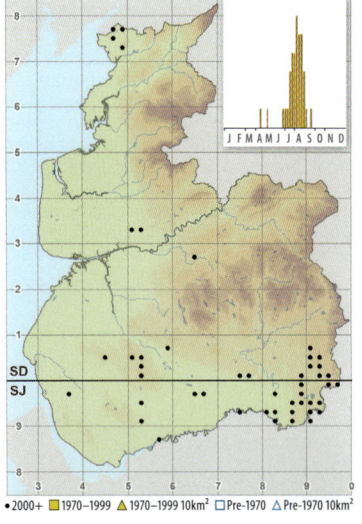

ROESLERSTAMMIIDAE

14.001 *Bucculatrix cristatella*

■ **LANCS STATUS** Occasional / local resident 1912–2019

First recorded in the Formby area in 1912 by W.A. Tyerman, and next at Crosby by W. Mansbridge, who described the species as 'scarce and local' (1940). More recently recorded on 21 April 2004 by K. McCabe, who found mines, larvae, and moulting cocoonets on Yarrow at Flixton. Subsequent records all from southern areas of the county, other than a moth at Fairhaven on 28 July 2015 (D. & J. Steeden).
■ **HABITAT** Rough grassland in lowland areas.
■ **FOODPLANTS** Yarrow. Larva mines the leaf initially; then feeds externally.
Double brooded. Most records are of the early stages. The Fairhaven moth was observed during daytime fieldwork. Once recorded at light, Flixton 15 July 2005.

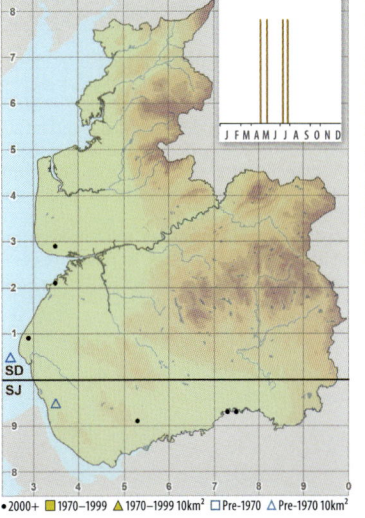

BUCCULATRICIDAE

The Moths of Lancashire

BUCCULATRICIDAE

14.002 *Bucculatrix nigricomella*

■ **LANCS STATUS** Frequent / widespread resident   c.1889–2022

First recorded in the late 19th century at Preston by J.H. Threlfall (Ellis, 1890), but not again until 1998 in Flixton. There have been around 100 records since, mostly of the adults. Early stages were first noted at Flixton in 2003 with at least five mines in a single leaf (K. McCabe). Fairly widespread over the county, mainly in lowland areas, but also at a few upland sites such as Rishton and Briercliffe.
■ **HABITAT** Verges, scrub, grassland.
■ **FOODPLANTS** Oxeye Daisy; initially mining the leaf before feeding externally. Double brooded. Tenanted mines, grazing larvae and moulting cocoonets found on leaves of the foodplant. Comes to light; occasionally disturbed during the day.

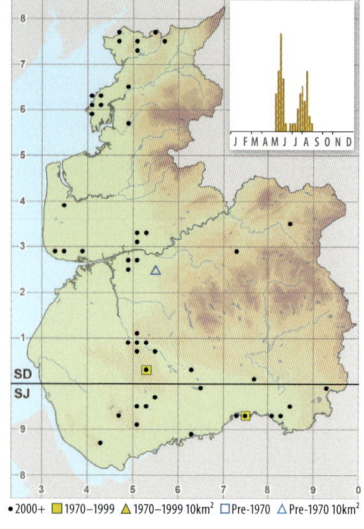

• 2000+   □ 1970–1999   △ 1970–1999 10km²   □ Pre-1970   △ Pre-1970 10km²

14.003 *Bucculatrix maritima*

■ **LANCS STATUS** Occasional / local resident   1872–2022

Larvae recorded on the Wyre estuary, 7 August 1872, by J.B. Hodgkinson and J.H. Threlfall. Later, at Stalmine (Ellis, 1890), then Bolton-le-Sands and Southport (Mansbridge, 1940). Unrecorded again until noted at Lytham by D.&J. Steeden in 1997. Further coastal records from VC59 at Hale, Marshside and Southport, and in VC60 from Bolton-le-Sands, Fluke Hall, Glasson, Sunderland Point, etc.
■ **HABITAT** Saltmarshes.
■ **FOODPLANTS** Sea Aster. The larva mines the leaves, readily switching to a fresh leaf. Double brooded. The adult moth has been recorded on twelve occasions, attracted to light on five of these. Tenanted leaf-mines recorded in April, May and August.

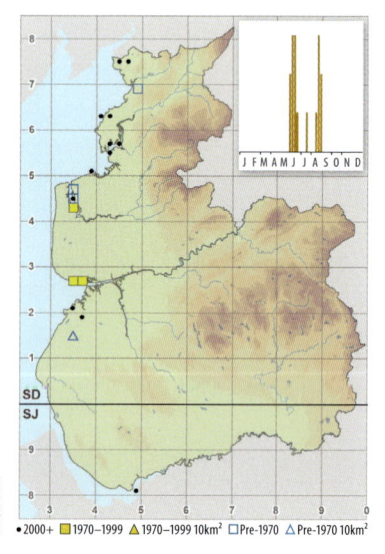

• 2000+   □ 1970–1999   △ 1970–1999 10km²   □ Pre-1970   △ Pre-1970 10km²

14.007 *Bucculatrix albedinella*

■ **LANCS STATUS** Scarce / very local resident   2022–2023

Over 30 vacated mines found in an elm hedgerow at Ravenmeols, Formby on 26 August 2022 (B. Smart). The mines showed the characteristic appearance of the species, with 2–4 diverticulae (blind endings). Smaller numbers of vacated mines later found at Ainsdale on 22 September 2022. Unclear if these are long-standing or more recent arrivals. Cheshire seems an unlikely source, as only recorded once. Rare in Yorkshire too, and limited to the south-east of the county, far from the Lancs border.
■ **HABITAT** Hedgerows.
■ **FOODPLANTS** Elm. Larvae, mining in early instars and later feeding on the underside of the leaves, were noted at Ravenmeols on 4 July 2023.

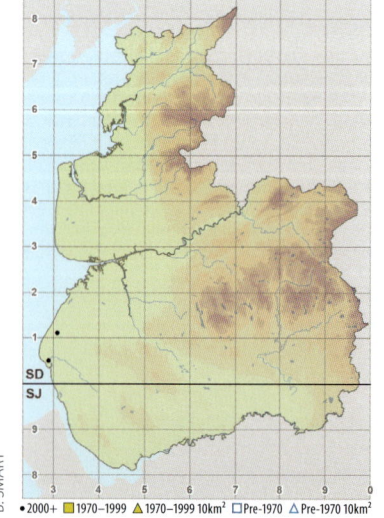

• 2000+   □ 1970–1999   △ 1970–1999 10km²   □ Pre-1970   △ Pre-1970 10km²

14.008 *Bucculatrix cidarella*

■ **LANCS STATUS** Frequent / widespread resident   c.1859–2022
Described as abundant in Manchester (Stainton, 1859), although it was 143 years before the next record, with leaf-mines in Flixton, 2002 (K. McCabe). Most of the 50+ records are of leaf-mines and mostly in VC59. Numerous records from Tameside, Oldham and Stockport from 2021 to 2022 by S.H. Hind. Sites in VC60 include Gait Barrows, Birk Bank and Preston. Absent from Sefton and Fylde coasts.
■ **HABITAT** Wetland sites, woodland edges where Alder is present.
■ **FOODPLANTS** Alder. Larva mines the leaf initially; then feeds externally.
Single brooded. Larva beaten from Alder at Chorlton 28 September 2014, with adult emerging May 2015. Adult records are of moths at light or noted during the day.

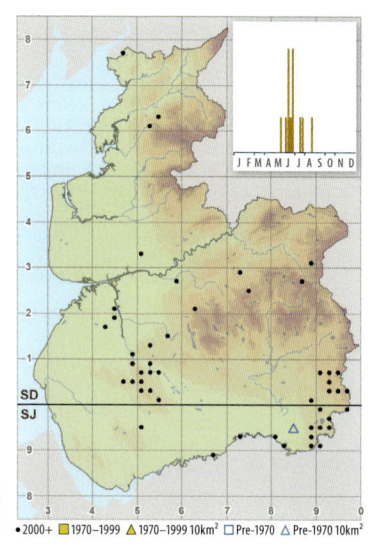

14.009 *Bucculatrix thoracella*

■ **LANCS STATUS** Frequent / widespread resident   c.1859–2022
First recorded in Manchester during the mid-19th century (Stainton, 1859). It was not until 1999 that it was further recorded in the county, when trapped at Flixton by K. McCabe, with mines found at Burtonwood by S.H. Hind in the same year. The subsequent 200 records demonstrate significant range expansion in the south of the county, less so in VC60, as unrecorded from Preston to Morecambe Bay.
■ **HABITAT** Woodland edges, gardens, roadside trees.
■ **FOODPLANTS** Lime, incl. Small-leaved Lime and hybrids; larva initially mining the leaf. Double brooded. Records of larvae, mines and moulting cocoonets on leaves, and pupal cocoons on lime trunks. Adults recorded at light and on lime during the day.

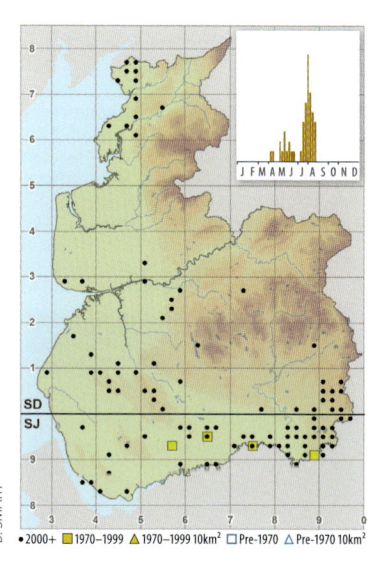

14.010 *Bucculatrix ulmella*

■ **LANCS STATUS** Frequent / widespread resident   c.1859–2022
First reported in Manchester district (Stainton, 1859) and by J.H. Threlfall at Preston (Ellis, 1890). As with many other micro-moths, it was then unrecorded for over 100 years. Noted by A.M. Emmet at Gait Barrows in 1995. 120 further records, mainly from lowland areas in the south, but also on northern limestone, e.g., Dalton Crags.
■ **HABITAT** Woodland, hedgerows, parks.
■ **FOODPLANTS** Oak. Larva initially mines, then feeds externally beneath the leaf. Double brooded. Most records are of mines. Adult recorded over 35 times, at light and netted by day. A couple have been dissected. Caution is required as the very similar elm-feeding species *B. ulmifoliae* is spreading from the southern counties.

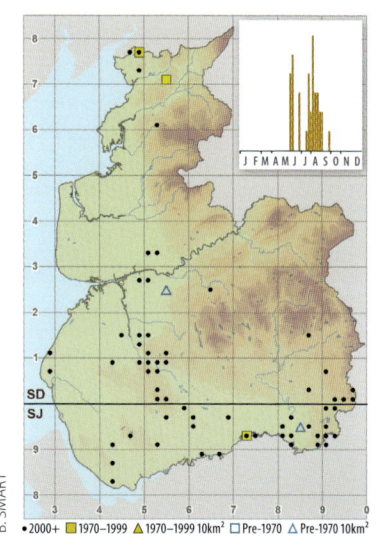

BUCCULATRICIDAE

The Moths of Lancashire   **89**

BUCCULATRICIDAE

14.012 *Bucculatrix bechsteinella*

■ **LANCS STATUS** Scarce / very local resident   c.1857–2023

Reported from Liverpool district at Olive Mount and Mosley Hill (presumably Mossley Hill) by C.S. Gregson (1857). A Fleetwood specimen is in the NHM London, recorded by F.W. Whittle c.1900. Next, noted when adult male beaten from Dog Rose growing beneath Hawthorn at Formby by B. Smart on 3 June 2022. Around 40 vacated mines, cocoonets and a pupal cocoon were found on the same Hawthorn, 26 August 2022.
■ **HABITAT** Hedgerows.
■ **FOODPLANTS** Hawthorn. Larvae, mining in early instars and later feeding on the upperside of the leaves, were noted at Formby on 4 July 2023.
Likely double brooded, although the pattern locally has not yet been confirmed.

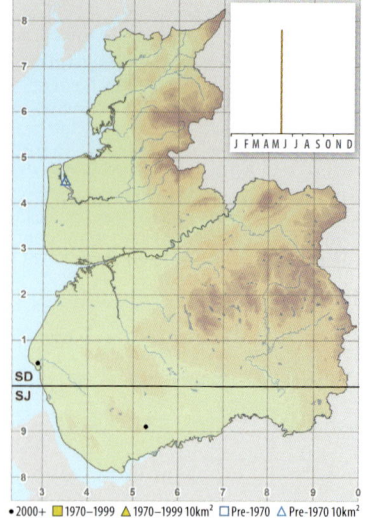

14.013 *Bucculatrix demaryella*

■ **LANCS STATUS** Scarce / very local resident   2004–2022

A fairly recent discovery, but whether previously overlooked or recently arrived is unclear. The first record was of an adult netted by S. Palmer at Gait Barrows on 22 May 2004. A few more adult records were followed in 2009 by one of leaf-mines, also at Gait Barrows (S. Tomlinson). All records are from VC60, with sites including Warton Crag, Leighton Moss, Yealand Hall Allotments, Trowbarrow Quarry and Docker Moor.
■ **HABITAT** Lowland limestone woods and scrub, and upland woodland.
■ **FOODPLANTS** Birch. The larva mines the leaf initially; then feeds externally.
Single brooded. Larvae and leaf-mines noted from July and August. Most records are of adults, found during the day and at dusk, and also recorded at light.

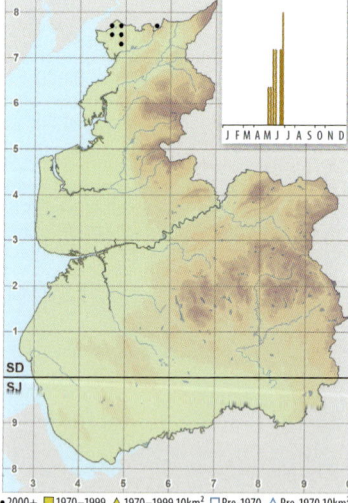

GRACILLARIIDAE

15.002 *Caloptilia cuculipennella*

■ **LANCS STATUS** Scarce / very local resident   2013–2021

First found on 20 September 2013, when single moths were netted at Gait Barrows and Yealand Storrs (S. Palmer & M. Young). It was then noted at Warton Crag on a few occasions between 2014 and 2021 (J. Patton) during the day and at light. A pupa, with old mines nearby, was found at Hawes Water in 2016 (B. Smart).
■ **HABITAT** Woodland and scrub on the lowland limestone of north Lancashire.
■ **FOODPLANTS** Ash.
Comes sparingly to actinic light. Readily disturbed during the day in small numbers. With the nearest UK colony in Cumbria having been there before 1890, it has either been overlooked or has recently started expanding its range southwards.

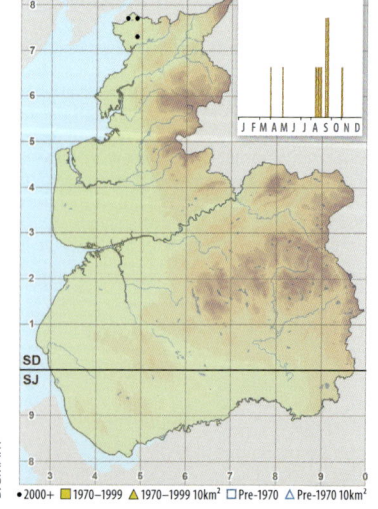

15.003 *Caloptilia populetorum*

■ **LANCS STATUS** Occasional / local resident   2013–2022

First noted in Cheshire in 2013, it reached Flixton in the same year, on 3 May (K. McCabe) and Chorlton in September. Since then, it has spread east along the Mersey valley, reaching Mossley in 2021, and north along the Sefton Coast, to Southport by 2020. In September 2022, a larva was found (and bred) in Trowbarrow, north VC60.

■ **HABITAT** Open woodland, scrub, mosses and brownfield sites.

■ **FOODPLANTS** Birch spp., including Silver Birch.

Double brooded, the second brood overwintering as an adult. Comes to light. Early stages have been found in late May, and August to September. Generally found in small numbers; seven came to light in Parr, St Helens on 19 August 2018 (R. Banks).

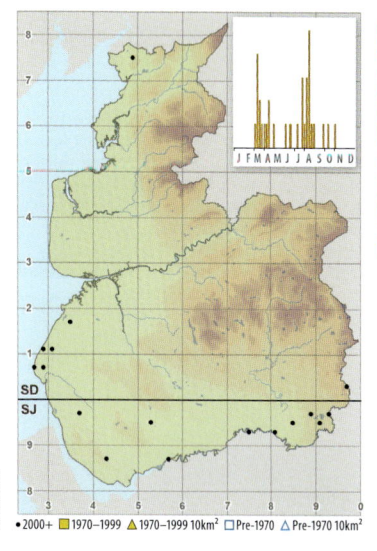

15.004 *Caloptilia elongella*

■ **LANCS STATUS** Frequent / widespread resident   1950–2022

First confirmed by M. Leech & H. Michaelis in 1950, who found it commonly among Alder in Formby. Since then, it has been noted at many well scattered sites in lowland and upland areas. Larval feeding cones can be readily found in summer and autumn.

■ **HABITAT** Damp woodland, scrub, gardens, parks and brownfield sites.

■ **FOODPLANTS** Alder and Grey Alder.

Double brooded, the second brood overwintering as an adult. Comes to light. Listed under *G. elongella* by Gregson (1857), Ellis (1890) and Mansbridge (1940), but none detail the larval foodplant. The very similar *C. betulicola* was not separated from *C. elongella* until 1927, so none of the early records can be attributed to species level.

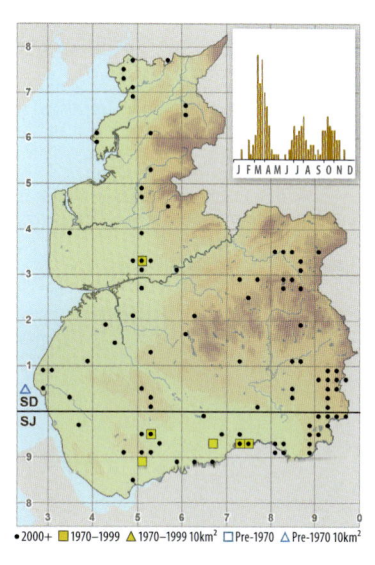

15.005 *Caloptilia betulicola*

■ **LANCS STATUS** Occasional / widespread resident   1957–2022

Not recognised as a Lancashire species (cf. *C. elongella*) until 1957, when it was bred from a mine on birch in Formby by H.N. Michaelis (Leech & Michaelis, 1957). Most frequently encountered in lowland areas, such as Astley Moss, Billinge, Flixton, Gait Barrows, Orrell, Longton and Winmarleigh Moss, usually with native birch growing nearby as self-seeding saplings to mature trees. Less frequent in upland areas.

■ **HABITAT** Mosses, moorland edge, scrub, brownfield sites and gardens.

■ **FOODPLANTS** Birch spp., including Downy Birch.

Double brooded, the second brood overwintering as an adult. Comes to light. The two broods of larval mines (tenanted) can be found in early and late summer.

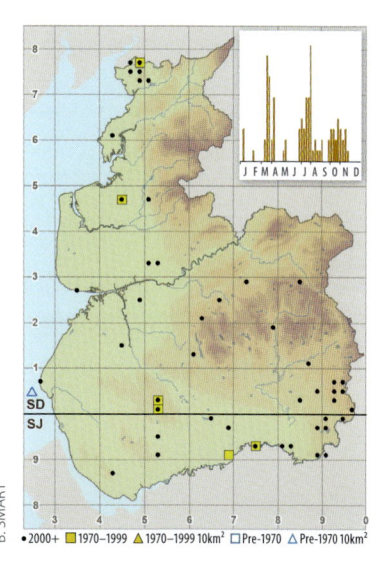

The Moths of Lancashire   **91**

GRACILLARIIDAE

15.006 *Caloptilia rufipennella*

■ **LANCS STATUS** Frequent / widespread resident   1994–2022

It first appeared in Britain in 1970, in Essex, and a few years later in East Scotland, which may explain why it was noted in VC59, on 6 August 1994, in Reddish Vale (E. Kearns), a year before it was first found in Cheshire. Occurs in both lowland and upland areas, with regular, annual sightings in areas such as Billinge, Fazakerley, Flixton, Ormskirk, Victoria Park, Parr and Preston, usually in single figures.

■ **HABITAT** Wherever Sycamore occurs, in woods, hedges, gardens and parks.

■ **FOODPLANTS** Sycamore.

Double brooded, the second brood overwintering as an adult. Larval cones are found in summer and autumn, but tenanted cones are rarely recorded.

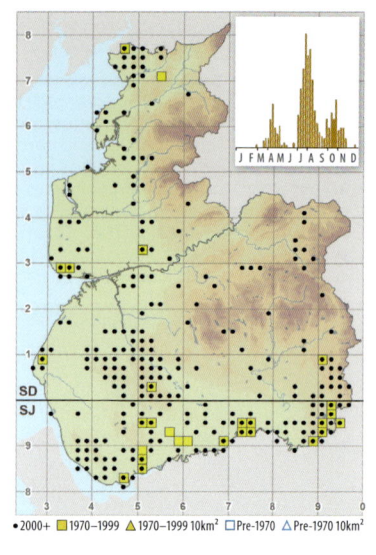

15.007 Azalea Leaf Miner *Caloptilia azaleella*

■ **LANCS STATUS** Occasional / local resident   1999–2022

An accidental introduction on imported Azaleas, it adapted to outdoor plants across southern England, spreading north and reaching Cheshire in 1994. First noted in Lancashire, at light, in Pennington on 8 May 1999 (P. Pugh). Since then, it has spread across much of southern, lowland VC59. Subsequently, the expansion rate slowed considerably, but it reached Heysham in 2009 and Brierclffe in 2016 (G. Turner).

■ **HABITAT** Gardens, parks and churchyards.

■ **FOODPLANTS** Azalea.

Appears to be triple brooded and comes readily to light. Mines or cones have been found from January to June and August to October, and a pupa in mid-March.

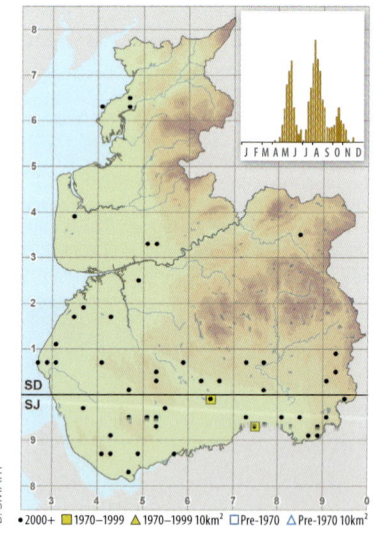

15.008 *Caloptilia alchimiella*

■ **LANCS STATUS** Scarce / very local resident   2003–2011

Originally listed as common and widespread, any noted prior to 1972 when the very similar *C. robustella* was identified as new to science, and without a current specimen to support the record, could refer to either species. Variations in forewing markings have continued to cause identification problems. Confirmed records are from Billinge on 15 July 2003 (C. Darbyshire), Longridge Fell woodland in 2004, Gait Barrows in 2004 and 2008, Astley Moss in 2005 and Baines Crag in 2011, the last by J. Girdley.

■ **HABITAT** Mature deciduous woodland and copses.

■ **FOODPLANTS** Oak sp.

Apparently single brooded. Comes to light. Larva and mines the same as *C. robustella*.

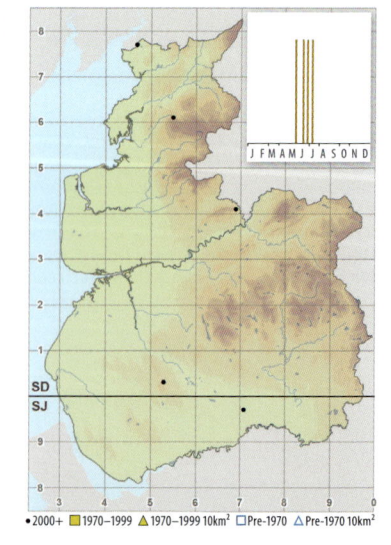

92   *The Moths of Lancashire*

15.009 *Caloptilia robustella*

■ **LANCS STATUS** Occasional / widespread resident   2000–2022

Discovered, new to science in 1972 (cf. *C. alchimiella*). First confirmed in Parbold, on 31 May 2000 (S. Palmer), but it remains primarily a VC59 species. It seems to be slowly expanding its range in the south and is found regularly in Billinge and Flixton. Much less frequent in VC60, although found in the north, at Botton Mill, in 2009.
■ **HABITAT** Woodland, copses and hedgerow standards.
■ **FOODPLANTS** Oak.

Probably double brooded. Comes to light. Larvae have only been found in October. The variability in forewing markings between this and *C. alchimiella* often cause identification problems and occasional sampling for dissection is recommended.

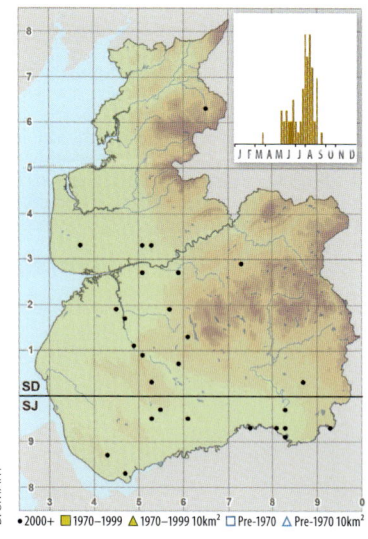

• 2000+   ▪ 1970–1999   ▲ 1970–1999 10km²   □ Pre-1970   △ Pre-1970 10km²

15.010 *Caloptilia stigmatella*

■ **LANCS STATUS** Frequent / widespread resident   1846–2022

Found at Cheetham Hill in June 1846 (Edleston, 1846), followed by Chat Moss, Preston, Formby and St Annes (Mansbridge, 1940). Occasionally recorded during the remainder of the 20th century at sites such as Carr House Common, Silverdale and Woolston. Thereafter, found widely across the whole county, mostly in single figures, at light or as a leaf mine. Very regular at many sites, particularly Flixton and Preston.
■ **HABITAT** Damp woodland, copses, sand dunes, scrub, gardens and brownfield sites.
■ **FOODPLANTS** Grey, Goat & Crack Willows, Osier, Aspen, Black-, Grey & White Poplar.

Double brooded, the second overwintering as an adult. Cones can be found from May to November, but are probably only tenanted in early summer and autumn.

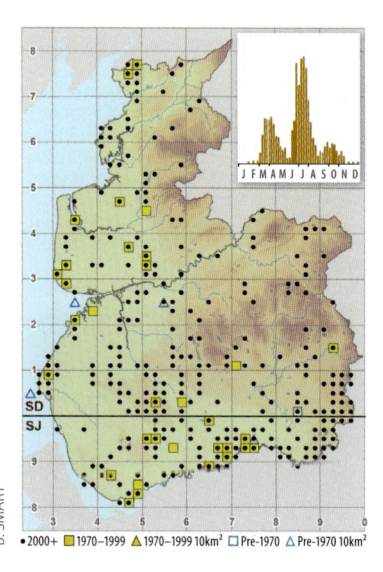

• 2000+   ▪ 1970–1999   ▲ 1970–1999 10km²   □ Pre-1970   △ Pre-1970 10km²

15.012 *Caloptilia semifascia*

■ **LANCS STATUS** Scarce / local resident   2020–2022

A recent addition to the Lancashire fauna, with the first found at light in Woolton, Liverpool, on 30 July 2020. Subsequently seen during 2021 and 2022 (all S.J. McWilliam). Elsewhere, it was found in Liverpool's Festival Gardens on 13 July 2022 (recorder unknown) and Parr, St Helens, on 13 August 2022 (R. Banks). Markings very variable within and between each of the two broods. Nationally, *C. honoratella* is expanding northwards and will need to be considered when recording *C. semifascia*.
■ **HABITAT** Parks, hedgerows and gardens.
■ **FOODPLANTS** Not recorded in Lancashire. Vacated cones, possibly of this sp., found on Field Maple in the Liverpool area a few years beforehand. Breeding through required.

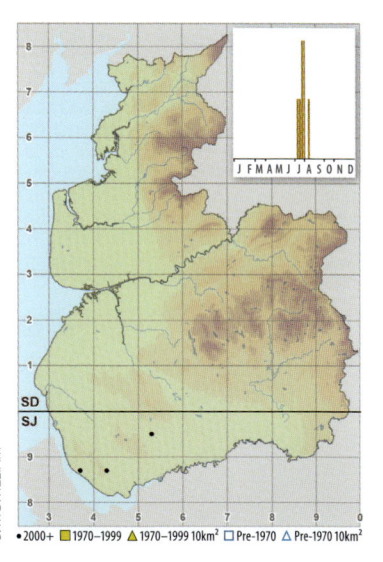

• 2000+   ▪ 1970–1999   ▲ 1970–1999 10km²   □ Pre-1970   △ Pre-1970 10km²

GRACILLARIIDAE

The Moths of Lancashire   **93**

GRACILLARIIDAE

15.014 *Gracillaria syringella*

■ **LANCS STATUS** Frequent / widespread resident   c.1857–2022

First noted in the Liverpool district, wherever Lilac grows, by Gregson (1857) and later in Sefton Park, Formby, Daubhill and St Helens. From the late 20th century onward, recorded widely across the whole county, at light and by its distinctive feeding signs, at sites such as Newton-with-Scales in 1987 (M. Evans), St Annes in 1991, Silverdale in 1994 and Lane Ends in 1996. Recent upland records have included those from Brierfield, Rochdale, Oldham Edge and Towneley Park.
■ **HABITAT** Woodland, hedgerows, gardens and limestone pavement with scrub growth.
■ **FOODPLANTS** Ash, Lilac and Privet.

Double brooded and has occasionally appeared in October. Comes to light.

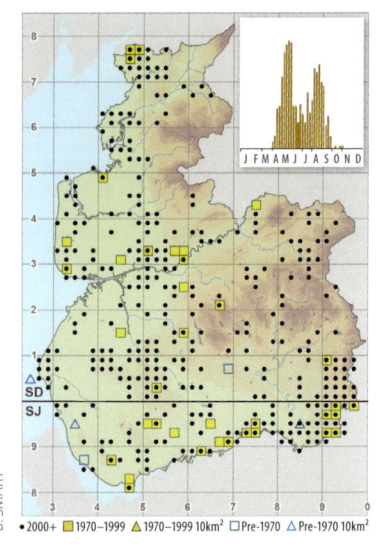

15.015 *Aspilapteryx tringipennella*

■ **LANCS STATUS** Frequent / widespread resident   c.1859–2022

First noted in Withington by J. Chappell c.1859 and later, from Fleetwood, Formby and Silverdale (Mansbridge, 1940). Since the mid-1990s, recorded widely at many lowland sites, as an adult to light and as a leaf-mine, the latter sometimes in moderate numbers. Most regular at sites such as Billinge, Chorlton, Flixton, Preston and St Helens. More local and less frequent further inland and approaching higher ground, in areas such as Adlington, Briercliffe, Rishton, Rochdale and Worsaw Hill.
■ **HABITAT** Rough grassland, on verges, banks, waste ground and brownfield sites.
■ **FOODPLANTS** Ribwort Plantain.

Double brooded. Comes to light and readily disturbed by day in sunshine.

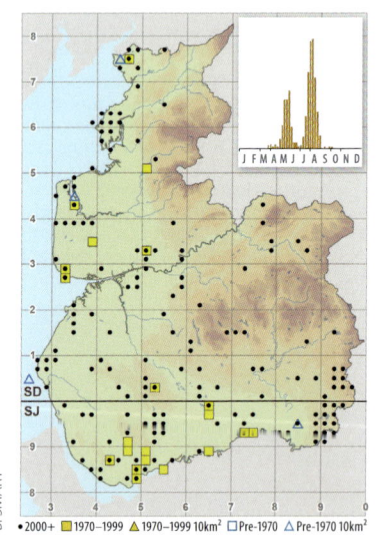

15.016 *Euspilapteryx auroguttella*

■ **LANCS STATUS** Occasional / local resident   c.1857–2022

Found in Woolton (Gregson, 1857), Chorlton in 1859, Pilling Moss in 1865 and Scorton c.1889. The next was not until 1995, at Gait Barrows (A.M. Emmet), a site with a few other reports to 2000. Elsewhere, in the north, it was regular on Heysham N.R. and Middleton N.R. from 2007 to 2022. In the south, it occurs mainly in the Mersey valley at sites such as Rixton Clay Pits, Flixton, Chorlton and Hollins Green.
■ **HABITAT** Rough grassland on lowland limestone, banks and brownfield sites.
■ **FOODPLANTS** St John's-wort, including Perforate St John's-wort.

Double brooded, with larvae found from late May to early July and end of August to mid-November. Rare at light.

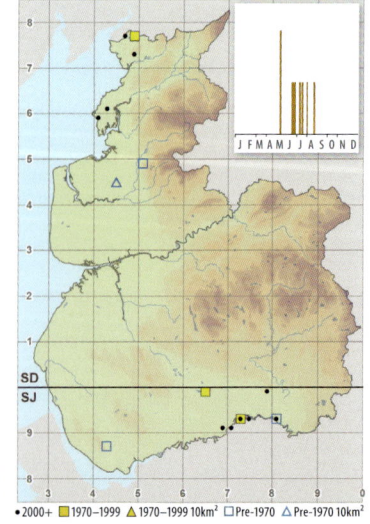

94   *The Moths of Lancashire*

GRACILLARIIDAE

15.017 *Calybites phasianipennella*

■ **LANCS STATUS** Occasional / local resident   1923–2022

Mansbridge (1940) noted it as scarce, with two records from Ainsdale in 1923. Not seen again until 14 August 2004 in St Annes, in Ormskirk in 2005, Roby in 2006 and Haydock in 2007. Recorded regularly at light in Hale on 14 occasions between 2008 and 2021 (C. Cockbain). The population appears stable, but is mainly restricted to VC59 in the Mersey valley, one or two brownfield sites and a few coastal areas.
■ **HABITAT** Rough grassland, riverbanks, waste ground and brownfield sites.
■ **FOODPLANTS** Sheep's Sorrel, Redshank and Common Bistort.
Probably double brooded. Comes to light. Larval cones found from August to October, sometimes in numbers, such as 50 in Ainsdale on 26 August 2021 (B. Smart).

• 2000+   ▢ 1970–1999   ▲ 1970–1999 10km²   ▢ Pre-1970   △ Pre-1970 10km²

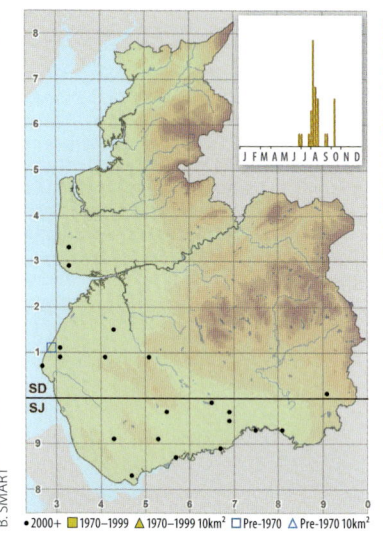

15.018 *Povolnya leucapennella*

■ **LANCS STATUS** Considered Extinct   c.1857

The sole record relates to an undated discovery by C.S. Gregson. In Gregson (1857), under *Coriscium sulphurellum* Haw., he reports that 'Boors Wood, Hale, is the only place I have met with this pretty insect in our district. August'.
■ **HABITAT** Deciduous woodland.
■ **FOODPLANTS** Not recorded in Lancashire.
Elsewhere in the British Isles, it feeds on oak (evergreen and deciduous) and possibly has a preference for young trees (Heath & Emmet, 1985). It is double brooded, with the adult of the second brood overwintering and with its long flight period the moth could be searched for in any month of the year.

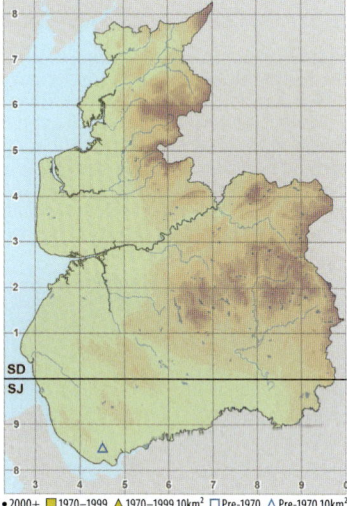

• 2000+   ▢ 1970–1999   ▲ 1970–1999 10km²   ▢ Pre-1970   △ Pre-1970 10km²

15.019 *Acrocercops brongniardella*

■ **LANCS STATUS** Occasional / widespread resident   2001–2022

This species had been spreading northwards nationally and was first found, when attracted to light in St Helens town centre, on 25 July 2001 (D. Owen). By 2004, it had moved into Billinge, Chorlton, Dalton (nr. Wigan), Flixton and Rufford, with larval mines sometimes abundant. Since then, it appears its range expansion has slowed considerably, reaching its most northerly point, Euxton, in 2007. It has continued eastwards along the Mersey Valley, reaching Ashton-under-Lyne in 2022 (S.H. Hind).
■ **HABITAT** Woodland and isolated oaks in parks and hedgerows.
■ **FOODPLANTS** Oak, including Pedunculate Oak.
Single brooded, overwintering as an adult. Tenanted mines found in May and June.

• 2000+   ▢ 1970–1999   ▲ 1970–1999 10km²   ▢ Pre-1970   △ Pre-1970 10km²

The Moths of Lancashire

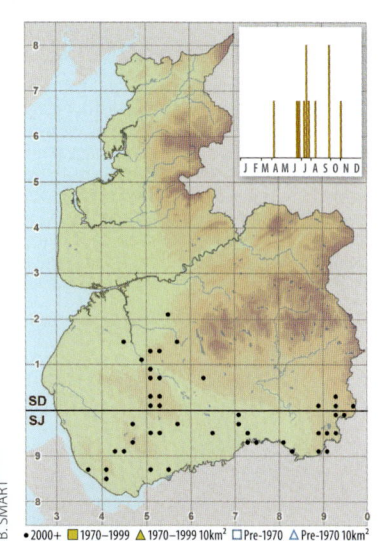

GRACILLARIIDAE

15.022 *Callisto denticulella*

■ **LANCS STATUS** Occasional / widespread resident c.1889–2022

First noted as 'local on Apple' in the Preston area by J.H. Threlfall (Ellis, 1890), later in the Formby and Silverdale areas (Mansbridge, 1940). Since 1998, it has been recorded widely, but in low numbers, wherever Apple or Crab Apple trees are encountered in the wider lowland countryside. Infrequently found in upland areas. Sites with more regular reports include Chorlton, Flixton and Sunderland Point.

■ **HABITAT** Edges of woodland, hedgerows, parks and gardens.
■ **FOODPLANTS** Apple, including Crab Apple.

Apparently single brooded, but not clear from the available data. Rarely comes to light. Nearly all records are of leaf mines or folds, with very few checked for larvae.

● 2000+ ☐ 1970–1999 △ 1970–1999 10km² ☐ Pre-1970 △ Pre-1970 10km²

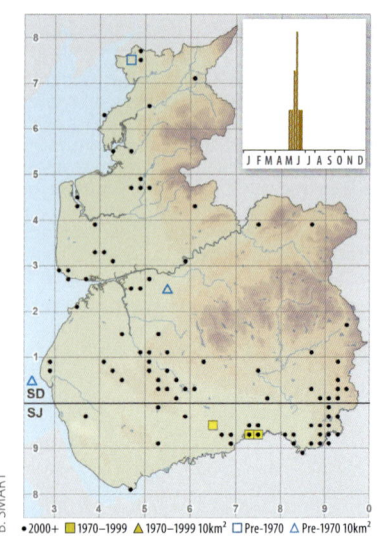

15.025 *Parornix betulae*

■ **LANCS STATUS** Occasional / widespread resident c.1857–2022

Noted as common amongst young birch on mosses (Gregson, 1857) and Mansbridge (1940). Mosses with recent reports include Astley, Rixton, Cockerham, Winmarleigh and Heysham, as well as from birch scrub elsewhere, e.g. Gait Barrows and Flixton. To date all adults, across many sites, have proved to be *P. betulae*, but larval feeding signs on their own, although currently accepted, may be less reliable. More breeding through of larvae would be advisable to ensure *P. loganella* is not being overlooked.

■ **HABITAT** Mosses, scrub, rough ground and brownfield sites.
■ **FOODPLANTS** Birch spp., including Downy and Silver Birch.

Double brooded. Comes to light.

● 2000+ ☐ 1970–1999 △ 1970–1999 10km² ☐ Pre-1970 △ Pre-1970 10km²

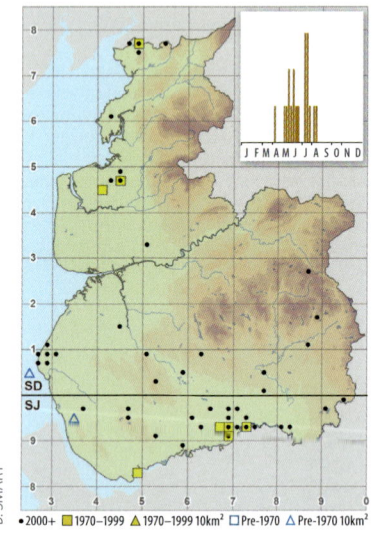

15.028 *Parornix anglicella*

■ **LANCS STATUS** Abundant / ubiquitous resident c.1857–2022

First recorded in the Liverpool district (Gregson, 1857) and subsequently noted as common and widespread amongst Hawthorn throughout (Mansbridge, 1940). Since then, the distinctive larval cones have enabled recording of this species wherever hawthorns have been searched, in both lowland and upland parts of the county. Sites with a long series of annual records include Billinge, Flixton, Parr and Preston.

■ **HABITAT** Woodland edges, hedgerows, gardens, scrub and brownfield sites.
■ **FOODPLANTS** Hawthorn. Rarely on Rowan and once on Wild Service Tree.

Double brooded. Comes to light in small numbers. The larval cones can often be abundant, with vacated cones readily found as long as leaves remain on the trees.

● 2000+ ☐ 1970–1999 △ 1970–1999 10km² ☐ Pre-1970 △ Pre-1970 10km²

96 *The Moths of Lancashire*

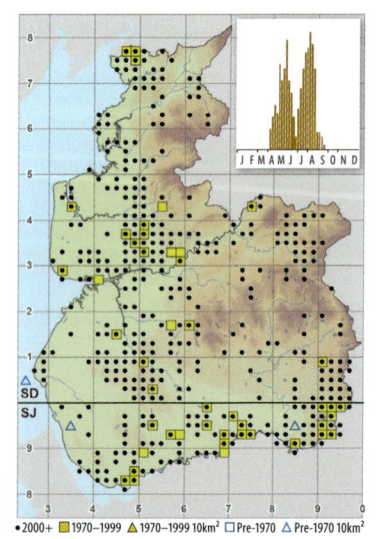

15.029 *Parornix devoniella*

■ **LANCS STATUS** Frequent / widespread resident   c.1857–2022
First noted in Croxteth Woods (Gregson, 1857), followed by reports from Manchester area (c.1859) and Preston district (c.1890). Mansbridge (1940) reported it to be local, but common among Hazel and this has continued to be the case to the present day. Nearly all records are of its distinctive larval leaf folds, which can be found abundantly at times. Regular sites include Chorlton, Flixton and Gait Barrows.
■ **HABITAT** Open woodland, scrub, hedgerows and brownfield sites.
■ **FOODPLANTS** Hazel.
Double brooded. Rare at light. A very late individual was noted, at light, on 30 October 2022 in Preston (S. Palmer), the identity confirmed by dissection.

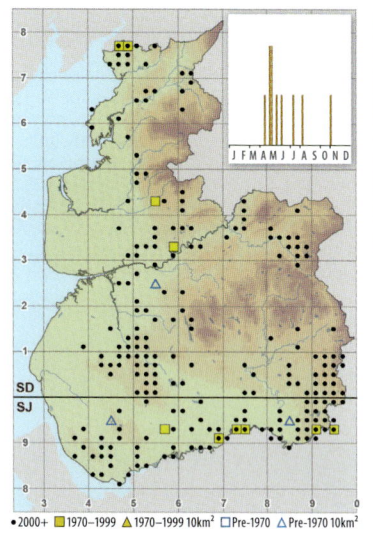

15.030 *Parornix scoticella*

■ **LANCS STATUS** Occasional / local resident   1995–2022
Although noted as very local historically in adjacent parts of Cheshire and Cumbria, the first record for Lancashire was not until 26 October 1995 at Anglezarke (S. Palmer), when a leaf fold was found on Rowan. It was then noted at Gait Barrows in 1997 and, since then, at several well-scattered sites. Usually associated with Rowan, in lowland and upland areas, but folds were found on Apple in Chorlton in October 2022 (B. Smart). Recent searches in south-east VC59 found it to be rare (S.H. Hind).
■ **HABITAT** Woodland, including adjacent to moorland, scrub and parks.
■ **FOODPLANTS** Rowan, Apple (bred to confirm) and Whitebeam.
Appears to be single brooded. Seen once at light. Larval folds, August to October.

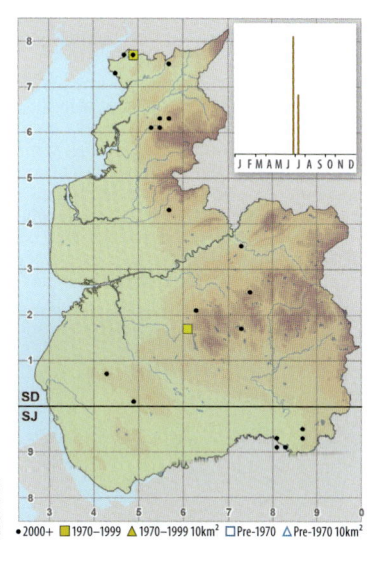

15.032 *Parornix finitimella*

■ **LANCS STATUS** Occasional / local resident   2000–2022
Not identified as a British species until 1917 and any previous records would have been mixed up with *P. torquillella*. The first confirmed record was not until 21 June 2000 at Gait Barrows, when a few tenanted folds were found (J.R. Langmaid). This site has since produced several additional records. Elsewhere it is regular in Chorlton (B. Smart) and has been noted at seven other widespread, mainly lowland, localities.
■ **HABITAT** Hedges and scrubland, where blackthorn thickets occur.
■ **FOODPLANTS** Blackthorn.
Double brooded. Comes occasionally to light. Only dissected adults and tenanted larval folds, where larval leg and abdomen colour were checked, have been accepted.

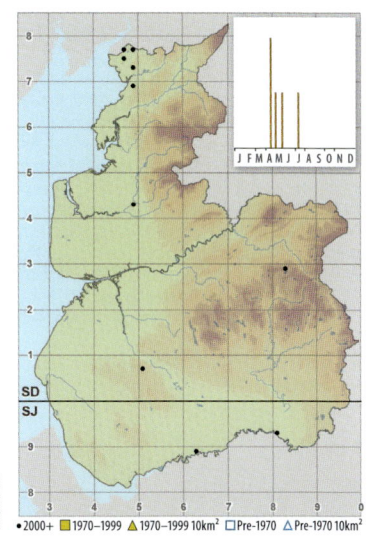

GRACILLARIIDAE

The Moths of Lancashire   97

GRACILIARIIDAE

15.033 *Parornix torquillella*

■ **LANCS STATUS** Occasional / widespread resident   1997–2022

More widespread than previous species, but the confusion over identification of pre-1917 records still apply. Since then, only records of tenanted folds, where the larva could be examined, and dissected adults are included. Sites with regular reports include Bay Horse, Chorlton, Flixton and Gait Barrows. Usually recorded in small numbers, but 20 tenanted mines were found in Parbold in 2002 (C.A. Darbyshire). Present in upland sites where *P. finitimella* was absent (S.H. Hind pers. comm.).
■ **HABITAT** Hedgerows, scrubland, brownfield sites and gardens.
■ **FOODPLANTS** Blackthorn and occasionally plum sp.
Double brooded, the second brood less numerous than first. Comes to light.

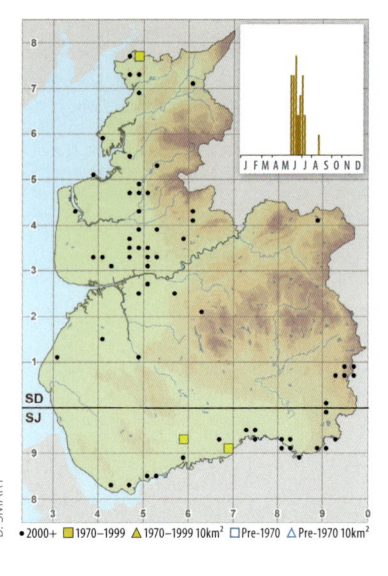

15.034 *Phyllonorycter harrisella*

■ **LANCS STATUS** Frequent / widespread resident   c.1857–2022

Described by Gregson (1857) as 'plentiful in oak woods' and by Mansbridge (1940) as 'abundant amongst oaks everywhere'. Remains numerous with status likely unchanged, although distribution is somewhat patchy north of the Ribble. Recorded in lowland areas, e.g., Sefton Coast, and Pennine fringes, e.g., Shaw, Briercliffe, etc.
■ **HABITAT** Oak woodland and hedgerows.
■ **FOODPLANTS** Pedunculate Oak; likely other oak species too.
At least double brooded, possibly with occasional third generation. Over 100 records of adult at light and in the field; far fewer of the leaf-mine, as moth should be bred, or pupa examined, to separate the mine from other oak-feeding *Phyllonorycter*.

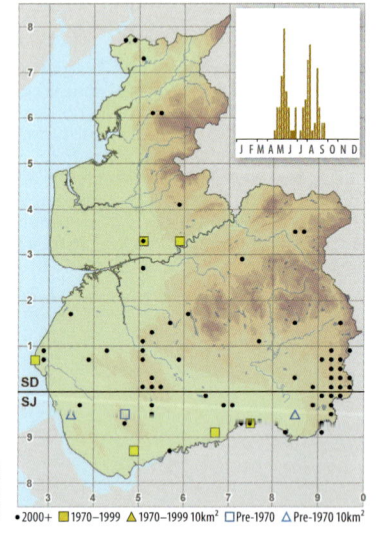

15.036 *Phyllonorycter heegeriella*

■ **LANCS STATUS** Rare / very local resident   1941

The sole accepted Lancashire record is of a moth netted by L.T. Ford at Warton Crag in 1941 (Lancs and Cheshire Fauna Society Annual Report, 1940 to 1942). There have been a few records of leaf-mines submitted, but without adult emergence. Whilst the mines of *P. heegeriella* are typically small (less than 1cm wide), larvae of other Oak-feeding *Phyllonorycter* have occasionally been noted to form similar mines. Therefore, records of mines, where the moth has not been bred, are considered unreliable. This is quite a local species in the neighbouring counties of Yorkshire and Cheshire, with records from both during 2022 (yorkshiremoths.co.uk; cheshire-chart-maps.co.uk, both accessed 1 September 2023). Probably under-recorded in the county. The rearing through of potential *P. heegeriella* leaf mines on oak, collected in July or September and October, would be a worthwhile project.
■ **HABITAT** Limestone woodland.
■ **FOODPLANTS** Not recorded in Lancashire.

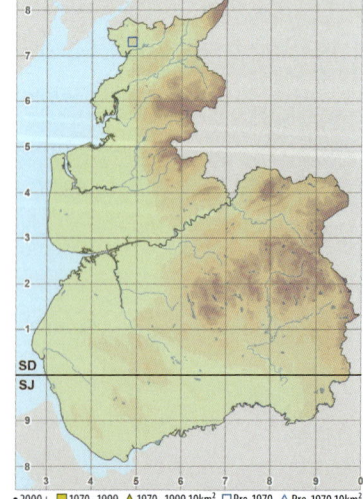

The Moths of Lancashire

15.039 *Phyllonorycter quercifoliella*

■ LANCS STATUS Frequent / widespread resident   c.1857–2022

First noted in the Liverpool district 'where oaks grow' (Gregson, 1857), and later, as common (Ellis, 1890). Mansbridge (1940) described the moth as abundant. Present in upland and lowland areas although distribution is patchy, perhaps due to difficulty in recording the mines rather than absence. Probably the most common *Phyllonorycter* species mining deciduous oak, certainly the most reared, with over 40 such records.
■ HABITAT Oak woodland, hedgerows, parks, gardens and isolated trees.
■ FOODPLANTS Oak.
At least double brooded, although graph suggests an occasional triple brood. Adults noted at light and in the field. Mine identification requires the adult to be bred.

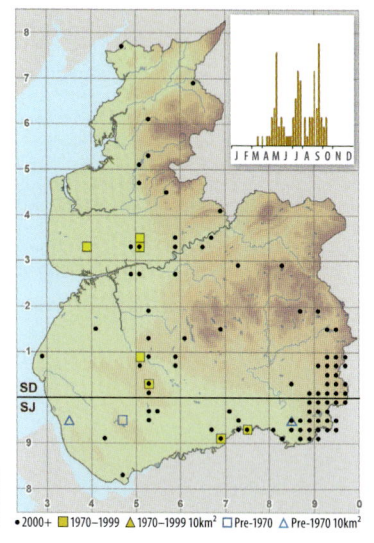

● 2000+   □ 1970–1999   △ 1970–1999 10km²   □ Pre-1970   △ Pre-1970 10km²

15.040 *Phyllonorycter messaniella*

■ LANCS STATUS Frequent / widespread resident   c.1857–2022

Present in Liverpool district around evergreen oaks (Gregson, 1857). Later, noted at Agecroft by J. Chappell and at Preston by J.H. Threlfall (Ellis, 1890). Unrecorded for over 100 years before recorded at Lightfoot Green in 1995 by S. Palmer. Widely distributed throughout Lancs; primarily, but not exclusively, in lowland areas.
■ HABITAT Woodland, hedgerows, parks, gardens.
■ FOODPLANTS Holm Oak, Pedunculate Oak, Beech, Sweet Chestnut, Hornbeam. Possibly triple brooded with an extended flight period; recorded up to mid-November. Adults recorded at light and occasionally flying around the foodplants. Moths bred from leaf-mines on oak, Beech and Sweet Chestnut.

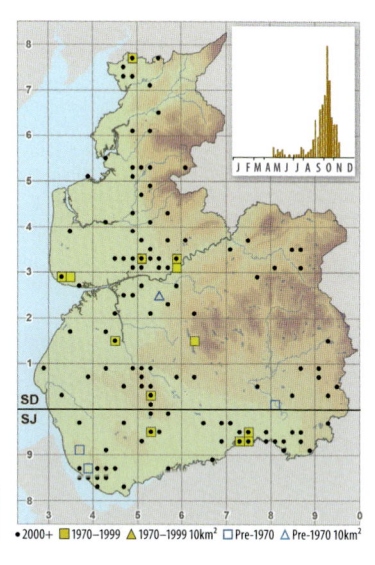

● 2000+   □ 1970–1999   △ 1970–1999 10km²   □ Pre-1970   △ Pre-1970 10km²

15.041 *Phyllonorycter platani*

■ LANCS STATUS Frequent / local resident   2006–2022

Leaf-mines first noted at Victoria Park, Stretford, by K. McCabe on 23 October 2006, since when the range has expanded throughout the south of the county wherever the foodplant is found. The first VC60 record was of a single mine on a young, newly planted tree in Preston 2019, with a further record nearby in 2022 (C.A. Palmer).
■ HABITAT Parks and streets, where the host tree has been planted.
■ FOODPLANTS London Plane.
Double brooded. Mines usually plentiful wherever they occur, with a maximum count of 60 in Chorlton in 2022. Mines mainly on the underside of the leaf; occasional on the upperside. Over 80 leaf-mine records. Adult recorded at light, Chorlton 2014.

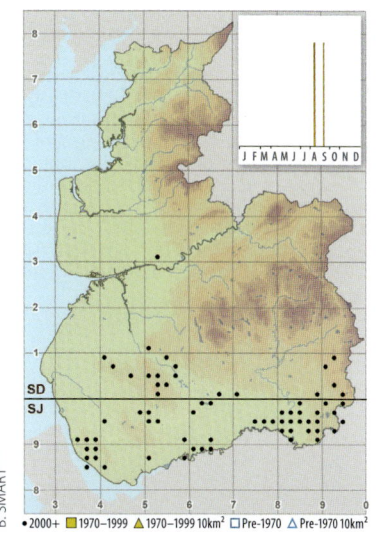

● 2000+   □ 1970–1999   △ 1970–1999 10km²   □ Pre-1970   △ Pre-1970 10km²

GRACILLARIIDAE

15.043 *Phyllonorycter oxyacanthae*

■ **LANCS STATUS** Abundant / widespread resident c.1857–2022

C.S. Gregson (1857) described a moth as plentiful in thorn hedges. Named only as '*Lithocolletis -----?*', it is likely this referred to *P. oxyacanthae*, only recognised as a species in 1856. Subsequently, noted at Hale, Woolton, Knowsley (Mansbridge, 1940), and at Didsbury by H.N. Michaelis in 1948. Not recorded again until 1977, but over 700 times since, and widely distributed throughout lowland and upland areas.
■ **HABITAT** Hedgerows, gardens, parks, woodland, moorland.
■ **FOODPLANTS** Hawthorn, Pear.

Double brooded. The vast majority of records are of leaf-mines; easily differentiated from other Hawthorn-feeding *Phyllonorycter*, whereas adults require dissection.

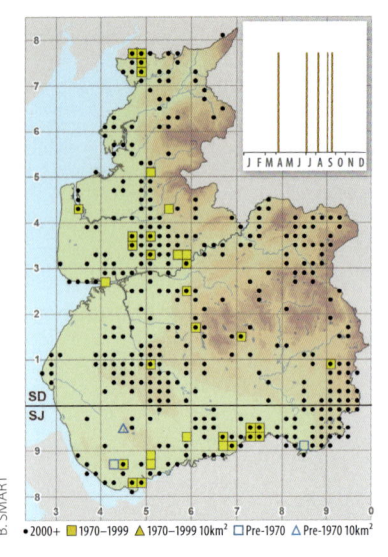

15.044 *Phyllonorycter sorbi*

■ **LANCS STATUS** Frequent / widespread resident c.1889–2022

The earliest records were by J.B. Hodgkinson from Preston and Longridge (Ellis, 1890). Mansbridge (1940) describes the moth as common on the foodplant, noting records from Burnley, Woolton, and Sales Wood. Distribution is largely confined to inland sites, with absence from Fylde coast and a single Sefton Coast record. Frequent in upland areas, probably reflecting the host tree's distribution.
■ **HABITAT** Moorland, woodland, hedgerows, mosses.
■ **FOODPLANTS** Rowan.

Double brooded. As with the previous species, almost all records are of leaf-mines, although there are a couple of records of moths at light.

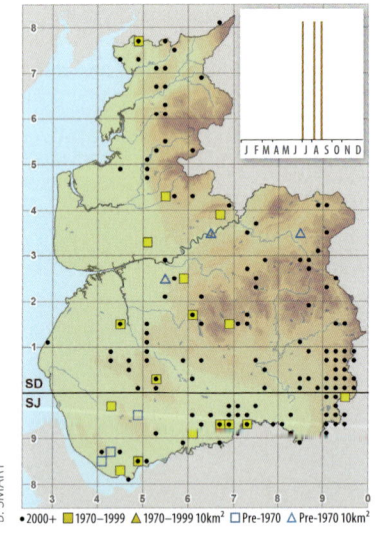

15.046 *Phyllonorycter blancardella*

■ **LANCS STATUS** Frequent / local resident 1919–2022

The first confirmed record is by W. Mansbridge of a moth at Sales Wood, Prescot in June 1919. He also notes records from Formby and Silverdale (1940); less certain due to name changes and potential confusion with *P. hostis*, a species unconfirmed in the county. Reasonably well distributed in VC59, less so in VC60. However, as few have been reared and dissected, any conclusions should be treated with caution.
■ **HABITAT** Orchards, gardens, hedgerows, woodland edges.
■ **FOODPLANTS** Apple, Crab Apple. (These foodplants are also used by *P. hostis*).

Double brooded. The vast majority of records are of leaf-mines. There is a single record of a moth attracted to light and subsequently dissected.

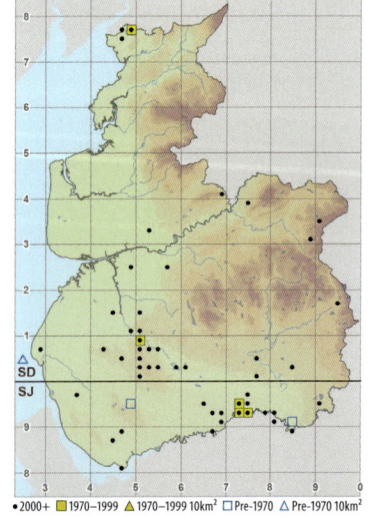

100 *The Moths of Lancashire*

15.048 *Phyllonorycter junoniella*

■ **LANCS STATUS** Rare / very local resident   **1856–1879**

Leaf-mines were recorded on 21 April 1856 by C.S. Gregson on high ground near Chorley. Referring to both this species and *Fomoria weaveri*, he later wrote 'every mountain on which the food-plant grows seems to produce the larvae… The reason they are so bad to find is, that as soon as they are nearly full-fed, the leaf falls off and must be searched for on the ground' (Gregson, 1856a). Subsequently, moths and larvae were collected by J.B. Hodgkinson on 28 June 1879 at Green Thorn Fell, Ribble Valley. He reported the foodplant as Bearberry, presumably a misidentification of Cowberry (Hodgkinson, 1879c). The lack of subsequent records suggests the possibility that the moth is now extinct within the county. However, the relative inaccessibility of potential sites and the moth's continued presence on Cheshire and Yorkshire moorland suggests that searches of suitable Cowberry sites on the Lancashire fells and moors may yet yield positive results.

■ **HABITAT** Moorland, where the foodplant grows.
■ **FOODPLANTS** Cowberry.

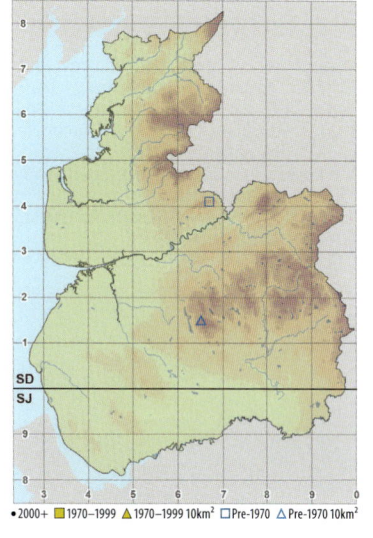

• 2000+   ▢ 1970–1999   ▲ 1970–1999 10km²   ☐ Pre-1970   △ Pre-1970 10km²

15.049 *Phyllonorycter spinicolella*

■ **LANCS STATUS** Occasional / local resident   **c.1859–2022**

First recorded in the Manchester district (Stainton, 1859); later, by J.H. Threlfall in Preston and by J. Chappell in Withington (Ellis, 1890). All early authors reported it as common or abundant where the foodplant grows. Modern records are distributed mainly across lowland areas of the county. Scarce in the south-eastern corner of Greater Manchester (S.H. Hind, 2021 pers. comm.) and in upland areas.

■ **HABITAT** Woodland edges, hedgerows, gardens. Hind also reported that mines were often associated with sheltered sucker growth on the warm side of wilder hedgerows.
■ **FOODPLANTS** Blackthorn, Wild Plum.

Double brooded. Most records are of mines. Adult once at light, Longton, May 2020.

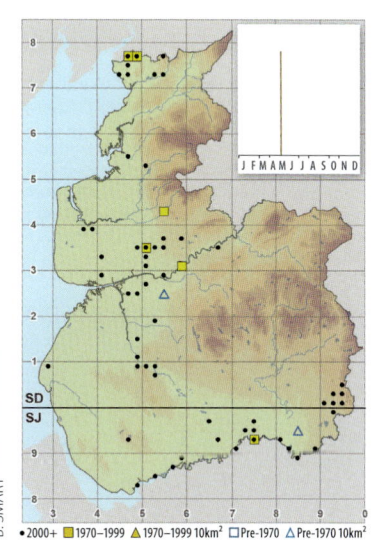

• 2000+   ▢ 1970–1999   ▲ 1970–1999 10km²   ☐ Pre-1970   △ Pre-1970 10km²

15.050 *Phyllonorycter cerasicolella*

■ **LANCS STATUS** Occasional / local resident   **2001–2022**

Likely to be a recent colonist, rather than a previously undetected presence. Leaf-mines first recorded on 28 October 2001, by C.A. Darbyshire at Billinge Hill, then the following day at Flixton by K. McCabe. Can be expected wherever cherry grows in the south of the county, but remains unrecorded further north than Mere Sands Wood, where it was noted in 2004. Records from 26 sites by S.H. Hind in 2021–2022 showed the species to be frequent in south-east Manchester, Tameside and Oldham.

■ **HABITAT** Gardens, hedgerows, woodland edges, amenity planting.
■ **FOODPLANTS** Cherry.

Double brooded. Recorded on 88 occasions, all in the leaf-mining stage.

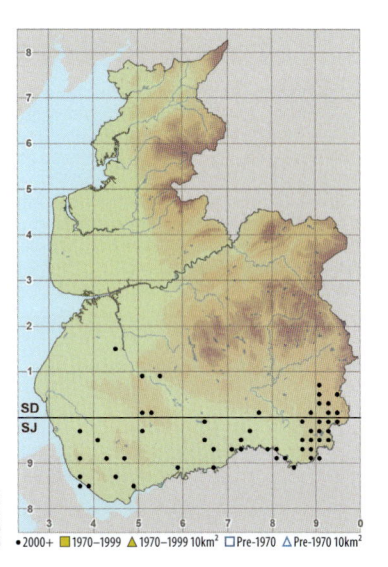

• 2000+   ▢ 1970–1999   ▲ 1970–1999 10km²   ☐ Pre-1970   △ Pre-1970 10km²

GRACILLARIIDAE

15.051 *Phyllonorycter lantanella*

■ **LANCS STATUS** Occasional / local resident   2019–2022

Another recent arrival in the county, leaf-mines found in abundance by B. Smart on Laurustinus at a supermarket car-park in Salford, 2019. Subsequently, recorded in Manchester, Irlam, Oldham and Rochdale. Also, noted from three sites on Fylde coast in 2022. The absence of records between these two geographical areas suggests the populations appeared independently from each other; likely due to use of the foodplant in landscaping, potentially introducing the moth into new localities.

■ **HABITAT** Amenity planting, parks, gardens.
■ **FOODPLANTS** Laurustinus.
Double brooded. All records are of leaf-mines, frequently parasitised.

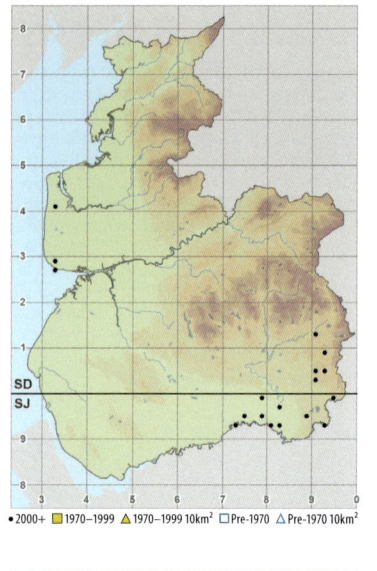

● 2000+   ☐ 1970–1999   △ 1970–1999 10km²   ☐ Pre-1970   △ Pre-1970 10km²

15.052 *Phyllonorycter corylifoliella*

■ **LANCS STATUS** Occasional / local resident   c.1857–2022

Reported amongst thorn hedges in Liverpool and Manchester districts (Gregson, 1857; Stainton, 1859). Ellis (1890) notes it is 'common among Hazel', a strange comment as Hazel is not a recognised foodplant. Widespread in much of VC59, and possibly increasing in frequency. First Sefton Coast record in 2022. The moth appears to be absent from upland areas and around south-east Manchester (S. H. Hind, pers. comm.). Also, absent from much of VC60, other than a few northern limestone sites.

■ **HABITAT** Woodland edges, hedgerows.
■ **FOODPLANTS** Hawthorn, Rowan, Apple, Whitebeam. Once on birch (*f. betulae*).
Double brooded. All records are of the leaf-mines.

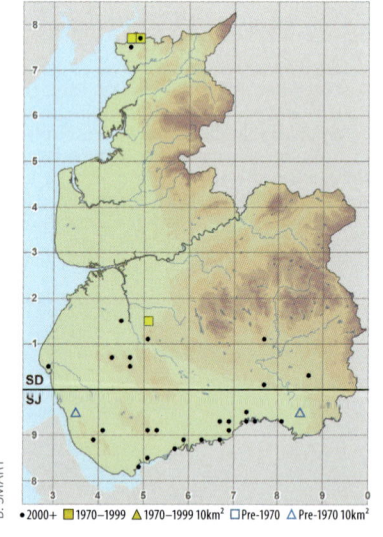

● 2000+   ☐ 1970–1999   △ 1970–1999 10km²   ☐ Pre-1970   △ Pre-1970 10km²

15.053 Firethorn Leaf Miner *Phyllonorycter leucographella*

■ **LANCS STATUS** Abundant / widespread resident   1999–2022

This initially adventive species is now resident over almost the whole county, its spread helped by widespread planting of the main host, and by a degree of polyphagy. First recorded on Firethorn at Salford by K. McCabe on 12 February 1999, reaching Preston in 2001 and Silverdale in 2003. Numbers can be huge, with a high count of 5000 mines noted on Firethorn in Wigan, 2004 by C.A. Darbyshire.

■ **HABITAT** Parks, gardens, amenity planting, hedgerows.
■ **FOODPLANTS** Firethorn, Hawthorn, Apple, Crab Apple, Rowan, Whitebeam, Swedish Whitebeam, Pear, Quince, cherry, Cherry Laurel, Beech, Cotoneaster, Sycamore.
Double brooded. Recorded at light on over 500 occasions.

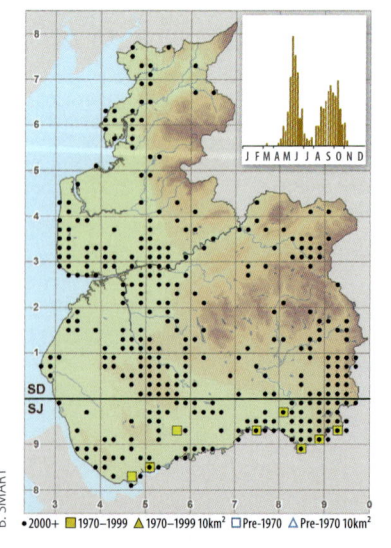

● 2000+   ☐ 1970–1999   △ 1970–1999 10km²   ☐ Pre-1970   △ Pre-1970 10km²

The Moths of Lancashire

15.054 *Phyllonorycter viminiella*

■ **LANCS STATUS** Frequent / local resident   c.1859–2022

First recorded in Manchester (Stainton, 1859). Mansbridge described status as not uncommon, but local, and reported presence at Formby, Knowsley and Simonswood Moss (1940). It is now quite widespread throughout the county, but likely to be under-recorded, as breeding is required to confirm identity. Initially seen as primarily a miner of narrow smooth-leaved willows but will also use rough-leaved willows.

■ **HABITAT** Wetland sites and woodland edges in lowland parts of the county.

■ **FOODPLANTS** Crack-willow, Osier, Grey Willow, Goat Willow.

Double brooded. The majority of the 85 records are of leaf-mines. Adults have been recorded on 36 occasions; at light and netted around the foodplants.

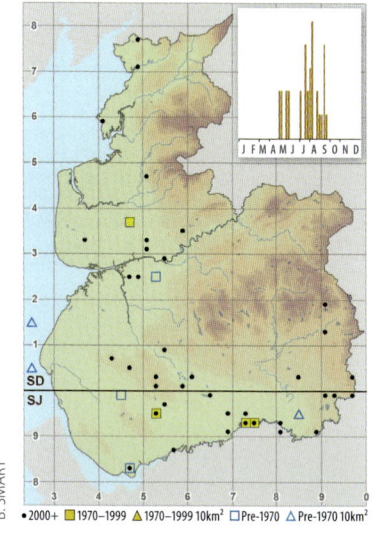

• 2000+   ■ 1970–1999   ▲ 1970–1999 10km²   □ Pre-1970   △ Pre-1970 10km²

15.055 *Phyllonorycter viminetorum*

■ **LANCS STATUS** Considered Extinct   c.1857–1940

Gregson (1857) states the moth occurs in the Liverpool district where Osiers grow. The species was also described as abundant in the Manchester district by Stainton (1859). Formby was later added to the list of sites, with the moth described as local among Osier (Mansbridge, 1940). There have been no further records of the species in the county, and its continued presence must be considered doubtful. The species has only once been recorded in Cheshire, just south of the Mersey in the Warrington area in 1989 (cheshire-chart-maps.co.uk, accessed 1 September 2023). The two Yorkshire records, from Selby and Scarborough, both date from the late 1800s (yorkshiremoths.co.uk, accessed 1 September 2023).

■ **HABITAT** Freshwater sites.

■ **FOODPLANTS** Osier. The plant is widespread in damp places, by streams and rivers, and is also widely planted. The leaf-mines are worth looking for wherever the foodplant occurs but will need to be bred through to differentiate them from *P. viminiella*.

Double brooded. Gregson gave the phenology as May and August, although it is unclear whether he was referring to adults or leaf-mines.

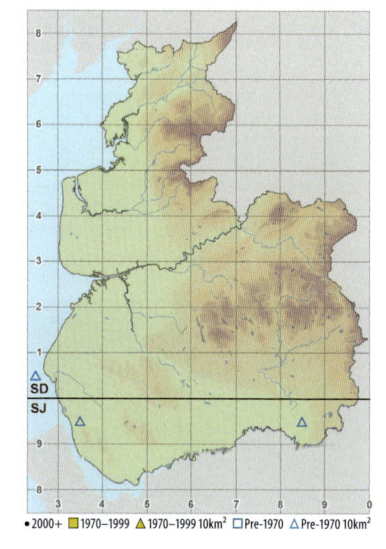

• 2000+   ■ 1970–1999   ▲ 1970–1999 10km²   □ Pre-1970   △ Pre-1970 10km²

15.056 *Phyllonorycter salicicolella*

■ **LANCS STATUS** Occasional / local resident   c.1857–2021

Only 20 or so county records. Not an easy species to identify, either in the leaf-mining or the adult form, suggesting that under-recording may play a part in this low total. Historical records show the moth to have been present in the Liverpool and Manchester districts (Gregson, 1857; Stainton, 1859). However, Mansbridge (1940) noted the moth was unrecorded since 1890. This drought persisted, and it was not until 2001 that the moth was again recorded, at Kearsley and at Flixton.

■ **HABITAT** Woodland edges in upland and lowland areas.

■ **FOODPLANTS** Goat Willow, Grey Willow.

Double brooded. Most records are of bred moths. Occasionally at light.

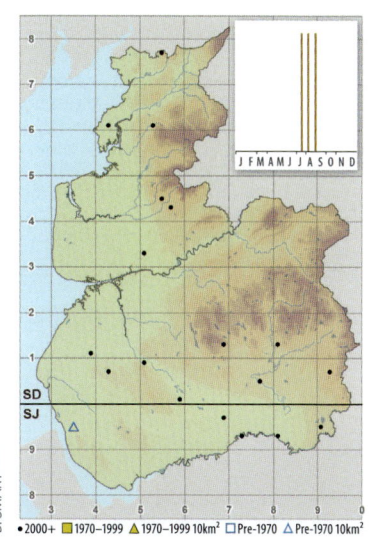

• 2000+   ■ 1970–1999   ▲ 1970–1999 10km²   □ Pre-1970   △ Pre-1970 10km²

GRACILLARIIDAE

The Moths of Lancashire   103

GRACILLARIIDAE

15.057 *Phyllonorycter dubitella*

■ **LANCS STATUS** Occasional / local resident   1986–2021

Recorded at light on a couple of occasions. One of those, in 1986 at Gait Barrows by L.W. Hardwick, was a first for Lancashire. Since then, records have shown the moth to be reasonably widespread in the south-east of the county, from Denton, Abbey Hey, Chorlton, Flixton etc. Other than Gait Barrows, the only VC60 records are from Lightfoot Green and Dobcroft N.R. by S. & C. Palmer in 2019. Only 19 records in total, and likely under-recorded due to difficulty in identifying leaf-mine and moth.

■ **HABITAT** Woodland edges, reclaimed industrial sites.
■ **FOODPLANTS** Goat Willow, Grey Willow.

Double brooded. Bred from eight tetrads in south-east of VC59, 2021 (S.H. Hind).

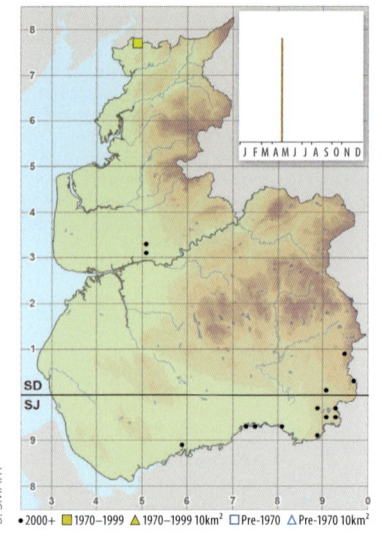

• 2000+   ☐ 1970–1999   △ 1970–1999 10km^2 ☐ Pre-1970   △ Pre-1970 10km^2

15.058 *Phyllonorycter hilarella*

■ **LANCS STATUS** Occasional / local resident   c.1857–2022

First reported from Liverpool (Gregson, 1857), with other 19th century records from Chat Moss, Preston, and the Manchester area where it was considered abundant (Stainton, 1859). Later, described as local and uncommon (Mansbridge, 1940). From Formby in 1957 and Warton Crag in 1974. In 2021–2022, bred from leaf-mines in 17 south-eastern tetrads of the county, one-third of all records (S.H. Hind). Noted in lowlands and at more elevated sites, e.g., Mossley, Oswaldtwistle, Docker Moor.

■ **HABITAT** Woodland edges, brownfield sites, mosses.
■ **FOODPLANTS** Goat Willow, Grey Willow.

Double brooded. Adults recorded at light or netted around the foodplant.

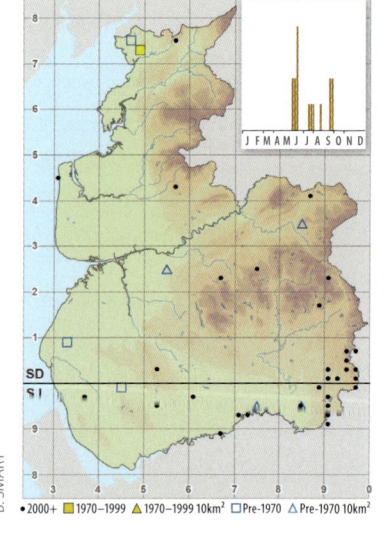

• 2000+   ☐ 1970–1999   △ 1970–1999 10km^2 ☐ Pre-1970   △ Pre-1970 10km^2

15.060 *Phyllonorycter ulicicolella*

■ **LANCS STATUS** Rare / very local resident   c.1889–2022

Ellis (1890) described the moth's distribution as local among Gorse, noting records by J.B. Hodgkinson at Fleetwood and by J.H. Threlfall at Stalmine. Also, at Hale (Mansbridge, 1940). Subsequently, unrecorded until 2004, when netted at Freshfield Dune by G. Jones and K. McCabe. Tenanted mines on Gorse twigs were found at Ainsdale Dunes on 18 March 2022 by B. Smart, with a couple of moths beaten from the same bush on 10 June 2022, probably the easiest way to look for this moth.

■ **HABITAT** Fixed sand-dunes. Possibly elsewhere where the foodplant occurs.
■ **FOODPLANTS** Gorse.

Single brooded. Under-recorded due to the difficulty in detecting the early stages.

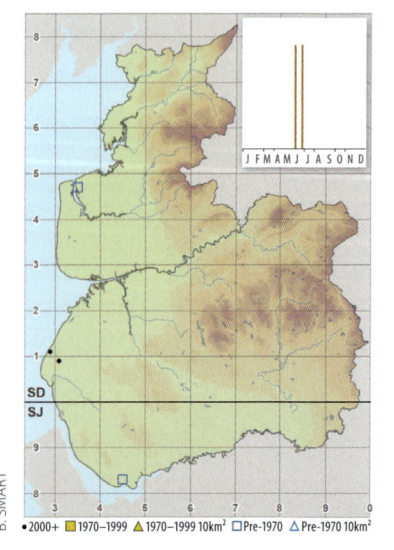

• 2000+   ☐ 1970–1999   △ 1970–1999 10km^2 ☐ Pre-1970   △ Pre-1970 10km^2

15.061 *Phyllonorycter scopariella*

■ **LANCS STATUS** Rare / very local resident   1879–2022

Only recorded on four occasions. The first was from Dutton, nr. Ribchester by J.B. Hodgkinson in 1879 (Hodgkinson, 1880a), and then at Formby in 1939, when W. Mansbridge found it occurring on Broom bushes in his garden (Mansbridge, 1940). Next, noted in 1953 by H.N. Michaelis at Freshfield. Thereafter, not recorded until 2022, when the moth was bred from pale brown mines at Freshfield by B. Smart. The mines were collected on 14 April 2022, with the moth emerging on 9 May 2022.
■ **HABITAT** Worth looking for wherever foodplant occurs, especially at coastal sites.
■ **FOODPLANTS** Broom.
Single brooded.

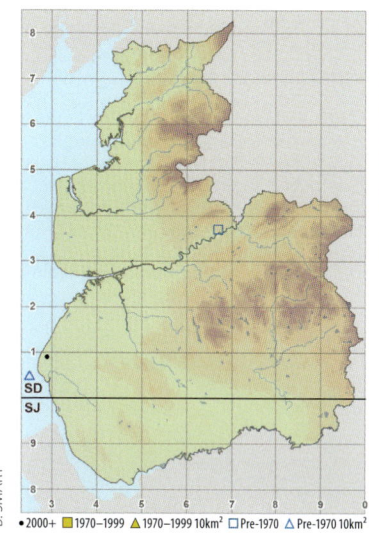

• 2000+   ■ 1970–1999   ▲ 1970–1999 10km²   □ Pre-1970   △ Pre-1970 10km²

15.063 *Phyllonorycter maestingella*

■ **LANCS STATUS** Abundant / widespread resident   c.1857–2022

First documented by C.S. Gregson (1857) who noted it to be present in the Liverpool district wherever Beech hedges occur. Described as common around Beech by Ellis (1890) and Mansbridge (1940). Leaf-mines of the species are likely to be found wherever the host occurs; at coastal and inland sites, lowland and upland areas, and throughout the north and south of the county. Recorded on 620 occasions in total.
■ **HABITAT** Woodland, hedgerows, parks, gardens, etc.
■ **FOODPLANTS** Beech.
Double-brooded. Most records are of the long, thin mines, differing from the rounder mines of *P. messaniella* on Beech. Occasionally recorded as an adult on Beech leaves.

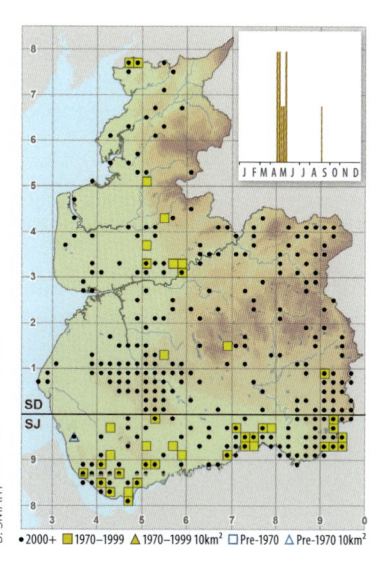

• 2000+   ■ 1970–1999   ▲ 1970–1999 10km²   □ Pre-1970   △ Pre-1970 10km²

15.064 Nut Leaf Blister Moth *Phyllonorycter coryli*

■ **LANCS STATUS** Frequent / widespread resident   c.1859–2022

Reported as abundant in the Manchester district by Stainton (1859) and later, at Preston by J.H. Threlfall (Ellis, 1890). Mansbridge (1940) described the moth as local. No further records until 1995 at Gait Barrows by A.M. Emmet and at Brock Bottom by S. Palmer. Possibly increased in abundance since, with the 450 records distributed over most of the county, although remains absent from the Sefton and Fylde coasts.
■ **HABITAT** In inland areas where the foodplant occurs.
■ **FOODPLANTS** Hazel; on mature trees, saplings and hedges.
Double brooded. Other than a single dissected moth, all records are of leaf-mines. The white, upperside mines, distinct from any other Hazel mine, are easily identified.

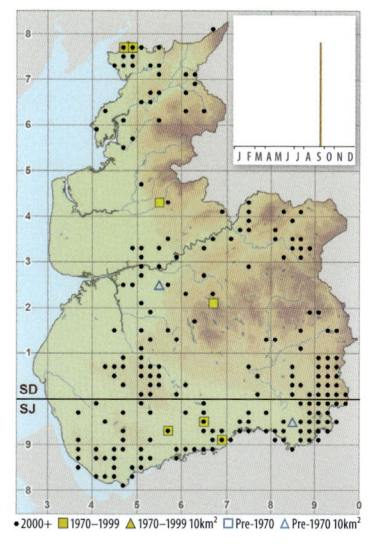

• 2000+   ■ 1970–1999   ▲ 1970–1999 10km²   □ Pre-1970   △ Pre-1970 10km²

The Moths of Lancashire   **105**

GRACILLARIIDAE

15.065 *Phyllonorycter esperella*

■ **LANCS STATUS** Occasional / local resident   1995–2022

First noted on 22 August 1995 by L.W. Hardwick at Dumplington. Next, recorded at Flixton and Kearsley in 2001. It has been recorded on over 70 occasions since but appears restricted to the south. A high count of 50 mines was in 2003 on Hornbeams planted by the East Lancs Road in St Helens, recorded by C.A. Darbyshire. Noted in 23 tetrads of south-eastern Greater Manchester by S.H. Hind during 2021 and 2022.

■ **HABITAT** Parks, gardens, woodland edges, street and amenity planting.

■ **FOODPLANTS** Hornbeam. Leaf-mines found in July and from August to October. Double brooded. All records are of the easily recognisable upper surface mines, sometimes causing the leaf-edge to fold over.

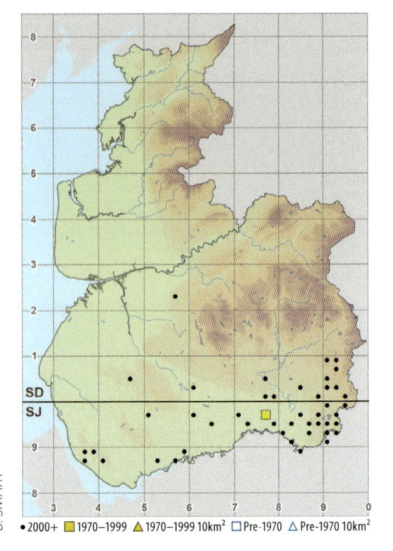

• 2000+   □ 1970–1999   △ 1970–1999 10km²   □ Pre-1970   △ Pre-1970 10km²

15.066 *Phyllonorycter strigulatella*

■ **LANCS STATUS** Occasional / local resident   1997–2022

A recent arrival with leaf-mines first noted at Flixton on 19 July 1997 by S.H. Hind and K. McCabe. The first VC60 record was from Scorton 2003. Bred from Italian Alder at Lundsfield Quarry in 2012 (S. Palmer). Mostly from lowland areas, but also from more upland areas, e.g., Oldham, Stalybridge, Whitworth, Littleborough, etc. More frequent in the south of the county, with approximately 90% of records from VC59.

■ **HABITAT** Amenity planting, hedgerows, woodland edges.

■ **FOODPLANTS** Grey Alder. Once on Italian Alder.

Double brooded. Most records are of the long, thin, underside mines. The moth is relatively easily identifiable and has been recorded at light and on Grey Alder.

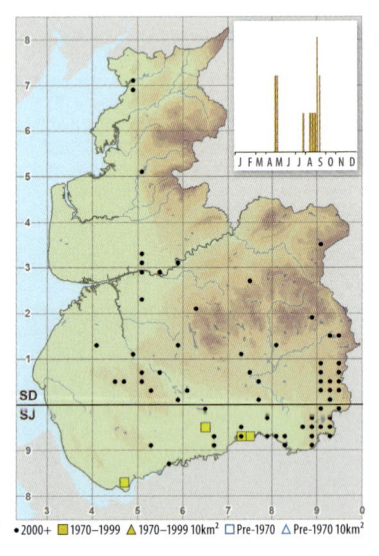

• 2000+   □ 1970–1999   △ 1970–1999 10km²   □ Pre-1970   △ Pre-1970 10km²

15.067 *Phyllonorycter rajella*

■ **LANCS STATUS** Abundant / widespread resident   c.1857–2022

Noted as profuse in Alder plantations around Formby (Gregson, 1857) and abundant everywhere among Alder (Mansbridge, 1940). This status appears little changed with leaf-mines readily encountered on their foodplants throughout the county, whether in upland locations such as Birk Bank or lowland locations such as Heysham Moss. Easily the most recorded Alder-feeding *Phyllonorycter*, with 582 records.

■ **HABITAT** Likely to be found wherever the foodplants occur.

■ **FOODPLANTS** Alder, Italian Alder, Grey Alder.

Probably triple brooded, at least in favourable years. Most records are of the mine, although adult also recorded on over 100 occasions, at light and around Alder.

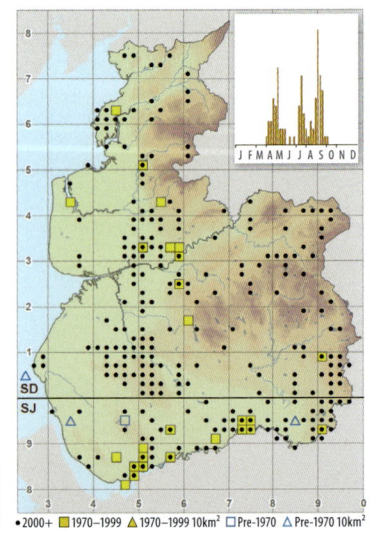

• 2000+   □ 1970–1999   △ 1970–1999 10km²   □ Pre-1970   △ Pre-1970 10km²

15.069 *Phyllonorycter anderidae*

■ **LANCS STATUS** Scarce / very local resident   1950–2022

First recorded by H.L. Burrows at Chat Moss in 1950, with adults reared from leaf-mines. More recently, tenanted mines were recorded during October on birch seedlings at Astley Moss in 2008 and Cadishead Moss in 2019 and 2022, with moths bred on each occasion. A moth was also netted during the daytime at the latter site on 31 July 2007 (K. McCabe). The only other record is of a moth at light at Heysham Moss on 27 August 2014 (J. Patton).
■ **HABITAT** Mosses; preferring plants growing on the edge of the wetter areas.
■ **FOODPLANTS** Birch, with a preference for seedlings.
Double brooded.

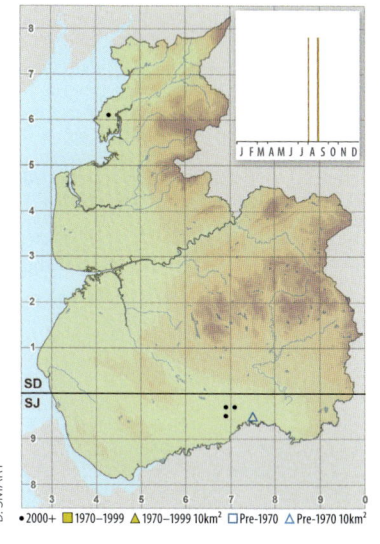

• 2000+   ☐ 1970–1999   △ 1970–1999 10km²   ☐ Pre-1970   △ Pre-1970 10km²

15.070 *Phyllonorycter quinqueguttella*

■ **LANCS STATUS** Occasional / local resident   c.1857–2022

Limited to sites on the Sefton and Fylde coasts where the foodplant occurs. First noted 'among the small sallows on the sand hills' around Liverpool (Gregson, 1857). Later, reported from Lytham by J.B. Hodgkinson (Ellis, 1890), Crosby and Formby, with Mansbridge (1940) describing it as locally common. Population appears stable in the dunes, with many mines recorded in the dunes at Lytham and Ainsdale in 2022.
■ **HABITAT** Fixed dunes and along paths through the dune edges.
■ **FOODPLANTS** Creeping Willow.
Double brooded. All records are of the leaf-mines, with moths emerging from mines collected in July and October.

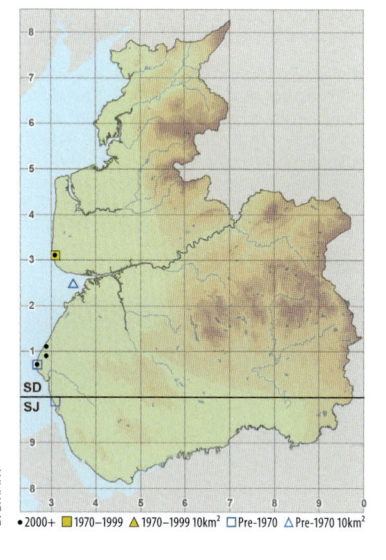

• 2000+   ☐ 1970–1999   △ 1970–1999 10km²   ☐ Pre-1970   △ Pre-1970 10km²

15.073 *Phyllonorycter lautella*

■ **LANCS STATUS** Occasional / local resident   1995–2022

As the mines are distinctive, the lack of historic records suggests this is a recent arrival in the county. First noted at Haighton, nr. Preston on Pedunculate Oak by S. Palmer on 11 November 1995, then at Gait Barrows in 1997. Range expansion noted during next few years, although increased recording likely also played a part, with records from Parbold, Flixton, Hindle, Skelmersdale, Billinge during 2001. Now appears more prevalent in the south. Unrecorded from Fylde and Sefton coasts.
■ **HABITAT** Woodland edges, scrub, hedgerows.
■ **FOODPLANTS** Oak. Usually on saplings; more mature trees occasionally used.
Double brooded. Almost all the 112 records are of the large, underside mines.

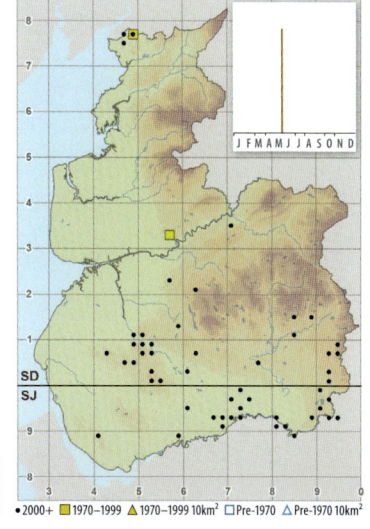

• 2000+   ☐ 1970–1999   △ 1970–1999 10km²   ☐ Pre-1970   △ Pre-1970 10km²

The Moths of Lancashire   **107**

GRACILLARIIDAE

15.074 *Phyllonorycter schreberella*

■ **LANCS STATUS** Scarce / local resident   c.1859–2021

First recorded in Manchester (Stainton, 1859), and at Preston (Ellis, 1890). Mansbridge (1940) noted the lack of any further records. Not recorded again until 2002, when noted at Bluebell Wood, nr. Rixton by K. McCabe. Current distribution appears restricted to the southern districts bordering Cheshire, such as Flixton, Stretford, Widnes, Speke. Unrecorded from upland sites.

■ **HABITAT** Woodland edges, hedgerows.
■ **FOODPLANTS** Elm.

Double brooded. Only 13 records, all of leaf-mines, with adults emerging from tenanted mines collected in October and November.

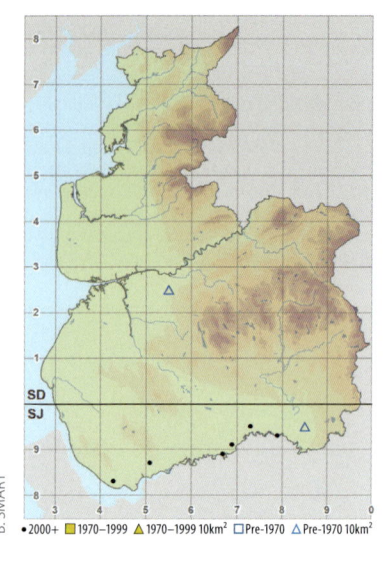

● 2000+ ☐ 1970–1999 △ 1970–1999 10km² ☐ Pre-1970 △ Pre-1970 10km²

15.075 *Phyllonorycter ulmifoliella*

■ **LANCS STATUS** Frequent / widespread resident   c.1857–2022

First recorded at Croxteth Park, Liverpool (Gregson, 1857). Historic status ranged from locally common (Ellis, 1890) to 'common throughout' (Mansbridge, 1940). It is likely the status has changed little over the last two centuries. Well distributed, but with an absence of records on the Fylde coast, a well-recorded area, although one where birch is uncommon. Also, noted in upland areas, such as Dalton Crags.

■ **HABITAT** Woodland edges, hedgerows, heaths, mosses, dunes.
■ **FOODPLANTS** Silver Birch, Downy Birch.

Double brooded. Almost all the 400 plus records are of the leaf-mines, with the moth recorded a few times at light and in the field.

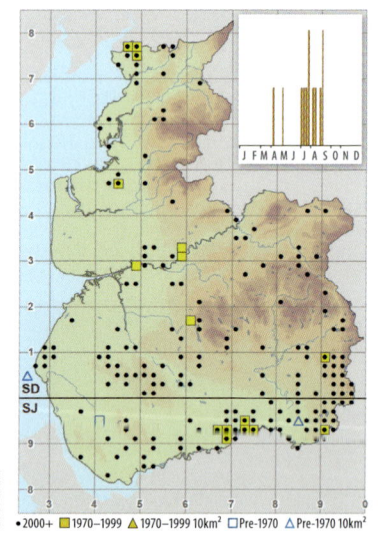

● 2000+ ☐ 1970–1999 △ 1970–1999 10km² ☐ Pre-1970 △ Pre-1970 10km²

15.076 *Phyllonorycter emberizaepenella*

■ **LANCS STATUS** Occasional / local resident   c.1857–2022

First noted at Hale (Gregson, 1857), with other 19th century records from Manchester and Preston. Described as common at the latter site (Ellis, 1890). Mansbridge (1940) noted the absence of further records, and it was 1995 before it was next reported, at Gait Barrows by A.M. Emmet. Appears to be a lowland species, with recent records from Billinge, Skelmersdale, Flixton, Chorlton, etc., although scarce at coastal sites.

■ **HABITAT** Parks, gardens, hedgerows.
■ **FOODPLANTS** Snowberry, Honeysuckle, Himalayan Honeysuckle.

Double brooded, although the autumn generation of mines is far more numerous. Commonest on Snowberry. Adults at light and netted at dusk around the foodplants.

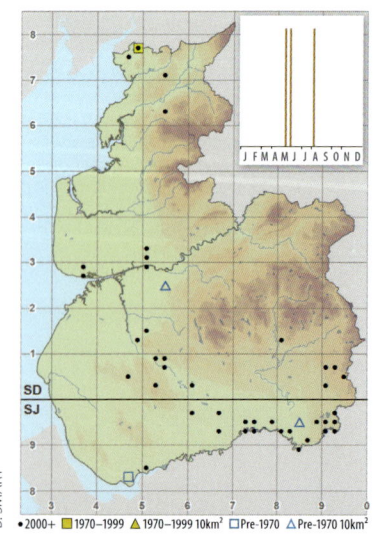

● 2000+ ☐ 1970–1999 △ 1970–1999 10km² ☐ Pre-1970 △ Pre-1970 10km²

15.078 *Phyllonorycter tristrigella*

■ **LANCS STATUS** Frequent / widespread resident   1846–2022
First recorded at Cheetham Hill (Edleston, 1846). Also, from Prescot (Gregson, 1857) and Preston (Ellis, 1890). Described as locally abundant by Mansbridge (1940), although no further records given. Recorded at Formby by H.N Michaelis in 1953, noting mines to be scarce on elm. Despite the devastation of Dutch elm disease, this moth has survived and appears widespread, primarily in lowland areas, but also from more upland areas such as Whitworth, Burnley and Oldham.
■ **HABITAT** Woodland edges, hedgerows.
■ **FOODPLANTS** Wych Elm. Probably other elm species also.
Double brooded. All records are of leaf-mines, other than once at light in 2015.

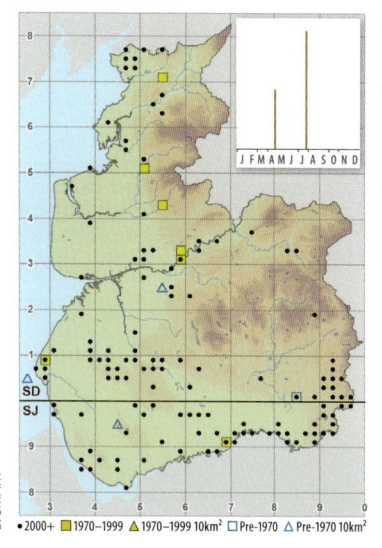

● 2000+   ■ 1970–1999   ▲ 1970–1999 10km²   □ Pre-1970   △ Pre-1970 10km²

15.079 *Phyllonorycter stettinensis*

■ **LANCS STATUS** Occasional / local resident   1856–2022
The 'peculiar mine' was noted on Alders near Formby in July 1856 (Gregson, 1857). Mansbridge (1921) reported 1,000s present in late May 1919 at Eccleston Mere and Sales Wood, Prescot. He later described distribution as very local (Mansbridge, 1940). The range and quantity of recent records, around 100 in 21st century, suggest a slight widening of distribution, possibly related to increased study of this genus. Appears restricted to lowland areas. Moths have been recorded at light on a few occasions.
■ **HABITAT** In damp, lowland habitat where Alder occurs.
■ **FOODPLANTS** Alder; single record on Grey Alder. Mines on upperside of leaves.
Double brooded. Second generation mines vastly outnumber the first generation.

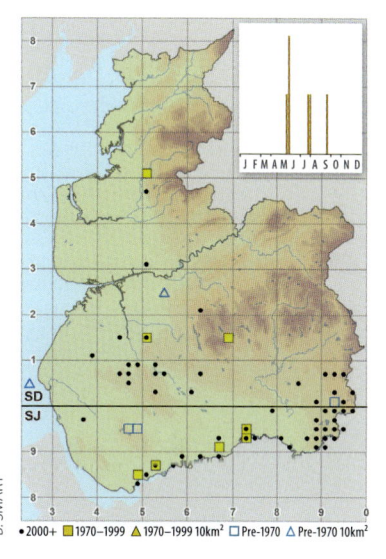

● 2000+   ■ 1970–1999   ▲ 1970–1999 10km²   □ Pre-1970   △ Pre-1970 10km²

15.080 *Phyllonorycter froelichiella*

■ **LANCS STATUS** Occasional / local resident   c.1857–2022
Noted in May 'among the alders beyond Huyton Quarry' (Gregson, 1857); presumably an adult record as an unusual time of year to record the leaf-mines. Reported from Preston as local (Ellis, 1890). No further records until 1995, when mines were noted at Scorton by S. Palmer. Since then, over 90 records, in lowland and upland areas. Remains unrecorded from Sefton and Fylde coasts.
■ **HABITAT** Freshwater sites, damp woodlands, rough grassland with scattered trees.
■ **FOODPLANTS** Alder, Grey Alder.
Likely single brooded. The long, uncreased, underside mines noted from September to November on over 100 occasions. Twice at light, in late June and early July.

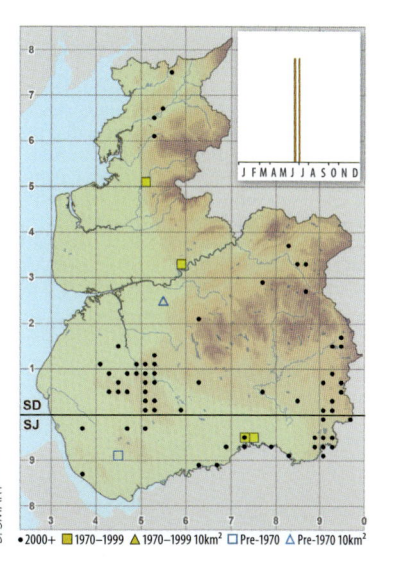

● 2000+   ■ 1970–1999   ▲ 1970–1999 10km²   □ Pre-1970   △ Pre-1970 10km²

GRACILLARIIDAE

The Moths of Lancashire   **109**

GRACILLARIIDAE

15.081 *Phyllonorycter nicellii*

■ **LANCS STATUS** Frequent / widespread resident   c.1857–2022

Recorded in 19th century at Croxteth Country Park (Gregson, 1857) and Preston (Ellis, 1890). Status was later described as scarce (Mansbridge, 1940). No further records until noted at Risley Moss in 1995 (L.W. Hardwick). Since when, the moth appears to have significantly increased in distribution and abundance. It is now found over most of the county, although remains unrecorded from the Fylde coast and much of the West Pennines. Adults have been recorded at light and netted around the host tree.
■ **HABITAT** Woodland edges, hedgerows, parks, gardens.
■ **FOODPLANTS** Hazel.
Double brooded. Most records are of the underside mines between parallel veins.

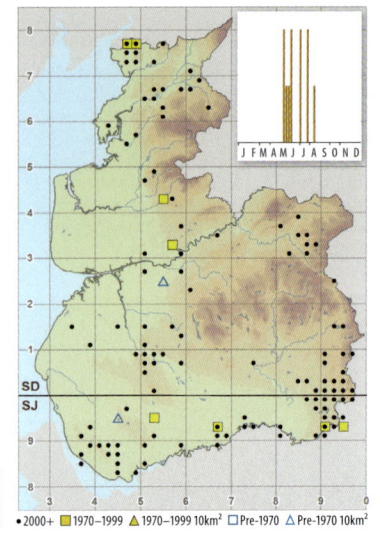

● 2000+   ▢ 1970–1999   △ 1970–1999 10km²   ▢ Pre-1970   △ Pre-1970 10km²

15.082 *Phyllonorycter klemannella*

■ **LANCS STATUS** Occasional / local resident   c.1856–2022

Noted by Gregson (1857) at Formby, mining 'the same trees' as *P. stettinensis*. Other 19th century records from Preston and Manchester. The early writers did not comment on status. It was over 100 years before the next record, at Scorton in 1995 (S. Palmer). Distribution shows a southern bias with only seven of 63 records from VC60. Mainly from lowland sites, e.g., Sefton Park, Spike Island, Rainford, etc.
■ **HABITAT** Woodland edges, hedgerows. Favours damp habitats where Alder grows.
■ **FOODPLANTS** Alder, Italian Alder.
Single brooded. Once at light. All other records are of the rounded, hardly creased, underside mines, noted from late August to November.

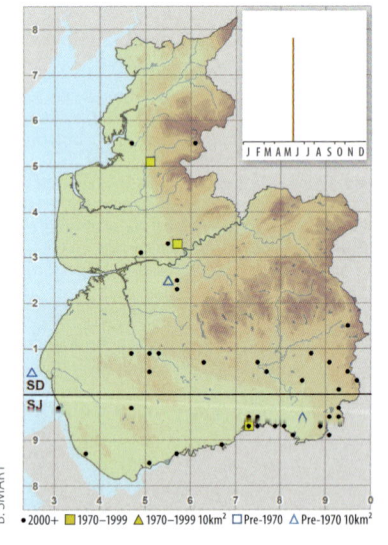

● 2000+   ▢ 1970–1999   △ 1970–1999 10km²   ▢ Pre-1970   △ Pre-1970 10km²

15.083 *Phyllonorycter trifasciella*

■ **LANCS STATUS** Occasional / local resident   c.1857–2022

Reported from the Liverpool district 'in woods and old hedges where honeysuckle is plentiful' (Gregson, 1857). Later, described as abundant (Stainton, 1859), generally distributed (Ellis, 1890), and not common (Mansbridge, 1940). Next recorded at Flixton 1998. Distribution remains patchy. More frequent in the south of the county; at Chorlton, Formby, Briercliffe, etc. From Fluke Hall, Gait Barrows, etc. in the north.
■ **HABITAT** Woodland edges, hedgerows, gardens.
■ **FOODPLANTS** Honeysuckle. Less frequently on Snowberry, Himalayan Honeysuckle.
Probably triple brooded. Most records are of the underside mines which cause the leaves to twist. Moths have also been recorded at light and flying around foodplant.

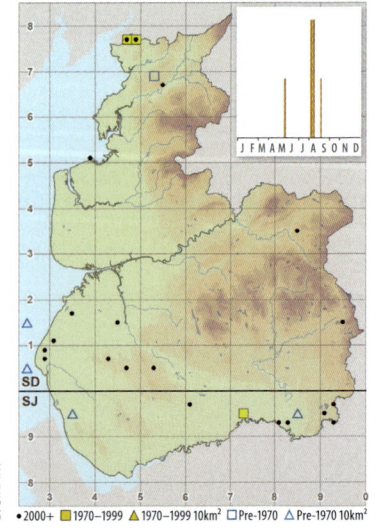

● 2000+   ▢ 1970–1999   △ 1970–1999 10km²   ▢ Pre-1970   △ Pre-1970 10km²

110 *The Moths of Lancashire*

15.084 *Phyllonorycter acerifoliella*

■ **LANCS STATUS** Occasional / local resident   **1995–2022**

Noted from the Liverpool district (Byerley, 1854). However, this is likely to have been from the Wirral and is considered unconfirmed for Lancs. The species was absent from Gregson, Stainton, Ellis and Mansbridge's early lists of the county moth fauna, and so the first record was by A.M. Emmet on 18 October 1995, with mines noted at Gait Barrows. This find appears to be somewhat of an outlier, as the 139 subsequent Lancashire records are all from southern and central parts of the county.
■ **HABITAT** Woodland edges, hedgerows, parks.
■ **FOODPLANTS** Field Maple.
Double brooded. Most records are of the leaf-mines. At light on nine occasions.

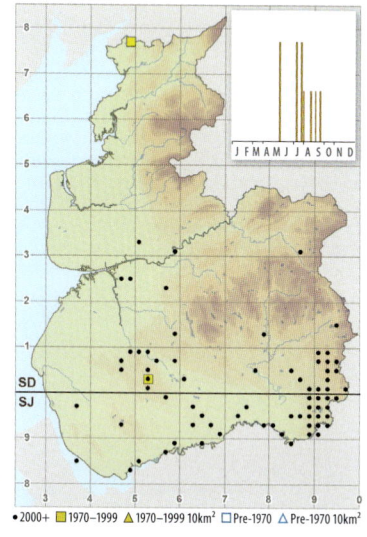

● 2000+   □ 1970–1999   △ 1970–1999 10km²   □ Pre-1970   △ Pre-1970 10km²

15.085 *Phyllonorycter joannisi*

■ **LANCS STATUS** Occasional / local resident   **1999–2022**

Although not recorded in the county until 12 October 1999, the mines have since been noted on over 220 occasions. The first record was of leaf-mines at Flixton by K. McCabe, where mines were frequent on the host tree. Most subsequent records are from south and central Lancashire, presumably connected to foodplant distribution, with a few upland and northern records from Scorton, Clitheroe and Briercliffe.
■ **HABITAT** Parks, streets, amenity planting.
■ **FOODPLANTS** Norway Maple.
Double brooded. Whilst most records are of the underside mines, the adult has been recorded at light on 49 occasions.

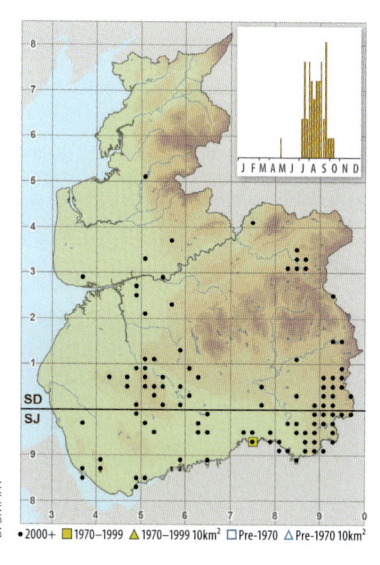

● 2000+   □ 1970–1999   △ 1970–1999 10km²   □ Pre-1970   △ Pre-1970 10km²

15.086 *Phyllonorycter geniculella*

■ **LANCS STATUS** Frequent / widespread resident   **1948–2022**

First recorded in Didsbury in 1948 by H.N. Michaelis, who also noted the species at Formby in 1949 and Silverdale in 1953. The geographical spread of these records suggest it could have been here for some time, despite the moth's absence from Mansbridge's 1940 list. The next records were from Billinge in 1989, Speke and St Helens in 1994, and Gait Barrows in 1995. Now, well distributed and likely to be encountered in lowland and upland areas wherever the foodplant occurs.
■ **HABITAT** Woodland edges, hedgerows, parks.
■ **FOODPLANTS** Sycamore.
Double brooded. Of the 522 records, over 80 are of the moth to light.

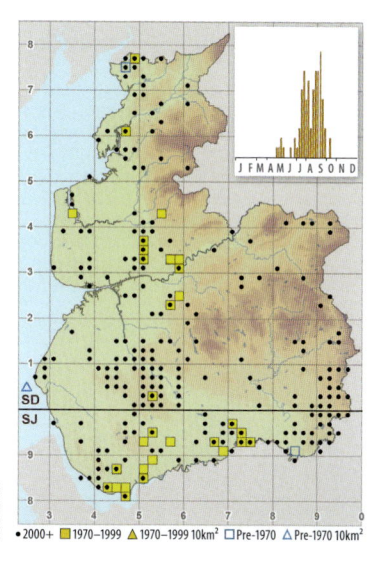

● 2000+   □ 1970–1999   △ 1970–1999 10km²   □ Pre-1970   △ Pre-1970 10km²

GRACILLARIIDAE

15.089 Horse-chestnut Leaf Miner *Cameraria ohridella*

■ **LANCS STATUS** Initially adventive. Now abundant / widespread resident 2007–2022
A recent arrival in the UK, this moth reached Chorlton in 2007, with mines noted by B. Smart. The first VC60 mines were found two years later. Now extremely common, occurring in huge numbers with a highest count of 1000 mines at Ellerbeck in 2018. Many trees are clearly even more infested than this, with uncertain long-term effects on the health of the host plant. Abundance noted on Horse-chestnut at Reddish, 15 October 2022, with at least 200 larval mines on nearby Sycamore (S.H. Hind).
■ **HABITAT** Wherever the host tree occurs.
■ **FOODPLANTS** Horse-chestnut, Sycamore.
Double brooded. Moths frequent at light and flying around foodplants.

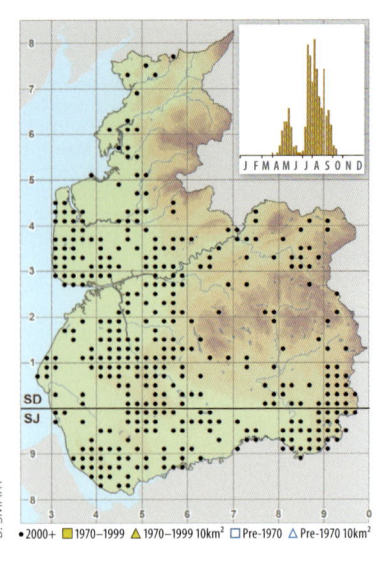

15.090 *Phyllocnistis saligna*

■ **LANCS STATUS** Scarce / very local resident 2019–2022
First recorded in Chorlton, as part of national northwards range expansion, by B. Smart on 15 September 2019, when mined twigs and leaves were noted on a single Crack-willow. Significant spread not detected until 2022, when mines noted at new sites in Chorlton and in Flixton, with increased density of mines on the affected trees.
■ **HABITAT** Woodland edges, close to lakes and rivers. So far, all records have been found in the Mersey Valley, within 200m or so of the river.
■ **FOODPLANTS** Crack-willow. Most records are of mined twigs and leaves. Bred from pupal cocoons on mined leaves, collected in Chorlton on 21 August 2022.
Double brooded. Once at light, on 15 July 2022 at Flixton (K. McCabe).

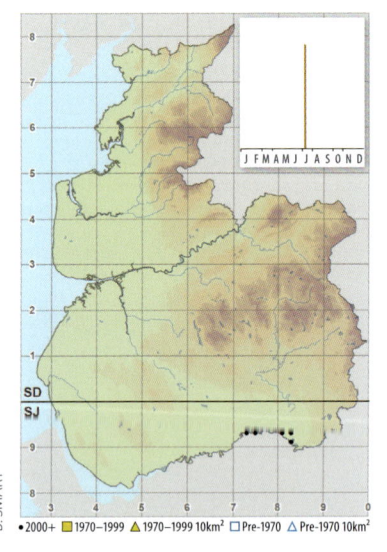

15.092 *Phyllocnistis unipunctella*

■ **LANCS STATUS** Frequent / local resident c.1857–2022
Recorded near poplars in Liverpool (Gregson, 1857). Strangely, not mentioned by Ellis (1890) or Mansbridge (1940). It was therefore over 130 years before the next record, an adult at light in Preston 1995. Whether due to genuine spread or increased effort, the next few years showed the moth to be widespread in southern and central Lancashire. Records remain sparse north of the Ribble. Primarily from lowland sites.
■ **HABITAT** Woodland edges, hedgerows, fixed dunes, parks.
■ **FOODPLANTS** Poplars, including Black-poplar and hybrids, Balsam Poplar.
Double brooded. The silvery gallery mine, on either side of the leaf, is distinctive and accounts for most records. Adults frequent at light and in the field.

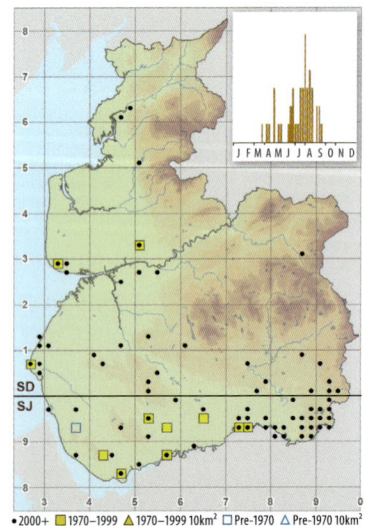

The Moths of Lancashire

16.001 Bird-cherry Ermine *Yponomeuta evonymella*

■ **LANCS STATUS** Abundant / widespread resident   c.1857–2022

'Taken by Mr Nixon at Hale, upon cherry trees' (Gregson, 1857), and noted to be locally common (Ellis, 1890). The moth was bred from larvae in Liverpool by W.A. Tyerman, sometime around 1917. Other pre-1940 records came from Reddish, Longworth Clough and Nelson. The species has been regularly recorded since, and apparently in increasing numbers. Reported on over 5,500 occasions in total throughout the county, which includes over 80 records of the larval webs on Bird Cherry. The larvae feed gregariously; occasionally in huge numbers, completely defoliating the host tree. The highest recorded count is of approximately 1,000 newly emerged moths on a Bird Cherry tree and on surrounding vegetation, all of which were covered in webbing, noted at Ainsworth, Bury on 29 June 2022 (P. Turner).

■ **HABITAT** Woodland edges, scrub, brownfield sites, amenity and street trees.

■ **FOODPLANTS** Bird Cherry. Presumably these were the cherry trees noted by Mr Nixon. Tenanted webs have been recorded from May to early July. The highest larval count was of approximately 100, feeding within a web at Springs Wood, Leck on 11 June 2006 (J.M. Newton).

Single brooded. Attracted to light. Unlike most of the genus, adults can be identified without knowledge of the larval foodplant, as the black dots are far more numerous than in other *Yponomeuta* species.

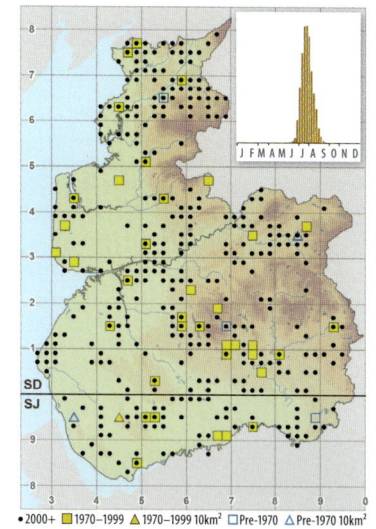

• 2000+   ■ 1970–1999   ▲ 1970–1999 10km²   □ Pre-1970   △ Pre-1970 10km²

16.002 Orchard Ermine *Yponomeuta padella*

■ **LANCS STATUS** Frequent / widespread resident   c.1859–2022

The species was noted by I. Byerley in *The Fauna of Liverpool* (1854). However, as most of the records within that work originate from south of the Mersey, i.e., in VC58, the first Lancashire record is considered to be J. Chappell's, from the Manchester area, c.1859 (Ellis, 1890). Early authors described the moth as locally common. It was noted quite extensively across Fylde during the 1980s and 1990s, with many records by M. Evans, and also by D. & J. Steeden and S. Palmer. It is likely still present in much of this area, but without many recent records to confirm this. The species has been primarily noted from lowland areas, but with a few upland records too, such as webs on Hawthorn scrub at Worsaw Hill on 15 June 2021 (S. Palmer). Webs can be extensive; one such at Clayton-le-Woods, nr. Chorley on 11 July 2012, affected 800m of hedge and contained 1,000s of larvae (G. Dixon).

■ **HABITAT** Woodland edges, hedgerows, scrub, brownfield sites.

■ **FOODPLANTS** Hawthorn, Blackthorn, Bullace, Cotoneaster.

Single brooded. Moths come to light, although the identity of this and the following two species cannot be safely determined without knowledge of the larval host plant. The map is therefore entirely based on recording of the larval and pupal webs and so is likely to be an underestimate of the moth's actual distribution.

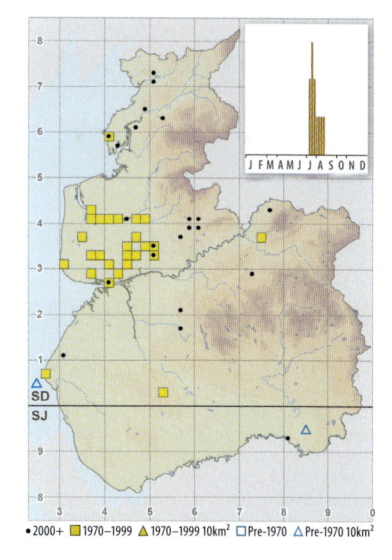

• 2000+   ■ 1970–1999   ▲ 1970–1999 10km²   □ Pre-1970   △ Pre-1970 10km²

The Moths of Lancashire   113

YPONOMEUTIDAE

16.003 Apple Ermine *Yponomeuta malinellus*

■ **LANCS STATUS** Occasional / local resident   1990–2022

A few moths were bred from a larval web at Newton-with-Scales in 1990 by M. Evans. Further records came in 2006 from Chorlton Water Park (B. Smart) and from Hutton (A. Barker), with webs noted by Manchester Ship canal, Flixton in 2015 (B. Smart). 2022 records of webs came from Green Bank, nr. Thrushgill by S. Palmer, from Reedley (M. Jackson) and Hardy Farm, Chorlton. Almost certainly under-recorded.

■ **HABITAT** Hedgerows, gardens.
■ **FOODPLANTS** Apple.

Single brooded. All confirmed records are of the early stages. As with the previous species, identity cannot be determined by external markings or by dissection.

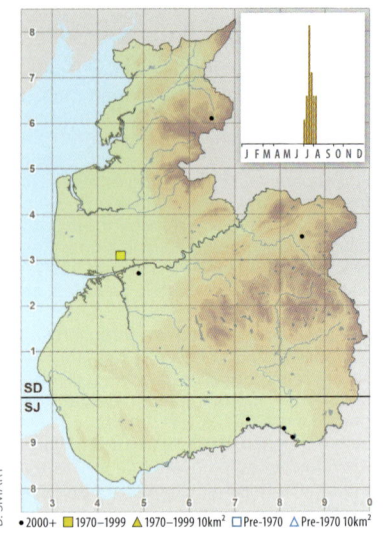

16.004 Spindle Ermine *Yponomeuta cagnagella*

■ **LANCS STATUS** Occasional / local resident   c.1859–2022

First recorded in Manchester with larvae on Spindle noted to be common (Stainton, 1859). Also, from Longridge by J.H. Threlfall (Ellis, 1890). Described by Mansbridge (1940) as common and well distributed without listing sites. Not noted again until 1986, when P.B. Hardy found defoliated Spindle bushes at Chorlton. Next, at Layton, Blackpool in 1987 (M. Evans). Recorded from approx. 25 sites, including six from 2021–2022. Primarily a lowland species, with garden records on Evergreen Spindle.

■ **HABITAT** Gardens, country parks, hedgerows.
■ **FOODPLANTS** Spindle, Evergreen Spindle.

Single brooded. Confirmed records are of larval webs on known foodplant.

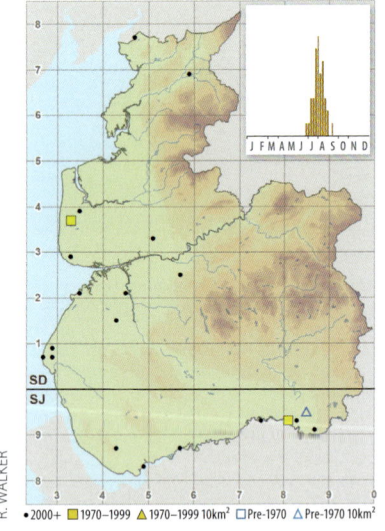

16.005 Willow Ermine *Yponomeuta rorrella*

■ **LANCS STATUS** Occasional / local resident   1992–2022

Larval webs first noted on willow at Carr House Green Common on 14 June 1992 (M. Evans). A second web was recorded on White Willow, on the north bank of the Mersey in Chorlton, 2010 (B. Smart). All other records have been from 2020 onwards, suggesting a gradual, northwards range expansion, including webs at Rimrose Valley Country Park and moths at Heaton Moor, Hutton, Tarleton and Great Harwood.

■ **HABITAT** Woodland edges, hedgerows.
■ **FOODPLANTS** White Willow, and likely other narrow-leaved willows.

Single brooded. Larval webs recorded on three occasions. Moths come to light and can be carefully identified based on external characteristics.

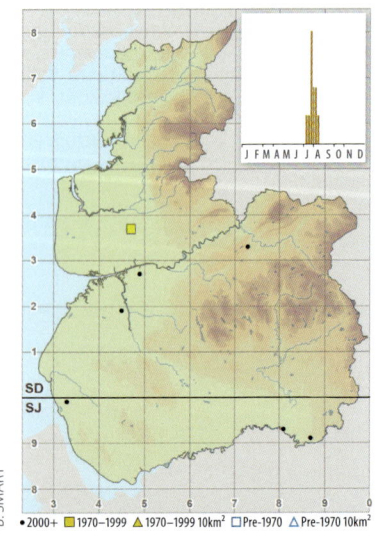

114   *The Moths of Lancashire*

16.007 *Yponomeuta plumbella*

■ **LANCS STATUS** Scarce / very local resident   1986–2023
First recorded in 1986 by L.W. Hardwick at Gait Barrows, then by M. Dempsey at Leighton Moss, 1992. All six subsequent records are from Gait Barrows and Hawes Water. The latter site produced the only larval record, on 21 May 2016, with a single early instar larva noted to be feeding within a shoot, causing it to droop; adult subsequently emerged (B. Smart). The species was not recorded again until 2023.
■ **HABITAT** Woodland edges, hedgerows.
■ **FOODPLANTS** Spindle. Distribution limited by foodplant.
Single brooded. Comes to light. Larval stages less obvious than preceding members of the genus. Later instars feed in small webs, usually only a few in each web.

• 2000+   ▢ 1970–1999   ▲ 1970–1999 10km²   ☐ Pre-1970   △ Pre-1970 10km²

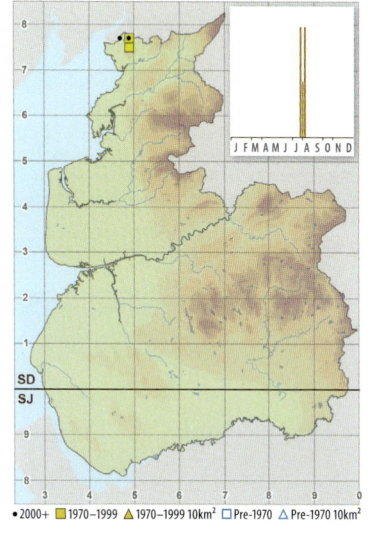

16.008 *Yponomeuta sedella*

■ **LANCS STATUS** Rare / very local resident   2018
This species has only been recorded on two occasions, both from the same garden in Great Sankey, Warrington, by J. Mitchell-Lisle in 2018. The moths, singletons on each occasion, were attracted to the Robinson MV trap on 6 May and 10 May, part of the first generation (this is a double brooded species nationally). The species may possibly have been introduced via *Sedum* from a nearby garden, or perhaps arrived as part of a northward expansion in range, with the first Yorkshire record occurring two years later. As yet, the moth is unrecorded in Cheshire or Cumbria.
■ **HABITAT** Not known, as both records are of individuals attracted to light.
■ **FOODPLANTS** Unrecorded within the county.

• 2000+   ▢ 1970–1999   ▲ 1970–1999 10km²   ☐ Pre-1970   △ Pre-1970 10km²

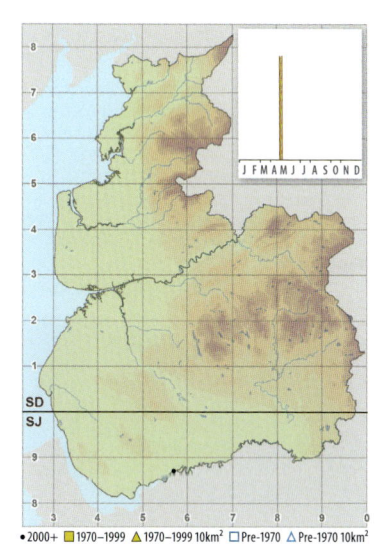

16.010 *Zelleria hepariella*

■ **LANCS STATUS** Occasional / local resident   1914–2022
Reported from Silverdale by P.F. Tinne in 1914, but not again until 1997 when noted by K. McCabe at Flixton. Next, from Gait Barrows in 1998. Further limestone records at Warton Crag, Eaves Wood, Yealand Conyers. Also, noted in Pennine districts, such as Burnley and Briercliffe, and some lowland areas away from Morecambe Bay, e.g., Parr, Chorlton, Lightfoot Green, etc. It is unclear whether the 62 records post-1914 represent movement into the county or the belated discovery of resident colonies.
■ **HABITAT** Limestone woodland, rough grassland.
■ **FOODPLANTS** Ash. Bred from larva found at Warton Crag 25 May 2018 (C. Palmer).
Single brooded. Records are of moths to light and netted during daytime fieldwork

• 2000+   ▢ 1970–1999   ▲ 1970–1999 10km²   ☐ Pre-1970   △ Pre-1970 10km²

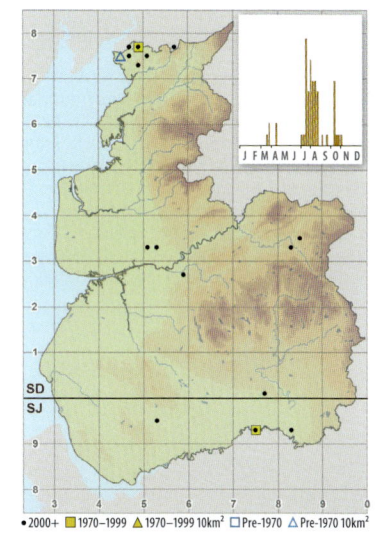

YPONOMEUTIDAE

16.011 *Zelleria oleastrella*

■ **LANCS STATUS** Adventive 2021

Only a single record is known from the region. Attracted to MV light at Lightfoot Green, Preston on 8 September 2021, the moth was recorded by C. & S. Palmer. Many new housing and amenity developments, with accompanying landscaping, were being constructed nearby, prior to and around this date, and it is possible that the moth emerged from an egg or larva introduced with an Olive tree planted in the locality, rather than representing part of a local breeding population.

■ **HABITAT** Not known.

■ **FOODPLANTS** Unrecorded within the county.
Nationally, double brooded.

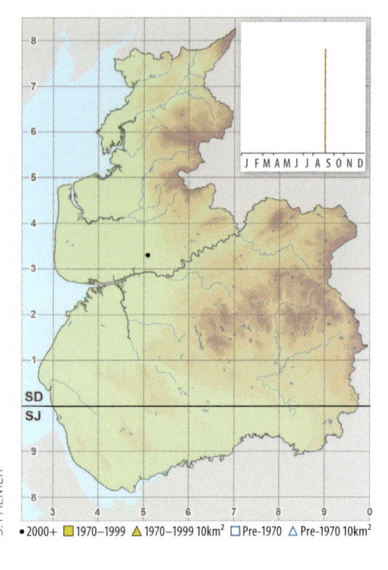

16.014 *Pseudoswammerdamia combinella*

■ **LANCS STATUS** Scarce / local resident c.1857–2018

Reported from the Liverpool district from 'mixed hedges, especially where blackthorn grows' (Gregson, 1857). Later, from Manchester (Stainton, 1859), Lancaster and Simonswood Moss (Mansbridge, 1940). Not noted again until 2011 when a moth came to light in Rainford (S. Williams). Subsequently, two in 2013, and one in 2015, all from Warton Crag, and one from Liverpool in 2018. The scattered nature of records suggests the continued presence of a few low-density populations.

■ **HABITAT** Woodland edges, hedgerows.

■ **FOODPLANTS** Unrecorded within the county.
Single brooded. Moths recorded at light, during day and at dusk.

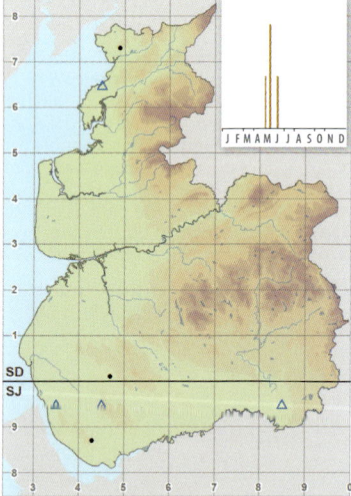

16.015 *Swammerdamia caesiella*

■ **LANCS STATUS** Frequent / local resident c.1857–2022

Mid- to late 19th century records from Rainford (Gregson, 1857), Manchester (Stainton, 1859) and Chat Moss (Ellis, 1890), and later noted to be 'abundant in birch woods' (Mansbridge, 1940). Primarily a lowland species, with more recent records from Little Woolden Moss, Heysham Moss, Ainsdale Dunes, Freshfield, Rixton Claypits and Docker Moor. Recorded on 76 occasions, all but 17 from this century.

■ **HABITAT** Mosses, woodland edges.

■ **FOODPLANTS** Birch, including Downy Birch.
Double brooded. Over half the records are of moths to light or netted during daytime or dusk. The others are of larvae and their webs, found on the upperside of a leaf.

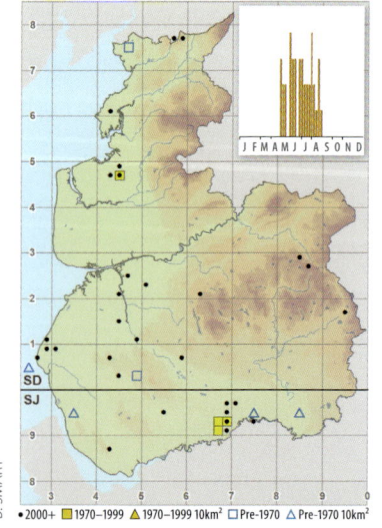

16.017 *Swammerdamia pyrella*

■ **LANCS STATUS** Frequent / widespread resident   c.1857–2022

Described by Mansbridge (1940) as a common and generally distributed species. Largely still the case, with over 300 21st century records, although relatively scarce in eastern and upland areas. First noted from Liverpool and Manchester districts (Gregson, 1857; Stainton, 1859). No further records until 1994, when recorded from Warton Bank by S. Palmer. Since then, regular at garden traps, including 178 records from Flixton since 1998 by K. McCabe, with a high count of nine in 2000.
■ **HABITAT** Woodland edges, hedgerows, gardens.
■ **FOODPLANTS** Hawthorn, Cherry, Blackthorn, Apple, Crab Apple.
Double brooded. Comes to light. 13 larval records, from June to late October.

● 2000+   ☐ 1970–1999   △ 1970–1999 10km²   ☐ Pre-1970   △ Pre-1970 10km²

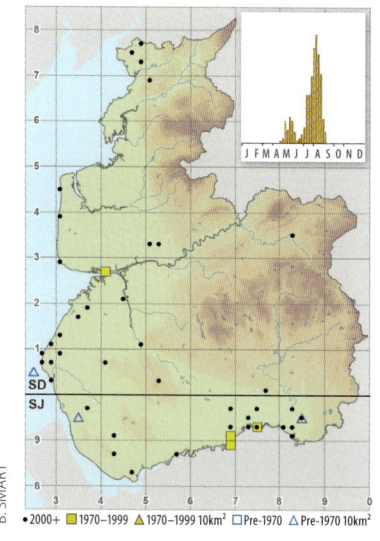

16.019 *Paraswammerdamia albicapitella*

■ **LANCS STATUS** Occasional / local resident   c.1859–2022

First noted in the Manchester district (Stainton, 1859). The foodplant was listed as Hawthorn, which is doubtful. Neither Ellis nor Mansbridge list the moth as a Lancashire species, suggesting they may have had doubts re: Stainton's report. Next, recorded from Storrs Moss in 1951 by N.L. Birkett, and in 1995 by S. McWilliam at Martin Mere. Most records from the Morecambe Bay area, with a few from Liverpool.
■ **HABITAT** Woodland edges, hedgerows.
■ **FOODPLANTS** Blackthorn. Early instar leaf-mine noted October 2018 (J. Patton) with moth bred from larva beaten from Blackthorn May 2022; both from Middleton N.R. Single brooded. Comes to light.

● 2000+   ☐ 1970–1999   △ 1970–1999 10km²   ☐ Pre-1970   △ Pre-1970 10km²

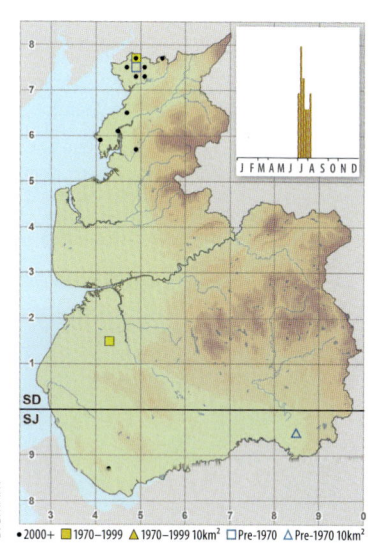

16.020 *Paraswammerdamia nebulella*

■ **LANCS STATUS** Abundant / widespread resident   c.1857–2022

With over 1,300 records, and distribution in lowland and upland areas, this moth and its larval webs can be expected wherever Hawthorn grows. First recorded in Liverpool (Gregson, 1857), and in Manchester (Stainton, 1859). Described as plentiful and well distributed (Mansbridge, 1940). Also, from Formby 1957, Ince Moss 1984 and Gait Barrows 1989. High count of 40 at Warton Bank in 2007. Population appears stable.
■ **HABITAT** Woodland edges, hedgerows.
■ **FOODPLANTS** Hawthorn.
Single brooded. Moths recorded at light, during day and at dusk. Tenanted leaf-mines noted in autumn, with larval webs recorded during May and June.

● 2000+   ☐ 1970–1999   △ 1970–1999 10km²   ☐ Pre-1970   △ Pre-1970 10km²

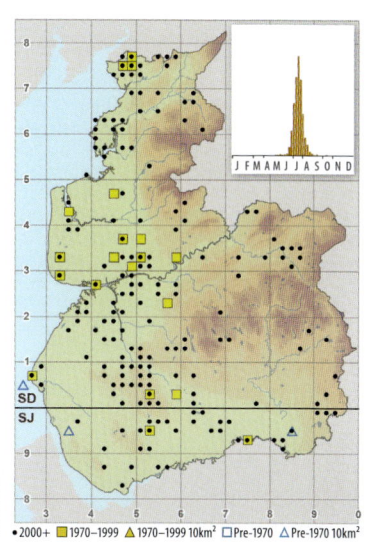

16.021 *Cedestis gysseleniella*

■ **LANCS STATUS** Scarce / local resident   2000–2022

A recent arrival in the county, with the first report on 12 August 2000 at Ainsdale, when the moth was recorded on a Lancashire Moth Group field trip. Noted on three occasions the following year, from Flixton, Littleborough and Parr. By 2011 it had been recorded on eleven occasions from ten sites. However, there has been an apparent decline since, with the species subsequently unrecorded until 1 July 2022, when a single moth was beaten from pine at Ainsdale Dunes.

■ **HABITAT** Pine woods. Probably from isolated pines in woodland, gardens, etc.
■ **FOODPLANTS** Unrecorded. 2022 record conforms to expected association with pine.
Single brooded. Comes to light and beaten from pines.

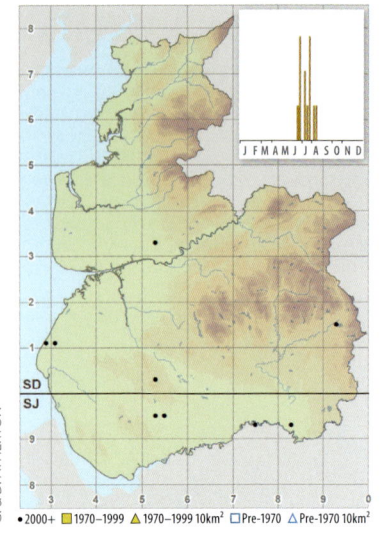

16.022 *Cedestis subfasciella*

■ **LANCS STATUS** Occasional / local resident   c.1859–2022

First reported from Manchester (Stainton, 1859), then at Chat Moss and Longridge (Ellis, 1890). Later, at Silverdale 1912 and Formby 1950. The next record was from Longridge Fell on 26 July 1996 (S. Palmer), with subsequent reports largely from lowland areas of VC59. The strongest populations are in the pine woods of the Sefton Coast, with over half of all sightings from Formby, Freshfield and Ainsdale. Other records from St Helens, Chorlton, Lightfoot Green, Billinge, Fazakerley, Westby, etc.

■ **HABITAT** Pine woods, mixed woodland, country parks.
■ **FOODPLANTS** Pine.
Larvae recorded mining needles during February and March.

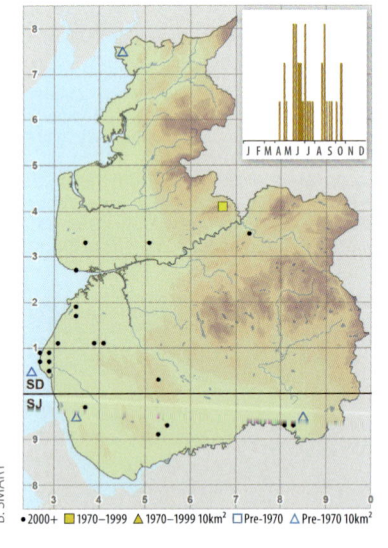

16.023 *Ocnerostoma piniariella*

■ **LANCS STATUS** Scarce / local resident   2002–2015

Mansbridge (1940) reports finding *O. piniariella* freely in pine woods throughout. Unfortunately, pre-1966 records may refer to this or the following, as *O. friesei* was only recognised as separate from this point onwards. Subsequently, there have been few confirmed records. The first of these was on 25 June 2002, a moth found resting on a birch leaf at White Moss Tip, Skelmersdale by C.A. Darbyshire. Further records are of a moth at light in Southport and pupal exuvia noted between spun needles at Freshfield, both from 2015. The Skelmersdale and Southport moths were both dissected. The pupal record was identified on the basis of differences in pupal morphology and confirmed by R. Edmunds. Whilst the map suggests the species is limited to south-west Lancashire, difficulties in identification mean it may potentially be rather more widespread.

■ **HABITAT** Pine woods, mixed woodland
■ **FOODPLANTS** Pine.
Single brooded. Comes to light. Moths from June and August.

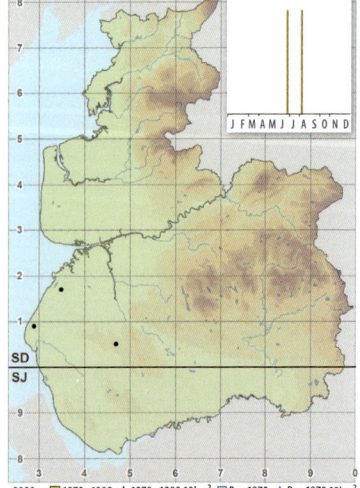

16.024 *Ocnerostoma friesei*

■ **LANCS STATUS** Occasional / local resident 2002–2022

The first confirmed record was of two males to light at Astley Moss on 14 April 2002 (I. Walker and G. Riley), with further records from Parr, Swinton, Southport, Flixton, Lightfoot Green, etc. The sole record from the far north of the county was on 28 March 2022, when a mating pair were noted on Scots Pine needles at Clougha by S. Palmer & B. Smart. High count of 13 to light at Southport, 15 April 2022 (R. Moyes).

■ **HABITAT** Pine woods, mixed woodland.
■ **FOODPLANTS** Pine.

Double brooded. Most records are of moths at light. Spun needles containing pupa were noted 17 February 2019 at Freshfield, with subsequent dissection of bred moth.

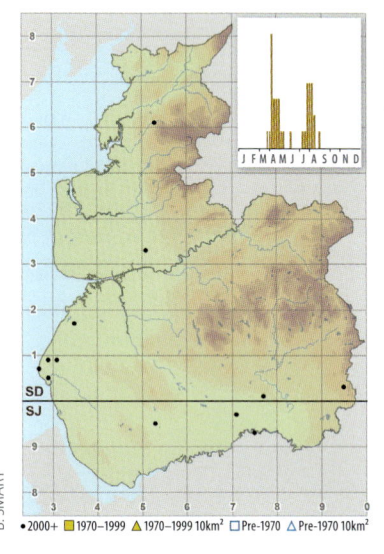

YPONOMEUTIDAE

17.002 *Ypsolopha nemorella*

■ **LANCS STATUS** Scarce / local resident c.1859–2020

First noted in the Manchester district (Stainton, 1859) and then in Preston and Cleveleys (Ellis, 1890). Recorded from Burnley around 1910 by A.E. Wright, but not again until 1986 at Gait Barrows, and 1987 from Bolton. This latter find was the last from VC59, with all subsequent records from northern sites, e.g., Wray, Jack Scout, Myers Allotment, Silverdale, etc. Only recorded on 13 occasions in Lancashire.

■ **HABITAT** Limestone woodland edges.
■ **FOODPLANTS** Not known in Lancashire.

Single brooded. Comes to light, with a few netted in the field. All from early July to 1 September; the latter, at Lord's Lot Wood in 2020, being the most recent (Ga. Jones).

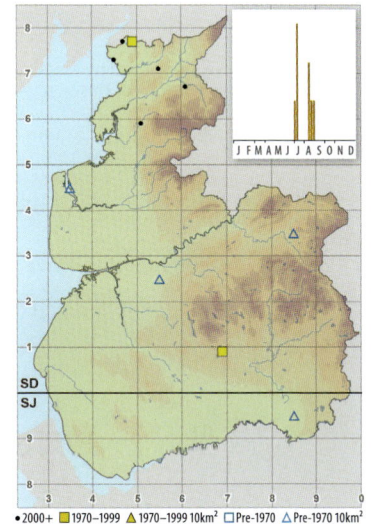

YPSOLOPHIDAE

17.003 Honeysuckle Moth *Ypsolopha dentella*

■ **LANCS STATUS** Frequent / widespread resident c.1857–2022

First reported from the Liverpool district (Gregson, 1857), and then in the Manchester area (Stainton, 1859). During the early 20th century, it was recorded from Burnley, Old Trafford and Hale, and described as 'local among Honeysuckle' (Mansbridge, 1940). 1,280 records post-1979, from a mixture of lowland sites, e.g., Lytham, Parr, and upland sites, e.g., Littleborough and Briercliffe. Continued high counts (21 in 1999, 20 in 2020) suggest the population remains healthy.

■ **HABITAT** Woodland edges, hedgerows, gardens.
■ **FOODPLANTS** Honeysuckle. Once on Snowberry (Chorlton, 18 May 2020, B. Smart).

Single brooded. Frequently to light. Larval spinnings from May to mid-June.

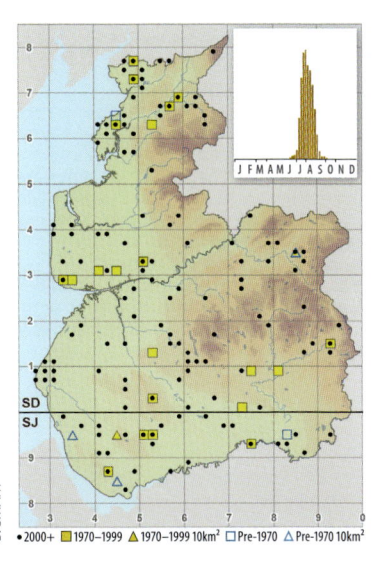

The Moths of Lancashire **119**

YPSOLOPHIDAE

17.005 *Ypsolopha scabrella*

■ **LANCS STATUS** Frequent / widespread resident   1953–2022

Unrecorded in the historic literature, suggesting possible arrival in the county during the mid-20th century. Recorded from Didsbury by H.N. Michaelis from 1953 to 1956, but not again until 1975 at Croxteth. Gradual northwards spread with records from Newton-with-Scales 1985 and Gait Barrows 1989. Now widespread, although less so in upland areas. High count of six, at Littleborough on 15 August 2000 (I. Kimber). Other recent records from Flixton, Parr, Hale, Westby, Fazakerley, Hindley, etc.
■ **HABITAT** Woodland edges, hedgerows, scrub.
■ **FOODPLANTS** Hawthorn.
Single brooded. Frequent at light. Five larval records, all between 14 May and 2 June.

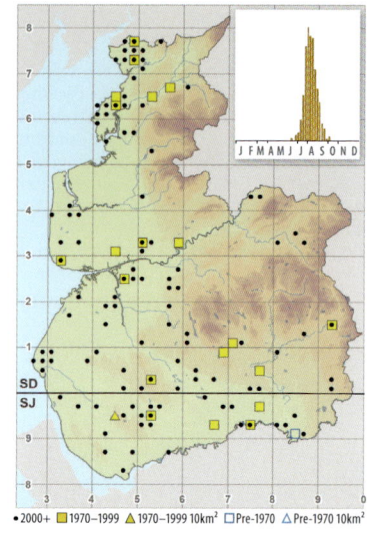

17.006 *Ypsolopha horridella*

■ **LANCS STATUS** Rare / very local resident   2021

There is just a single record of this species in the county, trapped at MV light in a garden at Tarleton by R. Pyefinch on 9 August 2021. This may possibly be a harbinger of further records as the moth moves northwards through the country. This process is already ongoing in Yorkshire where the previously unrecorded species has been noted on 22 occasions since 2011, and in Cheshire where 3 of the 4 records have occurred since 2019 (cheshire-chart-maps.co.uk; yorkshiremoths.co.uk, both accessed 1 September 2023).
■ **HABITAT** Unknown in the county.
■ **FOODPLANTS** Not recorded in Lancashire.

17.007 *Ypsolopha lucella*

■ **LANCS STATUS** Considered Extinct   c.1859

The only record we have for this species is from the Manchester district (Stainton, 1859). As previously discussed, the Manchester district is primarily VC59, but some of the southern and eastern edges belong to VC58 (Cheshire) and VC63 (south-west Yorkshire). The species remains unrecorded in Cheshire, and in Cumbria to the north. Recorded on just three occasions in Yorkshire, two of these being in VC63. These are from Wakefield 2006 and Barnsley 2012 (yorkshiremoths.co.uk, accessed 1 September 2023).
■ **HABITAT** Unknown in the county.
■ **FOODPLANTS** Not recorded in Lancashire.
Nationally, this species is associated with oak. The male is rare as the species is partially parthenogenetic (Langmaid *et al.*, 2018), a form of reproduction which does not involve fertilisation by males.

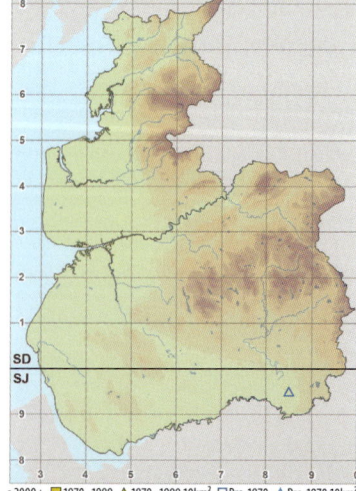

120 *The Moths of Lancashire*

17.008 *Ypsolopha alpella*

■ **LANCS STATUS** Occasional / local resident   2005–2022

Another recent arrival into the county, spreading northwards from Cheshire. All records so far have been confined to southern areas of the county. First recorded from Blackbrook, St Helens by G. & D. Atherton on 13 July 2005, with the same recorders finding the species nearby at Haydock a month later. St Helens seems a stronghold for the species with more recent records from Rainford and Parr. Also, noted from Roby, Flixton, Liverpool and Billinge.
■ **HABITAT** Not known. All records are of moths from garden light traps.
■ **FOODPLANTS** Unrecorded within the county.
Single brooded. High count of three at Rainford, 30 August 2018 (C. Daly, R. Moyes).

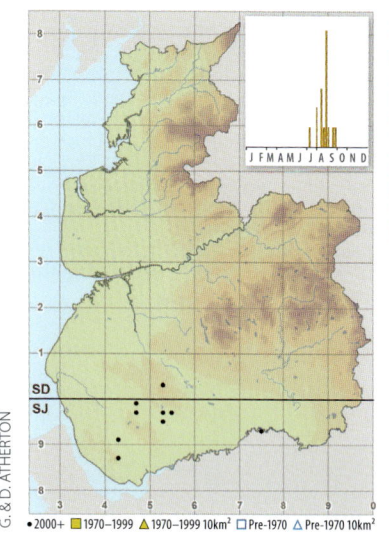

17.009 *Ypsolopha sylvella*

■ **LANCS STATUS** Occasional / local resident   c.1859–2020

First reported from the Manchester district (Stainton, 1859). However, all subsequent 12 records have come from the 21st century, suggesting northwards range expansion nationally. As with the previous species, all these records are restricted to the south of the county. The first was at Flixton on 28 September 2000 (K. McCabe), with others from Chorlton, St Helens, Billinge and Woolton.
■ **HABITAT** Oak woodland.
■ **FOODPLANTS** Oak.
Single brooded. Most records are of moths at light. Larva beaten from oak at Barlow Eye Tip, Chorlton on 12 June 2011, with moth emerging on 7 July 2011 (B. Smart).

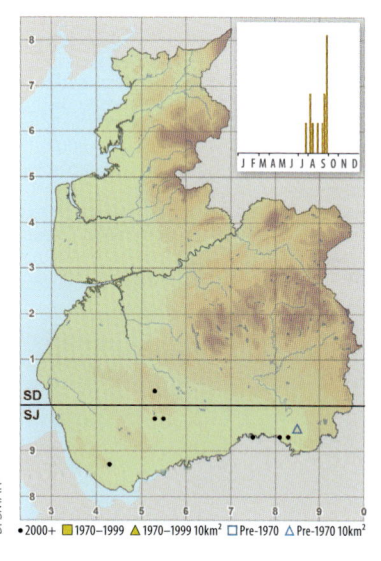

17.010 *Ypsolopha parenthesella*

■ **LANCS STATUS** Frequent / widespread resident   c.1857–2022

First noted from woodland in the Liverpool district (Gregson, 1857), and in Manchester (Stainton, 1859), but not recorded again until Crosby, 1951 (R. Pritchard). The fourth record was from Risley Moss in 1984, since when it has been regularly recorded from most lowland and upland areas. Noted nectaring on ragwort flowers at Holcroft Moss by K. McCabe, 2001. High count of 20 at St Helens, 2006 (R. Banks). Possibly in decline since, with no counts above three anywhere in the last five years.
■ **HABITAT** Woodland, country parks, scrub.
■ **FOODPLANTS** Oak, Hazel, birch, Goat Willow.
Single brooded. 700+ records, most of moths at light. Larvae noted May and June.

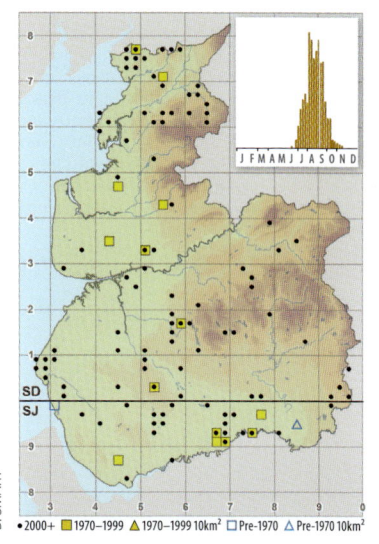

The Moths of Lancashire

YPSOLOPHIDAE

17.011 *Ypsolopha ustella*

■ **LANCS STATUS** Frequent / widespread resident   c.1857–2022

Early records from Croxteth Woods (Gregson, 1857), Manchester district (Stainton, 1859) and by J.B. Hodgkinson in Preston (Ellis, 1890). Reported as local and not common (Mansbridge, 1940). Not further recorded until 1991, when it was trapped at Holcroft Moss by S. McWilliam. There have been 264 records in total, all but 28 from this century. 21st century records include those from lowland areas such as Heysham Moss, Ainsdale, Ormskirk, and upland areas e.g., Wray, Littledale, Clougha.

■ **HABITAT** Oak woodland.

■ **FOODPLANTS** Oak. Larvae noted from late May to mid-June.

Single brooded. Records mostly of moths at light, even during the winter months.

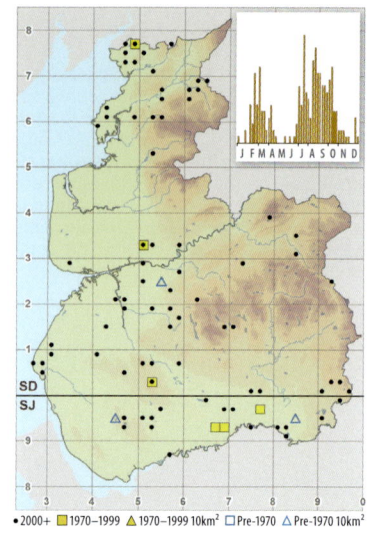

• 2000+   ▪ 1970–1999   ▲ 1970–1999 10km²   ▫ Pre-1970   △ Pre-1970 10km²

17.012 *Ypsolopha sequella*

■ **LANCS STATUS** Frequent / widespread resident   c.1859–2022

First reported from the Manchester district (Stainton, 1859), but not seen in the county again until 12 August 1975, at Fulwood Park, Liverpool by I.D. Wallace. Next, recorded at Chorley in 1983. Over 800 subsequent records confirm the moth to be thriving throughout lowland and upland areas, with a maximum count of 13 from Briercliffe, 24 September 2019 (G. Turner).

■ **HABITAT** Woodland edges, isolated trees in other locations, e.g., graveyards.

■ **FOODPLANTS** Sycamore (probable). Cocoons found on or beneath the foodplant.

Single brooded. Comes to light. Moth emerged 9 July 2004, from a cocoon found on the upperside of a Sycamore leaf at Billinge Hill three days earlier by C. Darbyshire.

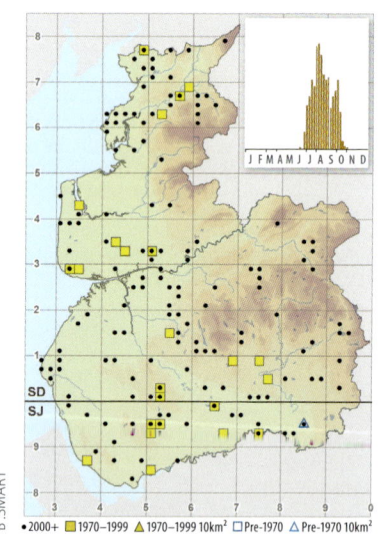

• 2000+   ▪ 1970–1999   ▲ 1970–1999 10km²   ▫ Pre-1970   △ Pre-1970 10km²

17.013 *Ypsolopha vittella*

■ **LANCS STATUS** Occasional / local resident   c.1859–2022

First reported from the Manchester district (Stainton, 1859), and next from Preston by J.B. Hodgkinson (Ellis, 1890). Status later noted as 'generally distributed and common on elm' (Mansbridge, 1940). Recorded from Didsbury in 1953 and St Helens in 1998. The 12 subsequent 21st century records show scattered distribution throughout lowland areas of the county, with recent records from Lytham St Annes, Pennington, Wigan, Rainford, Morecambe, Fulwood and Lightfoot Green.

■ **HABITAT** Woodland, gardens.

■ **FOODPLANTS** Elm. Unclear if Mansbridge's comments refer to larvae, adults, or both.

Single brooded. All modern records relate to moths trapped at light.

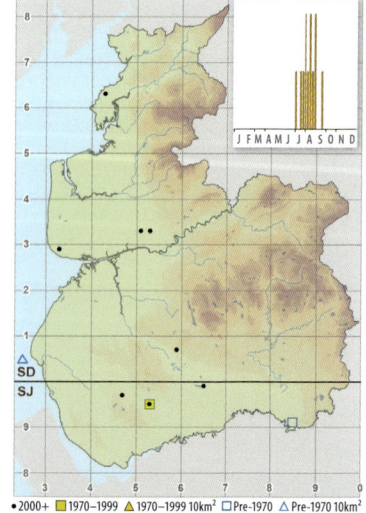

• 2000+   ▪ 1970–1999   ▲ 1970–1999 10km²   ▫ Pre-1970   △ Pre-1970 10km²

17.014 Ochsenheimeria taurella

■ **LANCS STATUS** Scarce / local resident   1859–2015
The 13 records are widely scattered with no evidence locally of attraction to light, suggesting possible under-recording. First reported in August 1859 under *birdella*, when 36 moths beaten from gorse in Blackpool by R.S. Edleston. Other records from Preston (Ellis, 1890) and Formby (Mansbridge, 1940), with latter describing the moth as local and uncommon. 21st century records from Salthill Quarry, Middleton N.R. and Dalton, nr. Wigan. Swept from grasses at latter site by C.A. Darbyshire in 2002.
■ **HABITAT** Rough grassland, primarily in lowland areas.
■ **FOODPLANTS** Unrecorded within the county.
Single brooded. Moths recorded during daytime fieldwork from 11 July to 21 August.

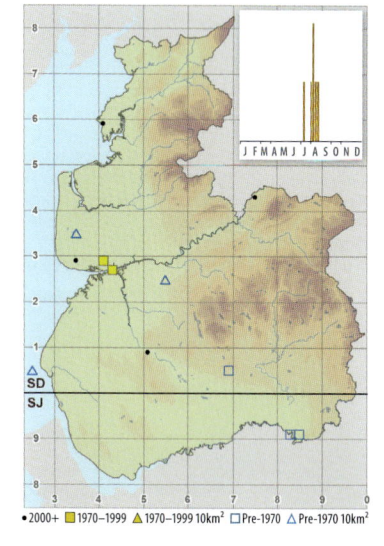

17.015 Ochsenheimeria urella

■ **LANCS STATUS** Scarce / local resident   c.1859–2021
First noted under *bisontella* in the Manchester district (Stainton, 1859). Later, from Lees, Longridge, Ribchester (Ellis, 1890), Rochdale and Burnley (Mansbridge, 1940). Mansbridge noted it to be 'common in the Pennine woods, scarce elsewhere.' Primarily an upland species, with recent records from Longridge Fell, Colne, and Pendleton Moor, but also from Middleton N.R. in 2014. A moth netted in daytime at Grains Bar, nr. Oldham by S.H. Hind on 25 July 2021 is the most recent of 14 records.
■ **HABITAT** Rough grassland.
■ **FOODPLANTS** Unrecorded within the county.
Single brooded. No records of moths at light. All netted or swept during day.

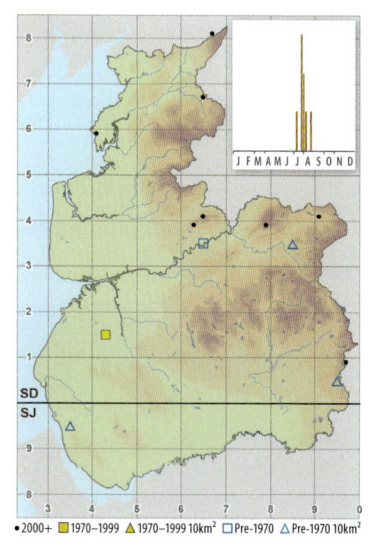

17.016 Cereal Stem Moth *Ochsenheimeria vacculella*

■ **LANCS STATUS** Considered Extinct   c.1859
The sole record of this species was noted to have been found indoors in the Manchester district (Stainton, 1859). It has rarely been recorded in neighbouring counties either, with a single Yorkshire record from 1946 (yorkshiremoths.co.uk, accessed 1 September 2023), three from Cheshire between 1917 and 1951 (cheshire-chart-maps.co.uk, accessed 1 September 2023), and not at all from Cumbria.
■ **HABITAT** Rough grassland, but occasionally found in sheds and houses, as with the Manchester example.
■ **FOODPLANTS** Not recorded in Lancashire.
Nationally, single brooded.

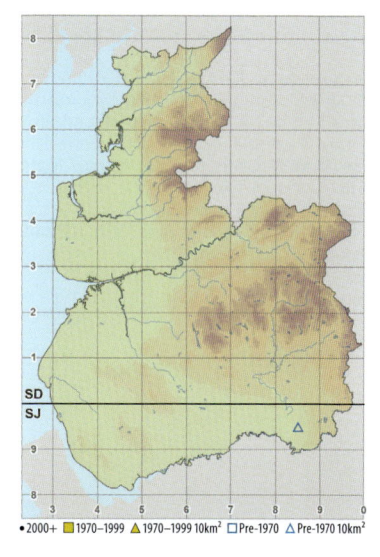

PLUTELLIDAE

18.001 Diamond-back Moth *Plutella xylostella*

■ **LANCS STATUS** Abundant / ubiquitous migrant and temporary resident   c.1857–2022
Reported from the Liverpool district by Gregson (1857), who noted it as 'more frequent in some years than others.' Described as abundant by Ellis (1890) and Mansbridge (1940). Over 6,000 records since, often in huge numbers. Estimated counts of 1,000 from Birkdale Green Beach, 900 from Formby and 700 from Fluke Hall, all from 2016, the last year of huge abundance of this migratory species.
■ **HABITAT** Rough grassland, gardens, arable fields, riverbanks.
■ **FOODPLANTS** Wild Mustard, Dame's-violet, *Brassica* sp. (inc. shop-bought broccoli). Double brooded, at least. Moths at light and easily disturbed during day. Larvae on crucifers, sometimes in high numbers e.g., 100s from Chorlton, July 2006 (B. Smart).

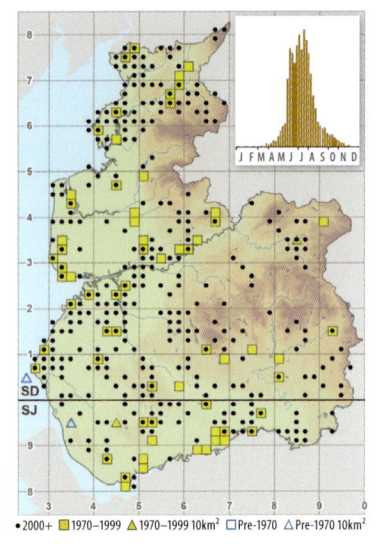

18.003 *Plutella porrectella*

■ **LANCS STATUS** Occasional / local resident   c.1857–2022
C.S. Gregson (1857) noted the larvae in the flower buds of 'white rock' in his own and other Liverpool gardens. Recorded in Preston by J.B. Hodgkinson and described as very local (Ellis, 1890). Later, from Formby (Mansbridge, 1940), Warton Crag in 1959 (C.J. Goodall), then Flixton in 1997 (K. McCabe). Over 400 subsequent records, mostly in VC59, but also 100 from the Morecambe Bay area around Heysham, Morecambe, Bolton-le-Sands, Yealand Conyers, etc. Larvae from Flixton, Rixton, Denton, Billinge.
■ **HABITAT** Rough grassland, parks, riverbanks etc. wherever foodplant occurs.
■ **FOODPLANTS** Dame's-violet. Identity of Gregson's 'white rock' unclear. Double brooded. Moths recorded at light and easily disturbed from vegetation.

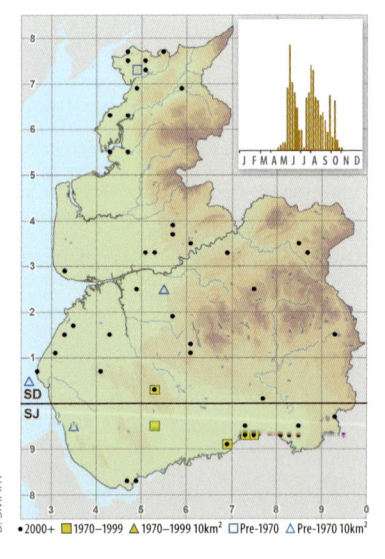

18.005 *Rhigognostis annulatella*

■ **LANCS STATUS** Rare / very local resident   1877–2006
First recorded on 8 June 1877, when larvae were found on a scurvygrass sp. at Morecambe by J.H. Threlfall, with adults subsequently reared (Threlfall, 1878). There is a single recent record of one attracted to MV light near the coast at Heysham N.R. on 15 April 2006 by P. Marsh and J. Roberts.
The species is rare in Cumbria, with just a single, confirmed record from Eaglesfield, nr. Cockermouth in 2018 (S. Colgate, pers. comm.). There are historical records from Cheshire and Yorkshire, but no recent sightings.
■ **HABITAT** Coastal locations.
■ **FOODPLANTS** Scurvygrass sp.
Single brooded. Comes to light. Larvae have not been recorded since 1877.

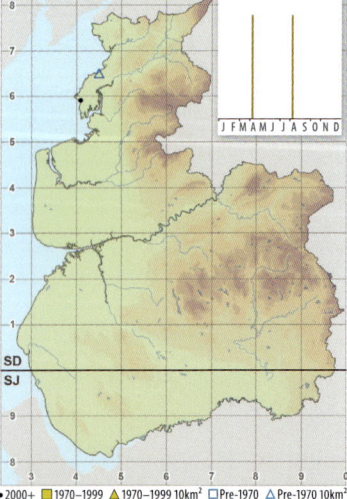

124 *The Moths of Lancashire*

18.007 *Eidophasia messingiella*

■ **LANCS STATUS** Occasional / local resident   c.1859–2022
First report from Manchester district (Stainton, 1859). Later, scores of larvae recorded on Large Bitter-cress at Brockholes Wood, nr. Preston on 13 May 1876 (Threlfall, 1877). Recorded at Flixton in 1997, the first record for over 100 years, with a moth trapped by K. McCabe. This area is a stronghold for the species, producing over half of the 46 county records. It also gave the high count of three adults on 25 June 2009. Other sites include Belmont, Rishton, Longton and Tyldesley.
■ **HABITAT** Rough grassland, scrub. From upland and lowland areas.
■ **FOODPLANTS** Large Bitter-cress.
Single brooded. Virtually all records are of moths at light, with a single larval record.

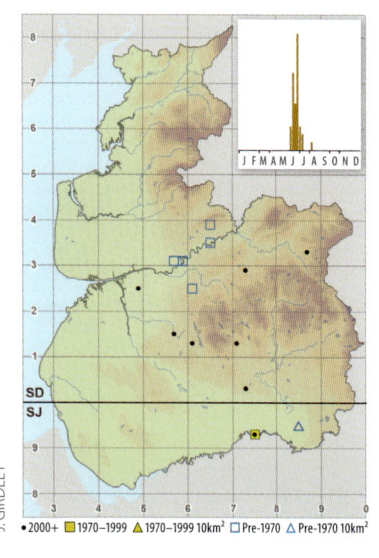

PLUTELLIDAE

19.001 *Orthotelia sparganella*

■ **LANCS STATUS** Occasional / local resident   c.1859–2022
First noted in the Manchester area (Stainton, 1859). A few scattered records are known between then and 2000, including from Pendleton in 1890, Didsbury in 1949, Heaton Mersey in 1955 and ten seen at Westby Clay Pits in 1980 (D. & J. Steeden). Since then, has been found infrequently at Flixton, Leighton Moss, Martin Mere and Silverdale Moss and, singly, at twelve other mainly lowland sites. Rarely wanders far from water, but one at MV light in Yealand Conyers on 24 July 2016 (B. Hancock).
■ **HABITAT** Larger wetland areas, ponds and brownfield sites with open water.
■ **FOODPLANTS** Not recorded in Lancashire.
Single brooded. Comes to light. Can be found resting on pond-side plants.

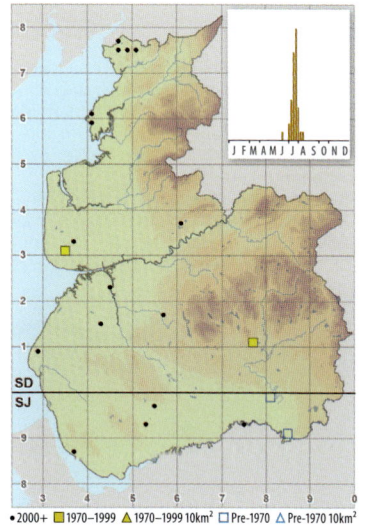

GLYPHIPTERIGIDAE

19.002 *Glyphipterix thrasonella*

■ **LANCS STATUS** Frequent / local resident   c.1857–2022
First noted in the Liverpool area (Gregson, 1857), followed by Preston c.1889 and Formby in 1950. Although present at Scorton in 1983 and near Inskip in 1986, it was not until the late 1990s that records increased considerably. Regular sites included Bold Moss, Chorlton, Docker Moor and Scorton. Daytime counts of up to 30 or so are regular, but larger counts can occur, with 100 at Chipping on 15 June 2021 (B. Dyson) and around 1,600 in Lamberts Meadow, Silverdale on 16 June 2009 (S. Palmer).
■ **HABITAT** Damp grassland, on sites such as commons, moors, mosses and dune slacks.
■ **FOODPLANTS** Not recorded in Lancashire.
Single brooded. Occasional at light; readily recorded during the day.

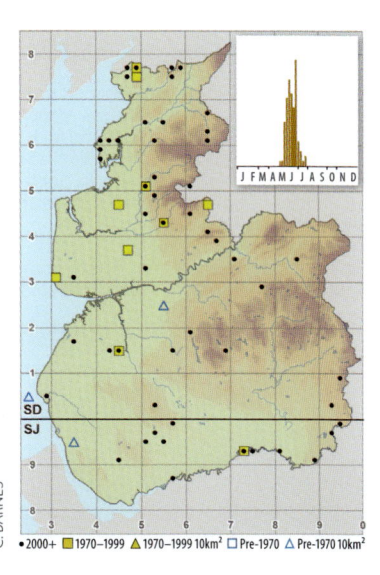

The Moths of Lancashire

GLYPHIPTERIGIDAE

19.003 *Glyphipterix fuscoviridella*

■ **LANCS STATUS** Frequent / local resident c.1857–2022

Noted on Crosby sandhills by Gregson (1857), with a few other records, from Chat Moss, Formby, Inskip area, Preston, Silverdale and St Annes, all prior to the 1990s. Thereafter, found regularly at several sites on vegetated coastal dunes including Ainsdale, Formby and St. Annes, with counts of 100 not unusual. More localised elsewhere, Ash Hill, Flixton being a regular site with good numbers present.

■ **HABITAT** Grassland, mainly on dune slacks, brownfield sites and limestone.
■ **FOODPLANTS** Field Wood-rush.

Single brooded. Occasional at light. Readily recorded in numbers during the day. Larvae have been found in roots and stems of the foodplant in February and March.

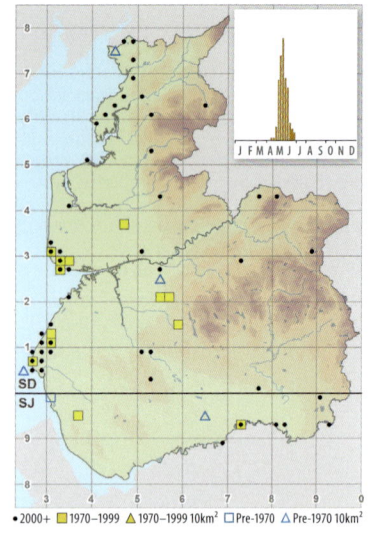

19.004 *Glyphipterix equitella*

■ **LANCS STATUS** Rare / vagrant 1953–1958

First noted in Heaton Mersey by W.D. Hincks on 2 July 1953 (Michaelis, 1954), but considered an introduction with plant material from out of region by Michaelis. The specimen is in Manchester Museum and, during a visit in 2022, a further specimen was found, also from Heaton Mersey by W.D. Hincks, dated 5 July 1958.

■ **HABITAT** Unknown.
■ **FOODPLANTS** Not recorded in Lancashire.

On 6 July 2001 and 10 June 2003, specimens identified as this species, came to light in Billinge (C.A. Darbyshire) with one checked by S. Palmer. Neither were retained and subsequent concerns about the identifications have not been possible to resolve.

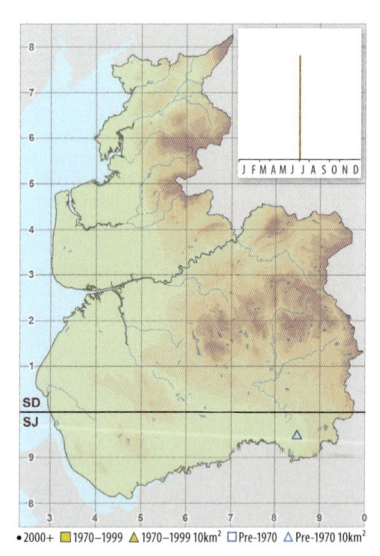

19.005 *Glyphipterix haworthana*

■ **LANCS STATUS** Occasional / very local resident 1846–2020

Reported as scarce on White Moss, Manchester in May 1846 (Edleston, 1846), it was later found on Chat Moss (1859) and near Bury (1889). It was not seen again until 2000, on Astley Moss, by S. McWilliam and this, together with Cadishead Moss, have remained the main sites since then. Seen twice on Heysham Moss in 2016. Up to 20 adults have been encountered, such as at Astley Moss on 21 April 2011.

■ **HABITAT** Lowland raised mosses.
■ **FOODPLANTS** Cottongrass sp.

Single brooded. Rare at light. Most sightings of adults were in flight during the day, or of larval feeding signs, where they spin cottongrass heads together.

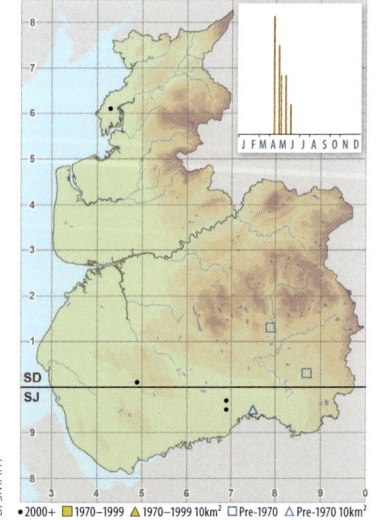

126 *The Moths of Lancashire*

19.007 Cocksfoot Moth *Glyphipterix simpliciella*

■ **LANCS STATUS** Abundant / widespread resident c.1857–2022

First reported from the Liverpool district (Gregson, 1857) and noted as locally abundant by most early authors. From mid-1980s onwards, it has been recorded from many locations across the county in both lowland and upland areas. Regularly monitored sites often produce counts of up to 200, but over 10,000 were estimated as present in Walthew Park, Upholland on 23 May 2004 (C.A. Darbyshire).
■ **HABITAT** Rough grassland, on wasteland, verges and hedgerows across the county.
■ **FOODPLANTS** Cock's-foot.

Single brooded. Rare at light. By day, on grasses and flowerheads, e.g., buttercup. Old larval feeding signs can be found at any time. One late August adult record in 2008.

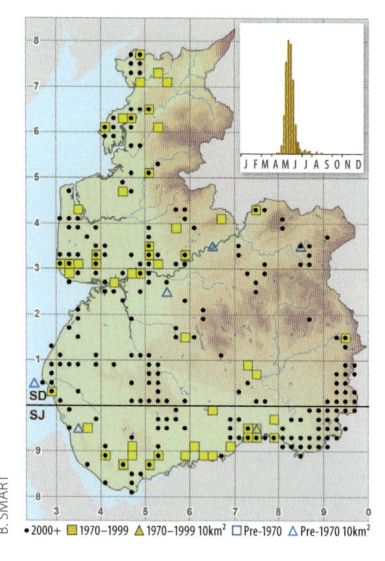

19.008 *Glyphipterix schoenicolella*

■ **LANCS STATUS** Rare / very local resident c.1914–2009

The only historical record comes from the Silverdale area c.1914 by W. Mansbridge and is mentioned, in passing, in the Manchester Entomological Society Annual Reports of that time. Despite this, the species is not documented in Mansbridge (1940). However, the moth was found again, in the same area, almost one hundred years later, at Hawes Water on the edge of Gait Barrows, on 2 June 2009. Three were netted during the day amongst Black Bog-rush (S. Palmer).
■ **HABITAT** Meadow on shell marl deposit by freshwater lake.
■ **FOODPLANTS** Not recorded in Lancashire.

Single brooded. A search on 5 April 2018 found possible feeding signs, but no larvae.

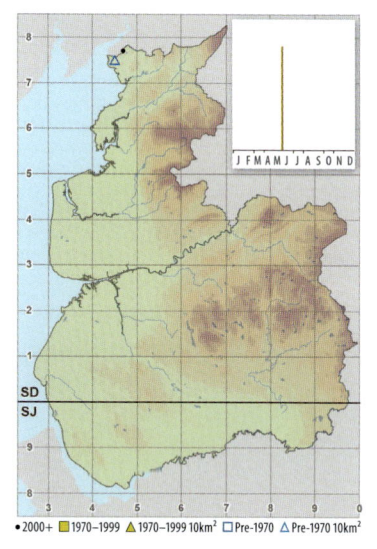

19.010 *Digitivalva pulicariae*

■ **LANCS STATUS** Considered Extinct pre-1890

The only records, all of them historical, are listed in Ellis (1890), with no details other than the recorders initials and a general location provided. The records could have been made at any time between the late 1840s to late 1880s, when these two recorders were active. The locations listed are Pendleton and Irlam for those reported by J. Chappell, and Preston and Cleveleys for those made by J.H. Threlfall.
■ **HABITAT** Unknown.
■ **FOODPLANTS** Not recorded in Lancashire.

The main national foodplant, Common Fleabane, is still present to the east of Cleveleys, on the Wyre estuary but searches for larvae have, so far, drawn a blank. It is also occasionally known to utilise Hemp Agrimony elsewhere and searches for the larvae on either plant should be carried out during June and July where it mines a leaf making a whitish blotch. May be overlooked.

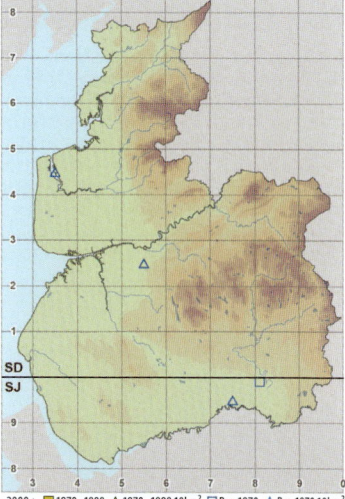

GLYPHIPTERIGIDAE

19.011 Leek Moth *Acrolepiopsis assectella*

■ **LANCS STATUS** Occasional / local resident   2012–2022

Expanding its range northwards in England, it was first recorded on 25 June 2012 in Crosby by J. Donnelly, the same year as the first for Cheshire. It was then noted in Fazakerley and Woolton in 2018 and has since slowly expanded its range eastwards. Recent sightings were from Brighton Grove Allotment, Manchester and Chorlton in 2020, and Fallowfield, Hindley and Victoria Park (Manchester) in 2022, all at light.
■ **HABITAT** Gardens and allotments.
■ **FOODPLANTS** Not recorded in Lancashire.

Early indications are that two broods have occurred in the county, with probable overwintering of the second brood as an adult. Comes to light.

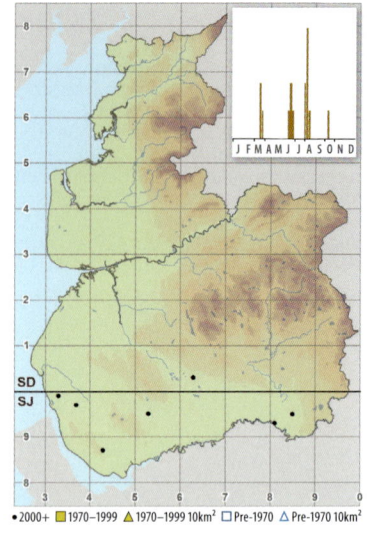

● 2000+   ▫ 1970–1999   ▵ 1970–1999 10km²   ▫ Pre-1970   △ Pre-1970 10km²

19.014 *Acrolepia autumnitella*

■ **LANCS STATUS** Occasional / widespread resident   1975–2022

First found as a larval mine in Formby during 1975 (M. Hull), but was not noted again until an adult was found in a spider's web in Warton, nr. Lytham on 10 December 2003 (S. Palmer). It has since spread rapidly throughout lowland parts, edging towards higher ground, such as Wrightington, in 2005. It is still scarce in east and north-east Lancashire, with a few records, e.g., from Rishton in 2020 and 2021.
■ **HABITAT** Hedgerows, gardens, scrubland and brownfield sites.
■ **FOODPLANTS** Bittersweet.

Double brooded, with second brood adults overwintering. Comes to light and can be found indoors. The large, bright white leaf-mines are easily seen in dense vegetation.

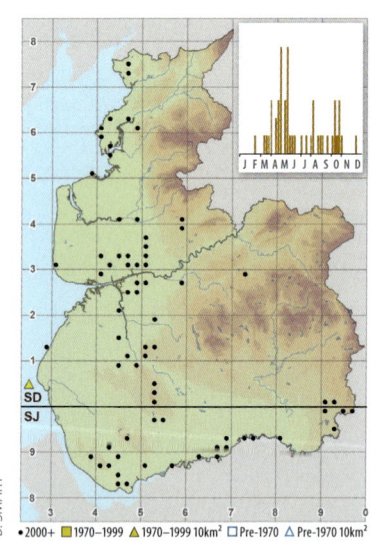

● 2000+   ▫ 1970–1999   ▵ 1970–1999 10km²   ▫ Pre-1970   △ Pre-1970 10km²

ARGYRESTHIIDAE

20.001 *Argyresthia laevigatella*

■ **LANCS STATUS** Scarce / local resident   1997–2012

The first confirmed record was from a Longridge Fell conifer plantation on 6 July 1997 (S. Palmer). Following this, moths have been recorded from seven other widely scattered sites, but never on more than one occasion. The most recent record was from an upland garden site at Botton Mill on 8 July 2012 (J. Girdley). It is suspected that the moth is rather under-recorded in the county.
■ **HABITAT** Conifer plantations and gardens with nearby Larch.
■ **FOODPLANTS** Not recorded in Lancashire.

Single brooded. Comes sparingly to light. Due to early confusion with *A. glabratella*, only those records with supporting specimens have been included.

● 2000+   ▫ 1970–1999   ▵ 1970–1999 10km²   ▫ Pre-1970   △ Pre-1970 10km²

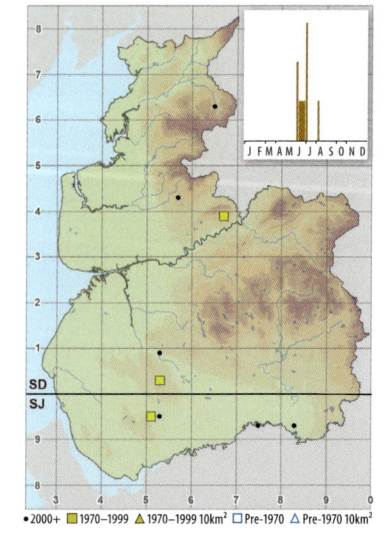

20.002 *Argyresthia glabratella*

■ **LANCS STATUS** Rare / very local resident 2005–2011
There have only been three records of this species within the county. The first of these was at Bold Moss, St. Helens on 7 June 2005 (R. Banks), the same recorder also finding one in his garden light trap at Parr, St. Helens on 23 June 2006. The final one was from Hardy Grove, Swinton on 25 May 2011 (G. Riley). A potentially under-recorded species which should be searched for in Norway Spruce plantations.
■ **HABITAT** Conifer plantations and gardens.
■ **FOODPLANTS** Not recorded in Lancashire.
Single brooded. Comes sparingly to light. Due to potential confusion with *A. laevigatella*, only those records with supporting specimens have been included.

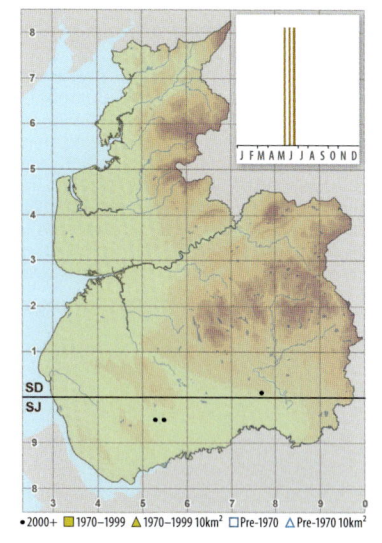

20.004 *Argyresthia arceuthina*

■ **LANCS STATUS** Rare / very local resident 2000–2009
All records are from an extensive area of lowland limestone pavement at Gait Barrows and were closely associated with the relatively few and well scattered Common Juniper bushes. The first two sightings, on 22 May 2000 (S. Palmer) and 21 May 2008 (S. Palmer & C. Barnes), were singletons tapped from the juniper bushes in the same area. Four were then disturbed from juniper bushes at Hawes Water, about 800m south-west of the previous site, but still within Gait Barrows N.N.R, on 2 June 2009 (S. Palmer).
■ **HABITAT** Limestone pavement.
■ **FOODPLANTS** Not recorded in Lancashire.

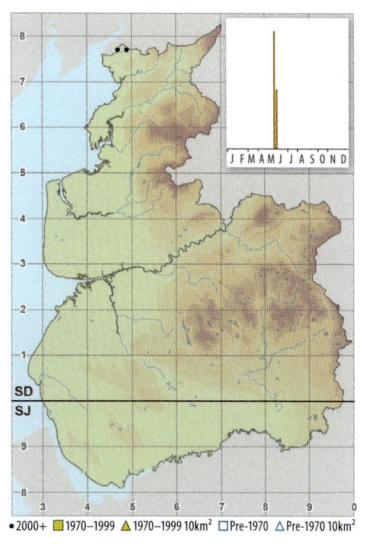

20.005 *Argyresthia trifasciata*

■ **LANCS STATUS** Frequent / widespread resident 2000–2022
Has spread rapidly northwards since arriving in Britain in 1982. First noted in Flixton on 12 May 2000 (K. McCabe) and reached VC60, in St Annes, a year later. Has since been found in many gardens across both lowland and upland areas. Frequently recorded in high numbers, with 71 in one garden light trap at Briercliffe on 2 June 2020 (G. Turner), although has declined at one or two sites in the last year or two.
■ **HABITAT** Gardens.
■ **FOODPLANTS** Cypress sp.
Single brooded, with a very small second brood occurring annually since 2012. The first indication of this second brood was at Billinge in 2003. Comes readily to light.

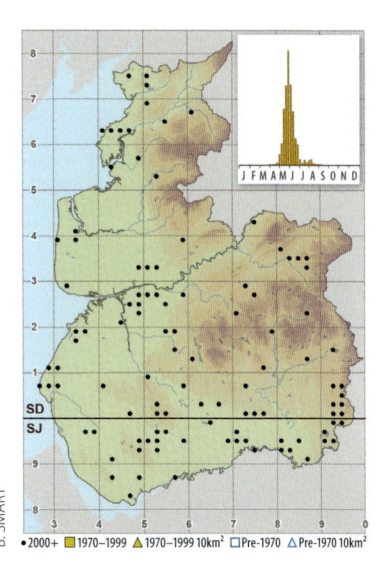

ARGYRESTHIIDAE

20.006 *Argyresthia dilectella*

■ **LANCS STATUS** Scarce / very local resident 1998–2022

Although common at Arnside (Cumbria) pre-1940, not reported in Lancashire until 16 July 1998 at Gait Barrows (S. Palmer). About this time, it also started appearing in a range of lowland gardens, such as in Billinge, Flixton, Preston and St Helens, with numbers peaking in 2006. Since then, a marked decline has occurred, resulting in its disappearance from gardens by 2014. The last records in the county were three from the Silverdale area (2016–17) and a singleton found at Little Woolden Moss in 2022.
■ **HABITAT** Limestone pavement and, formerly, in gardens.
■ **FOODPLANTS** Juniper. Beaten from Cypress bush on a moss in June 2022 (B. Smart).
Single brooded. Comes to light. Can be disturbed from wild Junipers during the day.

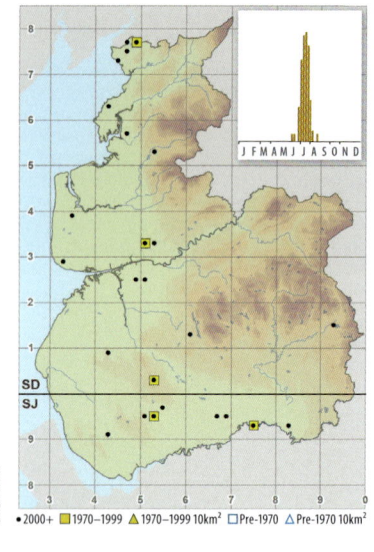

20.007 Cypress Tip Moth *Argyresthia cupressella*

■ **LANCS STATUS** Frequent / widespread resident 2007–2022

Despite being new to Britain in 1997 (in Suffolk), it didn't take long for it to reach Lancashire, coming to light on 15 May 2007 at Huyton (S. Tomlinson). Within two years it was found at Leighton Moss at the opposite end of the county. There are still some quite noticeable gaps in occurrence, suggesting some areas have benefitted from localised introductions via the horticultural trade, or accidental transportation. Counts of 20 or more in light traps are regular, such as in Hutton and St Helens.
■ **HABITAT** Gardens.
■ **FOODPLANTS** Cypress.
Single brooded. Comes to light. Larvae have been found during March and April.

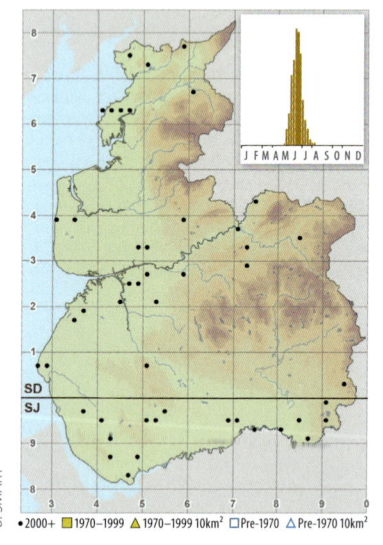

20.010 *Argyresthia ivella*

■ **LANCS STATUS** Considered Extinct 1939

The only known record of this species in the county was found in an unpublished manuscript of Lancashire and Cheshire Microlepidoptera (Hardwick, 1990). It refers to a 1939 report from the Silverdale area by H.N. Michaelis, with no further details or a reference source. N.L. Birkett also refers to this record in his card index of records held in Tullie House Museum, Carlisle, mainly covering Cumbria and nearby portions of VC60. This could be the source of the Hardwick note. Alongside this record Birkett adds '*do. J. Briggs*'. Briggs was a contemporary of Birkett's based in VC60 but Brigg's records, of which there are many, do not include any of this species.
■ **HABITAT** Unknown.
■ **FOODPLANTS** Not recorded in Lancashire, but elsewhere associated mainly with old apple orchards or, occasionally, isolated old apple trees in the wild. Examination of H.N. Michaelis specimens in various Museums, would benefit from more detailed research, not possible during the timescale of this project.

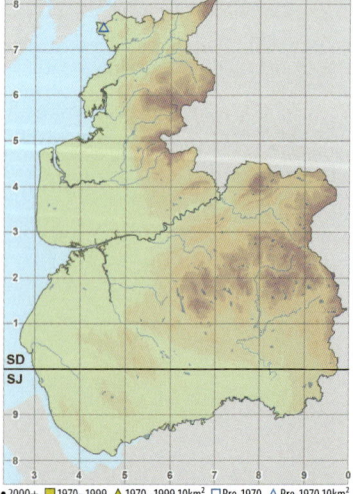

20.011 *Argyresthia brockeella*

■ **LANCS STATUS** Frequent / widespread resident   1846–2022

Noted at Chat Moss in July 1846 by R.S. Edleston and at Withnell Moor in April 1856, as a larva, by C.S. Gregson. Early authors noted it as very common to abundant. In recent years it has remained widely reported but large counts are now uncommon, exceptions being 300 at Herring Head Wood in 2006 and 100 at Astley Moss in 2011. Since 2015, the largest count was ten noted at Oldham Edge in July 2021 (S.H. Hind). Found once on 27 September 2004. An occasional form lacks the white marks.

■ **HABITAT** Woodland, moorland edge, mosses, gardens and brownfield sites.
■ **FOODPLANTS** Birch sp.

Single brooded. Comes to light. Larvae found in catkins during March and April.

● 2000+   □ 1970–1999   △ 1970–1999 10km²   □ Pre-1970   △ Pre-1970 10km²

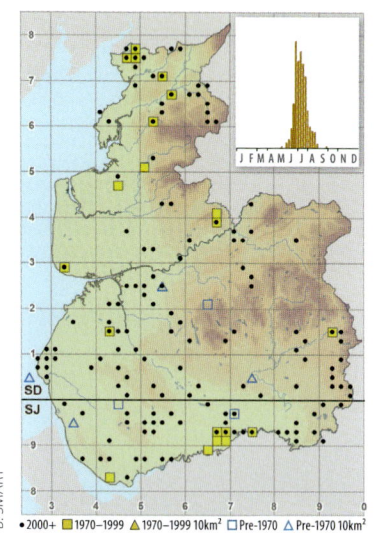

20.012 *Argyresthia goedartella*

■ **LANCS STATUS** Frequent / widespread resident   1846–2022

First found at Chat Moss in July 1846 by R.S. Edleston, with early authors noting it as very common in birch woods. In the 1950s, Leech & Michaelis (1957) also found it to be 'common amongst birch sp. and Alder in Formby'. Since then, found throughout the county in both lowland and upland areas but, in recent years, has not been recorded in quite as high numbers as previously. Counts above 20 are very unusual.

■ **HABITAT** Woodland, moorland edge, mosses, dunes, gardens and brownfield sites.
■ **FOODPLANTS** Birch sp.

Single brooded. Comes readily to light in moderate numbers. Larva found in birch catkins from late February to April.

● 2000+   □ 1970–1999   △ 1970–1999 10km²   □ Pre-1970   △ Pre-1970 10km²

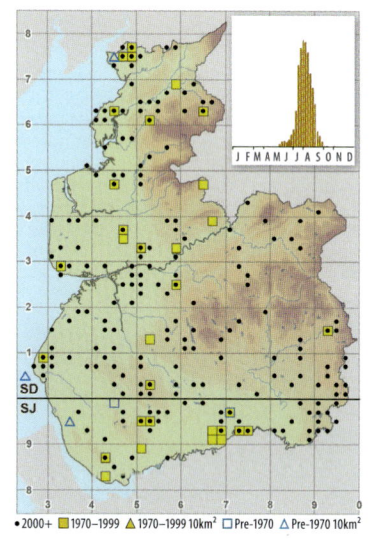

20.013 *Argyresthia pygmaeella*

■ **LANCS STATUS** Occasional / widespread resident   c.1857–2022

First reported from the Liverpool district (Gregson, 1857) and later, from Chat Moss c.1859 and Preston (Ellis, 1890). In Mansbridge (1940) it was noted as abundant amongst sallows, especially on sandhills. More recent records relate mostly to single moths coming to light or finds of slightly higher numbers of larvae. Generally erratic in occurrence and by no means annual, although seen in most years in Flixton.

■ **HABITAT** Woodland, dunes, mosses, brownfield sites, damp commons and gardens.
■ **FOODPLANTS** Creeping Willow, Grey Willow and possibly other willow spp.

Single brooded, with an extended flight period; one late individual found on 10 September 2015. Comes to light. The larvae have been found in April and May.

● 2000+   □ 1970–1999   △ 1970–1999 10km²   □ Pre-1970   △ Pre-1970 10km²

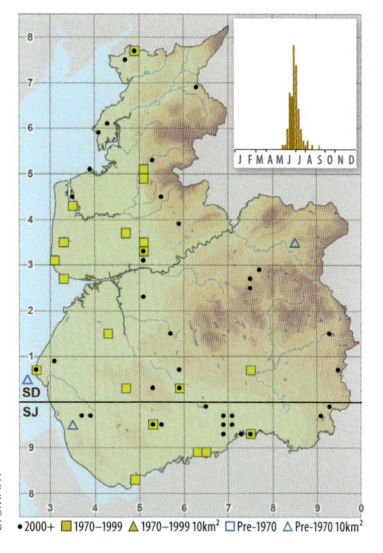

20.014 *Argyresthia sorbiella*

■ **LANCS STATUS** Scarce / very local resident   c.1857–2015

First reported between Knowsley and Kirkby (Gregson, 1857), then from Chat Moss c.1859 by J. Chappell. No more records are known until singletons were noted at Gait Barrows in 2007 and 2015 and two at light on Freshfield Dune Heath on 4 July 2015 (G. Jones). It is unexpected that all of the records relate to lowland areas where rowan, its main foodplant nationally, is less frequent, suggesting that its alternative national foodplant (whitebeam) might be worth examining in some areas.

■ **HABITAT** Mosses and limestone woodland.
■ **FOODPLANTS** Not recorded in Lancashire.
Single brooded. Comes to light.

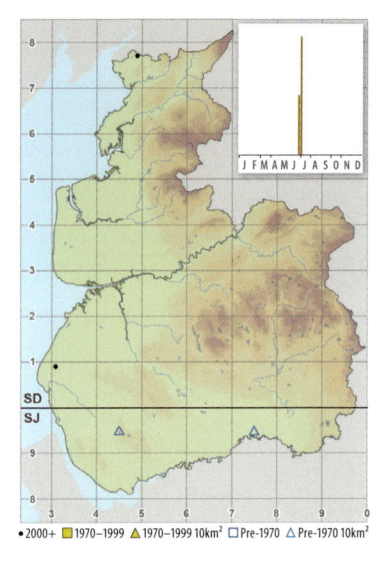

20.015 *Argyresthia curvella*

■ **LANCS STATUS** Occasional / local resident   c.1890–2022

The first was in Preston by J.H. Threlfall (Ellis, 1890). In June 1917, W. Mansbridge found it at Eccleston Mere and later, reported it as generally distributed in old orchards (Mansbridge, 1940). It was noted in Didsbury in 1959, but the next was not until 1996, in St Annes. Since then, it has been occasional and scattered in occurrence at over 30 sites but, rarely, with more than one or two records at each.

■ **HABITAT** Orchards and gardens.
■ **FOODPLANTS** Apple sp. Larvae April, May on flowering shoot in developing blossom. Single brooded. Comes to light. Between 1972 and 1987 records of *A. curvella* are likely to refer to *A. bonnetella* (Agassiz et al., 2013).

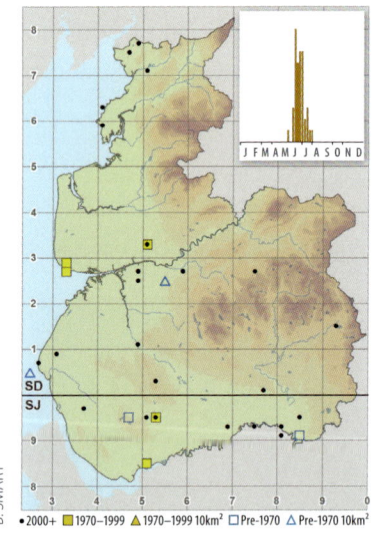

20.016 *Argyresthia retinella*

■ **LANCS STATUS** Frequent / widespread resident   c.1857–2022

First found in the Liverpool district (Gregson, 1857) and was later noted as common, especially on the mosses (Mansbridge, 1940). Leech & Michaelis (1957) also reported it to be common on birch in the Formby area and it has since been noted in most areas where birch grows naturally. Usually present in single figures, when it is easily disturbed from birch trees. It can occasionally be more numerous; 30 found during the day on Cadishead Moss in 2011 and 80 at light in Rainford in 2005 (R. Banks).

■ **HABITAT** Mosses, woodland, birch scrub, including on brownfield sites, and gardens.
■ **FOODPLANTS** Birch sp.
Single brooded. Comes to light. Larvae found in shoots in late April and May.

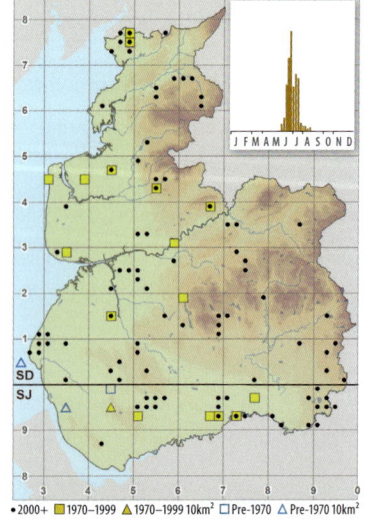

20.017 *Argyresthia glaucinella*

■ **LANCS STATUS** Rare / very local resident   1865–2006

Hodgkinson (1865) reports the first discovery of this moth in Lancashire whilst visiting 'Wildbottoms' (Hoghton Woods) in the second week of June 1865, where one was flying about an oak tree. Since then, there have been only two other records, when one was netted during the day next to an oak at Eaves Wood on 25 June 2004 (S. Palmer) and one came to light in Knowsley Safari Park on 19 June 2006 (K. McCabe).
■ **HABITAT** Oak woodland.
■ **FOODPLANTS** Not recorded in Lancashire. Looking for frass in the bark grooves of old oaks is a useful way to record the larva (M.R. Young pers. comm.).
Single brooded. Once at light. Stainton's (1859) Manchester record is in Cheshire.

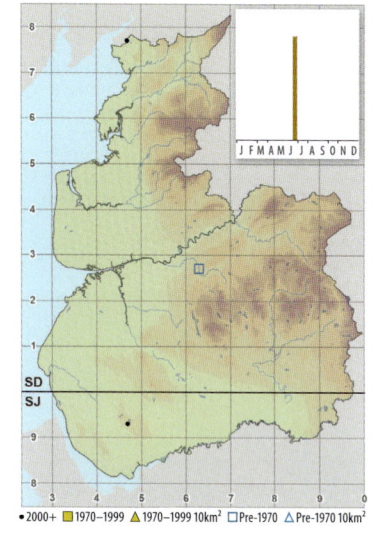

20.018 *Argyresthia spinosella*

■ **LANCS STATUS** Occasional / local resident   1846–2022

First reported from Chat Moss in June 1846 (Edleston, 1846). There were then two other historic reports from the Manchester district and one at Withington, all c.1859. It was not until 1995 that it was next noted, in Preston, and since then at several other scattered locations across the county; the common theme usually being the presence nearby of extensive Blackthorn scrub or hedgerows. Mostly found in ones and twos, the largest count, of six, was at light in Parr, St Helens on 6 June 2018.
■ **HABITAT** Hedgerows, limestone sites with scrub growth, brownfield sites and gardens.
■ **FOODPLANTS** Not recorded in Lancashire.
Single brooded. Comes to light. Can be disturbed by day from Blackthorn.

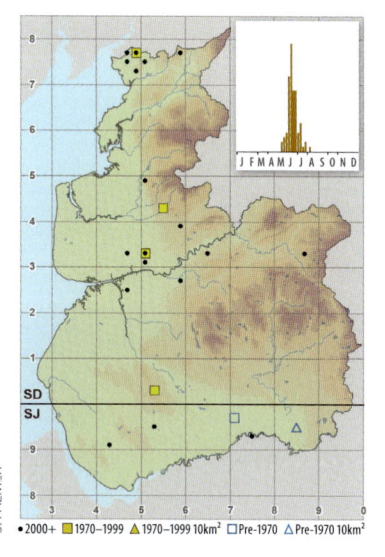

20.019 Apple Fruit Moth *Argyresthia conjugella*

■ **LANCS STATUS** Frequent / widespread resident   c.1857–2022

First recorded from the Liverpool district (Gregson, 1857), then Chat Moss c.1889 and Burnley (Mansbridge, 1940). Not noted again until 1991, at Holcroft Moss and has since been found wherever Rowan occurs naturally. Moorland edge woods or copses are favoured, where it can occur in good numbers; 60 were found by day at Hodge Clough in 2022 (S.H. Hind) and 120 at Clougha in 2008 (S. Palmer).
■ **HABITAT** Upland and limestone woodland, moorland edge, hedgerows and gardens.
■ **FOODPLANTS** Not recorded in Lancashire. Bred in 1879 without the foodplant being documented, but noted as occurring amongst Rowan.
Single brooded. Comes to light and readily disturbed from Rowan by day.

ARGYRESTHIIDAE

The Moths of Lancashire   **133**

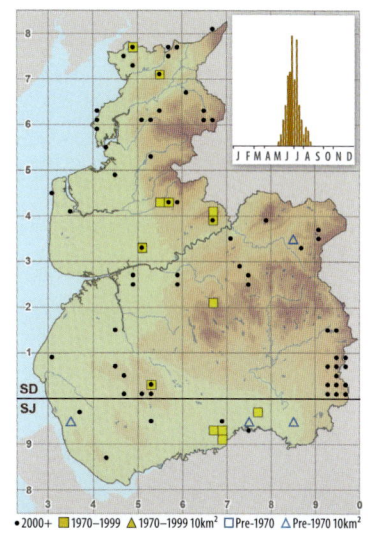

ARGYRESTHIIDAE

20.020 *Argyresthia semifusca*

■ **LANCS STATUS** Occasional / widespread resident   c.1857–2022

First recorded from Simonswood Moss (Gregson, 1857), later from Burnley, Croxteth and Preston, when considered to be local and uncommon (Mansbridge, 1940). Most post-1990 records have been from lowland garden light traps or occasionally when beaten from hedgerows. Sites where it is most regularly recorded at light have adjacent older hedgerows, such as in Flixton, Hale and Preston. Rare in upland areas, with only two sites with more than single records, in Adlington and Briercliffe.
■ **HABITAT** Hedgerows.
■ **FOODPLANTS** Not recorded in Lancashire.
Single brooded. Comes to light in small numbers.

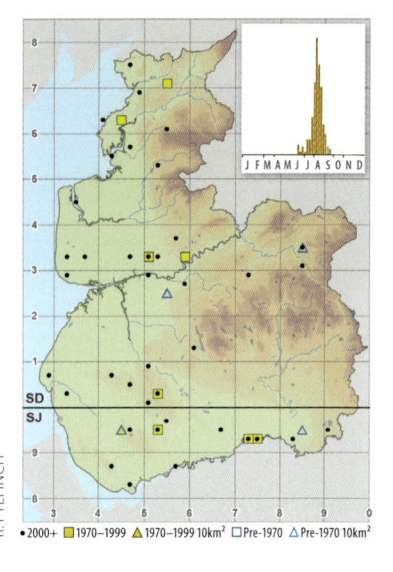

●2000+ ☐1970–1999 △1970–1999 10km² ☐Pre-1970 △Pre-1970 10km²

20.021 Cherry Fruit Moth *Argyresthia pruniella*

■ **LANCS STATUS** Occasional / widespread resident   c.1859–2022

First recorded from the Manchester district (Stainton, 1859) and later, reported as scarce and local (Mansbridge, 1940). Next found in 1949 where it was noted as common in a garden in Didsbury by H.N. Michaelis and, following that, in St Annes by D. & J. Steeden in 1995. Thereafter records increased markedly, with some sites having several sightings over the following years; these included Billinge, Briercliffe, Jackhouse N.R., Parr, Rochdale, Longton and Wigan. Usually occurs in single figures.
■ **HABITAT** Gardens, parks and other localities with planted cherry.
■ **FOODPLANTS** Cherry sp.
Single brooded. Comes to light. Larvae in leaf-buds or blossom in April and May.

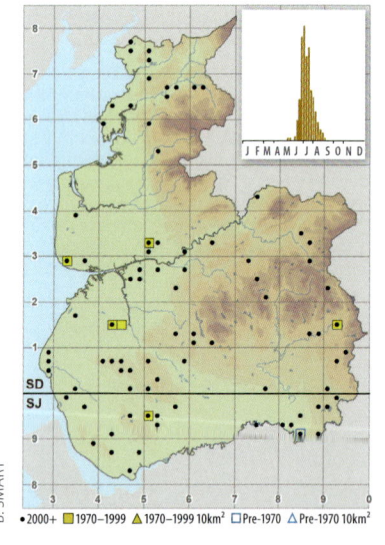

●2000+ ☐1970–1999 △1970–1999 10km² ☐Pre-1970 △Pre-1970 10km²

20.022 *Argyresthia bonnetella*

■ **LANCS STATUS** Frequent / widespread resident   c.1857–2022

Found in the Liverpool area (Gregson, 1857) and noted as 'abundant throughout' by other early authors. From the late 1980s, reported from several sites between Blackpool and Preston (M. Evans). An increase in the use of light traps across the county from 1995 led to many more records, with the moth becoming annual at a good number of sites, in small numbers. Larger counts are infrequent, but 50 were disturbed from an uncut Hawthorn hedge at Shaw Green on 30 June 2022.
■ **HABITAT** Hedgerows, gardens and scrubby areas, including brownfield sites.
■ **FOODPLANTS** Hawthorn.
An extended single brood. Comes readily to light. Larvae observed in April and May.

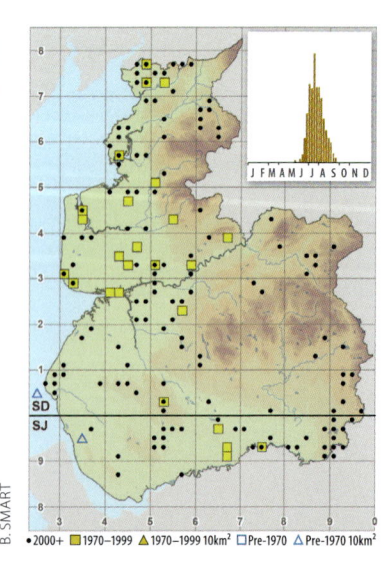

●2000+ ☐1970–1999 △1970–1999 10km² ☐Pre-1970 △Pre-1970 10km²

134   *The Moths of Lancashire*

20.023 *Argyresthia albistria*

■ **LANCS STATUS** Occasional / widespread resident   c.1859–2022
First noted from Irlam by J. Chappell c.1859 and Withington (Ellis, 1890). It was subsequently noted at Gait Barrows in 1986 and, from 1994 onwards, was encountered over an increasing range of lowland sites, mostly in garden light traps. Many of these involved just one or two sightings, but some reported it more regularly, such as in Dolphinholme and Preston.
■ **HABITAT** Hedgerows and any scrubby areas with Blackthorn thickets.
■ **FOODPLANTS** Blackthorn.
Single brooded. Comes to light and can be disturbed by day from Blackthorn thickets and older hedgerows. A larva was beaten from Blackthorn on 22 April 2022.

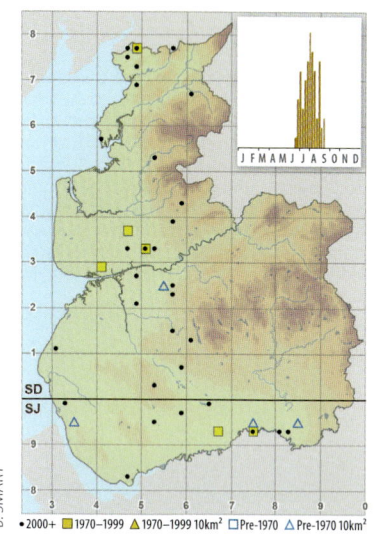

• 2000+   □ 1970–1999   ▲ 1970–1999 10km²   □ Pre-1970   △ Pre-1970 10km²

20.024 *Argyresthia semitestacella*

■ **LANCS STATUS** Occasional / local resident   c.1857–2021
C.S. Gregson first found this, among beeches, in the Croxteth area (Gregson, 1857), it then being reported from Irlam in 1859 and Burnley in 1910. The next was not noted until 22 August 1992, this from Holcroft Moss (S. McWilliam). Thereafter it was discovered in Chorlton, Fulwood (Preston), Shedden Clough, Rochdale, Ainsdale and north Preston, this last site not until 2017. Over half of all records are from Fulwood, where the garden light trap is run beneath mature Beech trees (A. Powell).
■ **HABITAT** Woodland and parkland.
■ **FOODPLANTS** Not recorded in Lancashire; adults associated with mature Beech trees.
Single brooded. Comes to light.

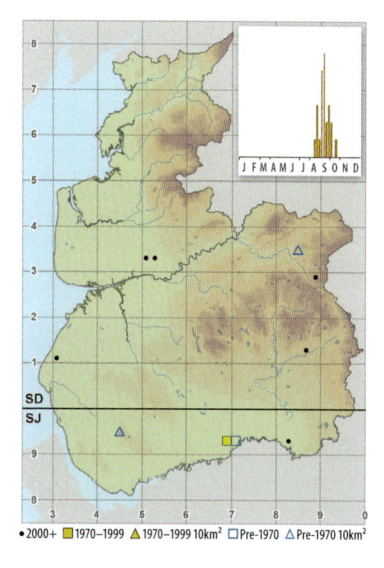

• 2000+   □ 1970–1999   ▲ 1970–1999 10km²   □ Pre-1970   △ Pre-1970 10km²

21.001 Apple Leaf Miner *Lyonetia clerkella*

■ **LANCS STATUS** Abundant / widespread resident   c.1857–2022
19th century records from Liverpool, 'near old orchards and gardens' (Gregson, 1857), Manchester and Silverdale. Described as 'local but fairly common' by Mansbridge (1940). Over 3,000 records since, from upland and lowland areas.
■ **HABITAT** Woodland, hedgerows, parks, gardens, orchards, etc.
■ **FOODPLANTS** Apple, Hawthorn, Wild Cherry, Blackthorn, Silver Birch, Downy Birch, Wild Plum, Quince, Rowan, Whitebeam, Swedish Whitebeam, Cherry Laurel, rose, cotoneaster, Bramble, Sycamore, Wych Elm and Buckthorn.
2 or 3 broods. Mainly on rosaceous trees; occasionally other hosts, e.g., Buckthorn from Eaves Wood 2022 (S. Palmer). Frequent at light and around foodplants.

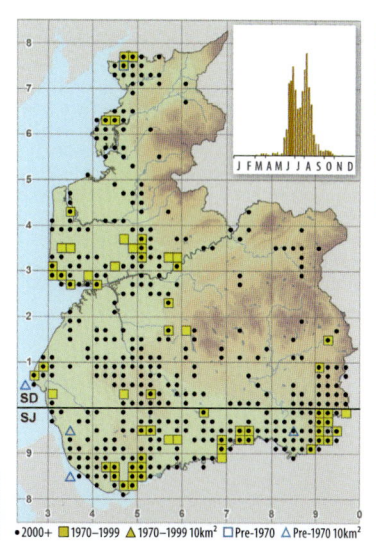

• 2000+   □ 1970–1999   ▲ 1970–1999 10km²   □ Pre-1970   △ Pre-1970 10km²

The Moths of Lancashire

LYONETIIDAE

21.004 Laburnum Leaf Miner *Leucoptera laburnella*

■ **LANCS STATUS** Occasional / local resident   c.1857–2022

Leaf-mines noted 'where Laburnum is plentiful' (Gregson, 1857). From Green Thorn Fell, 28 June 1879 by J.B. Hodgkinson on Dyer's Greenweed. Other early records associated with Dyer's Greenweed from Samlesbury, Simonswood Moss and Dutton, nr. Ribchester (Mansbridge, 1940). Subsequent leaf-mine records all on Laburnum, from lowland and upland sites where the tree has been planted, often in gardens.
■ **HABITAT** Parks, gardens, grassland.
■ **FOODPLANTS** Laburnum and Dyer's Greenweed.
Double brooded. Occasionally netted around foodplant and at light. Most records are of leaf-mines; moths occasionally bred. High count of 100 mines at Parbold in 2003.

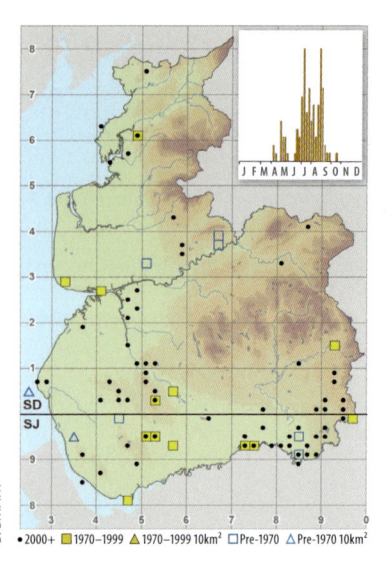

● 2000+   ☐ 1970–1999   △ 1970–1999 10km²   ☐ Pre-1970   △ Pre-1970 10km²

21.005 *Leucoptera spartifoliella*

■ **LANCS STATUS** Scarce / local resident   c.1889–2008

First noted from Preston by J.H. Threlfall and from Longridge by J.B. Hodgkinson (Ellis, 1890). Later reported as 'common where broom grows freely' (Mansbridge, 1940). Moths were noted on five occasions at Lightfoot Green from 1995 to 1999 but have not been seen since, due to local loss of foodplant. Subsequently, recorded at Stanah, Cuerden, Formby, Littleborough, Billinge, Scorton, Accrington, etc. However, the moth has declined throughout, with the last record, from Upholland, in 2008.
■ **HABITAT** Rough grassland, country parks and gardens.
■ **FOODPLANTS** Broom.
Double brooded. Comes to light. Cocoons on twigs at Hollingworth Lake, May 2003.

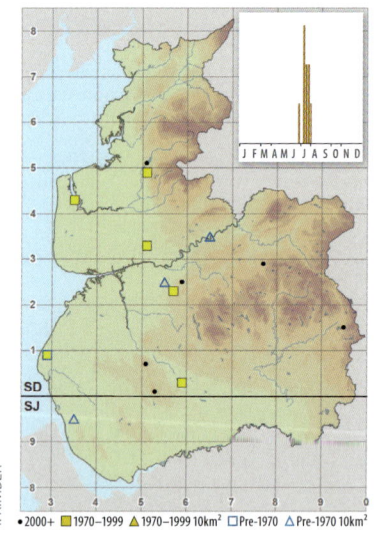

● 2000+   ☐ 1970–1999   △ 1970–1999 10km²   ☐ Pre-1970   △ Pre-1970 10km²

21.008 Pear Leaf Blister Moth *Leucoptera malifoliella*

■ **LANCS STATUS** Occasional / local resident   c.1857–2021

Primarily from lowland areas, occurrence of this moth tends to be rather erratic, suddenly appearing at a particular site before disappearing again for a few years. First records from Liverpool (Gregson, 1857), Flixton, Glazebrook and Preston (Ellis, 1890). Reported from Didsbury in 1949 by J. Harding, but then a long wait for the next record, from Great Moss, nr. Wigan by C.A. Darbyshire in 2002. Subsequently, recorded from St Helens, Flixton, Medlock Valley, Longton, Heysham Moss, etc.
■ **HABITAT** Hedgerows, woodland edges.
■ **FOODPLANTS** Hawthorn, Apple, Crab Apple, Rowan, Pear, Bullace.
Single brooded. Once beaten from Hawthorn. All other records are of leaf-mines.

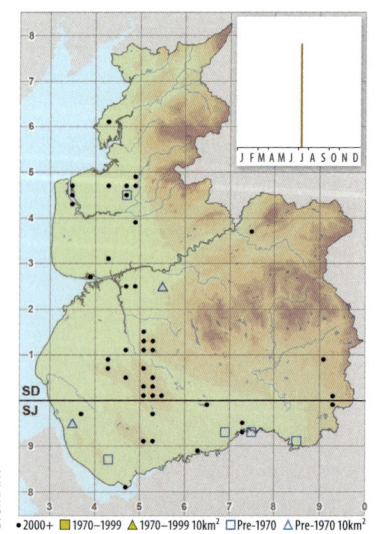

● 2000+   ☐ 1970–1999   △ 1970–1999 10km²   ☐ Pre-1970   △ Pre-1970 10km²

22.001 Atemelia torquatella

■ **LANCS STATUS** Considered Extinct 1937

A record of this species appears in Mansbridge (1940), contained within a supplementary list on page 262 of that document which does not appear to have been previously picked up at a national data level. Under the heading 'New to Lanc.', it relates to one caught in the Yealand area in 1937 by A.E. Wright. Mansbridge (loc. cit.) earlier, in the same publication (assessed as probably written about 20 years previously), notes it as a scarce species and only taken in Witherslack (Cumbria) by A.E. Wright and himself. This appears to be the most southerly known British record.

■ **HABITAT** Unknown.
■ **FOODPLANTS** Not recorded in Lancashire.

The moth is otherwise found in northern England and across Scotland. It is most frequently associated with Downy Birch seedlings, in areas of birch regeneration. The larvae make large blotch mines, often with two or three larvae in a blotch, with a light web spun beneath and with silken ties connecting adjacent leaves (Emmet, 1996; M.R. Young, pers. comm.).

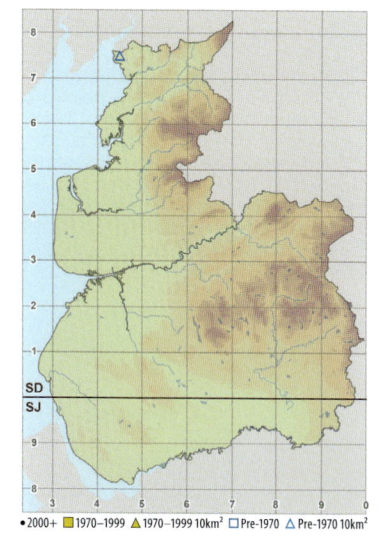

• 2000+ ■ 1970–1999 ▲ 1970–1999 10km² □ Pre-1970 △ Pre-1970 10km²

22.002 Ash Bud Moth *Prays fraxinella*

■ **LANCS STATUS** Occasional / widespread resident c.1857–2022

Occurs in two forms, the nominate black and white, and a dark form, f. *rustica*. Both forms were reported as common and generally distributed in the 19th and early 20th centuries. First noted in Liverpool area by Gregson (1857). Both forms remain widely distributed, but modern records are usually of low single figures. Annual, or almost so, at only a few sites, such as Flixton, Parr and Preston.

■ **HABITAT** Woodland, hedgerows and gardens.
■ **FOODPLANTS** Ash. The potential effect of Ash die-back will need to be monitored.

Double brooded. Comes to light. Only 1st brood larvae found, in mid-September, and after wintering, in May. See under *P. ruficeps* for further comment on f. *rustica*.

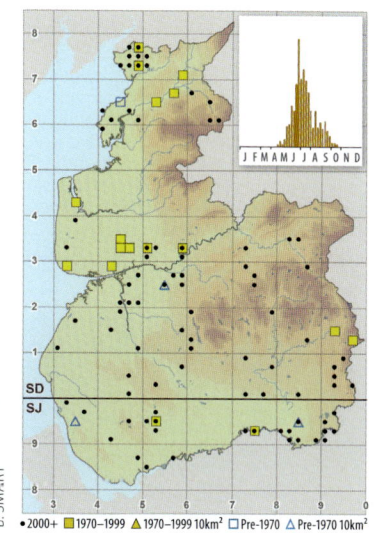

• 2000+ ■ 1970–1999 ▲ 1970–1999 10km² □ Pre-1970 △ Pre-1970 10km²

22.003 *Prays ruficeps*

■ **LANCS STATUS** Occasional / local resident 2013–2022

Added to the British list in Agassiz, *et al.*, (2013), it had been flagged up, in Sterling and Parsons (2012), as deserving of addition to the British list; identification criteria were noted therein. Using these, the moth, previously confused with some *P. fraxinella* f. *rustica*, was seen in Preston on 8 August 2013 by S. Palmer. Since then, it has been found at 14 other scattered locations but is probably under-recorded.

■ **HABITAT** Woodland, hedgerows and gardens.
■ **FOODPLANTS** Not recorded in Lancashire.

Double brooded. Comes to light. It appears to be unknown if *P. ruficeps* was overlooked amongst *P. fraxinella* f. *rustica* specimens prior to 2013.

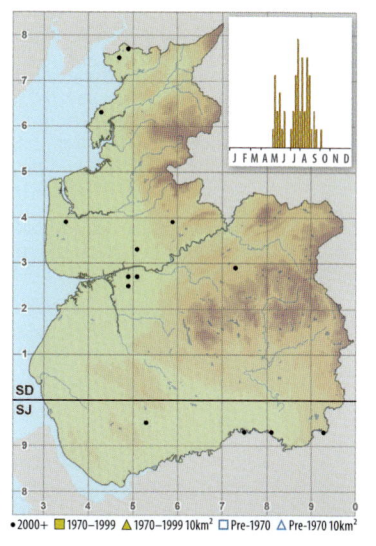

• 2000+ ■ 1970–1999 ▲ 1970–1999 10km² □ Pre-1970 △ Pre-1970 10km²

PRAYDIDAE

The Moths of Lancashire **137**

BEDELLIIDAE

24.001 *Bedellia somnulentella*

■ **LANCS STATUS** Occasional / widespread resident   1998–2022

First noted in Cheshire in 1998, which prompted successful searches in Flixton on 9 September 1998 (K. McCabe) and, two days later, at Hale (B.T. Shaw). Reported from Skelmersdale in 2004 and other lowland parts of southern VC59 to 2022. It is very local elsewhere, with the most northerly records from Stanah and Winmarleigh. Numbers fluctuate annually, with 2022 being a good year; possibly overlooked.
■ **HABITAT** Hedges, field margins, scrub, brownfield sites and waste ground.
■ **FOODPLANTS** Hedge Bindweed and Field Bindweed.
Double brooded. Occasionally at light. Readily found as a larva or pupa in its obvious spinnings. Two overlapping larval broods from mid-August to the end of October.

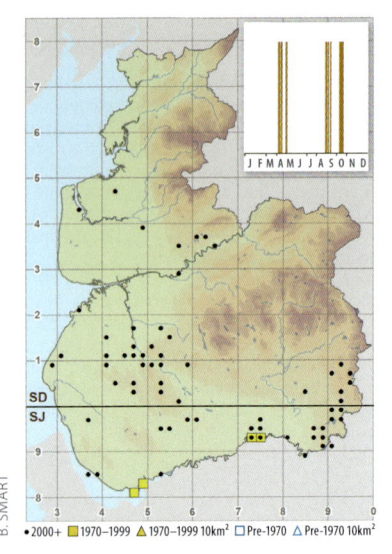

• 2000+   ☐ 1970–1999   △ 1970–1999 10km²   ☐ Pre-1970   △ Pre-1970 10km²

SCYTHROPIIDAE

25.001 **Hawthorn Moth** *Scythropia crataegella*

■ **LANCS STATUS** Occasional / local resident   c.1859–2022

First reported in Manchester (Stainton, 1859), then at Stretford (1889) and Didsbury (1949). Next, at light in Billinge on 20 June 2000 (C.A. Darbyshire), with a few other southern sites added up to 2014. It remains, primarily, a species of the Mersey Valley area, with most sightings from Chorlton, Flixton and Woolton (Liverpool). In 2022, there were signs of a slight range extension occurring, into Southport and Longridge.
■ **HABITAT** Hedges, scrub, brownfield sites and waste ground.
■ **FOODPLANTS** Hawthorn.
Double brooded. Comes to light. Larvae have been seen mining leaves in August and September, then making a conspicuous web in May and June.

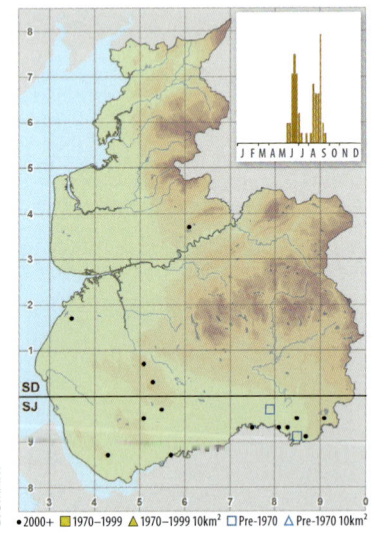

• 2000+   ☐ 1970–1999   △ 1970–1999 10km²   ☐ Pre-1970   △ Pre-1970 10km²

AUTOSTICHIIDAE

27.001 *Oegoconia quadripuncta*

■ **LANCS STATUS** Occasional / widespread resident   1990–2022

A gradual range extension in Britain led to its discovery in St Annes on 14 July 1990 (D. & J. Steeden). It was not reported from VC59 until 1999, at Flixton, Pennington and Parr. It has since spread erratically into other lowland parts, reaching Preston in 2008, Morecambe in 2009 and Formby in 2010. The only upland record has been from Briercliffe in 2022 (G. Turner). Only records of dissected moths are included.
■ **HABITAT** Gardens.
■ **FOODPLANTS** Not recorded in Lancashire.
Single brooded. Comes to light. An externally identical species, *O. deauratella*, is present in Yorkshire and could occur in Lancashire at any time.

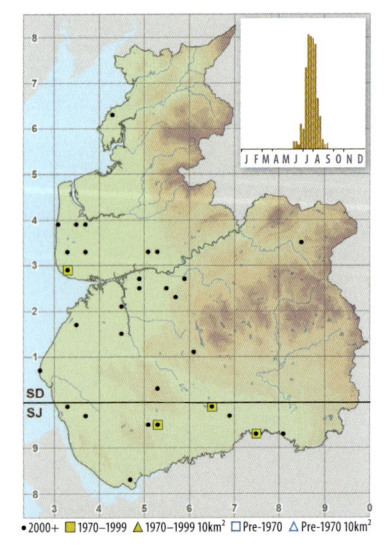

• 2000+   ☐ 1970–1999   △ 1970–1999 10km²   ☐ Pre-1970   △ Pre-1970 10km²

28.004 *Denisia similella*

■ **LANCS STATUS** Rare / very local resident   c.1859–2022

Reported from the Manchester district (Stainton, 1859) and in 'Wildbottoms' (Hoghton Wood), west of Blackburn, in June 1865 by J.B. Hodgkinson (Hodgkinson, 1865) and has remained a very elusive species since then. The only other reports were in 2022, with one attracted to a 20w light in Longridge on 22 May (D. Lambert) and one found by day, settled on a stone wall, at Green Bank on 29 May (S. Palmer).

■ **HABITAT** Moorland edge and a lowland meadow.
■ **FOODPLANTS** Not recorded in Lancashire.

Single brooded. Once at light and twice found resting during the day. On one occasion, found in an old hut, in woodland; the other on an old stone wall.

•2000+   ■1970–1999   ▲1970–1999 10km²   □Pre-1970   △Pre-1970 10km²

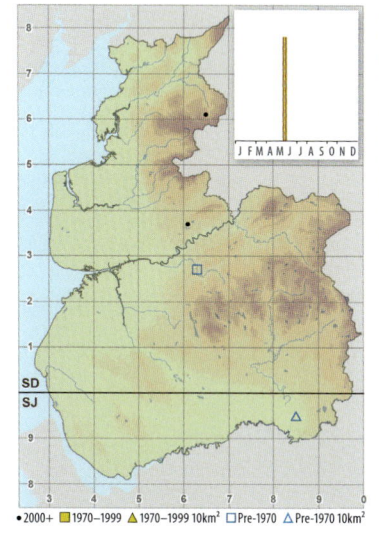

28.005 *Denisia albimaculea*

■ **LANCS STATUS** Rare / very local resident   2002

Although recorded on only two occasions in the county and from two well separated sites, it was remarkable that they appeared within three days of each other. The first was at Chorlton, on a tree trunk, on 28 May 2002 (B. Smart) and the next, 7km east of the first, came to light at Flixton on 31 May 2002 (K. McCabe). It is considered likely that this is a long-established, but overlooked, species in the county.

■ **HABITAT** Old established suburban areas with mature trees.
■ **FOODPLANTS** Not recorded in Lancashire.

Single brooded. Comes to light and rests on tree trunks during the day.

•2000+   ■1970–1999   ▲1970–1999 10km²   □Pre-1970   △Pre-1970 10km²

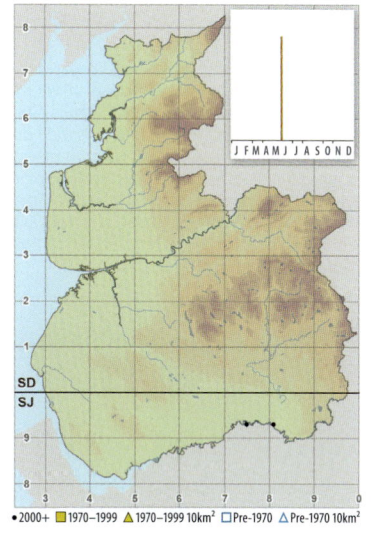

28.007 *Denisia subaquilea*

■ **LANCS STATUS** Rare / very local resident   2011–2019

There are two records of this species in the county, the first was netted during a daytime visit to Freshfield Dune Heath (Lancashire Wildlife Trust Reserve) on 11 May 2011 by G. Jones. The second, in 2019, was found by C. Fletcher somewhere in VC59, but the location and date were not documented. The moth was later examined by S. Palmer, confirming the identification. It is possible it was from the same site as the 2011 specimen, but this has not been possible to ascertain with certainty.

■ **HABITAT** Dune heath.
■ **FOODPLANTS** Not recorded in Lancashire.

Single brooded. Comes to light.

•2000+   ■1970–1999   ▲1970–1999 10km²   □Pre-1970   △Pre-1970 10km²

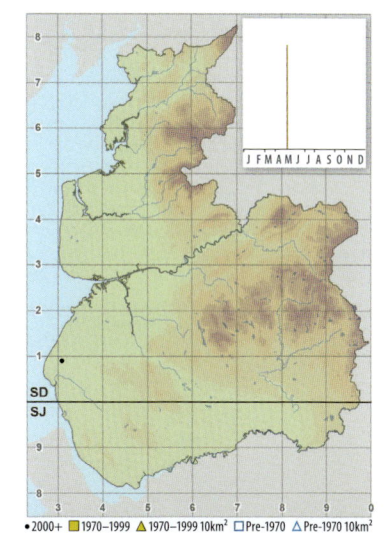

OECOPHORIDAE

28.008 *Metalampra italica*

■ **LANCS STATUS** Rare / very local resident   2019–2021

This species has expanded its range rapidly northwards since its first British record in Devon in 2003 and reached Cheshire in 2013. The first documented record in Lancashire was from near Clitheroe on 15 July 2019 by B. Honeywell running an 8w actinic light trap. The only other record, for which details were submitted, relates to one on the 16 July 2021, again at actinic light (twin 30w), in Brockhall Village by J. Jones.

■ **HABITAT** Gardens and parks.
■ **FOODPLANTS** Not documented in Lancashire.

Single brooded. Comes to light.

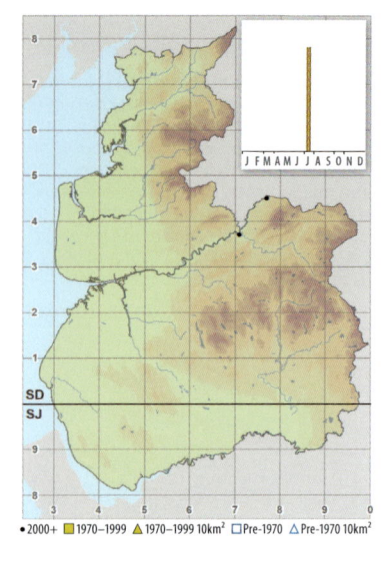

28.009 White-shouldered House-moth *Endrosis sarcitrella*

■ **LANCS STATUS** Frequent / widespread resident   c.1857–2022

First noted in Liverpool (Gregson, 1857) with reports since then, from most major conurbations and human settlements. Strongly associated with Liverpool Docks prior to containerisation. Reports across the county refer to its annual occurrence indoors, with barely a year passing without it being seen. Despite this, it is rarely found in double figures, the largest count of 21 being at light in Parr in 2008 (R. Banks).

■ **HABITAT** Within houses, outbuildings, dense hedgerows and once, debris in a ditch.
■ **FOODPLANTS** Detritus, stored and spilt bird-food and bird nests.

Although recorded in every month of the year, it appears to be double brooded in most locations. Found regularly indoors and comes readily to light outside.

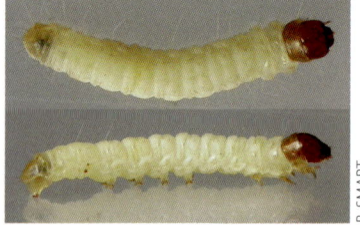

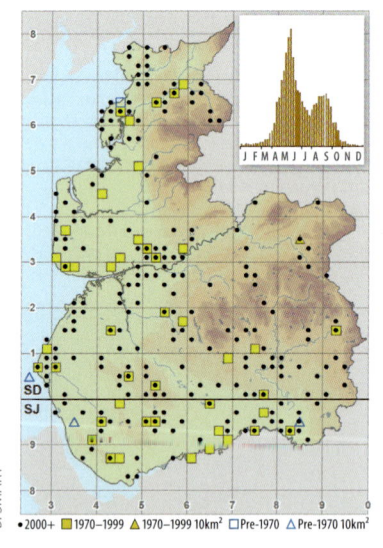

28.010 Brown House-moth *Hofmannophila pseudospretella*

■ **LANCS STATUS** Abundant / widespread resident   c.1857–2022

First noted in grain houses around Liverpool Docks (Gregson, 1857), and, as with the previous species, reports have since been received from many towns, cities and other human settlements, as well as the wider countryside. Reports from many houses across the county indicate an annual occurrence in small to moderate numbers. In Chorlton in 2006, approx. 100 larvae and pupae were found in a drawer within a garden shed (B. Smart), the same recorder noted larvae eating through plastic bags.

■ **HABITAT** Within houses, outbuildings, woodlands and hedgerows.
■ **FOODPLANTS** Stored seed, wood, bird nest material, scraps of foodstuffs etc.

Found regularly indoors and outdoors and comes readily to light outside.

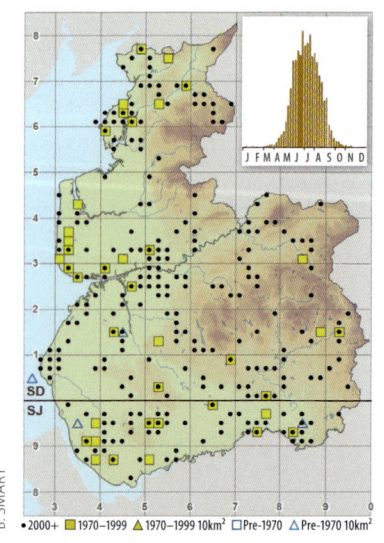

The Moths of Lancashire

28.011 *Borkhausenia minutella*

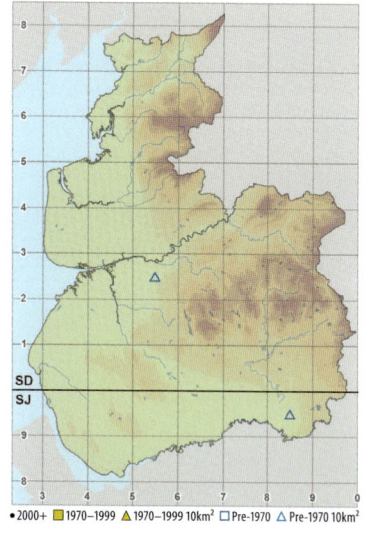

■ **LANCS STATUS** Considered Extinct.   c.1859–c.1889

Reported from the Manchester area by Stainton (1859), and also, possibly around the same time, by J.B. Hodgkinson (Ellis, 1890). It seems likely that Hodgkinson's record was the source of Stainton's report, but this has not been possible to prove. Ellis (loc. cit.) reported the species as local, and added an additional record from the Preston area, also by J.B. Hodgkinson. None of these had dates or more precise locations documented by Ellis. Mansbridge (1940), reported it to be 'a scarce and local species' but added no additional records to those already known.

■ **HABITAT** Unknown.

■ **FOODPLANTS** Not recorded in Lancashire.

It was considered to be extinct in the British Isles (Emmet & Langmaid, 2002) due to changes in farming practises and modernisation. However, it has since been noted on three occasions in recent years (G. Tordoff pers. comm.), twice in England (Lincolnshire and Shropshire) and once in Wales (Flintshire).

● 2000+   ■ 1970–1999   ▲ 1970–1999 10km²   □ Pre-1970   △ Pre-1970 10km²

28.012 *Borkhausenia fuscescens*

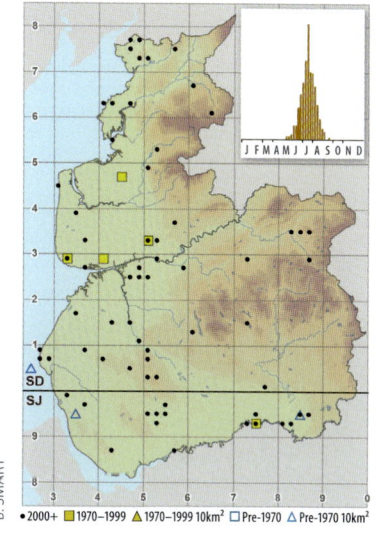

■ **LANCS STATUS** Frequent / widespread resident   c.1859–2022

Noted as abundant in the Manchester area (Stainton, 1859) and, also from the same area by J. Chappell (Ellis, 1890). In Formby, during the 1950s, recorded as common, but with no further records until 8 July 1994, when it was found at Warton, nr. St Annes (S. Palmer). Since then, records have been received from across the county, mainly in ones and twos. The largest single counts were of eight in light traps at Hale in 2012 (J. Mitchell-Lisle) and in Parr, St Helens in 2015 (R. Banks).

■ **HABITAT** Gardens, outbuildings and hedgerows, sometimes coming into houses.

■ **FOODPLANTS** Not recorded in Lancashire.

An extended single brood. Comes to light.

● 2000+   ■ 1970–1999   ▲ 1970–1999 10km²   □ Pre-1970   △ Pre-1970 10km²

28.014 *Crassa unitella*

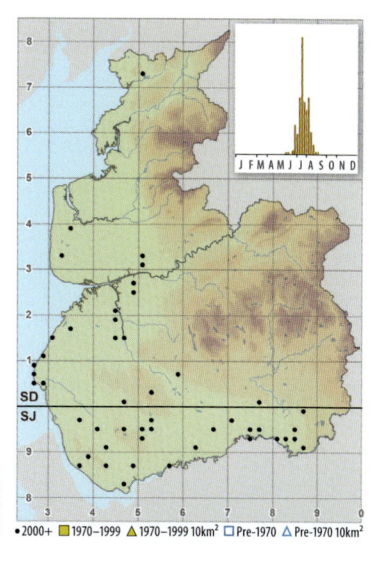

■ **LANCS STATUS** Occasional / local resident   2006–2022

The species moved north into Cheshire in 1989, but the first Lancashire records were not until 20 July 2006, in Roby (S. Tomlinson) and Withington (M. Earlam). Within a few years it had colonised much of the Mersey Valley and Sefton Coast, appearing to consolidate its south Lancashire position, before moving north again. First noted in VC60 at Hardhorn in 2016 and since then, has been seen occasionally at Marton (Blackpool), Preston (few sites) and Warton, Silverdale on 14 Aug. 2020 (P. Stevens).

■ **HABITAT** Unknown. Nearly all records are from garden light traps.

■ **FOODPLANTS** Not recorded in Lancashire.

Single brooded. Comes to light. Seven were in a Woolton light trap on 22 July 2019.

● 2000+   ■ 1970–1999   ▲ 1970–1999 10km²   □ Pre-1970   △ Pre-1970 10km²

OECOPHORIDAE

The Moths of Lancashire   **141**

OECOPHORIDAE

28.015 *Batia lunaris*

■ **LANCS STATUS** Occasional / local resident   1998–2022

New to Cheshire in 1997, it was only a year later that it arrived in Flixton, on 9 August 1998 (K. McCabe). This remained the sole site until 2002, when it came to light in Chorlton (B. Smart). Since then, it has spread slowly northwards, up the coast and infilled over much of the Mersey Valley area. Examples of its erratic expansion include the following first records; Marshside in 2004, Parr in 2005, Roby in 2006, Formby in 2007, Hough Green and Swinton in 2009 and Hale in 2016.

■ **HABITAT** Gardens, scrub and brownfield sites.
■ **FOODPLANTS** Not recorded in Lancashire.
Single brooded. Comes to light in low numbers.

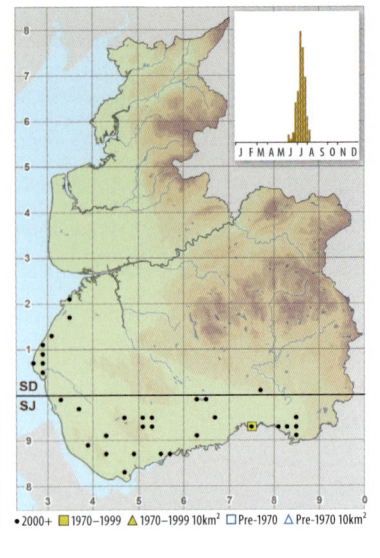

28.019 *Esperia sulphurella*

■ **LANCS STATUS** Frequent / widespread resident   1846–2022

First noted at Cheetham Hill in June 1846 by R.S. Edleston and all early authors reporting it as abundant where decayed trees with bark were found. In recent decades, it has remained regularly and widely reported in both lowland and upland areas, often seen flying in sunny conditions around dead wood. A group of eight moths were noted in flight near a thick hedge at Sunderland Point on 27 May 2016.

■ **HABITAT** Hedgerows, gardens and woodland.
■ **FOODPLANTS** Rotten wood.
Single brooded, with two late July records. Comes to light. Larvae have been found, in January and April, in a holly tree stump, a dying birch trunk and a rotten bird table.

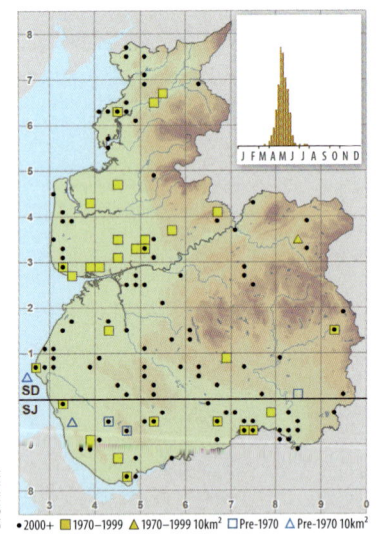

28.022 *Alabonia geoffrella*

■ **LANCS STATUS** Considered Extinct   c.1857–c.1920

Only three historic records are known, with both Ellis (1890) and Mansbridge (1940) give its status as 'very local and scarce'. The first report was of a few found near the old mill dam at Garston, Liverpool by C.S. Gregson, with no date specified, (Gregson, 1857). It was also noted as being present in the Manchester district (Stainton, 1859), an area which could cover some parts of adjacent vice counties. The final sighting, again without a specific date, but possibly around 1920, was by W.G. Clutten where he noted it as rare in the Burnley area (Mansbridge, loc. cit.). Best searched for alongside old hedgerows in sunny conditions in May and early June.

■ **HABITAT** Unknown.
■ **FOODPLANTS** Not recorded in Lancashire.

Elsewhere, it has been noted on four occasions in Cheshire, the last in 2018 (Cheshire-moth-charts.co.uk – accessed 30 August 2023) and on a few occasions in Yorkshire up to 2022 (yorkshiremoths.co.uk – accessed 30 August 2023). The data supporting a VC60 distribution dot in *The Moths and Butterflies of Great Britain and Ireland, Vol. 4 (1)* has not been discovered.

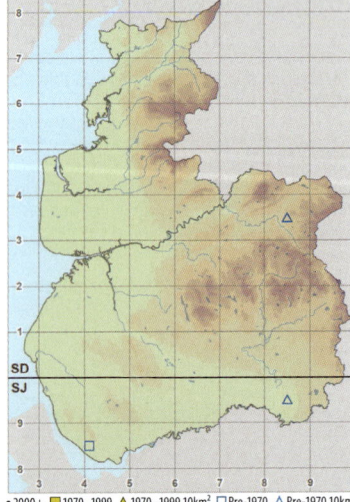

142 *The Moths of Lancashire*

28.024 *Tachystola acroxantha*

■ **LANCS STATUS** Frequent / widespread resident   1996–2022
After arriving in Britain (Devon) in 1908, a geographically isolated population became established in north-west England in 1986. The first Lancashire record was in Flixton on 18 October 1996 (K. McCabe) and a strong population developed in this area. In 1999 there were signs it was on the move and it reached Rochdale by 2007, Torrisholme in 2008, Silverdale area in 2014 and Briercliffe in 2019. Double digit counts are infrequent, but 52 were found at light in Flixton on 16 May 2002.
■ **HABITAT** Gardens in urban, suburban and rural areas.
■ **FOODPLANTS** Old nesting material from a nest box. Pupa found spun in dead Ivy leaves. Double brooded over an extended period or, perhaps, a succession of broods.

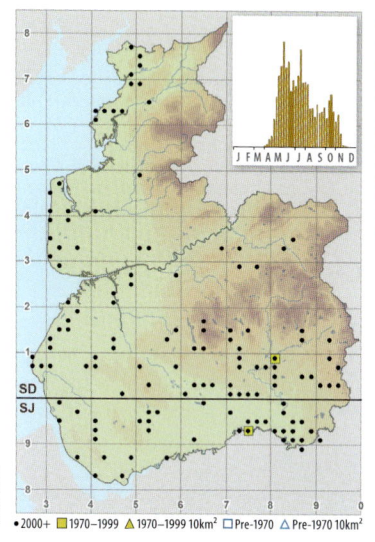

OECOPHORIDAE

28.025 *Pleurota bicostella*

■ **LANCS STATUS** Occasional / local resident   1844–2022
First found at White Moss (Edleston, 1844), where it was also abundant in 1846. Reported as common on heaths and mosses by Mansbridge (1940) and present on Freshfield Dune Heath c.1950. Winmarleigh, Cockerham and Astley Mosses were regular sites prior to 2014, but records have decreased across the county since then. A few recent moorland reports suggest it is under-recorded and, on rare occasions, it comes to garden light traps, such as at Worsthorne (2019) and Briercliffe (2022).
■ **HABITAT** Mosses and damp moorland.
■ **FOODPLANTS** Not recorded in Lancashire.
Single brooded. Comes rarely to light. Easily disturbed by day in sunny conditions.

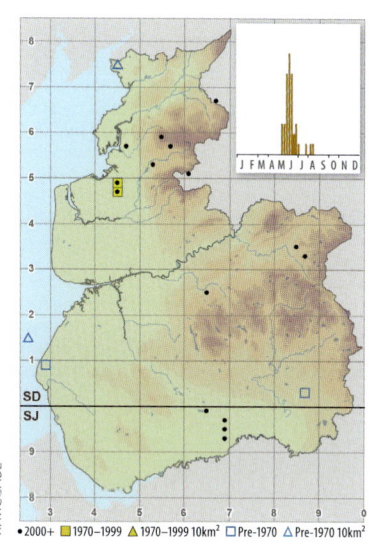

29.001 *Diurnea fagella*

■ **LANCS STATUS** Frequent / widespread resident   c.1857–2022
First noted in the Liverpool area (Gregson, 1857) and subsequently, as very common by Ellis (1890) and Mansbridge (1940). The dark form was frequent in the early 20th century. From the 1960s onwards, noted regularly from several sites, with light trapping later adding records from across the region, usually in single figures.
■ **HABITAT** Woodland, gardens, scrub, brownfield sites and mature hedgerows.
■ **FOODPLANTS** Oak sp., Hazel, Grey Willow, Osier, Small-leaved Lime, birch sp., Beech, Hawthorn, Bramble, Blackthorn, Rowan, Sycamore and Dog Rose.
Single brooded. Comes to light. The wingless females can be found on tree trunks. Larvae have been found from June to October and pupae in November.

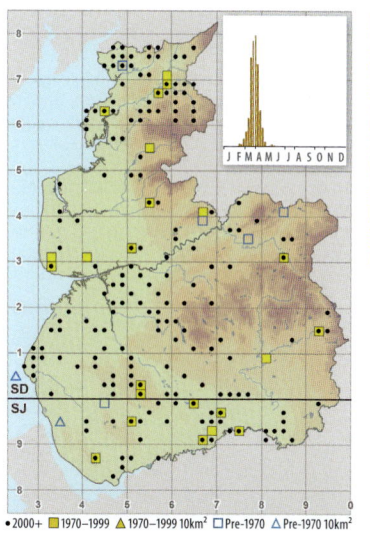

CHIMABACHIDAE

The Moths of Lancashire

CHIMABACHIDAE

29.002 *Diurnea lipsiella*

■ **LANCS STATUS** Occasional / very local resident   c.1857–2022

Noted in the Liverpool area (Gregson, 1857) and in 1910 and 1917 by W.A. Tyerman, then from Burnley in 1920 (W.G. Clutten). Since then, most records have been from its present stronghold in Silverdale's limestone woodland. The few other sites were at Woolton in 1940, Preston in 1995, Churn Clough in 2006 and Burton-in-Kendal in 2007. Around half of the county's records are from Warton Crag, mostly in single figures, but 15 came to actinic light there on 24 October 2013 (J. Patton).
■ **HABITAT** Long-established, deciduous woodland.
■ **FOODPLANTS** Not recorded in Lancashire.
Single brooded. Comes to light. Two observed in daytime flight, at Gait Barrows.

• 2000+   □ 1970–1999   ▲ 1970–1999 10km²   □ Pre-1970   △ Pre-1970 10km²

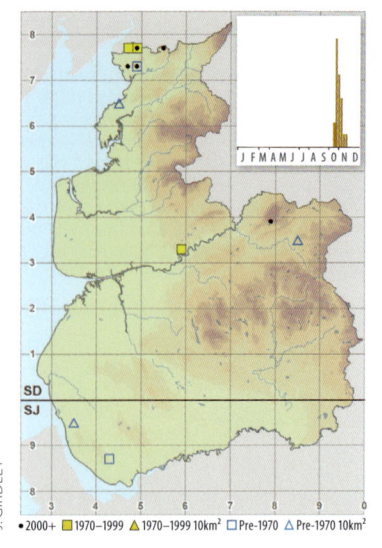

29.003 *Dasystoma salicella*

■ **LANCS STATUS** Considered Extinct   c.1857–c.1890

Ellis (1890) reports this species to be of very rare occurrence in the region and documents records from Huyton by C.S. Gregson and the Manchester district by Stainton (1859). In Gregson's own work of 1857, he provides a little extra information, noting, 'in old mixed hedges at the end of April'. This could refer to the Huyton record, or another he lists from Cheshire, but he fails to mention any dates. There are no additional records listed in the coverage of this species in Mansbridge (1940).
■ **HABITAT** Old mixed hedgerows.
■ **FOODPLANTS** Not recorded in Lancashire.
The male and wingless female can be found, nationally, in April, with the male flying in warm sunshine at about midday. Always local and scarce nationally, it was last noted in Cheshire in 1959 (cheshire-moth-charts.co.uk – accessed 30 August 2023) and in Yorkshire in 1975 (yorkshiremoths.co.uk – accessed 30 August 2023).

• 2000+   □ 1970–1999   ▲ 1970–1999 10km²   □ Pre-1970   △ Pre-1970 10km²

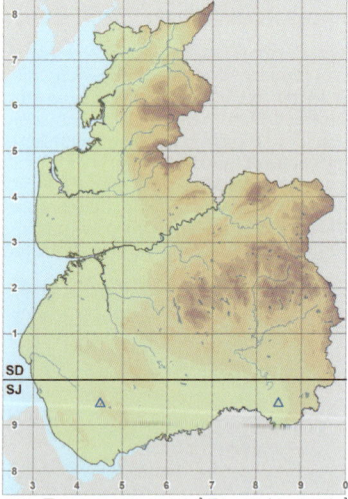

LYPUSIDAE

30.003 *Agnoea josephinae*

■ **LANCS STATUS** Scarce / very local resident   1997–2023

Although not found in the county until relatively recently, its earlier presence in the 19th century in south Cumbria, suggest it had been overlooked. First noted at Gait Barrows on 28 June 1997 (S. Palmer), with four further records from this site, in 1999, 2008, 2009 and 2015. It has only otherwise been found in a small oakwood on the edge of Birk Bank in 2002 and 2021 and at Trowbarrow Quarry in 2023.
■ **HABITAT** Deciduous woods on limestone and a small oakwood on moorland edge.
■ **FOODPLANTS** Not recorded in Lancashire.
Single brooded. Comes sparingly to light. Can be disturbed during the daytime. Microscopic examination usually required to separate other *Agnoea* species.

• 2000+   □ 1970–1999   ▲ 1970–1999 10km²   □ Pre-1970   △ Pre-1970 10km²

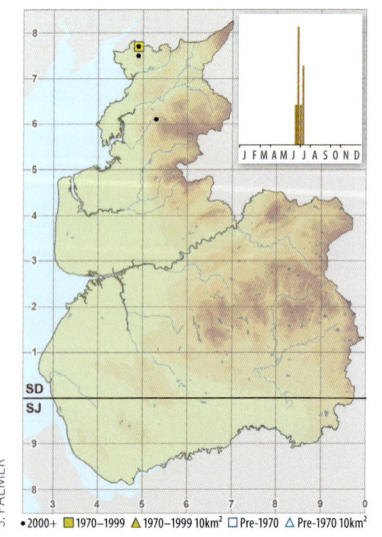

30.004 Amphisbatis incongruella

LANCS STATUS Scarce / very local resident   2003–2012
The moth will almost certainly have been overlooked in the past due to its small size and early flight period. First recorded on Astley Moss on 3 April 2003, when swept from heather by K. McCabe and also noted here on four further occasions, all in low numbers. Otherwise, only seen on the Cockerham and Winmarleigh Moss complex, between 2005 and 2012 (S. Palmer). Twenty-one were noted on 28 March 2012.
HABITAT Lowland raised bogs.
FOODPLANTS Not recorded in Lancashire.
Single brooded. Has only been found whilst searching or sweeping mosses during sunny conditions from late morning to early afternoon.

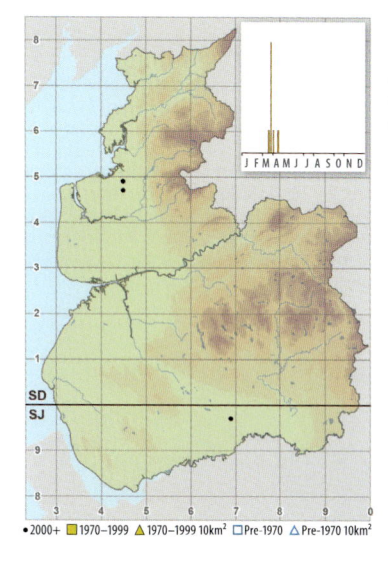

LYPUSIDAE

31.001 Carcina quercana

LANCS STATUS Frequent / widespread resident   c.1857–2022
Noted in the Liverpool district (Gregson, 1857), Manchester in 1859 and Formby c.1940. At that time, considered generally distributed, a situation that remains unchanged to the present day. The largest count was 14 at Rishton on 18 July 2022.
HABITAT Woodland, gardens, scrub, hedgerows, dunes and brownfield sites.
FOODPLANTS Apple, Aspen, Beech, Blackthorn, Bramble, Hawthorn, Norway Maple, oak sp., Pyracantha and Sycamore.
Single brooded, but with a small but increasing second brood noted since 1995. Comes to light. Larvae found late September to mid-May, with a few, possible second brood individuals, found during the summer months and bred the same year.

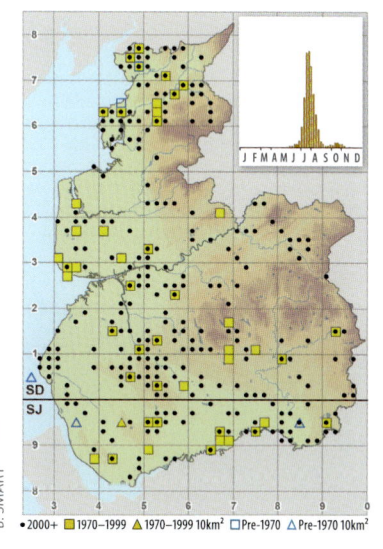

PELEOPODIDAE

32.001 Semioscopis avellanella

LANCS STATUS Scarce / very local resident   2000–2011
Considered an overlooked species, as it was present in south Cumbria in the early 20th century. First noted at Gait Barrows on 22 April 2000 (S. Palmer) and also again in 2004 and 2010, the latter by C. Barnes. The only other relatively regular location is Eaves Wood, a Small-leaved Lime wood nr. Silverdale. It was found there in 2004, and twice in 2011. One came to a lighted window in Silverdale on 20 April 2006 (T. Riden).
HABITAT Deciduous woodland on limestone.
FOODPLANTS Not recorded in Lancashire.
Single brooded, Comes to light in small numbers. There is a VC59 dot in *The Moths and Butterflies of Great Britain and Ireland Vol. 4 (1)*, but with no supporting data.

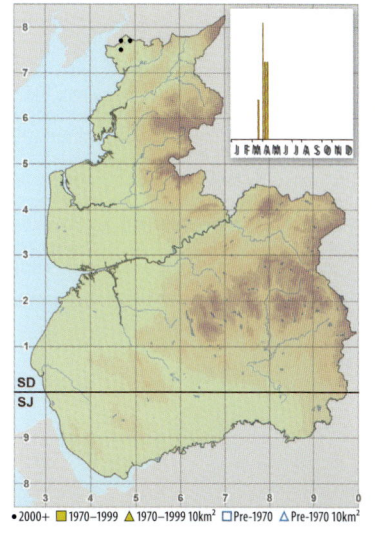

DEPRESSARIIDAE

The Moths of Lancashire   **145**

DEPRESSARIIDAE

32.002 *Semioscopis steinkellneriana*

■ **LANCS STATUS** Occasional / local resident   1985–2018

Noted at Winmarleigh Moss on 17 May 1985 at dusk and also at Hardhorn in 1986 (both M. Evans). In 2001, it was found at Gait Barrows, Marton (Blackpool) and Burrow Heights, with records infrequent over the next few years. In 2008, the first upland area report came from Herring Head Wood, with Lord's Lot Wood and Docker Moor following a few years later. It remains most frequent on the limestone, and ten came to light in Eaves Wood on 7 April 2012 (G. Jones).
■ **HABITAT** Deciduous woods and copses, on limestone, mosses and moors.
■ **FOODPLANTS** Not recorded in Lancashire.
Single brooded. Comes to light in small numbers and has been disturbed at dusk.

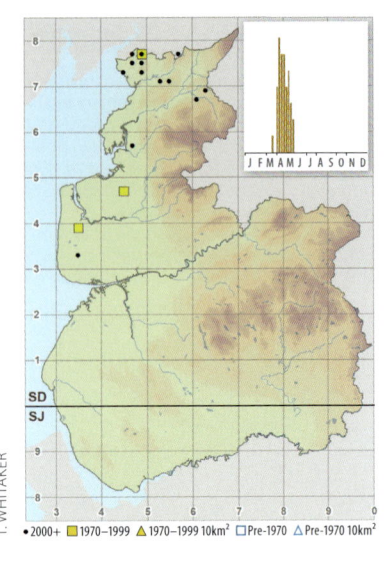

32.006 *Exaeretia allisella*

■ **LANCS STATUS** Occasional / local resident   c.1857–2010

Found in the Liverpool area where Mugwort abounds (Gregson, 1857), Warrington c.1889 (J.B. Hodgkinson) and Southport in 1910 by W.G. Clutten. Reports from Formby and Chorlton followed in the 1950s, while in the 1960s it was bred from larvae in Didsbury. More recently, one came to light in Billinge in 1989 and two were seen in Flixton in 1998. The last record was of two on Cadishead Moss on 26 July 2010 (K. McCabe). A dot on the national VC maps for VC60 has no supporting data.
■ **HABITAT** Grassland, mosses, dunes and waste ground, including brownfield sites.
■ **FOODPLANTS** Mugwort.
Single brooded. Comes to light. Once, plentiful as larvae in Mugwort roots.

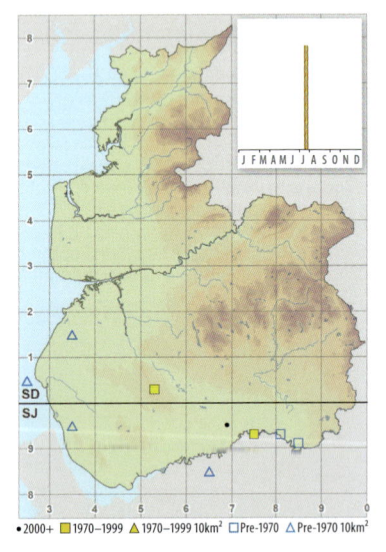

32.007 *Agonopterix ocellana*

■ **LANCS STATUS** Frequent / widespread resident   c.1857–2022

Found in Liverpool district (Gregson, 1857), Lytham (1865) and Preston (pre-1890). It was later described as general and common throughout (Mansbridge, 1940). The first known upland record was not until 1955, in Oldham, and it has remained local on higher ground since then. In lowland areas it is recorded widely, but infrequently. Flixton has by far the largest number of records, almost half of the 480 on the county database, the largest single count being seven on 29 March 2011 (K. McCabe).
■ **HABITAT** Woodland, gardens, scrub on mosses and limestone, and brownfield sites.
■ **FOODPLANTS** Creeping, Goat and Grey Willows.
Single brooded, overwintering as an adult. Comes to light in small numbers.

146   The Moths of Lancashire

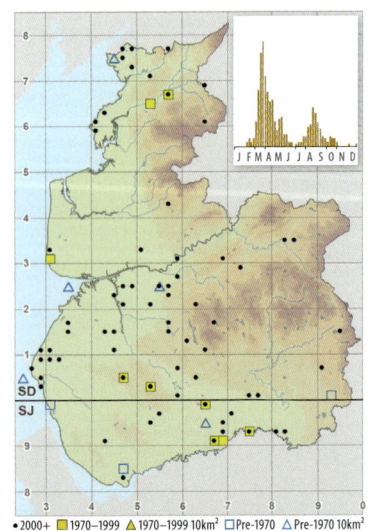

32.008 *Agonopterix liturosa*

■ **LANCS STATUS** Rare / very local resident   c.1857–2012
First recorded, under *Depressaria hypericella* Hübn., at Childwall near Liverpool in July, with no year given (Gregson, 1857). Since then, there have been only two records, the first at Gait Barrows in 1986 by S. McWilliam and one at Heysham N.R. (below) on 30 July 2012. The latter came to a 160w blended bulb moth trap and was recorded by A. Draper.
■ **HABITAT** Dry grassland on limestone and a brownfield site.
■ **FOODPLANTS** Not documented in Lancashire, but it is reported as having been found amongst Perforate St. John's-wort, this being in the 19th century.
Single brooded. Has come to light on one occasion.

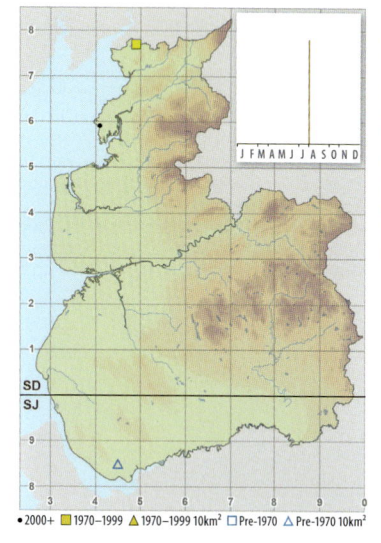

32.009 *Agonopterix purpurea*

■ **LANCS STATUS** Considered Extinct   1872–?
The only records relating to this species being found in the county were by J.B. Hodgkinson in the Blackpool area in about July 1872 (Hodgkinson, 1872) and by Hodgkinson and J.H. Threlfall at Lytham, undated, (Ellis, 1890). The former site was described as a location 'about six miles from Blackpool'. It is possible that both reports refer to the same site, visited on different occasions. Larvae were found and moths were bred in 1872.
■ **HABITAT** Coastal grassland.
■ **FOODPLANTS** Not documented in Lancashire.
Extensive searches for larvae on Wild Carrot took place in estuarine and coastal localities, as well as slightly inland, from the docks at Preston round to Rossall Point and on the Wyre estuary during 2022, but no *A. purpurea* larvae were located (J. Steeden and S. Palmer).

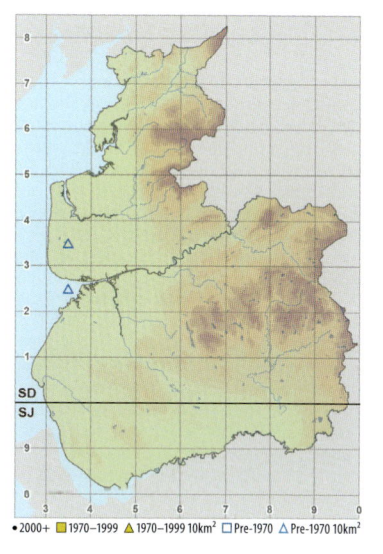

32.010 *Agonopterix conterminella*

■ **LANCS STATUS** Occasional / widespread resident   1865–2022
First found as larvae on sallows at Pilling Moss in June 1865 (Hodgkinson, 1865). Records from 16 generally coastal and lowland sites followed, including Crosby (1900), Formby (1950), Leighton Moss (1952) and St Annes (1995). Since then, it has been widely, but infrequently, recorded in lowland areas as singletons, and mostly at light. More frequent reports have come from Billinge, Flixton and north Preston. Few upland records are known, but include Briercliffe, Docker Moor and High Tatham.
■ **HABITAT** Scrub, mosses, woodland, dune slacks, hedgerows and brownfield sites.
■ **FOODPLANTS** Sallows, including Creeping, Goat and Grey Willow and Osier.
Single brooded. Comes to light. Larvae found in May and June.

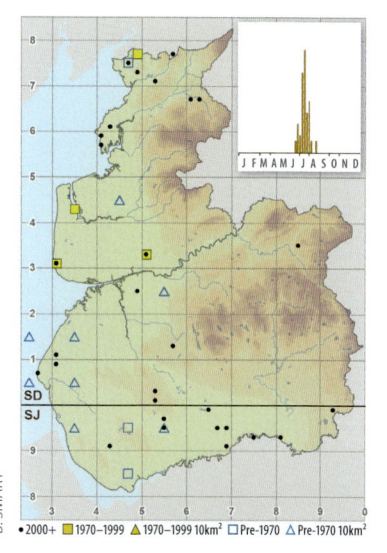

DEPRESSARIIDAE

32.011 *Agonopterix scopariella*

■ **LANCS STATUS** Occasional / local resident   **2000–2019**

The first came to light in Billinge on 7 March 2000 (C.A. Darbyshire) and was quickly followed by another, two days later, in Ormskirk (G. Jones). In 2003, it was found in St Annes and, over the following years, it was noted at twelve further lowland sites, usually as singletons at light. Four were found in a Southport trap on 20 April 2011.
■ **HABITAT** Coastal grassland, rough ground and brownfield sites.
■ **FOODPLANTS** Broom.
Single brooded, overwintering as an adult. Comes to light. Larvae found May to June on spun leaves and shoots. The records seem to indicate this to be a recent arrival in the county, which is backed up by Cheshire's first not being until 2004.

● 2000+   ☐ 1970–1999   △ 1970–1999 10km²   ☐ Pre-1970   △ Pre-1970 10km²

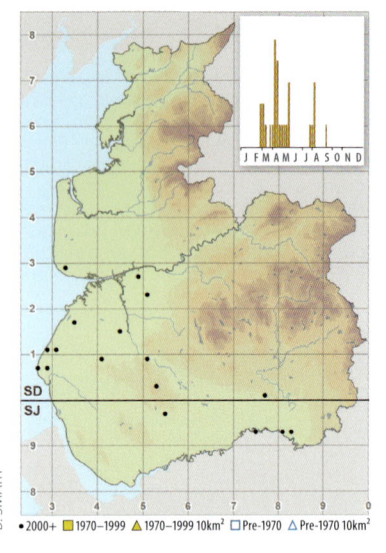

32.012 *Agonopterix atomella*

■ **LANCS STATUS** Considered Extinct   **1860–c.1879**

Emmet & Langmaid (2002) report the first British record as being by J.B. Hodgkinson and give the reference, Hodgkinson (1879a). The location was not mentioned until later, when larvae were reported from Dutton, near Longridge on 28 June 1879 (Hodgkinson, 1879b), with further details in Hodgkinson (1878 and 1879c). As well as Dutton, the various reports mention Preston, Longridge and Salmesbury areas with J.H. Threlfall reporting larvae on *Genista tinctoria* near Preston in June 1876 (Threlfall, 1877). Ellis (1890) adds a record from Longridge by Hodgkinson and from Salmesbury, near Preston by Threlfall. The only other location mentioned is from near Stoneyhurst in Hodgkinson (1879c). It has not been possible to confirm how many sites were involved, but it is most likely to have been a series of small sites, with Dyer's Greenweed occurring in that area on old hay meadows on south facing slopes at that time.
■ **HABITAT** Herb-rich meadows.
■ **FOODPLANTS** Dyer's Greenweed.
Only larvae were reported, in late June, although many appear to have been bred through. Dyer's Greenweed has, since then, declined considerably in the county (Greenwood, 2012 and pers. comm.). Records under the name *Depressaria atomella*, in Stainton (1859) were made prior to its separation from *A. scopariella*.

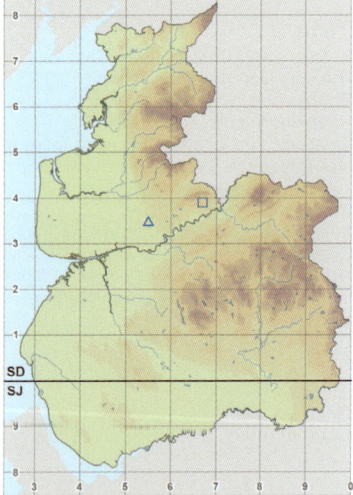

● 2000+   ☐ 1970–1999   △ 1970–1999 10km²   ☐ Pre-1970   △ Pre-1970 10km²

32.013 *Agonopterix carduella*

■ **LANCS STATUS** Occasional / very local resident   **1917–2018**

W. Mansbridge found two larvae on the Crosby sandhills in 1917 (Leech and Michaelis, 1957), but it was not until 1941 that it was next noted, as very common on Warton Crag by L.T. Ford. A single larva was discovered in Freshfield in 1955 and in 2005 three imago came to light in Flixton (K. McCabe). All other records, to the present day, have come from Warton Crag, including counts of around 15 larvae in 2003 (S. Palmer), and six found there in 2014 (B. Elliot *et al.*).
■ **HABITAT** Limestone grassland, a single brownfield site and sand dune areas.
■ **FOODPLANTS** Black Knapweed. Leech and Michaelis (1957) list *'Carduus'*.
Single brooded, overwinters as an adult. Larvae have been found in June.

● 2000+   ☐ 1970–1999   △ 1970–1999 10km²   ☐ Pre-1970   △ Pre-1970 10km²

148   *The Moths of Lancashire*

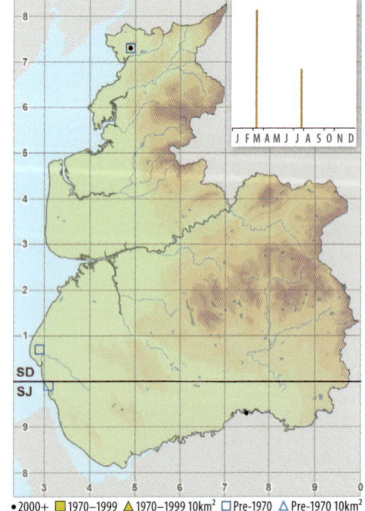

32.015 *Agonopterix subpropinquella*

■ **LANCS STATUS** Scarce / very local resident c.1857–2022
The first was noted on Crosby sandhills (Gregson,1857), while Edleston (1859) discovered 28 specimens at Blackpool in the middle of August 1859. The form *rhodochrella* was found on the west bank of R Wyre in September 1872 and another, much later, in Flixton in 2004. Only two other records are known, a second from Flixton on 4 August 2018 (K. McCabe) and one in Formby on 1 May 2022 (R. Walker).
■ **HABITAT** Coastal areas and brownfield sites.
■ **FOODPLANTS** Not recorded in Lancashire.
Single brooded, overwinters as an adult. The moth appears in two forms, one of which, f. *rhodochrella*, has the thorax, and sometimes the head, blackish in colour.

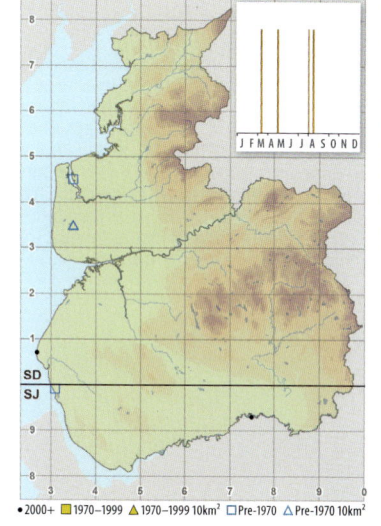

32.016 *Agonopterix propinquella*

■ **LANCS STATUS** Frequent / widespread resident c.1857–2022
First noted on Crosby sandhills (Gregson, 1857) and later, at Lytham. Early authors, including Mansbridge (1940), considered it chiefly a coastal sandhills species. Since the mid-1990s it has been found in many lowland areas, and from around 2001, commenced a spread into the uplands, where it remains rather local to the present day. Regular in small numbers at several lowland sites, including Flixton, Parr and Preston. The largest count was of five to light in Flixton on 15 May 2022 (K. McCabe).
■ **HABITAT** Rough grassland, gardens, waste ground and brownfield sites.
■ **FOODPLANTS** Creeping Thistle.
Single brooded, overwinters as an adult. The sole larval record was in early August.

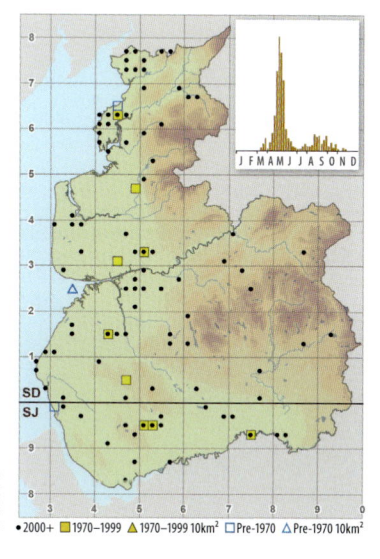

32.017 *Agonopterix arenella*

■ **LANCS STATUS** Frequent / widespread resident c.1857–2022
First recorded in the Liverpool district (Gregson, 1857), it was later considered local but not uncommon by Mansbridge (1940). Since then, has been found regularly at many sites across the county, in both lowland and upland regions. It is not unusual for an overwintering adult to be disturbed from vegetation or outhouses during the winter months. Seen mostly in single figures as adults but can be abundant as a larva.
■ **HABITAT** Rough grassland, gardens, waste ground, open scrub and brownfield sites.
■ **FOODPLANTS** Black Knapweed, Creeping Thistle, Greater Burdock and Spear Thistle.
Single brooded, overwinters as an adult. Tenanted larval spinnings have been found from June to August and it has once utilised a garden-planted Greater Knapweed.

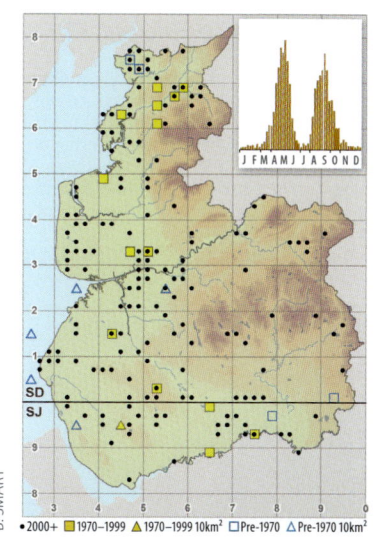

The Moths of Lancashire

DEPRESSARIIDAE

32.018 *Agonopterix heracliana*

■ **LANCS STATUS** Abundant / ubiquitous resident   c.1857–2022

Reported as common everywhere, since the earliest record from the Liverpool area (Gregson, 1857). Counts of more than single figures of adults are unusual, with the 23 at Flixton in 2010 (K. McCabe) exceptional. Larvae, however, are regularly abundant.

■ **HABITAT** Across all habitats, wherever umbellifers grow.

■ **FOODPLANTS** Angelica, Cow Parsley, Fennel, Ground Elder, Hedge Parsley, Hemlock, Hemlock Water Dropwort, Hogweed, Chervil, Sweet Cicely, Wild Carrot, Wild Parsnip.

Single brooded; overwinters as an adult and readily found in outbuildings at that time of year. Tenanted larval spinnings in the leaves and, much less frequently, in the flowers, have been found from mid-May to early August.

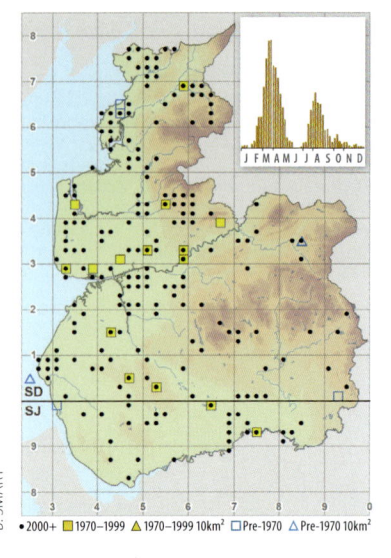

● 2000+ ☐ 1970–1999 △ 1970–1999 10km² ☐ Pre-1970 △ Pre-1970 10km²

32.019 *Agonopterix ciliella*

■ **LANCS STATUS** Occasional / local resident   c.1859–2022

First recorded c.1859 in the Stretford area by J. Chappell (Ellis, 1890) and later noted at several sites, such as Gathurst, Lytham and Preston, up to the mid-20th century. Records since have been more localised, with at least half from Flixton to light between 1998 and 2015 (K. McCabe). Probably under-recorded elsewhere.

■ **HABITAT** A range of habitats, generally in shady or damp localities.

■ **FOODPLANTS** Angelica, Cow Parsley and Hogweed.

Single brooded, overwintering as an adult. Tenanted larval spinnings in leaves found from June to early September, occasionally in numbers. 40 larvae found in one area near Longridge, on all three of the listed foodplants, on 30 July 2012 (S. Palmer).

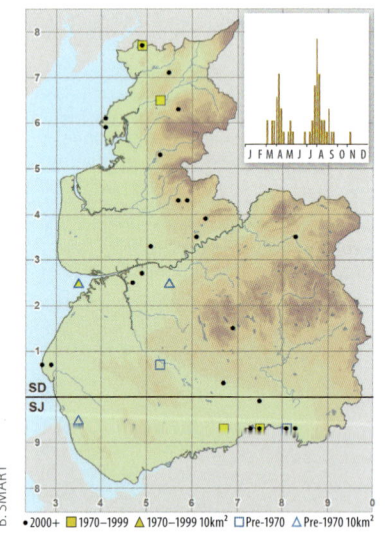

● 2000+ ☐ 1970–1999 △ 1970–1999 10km² ☐ Pre-1970 △ Pre-1970 10km²

32.024 *Agonopterix assimilella*

■ **LANCS STATUS** Occasional / local resident   c.1857–2022

First recorded in Broadgreen (Liverpool) and Roby (Gregson, 1857). Found later in the Formby and Freshfield areas (Mansbridge, 1940) and in Didsbury in 1955. Records since have been sporadic, including from Ashton Moss, Blackpool, Fazakerley, Martin Mere and Woodvale. Most records are from the sandy grassland areas on the Sefton Coast, where Broom occurs in good quantity.

■ **HABITAT** Coastal sandy grassland, rough ground and brownfield sites.

■ **FOODPLANTS** Broom.

Single brooded, with an extended flight period. Comes sparingly to light. Larvae found between spun stems from February to mid-April. Two early adults in March.

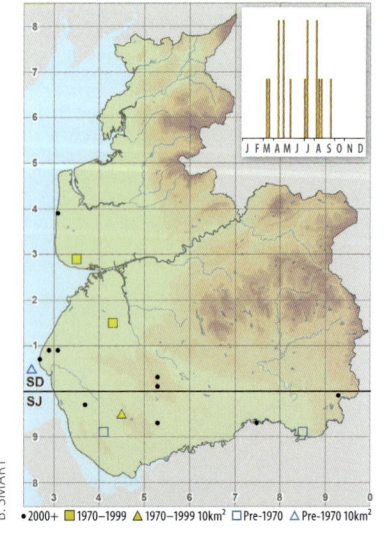

● 2000+ ☐ 1970–1999 △ 1970–1999 10km² ☐ Pre-1970 △ Pre-1970 10km²

150   *The Moths of Lancashire*

32.025 *Agonopterix nanatella*

■ **LANCS STATUS** Rare / very local resident 1865–2014

First recorded as an adult at Lytham dunes in late July 1865 (Hodgkinson, 1865), with the larvae found there in 1866 and 1876, the last by J.H. Threlfall. In Mansbridge (1940) it was noted as having been found on Crosby sandhills and found commonly at Formby (no further details given) but since then, not noted from those areas. Searches for the larvae at the Lytham St Annes NR on 25 May 2009 produced three spinnings, with the same again in 2013 and 2014 (S. & C. Palmer and J.R. Langmaid).
■ **HABITAT** Coastal sandy grassland.
■ **FOODPLANTS** Carline Thistle.
Single brooded. Larvae found in spun leaves from mid-May to late June.

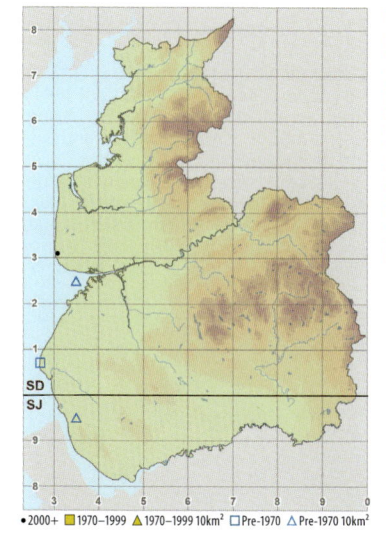

32.026 *Agonopterix kaekeritziana*

■ **LANCS STATUS** Occasional / very local resident c.1940–2015

Early name confusion meant the records in Ellis (1890), under *A. liturella*, and also probably by Gregson (1857), referred to a different species. This was corrected in Mansbridge (1940) who listed the following sites; Blackpool, Burnley, Formby and Wavertree, but without dates. Most of the nine later records related to larvae found on coastal or limestone sites, particularly at Warton Crag. Adults were attracted to light at Flixton in 1997, High Tatham in 2006 and at Warton Crag in 2014 (J. Patton).
■ **HABITAT** Grassland on dunes, limestone and, rarely, other open grassy areas.
■ **FOODPLANTS** Black Knapweed.
Single brooded. Comes sparingly to light. Single larvae found in late May and June.

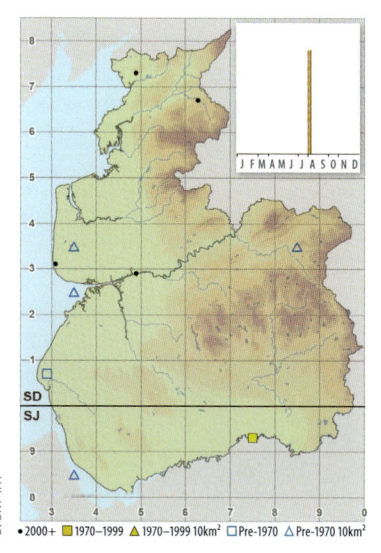

32.029 *Agonopterix umbellana*

■ **LANCS STATUS** Occasional / local resident c.1857–2022

First recorded in the Liverpool district (Gregson, 1857), where it was noted as common among gorse bushes. More recently, there have been occasional records from several sites, such as Docker Moor, Heysham, Martin Mere, Preesall Hill and Warton Crag, all sites with good stands of gorse. Freshfield Dune Heath is the most regular site for this moth, with records in most years between 2009 and 2022.
■ **HABITAT** Rough ground on sandy soil, limestone, moorland edge and brownfield sites.
■ **FOODPLANTS** Gorse.
Single brooded, overwintering as an adult. Comes to light in small numbers. A tenanted spinning was found on gorse in Freshfield on 10 June 2022 (B. Smart).

The Moths of Lancashire

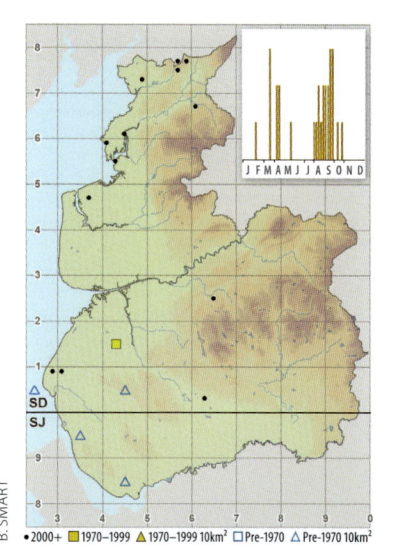

DEPRESSARIIDAE

32.030 *Agonopterix nervosa*

■ **LANCS STATUS** Occasional / widespread resident   1844–2022
First noted at Kersal Moor in August 1844 (Edleston, 1844), with further records from Liverpool and Manchester areas following. It was found in Rochdale in 1948 and Morecambe in 1959 (C.J. Goodall) and has since been widely reported from lowland and upland areas, most frequently in areas with gorse stands, such as at Heysham. Largest counts involved larvae, with 20 on gorse at Sunderland Point in 2011.
■ **HABITAT** Hedgerows, rough ground, moorland edges, gardens and brownfield sites.
■ **FOODPLANTS** Gorse, Broom and Laburnum.
Single brooded; rarely overwinters as an adult. Comes to light. Larvae found from mid-April to early June in flower heads on gorse and in spun leaves and stems.

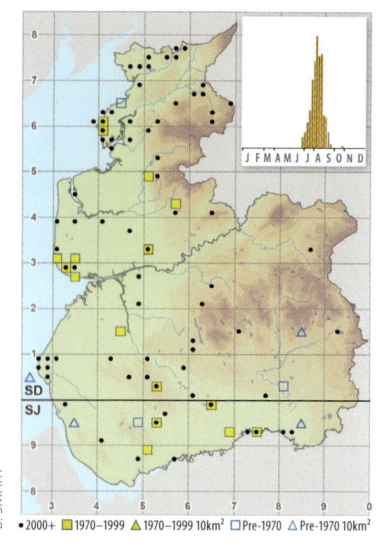

● 2000+   ◻ 1970–1999   ▲ 1970–1999 10km²   ◻ Pre-1970   △ Pre-1970 10km²

32.031 *Agonopterix alstromeriana*

■ **LANCS STATUS** Occasional / local resident   c.1889–2022
First reported at Lytham pre-1890 by J.B. Hodgkinson (Ellis, 1890) and c.1900 by W.G. Clutten in the Blackpool and St Annes areas. Since the mid-1990s, it has been reported fairly regularly from Flixton and Southport, and more intermittently from several other mainly coastal and Mersey Valley sites. Upland reports are rare, with the only ones from Worsthorne in 2002 (G. Gavaghan) and Royton in 2009 (R. Hart). Found as a larva on many Hemlock plants in the Blackpool area in 2022 (J. Steeden).
■ **HABITAT** Disturbed areas and rough ground where Hemlock occurs.
■ **FOODPLANTS** Hemlock.
Single brooded. Comes to light. Larvae found in leaf spinnings from June to mid-July.

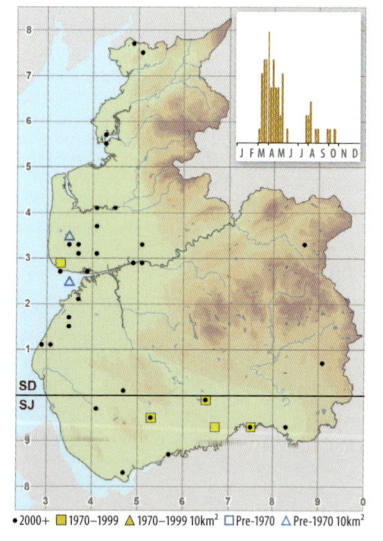

● 2000+   ◻ 1970–1999   ▲ 1970–1999 10km²   ◻ Pre-1970   △ Pre-1970 10km²

32.032 *Agonopterix angelicella*

■ **LANCS STATUS** Occasional / local resident   c.1859–2022
Noted in the Manchester area (Stainton, 1859) and then, bred from a larva on Angelica at Pilling Moss in 1865 (Hodgkinson, 1865). Subsequently recorded from Lytham, Preston, Salwick, Burnley and Formby, all prior to 1937. From the late 1990s it has been found with great regularity at light in Flixton (K. McCabe), this constituting well over three quarters of the 120 plus county records. All other sightings have been of singles in lowland locations such as Abram and Heysham.
■ **HABITAT** Damp woodland edges, mosses and verges where Wild Angelica occurs.
■ **FOODPLANTS** Wild Angelica and, occasionally, Hogweed.
Single brooded. Comes to light. Larvae found from mid-May to late June.

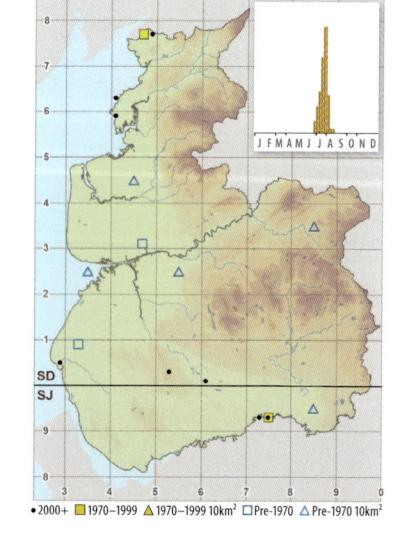

● 2000+   ◻ 1970–1999   ▲ 1970–1999 10km²   ◻ Pre-1970   △ Pre-1970 10km²

32.035 *Agonopterix yeatiana*

■ **LANCS STATUS** Occasional / local resident   1865–2022

In July 1865, larvae were found commonly on Wild Carrot at Lytham (Hodgkinson, 1865), the same recorder noting an adult in Blackpool on 12 August 1872. The moth was not recorded again until 1949 when it was located at Freshfield and, after that, in 1995 on the Lytham St Annes N.R. Since then, it has been found within a few miles of the coast at nine sites, mainly to light. Records have increased slightly since 2019.
■ **HABITAT** Coastal dune grassland.
■ **FOODPLANTS** Wild Carrot and Hemlock Water Dropwort.
Single brooded, overwintering as an adult. Comes to light. Larvae found from early June and during July. Inland light trap reports may relate to wanderers.

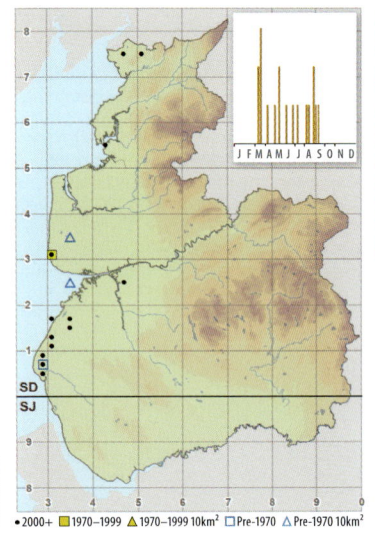

32.036 Parsnip Moth *Depressaria radiella*

■ **LANCS STATUS** Frequent / widespread resident   c.1859–2022

First noted in the Withington, Chat Moss and Glazebrook areas by J. Chappell (Ellis, 1890), amongst others. In the late 1980s, noted regularly in Morecambe and Billinge, with records from many more sites during the 1990s, including Claughton, Flixton, and Heysham. From 2012, inland and upland records started to decrease a little, this continuing to the present day. In contrast, coastal populations appear more stable.
■ **HABITAT** Rough grassland, verges, open scrubland, moss edges and brownfield sites.
■ **FOODPLANTS** Hogweed and Wild Parsnip. Possibly also Stone Parsley and Cow Parsley.
Single brooded, overwintering as an adult. Comes to light. Larvae found from late May to late July amongst spinnings in the flowerheads.

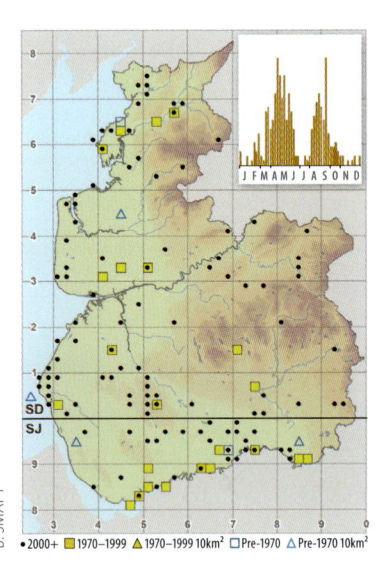

32.038 *Depressaria badiella*

■ **LANCS STATUS** Occasional / local resident   c.1857–2021

Only two historic records are known, from the Liverpool district (Gregson, 1857) and from Lytham, undated, by J.B. Hodgkinson (Ellis, 1890). Subsequently seen at Gait Barrows in 1984 and 1986, Bold Moss in 2000, and the first of a few at Flixton in 2003. From 2004 onwards, it was recorded once or twice from ten scattered sites and also at Heysham N.R., almost annually from 2009 to 2015. In 2021, it was noted at two upland limestone sites, at Worsaw Hill and Dalton Crags, the last by J. Patton.
■ **HABITAT** Limestone and sandy grassland, and flower-rich, grassy brownfield sites.
■ **FOODPLANTS** Not recorded in Lancashire.
Single brooded. Comes to light.

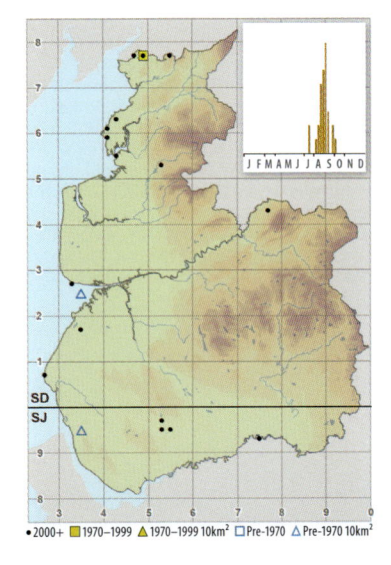

DEPRESSARIIDAE

The Moths of Lancashire   **153**

DEPRESSARIIDAE

32.039 *Depressaria daucella*

■ **LANCS STATUS** Abundant / widespread resident   c.1889–2022

Found in Lytham by J.B. Hodgkinson and Penwortham by J.H. Threlfall (Ellis, 1890), and later noted as local, but fairly common, by Mansbridge (1940). In the 1960s to '80s, recorded from Freshfield, Martin Mere, Morecambe and Swillbrook, amongst other sites. Records increased from 1998, it being regularly found at both light or as a larva at many, mainly lowland sites, including Bay Horse, Flixton and Southport. Larvae often abundant, with 300 found on Warton Marsh on 8 June 2002 (S. Palmer).
■ **HABITAT** Damp meadows, ditches, canal edges and other wet ground.
■ **FOODPLANTS** Hemlock Water Dropwort.
Single brooded, overwintering as an adult. Comes to light. Larva seen June and July.

● 2000+   ■ 1970–1999   ▲ 1970–1999 10km²   □ Pre-1970   △ Pre-1970 10km²

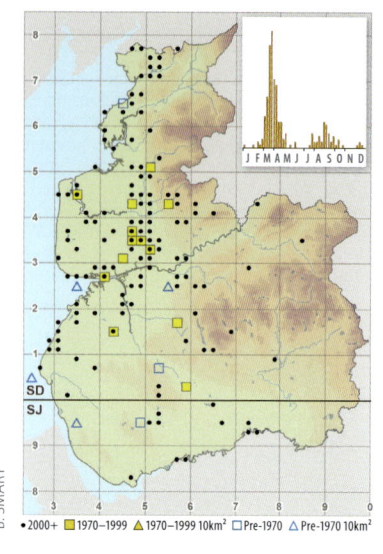

32.040 *Depressaria ultimella*

■ **LANCS STATUS** Rare / very local resident   2000–2008

There are only three records of this species in Lancashire, all of them found as singletons at light, the first to a 160w blended bulb and the other two being 125w Mercury Vapour type traps. First discovered at Heysham N.R. on 30 August 2000 by J. Roberts and confirmed by J.R. Langmaid and the next at Tinkers Lane, Bay Horse, near Dolphinholme on 2 June 2007 (N.A.J. Rogers). The final one was in a garden on Heysham Road, Morecambe on 9 May 2008 (D.J. Holding).
■ **HABITAT** Brownfield sites with wetland areas and sites close to wet ditches.
■ **FOODPLANTS** Not recorded in Lancashire.
Single brooded, overwintering as an adult. Comes to light.

● 2000+   ■ 1970–1999   ▲ 1970–1999 10km²   □ Pre-1970   △ Pre-1970 10km²

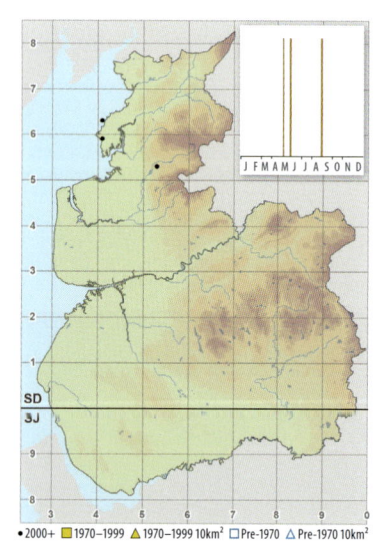

32.042 *Depressaria pulcherrimella*

■ **LANCS STATUS** Scarce / very local resident   1865–2022

First recorded by J.B. Hodgkinson near Pilling Moss, with 'nine smoked from old grass around a tree root', in late June 1865 (Hodgkinson, 1865). The next was in the Preston area by J.H. Threlfall (Ellis, 1890). Since then, only reported from Green Bank, by P.J. Marsh with singles at light on 10 July 2021 and 11 July 2022. The farmstead where the light was run has meadows nearby where Pignut is plentiful.
■ **HABITAT** Upland meadows with extensive Pignut growth.
■ **FOODPLANTS** Not recorded in Lancashire.
Single brooded. Comes to light. The habitat associated with the historic records is unknown. Probably overlooked, but searches for larvae in 2022 failed to locate any.

● 2000+   ■ 1970–1999   ▲ 1970–1999 10km²   □ Pre-1970   △ Pre-1970 10km²

154   The Moths of Lancashire

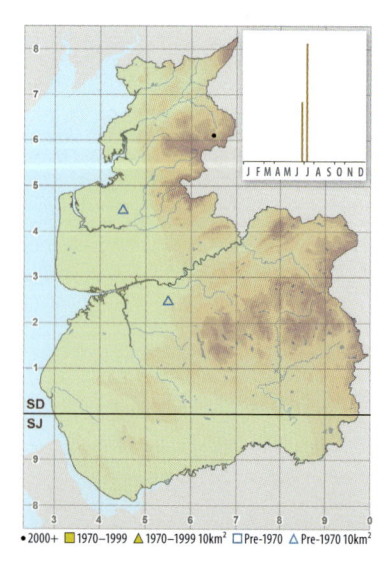

32.043 *Depressaria sordidatella*

■ **LANCS STATUS** Rare / very local resident   c.1857–2011

First recorded by C.S. Gregson in the Liverpool area by 'smoking' vegetation in the vicinity of the larval foodplant (Gregson, 1857). The next record was from near Pilling Moss, where a number of larvae were found on Cow Parsley in June 1865 (Hodgkinson, 1865). Undated records from Penwortham and Lytham are listed in Ellis (1890). The only recent report was of a singleton, attracted to MV light, at Flixton (K. McCabe) on 20 July 2011.
■ **HABITAT** Rough, damp grassland, including on a brownfield site.
■ **FOODPLANTS** Cow Parsley.
Single brooded. Comes to light. Historically, larvae found in June.

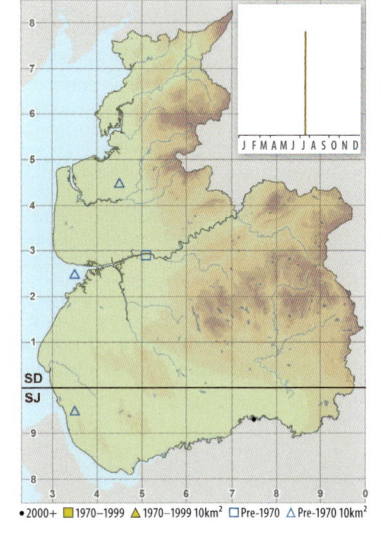

S. PALMER

32.044 *Depressaria douglasella*

■ **LANCS STATUS** Considered Extinct   c.1857–1918

First reports were of four found on the Crosby sandhills (Gregson, 1857). Later, many larvae were reported from an unspecified location six miles from Blackpool (possibly near Fleetwood) in early June 1872; twelve of these were bred (Hodgkinson, 1872). A little later that year, it was also reported on the west bank of R Wyre (near Fleetwood) by the same recorder. Ellis (1890) reports it to be confined to the Lancashire coast but adds an undated record from 'Wardless' by J.H. Threlfall. No locality with this name has been found in Lancashire, the nearest being an upland site in Cumbria. Subsequently, the only other report came from Burnley in 1918 by W.G. Clutten (Mansbridge, 1940), which gives the moth's status by then as 'another rare species with us'.
■ **HABITAT** Sand dunes and an estuarine bank.
■ **FOODPLANTS** Not documented in Lancashire despite many being bred. Elsewhere, it feeds mainly on Wild Carrot or Wild Parsnip; occasionally on Upright Hedge-parsley (Emmet and Langmaid, 2002).
Single brooded. The Burnley record is unusual in being so far inland. This species has shown a significant decline nationally.

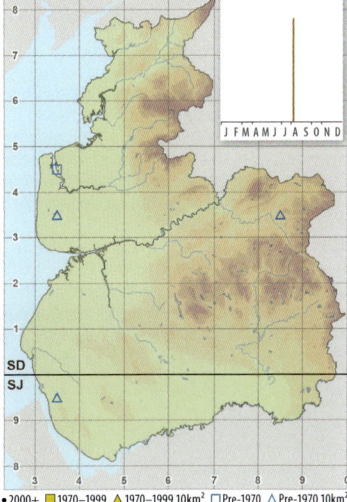

32.045 *Depressaria albipunctella*

■ **LANCS STATUS** Uncertain / considered Extinct   c.1857–c.1890

First referred to in Gregson (1857) where he writes, '*albipunctella* Hübn., amongst pimpernel….at Hightown. August.' Later, Ellis (1890) reports it to be scarce, reported only from the coast and Mansbridge (1940) as 'another rare species'. Both of these authors document Gregson's record and add one other, 'Cleveleys, J.H. Threlfall', without a date or whether the record refers to a larva or adult. These records have been widely accepted, by the aforementioned Lancashire authors and later, by Emmet & Langmaid (2002). No specimens have been located.
■ **HABITAT** Coastal areas.
■ **FOODPLANTS** Not recorded in Lancashire. It is unknown which plant Gregson's 'pimpernel' refers to; possibly a *Pimpinella* sp., but none of this genus are reported as foodplants of *P. albipunctella*. Nationally, the larvae feed on a range of umbellifers, but mainly Wild Carrot or Upright Hedge-parsley.
The species shows some evidence of a national decline (Emmet & Langmaid, loc. cit.).

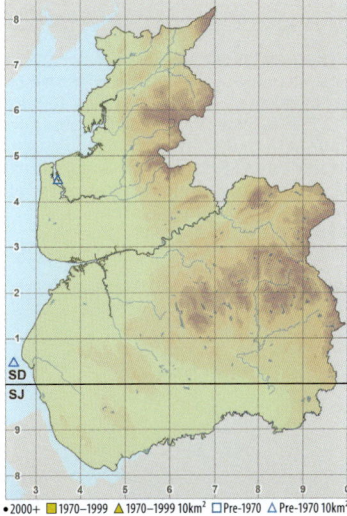

DEPRESSARIIDAE

The Moths of Lancashire

DEPRESSARIIDAE

32.047 *Depressaria chaerophylli*

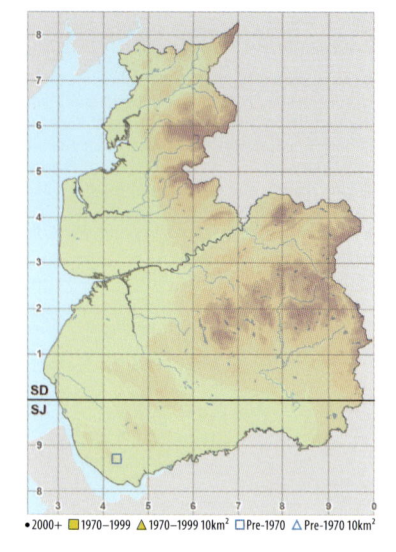

■ **LANCS STATUS** Considered Extinct   c.1857

The sole Lancashire record is an undated report from Woolton, Liverpool mentioned in Gregson (1857), who adds that it occurs 'near farm houses and dirty gardens in August'. No further information or records are added in Ellis (1890) or Mansbridge (1940). Gregson (loc. cit.) also mentions that it was found at Prenton in Cheshire, the only record known in that county (cheshire-moth-charts.co.uk – accessed 31 August 2023).

■ **HABITAT** Nationally, the habitat is noted as old hedgerows, (etc), especially sunken lanes, where the foodplant (*Chaerophyllum*) appears regularly, year after year (Emmet & Langmaid, 2002). If these were the conditions associated with the farm houses and gardens above, then they will have been long lost as Liverpool expanded its urban influence over this area.

■ **FOODPLANTS** Not recorded in Lancashire.

Nationally, the moth has not been noted elsewhere in north-west England although it does extend further north on the east side of the country.

● 2000+   ▪ 1970–1999   ▲ 1970–1999 10km²   ▫ Pre-1970   △ Pre-1970 10km²

32.050 *Telechrysis tripuncta*

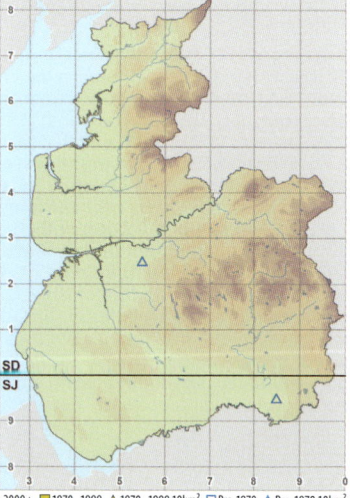

■ **LANCS STATUS** Considered Extinct   c.1859–c.1889

Documented as scarce and local in the north–west region by Mansbridge (1940), the listed records have not been possible to map with any accuracy. The species map therefore has an approximated blue 10km triangle for each of the two records. The first is listed in Stainton (1859) as from the Manchester district, but it has not been possible to establish, with any certainty, if this was within VC59 or not. The other record is noted in Ellis (1890), with no date, from the Preston area by J.B. Hodgkinson which could place the record in either VC59 or VC60.

■ **HABITAT** Unknown in the county.

■ **FOODPLANTS** Not recorded in Lancashire.

Nationally, the species is close to the northern limit of its range in Lancashire and frequents woodland and hedgerows. It is probably associated with dead wood in such locations and has a flight period mainly in June with no indication of a strong attraction to light, the main flight times noted as around dusk and dawn (Emmet & Langmaid, 2002).

● 2000+   ▪ 1970–1999   ▲ 1970–1999 10km²   ▫ Pre-1970   △ Pre-1970 10km²

COSMOPTERIGIDAE

34.001 *Pancalia leuwenhoekella*

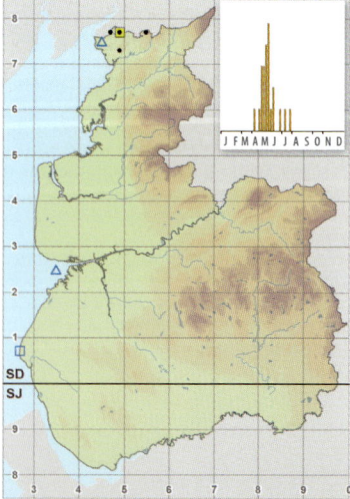

■ **LANCS STATUS** Occasional / very local resident   1866–2022

First recorded at Lytham St Annes on 22 May 1866 by J.B. Hodgkinson, and then at Silverdale by J.H. Threlfall (Ellis, 1890). Next, seen 'on the sand-hills at Formby in 1936' (Mansbridge, 1940). All subsequent records have come from the northern limestone areas, with sites including Gait Barrows, Warton Crag and Dalton Crags. Recorded on 25 occasions in total, with flight period from 26 April to 25 July. Maximum count of eight moths, noted at Gait Barrows on 19 May 2018 by J. Patton.

■ **HABITAT** Limestone grassland.

■ **FOODPLANTS** Not known in Lancashire.

Single brooded. Twice at light. Most records are of moths netted during day or dusk.

● 2000+   ▪ 1970–1999   ▲ 1970–1999 10km²   ▫ Pre-1970   △ Pre-1970 10km²

156 *The Moths of Lancashire*

34.003 *Euclemensia woodiella*

■ **LANCS STATUS** Adventive / temporary resident / Extinct   1829

In mid-June 1829, Robert Cribb, an apothecary's assistant living in Ancoats, Manchester, discovered a colony of this little orange moth flying around a hollow oak tree on Kersal Moor. He collected all that he could find and it was not seen again, there or anywhere else, for nearly 200 years. The moth was generally considered to be extinct until photographed in 2013 at the Lady Bird Johnson Wildflower Centre in Austin Texas (Ridout, 2016).

Kersal Moor was a small oasis of countryside near the centre of Manchester, which at that time was an overcrowded and heavily populated industrial city supplied with cotton and other goods from America via Liverpool and the linked rivers Mersey and Irwell.

Surprisingly the harsh working conditions promoted enquiry into the natural sciences (Cash, 1873). Societies were formed by self-taught workers and artisans who made important contributions to their chosen subjects. Entomology was studied by the Banksian Society, named after Sir Joseph Banks. Members were required to provide three perfect specimens of plants or insects each year for the Societies collections and subscription money was used to purchase books and equipment. Cribb gave two specimens of his moth to Sam Carter who was curator of the Societies' insect cabinet, and the third was given to R. Wood, his employer and fellow society member, to send for identification to John Curtis, then working on his magnificent book British Entomology.

• 2000+ ☐ 1970–1999 ▲ 1970–1999 10km² ☐ Pre-1970 △ Pre-1970 10km²

Cribb seems to have maintained a harmonious friendship with his fellow entomologists until the Society received the April 1830 part of Curtis's volume six containing an illustration and description of *Pancalia* (now *Euclemensia*) *woodiella*. Cribb, furious that the moth had been named after Wood, abandoned entomology (Sidebotham, 1884).

He refused to part with any more specimens, which he kept in a store box. One day, Sam Carter met him and offered ten shillings for the box, Cribb agreed but stated that it was deposited in a Beer House in the Oldham Road as security for an unpaid five-shilling bar bill. Carter gave the money to redeem the pledge but Cribb did not return. The next time they met Carter repeated his offer and also agreed to pay the bar bill. This time he accompanied Cribb to the Beer House. Unfortunately, the landlady had grown tired of waiting and put the box on the fire. The money on offer here is put into context by noting that a man operating two spinning mules at that time, could produce a sufficient weight of yarn to earn 20 shilling in a 58-hour week (Chadwick, 1860). Carter was really keen to obtain the box and Cribb was drinking too much beer!

So only three specimens remained. In 1862 Curtis died, and his wife sold his collections, including the *woodiella* type, to the National Museum of Victoria (now the Museums Victoria) in Melbourne, Australia. The two Carter specimens were sold with Carter's collection to the Manchester Society of Natural History and on to the Manchester Museum. Lord Walsingham exchanged one of these specimens for over two thousand British micro-lepidoptera and it is now in the Natural History Museum in London (Cook, 2018).

Walsingham's entomological assistant Durrant understood where the moths probably came from (Cook, 2018). He identified them as *Euclemensia* and noted that the larvae of the other two species in the genus lived in the hard gall-like female scale

Kersal Moor by William Wyld 1852

COSMOPTERIGIDAE

Euclemensia woodiella ...continued

insects (Kermesidae) on the twigs and branches of oak trees along the eastern coast of America.

Presumably, *E. woodiella* had been accidentally introduced and formed a colony where there was a tree infested with Kermes scale insects on Kersal Moor. Scale insect biology and Liverpool shipping records indicate that the most likely importation source was an infested oak sapling. This may have been imported for a landscaping scheme or for the Manchester Royal Botanical Gardens that was planned in 1827 and opened in 1831.

Brian Ridout

34.004 *Limnaecia phragmitella*

■ **LANCS STATUS** Frequent / widespread resident   1932–2022

Larvae noted by A.E. Wright in Reedmace heads at Silverdale in 1932. Next, recorded from Leighton Moss in 1951 by N.L. Birkett, then at Risley Moss by M. Hull in 1982. Subsequently, noted at freshwater sites such as Martin Mere, Little Woolden Moss, Rixton Claypits, Stretford Ees and Hollinwood Canal. The moth remains widespread, although maximum counts of adults at light are down from a decade or two ago.

■ **HABITAT** Lakes, ponds, rivers, streams, canals.

■ **FOODPLANTS** Reedmace (Bulrush) and Lesser Bulrush.

Single brooded. Most records are of moths recorded at light. Larvae noted feeding gregariously in the seed-heads of the foodplants.

● 2000+ □ 1970–1999 ▲ 1970–1999 10km² □ Pre-1970 △ Pre-1970 10km²

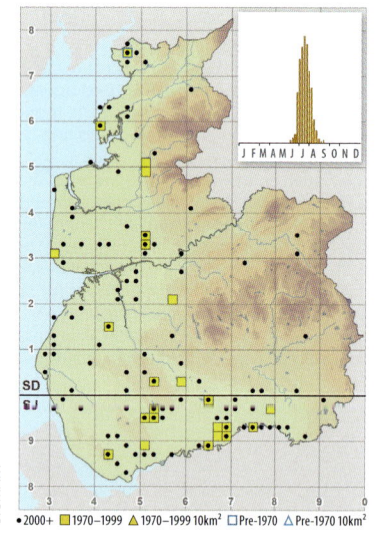

34.0111 *Anatrachyntis badia*

■ **LANCS STATUS** Adventive   2012

This moth has only been recorded on a single occasion within the county, when a moth was found indoors at Bootle by C. Fletcher on 28 October 2012. The moth was noted in a bag of grapes, thought to originate from Spain. Subsequently dissected, the moth, a male, was confirmed as this species.

■ **HABITAT** Unknown in the county.

■ **FOODPLANTS** Not recorded in Lancashire.

● 2000+ □ 1970–1999 ▲ 1970–1999 10km² □ Pre-1970 △ Pre-1970 10km²

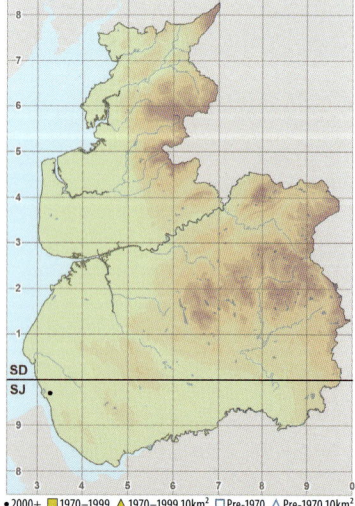

34.014 *Sorhagenia janiszewskae*

LANCS STATUS Scarce / very local resident   2000–2022

Larval workings noted on Buckthorn at Gait Barrows in 1999, although rearing unsuccessful. The first confirmed record is of two moths tapped from the foodplant and netted at Gait Barrows by R.M. Palmer and S. Palmer on 27 July 2000, a significant range extension for this species of southern England (Palmer, Palmer and Langmaid, 2001). Larval workings also found nearby at Hawes Water in 2016.
- **HABITAT** Limestone woodland edges.
- **FOODPLANTS** Buckthorn.

Single brooded. A tenanted shoot was located at Gait Barrows by S. Palmer on 31 May 2022. The adult, a female, subsequently emerged on 29 June 2022.

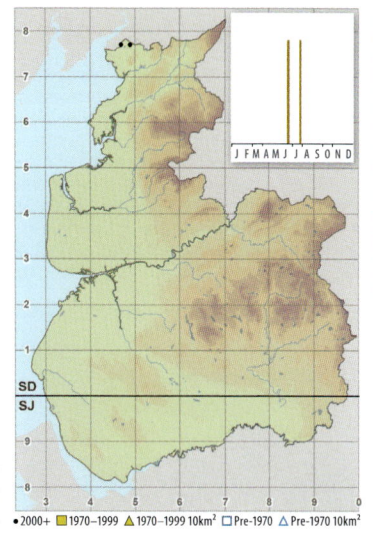

COSMOPTERIGIDAE

35.001 *Aproaerema sangiella*

LANCS STATUS Rare / very local resident   2021–2022

First discovered on Dalton Crags in small numbers during a wider survey of Hutton Roof on 8 and 19 July and 9 August 2021, by J. Patton. A few larvae were located in this area on 26 April 2022 on Common Bird's-foot-trefoil adjacent to, or growing over, bare limestone. Not located on lowland limestone areas, such as Gait Barrows, despite regular fieldwork and light trapping between 1995 and 2022.
- **HABITAT** Upland limestone, away from invasive scrub and denser vegetation.
- **FOODPLANTS** Common Bird's-foot-trefoil.

The larva spins the leaves at the tip of a stem into a ball-shaped structure. Single brooded. Attracted to actinic light.

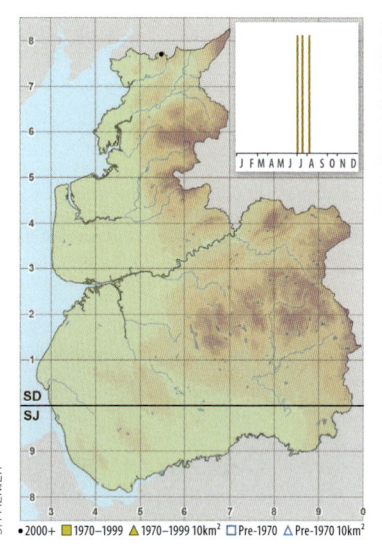

GELECHIIDAE

35.002 *Aproaerema cinctella*

LANCS STATUS Rare / very local resident   2004

The only record of this species relates to one of the form with an obsolescent white fascia attracted to a garden MV light trap on 28 June 2004 (K. McCabe). It probably originated from an adjacent vegetated fly ash tip where its national, larval foodplant is common. The lack of further records to light may be associated with open parts of the site now being shielded from the light trap by developing woodland. Searches of Common Bird's-foot-trefoil failed to produce any sign of larval feeding.
- **HABITAT** Brownfield site.
- **FOODPLANTS** Not recorded in Lancashire.

Single brooded. Attracted to light.

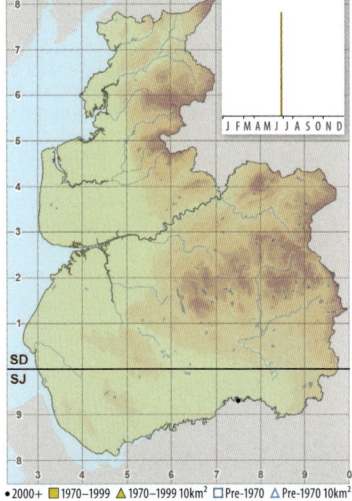

The Moths of Lancashire **159**

GELECHIIDAE

35.003 *Aproaerema larseniella*

■ **LANCS STATUS** Rare / very local resident   2020

The single county record relates to one attracted to MV light at Freshfield Dune Heath on 12 July 2020 (R. Walker), an identification confirmed by dissection. This is on the northern limit of its range on the west side of England and may possibly be part of a northward extension in its range nationally.

■ **HABITAT** Lowland heath. There are freshwater ditches nearby to the trapping site which may contain suitable damp habitat for the national, larval foodplant.
■ **FOODPLANTS** Not recorded in Lancashire.
Single brooded; attracted to light. Searches of Greater Bird's-foot-trefoil adjacent to the trap site would be worthwhile to establish if the moth is breeding.

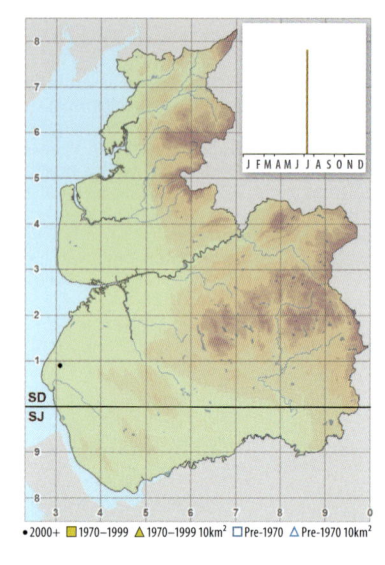

35.004 *Aproaerema taeniolella*

■ **LANCS STATUS** Occasional / very local resident   c.1859–2022

First noted in the Manchester district (Stainton, 1859). All subsequent records, from Lytham (1870) to Mere Sands Wood (2022) are of small numbers from grassland in coastal or open, lowland areas within 10km of the coast. The larva is found in May but, on one occasion, spinnings with small larvae similar to this species were found in October at a known site for the moth; the larvae failed to make it through the winter.

■ **HABITAT** Short, often sparsely vegetated grassland on sandy coasts, lowland limestone and lightly vegetated parts of brownfield sites, including quarries.
■ **FOODPLANTS** Common Bird's-foot-trefoil.
Single brooded. Readily disturbed in sunny conditions; occasionally attracted to light.

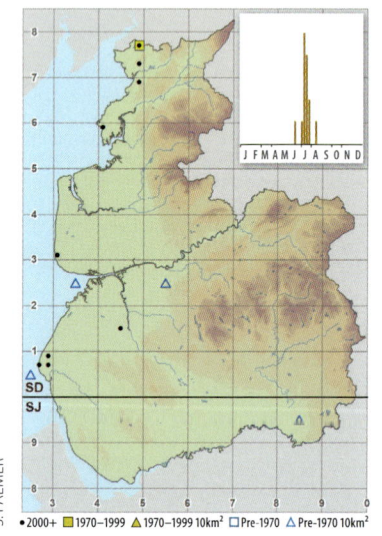

35.010 *Aproaerema anthyllidella*

■ **LANCS STATUS** Occasional / local resident   c.1859–2022

Found in the Manchester district c.1859, but recorder unknown (Stainton, 1859), with few subsequent records until the mid-1990s. Since then, recorded widely from a range of scattered lowland sites in small numbers, mostly in garden moth traps in southern lowland Lancashire. Locally common on sand dunes as a larva.

■ **HABITAT** Short grassland on sandy coasts, lowland limestone and brownfield sites.
■ **FOODPLANTS** Kidney Vetch. Has also been found well away from known Kidney Vetch sites, suggesting it also utilises other Fabaceae in the county.
Double brooded, with a stronger second brood. Readily disturbed in sunny conditions and attracted to light.

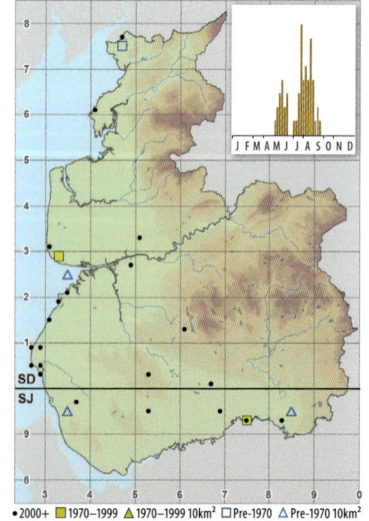

160 *The Moths of Lancashire*

35.011 *Anacampsis populella*

- **LANCS STATUS** Occasional / local resident c.1857–2022

First recorded on the Crosby Sandhills (Gregson, 1857) and then, from Formby in 1904 and Freshfield in 1946. Noted exclusively from the Sefton Coast until 1995 and the larger dunes remain its stronghold (Leech & Michaelis, 1957 and Smart, 2022). Away from the coast in VC59, it was discovered on Risley Moss (L.W. Hardwick) in 1995 and since then, in Flixton from 1999 to the present day (K. McCabe), Chorlton, Dalton, Halewood, Little Woolden Moss, Parr, Prescot Reservoir and Rixton. First noted in VC60 from the St Annes area in 1996 (J. Steeden), where it remains well established on the dunes. Otherwise only single reports are known, from Heysham Moss in 2018 and Gait Barrows in 2019 (both S. Garland). Possibly under-recorded. Due to potential confusion with *A. blattariella*, only bred or dissected specimens are included in this text. The dark form, f. *fuscatella*, is known to occur in dune areas.

- **HABITAT** Well vegetated sand dunes, wetland sites, gardens and brownfield sites.
- **FOODPLANTS** Creeping Willow, Aspen, White Poplar, Goat Willow, Grey Willow and once on a narrow-leaved willow species.

Single brooded. Attracted to light and can be found resting on trunks of larger poplar and willow trees. The larvae have been found between 2 May and 25 June with most from mid-May to early June. On Creeping Willow, the larvae spin leaves up against the stem, whereas on larger leaved foodplants they roll the leaf.

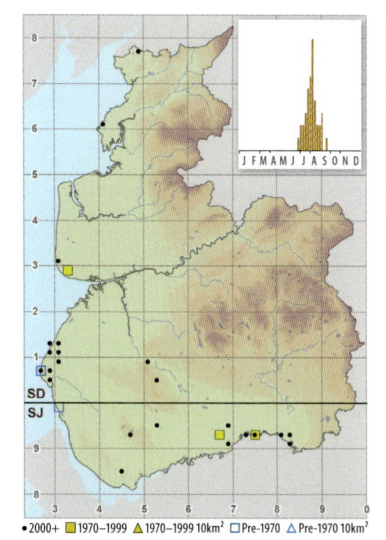

• 2000+ ▫ 1970–1999 ▲ 1970–1999 10km² ☐ Pre-1970 △ Pre-1970 10km²

35.012 *Anacampsis blattariella*

- **LANCS STATUS** Occasional / very local resident 2001–2022

This species was not separated from *A. populella* until 1947 and it is considered likely that reports of that moth from mosses in the 19th century will have included *A. blattariella*. No early specimens of *A. populella* from mossland sites have been located to enable further investigation. The first confirmed record was of two which came to light at Silverdale Moss on 24 July 2001 (S. Palmer et al.), a site with many mature birch trees. Since then, it has been attracted to light at Sidings Lane, St Helens on 28 July 2001 (R. Banks), Astley Moss in 2004, Holcroft Moss in 2005, Gait Barrows and Mere Sands Wood in 2007, Swinton in 2010 and Warton Crag in 2013, with a few other subsequent sightings elsewhere, mostly of singletons. Sites where larvae have been found include Holcroft Moss, Chorlton, Formby, Astley Moss and Ainsdale. Although most locations are in VC59, with a few outliers in the Silverdale area, it seems likely that this species is under-recorded in areas with mature birch.

- **HABITAT** Lowland raised bogs and other lowland sites where mature birch is common.
- **FOODPLANTS** Birch sp.

Single brooded. Attracted to light and can be found resting on mature birch trunks. The larva can be found from late April to early June, spinning a leaf into a roll.

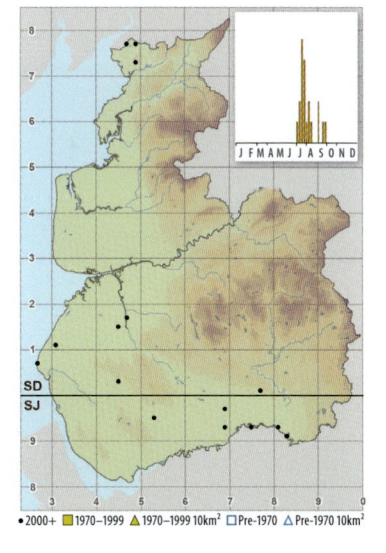

• 2000+ ▫ 1970–1999 ▲ 1970–1999 10km² ☐ Pre-1970 △ Pre-1970 10km²

GELECHIIDAE

35.013 *Anacampsis temerella*

■ **LANCS STATUS** Scarce / very local resident c.1857–2022

This is a specialist of fixed dunes and has Nationally Notable (Scarce) A status. Its survival on the Lancashire coast has been of great concern in recent years.

The earliest Lancashire records were noted during the 1850s from the Crosby sand-hills (Gregson, 1857). Later, found at Lytham in July 1865 by J.B. Hodgkinson, with the same recorder finding larvae on the only known British foodplant, Creeping Willow, the following May. Recorded by W. Mansbridge at Formby in 1922 (bred specimens are in World Museum). He also described the moth as 'common on the Lancashire sandhills,' also at Ainsdale, Lancaster and Burnley (Mansbridge, 1940). The record from Burnley by A.E. Wright, approximately 40 miles from any dunes, seems extremely unlikely, with the Lancaster record also considered unconfirmed.

Records from 1950 describe the larvae as common at Ainsdale, Crosby and Formby, all on the Sefton Coast. Also found at Freshfield in 1956 and 1963, since when records have been few. Larvae were reported from Ainsdale in 1976, and a few moths in 1984 at Formby. Until 2021, this was the last sighting from the Sefton Coast, with a single moth netted at Lytham by D. & J. Steeden in 2013 the last noted in the county.

In 2021, The Tanyptera Trust sponsored a survey to assess whether the moth remained in the Sefton Coast dune systems, by searching for the larval stage in spinnings of Creeping Willow during late spring (Smart, 2022). Due to the similarity of the larva and its spinnings to that of *Anacampsis populella*, larvae had to be collected and reared to imago before identity could be confirmed.

In total, 20 moths of *A. temerella* emerged between 15 June 2021 and 9 July 2021 from spinnings collected from Ainsdale Dunes NNR and Ainsdale and Birkdale Sandhills LNR between 28 May 2021 and 22 June 2021. This confirmed that *A. temerella* remains present on the Sefton Coast, seemingly in reasonable numbers.

The larvae were found in the fixed dunes, in extensive stands of Creeping Willow amongst the dune slacks. Vegetation was short, aided by the presence of rabbits and also by the sheep and cattle introduced over winter.

Following the successful emergences, a few consistent differences between larvae of *A. temerella* and *A. populella* were described, thus aiding future recording (Smart, loc. cit.). Unfortunately, larval searches in 2021 and 2022 at Starr Hills Dunes, Lytham, at Hightown and at Cabin Hill, Formby were unsuccessful, but should ideally be repeated.

• 2000+ ☐ 1970–1999 △ 1970–1999 10km² ☐ Pre-1970 △ Pre-1970 10km²

162 *The Moths of Lancashire*

35.017 *Neofaculta ericetella*

■ **LANCS STATUS** Abundant / widespread resident   1842–2022

Found abundantly on Chat Moss (Edleston, 1846) and has subsequently been noted, usually at least 'commonly', wherever Heather grows in profusion. Well over 1,000 adults were noted on Astley Moss on 22 May 2004 (K. McCabe). Occasionally attracted to light traps in suburban gardens, away from heathery areas.

■ **HABITAT** Lowland raised bogs, heather moors and, more generally, wherever heather grows commonly in the wild.

■ **FOODPLANTS** Heather.

Single brooded over an extended period. Found once in September and on 10 October 2009. Larvae noted from November to April and a pupa in late March.

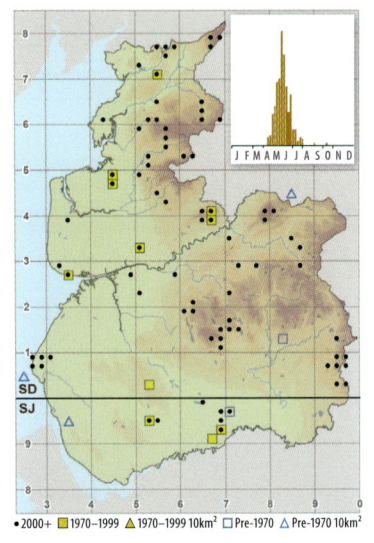

35.018 *Hypatima rhomboidella*

■ **LANCS STATUS** Occasional / widespread resident   c.1859–2022

First noted in the Chat Moss area c.1859 by J. Chappell (Ellis, 1890), followed by a small number of records until the 1950s, although Mansbridge (1940) noted it as 'common and widespread amongst birch'. In more recent years, recorded widely, but most frequently in lowland parts. Overall, it is considered that the status has changed little since the mid-19th century. Mostly found as singletons, but six came to MV light in Brinscall (a mature birch woodland site) on 19 September 2012 (G. Jones).

■ **HABITAT** Woods, mosses, hedgerows, gardens and scrubland.

■ **FOODPLANTS** Birch and Hazel.

Single brooded. The few larval records were noted between late May and early July.

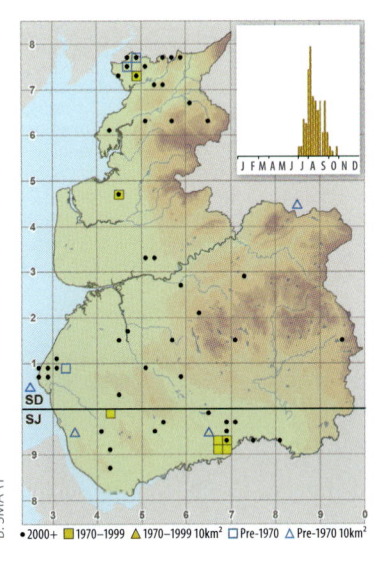

35.020 *Anarsia spartiella*

■ **LANCS STATUS** Occasional / local resident   1877–2022

Found as larvae spinning the shoots of Dyer's Greenweed on 8 June 1877 by J.H. Threlfall on the 'cliffs' at Morecambe (Threlfall, 1878). Since then, records have been few and scattered across mainly lowland and coastal areas, including Billinge, Formby, Freshfield and Warton Crag. Recently, also found on the edge of moorland in Dolphinholme, at Docker Moor and, in the east of the county, in Briercliffe.

■ **HABITAT** Heathland, dunes and adjacent to or on brownfield sites. Dyer's Greenweed has disappeared from most, if not all, of its former sites.

■ **FOODPLANTS** Broom, Gorse and Dyer's Greenweed.

Single brooded. Attracted to light. Larvae have been found in June.

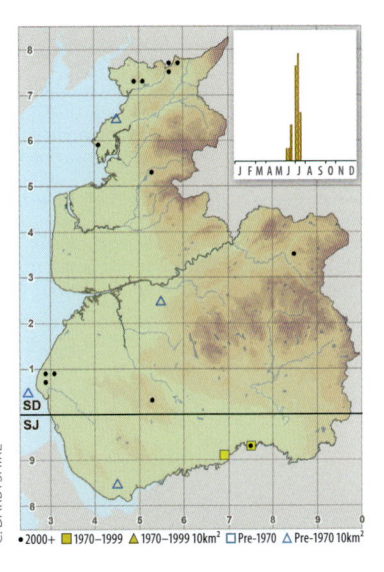

The Moths of Lancashire   **163**

GELECHIIDAE

35.022 Juniper Webber *Dichomeris marginella*

■ **LANCS STATUS** Rare / very local resident 1955–2022

Although noted at Arnside (Cumbria) from the late 19th century, it was not found in Lancashire until 30 July 1955 by R. Fairclough, in the Silverdale area. This locality, particularly Gait Barrows, has remained its core area in the county, but always in low numbers. From 1995–2005 it was noted in St Annes (19 records), suggestive of breeding on garden juniper cultivars. The first larva found in Lancashire was at Gait Barrows on 31 May 2022 (E.M. Riley), in a dense spinning amongst needles.

■ **HABITAT** Limestone pavement and, for a few years, in near-coastal urban gardens.
■ **FOODPLANTS** Juniper.
Single brooded. Attracted to light.

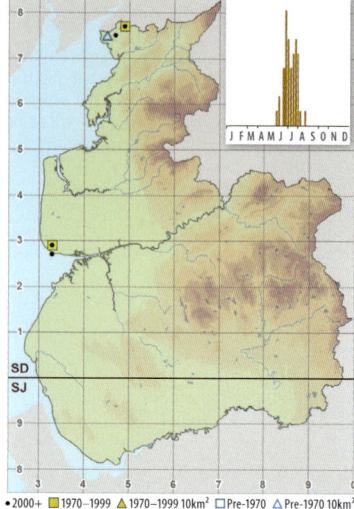

35.026 *Acompsia cinerella*

■ **LANCS STATUS** Scarce / very local resident c.1859–2021

First noted in Stretford c.1859 by J. Chappell (Hardwick, 1990), followed by a sighting at Lytham in 1865. Possibly overlooked in the past, as a 1924 specimen from the Formby area was found, misidentified, within specimens of *Bryotropha politella*. Most subsequent records are from the Silverdale area apart from a singleton at Heysham in 2009 and on the upland limestone of Dalton Crags in 2021.

■ **HABITAT** Grassland with scrub or near woodland, mainly on limestone or sandy sites.
■ **FOODPLANTS** Not recorded in Lancashire.
Single brooded. Records mostly relate to singletons attracted to light or, much less frequently, disturbed during the day.

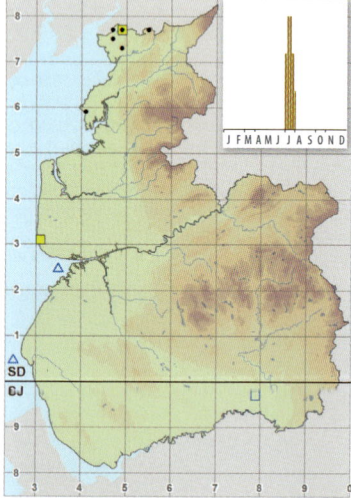

35.028 *Brachmia blandella*

■ **LANCS STATUS** Occasional / local resident 1997–2022

A recent arrival in Lancashire from the south. The first was in Flixton on 8 July 1997 (K. McCabe), this site producing over half the subsequent 290 county records. Elsewhere the spread has been slow, patchy and erratic and restricted to lowland parts of VC59. It was not until 2016 that it reached a site near Longton (J. Girdley), and this remains its most northerly location in the county.

■ **HABITAT** Brownfield sites, gardens, mosses, scrubby grassland and vegetated dunes.
■ **FOODPLANTS** Not recorded in Lancashire.
Single brooded, but with very late records on 29 September 2014 (B. Smart) and 26 September 2019 in Parr (R. Banks). Comes readily to light in low single figures.

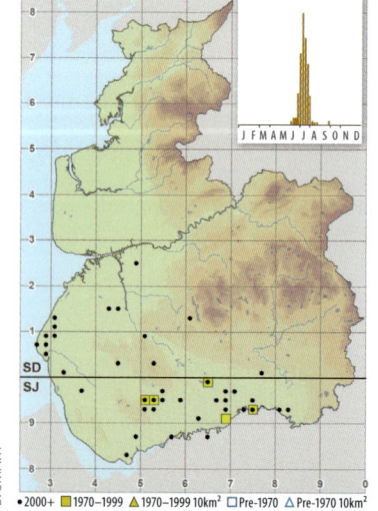

164   The Moths of Lancashire

35.031 Helcystogramma rufescens

■ **LANCS STATUS** Occasional / widespread resident   c.1857–2022

First recorded on 'roadsides and clay banks near the coast' in the Liverpool area (Gregson, 1857), with subsequent reports from many well scattered lowland sites to the present day. A notable increase in records took place between 1998 and 2012, mainly in the Flixton area but, to a lesser extent, in other lowland parts of VC59. Since then, it has maintained its distribution, but numbers of records have decreased.
■ **HABITAT** Gardens, grassland and waste ground, including brownfield sites.
■ **FOODPLANTS** Grasses, including a *Festuca* sp.
Single brooded, with a second brood on two occasions. Comes readily to light. Larvae have been found from late March to mid-June and, once, in mid-August.

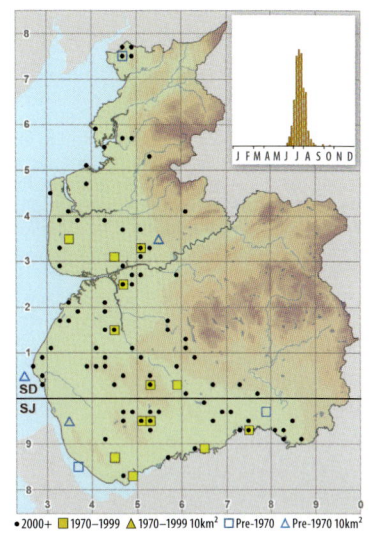

35.032 Hollyhock Seed Moth *Pexicopia malvella*

■ **LANCS STATUS** Considered Extinct   1844–c.1859

First noted at Chat Moss on 2 July 1844 (Edleston, 1844) and later, from the Manchester area (Stainton, 1859), where it was reported as common. The Manchester record was repeated in Ellis (1890) and Mansbridge (1940), but with no additional records and it was designated as scarce by the latter author. No reference is made in either of these publications to the report by Edleston; was it over-looked or considered unconfirmed? The location, Chat Moss, is a very unlikely habitat for this species. However, Eddleston published his records regularly in the journals of the time, which were edited by well-known and respected entomologists. They would comment if records were considered particularly unlikely, which did not occur in this case. The fact that Stainton (loc. cit.) also gave it as common in the area suggests the record was accepted more widely and perhaps overlooked in Lancashire.
■ **HABITAT** Unknown.
■ **FOODPLANTS** Not documented in Lancashire. Stainton (loc. cit.) made reference to use of gardens nationally by this species and as the main foodplant, Marsh-mallow (*Althea officinalis*) is not known in north-west England, Hollyhock would seem to be the more likely contender as its regional pabulum.

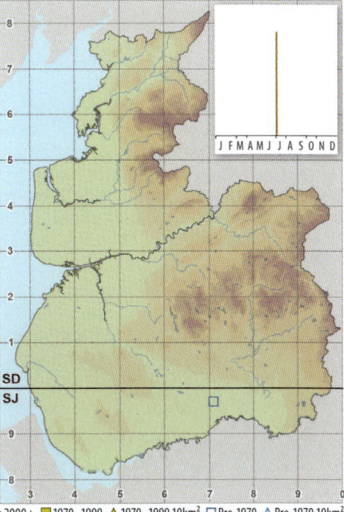

35.034 Angoumois Grain Moth *Sitotroga cerealella*

■ **LANCS STATUS** Adventive   c.1890–1958

Until the late 1950s, commonly imported with grain shipments in all stages of its life cycle. Noted as particularly abundant in the mid- to late 19th century by C.S. Gregson (Ellis, 1890) and in 1917 (Mansbridge, 1940), in grain warehouses at Liverpool Docks; also found outdoors under window-sills of nearby buildings. Since the 1950s, modern import methods and food hygiene procedures appear to have significantly reduced its occurrence, but no recent data on this appears to be readily accessible.
■ **HABITAT** Inside buildings used for grain storage.
■ **FOODPLANTS** On imported Indian corn and wheat.
Continuously brooded in suitable indoor conditions.
This species has, on rare occasions, been noted coming to garden light traps in southern counties of England. More usually though it continues to be found in imported, packaged grain products which can originate from a range of continents.

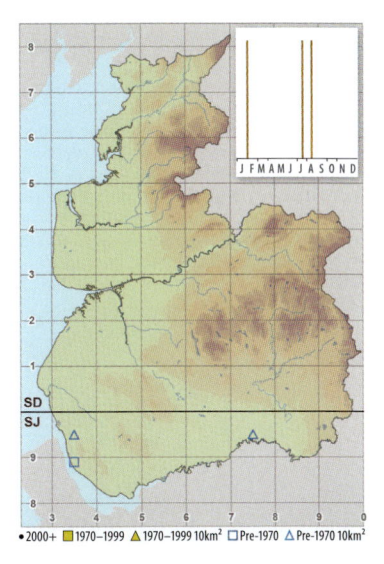

GELECHIIDAE

35.035 *Chrysoesthia drurella*

■ **LANCS STATUS** Occasional / local resident   c.1889–2022

Formerly considered local, but with only two documented pre-2000 records, from Lytham by J.B. Hodgkinson (Ellis, 1890) and Crosby in 1915 (Mansbridge, 1940). Recent records have been restricted to lowland parts of south Lancashire and Greater Manchester. All relate to the second brood larval mines, mostly tenanted, found on plants on the edges of arable farmland.

■ **HABITAT** Disturbed ground on peaty soil.

■ **FOODPLANTS** Orache sp. and *Chenopodium* spp., including Fat Hen.

The larvae can be quite numerous at sites where the foodplants are common on field edges. No records are documented of the moth coming to light.

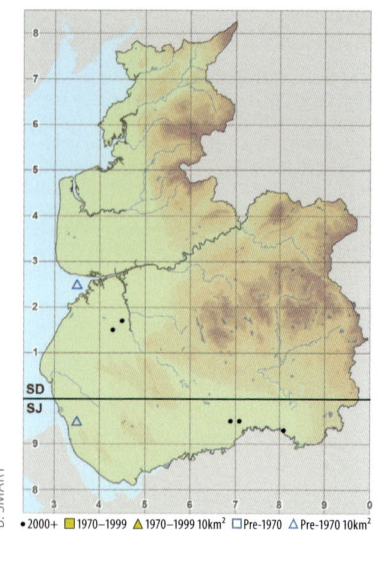

35.036 *Chrysoesthia sexguttella*

■ **LANCS STATUS** Frequent / local resident   c.1857–2022

Variously described as 'not abundant' in the Liverpool district (Gregson, 1857), at Stanley, common in the Manchester area (Stainton, 1859), but scarce by Mansbridge (1940), with no post-1890 records. It was not noted again until Didsbury in 1947. More recently, noted sparingly in lowland areas, but mainly as a larva on the edges of saltmarshes where foodplants proliferate and disturbed ground in the Mersey valley.

■ **HABITAT** Saltmarsh edges, disturbed ground on edges of arable fields and mosses.

■ **FOODPLANTS** Fat-hen, Orache, Grass-leaved Orache and goosefoot spp.

Double brooded. Occupied larval mines have been found in early May, mid-July to mid-August and late August through September. Rarely comes to light.

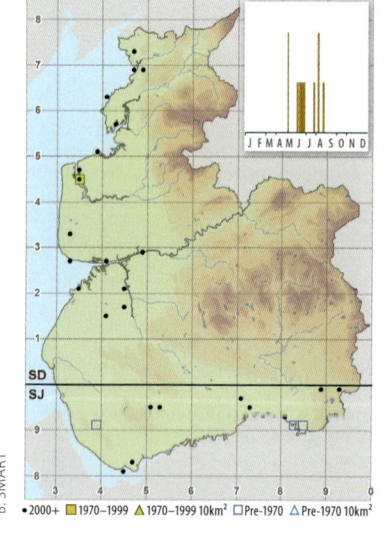

35.038 *Bryotropha domestica*

■ **LANCS STATUS** Frequent / widespread resident   c.1857–2022

First recorded in Liverpool c.1857 (Gregson, 1857) and noted as local but fairly common throughout the region in the early 20th century (Mansbridge, 1940). In the mid-1990s there was a notable increase in records, which may be partly down to an increased use of light traps and an expanding interest in micro-moths. However, the moth had also started expanding its range nationally around this time, suggesting the increase was occurring naturally across the region.

■ **HABITAT** Urban and suburban areas and gardens.

■ **FOODPLANTS** Not recorded in Lancashire.

Single brooded over an extended period. Readily attracted to light in small numbers.

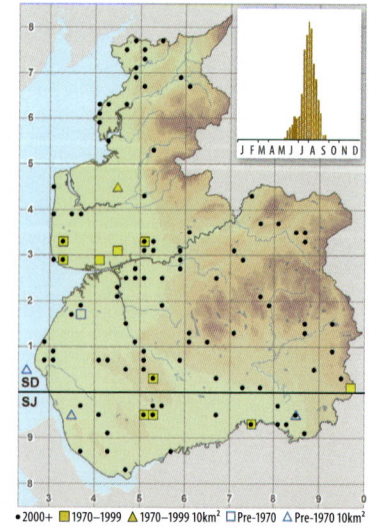

166   *The Moths of Lancashire*

35.039 *Bryotropha politella*

■ **LANCS STATUS** Occasional / local resident 2001–2022

The first confirmed record was of two at Leck Fell on 23 June 2001 (S. Palmer), with up to 80 noted here in dusk flight between 2005 and 2015. Elsewhere, 30 seen in evening flight on Whit Moor, Claughton, in 2009. Single figure counts at several, scattered, upland sites by day or at light suggest it is overlooked in the uplands. A historic record from Formby dunes proved erroneous on examining the specimen.
■ **HABITAT** Upland, flower rich, limestone grassland and moorland areas.
■ **FOODPLANTS** *Rhytidiadelphus squarrosus*.

Single brooded. Flies at dusk in suitable habitat and occasionally comes to light. A larva was found at Gannow Fell on 7 April 2015 (R. Heckford & S. Beavan) and bred.

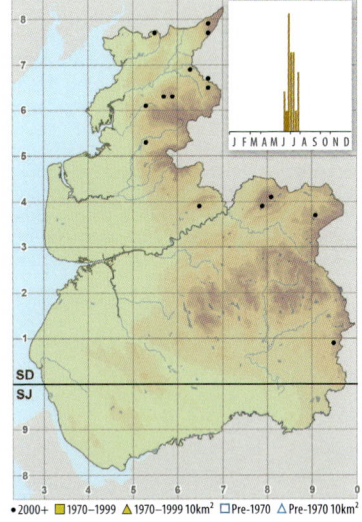

35.040 *Bryotropha terrella*

■ **LANCS STATUS** Frequent / widespread resident c.1857–2022

All former Lancashire authors note it as widespread and abundant and the first such was from the Liverpool district (Gregson, 1857). In recent decades, it has been noted widely across the county, but overall numbers reported have declined. Daytime counts in some prime habitats are still high, e.g., 70 at Heysham N.R. on 15 July 2014.
■ **HABITAT** Short, mossy grassland across the county, particularly in coastal dunes, limestone areas and brownfield sites. Also found on road verges and in gardens.
■ **FOODPLANTS** Moss, believed to be *Rhytidiadelphus squarrosus*.

Single brooded, over an extended period, with one late record from Great Sankey on 25 September 2021. Nectars on Common Ragwort; less strongly attracted to light.

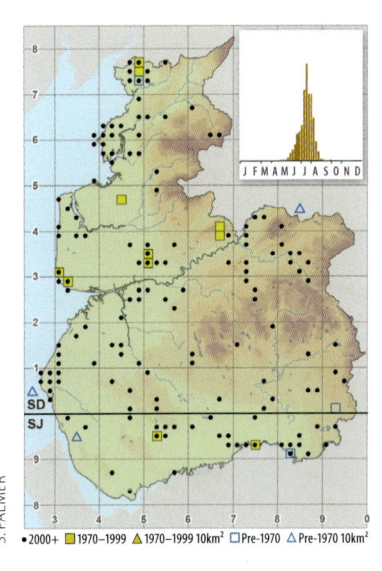

35.041 *Bryotropha desertella*

■ **LANCS STATUS** Scarce / very local resident c.1857–2021

Formerly reported on the sand hills of the Liverpool area (Gregson, 1857) and was noted as common in the Formby dunes during the 1920s (W. Mansbridge). Since then, seen once at Crosby (1956), occasionally in the dunes at Formby and Ainsdale (2011–2021) and in small numbers at St Annes during daytime searches. The species has shown no propensity to wander away from the dune systems.
■ **HABITAT** Short, broken grassland in the larger coastal dune systems.
■ **FOODPLANTS** Not recorded in Lancashire.

Single brooded with an extended flight period. Appears poorly attracted to light, which might explain its apparent scarcity in the Formby area in recent decades.

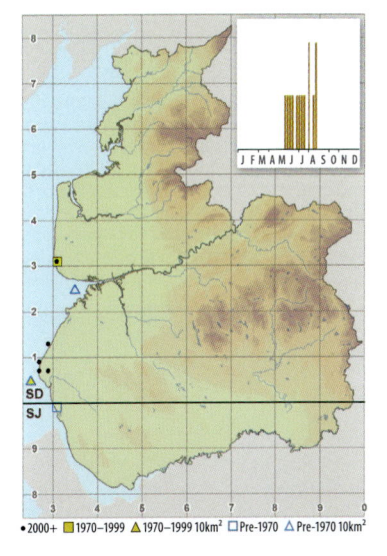

GELECHIIDAE

35.042 *Bryotropha boreella*

■ **LANCS STATUS** Scarce / very local resident   1996–2015

First recorded, when netted during the day, at Jeffrey Hill, Longridge on 11 August 1996 (S. Palmer). Its subsequent regular presence there prompted a successful search for the larva on 8 April 2015 (Heckford, Beavan & Palmer, 2015).

■ **HABITAT** Moorland, with damp to wet mossy areas or herb-rich upland slopes.

■ **FOODPLANTS** Mosses, specifically *Hypnum jutlandicum*, *Rhytidiadelphus squarrosus* and *Aulacomnium palustre*.

Single brooded. Readily disturbed in small numbers on sunny, warm days and has once been observed flying in good numbers at dusk on a northwest-facing, grassy, mossy, herb-rich, upland slope at Leck Fell.

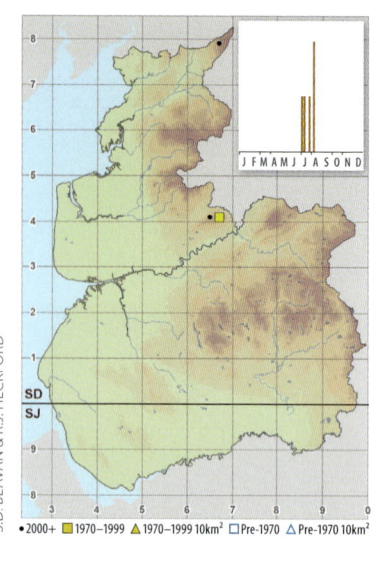

35.045 *Bryotropha basaltinella*

■ **LANCS STATUS** Rare / very local resident   1941

The specimen supporting this record was netted on Warton Crag in late June 1941 by L.T. Ford and is in the Natural History Museum. At the time, this was well north of all other records in the British Isles. It was not until 1958 that it was located anywhere else in Britain north of a line between the R Severn and the Wash; this at Spurn, on the coast of south-east Yorkshire. Despite considerable daytime fieldwork and light trapping on and near to Warton Crag in recent decades, it has not been re-found in that area or elsewhere. Possibly extinct in the county, but see note below.

■ **HABITAT** Limestone grassland near the coast.

■ **FOODPLANTS** Not recorded in Lancashire.

Single brooded with an extended flight period nationally, it comes readily to light. In recent decades it has expanded its range northwards across the Midlands of England and was found in Cheshire in 1997. It has also been noted in Tynemouth in 2008 and for the first time, in Scotland, in 2020. Any small, stocky and dark *B. domestica* look-alike should be carefully examined as it may well prove to be this species.

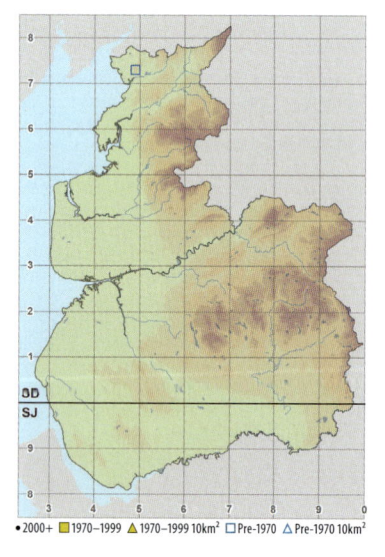

35.046 *Bryotropha senectella*

■ **LANCS STATUS** Occasional / local resident   c.1857–2022

Formerly common on all sand hills (Gregson, 1857 & Mansbridge, 1940). It is still mainly a coastal species, but increased light-trapping from the 1990s onwards demonstrated that it also inhabits parts of the Mersey Valley and surrounding areas, such as Billinge and Flixton. Rarely found in upland areas, exceptions being at Rishton, near Blackburn and on the northern upland limestone of Dalton Crags.

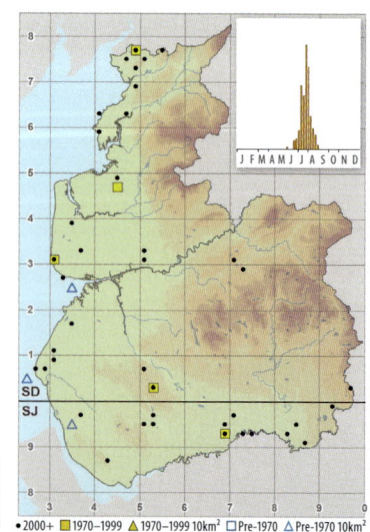

■ **HABITAT** Coastal dunes, herb-rich grassland, brownfield sites and limestone grassland.

■ **FOODPLANTS** Found once as a pupa in rooftop moss.

Single brooded with one early record in late May. Comes to light in small numbers. Found at dusk in dunes, sometimes in double figures, nectaring on Common Ragwort.

168 *The Moths of Lancashire*

35.047 *Bryotropha affinis*

■ **LANCS STATUS** Frequent / widespread resident   c.1857–2022

Early records from the Liverpool district (Gregson, 1857) and Manchester c.1859, report it to be of regular occurrence. However, by the late 19th century, a decline had set in, with it being described as local by 1890 and rare by 1940. There are no known records in the county between 1952 and 1983. Reappeared in 1984 at Lytham Hall (M. Evans), moving into lowland south, and coastal north Lancashire by 1999. Since then, found at light across much of the county, in both lowland and upland areas.

■ **HABITAT** A wide range of urban and rural habitats, including gardens.
■ **FOODPLANTS** Not recorded in Lancashire.

Probably single-brooded, with an extended flight period. Comes readily to light.

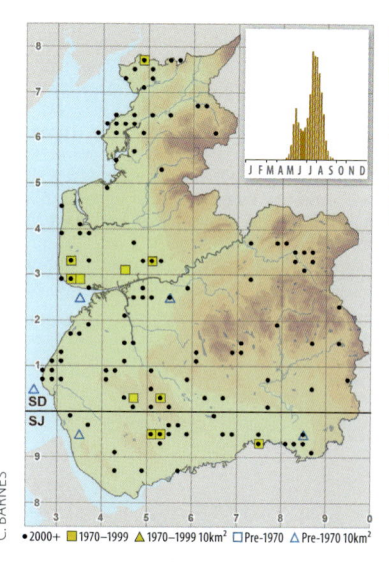

35.048 *Bryotropha umbrosella*

■ **LANCS STATUS** Scarce / very local resident   1866–2022

First noted in the dunes at Lytham on 22 May 1866 (Hodgkinson, 1866) and since, reported sparingly from most of the larger dune systems, north to St Annes. Last known at Crosby in 1914, where habitat loss caused its demise. Since 2000, found at Birkdale, Fairhaven, Formby and St Annes, always as singletons. Appears to be declining at St Annes with development and footfall degrading the site.

■ **HABITAT** Larger coastal sand dunes.
■ **FOODPLANTS** Not recorded in Lancashire.

Single brooded. Can be disturbed during the day and will occasionally come to light. The nominate form occurs most frequently, the paler f. *mundella* much less so.

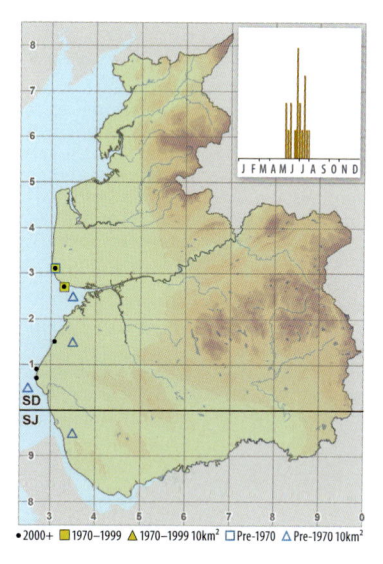

35.049 *Bryotropha similis*

■ **LANCS STATUS** Occasional / local resident   c.1859–2022

Stainton (1859) mentions its presence in the Manchester area, but with no other details. In 1961 H.N. Michaelis recorded one in the Bury area, but it was not until after 1994 that it was reported more widely across the county. The following quarter century or so produced 36 widely scattered records from coast to upland, but with a bias toward moorland edge sites. The main concentration was from Bay Horse, near Dolphinholme, with 16 noted between 2005 and 2018 (N.A.J. Rogers).

■ **HABITAT** Moorland edge; occasionally gardens, limestone sites, dunes and mosses.
■ **FOODPLANTS** Not recorded in Lancashire.

Single brooded. Comes weakly to light and once seen nectaring on Common Ragwort.

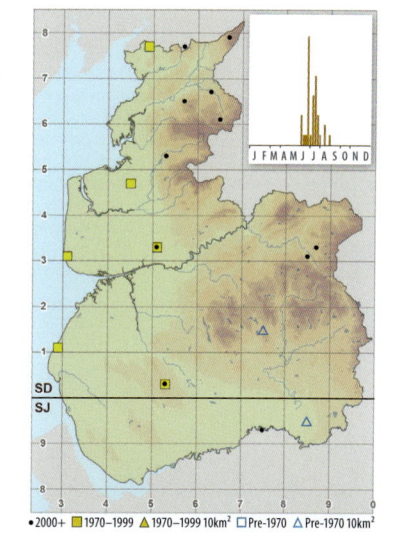

GELECHIIDAE

35.050 *Aristotelia ericinella*

■ **LANCS STATUS** Occasional / local resident c.1857–2019

Noted in the Liverpool district (Gregson, 1857), at Chat Moss c.1859 by J. Chappell and Farington Moss (c.1889) by J.H. Threlfall. Lowland mosses in these areas have since suffered considerable peat extraction, drainage for agriculture or have been developed. It was not recorded again until 1996, on Winmarleigh Moss, and subsequently on Bold Moss, Cockerham Moss and Freshfield Dune Heath.

■ **HABITAT** Lowland raised bog and dune heathland.
■ **FOODPLANTS** Heather.

Single brooded, with an extended flight period and a single September record. Comes to light and readily disturbed on sunny days. Larvae have been found in June.

● 2000+ ☐ 1970–1999 ▲ 1970–1999 10km² ☐ Pre-1970 △ Pre-1970 10km²

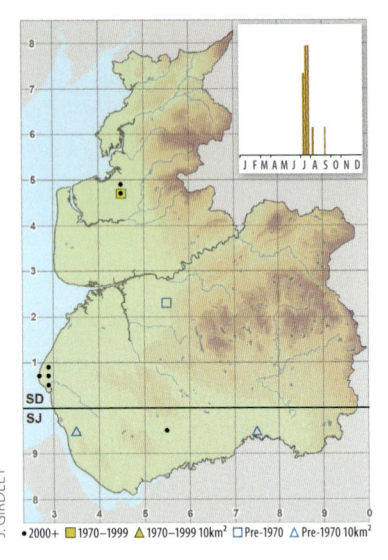

35.052 *Aristotelia brizella*

■ **LANCS STATUS** Considered Extinct 1940

The only record of this coastal species came in 1940, when L.T. Ford spent some time in north-west England, including in parts of Lancashire. On 5 June 1940, on a warm evening at 8pm, he visited the saltmarsh at Bolton-le-Sands. Here, amongst other saltmarsh micros, which he commented were on the wing in countless numbers, he found *A. brizella* (Ford, 1941). Although the habitat still looks suitable for this moth, searches of this and other areas of saltmarsh, have failed to find any signs of this species. As well as looking for adults, the visits involved examination of Thrift and Sea Lavender plants for larval activity (the national larval foodplants for this species).

■ **HABITAT** Saltmarsh.
■ **FOODPLANTS** Not recorded in Lancashire.

Single brooded.

Even in 1940, this was a very isolated colony nationally, and if it has been lost there seems to be little chance of it naturally repopulating from elsewhere. There are no obvious signs that it is expanding its range nationally.

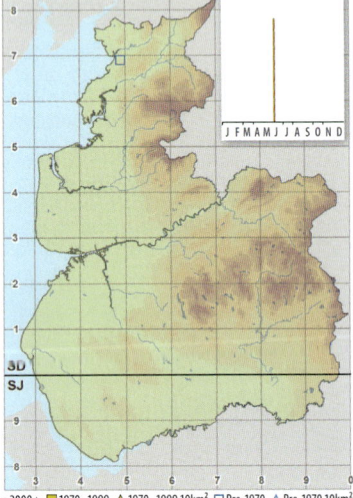

● 2000+ ☐ 1970–1999 ▲ 1970–1999 10km² ☐ Pre-1970 △ Pre-1970 10km²

35.053 *Isophrictis striatella*

■ **LANCS STATUS** Rare / very local resident 2006–2022

Only three sightings are known, all recent. The first in Flixton on 25 July 2006 (on a Tansy flowerhead) and the next, at light, was close to the first site, on 31 July 2012 (both K. McCabe). The third came to light in Parr, St Helens on 18 July 2022 (R. Banks). Specimens from 2006 and 2022 were dissected to exclude *I. anthemidella*.

■ **HABITAT** Gardens, rough ground and brownfield sites.
■ **FOODPLANTS** Not recorded in Lancashire.

Single brooded. Tansy plants were examined for larvae in St Helens around 1960 (Uffen, 1960) and in Flixton c.2007, but no larval feeding signs were found. This suggests the species is a new arrival, rather than an overlooked long-term resident.

● 2000+ ☐ 1970–1999 ▲ 1970–1999 10km² ☐ Pre-1970 △ Pre-1970 10km²

The Moths of Lancashire

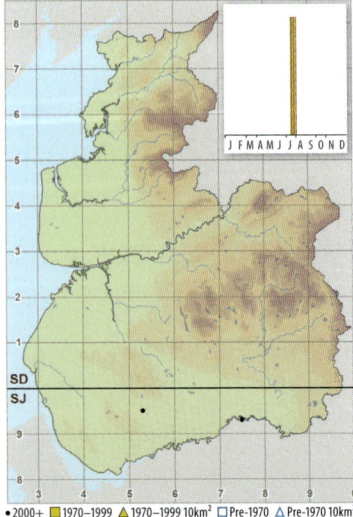

35.056 *Metzneria lappella*

■ **LANCS STATUS** Occasional / widespread resident　　　　　1955–2022
First noted in the Silverdale area in 1955 and 1956 by H.N. Michaelis, where it was bred. It was 1999 before the next record, in Flixton to light. Between 2002 and 2013 a large number of records were provided by C.A. Darbyshire who collected burdock seedheads from across central, lowland VC59. Infrequently found in upland areas and VC60, which may in part be down to a lack of fieldwork, searching for larvae.
■ **HABITAT** Disturbed ground, sand dunes, brownfield sites and edges of arable land.
■ **FOODPLANTS** Lesser Burdock. The recorded use of Greater Burdock as a larval foodplant in Lancashire is probably based on a misidentification of the plant.
Single brooded. Larvae, and later pupae, occupy seed-heads from September to April.

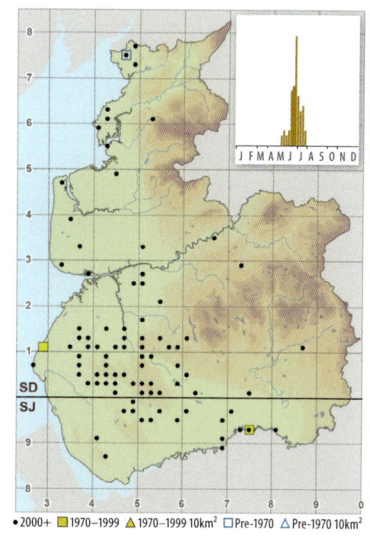

35.058 *Metzneria metzneriella*

■ **LANCS STATUS** Occasional / widespread resident　　　　　c.1859–2022
First noted in the Manchester area (Stainton, 1859), but later considered scarce and local in the county for many years (Mansbridge, 1940). There was a single record from Holden Clough in 1955, but it was not until after 1994 that reports increased in frequency. A combination of checking seedheads for larvae and light trapping soon established it occurred widely in lowland parts and was sometimes locally common. It remains rather infrequently reported from higher areas, but may be overlooked.
■ **HABITAT** Meadows, roadside verges, brownfield sites and dune grassland.
■ **FOODPLANTS** Common Knapweed.
Single brooded. The larvae or pupae are found in seedheads from September to May.

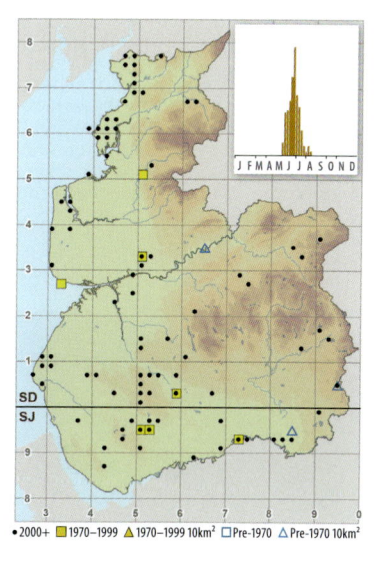

35.060 *Apodia martinii*

■ **LANCS STATUS** Occasional / local resident　　　　　1996–2022
A recent arrival in the county when one came to light in Preston on 13 August 1996 (S. Palmer); an accidental introduction via purchased pond plants cannot be ruled out. It was next noted, commonly, at Rixton Clay Pits on 19 August 1998 and has since been found at several lowland sites where fleabane occurs in quantity. Recent records in the Silverdale area suggest it may be using Ploughman's Spikenard in that area. There is one upland record, from Robert Hall Moor on 19 July 2013 (S. Garland).
■ **HABITAT** Damp meadows, coastal marshes and wet brownfield sites.
■ **FOODPLANTS** Common Fleabane.
Single brooded. Larvae & pupae from Oct. to March. Formerly named *A. bifractella*.

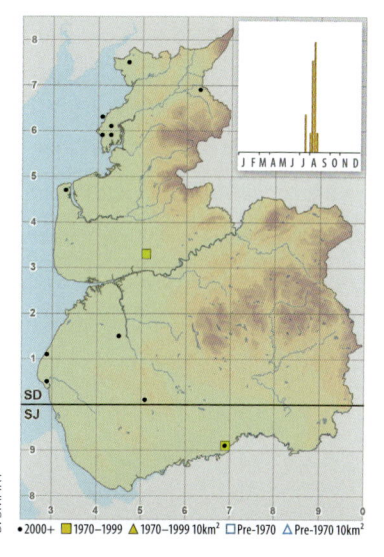

The Moths of Lancashire　171

GELECHIIDAE

35.061 *Ptocheuusa paupella*

■ **LANCS STATUS** Considered Extinct　　　　　　　　　　pre-1890

A single record of this distinctive species is mentioned in Ellis (1890) with the location noted as Cleveleys and the recorder J.H. Threlfall, a well-known and respected lepidopterist of that time. No specific date is mentioned, but the main larval foodplant of this species nationally, Common Fleabane, does occur in parts of the Cleveleys area. Up to 1890, the nearest documented British record was in Norfolk and more recently it was not until 2006 that it was noted in Caernarvonshire, North Wales and 2020 in south-west Yorkshire.

■ **HABITAT** Not known.

■ **FOODPLANTS** Not recorded in Lancashire.

The record was accepted within the county by Ellis (loc. cit.) and later, by Mansbridge (1940). It was similarly accepted and published in Emmet and Langmaid (2002a). Despite this, further supporting details or the examination of a specimen, if extant, would be desirable.

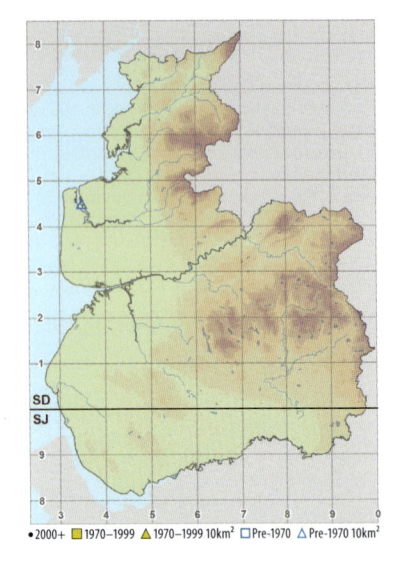

35.065 *Monochroa cytisella*

■ **LANCS STATUS** Occasional / local resident　　　　　　　c.1859–2022

Although first reported from the Manchester area (Stainton, 1859), it was not until 1986 that it was rediscovered, near Dolphinholme, by M.W. Harper. In 2002, ten larval galls were found near Belmont and it has subsequently been attracted to light, or found as a larva, at twelve other well scattered lowland and upland sites.

■ **HABITAT** The drier parts of moorland edge, limestone grassland, mosses and woods, where Bracken abounds.

■ **FOODPLANTS** Bracken.

Single brooded. Larval galls have been found from late April to mid-June. Probably under-recorded as an adult with larval galls found to be frequent on occasions.

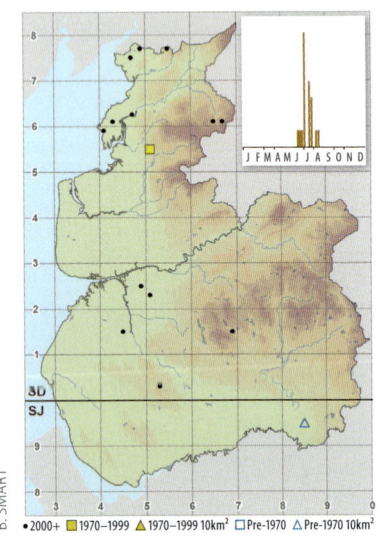

35.066 *Monochroa tenebrella*

■ **LANCS STATUS** Occasional / local resident　　　　　　　c.1857–2022

First found in Liverpool Botanic Garden (Gregson, 1857) and, later that century, in Manchester, Farington and Lytham. Considered local by Mansbridge (1940), it has since been found to be present in reasonable numbers at favoured sites which include Longridge Fell and Cockerham Moss. Males can be confused with *Oxypteryx unicolorella*, but the white tips to the antennae on females readily separate the two.

■ **HABITAT** Drier parts of moorland, mosses and other areas of acidic grassland.

■ **FOODPLANTS** Not recorded in Lancashire.

Single brooded. Seen flying in numbers over a large patch of Sheep's Sorrel on an un-grazed grassy bank in warm, sunny conditions.

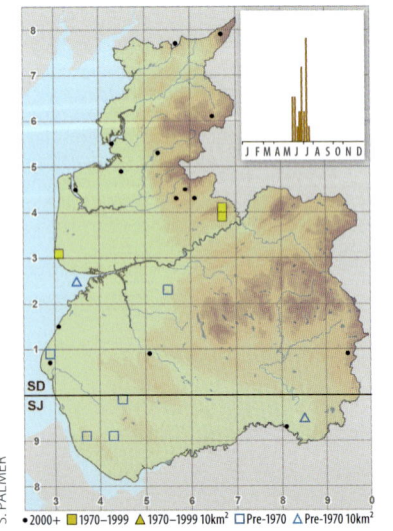

172　*The Moths of Lancashire*

35.068 *Monochroa tetragonella*

■ **LANCS STATUS** Scarce / very local resident   2010–2021

First noted in Britain in 1874, in Norfolk, it has subsequently been found at 18 widely scattered coastal saltmarsh sites in England and Wales. Since 2000, it has only been recorded at ten sites nationally but is possibly under-recorded. It was not until a MV light trap was run in a garden on the edge of the R Lune estuary at Sunderland Point, that it was added to the Lancashire fauna, on 30 June 2010 (J. Girdley). Since then, it has also been found in suitable habitat to the north of Sunderland Point and on the north coast of the Fylde, near Fluke Hall. There are also a few records of the moth coming to light a kilometre or so inland, at Heysham Nature Reserve in 2014 and Silverdale Moss in 2015, suggesting it occasionally disperses from its breeding sites.

In Lancashire, the moth occupies the upper saltmarsh zone where the larvae can be found feeding in the rootstock and lower stem of Sea-milkwort causing a distinctive browning of the leaves and wilting of the plant, sometimes killing it. Searches for tenanted plants during April and May suggest that, in Lancashire at least, it is present only in low numbers. However, infected plants can be difficult to locate and examine for larval presence. Similarly, the adult is rarely recorded in more than ones and twos, with the majority found at light (including to actinic light) or flying at dusk, low over the saltmarsh. The dusk flight is difficult to monitor as other species of a similar size fly at the same time in much greater abundance (e.g. *Coleophora adjunctella* and *Elachista scirpi*). The combination of these factors mean it has not been possible to assess how the moth is faring since its original discovery.

It is considered likely to be an overlooked, long-term resident which may have been missed in other areas of extensive saltmarsh in the county. Coastal sites, such as at Bolton-le-Sands, Carnforth and Marshside (north of Southport) would be worth investigating. It is unknown whether it overwinters as an ovum or a small larva, but whichever is the case, they will have to deal with regular tidal inundations.

The moth is accorded pRDB1 status in Davis (2012), but subsequent additional records in England and Wales suggest it is a little more widespread than previously known and that Nationally Scarce A might now be a more appropriate designation.

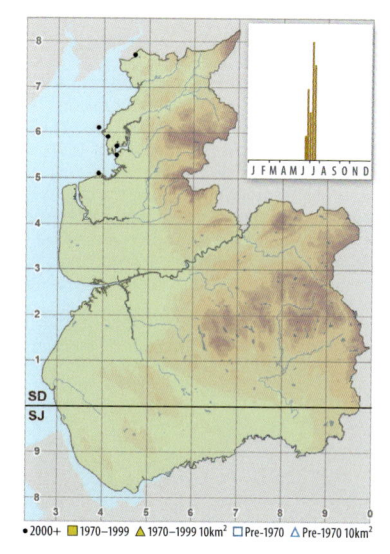

● 2000+   ■ 1970–1999   ▲ 1970–1999 10km²   □ Pre-1970   △ Pre-1970 10km²

GELECHIIDAE

The Moths of Lancashire   173

GELECHIIDAE

35.071 *Monochroa lucidella*

■ **LANCS STATUS** Occasional / local resident   1855–2022

Recorded historically on four occasions between July 1855, when found in old clay pits by N. Cook in the Liverpool district (Gregson, 1857), and Freshfield in 1952, by H.L. Burrows. Since the summer of 1999, records have increased markedly, including singletons appearing at light well away from suitable habitat. On 26 June 2007 around 30 were found during the day in the Freshfield area (G. & D. Atherton).

■ **HABITAT** Shallow edges of freshwater pools and lakes with emergent rush growth.

■ **FOODPLANTS** Not recorded in Lancashire.

Single brooded. Searches at Middleton N.R., where the moth was swept from amongst Common Spike-rush, have failed to locate any larvae.

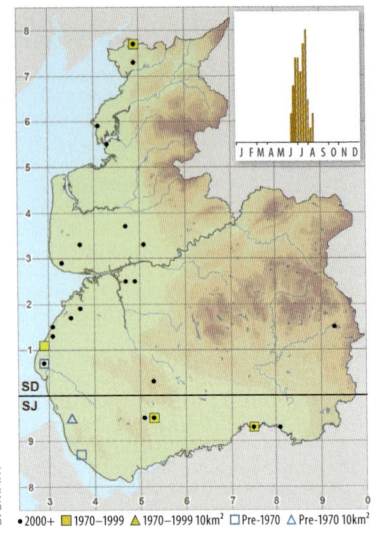

35.076 *Monochroa suffusella*

■ **LANCS STATUS** Rare / very local resident   2007–2022

A recent discovery and considered an overlooked, long-term resident. The first two sightings were of singletons, both attracted to light at Astley Moss on 17 May 2007 (actinic light) and 27 June 2009 (125w MV) by K. McCabe. A few searches were made at this site for larvae, but were unsuccessful. Further larval searches at Little Woolden Moss on 11 March 2022 (B. Smart) produced a tenanted leaf mine.

■ **HABITAT** Lowland raised bogs.

■ **FOODPLANTS** Cottongrass.

Single brooded. Possibly overlooked at Cockerham / Winmarleigh Mosses, on the Fylde, where conditions and foodplant availability appear suitable for this species.

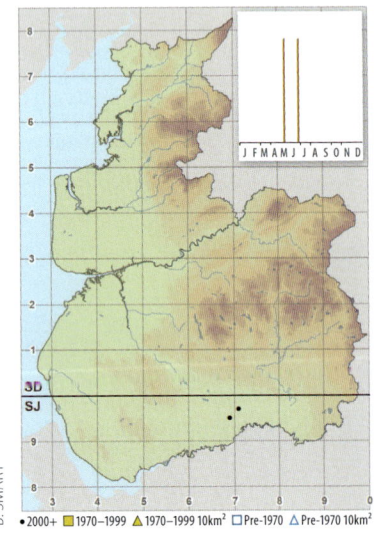

35.077 *Monochroa hornigi*

■ **LANCS STATUS** Rare / vagrant   2022

A very recent and unexpected discovery in the county, the nearest known record nationally being 100km east in Yorkshire. It was discovered during an examination of unidentified specimens retained during a six-month light trap survey at Maple Farm Nursery, near Westby, by J. Steeden. The moth was caught on 16 June 2022 and confirmed by dissection (a male). An adventive origin seems very unlikely. As to whether it was a vagrant or an overlooked resident, only time may tell.

■ **HABITAT** Unknown.

■ **FOODPLANTS** Not recorded in Lancashire.

Single brooded.

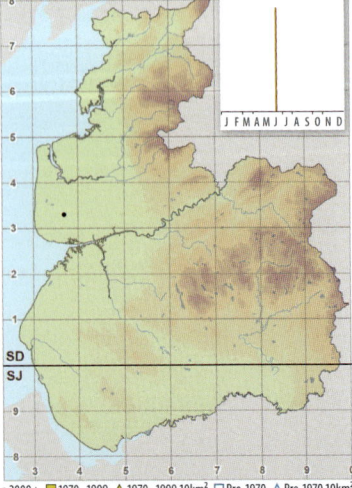

174 *The Moths of Lancashire*

35.079 *Oxypteryx wilkella*

■ **LANCS STATUS** Rare / very local resident   c.1857–2016

There are only two records in Lancashire, both from roughly the same area of coastal dunes, but over 150 years apart. The first was reported as being disturbed from herbage near the mouth of the R Alt, near Hightown (Gregson, 1857). The second came when three were attracted to MV light run next to a gently sloping, south-facing, sandy bank, on MOD land at Altcar on the 26th June 2016, by R. Walker. It may be an overlooked species elsewhere on the Sefton Coast and St Annes dunes.
■ **HABITAT** Dry broken grassland on large coastal sand dunes.
■ **FOODPLANTS** Not recorded in Lancashire.
Single brooded. Comes to light.

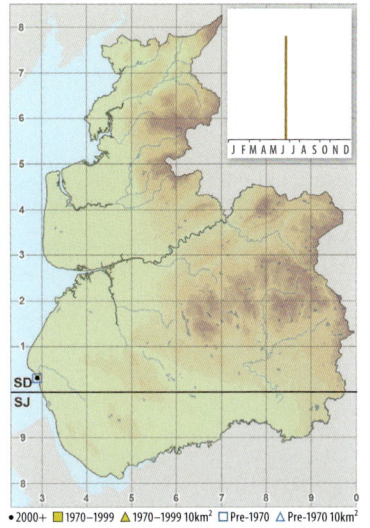

35.080 *Oxypteryx unicolorella*

■ **LANCS STATUS** Scarce / very local resident   1914–2015

Due to potential confusion with the similar *Monochroa tenebrella*, early records from Wardless by J.H. Threlfall (Ellis, 1890) and Woodvale (Mansbridge, 1940) are unconfirmed. One specimen was available for examination, allowing confirmation, from Silverdale on 20 June 1914, by W. Mansbridge. All other confirmed records are more recent and also from the Silverdale area; at Gait Barrows from 1999 to 2015 (including five on 3 June 2008), and one at Jack Scout on 2 July 2008 (S. Palmer).
■ **HABITAT** Lowland limestone grassland.
■ **FOODPLANTS** Not recorded in Lancashire.
Single brooded. Comes to light and occasionally disturbed during the day.

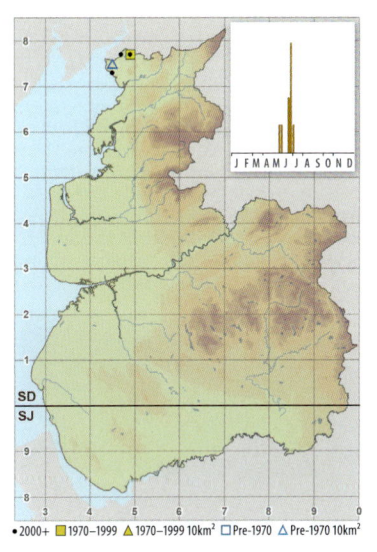

35.081 *Oxypteryx atrella*

■ **LANCS STATUS** Occasional / local resident   c.1889–2021

First noted in Lytham by J.B. Hodgkinson, date unknown (Ellis, 1890). All other records are more recent and from the northern limestone areas. The first of these was in 1986 at Gait Barrows (Hardwick, 1990), the main site for this species in the county. The other locations, with fewer records, are Leighton Moss, from 2006 to 2010, Butterfly Conservation's reserve at Myers Allotment on 5 August 2011 and Crag House Allotment, Hutton Roof, on 9 August 2021 (J. Patton).
■ **HABITAT** Mostly on, or adjacent to limestone grassland, in lowland and upland areas.
■ **FOODPLANTS** Not recorded in Lancashire.
Single brooded. Comes to light and can be swept by net during the daytime.

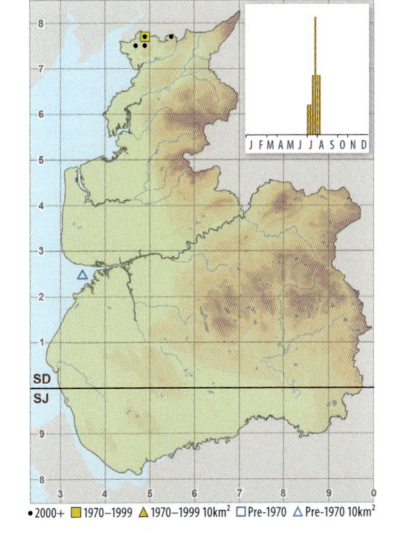

The Moths of Lancashire   **175**

GELECHIIDAE

35.085 *Athrips mouffetella*

■ **LANCS STATUS** Occasional / local resident c.1859–2022

First recorded at Withington and in Glazebrook c.1859 by J. Chappell and later, c.1889 in Cleveleys (J.H. Threlfall). Since 1985 it has been found at many scattered sites, mainly in lowland VC59, less often in VC60 and in some upland gardens. Sites include Billinge, Briercliffe, Flixton, Heysham, Parbold and Preston. It has shown a tendency to arrive, stay for a few years and then disappear, without obvious cause.
■ **HABITAT** Hedgerows, gardens, scrubland and edges of woodland.
■ **FOODPLANTS** Honeysuckle.
Single brooded, with a very early record on 3 June 2018. Comes to light. The larva has been found twice, in a garden in Chorlton, in mid-May 2007 and 2008 (B. Smart).

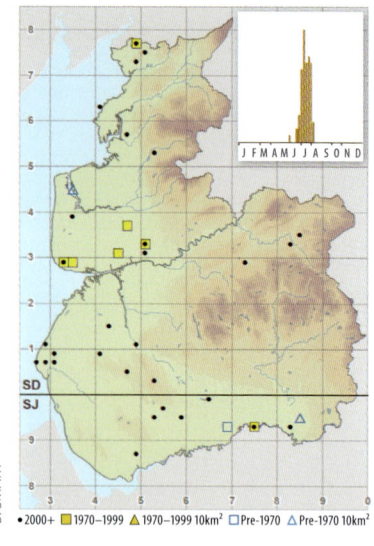

35.089 *Prolita sexpunctella*

■ **LANCS STATUS** Occasional / local resident 1844–2021

Reported by Edleston (1844) as occurring in the Manchester area and, by the same recorder in 1846, as abundant on Chat Moss. Thereafter, found at five mainly upland locations until 1918. It was discovered on Winmarleigh Moss in 1992, a site which combined with nearby Cockerham Moss, has produced most of the recent records. Other recent locations include Jeffrey Hill, Pendle Hill and Bishops Park, near Oldham.
■ **HABITAT** Lowland raised bogs and damp moorland.
■ **FOODPLANTS** Not recorded in Lancashire.
Single brooded. Readily disturbed from amongst low-growing mosses, Bell Heather and Cranberry on sunny days.

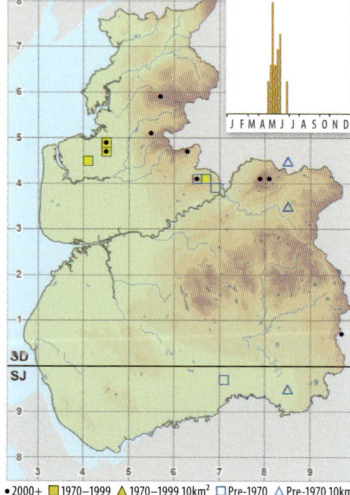

35.091 *Sophronia semicostella*

■ **LANCS STATUS** Considered Extinct c.1859–1953

The earliest record was reported in Stainton (1859), where it had been found in the Manchester district, but with no other details. It is possible this record may refer to smaller parts of this region that are outside the Lancashire VC59 area. The only other records are from Didsbury by H.N. Michaelis in 1952 and 1953. The 1952 record was documented as follows: 'Two specimens were taken flying over grass (mainly *Dactylis glomerata*) at sunset on 3 July 1952, near the River Mersey at Didsbury' (Michaelis, 1953). This is on the northern limit of the species national range and loss or degradation of habitat may well be the reason for its apparent demise.
■ **HABITAT** Dry grassland.
■ **FOODPLANTS** Not recorded in Lancashire.
Single brooded. Has been observed flying at dusk.
Elsewhere, it has been reported in North-west Yorkshire as recently as 2011 and Cheshire in 2012. It may therefore still be present in Lancashire and is best searched for flying over, or disturbed from, areas of broken ground with short grass in afternoon sunshine.

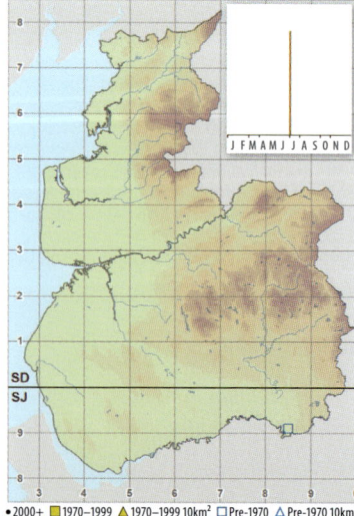

35.092 *Mirificarma lentiginosella*

■ **LANCS STATUS** Considered Extinct **1879**

J.B. Hodgkinson noted finding larvae on *Genista tinctoria* on 28 June 1879 (Hodgkinson, 1879) in his report of entomological activities in the Dutton area, near Longridge. This locality is north of the R Ribble with extensive meadows, leading upwards to the moorland at Longridge Fell. Greenwood (2012) documents historic records of Dyer's Greenweed in three nearby tetrads, but the precise location for this colony of *M. lentiginosella* has not been possible to establish.
■ **HABITAT** Dry grassland.
■ **FOODPLANTS** Dyer's Greenweed.
Single brooded. The plant is now very rare in VC60 (Greenwood, pers. comm., 2021). Although the moth was last noted in Cheshire in 1946, if any areas of the foodplant are located in Lancashire, a search for the moth or its larvae would be recommended. In such circumstances, it was discovered in south Northumberland in 2013.

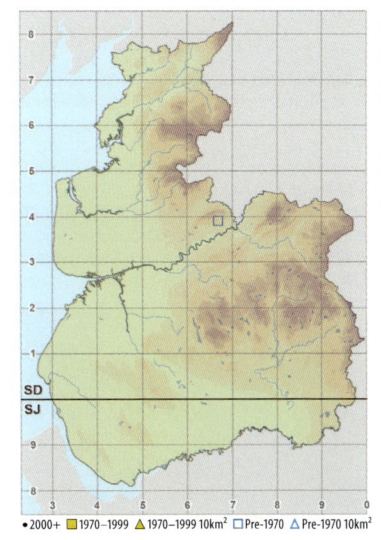

• 2000+ ▢ 1970–1999 ▲ 1970–1999 10km² ▢ Pre-1970 △ Pre-1970 10km²

35.093 *Mirificarma mulinella*

■ **LANCS STATUS** Occasional / widespread resident **c.1857–2022**

Noted as 'common wherever gorse occurs' (Gregson, 1857), this repeated by Ellis (1890), Mansbridge (1940) and Leech & Michaelis (1957). Next noted at Altcar in 1987 and Preston in 1995, it has since been regularly recorded across both lowland and upland parts of the county, most commonly in garden light traps close to sites with extensive gorse. Such locations include Billinge, Flixton and Heysham N.R.
■ **HABITAT** Moorland, rough pasture, sandy areas, roadside verges and brownfield sites.
■ **FOODPLANTS** Gorse and Broom.
Single brooded. Can be disturbed from the larval foodplants and comes to light. Readily found as a larva in gorse or Broom flowers from late March to mid-May.

• 2000+ ▢ 1970–1999 ▲ 1970–1999 10km² ▢ Pre-1970 △ Pre-1970 10km²

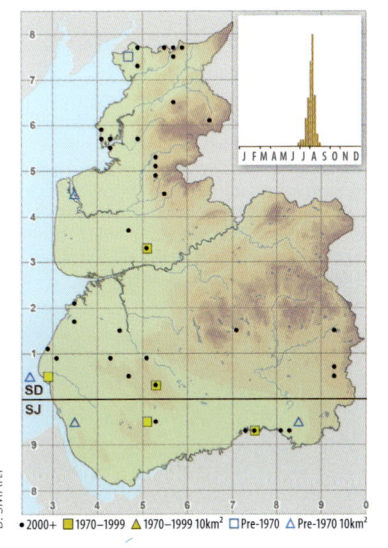

35.094 *Aroga velocella*

■ **LANCS STATUS** Occasional / local resident **c.1859–2022**

Reported from the Manchester area (Stainton, 1859) and then from Bolton-le-Sands by L.T. Ford in 1940. Not noted again until 1995, when seen in Preston and in 1997 and 1998, when found in good numbers on a clear-felled forestry site at Longridge Fell, with Sheep's Sorrel well established. Since then, recorded at six sites including Freshfield Dune Heath where twelve came to light on 12 July 2020 (J. Girdley).
■ **HABITAT** Moorland edge, heathland, a brownfield site and grassy banks.
■ **FOODPLANTS** Not recorded in Lancashire.
Double brooded. Attracted to light. Appears to wander, on occasions, to presumably locate newly established populations of the larval foodplant.

• 2000+ ▢ 1970–1999 ▲ 1970–1999 10km² ▢ Pre-1970 △ Pre-1970 10km²

The Moths of Lancashire **177**

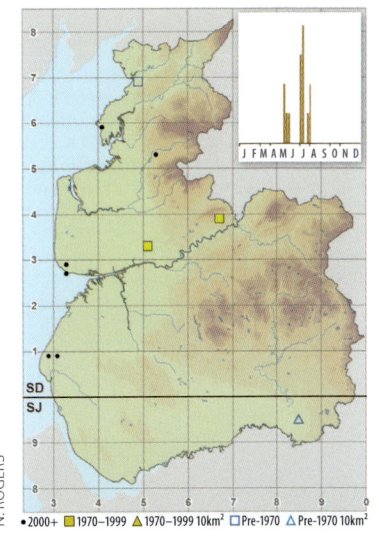

GELECHIIDAE

35.095 *Chionodes distinctella*

■ **LANCS STATUS** Scarce / very local resident   1865–2020

First seen at Lytham in July 1865 (Hodgkinson, 1865) and next, in Formby by W. Mansbridge in 1915. It was not seen again until a light trapping session at St Annes N.R. on 22 July 1997 (S. Palmer *et al.*) and thereafter, has only been noted on seven occasions on the Altcar, Birkdale, Formby and Freshfield dune systems between 1999 and 2020, always in low numbers. The largest counts were of three at light at Freshfield Dune Heath on 30 August 2016 and Formby on 23 July 2018 (R. Walker).
■ **HABITAT** Larger coastal sand dunes and dune heath.
■ **FOODPLANTS** Not recorded in Lancashire.
Single brooded. Attracted to light.

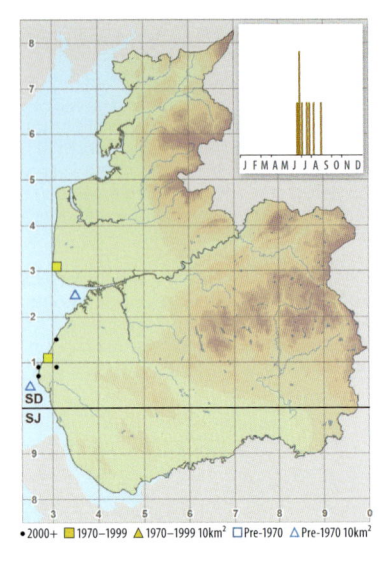

35.096 *Chionodes fumatella*

■ **LANCS STATUS** Occasional / local resident   1952–2019

First noted on the 18 August 1952 by H.N. Michaelis, when a few were smoked from an overhanging bank in the sand dunes at Freshfield (Michaelis, 1953). Not seen again until 1997, when it was discovered on Lytham St. Annes N.R. Since then, found at eight lowland sites, most of these being coastal or near-coastal. The remainder were at inland garden light traps, in Flixton, Parr and Swinton. This follows a national trend, in recent decades, for the species to spread inland to suitable areas.
■ **HABITAT** Sandy coasts, dune heath, brownfield sites and, rarely, gardens.
■ **FOODPLANTS** Not recorded in Lancashire.
Single brooded and comes to light.

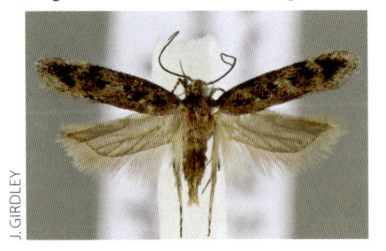

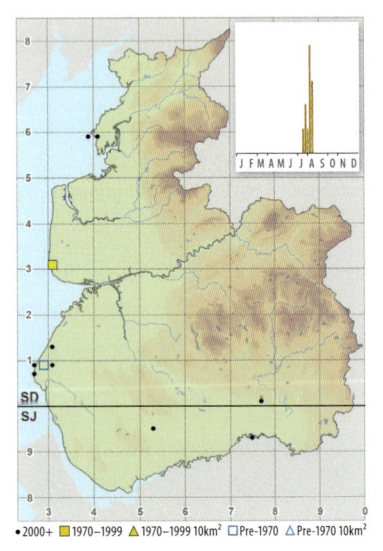

35.097 *Gelechia rhombella*

■ **LANCS STATUS** Rare / local resident   1845–2005

Found in Cheetham Hill, Manchester in June 1845 by R.S. Edleston (Edleston, 1846) and Irlam pre-1889 by J. Chappell. It was next noted as a larva, on apple, in Didsbury by H.N. Michaelis in 1954 and 1956, the latter emerging on 14 August 1957. In September 2001, it was reported from Gait Barrows by R. Petley-Jones, the only known record from VC60. During research for this book in 2022, a photo was discovered of this species, taken in Haydock, on 17 July 2005 (G. & D. Atherton).
■ **HABITAT** Gardens.
■ **FOODPLANTS** Apple.
Single brooded. Comes to light.

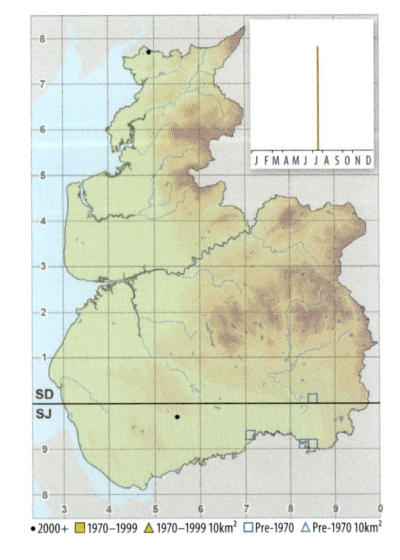

178 *The Moths of Lancashire*

35.101 *Gelechia sororculella*

■ **LANCS STATUS** Occasional / local resident   c.1857–2021

Gregson (1857) found it commonly amongst the small willows on sand dunes of the Liverpool area, as did Hodgkinson, in 1865, at Lytham. Apart from reports by W.G. Clutten of it being scarce in the Burnley and Pendle areas in 1920, and one at Spring Wood, Chorley in 1996, all other records have been from coastal sites, mosses, or gardens in lowland areas. Mostly found as singletons, but at least five were disturbed from a sea buckthorn thicket during the day at Lytham St Annes N.R. on 11 July 2008.
■ **HABITAT** Dune slacks, lowland raised bogs, brownfield sites and rarely gardens.
■ **FOODPLANTS** Creeping Willow and Grey Willow.
Single brooded. Comes to sugar, occasional at light and can be swept during the day.

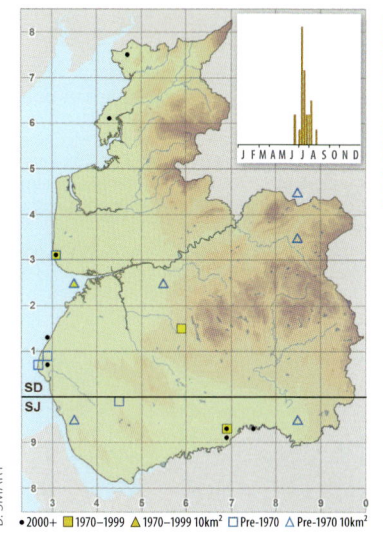

35.103 *Gelechia cuneatella*

■ **LANCS STATUS** Scarce / very local resident   2001–2021

A singleton was noted in a garden light trap in Flixton on 29 July 2001 by K. McCabe, the first time this species had been found in north-west England. It was subsequently seen at the same site in 2014, 2017 and 2021. Elsewhere, it has also been noted at Marton, Blackpool, coming to light on 24 August 2002, at Leighton Moss on 21 September 2009, and Chorlton on 4 September 2014 (B. Smart).
■ **HABITAT** River valleys and wetland sites with large, old willows present.
■ **FOODPLANTS** Not recorded in Lancashire.
Single brooded. Attracted to light. Larval searches have taken place at Flixton and Leighton Moss, without any success. Considered an overlooked, long-term resident.

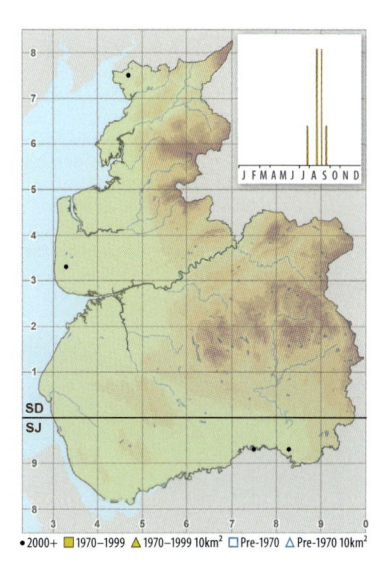

35.105 *Gelechia nigra*

■ **LANCS STATUS** Rare / vagrant; possible resident   2011–2023

Status uncertain, with a few reports mostly from different sites and initially north of other national sightings. Singletons found at Formby on 21 July 2011 (R. Walker), Swinton on 31 July 2011 (G. Riley) and at Widnes on 5 July 2019 (R. Kinsella). A second Formby specimen was attracted to light on 27 July 2023 (R. Walker) suggesting it may now be resident in the county. Searches for larvae, or adults on tree trunks along the Sefton Coast poplars would be worthwhile.
■ **HABITAT** Unknown as all records are from garden light traps.
■ **FOODPLANTS** Not recorded in Lancashire.
Single brooded. Attracted to light.

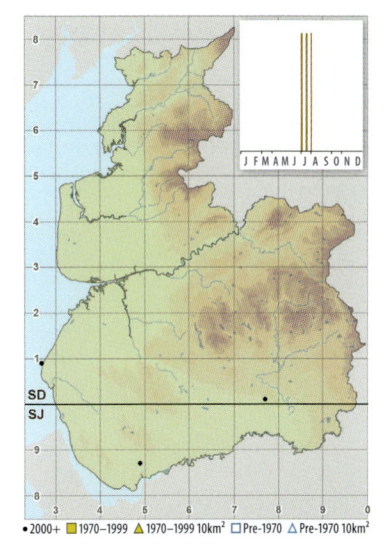

GELECHIIDAE

35.107 *Psoricoptera gibbosella*

■ **LANCS STATUS** Rare / local resident c.1859–2022

First recorded in the Manchester area during the mid-19th century (Stainton, 1859). It was not until 2000 that it was rediscovered in the county, when it came to MV light in a Flixton garden on 22 August (K. McCabe). Despite regular light trapping at this site, it has not been seen there again. The only other records were nr. Longton in 2016, Rainford in 2018, Chorlton in 2019 and Hoghton in 2022 (G. Dixon).
■ **HABITAT** Mature oaks in hedgerows and deciduous woodland.
■ **FOODPLANTS** Oak.

Single brooded. Comes to light, as singletons. Larvae were found on one occasion at Chorlton, with two spinnings on oak leaves on 19 May 2019 (B. Smart).

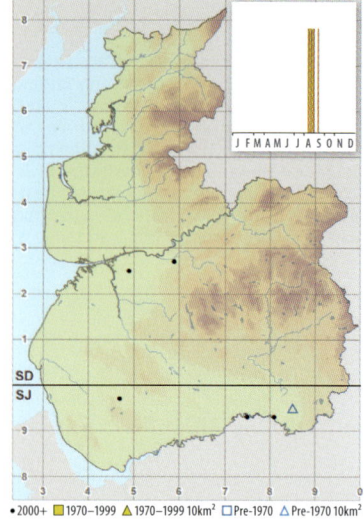

35.109 *Scrobipalpa acuminatella*

■ **LANCS STATUS** Occasional / widespread resident c.1889–2022

An undated record from Lytham by J.B. Hodgkinson is listed in Ellis (1890), with a few further records up to 1998, from Silverdale, Freshfield, Didsbury and Preston. Since 1999, records have increased considerably (partly due to increased light trapping), with a scattered range across lowland parts of the region. Upland records are less frequent and most are post 2010, including a mix of larval finds and light trapping.
■ **HABITAT** Grassy banks, field edges and verges, brownfield sites and gardens.
■ **FOODPLANTS** Creeping Thistle, Spear Thistle and a *Carduus* thistle sp.

Double brooded, first brood the stronger of the two. Attracted to light. Larvae can be found from mid-June to late July and late September to mid-November.

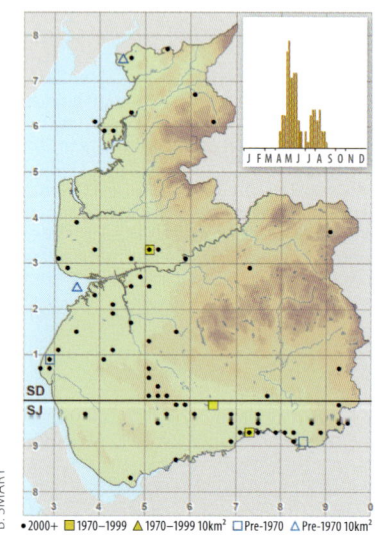

35.113 *Scrobipalpa salicorniae*

■ **LANCS STATUS** Scarce / very local resident 2004–2018

First found as a larva in 2004 at Glasson and Marshside (S. Palmer), the latter bred. Since then, all but three of the twelve other records involved light trapped individuals at Sunderland Point. Of the remainder, two were from coastal sites, while the other occurred well inland (as a vagrant) at Rishton on 25 June 2018 (D. Bickerton).
■ **HABITAT** Saltmarshes.
■ **FOODPLANTS** Sea Aster (possibly not its main foodplant in the region).

Single brooded. Comes to light. The larvae have been found in late April and May. Probably an overlooked long-term resident, this species was known as *S. salinella* until 2010, the name change coming with a proviso (Agassiz *et al.*, 2013).

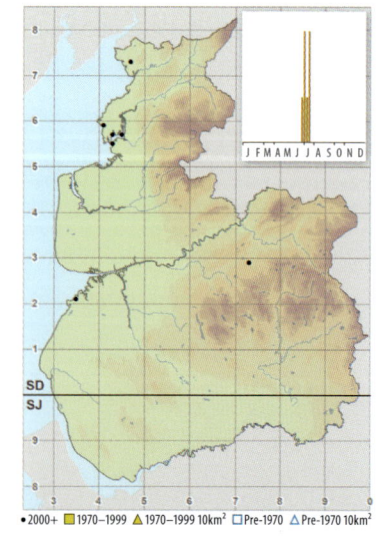

35.114 *Scrobipalpa instabilella*

■ **LANCS STATUS** Frequent / local resident   1996–2022

Early records are considered unverified due to identification issues and no historic specimens have been located, so far, to enable dissection. Since the first confirmed record, on 25 July 1996 at Stanah (S. Palmer), it has been noted on all of the larger Lancashire saltmarshes, sometimes in considerable numbers; e.g., about 150 at Burrows Marsh in July 2004. One vagrant was noted inland, in north Preston, in 2001.
■ **HABITAT** Saltmarshes.
■ **FOODPLANTS** Sea-purslane.
Single brooded. Comes to light and easily disturbed by day. Larvae have been found from late March to mid-May. The forewing markings can be very variable.

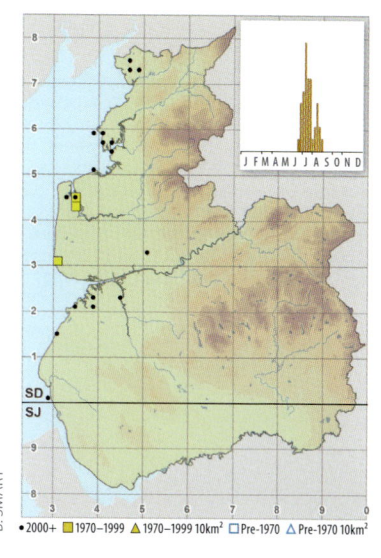

35.115 *Scrobipalpa nitentella*

■ **LANCS STATUS** Occasional / very local resident   2003–2022

As with other *Scrobipalpa* spp., early records are confused and considered unverified due to identification issues. The first confirmed record was from Heaton Marsh on 13 July 2003 (S. Palmer) and, since then, it has been noted at five other sites, mostly along the edges of the coastal estuaries of the R Ribble and R Wyre. In 2022, a small, dark, male was light-trapped well inland, north-east of Longridge; perhaps the very warm weather conditions had drifted it away from its usual saltmarsh haunts.
■ **HABITAT** Drier edges of saltmarshes.
■ **FOODPLANTS** *Atriplex* sp.
Single brooded. Comes to light. Larvae found in September at Marshside (B. Smart).

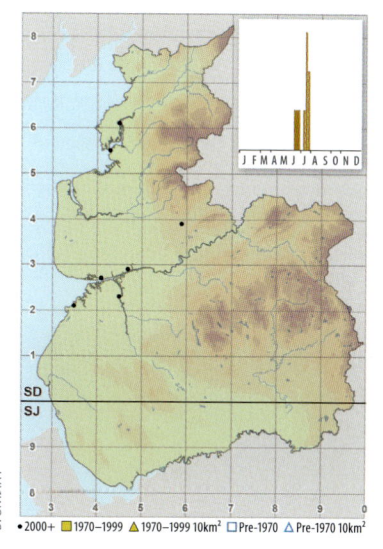

35.116 *Scrobipalpa obsoletella*

■ **LANCS STATUS** Rare / very local resident   2014–2017

Recorded on the coast at Middleton, as singletons, on 30 August 2014 and 13 August 2017 by J. Patton using an actinic light trap. Seven larvae were subsequently found on the same stretch of coastline, inside orache stems, on 14 September 2017 by K. McCabe, S. Palmer and J. Patton.
■ **HABITAT** Upper reaches of a pebble beach, bordered by a raised earth bank.
■ **FOODPLANTS** Babington's Orache.
The number of broods are not possible to assess due to the limited number of records. Comes to light. Larvae have been found in September, adults emerging the following May and June. Considered to be a previously overlooked species.

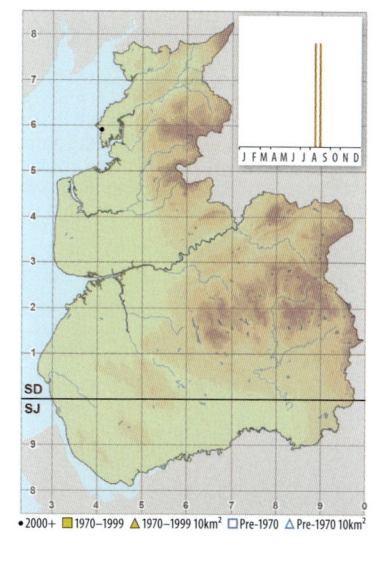

The Moths of Lancashire   **181**

GELECHIIDAE

35.117 *Scrobipalpa atriplicella*

■ **LANCS STATUS** Rare / very local resident c.1889–2022

A record, without date, from the Fleetwood area by J.B. Hodgkinson is noted in Ellis (1890). It was next found in Formby, in 1922, by W. Mansbridge (Mansbridge, 1940). Since then, there have only been three records, of singletons, at Risley Moss (M. Hull) on 28 July 1982, at Cadishead Moss on 28 April 2010 (K. McCabe) and at Birkdale Green Beach on 2 September 2022 (R. Walker).
■ **HABITAT** Arable field borders adjacent to mosses and vegetated coastal dunes.
■ **FOODPLANTS** Not recorded in Lancashire.
Double brooded. Comes to light. Unsuccessful searches for the larvae have been carried out on field edges in peaty areas on a few occasions.

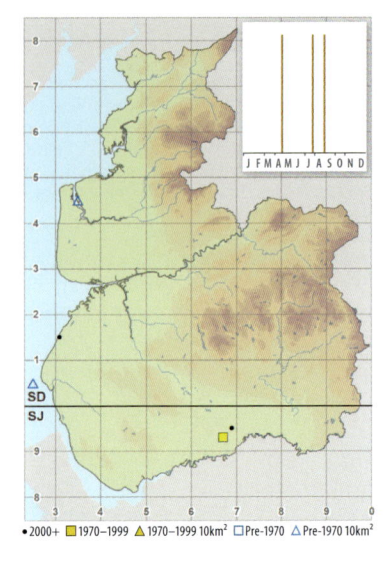

35.118 *Scrobipalpa ocellatella*

■ **LANCS STATUS** Vagrant 2022

During the summer of 2022, this species bred on inland Beet crops in unprecedented numbers in parts of south-east England. A major eruptive dispersal then took place, reaching as far as Lancashire and beyond. The first records were from Flixton on 12 August (K. McCabe) and Victoria Park, Manchester on 17 August (P. Pemberton). It was then reported from five locations between the 2 September and 7 September, with four found in Parr and Silverdale during that period and three in Preston.
■ **HABITAT** Unknown.
■ **FOODPLANTS** Not recorded in Lancashire.
Comes to light. Larval searches on Sea Beet during October 2022 were unsuccessful.

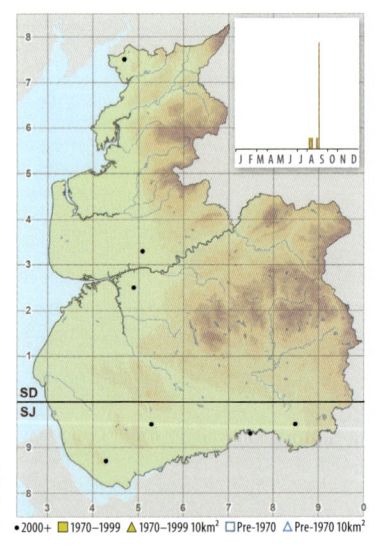

35.119 *Scrobipalpa samadensis*

■ **LANCS STATUS** Scarce / local resident 1877–2019

In 1877, larvae were discovered in roots of sea plantain on the banks of the R. Wyre by J.H. Threlfall (Threlfall, 1878), but misidentified as *Gelechia instabilella*. The true identity was later established by Stainton (1882), under *G. plantaginella*. Not seen again until 21 July 1997 (S. Palmer) at Glasson and has since been found at six other sites, mostly at light. Few larvae found on 28 April 2019, at Fluke Hall marshes, mining the foodplant's leaves and root crown (J. Girdley, S. Palmer & B. Smart).
■ **HABITAT** Saltmarshes.
■ **FOODPLANTS** Sea Plantain.
Single brooded. Comes to light. Larvae have been found from late April to mid-May.

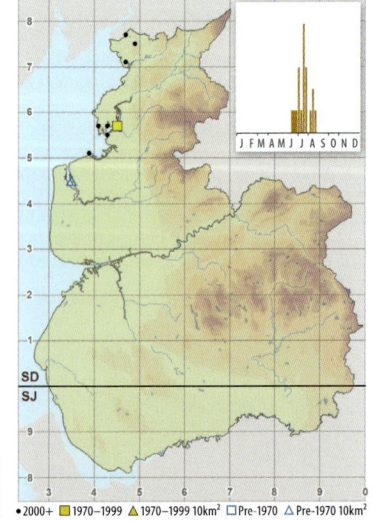

182 *The Moths of Lancashire*

35.120 Thyme Moth *Scrobipalpa artemisiella*

■ **LANCS STATUS** Occasional / local resident 1865–2022
First noted, in abundance, at Lytham St Annes in July 1865 (Hodgkinson, 1865), but not reported again until 1986 at Gait Barrows (L. Hardwick). The latter remains the most regular site for this species, with about 60 seen on 11 July 2021. Elsewhere, up to 30 were found on Worsaw Hill, near Clitheroe, between mid-June and late July 2021 and two on upland limestone at Dalton Crags on 8 July 2022 (J. Patton *et al.*).
■ **HABITAT** Limestone and, formerly, dune grassland, where Thyme occurs.
■ **FOODPLANTS** Not recorded in Lancashire.
Single brooded. Readily disturbed on sunny days, sometimes in good numbers, from larger clumps of Thyme. Occasionally comes to light.

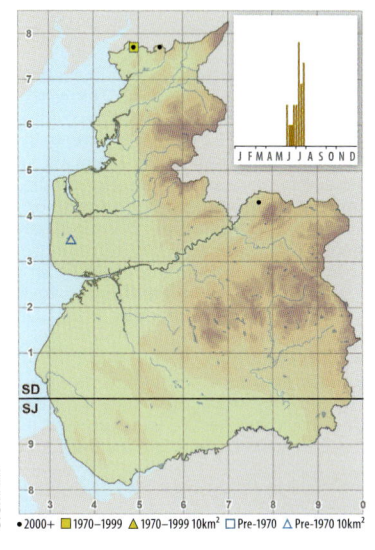

35.123 *Scrobipalpa costella*

■ **LANCS STATUS** Occasional / widespread resident c.1857–2022
First recorded in Liverpool (Gregson, 1857), but not noted again until 1921, in Formby by W.A. Tyerman (Mansbridge, 1940). Infrequently reported until the mid-1990s, since when records and its geographical range have increased considerably, reaching the edges of the uplands by the end of the 20th century. Despite this, the only reports from upland areas have been at light in Littleborough and Rishton. Searches for the larval feeding signs are an effective method of locating this species.
■ **HABITAT** Hedgerows, scrub and gardens.
■ **FOODPLANTS** Bittersweet, mining the leaves - cf. *Acrolepia autumnitella*.
At least double brooded and recorded in every month except January. Comes to light.

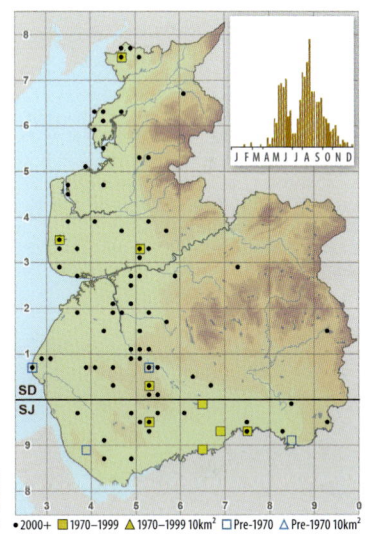

35.126 Potato Tuber Moth *Phthorimaea operculella*

■ **LANCS STATUS** Adventive 1996
A single larva discovered in a supermarket-sourced potato in Preston on 8 June 1996, was subsequently bred to confirm the identity (S. Palmer). The origin of the potato in question was not established.
■ **HABITAT** Imported food material.
■ **FOODPLANTS** Potato.
Elsewhere, it has also been found on tomato, but larvae have not been located in the wild. It was first noted in the British Isles in 1935 at London's East India Dock on imported potatoes from Malta. It has been found at light on three occasions in southern England, once as a possible migrant (Emmet & Langmaid, 2002).

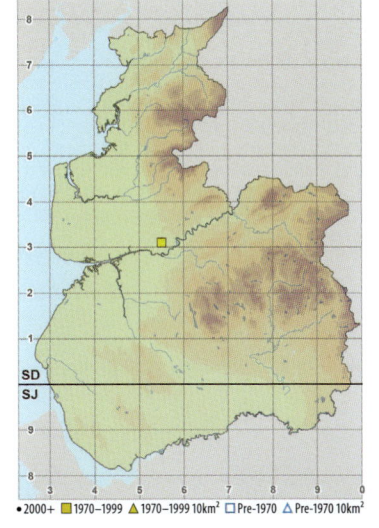

The Moths of Lancashire **183**

GELECHIIDAE

35.127 *Tuta absoluta*

■ **LANCS STATUS** Adventive / possible rare migrant   2009–2023

Originally known from DEFRA intercepts in Lancashire during 2009, the location of which has not been possible to confirm. For mapping purposes, they are shown as Liverpool Docks, but could potentially relate to any locations where tomatoes are imported into the county from abroad. Nationally, from being exclusively an adventive species, there are strong indications that it is now a regular migrant in small numbers in southern and eastern England. It has also become established as a pest in some tomato-growing glasshouses. As a result, the origin of the single moth found in a light trap at Ainsdale on 14 September 2022 (R. Moyes) could relate to a migrant or be of British origin from any established colonies in greenhouses. The same will be the case for the male that appeared at Southlands, near Longton on 7 September 2023 (J. Girdley), on a night when other migrant species were noted.

■ **HABITAT** Unknown.

■ **FOODPLANTS** Imported tomato.

The small size of this moth means it is easily overlooked, but the larval feeding signs could be checked for on greenhouse tomato plants. These include mines in the leaves and borings into the buds, stalks and fruits.

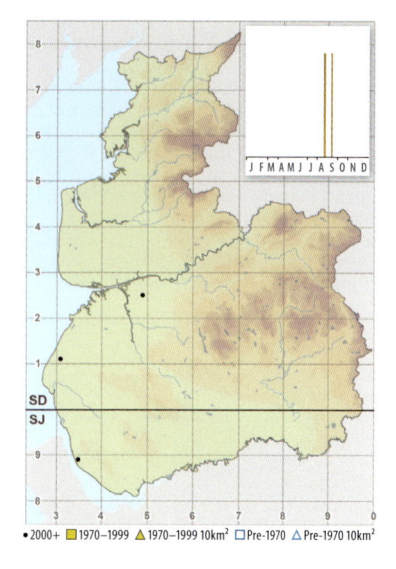

35.128 *Caryocolum alsinella*

■ **LANCS STATUS** Scarce / very local resident   c.1889–2016

First noted in the Lytham area by J.H. Threlfall, as a pupa amongst *Cerastium* sp., and in the same area by J.B. Hodgkinson (Ellis, 1890). Two from Formby dated 18 June 1952 were found in the H.N. Michaelis coll., in Manchester Museum. Since 1999, it has been attracted to light at both Ainsdale and Lytham St Annes N.R., but most success has been had in locating larvae. Six tenanted spinnings were found at Formby in 2006 (K. McCabe *et al.*) and eleven at St Annes on 9 May 2016 (S. Palmer).

■ **HABITAT** Larger coastal dunes.

■ **FOODPLANTS** Mouse-ear sp. (either *Cerastium diffusum* or *C. semidecandrum*).

Single brooded. Rarely comes to light. Larvae have been found during May.

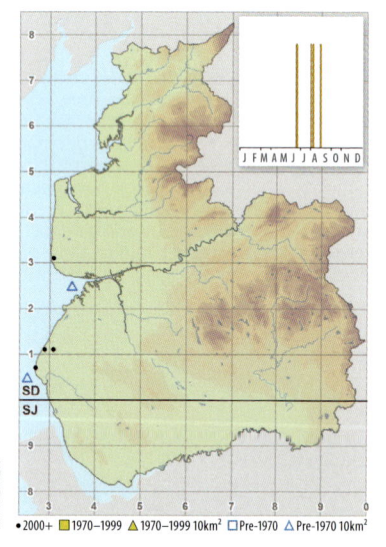

35.129 *Caryocolum viscariella*

■ **LANCS STATUS** Occasional / local resident   1877–2022

Larvae found, in abundance, in Preston and Thornton in 1877 by J.H. Threlfall (1878) and, in lesser numbers, at Preston in 1894, Bare (Morecambe) in 1940 and Crosby in 1951. From 2001, it was light-trapped in ones or twos at various locations, including Burrow Heights and Flixton and, more frequently, at Bay Horse. Larval searches for this mainly lowland species, have proved successful at several sites including St Helens in 2015, Fluke Hall in 2016, Sunderland Point in 2021 and Bare in 2022.

■ **HABITAT** Shady hedgerows, verges and woodland edges.

■ **FOODPLANTS** Red Campion.

Single brooded. Irregularly at light. Larvae have been found in April and May.

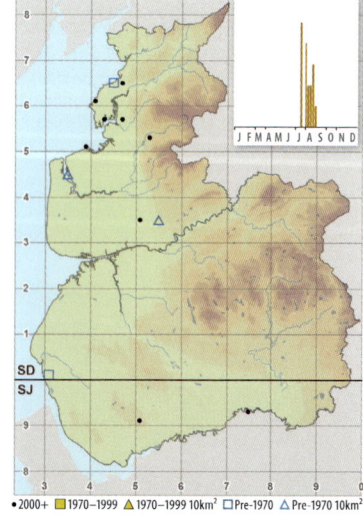

184 *The Moths of Lancashire*

35.131 *Caryocolum marmorea*

■ **LANCS STATUS** Occasional / local resident c.1857–2021
First found on the sand dunes near Liverpool by Gregson (1857), followed by Lytham in 1865 and later noted as abundant on sand dunes by Ellis (1890), Mansbridge (1940) and, at Formby, by Leech and Michaelis (1957). Since then, only seven sand dune records are known, from Altcar, Formby and Lytham St Annes N.R., up to 2021. Unusually, three have been found further inland; from Yealand Storrs in 1977 (M. Hull) and two from Flixton, on 5 September 2006 and 17 August 2012 (K. McCabe).
■ **HABITAT** Larger coastal sand dunes.
■ **FOODPLANTS** Not recorded in Lancashire.
Single brooded. Comes at light. Possibly overlooked in recent years.

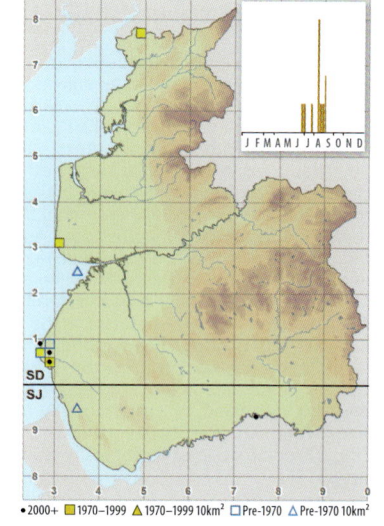

35.132 *Caryocolum fraternella*

■ **LANCS STATUS** Occasional / local resident c.1859–2018
First recorded in Withington c.1859 by J. Chappell and then near Pilling Moss in June 1865 (J.B. Hodgkinson). Identification issues may have confused matters to some extent in the early 1900s, with, for example, a specimen labelled *C. junctella* in the Mansbridge coll., later found to be *C. fraternella* on dissection. Since 2002, noted from seven lowland sites, the last at Middleton N.R. on 22 July 2018 (P.J. Marsh).
■ **HABITAT** Dune and sandy grassland, brownfield sites and edges of mosses.
■ **FOODPLANTS** Common Mouse-ear.
Single brooded. Comes sparingly to light and occasionally nectars on Common Ragwort. Larvae have been found from late April to early May.

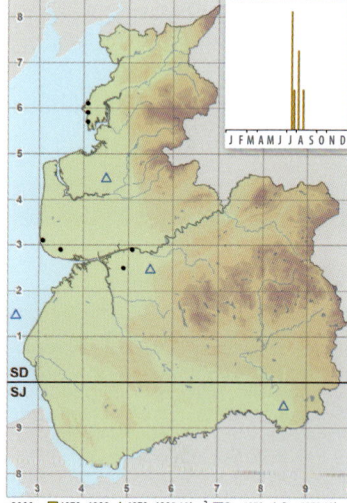

35.133 *Caryocolum blandella*

■ **LANCS STATUS** Considered Extinct c.1859–1889
Stainton's Manual of 1859 records this species from the Manchester area. It is possible that this reference related to a record from Flixton, by J. Chappell, as well as one from Carrington Moss (Cheshire) by the same recorder. With no dates detailed for these in any publications, it has not been possible to confirm this. In June 1865, J.B. Hodgkinson reported it from the Pilling Moss area (Hodgkinson, 1865), a slightly early month for this species (unless a larva). Later, Ellis (1890) mentions a record from Lytham by Hodgkinson and also refers to its discovery in the Preston area, by J.H. Threlfall, both without dates. No further records are added in Mansbridge (1940). In adjacent counties, the moth has been noted more recently, in Cheshire up to 1953, just across the border in mid-west Yorkshire in 2021 and in more northerly parts of Cumbria in 2022.
■ **HABITAT** Unknown.
■ **FOODPLANTS** Not recorded in Lancashire.
Single brooded nationally. The recent records in some adjacent counties suggest searches in old lanes with Greater Stitchwort would be of value, in spring for the larva and summer for the adult.

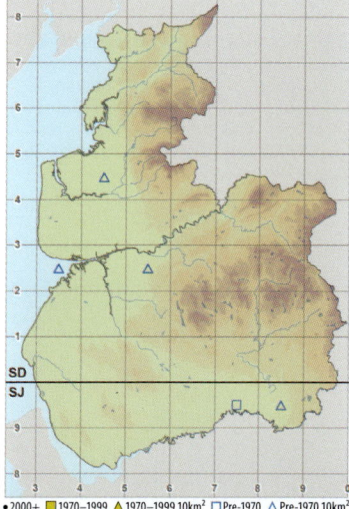

GELECHIIDAE

35.137 *Caryocolum tricolorella*

■ **LANCS STATUS** Scarce / local resident　　　　　c.1859–2016

First reported in the Manchester area as common, but without any specific details (Stainton, 1859). This area includes small portions of adjacent vice counties. Later, it was noted in Ellis (1890) from the Preston area by J.B. Hodgkinson and J.H. Threlfall, again with no specific site or date listed. Mansbridge (1940) repeats the details as listed in Ellis (loc. cit.) with both authors detailing it as local and not common. In 2005 (noted as in July or August), it was attracted to an MV light trap run by N.A.J. Rogers, in Bay Horse, near Dolphinholme. A further five singletons were recorded at the same site from 5 August 2006 to 18 July 2013. This is a moorland edge site with shaded, herb-rich verges and deciduous woodland nearby, with plenty of Greater Stitchwort present. Elsewhere, it came to light at Mill Houses on 11 August 2010 and Herring Head Wood on 22 August and 10 September 2015 (all P.J. Marsh). In the same year, on 4 April, six larvae were located feeding in spun heads of the foodplant along a 30-yard stretch of shaded verge near Wyresdale Park, Scorton (S. Palmer) and at the same site on 30 March 2016 (S. Palmer & B. Smart). Almost certainly under-recorded.

■ **HABITAT** Woodland, copses and grassy verges adjacent to scrub or hedgerows.
■ **FOODPLANTS** Greater Stitchwort.

Single brooded. Comes occasionally to light. Larvae found from late March into April.

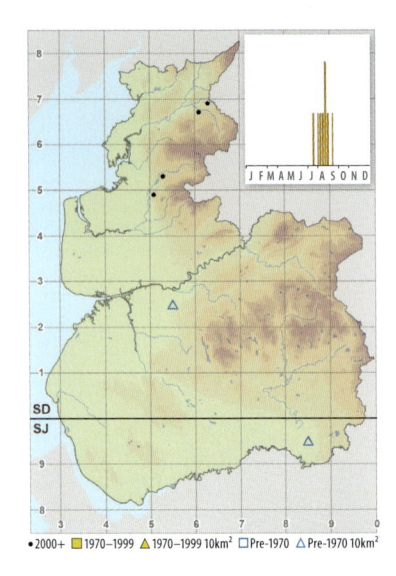

35.141 *Teleiodes vulgella*

■ **LANCS STATUS** Occasional / widespread resident　　　　　c.1857–2022

Found commonly in the Liverpool area (Gregson, 1857) and later, as locally common (Ellis, 1890), with Irlam and the Preston district listed as known sites. Mansbridge (1940) noted it as fairly common but local and added Childwall and Wavertree, while H.N. Michaelis reported it from Formby in 1953 and Didsbury on 19 July 1958. Not noted again until 1995, when increased and more widespread recording effort produced sightings from a wide range of sites including Flixton, Lightfoot Green, Marton, Parr, Rixton, Billinge and Orrell. From 2000 onwards, as well as increasing in many southern and western parts of the county, records were received from a wider range of locations into the east and north, suggesting a genuine range expansion was occurring. These included two from Littleborough in June 2000, Jack Scout, nr. Silverdale in 2004, south of Oldham in 2007, Briercliffe in 2013 and reaching Wray, in north-east Lancashire, on 14 June 2020 (Ga. Jones).

■ **HABITAT** Woodland, mature hedgerows and gardens.
■ **FOODPLANTS** Hawthorn and Rowan.

Single brooded, with an occasional second brood in some years, this first noted on 8 October 2003 in Preston. Comes readily to light. Larvae found from September to early May and also, in years with a second brood, from mid-July to early August.

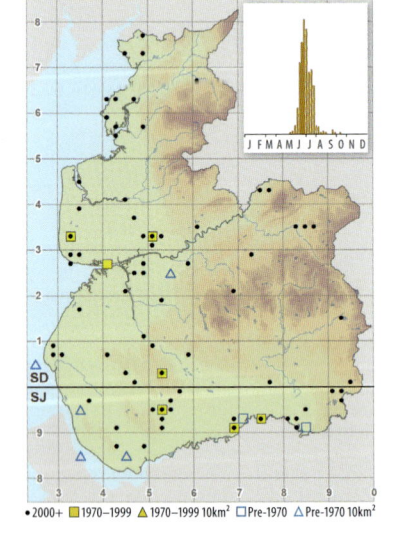

The Moths of Lancashire

35.143 *Teleiodes luculella*

■ **LANCS STATUS** Scarce / local resident   c.1859–2023

Reported as common in Manchester area (Stainton, 1859), but later considered local and not very common by Ellis (1890). In 1920, it was noted as scarce in Pendle and was not reported again in the county until 1995. From then, single reports were received from several lowland sites, the exception being in north Preston, where nine sightings were made, the last in 2015. Since then, only recorded as a singleton at light in Hale on 7 July 2023 (C. Cockbain).

■ **HABITAT** Woodland and hedgerows with Oak standards.

■ **FOODPLANTS** Oak.

Single brooded. Occasional at light. A larva was found at Rixton Clay Pits in October.

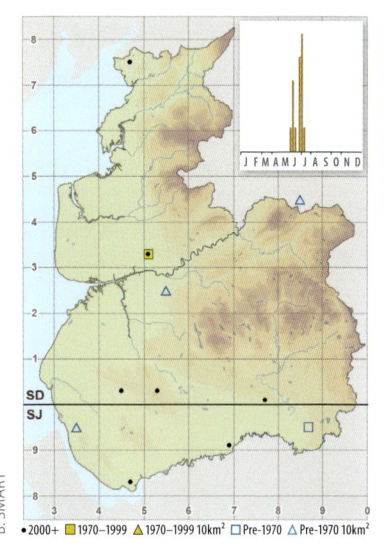

35.145 *Neotelphusa sequax*

■ **LANCS STATUS** Scarce / very local resident   1997–2022

Although not noted until 23 June 1997, when vacated larval spinnings were found on Warton Crag (S. Palmer), the species was present in nearby Cumbria from at least the late 19th century. It seems likely, therefore, that it had been overlooked in Lancashire. Warton Crag remains its main stronghold in the county, although it has also been found at Gait Barrows on several occasions between 1997 and 2016. Most records relate to larval feeding signs, with counts of up to 15 spinnings not unusual.

■ **HABITAT** Limestone grassland.

■ **FOODPLANTS** Common Rock-rose.

Single brooded. Occasional at light. Larvae can be found between mid-April and June.

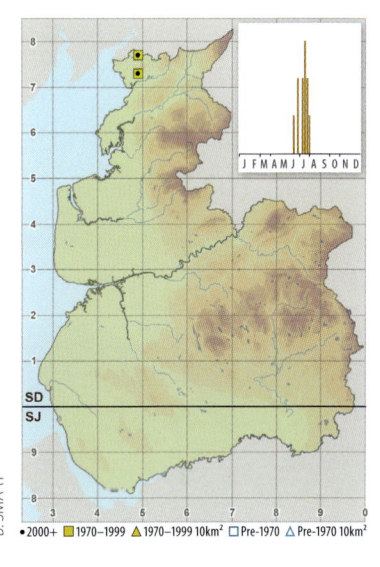

35.146 *Teleiopsis diffinis*

■ **LANCS STATUS** Occasional / widespread resident   c.1857–2022

Noted in the Liverpool (Gregson, 1857) and Manchester districts (Stainton, 1859) with a few other scattered records, such as Burnley, 20 June 1919 (A.E. Wright), received from across the county into the 1970s. It is suspected that light trapping has been the main driver behind a more recent increase and spread of records, particularly in the uplands. Probably under-recorded in moorland areas. Numbers attracted to light traps are always in single figures and mostly only ones or twos.

■ **HABITAT** Dry, mainly acidic grassland on moors, heaths, banks and, rarely, gardens.

■ **FOODPLANTS** Not recorded in Lancashire.

Double brooded. Comes readily to light.

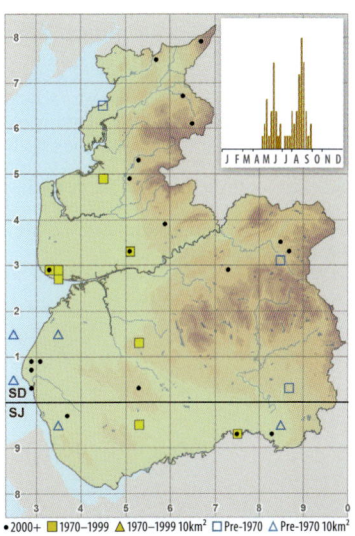

GELECHIIDAE

35.147 *Carpatolechia decorella*

■ **LANCS STATUS** Rare / very local resident   2003–2021

Known to be present in nearby parts of Cumbria in the 1920s, so was probably overlooked before the first was noted on 12 April 2003 on Gait Barrows, at light, by L. Sivell. Of the four other county records, three were also from the same site, on 6 May 2008, 22 April 2012 and 25 July 2014. The only record away from the northern limestone, was on National Trust land in Formby, where a regularly run light trap attracted one on 10 August 2021 (R. Walker & T. Davenport).

■ **HABITAT** Lowland limestone woodland and once, from a wood on coastal dunes.
■ **FOODPLANTS** Not recorded in Lancashire.

Single brooded; adult overwinters. Comes to light. Forewing markings very variable.

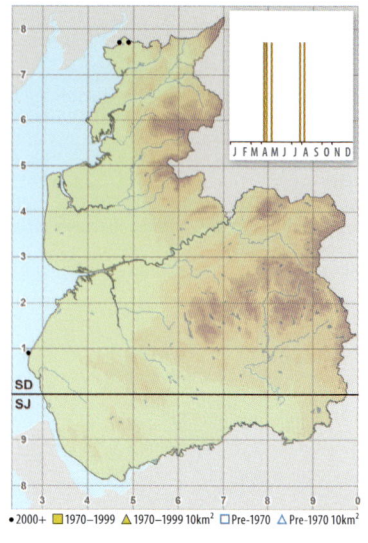

35.148 *Carpatolechia fugitivella*

■ **LANCS STATUS** Occasional / widespread resident   c.1857–2022

First recorded from Croxteth, Liverpool (Gregson, 1857) but then, not until 1938 when a larva was found (and bred) in Formby (Leech & Michaelis, 1957). Considered scarce by early recorders. Since the late 1990s, has been recorded fairly regularly at light traps across lowland parts but always in small numbers. Sites where it has appeared most frequently, although not annually, include Flixton, Parr and north Preston. Does not appear to have been adversely affected by Dutch elm disease.

■ **HABITAT** Woodland, parkland, hedgerows and gardens.
■ **FOODPLANTS** Elm sp.

Single brooded. Comes readily to light.

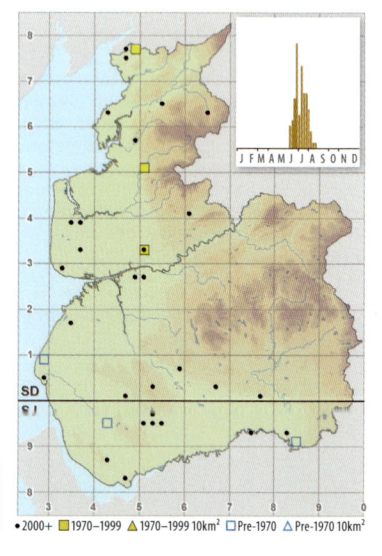

35.149 *Carpatolechia alburnella*

■ **LANCS STATUS** Occasional / local resident   1998–2022

It was 1935 before this species was recognised in Britain, and 1998 before it was recorded in Lancashire, this at Martin Mere, by R. Underwood and identified by M. Hull. The next was in Orrell in 2001 and thereafter, observed irregularly, mainly in VC59, in lowland sites where birch occurs naturally. It has only been found on three occasions north of the R. Ribble, in the Silverdale area in 2008 and 2009 and at Cockerham Moss in 2010, where six were disturbed from birch during the day.

■ **HABITAT** Open woodland, meres, mosses, brownfield sites and, occasionally, gardens.
■ **FOODPLANTS** Birch sp.

Single brooded. Comes to light in small numbers. Larvae found in May and early June.

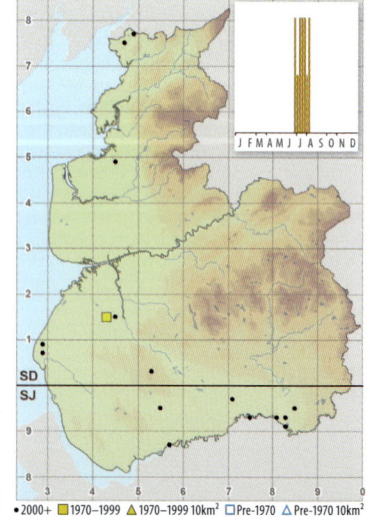

35.150 *Carpatolechia notatella*

■ **LANCS STATUS** Occasional / local resident c.1859–2022

First noted in the Manchester district, although recorder and location unknown (Stainton, 1859); then in the Preston area by J.H. Threlfall (Ellis, 1890). Larvae were found in the Oldham area in 1955 and an adult on Risley Moss in 1982. Since then, recorded at a variety of inland and, occasionally, upland sites to the present day, many in the larval stage. Rarely found as an adult in more than ones or twos, with most sightings of individuals disturbed from the larval foodplant during the day.
■ **HABITAT** Damp areas in woodland, brownfield sites and alongside rivers and streams.
■ **FOODPLANTS** Grey and Goat Willow.
Single brooded. Rare at light. Larvae can be found in September.

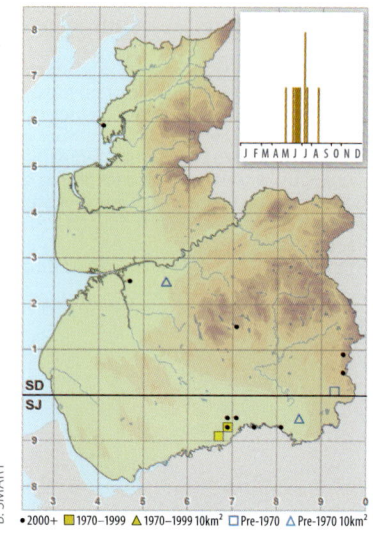

35.151 *Carpatolechia proximella*

■ **LANCS STATUS** Occasional / widespread resident c.1857–2022

Noted on a range of the county's southern mosses (Gregson, 1857 and Ellis, 1890). Elsewhere, near Pendle Hill in 1920 and Formby in 1950, where it was noted as common among birch. Since the mid-1990s, found on lowland raised bogs such as Astley, Cockerham and Heysham Moss, at upland wooded sites, such as Bay Horse, Beacon Fell, Docker Moor and White Coppice, amongst birch on limestone at Gait Barrows and within a few gardens, such as in Flixton and Rishton.
■ **HABITAT** Mosses, heaths, woods, moorland edges and sometimes gardens.
■ **FOODPLANTS** Birch sp.
Single brooded. Comes to light. Larvae can be found in August and September.

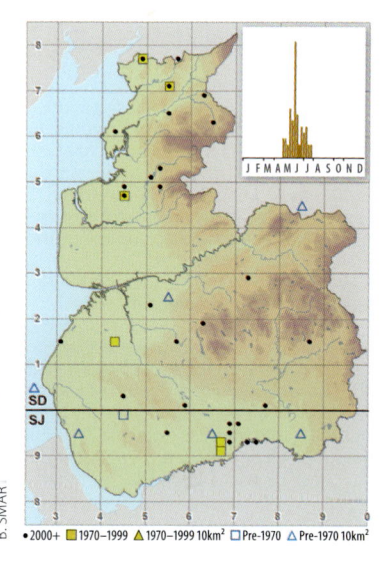

35.152 *Pseudotelphusa scalella*

■ **LANCS STATUS** Considered Extinct 1915

The *Lancashire and Cheshire Fauna Committee Annual Report* by W. Mansbridge (1916) states, 'Among the specimens sent by the latter (W.G. Clutten) was a specimen of *Gelechia scalella* which is a new record for our area and curiously enough the same species was recorded for the first time for Yorkshire in 1915'. Brindle (1939) lists it from Pendle Hill as scarce. It seems likely that these refer to the same record. However, Mansbridge (1940) makes no reference to *P. scalella*. It is not known whether Mansbridge subsequently considered the record incorrect, or if it was overlooked when compiling the updated text in his supplement list.
■ **HABITAT** Unknown.
■ **FOODPLANTS** Not recorded in Lancashire.
In adjacent counties, it was noted once in Cheshire during 1935, on the southern edge of that county. In Yorkshire, it has been noted in VC63 on several occasions up to 2004 (cheshire-moth-charts.co.uk and yorkshiremoths.co.uk accessed September 2023). Searches for adults on oak trunks in mature deciduous woodland during May and June would be useful if trying to relocate this species in Lancashire.

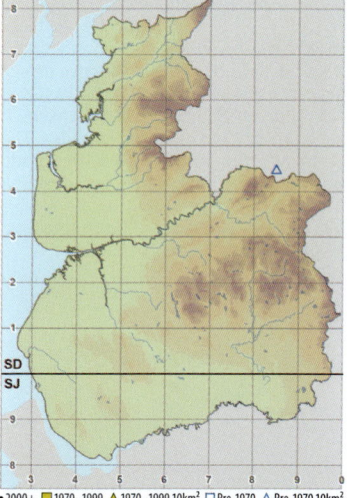

The Moths of Lancashire **189**

GELECHIIDAE

35.153 *Pseudotelphusa paripunctella*

■ **LANCS STATUS** Scarce / very local resident   c.1857–2011

First reported from 'fallows around mosses' in the Liverpool district (Gregson, 1857) and then from the Manchester district, by Stainton (1859). A single larva was found on birch at Astley Moss on 1 September 2001 (I. Smith) and subsequently bred. The moth continued to be recorded regularly from this Moss, at light, until May 2011. It is unknown whether reduced recording or changes in site management led to the lack of more recent records. Only one other record is known, from Swinton in 2011.
■ **HABITAT** Moss edges with scrub and tree growth.
■ **FOODPLANTS** Birch.
Single brooded. Usually seen in low numbers but seven came to MV light in 2010.

• 2000+   ▢ 1970–1999   ▲ 1970–1999 10km²   ▢ Pre-1970   △ Pre-1970 10km²

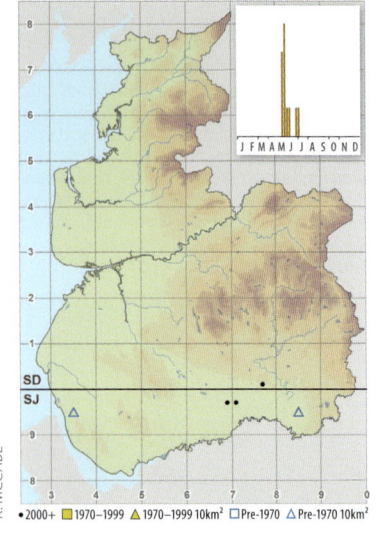

35.154 *Xenolechia aethiops*

■ **LANCS STATUS** Considered Extinct   c.1857–c.1890

The earliest records paint a rosy picture, with it being noted as plentiful on the mosses of the Liverpool district, where burnt (Gregson, 1857). Shortly afterwards, it was found on Chat Moss by J. Chappell, probably around 1859. Despite the apparent mid-19th century profusion, the only other sighting, undated but prior to 1890, was by J.H. Threlfall in Longridge (Ellis, 1890). Despite this, it was still being reported as local on heaths and mosses. The rapid decline and possible extinction of this species appeared complete by the time Mansbridge (1940) noted that there were no additional records since those published in 1890.
■ **HABITAT** Mosses and heather moorland.
■ **FOODPLANTS** Not recorded in Lancashire.
Recent searches in early spring sunshine on relatively recently burnt moorland, its apparent favoured habitat nationally, have drawn a blank. Nationally, records have declined over most of its range, particularly in the southern half of Britain, where the last record anywhere south of Lancashire, was in 1977 in Shropshire. It has, however, been noted more recently in moorland habitat, and once as a wanderer, not that far to the east, into Yorkshire. Further daytime searches in March and April would be recommended.

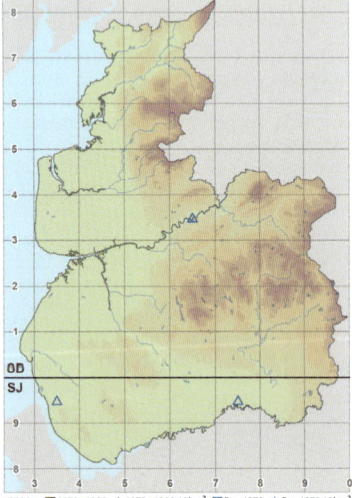

• 2000+   ▢ 1970–1999   ▲ 1970–1999 10km²   ▢ Pre-1970   △ Pre-1970 10km²

35.157 *Recurvaria leucatella*

■ **LANCS STATUS** Vagrant   2006–2010

Only two records of this distinctive gelechiid species have been received, with the first a single moth in Hale on 16 July 2006 (C. Cockbain) and two at light in Adlington on 21 July 2010 (P. Krischkiw). This, initially, seemed to be in line with a national northern extension of range. The species had started moving into lowland, north-west Cheshire in 2002, with a gradual but slow increase, at least until 2020. However, it has not been recorded again in Lancashire since 2010.
■ **HABITAT** Unknown.
■ **FOODPLANTS** Not recorded in Lancashire.
Single brooded. Comes to light.

• 2000+   ▢ 1970–1999   ▲ 1970–1999 10km²   ▢ Pre-1970   △ Pre-1970 10km²

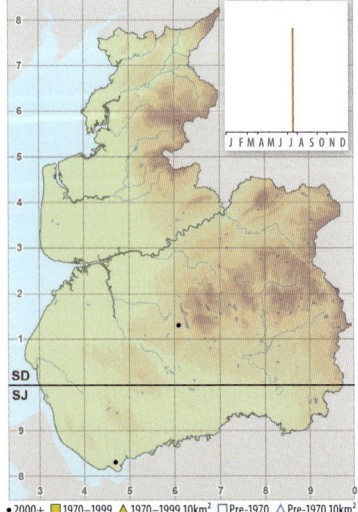

35.159 *Exoteleia dodecella*

■ **LANCS STATUS** Occasional / widespread resident   c.1857–2022
First reported in the Liverpool area by Gregson (1857), then on Chat Moss (Ellis, 1890) and Simonswood Moss in 1907 and 1915, by W. Mansbridge. It was not until the early 2000s, that an increase in sightings took place, across various parts of the county. This was mainly in southern areas, such as Billinge, Formby and Southport, but also in Preston and the Fylde. Infrequently recorded in much of VC60, with a few exceptions, but it seems likely to be under-recorded.
■ **HABITAT** Where pines are found, including woodland, moorland edge and gardens.
■ **FOODPLANTS** Pine sp., including, probably, Scots and Corsican Pine.
Single brooded. Comes to light. Larvae have been found in February and March.

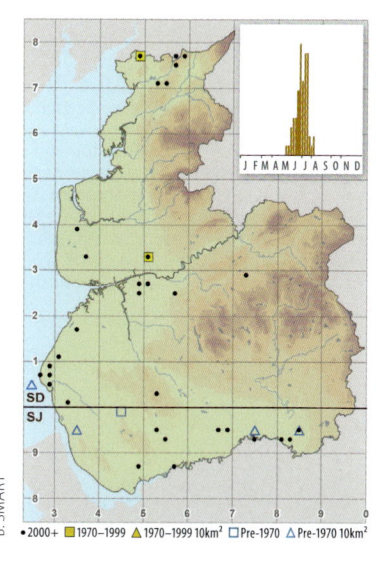

35.160 *Stenolechia gemmella*

■ **LANCS STATUS** Occasional / local resident   c.1859–2022
First noted in Manchester (Stainton, 1859) and then, at a few other lowland sites (e.g., Prescot in 1919), prior to the 1990s. Since then, reported regularly in Billinge (C. Darbyshire), Flixton and north Preston, and on a few occasions at 13 other mainly lowland sites. Upland site records restricted to Bay Horse in 2008 and Rishton in 2020. Removal of old hedgerow oak standards during housing developments are affecting local populations, with reduced reports in some areas.
■ **HABITAT** Woodland and hedgerows with mature oaks.
■ **FOODPLANTS** Not recorded in Lancashire.
Single brooded. Comes occasionally to light in ones and twos.

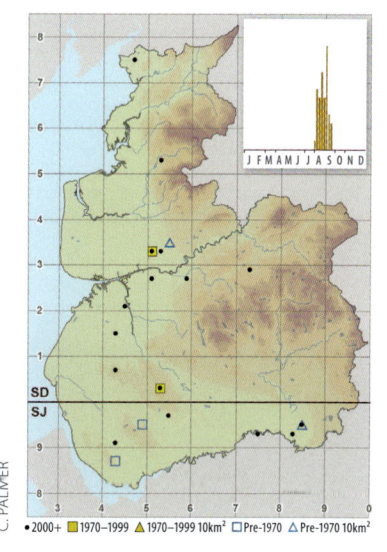

36.001 *Batrachedra praeangusta*

■ **LANCS STATUS** Occasional / widespread resident   c.1857–2022
First noted in the Liverpool area on the trunks of poplars (Gregson, 1857) and locally elsewhere (Ellis, 1890). Mansbridge (1940) noted it as common amongst Aspen and Poplar. In May 2004, a larva was seen descending on a silk thread, but without a foodplant being obvious (B. Smart) - willow was nearby. Is now annual at many sites in the lowlands, usually in small numbers. Since 2000, has also been seen in upland sites, such as Briercliffe, Littleborough, Oswaldtwistle, and Worsthorne.
■ **HABITAT** Sand dunes, mosses, brownfield sites, scrubby woodland and gardens.
■ **FOODPLANTS** Not confirmed in Lancashire.
Single brooded. Comes to light; 14 were in a MV trap at Gait Barrows in 2009.

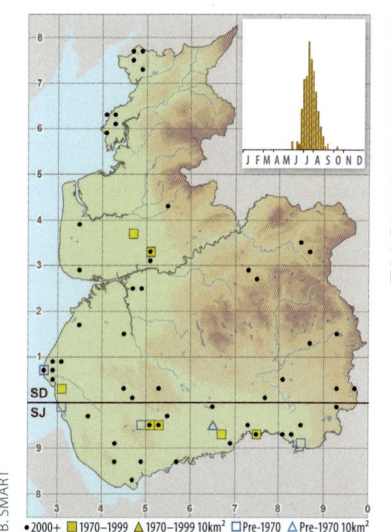

The Moths of Lancashire   **191**

BATRACHEDRIDAE

36.0019 *Batrachedra confusella*

■ **LANCS STATUS** Scarce / local resident   2009–2022

Originally recorded under *B. pinicolella* in Preston on 4 July 2009 by S. Palmer, this species was later split into *B. pinicolella* and *B. confusella* (Berggren *et al.*, 2022). All kept specimens have been dissected and redetermined, proving to be *B. confusella*. These included the first '*B. pinicolella*' above, one from Formby in 2014 and one from Parr, St Helens in 2018. Specimens from Formby (2013), Penketh (2015) and Chorlton (2020) were not retained; a photo of the latter moth is shown here (right).

■ **HABITAT** A large coastal Corsican Pine plantation and sites with isolated mature pines.

■ **FOODPLANTS** Not recorded in Lancashire.

Too few records are available to establish voltinism. Attracted to light.

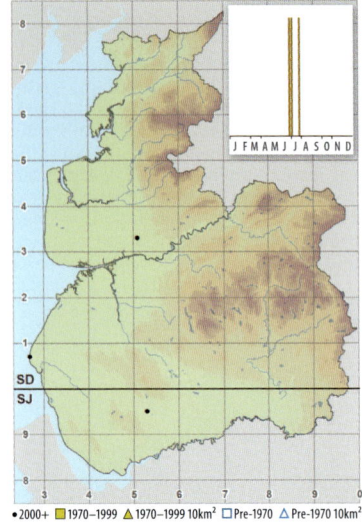

COLEOPHORIDAE

37.005 *Coleophora lutipennella*

■ **LANCS STATUS** Frequent / widespread resident   1995–2022

An early larval case of this species was found by A.M. Emmet at Gait Barrows on 18 October 1995. The next record is of an adult trapped at light at Lightfoot Green on 19 July 1996 by S. Palmer. Most of the 100+ records have been from lowland areas, such as a larval case feeding on oak at Flixton on 24 May 2019, with dissection of resultant adult. There are some upland records too, from Holden Clough, Rishton, Burnley etc.

■ **HABITAT** Oak woodland, isolated trees at brownfield sites, country parks etc.

■ **FOODPLANTS** Oak.

Single brooded. Almost all records of moths to light. Reared on a couple of occasions from cases found during May. Dissection required to separate from *C. flavipennella*.

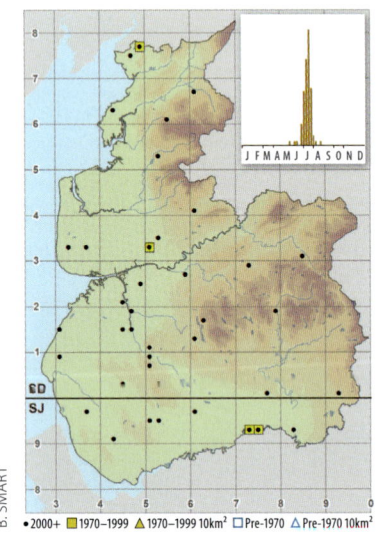

37.006 *Coleophora gryphipennella*

■ **LANCS STATUS** Occasional / local resident   c.1857–2022

First noted from the Liverpool district by Gregson (1857), and from Irlam and Withington by J. Chappell (Ellis, 1890). A well distributed species with the 100 or so subsequent records evenly divided between VC59 and VC60. Mainly found in lowland areas such as Pennington Flash and Rixton Claypits, and on limestone sites such as Gait Barrows and Warton Crag. Has also been recorded in more upland areas, e.g., Colne, Clitheroe, Dalton Crags, etc. Absent from much of the Mersey Valley.

■ **HABITAT** Woodland edges, hedgerows.

■ **FOODPLANTS** Rose, including Dog Rose and Burnet Rose.

Single brooded. At light on a handful of occasions. Most records are of larval cases.

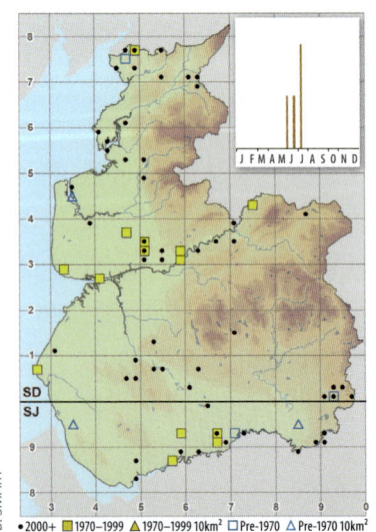

The Moths of Lancashire

37.007 *Coleophora flavipennella*

■ **LANCS STATUS** Occasional / local resident   1999–2022

It has not been possible to allocate the historic oak-feeding *Coleophora* records from Ellis (1890) to this or *C. lutipennella*. The first confirmed records were of moths to light at Gait Barrows on 22 June 1999, and Cuerden Valley a month later. Both of these were trapped by S. Palmer and subsequently dissected. Other sites recorded include Parr, Docker Moor, Mill Houses, Freshfield Dune Heath, Mere Sands Wood.
■ **HABITAT** Oak woodland, isolated trees at brownfield sites, country parks, etc.
■ **FOODPLANTS** Oak

Single brooded. Small late autumn cases bear a dorsal hump from a small fragment of mined leaf, absent in *C. lutipennella*. Spring cases of the two are indistinguishable.

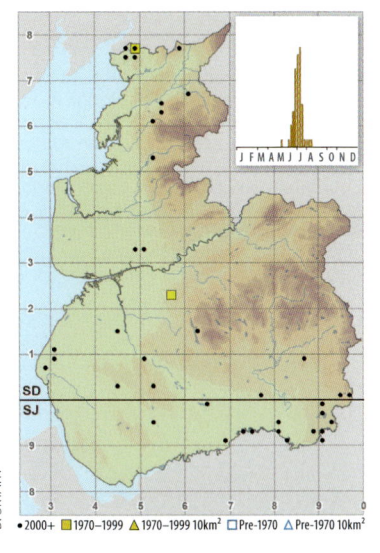

37.009 *Coleophora milvipennis*

■ **LANCS STATUS** Scarce / very local resident   2001–2022

Appears to be a recent arrival into the county, as absent from historical literature. First recorded at Astley Moss as a larval case on birch by I.F. Smith on 9 September 2001. The highest count for the species was at the same site in 2004, when ten cases were recorded by K. McCabe. A moth netted during daytime field work at nearby Cadishead Moss on 10 June 2010 was dissected and revealed as this species. All subsequent records have come from Cadishead Moss and Little Woolden Moss.
■ **HABITAT** Mosses.
■ **FOODPLANTS** Birch, including Downy Birch.

Single brooded. Almost all records of larval cases. Once netted during day.

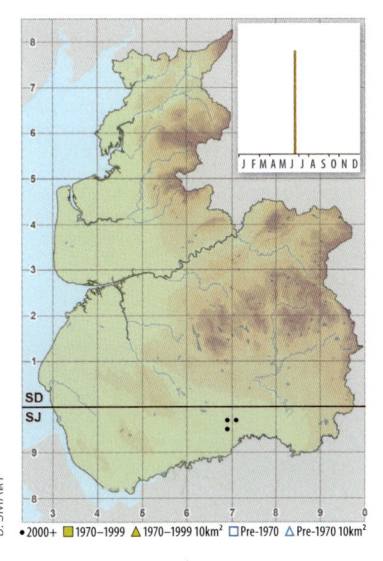

37.012 *Coleophora limosipennella*

■ **LANCS STATUS** Rare / very local resident   2021

There is only a single record of this species, with the very distinctive case noted on an elm sapling on Montague Road, Freshfield on 28 May 2021 by B. Smart. The larva was active, with a number of feeding windows evident on the nearby leaves. Subsequent searches for other cases nearby, and at Formby and Ainsdale, have so far been unsuccessful. The record may be part of a northwards range expansion of this species, evidenced by increased reporting of the cases from Yorkshire and Cheshire this century.
■ **HABITAT** Elm hedgerows.
■ **FOODPLANTS** Elm.

Nationally, single brooded.

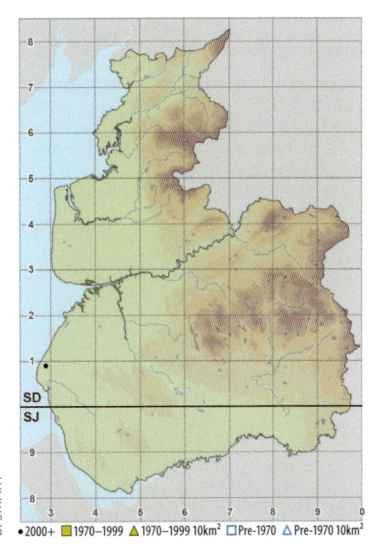

COLEOPHORIDAE

The Moths of Lancashire   193

COLEOPHORIDAE

37.013 *Coleophora siccifolia*

■ **LANCS STATUS** Scarce / local resident   c.1889–2021

First noted in Preston by J.B. Hodgkinson and J.H. Threlfall (Ellis, 1890). It was over 100 years before the next record, a case on Hawthorn at Charnock, recorded by S. Palmer on 12 June 1999, with an adult at light in Pennington three days later. Subsequent adult records from Flixton and Rainford, with larval cases found at Chorlton (twice), Stockport, Catterall and Pepper Hill, nr. Kirkham.

■ **HABITAT** Woodland edges, hedgerows.
■ **FOODPLANTS** Apple, Hawthorn, Rowan.

Single brooded, although appears largely biennial due to two-year life-cycle. All records of the moth are from odd years, with four of six larval records from even.

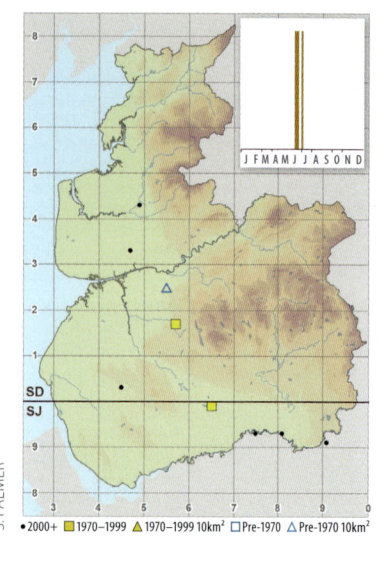

● 2000+   ▢ 1970–1999   ▲ 1970–1999 10km²   ▫ Pre-1970   △ Pre-1970 10km²

37.014 *Coleophora coracipennella*

■ **LANCS STATUS** Occasional / local resident   1919–2022

Only separable from *C. spinella* and *C. prunifoliae* by dissection. The first of 18 confirmed records was of a moth bred from Hawthorn by W. Mansbridge in 1919 at Wavertree, dissected in 2022 by S. Palmer. The next two records were both on 6 July 2001, of moths to light at Lightfoot Green and Marton. Recorded as a larval case on Blackthorn from Chorlton in 2004, with adult bred and examined. Other larval records with moths reared, from Flixton, Denton, Marshside, Broughton and Lea.

■ **HABITAT** Woodland edges, hedgerows.
■ **FOODPLANTS** Hawthorn, Blackthorn, Apple.

Single brooded. Comes to light. Larval cases recorded from April to June.

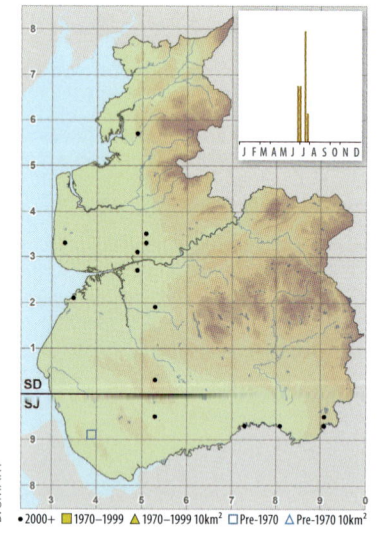

● 2000+   ▢ 1970–1999   ▲ 1970–1999 10km²   ▫ Pre-1970   △ Pre-1970 10km²

37.015 *Coleophora serratella*

■ **LANCS STATUS** Frequent / widespread resident   c.1857–2022

First noted from Liverpool, in areas 'where alders grow, always plentiful' (Gregson, 1857). Later, from Preston and Chat Moss (Ellis, 1890). Other pre-1970 records from Didsbury and Haskayne. Far more widely recorded post-1970 in lowland and upland areas, with two counts of 200 cases by S. Palmer. The first was on Alder by the north bank of the Ribble, Preston in 2001, the second on birch at Cockerham Moss in 2011.

■ **HABITAT** Woodland, hedgerows, gardens, isolated trees.
■ **FOODPLANTS** Silver Birch, Downy Birch, Alder, Italian Alder, Grey Alder, Hazel, elm, Apple, Crab Apple. Once on Rowan, Bramble, sallow and Norway Maple.

Single brooded. Comes to light, although dissection required. Most records of cases.

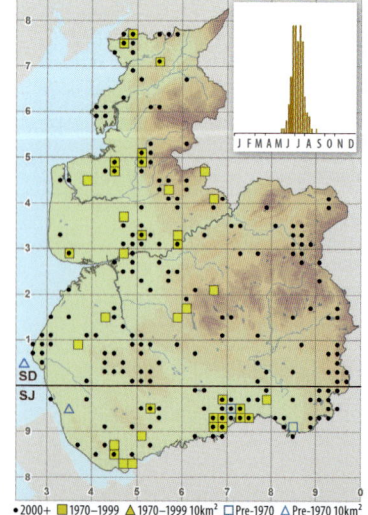

● 2000+   ▢ 1970–1999   ▲ 1970–1999 10km²   ▫ Pre-1970   △ Pre-1970 10km²

The Moths of Lancashire

37.016 Apple & Plum Case-bearer *Coleophora spinella*

■ **LANCS STATUS** Frequent / widespread resident   1998–2022
The 68 records underestimate actual abundance, as even bred moths require dissection to confirm, but also from doubt cast on historical records of *C. spinella* by the subsequent recognition of *C. prunifoliae* and *C. coracipennella* as separate species. The first confirmed record was of a case on Apple at Woodplumpton on 22 June 1998 by S. Palmer, with the moth bred. Later from Parr in 1999, Flixton in 2000 and Crossgill in 2001. Mainly from lowland areas with just a few upland records.
■ **HABITAT** Woodland, hedgerows, parks, gardens.
■ **FOODPLANTS** Hawthorn, Apple and cotoneaster.
Single brooded. Frequently to light. Larval cases noted and bred on 18 occasions.

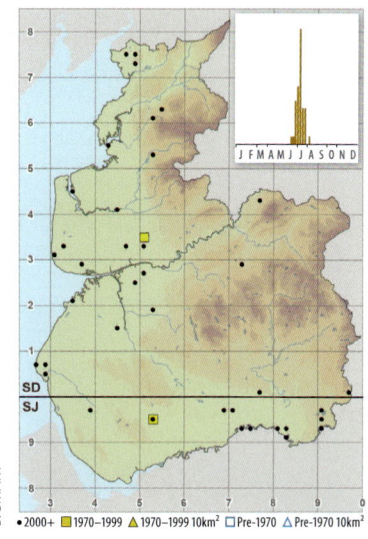

● 2000+   ▪ 1970–1999   ▲ 1970–1999 10km²   □ Pre-1970   △ Pre-1970 10km²

37.017 *Coleophora prunifoliae*

■ **LANCS STATUS** Rare / very local resident   2008
The only Lancashire record to date is of a larval case found on 21 May 2008 at Gait Barrows by C. Barnes and S. Palmer. The moth, a male, emerged on 20 June 2008 and was dissected to confirm identity. As with *C. spinella* and *C. coracipennella*, the similarity of cases, shared foodplants, and lack of external differences mean that dissection is required to separate individuals of the three species.
■ **HABITAT** Woodland.
■ **FOODPLANTS** Blackthorn.
Single brooded.

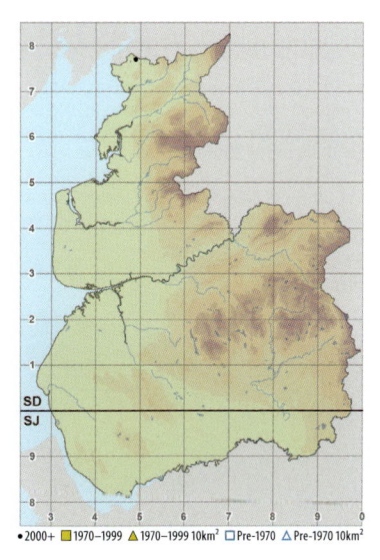

● 2000+   ▪ 1970–1999   ▲ 1970–1999 10km²   □ Pre-1970   △ Pre-1970 10km²

37.020 *Coleophora fuscocuprella*

■ **LANCS STATUS** Scarce / very local resident   1998–2022
Of the 20 records, two are from Warton Crag. All others are from Gait Barrows. This includes the first county record, a larval case found on Hazel on 16 July 1998 by S. Palmer, and the highest count of approximately 100 cases across the reserve in both sheltered and open areas, in small and larger bushes, on 11 July 2021 by M. Young, C. Palmer & S. Palmer. Accompanying the cases have been the distinctive feeding signs, where the leaves are peppered with many tiny holes.
■ **HABITAT** Limestone woodland.
■ **FOODPLANTS** Hazel.
Single brooded. All records are of the early stages, with the moth reared once.

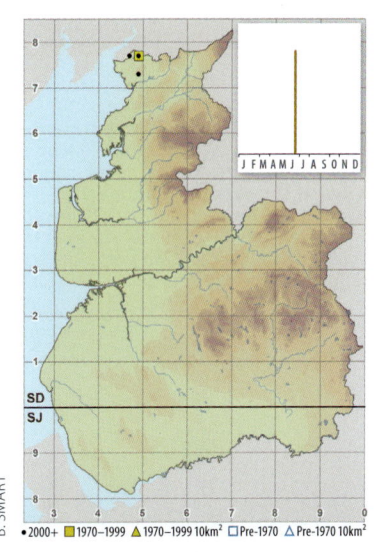

● 2000+   ▪ 1970–1999   ▲ 1970–1999 10km²   □ Pre-1970   △ Pre-1970 10km²

COLEOPHORIDAE

37.022 *Coleophora lusciniaepennella*

■ **LANCS STATUS** Frequent / widespread resident   c.1857–2022

First reported from the Liverpool district, 'on sallows in the lane leading from Mosley Hill to Allerton; also on the sweet gale at Simonswood Moss.' (Gregson, 1857). Also from Chat Moss, Farington, Preston and Longridge (Ellis, 1890). Primarily found in lowland areas, e.g., Formby, Cockerham Moss and Ash Hill, Flixton. High counts of 20 cases from Lytham on Creeping Willow, and Winmarleigh Moss on Bog-myrtle.

■ **HABITAT** Woodland, dune systems, mosses, brownfield sites.

■ **FOODPLANTS** Grey Willow, Goat Willow, Creeping Willow, Bog-Myrtle. Also noted feeding on birch and Meadowsweet, growing amongst extensive Creeping Willow. Single brooded. Often to light with 39 confirmed records. Most records are of cases.

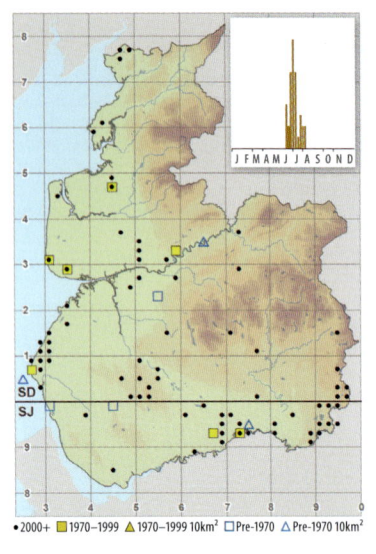

37.027 *Coleophora potentillae*

■ **LANCS STATUS** Rare / very local resident   2006

Possibly under-recorded. However, the larval cases have only been found on one occasion in the county, with three tenanted cases noted at Davyhulme Country Park on 20 October 2006. The larvae, noted by K. McCabe, were feeding on Bramble. Rearing unsuccessful, but identity subsequently confirmed from the photographs by P.H. Sterling. Neighbouring counties show no records from Cheshire and Cumbria, and seven records from Yorkshire, all from South-west Yorkshire (VC63).

■ **HABITAT** Woodland edges.

■ **FOODPLANTS** Bramble.

Single brooded. Unrecorded at light.

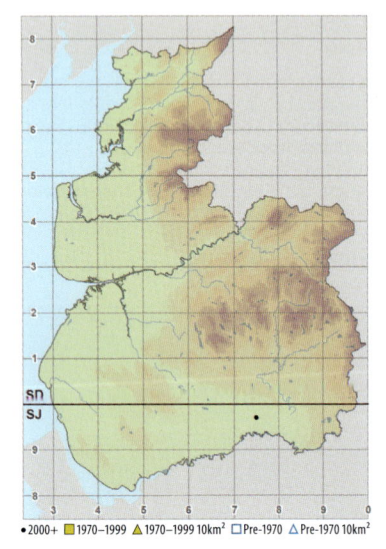

37.028 *Coleophora juncicolella*

■ **LANCS STATUS** Occasional / local resident   c.1859–2022

First recorded in Manchester (Stainton, 1859), and at Farington and Longridge by J.H. Threlfall (Ellis, 1890). The next record was of cases on Heather at Winmarleigh Moss on 25 June 1996 by S. Palmer. Nine cases were swept from Astley Moss in 2003 by K. McCabe. Also, from Birk Bank, Longridge Fell, Smithills Moor, Mossley, etc.

■ **HABITAT** Moorland, lowland mosses.

■ **FOODPLANTS** Heather, Cross-leaved Heath.

Single brooded. Twice at light. Cases obtained by sweeping the heather on nine occasions. Larvae have also been obtained by collecting debris from beaten vegetation in a container and waiting for them to crawl up the sides.

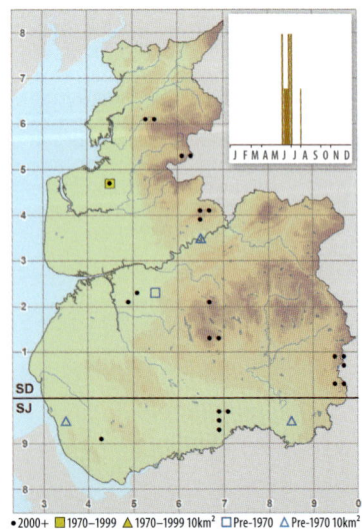

The Moths of Lancashire

37.029 *Coleophora orbitella*

■ **LANCS STATUS** Scarce / local resident   1951–2020
First recorded in 1951 by H.N. Michaelis from Heaton Norris, Stockport. Cases next found at Winmarleigh Moss on Downy Birch by S. Palmer on 9 October 2004. Moths trapped at Parr, St Helens in 2006 (R. Banks) and Billinge in 2013 (C.A. Darbyshire), both dissected to confirm identity. The only other record is of two final instar larval cases found at Cragg Wood, Littledale by S. Palmer on 4 September 2020. One was fixed for overwintering, with the other still feeding.
■ **HABITAT** Woodland, mosses.
■ **FOODPLANTS** Birch, including Downy Birch.
Single brooded. Twice at light. All other records are of larval cases.

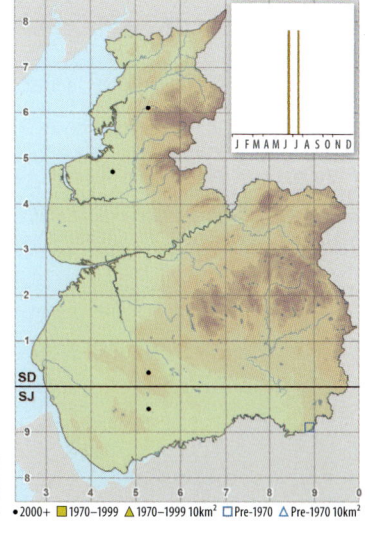

37.030 *Coleophora binderella*

■ **LANCS STATUS** Occasional / local resident   1996–2022
Two cases found on Alder at Scorton in May 1996 by S. Palmer, with adults bred. Further recorded in May 1998 at nearby Dolphinholme. Other larval records from Astley Moss, Cadishead Moss, Sutton Manor and Shaw, with adults recorded from Lightfoot Green, Flixton, Swinton, Freshfield Dune Heath and Parr, St Helens.
■ **HABITAT** Woodland, mosses.
■ **FOODPLANTS** Alder, birch. Case once found on Sweet Chestnut at Cadishead Moss 14 September 2009 by K. McCabe, with evidence of mining. Birch was growing nearby.
Single brooded. Eight records of cases, and eight of moths to light. Once netted during daytime field work. Larval cases recorded in autumn and again in May-June.

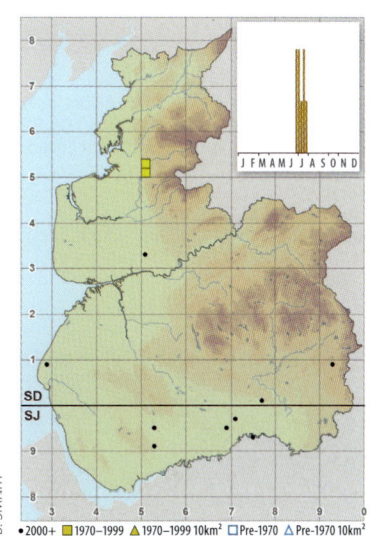

37.033 *Coleophora trifolii*

■ **LANCS STATUS** Occasional / local resident   1922–2022
First recorded at Woodvale, Freshfield by W.A. Tyerman in 1922. A long absence before the next record in 1984 of a case on the foodplant at Ince Moss by L.W. Hardwick. Moths to light at Flixton 1998, and Parr and Billinge 1999. Most records are from lowland sites, although with a few from eastern areas below the Pennines, such as Briercliffe, Rishton and Littleborough. High count of 15 moths from Heysham N.R. in 2003 (S. Palmer). 2022 records from St Helens, Denton, Middleton N.R., etc.
■ **HABITAT** Brownfield sites, rough grassland.
■ **FOODPLANTS** Melilot, including Ribbed Melilot. Cases noted on 18 occasions.
Single brooded. Moths noted around foodplant in day and also at light.

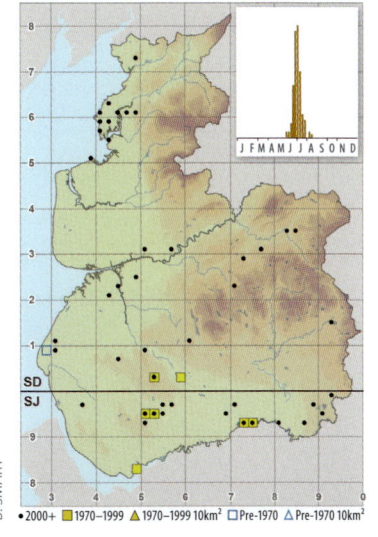

COLEOPHORIDAE

The Moths of Lancashire   197

COLEOPHORIDAE

37.034 *Coleophora frischella*

■ **LANCS STATUS** Rare / very local resident   2011–2012

Only recorded on two occasions in the county. The first was of a moth netted during the day at Upper Coldwell Reservoir, Nelson by K. McCabe on 19 May 2011. The second was at Hawes Water on 25 May 2012 by G. Jones, again of a moth netted during the day. As with almost all members of this genus, dissection was required to confirm identity. The early stages have not been found.
■ **HABITAT** Rough grassland.
■ **FOODPLANTS** Not recorded in Lancashire.
Moths recorded during daytime field work.

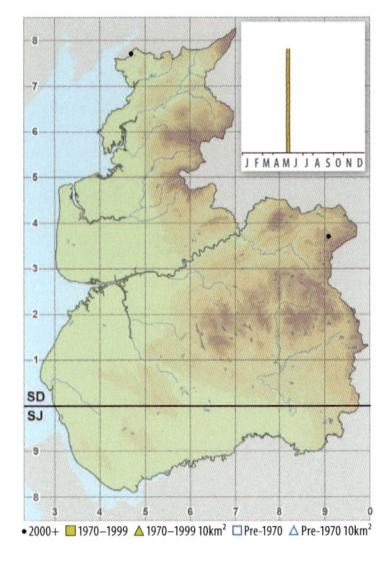

S. PALMER

37.035 Clover Case-bearer *Coleophora alcyonipennella*

■ **LANCS STATUS** Rare / local resident   2010–2023

The first record we have for this species is of a moth trapped at light at Bay Horse, Dolphinholme on 21 May 2010 by N.A.J. Rogers, with the second eight years later at the same site. Subsequently, recorded from Clayton-le-Dale by D. Bickerton on 4 September 2022, the first VC59 record, and the following day by J. Steeden at Westby. Recorded on a single occasion in 2023, a male trapped at light in Chorlton on 17 June 2023 (B. Smart).
■ **HABITAT** Unknown. All moths have been recorded at light, mainly in gardens.
■ **FOODPLANTS** Not known in Lancashire.
Double brooded. Dissection essential to separate this species from *C. frischella*.

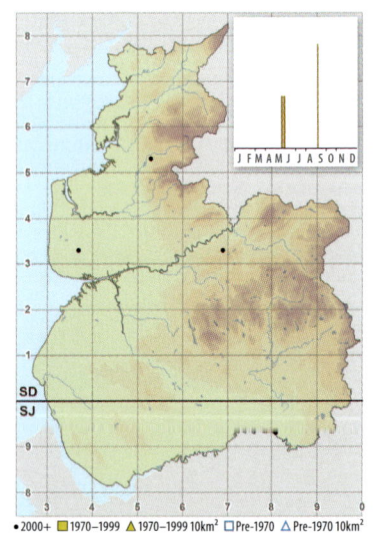

B. SMART

37.038 *Coleophora lineolea*

■ **LANCS STATUS** Occasional / local resident   2000–2020

First recorded at Parr, St Helens on 19 July 2000 by R. Banks, and again two days later at Flixton. Both were of moths at light. Moths also recorded at Bispham, Sunderland, Morecambe and Swinton. Larval cases first noted on Hedge Woundwort at Ash Hill, Flixton in May 2011 by K. McCabe and R. Hilton. Noted at the same site in 2015 on Field Woundwort. Maximum count is of ten cases on 14 September 2017 at Middleton N.R., the only other site where the early stages have been found.
■ **HABITAT** Brownfield sites, rough grassland.
■ **FOODPLANTS** Hedge Woundwort, Field Woundwort.
Single brooded. Comes to light and netted once at dusk.

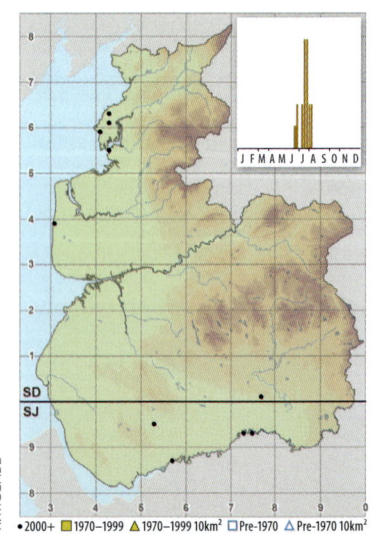

B. SMART   *K. MCCABE*

198   *The Moths of Lancashire*

37.044 *Coleophora discordella*

■ **LANCS STATUS** Frequent / widespread resident   c.1857–2022
First reported as 'plentiful … on Crosby sand-hills' (Gregson, 1857). Other 19th century records were from Pilling Moss, Preston, Manchester and Lytham, with the moth later described as common (Mansbridge, 1940). 21st century records of cases from Docker Moor, Chorlton, Rixton, Middleton N.R., Potts Corner, Marshside, etc. Likely to be found throughout lowland and upland areas where the foodplants occur.
■ **HABITAT** Rough grassland, brownfield sites, coastal meadows.
■ **FOODPLANTS** Common Bird's-foot-trefoil. Less often on Greater Bird's-foot-trefoil. Single brooded. Comes to light. Also, found amongst foodplant during day, and by sweeping. 100 of the 160 records are of cases, found from April to October.

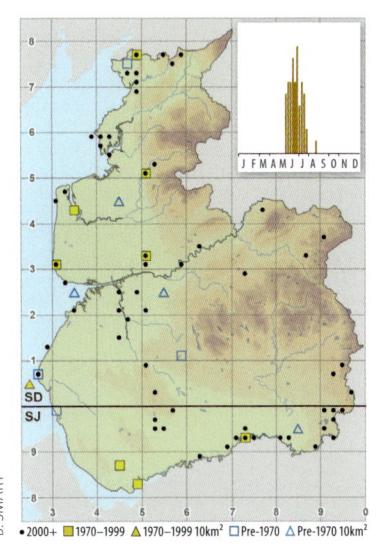

37.046 *Coleophora deauratella*

■ **LANCS STATUS** Occasional / local resident   2002–2022
Seems likely to be a new arrival in the county, as no Lancashire sites were mentioned by the early authors. A moth to light at Billinge in 2002 (C.A. Darbyshire) and one the following year in Chorlton were the first confirmed records of this species. The highest counts are of 18 cases at Great Moss, nr. Wigan by C.A. Darbyshire in 2007, and of 16 adults netted at Ashton Moss by S.H. Hind in 2022. Cases also noted at Upholland, Roby Mill, Billinge Hill and Broadfield Park, Rochdale, etc.
■ **HABITAT** Rough grassland, brownfield sites.
■ **FOODPLANTS** Red Clover. Cases recorded in late summer and early autumn. Single brooded. Most records are of moths at light or netted during the day.

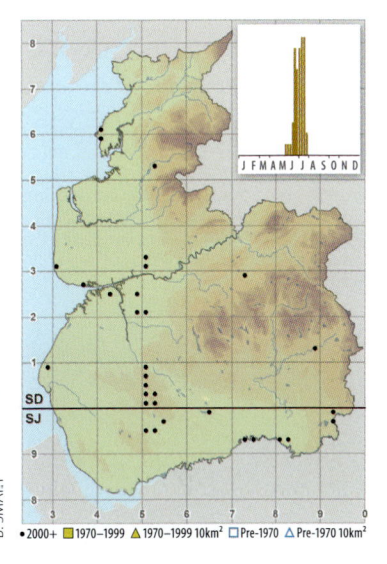

37.048 *Coleophora mayrella*

■ **LANCS STATUS** Frequent / widespread resident   c.1889–2022
First recorded in Preston by J.H. Threlfall with status described as local (Ellis, 1890). Next, from Silverdale in 1914 by W. Mansbridge. Recorded from Scorton, Formby and Martin Mere in 1980s, and in 2022 from Lightfoot Green, Birk Bank, Freshfield Dune Heath, Birkdale, Wray, Morecambe and Warton Crag. Over 198 records, primarily lowland. Also, a few upland records from Worsaw Hill, Wray, etc.
■ **HABITAT** Rough grassland, brownfield sites.
■ **FOODPLANTS** White Clover.
Single brooded. Adults come readily to light and have been netted during day. Larval cases noted among collected flower-heads in July and August, first in 2005.

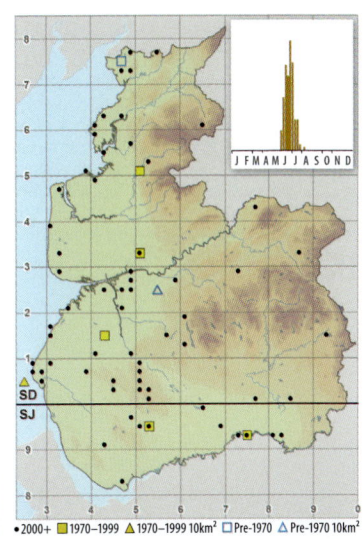

COLEOPHORIDAE

37.049 Pistol Case-bearer *Coleophora anatipennella*

■ LANCS STATUS Scarce / local resident 2000–2018

Historical records from Irlam, Manchester, Formby and Preston remain unconfirmed due to confusion at the time with *C. albidella* and the fact that the foodplant was not documented. The first confirmed record therefore, is of larval feeding signs, with over 20 small holes noted on a Hawthorn leaf at Stretford, 11 November 2000 by K. McCabe. Similar signs, but with cases also present, were noted on Blackthorn at Flixton on 6 October 2018. Again, noted on Blackthorn, although without presence of cases, at Spike Island, Widnes in October 2006 (K. McCabe). A moth trapped at Woodvale on 30 June 2018 by J. Girdley was dissected and confirmed as this species. Remains unrecorded from VC60. More frequently recorded in Cheshire and Yorkshire, with 93 records from the former county and 4 from 2022 alone in Yorkshire (cheshire-chart-maps.co.uk; yorkshiremoths.co.uk, both accessed 1 September 2023).
■ HABITAT Woodland edges, hedgerows.
■ FOODPLANTS Blackthorn, Hawthorn.
Single brooded. Most records are of larval feeding signs. Once recorded at light.

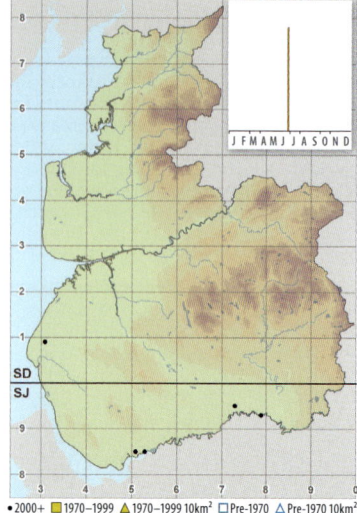

37.050 *Coleophora albidella*

■ LANCS STATUS Occasional / local resident 1928–2022

Due to early confusion with *C. anatipennella*, only records on willow or subsequently dissected have been accepted. The first was collected by W.G. Clutten at Burnley in 1928, dissected by S. Palmer in 2022. All other records from lowland areas, mainly coastal sites where Creeping Willow is utilised. Cases recorded by H.N. Michaelis in 1958 from Formby. High counts include Heysham Moss (30 moths, Aug 2015), Lytham St Annes (10 moths, August 2007) and Cadishead Moss (6 cases, April 2019).
■ HABITAT Sand dunes, woodland edges, brownfield sites.
■ FOODPLANTS Creeping Willow, Grey Willow, Goat Willow.
Single brooded. Comes to light. Tenanted larval cases from April to July.

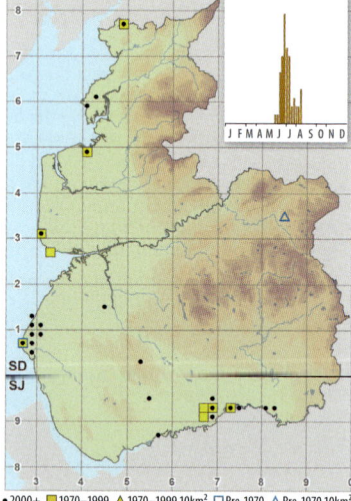

37.051 *Coleophora kuehnella*

■ LANCS STATUS Rare / very local resident 1945–1946

A moth was bred from a larva found at Formby by S. Charlson in 1945, presumably collected on oak as per the nationally known host, although the foodplant is not documented. He found two further larval cases in 1946 (Mansbridge, 1947). There have been no subsequent records within the county. Records are also extremely sparse in surrounding counties, with none from Cumbria, one from Cheshire in the mid-20th century, and a handful from Yorkshire, the most recent in 2017 (cheshire-chart-maps.co.uk; yorkshiremoths.co.uk, both accessed 1 September 2023).
■ HABITAT Unknown in the county.
■ FOODPLANTS Not recorded in Lancashire.

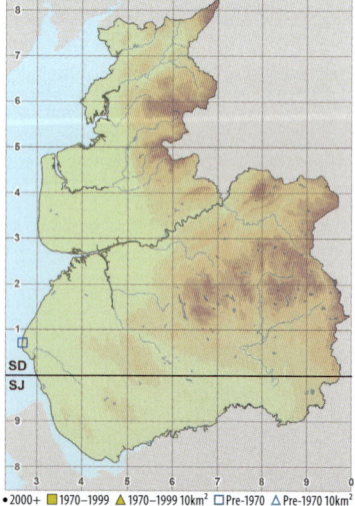

37.052 *Coleophora ibipennella*

■ **LANCS STATUS** Rare / very local resident   2007
The sole record is of two cases found on oak by K. McCabe at Highfield Moss, nr. Lowton on 13 October 2007, confirmed as this species by J.R. Langmaid and S. Palmer. Slightly more frequent in the surrounding counties than the previous species, with nine Cheshire records and twelve from Yorkshire. Unrecorded in Cumbria.
■ **HABITAT** Mosses.
■ **FOODPLANTS** Oak.

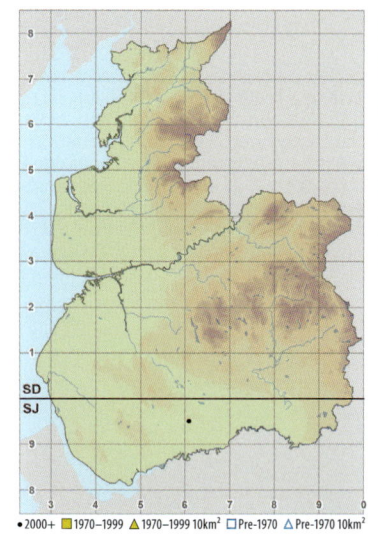

37.053 *Coleophora betulella*

■ **LANCS STATUS** Occasional / local resident   2000–2022
Recorded on 21 occasions, all from this century, and all from lowland sites in the south of the county. First noted by K. McCabe in 2000 with a larval case found on birch at Flixton. Cases also recorded from Cadishead Moss and Hardy Farm, Chorlton. Moths trapped at light in Flixton, Chorlton, Rainford, Holcroft Moss, Billinge and Swinton. Highest count is of three moths to light (one female, two males) at Holcroft Moss, 9 July 2005 (K. McCabe).
■ **HABITAT** Mosses, brownfield sites.
■ **FOODPLANTS** Birch.
Single brooded. 12 records of adults to light. The other records are of larval cases.

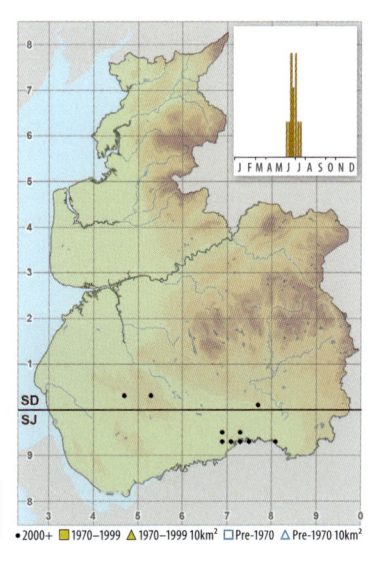

37.054 *Coleophora currucipennella*

■ **LANCS STATUS** Rare / very local resident   2018
Another apparently rare coleophorid with just a single county record. Trapped at MV light in a Flixton garden by K. McCabe on 12 June 2018. The moth, a male, was subsequently dissected and identity confirmed. The records from surrounding counties (six Cheshire records – none since 1990, none in Cumbria, and none in Yorkshire since 1885) suggest this is an unlikely candidate to have spread into the area. Possibly there is a small, extant Flixton colony, previously unrecorded.
■ **HABITAT** Unknown in the county.
■ **FOODPLANTS** Not recorded in Lancashire.

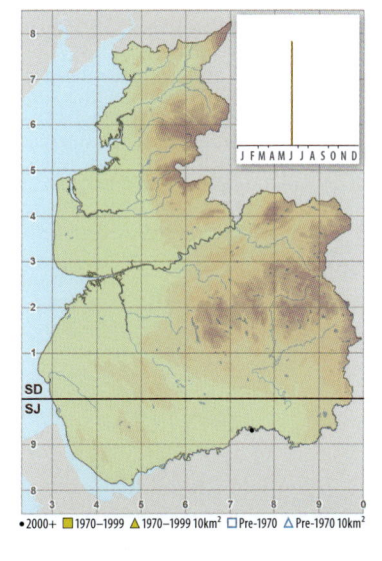

37.055 *Coleophora pyrrhulipennella*

■ **LANCS STATUS** Scarce / very local resident   c.1857–2019

First noted near Liverpool (probably on the adjacent mosses) by Gregson (1857) and in the Manchester district (Stainton, 1859). Since then, only recorded from a lowland raised bog complex at Cockerham and Winmarleigh Mosses from 1996 to 2012, towards the higher ground at Longridge Fell, 2003 and 2019 and unusually, at light, in Billinge in 1999 and 2002 (C. Darbyshire). Probably under-recorded in the county.

■ **HABITAT** Mosses and moorland.

■ **FOODPLANTS** Heather and Heath sp. (*Erica*), preferring damper areas.

Single brooded. Rare at light. Has been swept, as tenanted larval cases, from March to May, with 50 found in 2005 at Winmarleigh (S. Palmer), and empty cases in June.

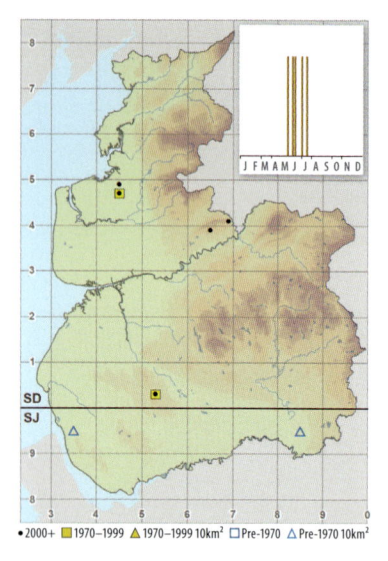

37.063 *Coleophora albicosta*

■ **LANCS STATUS** Frequent / widespread resident   c.1853–2022

First noted in the Manchester area (Stainton, 1859) and later, as 'common among Gorse throughout' (Mansbridge, 1940). Since then, recorded in lower numbers, but it remains well distributed over lowland and upland areas. Some sites do have larger counts, with 35 to light at Freshfield Dune Heath in 2016 and 12 at Docker Moor in 2021. Sites with regular reports are limited, but include Flixton and Bay Horse.

■ **HABITAT** Rough ground, brownfield sites, edges of upland grassland and moors.

■ **FOODPLANTS** Gorse.

Single brooded. Comes to light. A few late-season adults have been attracted to light between late July and early August in the Altcar and Formby area, post 2000.

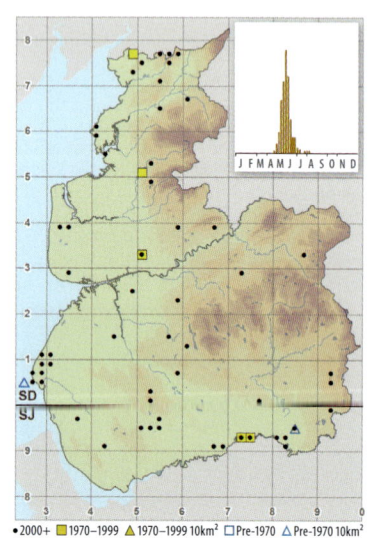

37.066 Larch Case-bearer *Coleophora laricella*

■ **LANCS STATUS** Occasional / widespread resident   c.1857–2022

Found in the Liverpool area (Gregson, 1857) and Manchester districts (Stainton,1859) and then, in 1995, cases were noted at Longridge Fell (S. Palmer). An adult was found at St Annes, and cases in Reddish Vale, both in 1999 and it has subsequently been located across much of inland Lancashire, but less frequently near the coast. Counts of cases are usually in low numbers, but 30 were seen at Chorlton Water Park on 9 May 2009. Not attracted to light with any regularity, unless larch trees are close by.

■ **HABITAT** Planted woodland, copses, shelter belts and gardens.

■ **FOODPLANTS** Larch.

Single brooded. Comes to light. Most often found as a larval case, September to May.

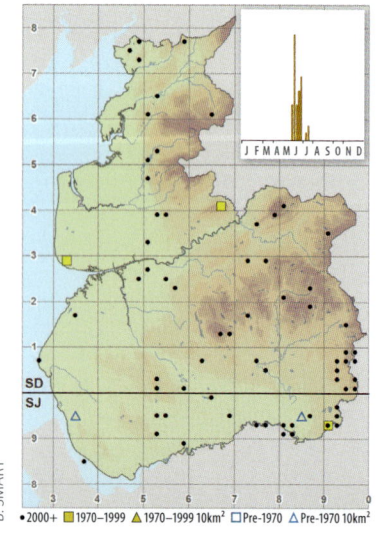

37.068 *Coleophora adjunctella*

■ **LANCS STATUS** Frequent / local resident   c.1889–2020

J.B. Hodgkinson first located it in the Preston area, sometime between 1882 and 1889 (Ellis, 1890) and, in 1940, L.T. Ford noted larval cases and adults at Bolton-le-Sands (Ford, 1941). More recently, found in St Annes (1996) and Burrow Heights Farm (2006). Low numbers came to light at Sunderland Point in 2011 and seen by day at Carnforth in 2014. Searches of saltmarsh at Fluke Hall in 2016 and 2019 showed it to be locally abundant as an adult flying at dusk, with 500 plus on 5 June 2016.

■ **HABITAT** Saltmarshes.
■ **FOODPLANTS** Saltmarsh Rush.

Single brooded. Rare at light. A wanderer was seen inland, south of Preston, in 2020.

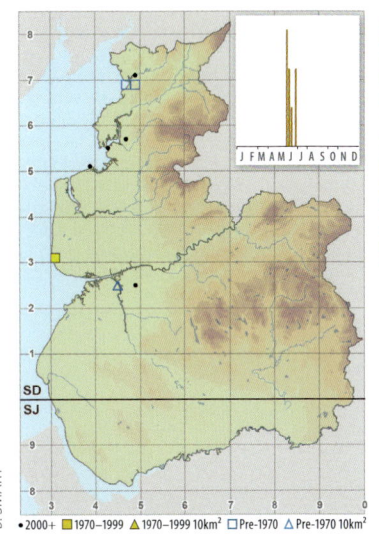

● 2000+   ▪ 1970–1999   ▲ 1970–1999 10km²   ▫ Pre-1970   △ Pre-1970 10km²

37.069 *Coleophora caespititiella*

■ **LANCS STATUS** Frequent / widespread resident   1984–2022

Records, including historic, of the very similar rush-feeding species are considered unconfirmed unless subsequently dissected (Emmet, 1996). First noted in Kirkham on 8 June 1984 (M. Evans) and in 1991 at Carr House Common. Since 1996 has been recorded widely, usually in small numbers, in lowland localities, including Catterall, Cuerden, Flixton, Holcroft Moss, and Scorton. Local in upland areas, including Besom Hill, Briercliffe, Mossley and Rishton. 114 came to light in Preston on 2 June 2020.

■ **HABITAT** Mosses, damp meadows, wet commons, canals and lower moorland edges.
■ **FOODPLANTS** Soft-rush and Jointed Rush.

Single brooded. Comes to light. Case found in a Hedge Woundwort stem in winter.

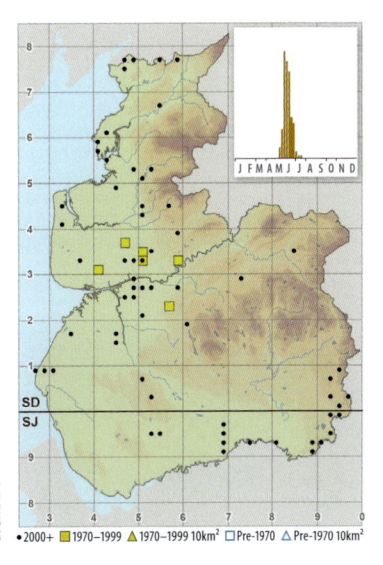

● 2000+   ▪ 1970–1999   ▲ 1970–1999 10km²   ▫ Pre-1970   △ Pre-1970 10km²

37.070 *Coleophora tamesis*

■ **LANCS STATUS** Occasional / local resident   1999–2022

Records, including historic ones, of the very similar rush-feeding species are considered unconfirmed unless subsequently dissected (Emmet, 1996). First seen in Formby on 1 July 1999 (S. Palmer), this followed by records from Scorton, Flixton, Littleborough and Gait Barrows. Usually present in small numbers, records have been generally scattered and infrequent, mostly in lowland areas. Sites with more regular sightings are Preston (2000–2014) and Bay Horse from 2006–2018 (N. Rogers).

■ **HABITAT** Wetlands, including ponds, edges of moorland and damp meadows.
■ **FOODPLANTS** Jointed Rush.

Single brooded. Comes to light. Cases seen in Chorlton and Freshfield in August 2015.

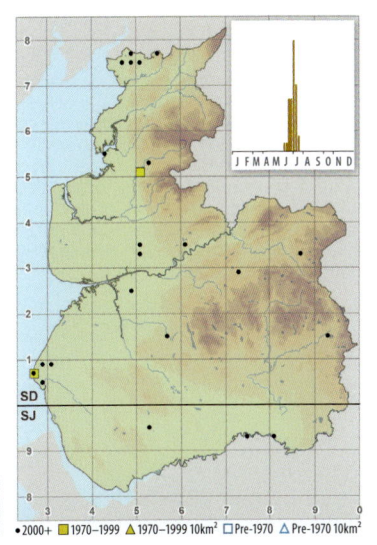

● 2000+   ▪ 1970–1999   ▲ 1970–1999 10km²   ▫ Pre-1970   △ Pre-1970 10km²

COLEOPHORIDAE

37.071 *Coleophora glaucicolella*

■ **LANCS STATUS** Occasional / widespread resident   1934–2022

Records, including historic, of the very similar rush-feeding species are considered unconfirmed unless subsequently dissected (Emmet, 1996). Early W. Mansbridge material, dissected by F.N. Pierce, has provided records from Formby in 1934 and Ainsdale in 1940. More recently it was noted on Gait Barrows in 1997, Billinge and Longshaw in 1999 and Hollingworth Lake in 2001. Sites with more regular records include Bay Horse, Dobcroft N.R., Flixton, Gait Barrows, Preston and near Longton.
■ **HABITAT** Damp meadows, wetland sites, gardens and edges of moorland.
■ **FOODPLANTS** Rush spp., including Hard Rush.
Single brooded. Comes to light. Larval cases common at Mere Sands Wood in 2003.

● 2000+   ☐ 1970–1999   ▲ 1970–1999 10km²   ☐ Pre-1970   △ Pre-1970 10km²

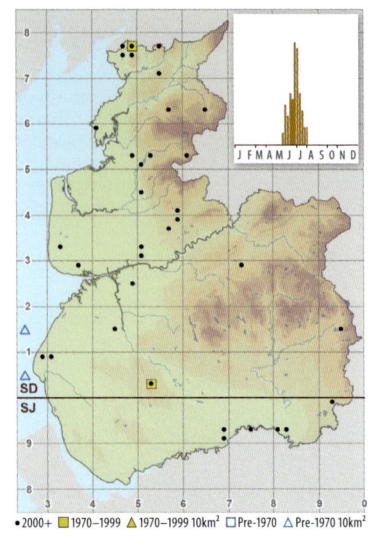

37.072 *Coleophora otidipennella*

■ **LANCS STATUS** Occasional / local resident   c.1859–2021

First noted in the Manchester district (Stainton, 1859), then at Pilling Moss in 1865 (Hodgkinson, 1865) and Crosby (Mansbridge (1940). Since then, most records have come from the old fly-ash tip at Flixton between 1999 and 2019, the largest count being a daytime one of 30 on 20 April 2009 (K. McCabe). Seen on five occasions in Chorlton (2003–2010), and a few times at other sites including Bold Moss, Heysham N.R., Marshside, Parr, Preston and Tame Valley, this last on 19 May 2021 (S.H. Hind).
■ **HABITAT** Dune grassland, lowland meadows, gardens and brownfield sites.
■ **FOODPLANTS** Field Wood-rush.
Single brooded. Comes to light. A lowland species, considered overlooked.

● 2000+   ☐ 1970–1999   ▲ 1970–1999 10km²   ☐ Pre-1970   △ Pre-1970 10km²

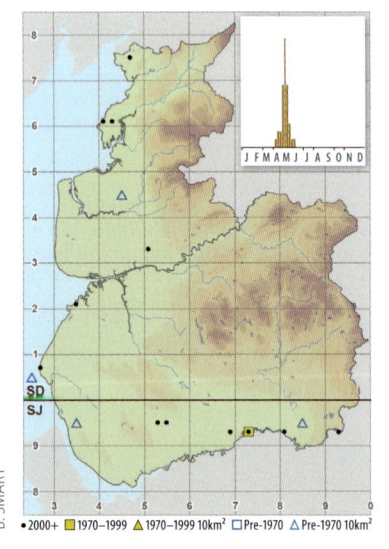

37.073 *Coleophora alticolella*

■ **LANCS STATUS** Abundant / widespread resident   1984–2022

Records, including historic, of the very similar rush-feeding species are considered unconfirmed unless subsequently dissected (Emmet, 1996). The first was noted on 8 June 1984 at Kirkham (M. Evans) and it was subsequently seen at Roddlesworth Reservoir, Longridge Fell, Lord Lot's Wood, Scorton and Littleborough. It can be extremely abundant at some upland sites, such as on Leck Fell, but also occurs widely, if less abundantly, in lowland areas. It is uncommon in most coast areas.
■ **HABITAT** Moorland, mosses, damp rough grassland and inland brownfield sites.
■ **FOODPLANTS** Rush spp., Soft-rush and Heath-rush.
Single brooded. Comes to light.

● 2000+   ☐ 1970–1999   ▲ 1970–1999 10km²   ☐ Pre-1970   △ Pre-1970 10km²

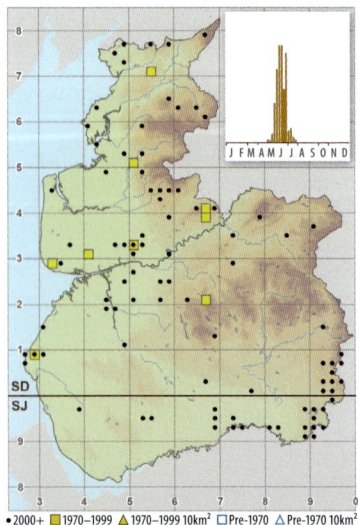

37.074 *Coleophora taeniipennella*

■ **LANCS STATUS** Occasional / local resident   1983–2022

Records, including historic ones, of the very similar rush-feeding species are considered unconfirmed unless subsequently dissected (Emmet, 1996). First found at Scorton in 1983 by L.W. Hardwick, next at Formby in 1999, Mere Sands Wood in 2003, Bold Moss in 2005, Gait Barrows in 2006 and Bay Horse in 2007. Appears to be one of the least common rush-feeding *Coleophora*, with nearly all sites in lowland areas, but could be overlooked amongst the more common species in the uplands.
■ **HABITAT** Brownfield sites, damp acidic grassland and lower moorland edges.
■ **FOODPLANTS** Jointed Rush.
Single brooded. Comes to light. Several larval cases have been found in October.

● 2000+   ■ 1970–1999   ▲ 1970–1999 10km²   □ Pre-1970   △ Pre-1970 10km²

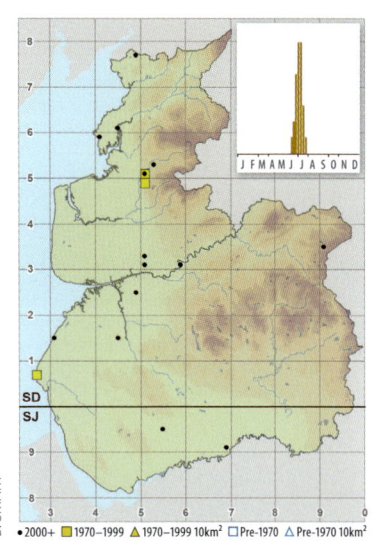

37.078 *Coleophora maritimella*

■ **LANCS STATUS** Occasional / local resident   1940–2021

First recorded in November 1940 by L.T. Ford who reported considerable numbers of larvae on the seeds of *Juncus maritimus* growing on the saltmarsh at Carnforth (Ford, 1941). It was not seen again until 2000 at Lane Ends, followed by 2003 at Potts Corner and 2004 at Burrows Marsh, while 20 larvae were found on a second visit to Potts Corner on 2 May 2006. A further site, Sunderland Point, was added on 3 June 2011 when one came to a light trap run nearby to the saltmarsh (J. Girdley).
■ **HABITAT** Saltmarsh.
■ **FOODPLANTS** Sea-rush.
Single brooded. Unusual at light. Tenanted larval cases found in November and May.

● 2000+   ■ 1970–1999   ▲ 1970–1999 10km²   □ Pre-1970   △ Pre-1970 10km²

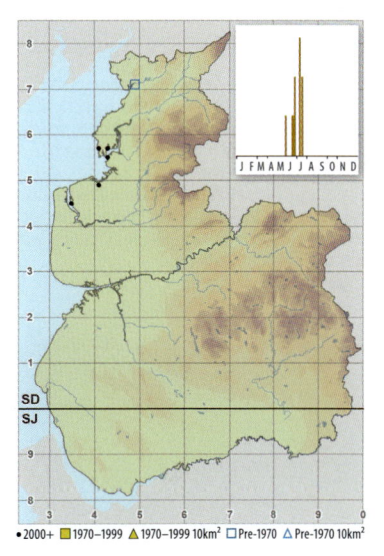

37.080 *Coleophora virgaureae*

■ **LANCS STATUS** Occasional / local resident   1997–2022

The first was netted by day at Stanah on 15 August 1997 (S. Palmer). Since then, records were mainly split between saltmarsh and limestone, with two further inland. Saltmarsh sites include Hale, Marshside and Sunderland and the limestone sites are Gait Barrows and Warton Crag. Adults and cases are often present in double figures at Gait Barrows and 30 adults were seen at Marshside on 8 August 2013.
■ **HABITAT** Limestone pavement and saltmarsh.
■ **FOODPLANTS** Goldenrod, Sea Aster.
Single brooded. Comes to light. Cases found from September to November. A very late adult was netted by day at Gait Barrows on 16 September 2008 (S. Palmer).

● 2000+   ■ 1970–1999   ▲ 1970–1999 10km²   □ Pre-1970   △ Pre-1970 10km²

The Moths of Lancashire   205

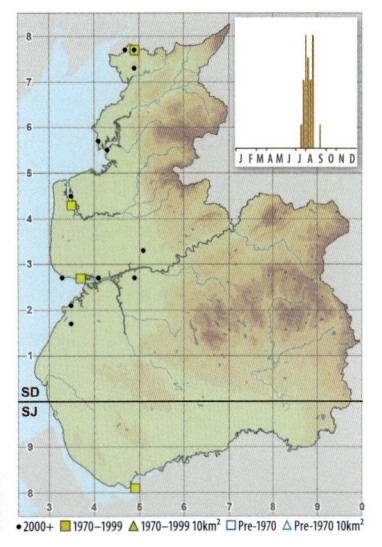

37.082 *Coleophora asteris*

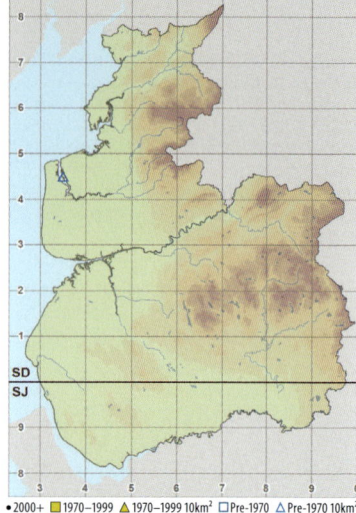

■ **LANCS STATUS** Considered Extinct 1875

J.B. Hodgkinson reported finding three colonies, under the name *C. tripoliella*, a known synonym of *C. asteris*, on Sea Aster near Fleetwood in 1875 and was listed as new to Britain based on these discoveries (Hodgkinson, 1875). It has since been found to be locally common on saltmarshes in the east and south of England and is also present at a site in North Wales and North Scotland (M. Young pers. comm.).
■ **HABITAT** Saltmarsh.
■ **FOODPLANTS** Sea Aster.

Searches in 2021 and 2022 for larvae at a few of the larger saltmarsh sites in the county with plenty of Sea Aster only produced cases of *C. virgaureae* (bred and dissected to confirm). The cases of these two species can be found in the seedheads of the foodplant and are reported to be separable (Emmet, 1996: 302), although breeding and dissection is considered necessary to confirm. Further searches of Sea Aster on the east and west banks of R Wyre are recommended, as well as at other saltmarsh sites in the county as it could be easily overlooked.

• 2000+ ■ 1970–1999 ▲ 1970–1999 10km² ☐ Pre-1970 △ Pre-1970 10km²

37.083 *Coleophora saxicolella*

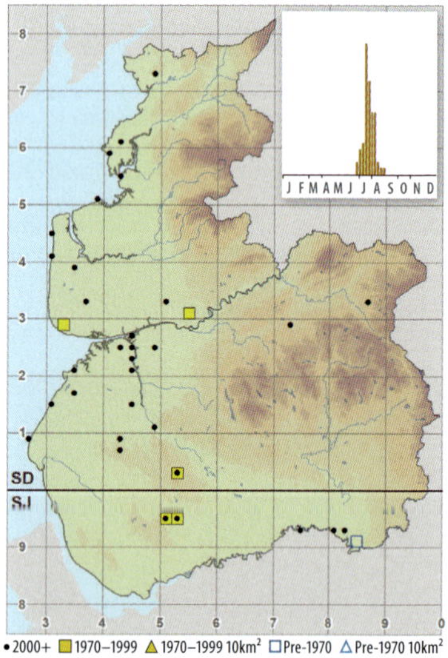

■ **LANCS STATUS** Occasional / widespread resident 1951–2022

First noted in Didsbury in 1951 by H.N. Michaelis, and next in 1995 at Ribbleton. Since then, usually as singletons at light in coastal localities and at some inland sites, such as Billinge, Flixton, Parr and Southport. Only recorded twice from higher ground, at Rishton and Worsthorne. Those sites with larger counts include Hesketh Out Marsh, Westhead old railway embankment at Ormskirk and Marshside. Probably under-recorded throughout. Only dissected specimens are included in the text.
■ **HABITAT** Rough ground, edges of saltmarshes and estuaries and edges of arable fields.
■ **FOODPLANTS** *Atriplex* sp.

Single brooded. Comes to light. Tenanted larval cases have been found in September.

B. SMART

• 2000+ ■ 1970–1999 ▲ 1970–1999 10km² ☐ Pre-1970 △ Pre-1970 10km²

37.086 *Coleophora versurella*

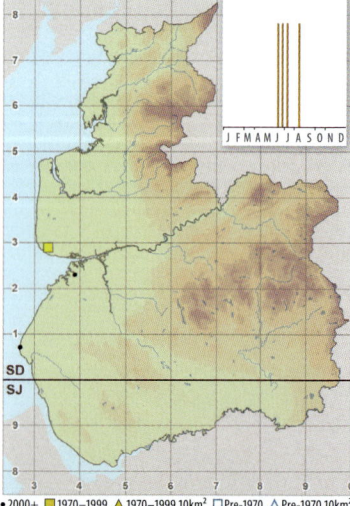

■ **LANCS STATUS** Rare / very local resident 1995–2021

Only recorded on four occasions, but probably overlooked. Two specimens were discovered, in 2022, in a large selection of retained micro-moths from the last 30 years or so, and following dissection (by S. Palmer) produced the earliest known records of this rare species in the county. These were attracted to light in St Annes on 1 July 1995 and 11 July 1997 (D. & J. Steeden). Prior to these, the first documented record refers to one at light in Banks on 16 August 2005 (A. Barker) and the other was light trapped at Lifeboat Rd., Formby on 16 June 2021 (R. Walker).
■ **HABITAT** Rough ground in coastal area, including both sandy and saltmarsh sites.
■ **FOODPLANTS** Not recorded in Lancashire. Nationally, it is associated with orache or goosefoot.

Single brooded. Comes to light. Only dissected specimens are included in the text. Elsewhere, the moth is found on waste ground, root crop field edges and saltmarshes. A few Coleophora species feed on such plants in these habitats, so breeding and dissection will be necessary to confirm identity.

• 2000+ ■ 1970–1999 ▲ 1970–1999 10km² ☐ Pre-1970 △ Pre-1970 10km²

37.087 *Coleophora vestianella*

■ **LANCS STATUS** Rare / very local resident   1951–2022

First recorded in Didsbury by H.N. Michaelis on 2 July 1951 and was noted as occasional there to 1954 (Michaelis, 1953a). The only other confirmed reports were from Formby on 14 August 2020 and Birkdale Green Beach on 19 July 2022 (both R. Walker). Emmet (1996) comments, found on the Lancashire sandhills, but with no details. There were additional, but unconfirmed, records from Fleetwood, Liverpool and Preston (Mansbridge, 1940) and a search for historic specimens is advised. Only specimens determined by dissection have been included in text.
■ **HABITAT** Coastal dunes. The inland site habitat is unknown.
■ **FOODPLANTS** Not recorded in Lancashire. Single brooded. Comes to light.

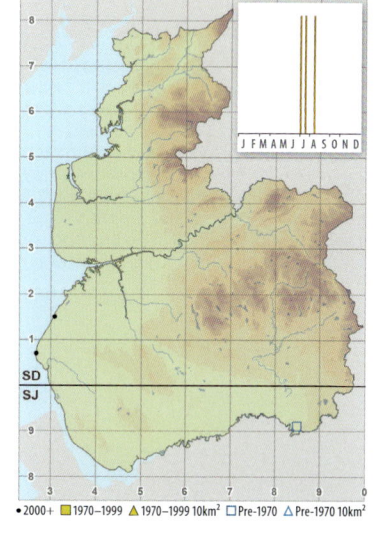

37.088 *Coleophora atriplicis*

■ **LANCS STATUS** Occasional / local resident   1996–2021

First confirmed from St Annes on 29 June 1996 (D. & J. Steeden) and from Glasson on 20 June 1997 (S. Palmer). Subsequently recorded from a range of saltmarshes, mostly during daytime visits, including Bazil Point, Burrows Marsh, Fairhaven, Marshside, Stalmine, Stanah and Sunderland. Usually seen as singletons, but ten came to light at Sunderland Point in 2006. Only dissected specimens are included and hence historic records without specimens have not been possible to confirm.
■ **HABITAT** Saltmarsh.
■ **FOODPLANTS** *Atriplex* sp.
Single brooded. Comes to light. A wanderer came to light at Bay Horse in 2018.

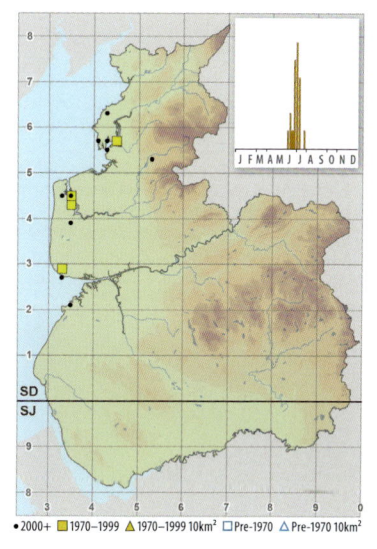

37.090 *Coleophora artemisicolella*

■ **LANCS STATUS** Frequent / local resident   1867–2022

First located in the Warrington district in 1867 by N. Greening. The next was not until October 1998, and annually since then, in Flixton (K. McCabe) at light and as larval cases. Most records in lowland VC59 are of larval cases at sites such as Bickerstaffe, Holcroft Moss, Irlam Moss, Ormskirk, Rainford and Skelmersdale. Appears to be very local in VC60, with an adult to light in Preston on 29 July 2002 and cases at Middleton N.R. in 2018 and 2019. Cases can be common; 16 were found in Southport in 2022.
■ **HABITAT** Rough and disturbed ground in lowland areas, mosses and brownfield sites.
■ **FOODPLANTS** Mugwort.
Single brooded. Comes occasionally to light. Feeding signs found August to April.

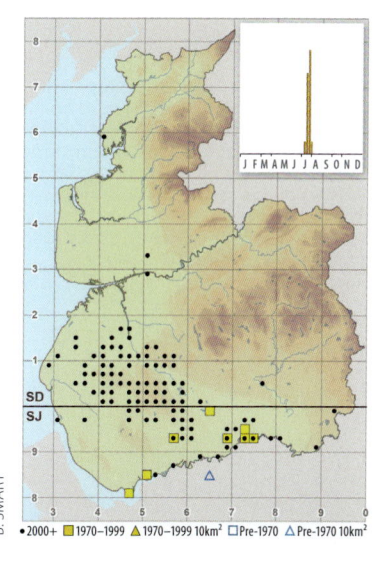

COLEOPHORIDAE

The Moths of Lancashire   207

COLEOPHORIDAE

37.093 *Coleophora peribenanderi*

■ **LANCS STATUS** Frequent / widespread resident   c.1889–2022

First found in Preston (J.B. Hodgkinson), pre-1890, under *C. therinella*, and then at Ainsdale. It was noted as a scarce and local species by Mansbridge (1940). Thereafter found in Didsbury in 1952, Freshfield in 1953 and Preston in 1995, where it has been a regular to the present day. Although it comes to light in small numbers, hunting for the larval case is the most productive way to record this species. Searches in autumn, in lowland and upland areas, have produced records across much of the county.
■ **HABITAT** Rough grassland, banks, verges, meadows, gardens and brownfield sites.
■ **FOODPLANTS** Creeping Thistle and once on a *Carduus* sp.
Single brooded. Comes to light. Found as cases Sept. to Nov. and May; rarely in June.

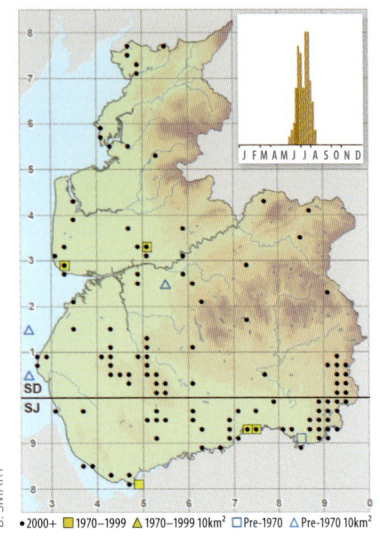

37.098 *Coleophora inulae*

■ **LANCS STATUS** Rare / very local species   2021

The only Lancashire record of this species was of one to MV light on 9 July 2021 at Mere Sands Wood, by J. Girdley. The nearest known records are in Yorkshire (yorkshiremoths.co.uk, accessed 13.2.2023), with the first there in 1995. It remains scarce and very local in that county, but has spread a little since then, suggesting this could be a new arrival in Lancashire, rather than an overlooked resident.
■ **HABITAT** Damp, herb rich meadow.
■ **FOODPLANTS** Not recorded in Lancashire.
Nationally, this species is single brooded. The moth was trapped adjacent to an extensive area of Common Fleabane, but searches in 2023 failed to locate any cases.

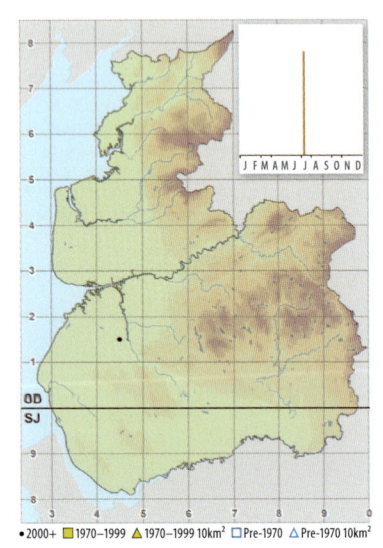

37.099 *Coleophora striatipennella*

■ **LANCS STATUS** Occasional / widespread resident   1997–2022

There is no indication of this species having been recorded historically, but it was probably overlooked. The first was netted by day at Gait Barrows on 28 June 1997 (S. Palmer) and, in the same year, it came to light in St Annes on 15 July (D. & J. Steeden). Most of the 40 records are from coastal and other lowland sites such as Chorlton and Parr, but do include a few upland sites, such as Bay Horse, Dalton Crags and Worsthorne. Only specimens which have been dissected are included in this text.
■ **HABITAT** Grassland on dunes, banks, rough areas and brownfield sites.
■ **FOODPLANTS** Common Mouse-ear.
Single brooded. Comes to light. Cases have been found from mid-July into August.

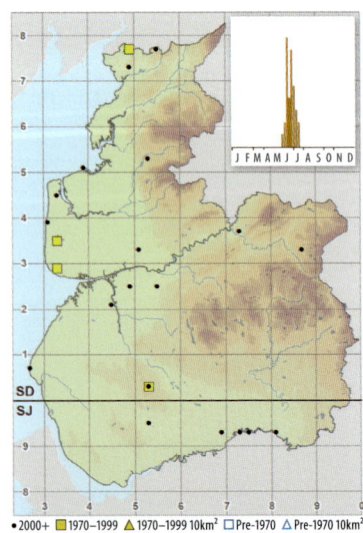

37.102 *Coleophora argentula*

■ **LANCS STATUS** Frequent / widespread resident **1997–2022**

Suspected to have spread north into the county in recent decades; not found in Cheshire until 1996 (see also map in Emmet, 1996). First recorded in Lancashire as a case on Yarrow in Flixton, October 1997 (K. McCabe), a site with near annual records since then. Mostly a species of lowland areas where Yarrow grows commonly. Not recorded in VC60 until 2006 in Lytham. One upland record, from Rishton, in 2022.
■ **HABITAT** Rough grassland, banks, verges and brownfield sites.
■ **FOODPLANTS** Yarrow.
Single brooded. Uncommon at light. Most frequently seen as larval cases, late August to March, sometimes abundantly. 147 cases seen on Middleton N.R. in October 2019.

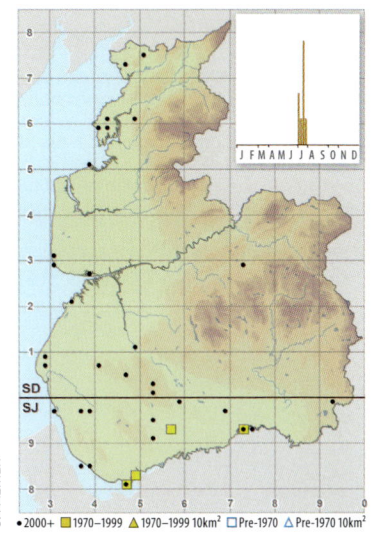

37.104 *Coleophora adspersella*

■ **LANCS STATUS** Rare / very local resident **1997–2023**

Dissection of a retained *Coleophora* specimen in 2022 by S. Palmer, produced the earliest record of this species in the county, from Fleetwood on 8 August 1997 (D. & J. Steeden). Prior to this, the first documented record related to one found during the day at Warton bank, by the Ribble estuary saltmarsh, on 21 June 2007. Later, one came to light on the south side of the same estuary (A. Barker) on 21 July 2007 and one was found in Southport on 20 July 2012 (R. Moyes). Just prior to publication, a male came to light in north Preston on 27 July 2023 (S. Palmer) and is considered a wanderer from its more usual habitats.
■ **HABITAT** Waste ground and saltmarshes.
■ **FOODPLANTS** Not recorded in Lancashire.
Single brooded. Has come to light on three occasions, the remainder being netted during the daytime adjacent to saltmarsh. All records have been confirmed by dissection.

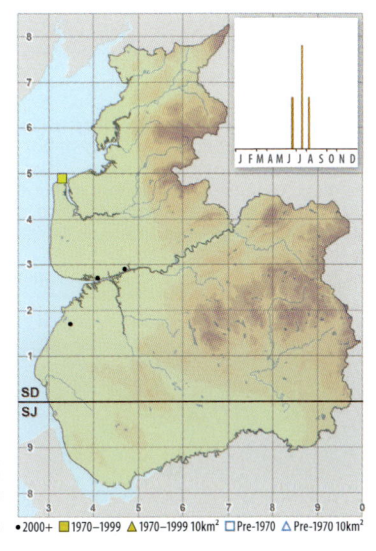

37.106 *Coleophora paripennella*

■ **LANCS STATUS** Occasional / local resident **c.1956–2022**

Despite being known from Grange, Cumbria in 1884, it was not recorded in Lancashire until a few cases were found on Formby Moss by H.N. Michaelis (Leech & Michaelis, 1957). From the late 1990s, recorded widely in coastal lowland areas and along the Mersey Valley. Further inland, encountered much less frequently, but seen twice in Clitheroe, in a disused quarry. The only upland record was of a case found at Snipe Clough on 17 April 2022 (S.H. Hind). 30 larval cases found at Flixton in 2011.
■ **HABITAT** Rough grassland, verges, edges of fields, waste ground and brownfield sites.
■ **FOODPLANTS** Black Knapweed and once on Creeping Thistle; makes a black case.
Single brooded. Once at light. Cases, in small numbers, occur from August to May.

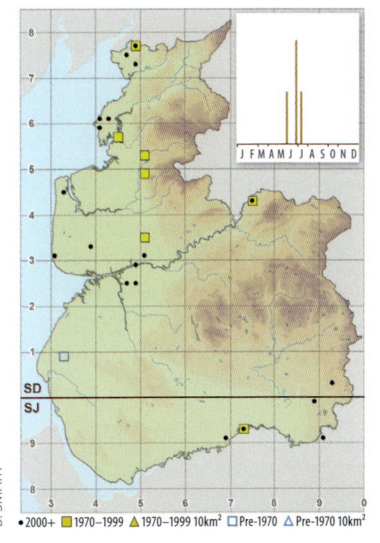

COLEOPHORIDAE

37.108 *Coleophora salicorniae*

■ **LANCS STATUS** Scarce / local resident; occasional vagrant 2006–2022
First noted, away from its saltmarsh habitat, at Bay Horse (N.A.J. Rogers) and Leighton Moss (S. Palmer), on 22 July 2006. The former site also noted another on 5 August 2006, all are considered vagrants. It was then found in its more usual habitat, at Sunderland Point, between 2011 and 2014 (J. Girdley & P.J. Marsh), including twelve on 31 July 2011. A vagrant came to light near Longton on 12 July 2022 and a probable resident was seen at light on Birkdale Beach on 19 July (both J. Girdley).
■ **HABITAT** Saltmarshes.
■ **FOODPLANTS** Not recorded in Lancashire.
Single brooded. Comes to light. Shows a national tendency to disperse inland.

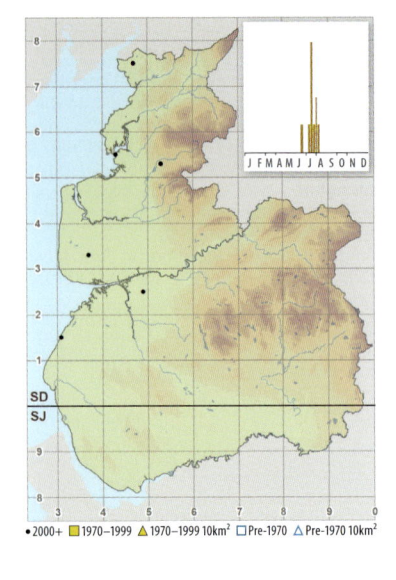

ELACHISTIDAE

38.001 *Perittia obscurepunctella*

■ **LANCS STATUS** Rare / very local resident c.1857–2018
Noted as present at Edge Lane and Fletcher Grove, Liverpool (Gregson, 1857) and later described as scarce on Honeysuckle, without documentation of further sites or records (Mansbridge, 1940). It was not until 23 April 2014 that the moth was rediscovered in the county, when a day-flying adult was netted at Warton Crag by J. Girdley. Subsequent searches for mines at the same site revealed a vacated mine in 2017, and a single leaf, tenanted by two larvae on 24 June 2018 (Smart, 2021).
■ **HABITAT** Semi-open woodland on limestone.
■ **FOODPLANTS** Honeysuckle.
Single brooded. The moth flies in sunshine; unrecorded at light.

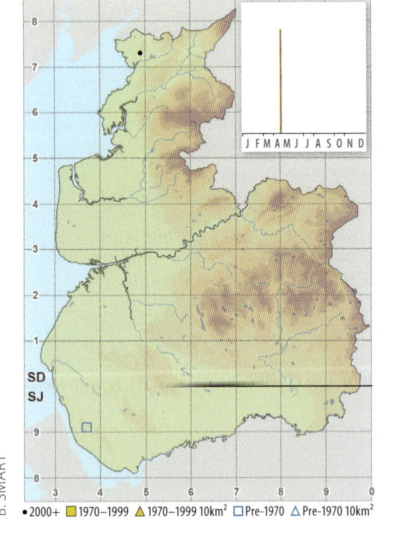

38.004 *Elachista argentella*

■ **LANCS STATUS** Frequent / widespread resident c.1857–2022
First noted in the Liverpool district on all waste lands (Gregson, 1857), and later, as abundant everywhere (Ellis, 1890). This is still a common species of lowland and upland grassland, with almost 700 records. Around 50 were swept at Wigan Flashes on 2 June 2002 by C.A. Darbyshire. Attracted to light in small numbers and readily disturbed from low vegetation during the day, occasionally flying in sunshine.
■ **HABITAT** Most grassland habitats, woodland edges, meadows, fixed dunes, etc.
■ **FOODPLANTS** Tufted Hair-grass, Red Fescue, Cock's-foot, Creeping Soft-grass, etc.
Single brooded. Tenanted mines recorded during spring on various narrow and broad-leaved grasses, with adults subsequently reared on at least six occasions.

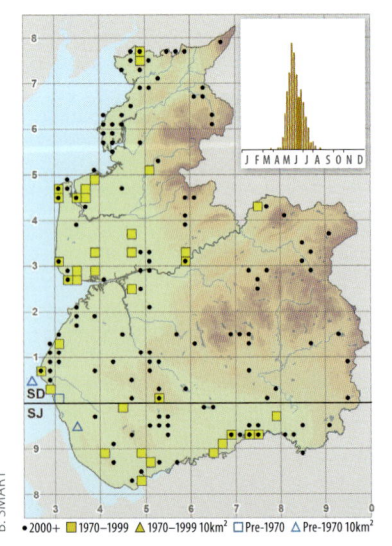

38.005 *Elachista triatomea*

■ **LANCS STATUS** Occasional / local resident   **c.1889–2022**

First recorded in the Morecambe area by J.H. Threlfall (Ellis 1890) and at Lostock by C.E. Stott (Mansbridge, 1940). The third record, at Lytham 1988, was the first of 50 modern records. Distribution in the north of the county shows a coastal bias, less evident in south Lancashire, where the moth can be found at many inland sites.

■ **HABITAT** Meadows, scrub and coastal grassland.

■ **FOODPLANTS** Red Fescue.

Single brooded. Recorded at light, and during the day. The leaf-mine pictured was found in May 2018, by checking an area where two females had been observed displaying to a swarm of amorous males on a late June morning the previous year.

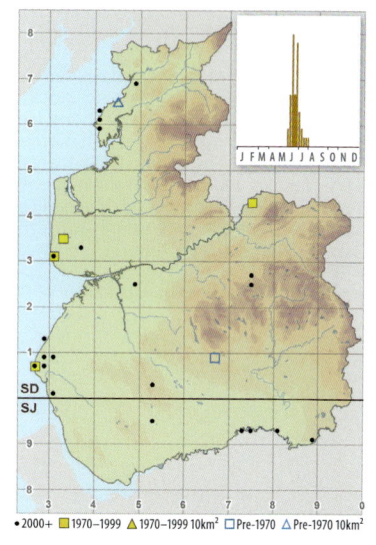

• 2000+   ▪ 1970–1999   ▲ 1970–1999 10km²   ▫ Pre-1970   △ Pre-1970 10km²

38.007 *Elachista subocellea*

■ **LANCS STATUS** Considered Extinct   **c.1891**

In 1915, C.E. Stott sent a number of micro-moths to W. Mansbridge for determination. These were collected in the Lostock area during the previous 25 years. Amongst them was a specimen of this elachistid. The moth is reasonably distinctive and so the lack of any subsequent records suggests the probability that it is now extinct within the county. Mansbridge (1940) states the record was from 1891, although this may be an approximate date. As to neighbouring counties, there is a single Cheshire record, from Alsager in 2014 (cheshire-chart-maps.co.uk, accessed 1 September 2023). It is considered a rare and very local resident in Yorkshire, although there are three records from 2022 (yorkshiremoths.co.uk, accessed 1 September 2023). VC69 (south Cumbria) is circled on the hand-written record on the Micro-moth Distribution Maps kept by A.M. Emmet, although there are no records in the Cumbrian database (S. Colgate, pers. comm.).

■ **HABITAT** Not known within the county. However, if present, woodland edges and areas of grassland containing False Brome are the likeliest habitats.

■ **FOODPLANTS** Not known in Lancashire.

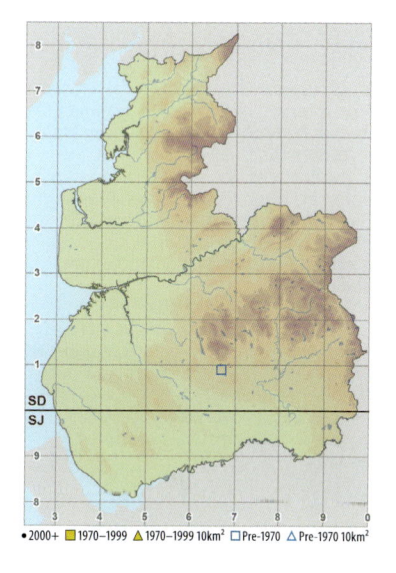

• 2000+   ▪ 1970–1999   ▲ 1970–1999 10km²   ▫ Pre-1970   △ Pre-1970 10km²

38.008 *Elachista triseriatella*

■ **LANCS STATUS** Occasional / very local resident   **2000–2022**

A few moths noted during daytime field work at Gait Barrows on 21 June 2000 by J.R. Langmaid and R.M. Palmer were the first records of this species. Unrecorded for a further 21 years until a large colony was discovered at Dalton Crags by J. Patton, with adults netted in the field and trapped at light in good numbers; 35 on 8 July 2021. A subsequent early morning visit to the site in July 2022 revealed seven moths at overnight traps, but many more flying and easily kicked up from the grasses.

■ **HABITAT** Limestone grassland in upland and lowland habitats.

■ **FOODPLANTS** Not recorded in Lancashire.

Single brooded. Readily disturbed on sunny mornings. Also, attracted to light.

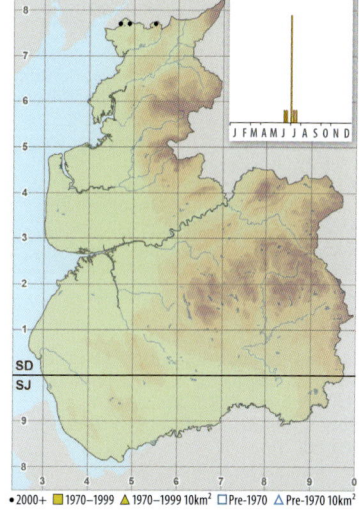

• 2000+   ▪ 1970–1999   ▲ 1970–1999 10km²   ▫ Pre-1970   △ Pre-1970 10km²

The Moths of Lancashire

ELACHISTIDAE

38.013 *Elachista cingillella*

■ **LANCS STATUS** Rare / very local resident   2000–2015

As far as we can find, there are no Lancashire records of this moth in any of the historical entomological publications. However, it has twice been recorded this century, on both occasions at Gait Barrows. The first was netted during the day on 23 May 2000. The second was recorded at light on 23 June 2015. Both records were of males, recorded and subsequently dissected by S. Palmer, with identity determined by J.R. Langmaid. The early stages have not been identified within the county. Unrecorded in Cheshire and Yorkshire. There is a 19th century record from Grange-over-Sands in Cumbria (Emmet, 1996).

■ **HABITAT** Limestone grassland.

■ **FOODPLANTS** Not known in Lancashire.

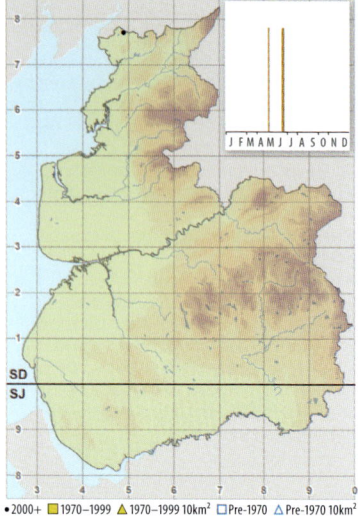

38.015 *Elachista gangabella*

■ **LANCS STATUS** Occasional / local resident   c.1940–2022

First recorded by L.T. Ford at Yealand Conyers c.1940; specimen is in the Natural History Museum. Next, at Leighton Moss 1998, then Gait Barrows 2004, with 18 records in total, all from northern limestone sites. Tenanted mines noted at Dalton Crags, Warton Crag, Gait Barrows, with mines particularly numerous at the latter.

■ **HABITAT** Limestone woodland edges and grassland.

■ **FOODPLANTS** False Brome.

Double brooded. Adults recorded from May to early July, during the day, and once at light. Tenanted mines recorded from September to November, and once in April, having overwintered. Mines containing early instar larvae noted in late July.

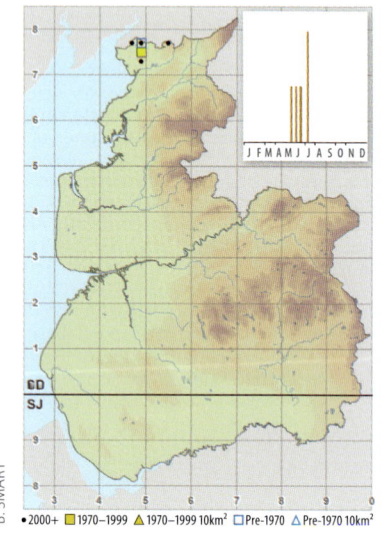

38.016 *Elachista subalbidella*

■ **LANCS STATUS** Occasional / widespread resident   c.1889–2022

First recorded at Pilling Moss by J.B. Hodgkinson (Ellis, 1890). Later, noted as common at Formby Moss by H.W. Wilson in 1953. Twice recorded in the 1980s, at Winmarleigh Moss and Risley Moss, and on a further 20 or so occasions this century.

Mainly from lowland sites, with a few at more upland sites, e.g., Oswaldtwistle and Docker Moor. The few garden records may relate to wanderers from their usual habitats.

■ **HABITAT** Mosses and moorland. Occasional in gardens.

■ **FOODPLANTS** Purple Moor-grass.

Single-brooded. Moths netted and swept during day, and also at light. Leaf-mines noted at Little Woolden Moss September 2015 and twice since; adults reared.

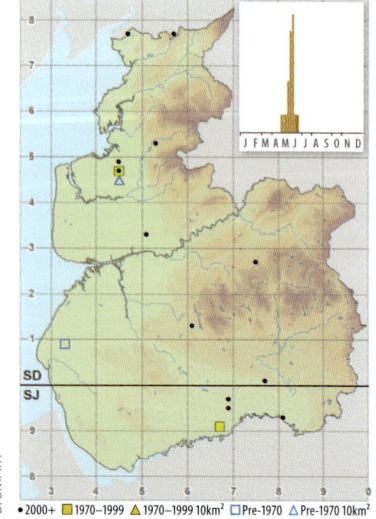

212   *The Moths of Lancashire*

38.017 *Elachista adscitella*

■ **LANCS STATUS** Occasional / local resident   c.1940–2022

The esteemed entomologist, L.T. Ford, occasionally visited Lancashire during the 1940s, and was responsible for many good records, including the first of this moth, at Halton, Lancaster. Next, recorded in 1997, then 2006, both at Gait Barrows. 36 records in total, with 14 of these from intensive work at Dalton Crags in 2021 by J. Patton, including a high count of 108 moths from 4 traps across the site on 9 August.

■ **HABITAT** Limestone grassland, on lowland and upland sites.

■ **FOODPLANTS** Blue Moor-grass. Double brooded. Moths recorded during the day and at light traps. Tenanted mines were noted at Gait Barrows on 5 April 2018 and at Dalton Crags on 26 April 2022.

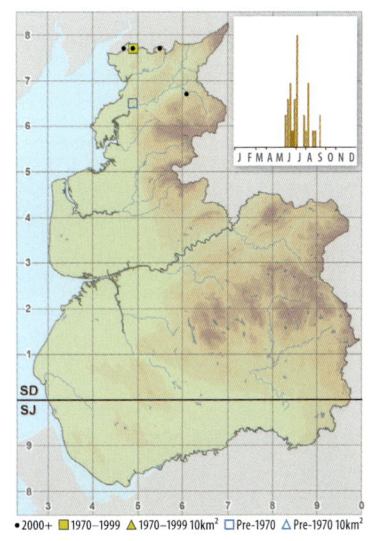

● 2000+   ☐ 1970–1999   ▲ 1970–1999 10km²   ☐ Pre-1970   △ Pre-1970 10km²

38.018 *Elachista bisulcella*

■ **LANCS STATUS** Scarce / local resident   c.1889–2022

First recorded in the late 19th century by J.H. Threlfall from the Preston district (Ellis, 1890), but not again until the 1940s at Halton and then in 2006. The four 21st century records, from Burnley, Colne, Docker Moor and Middleton N.R., suggest the moth is thinly but widely distributed. All were from the period between 13 July to 11 August.

■ **HABITAT** Woodland edges and grassland, at lowland and upland sites.

■ **FOODPLANTS** Unrecorded. Moths were bred from larvae on an undocumented foodplant at Halton, Lancaster by L.T. Ford, around 1940. The specimens produced were later dissected by E.C. Pelham-Clinton to confirm the species involved. Single brooded. The four recent records were of moths at light or netted during day.

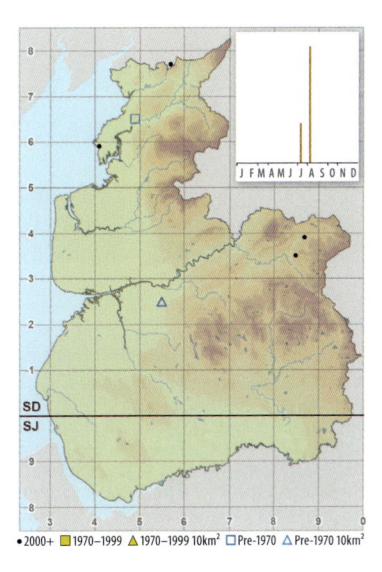

● 2000+   ☐ 1970–1999   ▲ 1970–1999 10km²   ☐ Pre-1970   △ Pre-1970 10km²

38.022 *Elachista gleichenella*

■ **LANCS STATUS** Occasional / local resident   1999–2021

Recent examination has shown that some of the earliest specimens purporting to be this moth were actually female *E. apicipunctella*. As a result, the historical records have been considered unconfirmed. The earliest definite *E. gleichenella* record is of four moths at Clitheroe by S. Palmer on 10 June 1999. All subsequent records have been in the north of the county, at Warton Crag, Gait Barrows and Myers Allotment.

■ **HABITAT** Limestone woodland and grassy banks.

■ **FOODPLANTS** Sedge (probably Wood Sedge).

Single brooded. Three moths were reared from sedge mines found at Gait Barrows on 10 April 2021 by S. Palmer. All other records are of moths netted during the day.

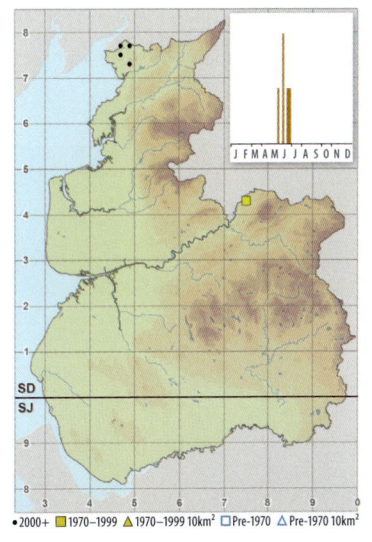

● 2000+   ☐ 1970–1999   ▲ 1970–1999 10km²   ☐ Pre-1970   △ Pre-1970 10km²

ELACHISTIDAE

38.023 *Elachista biatomella*

■ **LANCS STATUS** Scarce / local resident c.1889–2021

First reported at Silverdale in the late 19th century by J.H. Threlfall (Ellis, 1890) and at Freshfield by W. Mansbridge on 25 September 1922, with its status considered to be 'local and infrequent' (Mansbridge, 1940). The moth was next recorded at Starr Hills, Lytham in 1997 by D. & J. Steeden. More recent sightings have come from Warton Crag, where trapped at light on 9 July 2005 by M. Tordoff, and at Gait Barrows, with the species noted on five occasions between 2008 to 2015, all at light. The moth was also recorded by J. Patton at Heysham Moss at dusk on 17 August 2017.

■ **HABITAT** Woodland edges, exposed grassland, hillsides.

■ **FOODPLANTS** Glaucous Sedge. Eight tenanted mines containing early instar larvae were recorded at Worsaw Hill in October 2021 (S. Palmer & B. Smart). Two of the mines were successfully overwintered and gave rise to adults the following spring. Single brooded. Most records are of moths taken at light. Once netted at dusk.

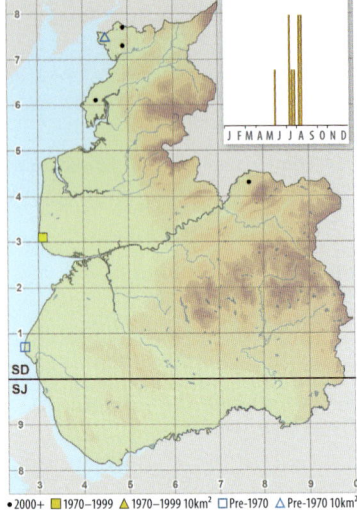

• 2000+   □ 1970–1999   ▲ 1970–1999 10km²   □ Pre-1970   △ Pre-1970 10km²

38.024 *Elachista poae*

■ **LANCS STATUS** Scarce / local resident 1961–2023

First noted at Reddish in May 1961 by H.N. Michaelis. The moth was also recorded at Orrell on 4 July 2001 by P. Alker and at Lightfoot Green on 14 May 2022 by S. Palmer, both records from garden light traps. Leaf-mines were noted on Reed Sweet-grass during 2021 at four sites in Rochdale from 20 March to 25 April, three along the Rochdale Canal and once at a pond in Mandale Park, a reclaimed tip (Smart, 2021). The pupa was also found spun to a blade of the foodplant at the latter site on 10 April 2021. The first of the adults from the canalside mines emerged on 27 April 2021. The early stages were also recorded by B. Smart at Stretford Ees in 2023, with adults reared.

■ **HABITAT** Borders of canals, streams, ponds, etc., where the foodplant occurs.

■ **FOODPLANTS** Reed Sweet-grass. Larvae mine the leaves in early spring. Appears to be double brooded, with moths recorded in May and July.

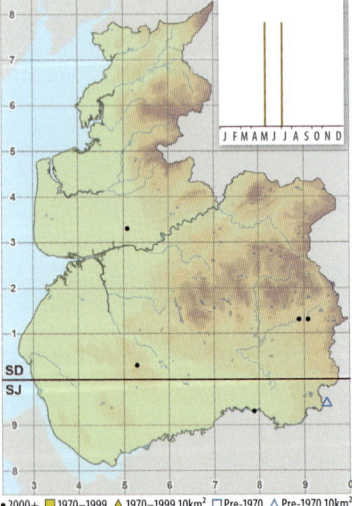

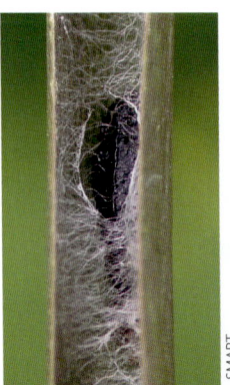

• 2000+   □ 1970–1999   ▲ 1970–1999 10km²   □ Pre-1970   △ Pre-1970 10km²

214 *The Moths of Lancashire*

38.025 *Elachista atricomella*

■ **LANCS STATUS** Frequent / widespread resident   c.1889–2022

Noted at Preston by J.H. Threlfall, with the moth described as local (Ellis, 1890). Recorded a few times in the 1940s and 1950s, then not until 1986. Subsequent trapping has shown this moth to be well distributed throughout. Very frequent in K. McCabe's Flixton garden, with over half of the 923 county records of this moth.

■ **HABITAT** Rough grassland, woodland edges, hillsides.
■ **FOODPLANTS** Cock's-foot.

Single brooded. Attracted to light, sometimes in numbers (40 to actinic light at Worsaw Hill, 20 July 2021). Occasionally disturbed from vegetation during day. Tenanted mines common where foodplant present, with moths first reared 2018.

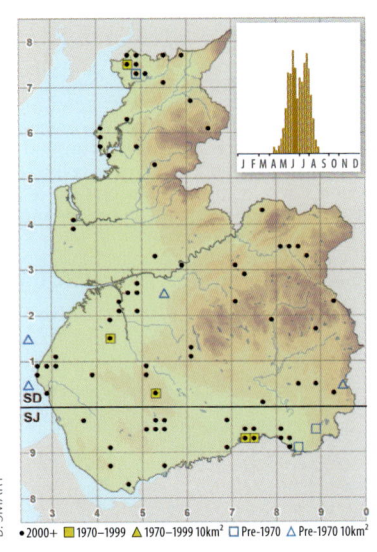

38.026 *Elachista kilmunella*

■ **LANCS STATUS** Occasional / local resident   c.1857–2021

Recorded in the 19th century at Kirkby Moss (Gregson, 1857) and at Longridge Fell in 1879 by J.B. Hodgkinson. Later, described as locally abundant (Mansbridge, 1940). This may still be the case in the correct habitat, as suggested by a daytime count of 50 moths at Winmarleigh Moss on 6 May 2003. The most recent record is of two males to light at Docker Moor on 29 June 2021. Other 21st century sites include Blaze Moss, Astley Moss, Bay Horse, Cockerham Moss and Jeffrey Hill.

■ **HABITAT** Moorland and mosses.
■ **FOODPLANTS** Leaf-mines unrecorded in Lancashire, or elsewhere.

Single-brooded. Most records are of moths flying in daytime. Twice trapped at light.

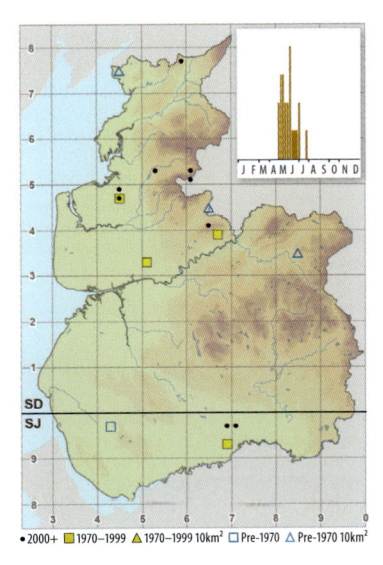

38.028 *Elachista alpinella*

■ **LANCS STATUS** Scarce / very local resident   c.1854–2015

The type locality given by Stainton (1854) was the Manchester district; probably VC59, although a small portion of Manchester extends into adjacent Vice-counties. The report was based upon a moth 'taken by Mr Edleston, on moors near Manchester, in August'. Good numbers of the species were reported at Dutton, nr. Ribchester in 1879 by Hodgkinson (1880a). He reported that 'one little wet sheltered corner on the moors yielded me a couple of hundred of the hitherto rare *Elachista alpinella* of Edleston', further noting, 'the female of this species is rather the handsomer'. The moth was later listed by Mansbridge (1940), with the entry for Lancashire noting, 'recorded from Lanc. in Meyrick's *Revised Handbook of British Lepidoptera*, but I do not know on what authority'. The moth was next recorded on 1 August 2007, with a singleton trapped at light at Dalton, nr. Parbold by C.A. Darbyshire. The only two subsequent records of this species were also noted at light; at Gait Barrows on 25 July 2014 and at Lightfoot Green, Preston on 22 August 2015, both trapped by S. Palmer. These scattered locations suggest the possibility that the moth may be under-recorded. The three 21st century specimens, all male, were dissected by S. Palmer to confirm their identity.

■ **HABITAT** Damp moorland and grassland.
■ **FOODPLANTS** Early stages and foodplant unrecorded locally. Elsewhere, long, linear mines are made in various sedges, such as Lesser and Greater Pond-sedge, and are best looked for in late spring. Single brooded.

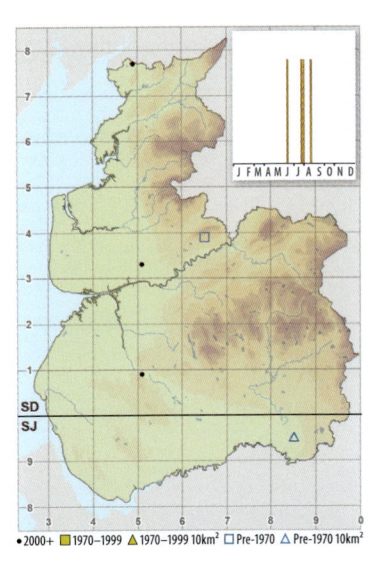

The Moths of Lancashire **215**

ELACHISTIDAE

38.029 *Elachista luticomella*

■ **LANCS STATUS** Scarce / local resident   c.1857–2022

First reported after being bred freely from Cock's-foot, presumably in the Liverpool district, although exact location unclear (Gregson, 1857). Further pre-1950 records at Burnley, Preston and a number around Liverpool. Widely distributed but possibly with declining frequency. There are only five 21st century records, from Holcroft Wood, Cockerham Moss, Warton Crag, Middleton N.R. and Garstang.
■ **HABITAT** Rough grassland on mosses and woodland edges.
■ **FOODPLANTS** Cock's-foot.

Single brooded. Larvae unrecorded since noted by Gregson. Recent records relate to moths trapped at light or disturbed from vegetation during daytime fieldwork.

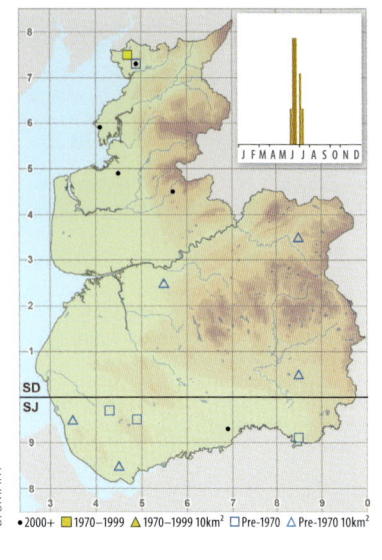

38.030 *Elachista albifrontella*

■ **LANCS STATUS** Frequent / widespread resident   c.1857–2022

First reported from Liverpool, with abundance noted 'under hedges in sheltered places' (Gregson, 1857). Status was later described as common and generally distributed by Ellis (1890). This appears to still be the case, with the moth having been recorded over 70 times in the county. Mainly noted at lowland sites, such as the Mersey Valley, but also at a few upland locations, such as Leck Fell and Parlick Fell.
■ **HABITAT** Rough grassland, meadows, woodland edges.
■ **FOODPLANTS** Cock's-foot and possibly other grasses. Leaf-mines noted April and May.

Single brooded, although a late record from 8 September 2013 may represent a partial second generation. Easily disturbed during the day; occasional at light.

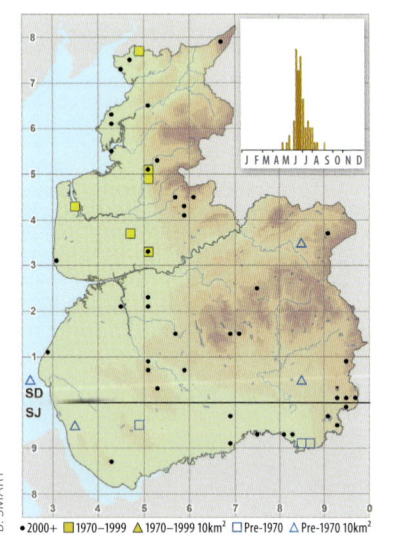

38.032 *Elachista apicipunctella*

■ **LANCS STATUS** Occasional / local resident   1850–2022

First recorded in July 1850 'on the fallows at Simonswood Moss' (Gregson, 1857). Later, described as 'local and not common' (Mansbridge, 1940). Recent records suggest it is fairly common in VC59; noted from St. Helens, Flixton, Chorlton, etc. Much less so in VC60 with only five records since 1910, all from 2003 to 2009, and none since. Primarily a lowland species, although with one or two upland records.
■ **HABITAT** Rough grassland, woodland edges.
■ **FOODPLANTS** Unknown locally, although feeding in grass stems noted (Gregson, 1857).

Single brooded, with possible, occasional second brood in August. Attracted to light. Also, noted flying during the day and easily disturbed from vegetation.

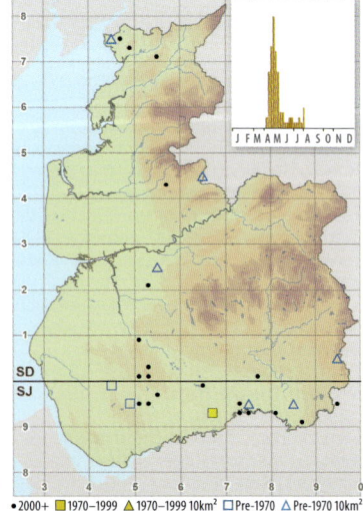

216 *The Moths of Lancashire*

38.033 *Elachista subnigrella*

■ **LANCS STATUS** Rare / very local resident   **1919–2017**
First recorded by Mansbridge at Sales Wood, Prescot in 1919, later describing the moth as 'a common species and probably more widespread than the records suggest' and noting the moth's presence at Formby and Liverpool (Mansbridge, 1940). It certainly appears far less common now, with only two post-1940 records, both of moths netted in the field at Warton Crag in 2014 and 2017 by S. Palmer.
■ **HABITAT** Dry, open grassland.
■ **FOODPLANTS** Not known in Lancashire.
Double brooded nationally, although the county records are too few to clarify the local picture. Both recent records were encountered during mid-June.

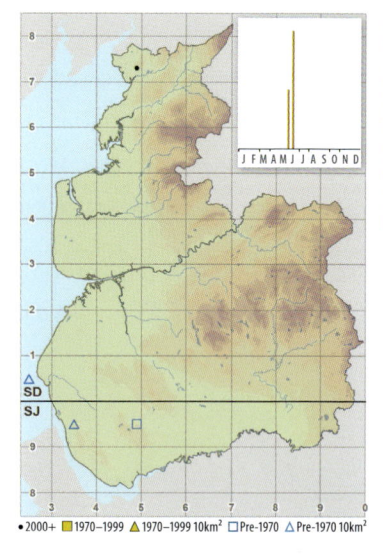

38.036 *Elachista humilis*

■ **LANCS STATUS** Occasional / local resident   **c.1865–2022**
Recorded at Brockholes Wood near Preston in the 1860s by J.B. Hodgkinson. A handwritten note in S. Palmer's copy of Mansbridge (1940) states the moth was noted by L.T. Ford at Halton in 1941, describing it to be common. Bred from mines at Formby in 1955, then absent for over 60 years. Records from Chorlton, Grize Dale and Tameside since 2018, suggest the moth is reasonably well-established.
■ **HABITAT** Sheltered areas of rough grassland.
■ **FOODPLANTS** Tufted Hair-grass.
Single brooded. Disturbed from vegetation on three occasions during daytime fieldwork in June (S.H. Hind). Tenanted mines found from late March to late May.

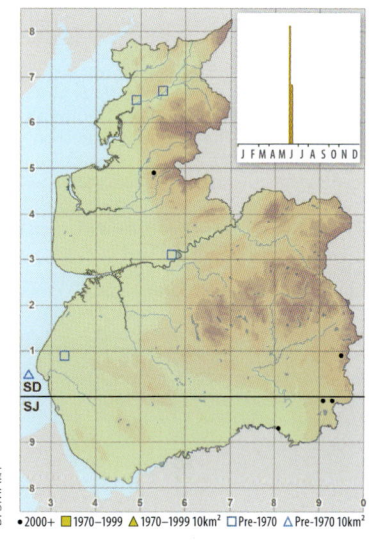

38.037 *Elachista canapennella*

■ **LANCS STATUS** Frequent / widespread resident   **c.1889–2022**
Reported as common by J.H. Threlfall at Preston (Ellis, 1890) with pre-1960 records from Crosby, Formby, Burnley, Silverdale, Oldham, Wavertree. Unrecorded from 1955 to 1989, but on many occasions since. Vast majority of the 450 records are from lowland habitats. Occasionally from upland areas, such as by Docker Moor bothy in 2021, and on Dalton Crags 26 April 2022, when a parasitised larva was found mining a blade of Sheep's Fescue, with identity confirmed by DNA barcoding.
■ **HABITAT** Most grassland habitats.
■ **FOODPLANTS** Sheep's Fescue.
Double brooded. Most records at light; occasionally netted during day.

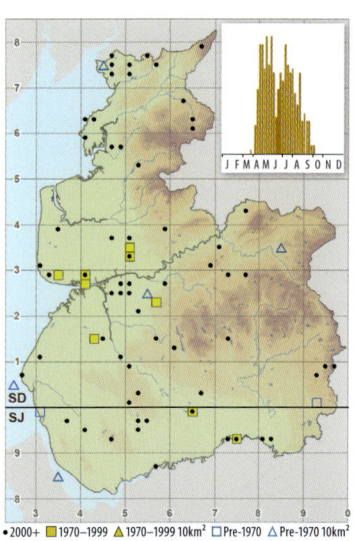

The Moths of Lancashire   **217**

ELACHISTIDAE

38.038 *Elachista rufocinerea*

■ **LANCS STATUS** Frequent / widespread resident   c.1859–2020
First recorded in the Manchester area (Stainton, 1859). Described as 'abundant everywhere' by Ellis (1890) and Mansbridge (1940). Remains widely distributed, although 25 Lancashire records in the last decade (and none since 2020), compared to 45 in the ten years prior, suggest decreasing frequency. Almost all records are from lowland sites. Approx. 300 moths noted at dusk, including numerous mating pairs, at Dalton, near Parbold 20 April 2007 (C.A. Darbyshire).
■ **HABITAT** Rough grassland.
■ **FOODPLANTS** Unrecorded. Leaf-mines recorded elsewhere on Creeping Soft-grass, etc. Single brooded. Moths recorded at light and on warm late spring evenings.

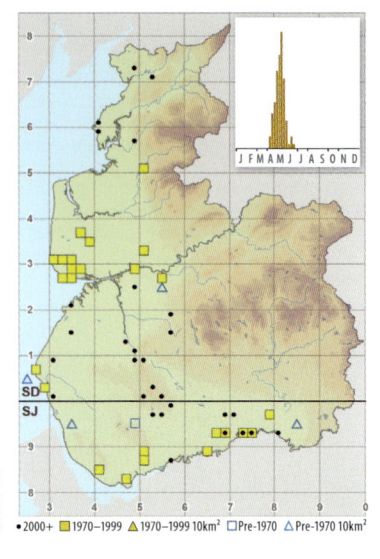

● 2000+   ■ 1970–1999   ▲ 1970–1999 10km²   □ Pre-1970   △ Pre-1970 10km²

38.039 *Elachista maculicerusella*

■ **LANCS STATUS** Frequent / widespread resident   c.1846–2022
Recorded by R.S. Edleston on White Moss in Manchester, with incidence described as scarce (Edleston, 1846). Later, noted as plentiful in wet parts of the mosses (Gregson, 1857), then as locally common (Mansbridge, 1940). Remains plentiful in lowland areas with over 400 records. A frequent visitor to garden traps.
■ **HABITAT** River and canal banks, wet meadows, marshes and mosses.
■ **FOODPLANTS** Reed Canary–grass, Common Reed and Meadow Foxtail. Phenology graph suggests bivoltinism. Tenanted mines April to May, and occasionally July to August. Reared from Meadow Foxtail in 2021, nationally a previously unrecorded foodplant. Moths easily disturbed from low vegetation during the day.

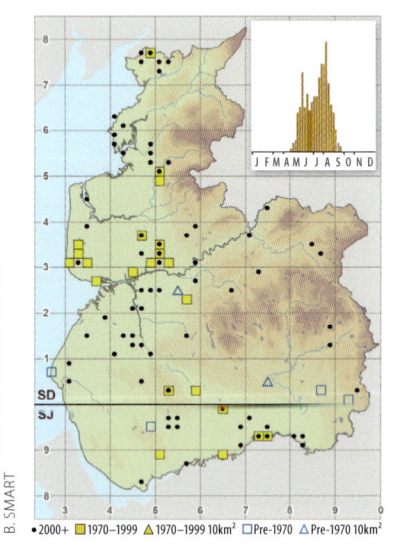

● 2000+   ■ 1970–1999   ▲ 1970–1999 10km²   □ Pre-1970   △ Pre-1970 10km²

38.040 *Elachista trapeziella*

■ **LANCS STATUS** Rare / very local resident   c.1889–2012
First recorded in the county in the 1880s by J.H. Threlfall at Caton, near Lancaster (Ellis, 1890). There were no further records until 27 April 2012, when a tenanted mine was found at woodland in Brock Valley, near Claughton by S. Palmer. The larva quickly pupated on 29 April with emergence of the moth in late May 2012. A subsequent search around the same locality in 2018 was unsuccessful, and there have been no further records elsewhere in the county.
■ **HABITAT** Shaded woodland and riverbanks.
■ **FOODPLANTS** Hairy Wood-rush.
Single brooded.

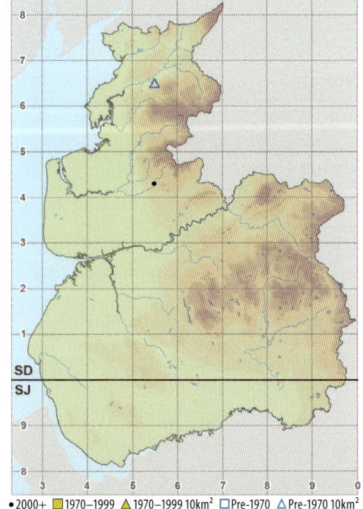

● 2000+   ■ 1970–1999   ▲ 1970–1999 10km²   □ Pre-1970   △ Pre-1970 10km²

218   *The Moths of Lancashire*

38.041 *Elachista cinereopunctella*

■ **LANCS STATUS** Scarce / very local resident   c.1940–2022

There is a 1940s record of this moth from Bolton-le-Sands, with a specimen in L.T. Ford's collection. No further records until 2 June 2009, when two adults were netted during the day at Hawes Water (S. Palmer). Distribution is limited to the Silverdale area with other records from Challan Hall Allotment, Gait Barrows and Warton Crag.
■ **HABITAT** Woodland edges.
■ **FOODPLANTS** Glaucous Sedge, possibly other sedges.
Single brooded. Leaf-mines fairly numerous beneath the trees at Hawes Water in 2016 and 2017, with adults reared (S. Palmer, B. Smart). Larvae very distinctive, thus allowing mines to be recorded elsewhere without requirement to breed through.

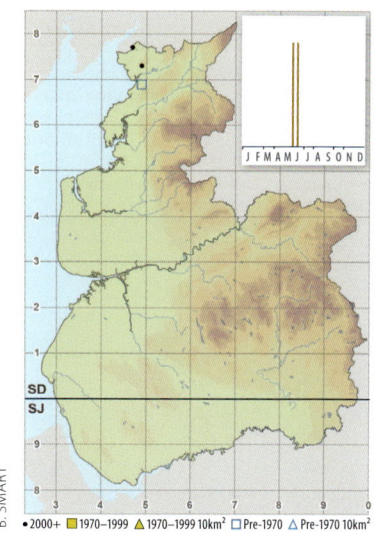

38.042 *Elachista serricornis*

■ **LANCS STATUS** Scarce / very local resident   c.1889–2021

First recorded from Pilling Moss by J.B. Hodgkinson (Ellis, 1890). No further records until 18 May 2005, when a single moth was netted during the day by S. Palmer at Winmarleigh Moss. All subsequent records have been from the north of the county, with upland records from Longridge Fell, Docker Moor and Whit Moor.
■ **HABITAT** Mosses and moorland.
■ **FOODPLANTS** Unrecorded locally. The moth is unrecorded at the Chat Moss complex, perhaps surprising given the high density of the usual foodplant, cottongrass.
Single brooded. Unrecorded at light. The five 21st century records are all of moths netted during the day, from 18 May to 24 July.

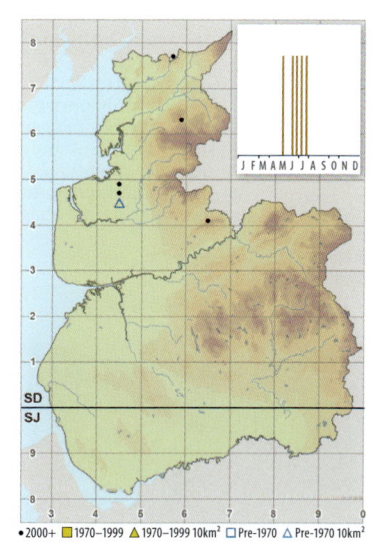

38.043 *Elachista scirpi*

■ **LANCS STATUS** Scarce / very local resident   c.1940–2019

Distribution is limited to the saltmarshes around Morecambe Bay. First recorded around 1940, with larvae common, although many parasitised, at Carnforth saltmarsh by L.T. Ford. However, not seen in the county again until 16 July 2013 at Sunderland Point by S. Palmer. Moths later recorded at Fluke Hall and Heysham Moss, the latter found away from the marsh following a very high tide.
■ **HABITAT** Saltmarshes.
■ **FOODPLANTS** Saltmarsh Rush.
Single brooded. Moths recorded by dusk and daytime fieldwork. Tenanted mines from saltmarsh at Fluke Hall, near Pilling, on 28 April 2019, with adults reared.

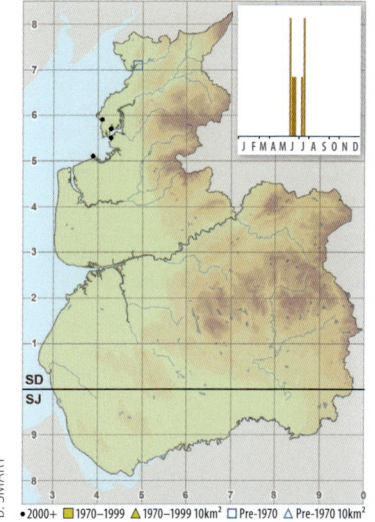

The Moths of Lancashire   **219**

ELACHISTIDAE

38.044 *Elachista eleochariella*

■ **LANCS STATUS** Rare / very local resident **2013**

Only recorded once in the county, on 25 June 2013, at Lytham St Annes Nature Reserve by J.R. Langmaid and S. Palmer, with dissection confirming identification. The moth, a male, was netted from amongst the wet area in the southern dune slack.

■ **HABITAT** Dune slacks.

■ **FOODPLANTS** Not known in Lancashire.

Nationally, single brooded.

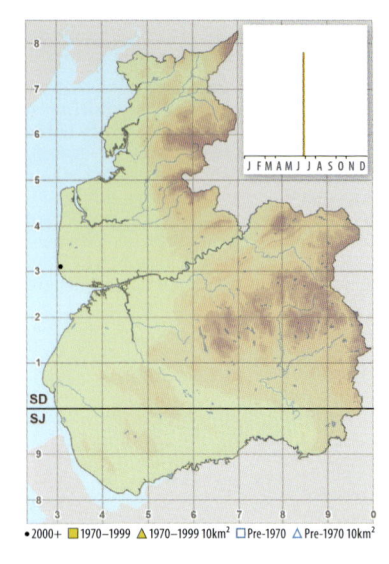

38.045 *Elachista utonella*

■ **LANCS STATUS** Rare / very local resident **1915–2017**

First recorded in June 1915 at Simonswood Moss, Knowsley by W. Mansbridge, although strangely this record is omitted from his 1940 list. 'Probably others will be found this year', he predicted (Mansbridge, 1917). Unfortunately, this was incorrect, as the only subsequent record was over 100 years later. This was a moth trapped at MV light in Flixton by K. McCabe on 16 August 2017; presumably a wanderer from the Chat Moss complex. The moth, a male, was dissected to confirm identity.

■ **HABITAT** Mosses.

■ **FOODPLANTS** Not known in Lancashire.

Nationally, single brooded.

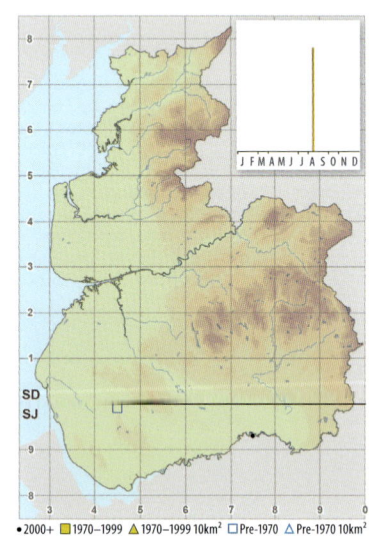

38.046 *Elachista albidella*

■ **LANCS STATUS** Occasional / local resident **c.1857–2022**

Abundance noted amongst cottongrass on mosses (Gregson, 1857). Recorded at Longridge Fell in 1871 by J.B. Hodgkinson, and at Simonswood Moss 1908 to 1915 by W. Mansbridge. Later, described as common on the mosses (Mansbridge, 1940). A dusk count of 30 at Astley Moss by K. McCabe in June 2007 suggests it may still be.

■ **HABITAT** Mosses and bogs in lowland and upland areas.

■ **FOODPLANTS** Cottongrasses.

Single brooded. Tenanted, but parasitised, mines at Little Woolden Moss in 2018, 2019, 2022 (B. Smart). Adults recorded at light, and in flight during afternoon and dusk. Occasionally trapped in gardens, likely as vagrants from the mosses and hills.

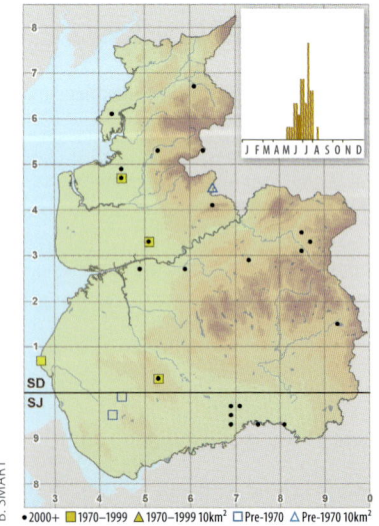

220 *The Moths of Lancashire*

38.047 *Elachista freyerella*

■ **LANCS STATUS** Occasional / widespread resident   1854–2017
Bred from leaf-mines found nr. Liverpool by C.S. Gregson in March 1854 (Stainton, 1855). Later, described as 'a common and variable species' (Mansbridge, 1940). Records suggest possible recent decline, with just four post-2011 records, compared to 15 in the decade prior. Whilst the map suggests it is widely distributed, all recent records are from VC60, including a high count of 23 moths, found in a water trough at Warton Crag on 31 May 2013 by S. Palmer. Unrecorded in VC59 since 2005.
■ **HABITAT** Rough grassland in lowland and upland areas, mainly the former.
■ **FOODPLANTS** *Poa* sp. The 1854 larvae were noted as black-headed and greenish-grey. Double brooded. Occasionally to light. More commonly netted or swept during day.

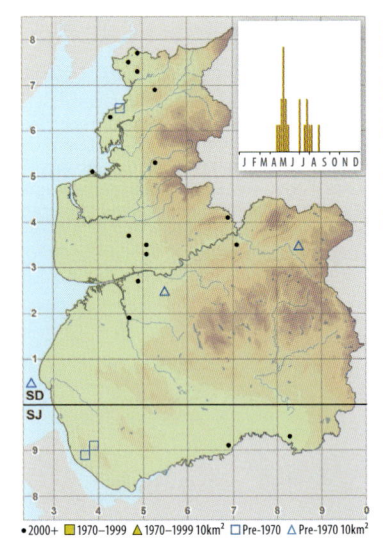

39.001 *Blastodacna hellerella*

■ **LANCS STATUS** Frequent / widespread resident   1912–2022
Because of name changes, it is difficult to determine which species, this or *B. atra*, is referred to in some historic records. A.W. Boyd's 1912 record from Boggart Hole Clough is believed to refer to this species, later noted as common (Mansbridge, 1940). Abundance seems unchanged. Recorded throughout upland and lowland areas wherever foodplant occurs. Maximum count of 31 to light at Flixton, 22 June 2003.
■ **HABITAT** Woodland edges, scrub, hedgerows, etc.
■ **FOODPLANTS** Hawthorn berries (haws).
Single brooded, with occasional, small second generation. Moths to light, swept or netted in day. Records of small holes in haws only reliable if bred, or larva checked.

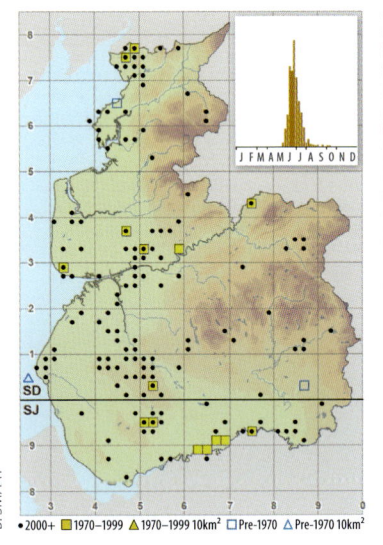

39.002 Apple Pith Moth *Blastodacna atra*

■ **LANCS STATUS** Rare / very local resident   c.1889–2002
First recorded by J.B. Hodgkinson from Preston (Ellis, 1890). Thereafter, noted by H.N. Michaelis at Didsbury, Manchester in 1953 and on 4 August 1956. The only subsequent record is of a male trapped at light on 27 July 2002 at Billinge by C.A. Darbyshire. Dissections of the 1956 and 2002 specimens were carried out by S. Palmer in 2022 as part of the preparation for this book and confirmed the identity of both as this species.
■ **HABITAT** Unknown in the county.
■ **FOODPLANTS** Not recorded in Lancashire.
Nationally, single brooded.

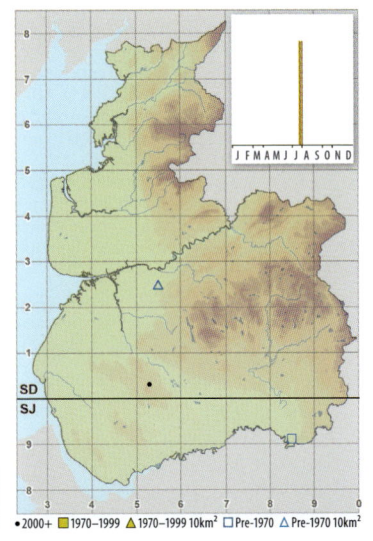

The Moths of Lancashire   **221**

PARAMETRIOTIDAE

39.003 *Spuleria flavicaput*

■ **LANCS STATUS** Occasional / widespread resident   c.1857–2022

Recorded 'in thorn hedges' in Liverpool (Gregson, 1857), and noted as common by other early authors. However, unrecorded from 1950 to 1989. Over 35 records since, but some concern that annual totals of three to four in early 2000s, have dwindled to one or two. Mainly from lowland areas such as Flixton, Halewood, Hoghton, Astley Moss, etc., with upland records limited to a singleton at Birk Bank in 2003.

■ **HABITAT** Woodland edges, scrub, hedgerows.
■ **FOODPLANTS** Hawthorn.

Single brooded. Moth at light and netted during day. Larvae feed within a twig, pupating within after forming exit-hole; recorded on a few occasions at Chorlton.

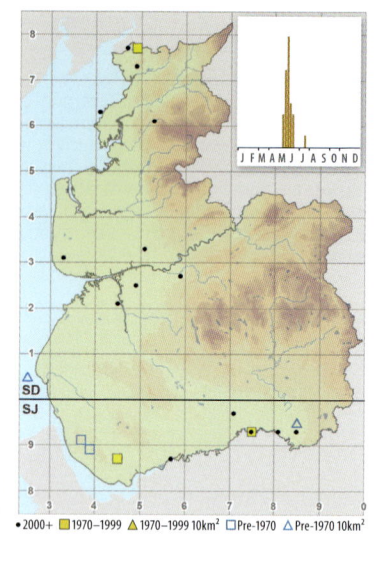

39.005 *Chrysoclista linneella*

■ **LANCS STATUS** Rare / very local resident   2022–2023

Trapped at light by P. Pemberton at Victoria Park, Manchester on 14 August 2022, the first county record. The finder noted mature limes about 200m from the garden, possibly the source of the record. Alerted by this find, B. Smart noted feeding signs of reddish-brown frass extruding from the trunks of three lime trees in Chorlton on 4 May 2023, with identity of the causer confirmed by P.H. Sterling. Searches of around 40 mature limes in Preston (2023) failed to locate feeding signs (S. Palmer).

■ **HABITAT** Tree-lined streets and parks.
■ **FOODPLANTS** Larvae feed beneath bark on the trunks of mature limes.

Single brooded.

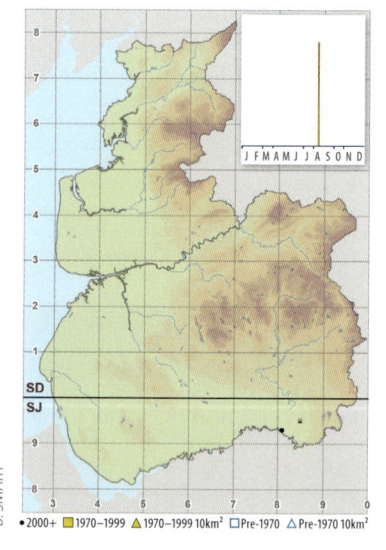

39.006 *Chrysoclista lathamella*

■ **LANCS STATUS** Rare / very local resident   2005–2011

First recorded at St Helens by G. & D. Atherton on 18 June 2005. The only subsequent record was of a moth noted during the middle of the day on 20 June 2011 by K. McCabe. This was netted whilst flying around willows in sunshine at Cadishead Moss. It seems likely that this species is under-recorded, perhaps similar to the Yorkshire situation where records have been very sparse, with only six from 1882 to 2022, rather than a recent arrival into the county.

■ **HABITAT** In woods and bogs where willows occur.
■ **FOODPLANTS** Unrecorded within the county.

Nationally, single brooded.

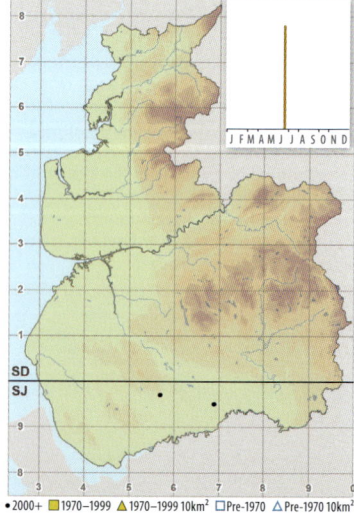

222   The Moths of Lancashire

40.001 *Mompha conturbatella*

■ **LANCS STATUS** Occasional / local resident   c.1927–2022

Whilst absent from other historical literature, noted by Meyrick (1927) to be local in the county. Next, recorded at Didsbury in 1950 by H.N. Michaelis. He also noted the species in 1953 at Chat Moss, Worsley and Formby Moss, after commenting that it was now widely distributed in the north-west due to spread of the foodplant (Michaelis, 1962). Larval spinnings found at Longworth Clough, Longton, Astley Moss, Chorlton Ees, Gait Barrows, and from Ainsdale and Birkdale Dunes, the latter in 2021.

■ **HABITAT** Scrub, brownfield sites, disturbed ground, fixed dunes.

■ **FOODPLANTS** Rosebay Willowherb. Larvae feed in spinnings of the terminal leaves. Single brooded. Comes to light. The larvae have been recorded from April to June.

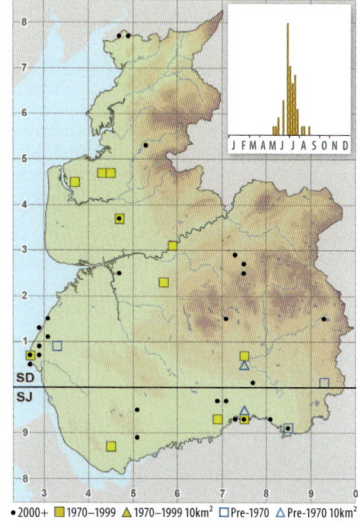

40.002 *Mompha ochraceella*

■ **LANCS STATUS** Frequent / widespread resident   c.1889–2022

First from Preston by J.B. Hodgkinson (Ellis, 1890), then Woodvale by W.A. Tyerman in 1922. Both Ellis and Mansbridge (1940) described it as very local. H.N. Michaelis made a number of finds of this species during the 1950s. These included the moth at Didsbury, a cocoon at Holden Clough, Oldham, and larvae in stems of the foodplant at Silverdale and Formby Moss. Recorded from Newton-with-Scales by M. Evans in 1984. Over 200 records since; less frequent on the higher ground of eastern Lancs.

■ **HABITAT** Rough grassland, usually in damp situations.

■ **FOODPLANTS** Great Willowherb. Larval records from May.

Single brooded. Almost all recent records are of moths to light.

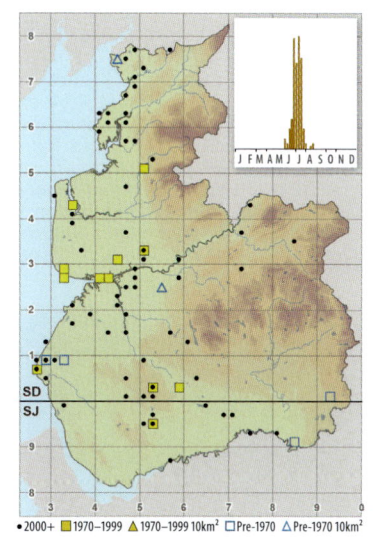

40.003 *Mompha lacteella*

■ **LANCS STATUS** Scarce / very local resident   1879–2023

From Dutton, nr. Ribchester by J.B. Hodgkinson in 1879 (Hodgkinson, 1880a). Later, described as very local, with no other sites noted (Mansbridge, 1940). No further reports until netted at dusk on 7 June 2003 and on 29 June 2009, both by S. Palmer at Gait Barrows. Four subsequent records of moths at light at Worsthorne in 2012 (G. Gavaghan, G. Turner), Silverdale Moss in 2015 (J. Patton), Bolton-le-Sands in 2018 (S. Garland) and nr. Longton on 11 July 2023 (J. Girdley).

■ **HABITAT** Scrub, rough grassland.

■ **FOODPLANTS** Unrecorded within the county.

Single brooded. Only six records, three of which at light. No larval records.

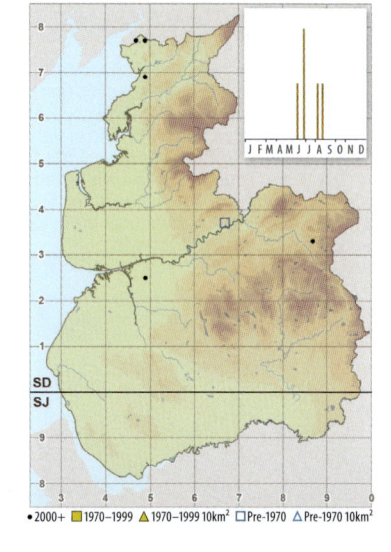

The Moths of Lancashire   **223**

MOMPHIDAE

40.004 *Mompha propinquella*

■ **LANCS STATUS** Frequent / widespread resident   1865–2022

The only historical records are from Pilling Moss in 1865 (Hodgkinson, 1865) and from Preston (Ellis, 1890). Considered a rare species by Mansbridge (1940). Unrecorded from 1890, followed by 130 well distributed records since 1984, suggesting significant range expansion. Records from 2022 from diverse locations, e.g., Flixton, Silverdale, Parr, Crosby, Worsthorne, Longton, Briercliffe, etc. Leaf-mines also noted by B. Smart, with adults reared, at Wheatacre, St Helens 2017 and at Hardy Farm, Chorlton 2019.
■ **HABITAT** Scrub, rough grassland.
■ **FOODPLANTS** Great Willowherb.
Single brooded. Leaf-mines recorded from March to April. Moth frequently to light.

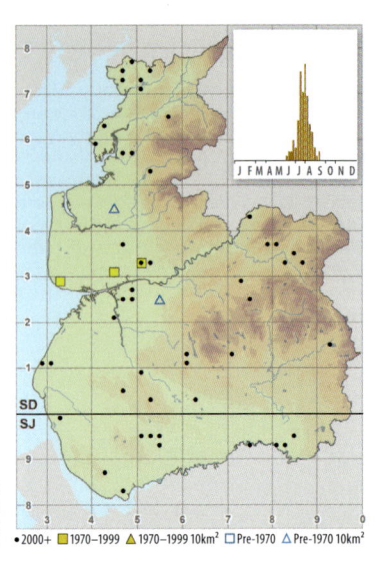

• 2000+ ☐ 1970–1999 ▲ 1970–1999 10km² ☐ Pre-1970 △ Pre-1970 10km²

40.006 *Mompha jurassicella*

■ **LANCS STATUS** Occasional / local resident   2011–2022

One of a number of momphids to have extended its range north-west during this century. Perhaps surprising, therefore, that the first records came from the north of the county. Recorded at Heysham N.R. on 4 September 2011 by J. Girdley, and in Morecambe the following March. The first VC59 records came from Flixton and Chorlton in 2015 and 2016. Has subsequently become very common in the Mersey Valley, with many stems of the foodplant containing numerous larvae.
■ **HABITAT** Scrub, rough grassland, often in damp situations.
■ **FOODPLANTS** Great Willowherb. Larvae noted in stems from June to August.
Single brooded. Moths found indoors, netted in daytime, and at light.

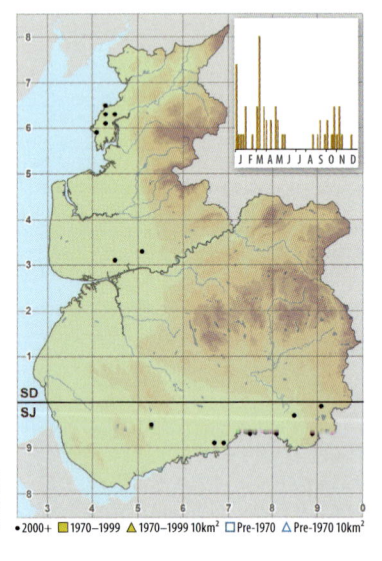

• 2000+ ☐ 1970–1999 ▲ 1970–1999 10km² ☐ Pre-1970 △ Pre-1970 10km²

40.007 *Mompha bradleyi*

■ **LANCS STATUS** Occasional / local resident   2011–2022

Another recent arrival in the county, first recorded from Flixton by K. McCabe on 5 May 2011, and from Great Sankey by J. Mitchell-Lisle later the same year. Became a frequent visitor at both sites, and at Parr and Lightfoot Green, in the years since. Moths recorded from 12 Tameside and south-east Manchester tetrads by S.H. Hind in 2021–2022. Tenanted larval galls noted in shoots of the foodplant at Chorlton, Blackley, Ashton-on-Ribble and Risley Moss between July and September.
■ **HABITAT** Scrub, rough grassland, brownfield sites.
■ **FOODPLANTS** Great Willowherb.
Single brooded. Moths recorded at light and at rest, often indoors.

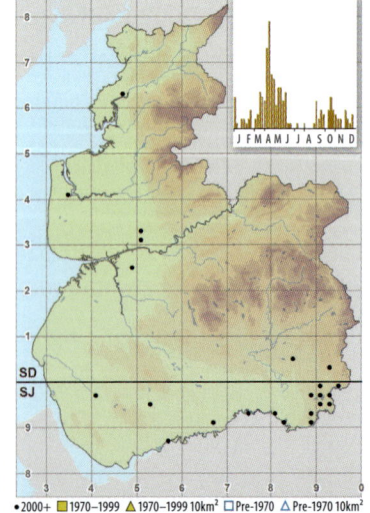

• 2000+ ☐ 1970–1999 ▲ 1970–1999 10km² ☐ Pre-1970 △ Pre-1970 10km²

40.008 Mompha subbistrigella

■ **LANCS STATUS** Frequent / widespread resident   1955–2022

Noted at Didsbury in 1955 by H.N. Michaelis. Next, recorded in 1990's from Lytham, Flixton, Lightfoot Green, and Lostock Hall. High counts of 23 from 2012 at St Helens, and 18 in 2008 and 2014. Our most recorded momphid with over 1,800 records. However, numbers appear reduced, with maximum counts of four and six in the last two years. The larvae feed in curved seed-pods of the foodplant in July and August.
■ **HABITAT** Scrub, gardens, disturbed ground, brownfield sites.
■ **FOODPLANTS** Broad-leaved Willowherb; possibly other smaller willowherb species. Single brooded. Larvae recorded on five occasions. Comes to light, but also noted indoors, often resting in corners of window frames, etc., especially over winter.

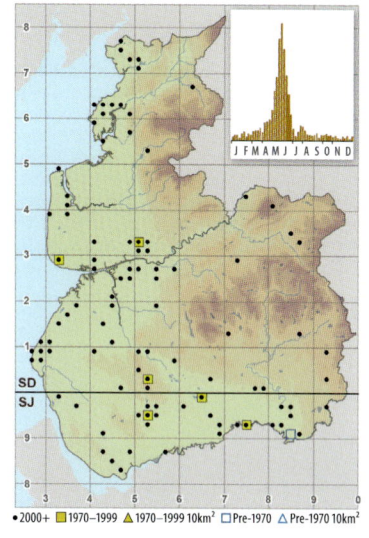

40.009 Mompha sturnipennella

■ **LANCS STATUS** Occasional / local resident   2002–2022

Over 140 records, all from VC59. First recorded 20 April 2002 by K.McCabe at Flixton. Probably new to the county or overlooked due to similarity with the previous species. Regular at the Flixton trap for four years before noted elsewhere. From St Helens in 2006, Withington 2007, and from Freshfield and Formby on the Sefton Coast in 2007. The latter area responsible for almost all records of the larval galls, fairly common among stands of the foodplant growing along paths through the dunes and woods.
■ **HABITAT** Disturbed ground, dune edges.
■ **FOODPLANTS** Rosebay Willowherb. Tenanted galls noted in stems from May and June. Double brooded. Moths recorded at light, occasionally indoors.

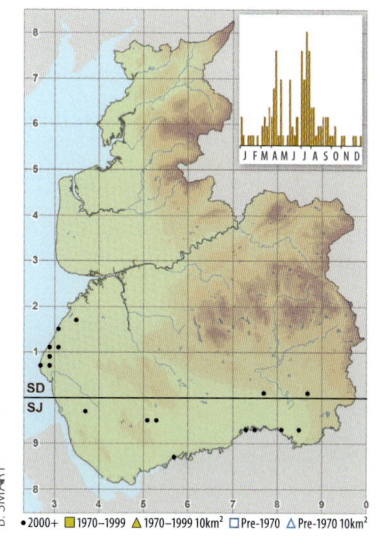

40.010 Mompha epilobiella

■ **LANCS STATUS** Frequent / widespread resident   1914–2022

First recorded in 1914 by W. Mansbridge, with moths bred from pupae collected at Speke (Mansbridge, 1915). Thereafter, from Formby 1925, Didsbury 1948 and Morecambe 1960, but not again until 1995, when noted at Bryning, Lightfoot Green, Dumplington and Risley Moss. Largely absent from the higher ground in the north-east of the county. A quarter of the 600 plus records are of the early stages.
■ **HABITAT** Rough grassland, damp meadows, disturbed ground.
■ **FOODPLANTS** Great Willowherb.
Double brooded. Frequent at light. Also, found indoors, and flying around foodplant. Larvae and pupae in spun flowering shoots from late May to September.

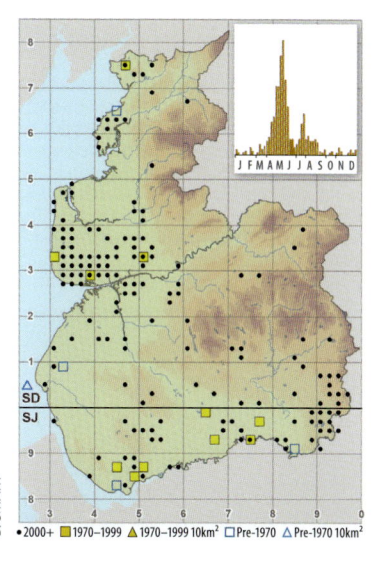

The Moths of Lancashire   **225**

MOMPHIDAE

40.011 *Mompha langiella*

■ **LANCS STATUS** Occasional / local resident   1984–2022

Appears to have spread into Lancashire from counties to the north and south during the last few decades. First recorded at Risley Moss in 1984 by M. Diamond; then in Halewood in 1998 by S. McWilliam, Woolton in 2011 and Didsbury in 2013. The first north Lancs. record was at Torrisholme in 2015 (A. Draper). Spread through the county accompanied by an apparent increase in larval use of willowherbs, rather than being restricted to the former main foodplant, Enchanter's Nightshade.
■ **HABITAT** Woodland disturbed ground, brownfield sites, along paths and verges.
■ **FOODPLANTS** Enchanter's-nightshade, Great Willowherb, Rosebay Willowherb. Single brooded. Moths come to light. Tenanted mines noted from May to August.

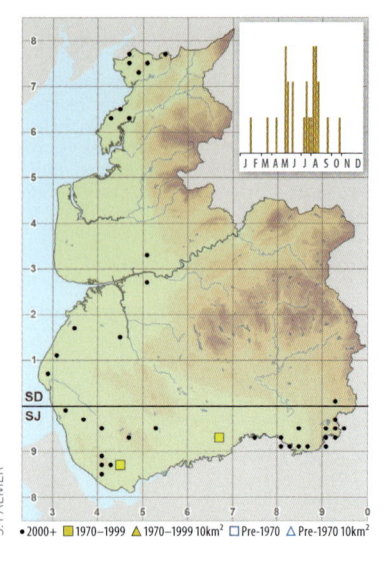

40.012 *Mompha miscella*

■ **LANCS STATUS** Scarce / very local resident   c.1920–2022

Recorded from Silverdale in the early 20th century (Mansbridge, 1940), and described by the same author as scarce and local. Restricted foodplant distribution ensures the moth remains restricted to the limestone sites of Morecambe Bay. The first modern record was of a vacated mine at Gait Barrows on 15 May 1998 (S. Palmer). Also recorded from Warton Crag, Jack Scout, Leighton Moss, Myers Allotment, etc.
■ **HABITAT** Limestone grassland.
■ **FOODPLANTS** Common Rock-rose. 20 of the 38 records are of the leaf-mines. Double brooded. Moths noted in day around foodplant. Also to light, although no evidence of the moth reaching light-traps in nearby gardens.

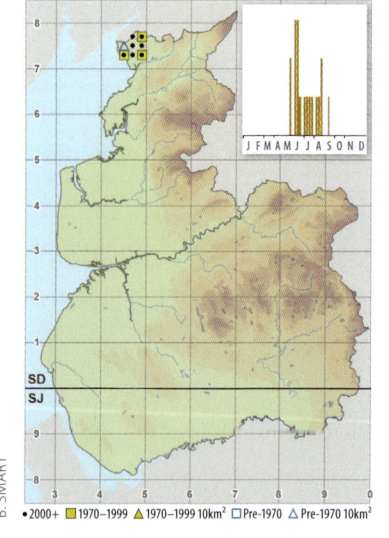

40.013 *Mompha locupletella*

■ **LANCS STATUS** Occasional / widespread resident   c.1857–2022

First reported from the Liverpool district (Gregson, 1857). Later, from Pilling Moss and Brockholes Quarry by J.B. Hodgkinson, and Salwick by J.H. Threlfall (Ellis, 1890). No further records until 1950, when recorded in Irlam by H.N. Michaelis. Next, from Holden Clough in 1955 and Scorton in 1995. Subsequent records show moth to be quite well distributed in lowland and upland locations, examples of the latter being Oswaldtwistle, Edenfield and Docker Moor. Larval mines reported on ten occasions.
■ **HABITAT** Scrub, particularly in damp habitats.
■ **FOODPLANTS** Willowherbs, including Broad-leaved Willowherb and Marsh Willowherb. Double brooded. Most of the 169 records are at light. Occasionally netted in the field.

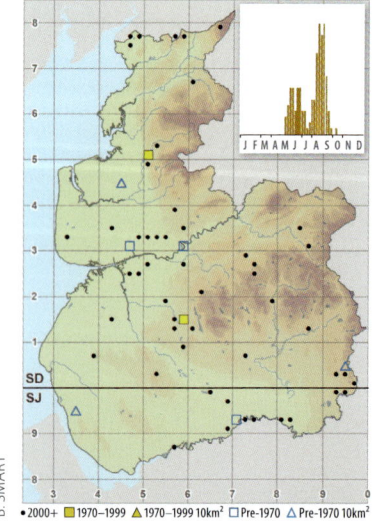

226 The Moths of Lancashire

40.014 *Mompha terminella*

■ **LANCS STATUS** Scarce / very local resident c.1859–2022

Whilst first reported from Manchester (Stainton, 1859), most subsequent records have come from limestone sites around Silverdale. The first was of vacated leaf-mines at Dalton, nr. Burton-in-Kendal 2008 (S. Palmer). Next, recorded at Silverdale Moss in 2014, with a moth coming to light (J. Patton). Other records from Challan Hall Allotment, Lords Lot, Gait Barrows and Warton Crag. There are eleven records of the leaf-mines, with the moth reared from a mine collected at Warton Crag in 2018.
■ **HABITAT** Clearings and paths in limestone woodland.
■ **FOODPLANTS** Enchanter's-nightshade. Tenanted mines recorded in July and August. Single brooded. Most records are of leaf-mines. Once at light and once netted in day.

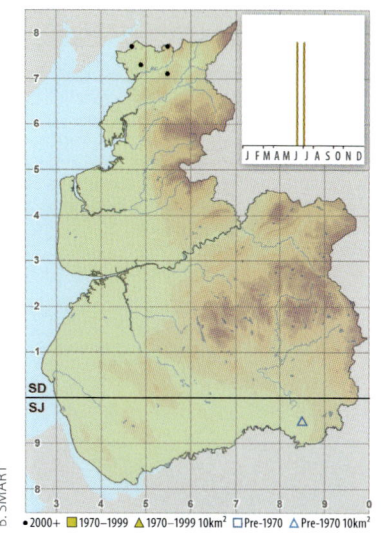

40.015 *Mompha raschkiella*

■ **LANCS STATUS** Frequent / widespread resident 1953–2022

Absent from older historical literature, the subsequent arrival and spread through the county mirrors the spread of the foodplant (Michaelis, 1962). Recorded in 1953 by H.N. Michaelis at Formby. First noted away from Sefton Coast in 1988 when recorded at Winmarleigh Moss by M. Evans. The following decade saw significant range expansion with records from Great Plumpton, Woolston Eyes, Flixton, Lytham, Daisy Nook, Stenner Woods, Beacon Fell, etc. Almost half the 947 records are of leaf-mines.
■ **HABITAT** Brownfield sites, scrub, woodland clearings, paths and verges.
■ **FOODPLANTS** Rosebay Willowherb. Tenanted mines noted from June to October. Double brooded. Moths recorded at light and in day, flying around the foodplant.

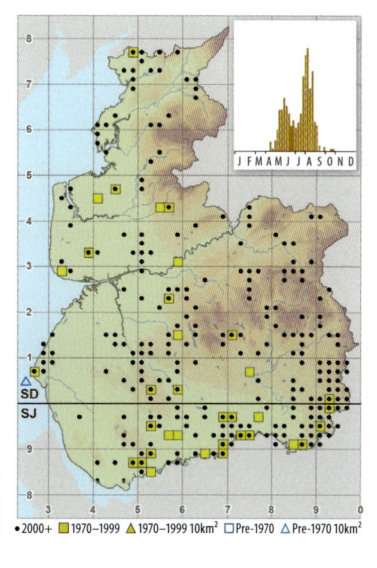

41.002 *Blastobasis adustella*

■ **LANCS STATUS** Abundant / widespread resident 1973–2022

First recorded on 21 July 1973 by N.L. Birkett from Leighton Moss, then at Yealand Storrs by M. Hull, 1977. 1980s records restricted to Newton-with-Scales, Lowton, Gait Barrows. Increasingly frequent from 1990s onwards and now found across the county in all habitats, including upland sites such as Dalton Crags, Wray and Docker Moor. High count of 150 moths to light at Blackburn on 30 July 2022 by D. Bickerton.
■ **HABITAT** Gardens, woodland, scrub, etc.
■ **FOODPLANTS** Pine, Yew, Rhododendron, Wild Teasel, Hart's-tongue Fern, Gorse, Larch, Yarrow, burdock, knapweed, etc., with a preference for dead plant material. Single brooded. Moths at light. Also reared from old Magpie nest, Chorlton in 2019.

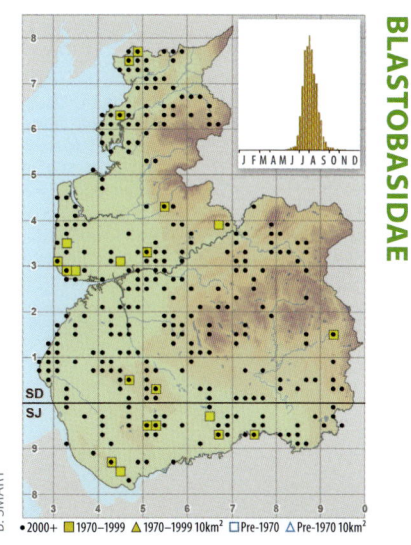

MOMPHIDAE

BLASTOBASIDAE

The Moths of Lancashire 227

BLASTOBASIDAE

41.003 *Blastobasis lacticolella*

■ **LANCS STATUS** Abundant / widespread resident   1995–2022

A Madeiran native, this species was first recorded in England in 1946, and by 1995 had reached Lancs, with moths taken at Risley Moss by L.W. Hardwick and Lightfoot Green by S. Palmer. From Trafford Park in 1997, Skelmersdale in 1998, and Parr, Billinge and Littleborough in 1999, and subsequently throughout. High count of 25 moths attracted to light in a tunnel at a Skelmersdale industrial estate in 2000.
■ **HABITAT** Likely to be found wherever foodplants occur.
■ **FOODPLANTS** Rose hips, haws, acorns, Ash seeds, Honeysuckle flowers, oak leaves and Hogweed seeds.
Likely double brooded. Moths at light. Larvae noted August to October.

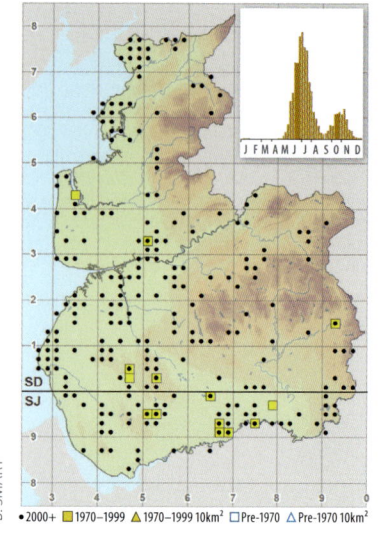

● 2000+   ☐ 1970–1999   ▲ 1970–1999 10km²   ☐ Pre-1970   △ Pre-1970 10km²

41.005 *Blastobasis rebeli*

■ **LANCS STATUS** Occasional / local resident   2010–2022

Now recorded on 63 occasions, this recent arrival was first noted at Martin Mere on 11 July 2010 by A. Bunting, and from Cuerden Valley a month later by S. Palmer. Trapped at Wigan in 2013, 2014 by G. & B. Wynn. Next from Roby Mill, Southport, St Helens, Rufford, Freshfield, Burscough, Longton etc. Highest counts of five from Mere Sands Wood and Southport, both in 2022. Records so far limited to the south-west quarter of the county, with the sole VC60 report from Lytham in 2019.
■ **HABITAT** Not known. Almost all records from garden light traps.
■ **FOODPLANTS** Unrecorded within the county.
Likely single brooded. All records are of moths at light.

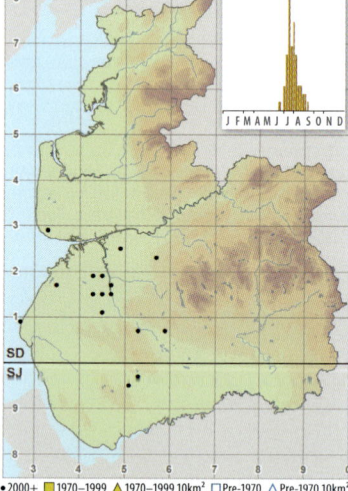

● 2000+   ☐ 1970–1999   ▲ 1970–1999 10km²   ☐ Pre-1970   △ Pre-1970 10km²

STATHMOPODIDAE

42.002 *Stathmopoda pedella*

■ **LANCS STATUS** Occasional / local resident   2001–2022

Believed to be a recent arrival in the county, it was first noted in Flixton on 21 July 2001 (K. McCabe), a site where it has remained almost annual to the present day. It was next seen in Chorlton in 2003, Martin Mere in 2004 and Freshfield, Bold Moss and Formby in 2007. Records suggest a very slow movement northwards, with it reaching Rishton in 2016 (D. Bickerton) and Hoghton (G. Dixon) in 2018.
■ **HABITAT** Damp areas with Alder, edges of Meres and scrubby brownfield sites.
■ **FOODPLANTS** Not recorded in Lancashire.
Single brooded. Comes to light in low single figures. Has been found resting on, or swept from, Alder by day and was once found indoors.

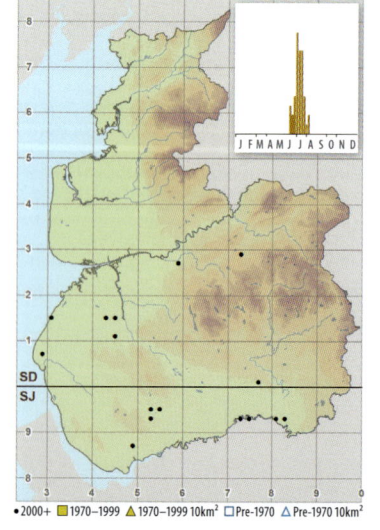

● 2000+   ☐ 1970–1999   ▲ 1970–1999 10km²   ☐ Pre-1970   △ Pre-1970 10km²

43.001 *Scythris fallacella*

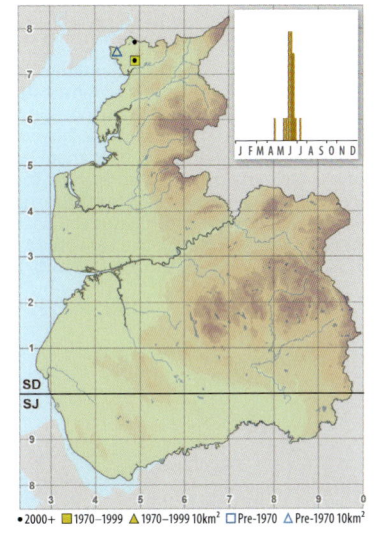

■ **LANCS STATUS** Occasional / very local resident   1914–2022

In Britain, this pRDB1 moth is restricted to northern England, where it is a species of open limestone slopes.

The earliest British records are from Yorkshire in the mid- to late 19th century and it was later found to be present in Cumbria in the early 20th century. The first confirmed records in Lancashire came as a result of genitalia examination of specimens, by F.N. Pierce, collected in the Silverdale area (Mansbridge, 1940). Although it was noted as very common, no further details were given, but the most likely recorder was A. E. Wright. It was not until 21 June 1914 that a full date was documented in Lancashire (*Proceedings of the Manchester Entomological Society*, 1914) but without the recorder being mentioned.

The next report was on 23 June 1997 from Warton Crag (S. Palmer), and this site remains its stronghold in Lancashire, with counts of a dozen or so not unusual. It has also been observed, less frequently, at Gait Barrows between 2004 and 2021.

A marked decline in numbers has taken place since the early 20th century, when it was noted as very common in the Silverdale area. It can be speculated that habitat loss and scrub encroachment are major factors in this decline. Climate change may also now be a contributing factor, with extremes of temperature and, particularly, periods of spring drought stressing the vegetation on the exposed slopes.

Based on observations at Warton Crag on 17 May 2012 (Heckford & Beavan, 2013), the larva was described for the first time in Britain. The larvae were found on Common Rock-rose, appearing to favour plants amongst mosses, growing over limestone outcrops. They fed from within a silken spinning or tube amongst the leaves and stems, and one pupa was found within a cocoon attached to leaves of the foodplant. Several larvae were located and this is the only known occasion when larvae have been observed in the county, despite more recent searches.

The moth flies in warm, often sunny conditions close to the ground and has been observed on the flowers of Common Rock-rose as well as Mouse-ear-hawkweed. Its usual flight period is from late May to early July but it has been noted on one occasion as early as the last week of April, in 2021 (J. Patton), when other species were also noted on the wing earlier than usual.

The two main sites for this species, Warton Crag and Gait Barrows, are currently managed with occasional scrub clearance taking place in areas where the moth occurs. It will be important to continue this practice, to maintain suitable open areas for rock-rose to thrive, particularly in areas where it overlies the limestone outcrops.

• 2000+   ■ 1970–1999   ▲ 1970–1999 10km²   □ Pre-1970   △ Pre-1970 10km²

SCYTHRIDIDAE

The Moths of Lancashire   **229**

SCYTHRIDIDAE

43.002 *Scythris grandipennis*

■ **LANCS STATUS** Rare / very local resident   c.1859–2015

The first report came from the Manchester district (Stainton, 1859), where it was noted as abundant. The precise location of this could not be assessed. The only other historic record was from the Silverdale area around 1910 by A.E. Wright, listed in Mansbridge (1940). Since then, there has only been one recent record, that being of a single moth disturbed from gorse, found in the central part of Heysham Moss, by day, on 24 June 2015 by J. Patton. Probably under-recorded.

■ **HABITAT** Rough ground, mosses and coastal areas, where old gorse stands occur.
■ **FOODPLANTS** Not recorded in Lancashire.

Single brooded. Searches for larval webs on gorse would be worthwhile.

• 2000+   ▪ 1970–1999   ▲ 1970–1999 10km²   ▫ Pre-1970   △ Pre-1970 10km²

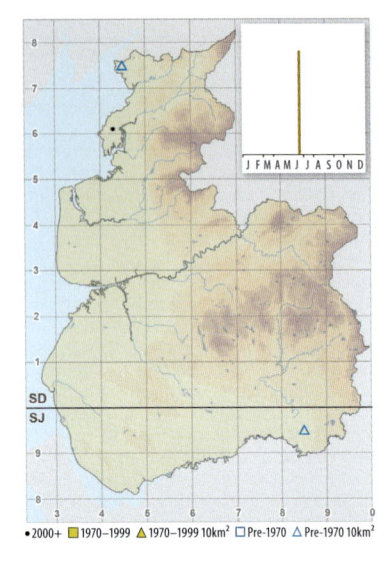

43.004 *Scythris picaepennis*

■ **LANCS STATUS** Occasional / very local resident   1997–2022

First noted at Fairhaven on 25 May 1997 (D. & J. Steeden), with many flying in the evening. Next from Gait Barrows in 1998 and Freshfield in 2007. Since then, regularly reported, sometimes in low double figures, from St Annes N.R. and Gait Barrows. Five seen by day amongst Common Bird's-foot-trefoil on Dalton Crags on 8 July 2022.

■ **HABITAT** Short sward in sand dunes, and lowland and upland limestone grassland.
■ **FOODPLANTS** Not recorded in Lancashire.

Single brooded; one seen at light on 19 September 2014 in Formby (R. Walker). Rare at light, flies in sunshine. A Manchester district record (Stainton, 1859), is considered to be from outside the VC59 portion of this area due to its habitat requirements.

• 2000+   ▪ 1970–1999   ▲ 1970–1999 10km²   ▫ Pre-1970   △ Pre-1970 10km²

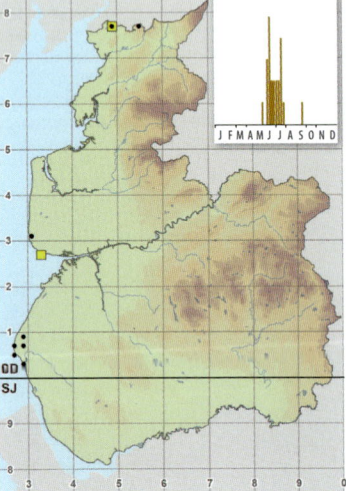

ALUCITIDAE

44.001 Twenty-plume Moth *Alucita hexadactyla*

■ **LANCS STATUS** Frequent / widespread resident   c.1857–2022

First recorded in the Liverpool and Manchester districts (Gregson, 1857; Stainton, 1859). Next, from Formby and Wavertree (Mansbridge, 1940), Morecambe 1965 and Burnley 1979. 12 records during the 1980s and an explosion since with over 3800 records from across the county. Likely to be encountered wherever the foodplant occurs. High count of 25 noted at dusk by G. Jones at Longton on 3 June 2016.

■ **HABITAT** Hedgerows, gardens, woodland edges.
■ **FOODPLANTS** Honeysuckle. Larvae found in the flowers on numerous occasions.

Double brooded, with moths found throughout the year, often indoors during winter. Comes to light. Frequently noted flying around foodplant at dusk.

• 2000+   ▪ 1970–1999   ▲ 1970–1999 10km²   ▫ Pre-1970   △ Pre-1970 10km²

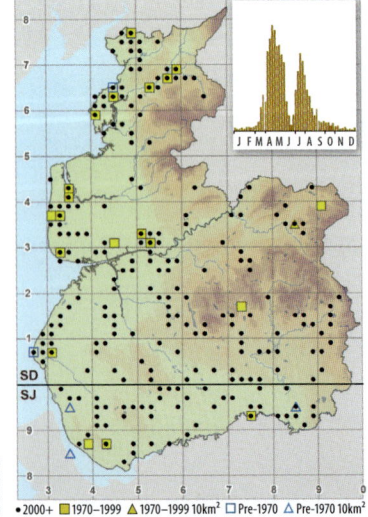

45.004 Triangle Plume *Platyptilia gonodactyla*

■ **LANCS STATUS** Frequent / widespread resident 1846–2022

First recorded in July 1846 from Pendlebury Wood, and described as scarce (Edleston, 1846). Also, from Liverpool and Manchester districts (Gregson, 1857; Stainton, 1859). From Trawden and Nelson in 1916 by I. Hartley, and then by H.N. Michaelis at Didsbury and Formby in 1953 and 1957. Over 400 records since 1950s. Particularly frequent in lowland areas but also occurs in districts on the moorland edges, e.g., Shaw, Briercliffe, Rishton, Longridge, etc. High count of 30 at Middleton N.R., 2016.
■ **HABITAT** Rough grassland, waste ground.
■ **FOODPLANTS** Coltsfoot. Larvae in stems; pupae in seed-heads.
Double brooded. At light and easily disturbed from vegetation.

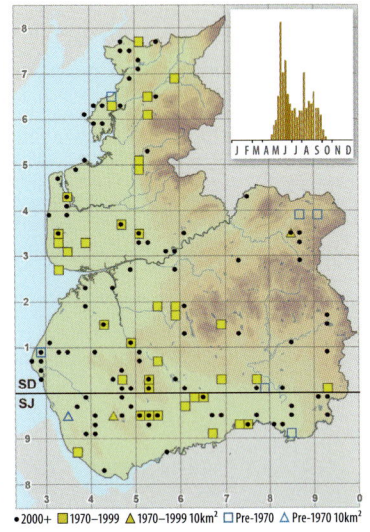

45.008 Yarrow Plume *Gillmeria pallidactyla*

■ **LANCS STATUS** Occasional / local resident c.1859–2022

Due to confusion with the following species, some historical records cannot be assigned to either. An exception is the report of larvae feeding on Yarrow in the Manchester district (Stainton, 1859). Next confirmed record from Didsbury, 1949 by H.N. Michaelis. Recorded in lowland and upland areas. Appears to be increasing in frequency and abundance with over 40 records from 2021–2022, including a high count of eight from Hazelrigg Weather Station, Lancaster 11 July 2022 by J. Patton.
■ **HABITAT** Rough grassland, verges.
■ **FOODPLANTS** Yarrow.
Single brooded. Moths to light. Occasionally netted in field around foodplant.

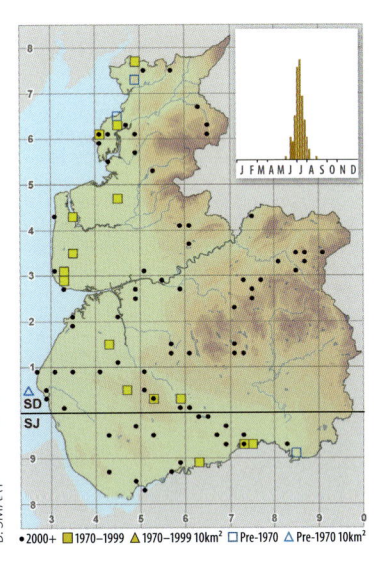

45.009 Tansy Plume *Gillmeria ochrodactyla*

■ **LANCS STATUS** Occasional / local resident 1997–2022

The same qualifications apply to historical records as noted re: *G. pallidactyla*. The first confirmed record, therefore, was of a moth netted at Flixton by K. McCabe on 19 July 1997, a real hotspot for the species with 125 of 139 records. Also, recorded at Brockhall, Lytham, Oswaldtwistle, Parr, Heysham Moss and Gillmoss, nr. Knowsley. Larvae recorded in Tansy stems at Flixton and Middleton N.R. during May. Maximum adult counts of four from Flixton on three occasions between 2006 and 2019.
■ **HABITAT** Rough grassland, brownfield sites, woodland edges.
■ **FOODPLANTS** Tansy.
Single brooded. Comes to light. Occasionally found resting on Tansy.

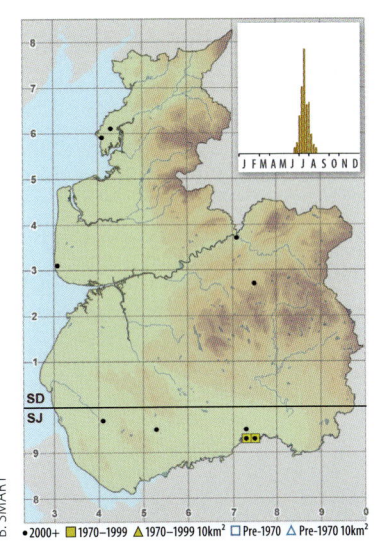

PTEROPHORIDAE

45.010 Beautiful Plume *Amblyptilia acanthadactyla*

■ **LANCS STATUS** Frequent / widespread resident   c.1857–2022

First records from Hale Marsh (Gregson, 1857), Dutton (Ellis, 1890), and Formby (Mansbridge, 1940), with the latter author describing a 'very scarce species'. Only recorded on four further occasions in 20th century. Population has since exploded, moving into gardens, and with over 2,000 21st century records. Found on lowland areas and also on moorland edges at Clitheroe, Rishton, Wray, Longridge, etc.

■ **HABITAT** Rough grassland, gardens, woodland clearings, hedgerows, moorland.

■ **FOODPLANTS** Herb Robert, Cranesbill. Pupal exuvia on Hedge Woundwort. Emerged October 2022 from collected Pot Marigold flowers, Bolton-le-Sands (S. Garland). Double brooded. Frequent at light. Easily disturbed from vegetation.

● 2000+   ■ 1970–1999   ▲ 1970–1999 10km²   □ Pre-1970   △ Pre-1970 10km²

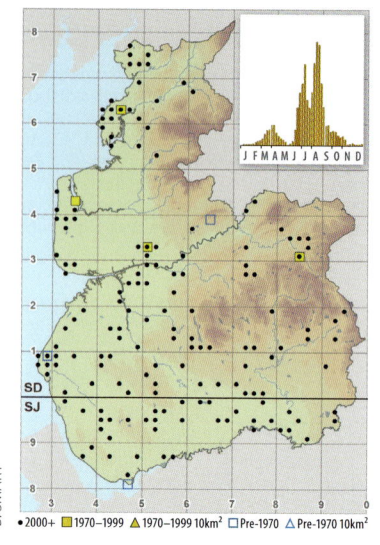

45.011 Brindled Plume *Amblyptilia punctidactyla*

■ **LANCS STATUS** Scarce / local resident   1842–2021

First 'taken by Mr Nixon upon the window curtains in his house at Hale,' on 21 June 1842 (Gregson, 1857). Not noted again until 2 April 2001, when recorded by B. Brigden at Bispham. Whilst far less frequent than the previous species, the moth has been reported in most subsequent years, all singletons. Sites include Sunderland, Great Sankey, Leyland, Walmer Bridge and Leighton Moss. As the map shows, all records have come from lowland areas, rarely far from the coast or major estuaries.

■ **HABITAT** Damp woodland, banks of streams and ditches.

■ **FOODPLANTS** Vacated pupal exuviae found on Hedge Woundwort at Eaves Wood. Double brooded. Most of the 21 records are of moths at light. Larvae unrecorded.

● 2000+   ■ 1970–1999   ▲ 1970–1999 10km²   □ Pre-1970   △ Pre-1970 10km²

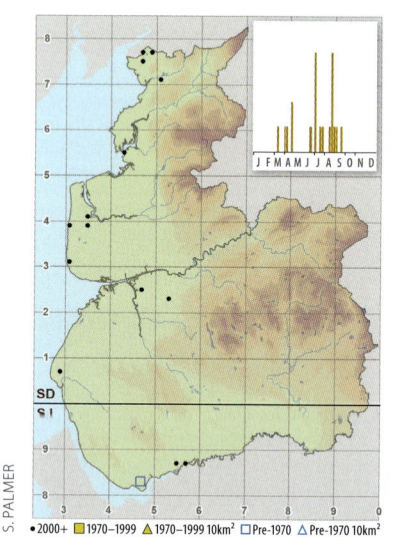

45.012 Brown Plume *Stenoptilia pterodactyla*

■ **LANCS STATUS** Occasional / local resident   1846–2022

First recorded from Pendlebury Wood (Edleston, 1846). Other 19th century records from Warbreck Moor, Liverpool (Gregson, 1857) and Manchester (Stainton, 1859). Described by Mansbridge (1940) as locally common. Recorded regularly during first half of 20th century, then unrecorded between 1956 and 1991. Recorded from Caton, Lancaster and Lytham from 1992–1994. Subsequently, from Formby, Flixton, Gait Barrows, Warton Crag, Docker Moor, Holcombe, Fairhaven, Carnforth, etc.

■ **HABITAT** Woodland edges, limestone grassland, fixed dunes.

■ **FOODPLANTS** Not recorded in Lancashire.

Single brooded. Comes to light. Netted at dusk and in the day.

● 2000+   ■ 1970–1999   ▲ 1970–1999 10km²   □ Pre-1970   △ Pre-1970 10km²

232   *The Moths of Lancashire*

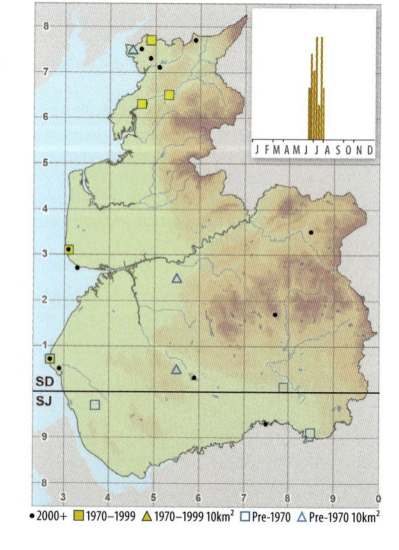

45.013 Twin-spot Plume *Stenoptilia bipunctidactyla*

■ **LANCS STATUS** Occasional / local resident   **c.1859–2022**

First recorded at Manchester (Stainton, 1859), then at Kenyon, Warrington by J. Chappell (Ellis, 1890). Noted from Crosby sand-hills, Formby Moss and Morecambe during 20th century. Increased recording this century may represent actual increase in distribution with more inland records; from Haydock, Rivington, Wigan Flashes, Lightfoot Green, Briercliffe, etc. Noted flying around foodplant (Devil's-bit Scabious) from four Tameside tetrads by S.H. Hind in 2021 and 2022. Larva unrecorded.
■ **HABITAT** Rough grassland, woodland edges.
■ **FOODPLANTS** Not recorded in Lancashire.
Single brooded. Moths recorded at light and noted amongst foodplant during day.

• 2000+   ■ 1970–1999   ▲ 1970–1999 10km²   ☐ Pre-1970   △ Pre-1970 10km²

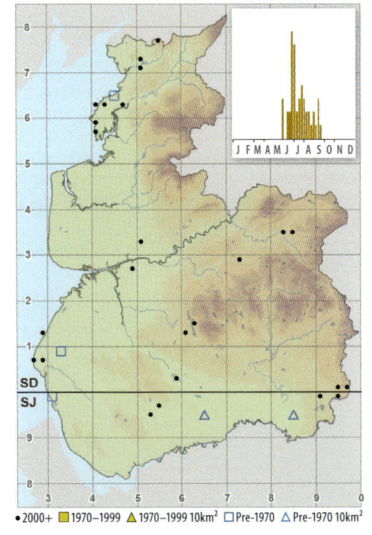

45.019 Saxifrage Plume *Stenoptilia millieridactyla*

■ **LANCS STATUS** Scarce / local resident   **2000–2023**

Thought to have reached Britain as an accidental introduction via Irish plants (Hart, 2011). The moth spread widely since the first British record at Chesterfield in 1969, reaching Lancashire in 2000, when recorded at Littleborough by I. Kimber and at Worsthorne by G. Gavaghan. In Lancs, initially thrived for 15 years or so at a limited number of sites, suggesting local populations related to garden plants. Just three post-2018 records: Rishton 2020, Leighton Moss 2021, Silverdale 2023.
■ **HABITAT** Gardens.
■ **FOODPLANTS** Mossy Saxifrage. Larvae noted in a Blacko garden 15 May 2006 (C. Hart). Has a partial second brood. Almost all records are of moths at light.

• 2000+   ■ 1970–1999   ▲ 1970–1999 10km²   ☐ Pre-1970   △ Pre-1970 10km²

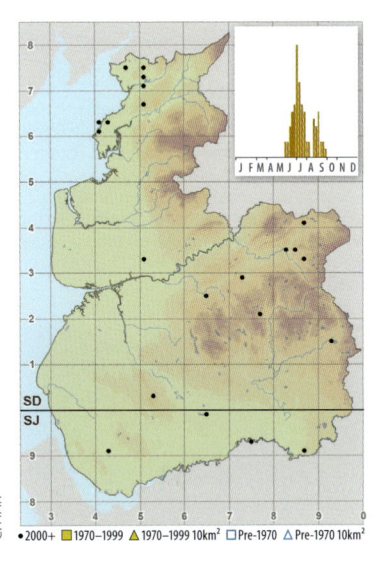

45.021 Dowdy Plume *Stenoptilia zophodactylus*

■ **LANCS STATUS** Scarce / local resident   **c.1857–2015**

First British record at Southport (Gregson, 1857). Next, recorded at Ainsdale in 1920. Larvae found on centaury at Birkdale dunes on 29 July 1923 by H. Britten. The first inland records were from Flixton 1998 and Pickering's Pasture, Hale in 1999, St Helens 2000, Widnes 2001, Chorlton 2003 and Roby Mill 2004, but not since. Subsequent records from Heysham N.R. (on eight occasions), Ainsdale and Freshfield, but concerningly, none since 2015.
■ **HABITAT** Dunes, dry grassland.
■ **FOODPLANTS** Centaury. Twice on Yellow-wort, from Formby 1950 and Ainsdale 1953.
Double brooded. Of the 32 records, four are larval. 17 are of moths to light.

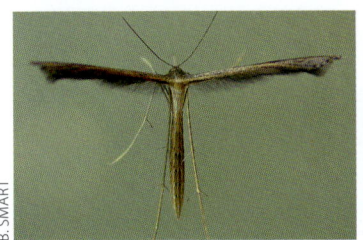

• 2000+   ■ 1970–1999   ▲ 1970–1999 10km²   ☐ Pre-1970   △ Pre-1970 10km²

The Moths of Lancashire **233**

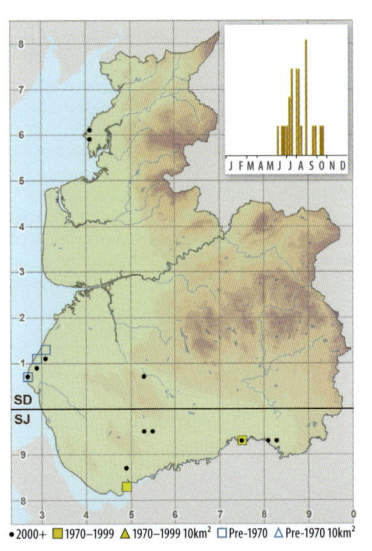

45.023 Crescent Plume *Marasmarcha lunaedactyla*

■ **LANCS STATUS** Rare / vagrant   2020

A rather peculiar record was the appearance of this moth, to MV light in a garden trap in Chorlton, Manchester by B. Smart on 24 June 2020. The specimen was checked by S. Palmer to ensure correct identification. The foodplant, rest-harrow, is not known from the locality but may occur in nearby gardens. There have been no further records. Unrecorded from Cheshire and Cumbria but recorded from south-west and mid-west Yorkshire (VCs 63 and 64).

■ **HABITAT** Unknown in the county.
■ **FOODPLANTS** Not recorded in Lancashire.
Nationally, single brooded.

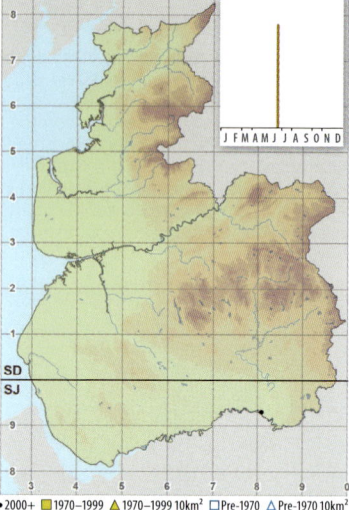

B. SMART

45.027 Scarce Light Plume *Oxyptilus laetus*

■ **LANCS STATUS** Rare / migrant   2019

The only two records of this species occurred within a fortnight of each other. Despite the lack of geographical proximity, they seem likely to be part of the same migratory movement, which also resulted in the first Yorkshire record and the second from Cheshire during the same period. The first moth was recorded in a Silverdale garden by J. Patton on 4 July 2019, followed by a second on 15 July 2019, at Houghton Green, nr. Warrington, noted by G. Dunbar.

■ **HABITAT** Unknown in the county.
■ **FOODPLANTS** Not recorded in Lancashire.
Moths recorded at MV light.

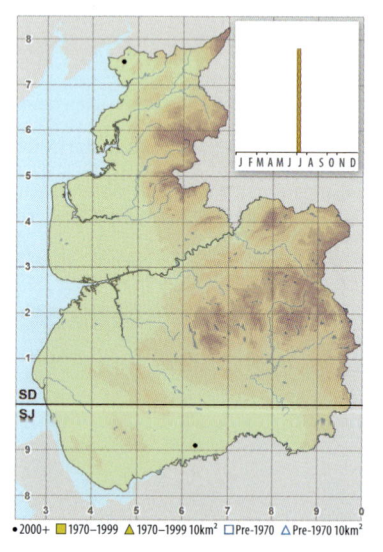

J. GIRDLEY

45.028 Wood Sage Plume *Capperia britanniodactylus*

■ **LANCS STATUS** Rare / very local resident   1914–2006

First recorded at Trowbarrow Quarry on 20 June 1914, during a Manchester Entomological Society field meeting. Reported from the Formby sand dunes (Mansbridge, 1940), and later described as common at the site (Leech and Michaelis, 1957). The only other record is of a singleton trapped at Gait Barrows on a Moth Night event run by the Lancashire Moth Group on 30 June 2006. Searches for larvae on Wood Sage at the latter site have so far been unsuccessful.

■ **HABITAT** Woodland edges, shady areas of dry scrub.
■ **FOODPLANTS** Unrecorded within the county.
Single brooded. At MV light in 2006. Other finds likely to be daytime records.

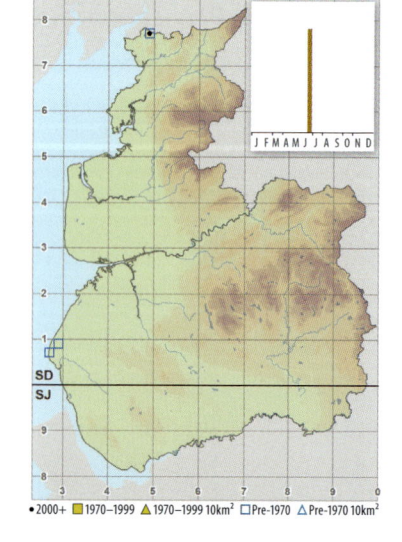

S. PALMER

45.030 White Plume *Pterophorus pentadactyla*

■ **LANCS STATUS** Frequent / widespread resident   **c.1857–2022**
First noted in the Liverpool district (Gregson, 1857). Other 19th century records from Withington, Irlam and Patricroft by J. Chappell (Ellis, 1890). Recorded at Old Trafford and Middle Hulton in 1925 and reported as 'common and general' (Mansbridge, 1940). Appears well distributed throughout lowland and more upland areas, such as Briercliffe, Colne and Mill Houses. Recorded on 438 occasions, with 22 from 2022.
■ **HABITAT** Grassland, brownfield sites, gardens.
■ **FOODPLANTS** Bindweed species. Larvae were recorded by L.W. Hardwick on bindweed at Ince Moss, Wigan in August 1993 and at Pickering's Pasture, Hale in June 1994.
Single brooded. Moths netted, trapped at light, swept, and noted around bindweed.

• 2000+   ▪ 1970–1999   ▲ 1970–1999 10km²   ▫ Pre-1970   △ Pre-1970 10km²

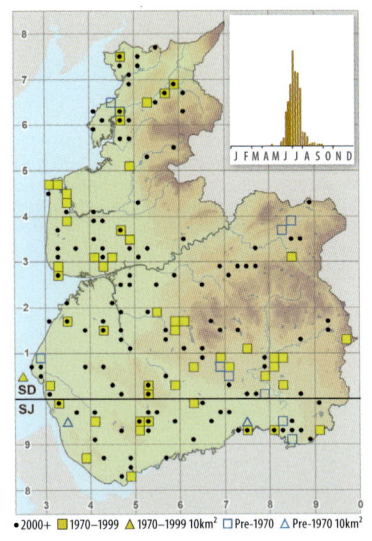

45.033 Thyme Plume *Merrifieldia leucodactyla*

■ **LANCS STATUS** Scarce / very local resident   **c.1910–2022**
The earliest records were not critically examined. However, the rediscovery of the species at Dalton Crags / Hutton Roof complex by J. Patton, with records on 17 June 2021 and 7 July 2022, provided a specimen. The confirmation via dissection proved the moth's presence in the area, and so the earlier records from Silverdale by A.E. Wright c.1910 and from the Lancaster district by M. Dempsey in 1986 were assigned to this species, with the very similar *M. tridactyla* considered extremely unlikely.
■ **HABITAT** Limestone grassland.
■ **FOODPLANTS** Not recorded in Lancashire.
Single brooded. Moths at light or flying around Wild Thyme, the foodplant nationally.

• 2000+   ▪ 1970–1999   ▲ 1970–1999 10km²   ▫ Pre-1970   △ Pre-1970 10km²

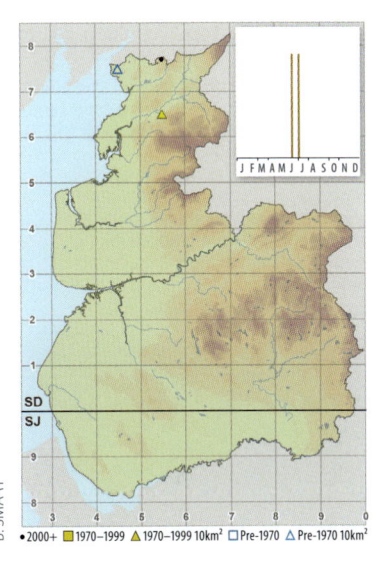

45.034 Dingy White Plume *Merrifieldia baliodactylus*

■ **LANCS STATUS** Considered Extinct   **c.1940**
The single record of this species is from Silverdale, recorded by A.E. Wright in the early 20th century (Mansbridge, 1940). The text describes the moth as 'very local but fairly common,' although the only other 'Lancashire' site given, Grange-over-Sands, is now part of Cumbria, in VC69. Unrecorded in Yorkshire and Cheshire. It is at the northern limit of its national range in Cumbria.
■ **HABITAT** Limestone grassland.
■ **FOODPLANTS** Not recorded in Lancashire.
Elsewhere, the species is associated with Marjoram, a plant present in the limestone grassland of the Silverdale area (Greenwood, 2012). The larva could be searched for in May to early June feeding on the shorter stems hidden by longer growth, the wilting of such stems a useful feeding sign (Hart, 2011).

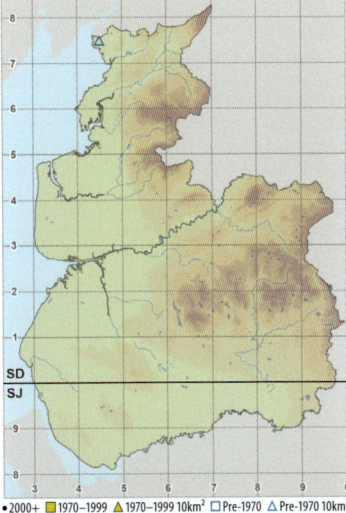

• 2000+   ▪ 1970–1999   ▲ 1970–1999 10km²   ▫ Pre-1970   △ Pre-1970 10km²

PTEROPHORIDAE

45.037 Dusky Plume *Oidaematophorus lithodactyla*

- **LANCS STATUS** Occasional / local resident   c.1859–2022

First recorded from the Manchester district (Stainton, 1859), and from Freshfield and Formby, with larvae on Common Fleabane at the latter site (Mansbridge, 1940). The first modern record was from Clayton Green in 1995 by S. Palmer. Larvae further recorded at Middleton N.R., Heysham N.R., Rixton Claypits and Tame Valley, Denton. Moths recorded on more than one occasion from Billinge Hill, Dobcroft N.R., Mere Sands Wood, Sunderland, Yealand Conyers and Stannings Folly Wood, nr. Wymott.
- **HABITAT** Grassland.
- **FOODPLANTS** Common Fleabane. Larvae from late May to early July.

Single brooded. Comes to light. Easily disturbed from vegetation and foodplant.

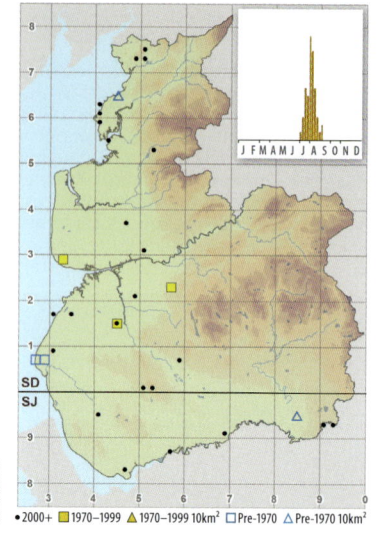

• 2000+ ☐ 1970–1999 △ 1970–1999 10km² ☐ Pre-1970 △ Pre-1970 10km²

45.038 Plain Plume *Hellinsia tephradactyla*

- **LANCS STATUS** Scarce / very local resident   c.1859–2018

Almost all records are from the Morecambe Bay limestone, the only exception being the first, noted in Manchester (Stainton, 1859). Later, from Silverdale by A.E. Wright (Mansbridge, 1940), with the latter author describing the moth as 'local and not common.' The next record was at Gait Barrows on 28 June 1997 by S. Palmer. The moth has been noted here on eight further occasions, but not since 2014. The most recent records have come from Warton Crag, in June and July 2018 by J. Patton.
- **HABITAT** Limestone woodland.
- **FOODPLANTS** Larvae unrecorded within the county.

Single brooded. Most records are of moths at light.

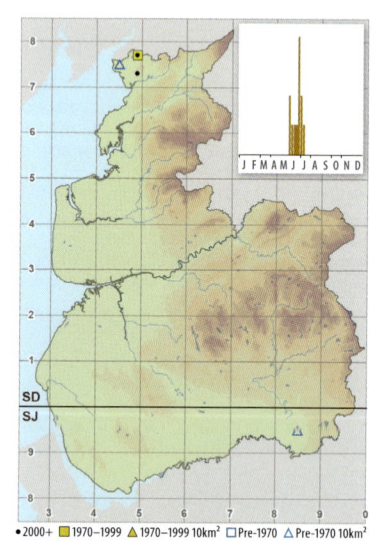

• 2000+ ☐ 1970–1999 △ 1970–1999 10km² ☐ Pre-1970 △ Pre-1970 10km²

45.040 Mugwort Plume *Hellinsia lienigianus*

- **LANCS STATUS** Occasional / local resident   2001–2021

Appears a very recent arrival in the county. All 18 records to date have been limited to the south of the county. The first was of a moth trapped at light in Flixton on 23 July 2001 by K. McCabe. The following three years saw further records from Woolston Eyes, Widnes, St Helens and Skelmersdale. First noted at Huyton in 2009 and at Hunts Cross, Liverpool in 2018. Most recently recorded on 16 June 2021 at light and sheet in the Lifeboat Road Car Park, Formby by J. Girdley.
- **HABITAT** Dry grassland, scrub, disturbed ground.
- **FOODPLANTS** Unrecorded within the county.

Single brooded. All records are of moths attracted to light.

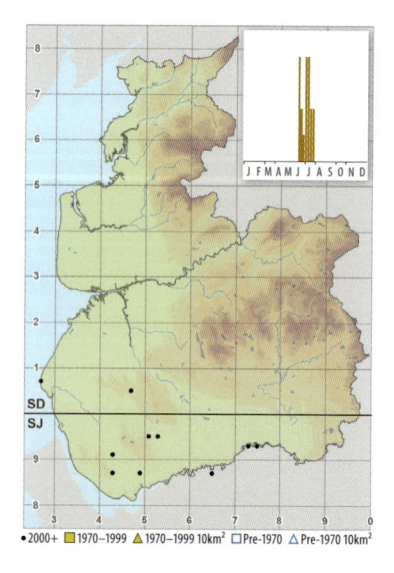

• 2000+ ☐ 1970–1999 △ 1970–1999 10km² ☐ Pre-1970 △ Pre-1970 10km²

45.041 Small Goldenrod Plume *Hellinsia osteodactylus*

■ **LANCS STATUS** Frequent / widespread resident c.1910–2022
Recorded around 1910 from Silverdale by A.E. Wright (Mansbridge, 1940), and at Crosby by W. Mansbridge in 1913 (Leech and Michaelis, 1957). Noted as common among Goldenrod in north Lancs (Mansbridge, 1940). 255 subsequent records; the majority from Sefton Coast, Morecambe Bay limestone, and from gardens, e.g., 55 times in K. McCabe's Flixton garden. Larvae recorded in flower-heads of the two foodplants, with moth bred from Goldenrod larva at Gait Barrows 2009 by S. Palmer.
■ **HABITAT** Woodland clearings, rough grassland, fixed dunes.
■ **FOODPLANTS** Common Ragwort, Goldenrod.
Single brooded. Comes to light. Moths recorded at rest on foodplant.

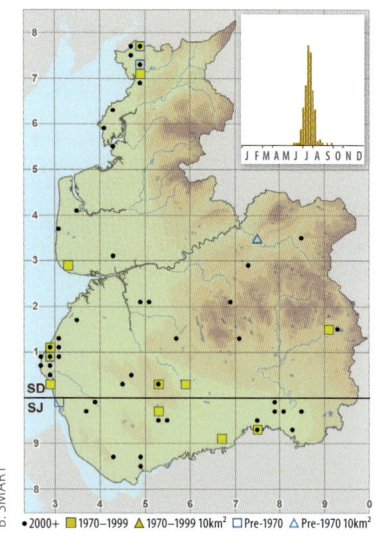

45.043 Hemp Agrimony Plume *Adaina microdactyla*

■ **LANCS STATUS** Occasional / local resident 1953–2022
A flourishing colony at Holden Clough, Oldham was first recorded by L.N. Kidd in June 1953. Noted from Hawes Water in 1957 and 1960 by H.N. Michaelis. Larval galls in stems recorded by L.W. Hardwick from Ince Moss in 1984. Moths bred from tenanted stems collected in January 2015 at Rixton by B. Smart. This lowland species is most frequently recorded at light. The widespread nature of subsequent records suggests that absence from 19th century reports is likely due to the moth being overlooked.
■ **HABITAT** Woodland edges, scrub, hedgerows, paths.
■ **FOODPLANTS** Larvae in galls in stems of Hemp Agrimony.
Single brooded; September adults may represent a partial second generation.

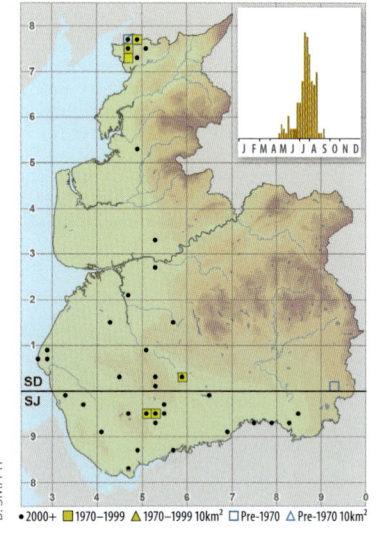

45.044 Common Plume *Emmelina monodactyla*

■ **LANCS STATUS** Abundant / widespread resident c.1910–2022
Recorded around 1910 by A.E. Wright from Burnley, with contemporaneous records documented on H. Britten's card index from Wigan, Burnley, Silverdale and the Fylde area (1950s). Described as infrequent by Mansbridge (1940), suggesting a significant upturn since in fortunes. Now, likely wherever foodplants occur. High count of 47 at dusk in a Morecambe garden on 14 September 2009 by J. Girdley.
■ **HABITAT** Woodland edges, scrub, brownfield sites, gardens, etc.
■ **FOODPLANTS** Bindweeds, including Hedge Bindweed. Larvae found every month from June to October, including ten at Ashton Moss on 12 August 2003 by I.F. Smith.
Double brooded. Moths frequent at light. Easily disturbed from vegetation.

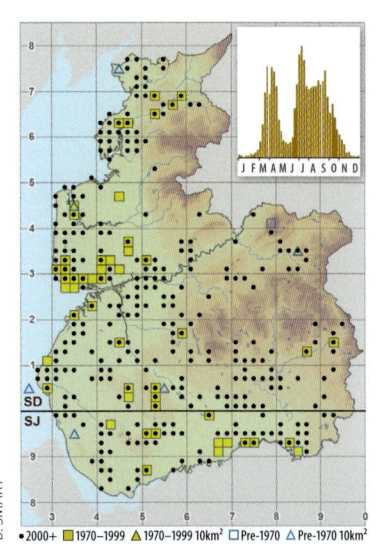

The Moths of Lancashire

SCHRECKENSTEINIIDAE

46.001 *Schreckensteinia festaliella*

■ **LANCS STATUS** Occasional / local resident   c.1859–2021

First reported from Chat Moss c.1859 by J. Chappell. Next, in Dutton c.1889 by J.B Hodgkinson, Pendle Hill and Preston (Ellis, 1890), Caton Moor in 1940 and Risley Moss in 1982. Since then, found regularly at Flixton and Warton Crag, but much less frequently elsewhere and mainly in lowland areas. Noted, as singletons, from two upland sites at Miles Hill in 2006, and Bishops Park on 13 June 2021 (S.H. Hind).

■ **HABITAT** Scrubland and open woodland, in sheltered patches of course vegetation.
■ **FOODPLANTS** Not recorded in Lancashire.

Records suggest it may be double brooded, or perhaps continuously brooded from Spring to early Autumn. Comes to light and sometimes seen in flight during the day.

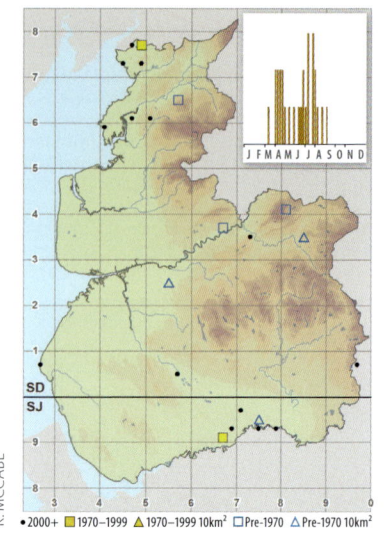

EPERMENIIDAE

47.001 *Phaulernis fulviguttella*

■ **LANCS STATUS** Rare / very local resident   c.1857–2016

First noted in the Liverpool district (Gregson, 1857) and in Barton, nr. Manchester c.1859, by J. Chappell. It was then found in the Preston area by J.B. Hodgkinson (Ellis, 1890) without any date, and in the Burnley area (1916) by W.G. Clutten (Mansbridge, 1916a). In 1953, it was seen at Holden Clough, Oldham by L. N. Kidd, this reported by H.N. Michaelis in the Lancashire and Cheshire Fauna Committee Report of that year. The sole modern record relates to an adult photographed on an unnamed umbellifer flowerhead at Gait Barrows on 6 August 2016 (N. Garnham).

■ **HABITAT** Unknown in the county.
■ **FOODPLANTS** Not recorded in Lancashire.

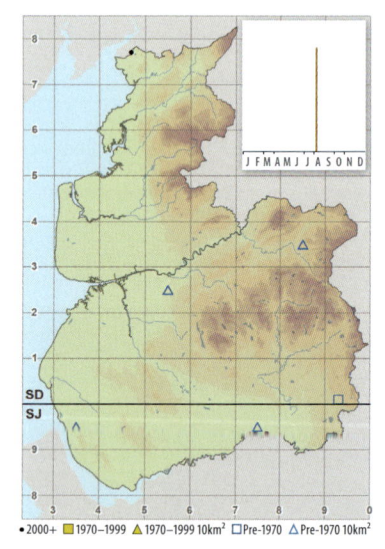

47.004 *Epermenia aequidentellus*

■ **LANCS STATUS** Rare / vagrant   2019

The sole Lancashire record came when a singleton was attracted to MV light on National Trust land at Formby on 25 June 2019 (J. Girdley). The area is made up of stabilised sandy grassland, about 700m east of the coastline, enclosed on two sides by mature pine tree growth. The presence nearby of potential larval foodplants, such as Wild Carrot, has not been investigated. This record has been difficult to assess in a national context. The moth is not known to be migratory or dispersive and the nearest known location was a singleton in Harlech, Wales in 2006.

■ **HABITAT** Stabilised dune grassland, with surrounding pine woodland.
■ **FOODPLANTS** Not recorded in Lancashire.

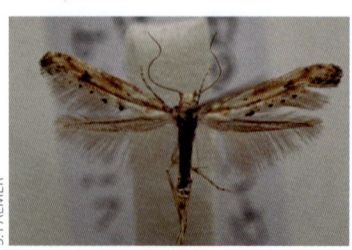

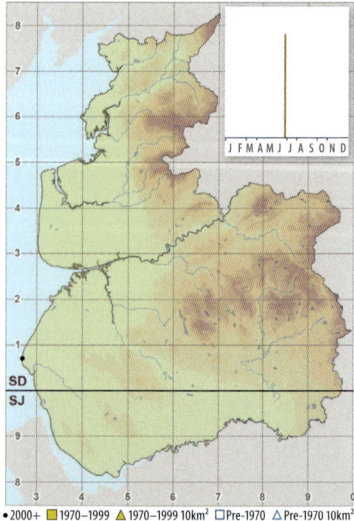

238 *The Moths of Lancashire*

47.005 *Epermenia chaerophyllella*

■ **LANCS STATUS** Frequent / widespread resident   c.1889–2022
A single early record, from Cleveleys by J.H. Threlfall (Ellis, 1890). Since 1995, it has been reported widely in good numbers, mainly as a larva in lowland and some upland areas. A survey of south-east VC59 in 2021–22 (S.H. Hind) found it regularly, including a count of 67 larvae on a single Hogweed plant. Most regularly noted at Flixton, Parr, Chorlton, Lunt Meadows, Longton, Billinge, Rixton Claypits and Lightfoot Green.
■ **HABITAT** Rough grassland, verges etc, usually wherever the larval foodplants grow.
■ **FOODPLANTS** Hogweed and Angelica.
Double brooded, the adult overwintering. Comes to light. Larvae feed gregariously on the underside of leaves, from May to August.

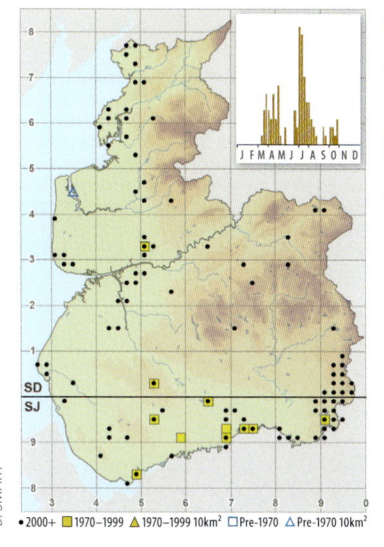

47.006 *Epermenia falciformis*

■ **LANCS STATUS** Rare / very local resident   2021–2022
This species was new to Cheshire in 2020 and the subsequent arrival in Lancashire fits the profile of a species that is expanding its range northwards in Britain. The first record was on 4 July 2021 at Flixton by K. McCabe. Later that year, on 20 August, a probable second brood individual was found in the same Flixton garden. Although not noted in Flixton in 2022, the species did come to light at Parr, St Helens, 20km to the east, on 11 August 2022 (R. Banks). All three records involved singletons.
■ **HABITAT** Unknown.
■ **FOODPLANTS** Not recorded in Lancashire. Associated with Angelica elsewhere.
Probably double brooded. Comes to light.

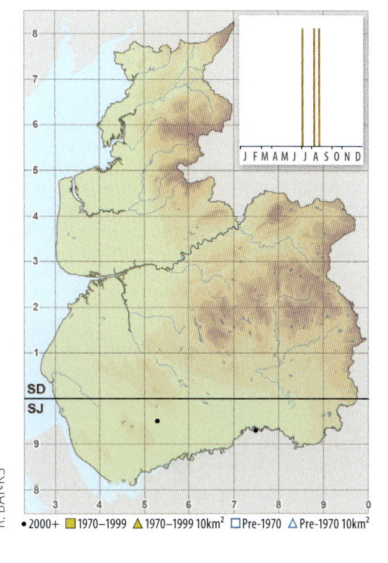

EPERMENIIDAE

48.001 *Anthophila fabriciana*

■ **LANCS STATUS** Abundant / ubiquitous resident   c.1857–2022
Recorded from Liverpool and Manchester districts (Gregson, 1857; Stainton, 1859). Later, described as abundant everywhere (Mansbridge, 1940), with additional records from Knowsley Park 1907 and Crosby 1908. Clearly there is no shortage of the foodplant, and this species continues to thrive throughout all parts of the county. There are 18 counts of one hundred moths or more, the highest being an estimate of 1,000 at Orrell Water Park on 1 June 1999 by C.A. Darbyshire.
■ **HABITAT** Scrub, rough grassland, woodland edges, paths and verges, etc.
■ **FOODPLANTS** Common Nettle. Tenanted spinnings recorded from April to September.
Double brooded. Most records are of moths flying in sunshine. Occasional at light.

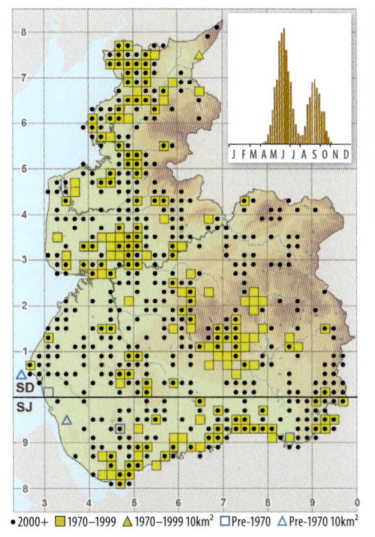

CHOREUTIDAE

The Moths of Lancashire   239

CHOREUTIDAE

48.002 *Prochoreutis myllerana*

■ **LANCS STATUS** Scarce / local resident c.1985–2022

Historic records cannot be confirmed, as up to 1939 this species and *P. sehestediana* were mostly considered a single species (Heath *et al.*, 1985). The map in the latter text contains a dot for VC59, signifying a previous record; details unknown. The next record was of around 30 moths flying and easily disturbed from their foodplant on the edge of the lake at Chorlton Water Park by B. Smart on 27 July 2004. Also noted at Heysham Moss, Silverdale Moss and Challan Hall Allotment from 2014 to 2022.

■ **HABITAT** Edges of freshwater bodies.

■ **FOODPLANTS** Skullcap. Moths bred from larvae, Chorlton 2007; dissected by S. Palmer. Double brooded. Moths at light and netted by day. Dissection required to confirm.

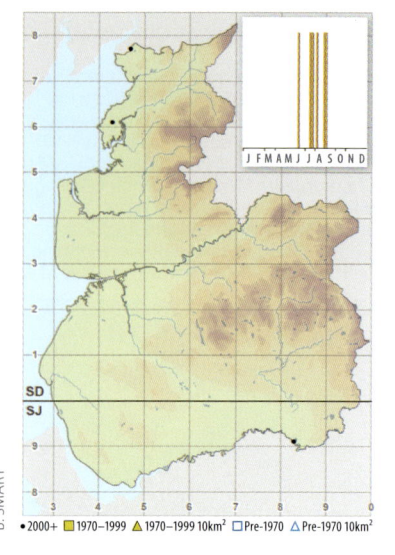

● 2000+   ☐ 1970–1999   ▲ 1970–1999 10km²   ☐ Pre-1970   △ Pre-1970 10km²

48.005 *Tebenna micalis*

■ **LANCS STATUS** Rare / migrant 2021–2022

Only recorded twice within the county, and both in the last few years. First reported from Southport on 10 August 2021 by D. Nickeas. The second moth was netted at dusk at Heysham Moss by J. Patton on 28 August 2022. Whilst this is a migrant, it can also be a temporary resident, with the gregarious larvae feeding in webs on Common Fleabane. As yet, there is no indication of this within Lancashire. There is a single Yorkshire record of a *Tebenna* sp., almost certainly *T. micalis*, from 2021. Unrecorded in Cheshire.

■ **HABITAT** Unknown in the county.

■ **FOODPLANTS** Not recorded in Lancashire.

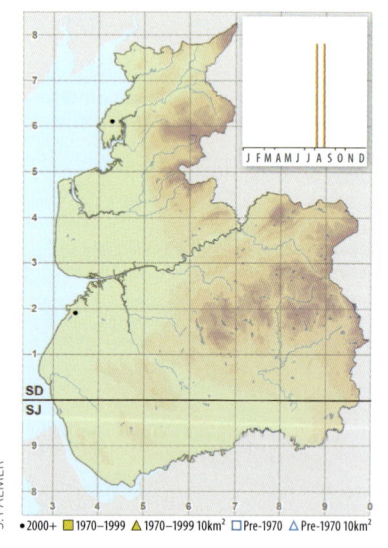

● 2000+   ☐ 1970–1999   ▲ 1970–1999 10km²   ☐ Pre-1970   △ Pre-1970 10km²

48.007 Apple Leaf Skeletonizer *Choreutis pariana*

■ **LANCS STATUS** Occasional / local resident 1936–2022

First recorded from Didsbury in 1936 by H.N. Michaelis, with adults reared from larvae, foodplant unrecorded. Next, noted at Armitage Gardens, Liverpool in 1993. Larvae on Crab Apple at Chorlton in 2005, Apple at Flixton 2007 and Hawthorn at St Helens in 2008. Moth noted by S. Tomlinson during the day at Prescot Reservoirs in 2007. Other records of this mainly lowland species have come from Aintree, King's Moss, Birkdale Hills, Poulton-le-Fylde, Westby, Littleborough, Adlington and Longton.

■ **HABITAT** Woodland edges, hedgerows, scrub.

■ **FOODPLANTS** Apple, Crab Apple, Hawthorn. Tenanted larval spinnings noted June-July. Double brooded. Comes to light. Occasionally, netted in day.

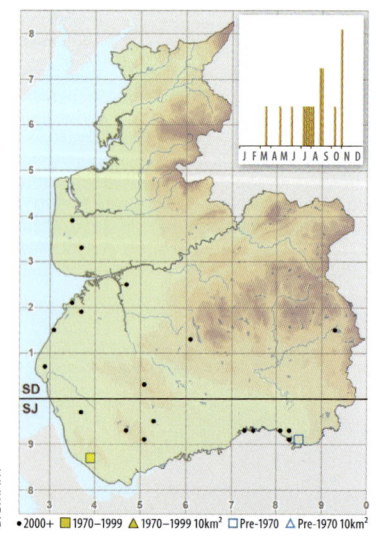

● 2000+   ☐ 1970–1999   ▲ 1970–1999 10km²   ☐ Pre-1970   △ Pre-1970 10km²

240 *The Moths of Lancashire*

49.001 *Olindia schumacherana*

■ **LANCS STATUS** Occasional / local resident   c.1890–2022

Noted as 'taken in woods' by N. Cooke (Byerley, 1854), but this may relate to VC58. The first confirmed Lancashire record was at Brockholes Wood, Preston (Ellis, 1890). Apart from an isolated 2016 report at Far Banks (VC59), all other recent records have been in VC60 in or adjacent to deciduous woods. These include Brock Bottoms, Mill Houses, Silverdale area, Botton Mill (12 on 4 July 2014) and at Leighton Moss, its main site, where a possible 2nd brood was noted. Mainly occurs as singletons at light.
■ **HABITAT** Mature deciduous woodland; rarely in more open areas.
■ **FOODPLANTS** Not recorded in Lancashire.
Single brooded; a possible, localised, second brood occurred in 2021 and 2022.

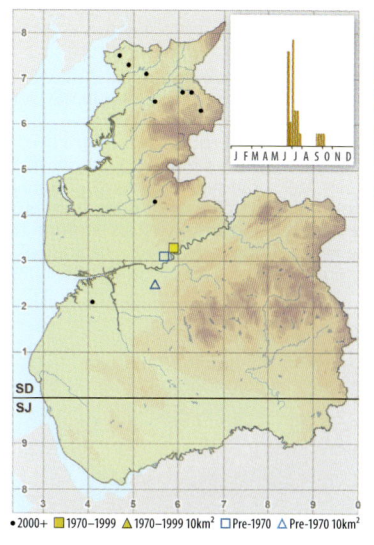

49.002 *Isotrias rectifasciana*

■ **LANCS STATUS** Occasional / widespread resident   c.1859–2022

Noted in the Manchester district by J. Chappell c.1859, the Crosby sandhills in 1885 (C.S. Gregson) and Preston, Burnley and Formby areas, all prior to 1927. Few records known from then to the late 1990s, mostly from limestone and sand dune sites. Since then, most records have been from Flixton and Gait Barrows, but with scattered, occasional lowland records elsewhere. The very few reports from higher parts of the county are historic and some may be misidentifications of *Cnephasia* sp.
■ **HABITAT** Sand dunes, hedgerows and scrub on limestone and brownfield sites.
■ **FOODPLANTS** Not recorded in Lancashire or nationally, but possibly hawthorn.
Single brooded. Comes occasionally to light mostly in low single figures.

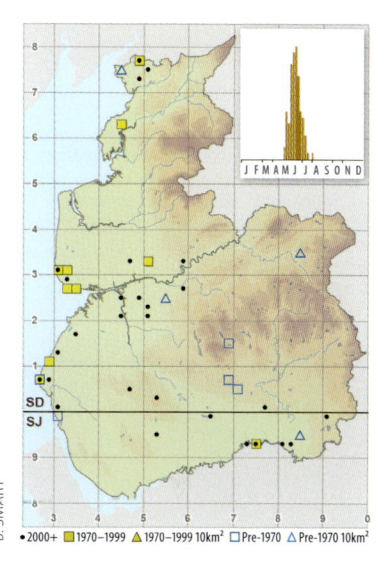

49.004 Red-barred Tortrix *Ditula angustiorana*

■ **LANCS STATUS** Frequent / widespread resident   c.1857–2022

First noted in the Liverpool area (Gregson, 1857), occurring in old gardens where pear trees grow (Gregson, 1857). Later, seen in Chorlton and Withington in 1859 and Preston c.1890. Since the 1980s, it has been reported from most lowland light trap sites, usually in small numbers. Sightings in higher parts were very few until Caton in 1990 and Littleborough in 1999, since when it has been noted widely.
■ **HABITAT** Wide range, but mostly gardens, parks and amenity planted areas.
■ **FOODPLANTS** Gorse, Broom, Pyracantha, Rowan, pine sp., cedar, Ivy, Bramble, oak, elm, Ash, birch, Burnet Rose and Oak Marble Gall. Spins leaves, September to May. Single brooded, with an occasional, small, second brood in autumn. Comes to light.

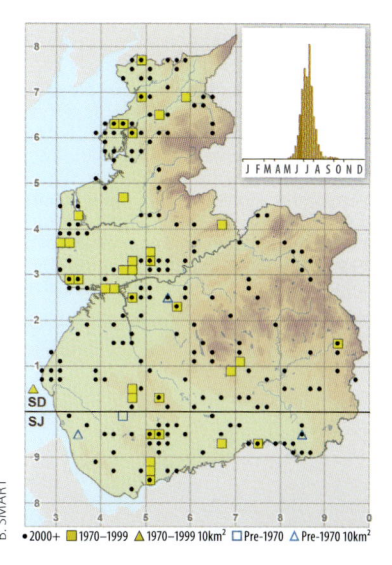

The Moths of Lancashire   **241**

TORTRICIDAE

49.005 *Epagoge grotiana*

■ **LANCS STATUS** Rare / very local resident   c.1853–2021

First reported from the mosses in the Liverpool and Warrington district by N. Cooke (Byerley, 1854), with Gregson (1857) adding it occurs on the drier parts of the mosses near Liverpool (i.e., within a twelve-mile radius). It was also reported from Irlam and Barlow Moor by J. Chappell and in the Preston district by J.B. Hodgkinson (Ellis, 1890). Not seen again until 27 June 2021 (two) and 9 July 2021 (one), when they were attracted to MV light run at Mere Sands Wood N.R. (J. Girdley, *et al.*).

■ **HABITAT** Woodland and scrubland around mosses and meres.
■ **FOODPLANTS** Not recorded in Lancashire.
Single brooded. Comes sparingly to light.

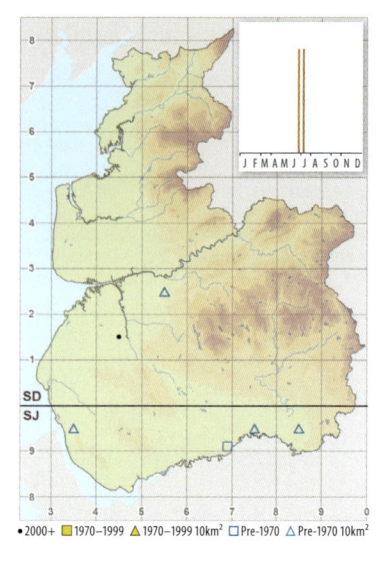

49.008 *Philedone gerningana*

■ **LANCS STATUS** Occasional / local resident   1846–2021

Noted as rare on Chat Moss by R.S. Edleston in July 1846, abundant on Risley Moss by Gregson (1857) and common on moors and mosses of the region by Ellis (1890). In 1954, it was reported from Formby Moss, but then, not until after the mid-1980s, on Winmarleigh Moss and at Birk Bank. Subsequently noted widely, but infrequently, on the moors of VC60, and, less so, in VC59. Reports from the mosses have been few in recent years, but include Winmarleigh Moss, up to 1999 and Astley Moss, up to 2009.

■ **HABITAT** Moorland and mosses.
■ **FOODPLANTS** Heather.
Single brooded. Comes occasionally to light, but most sightings are during the day.

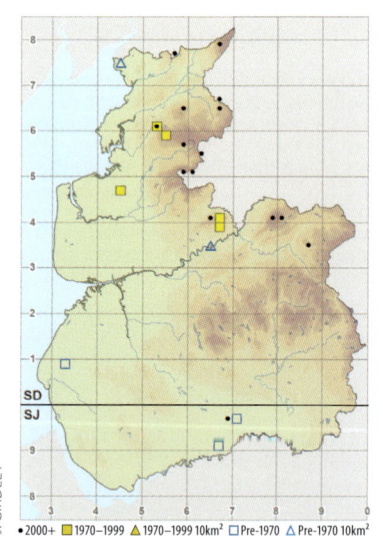

49.009 *Capua vulgana*

■ **LANCS STATUS** Occasional / widespread resident   c.1857–2022

There are only two historic records of this species, from Boor's Wood, Hale (Gregson, 1857) and Burnley in 1920, by W.G. Clutten. It was not reported again until 1990, at Crossgill. Since 1996, it has been noted almost annually at a range of scattered sites including Flixton, Ainsdale, Docker Moor, Gait Barrows, Lord's Lot Wood, Silverdale Moss and Warton Crag. The largest counts, of between six and eleven individuals, have all been to light at Gait Barrows, from mid-May to early June.

■ **HABITAT** Woodland on limestone, moorland edge, near sand dunes and mosses.
■ **FOODPLANTS** Not recorded in Lancashire.
Single brooded. Comes to light in small numbers, but can be disturbed by day.

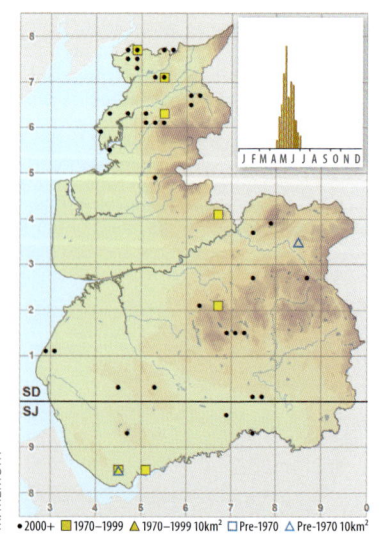

49.010 *Philedonides lunana*

■ **LANCS STATUS** Occasional / local resident   1856–2022
Noted at Withnell Birch Clough on 20 April 1856 (Gregson, 1856), at the same site in 1865 by J.B. Hodgkinson with over 100 noted, and also near Longridge c.1889. Not reported again until 2002, at Blackstone Edge, Darwen in 2012 and on a few occasions at Pendle Hill during April 2014, with up to 14 seen (G. Turner). Since then, found at only five moorland sites, but considered under-recorded.
■ **HABITAT** Moorland.
■ **FOODPLANTS** Not recorded in Lancashire.
Single brooded. All reports were in flight on sunny, early spring days. Reported once, by day, at Gait Barrows in 1987 (E. Emmett), well away from its usual habitat.

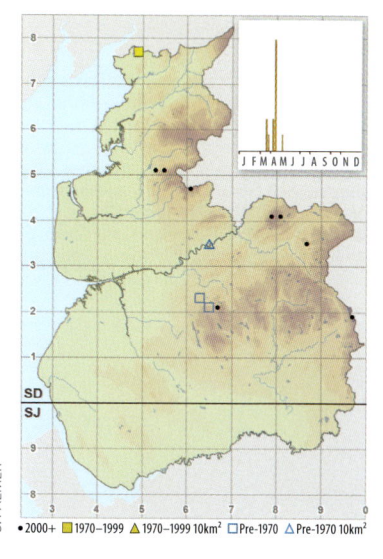

49.013 Large Fruit-tree Tortrix *Archips podana*

■ **LANCS STATUS** Frequent / widespread resident   c.1857–2022
First noted in the Liverpool area (Gregson, 1857) and thereafter considered as plentiful, occasionally abundant, and generally distributed. Since the 1960s, it has been regular at many lowland sites, usually in low single figures, although 20 were seen in an actinic trap at Billinge on 27 June 1992 (C.A. Darbyshire). Not noted from more upland parts until the late 1990s, where it remains local and infrequent.
■ **HABITAT** Hedgerows, gardens, woodland, scrub and brownfield sites.
■ **FOODPLANTS** Birch, including Downy Birch, oak, Hawthorn, Apple and rose.
Mainly single brooded, with a small, occasional, second brood in the autumn since 1989. Comes to light. Larvae have been found spinning leaves in May and June.

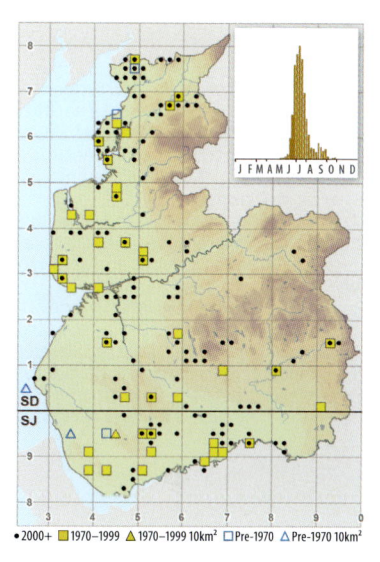

49.015 Variegated Golden Tortrix *Archips xylosteana*

■ **LANCS STATUS** Occasional / local resident   c.1857–2022
Noted as plentiful in a nut hedge in Croxteth (Gregson, 1857) and, by other early authors, as common throughout, including Formby in 1950. Between then and 1999 there were very few records. From then, until 2015 it was found in small numbers at about 20 widely scattered sites, with Mill Houses as its stronghold. Upland reports were few in number with most of these from north-east VC60. Since 2015, it has only been recorded at nine sites and nowhere has had more than four records per year.
■ **HABITAT** Scrub, hedgerows, gardens and brownfield sites.
■ **FOODPLANTS** Silver Birch.
Single brooded. Comes to light in low single figures. Showing a steady decline.

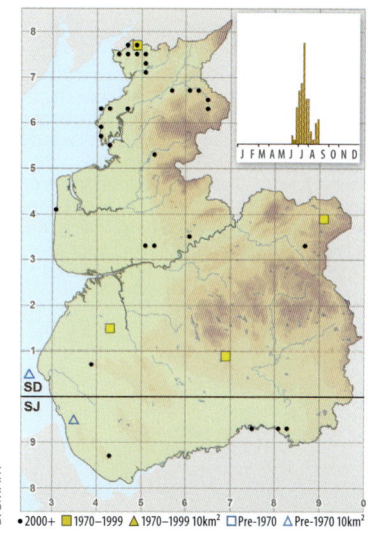

TORTRICIDAE

49.016 Rose Tortrix *Archips rosana*

■ **LANCS STATUS** Occasional / local resident   c.1857–2021

Noted as abundant everywhere within the Liverpool district (Gregson, 1857), with Mansbridge (1940) reporting it as a very common species in the region. From the 1940s to '80s, only noted in the Formby area on dunes containing Creeping Willow. This apparent reduction in range has since been confirmed, with records from few sites such as Flixton, Formby and St Annes. Scarce in upland areas, with the only recent record from Rishton on 23 June 2020 (D. Bickerton).

■ **HABITAT** Hedges, woodland and scrub, particularly adjacent to sand dunes.

■ **FOODPLANTS** Sallows, including Creeping Willow, Apple and birch.

Single brooded. Infrequent at light. Larvae have been found in May and June.

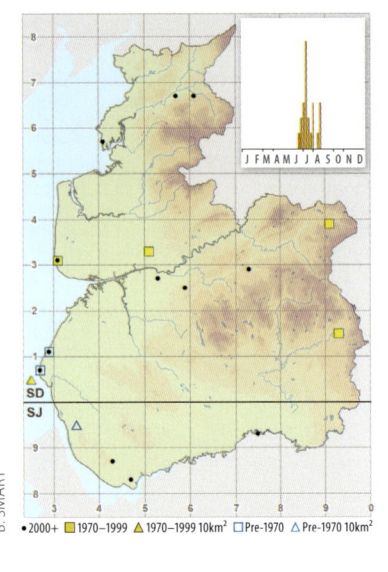

● 2000+   ▪ 1970–1999   ▲ 1970–1999 10km²   □ Pre-1970   △ Pre-1970 10km²

49.018 *Choristoneura hebenstreitella*

■ **LANCS STATUS** Rare / very local resident   c.1857–1990

Four specimens were taken at Little Britten Wood, a site said to have been between the Little Britain Pub and Simonswood Moss, near Kirkby (Gregson, 1857). It was next reported in the Preston area and at Longridge, without any dates, by J.B. Hodgkinson (Ellis, 1890). The final record relates to a June 1990 sighting at Birk Bank by M. Dempsey.

■ **HABITAT** Deciduous woodland, near mosses and moors.

■ **FOODPLANTS** Not recorded in Lancashire. Elsewhere it is polyphagous on deciduous trees and shrubs and associated with large woods and moorland areas where it is known to utilise Bilberry.

Believed to be single brooded. Byerley (1854) noted that it was found in gardens in the Warrington area c.1853 by N. Cooke, but these records are not listed by any subsequent authors and are considered unconfirmed. Its apparent absence in the county for long periods is difficult to explain and with a national distribution extending north to the Scottish border it may be overlooked, particularly in moorland areas where Bilberry occurs.

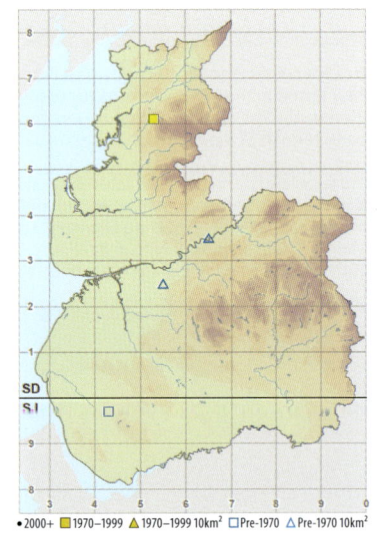

● 2000+   ▪ 1970–1999   ▲ 1970–1999 10km²   □ Pre-1970   △ Pre-1970 10km²

49.020 *Argyrotaenia ljungiana*

■ **LANCS STATUS** Occasional / local resident   1842–2014

First noted on White Moss, Manchester in 1842 by R.S. Edleston and on Rixton Moss in 1853, by N. Cooke. Early authors report it as common on the mosses and heaths, among young firs, at sites such as Rixton Moss, Chat Moss, Pilling Moss, Longridge and Pendle Hill. From 1994 to 2012, noted regularly on Cockerham and Winmarleigh Mosses, with 20 seen at the former site on 23 July 2010. Seen, less frequently, at several other sites, the last being at Longridge Fell on 18 May 2014 (G. Dixon).

■ **HABITAT** Lowland raised bogs and moorland.

■ **FOODPLANTS** Not recorded in Lancashire.

Double brooded. Mainly recorded by day, sometimes at light. Probably overlooked.

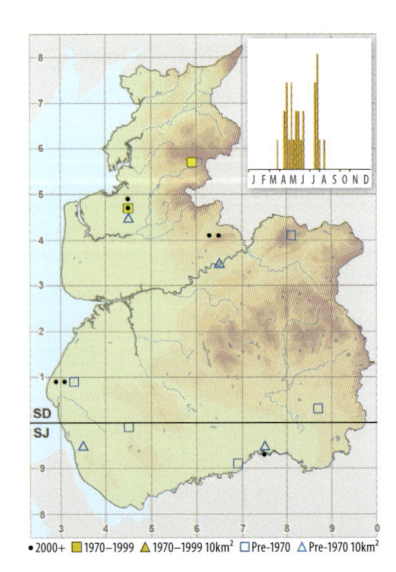

● 2000+   ▪ 1970–1999   ▲ 1970–1999 10km²   □ Pre-1970   △ Pre-1970 10km²

244 *The Moths of Lancashire*

49.021 *Ptycholomoides aeriferana*

■ **LANCS STATUS** Occasional / local resident   1999–2020

This species has slowly been expanding its range northwards after arriving in Britain in 1951, reaching Cheshire in 1995 and Lancashire, at Wigan Flashes, on 26 June 1999 (S. Palmer *et al.*). It was then noted in Parr and Walmer Bridge in 2001, Chorlton in 2002, Flixton in 2003, Martin Mere in 2009 and Rochdale in 2010. Since then, the expansion rate has slowed, with new sites in Rishton and Swinton in 2013 and, within VC60, at Fulwood, Preston on 28 June 2014 (A. Powell) and north Preston in 2018.
■ **HABITAT** Gardens. Nationally, associated with Larch plantings.
■ **FOODPLANTS** Not recorded in Lancashire.
Single brooded. Comes to light mainly as singletons. Only reported once since 2018.

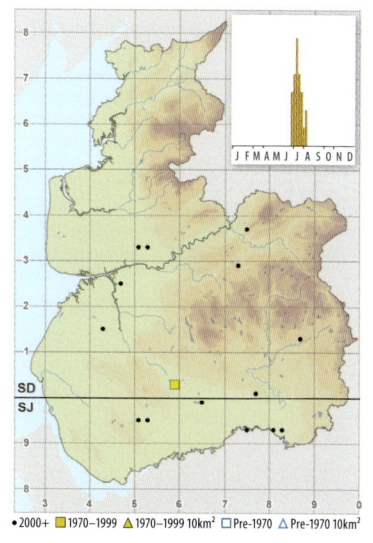

● 2000+ ☐ 1970–1999 △ 1970–1999 10km² ☐ Pre-1970 △ Pre-1970 10km²

49.022 *Ptycholoma lecheana*

■ **LANCS STATUS** Occasional / widespread resident   c.1857–2022

Originally noted as plentiful in woods in the Liverpool area (Gregson, 1857). Later, considered local and infrequent by Mansbridge (1940), with records up to 1940 from Prestwich, Preston, Balderstone, Southport and Burnley. By the late 1990s, it had also been seen at Formby Moss, Morecambe and Billinge. Thereafter, scattered reports were received from across the county, but always in small numbers. Rarely seen with any regularity, but it did appear in Billinge on nine occasions from 1989 to 2004.
■ **HABITAT** Woodland, gardens and brownfield sites with mature scrub.
■ **FOODPLANTS** Goat Willow, oak, elm, White Poplar, birch, Cherry Plum and lime.
Considered single brooded. Comes sparingly to light. Larvae found in April and May.

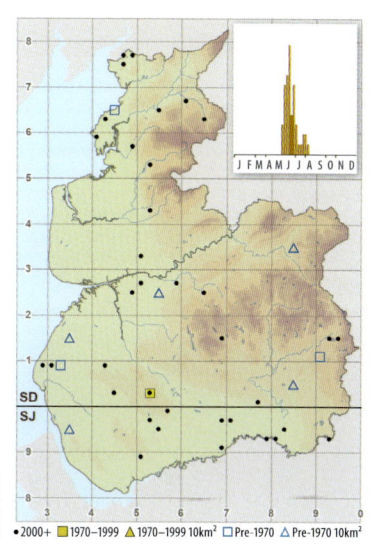

● 2000+ ☐ 1970–1999 △ 1970–1999 10km² ☐ Pre-1970 △ Pre-1970 10km²

49.023 *Pandemis cinnamomeana*

■ **LANCS STATUS** Occasional / local resident   1907–2022

First noted on Kirkby Moss in 1907 and then Simonswood Moss (Mansbridge, 1940). Not seen again until the 1980s, at Mount Vernon, Caton and Warton Crag, by E. Emmett. Since then, most reports are from northern VC60, occasionally elsewhere, such as Flixton, Hale, Churn Clough Res., Pendleton Moor, Rishton and Rochdale. In the north, its main sites are at Gait Barrows, Docker Moor and Mill Houses.
■ **HABITAT** Deciduous woods near moors, mosses and on limestone, and hedgerows.
■ **FOODPLANTS** Larch is the only foodplant noted in the county.
Double brooded. Comes to light, usually in low single figures. 17 came to light at Docker Moor on 3 July 2014 (P.J. Marsh & J. Roberts).

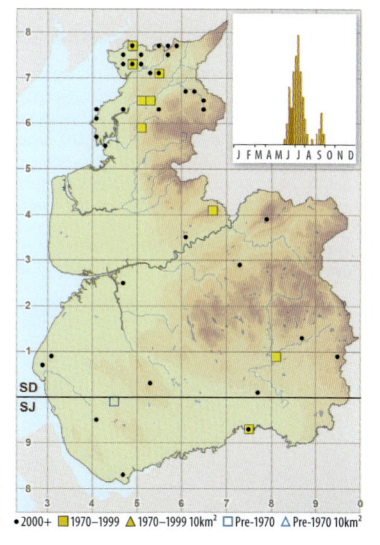

● 2000+ ☐ 1970–1999 △ 1970–1999 10km² ☐ Pre-1970 △ Pre-1970 10km²

49.024 Chequered Fruit-tree Tortrix *Pandemis corylana*

■ **LANCS STATUS** Frequent / widespread resident   c.1859–2022
First at Barton Moss c.1859 J. Chappell and Preston (Ellis, 1890), where it was noted as local; later, 'generally distributed' (Mansbridge, 1940). In recent decades, noted widely in lowland parts of VC59 and northern VC60; less frequently in upland areas. Found at Leighton Moss in 1960 and now regular at this site. Infrequent in central and southern VC60, although regular in small numbers in the Preston area.
■ **HABITAT** Woodland, hedgerows, gardens and brownfield sites.
■ **FOODPLANTS** Elm, Grey Willow, Creeping Thistle and Alder Buckthorn.
Single brooded. Comes to light in low numbers; 24 were seen at light in Parr in 2018. Larvae found May and June. Early June records are unusual and require verification.

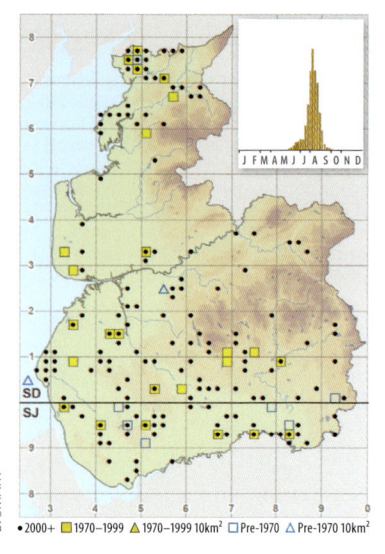

● 2000+   ■ 1970–1999   ▲ 1970–1999 10km²   □ Pre-1970   △ Pre-1970 10km²

49.025 Barred Fruit-tree Tortrix *Pandemis cerasana*

■ **LANCS STATUS** Abundant / widespread resident   c.1857–2022
First noted in the Liverpool district (Gregson, 1857) and considered abundant throughout by early authors. Few early records were documented, but these did include Kirkby Moss, Formby, Croxteth, Silverdale, Burnley and Crosby, up to the 1970s. Thereafter reported widely from both lowland and upland regions.
■ **HABITAT** Woodland, hedgerows, gardens, scrubland and brownfield sites.
■ **FOODPLANTS** Birch sp., Downy Birch, Beech, Sycamore, sallows, Larch, Burnet Rose, Honeysuckle, Lilac, Buckthorn and oak sp.
Single brooded, with an extended flight period. Comes to light, sometimes in double figures, with 40 at Fulwood 14 June 2014. Larvae seen from November to May.

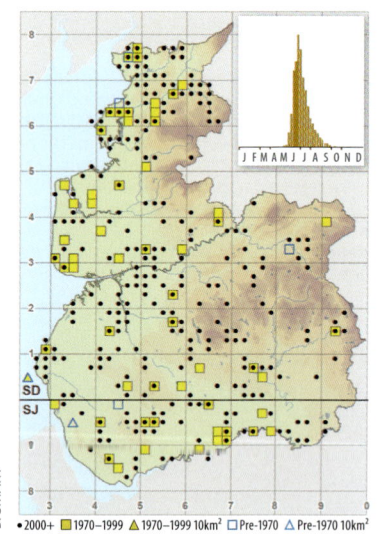

● 2000+   ■ 1970–1999   ▲ 1970–1999 10km²   □ Pre-1970   △ Pre-1970 10km²

49.026 Dark Fruit-tree Tortrix *Pandemis heparana*

■ **LANCS STATUS** Frequent / widespread resident   c.1857–2022
First noted in the Liverpool district (Gregson, 1857) and reported by all other early authors as, variously, abundant in hedgerows or common everywhere, including from Burnley between 1917 and 1920. Since 1980, it has been seen regularly at many sites, including Billinge, Flixton, Morecambe, Rishton, Briercliffe, Burrow Heights, Hale, Rochdale, Fulwood, Leighton Moss, Walmer Bridge, Longton and Sunderland Point.
■ **HABITAT** Woodland, gardens, hedgerows, scrub and brownfield sites.
■ **FOODPLANTS** Birch, sallow, Blackthorn, Ash, Aspen and rose.
Single brooded. Comes to light with 21 in Flixton on 24 June 2005. Since about 2015, double digit counts have all but disappeared. Larvae found from March to June.

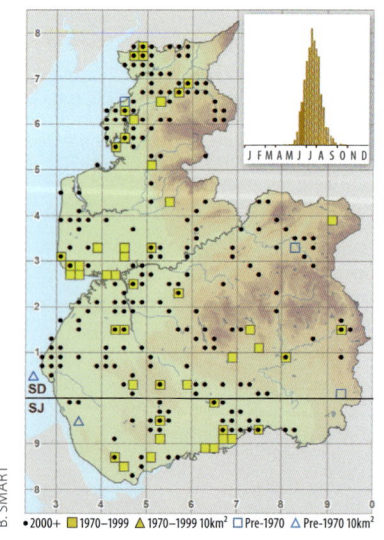

● 2000+   ■ 1970–1999   ▲ 1970–1999 10km²   □ Pre-1970   △ Pre-1970 10km²

49.028 *Syndemis musculana*

- **LANCS STATUS** Frequent / widespread resident   c.1857–2022

First noted in the Liverpool district (Gregson, 1857), then Yealand Conyers in 1910. Mansbridge (1940) reported it as common in woodlands, and Leech & Michaelis (1957) as common in the Formby area. Since the mid-1980s it has been noted widely across the county, more frequently, but not exclusively, in lowland areas. Sites with long-term record sets include Ainsdale, Billinge, Flixton, Preston and Warton Crag.
- **HABITAT** Woodland, moorland, mosses, scrub and, less frequently, in gardens.
- **FOODPLANTS** Blackthorn, birch sp., willow sp., Bilberry, Broom, Heather, Honeysuckle and Angelica.

Single brooded. Comes to light in single figures. Larvae found August to November.

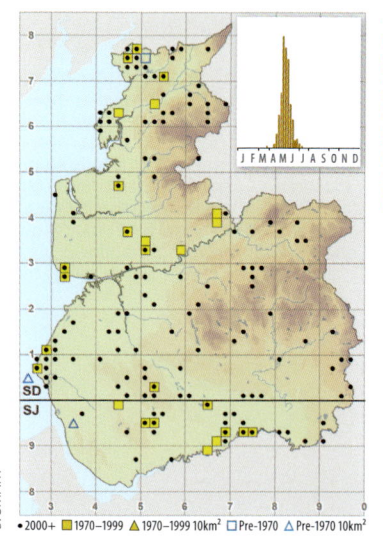

• 2000+   ■ 1970–1999   ▲ 1970–1999 10km²   □ Pre-1970   △ Pre-1970 10km²

49.029 *Lozotaenia forsterana*

- **LANCS STATUS** Frequent / widespread resident   1848–2022

First noted, as a larva, at Longridge Fell by J.B. Hodgkinson in 1848. Thereafter, found in the Liverpool district in 1857, Simonswood Moss in 1908 and Belmont in 1921. Since the late 1980s, it has been reported regularly from both moorland areas and garden light traps at inland sites, rarely coastal. Sites with larger datasets include Adlington, Billinge, Briercliffe, Hoghton, Orrell, Preston, Rishton and Walmer Bridge.
- **HABITAT** Moorland, mosses, gardens, scrub, hedgerows and brownfield sites.
- **FOODPLANTS** Ivy, Heather, Bilberry, Cowberry, Black Knapweed, Quince, Bay, Holly.

Single brooded. Comes to light. Larvae found from late March to May. An MV light run on Longridge Fell on 24 July 2012 attracted at least 31, but probably many more.

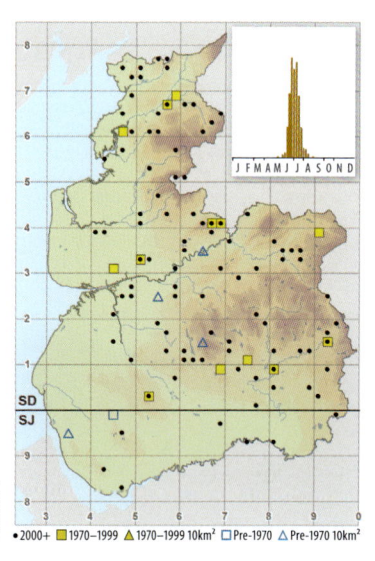

• 2000+   ■ 1970–1999   ▲ 1970–1999 10km²   □ Pre-1970   △ Pre-1970 10km²

49.030 Carnation Tortrix *Cacoecimorpha pronubana*

- **LANCS STATUS** Occasional / local resident   1981–2022

First noted at Sharples on 14 July 1981 by E.G. Hancock, following an expansion of range northwards in Britain. Next, seen at Claughton in 1987 and then, throughout the 1990s, at St Annes, Hornby and Morecambe, with 79 noted in 1995 at the last site. Elsewhere, it was reported from Flixton in 1996, Lytham in 1997 and St Helens in 1999. Despite scattered, lowland records since then, including Radcliffe in 2020 and Crosby in 2022, records are now rare, with no more than two a year since 2012.
- **HABITAT** Gardens and hedgerows.
- **FOODPLANTS** Not recorded in Lancashire.

Up to three broods. Comes to light. The bright orange hindwing is definitive.

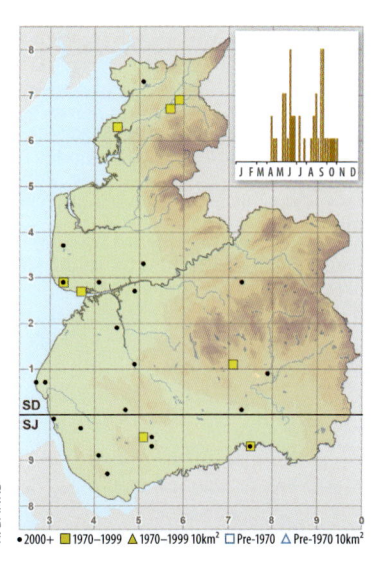

• 2000+   ■ 1970–1999   ▲ 1970–1999 10km²   □ Pre-1970   △ Pre-1970 10km²

The Moths of Lancashire

49.031 Timothy Tortrix *Zelotherses paleana*

■ **LANCS STATUS** Frequent / widespread resident   c.1857–2022
Noted in Liverpool (Gregson, 1857), then as widely distributed, which still applies to the present day. First reported from higher parts in Burnley c.1974 but remains less frequent in the uplands. 43 seen at Myers Allotment on 14 July 2012 (A. Barker).
■ **HABITAT** Rough grassland in a wide range of habitats, particularly brownfield sites.
■ **FOODPLANTS** Cock's-foot Grass, Purple Moor-grass, Creeping Thistle, Spear Thistle, Marsh Thistle, Black Knapweed, Greater Knapweed, Ribwort Plantain, Hogweed, Sea Aster, mouse-ear sp., vetch sp., Yellow Vetchling, Butterbur, Common Ragwort, Colt's-foot, Tall Melilot, Cuckooflower, *Chenopodium* sp. and possibly Mullein.
Single brooded. Comes to light. Larvae found from late October to early July.

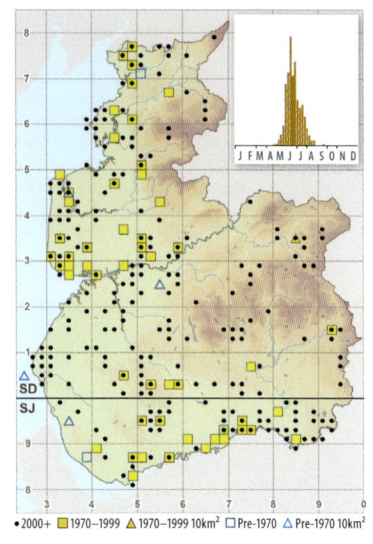

49.033 Bilberry Tortrix *Aphelia viburnana*

■ **LANCS STATUS** Occasional / widespread resident   1846–2022
Found on Chat Moss in July 1846 (Edleston, 1846) and later, at a scattered range of mosses and moors and reported as common (Mansbridge, 1940). Since then, regular sites include Astley Moss, Oswaldtwistle, Thrushgill and Winmarleigh Moss. Wanders on occasions and, as such, has been seen at Birkdale and Bispham and a few other locations, all as singletons. Larger counts include 50 at Astley Moss in July 2003.
■ **HABITAT** Mosses and moorland.
■ **FOODPLANTS** Bilberry, Cross-leaved Heath, Bog Myrtle and Bog Rosemary.
Single brooded. Comes to light. Larvae found May, June and July. A report of 20 on 6 May 2018 was exceptional, but was from a reliable recorder and amongst Bilberry.

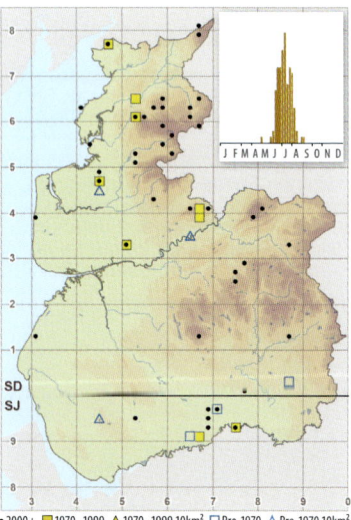

49.035 *Clepsis senecionana*

■ **LANCS STATUS** Occasional / local resident   c.1853–2022
First noted on Liverpool district mosses by N. Cooke (Byerley, 1854), then Chat Moss c.1859 (J. Chappell), Longridge c.1888, Simonswood Moss c.1900, Kirkby Moss in 1907 and Burnley in 1910. Ellis (1890) reported it as scarce, while Mansbridge (1940) wrote 'locally common among Sweet Gale'. Since then, most records have been from Birk Bank, Longridge Fell, Beacon Fell and Pendle Hill. Probably under-recorded.
■ **HABITAT** Mosses and moorland.
■ **FOODPLANTS** Not recorded in Lancashire.
Single brooded. Best found by day. An August 1986 Caton specimen and dissection slide, in World Museum, has been examined and is a misidentification.

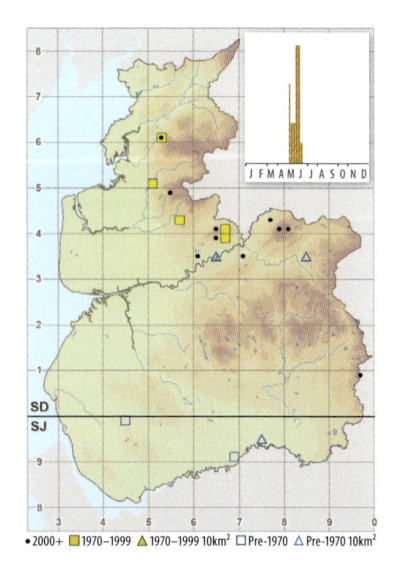

248 *The Moths of Lancashire*

49.037 Cyclamen Tortrix *Clepsis spectrana*

■ **LANCS STATUS** Frequent / widespread resident   c.1857–2022
Found in the Liverpool district (Gregson, 1857), Allerton in 1908, Burnley in 1911, Blackpool in 1925 and Middle Hulton in 1946. In the 1950s, noted from Formby and Morecambe. Since the 1980s, has been widely recorded in the lowlands, but is less frequent in upland regions. Regular at many sites, such as Billinge, Hale, Sutton Leach, Parr, Longton, Hoghton, Leighton Moss, Martin Mere and Sunderland Point.
■ **HABITAT** Wide range of habitats, particularly marshy places (Mansbridge, 1940).
■ **FOODPLANTS** Great Willowherb, Common Nettle, Bramble sp., dock sp., Autumn Hawkbit, Coltsfoot, Ivy, Iris, *Ranunculus* sp., bindweed and rose sp.
Double brooded. Comes to light. Larvae found from mid-March to mid-June.

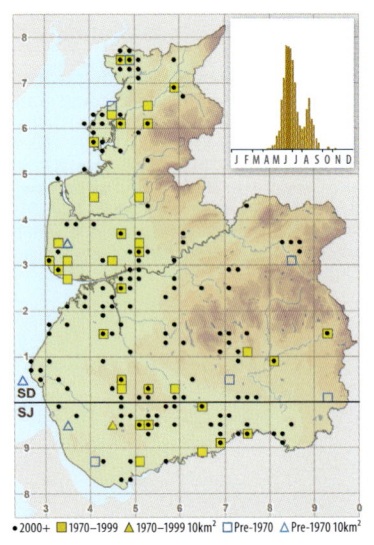

49.038 *Clepsis consimilana*

■ **LANCS STATUS** Frequent / widespread resident   c.1857–2022
Plentiful in the Liverpool district (Gregson, 1857). Thereafter, noted as common and widespread (Mansbridge, 1940), but no sites were listed. In the 1990s, found mainly in lowland west and south Lancashire, such as St Annes, Preston and Flixton. Since then, has expanded its range across the region, including into the uplands.
■ **HABITAT** Gardens, parks, hedgerows, amenity plantings and brownfield sites.
■ **FOODPLANTS** Privet.
Single brooded. A small second brood commenced in 2001, increasing in size from 2018 and, in the following year, was also noted in a few upland areas. Comes to light, mostly in single figures. 30 were seen at Middleton N.R. on 12 July 2015.

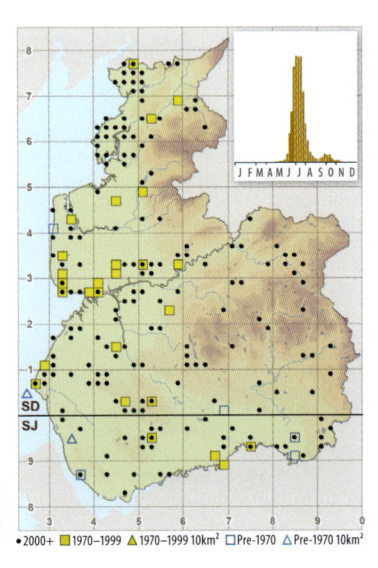

49.039 Light Brown Apple Moth *Epiphyas postvittana*

■ **LANCS STATUS** Abundant / widespread resident   1986–2022
Arrived in Cornwall, as an adventive from Australia, spreading rapidly north and reaching Lancashire on 2 September 1986 at Woolton, Liverpool (S. McWilliam). Since spread across most of the county; much less frequent in rural areas away from human habitation. Apparently absent from the highest areas. It reached Morecambe in 1996, Littleborough in 1999, Roeburndale in 2001 and Silverdale in 2002.
■ **HABITAT** Most frequent in urban or suburban areas, gardens, parks and hedgerows.
■ **FOODPLANTS** A very wide range of plants, with over 30 species noted in the county.
Two overlapping broods. Comes to light, the largest count being 150 near Middleton N.R. on 18 September 2015. Numbers vary according to how harsh the winters are.

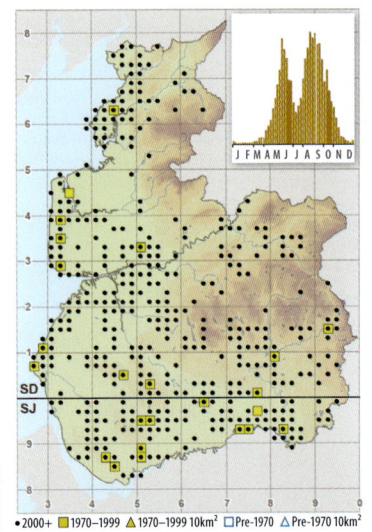

The Moths of Lancashire   **249**

TORTRICIDAE

49.040 *Lozotaeniodes formosana*

■ **LANCS STATUS** Occasional / widespread resident   1996–2022
Arrived in Britain in 1945 and moved gradually northwards, reaching Billinge on 17 July 1996 (C.A. Darbyshire). Since then, noted in Flixton in 1997, Pennington in 1998 and Mere Sands Wood in 1999. This slow rate of expansion has continued across lowland VC59 and into mainly coastal locations in VC60, reaching Thornton in 2003 and Carnforth in 2016. Upland records are few in number but include Worsthorne on 8 August 2006, Briercliffe from 25 July 2019 and Shaw on 18 July 2020 (G. Crowder).
■ **HABITAT** Locations with Scots Pine, but is prone to wandering.
■ **FOODPLANTS** Not recorded in Lancashire.
Single brooded. Comes to light, mostly in single figures.

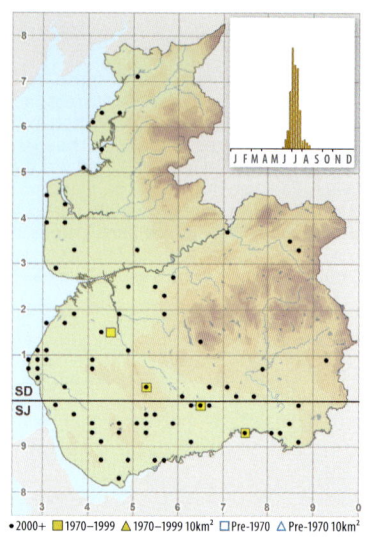

49.042 *Neosphaleroptera nubilana*

■ **LANCS STATUS** Occasional / very local resident   1987–2016
Considered an overlooked resident, first noted when five were encountered during the day at Bucks Moss Wood, Salwick on 12 July 1987 by M. Evans. Then, found at Clayton Green on 11 July 1998 and Bispham Hall, Billinge on 10 July 2000. It has been noted on eight further occasions, all in lowland regions, with larvae numerous near Forton M6 Services (north) on 19 May 2006 (S. Farrell) and 25 flying around a Hawthorn hedge at Sycamore House, Ormskirk on 4 July 2006 (C.A. Darbyshire).
■ **HABITAT** Hedgerows and scrubland.
■ **FOODPLANTS** Hawthorn.
Single brooded. Comes occasionally to light. Larvae found in May.

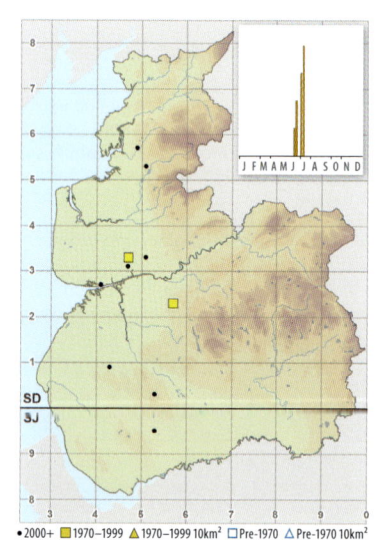

49.043 *Exapate congelatella*

■ **LANCS STATUS** Occasional / widespread resident   1852–2022
First noted in Rainhill in October 1852 (Gregson, 1857) and then, prior to 1960, from Burnley, Crosby, Formby, Fallowfield, Old Trafford, Morecambe and Warton Crag. Since the mid-1990s has been recorded in widely scattered localities in both lowland and upland regions. Usually recorded in single figures, although 45 were attracted to a Victoria Park, Manchester light trap on 7 November 2020 (P. Pemberton).
■ **HABITAT** Hedgerows, gardens, mosses, moorland, woodland and brownfield sites.
■ **FOODPLANTS** Not recorded in Lancashire.
Single brooded. Comes to light. A late-season moth, with a wingless female, which may be under-recorded due to reduced light trapping this late in the year.

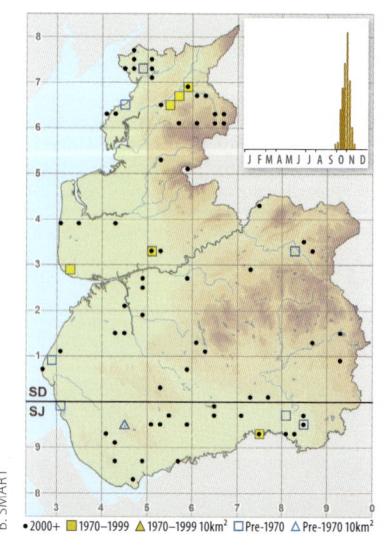

250   *The Moths of Lancashire*

49.044 *Tortricodes alternella*

■ **LANCS STATUS** Frequent / widespread resident   c.1857–2022

Reported to swarm in oak woods in the Liverpool district (Gregson, 1857) and later as 'abundant in oak woods' (Mansbridge, 1940). Documented sites were limited to the Burnley area. It wasn't until the late 1990s that it was noted again, at locations such as Dean Wood, Euxton, Cuerden, Billinge, Flixton and Preston. Larger counts occurred at Botton Mill, Worden Pk., Herring Head Wood and, particularly, at Mill Houses. 150 were attracted to a Chorley light trap on 10 March 2001 (G. Jones).

■ **HABITAT** Mainly deciduous oak woodland or hedgerows with oak standards.

■ **FOODPLANTS** Oak sp.

Single brooded. Comes to light mostly in single figures. Larvae found in May.

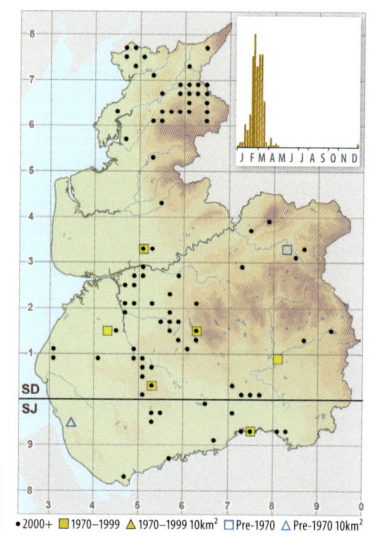

• 2000+   □ 1970–1999   △ 1970–1999 10km²   □ Pre-1970   △ Pre-1970 10km²

49.045 *Eana osseana*

■ **LANCS STATUS** Frequent / local resident   c.1859–2022

First noted on Chat Moss c.1859 by J. Chappell and then, Longridge c.1886, Silverdale c.1888, Burnley in 1910 and Storrs Moss in 1951. Since then, found regularly on grassy, upland sites, such as Caton, Birk Bank, Ward's Stone, Crossgill, Thrushgill, Bay Horse, Trough of Bowland, Pendle Hill and Leck Fell. It is unusual to see this species in lowland areas, but singletons were present at Heysham N.R. from 2000 to 2015.

■ **HABITAT** Mostly upland acidic and limestone grassland. Occasional on mosses.

■ **FOODPLANTS** Not recorded in Lancashire.

Single brooded. Comes to light, usually in low single figures. Over 200 were seen during the day on Tarnbrook Fell on 17 July 2021 by B. Dyson.

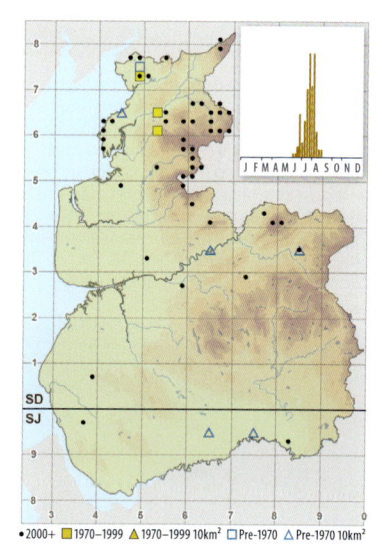

• 2000+   □ 1970–1999   △ 1970–1999 10km²   □ Pre-1970   △ Pre-1970 10km²

49.047 *Eana incanana*

■ **LANCS STATUS** Scarce / very local resident   c.1920–2015

A few were noted, over a period of time, in the Burnley area by W.G. Clutten (Mansbridge, 1940) but dates are unknown. Next reported from Risley Moss on 24 July 1982 (M. Hull) and, in 1986, it was found at Gait Barrows, the first of five records from this site, up to 2003. On 25 June 2004 it was seen in Eaves Wood, Silverdale and in Heysham in 2009, Whalley in 2012, Mill Houses in 2014 and nr. Aughton in 2015.

■ **HABITAT** Deciduous woodland with plentiful bluebell growth.

■ **FOODPLANTS** Not recorded in Lancashire.

Single brooded. Comes to light. Some well-marked forms of *Cnephasia stephensiana* can be superficially similar.

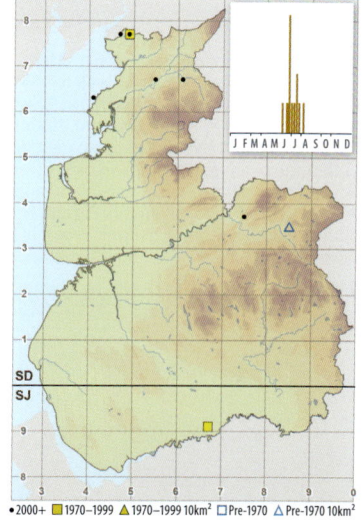

• 2000+   □ 1970–1999   △ 1970–1999 10km²   □ Pre-1970   △ Pre-1970 10km²

The Moths of Lancashire

TORTRICIDAE

49.048 *Eana penziana*

■ **LANCS STATUS** Rare / very local resident   1938–2015

First noted from Silverdale in 1938, per H.N. Michaelis in Hardwick (1990). Bradley, *et al.* (1973) include a drawing of f. *bellana* from Warton Crag which is thought to relate to one of those found by L.T. Ford in 1941. Next, reported from Gait Barrows, with singletons coming to MV light in 1993, 6 July 1998, 8 July 1999, 11 September 2000 and July 2001 (all R. Petley-Jones). A single f. *bellana* was noted from Leck Fell on 9 July 2015 (P.J. Marsh). It has not been possible to establish the forms seen at Gait Barrows with any certainty, but at least one may have been f. *colquhounana*.

■ **HABITAT** Limestone pavement in both coastal and upland areas.
■ **FOODPLANTS** Not recorded in Lancashire. Single brooded.

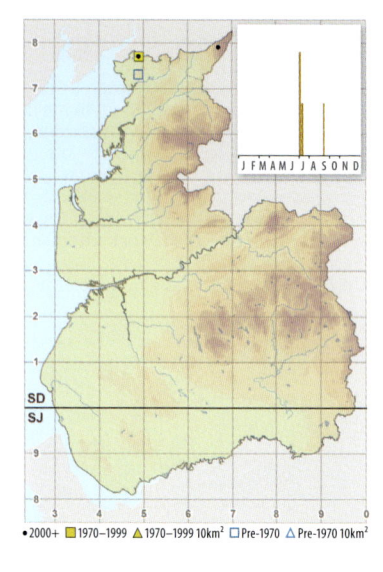

49.049 Light Grey Tortrix *Cnephasia incertana*

■ **LANCS STATUS** Frequent / widespread resident   c.1857–2022

First noted in Liverpool (Gregson, 1857) and later, from Simonswood Moss, Formby, Silverdale and Morecambe, up to the 1960s. From then, found widely across the county, most frequently in lowland areas. Regular at sites such as Flixton and Preston, and 85 attracted to light at Sunderland Point on 4 July 2012 (J. Girdley).

■ **HABITAT** Scrubland, hedgerows, gardens, brownfield sites and woodland.
■ **FOODPLANTS** Common Bird's-foot-Trefoil, mouse-ear sp., Goat Willow, Hogweed, Pignut, Creeping Thistle, Great Willowherb, Ribwort Plantain and Creeping Buttercup. Single brooded. Comes to light. Larvae found late August to early June. This species can be confused with other *Cnephasia*.

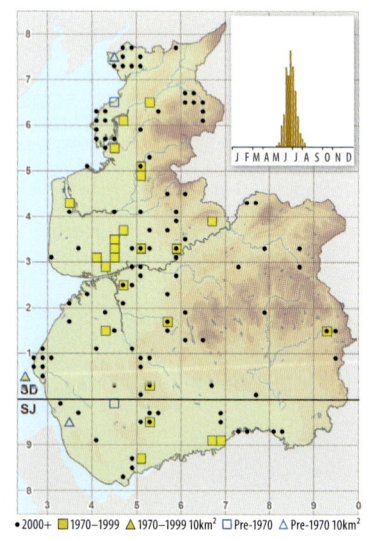

49.050 Grey Tortrix *Cnephasia stephensiana*

■ **LANCS STATUS** Occasional / local resident   1949–2022

First noted in Freshfield from 1949 to 1951 by C. de G. Fraser (det. H.N. Michaelis) and from Morecambe on 21 July 1959 by C.J. Goodall (det. S. Palmer). Since then, found at Crosby, St Annes, Preston, Flixton and Longridge Fell. Never a common species, records have declined steadily since 2006 and more markedly since 2015, with none received in 2021 and only three in 2022, twice near Rishton and at Parr.

■ **HABITAT** Gardens, hedges, scrubland and brownfield sites.
■ **FOODPLANTS** Ribwort Plantain, Pignut, Hogweed and Creeping Buttercup. Single brooded. Comes to light, in single figures. Larvae found in May. Early records have not been confirmed unless specimens were available for examination.

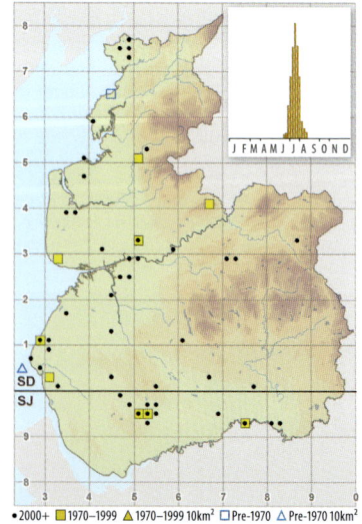

49.051 Flax Tortrix *Cnephasia asseclana*

■ **LANCS STATUS** Frequent / widespread resident   1976–2022

Early records under *C. virgaureana* and noted as common and general, probably refer to this sp. First confirmed at Crosby on 18 July 1976 by E.C. Pelham-Clinton, and then noted in many lowland areas, such as Flixton, Fulwood, Longton, Preston and Parr. Apparently more local in upland areas, such as Rishton, but probably under-recorded.
■ **HABITAT** Rough grassland, hedgerows, scrub, mosses, gardens and brownfield sites.
■ **FOODPLANTS** Bush Vetch, Bird's-foot-Trefoil, Tansy, Garlic Mustard, Creeping Thistle, Creeping Buttercup, Pignut, Ribwort Plantain, Mugwort, Marjoram and Wood Sage. Single brooded. Comes to light, usually in single figures. Dark and light forms occur. Larvae found May and June. Dissection is usually required to confirm identity.

● 2000+ ☐ 1970–1999 ▲ 1970–1999 10km² ☐ Pre-1970 △ Pre-1970 10km²

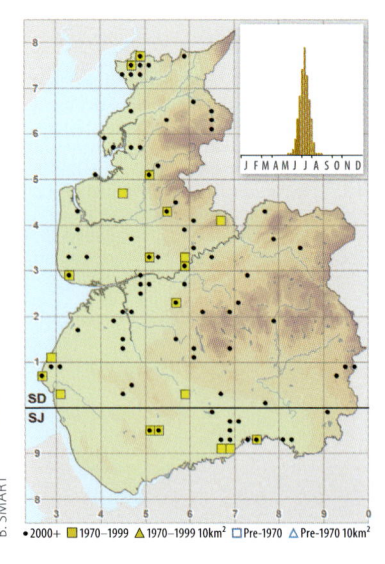

49.052 *Cnephasia pasiuana*

■ **LANCS STATUS** Rare / vagrant   2022

Several records under *C. pascuana*, a synonym of *pasiuana*, are noted from Preston, Longridge and Burnley in Mansbridge (1940), but none can be attributed with certainty to this species. Despite critical examination of many *Cnephasia* specimens since the 1990s, this species was not discovered in the county until a male was found at Small Rishton Reservoir on 8 July 2022 by D. Bickerton. Whether it is an overlooked resident or part of a wider expansion of range, is unknown.
■ **HABITAT** Unknown.
■ **FOODPLANTS** Not recorded in Lancashire.
Searches for extant historic specimens would be worthwhile.

A photograph of the dissection of the moth is included for this and the next species as the moths were worn and unrecognisable.

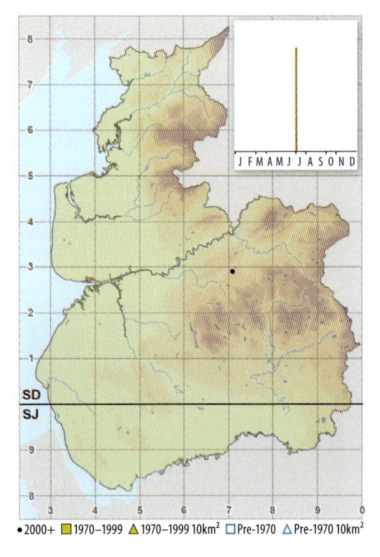

● 2000+ ☐ 1970–1999 ▲ 1970–1999 10km² ☐ Pre-1970 △ Pre-1970 10km²

49.054 *Cnephasia genitalana*

■ **LANCS STATUS** Rare / vagrant   2022

As with *C. pasiuana*, this species was an unexpected addition to the county fauna in 2022, when a single specimen came to light in the Tarleton area on the 29 July (A. Barker). There have been no other suspected specimens prior to the arrival of this individual. It is considered likely that its arrival was part of a wider expansion of range from the south, associated with the very warm weather of 2022; Cheshire had its first and only record in 2019. The moth closely resembles other *Cnephasia* species and could be overlooked amongst the more common *Cnephasia*.
■ **HABITAT** Unknown.
■ **FOODPLANTS** Not recorded in Lancashire.

● 2000+ ☐ 1970–1999 ▲ 1970–1999 10km² ☐ Pre-1970 △ Pre-1970 10km²

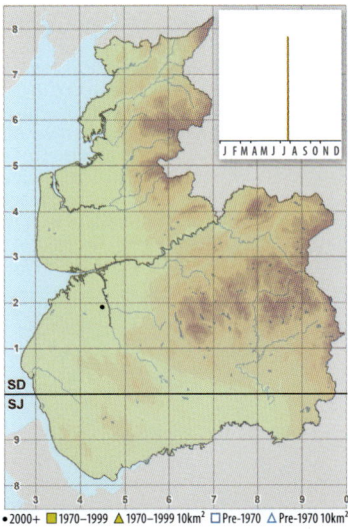

TORTRICIDAE

49.056 *Cnephasia conspersana*

■ **LANCS STATUS** Occasional / local resident c.1888–2022

First recorded in the Preston district by J.B. Hodgkinson and next by L.T. Ford, at Leighton Moss, in 1941. Mainly a coastal species, but occasionally found a little further inland, it was noted at St Annes dunes N.R. in 1995 and 1997, St Helens in 1999, Ainsdale and Formby in 2000 and Altcar in 2003. Most frequently found at Heysham N.R., associated with the flower-rich grassland, between 2004 and 2011.

■ **HABITAT** Coastal and limestone grassland, and a few lowland brownfield sites.
■ **FOODPLANTS** Oxeye Daisy, hawkweed sp., Common Bird's-foot-trefoil and Common Rock-rose, mainly in spun flowerheads and occasionally on spun terminal leaves.
Single brooded. Comes to light in single figures. Larvae found from mid-May to June.

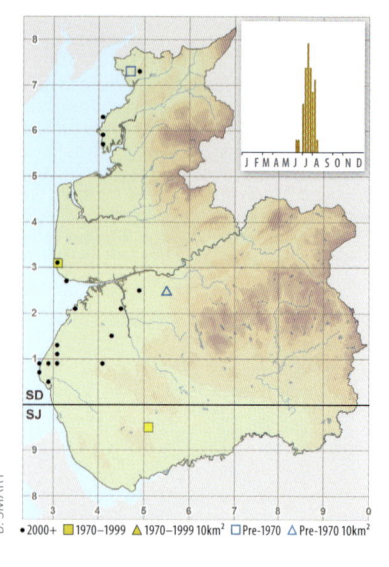

• 2000+   □ 1970–1999   ▲ 1970–1999 10km²   □ Pre-1970   △ Pre-1970 10km²

49.057 *Cnephasia longana*

■ **LANCS STATUS** Scarce / local resident c.1857–2022

Found in the Liverpool district (Gregson, 1857), the Formby area in 1904 (W. Mansbridge) and Ainsdale and Formby c.1950, where common on the sandhills. It was noted again at Formby in 1976, at St. Annes in 1995 and St Helens in 2004. Sporadically noted at a few low inland sites since then, these being at Martin Mere, Ormskirk, Woolton, Bickerstaffe, Adlington, Billinge, Eccleston and Hale. Occurs in low single figures at light and once netted on the edge of Cadishead Moss, in 2010.

■ **HABITAT** Coastal grassland, rough ground and brownfield sites.
■ **FOODPLANTS** Not recorded in Lancashire.
Single brooded. Comes to light.

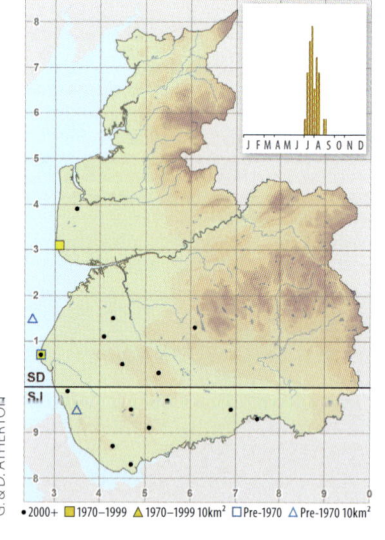

• 2000+   □ 1970–1999   ▲ 1970–1999 10km²   □ Pre-1970   △ Pre-1970 10km²

49.059 Green Oak Tortrix *Tortrix viridana*

■ **LANCS STATUS** Frequent / widespread resident c.1857–2022

First noted in the Liverpool district (Gregson, 1857) and on Simonswood Moss in 1904. It was later described as an abundant pest in oak woods (Mansbridge, 1940). Since then, it has been found wherever oak trees grow, with regular sites including Billinge, Botton Mill, Claughton, Flixton, Fulwood, Mill Houses and Preston.

■ **HABITAT** Deciduous woodland and hedgerows with oak standards.
■ **FOODPLANTS** Pedunculate Oak, oak sp. and once on elm.
Single brooded. Comes to light mostly in single figures. Larvae found from April to June. Significant leaf damage to oaks last reported in 1948, while there have been no counts of over 30 adults since 1986, when 500 were seen at Birk Bank (M. Dempsey).

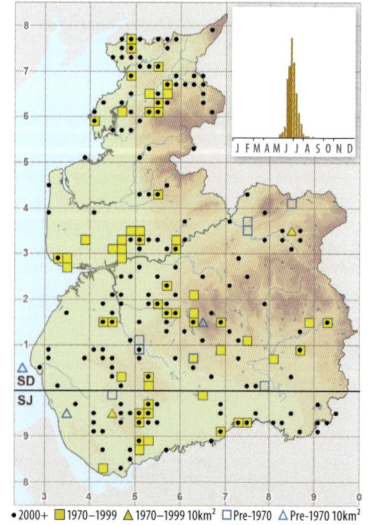

• 2000+   □ 1970–1999   ▲ 1970–1999 10km²   □ Pre-1970   △ Pre-1970 10km²

254   *The Moths of Lancashire*

49.060 *Aleimma loeflingiana*

■ **LANCS STATUS** Frequent / widespread resident   c.1857–2022

Reported as plentiful in the Liverpool area by C.S. Gregson and later, during the 19th century, from Barlow Moor, Irlam, Silverdale and Preston (Ellis, 1890). Mansbridge (1940) noted it as 'common and general'; since then, found regularly in many lowland areas, but excluding much of the Fylde. Less frequent in upland areas, where it is mainly associated with the northern, wooded, river valleys.

■ **HABITAT** Woodland and hedgerows with oak standards.
■ **FOODPLANTS** Oak sp.

Single brooded. Comes to light, usually in single figures. Largest count was 22 at MV light in Mill Houses on 4 July 2011 (P.J. Marsh). Larvae found in May.

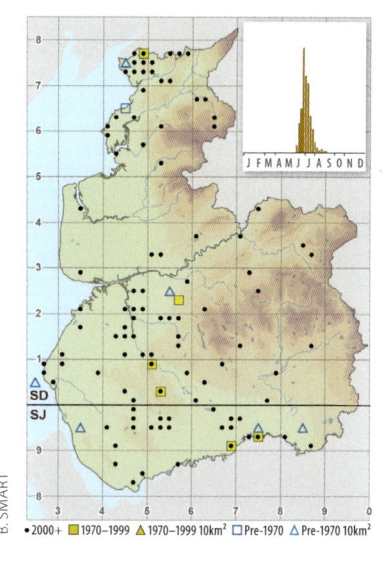

49.061 *Acleris holmiana*

■ **LANCS STATUS** Occasional, now Scarce / local resident   c.1859–2023

First noted in Irlam, Glazebrook and Barlow Moor by J. Chappell c.1859 and later, reported as local, but common where found (Ellis, 1890). Between the late 1990s and 2015 it was found at many scattered sites, mostly in the central and northern lowlands. Sites with regular reports in that period included Flixton, Gait Barrows, Leighton Moss, Sunderland Point and Wray. Since then, it has declined considerably and was last noted on 9 August 2020 at Yealand Conyers (B. Hancock).

■ **HABITAT** Woodland edges, scrubland, hedgerows and occasionally gardens.
■ **FOODPLANTS** Rowan and Blackthorn, the latter on 2 June 2023.

Single brooded. Comes to light in low single figures.

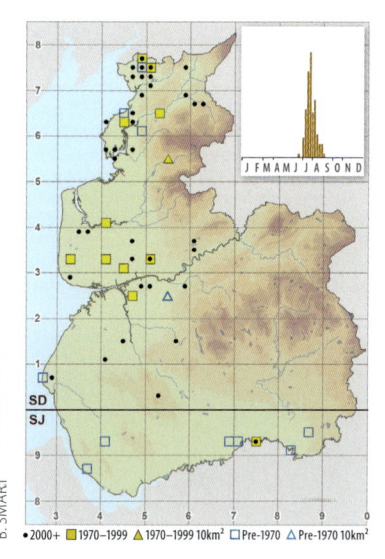

49.062 *Acleris forsskaleana*

■ **LANCS STATUS** Frequent / widespread resident   c.1857–2022

First noted in the Liverpool district (Gregson, 1857) and then, as common and general across the county by Mansbridge (1940). Prior to the 1990s, found at sites such as Prestwich, Crosby, Gait Barrows, Martin Mere and Morecambe. Records have increased in regularity and distribution since c.2000, at sites such as Billinge, Briercliffe, Carnforth, Flixton, Morecambe, Parr, Preston, Rishton and St Annes.

■ **HABITAT** Woodland, hedgerows, parks, scrubland and gardens.
■ **FOODPLANTS** Norway Maple and Field Maple. Adults common amongst Sycamore.

Single brooded; a possible, small second brood in October. Comes to light, mostly in single figures, 16 occurred at Mill Houses on 4 July 2011. Larvae found in May.

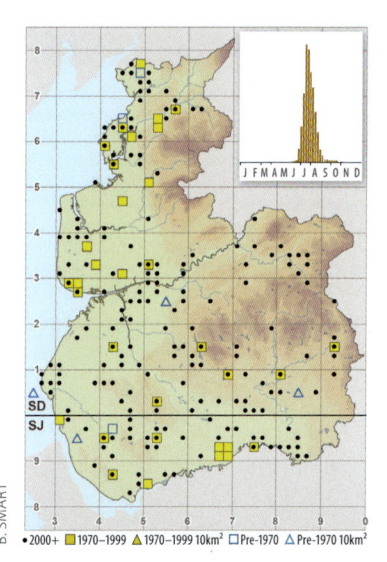

The Moths of Lancashire   **255**

49.063 *Acleris bergmanniana*

■ **LANCS STATUS** Frequent / local resident   c.1857–2022

Noted as common coastally among roses in the Liverpool area by C.S. Gregson and later, as common throughout by Mansbridge (1940). Since then, has become more scattered in distribution. Sunderland Point retains a strong colony on Burnet Rose and a large, roadside, rose bush at Parlick produced many moths in 2022. In contrast, the last sighting at Billinge was in 1999 and, in Preston, it was last noted in 2015.
■ **HABITAT** Hedgerows, scrubland, sand dunes, brownfield sites and gardens.
■ **FOODPLANTS** Rose spp., including Burnet Rose.
Single brooded. Comes to light; can be disturbed from hedgerow rose bushes by day and flies around them at dusk. Larvae found in May, occasionally in double figures.

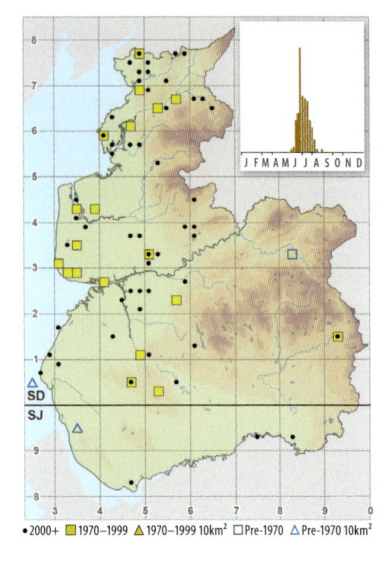

• 2000+   ■ 1970–1999   ▲ 1970–1999 10km²   □ Pre-1970   △ Pre-1970 10km²

49.064 *Acleris caledoniana*

■ **LANCS STATUS** Occasional / local resident   1857–2022

First noted at Cadishead Moss in 1857 by C.S. Gregson and Chat Moss c.1883, these being the only lowland records. Near Longridge c.1879, and the Burnley area in 1919. Not noted again until 2008 at Brinscall, followed by seven other very scattered upland sites to the present day. Largest count was 60 on Jeffrey Hill on 18 August 2011, by day. Considered under-recorded. See further comment under *A. comariana*.
■ **HABITAT** Moorland and, formerly, from lowland raised bogs.
■ **FOODPLANTS** Bilberry.
Single brooded. Readily disturbed by day from areas with plenty of heather and bilberry. Sole larval record was on 30 May 2002 at Jeffrey Hill, Longridge (S. Palmer).

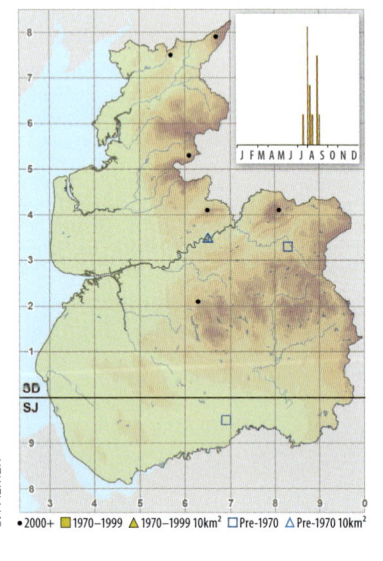

• 2000+   ■ 1970–1999   ▲ 1970–1999 10km²   □ Pre-1970   △ Pre-1970 10km²

49.065 Strawberry Tortrix *Acleris comariana*

■ **LANCS STATUS** Occasional / local resident   1924–2022

Noted as common near Woodvale in 1924 (Mansbridge, 1940) with the various forms found listed, as per Sheldon (1925). Since then, reported at scattered sites across lowland areas, often near the coast and in the Mersey Valley. Good numbers were reported at Flixton in 2002 (K. McCabe) and since noted here annually to 2022.
■ **HABITAT** Grassland in coastal areas, a few lowland gardens and brownfield sites.
■ **FOODPLANTS** Marsh Cinquefoil; once on Blackthorn, a very unusual foodplant.
Double brooded. Comes to light. Often misidentified due confusion with *A. laterana* with dissection required. Similar, both externally and in the genitalia, to *A. caledoniana* with current research attempting to clarify their specific status.

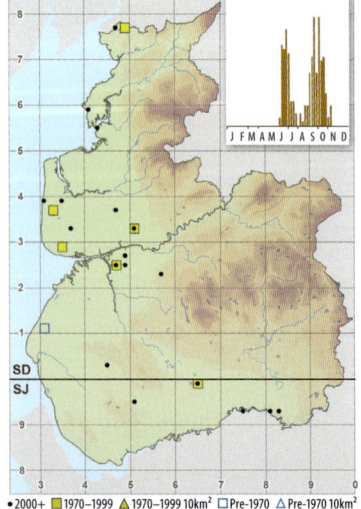

• 2000+   ■ 1970–1999   ▲ 1970–1999 10km²   □ Pre-1970   △ Pre-1970 10km²

256   *The Moths of Lancashire*

49.066 *Acleris laterana*

■ **LANCS STATUS** Frequent / widespread resident   1998–2022
Early records, under *Peronea latifasciana* (noted as fairly common and well distributed), as well as modern records, are not usually possible to allocate to this species without dissection to exclude *A. comariana*. First confirmed record from Brock Bottom in 1998 (S. Palmer). Since, found regularly at several sites including Leighton Moss, Flixton, Preston, Rishton and Bay Horse. Considered under-recorded.
■ **HABITAT** Gardens, woodland, parks, scrubland and brownfield sites.
■ **FOODPLANTS** Rowan, Meadowsweet and Blackthorn.
Single brooded. Comes to light, sometimes in double figures, e.g., 48 at Flixton in 2021 (K. McCabe). Larvae found in May.

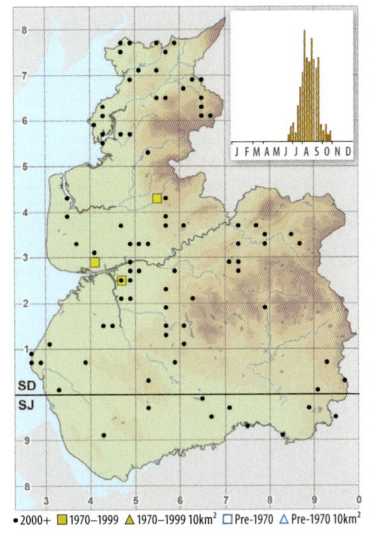

49.067 *Acleris abietana*

■ **LANCS STATUS** Rare / very local resident or vagrant   2001–2018
A relatively new arrival in the county, possibly expanding range southward following its arrival in Britain during 1965, in Perthshire (Sterling and Parsons, 2012). The first Lancashire record was from Burrow Heights, when one came to light on 23 October 2001 (B. Cockburn). Otherwise, it has only been seen at Docker Moor on 11 May 2015 and 18 April 2018, as singletons to light, by P.J. Marsh. These two records from a site which contains large conifer plantations, suggest it may be locally established.
■ **HABITAT** Coniferous woodland.
■ **FOODPLANTS** Not recorded in Lancashire.
Single brooded, overwintering as an adult.

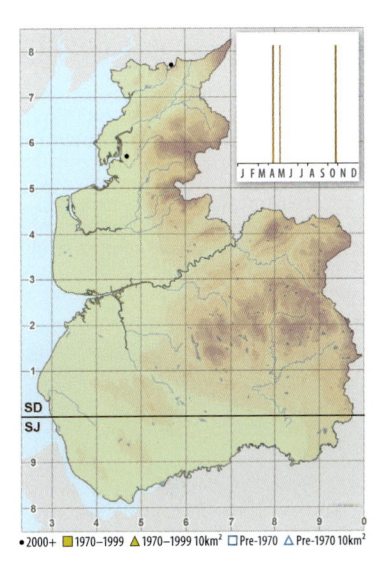

49.069 *Acleris sparsana*

■ **LANCS STATUS** Frequent / widespread resident   c.1857–2022
First noted in the Liverpool district (Gregson, 1857) as plentiful where Beech trees grow. Reported as common and general throughout the region (Mansbridge, 1940), and this has remained the case to the present day. Since 2000, it has been found regularly at many sites, including Parr, Preston, Rishton, Roby, St Annes and Wigan.
■ **HABITAT** Woodland, gardens, hedgerows and parks.
■ **FOODPLANTS** Sycamore, Field Maple, Beech, oak, Hawthorn & Hornbeam.
Single brooded, overwintering as an adult, but very infrequent after December. Comes to light, mostly in single figures; 24 came to light below a mature Beech on 2 October 2013 in Fulwood (A. Powell). Larvae found from mid-May to end of August.

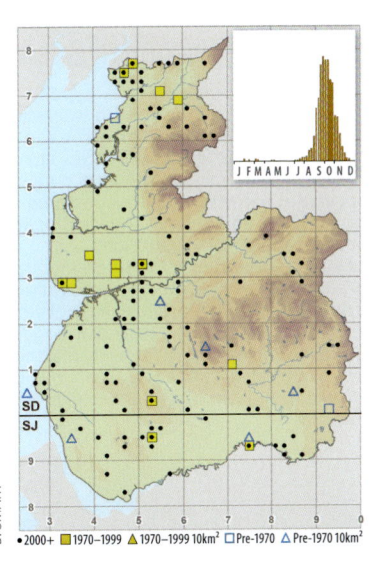

The Moths of Lancashire   **257**

TORTRICIDAE

49.070 Rhomboid Tortrix *Acleris rhombana*

■ **LANCS STATUS** Frequent / widespread resident   c.1857–2022

First noted in Liverpool, where abundant in hedgerows (Gregson, 1857). Early authors, up to 1940, reported it as abundant everywhere but with few supporting records. Since then, found widely in lowland parts of the county, but never abundantly and is rather infrequent in upland areas. Sites with long-term datasets include Billinge, Flixton, Heysham, Longton, Parr, Preston and Sunderland Point.

■ **HABITAT** Hedgerows, woodland, scrub, gardens, parks and brownfield sites.
■ **FOODPLANTS** Rowan, Hawthorn, Blackthorn, Cherry and Apple.

Single brooded. Comes to light, mostly in single figures; 20 were found at Sutton Leach on 28 August 2008 (S. Briers). Larvae found from late April to mid-June.

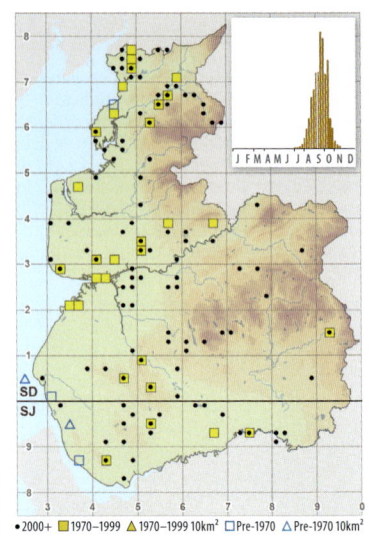

• 2000+   ■ 1970–1999   ▲ 1970–1999 10km²   □ Pre-1970   △ Pre-1970 10km²

49.071 *Acleris emargana*

■ **LANCS STATUS** Frequent / widespread resident   c.1857–2022

Found in Croxteth by Gregson (1857) and later described as not common, but fairly well distributed at Chat Moss, Preston, Formby, Bolton and Wigan (Mansbridge, 1940). Since then, noted widely across the county with sample specimens checked at many locations to exclude *A. effractana*. Regularly found, in small numbers, at sites such as Flixton, Fulwood, Haydock, Heysham, Leighton Moss and Oswaldtwistle. 15 attracted to light at Heysham Moss on 7 September 2016.

■ **HABITAT** Damp woodland, mosses, scrubland, hedgerows, and brownfield sites.
■ **FOODPLANTS** Willow spp., including Crack-willow and Grey Willow.

Single brooded. Comes to light. Larvae, May. See also comments under *A. effractana*.

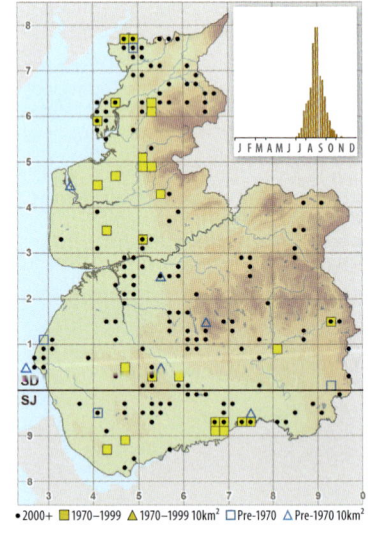

• 2000+   ■ 1970–1999   ▲ 1970–1999 10km²   □ Pre-1970   △ Pre-1970 10km²

49.072 *Acleris effractana*

■ **LANCS STATUS** Rare / very local resident   2011–2012

Not recognised as a distinct British species until 2005. This moth can be variable in markings and in the extent of the costal scallop and there is therefore potential confusion with other *Acleris* species such as *A. emargana* and *A. hastiana*. Despite microscopic examination of many potential candidates, *A. effractana* has only been located at one moorland edge site. Singletons were noted attracted to MV light at Bay Horse on 14 August 2011 and 18 August 2012, by N. Rogers, and both were males, confirmed by dissection.

■ **HABITAT** Associated with a single moorland edge area.
■ **FOODPLANTS** Not recorded in Lancashire.

The life history of this species has not yet been differentiated from that of *A. emargana*, with both feeding on a range of willow species. Any *Acleris* specimen encountered with weak forewing markings and a shallow costal scallop, would be worth checking and it should be borne in mind that confirmation of this species is currently only considered possible by microscopic examination of critical internal features.

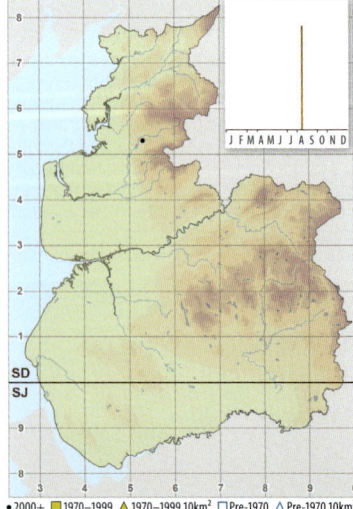

• 2000+   ■ 1970–1999   ▲ 1970–1999 10km²   □ Pre-1970   △ Pre-1970 10km²

258   *The Moths of Lancashire*

49.073 *Acleris schalleriana*

■ LANCS STATUS Rare / very local resident   1987–2016

This species was found in the 19th century in south Cumbria and is considered an overlooked, long-term resident in Lancashire. All records are from Gait Barrows, the first in 1987 by M. Dempsey. Thereafter, it was reported in 1989 by M. Shaw and at MV light on the 27 September 1994 and 11 September 2000 (R. Petley-Jones). On 12 June 2016, two larvae were found feeding on the leaves of Guelder-rose within the Gait Barrows boundary, at Hawes Water, by B. Smart, and emerged on 17 July 2016.
■ HABITAT Lowland limestone scrub.
■ FOODPLANTS Guelder-rose.
Double brooded, but only based on one larval and two full-dated adult records.

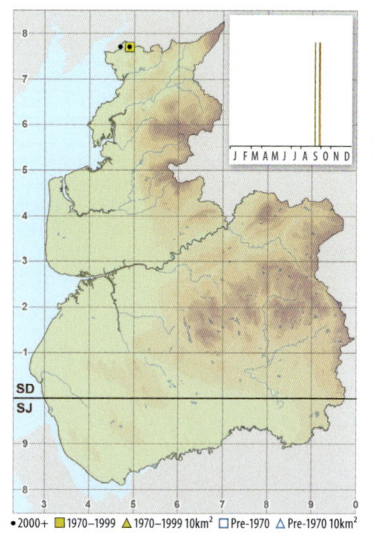

49.075 *Acleris umbrana*

■ LANCS STATUS Rare / very local resident   2009–2011

Recorded from Windermere, Cumbria, on Blackthorn, once during the late 19th century, suggesting it is an overlooked resident in Lancashire. First noted at Gait Barrows on 22 April 2009 by R. Petley-Jones, then at the same site during the day on 1 December 2009, when one was disturbed during scrub-clearance work. The next was found at Yealand Hall Allotment on 30 October 2010 by G. Jones and the last at Gait Barrows on 31 October 2011 by R. Petley-Jones at light.
■ HABITAT Scrub on lowland limestone.
■ FOODPLANTS Not recorded in Lancashire.
Considered single brooded, over-wintering as an adult. Comes to light.

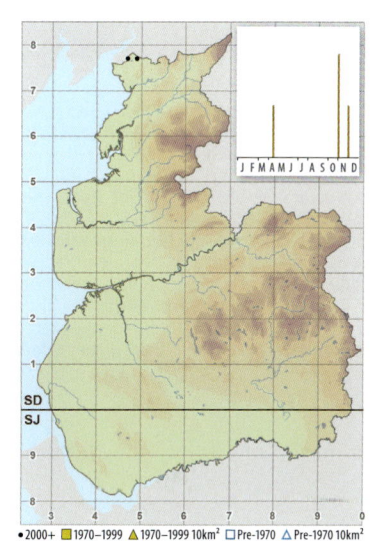

49.077 Garden Rose Tortrix *Acleris variegana*

■ LANCS STATUS Frequent / widespread resident   c.1857–2022

First found in Liverpool (Gregson, 1857) and thereafter noted by all early authors as abundant throughout the region, including in Crosby, Mossley Hill and Silverdale. In more recent decades recorded widely across much of the county, including upland sites such as Briercliffe, Rishton, Littleborough and Worsthorne. Long-term datasets are known from many sites such as Bispham, Flixton, Hale, Morecambe and St Annes.
■ HABITAT Scrubland, gardens, hedgerows, parks and brownfield sites.
■ FOODPLANTS Rose sp., Burnet Rose, Hawthorn, Blackthorn, Rowan and Apple.
Single brooded. Comes to light; 16 seen at Flixton on 1 September 1999 (K. McCabe). Larvae found from late May to mid-July. Occasionally, an all dark grey form occurs.

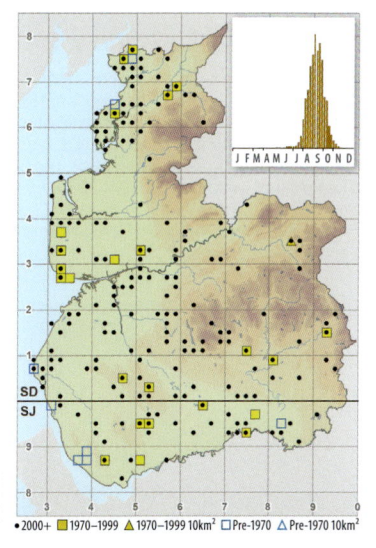

The Moths of Lancashire

49.078 *Acleris aspersana*

■ **LANCS STATUS** Occasional / local resident   c.1857–2022

First noted in the Liverpool district (Gregson, 1857) and then from Longridge in 1888, Silverdale in 1909 and the Formby area in the 1950s. For the remainder of the 20th century, it was recorded occasionally at sites such as Claughton, Gait Barrows, St Annes, Morecambe and Preston. Since 2000, records have increased to some extent and have been split equally between lowland, mostly coastal grassland, and herb-rich upland sites, including Ainsdale, Leighton Moss, Worsaw Hill and Thrushgill.

■ **HABITAT** Flower-rich meadows on dunes, limestone, hillsides and heaths.
■ **FOODPLANTS** Meadowsweet, cinquefoil sp. and bridewort sp.

Single brooded. Comes to light in low single figures. Larvae found in May and June.

● 2000+   ■ 1970–1999   ▲ 1970–1999 10km²   □ Pre-1970   △ Pre-1970 10km²

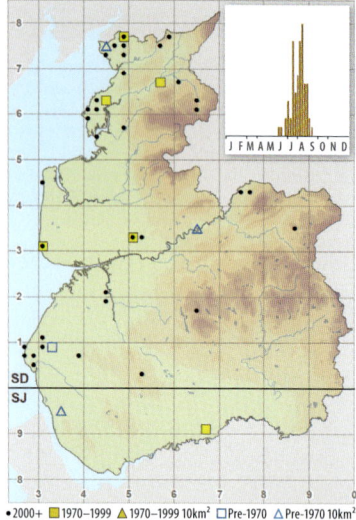

49.079 *Acleris shepherdana*

■ **LANCS STATUS** Considered Extinct   1858

Only recorded on one occasion, this by J.B. Hodgkinson who reports in the *Entomologist's Weekly Intelligencer* of 7 August 1858 that 'this week I have bred *Peronea shepherdana* from the larva which I found feeding on *Spiraea ulmaria*, at Lytham, in June.' (Hodgkinson, 1858).

■ **HABITAT** Possibly coastal dune slack. The precise site for this record is not known.
■ **FOODPLANTS** Meadowsweet.

Searches for an untidy spinning in the leaves of young shoots during May and June in wet coastal areas, where meadowsweet grows, would be worthwhile. To date no such feeding signs have been located. Nationally, the Lancashire record is a very isolated one with the main colonies being in south and east England. Although this record has been accepted widely in various national publications, it is considered advisable that further attempts are made to locate a specimen.

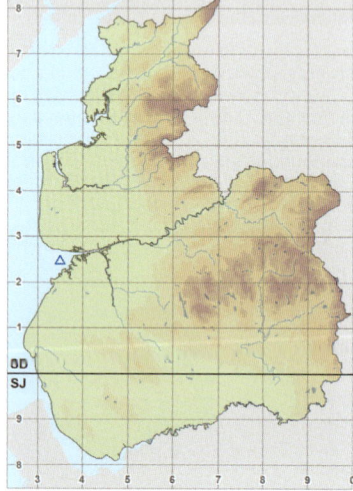

● 2000+   ■ 1970–1999   ▲ 1970–1999 10km²   □ Pre-1970   △ Pre-1970 10km²

49.080 *Acleris hastiana*

■ **LANCS STATUS** Frequent / local resident   c.1857–2022

Noted in the Liverpool area (Gregson, 1857), on Chat Moss in 1859, Lytham and Farrington c.1888. Since then, mostly on coastal dunes, the wetter parts of the Mersey Valley, Billinge, Heysham N.R. and Leighton Moss; infrequent elsewhere.

■ **HABITAT** Dune slacks, damp woods, mosses, riverside trees and scrub.
■ **FOODPLANTS** Grey Willow, Goat Willow, Creeping Willow, White Poplar and Black-poplar hybrid. Pupa found in trunk of Crack-willow.

Double brooded, 2nd brood overwintering as adult. Comes to light. Larvae have been found, in numbers, from April to mid-June and mid-July to September. Great numbers were observed flying in early evening sun at Formby in March 1929.

● 2000+   ■ 1970–1999   ▲ 1970–1999 10km²   □ Pre-1970   △ Pre-1970 10km²

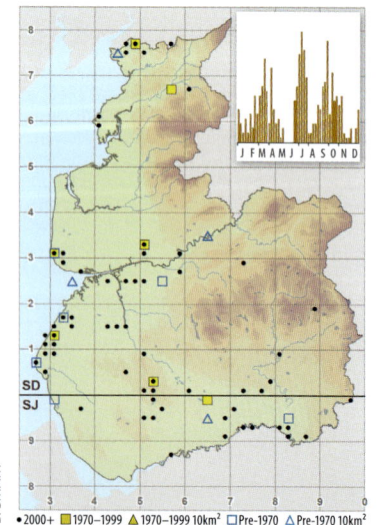

49.082 *Acleris hyemana*

■ **LANCS STATUS** Occasional / widespread resident   c.1857–2022
Recorded from mosses nr. Liverpool (Gregson, 1857) and later from Chat Moss, Farington, Longridge and Pilling. There were very few further records until c.1999 with possible wanderers found in Flixton and Pennington, although isolated areas of Heather, such as at Royal Lytham Golf Course, do support this species. Since then, 39 records have been received from 27 sites, usually where Heather is abundant.
■ **HABITAT** Moorland, mosses and heaths.
■ **FOODPLANTS** Not recorded in Lancashire.
Single brooded, overwintering as an adult. Comes to light in small numbers; readily disturbed from amongst Heather on sunny days. Considered very under-recorded.

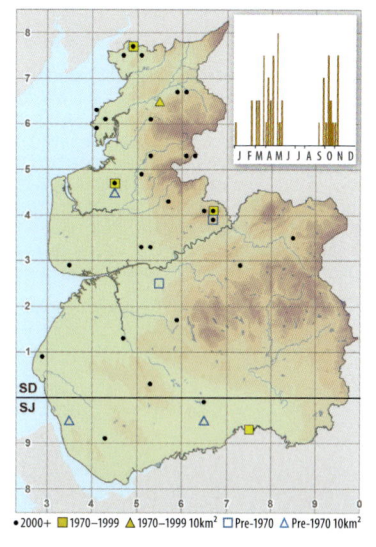

49.083 *Acleris ferrugana*

■ **LANCS STATUS** Occasional / local resident   2003–2022
Not recognised as a separate species in Britain until 1915, with Mansbridge (1940) recording it only on birch (= *A. notana*). Not separable from *A. notana* other than by dissection or breeding from oak. First confirmed records were from Chorlton on 8 & 23 February 2003 by B. Smart. Thereafter, found at several well scattered sites including Flixton, Cuerden, north Preston, Mill Houses, Rufford and Tame Valley.
■ **HABITAT** Deciduous woodland and hedgerows with oak standards.
■ **FOODPLANTS** Oak sp., with larvae found in May and August to September.
Double brooded, the second brood overwintering as an adult. Comes to light, in low single figures.

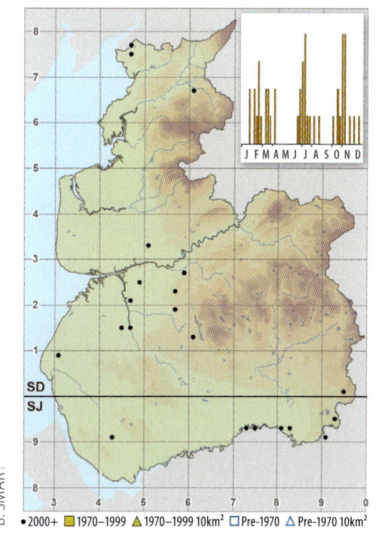

49.084 *Acleris notana*

■ **LANCS STATUS** Frequent / local resident   c.1940–2022
Earliest records fail to note foodplant, so could relate to *A. ferrugana* (see also note under that sp.). Mansbridge (1940), under *A. ferrugana*, notes it amongst birch on mosses, with no dates or sites. First record with data was from Formby (Leech & Michaelis, 1957). Since then, found amongst lowland birch scrub, such as at Little Woolden Moss, Flixton and Gait Barrows. 20 seen Winmarleigh Moss, October 2021.
■ **HABITAT** Woodland, scrubland, mosses and brownfield sites.
■ **FOODPLANTS** Birch sp., including Silver Birch.
Double brooded, the second brood overwintering as an adult. Comes to light. Larvae found in untidy leaf spinnings from May to June and from August to September.

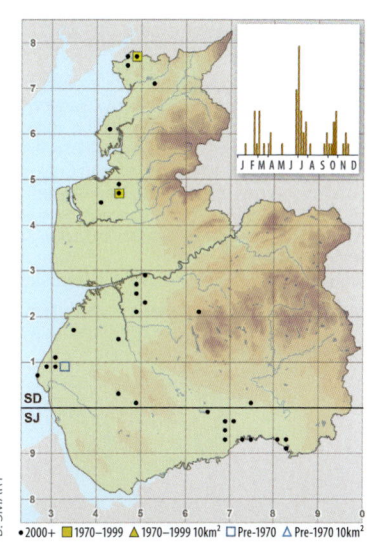

The Moths of Lancashire   261

TORTRICIDAE

49.086 *Acleris logiana*

■ **LANCS STATUS** Scarce / very local resident   2017–2022

Probably arrived in south-east Lancashire following a northern expansion of range on the east side of England. All records have been of singletons, the first in Flixton on 18 July 2017 (K. McCabe). Six of the seven other records also came from this site which has extensive birch tree growth nearby. Of the Flixton records, two came in 2021 and four in 2022, indicating local breeding is taking place. The final record relates to a (bred) pupa, found between spun birch leaves at Chorlton, 13 June 2022 (B. Smart).
■ **HABITAT** Scrubland and deciduous woodland edge with birch trees.
■ **FOODPLANTS** Pupa found between spun birch leaves.
Double brooded, the autumn brood overwintering as an adult. Comes to light.

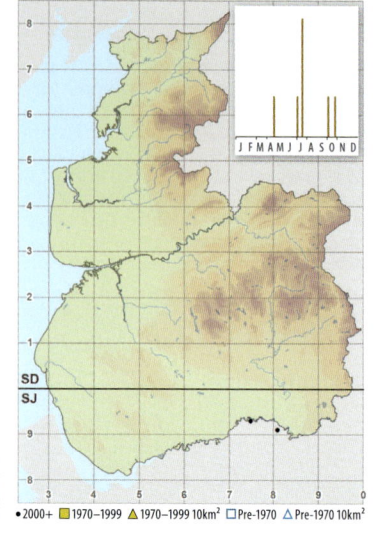

49.087 *Acleris literana*

■ **LANCS STATUS** Occasional / local resident   c.1857–2022

First reported in Liverpool, where its presence was noted in 'old woods' (Gregson, 1857). It was not seen again until 1996, at Gait Barrows, with further reports there until 2014. This coincided with a more general range increase, arriving at Dalton in 2006, Mill Houses from 2009 to 2015, Myers Allotment and Wray in 2011, Leighton Moss and Botton Head in 2015 and Silverdale in 2021. More widely, it was noted at Roby in 2006, Briercliffe in 2020 and Coppull, Fleetwood and Formby in 2022.
■ **HABITAT** Oak woodland and more isolated oaks in recent years.
■ **FOODPLANTS** Not recorded in Lancashire.
Single brooded, overwintering as an adult. Comes to light, mostly as singletons.

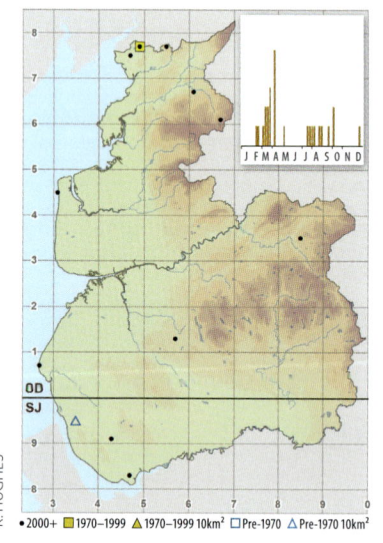

49.090 *Eulia ministrana*

■ **LANCS STATUS** Occasional / local resident   c.1857–2018

Noted as plentiful in Croxteth by Gregson (1857) and as common in Formby c.1950 (Leech & Michaelis, 1957). Since then, it has been lost from these areas, with most now found on lowland mosses, such as Risley and Astley Moss, limestone scrubland, such as at Gait Barrows and Warton Crag, and the northern woodlands of Mill Houses, Lord's Lot Wood and Herring Head Wood. 15 noted flying around birch at Astley Moss in 2004 (K. McCabe), but none seen here since 2011.
■ **HABITAT** Scrub, mosses, moorland edge and upland deciduous woods.
■ **FOODPLANTS** Birch.
Single brooded. Comes to light in low numbers. Larvae found August and September.

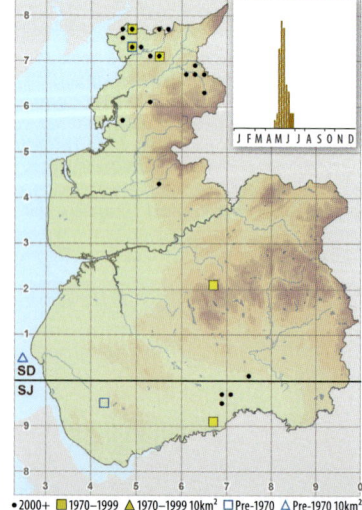

The Moths of Lancashire

49.091 *Pseudargyrotoza conwagana*

■ **LANCS STATUS** Frequent / widespread resident   c.1857–2022

First reported in the Liverpool district, where it was described as 'plentiful in lanes' (Gregson, 1857). Mansbridge recorded the moth at Silverdale in 1911, noting it to be 'common in N. Lancs, infrequent elsewhere' (Mansbridge, 1940). Noted at Parbold in 1973 and at Gait Barrows in 1986, with a subsequent increase in range and abundance, and likely to be found wherever Ash is present. As yet, it is unclear how the moth will be affected by the spread of Ash Dieback fungal disease in the county.
■ **HABITAT** Woodland edges, parks, gardens.
■ **FOODPLANTS** Ash seeds. Larvae recorded from September to November.
Single brooded. Comes readily to light. Moths also beaten from the foodplant.

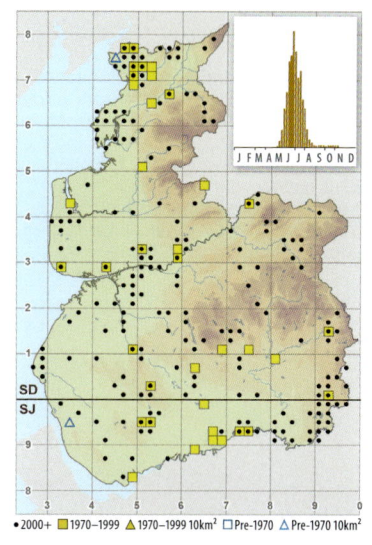

49.092 *Phtheochroa inopiana*

■ **LANCS STATUS** Occasional / local resident   c.1888–2022

This lowland species was first noted in the Preston district by J.B. Hodgkinson (Ellis, 1890). Next, from Gathurst 1935. Not seen again until 15 July 1989 when trapped at Woolston Eyes by S. McWilliam. Recorded at Wyreside Ecology Centre at Stanah in 1993, Wigan Flashes in 1999. All subsequent records have been from either Sefton Coast, e.g., Freshfield, Woodvale, Birkdale Green Beach, etc., or from north Lancs., e.g., Middleton N.R., Heysham N.R., Lancaster University, Sunderland Point, etc.
■ **HABITAT** Rough grassland, coastal scrub.
■ **FOODPLANTS** Not recorded in Lancashire.
Single brooded. 40 records, almost all of moths at light.

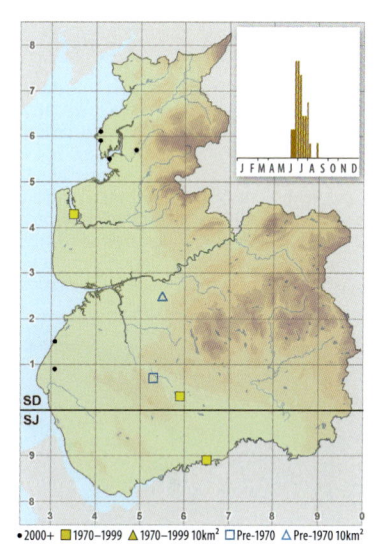

49.094 *Phtheochroa sodaliana*

■ **LANCS STATUS** Rare / very local resident   1994–2015

Whilst relatively new for the county, the moth may have been previously unnoticed, a theory supported by a pre-1940 record from Witherslack, VC69 (Mansbridge, 1940). First recorded on 8 July 1994 by R. Petley-Jones at Gait Barrows, with another at the same site on 20 June 2000 (S. Palmer). Next, trapped at light by K. McCabe and S.H. Hind at Leighton Moss in 2003, with the most recent record from Warton Crag in 2015 by J. Patton. All records are from limestone sites in the Silverdale area.
■ **HABITAT** Limestone woodland and hedgerows where Buckthorn grows.
■ **FOODPLANTS** Unrecorded within the county.
Single brooded. Comes to light.

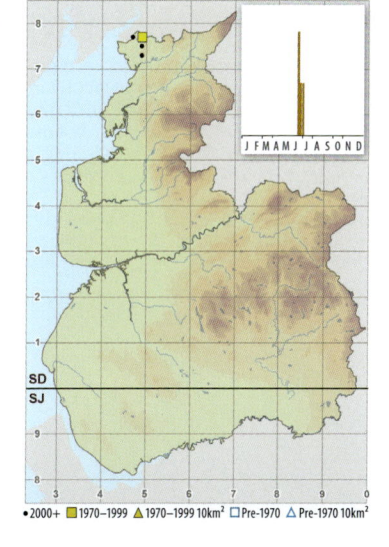

The Moths of Lancashire   263

TORTRICIDAE

49.095 *Phtheochroa rugosana*

■ **LANCS STATUS** Rare / very local resident   2000–2010

Both records of this moth are from garden light traps in the south of the county and are seemingly part of the range expansion from southern England that has also led to many new records in Cheshire and Yorkshire. First recorded at light in Flixton by K. McCabe on 18 June 2000. The only subsequent record was from Hale on 3 June 2010, recorded by C. Cockbain.

■ **HABITAT** Unknown in the county.
■ **FOODPLANTS** Not recorded in Lancashire.
Comes to light.

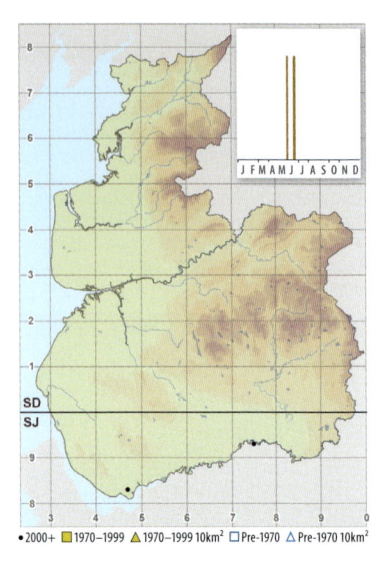

49.096 *Hysterophora maculosana*

■ **LANCS STATUS** Scarce / very local resident   c.1888–2020

From Brockholes Wood, Preston (Ellis, 1890), and then Knowsley Park, Burnley and Sales Wood, nr. Prescot, with status noted as 'sometimes common among Bluebells' (Mansbridge, 1940). The next record was in 2003 at Lord's Lot Wood by S. Palmer, with further records from Bay Horse and Lancaster. Since 2009, only noted at Warton Crag, with 12 records in total, including a high count of eleven on 29 May 2009.

■ **HABITAT** Deciduous, including limestone, woods.
■ **FOODPLANTS** Bluebells. Vacated seed-heads noted at Warton Crag 16 June 2017. Single brooded. Once at light. The other records are moths observed in the day, usually found among Bluebells. The larva remains unrecorded.

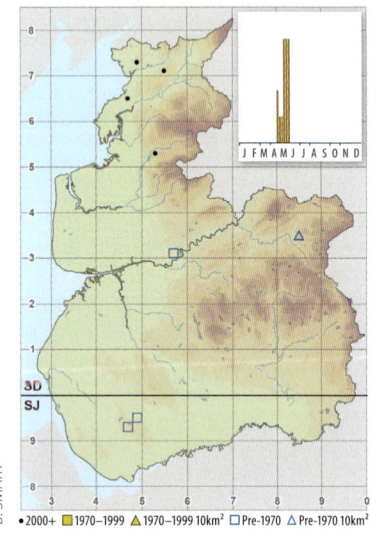

49.097 *Cochylimorpha straminea*

■ **LANCS STATUS** Frequent / widespread resident   c.1857–2022

First reported from Liverpool as 'abundant in grasslands' (Gregson, 1857). The moth went unrecorded from the mid-1950s until 1982, when it was noted at Risley Moss by M. Hull. Occasional in the uplands, but mainly known from lowland areas in the west, including 75 Heysham N.R. records. High count of 30 moths on knapweed at Hornby 20 August 2006 by J. Newton. Previously regular at Flixton, but just once since 2012.

■ **HABITAT** Rough grassland, scrub.
■ **FOODPLANTS** Bred by H.N. Michaelis in 1955. Although not documented, presumably from Black Knapweed, as any newly identified host would have been published. Double brooded. Comes to light and netted in day.

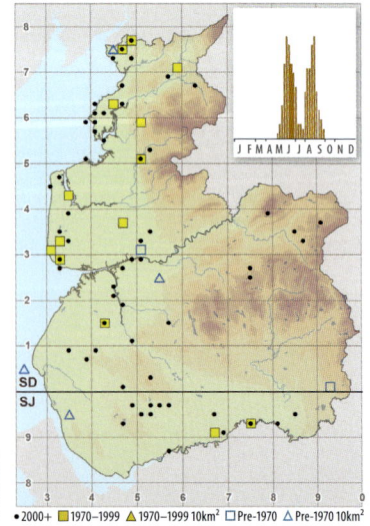

264 *The Moths of Lancashire*

49.100 *Phalonidia curvistrigana*

■ **LANCS STATUS** Scarce / very local resident 1906–2016
The first record is from Silverdale by A.E. Wright in 1906. Later, described as 'scarce among Goldenrod' (Mansbridge, 1940). Next reported from Gait Barrows in the 1980s by I. Rutherford. All seven subsequent records have been from the latter site, with singletons recorded in 1994, 1999, 2009 and 2016. Five were recorded in total during 2008, including three on the night of 3 July 2008 by R. Petley-Jones, C. Barnes and S. Palmer.
■ **HABITAT** Limestone woodland clearings.
■ **FOODPLANTS** Not recorded in Lancashire.
Single brooded. Moths recorded at light and at dusk.

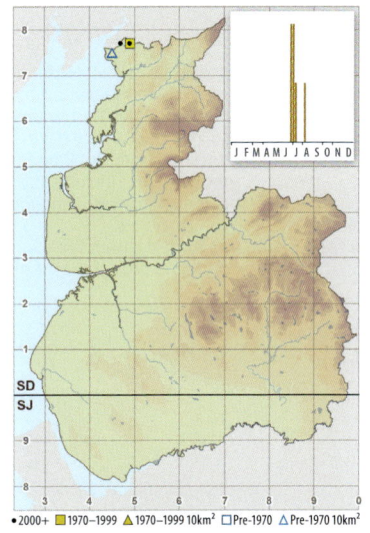

49.101 *Phalonidia manniana*

■ **LANCS STATUS** Occasional / local resident 2002–2022
Unfortunately, since *P. udana* was split from *P. manniana* in 2012, it has not been possible to confirm the historical records associated with this species. The first confirmed record was of a moth trapped in Flixton by K. McCabe on 16 June 2002 and subsequently re-examined following the split. Also recorded from Birkdale Green Beach, Mere Sands Wood, Little Hawes Water and Bold Moss, St Helens. Highest count of five at light from Mere Sands Woods by J. Girdley on 27 June 2021.
■ **HABITAT** Heaths, mosses and freshwater sites.
■ **FOODPLANTS** Not recorded in Lancashire.
Single brooded. Most records are of moths to light.

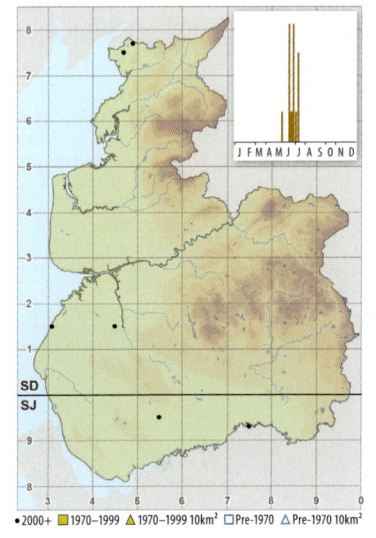

49.103 *Phalonidia affinitana*

■ **LANCS STATUS** Occasional / local resident c.1857–2022
First recorded from 'marshy places' around the Liverpool district (Gregson, 1857). Described as scarce (Ellis, 1890), and only in saltmarshes (Mansbridge, 1940). Regular sites are Sunderland Point (and a nearby garden), Marshside and Burrows Marsh. Unusual inland record at light from Billinge by C.A. Darbyshire 12 June 2000, subsequently dissected. Most recently from Silverdale 20 May 2022 by J. Stonehouse.
■ **HABITAT** Saltmarshes.
■ **FOODPLANTS** Unrecorded. Pupal cocoon amongst Sea Purslane stems 11 April 2002, with emergence 1 May (S. Palmer). The only known foodplant nationally is Sea Aster.
Single brooded. Moths recorded at light and netted by day and at dusk.

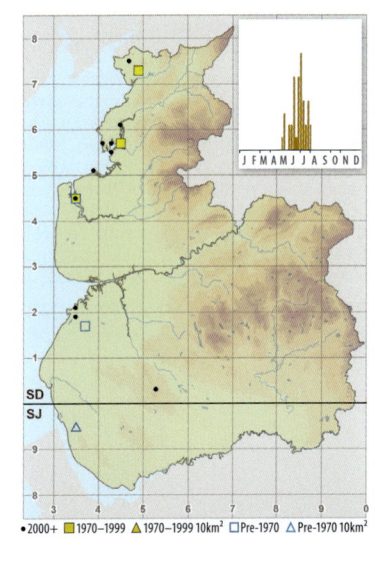

49.105 *Gynnidomorpha vectisana*

■ **LANCS STATUS** Occasional / local resident   1872–2022

Over 100 moths noted from Fleetwood by J.B. Hodgkinson and J.H. Threlfall in August 1872. Also, recorded at Marton Mere, Blackpool (Ellis, 1890). Reported from Bolton-le-Sands in 'countless numbers' in flight over the saltmarsh by L.T. Ford in 1940. 20 moths recorded from Out Rawcliffe, north of the Wyre, in August 2007 by D. & J. Steeden. Other VC60 sites include Sunderland Point, Fluke Hall, Pilling Marsh, etc. The few VC59 records include one from Bretherton, nr. Longton in 2022 by A. Barker.
■ **HABITAT** Coastal saltmarshes.
■ **FOODPLANTS** Unrecorded within the county.
Single brooded. Most records of moths netted during day. Occasional at light.

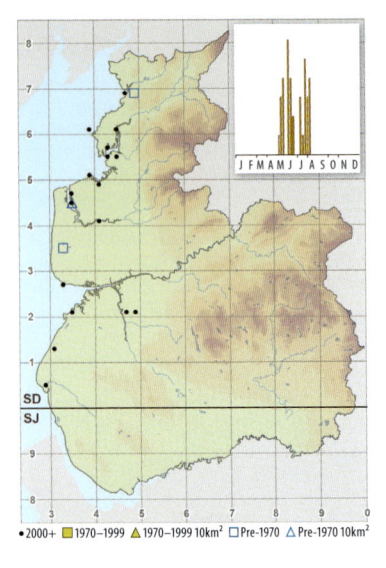

49.108 *Gynnidomorpha alismana*

■ **LANCS STATUS** Scarce / very local resident   1922–2009

Recorded by W. Mansbridge in 1922 from Formby, later describing the moth as 'scarce and local' (Mansbridge, 1940). Next recorded in 2000, with a record from K. McCabe's Flixton back garden on 16 June, and from Bold Moss, St Helens by R. Banks on 6 July. The next record was from Leighton Moss by S. Palmer in 2001, and again on 1 July 2009. There have been no subsequent reports. Known from Cheshire where last recorded in 2017 (cheshire-chart-maps.co.uk, accessed 1 September 2023), and from Yorkshire, with four records from 2022, and one of moths reared from larvae in stems of Water Plantain in 1994 (yorkshiremoths.co.uk, accessed 1 September 2023). Unrecorded in Cumbria.
■ **HABITAT** Freshwater sites where Water Plantain grows.
■ **FOODPLANTS** Not recorded in Lancashire.
Single brooded. Comes to light.

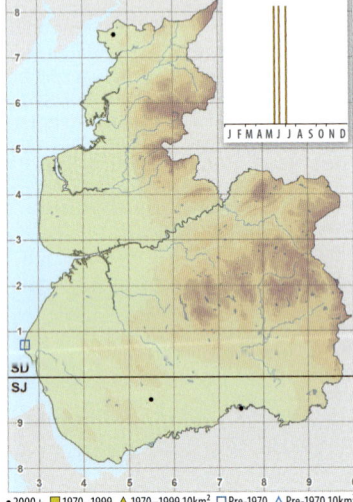

49.109 *Agapeta hamana*

■ **LANCS STATUS** Frequent / widespread resident   c.1857–2022

The most frequently recorded of the 'cochylids', with 1783 records, and very widely distributed. First reported from the Liverpool district (Gregson, 1857). Described as 'common and well-distributed' (Ellis, 1890), but later as 'apparently not common anywhere' (Mansbridge, 1940). The highest count is of 50 moths, flying at dusk by ponds at Scarth Hill, nr. Ormskirk on 5 June 2007 (C.A. Darbyshire). There are a few upland records, e.g., Turton, Leck Fell, Wray, etc., in addition to many lowland sites.
■ **HABITAT** Rough grassland, scrub.
■ **FOODPLANTS** Not recorded in Lancashire.
Single brooded. Recorded at light-traps and flying during the day.

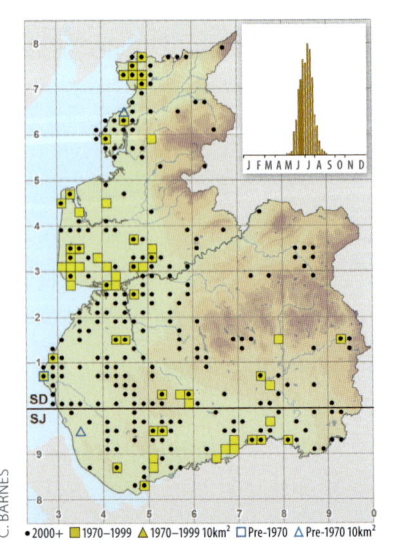

49.110 *Agapeta zoegana*

■ **LANCS STATUS** Frequent / local resident   c.1857–2022

Less common than the previous species, with 341 records, mainly in the west and the south. First record from Liverpool district (Gregson, 1857), then at Silverdale 1910 and Southport 1920. Described by Mansbridge as fairly common (1940). Recorded at Ainsdale in 1949 by H.N. Michaelis. High counts from Warton Crag, including one of 75 by J. Clifton & G. Powell on 3 August 2002. Regular at Leighton Moss, Sunderland, Heysham, etc. Decreased numbers noted in the Mersey Valley over the last decade.
■ **HABITAT** Meadows and rough ground where knapweed is found.
■ **FOODPLANTS** Not recorded in Lancashire.
Single brooded. Occasionally noted in the field during the day. Most records at light.

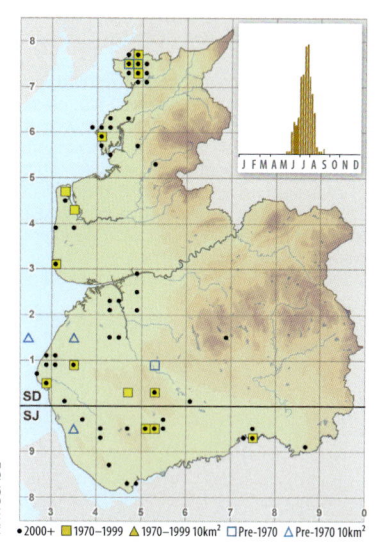

49.111 *Eupoecilia angustana*

■ **LANCS STATUS** Frequent / widespread resident   1846–2022

Found in great abundance at Chat Moss in July 1846 (Edleston, 1846). Noted 'where gorse grows' (Gregson, 1857), and later, as 'abundant on moors and heaths' (Mansbridge, 1940). Now found throughout lowland and upland areas. High count of 250 flying in evening, 28 August 2005, at Astley Moss. Four moths emerged in June 2021, from larvae found in seed-heads in September 2020 by B. Smart at Chorlton.
■ **HABITAT** Rough grassland, woodland edges, mosses and moorland.
■ **FOODPLANTS** Ribwort Plantain.
Probably single brooded with prolonged emergence period. Graph suggests a second peak in August. Moths trapped at light and disturbed from vegetation in the day.

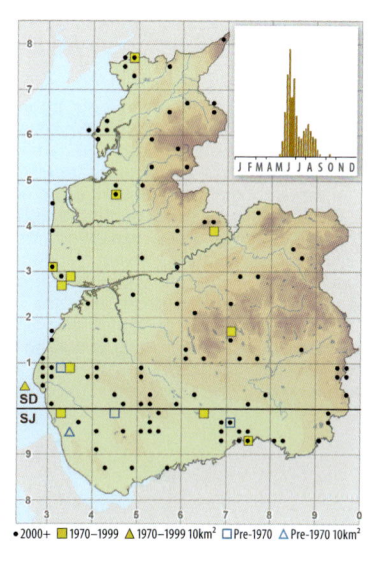

49.114 *Aethes hartmanniana*

■ **LANCS STATUS** Considered Extinct   c.1888–1953

First noted in the late 19th century at Longridge by J.B. Hodgkinson and at Preston by J.H. Threlfall, and described as local (Ellis, 1890). Recorded four times from 1950 to 1953 at Formby and Formby Moss, by H.N. Michaelis, R. Prichard and H.W. Wilson, but not since, here or elsewhere in the county. Potential confusion with *Aethes piercei* (noted by Sterling and Parsons, 2011), as well as absence of the nationally preferred habitat, chalk and limestone grassland, in the Formby area, casts some doubt on the previous records. There are historical records from Cumbria and Yorkshire, although it is worth noting that Bradley, Tremewan and Smith (1973) considered the more northerly English (and Scottish) records of *hartmanniana* as unconfirmed. Unrecorded in Cheshire.
■ **HABITAT** Unknown
■ **FOODPLANTS** Not recorded in Lancashire.
Nationally, single brooded. Further searches for extant specimens, particularly those from the Formby area, are recommended.

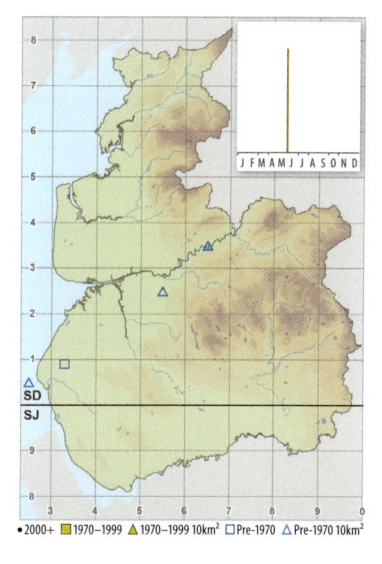

TORTRICIDAE

49.120 *Aethes smeathmanniana*

■ **LANCS STATUS** Frequent / widespread resident c.1857–2022

Gregson (1857) stated the moth was present in old, wet grasslands, although rarely seen in the Liverpool district. Next, recorded at Leighton Moss and Freshfield in 1949 and 1953. The 350+ records this century suggest the moth to be widespread, although with a southern and lowland bias, and in good numbers, with highest count of 24 to light on 6 August 2020 at Hale by C. Cockbain. Moth bred from tenanted Tansy seed-heads, collected at Ash Hill, Flixton in 2012, by K. McCabe and S. Palmer.
■ **HABITAT** Rough grassland, woodland edges.
■ **FOODPLANTS** Tansy, Yarrow.
Double brooded with a possible rare third or extended second brood to mid-October.

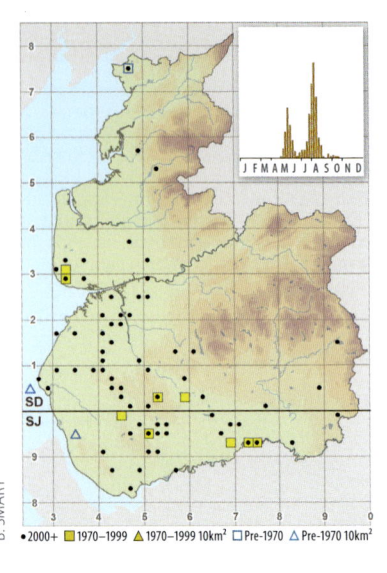

49.121 *Aethes tesserana*

■ **LANCS STATUS** Rare / very local resident c.1857–1988

A very infrequently recorded moth, although the 1988 record suggests it may still be present. The first records were from Woolton and Garston old pastures in the Liverpool district (Gregson, 1857). Next recorded from Didsbury in 1949 by H.N. Michaelis, with the recorder suggesting possible accidental introduction with larval foodplants. The final record was by M. Broomfield at Claughton in July 1988. Whilst no specimen or photograph was available for the latter record, on the basis of the recorder's experience, the presence of suitable habitat, and the reasonably distinctive forewing markings, the record was accepted. This is a very local species in Yorkshire with five records since 2005 from Middleton-on-the-Wolds (yorkshiremoths.co.uk, accessed 14 September 2023). Only recorded on a single occasion in Cheshire, at Whaley Bridge in 2019 (cheshire-chart-maps.co.uk, accessed 14 September 2023). Recorded historically in Cumbria.
■ **HABITAT** Dry grassland.
■ **FOODPLANTS** Unrecorded within the county.
Single brooded. Once to light, in 1988.

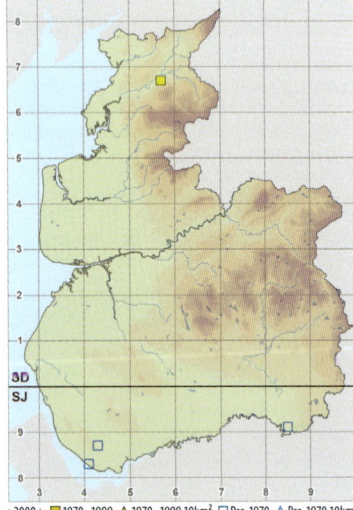

49.122 *Aethes dilucidana*

■ **LANCS STATUS** Considered Extinct c.1888

There are two 19th century records of this species, from Blackpool and Lytham by J.B. Hodgkinson; at the latter site with J.H. Threlfall (Ellis, 1890). Ellis considered the moth to be rare. With no subsequent records, the likelihood is that this mainly southern species is now extinct within the county. There is one single Yorkshire record from the late 19th century (yorkshiremoths.co.uk, accessed 1 September 2023), and a dot indicating at least one record for VC69 Cumbria on the map in *The Moths and Butterflies of Great Britain and Ireland, Volume 5:1* (Bland (ed.) 2014).
■ **HABITAT** Unknown.
■ **FOODPLANTS** Not recorded in Lancashire.
Nationally, this single brooded species occurs in mainly coastal areas, the larvae utilising Wild Parsnip or Hogweed. Searches for the feeding signs in the autumn or collection of old stems in coastal localities in the early spring may be worthwhile.

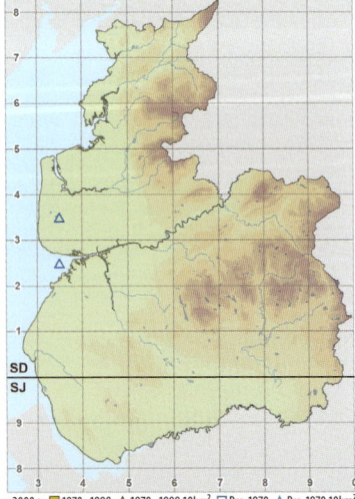

49.123 Aethes beatricella

■ **LANCS STATUS** Scarce / very local resident 1997–2021

This species, formerly restricted to southern counties, has expanded its range northwards, with the Lancs records accompanied by similar changes in Cheshire and Yorkshire. First recorded at Flixton on 23 June 1997 by K. McCabe, and again in 2012. Next, recorded by C. Cockbain at Hale on 27 June 2009 and has been fairly regular at this site since, with 13 moths trapped in total, high count of four on 10 August 2013. The only other records are from Martin Mere in 2009 and Southport in 2011.
■ **HABITAT** Unknown, as all but the Martin Mere moth have been trapped in gardens.
■ **FOODPLANTS** Not recorded in Lancashire.
Single brooded. All records are of moths to light.

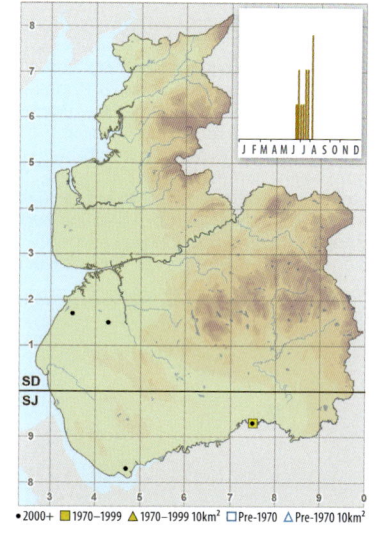

49.124 Aethes francillana

■ **LANCS STATUS** Rare / very local resident c.1888–2018

First recorded by J.H. Threlfall from Lytham St. Annes (Ellis, 1890). It was quite a surprise when the moth finally turned up again on 26 June 2018, to a 125w MV light above a sheet at Formby, recorded by J. Girdley. Possibly, even more of a surprise when a further moth was trapped at Woodvale four nights later, by the same recorder, with both moths dissected to confirm identity.
■ **HABITAT** Coastal grassland where Wild Carrot grows.
■ **FOODPLANTS** Not recorded in Lancashire.
Nationally, single brooded. Comes to light.

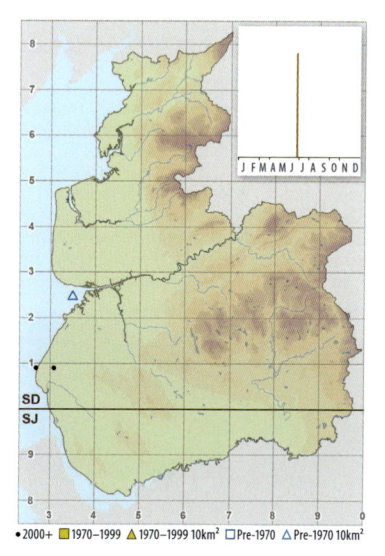

49.127 Aethes cnicana

■ **LANCS STATUS** Frequent / widespread resident c.1906–2022

Appears to be increasing in number, with half of the 186 records coming from the last five years. The highest count, 39 from Docker Moor in 2021, shows it is thriving in upland, as well as lowland habitats. First recorded in the early 1900s from Lancaster by C.H. Forsythe, and from Burnley by W.G. Clutten (Mansbridge, 1940). Has since been recorded throughout most of the county, although absent from Fylde coast.
■ **HABITAT** Rough grassland and moorland where Spear and Marsh Thistles grow.
■ **FOODPLANTS** Unrecorded within the county.
Single brooded. Comes to light. Bred from larva at Holden Clough, Oldham by H.N. Michaelis in 1955, although the foodplant, presumably thistle, was undocumented.

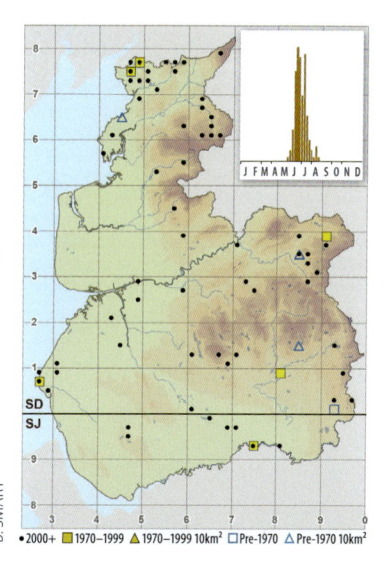

The Moths of Lancashire 269

49.128 Aethes rubigana

- **LANCS STATUS** Frequent / local resident   c.1888–2022

Recorded from Longridge by J.B. Hodgkinson (Ellis, 1890). Later reported to be less frequent than the previous species (Mansbridge, 1940). Noted from Rochdale in 1947 by J. Hardman and at Billinge in 1989. Frequently recorded since mid-1990s with three to 13 records a year, although perhaps concerningly, only one in 2022, from Leighton Moss. Of the 150 records, the highest count is eight from a Leighton Moss trap, positioned close to burdock, recorded by J. Beattie, 6 July 2009.
- **HABITAT** Grassland, scrub, woodland edges.
- **FOODPLANTS** Not recorded in Lancashire.

Single brooded. Almost all records are of moths at light.

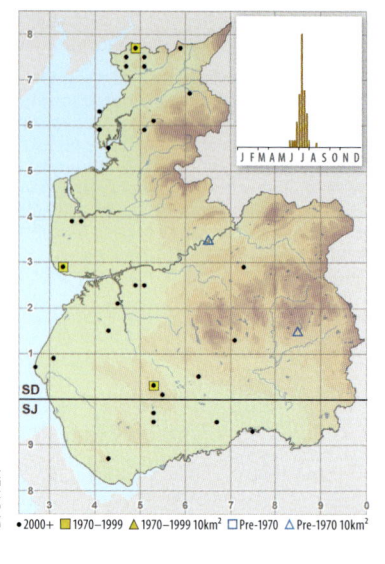

49.129 Cochylidia rupicola

- **LANCS STATUS** Scarce / local resident   c.1888–2020

First reported from Ditton Marsh, Widnes by C.S. Gregson, and from Brockholes Wood, nr. Preston by J.B. Hodgkinson, with status given as scarce (Ellis, 1890). Noted by L.T. Ford at Hawes Water in 1941. More recently, recorded at Gait Barrows from 1997 to 2001, Burscough in 2004, Rainford 2005, Bold Moss in 2005 and 2007, and at Hale in 2008 and 2020. The highest count is of three moths amongst Hemp Agrimony at Rufford Canal, Burscough by C.A. Darbyshire on 26 July 2004, at dusk.
- **HABITAT** Riverbanks and freshwater marshes.
- **FOODPLANTS** Unrecorded within the county.

Single brooded. Comes to light and netted at dusk.

49.132 Cochylidia implicitana

- **LANCS STATUS** Occasional / local resident   1984–2022

This species, previously limited to the south of the country, has expanded its range during the last few decades. First recorded in the county at Ince Moss in 1984 by L.W. Hardwick, then at Heysham N.R. in 1993 and at Lytham St. Annes 1996. Most frequently reported from the garden traps at Flixton by K. McCabe since 1997, and Hale by C. Cockbain since 2020. Of the 39 records, most are singletons, with records of two on just a handful of occasions. Confined to lowland areas of the county.
- **HABITAT** Rough wasteland and coastal areas.
- **FOODPLANTS** Not recorded in Lancashire.

Single brooded. Almost all records are of moths to light.

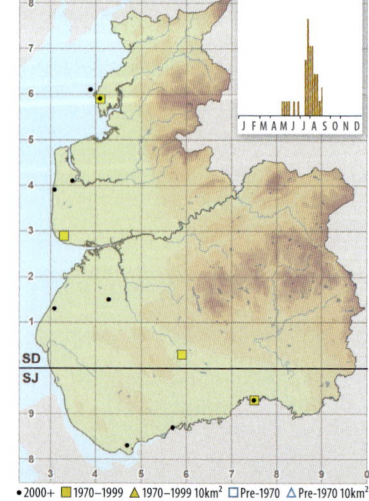

49.133 *Thyraylia nana*

■ **LANCS STATUS** Frequent / local resident   c.1859–2022

Reported from Chat Moss by J. Chappell in the latter half of the 19th century (Ellis, 1890). Later from Simonswood Moss and Knowsley Park, and described as 'common and generally distributed among birch' (Mansbridge, 1940). From Formby 1957, then Haskayne, Risley Moss and Woolston Eyes in 1980s. Recorded on 107 occasions, from lowland areas and on the edges of the moors, from Oldham, Briercliffe, etc. High count of ten moths at Holcroft Moss on 7 June 2005.
■ **HABITAT** Birch woodland, mosses.
■ **FOODPLANTS** Not recorded in Lancashire.
Single brooded. Comes to light, but also noted at rest on birch.

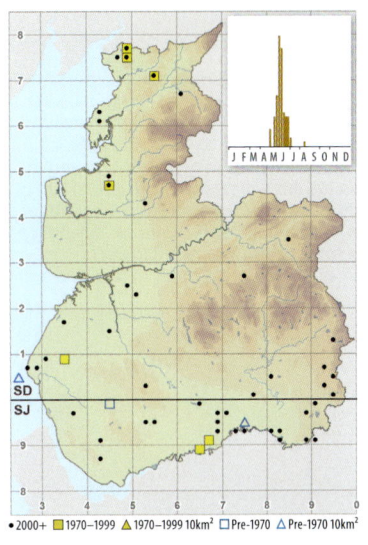

49.134 *Cochylis roseana*

■ **LANCS STATUS** Occasional / local resident   2004–2021

Reported as occurring 'as far north as…Lancashire' (Bradley *et al.*, 1973), although further details of any early records are unknown. The first confirmed record was of larvae in Wild Teasel seed-heads at Irlam, by K. McCabe on 28 March 2004, with emergence in June. Moths were trapped at light in Liverpool by S. McWilliam in 2018, and by J. Mitchell-Lisle at Great Sankey in 2019. Larvae recorded February 2019 at Sutton Manor, St Helens by B. Smart. Appears restricted to south Lancs.
■ **HABITAT** Rough grassland, brownfield sites.
■ **FOODPLANTS** Wild Teasel. Larvae feed by boring through the seeds.
Single brooded. Comes to light. Larvae found in seed-heads from February to May.

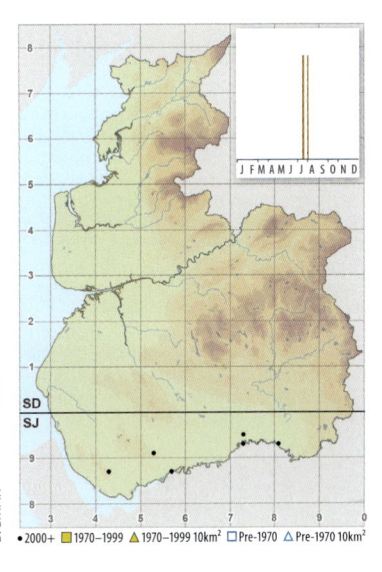

49.137 *Neocochylis dubitana*

■ **LANCS STATUS** Frequent / widespread resident   1914–2022

Bred from larvae by W. Mansbridge from the Crosby sand-hills in 1914, although foodplant not noted. He later noted the moth to be common at other Sefton Coast sites (Mansbridge, 1940). This is a lowland species, with distribution concentrated in the south and west. The first inland records were from Didsbury in 1948, Risley Moss 1982 and Martin Mere 1986. Larva on Common Ragwort at Chorlton in 2007. Noted frequently since, although only twice in 2022, which is potentially of some concern.
■ **HABITAT** Grassland, rough wasteland, coastal areas.
■ **FOODPLANTS** Common Ragwort.
Double brooded. Most records are of moths to light. Occasionally netted during day.

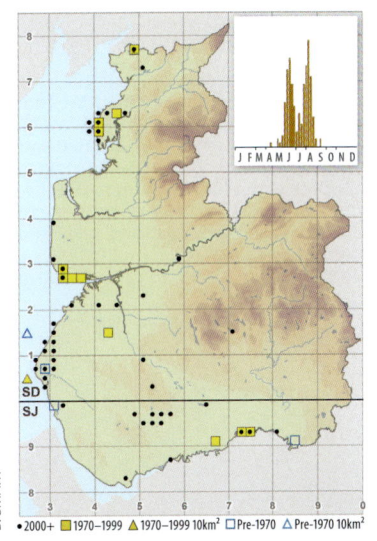

49.139 *Cochylichroa atricapitana*

■ **LANCS STATUS** Frequent / widespread resident   c.1857–2022

First recorded from the Liverpool district, where found 'freely around the sand-hills' (Gregson, 1857). Many early records from coastal districts, with first inland records from Knowsley c.1940 and Risley Moss in 1982. Similar distribution to the previous, with absence from north-eastern and upland areas. No ongoing concern regarding status, with 70 records from 2022. High count of 55 to light at Formby in 2014. Larvae noted within Common Ragwort stems at Chorlton in autumn 2020 (B. Smart).
■ **HABITAT** Grassland, rough wasteland, coastal areas.
■ **FOODPLANTS** Common Ragwort.
Double brooded. Comes to light. Occasionally, swept and netted during day.

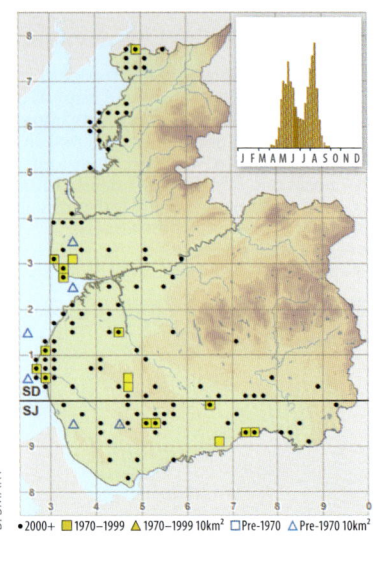

49.142 *Falseuncaria ruficiliana*

■ **LANCS STATUS** Considered Extinct   c.1888–1940

This species has only been reported in Lancashire on two occasions, with the second from over 80 years ago. First reported from Longridge by J.B. Hodgkinson (Ellis, 1890), with the author describing its status as 'local and not common.' The moth was later noted from Silverdale to be 'local among cowslip' (Mansbridge, 1940). The situation is similar in Cheshire where the moth was last recorded in 1936. There are also historical records from Grange-over-Sands and Witherslack in south Cumbria (Mansbridge, 1940). The moth appears to still be present in East Yorkshire with four 21st century records from VC61 and VC62 (yorkshiremoths.co.uk, both accessed 1 September 2023).
■ **HABITAT** Unknown in the county.
■ **FOODPLANTS** Not recorded in Lancashire.
Nationally, this species occurs across the whole of the British Isles and, as well as cowslips, the larvae are associated with Lousewort, Yellow Rattle and Goldenrod. This range of foodplants and its national distribution suggest it may still be present but overlooked in the county.

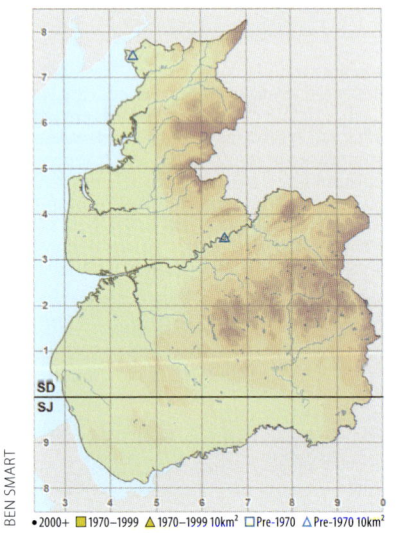

49.144 *Eudemis profundana*

■ **LANCS STATUS** Occasional / widespread resident.   c.1888–2022

First noted at Hoghton c.1888 by J.H. Threlfall (Ellis, 1890) and then the Silverdale area c.1934 (W. Mansbridge), these rare sightings in line with isolated records in adjacent counties. Next noted at Gait Barrows, Caton, Billinge and Leighton Moss between 1986 and 2002 with reports increasing further after 2006. Since then, it has been regular at a few sites such as Billinge, Flixton, Gait Barrows, Mill Houses and Swinton, and less frequently at several other locations.
■ **HABITAT** Woodland, scrubland, parks and gardens.
■ **FOODPLANTS** Oak.
Single brooded. Comes to light in small numbers. Larvae found late April to June.

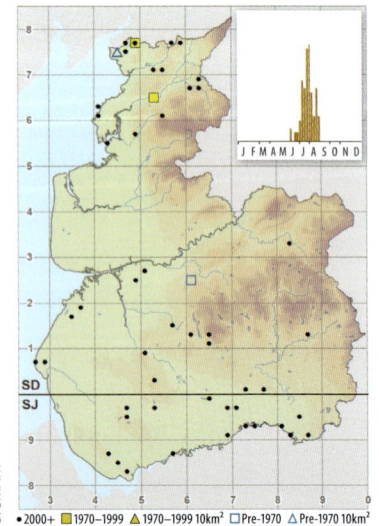

49.146 *Apotomis semifasciana*

- **LANCS STATUS** Scarce / local resident   c.1857–2021

Noted at Crosby, where bred (Gregson, 1857) and later, at Fleetwood c.1888 (J.B. Hodgkinson), Birkdale in 1912 and 1934 (W.G. Clutten), Storrs Moss in 1959 and Leighton Moss in 1961 (N. Birkett). Since then, there have been only 19 records from five near-coastal areas. These were at Heysham N.R. from 1995–2015, Leighton Moss from 2003 to 2018, Sunderland Point in 2012 and 2015, to the north of Potts Corner in 2014 and Middleton N.R. from 2009 to 2021 (J. Girdley and J. Patton).
- **HABITAT** Coastal slacks, wet woodland near open water and wet brownfield sites.
- **FOODPLANTS** Willow sp.

Single brooded. Comes to light in low numbers, with five at Sunderland Point in 2015.

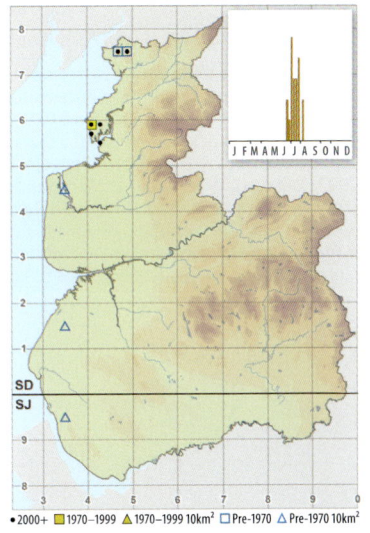

49.149 *Apotomis turbidana*

- **LANCS STATUS** Occasional / local resident   c.1857–2022

First noted at Kirkby Moss and Hale (Gregson, 1857), then Chat Moss in 1859 and the Preston area and Silverdale, c.1888. Apart from a few records c.1940 at Freshfield (Mansbridge, 1940) it was not reported again until 1988 at Woolston and, in 1989, at Martin Mere. Since then, it has been noted more widely, but still rather infrequently, from sites such as Flixton, Docker Moor, Gait Barrows and Littleborough (K. McCabe).
- **HABITAT** Birch woodland and scrub, including on mosses and moors.
- **FOODPLANTS** Birch. Larvae in May at Astley Moss 2013 & Little Woolden Moss 2016.

Single brooded. Comes to light, mostly in single figures. 15 attracted to light at Rainford in 2006 (R. Banks) and Freshfield in 2007 (G. & D. Atherton).

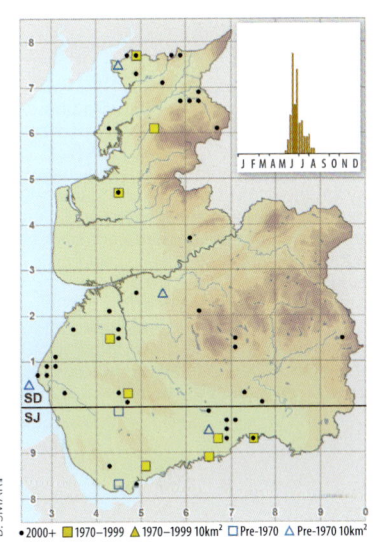

49.150 *Apotomis betuletana*

- **LANCS STATUS** Occasional / widespread resident   c.1857–2022

Noted in Kirkby Wood among birches (Gregson, 1857), Chat Moss c.1859 and Preston and Silverdale c.1888. Reported as common in Formby (Leech & Michaelis, 1957) but otherwise not seen with any regularity until the late 1980s. Since then, it has been found at many localities across VC59 and northern VC60, and most often at sites such as Ainsdale, Docker Moor, Flixton, Heysham Moss and Mill Houses.
- **HABITAT** Scrub, mosses, birch woodland, moorland edge and brownfield sites.
- **FOODPLANTS** Birch.

Single brooded. Comes to light, mostly in single figures; 20 were found on Docker Moor and Hindburndale in late July 2014. Larvae found in April and May.

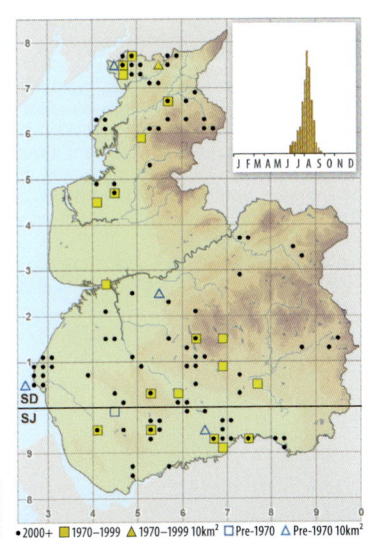

The Moths of Lancashire

TORTRICIDAE

49.151 *Apotomis capreana*

■ **LANCS STATUS** Scarce / very local resident   1999–2016

As a scarce species in northern Britain, it seems likely it was previously overlooked in Lancashire. Its close resemblance to other *Apotomis* spp. may have also been a factor. First reported from Gait Barrows on 15 June 1999 and in 2000 (R. Petley-Jones), followed by an isolated report from Hough Green, Widnes on 15 July 2002 (J. Clarke). Then found at Mere Sands Wood in 2007, Leighton Moss in 2010, Trowbarrow in 2012 and 2014 and Warton Crag on 10 August 2016 (J. Patton).
■ **HABITAT** Scrubby woodland, mainly in limestone areas.
■ **FOODPLANTS** Not recorded in Lancashire.
Single brooded. Occasionally comes to light in low single figures.

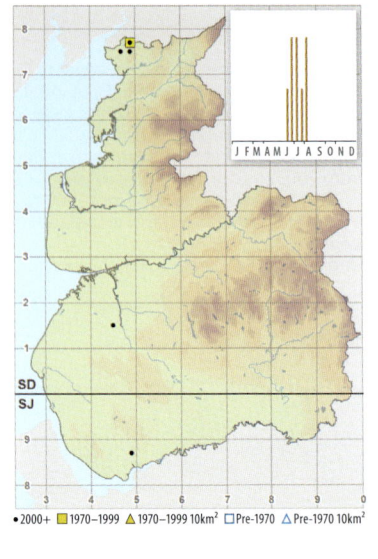

49.152 *Apotomis sororculana*

■ **LANCS STATUS** Rare / very local resident   c.1889–2014

Regarded as uncommon by early authors, it was first noted in the Silverdale and Preston areas prior to 1890 by J.B. Hodgkinson. It was then reported in the Lancaster area (C.H. Forsythe) and Hale Bank (W.A. Tyerman), prior to 1940, but specific dates were not documented. The next record was not until 1993 at Gait Barrows followed by another at the same site on 15 June 1999 (both R. Petley-Jones). More recently it has only been seen at Bay Horse on 25 July 2014 (N.A.J. Rogers).
■ **HABITAT** Scrub with birch present; once on the edge of moorland.
■ **FOODPLANTS** Not recorded in Lancashire.
Single brooded. Comes to light.
Although known to occur across the British Isles, the species is more frequently encountered on the hills and slopes of Scotland than in southern England, where it is considered scarce and local (Bradley *et al.*, 1979). Changes in climate may well be forcing this species to retreat northwards. The small number of Lancashire records overall, particularly in recent years, is maybe indicative of this.

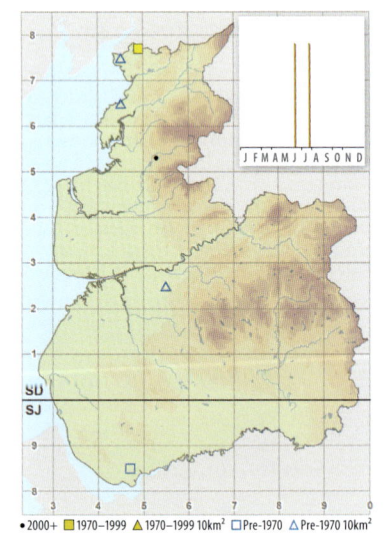

49.153 *Apotomis sauciana*

■ **LANCS STATUS** Occasional / local resident   c.1859–2022

Noted as plentiful on the Lancashire Mosses (Stainton, 1859) and abundant as larvae on Bilberry near Dutton in 1879 (J.B. Hodgkinson). Later, recorded widely across upland moors to the present day at sites such as Belmont, Burnley, Birk Bank, Ashworth Moor, Claughton, Thrushgill, Wolf Fell and Mossley. The Heysham record, on a sandy coastal headland in 2013, is considered a vagrant from elsewhere.
■ **HABITAT** Moorland and roadside verges where the foodplant grows.
■ **FOODPLANTS** Bilberry.
Single brooded. Comes to light in small numbers, but readily disturbed by day, with 80 seen at Longridge on 24 July 2012 (S. Palmer).

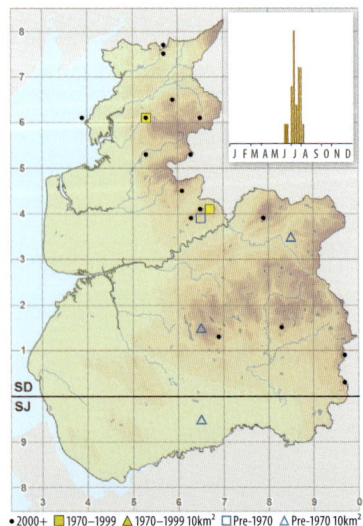

49.154 *Orthotaenia undulana*

■ **LANCS STATUS** Frequent / widespread resident   c.1857–2022
Common in old lanes in the Liverpool area (Gregson, 1857) and later, noted as general and abundant by early authors, up to 1914. Next recorded in 1983, at Winmarleigh Moss (M. Evans), where it was reported until 2017. Other recent sites include Flixton, Heysham, Gait Barrows, Scorton and Birk Bank and, more widely, on moors, mosses, brownfield sites and dry heaths; very infrequent elsewhere.
■ **HABITAT** Mosses, moors, scrubland; rare in gardens, unless close to these other areas.
■ **FOODPLANTS** Birch sp., Silver Birch, Apple, Bilberry and Heather. Larvae April to May. Single brooded; sometimes in August or September. Comes to light. Easily disturbed by day on drier parts of mosses and moors, with 110 at Clougha on 11 June 2008.

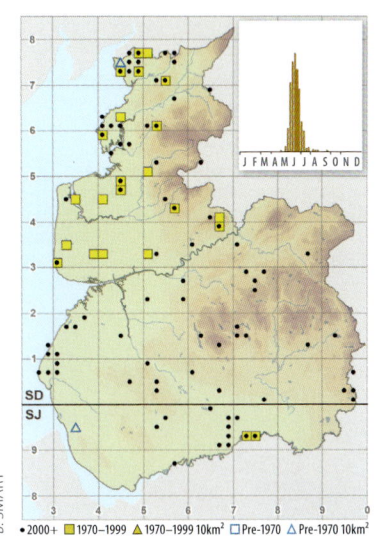

● 2000+   ■ 1970–1999   ▲ 1970–1999 10km²   □ Pre-1970   △ Pre-1970 10km²

49.155 *Hedya salicella*

■ **LANCS STATUS** Occasional / local resident   c.1857–2022
First seen at Huyton (Gregson, 1857), with Ellis (1890) noting it as not common and Mansbridge (1940) as scarce. Bred in numbers from Ainsdale and Formby c.1940–50 (B.B. Snell). Since 1991, it has spread, rather patchily, within the Mersey Valley and other parts of lowland VC59, but is infrequent and coastal elsewhere. Not annual at any sites, but in Flixton it has occurred as a singleton in most years (K. McCabe).
■ **HABITAT** Scrub on the coast, mosses, wetland and brownfield sites.
■ **FOODPLANTS** Willow and poplar sp.
Single brooded, with a small second brood in September on occasions. Comes to light mostly in single figures; 16 seen in a St Helens light trap on 1 July 2009 (R. Banks).

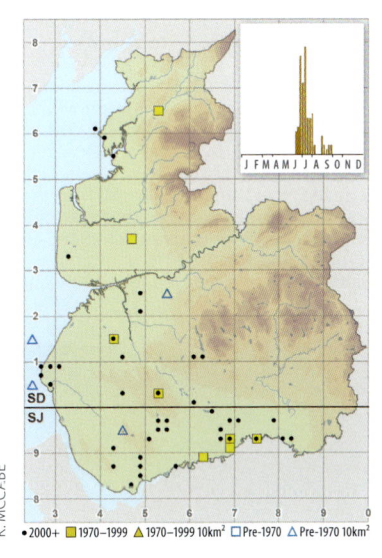

● 2000+   ■ 1970–1999   ▲ 1970–1999 10km²   □ Pre-1970   △ Pre-1970 10km²

49.156 Marbled Orchard Tortrix *Hedya nubiferana*

■ **LANCS STATUS** Abundant / widespread resident   c.1857–2022
First noted in the Liverpool district (Gregson, 1857) and later, by all early authors, as abundant throughout. This has remained the case to the present day with many regular sites across lowland and upland areas, including Billinge, Briercliffe, Longton, Martin Mere, Rishton, St Annes, St Helens area, Sunderland Point, Walmer Bridge and Yealand Conyers. The largest count was of 48 at Parr on 25 June 2001 (R. Banks).
■ **HABITAT** Hedgerows, gardens, scrubland, parks and brownfield sites.
■ **FOODPLANTS** Hawthorn, Apple, Crab Apple, Rowan, Pyracantha, Quince, Blackthorn, Wild Plum and Bramble. Pupa found in spun poplar leaves on one occasion.
Single brooded. Comes to light, often in double figures. Larvae September to May.

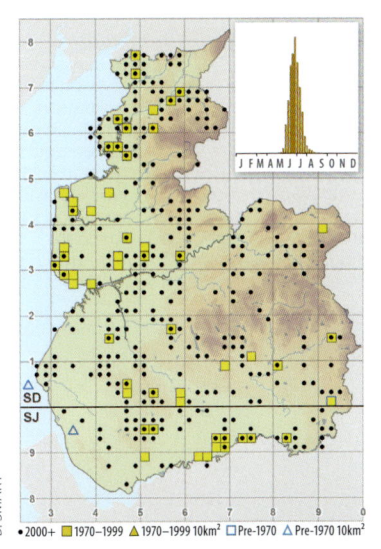

● 2000+   ■ 1970–1999   ▲ 1970–1999 10km²   □ Pre-1970   △ Pre-1970 10km²

TORTRICIDAE

49.157 Plum Tortrix *Hedya pruniana*

■ **LANCS STATUS** Frequent / widespread resident　　　　c.1857–2022
First noted in Liverpool (Gregson, 1857) and later, common everywhere (Ellis,1890). By 1940, W. Mansbridge wrote that it was much less common than previously, and this seems to have remained the case in urban areas. Elsewhere, particularly in lowland parts, it remains widespread, but less so in upland areas. Sites with regular sightings include Bay Horse, Flixton, Fulwood, Hale, Hoghton, Longton, Morecambe, Tarleton and Wray. Largest count was 20 seen by day in Ormskirk on 15 June 2005.
■ **HABITAT** Hedgerows, scrubland, gardens, parks and brownfield sites.
■ **FOODPLANTS** Blackthorn and Cherry Plum.
Single brooded. Comes to light, usually in single figures. Larvae found April and May.

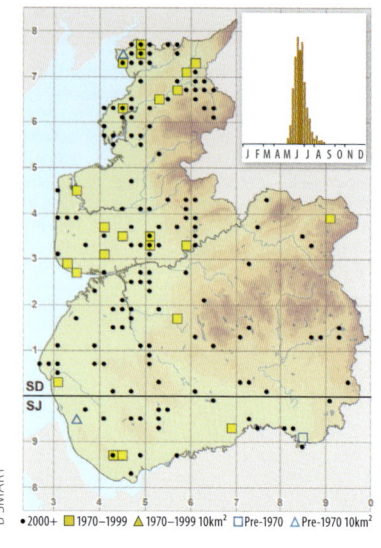

49.158 *Hedya ochroleucana*

■ **LANCS STATUS** Occasional / local resident　　　　c.1857–2022
Reported from Liverpool (Gregson, 1857), Irlam c.1859 and Preston c.1889 and noted as scarce by Mansbridge (1940). In 1997, one was found by day at Red Scar Wood, Preston, followed by 35 well scattered records to the present day. Sites with more than one report include High Tatham, Leighton Moss, Mill Houses and Brookhouse, with the most being nine records from Wray between 2011 and 2020. Larval spinnings were frequent on rose in an old hedgerow at Longridge on 5 May 2022.
■ **HABITAT** Hedgerows, scrub, woodland edge and brownfield sites.
■ **FOODPLANTS** Rose sp., in long-established hedgerows.
Single brooded. Comes to light in low single figures. Larvae found in May.

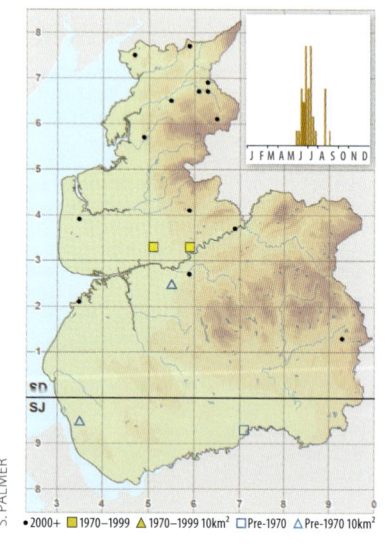

49.159 *Hedya atropunctana*

■ **LANCS STATUS** Occasional / very local resident　　　　c.1857–2022
First noted Simonswood Moss (Gregson, 1857), then Middleton, Manchester in 1859, Formby Moss in 1872 and again, on Simonswood Moss until 1908 (W. Mansbridge). Next, Winmarleigh Moss in 1991, this and nearby Cockerham Moss being regular sites until 2012. In 2010, seen twice in Rochdale and since then, has been regular at Little Woolden Moss. Larva found at Ainsdale in 2022. Probably under-recorded.
■ **HABITAT** Mosses, moorland and twice, surprisingly, in sandy coastal scrub.
■ **FOODPLANTS** Bog Myrtle and birch sp.
Single brooded. Comes to light mostly as singletons. 30 found by day on Cockerham Moss near Bog Myrtle on 8 July 2010 (S. Palmer). Larvae found July and September.

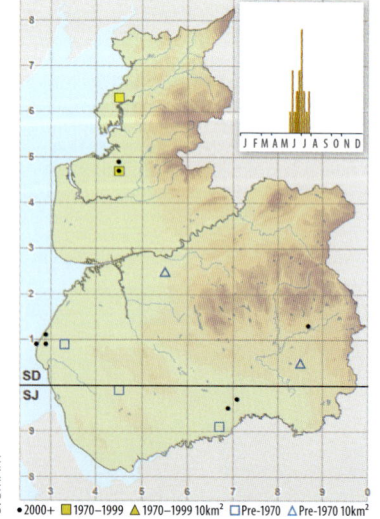

276　The Moths of Lancashire

49.160 Celypha rufana

■ **LANCS STATUS** Occasional / very local resident   c.1866–2017

Hodgkinson (1866) noted it in 'Lancashire, near the coast', the precise location of which is unknown and may refer to modern-day Cumbria. It was next reported in the Silverdale area in 1920 and at Gait Barrows in 1925 by W.G. Clutten. Since then, the majority of records have been at Gait Barrows, with 50 seen there on 7 Jun 2003. Elsewhere, noted very infrequently at Trowbarrow, Warton Crag and at Dock Acres.

■ **HABITAT** Shallow limestone grassland on a rocky substrate with moss present.
■ **FOODPLANTS** Wild Strawberry, Barren Strawberry, Mouse-ear-hawkweed and Common Ragwort (Heckford & Beavan, 2013).
Single brooded. Comes to light and flies by day. Larvae found in May.

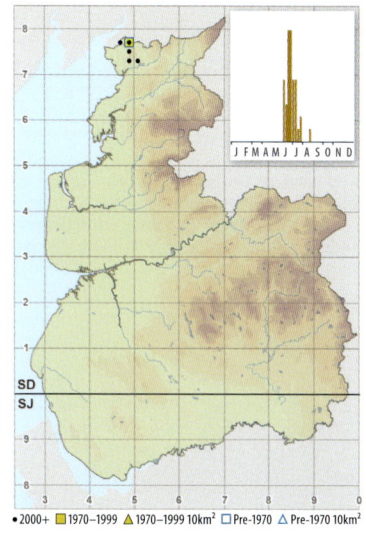

49.161 Celypha striana

■ **LANCS STATUS** Frequent / widespread resident   1857–2022

First seen in West Derby in 1857 (C.S. Gregson), then near Preston and the banks of the R Wyre c.1888 (J.H. Threlfall). Mansbridge (1940) reported it as occasional, mainly in near-coastal areas, with further sightings noted in Lytham, Morecambe and Silverdale up to the 1980s. It then commenced a wide-ranging but erratic expansion inland, reaching Preston in 1995, Littleborough in 1999 and Flixton in 2002. The range increase has continued to the present day, but it is infrequently seen in upland areas.

■ **HABITAT** Disturbed ground, grassland, verges, meadows and brownfield sites.
■ **FOODPLANTS** Not recorded in Lancashire.
Single brooded. Comes to light in single figures. 20 seen at Rainford on 22 June 2014.

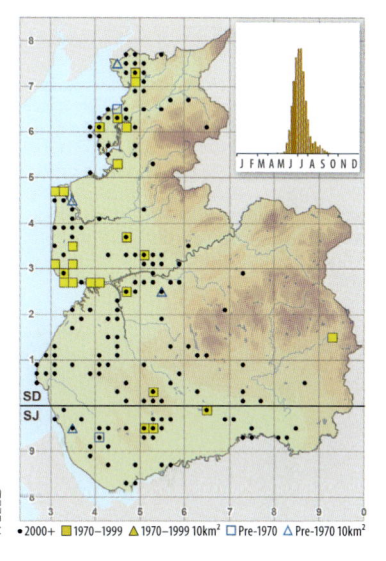

49.164 Celypha cespitana

■ **LANCS STATUS** Occasional / local resident   c.1857–2022

Noted in Crosby (Gregson, 1857) and then the Preston district c.1888 by J.H. Threlfall. In Mansbridge (1940), Formby was added to the list of sites, an area where it was also noted in 1976 (E.C. Pelham-Clinton). It was first reported at St Annes in 1995, but has only been seen there on a few other occasions since then. Although it has not been reported from Crosby for some time, most of the remaining 24 county records have been on the Sefton coastal dunes, from Altcar, north to Birkdale.

■ **HABITAT** Larger areas of stabilised sand dune grassland.
■ **FOODPLANTS** Not recorded in Lancashire.
Single brooded. Comes to light in single figures and can be disturbed on sunny days.

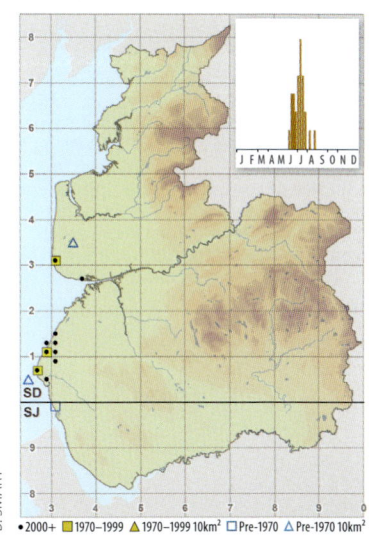

The Moths of Lancashire   **277**

TORTRICIDAE

49.166 *Celypha lacunana*

■ **LANCS STATUS** Abundant / ubiquitous resident   c.1857–2022

Noted in Liverpool (Gregson, 1857) and as abundant everywhere by all early authors. Since then, has been found throughout the county, often in double figures. A daytime count of over 200 was made in Skelmersdale on 16 June 2000 (C.A. Darbyshire).

■ **HABITAT** Grassland, scrub, rough ground, verges, hedgerows and brownfield sites.

■ **FOODPLANTS** Great Willowherb, Common Bird's-foot-Trefoil, Hogweed, tormentil, Silverweed, Honeysuckle, nettle, birch sp., Pignut, Creeping Buttercup, Spear Thistle, Cowberry, Cleavers, Bramble, mint, Dropwort, Ox-eye Daisy, vetch sp., forget-me-not, Ribwort Plantain, Rosebay Willowherb, Yarrow, Meadowsweet and Crosswort.

Single brooded over an extended period. Comes to light. Larvae found March to May.

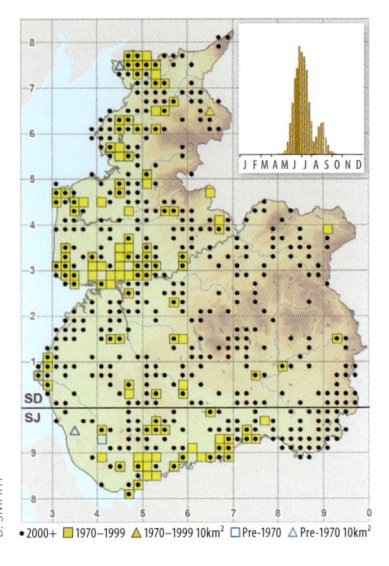

49.167 *Celypha rivulana*

■ **LANCS STATUS** Scarce / very local resident   c.1857–2019

Reported as abundant on Hale Marsh (Gregson, 1857), then on a saltmarsh near Preston (J.H. Threlfall) and in Lytham (J.B. Hodgkinson) c.1890. Next, reported from near Lancaster c.1940, by C.H. Forsythe. Since then, the few records have all been on sand dunes. At St Annes, two were noted in 1995, with singletons in 1997, 1999 and 2013. Three were seen at Ainsdale in 2016 and one there in 2019 (R. Moyes *et al.*).

■ **HABITAT** Sand dune grassland and slacks.

■ **FOODPLANTS** Not recorded in Lancashire.

Single brooded. Comes to light. A record from Longridge noted in Ellis (1890) seems unlikely and requires confirmation.

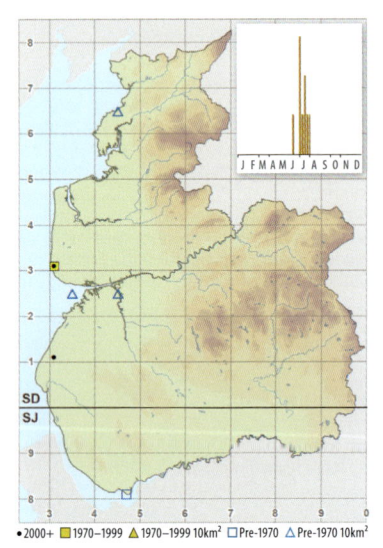

49.172 *Phiaris schulziana*

■ **LANCS STATUS** Occasional / local resident   1846–2021

Abundant on Chat Moss (Edleston, 1846) and wetter parts of mosses in the Liverpool area (Gregson, 1857). Also noted at Rixton (N. Cooke) and Simonswood (Ellis, 1890). Mansbridge (1940) reports it from Pilling Moss, Silverdale and the Burnley moors. Since then, most reports have been from Winmarleigh and Cockerham Mosses, but with several scattered, mainly upland reports from elsewhere, and a few wanderers.

■ **HABITAT** Mosses and, to a lesser extent (but may be under-recorded) on moorland.

■ **FOODPLANTS** Not recorded in Lancashire.

Single brooded. Frequently recorded in May, a month earlier than is widely noted nationally. 100 seen by day on Winmarleigh Moss on 17 May 2004 (B. Dyson).

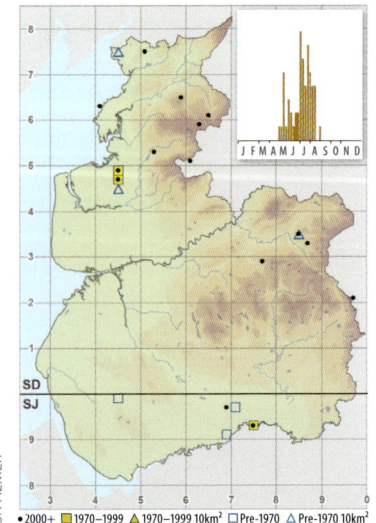

49.173 Phiaris micana

■ LANCS STATUS Considered Extinct pre-1888

The only reference to this species in the county was published in Ellis (1890) where it was reported, undated, as being found on Pilling Moss by J.B. Hodgkinson and J.H. Threlfall. It is possible that this could refer back to the mid-19th century, a time when Hodgkinson was particularly active in this area. The site name refers to an area of lowland raised bog, likely to have been more extensive at that time, but now reduced by drainage, peat extraction and agricultural use. Parts of this mossland still remain, as Cockerham Moss and Winmarleigh Moss, both Lancashire Wildlife Trust reserves. These sites are now being more extensively wetted by removal or blocking of drainage ditches.

■ HABITAT Mosses.

■ FOODPLANTS Not recorded in Lancashire.

This species has been found in wetland habitats in southern Scotland in the last decade and may therefore still be present on remaining mosses in very localised areas. Searches of the Cockerham and Winmarleigh Moss areas during July would be worth considering.

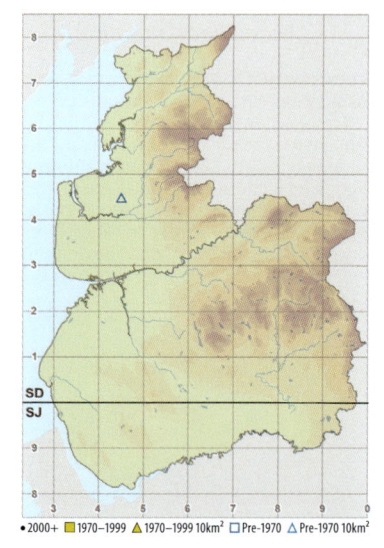

49.174 Phiaris palustrana

■ LANCS STATUS Frequent / local resident 1918–2021

First noted by W.G. Clutten near Burnley in 1918 (Mansbridge, 1920), but not again until 1990 at Birk Bank, by M. Dempsey. Came to light at Trough Summit in 2001 and since then has been noted at several upland sites, such as Blaze Moss, Holdron Moss, Green Bank, Leck Fell, Tarnbrook Fell and Worsthorne. Occasionally found well away from the moors, at sites such as Arkholme and Clitheroe. Larvae found at Gannow Fell and Jeffrey Hill in April 2015 and subsequently bred (S. Beavan & R.J. Heckford).

■ HABITAT Damper parts of moorland.

■ FOODPLANTS Moss spp., *Dicranum scoparium* and *Rhytidiadelphus squarrosus*.

Single brooded. Comes to light; readily disturbed by day; 30 on Jeffrey Hill in 2014.

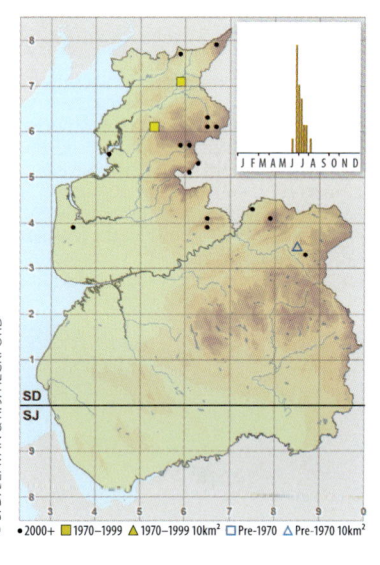

49.178 Stictea mygindiana

■ LANCS STATUS Rare / very local resident c.1889–2007

First noted in the Chat Moss complex by J.B. Hodgkinson (Ellis, 1890), but with Mansbridge (1940) reporting that there were no additional records since then. The only other older record was listed in Bradley *et al.* (1979), from near Quernmore (date and recorder unspecified). This sighting is not listed by the early authors, suggesting it was made sometime between 1940 and 1979. Although the recorder is unknown, Bradley (loc. cit.) regularly referred to records in this time frame made by L.T. Ford during his various site visits in north Lancashire (VC60) around 1941. Since then, it has only been reported twice in the county, when netted at Lord's Lot Wood on 3 June 1990, by M. Dempsey, and when one came to light at Briercliffe on 1 June 2007 (T. Lally). Probably still present in the county, overlooked, in very localised areas.

■ HABITAT On moorland and, formerly, lowland raised bogs.

■ FOODPLANTS Not recorded in Lancashire. Linked mainly with Cowberry elsewhere.

Single brooded. Searches for the moth in areas of moorland where Cowberry still proliferates would be worthwhile.

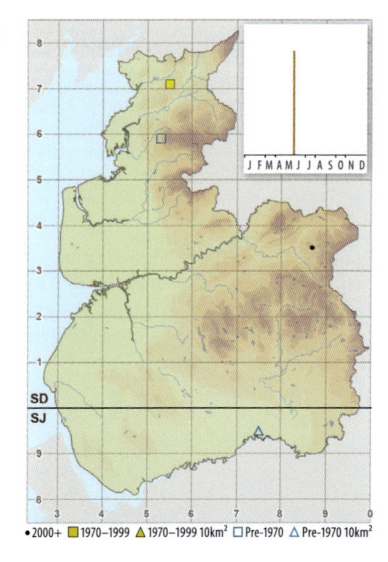

TORTRICIDAE

49.179 *Olethreutes arcuella*

■ **LANCS STATUS** Occasional / very local resident   2000–2022
Noted in the 19th century in south Cumbria and considered an overlooked resident in Lancashire. It was first found at Gait Barrows on 10 June 2000 (R. Petley-Jones), which remains its core site to the present day. Nearby, it has also been seen at Yealand Hall Allotment in 2010, Eaves Wood in 2015 and Warton Crag in 2021. A recently cleared woodland glade at Challan Hall Allotment held several on 12 June 2022.
■ **HABITAT** Sheltered, large, open clearings, with scrub, within deciduous woodland.
■ **FOODPLANTS** Not recorded in Lancashire.
Single brooded. Rare at light. Flies in sheltered conditions, resting on leaves in the sun. Once found in a garden light trap at Bispham on 9 July 2005 (B. Brigden).

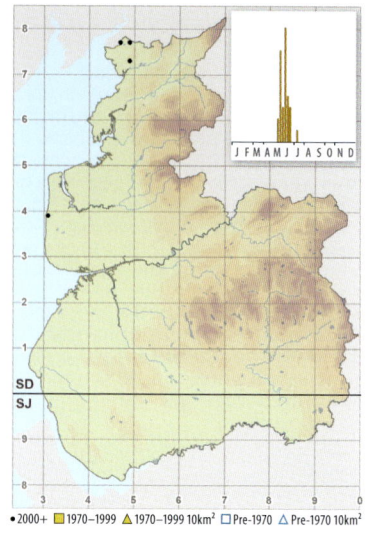

49.180 *Piniphila bifasciana*

■ **LANCS STATUS** Occasional / local resident   c.1859–2022
First noted in the Chat Moss area c.1859 by J. Chappell, then at Formby in 1924 and Simonswood Moss in c.1940. Not reported again until May 1990 (M. Dempsey) at Birk Bank, in 1999 at Gait Barrows and 2009 in Flixton and Little Crosby. Since then, its main stronghold has been in the areas where pines are common, at Formby (2010 to 2022) and Ainsdale (2012 to 2021). Elsewhere, it has been noted on single occasions at Wigan (2011), Billinge (2015), Southport (2017) and Briercliffe (2022).
■ **HABITAT** Localities with at least a few mature Pines nearby.
■ **FOODPLANTS** Not recorded in Lancashire.
Single brooded. Comes to light in low single figures.

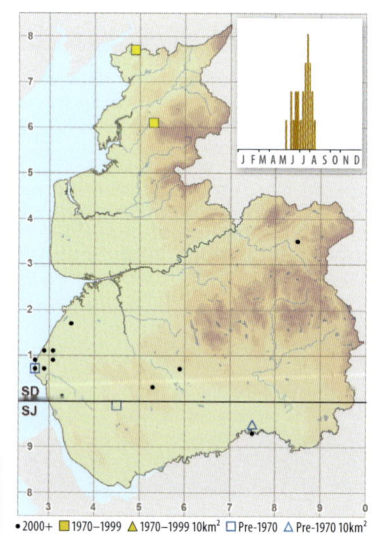

49.183 *Lobesia abscisana*

■ **LANCS STATUS** Frequent / widespread resident   1997–2022
Noted at Flixton (K. McCabe) and St Annes (D. & J. Steeden) in July 1997, as part of a national expansion of range. After a slow start, it spread rapidly, occupying many lowland sites, including St Helens and Billinge in 1999, Parbold and Rossall in 2001, Gait Barrows in 2008 and Rishton in 2014. Since then, local fluctuations in annual numbers are occurring with, for example, decreases at Parr and increases at Hale.
■ **HABITAT** Rough grassland, meadows, gardens and brownfield sites.
■ **FOODPLANTS** Creeping Thistle.
Double brooded, the first brood being smaller. 44 came to light at Ormskirk in 2005 (C.A. Darbyshire). Larvae found in spun shoots from June to early July and August.

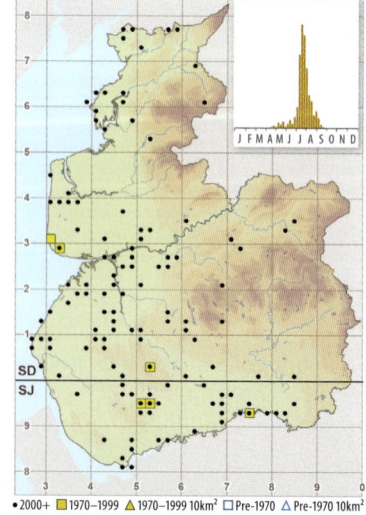

280   *The Moths of Lancashire*

49.184 *Lobesia reliquana*

■ **LANCS STATUS** Rare / very local resident   1994–2017
First recorded in the county at Gait Barrows on 24 June 1994 by R. Petley-Jones, with all but the last sighting from the same site. The next was on 21 May 2008, when three came to light (C. Barnes, *et al.*) and then on 29 May 2008 (R. Petley-Jones). It was last noted on 9 June 2017 when one came to light at White Moss, Silverdale (J. Patton). It is considered to be an overlooked long-term resident, having been found three times in parts of south Cumbria prior to 1890.
■ **HABITAT** Open deciduous woodland and scrub.
■ **FOODPLANTS** Not recorded in Lancashire.
Single brooded. Comes to light in small numbers.

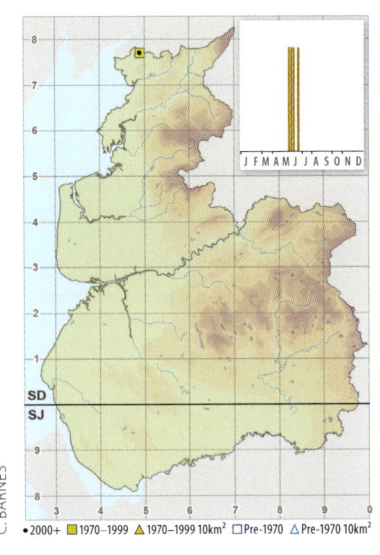

49.185 *Lobesia littoralis*

■ **LANCS STATUS** Occasional / widespread resident   c.1890–2022
First noted as local in the Fylde and Preston areas, and on the banks of R Wyre, by J.H. Threlfall and J.B. Hodgkinson, these all being linked with saltmarsh (Ellis, 1890). Further reports were received from Crosby, Southport, Bolton-le-Sands and Morecambe, up to 1960. It started appearing in gardens, where Thrift was planted, from 1997, with reports from Preston until 2008, and elsewhere, from Wray in the north to Flixton in the south and Briercliffe and Worsthorne in the east.
■ **HABITAT** Saltmarshes, rocky coasts and, while the foodplant remains, in gardens.
■ **FOODPLANTS** Thrift.
Double brooded. Comes to light and flies during the day; only found in single figures.

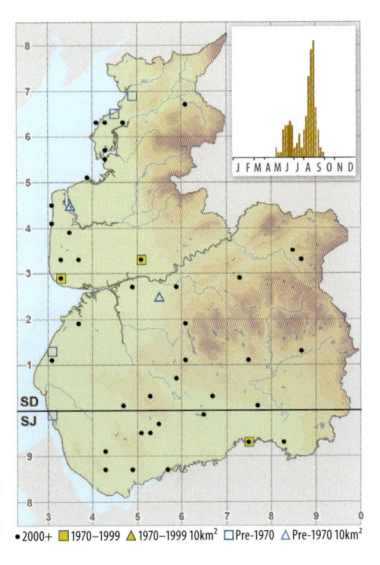

49.186 *Endothenia gentianaeana*

■ **LANCS STATUS** Occasional / local resident   2004–2022
Considered a recent arrival in the county, with most of the records recorded in the larval stage. First found as larvae in Wild Teasel heads at Irlam on 9 March 2004, with 30 seedheads collected and seven adults emerging (K. McCabe). Noted on the Canal towpath at Bridgewater in 2007, Barton upon Irwell in 2008, Woolston in 2012, Chorlton from 2015 to 2021, Sutton Manor, St Helens in 2019, Old Trafford in 2021 and Hodge Clough, Oldham in 2022. First found in VC60 at Middleton N.R. in 2019.
■ **HABITAT** Disturbed and waste ground, and brownfield sites.
■ **FOODPLANTS** Wild Teasel. Larva separable from that of *E. marginana* with care.
Double brooded. Tenanted seedheads found once in July and from October to May.

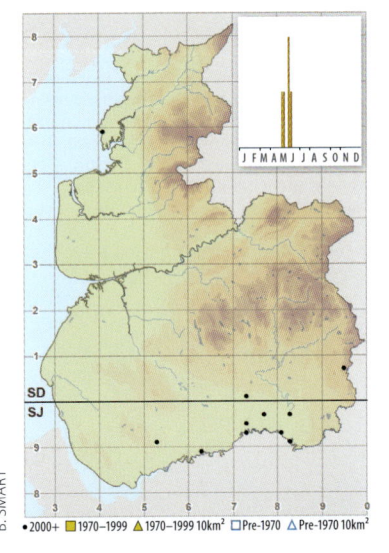

The Moths of Lancashire   **281**

49.187 Endothenia oblongana

■ **LANCS STATUS** Unknown   pre-1955

Reported, without a date, in Michaelis (1955) as found occasionally in the Formby area by W. Mansbridge. These records are not listed in Mansbridge (1940), which suggests that they were more recent (Mansbridge passed away in 1955) or that Michaelis had re-examined one or more Mansbridge specimens from earlier years and reidentified them as this species. As a result of Michaelis's publication of this record, it was documented in Bradley *et al.* (1979). Strangely, and to further muddy the waters, the record was not included in Leech and Michaelis (1957), an extensive and detailed list of Formby moths, in which it is believed that Michaelis coordinated and authored the micro-lepidoptera text. A search for an extant specimen is recommended. Ellis (1890) writes to report *'Penthina oblongana* Haw. (= *marginana* Haw.) at Simonswood Moss'. Despite the reference to *E. marginana*, Mansbridge (loc. cit.) lists this record under *E. oblongana* Haw. This record is considered unverified.

■ **HABITAT** Sandy grassland.

■ **FOODPLANTS** Not recorded in Lancashire. Recent research (Heckford, 2010), has shown that nationally the larva of this species feeds in the rootstock of Ribwort Plantain.

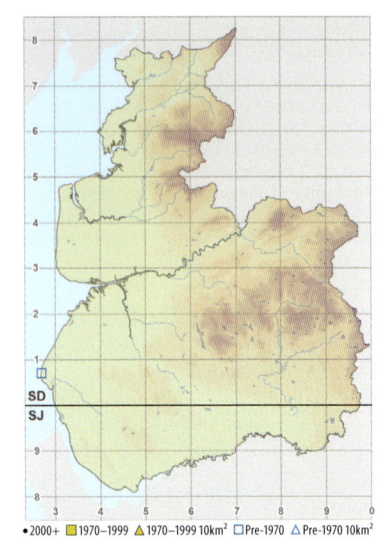

• 2000+ ▢ 1970–1999 ▲ 1970–1999 10km² ▢ Pre-1970 △ Pre-1970 10km²

49.188 Endothenia marginana

■ **LANCS STATUS** Occasional / widespread resident   2001–2022

There are a few records from 1857 to 1986 under different names which may well relate to this species, but for which supporting specimens are not available. The first confirmed record was from Flixton on 23 May 2001 (K. McCabe) and later noted in many lowland areas in VC59. It was not confirmed in VC60 until 2009 at St Annes (G. Jones) and later, at Fleetwood, Preston, Marton and Hyning Scout Wood.

■ **HABITAT** Disturbed ground, rough grassland, verges and brownfield sites.

■ **FOODPLANTS** Wild Teasel. Larva separable from that of *E. gentianaeana* with care. Double brooded. Comes to light. Teasel heads collected in December through to April, have produced moths. The hindwing colour in the male is diagnostic.

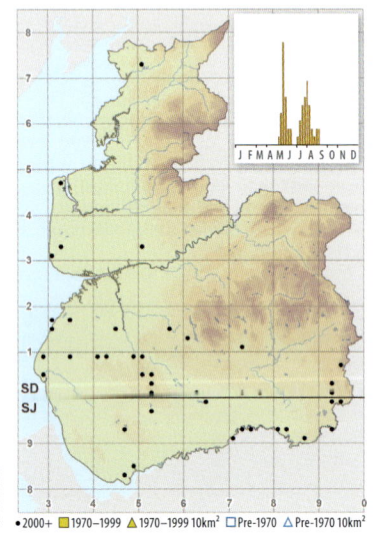

• 2000+ ▢ 1970–1999 ▲ 1970–1999 10km² ▢ Pre-1970 △ Pre-1970 10km²

49.190 Endothenia ustulana

■ **LANCS STATUS** Rare / very local resident   2014–2020

Likely an overlooked, long-term resident, the first record of which was a single female attracted to MV light at Gait Barrows on 25 July 2014 (S. & C.A. Palmer). The only other record relates to a female, this time attracted to actinic light, in Silverdale village on 13 August 2020 (J. Patton). Deciduous, limestone woodland is situated immediately behind the catch site.

■ **HABITAT** Woodland edge and rides, or open scrub, on limestone, where Bugle occurs.

■ **FOODPLANTS** Not recorded in Lancashire.

Single brooded. Comes to light.

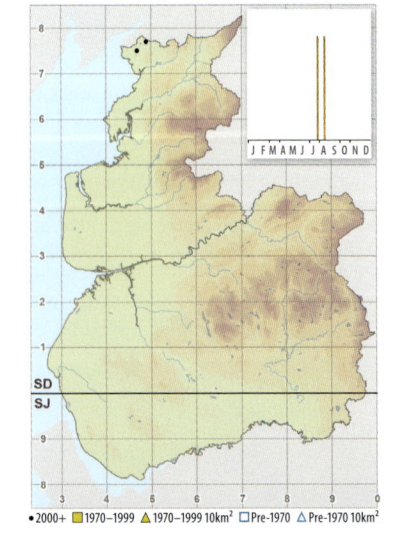

• 2000+ ▢ 1970–1999 ▲ 1970–1999 10km² ▢ Pre-1970 △ Pre-1970 10km²

49.191 *Endothenia nigricostana*

■ **LANCS STATUS** Scarce / very local resident   **c.1888–2019**
Recorded from Brockholes Wood, near Preston by J.B. Hodgkinson c.1888 (Ellis, 1890), this record not mentioned in Mansbridge (1940). Next, found by day alongside the Leeds - Liverpool Canal at Parbold on 13 June 2004 (C.A. Darbyshire), at Martin Mere to MV light on 7 June 2010 by A. Bunting, in Flixton on 14 June 2018 by K. McCabe and on two occasions as larvae at Rixton Clay Pits in January and February 2019 in the stem of the foodplant, subsequently bred, by B. Smart.
■ **HABITAT** Hedgerows and brownfield sites.
■ **FOODPLANTS** Hedge Woundwort.
Single brooded. Comes sparingly to light.

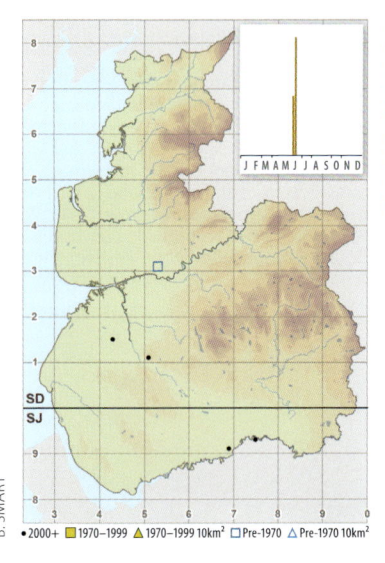

49.192 *Endothenia ericetana*

■ **LANCS STATUS** Occasional / local resident   **1920–2022**
The earliest record came from Formby in 1920, by W. Mansbridge, while H.N. Michaelis found it in Didsbury in July 1956. Since then, it has been reported from Sunderland Point in 1985, Martin Mere in 1991, north Preston in 1995, Risley Moss in 1999 and Burrow Heights in 2001. Eleven of the remaining 14 county records were in Flixton, from 2006–2022, the others being at Astley Moss on 5 August 2007, Warton Crag on 9 August 2013 and Silverdale Moss on 19 July 2016 (J. Patton).
■ **HABITAT** Rough ground and grassland, sometimes near wetlands.
■ **FOODPLANTS** Not recorded in Lancashire.
Single brooded. Comes to light in very low single figures.

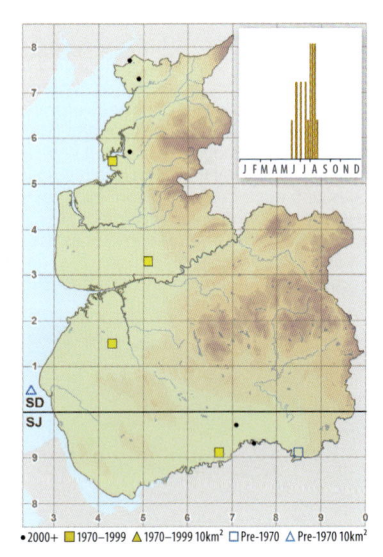

49.193 *Endothenia quadrimaculana*

■ **LANCS STATUS** Occasional / widespread resident   **c.1857–2022**
First noted in the Liverpool district (Gregson, 1857) and then in Preston c.1888 (Ellis 1890). Reported as scarce and local in Mansbridge (1940). Infrequently found on the Sefton Coast, Lytham, Storrs Moss and Leighton Moss, up to the mid-1990s. Since then, recorded regularly at several, mainly lowland, sites and usually in low single figures. These include Billinge, Martin Mere, Parr, Rishton and, particularly, Flixton.
■ **HABITAT** Marshes, mosses, wetlands, wet meadows and damp brownfield sites.
■ **FOODPLANTS** Not recorded in Lancashire.
Probably single brooded with an extended flight period. Comes to light. October records are not mentioned nationally and may relate to a small second brood.

The Moths of Lancashire   283

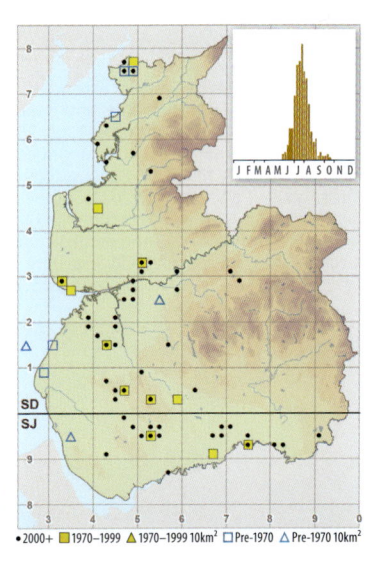

49.194 *Bactra lancealana*

■ **LANCS STATUS** Abundant / widespread resident   c.1857–2022

One of the county's most abundant species, it was first found in the Liverpool district (Gregson, 1857), with other early authors noting it as abundant. Sites with regular reports include Billinge, Flixton, Docker Moor, Thrushgill, Oswaldtwistle, Mill Houses, Leighton Moss, Leck Fell and many more. Numbers flying at dusk, at sites such as Dunscar Lodges and Sunderland Point, have been known to reach into the 1,000s.
■ **HABITAT** Wherever rushes grow, particularly on saltmarshes and damp moorland.
■ **FOODPLANTS** Not recorded in Lancashire.
Double brooded. Comes to light; readily disturbed from rushes. The considerable variability in forewing markings can lead to the misidentification of other *Bactra* sp.

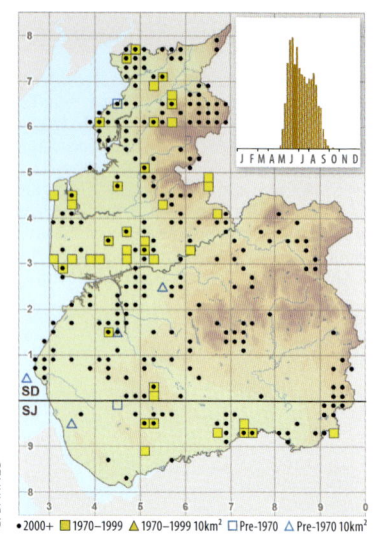

49.195 *Bactra furfurana*

■ **LANCS STATUS** Occasional / local resident   c.1857–2022

First noted in old clay pits in the Liverpool area (Gregson, 1857) and next, at Formby, where it was reported to be common (Mansbridge, 1940). Later, seen at Marton Mere in 1986 and on the banks of R Wyre at Stanah in 1996. Since 1999, it has been recorded at several scattered localities in lowland areas, usually in single numbers. Sites with more than one record include Heysham, Leighton Moss and Martin Mere.
■ **HABITAT** Wetland areas, including lakes, ponds and marshes. Occasionally wanders.
■ **FOODPLANTS** Not recorded in Lancashire.
Double brooded. Comes to light. 20 seen in dune slacks at Ainsdale by day in 2009. Some forms of *B. lancealana* look superficially similar.

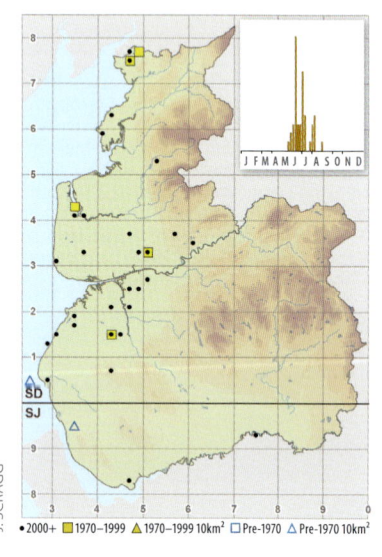

49.196 *Bactra lacteana*

■ **LANCS STATUS** Scarce / very local resident   2006–2017

An unrecognised British species until 1996, the first Lancashire specimens were attracted to light at Bay Horse on 1 July 2006 by N.A.J. Rogers. One of the two was dissected and confirmed by J.R. Langmaid. Subsequently recorded at the same site on 4 July 2006, 17 July 2007, 28 June 2009 and 7 July 2013. Three were found at White Moss, Yealand on 12 June 2020 (J. Girdley) and one in the centre of Southport on 30 June 2017 by R. Moyes (confirmed by N.A.J. Rogers). Comes occasionally to light.
■ **HABITAT** Moorland edge and damp acidic grassland. Once in an urban garden.
■ **FOODPLANTS** Not recorded in Lancashire.
Single brooded. Due to its similarity to *B. lancealana*, it requires dissection.

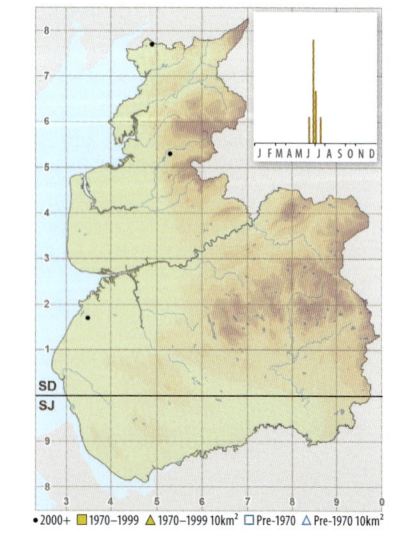

284   *The Moths of Lancashire*

49.197 *Bactra robustana*

■ **LANCS STATUS** Scarce / very local resident   **1941–2016**

First reported when several were found on Carnforth saltmarsh in 1941 by L.T. Ford (Mansbridge, 1944). Next, found as a singleton in a light trap at Sunderland Point on 3 July 2010 (J. Girdley). The only other known site for this species was the saltmarsh near Fluke Hall, when four were seen here during the day on 5 June 2016 and one, at the same site, on 14 June 2016 (S. & C. Palmer).

■ **HABITAT** Saltmarshes.
■ **FOODPLANTS** Not recorded in Lancashire.

Single brooded. Once attracted to light but more frequently found in the day. Could be an overlooked, longer-term resident. Confusion with *B. lancealana* is possible.

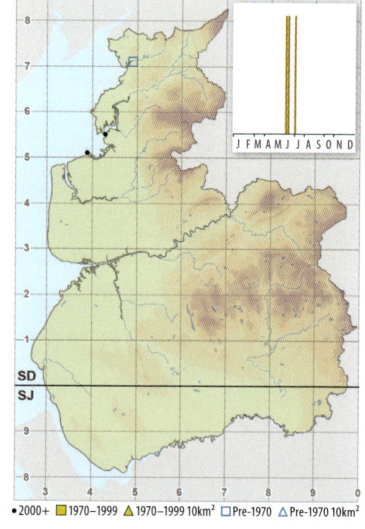

49.199 *Eucosmomorpha albersana*

■ **LANCS STATUS** Scarce / very local resident   **1987–2016**

An unusually isolated species, the nearest known colony being 80km east, across the Pennines. All records have been from the lowland limestone in the Silverdale area, the first being at Gait Barrows on 24 May 1987, by M. Dempsey. Except for two nearby finds, at Yealand Hall Allotment in 1987 and Warton Crag in 2015, all others were also from Gait Barrows (1989–2016), either to light or disturbed during the day. The largest count, of three, was on 22 May 2004 (S. Palmer). It has not been possible to assess if this is a recent arrival or an overlooked, longer-term resident.

■ **HABITAT** Scrubby areas on the lowland limestone.
■ **FOODPLANTS** Not recorded in Lancashire.

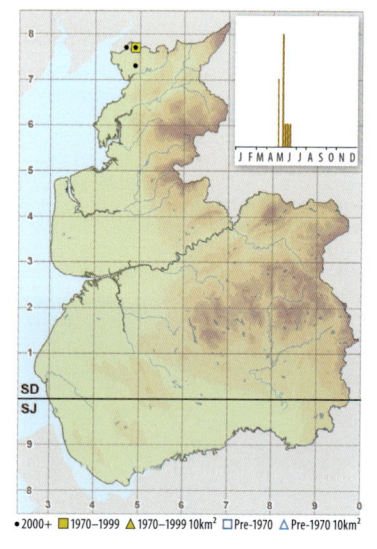

49.200 Cherry Bark Tortrix *Enarmonia formosana*

■ **LANCS STATUS** Frequent / widespread resident   **c.1887–2022**

Abundant, as pupae and adults, in a Liverpool garden, with no date given (Gregson, 1887), but where it is suggested to be an introduction on a Pear tree planted c.1846. Next, seen at Formby in 1912, St Annes in 1995 and Preston in 1996. Since 1999, it has gradually extended its range into many lowland gardens in VC59, as well as a few upland sites, at Briercliffe, Littleborough, Rishton and Rochdale. In 2003, it reached Leighton Moss, but has remained local in the lowlands of north VC60 since then.

■ **HABITAT** Gardens, orchards, parks and other areas with planted fruit trees.
■ **FOODPLANTS** In the bark of Cherry and Pear Trees.

Single brooded, with extended flight period. Comes to light; rests on fruit tree trunks.

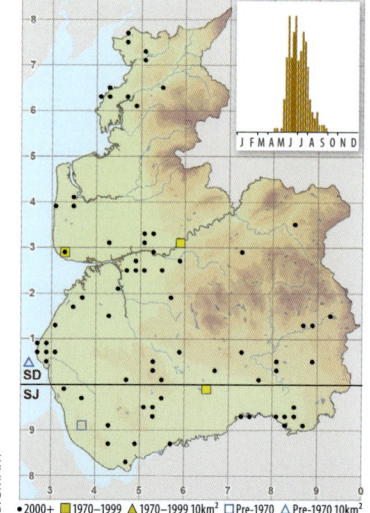

The Moths of Lancashire

TORTRICIDAE

49.201 *Ancylis unguicella*

■ **LANCS STATUS** Occasional / local resident   c.1853–2021

First noted on the mosses of the Liverpool area (un-mapped) by N. Cooke (Byerley, 1854) and next, in the Lancaster area in 1902. Not reported again until 1990, at Birk Bank and thereafter, on Longridge Fell in 1997 and Beacon Fell in 1999. Since then, it has occurred on moorland by Knots Wood in 2000, Winmarleigh Moss in 2004 and at seven other widely scattered sites, including Rochdale in 2012 and on Lee Fell on 14 May 2021 (R. Foster). Occasionally appears well away from its usual habitat.
■ **HABITAT** Mosses, moorland and other heathery sites.
■ **FOODPLANTS** Not recorded in Lancashire.
Single brooded. Comes to light and readily disturbed by day from Heather.

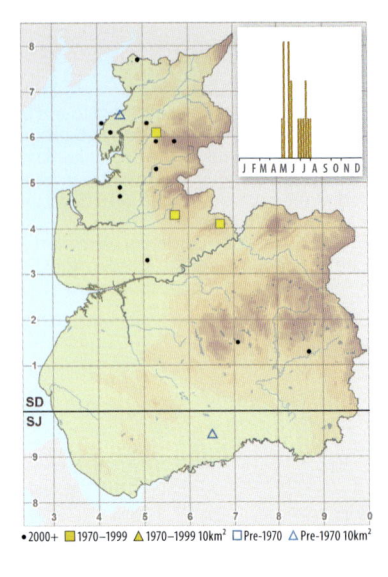

● 2000+   ■ 1970–1999   ▲ 1970–1999 10km²   □ Pre-1970   △ Pre-1970 10km²

49.202 *Ancylis uncella*

■ **LANCS STATUS** Occasional / local resident   c.1853–2011

First recorded 'on the mosses' of the Liverpool area (un-mapped) by N. Cooke (Byerley, 1854) and then from Longridge (Ellis, 1890). Next, noted at Risley Moss on 18 June 1983 (M. Hull), at Winmarleigh Moss from 1988 to 2007 and Holcroft Moss in 1998. Since 2000, found at a few additional sites, at Astley Moss from 2003 to 2010, Bold Moss in 2005, Cadishead Moss in 2010 and Cockerham Moss in 2011.
■ **HABITAT** Lowland raised bogs; once, historically, on moorland at lower altitude.
■ **FOODPLANTS** Not recorded in Lancashire.
Single brooded. Occasionally noted at light. Most frequently disturbed by day from amongst Heather, with 10 seen on 22 May 2004 on Astley Moss (K. McCabe).

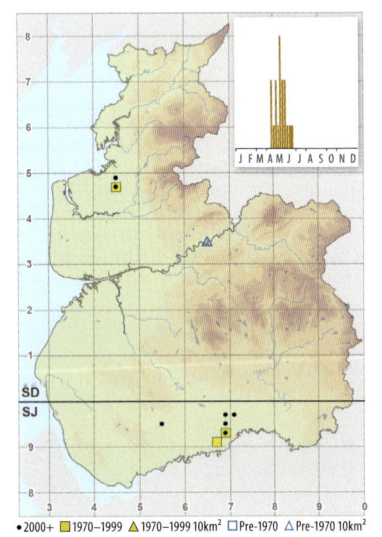

● 2000+   ■ 1970–1999   ▲ 1970–1999 10km²   □ Pre-1970   △ Pre-1970 10km²

49.204 *Ancylis obtusana*

■ **LANCS STATUS** Scarce / very local resident   2009–2017

Considered to be an overlooked resident as it was present in south Cumbria in the late 19th century. First noted, during the day, at Gait Barrows on 9 June 2009 by S. Tomlinson. It was later recorded from the same site, on 1 June 2011 (R. Petley-Jones and B. Hancock) when it came to light, and on 7 June 2017 (J.R. Langmaid and S. Palmer) during the day. Otherwise, only reported from Warton Crag, during the day, on 14 June 2017 (S. Palmer).
■ **HABITAT** Limestone scrub with Buckthorn or Alder Buckthorn present.
■ **FOODPLANTS** Not recorded in Lancashire.
Single brooded. Rare at light; can be disturbed during the day.

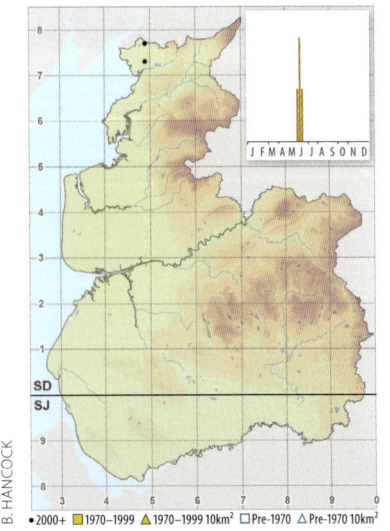

● 2000+   ■ 1970–1999   ▲ 1970–1999 10km²   □ Pre-1970   △ Pre-1970 10km²

286 *The Moths of Lancashire*

49.207 Ancylis geminana

■ **LANCS STATUS** Rare / very local resident 1994–2015

This sp., *A. diminutana* and *A. subarcuana* (the latter not known in Lancs.) have been the subject of taxonomic changes, and for some of this period, they were considered a single species (*A. geminana*) with two forms. In Lancs., they were historically noted under *A. biarcuana* Steph. and *A. diminutana* Haw. in Ellis (1890) and Mansbridge (1940). As a result, records from Crosby c.1857 and Pilling c.1886 were not possible to allocate to species level with certainty. The first confirmed record came from Gait Barrows on 30 June 1994 by R. Petley-Jones and it was subsequently found on three further occasions from this site, in 2001, 2002 and 2015, all as singletons. The last of these was on 13 August 2015, at light, by J. Patton. The only other confirmed record was of a singleton at Leighton Moss on 30 May 2009 (S. Palmer).
■ **HABITAT** Limestone scrub and wetland areas with willow present.
■ **FOODPLANTS** Not recorded in Lancashire.
Probably single brooded. Occasional at light.

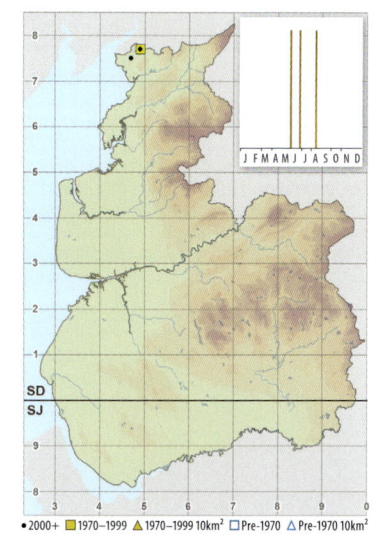

49.209 Ancylis diminutana

■ **LANCS STATUS** Scarce / very local resident 1994–2022

Historic Lancashire records were listed under *A. biarcuana* Steph. (now = *A. geminana* (Don.)) and *A. diminutana* Haw. For more details on the taxonomic position, see 49.207 *Ancylis geminana*. First noted from Gait Barrows in 1994 (R. Petley-Jones) and subsequently, on several occasions until 2015. Elsewhere, only seen at Leighton Moss, from 2003 to 2022.
■ **HABITAT** Limestone scrubland and wetlands where willows are present.
■ **FOODPLANTS** Not recorded in Lancashire.
With both *A. biarcuana* and *A. diminutana* noted from Pilling by J.B. Hodgkinson (Ellis, 1890), specimens would be needed to enable confirmation of these records.

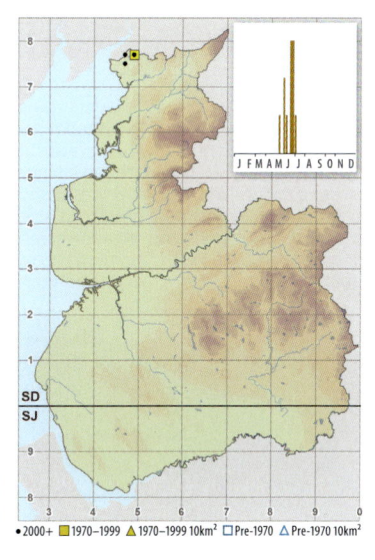

49.211 Ancylis myrtillana

■ **LANCS STATUS** Frequent / local resident 1842–2022

Noted on White Moss, Manchester in 1842 (Eddleston, 1844) and later, on Chat Moss c.1859 and Great Thorn Fell in 1879. Reported as common on moor and moss by Mansbridge (1940). Drainage of many lowland mosses has led to its loss from such areas since then. Still encountered regularly on moorland, such as at Longridge Fell, Birk Bank, Docker Moor and Pendle Hill. 50 were seen at Holcombe on 14 May 2000.
■ **HABITAT** Moors and Mosses.
■ **FOODPLANTS** Bilberry.
Single brooded. Comes to light; readily disturbed from bilberry by day. Sometimes comes to garden light traps well away from its usual habitat.

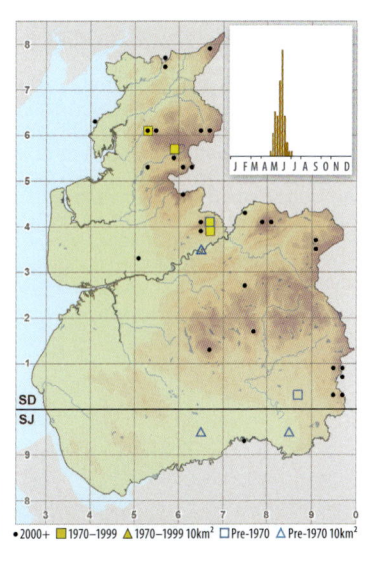

The Moths of Lancashire 287

49.212 Ancylis apicella

- **LANCS STATUS** Scarce / very local resident 1984–2022
First reported, under *Phoxopteryx siculana* Hb., from Simonswood Moss as bred from Bog-myrtle (Gregson, 1857), not a known foodplant. Record accepted by Ellis (1890), Mansbridge (1940) and Bradley *et al.* (1979) but is not mapped here. Only otherwise recorded from Gait Barrows on several occasions from 1984 to the present day. Both first and second brood moths have been found, mostly involving singletons. The largest count was of four at light on 29 May 2008 (R. Petley-Jones).
- **HABITAT** Limestone scrub.
- **FOODPLANTS** Buckthorn and Alder Buckthorn. Use of Bog-myrtle seems very unlikely. Double brooded. Comes to light. Larvae have been found in late July and mid-August.

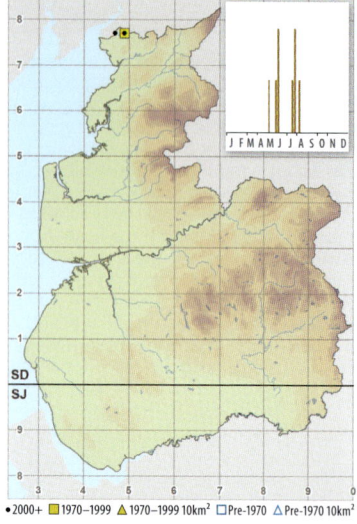

49.214 Ancylis badiana

- **LANCS STATUS** Abundant / widespread resident c.1857–2022
First noted in the Liverpool district (Gregson, 1857) with all early authors noting its abundance across the region and reported, since then, from many sites. The locations with most frequent records include Flixton, Adlington, Rishton, Oswaldtwistle, Parr, Briercliffe and Billinge. It can be equally common at both lowland and upland grassland sites. Numbers noted vary considerably, but an estimated 1,000 plus were present at Abbey Hey on 12 August 2021 (S.H. Hind).
- **HABITAT** Rough grassland, meadows, verges, gardens and brownfield sites.
- **FOODPLANTS** Bush Vetch, Common Vetch, Hairy Tare and Red Clover.
Double brooded. Comes to light and very active by day. Larvae May to November.

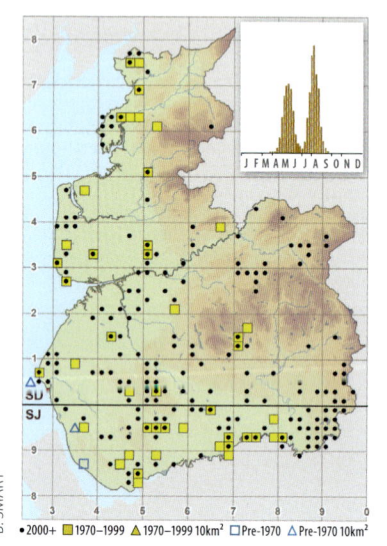

49.215 Ancylis achatana

- **LANCS STATUS** Occasional / widespread resident 1997–2022
Although first reported in Cheshire in 1986, it was not noted in Lancashire until 12 July 1997, when one came to light in Flixton (K. McCabe). This remained the sole county site until 1999, with records then received from Rixton, St Helens and Pennington. Its range expansion has continued at this slow pace to the present day, reaching Billinge in 2000, Chorlton in 2002, Parbold in 2007, near Lea (Preston) in 2008, Formby in 2010, Cockerham in 2011, Rishton in 2015 and Hoghton in 2021.
- **HABITAT** Scrub, gardens, hedgerows, parks and brownfield sites.
- **FOODPLANTS** Not recorded in Lancashire.
Single brooded. Comes to light, the largest count being 36 at Flixton on 3 July 2006.

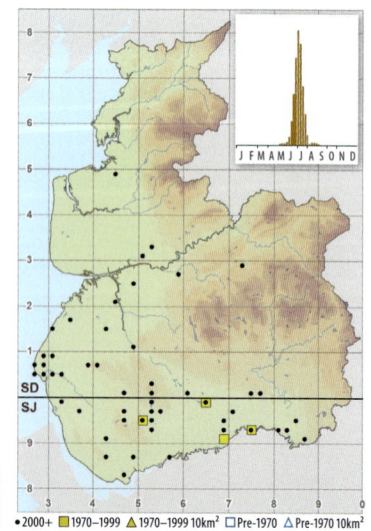

49.216 *Ancylis mitterbacheriana*

■ **LANCS STATUS** Scarce /very local resident   **c.1857–2020**

Reported as common in the Liverpool area (Gregson, 1857), with the only other historic records, Pendleton c.1859 (J. Chappell) and Burnley in 1910 by A.E. Wright & W.G. Clutten. Ellis (1890) reports it to be locally common, while Mansbridge (1940) notes it as local and infrequent. Next seen on 15 June 1999 at Gait Barrows, a site with four subsequent, mainly larval, records. Otherwise only found at Eaves Wood in 2009, White Moss, Yealand in 2012 and, as a larva, in Cragg Wood in 2020 (B. Smart).
■ **HABITAT** Deciduous woodland.
■ **FOODPLANTS** Mainly Oak sp., occasionally Beech.

Single brooded. Rare at light. Larvae have been found from August to October.

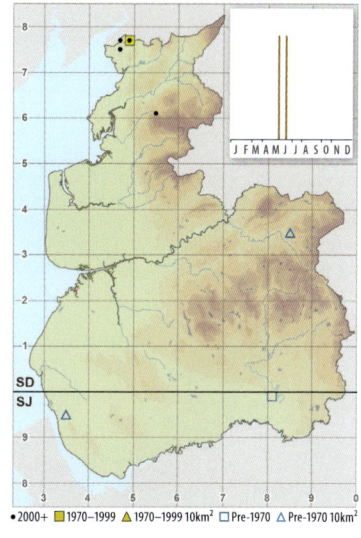

49.218 *Eriopsela quadrana*

■ **LANCS STATUS** Occasional / very local resident   **c.1940–2017**

Although found in south Cumbria in the late 19th and early 20th century, it was not noted in Lancashire until c.1940, by H.N. Michaelis; this, as with all later sightings, being at Gait Barrows. It was 1995 before the next report, at light, by R. Petley-Jones with regular sightings to 2017 in small numbers. Ten were seen on 3 May 2013.
■ **HABITAT** Vegetated limestone pavement.
■ **FOODPLANTS** Not recorded in Lancashire.

Single brooded. Comes to light in small numbers, but most frequently reported during the day when it can be readily disturbed, and flits low among Goldenrod. Possibly overlooked at other limestone sites.

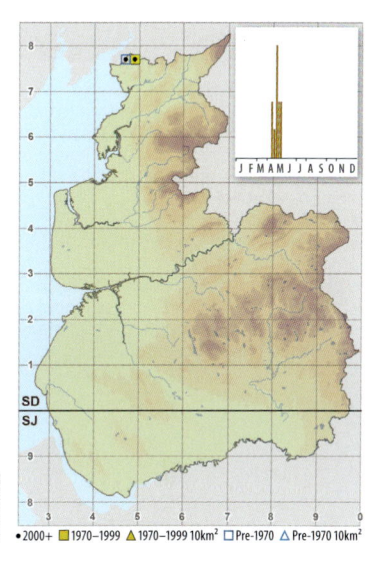

49.219 *Thiodia citrana*

■ **LANCS STATUS** Considered Extinct   **c.1859**

Reported from the Lytham area by J.B. Hodgkinson, this single but undated record is mentioned in Stainton (1859), Ellis (1890), Meyrick ([1928]) and Bradley *et al.* (1979). Stainton (loc. cit.) mentions 'amongst rest harrow' but it is not possible to say if this comment relates specifically to Hodgkinson's sighting. It is more likely to have been from the localised colonies in south-east England, where it was locally common. The original report of the discovery of this species by Hodgkinson has not been located and it is possible that none was written, despite his propensity to put pen to paper.
■ **HABITAT** Unspecified coastal area, possibly sand dune or rough coastal ground.
■ **FOODPLANTS** Not recorded in Lancashire.

Searches for spinnings in the flower-heads of Yarrow, in August or September and in shoots in May, would be worth carrying out. However, the considerable reduction in sandy, coastal grassland in the whole area along the coast from Fairhaven to Fleetwood, which continues apace to the present day with sea defences and housing development, strongly suggests it has been lost to the area.

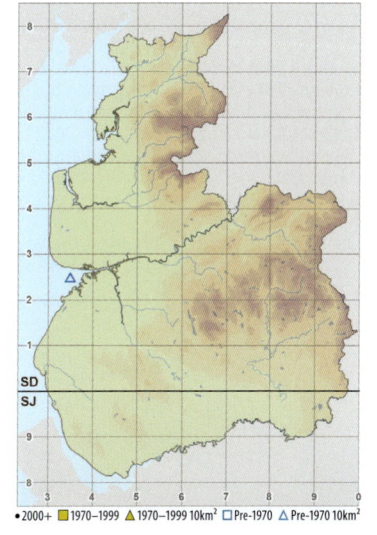

49.220 *Rhopobota myrtillana*

■ **LANCS STATUS** Occasional / local resident   c.1859–2022

Noted on Barton Moss c.1859 by J. Chappell, then from the Burnley area in 1910, Caton Moor c.1940 (present in large numbers) and the Bolton area in 1949. Next reported from Longridge Fell in 1997, Littleborough in 2002 and Barkin Bridge in 2006. Daytime surveys, on the south-east moors by S.H. Hind in 2021 and 2022, produced many records, with 17 seen at Higher Hartshead on 26 June 2021. This suggests the species is under-recorded in upland Lancashire.

■ **HABITAT** Moorland and, formerly, a lowland raised bog.
■ **FOODPLANTS** Not recorded in Lancashire.

Single brooded. Comes to light, but most often noted by day amongst Bilberry.

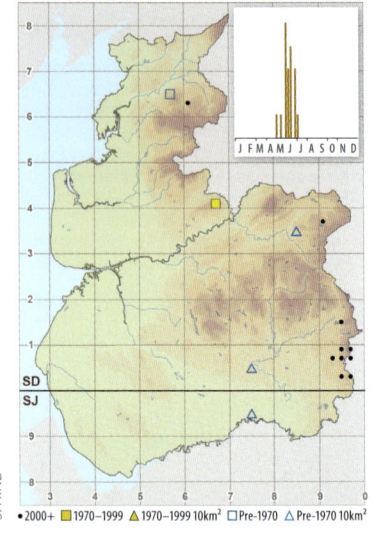

49.221 *Rhopobota stagnana*

■ **LANCS STATUS** Considered Extinct   c.1857

The only known records were from what may have formerly been boggy or marshy areas in the Liverpool area, at Ditton Marsh and Halewood Road. The records are mentioned in C.S. Gregson (1857) but, other than noting that they occurred in July, no further details are given and the sightings could have been some years prior to the above publication date.

■ **HABITAT** Unknown.
■ **FOODPLANTS** Not recorded in Lancashire. Nationally, dependent on the habitat in which the species occurs, it feeds on two different foodplants. The location will exclude the possibility of Small Scabious, while the potentially marshy nature in the locality would suggest that they would probably have been associated with Devil's-bit Scabious. What is certain, is that the two sites in this part of Liverpool, will have been lost to development.

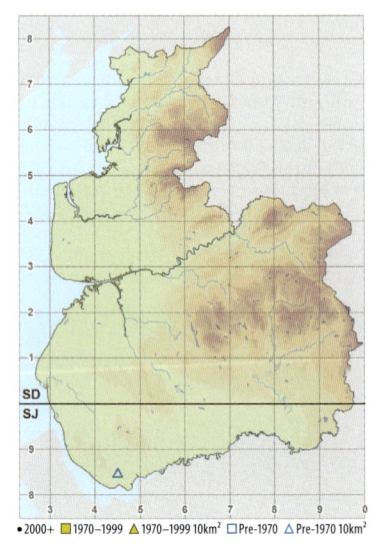

49.222 *Rhopobota ustomaculana*

■ **LANCS STATUS** Occasional / very local resident   2022

Considered to be a long established but overlooked resident species, it was first found when patches of Cowberry, known to be present in the Clougha area, were successfully searched on 28 March 2022 by B. Smart and S. Palmer. The site checked, Windy Clough, covered an area of about 100 square metres. Very many tenanted mines and larger spinnings were located and several bred through for confirmation.

■ **HABITAT** Un-grazed Heather moorland, maintained as a grouse moor.
■ **FOODPLANTS** Cowberry.

Considered to be single brooded, the larva makes an obvious brown mine, later exiting to feed between two spun leaves, one on top of the other.

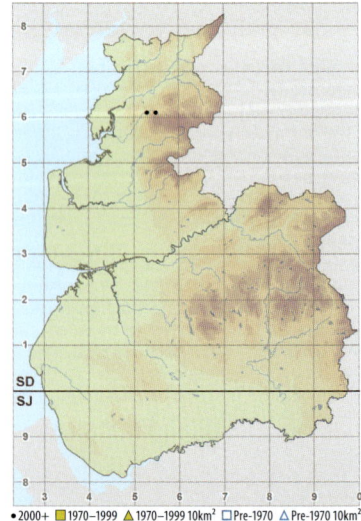

49.223 Holly Tortrix *Rhopobota naevana*

■ **LANCS STATUS** Frequent / widespread resident   1854–2022

First noted on Rivington Pike by Mr Buxton in 1854 and then, 'where holly grows' in Liverpool in 1857. (Mansbridge, 1940) reports it as very common. Noted widely across the county since the 1980s and regularly recorded at garden light traps at sites such as Bay Horse, Flixton, Fulwood, Heysham, Longton, Mill Houses, Morecambe, Preston and Sunderland Point. The few May records are exceptional and unusual.
■ **HABITAT** Woodland, gardens, parks, hedges, mosses and moorland.
■ **FOODPLANTS** Holly, Bilberry, Hawthorn and Blackthorn.
Single extended brood. Comes to light in single figures. Easily disturbed from Bilberry by day and 40 seen at Birk Bank at dusk on 8 July 2021. Larvae found May and June.

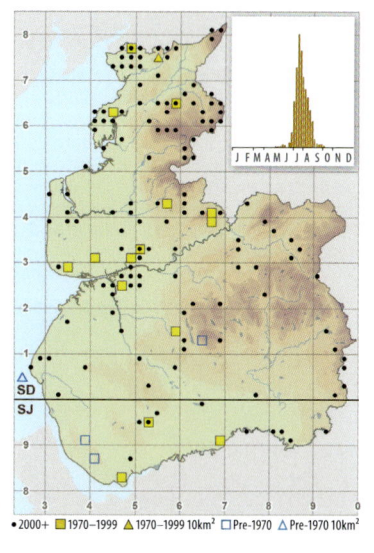

49.224 Bud Moth *Spilonota ocellana*

■ **LANCS STATUS** Frequent / widespread resident   c.1857–2022

First found in Liverpool (Gregson, 1857) and noted as fairly common throughout by Mansbridge (1940). Since the 1990s it has been reported regularly at many lowland sites, particularly in coastal areas and the major river valleys. It was not until around 2012 that records from higher ground increased, but it remains local in these areas.
■ **HABITAT** Woodland, hedgerows, scrub, gardens and brownfield sites.
■ **FOODPLANTS** Hawthorn, Blackthorn, Cherry, Rowan, Apple, oak sp., Alder Buckthorn and Sea Buckthorn.
Single extended brood. Comes to light. A larva found on 2 July 2021 emerged 24 July 2021, suggesting a second brood may be possible. Larvae usually September to May.

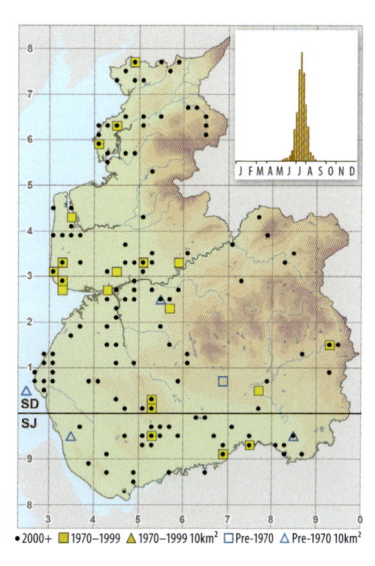

49.225 *Spilonota laricana*

■ **LANCS STATUS** Occasional / local resident   2005–2022

Up to at least 1979, this was considered a form of *S. ocellana* and, as a result, any earlier reports would have been overlooked within that common and widespread species. Despite the recent taxonomic split, modern records are likely to be further confused by the occasional occurrence of dark forms of *S. ocellana*. The similarity also applies with the larvae of these two but there are a few small differences and the larval foodplant will resolve this, it being associated with Larch, primarily, in other parts of the country. This species was first confirmed at Churn Clough reservoir on 3 September 2005 by A. Barker and K. McCabe (with 20 here in 2006), and at Miles Hill in 2006. Since then, found at nine further sites, including Docker Moor, Lord's Lot Bog, Bay Horse, Formby and Ainsdale, usually adjacent to Larch plantings.
■ **HABITAT** Mainly moorland edge and coastal plantations.
■ **FOODPLANTS** Not recorded in Lancashire.
Single brooded. Comes to light. The forewing is narrower than in *S. ocellana* and the nearly pure white ground colour is coarsely strigulate with blackish grey (Bradley, *et al.*, 1979).

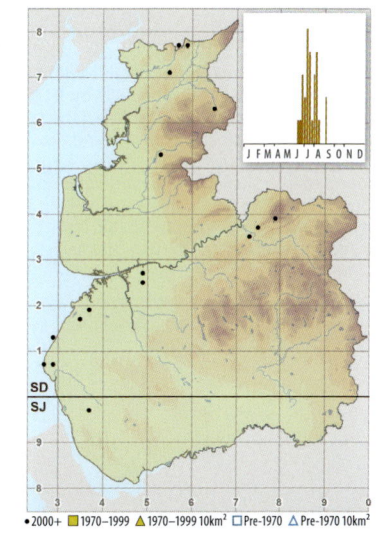

49.228 *Epinotia sordidana*

■ **LANCS STATUS** Occasional / local resident c.1888–2022

There is some doubt about the accuracy of the earliest records, in the 1850s, the first considered to be correct being from Longridge c.1888 by J.B. Hodgkinson. Reported from the Formby, Southport and Burnley areas in 1900, it was not noted again until 1951 at Leighton Moss and 1955 at Holden Clough, Oldham. From around 2000, it has been noted from a limited range of widespread sites, such as Scorton, St Annes, Flixton, Billinge, Chorlton and Rishton amongst others, but rarely with any regularity.

■ **HABITAT** Damp woodland, wetland and brownfield sites with Alders present.
■ **FOODPLANTS** Alder.

Single brooded. Comes to light, usually as singletons.

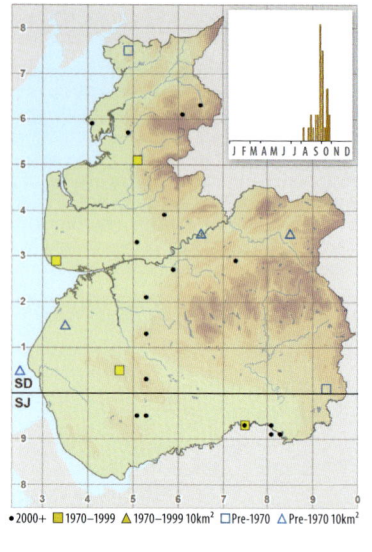

49.229 *Epinotia caprana*

■ **LANCS STATUS** Occasional / local resident c.1857–2021

Noted as abundant in the Liverpool area where willows grow (Gregson, 1857), it was subsequently found at St Annes, Birkdale and Formby, this continuing into the 21st century. Away from the coast, it is infrequent in occurrence. Sites with one or two such records include Billinge, Hic Bibi, Martin Mere, Mill Houses, Preston and the St Helens area. Most readily found as a larva in damp dune slacks, with nine recorded on Creeping Willow at St Annes N.R. on 25 May 2021 (B. Smart).

■ **HABITAT** Dune slacks and woodland or scrub adjacent to rivers or other wetland sites.
■ **FOODPLANTS** Willow spp., Creeping Willow.

Single brooded. Comes to light in low single figures. Larvae found from May to June.

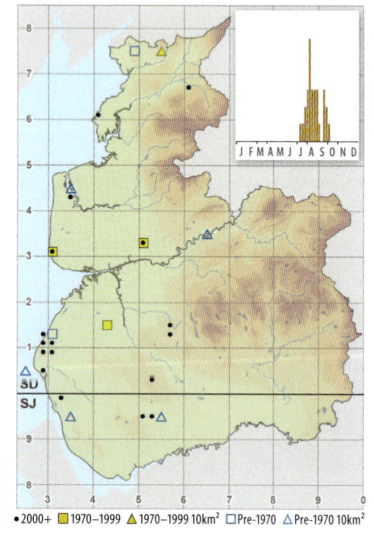

49.230 *Epinotia trigonella*

■ **LANCS STATUS** Occasional / local resident c.1859–2022

Found on Chat Moss around 1859 by J. Chappell, with the next in Longridge c.1888 and Simonswood Moss in 1905. For the next 90 years only a handful of records were received, these mostly from lowland mosses, some of which no longer exist. Since the mid-1990s records from mosses have been few in number but it was noted at a range of new sites. These included Flixton, Docker Moor, Gait Barrows and Warton Crag. These and other single site records indicate it may be rather under-recorded.

■ **HABITAT** Mosses, moorland edge, limestone and coastal scrubland with birch.
■ **FOODPLANTS** Not recorded in Lancashire.

Single brooded. Comes sparingly to light; 11 seen at Gait Barrows on 22 August 2008.

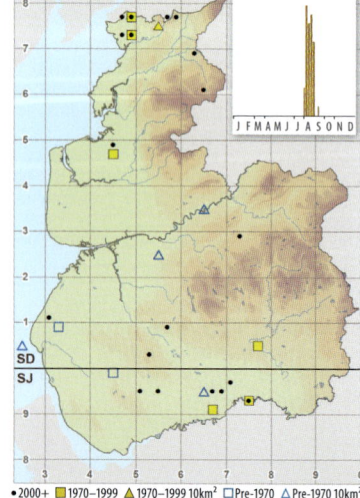

49.231 *Epinotia brunnichana*

■ **LANCS STATUS** Occasional / widespread resident   **1947–2022**
Reported (considered incorrectly) from the Liverpool area by Byerley (1854). Neither Ellis (1890) or Mansbridge (1940) mention any records under this name or *sinuana* (a synonym). Recorded from Daubhill in 1947 by M. Morris and later, at Yealand Storrs in 1977, the first of around 25 noted in the Silverdale area. It has also been seen regularly at sites such as Flixton, Mill Houses, Docker Moor, Ainsdale and Longton. Noted infrequently from several other sites across the county.
■ **HABITAT** Woodland, scrub, mosses, moorland edge and brownfield sites.
■ **FOODPLANTS** Birch.
Single brooded. Comes sparingly to light, but 30 were seen in Rainford on 26 July 2002.

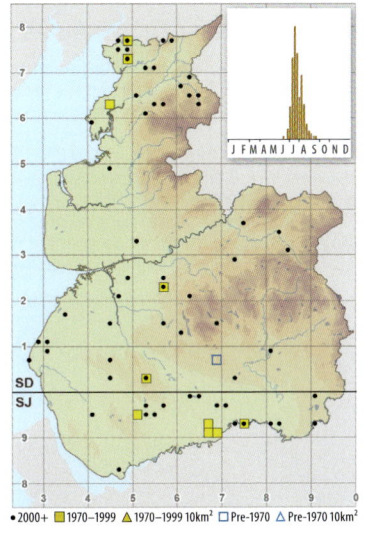

49.232 *Epinotia maculana*

■ **LANCS STATUS** Rare / very local resident   **c.1859–1984**
First noted at Chat Moss c.1859 by J. Chappell (Ellis, 1890) and then in the Dutton area, near Longridge, where many larvae were found and bred from young Aspen trees in 1879 (Hodgkinson, 1880a). J.H. Threlfall, a younger lepidopterist friend of Hodgkinson, also recorded it in the Longridge area (Ellis, 1890), this considered most likely to be from the Dutton site. Michaelis (1958) reports that 'two were beaten from *Populus alba* at Formby in September 1957'. The last one was observed on Risley Moss in 1984 by L. W. Hardwick.
■ **HABITAT** Where Aspen occurs; does not appear to use amenity planted trees.
■ **FOODPLANTS** Aspen. Two adults disturbed from White Poplar.
Probably single brooded. This distinctive, dark species with long, relatively narrow wings is mainly associated with Aspen nationally, but has been known to use other poplar species on occasions. It does come to light and is noted as being readily disturbed from tree trunks, when it flies a short distance to settle again lower down.

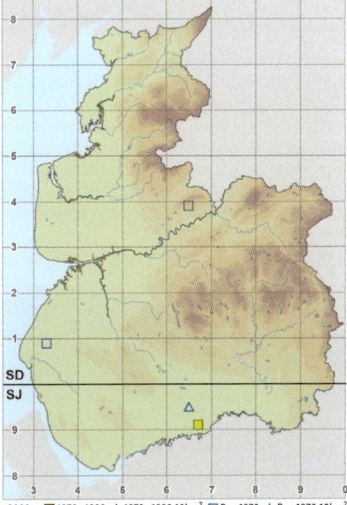

49.233 *Epinotia solandriana*

■ **LANCS STATUS** Occasional / widespread resident   **c.1857–2022**
Noted as abundant where birch grows in the Liverpool area (Gregson, 1857) and later, at Chat Moss in 1859 and Longridge c.1888. In the 1950s, noted regularly at Formby and, from the 1990s, found in small numbers at four sites in the Silverdale area. From 2000 onwards, records received from several additional lowland and upland sites, including Flixton, Mill Houses, Rochdale, Briercliffe and Herring Head Wood. The last site included 12 on one night in early September 2015 (J. Roberts).
■ **HABITAT** Woodland, scrub, mosses and moorland edge, where birch is common.
■ **FOODPLANTS** Birch sp.
Single brooded. Comes to light. Larval feeding signs noted from late May to early July.

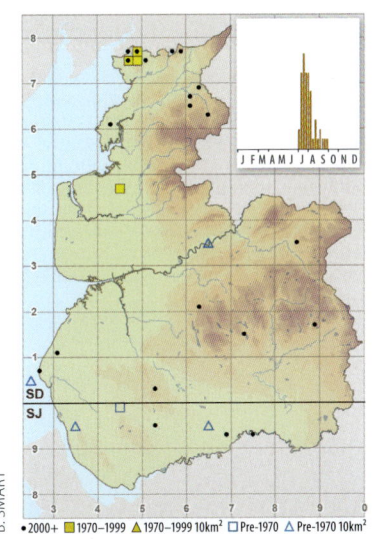

The Moths of Lancashire   **293**

TORTRICIDAE

49.234 *Epinotia abbreviana*

■ **LANCS STATUS** Occasional / widespread resident   c.1888–2022

First noted in the Liverpool district as locally common among elms by J.W. Ellis and later, at Formby (Mansbridge, 1940). Since the 1980s, it has been reported from a wide range of scattered localities, on an infrequent basis, in mainly lowland parts of the county. These included Flixton, Stretford, Preston, the Silverdale area and Stanah.
■ **HABITAT** Woodland and hedgerows.
■ **FOODPLANTS** Elm sp., Wych Elm.

Single brooded. Comes to light in small numbers. Tenanted larval spinnings found May to mid-June. A lack of documented historic records has made it impossible to assess what changes Dutch elm disease may have had on this species.

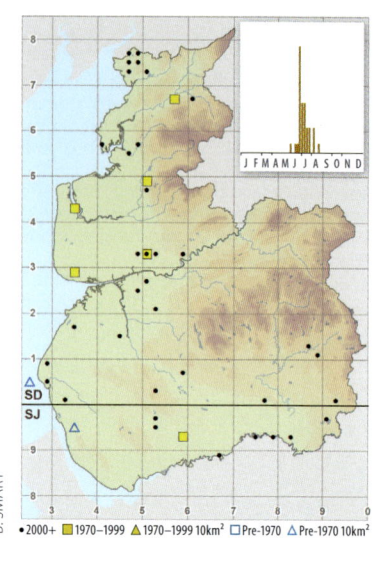

49.237 *Epinotia signatana*

■ **LANCS STATUS** Occasional / local resident   c.1859–2022

Early authors considered it to be very local, the first report being in Stainton (1859) from the Preston district. This could well refer to undated Salwick or Scorton records by J.B. Hodgkinson, later listed in Ellis (1890). Bradley, *et al.* (1979) mention its presence, without dates, in the Silverdale area, a locality that produced most modern reports and include Leighton Moss and Gait Barrows. It is rare elsewhere, with records from Mill Houses in 2018 and Chorlton (two sites) and Fitton Hill, in 2022.
■ **HABITAT** Scrub, particularly on limestone, hedgerows and brownfield sites.
■ **FOODPLANTS** Blackthorn.

Single brooded. Comes to light, mainly as singletons. Larvae late April (B. Smart).

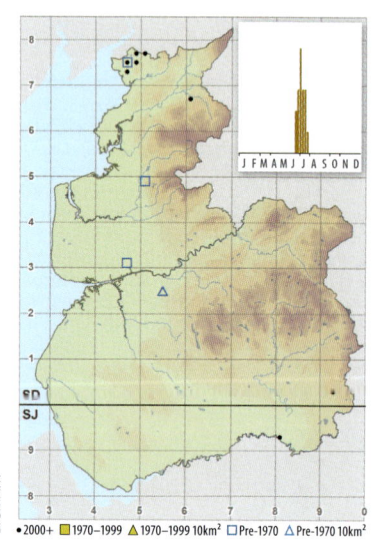

49.238 Willow Tortrix *Epinotia cruciana*

■ **LANCS STATUS** Occasional / local resident   c.1857–2021

Noted on Crosby sandhills in profusion (Gregson, 1857), at Yealand Storrs (1929) and several occasions at Formby (1947–50). Since then, the dunes from Altcar to Birkdale have been the major site for this species, with the dune remnants from Fairhaven to Blackpool also of importance. Occurred at Gait Barrows from 1996 to 2017, and elsewhere, sporadically across a range of inland habitats, including Astley Moss, Heysham Moss, Docker Moor and Scorton.
■ **HABITAT** Vegetated coastal dunes, scrub and brownfield sites, with willow growth.
■ **FOODPLANTS** Creeping Willow, once on Aspen.

Single brooded. Comes to light in single figures. Readily disturbed by day.

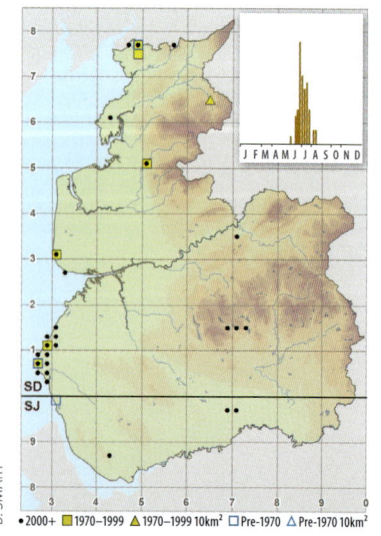

294 *The Moths of Lancashire*

49.239 *Epinotia mercuriana*

■ **LANCS STATUS** Considered Extinct   c.1910–c.1940

Records of this predominantly northern British species in the county are vague and lack detail. The first was reported by Mansbridge (1940) without any date, but is credited to A.E. Wright and W.G. Clutten, from the Burnley area. This record did not appear in Ellis (1890), suggesting it may well have been noted between 1890 and about 1920; at about this time it is believed that Wright moved to Cumbria. The only other record was published in the Lancashire and Cheshire Fauna Society report of 1940–42, written by W. Mansbridge. Here, he mentions the finding of the species by L.T. Ford, from Barn Fell, in 1940. Attempts to establish where this location is, have not been successful. As many of Ford's specimens are housed in the Natural History Museum in London, it is recommended that this would probably be a next stage in tracking down details of this upland species.
■ **HABITAT** Moorland.
■ **FOODPLANTS** Not recorded in Lancashire.
Bradley, *et al.* (1979) report that nationally, this species is rarely found below 200m and the larva is associated with plants such as heather and Bilberry.

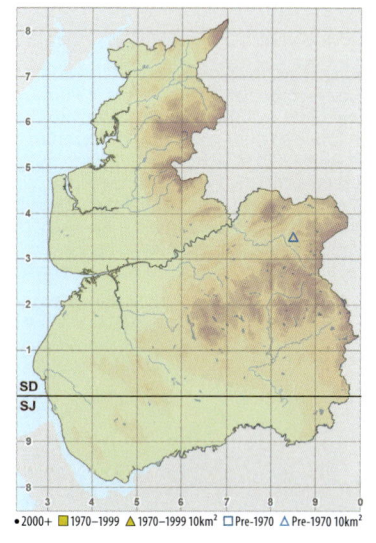

49.240 *Epinotia immundana*

■ **LANCS STATUS** Frequent / widespread resident   c.1857–2022

First recorded in Croxteth (Gregson, 1857) then, in Preston and Longridge c.1888 (Ellis, 1890) and Prescott in 1919. It was reported as common among Alder by Mansbridge (1940) and noted at Southport in 1943. Thereafter, found regularly at many sites across the county, including 20 on Alder trunks at Flixton on 12 May 2001 (K. McCabe) and 30 at light in Holmeswood on 28 April 2014 (S. Priestley).
■ **HABITAT** Woodland, scrub, mosses, riversides, gardens and brownfield sites.
■ **FOODPLANTS** Birch and Alder.
Double brooded. Comes to light, mostly in small numbers. Larvae feed in catkins in winter to early spring and in spun leaves at other times of the year.

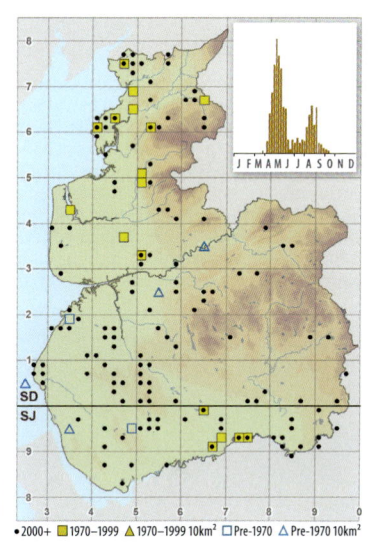

49.242 *Epinotia nanana*

■ **LANCS STATUS** Scarce / local resident   2006–2022

First recorded at light on Caton Moor on 4 July 2006 by J. Roberts and next, by G. Riley at Swinton on 3 and 27 June 2011 (different moths, a male and a female). Since then, recorded at Sunderland Point in 2013, Preston in 2015, Thrushgill in 2019 and, the last three, at Poulton on 2 and 14 July 2021 and 16 June 2022 (B. Dyson).
■ **HABITAT** A range of habitats from moorland to coastal lowlands at some of which there are nearby spruce trees.
■ **FOODPLANTS** Not recorded in Lancashire.
Single brooded. Comes to light as singletons. Bradley, *et al.* (1979) note it as scarce in Lancashire, but this relates to the old North Lancashire, in modern-day Cumbria.

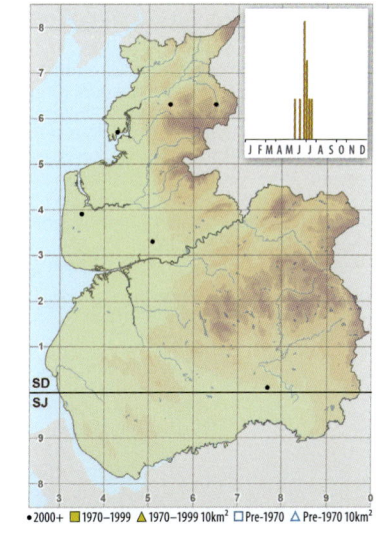

49.243 *Epinotia demarniana*

■ **LANCS STATUS** Occasional / local resident   1999–2022

First recorded at Risley Moss by P. Pugh on 2 June 1999, then at Holcroft Wood and Moss in 2002 and 2004. It was next seen at light in Flixton during 2005 and was netted during the day in 2006 on Astley Moss. In 2019 it was observed in flight around birch at Flixton on three occasions and once on Chat Moss. Flixton featured again in 2020, when one came to light, while daytime sightings were reported from Medlock Vale in June 2021 and near Oldham in June 2022, by S.H. Hind.

■ **HABITAT** Mosses, moorland edge and a brownfield site with birch growth.
■ **FOODPLANTS** Not recorded in Lancashire.

Single brooded. Comes to light. Can be found flying by day around birch trees.

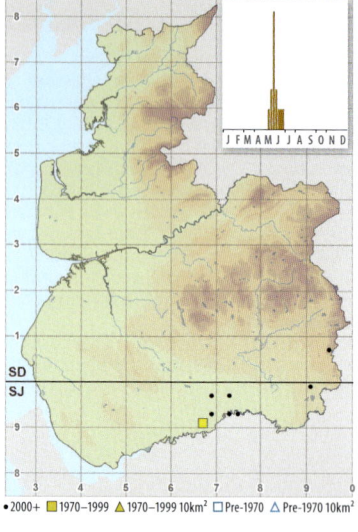

49.244 *Epinotia subocellana*

■ **LANCS STATUS** Occasional / local resident   1846–2021

First noted as abundant on Chat Moss (Eddleston, 1846), then in the Liverpool district (Gregson, 1857), at Preston (1888) and Burnley (c.1910). In the 1950s it was found on Formby Moss and at Holden Clough, Oldham. Since then, noted infrequently and usually as singletons, at several sites in both lowland and upland parts of the county. Such sites include Flixton (with most records, but mostly in the late 1990s), Risley Moss, Gait Barrows, Hough Green, Chorlton, Woolton, Besom Hill and Mossley.

■ **HABITAT** Mosses, scrub, moorland edge, limestone areas and brownfield sites.
■ **FOODPLANTS** Willow spp., including Goat Willow.

Single brooded. Comes to light and can be disturbed by day, both in low numbers.

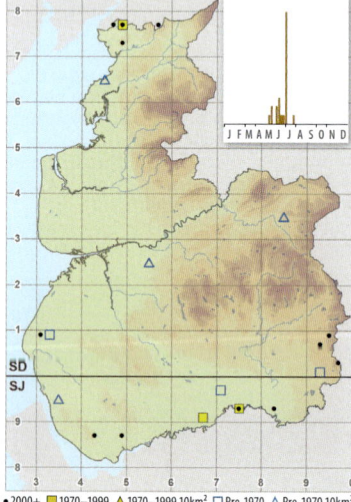

49.245 *Epinotia tetraquetrana*

■ **LANCS STATUS** Occasional / local resident   c.1857–2020

First recorded at Croxteth, Liverpool (Gregson, 1857) and near Longridge c.1888 (Ellis, 1890). Not seen again until 1950, on Formby Moss, where common on Alder, and in the Freshfield area. Since then, it has been observed on Risley Moss in 1985, Winmarleigh Moss in 1988 and several other mossland sites, often by day, to the present. Although less frequently than on the mosses, also found at Docker Moor, near birch scrub at Gait Barrows and in mature birch woodland at Lord's Lot Wood.

■ **HABITAT** Mosses, scrubland, birch woodland and moorland edges.
■ **FOODPLANTS** Birch and Alder.

Single brooded. Comes to light. 80 seen by day at Winmarleigh Moss on 4 June 2006.

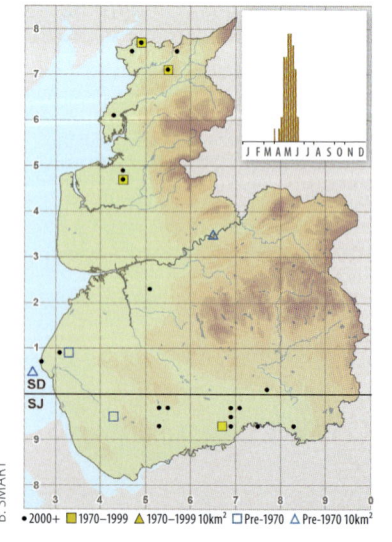

296 *The Moths of Lancashire*

49.246 *Epinotia pygmaeana*

■ **LANCS STATUS** Rare / very local resident   c.1910–1997
Noted as scarce and local within the North-west region by Mansbridge (1940), the only early Lancashire find was reported by A.E. Wright from the Burnley area, with the date estimated as sometime around 1910. It was not seen again until a singleton was netted in the daytime along the edge of a coniferous forestry plantation at Longridge Fell on 16 April 1997 (S. Palmer). A few other similar-sized tortricid moths were observed there in flight at the same time but none of these were netted.
■ **HABITAT** Conifer planation.
■ **FOODPLANTS** Not recorded in Lancashire.
Single brooded. Flies during sunny conditions. Bradley, *et al.* (1979) report it to be common in spruce plantations in various counties, including Lancashire. It is not certain if this refers to the Burnley record, or a further unknown record. It is very likely to be under-recorded in the county due to the early season flight period and lack of light trapping in conifer plantations. Daytime visits during April and May in sunny conditions are more likely to be productive.

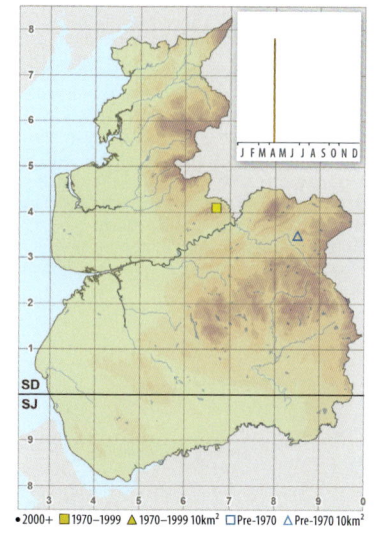

49.247 *Epinotia subsequana*

■ **LANCS STATUS** Rare / very local resident   2010
Seen in flight, in sunny conditions, on two occasions at the same site, Hudd Lee Wood, north-west of Hurst Green, near Longridge (D. Lambert). The first was on 28 April 2010, when three were seen, and one netted, flying around planted Noble Fir, a known national foodplant. The second report involved one being netted, out of two observed, on 3 May 2010.
■ **HABITAT** Conifer plantation.
■ **FOODPLANTS** Not recorded in Lancashire, but probably utilising the planted Noble Fir.
Single brooded. Flies in sunny conditions.
The extent of planting of this tree in the county is unknown. It is possible that the species was introduced with the initial planting, although it may have arrived naturally as it does have a wider national foodplant range than just this one species of fir. As far as is known, Lancashire is its currently most northerly outpost.

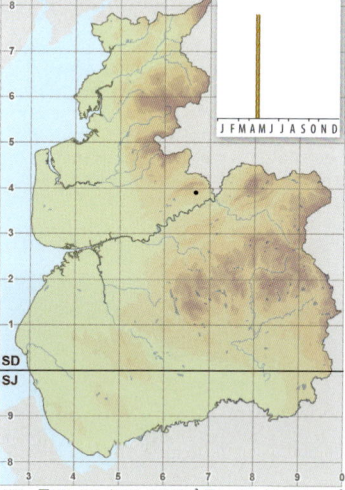

49.248 **Nut Bud Moth** *Epinotia tenerana*

■ **LANCS STATUS** Frequent / widespread resident   c.1857–2022
First noted in Liverpool area as 'a few in oak woods' (Gregson, 1857) and thereafter, up to 1950, in Preston, Silverdale, Burnley, Lancaster, Prescott and Formby. Since the mid-1990s it has been found widely across the county, with regular reports from sites such as Flixton, Gait Barrows, Leighton Moss, Oswaldtwistle, Preston and Rishton.
■ **HABITAT** Woodland, scrub, hedgerows, gardens and brownfield sites.
■ **FOODPLANTS** Hazel, in catkins and buds, and in a Birch catkin (bred to confirm).
A single extended brood, or quite possibly two broods, but larvae have only been found from late March to mid-April. Comes to light, mainly in single figures. Is readily disturbed by day, with 30 tapped from hedges at Osbaldeston Green on 1 July 2022.

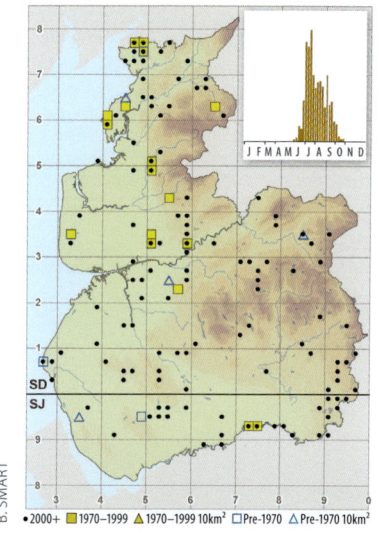

The Moths of Lancashire   **297**

TORTRICIDAE

49.249 *Epinotia ramella*

■ **LANCS STATUS** Frequent / widespread resident   1904–2022

Not mentioned as present in either Byerley (1854) or Ellis (1890). The first county record came from Simonswood Moss in 1904 by W. Mansbridge. There were a few records from Formby Moss and Storrs Moss in the 1950s, but by the late 1990s it was being reported much more widely, with regular reports from Billinge and Flixton for example. 50 came to light at Astley Moss on 27 August 2008 (K. McCabe).

■ **HABITAT** Woodland, mosses, scrubland and some brownfield sites where birch occurs.
■ **FOODPLANTS** Birch, in the catkins.

Single brooded. Comes to light, usually in single figures. Occurs in two forms, one of which is reminiscent of *E. bilunana* and with which it overlaps flight periods in July.

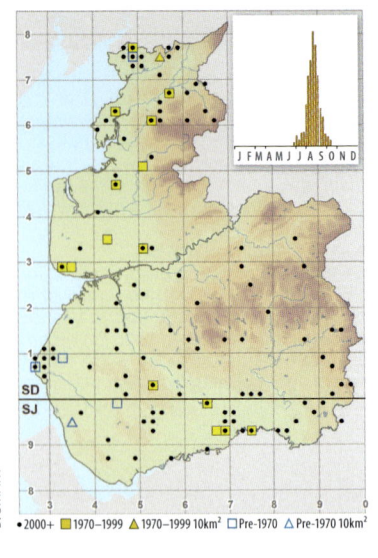

• 2000+   ☐ 1970–1999   ▲ 1970–1999 10km²   ☐ Pre-1970   △ Pre-1970 10km²

49.251 *Epinotia rubiginosana*

■ **LANCS STATUS** Occasional / local resident   1923–2020

First noted in Formby in 1923 (Mansbridge, 1924) and undated, at the same site, by W.A. Tyerman (Mansbridge, 1940). Records have continued from this area, including from Freshfield and Ainsdale, to the present day. First recorded away from the Sefton Coast in Preston on 28 June 1995, and since then scattered sightings have been noted across the county, such as on four occasions at Docker Moor.

■ **HABITAT** Localities with mature Pine sp.
■ **FOODPLANTS** Not recorded in Lancashire.

Single brooded. Comes to light, mostly in low single figures. Has occasionally been disturbed from Scots Pine during the daytime.

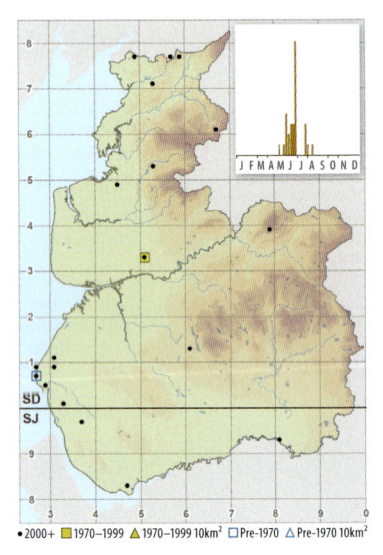

• 2000+   ☐ 1970–1999   ▲ 1970–1999 10km²   ☐ Pre-1970   △ Pre-1970 10km²

49.252 *Epinotia tedella*

■ **LANCS STATUS** Occasional / local resident   c.1857–2022

First recorded at Croxteth Park, Liverpool (Gregson,1857) who noted it as 'plentiful where silver firs grow'. It was next found in Burnley in 1910 and Formby in 1941, with no further record until the 1990s. Then, seen regularly at Longridge Fell during the daytime, between 1996 and 2012 and at Docker Moor, at light, from 2005 to 2021. There are scattered reports from many other conifer plantations and a count of 20 or more was made on a day visit to Thrushgill in June 2019. Probably under-recorded.

■ **HABITAT** Conifer Plantations.
■ **FOODPLANTS** Not recorded in Lancashire.

Single brooded. Comes to light. Readily disturbed by day in spruce plantations.

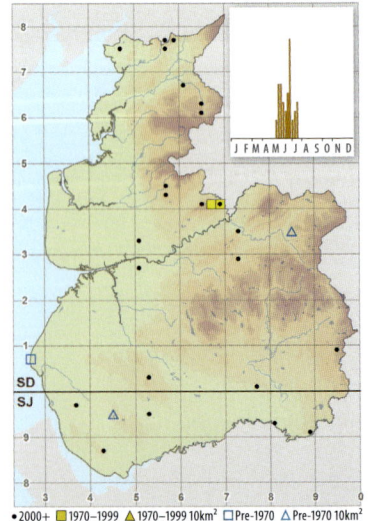

• 2000+   ☐ 1970–1999   ▲ 1970–1999 10km²   ☐ Pre-1970   △ Pre-1970 10km²

298 *The Moths of Lancashire*

49.254 *Epinotia bilunana*

■ **LANCS STATUS** Frequent / widespread resident   1846–2022

Noted as abundant on Chat Moss in June 1846 by R.S. Edleston, it was next seen in Croxteth in 1857, Longridge and Levenshulme in 1888 and Kirkby Moss in 1908, with Mansbridge (1940) reporting it as common among birch. It was not noted again until the 1980s at Martin Mere and Out Rawcliffe, but thereafter, across much of the county. Noted annually at some sites, such as Flixton, but elsewhere it has increased in regularity in recent years, such as in Hoghton, nr. Longton, Rishton and Southport.
■ **HABITAT** Mosses, woodland, scrub, gardens and brownfield sites, where birch grows.
■ **FOODPLANTS** Birch, in the catkins.
Single brooded. Comes to light in single figures; 30 seen at Rainford in June 2006.

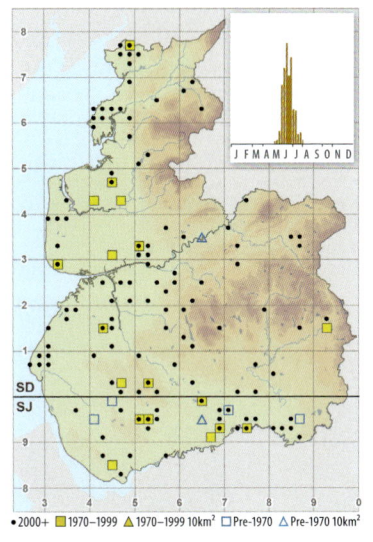

• 2000+ ■ 1970–1999 ▲ 1970–1999 10km^2 □ Pre-1970 △ Pre-1970 10km^2

49.255 *Epinotia nisella*

■ **LANCS STATUS** Frequent / widespread resident   c.1890–2022

First noted in Preston by J.B. Hodgkinson & J.H. Threlfall (Ellis, 1890) and later, as locally abundant in Formby (Mansbridge, 1940). Up until the early 1990s, the only additional sites were Crosby, Oldham and Yealand Storrs but, since then, records have been received from across the county and it is now regular at many sites.
■ **HABITAT** Woodland, scrubland, gardens, edges of wetland and brownfield sites.
■ **FOODPLANTS** Sallow spp.
Single brooded. Comes to light, sometimes in good numbers. *E. cinereana* was given specific status again in 2012, having been considered a form of *E. nisella* since c.1950. It is associated with Aspen nationally, but has no confirmed records in Lancashire.

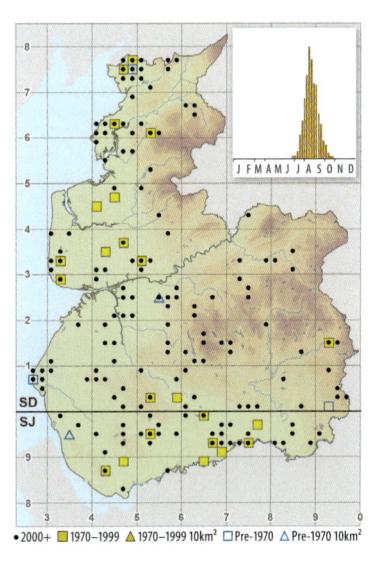

• 2000+ ■ 1970–1999 ▲ 1970–1999 10km^2 □ Pre-1970 △ Pre-1970 10km^2

49.257 Larch Tortrix *Zeiraphera griseana*

■ **LANCS STATUS** Scarce / local resident   c.1940–2018

First noted as fairly common among pines throughout (Mansbridge, 1940). Next from Yealand Storrs, 1977 by M. Hull, Clayton Green in 1998 and Lightfoot Green in 2004, the latter two by S. Palmer. Noted on three occasions each at Botton Mill, nr. Wray by J. Girdley and Churn Clough Reservoir, Sabden by A. Barker and K. McCabe. Other records from Bay Horse, Tatham, Lords Lot Wood, Baines Crag, Dalton Crags, Docker Moor and Pendleton Moor confirm this to be primarily an upland moth.
■ **HABITAT** Larch plantations, coniferous woodland.
■ **FOODPLANTS** Unrecorded within the county.
Single brooded. Most records are of moths at light.

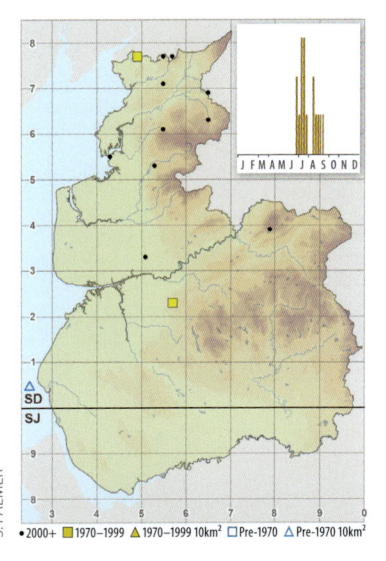

• 2000+ ■ 1970–1999 ▲ 1970–1999 10km^2 □ Pre-1970 △ Pre-1970 10km^2

The Moths of Lancashire

TORTRICIDAE

49.259 Spruce Bud Moth *Zeiraphera ratzeburgiana*

■ **LANCS STATUS** Occasional / local resident　　　　　　1986–2021

First noted by M. Dempsey on 7 August 1986 at Caton. Bred from larva collected on 29 June 1998 at Longridge Fell by S. Palmer. Further recorded in 1999 at Billinge and Flixton. Since then, recorded at Docker Moor on 12 occasions, at Crossdale Beck, nr. Lowgill and at several other moorland sites. Although less frequently, has also been found in the dunes at Formby near pine woodland, and also from gardens some way from coniferous woodland, suggesting populations may survive on isolated trees.

■ **HABITAT** Coniferous woods.
■ **FOODPLANTS** Sitka Spruce.
Single brooded. Almost all records are of moths to light.

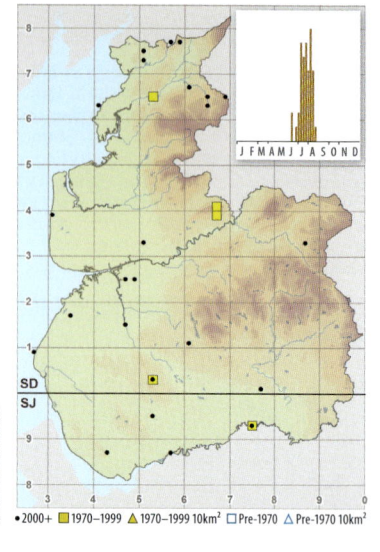

49.260 *Zeiraphera isertana*

■ **LANCS STATUS** Frequent / local resident　　　　　　c.1857–2022

Early authors considered it to be abundant on old oaks, with the first report coming from the Liverpool district (Gregson, 1857). The next records were over 100 years later, at Yealand Storrs in 1977 by M. Hull and at St Helens in 1980 by I.D. Wallace. Many subsequent garden records, with multiple sightings from Lightfoot Green, Flixton, Woolton, Haydock, Rishton, Wigan, Roby, Billinge, Chorlton, Huyton and Lytham Hall. Also, from Gait Barrows and Mill Houses in the north of the county.

■ **HABITAT** Oak woodland and isolated hedgerow oaks.
■ **FOODPLANTS** Oak. Larvae recorded in spun leaves during mid-May 2022 at Rixton.
Single brooded. Comes to light.

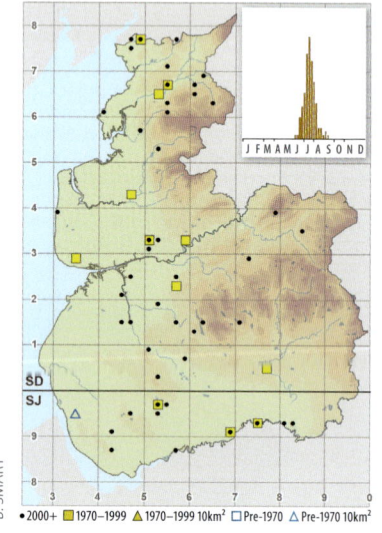

49.261 *Crocidosema plebejana*

■ **LANCS STATUS** Occasional / migrant or vagrant　　　　　　1998–2007

This occasional visitor to the county was first reported from Lytham St Annes by D. & J. Steeden on 1 July 1998, the first of five records at the site in two years, suggesting the possibility that the moth may have been temporarily resident. Three moths were recorded at Heysham by D.J. Holding during September 2005. The next records were of moths at Flixton and Bispham during 2006, and from Carnforth and Bay Horse, nr. Dolphinholme in 2007. Surprisingly, there have been no further records since.

■ **HABITAT** Coastal locations, gardens.
■ **FOODPLANTS** Not recorded in Lancashire.
Overlapping broods. Comes to light.

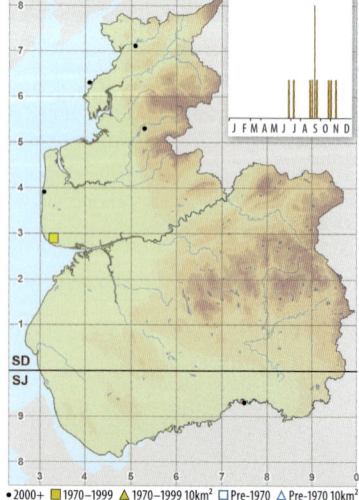

300　The Moths of Lancashire

49.264 *Eucosma obumbratana*

■ **LANCS STATUS** Occasional / local resident 1916–2022
The earliest known record is from August 1916, with about a dozen from the Crosby sand-hills, although the account suggests the moth was already known from the site. (Mansbridge, 1916a). Next, recorded at Freshfield Dune Heath in 1951 and at Flixton in 2000, the first of 17 Flixton records from a county total of 38. Further records from Chorlton, Hesketh Bank, Martin Mere, Crossens Marsh, Astley Moss, Longton, Rainford, Penketh and Lunt Meadows. Unrecorded in the north of the county.
■ **HABITAT** Sand dunes, saltmarshes, arable fields, waste ground.
■ **FOODPLANTS** Not recorded in Lancashire.
Single brooded. Comes to light. Occasionally encountered during day.

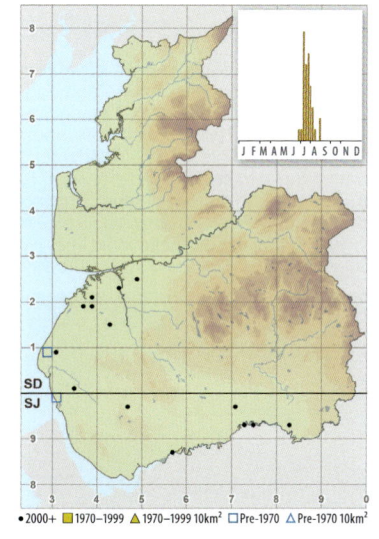

49.265 *Eucosma cana*

■ **LANCS STATUS** Abundant / widespread resident c.1857–2022
First documented from the Liverpool area as plentiful in old pastures (Gregson, 1857) and later as well-distributed (Mansbridge, 1940). Further recorded from Withington, Formby and Leighton Moss. Noted regularly throughout, from the 1980s onwards, including in upland areas, e.g., Leck Fell, Worsaw Hill, etc. High counts of 50 from Warton Crag 2002 (J. Clifton) and among thistles at Halton, nr. Lancaster in 2010.
■ **HABITAT** Rough grassland, scrub, brownfield sites.
■ **FOODPLANTS** Spear Thistle, Black Knapweed.
Single brooded. Comes to light. Frequently found around foodplants during day and at dusk. Larvae have been found from late July to September in seed-heads.

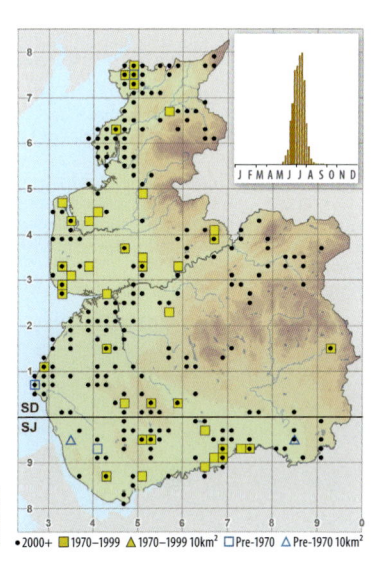

49.266 *Eucosma hohenwartiana*

■ **LANCS STATUS** Frequent / widespread resident c.1888–2022
Reported from Preston by J.H. Threlfall (Ellis, 1890) and described as scarce, with records from Silverdale, Lancaster, Formby and Bolton noted by Mansbridge (1940). Although numbers of records appear stable, counts are now predominantly of singletons. High count of 20 from Great Plumpton 29 July 1990 by M. Evans. Most frequently recorded at Flixton, Gait Barrows and Warton Crag.
■ **HABITAT** Dry grassland, scrub, brownfield sites.
■ **FOODPLANTS** Black Knapweed.
Single brooded. Readily to light. Easily disturbed from vegetation in meadows containing Knapweed. Larvae in seed-heads of the foodplant during August.

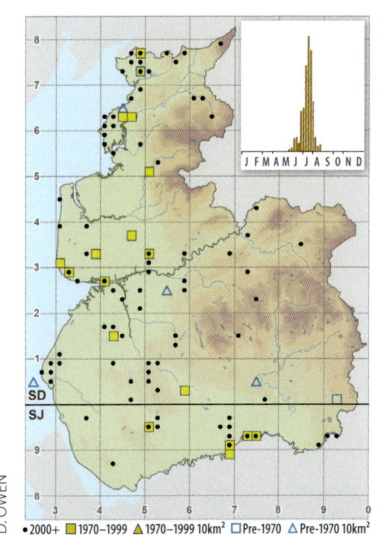

The Moths of Lancashire 301

TORTRICIDAE

49.269 *Eucosma campoliliana*

■ **LANCS STATUS** Frequent / widespread resident   1846–2022
First noted by R.S. Edleston (1846) among ragwort on Kersal Moor, Salford, and considered rare. Reported as 'abundant where rag weed grows' in Liverpool (Gregson, 1857). Noted from Lytham and Wyre valley by J.B. Hodgkinson (Ellis, 1890). Only three 20th century records prior to 1980s, but since recorded on 740 occasions, confirming increasing abundance. High count of ten at St Helens in 2003 (R. Banks).
■ **HABITAT** Rough grassland, scrub, brownfield sites.
■ **FOODPLANTS** Common Ragwort. Larvae found during August in flower-heads, including four at Cadishead Moss in 2019, with moths bred through.
Single brooded. Recorded during the day and at light.

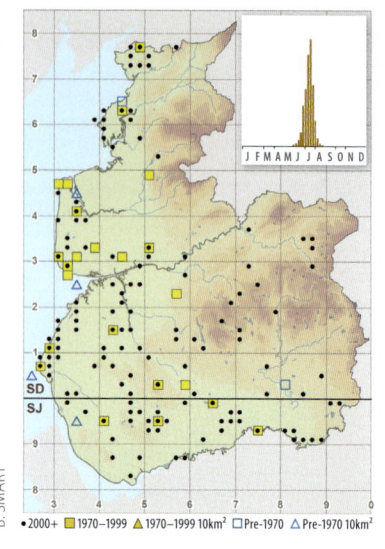

● 2000+   ☐ 1970–1999   △ 1970–1999 10km²   ☐ Pre-1970   △ Pre-1970 10km²

49.272 *Eucosma tripoliana*

■ **LANCS STATUS** Occasional / local resident   2003–2022
Seemingly, a recent arrival in the county, as first recorded on 1 August 2003 when trapped at light nr. Thurnham by B. Cockburn and L. Sivell. Further recorded at Oxcliffe Marsh, Heysham six days later, and at Marshside the following year. 20 to light at Birkdale Green Beach 2012 by G. Jones. Coastal sites such as these provide most records, although a few inland records, from Hoghton, Dolphinholme, Docker Moor, Preston and Longton, suggest a tendency to wander from the saltmarshes.
■ **HABITAT** Coastal sites and saltmarshes where foodplant present.
■ **FOODPLANTS** Sea Aster.
Single brooded. Comes to light. Larvae found in seed-heads, September to October.

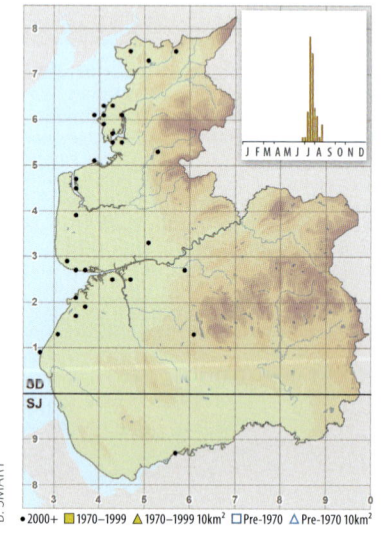

● 2000+   ☐ 1970–1999   △ 1970–1999 10km²   ☐ Pre-1970   △ Pre-1970 10km²

49.276 *Eucosma aspidiscana*

■ **LANCS STATUS** Occasional / very local resident   1984–2022
Recorded solely from the Morecambe Bay limestone area, with all records from last 40 years. Likely to have been previously missed rather than a new arrival. First noted on 25 May 1984 at Gait Barrows by E. Emmet, this site providing over half the 72 records. Further records from Leighton Moss 1993, Warton Crag 1994, Carnforth Ironworks 2008, Yealand Hall Allotments 2010, Trowbarrow Quarry 2012 and Yealand Conyers 2014. Larva found in Goldenrod stem, Gait Barrows by J. Patton, July 2022.
■ **HABITAT** Limestone grassland.
■ **FOODPLANTS** Goldenrod.
Single brooded. Rare at light. Netted in day, with 63 at Gait Barrows, S. Palmer 2010.

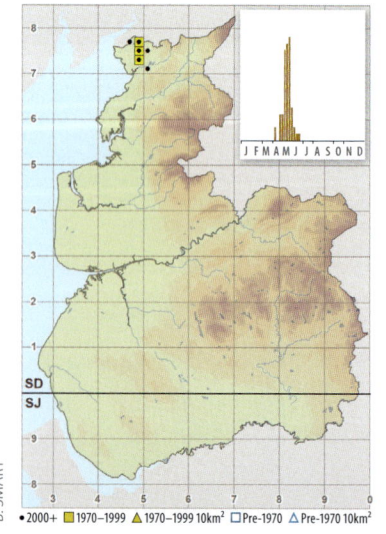

● 2000+   ☐ 1970–1999   △ 1970–1999 10km²   ☐ Pre-1970   △ Pre-1970 10km²

302   The Moths of Lancashire

49.279 *Gypsonoma dealbana*

■ **LANCS STATUS** Frequent / widespread resident c.1888–2022

First noted at Preston by J.B. Hodgkinson (Ellis, 1890), with the only other pre-1980 records coming from Formby in 1950 and at Rufford in 1958. Ellis had earlier described the species as local (1890), Mansbridge as common (1940). The moth was next recorded in 1980s at Risley Moss, Woolston Eyes, St Helens and Gait Barrows. Over 800 records since 1990 with a high count of 22 moths at Fowley Common, nr. Culcheth on 17 July 2006 (J.D. Wilson). Appears predominantly a lowland species.
■ **HABITAT** Woodland, hedgerows, parks, etc.
■ **FOODPLANTS** Oak, Hazel, willows, Hawthorn, Blackthorn, Plum, cherry, Crab Apple, poplar, Aspen, Small-leaved Lime, birch, Sycamore, Norway Maple, Dogwood.
Single brooded in an extended generation. Comes readily to light. Larval feeding frequently recorded in autumn as the larvae graze on underside of leaves from within a frass-covered tube, less so in spring when the larvae feed from within spun leaves.

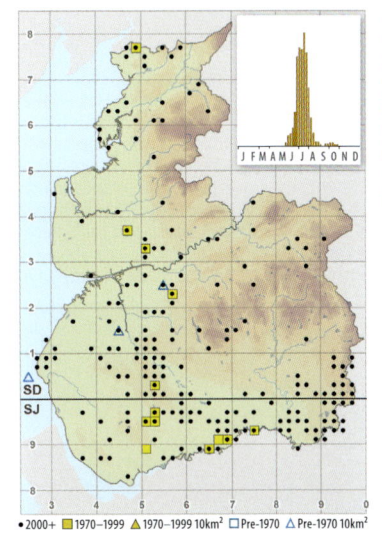

● 2000+ ■ 1970–1999 ▲ 1970–1999 10km² □ Pre-1970 △ Pre-1970 10km²

49.280 *Gypsonoma oppressana*

■ **LANCS STATUS** Occasional / local resident c.1908–2022

With just a single record from Simonswood Moss, nr. Kirkby (Mansbridge, 1940), it is possible the moth may have become temporarily extinct before its return to the county, when trapped at light in Chorlton on 5 July 2013 (B. Smart). Subsequently, recorded at Flixton in 2019, St Helens, Fazakerley and Stretford in 2020, Burscough, Oldham, Denton, etc. in 2021 and Briercliffe, Ashton, etc. in 2022. As yet, confined to southern and central areas of the county, but it does appear to be spreading north. A record of larval feeding at Lightfoot Green in 2021 represents the only VC60 sighting so far.
■ **HABITAT** Woodland edges and parks.
■ **FOODPLANTS** Black-poplar, Black-poplar hybrids, Lombardy Poplar.
Single brooded. Comes to light. Larval feeding noted in autumn, initially mining along the midrib of a leaf, then grazing on the underside from a frass-covered silken tube. Larvae are reddish-brown unlike the grey larvae of *G. dealbana*. Feeding signs have also been recorded in spring, as a frass-covered tube develops from tenanted buds.

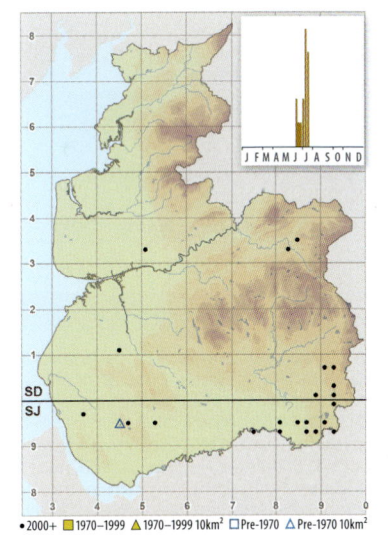

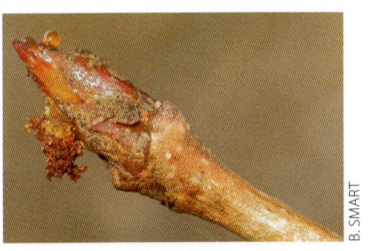

● 2000+ ■ 1970–1999 ▲ 1970–1999 10km² □ Pre-1970 △ Pre-1970 10km²

49.281 *Gypsonoma sociana*

■ **LANCS STATUS** Occasional / widespread resident   1846–2022

First recorded from Cheetham Hill, where considered common, by R.S. Edleston (1846). Nearly 100 years later, status listed as 'scarce in recent years, and local' (Mansbridge, 1940). Bred from poplar in 1939 at Ainsdale by H.N. Michaelis, with contemporaneous records from Preston, Southport, Prescot and Manchester. Over 250 records since 2000, with stable numbers of counts and individuals. High count of eight from Huyton 2009. Distribution rather patchy in VC60 and in upland areas.
■ **HABITAT** Woodland, parks and on isolated trees.
■ **FOODPLANTS** Poplar.
Single brooded. Moths readily to light. Once to pheromone lure at Hutton in 2018.

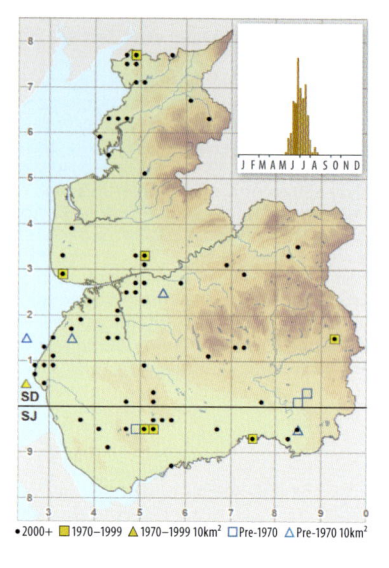

49.283 *Gypsonoma aceriana*

■ **LANCS STATUS** Occasional / local resident   c.1888–2022

Reported from St Annes by J.B. Hodgkinson (Ellis, 1890), with status given as very scarce. Abundance noted among Aspen on Crosby sand-hills in 1913 (Mansbridge, 1940). B.B. Snell and H.N. Michaelis found larvae common in poplar shoots at Freshfield, 1953. No further records until noted at Heysham N.R. in 1993. Subsequent records largely from garden traps. Primarily a lowland species, most frequent in the south and west, with a single record from Briercliffe (G. Turner) the only exception.
■ **HABITAT** Woodland edges, isolated trees.
■ **FOODPLANTS** Poplars, including Black-poplar, Balsam Poplar; feeding in shoots.
Single brooded. Comes to light. Tenanted shoots recorded from May to early June.

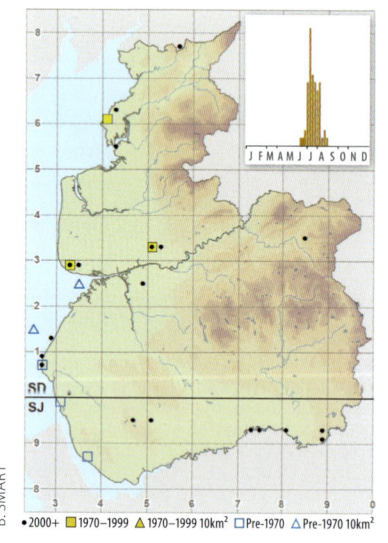

49.284 *Epiblema sticticana*

■ **LANCS STATUS** Occasional / local resident   c.1857–2022

Reported from the Liverpool district by Gregson (1857), describing the moth as common amongst Colt's-foot. From Stretford in 1919 and Formby and Ainsdale in 1950. Next, recorded from Bolton by S.P. Garland in 1986. Records infrequent, but do demonstrate a reasonably well-distributed species, present in lowland areas such as Middleton N.R. and upland areas such as Longridge Fell and Docker Moor.
■ **HABITAT** Rough ground, scrub, quarries, brownfield sites.
■ **FOODPLANTS** Not recorded in Lancashire.
Single brooded. Rarely to light. Ten of the 39 field records demonstrate an association with Colt's-foot, although early stages have not been detected.

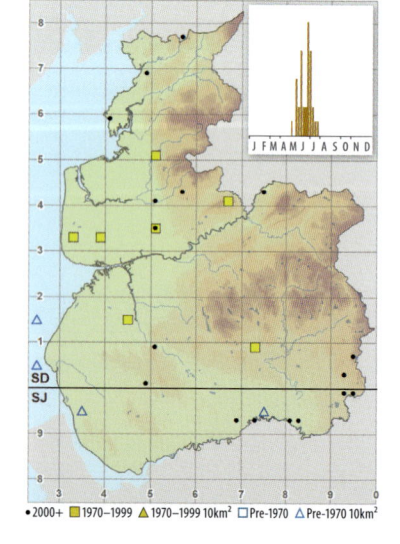

304 *The Moths of Lancashire*

49.285 *Epiblema scutulana*

■ **LANCS STATUS** Frequent / widespread resident   c.1857–2022
Noted among thistles in the Liverpool district (Gregson, 1857). Later, at Manchester by J. Chappell and at Longridge by J.B. Hodgkinson (Ellis, 1890), with Ellis describing it as 'tolerably common.' Frequently recorded since, save for an absence of records, as with many other species, from 1960s–1970s. The first post-1970 record was from Scorton by L.W. Hardwick, 1983, with 180 records since. Less frequent in VC59 and lowland areas, with suggestions of decline in last decade. Possibly under-recorded.
■ **HABITAT** Rough grassland, moorland, brownfield sites.
■ **FOODPLANTS** In stems of Creeping Thistle, and thistles of *Carduus* genus (1950).
Single brooded. At light and netted in field. Larvae noted in March in old stems.

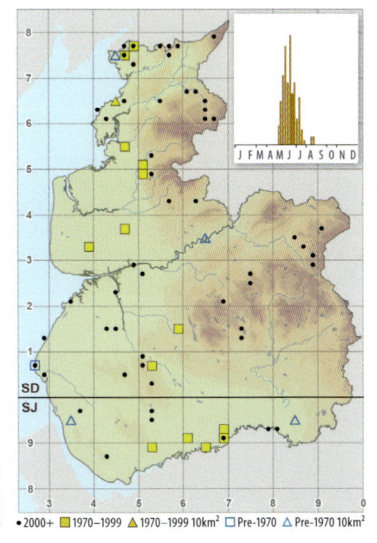

49.286 *Epiblema cirsiana*

■ **LANCS STATUS** Frequent / widespread resident   c.1910–2022
Not an easy moth to separate in the field from the previous, which combined with taxonomic changes, has likely meant both are under-recorded. First noted at Burnley by A.E. Wright, and from Silverdale, Formby and Crosby (Mansbridge, 1940). Next, from 1981 at Haskayne, and at Wesham in 1984 (M. Evans). Slightly more frequent in VC59 than VC60 and found in upland and in lowland areas. 48 records from 2013 to 2022, compares unfavourably to 87 the previous decade.
■ **HABITAT** Rough grassland, brownfield sites.
■ **FOODPLANTS** Black Knapweed; feeding in stems.
Single brooded. Rarely at light. Larvae recorded from January to May.

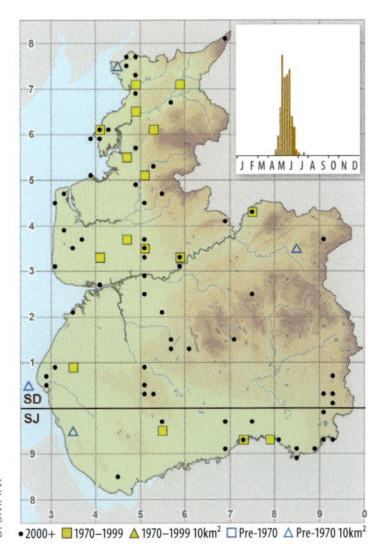

49.288 *Epiblema foenella*

■ **LANCS STATUS** Frequent / widespread resident   1903–2022
First reported from Southport by W.G. Clutten in 1903 (Mansbridge, 1940), and then in 1950s from Formby, Didsbury and Morecambe. Whilst frequency and range have increased in last 50 years, the moth remains absent from the fells, and is largely coastal in VC60, mirroring foodplant distribution. Recent counts suggest decline, with 79 records from 2013 to 2022, compared with 143 from 2003 to 2012. Almost all singletons since 2018, compared to a high of 20 at Altcar in 2013 (R. Walker).
■ **HABITAT** Rough grassland, brownfield sites.
■ **FOODPLANTS** Mugwort; feeding in rootstock.
Single brooded. Comes readily to light. Larvae recorded from March to May.

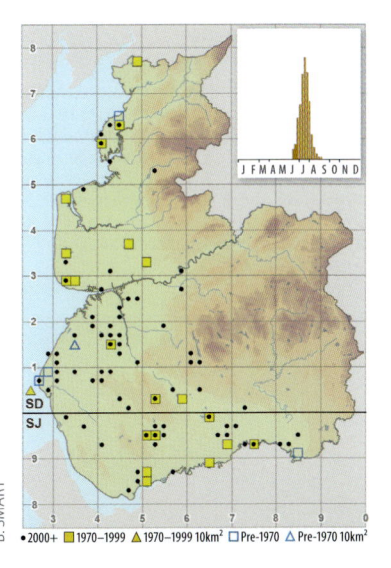

TORTRICIDAE

The Moths of Lancashire   **305**

TORTRICIDAE

49.289 *Epiblema costipunctana*

- **LANCS STATUS** Occasional / local resident   c.1888–2022

Reported from Lytham St Annes and the banks of the Wyre by J.B. Hodgkinson and from Preston by J.H. Threlfall (Ellis, 1890). Noted as 'common, especially on the sand-hills' (Mansbridge, 1940). From Gait Barrows and Great Plumpton in 1988, Morecambe in 1995, Jack Scout in 2001. Distribution shows quite a marked east / west split, with the only eastern records at Chorlton, Briercliffe (G. Turner), Medlock Vale and the Tame Valley (S.H. Hind), all from the last decade.
- **HABITAT** Rough grassland, scrub, brownfield sites, sand dunes.
- **FOODPLANTS** Common Ragwort. Larvae noted in rootstock, September to November. Double brooded. Comes to light. Moth reared from larvae, Chorlton 2021 (B. Smart).

● 2000+ ☐ 1970–1999 △ 1970–1999 10km^2 ☐ Pre-1970 △ Pre-1970 10km^2

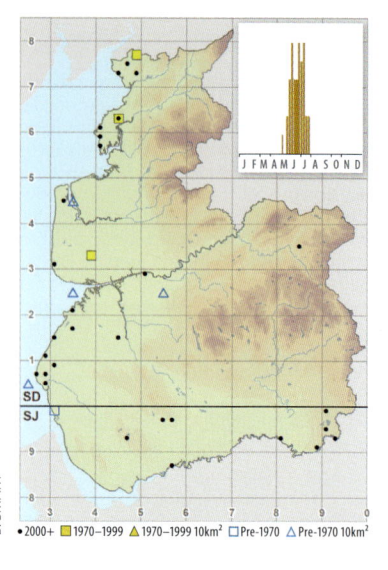

49.290 *Epiblema turbidana*

- **LANCS STATUS** Scarce / local resident   c.1859–2022

First noted from Stretford and Agecroft by J. Chappell around 1859, and in 1879 at Dutton. Described as local (Ellis 1890). Recorded just four times during the 20th century, at Heaton Moor, Burnley, Oldham and at Inskip, where a moth was netted around Butterbur on 17 June 1997 by D. & J. Steeden. Next, recorded at Chorlton in late afternoon of 21 June 2015. Further moths noted from Hoghton in 2016 (G. Dixon) and Bolton Green in 2022. Possibly under-recorded as rarely to light.
- **HABITAT** Damp meadows.
- **FOODPLANTS** Not recorded in Lancashire.
Single brooded. Only once at light. More often recorded in day, around Butterbur.

● 2000+ ☐ 1970–1999 △ 1970–1999 10km^2 ☐ Pre-1970 △ Pre-1970 10km^2

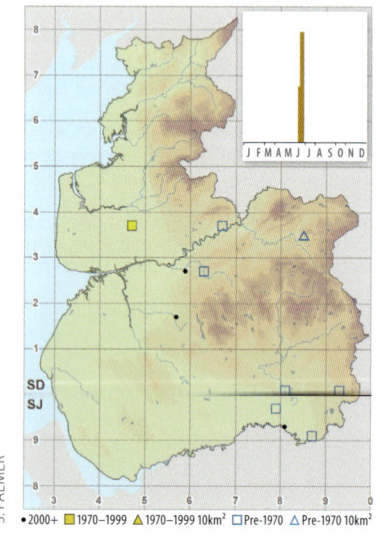

49.292 *Notocelia cynosbatella*

- **LANCS STATUS** Frequent / widespread resident   c.1857–2022

First reported as common on roses in Liverpool gardens (Gregson, 1857) and later, as common and generally distributed (Ellis, 1890). Noted widely across the county since the 1980s and regularly at light in gardens at Hale, Heysham, Clitheroe, Flixton, Swinton, Hoghton, Rishton, Lightfoot Green and many more. Occasionally as late as August, e.g., from Formby 17 August 2012. Possibly declining, with only 33 records of 39 moths from 2022, compared to 66 of 84 ten years earlier, in 2012.
- **HABITAT** Woodland, gardens, hedges.
- **FOODPLANTS** Rose.
Single brooded. Readily to light. Larvae recorded in April in shoots.

● 2000+ ☐ 1970–1999 △ 1970–1999 10km^2 ☐ Pre-1970 △ Pre-1970 10km^2

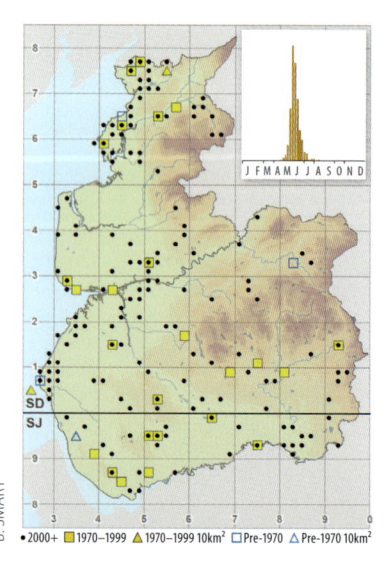

49.293 *Notocelia tetragonana*

■ **LANCS STATUS** Rare / very local resident   c.1888–2015

Recorded from Salwick and Brockholes Wood, nr. Preston by J.B. Hodgkinson (Ellis, 1890) with the author describing the species as very local. Next, from Silverdale, 1921 by W. Mansbridge (1940). Subsequently, remained unrecorded until 2 September 2014 (an unusually late date) from Gait Barrows by R. Petley-Jones, with identity confirmed by P.H. Sterling from the photo. Noted again at the same site the following year by R. Hemming, A. Simpson *et al.* on 29 June but has not been recorded since.
■ **HABITAT** Limestone woodland edges.
■ **FOODPLANTS** Unrecorded within the county.
Single brooded. Both recent records were of moths to light.

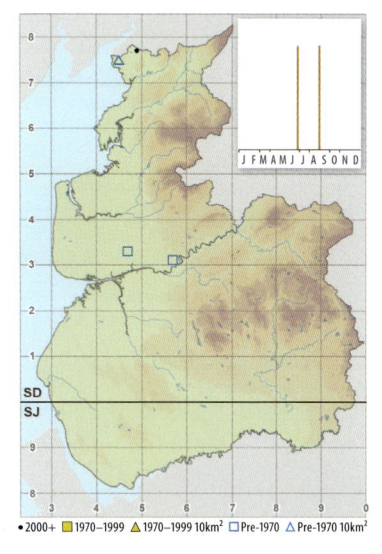

49.294 Bramble Shoot Moth *Notocelia uddmanniana*

■ **LANCS STATUS** Abundant / widespread resident   1844–2022

First recorded from Cheetham Hill on 15 July 1844 (Edleston, 1844) and Liverpool (Gregson, 1857). Later noted as common throughout (Mansbridge, 1940). With over 2500 records, found in all habitats throughout the county, its success no doubt linked to its ubiquitous foodplant. However, there are signs that whilst numbers of records remain high, moth counts have declined, with most recent records of singletons.
■ **HABITAT** Bramble patches in scrub, rough grassland, woodland edges, gardens etc.
■ **FOODPLANTS** Bramble.
There is a single extended generation. Comes readily to light. High count of 50 to light at Formby Point in 2000. Larvae in spun bramble leaves from April to mid-June.

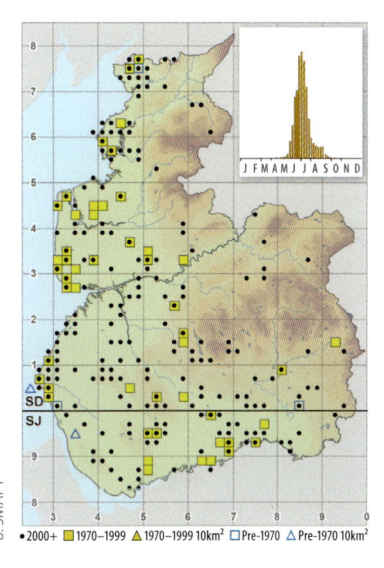

49.295 *Notocelia roborana*

■ **LANCS STATUS** Occasional / local resident   c.1857–2022

Reported from Liverpool district 'where roses grow' (Gregson, 1857). Described as generally distributed and tolerably common (Eliis, 1890), but by Mansbridge as rare (1940). Next, from Morecambe in 1960, Silverdale 1972 and St Annes 1988. Absent from much of east Lancs and infrequent in the south, e.g., just six Flixton records. More regular in the north, at Sunderland, Heysham, Mill Houses, etc., but declining throughout. Only 17 records from 2015–2022, with 69 in the eight years previous.
■ **HABITAT** Hedgerows, coastal scrub. Occasional on moorland.
■ **FOODPLANTS** Burnet Rose.
Single brooded. Comes to light. Larvae feeding in shoots; recorded late April to May.

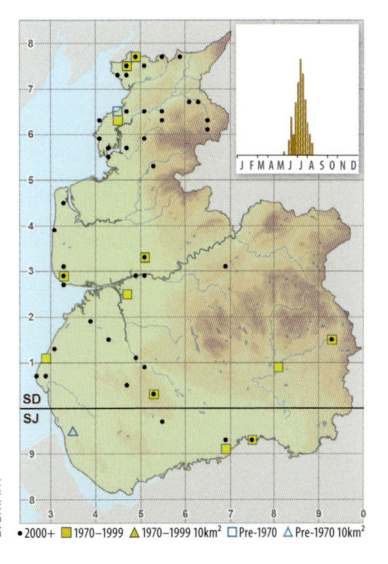

The Moths of Lancashire   **307**

TORTRICIDAE

49.296 *Notocelia incarnatana*

■ **LANCS STATUS** Scarce / very local resident   1909–2010

Bred from larvae found at Silverdale on 26 June 1909 (Mansbridge, 1914). The moth was later noted from Crosby and described as 'locally common' (Mansbridge, 1940), although it was not recorded again until 13 July 1995 when noted at Lytham St Annes by S. Palmer. The next records were from Gait Barrows in 2001, 2002 and 2006, all during June. The only other records are from Keer Falls, nr. Burton-in-Kendal on 26 July 2003 (J. Roberts) and at Bay Horse, nr. Dolphinholme on 18 and 20 July 2010 (N.A.J. Rogers). All counts of moths, where documented, have been of singletons. There have been no further records since 2010. Recorded in all the neighbouring counties, with 39 records from Cheshire, including 15 in the last 20 years, all from the Wirral (cheshire-chart-maps.co.uk, accessed 1 September 2023). Last recorded in Yorkshire during 2020 (yorkshiremoths.co.uk, accessed 1 September 2023).

■ **HABITAT** Open woodland, limestone areas, dune scrub.

■ **FOODPLANTS** Reared in 1909. Foodplant not documented but presumably rose. Single brooded. Most recent records have been of moths to light.

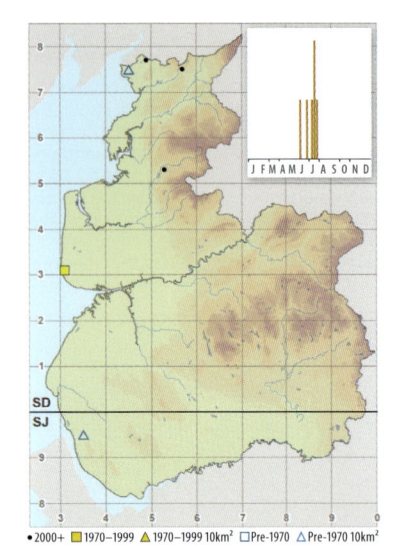

49.297 *Notocelia rosaecolana*

■ **LANCS STATUS** Frequent / widespread resident   c.1859–2022

Recorded from Irlam and Barlow Moor, Manchester by J. Chappell, with status described as local (Ellis, 1890) and later as 'a very common species' (Mansbridge, 1940). Recorded regularly since, although not at all from 1960 to 1986. Over 800 records divided equally between VC59 and VC60. Frequent in gardens, with moths bred from garden roses at Chorlton in 2006 and 2022. Counts appear stable in upland and lowland areas. Highest count is of ten, from Briercliffe 2020 (G. Turner).

■ **HABITAT** Open woodland, hedgerows, gardens.

■ **FOODPLANTS** Rose.

Single brooded. Virtually all modern adult records are at light. Larvae in April.

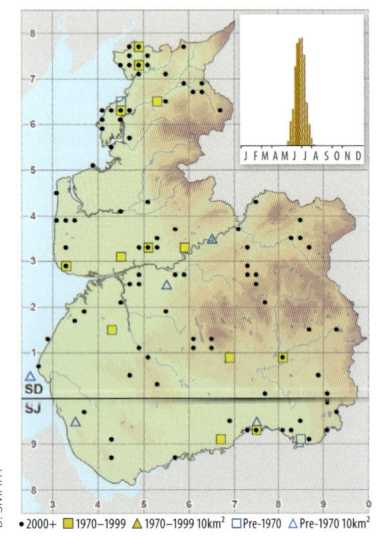

49.298 *Notocelia trimaculana*

■ **LANCS STATUS** Frequent / widespread resident   c.1857–2022

First recorded from the Liverpool district (Gregson, 1857). Ellis (1890) and Mansbridge (1940) noted the moth as common throughout, although no sites were mentioned. Not noted again until Caton in 1992, with records from Hornby, Gait Barrows and Flixton in 1997, the latter site proving a hot-spot, providing over 400 records, about one third of the total. Since noted from most of our lowland and upland habitats, with nine of the twelve counts above ten from the last five years.

■ **HABITAT** Open woodland, hedgerows, gardens.

■ **FOODPLANTS** Hawthorn.

Single brooded. Comes readily to light. Larvae recorded in April and early May.

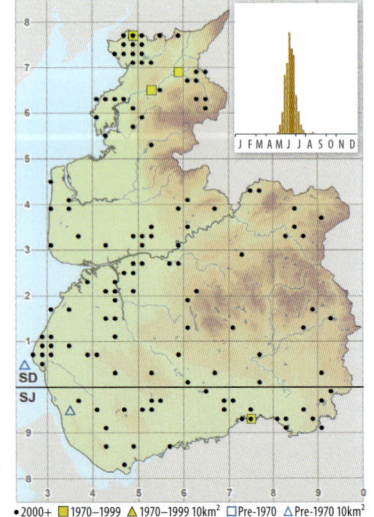

49.299 *Pseudococcyx posticana*

■ **LANCS STATUS** Rare / very local resident   1918–2010
The first record, taken from the Burnley area on 1 June 1918 by A.E. Wright, was initially thought to be *Clavigesta purdeyi*, with identity corrected in 1937 (Mansbridge, 1939). The only other record was of a moth netted during the day at Freshfield by G. Jones and S. Palmer on 21 May 2010. This moth, almost equally as scarce in neighbouring counties, could potentially be an under-recorded species with little attraction to light. Looking for the early stages on pine may be more productive.
Very local in neighbouring counties with two 21st century Yorkshire records and three from Cheshire during the same period, all of the latter from Alsager.
■ **HABITAT** Pine woodland.
■ **FOODPLANTS** Not recorded in Lancashire.
Nationally, single brooded, although unable to confirm locally from the two records.

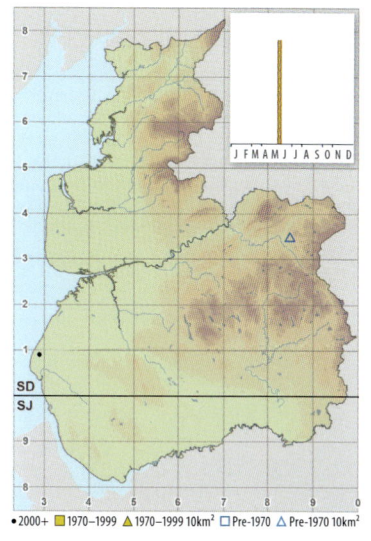

49.300 Pine Bud Moth *Pseudococcyx turionella*

■ **LANCS STATUS** Rare / Very local resident   1951–2018
There are only four records of this species, all from the Sefton Coast. The moth was first noted on 11 June 1951 at Formby by H.N. Michaelis (Michaelis, 1953a). The other three records all come from the last decade, suggesting the moth may be increasing in abundance. The second record came from National Trust land at Formby on 5 June 2016 by R. Walker. The third, on 12 June 2018, came from a Formby garden, again trapped by R. Walker. A fourth was on 27 June 2018 by R. Moyes, a little further north in Southport. The moth is a rare and very local resident of Yorkshire where it was last recorded in 2020 (yorkshiremoths.co.uk, accessed 1 September 2023). Also recorded from Cumbria. There have been no further records from Cheshire since reported from Rudheath (Ellis, 1890).
■ **HABITAT** Pine woodland.
■ **FOODPLANTS** Not recorded in Lancashire.
Single brooded. The three recent records are of moths to light.

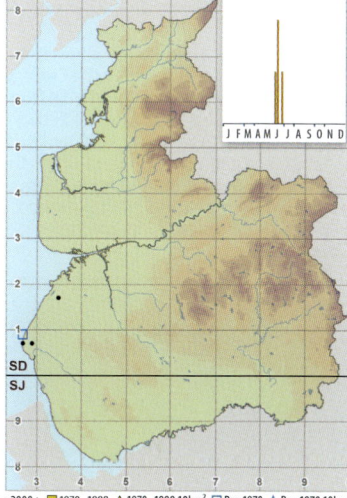

49.301 Pine Resin-gall Moth *Retinia resinella*

■ **LANCS STATUS** Rare / very local resident   2022
The Sefton Coast is well known for its extensive sand dunes, but inland and parallel to the coast are 260ha of pine trees, planted between 1890 and 1930 as a 'cash crop'. Following the first big storm of 2022, over 120 pines were lost on Formby National Trust lands. On 24 April 2022, R. Walker was inspecting the fallen pines and noted a gall containing a pupal exuvia among the pine needles. This was cut off and passed to S. Palmer, and further examined by M. Young, who was able to confirm the identification as this species.
■ **HABITAT** Pine woodland.
■ **FOODPLANTS** Pine.

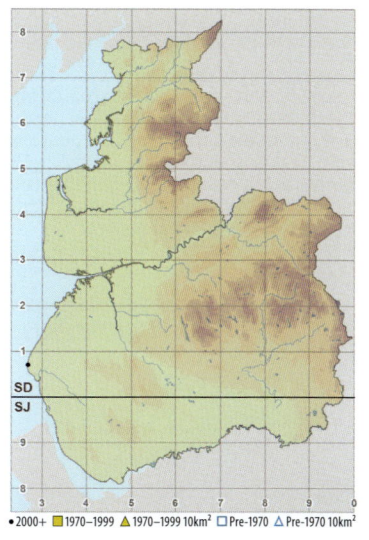

TORTRICIDAE

49.304 Pine Leaf-mining Moth *Clavigesta purdeyi*

■ **LANCS STATUS** Occasional / widespread resident 1950–2022

In view of the 1918 misidentification noted under *P. posticana*, the first record was on 31 July 1950 at Formby by S. Charlson. Spread has been rapid, with the first inland record from Didsbury in 1963 by H.N. Michaelis. Recorded on 145 occasions including some upland locations such as Shaw, Higher Tatham, Briercliffe. Moths reared by B. Smart from three larvae noted descending from pine on silken threads at Ainsdale 22 June 2021, and from two larvae beaten from pine nearby in June 2022.
■ **HABITAT** Pine woodland, parks, gardens.
■ **FOODPLANTS** Pine.
Single brooded. Comes to light. Moths and larvae beaten from pine.

● 2000+   ☐ 1970–1999   △ 1970–1999 10km²   ☐ Pre-1970   △ Pre-1970 10km²

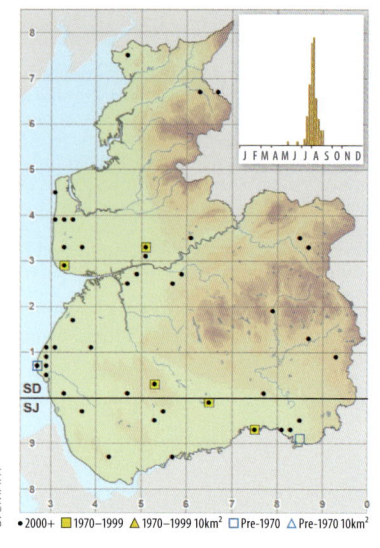

49.305 Pine Shoot Moth *Rhyacionia buoliana*

■ **LANCS STATUS** Occasional / local resident c.1857–2022

Half of the 64 records are from the Fylde and Sefton Coasts, with the remainder found inland, presumably using pine from gardens, parks and plantations. Reported from firs in the Liverpool district (Gregson, 1857). Later, noted as larvae on Corsican Pine at Avenham, Preston in 1872 by J.B. Hodgkinson. From Chat Moss and Formby in 1880s. 21st century records, of which there are 53, include moths at garden traps in Swinton, Bispham, Flixton, Hale, Billinge, St Helens, Heysham, Longton, etc.
■ **HABITAT** Pine woodland, parks, gardens.
■ **FOODPLANTS** Corsican Pine. Bred from larva in pine shoot, Freshfield 2022.
Single brooded. Almost all records are of moths at light. Larvae recorded in June.

● 2000+   ☐ 1970–1999   △ 1970–1999 10km²   ☐ Pre-1970   △ Pre-1970 10km²

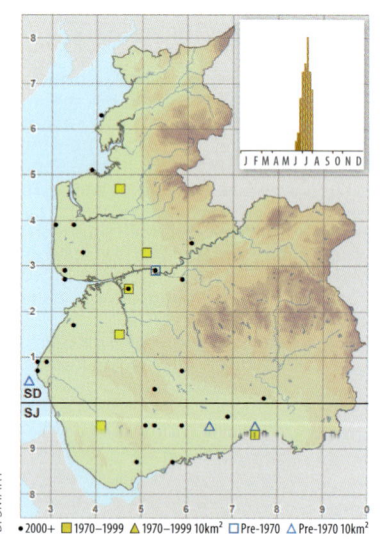

49.306 *Rhyacionia pinicolana*

■ **LANCS STATUS** Occasional / local resident c.1857–2022

The 62 records show a definite coastal bias, with inland populations restricted to the south of the county and the far north, including Docker Moor and Dalton Crags. First reported on 'firs' in the Liverpool district (Gregson, 1857). Next, recorded from Chat Moss and Formby in the late 1880s. Mansbridge added Sefton Park and Simonswood Moss to the known sites (1940). Recorded 50 times this century, with high counts of six noted at Formby in 2013 and 2015 by R. Walker.
■ **HABITAT** Pine woodland, gardens.
■ **FOODPLANTS** Pine. Bred from larva at Formby by R. Pritchard in 1953.
Single brooded. Comes to light. Less frequent in garden traps than *R. buoliana*.

● 2000+   ☐ 1970–1999   △ 1970–1999 10km²   ☐ Pre-1970   △ Pre-1970 10km²

The Moths of Lancashire

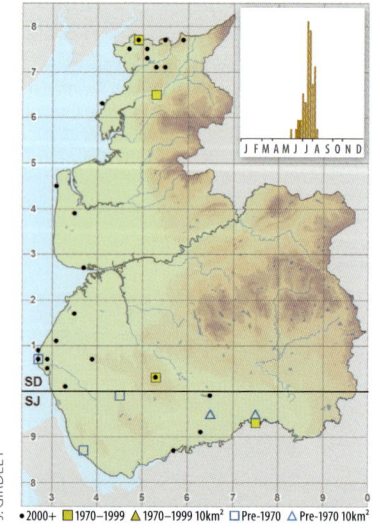

49.307 Spotted Shoot Moth *Rhyacionia pinivorana*

- **LANCS STATUS** Occasional / widespread resident   1844–2022

First recorded from Chat Moss on 2 June 1844 (Edleston, 1844), and from Silverdale in 1909 by W. Mansbridge. Noted as abundant by Mansbridge (1940), and as 'present in small numbers on pine at Freshfield and Formby' (Leech and Michaelis, 1957). Range and frequency increased since 2000 with 54 records, compared with just four in twentieth century. The post-2000 records include many from gardens away from pine woods and plantations, suggesting use of isolated pine trees.
- **HABITAT** Pine woodland, gardens.
- **FOODPLANTS** Pine.

Single brooded. Comes to light and netted at dusk.

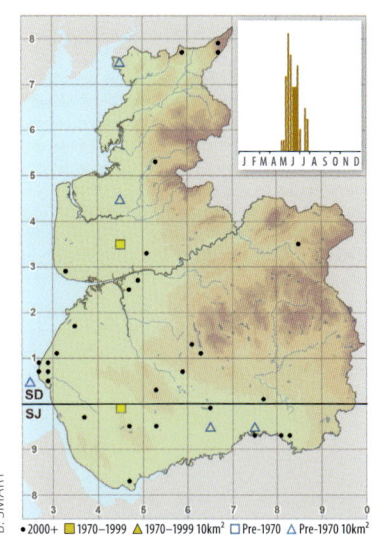

49.309 *Dichrorampha plumbana*

- **LANCS STATUS** Occasional / widespread resident   c.1910–2022

First recorded in early 20th century from Formby, Liverpool and Silverdale, and described as 'fairly common' (Mansbridge, 1940). Netted around Tansy from Flixton in 2000 and Rixton in 2001. Two moths, subsequently dissected, bred from Tansy roots by K. McCabe in Flixton 2012, a previously unrecorded foodplant. Nationally, known to use Yarrow and Oxeye Daisy. Mainly from lowland areas with occasional upland record, e.g., swept from Yarrow at Wray-with-Botton in 2022 by S. Palmer.
- **HABITAT** Rough grassland, scrub.
- **FOODPLANTS** Tansy.

Single brooded. Netted around foodplants. Only one of the 55 records at light.

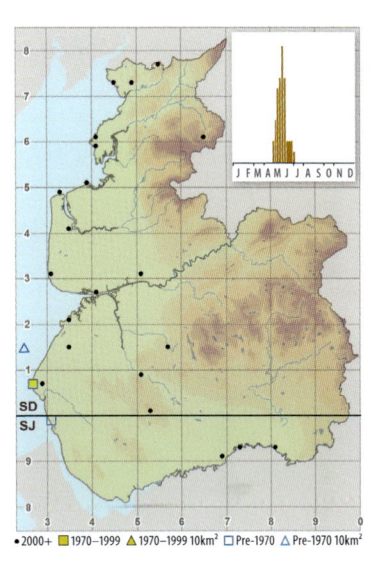

49.310 *Dichrorampha sedatana*

- **LANCS STATUS** Scarce / very local resident   1908–2014

First reported in 1908 on a garden wall by Sefton Park (Mansbridge, 1909). However, when writing of the species three decades later, the same author described the moth as scarce and known only from Lancaster (Mansbridge, 1940). Next recorded in 1999 at Flixton, and on 15 more occasions up to 2010, with up to 30 moths flying around or resting on Tansy, all recorded by K. McCabe. The only subsequent record was of a moth netted in the field on 16 May 2014 at Marshside by K. McCabe and S. Palmer.
- **HABITAT** Rough grassland, parks, gardens.
- **FOODPLANTS** Not recorded in Lancashire. Single brooded.

This is one of several Dichrorampha species requiring dissection to confirm identity.

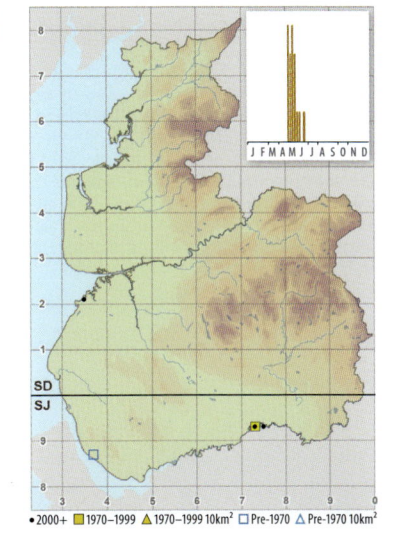

TORTRICIDAE

The Moths of Lancashire   311

TORTRICIDAE

49.311 Dichrorampha aeratana

■ **LANCS STATUS** Occasional / local resident   1909–2022

Early records were exclusively from Silverdale, with the first, by W. Mansbridge, in 1909, and the Sefton Coast, where noted at Crosby and Formby (Mansbridge, 1940). The next location noted was Great Plumpton, where a moth was recorded during June 1984 by M. Evans. Most subsequent records are from VC60, including Scorton, Warton Crag, Leighton Moss, Heysham N.R., etc., with VC59 records including Flixton, Longton, Bretherton and Ashton Moss. Dissection required to confirm identity.
■ **HABITAT** Rough grassland.
■ **FOODPLANTS** Unrecorded within the county.
Single brooded. Mainly recorded during the day. At light on seven occasions.

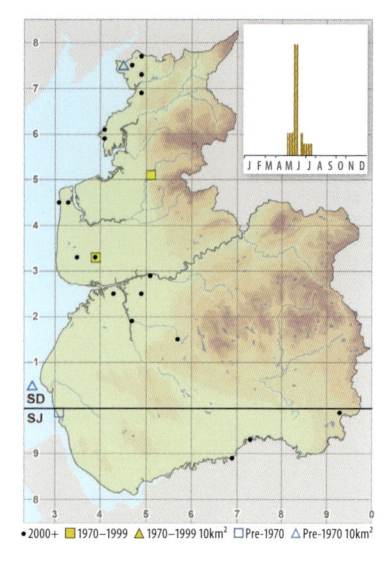

• 2000+   ◻ 1970–1999   △ 1970–1999 10km²   ◻ Pre-1970   △ Pre-1970 10km²

49.313 Dichrorampha acuminatana

■ **LANCS STATUS** Occasional / widespread resident   c.1910–2022

Reported from Crosby, Formby, Ainsdale and Silverdale in the early 20th century. Mansbridge considered the moth to be 'not uncommon but probably overlooked' (1940). Next recorded at Lytham St. Annes by D. & J. Steeden on 29 July 1995. 21st century records have shown the moth to be far more widespread, but with a bias towards coastal locations and southern parts of the county, such as the Mersey Valley. Dissection required to confirm identity.
■ **HABITAT** Dry grassland.
■ **FOODPLANTS** Unrecorded within the county.
Double brooded. Frequent at light, and in day around Tansy and Oxeye Daisy.

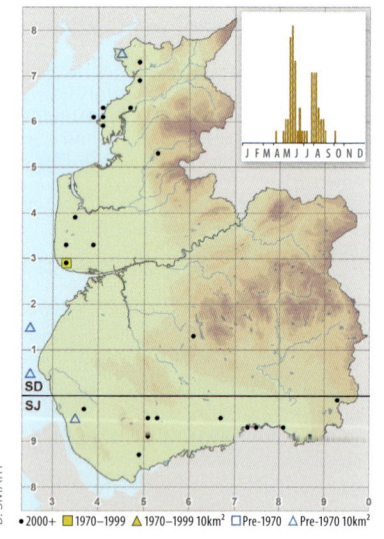

• 2000+   ◻ 1970–1999   △ 1970–1999 10km²   ◻ Pre-1970   △ Pre-1970 10km²

49.315 Dichrorampha simpliciana

■ **LANCS STATUS** Occasional / local resident   c.1910–2022

First confirmed from Wavertree, Simonswood Moss and Sales Wood, with the moth described as not uncommon (Mansbridge, 1940). It was 1996 before the next record, to light at St Annes Moss by D. & J. Steeden. 65 of the 97 records are from Flixton by K. McCabe, 53 of these at the garden light trap. Primarily a lowland species with only five VC60 records, including Sunderland Point, Poulton-le-Fylde and Lightfoot Green.
■ **HABITAT** Rough grassland, brownfield sites.
■ **FOODPLANTS** Unrecorded within the county.
Single brooded. Comes to light. Noted flying around Mugwort in good numbers, with high count of 50 at Flixton, 31 August 2010. Dissection required to confirm identity.

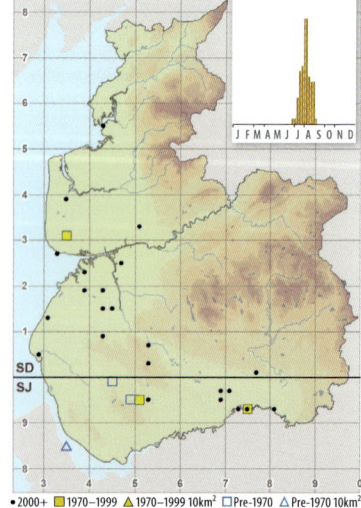

• 2000+   ◻ 1970–1999   △ 1970–1999 10km²   ◻ Pre-1970   △ Pre-1970 10km²

49.318 *Dichrorampha vancouverana*

■ **LANCS STATUS** Occasional / local resident   1989–2021
First recorded from Warton Crag on 5 July 1989 by M. Dempsey. Noted on a few occasions since 1995 at Great Plumpton, Lytham St Annes and Fairhaven by D. & J. Steeden, with a couple of males dissected. Mainly a coastal species, with further records from Formby, Freshfield, Southport, Marshside and Blackpool. The only inland records are from Flixton and St Helens.
■ **HABITAT** Dry grassland.
■ **FOODPLANTS** Not recorded in Lancashire.
Single brooded. Seven records are of moths at light, with others swept from vegetation, or flying around the national foodplants, Tansy and Yarrow.

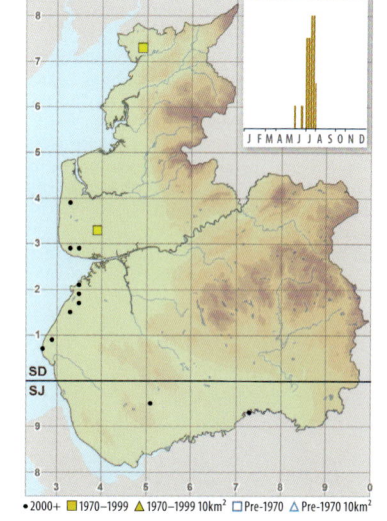

49.319 *Dichrorampha flavidorsana*

■ **LANCS STATUS** Occasional / local resident   2006–2022
Most records, including the first on 17 July 2006, are from Flixton, by K. McCabe. Only four are not, including one disturbed from Tansy at Chorlton in 2022. The other records are from Lytham St Annes in 2013, Heysham Moss in 2017 and by Tewitfield Canal, nr. Carnforth in 2022. The scattered nature of the records, from the north-west and south-east, and apparent reluctance to wander far from the foodplant, Tansy, suggest the species may be under-recorded. Dissection required to confirm identity.
■ **HABITAT** Rough grassland, scrub.
■ **FOODPLANTS** Feeding unrecorded, although moths frequently noted around Tansy.
Single brooded. No indication of attraction to light; all records of moths during day.

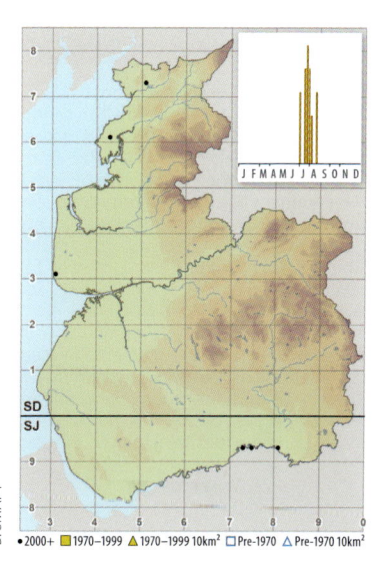

49.320 *Dichrorampha alpinana*

■ **LANCS STATUS** Occasional / widespread resident   c.1934–2022
First recorded at Silverdale by W. Mansbridge, with subsequent dissection by F.N. Pierce in 1934. Mansbridge later described the moth to be uncommon (1940). Next recorded by L.W. Hardwick, at Scorton in 1983, and at Hale in 1994, flying around the foodplant, Oxeye Daisy. Currently, appears to be faring quite well, with nine of the 55 records coming from 2022. The high count of nine is also from 2022 (7 July) by S.H. Hind at Ashton Moss. Likely to be present wherever Oxeye Daisy found.
■ **HABITAT** Rough grassland, gardens.
■ **FOODPLANTS** Not recorded in Lancashire.
Single brooded. To light on 23 occasions. Other records around Oxeye Daisy.

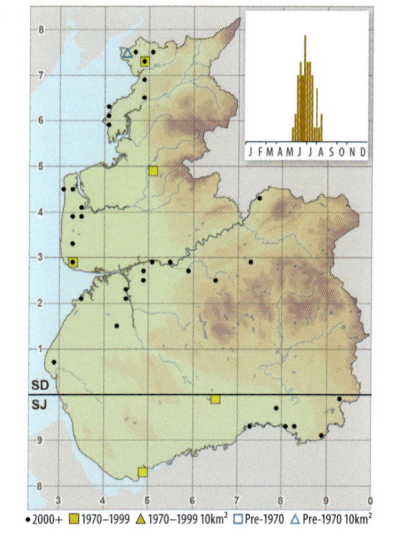

The Moths of Lancashire

49.321 *Dichrorampha petiverella*

■ **LANCS STATUS** Occasional / local resident 1914–2022

First reported from Silverdale by W. Mansbridge (1914), later noting the moth as common and well distributed, and adding a Formby record (1940). No other sites reported until 1988, when L.W. Hardwick noted at Woolston. The 1990s saw further records from Heysham N.R., Galgate, Glasson, Lancaster, Rossall Point and Mere Sands Wood. Bred in 2012 from Yarrow roots, Flixton, by K. McCabe. Population and range appear stable, although largely restricted to lowland, western locations.
■ **HABITAT** Rough grassland.
■ **FOODPLANTS** Yarrow.
Single brooded. Comes to light. Most records are from the day, often around Yarrow.

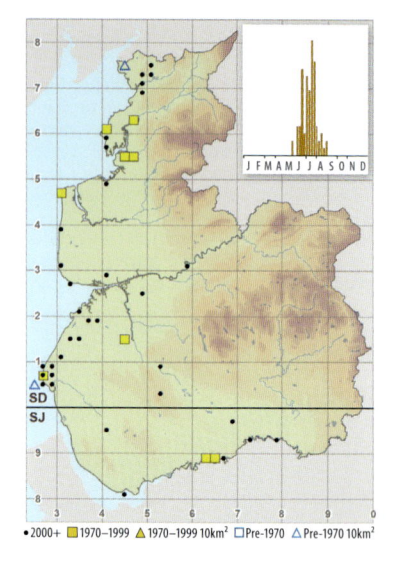

49.322 *Dichrorampha plumbagana*

■ **LANCS STATUS** Considered Extinct 1929

19th century records from Preston cannot be verified, as the moths were undissected or at least there is no record of such. W. Mansbridge later recorded a couple from Formby in 1929 (Mansbridge, 1940). Further clarification of this record came in The Lepidoptera of Formby, by Leech and Michaelis (1957), with the comment that 'the majority of specimens of *Hemimene* (now *Dichrorampha*) in the Mansbridge collection were checked by the late F.N. Pierce who examined the genitalia.' It has not been recorded in the county since. The species is unrecorded from Cheshire, but has been noted on four occasions in Yorkshire, with the most recent in 2020 (yorkshiremoths.co.uk, accessed 1 September 2023).
■ **HABITAT** Dunes.
■ **FOODPLANTS** Unrecorded within the county.
Single brooded, although unable to ascertain this from the few local records. The considerable difficulty in determining the group of very similar *Dichrorampha* species without dissection means that most of these are likely to be under-recorded in the county. This may well be the case with this Yarrow-feeding species, as the plant is widespread across the county.

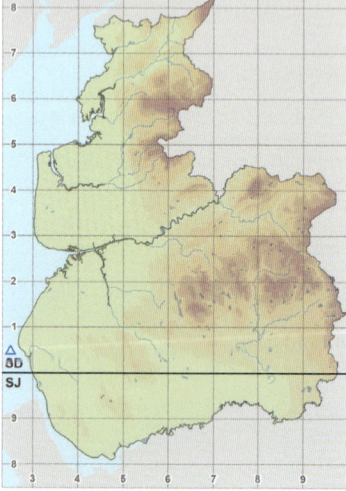

49.323 *Dichrorampha alpestrana*

■ **LANCS STATUS** Occasional / widespread resident 1927–2022

W. Mansbridge was the first to record this moth, noting it at Formby in 1927 (Leech and Michaelis, 1957), and also from Silverdale, Simonswood Moss, and Hale (Mansbridge, 1940), all confirmed by F.N. Pierce. Other dissected records from Marton Mere 1999, Flixton 2001, Ormskirk 2005, and subsequently, from Longton, Adlington, Billinge, Fulwood, Longridge, Thrushgill, Hest Bank, etc. A female, confirmed as this species, was bred in 2012 from Tansy roots, Flixton, by K. McCabe.
■ **HABITAT** Rough grassland, hillsides.
■ **FOODPLANTS** Tansy.
Single brooded. Of 79 records, 51 are of moths at light. Netted in day at foodplants.

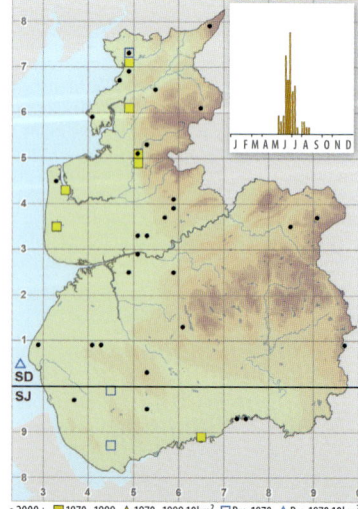

49.324 Pea Moth *Cydia nigricana*

■ **LANCS STATUS** Frequent / widespread resident   c.1888–2022

The first mention is of J.B. Hodgkinson recording the species among peas in the Wyre district (Ellis, 1890). From Ainsdale, Silverdale and Didsbury in the 1940s, Formby in 1984, and Stalmine, Fairhaven, Heysham N.R. and Flixton in the 1990s. These and subsequent records come predominantly from south and west, mainly in lowland areas, although a few on higher ground such as Oswaldtwistle and Rishton.
■ **HABITAT** Grassland, gardens, waste ground.
■ **FOODPLANTS** Pea and vetch pods. Larvae found in pods during July.
Single brooded, with occasional small second brood, e.g., 3 October 2018 at Longton. Noted flying around vetches, including 310 at Flixton in 2002. Infrequent at light.

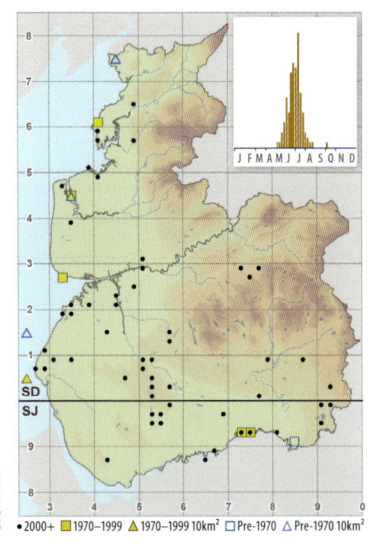

49.325 *Cydia ulicetana*

■ **LANCS STATUS** Frequent / widespread resident   c.1857–2023

Gregson reported the moth from Liverpool (1857), stating likely wherever Gorse occurs, a situation that remains true today. Noted a few times at Formby from early 20th century onwards. Next, from Preesall and Great Plumpton, 1984 by M. Evans. Recorded on over ten occasions each at Warton Crag, Flixton, Freshfield Dune Heath, etc., with upland records from Docker Moor, Longridge Fell, Birk Bank, etc.
■ **HABITAT** Dunes, scrub, moorland.
■ **FOODPLANTS** Gorse. Larvae noted in pods at Ainsdale on 4 July 2023 (B. Smart). Primarily single brooded, but a few later records suggest a small second brood. Moths frequently netted around Gorse. Occasional at light.

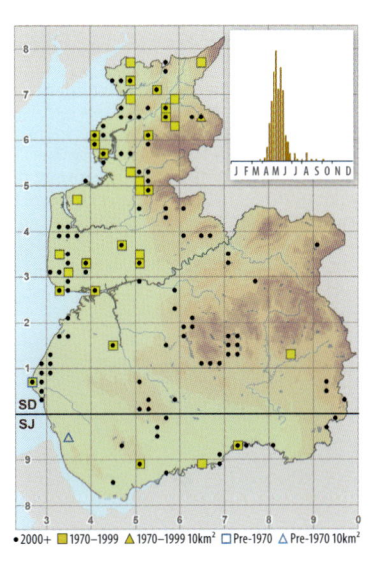

49.332 *Cydia coniferana*

■ **LANCS STATUS** Scarce / very local resident   1844–2020

Recorded from Manchester and Chat Moss by R.S. Edleston (1844; 1846), noting it to be very rare. Recorded in 1880s from Salwick, nr. Kirkham by J.B. Hodgkinson. Later from Formby and Simonswood Moss (Mansbridge, 1940), with the moth described as generally common among pines. Not noted again until 1997, when at Gait Barrows, and from Freshfield in 2007. Recorded 12 times by G. Turner in his Briercliffe garden since 2014. The highly scattered distribution suggests possibly under-recorded.
■ **HABITAT** Pine woodland.
■ **FOODPLANTS** Not recorded in Lancashire.
Single brooded. Attracted to light. Occasionally noted during the day.

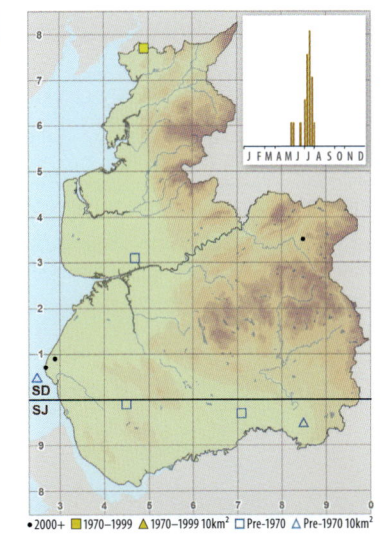

The Moths of Lancashire   **315**

TORTRICIDAE

49.334 *Cydia cosmophorana*

■ **LANCS STATUS** Rare / Very local resident   1914–2021

First noted by Mansbridge (1915) in his *Report on the Lepidoptera for 1914*. Outlining the year's finds, he wrote, 'beginning in the spring, I captured a fine specimen of *Coccyx* [now *Cydia*] *cosmophorana* in Gatebarrow Wood at Silverdale', i.e., at Gait Barrows. Subsequently, unrecorded until 11 June 2021, when a single moth was beaten from pine at Beacon Fell by S. Garland.

■ **HABITAT** Coniferous woodland.
■ **FOODPLANTS** Unrecorded. Elsewhere, associated with resinous nodules on pine bark.

Single brooded. The virtually imperceptible early stages and the habit of flying around tree-tops in early afternoon sunshine may partially explain the lack of records.

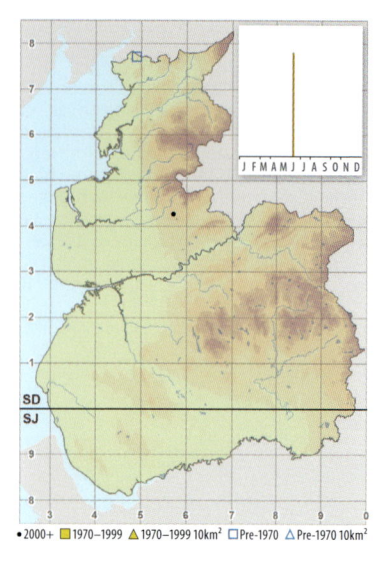

49.335 Spruce Seed Moth *Cydia strobilella*

■ **LANCS STATUS** Rare / very local resident   2016

There is just a single record of this moth; a singleton trapped at light in a Wigan garden by G. Wynn on 9 May 2016. Identity was confirmed from the photograph by C.A. Darbyshire, S. Palmer and R. Walker. The situation in Yorkshire and Cheshire is similar, with a 2015 record in the former (the first since 1916), and records from the latter in 2016 and 2020, the first Cheshire records.

■ **HABITAT** Unknown in the county.
■ **FOODPLANTS** Not recorded in Lancashire.

Single brooded. Collecting spruce cones after winter storms may yield the species.

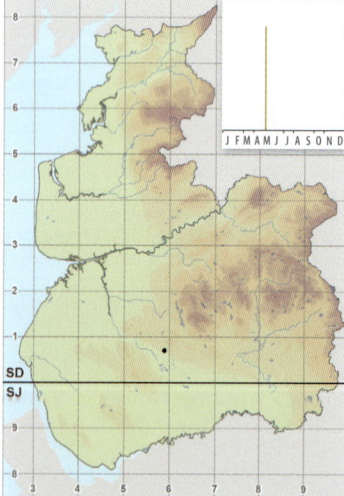

49.338 Codling Moth *Cydia pomonella*

■ **LANCS STATUS** Frequent / widespread resident   c.1859–2022

First recorded in Manchester by J. Chappell, followed by Silverdale and Preston (Ellis, 1890). Other pre-1940 records from Wavertree, Formby and Bolton. Reported as a common pest on apples by Mansbridge (1940). Thereafter, the moth has been recorded regularly at many sites across the county, including approx. 150 at a pheromone lure at Hutton, nr. Longton in June 2018 by A. Barker.

■ **HABITAT** Woodland, orchards, gardens, hedgerows, allotments.
■ **FOODPLANTS** Apple, Crab Apple.

An extended single generation. Readily to light, including 33 in Ormskirk 2016 by J. Watt. Larvae feed within apples and have been recorded from August to October.

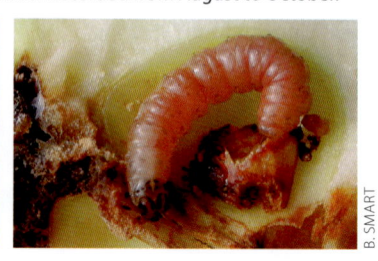

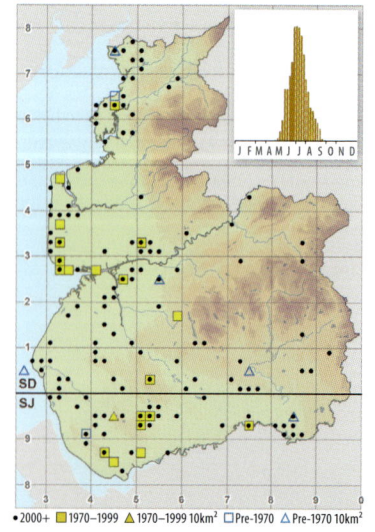

316   *The Moths of Lancashire*

49.341 *Cydia splendana*

■ **LANCS STATUS** Frequent / widespread resident   1905–2022
Recorded from Liverpool in 1905 by W. Mansbridge, from Lancaster the following year by C.H. Forsythe, and later from Childwall by Mansbridge, who noted the moth to be local in oak woods (1940). Records since 1990 have shown this moth to be widespread in lowland and upland areas with over 1,000 records. Comes readily to light, with high count of 20 from a Silverdale garden in August 2021 by J. Patton.
■ **HABITAT** Oak woodland.
■ **FOODPLANTS** Oak, feeding in acorns.
Single brooded. Bred from acorn collected at Risley Moss 10 September 2020 by K. McCabe and B. Smart, not emerging until 9 July 2022, nearly two years later.

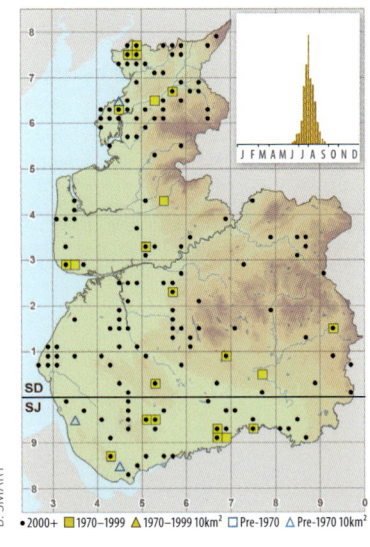

49.342 *Cydia fagiglandana*

■ **LANCS STATUS** Occasional / widespread resident   1999–2022
A recent addition to the county's fauna, first recorded from Pennington, Wigan by P. Pugh on 28 May 1999, followed by Gait Barrows in July 1999 and Lytham St. Annes in 2000. Recorded at light from Fulwood in 2006 by A. Powell, with over 90 subsequent records from Fulwood, over half the total county records. The majority of records are from lowland areas, although has also been noted on higher ground at Docker Moor, Clitheroe, Briercliffe, Rishton and Blackburn.
■ **HABITAT** Beech woodland.
■ **FOODPLANTS** Not recorded in Lancashire.
Single brooded. Comes readily to light, accounting for all 174 Lancashire records.

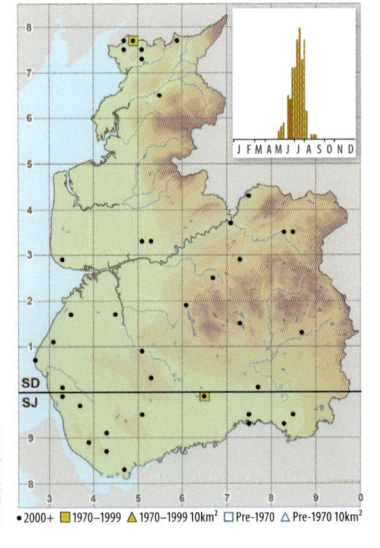

49.345 *Lathronympha strigana*

■ **LANCS STATUS** Frequent / widespread resident   c.1888–2022
First reported from Whitewell by J.H. Threlfall (Ellis, 1890) and from Silverdale on 4 July 1914 (Mansbridge, 1914). Not recorded again until 1983 at Scorton by L.W. Hardwick, with further 20th century records from Red Scar Wood, Heysham N.R., Flixton, Woolston Eyes, etc. Numbers of records greatly increased this century. Fairly widespread, with high count of 20 at Great Moss, nr. Wigan, 2004 by C.A. Darbyshire.
■ **HABITAT** Grassland, verges, woodland edges, gardens.
■ **FOODPLANTS** St John's-worts, including Perforated St John's-wort.
Single brooded, with occasional second generation. Comes to light. Larvae feed within spun shoots and recorded from April to June.

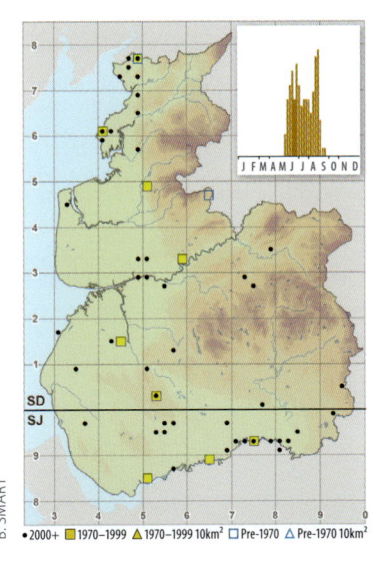

The Moths of Lancashire   **317**

49.347 *Grapholita compositella*

■ **LANCS STATUS** Frequent / widespread resident 1906–2022
Recorded at Standish 1906 by F.N. Pierce, this was the only record until 1984, when noted at Formby by L.W. Hardwick. With over 270 subsequent records, appears to have increased in range and abundance. Mainly a lowland species with over 100 records from the Mersey Valley. Emerged 16 August 2010 from White Clover heads collected one month earlier at Chorlton. High count of 40 from Heysham N.R. 2003.
■ **HABITAT** Rough grassland, meadows.
■ **FOODPLANTS** Red Clover, White Clover.
Double brooded, although first generation more numerous. Recorded around clover during day. Infrequent at light. Larvae on flower-heads, recorded in July and August.

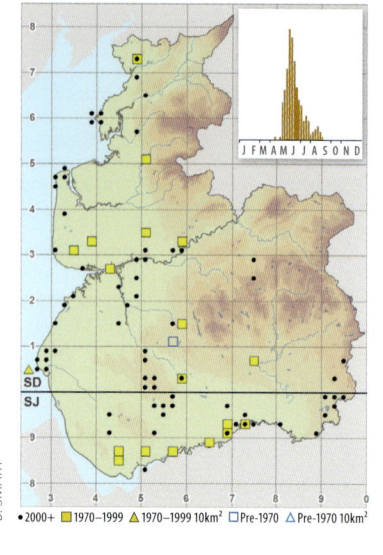

49.349 *Grapholita internana*

■ **LANCS STATUS** Occasional / local resident 1987–2019
M. Dempsey noted the moth at Gait Barrows in 1987, the only record prior to the 21st century. Next recorded at Hurst Green by D. Lambert on 21 May 2008, followed by records from Warton Crag 2009, Flixton 2010 and Birk Bank 2011. Noted at Docker Moor near gorse in 2015, with the most recent records from Warton Crag (a high count of ten moths) by B. Hancock, and Ash Hill, Flixton by K. McCabe, both in 2019.
■ **HABITAT** Scrub, moorland.
■ **FOODPLANTS** Not recorded in Lancashire.
Single brooded. Usually found flying around gorse bushes. Unrecorded at light.

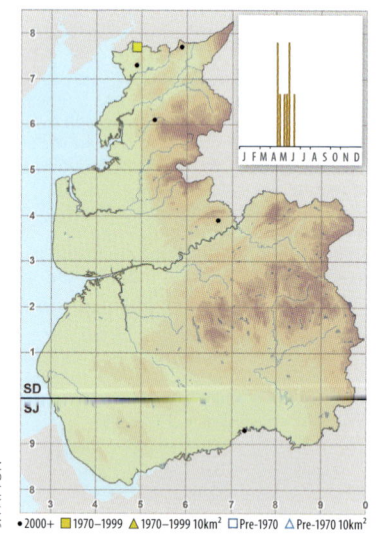

49.351 *Grapholita lunulana*

■ **LANCS STATUS** Occasional / local resident 1991–2022
First noted in the Blackpool area, from Marton 1991 and Squires Gate 1997, by D. & J. Steeden. Next recorded at Wigan Flashes in 2000 by C.A. Darbyshire, and at Parbold, Chorlton, Red Scar Wood and Lytham St Annes in 2002. Most records are from the coast and the south, although with a few from higher ground around Oldham, Bolton and Clitheroe. High count of 38 at Bispham Marsh on 17 May 2016 by B. Brigden.
■ **HABITAT** Rough grassland, brownfield sites.
■ **FOODPLANTS** Vetch pods.
Single brooded. Noted flying around foodplants from early morning to early evening. Very rarely to light. Larvae found during July in pods of unidentified vetch.

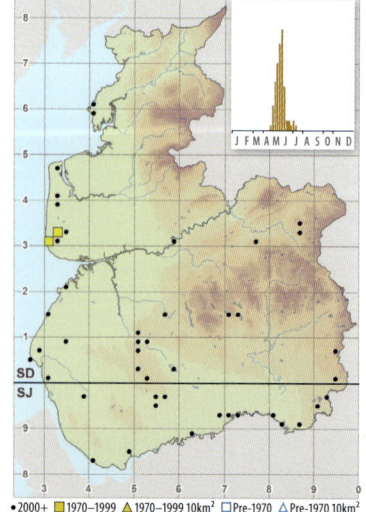

49.354 *Grapholita jungiella*

- **LANCS STATUS** Frequent / widespread resident   c.1920–2022

First recorded by C.H. Forsythe from Lancaster (Mansbridge, 1940). Noted at Ainsdale, with 'a few among *Vicia* in June 1949,' and larvae 'in pods of their food' (Leech and Michaelis, 1957). Further records from Didsbury 1964, Formby 1991, Slyne 1992, Silverdale 1995, Flixton 1997, over 160 times at latter site. Combined with Stretford, Chorlton and Didsbury records, the Mersey Valley is a stronghold for the species. Also from higher ground, such as Oswaldtwistle 2013 (Mark Memory).
- **HABITAT** Grassland, scrub, woodland edges, brownfield sites.
- **FOODPLANTS** Vetch pods.

Single brooded. Most records are of day-flying moths. Occasionally to light.

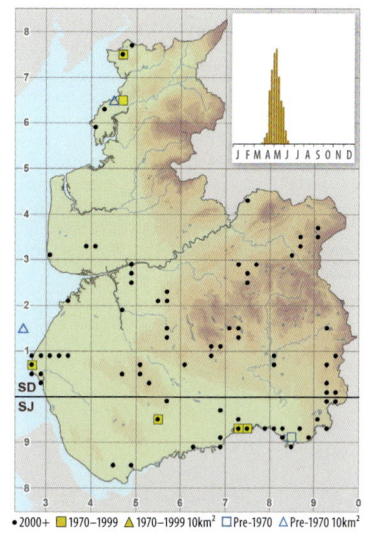

49.3541 Oriental Fruit Moth *Grapholita molesta*

- **LANCS STATUS** Adventive   2005

This moth, a native of China, occurs occasionally in the UK, usually associated with importation of fruit, such as peaches and nectarines. It has only been recorded on a single occasion in Lancashire, by B. Smart on 14 July 2005, when three moths emerged over a few days indoors in Chorlton, Manchester. All appeared in the dining room, close to a fruit bowl containing nectarines. A pot of Snapdragons was also nearby but was felt to be a far less likely source of the moths.
- **HABITAT** Unknown.
- **FOODPLANTS** Probably emerged from nectarines.

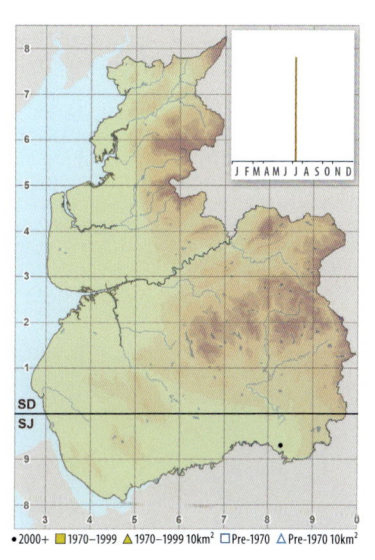

49.355 *Grapholita lathyrana*

- **LANCS STATUS** Considered Extinct   1879–1880

This species was recorded on two occasions by J.B. Hodgkinson at Dutton, nr. Ribchester. The first was in 1879 when larvae were collected, with adults emerging in April and May 1880. Unfortunately, the finder could not be certain of the foodplant (Hodgkinson, 1880d). The second was of larvae found in unexpanded flower-buds of Dyer's Greenweed on 8 July 1880. The finder noted, 'there were holes eaten at the base of the flowers, and snugly inside was another larva which had drawn a leaf up to the stem for shelter – they don't seem at all to eat the leaves' (Hodgkinson, 1880e). It is unclear whether the larvae were subsequently reared. The species is unrecorded in Cumbria, Cheshire and Yorkshire.
- **HABITAT** Grassland.
- **FOODPLANTS** Dyer's Greenweed, now a very scarce plant within the county (Greenwood, pers. comm.).

Single brooded, although unable to ascertain this from the two local records.

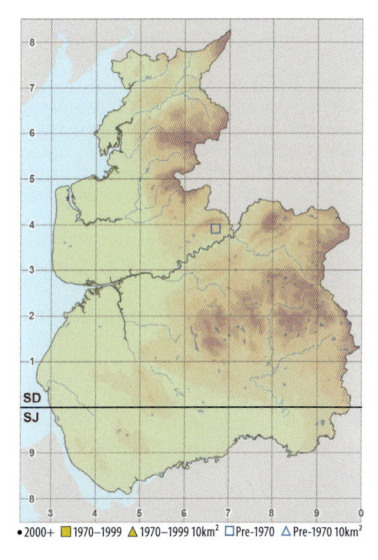

TORTRICIDAE

The Moths of Lancashire   319

TORTRICIDAE

49.357 Plum Fruit Moth *Grapholita funebrana*

■ **LANCS STATUS** Occasional / widespread resident c.1888–2022
Reported from Preston by J.B. Hodgkinson (Ellis, 1890). Subsequently, unrecorded until noted at Bay Horse by N.A.J. Rogers in 2007, followed by records from Hutton, Lightfoot Green, Flixton, Mere Sands Wood, Dalton Crags, etc., as part of northwards range expansion responsible for similar records in Cheshire and Yorkshire.
■ **HABITAT** Woodland, gardens, hedgerows.
■ **FOODPLANTS** Plum, Blackthorn; feeding in the fruit from late July to September. Single brooded, but with an extended flight period. Occasional at light, and to pheromone lures. The latter was responsible for the high count of 72 at Hunts Cross in 2018 by S. McWilliam.

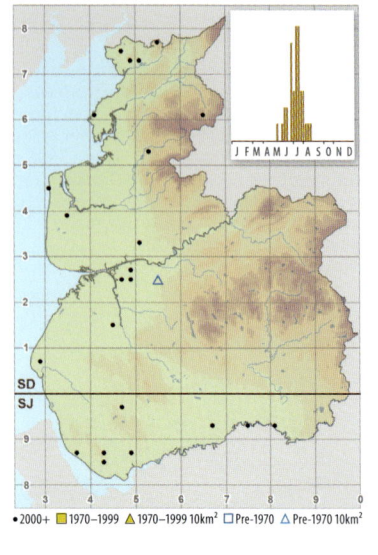

• 2000+   □ 1970–1999   ▲ 1970–1999 10km²   □ Pre-1970   △ Pre-1970 10km²

49.358 *Grapholita tenebrosana*

■ **LANCS STATUS** Occasional / local resident c.1888–2022
Recorded from Cleveleys by J.H. Threlfall and the Wyre district by J.B. Hodgkinson (Ellis, 1890), and from Formby, Ainsdale and Silverdale in early 1960s. Bred in numbers from rose hips at Formby in 1960 by H.N. Michaelis. 21st century records from Wray, Hoghton, Longton, Chorlton, Flixton, Freshfield, Formby and Oldham. Wider use of pheromones would likely confirm moth to be under-recorded.
■ **HABITAT** Woodland edges, hedgerows, gardens.
■ **FOODPLANTS** Roses, including Dog Rose, with larvae noted in September and October. Single brooded. Moths occasional at light. Attracted to pheromone lures for *G. funebrana* and *Pammene argyrana* at Chorlton, 2022.

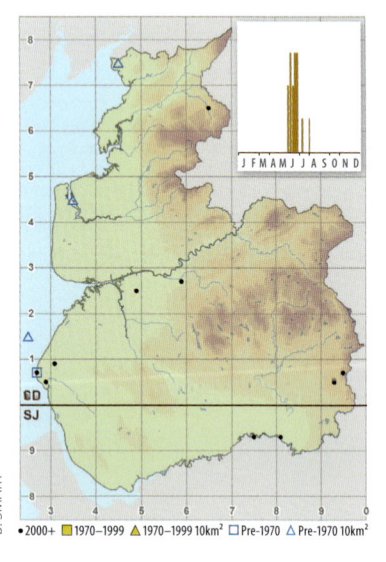

• 2000+   □ 1970–1999   ▲ 1970–1999 10km²   □ Pre-1970   △ Pre-1970 10km²

49.359 *Grapholita janthinana*

■ **LANCS STATUS** Frequent / widespread resident c.1888–2022
First noted at Fleetwood by J.B. Hodgkinson, and described as 'very local' (Ellis, 1890). Bred from larvae at Formby (Mansbridge, 1940). Further reported from Crosby 1976, Formby again in 1984 and Woodplumpton 1996. Many records since, probably due to combination of increased recording effort, awareness of early stages and perhaps genuinely increased frequency. Upland records include Worsaw Hill, 2021.
■ **HABITAT** Woodland, hedgerows.
■ **FOODPLANTS** Hawthorn; feeding in the haws.
Single brooded. Comes to light. Occasionally recorded during day by foodplant. Larvae feed within a spun pair of haws, and recorded from September to October.

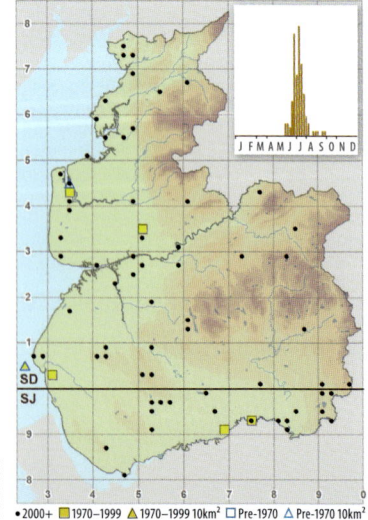

• 2000+   □ 1970–1999   ▲ 1970–1999 10km²   □ Pre-1970   △ Pre-1970 10km²

49.360 *Pammene splendidulana*

■ **LANCS STATUS** Rare / very local resident   c.1853–2016

First recorded by N. Cooke 'in the hedges about Warrington' (Byerley, 1854) and next by J.B. Hodgkinson from the Preston district (Ellis, 1890). Both Ellis and Mansbridge (1940) considered the moth to be uncommon. There was a long wait for the next record. This was of a moth netted during the day by J. Patton at Thrushgill Plantation, nr. Wray on 4 June 2015. Another turned up the following year, at Birk Bank, nr. Quernmore, recorded by J. Girdley on 27 May 2016.

■ **HABITAT** Moorland, oak woodland.
■ **FOODPLANTS** Unrecorded within the county.

Nationally, single brooded. Moths recorded during day.

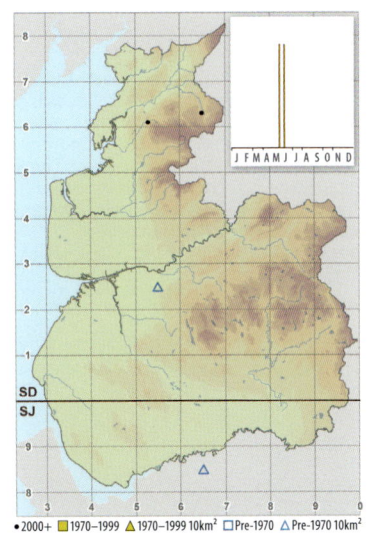

49.362 *Pammene giganteana*

■ **LANCS STATUS** Occasional / local resident   2021–2022

Pheromone lures have radically altered our understanding of the distribution of some of this genus, and this is probably the clearest example. First recorded at Birk Bank by J. Patton on 9 April 2021, and from Gait Barrows, Aughton Wood and Silverdale in the following days. Next from Rivington 20 April 2021 by J. Girdley. High count of nine lured in 45 minutes during a sunny morning in Chorlton on 25 March 2022 (B. Smart). Also recorded at Cottam, Preston and Fulwood in March 2022 by S. & C. Palmer.

■ **HABITAT** Oak woodland.
■ **FOODPLANTS** Unrecorded within the county.

Single brooded. Recorded using *Grapholita molesta* and *G. funebrana* lures.

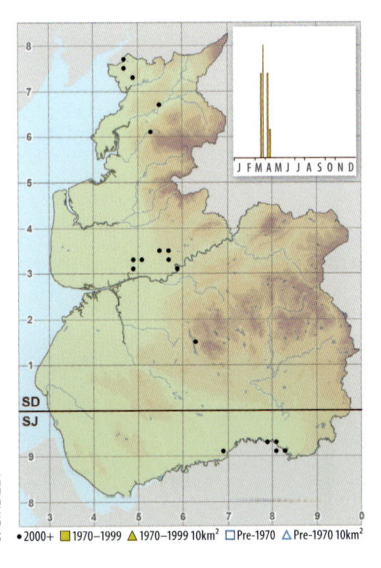

49.363 *Pammene argyrana*

■ **LANCS STATUS** Scarce / local resident   1846–2022

First noted by R.S. Edleston during June 1846 in abundance at Cheetham Hill (Edleston, 1846), with other 19th century records from Liverpool and Preston. Unrecorded in 20th century other than a comment that the moth is 'common in old oak woods and generally distributed' (Mansbridge, 1940). First recorded this century at Dalton, nr. Parbold by C.A. Darbyshire in 2003. More recently, from south Manchester, Haydock, Astley Moss and Herring Head Wood, nr. Tatham.

■ **HABITAT** Oak woodland, hedgerows.
■ **FOODPLANTS** Unrecorded within the county.

Single brooded. Post-2000 records include five at light and three on oak trunks.

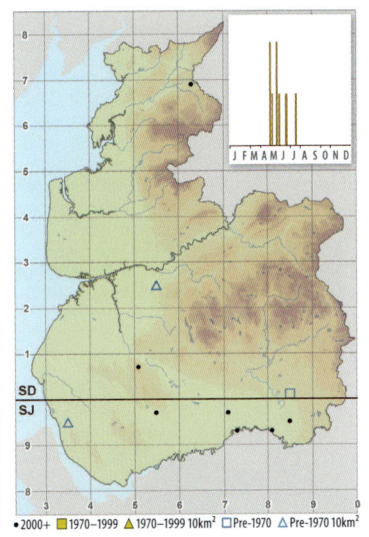

TORTRICIDAE

The Moths of Lancashire   **321**

49.364 *Pammene suspectana*

■ **LANCS STATUS** Scarce / very local resident   2022–2023

Another *Pammene* species with records solely based on moths attracted to pheromone lures. The *Grapholita funebrana* lure was again the attractant used. First recorded from a back garden in Chorlton by B. Smart around 4pm on 29 April 2022, with four moths attending the lure. The following day, four more were lured at nearby Hardy Farm. Two further records from the same back garden were noted on 7 and 9 May 2022, with a couple dissected by S. Palmer to confirm identity.
■ **HABITAT** Unknown in the county.
■ **FOODPLANTS** Not recorded in Lancashire.
Single brooded. Attracted to lure at Nether Kellett, VC60, 6 May 2023 (J. Mason).

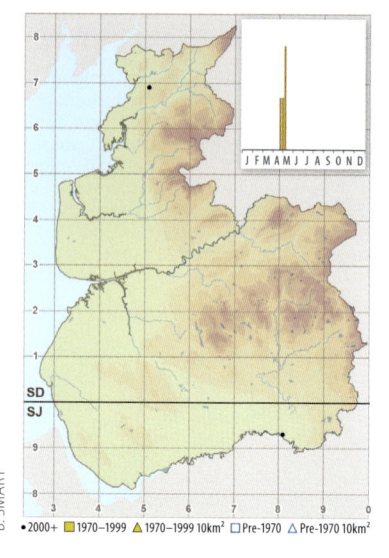

49.366 *Pammene obscurana*

■ **LANCS STATUS** Scarce / local resident   1909–2021

First reported from Simonswood Moss, nr. Kirkby on 18 June 1909 by W. Mansbridge (1940). The next record was of a male, resting on a fence post at Holcroft Moss on 5 May 2003, recorded by K. McCabe. Two years later, in the north of the county, a moth was noted flying among Heather at Winmarleigh Moss by S. Palmer, with another from Bold Moss, St Helens the same year. Further records from Flixton, Cockerham Moss, Longton and White Moss, nr. Yealand Redmayne.
■ **HABITAT** Birch woodland, mosses.
■ **FOODPLANTS** Not recorded in Lancashire.
Single brooded. The ten records include three moths at light, and seven during day.

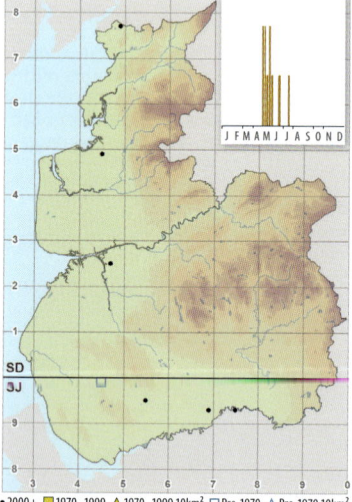

49.367 *Pammene fasciana*

■ **LANCS STATUS** Frequent / widespread resident   c.1859–2022

Recorded from Withington, Manchester by J. Chappell, Silverdale by J.B. Hodgkinson (Ellis, 1890) and Kirkby by W. Mansbridge in 1917. The next record was a moth trapped at Lightfoot Green in 1995 by S. Palmer. Light trapping has shown the moth likely to be found wherever oak occurs. Frequent sites include Dolphinholme, Hale, Billinge, Chorlton, Heaton Moor, etc. Also from upland areas, such as Leck Fell.
■ **HABITAT** Oak woodland, isolated oak trees.
■ **FOODPLANTS** Oak; feeding in acorns.
Single brooded. Occasional late record, e.g., 29 September 2001 at Lightfoot Green. Comes readily to light. Larval feeding noted from late August and September.

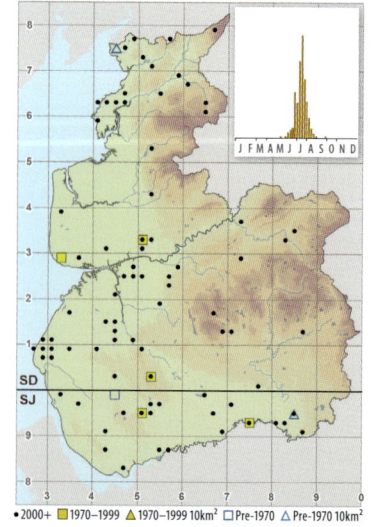

322 *The Moths of Lancashire*

49.368 *Pammene herrichiana*

■ **LANCS STATUS** Rare / very local resident   2011

This species, previously considered a form of *P. fasciana*, is mainly associated with southern English Beech woodland. Locally, the only record we have is of two moths trapped at Fulwood by A. Powell on 10 May 2011. The moths came to light under a large Beech tree at the bottom of his garden and based on external appearances were clearly *P. herrichiana*, a much darker moth than *P. fasciana*. A specimen was retained, but differences on dissection between the two are negligible, at best.
■ **HABITAT** Garden on edge of woodland with mature Beech trees.
■ **FOODPLANTS** Unrecorded within the county.
Nationally, single brooded.

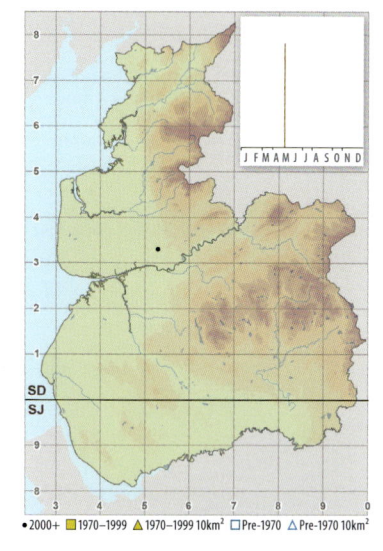

49.371 Fruitlet Mining Tortrix *Pammene rhediella*

■ **LANCS STATUS** Occasional / local resident   c.1857–2020

The first record is of moths flying in early April sunshine (Gregson, 1857). Further 19th century records from Preston and Cleveleys (Ellis, 1890). Noted at Silverdale and Lancaster, the latter by C.H. Forsythe, and described as rare (Mansbridge, 1940). From Formby 1949 and Glasson 1989. Primarily a lowland species, with 21 records during last 30 years; from Salthills Quarry, Haydock, Woolton, Rishton, etc.
■ **HABITAT** Woodland edges, hedgerows, scrub.
■ **FOODPLANTS** Not recorded in Lancashire.
Single brooded. Noted five times at light, but more commonly recorded during day, flying around Hawthorn. Probably under-recorded as rarely descends to lower levels.

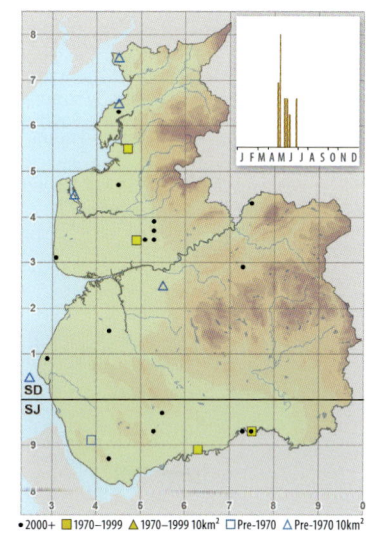

49.372 *Pammene populana*

■ **LANCS STATUS** Occasional / local resident   c.1857–2022

Noted to be 'abundant at Crosby on the sallows' by Gregson (1857), with moths bred from larvae collected at the same site in 1904 (Mansbridge, 1905). Recorded from Lytham St Annes by J.B. Hodgkinson in 1885 and more recently, from Fairhaven, by D. & J. Steeden in 1991. Further records from Lytham, Ainsdale, Formby, Middleton N.R. and Altcar demonstrate this to be a coastal species, mainly associated with Creeping Willow. High count of 50 at light at Formby Point on 12 August 2000.
■ **HABITAT** Sand dunes.
■ **FOODPLANTS** Creeping Willow. Once bred from Aspen at Formby, 2014.
Single brooded. Comes to light. Netted by day and at dusk. 14 larval records.

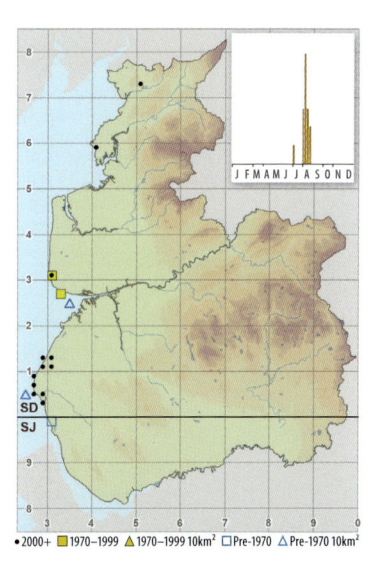

The Moths of Lancashire

TORTRICIDAE

49.373 *Pammene spiniana*

■ **LANCS STATUS** Rare / very local resident   2017

The sole Lancashire record is of a moth recorded during daytime field work at Gait Barrows on 8 August 2017 by J. Patton, and subsequently dissected by S. Palmer. It was netted in the early afternoon after being disturbed from a patch of well-developed Blackthorn. Recorded on four occasions in Cheshire and five in Yorkshire. Unrecorded in Cumbria.

■ **HABITAT** Blackthorn hedgerows.
■ **FOODPLANTS** Not recorded in Lancashire.
Single brooded.

● 2000+   ◻ 1970–1999   △ 1970–1999 10km²   ◻ Pre-1970   △ Pre-1970 10km²

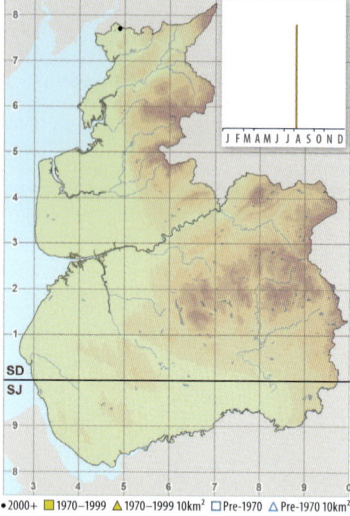

49.375 *Pammene regiana*

■ **LANCS STATUS** Widespread / frequent resident   c.1859–2022

Recorded by J. Chappell among Sycamores in Manchester district (Ellis, 1890). Further noted at Crosby, Formby, Ainsdale, Southport, Childwall, Huyton, Lancaster, and Boggart Hole Clough in north Manchester (Mansbridge, 1940), with other mid-1900s records from Morecambe, Oldham and Didsbury. Next from Silverdale and Heysham N.R. in 1990 and Orrell in 2000. 235 subsequent records, well-distributed throughout.

■ **HABITAT** Woodland, parks, gardens.
■ **FOODPLANTS** Sycamore; feeding in the seeds.
Single brooded. Frequently at light. Occasionally found on Sycamore trunks and foliage. Larval cocoons found beneath Sycamore bark during winter and early spring.

● 2000+   ◻ 1970–1999   △ 1970–1999 10km²   ◻ Pre-1970   △ Pre-1970 10km²

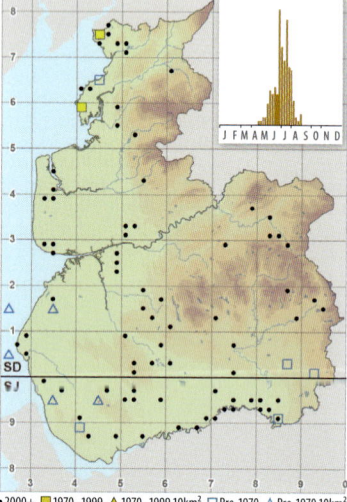

49.376 *Pammene aurita*

■ **LANCS STATUS** Frequent / widespread resident   2000–2022

A recent arrival in the county, the moth was first recorded at Flixton by K. McCabe on 5 August 2000. The following five years saw many garden records from Billinge, Orrell, St Helens, Ormskirk, Penketh, Lytham, etc., with range rapidly moving northwards to include Preston, Fluke Hall, Silverdale, Warton Crag, etc. Larva once recorded in leaf litter beneath Sycamore in a Chorlton garden in 2006; adult bred.

■ **HABITAT** Woodland, parks, gardens, hedgerows.
■ **FOODPLANTS** Unrecorded within the county.
Single brooded. Comes to light. Attracted to Hornet Moth, Currant Clearwing and Large Red-belted Clearwing lures, with ten to the latter on 1 August 2022, Chorlton.

● 2000+   ◻ 1970–1999   △ 1970–1999 10km²   ◻ Pre-1970   △ Pre-1970 10km²

324   The Moths of Lancashire

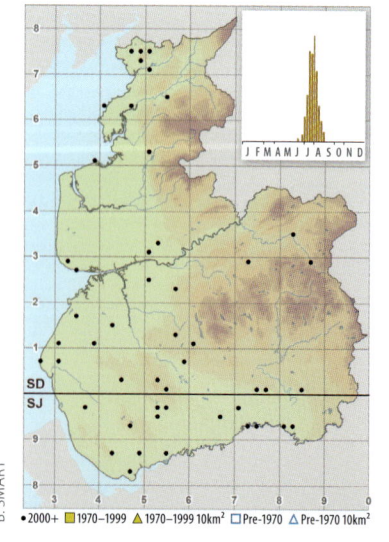

49.378 *Pammene ochsenheimeriana*

■ **LANCS STATUS** Rare / very local resident   2011

This moth, usually associated with coniferous woodland, has only been recorded on a single occasion within the county. A single moth was trapped at light by G. Riley in his Swinton garden in June 2011, and subsequently dissected by S. Palmer. There is a single record from Yorkshire and none from Cheshire or Cumbria.

■ **HABITAT** Unknown in the county.
■ **FOODPLANTS** Not recorded in Lancashire.
Nationally, single brooded.

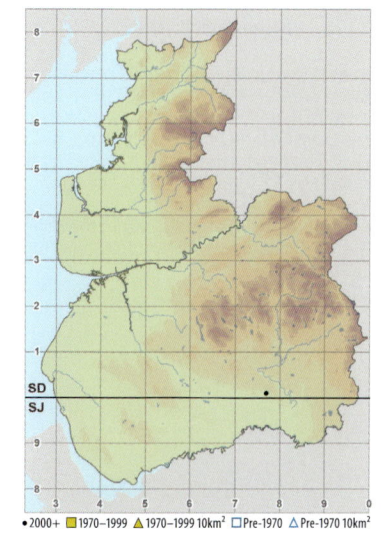

49.379 *Pammene aurana*

■ **LANCS STATUS** Occasional / local resident   c.1853–2022

The first record of this species, from Warrington by N. Cooke (Byerley, 1854), states that it is found on flowers. Searching flower-heads, specifically Hogweed, from May to July remains the easiest way to see the moth. Other 19th century records from Agecroft, Withington, Preston and Cleveleys, with status noted as scarce (Ellis, 1890). Fairly common in Mersey Valley, less so in other lowland areas. Remains absent from upland areas, and now also appears to be absent on Fylde coast and in Preston area.

■ **HABITAT** Rough grassland, scrub.
■ **FOODPLANTS** Hogweed; feeding signs on seeds noted from Yealand Redmayne, 2022.
Single brooded. Moth recorded in day; rarely at light. Once, to pheromone lure.

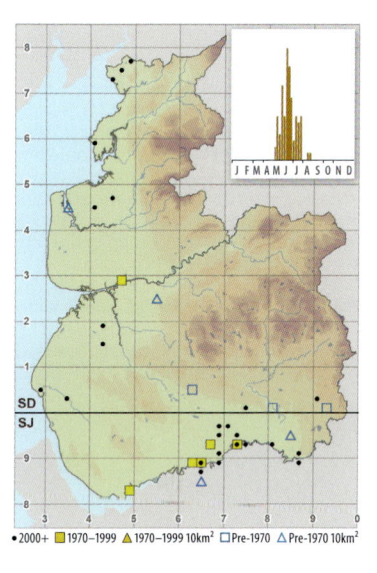

49.380 *Pammene gallicana*

■ **LANCS STATUS** Scarce / Local resident   c.1889–2022

First recorded at Fleetwood by J.B. Hodgkinson (Ellis, 1890), but not again until 28 July 2014, when a moth was noted on a Wild Carrot flower at Freshfield by B. Smart. In October 2017, larvae were found in seed-heads of the foodplant at Middleton N.R. by J. Patton. Identity was confirmed when an adult was bred from identical larvae at the same site in 2020. Further recorded at Formby and Fleetwood Dunes in 2022.

■ **HABITAT** Dry coastal grassland, waste ground, sand dunes.
■ **FOODPLANTS** Wild Carrot.
Single brooded. Moths noted on Wild Carrot flower-heads. Unrecorded at light. Larvae found in seed-heads, sometimes gregariously, in September and October.

TORTRICIDAE

The Moths of Lancashire   325

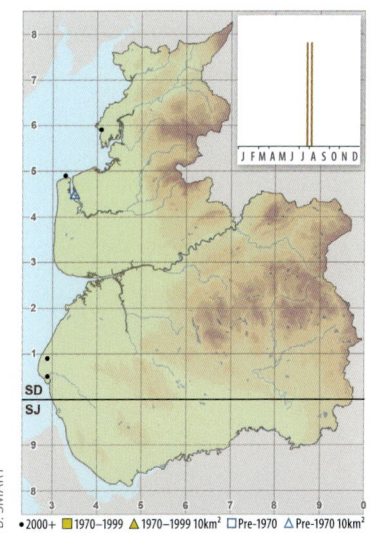

TORTRICIDAE

49.381 Strophedra weirana

■ **LANCS STATUS** Rare / very local resident   2015–2022

All three records are from Leek Hill Wood, an area of predominantly Beech woodland to the north and east of Warton Crag in the Morecambe Bay limestone area. The first, recorded by J. Patton, was of a male netted during the day on 1 July 2015. The two further records, noted by the same recorder, were of moths trapped at light on 22 June 2017 and 6 July 2022.

■ **HABITAT** Beech woodland.

■ **FOODPLANTS** Unrecorded within the county. Nationally, the foodplant is Beech, and it is worth looking in likely habitat for the distinctive larval feeding signs in autumn. Single brooded.

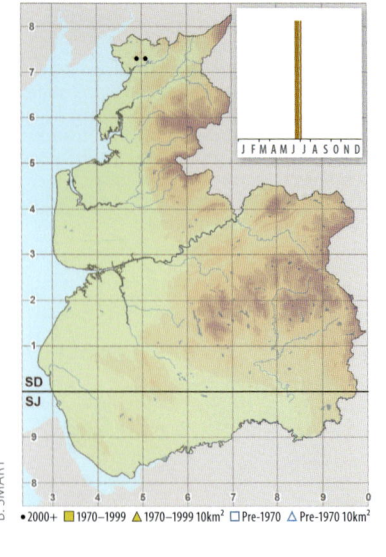

49.382 Strophedra nitidana

■ **LANCS STATUS** Considered Extinct   c.1888

Recorded by J.H. Threlfall and J.B. Hodgkinson from the Preston district during the 19th century, although the exact date(s) are unknown (Ellis, 1890). Neither is it clear whether the record(s) come from VC59 or VC60. There have been no subsequent records of the moth. It is rare in surrounding counties with just a few historical records from Cumbria and Yorkshire, and a single modern record from Biggin, nr. Selby, Yorkshire on 19 June 2012 (yorkshiremoths.co.uk, accessed 1 September 2023). Unrecorded in Cheshire.

■ **HABITAT** Unknown in the county.

■ **FOODPLANTS** Not recorded in Lancashire.

This single brooded species is fairly widespread in the British Isles. The larva feeds between two oak leaves spun together, from October onwards, which eventually fall to the ground. It is known to fly in late afternoon sunshine high up in oak trees, possibly partially explaining the lack of any further records (Langmaid, *et al.*, 2018).

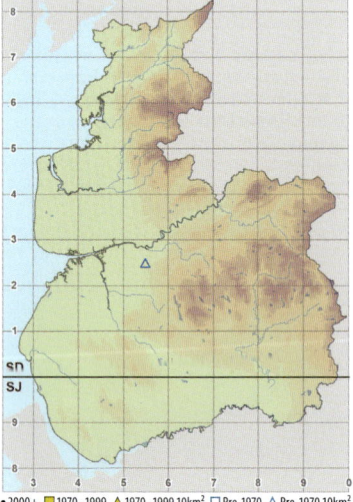

COSSIDAE

50.001 Goat Moth *Cossus cossus*

■ **LANCS STATUS** Scarce / very local resident   c.1857–2023

Found in much of the Liverpool district (Gregson, 1857) and abundant as larvae near Manchester (Chappell, 1861). Later, noted as 'generally distributed in wooded areas' (Ellis, 1890). Elsewhere it was noted at Bolton in 1920, Hoghton in 1939, St Annes (the only VC60 record) in 1939 and Patricroft in 1940. The last inland record was near Burnley c.1974 (Roberts & Wrightson, 1986). Recorded almost annually in small numbers in Altcar, Formby and Ainsdale areas; none in 2022; four in 2023.

■ **HABITAT** Sand dunes, planted with poplars. Unknown inland.

■ **FOODPLANTS** Balsam Poplar and *Salix* sp. pollards.

Single brooded. Comes to light. Most larvae have been seen in August to September.

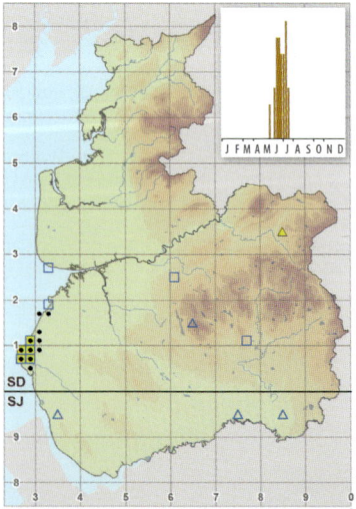

326 *The Moths of Lancashire*

50.002 Leopard Moth *Zeuzera pyrina*

■ **LANCS STATUS** Occasional / local resident   1880–2022

First recorded in the Rainhill area c.1880 by H.H. Higgins (Ellis, 1890) and then in Burnage (1907), Withington (1914), Huyton (1917) and Chorlton in 1920. Records have always been erratic and several years can pass with no reports. From 2008, it has become more regular in occurrence, but nearly always as singletons. Since 2019, noted at Coppull, Croft, Flixton, Hale, Haydock, Longton, Prescot and Tarleton.
■ **HABITAT** Woodland, gardens, parks and brownfield sites.
■ **FOODPLANTS** Ash. Pupa found in Ash trunk, Manchester, 26 January 1933 (H. Britten). Single brooded. Comes to light. A very late specimen was found at Beech Grove, the only VC60 record, on 18 September 2004 (M. Myerscough), seen alive by S. Palmer.

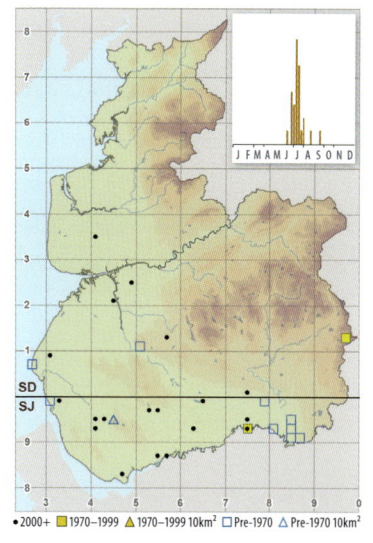

COSSIDAE

52.002 Hornet Moth *Sesia apiformis*

■ **LANCS STATUS** Considered Extinct   1907

The only record we have of this species relates to larval feeding signs from Blundellsands, nr. Crosby in 1907, noted by F.W. Gardener, and accepted by W. Mansbridge. The report is quoted by Mansbridge (1940) who clarified that it 'relates to the poplar feeding moth,' and further quotes Gardener's observation that 'the species appears to be less common than formerly, many localities having been destroyed, but there are still many trees on which the damage done by the larva is noticeable.' Mansbridge further commented that 'if looked for would probably be found more commonly than has hitherto been the case'. Unrecorded in Yorkshire. The last of the seven Cheshire records is from 1907 (cheshire-chart-maps.co.uk, accessed 1 September 2023).
■ **HABITAT** Unknown.
■ **FOODPLANTS** Poplar trunks.
Nationally, single brooded. Recent use of the pheromone lure developed for this species has been unsuccessful in south Manchester but is certainly worth trying in other areas where poplars are present.

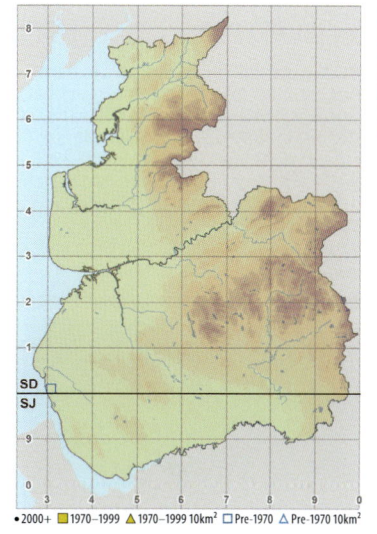

SESIIDAE

52.003 Lunar Hornet Moth *Sesia bembeciformis*

■ **LANCS STATUS** Frequent / widespread resident   1844–2022

Abundant on poplars at Middleton, nr. Rochdale (Edleston, 1844). Other 19th century records from Liverpool and Chat Moss. 215 widely distributed records since, including from upland areas. The LUN pheromone lure is responsible for 60 records since 2019, including L. Renshaw's ten records from Heysham N.R. and K. Haydock & J. Mills' 12 from Chorley. Trapped in mist nets at Knott End by C. Batty, 9 July 2021.
■ **HABITAT** Open woodland, hedgerows, damp woodland.
■ **FOODPLANTS** In trunks of Goat Willow, White Willow. Occasional on poplar.
Single brooded. Moth to pheromone lure and twice at light. Larval borings or pupal exuviae have been noted on 30 occasions.

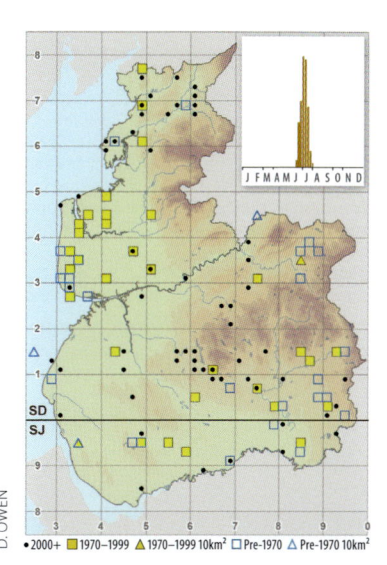

The Moths of Lancashire   **327**

SESIIDAE

52.006 White-barred Clearwing *Synanthedon spheciformis*

■ **LANCS STATUS** Considered Extinct   c.1845–c.1857

Reported from the Botany Bay Wood area of Chat Moss in Salford on 6 July 1845 by John Thomas, and determined by R.S. Edleston (Edleston, 1845). The moth was documented by Gregson (1857) from the same site, although this may have been based on the same record. Mansbridge (1940) later noted the moth to be 'of occasional occurrence only.' There are a few historic records from Cumbria, Yorkshire and Cheshire with the most recent being a record from Wilmslow, Cheshire in 1946 (cheshire-chart-maps.co.uk, accessed 1 September 2023).
■ **HABITAT** Alder woodland.
■ **FOODPLANTS** Not recorded in Lancashire. Elsewhere, the larvae are associated with Alder and birch trees.

Nationally, single brooded. It is possible that the use of pheromone lures may reveal surviving populations in the old Chat Moss area, although a single attempt in June 2022 at Rindle Wood was unsuccessful. The national picture though is not promising, as numbers of records are in decline despite ever wider use of such lures. The moth now appears to be limited to the Welsh borders, south-east England and the Norfolk Broads (Randle et al., 2019).

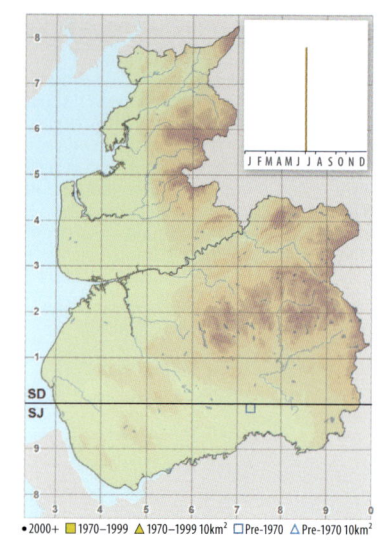

● 2000+   ☐ 1970–1999   ▲ 1970–1999 10km²   ☐ Pre-1970   △ Pre-1970 10km²

52.007 Large Red-belted Clearwing *Synanthedon culiciformis*

■ **LANCS STATUS** Rare / very local resident   c.1859–2004

First recorded in late 19th century at Chat Moss by J. Chappell and described as rare (Ellis, 1890). Noted by J. Collins and G.O. Day at Rixton Moss c.1900. Next, by J. Steeden in 1986 at Winmarleigh Moss. P. Cleary-Pugh recorded the moth with use of a pheromone lure at Astley Moss in 2001, with K. McCabe recording six pupal exuviae at Holcroft Moss in 2002. Unrecorded from the mosses since 2004; possibly linked to lack of suitable surveys, but perhaps also to changes in habitat management.
■ **HABITAT** Birch scrub on the mosses.
■ **FOODPLANTS** Birch. Bred from birch stumps at Rixton by J. Collins (Mansbridge, 1940).

Single brooded. Pupal exuviae noted protruding from birch stumps in 2002.

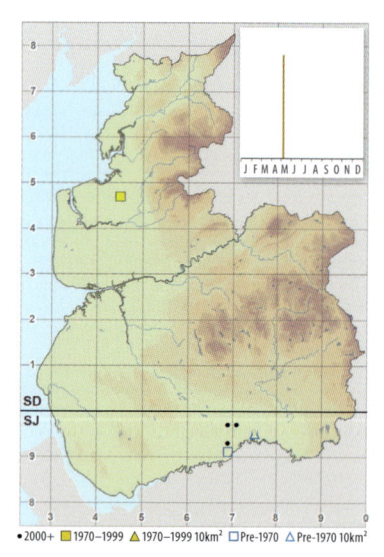

● 2000+   ☐ 1970–1999   ▲ 1970–1999 10km²   ☐ Pre-1970   △ Pre-1970 10km²

52.008 Red-tipped Clearwing *Synanthedon formicaeformis*

■ **LANCS STATUS** Occasional / local resident   1987–2022

Recorded at Altcar in 1987 by C.J. Palmer, and from Woolston Eyes in 2003 by S. McWilliam. Other than at Bold Moss, St Helens by R. Banks in 2017, all subsequent records are from the Sefton Coast, primarily Ainsdale. Surveying led by R. Walker with the FOR pheromone lure resulted in 58 records from the dunes since 2014, including a count of 73 in a half-mile survey on 22 June 2014. The moth has also been lured from Creeping Willow at Ainsdale and noted flying around the plant.
■ **HABITAT** Fixed dunes.
■ **FOODPLANTS** Creeping Willow likely to be the foodplant used at Ainsdale.

Single brooded. Early stages not recorded. 46 of the 65 records were at FOR lure.

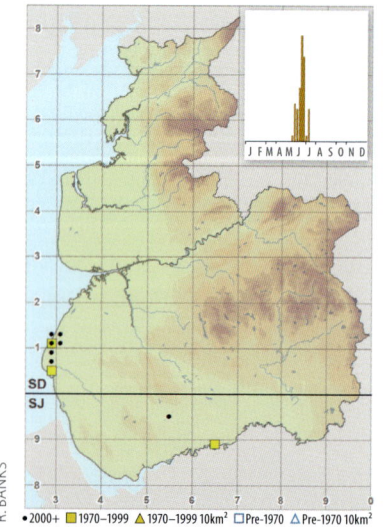

● 2000+   ☐ 1970–1999   ▲ 1970–1999 10km²   ☐ Pre-1970   △ Pre-1970 10km²

52.012 Yellow-legged Clearwing *Synanthedon vespiformis*

- **LANCS STATUS** Scarce / local resident　　　　　　　　　2015–2022

Probably a much under-recorded species, use of the VES pheromone lure offers the possibility of beginning to understand the true distribution of this moth within the county. First recorded in 2015 from Croft, Warrington by P. Brighton. The subsequent records were from two Chorlton gardens by B. Smart and S. Marsden in 2022, where use of the lure resulted in five records between 2 June and 13 July, the moths responding to the pheromone almost immediately, likely from nearby oaks.
- **HABITAT** Unknown.
- **FOODPLANTS** Not recorded in Lancashire.

Single brooded. Attracted to VES lure and once to the Hornet Moth lure.

● 2000+　☐ 1970–1999　▲ 1970–1999 10km²　☐ Pre-1970　△ Pre-1970 10km²

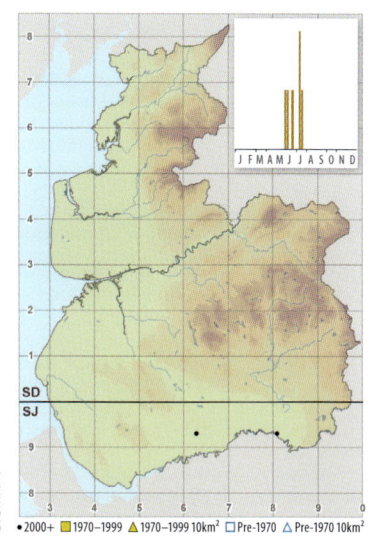

52.013 Currant Clearwing *Synanthedon tipuliformis*

- **LANCS STATUS** Frequent / local resident　　　　　　　　c.1853–2022

First report by N. Cooke from West Derby (Byerley, 1854). Other 19th century records from Hale, Warrington, Huyton, Crosby, Southport, Bolton. Mansbridge described the moth as well-distributed in south-west Lancs, but absent, or scarce, in the north (1940). Larval feeding was noted in currant bushes at Oldham in 1906 by R. Cottam. A frequently recorded Clearwing with 136 records, over 100 from this century, aided by lures. This primarily lowland species is largely unrecorded from north-east Lancs.
- **HABITAT** Gardens, allotments.
- **FOODPLANTS** Currant bushes.

Single brooded. Moths noted around foodplant and with pheromone use.

● 2000+　☐ 1970–1999　▲ 1970–1999 10km²　☐ Pre-1970　△ Pre-1970 10km²

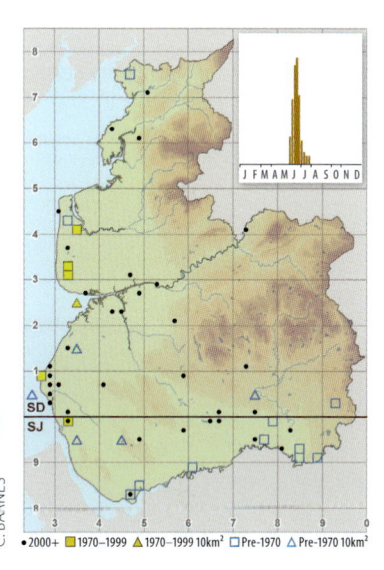

52.014 Six-belted Clearwing *Bembecia ichneumoniformis*

- **LANCS STATUS** Rare / very local resident　　　　　　　2019–2023

Recorded by P. Hopkins on 8 July 2019, using the API pheromone lure in a small patch of birds-foot trefoil in a disused area adjacent to Liverpool Festival Gardens. After ten minutes a single moth appeared, but after a further 20, no more moths lured. Five moths swept on the same site in 2020 by P. Brash, noting the habitat to be sparsely vegetated grassland … developed on what might have been crushed hardcore, with plenty of clover and bird's-foot trefoil. Unfortunately, much of the site has since been lost to Development. Further noted at Hightown, 2023 (S. Hartnett). Single brooded.
- **HABITAT** Short flower-rich grassland on brownfield site and coastal dune.
- **FOODPLANTS** Not recorded in Lancashire.

● 2000+　☐ 1970–1999　▲ 1970–1999 10km²　☐ Pre-1970　△ Pre-1970 10km²

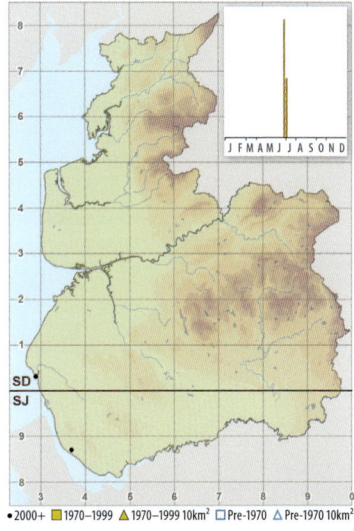

ZYGAENIDAE

54.002 Forester *Adscita statices*

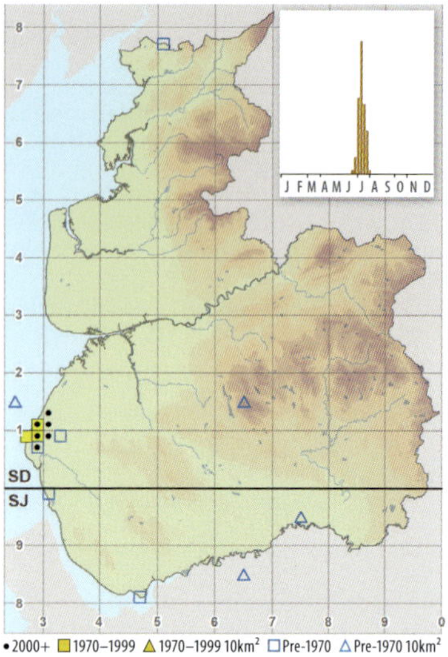

■ **LANCS STATUS** Scarce / very local resident   c.1857–2023

Although afforded the national status of Least Concern (Fox *et al.*, 2019), this species has declined at a national level and within Lancashire, where it is now restricted to a five km stretch of coastal grassland, with one primary colony, and three satellite areas with a few recent sightings.

It was formerly known from seven well scattered locations in lowland VC59, with a single site in north VC60. The first records were from Hale Marsh and Crosby (Gregson, 1857), followed by Chat Moss (1859), Warrington and Chorley (both 1885). It was then found at Ainsdale in 1912 and Formby in 1938. The single VC60 record came from White Moss, Yealand in 1929 (W.P. Stocks), the last record away from the Sefton Coast.

Since then, there have been occasional records from Ainsdale, Formby Golf Course, Formby (Larkhill Heath), Formby Moss, Freshfield Dune Heath and Freshfield (Massam's Slack). In 2008, annual recording was initiated on Ainsdale NNR, mainly Pinfold Meadow, and at Larkhill in Formby. Sightings since 2008 have been annual on Ainsdale NNR up to 2022, but the Larkhill colony appears lost, with no records since 2013. One seen on Freshfield Dune Heath in 2020, and up to seven in 2023.

The adults fly low to the ground from mid-morning to late afternoon in sunshine, sometimes visiting ragwort flowers in small groups; Bramble or Rosebay Willowherb can be used when ragwort is not available. Usually recorded in small numbers, double figures have been seen in some years, reaching a peak of 34, on Ainsdale NNR, in 2017. Forester pheromone trials took place in 2015–16 on Ainsdale with some success (a feature not always replicated in other trials elsewhere in England). The larvae, or its feeding signs, have not been found, but nationally it utilises Common and Sheep's Sorrel.

An aerial photo of Ainsdale in 1945 shows the, now named, Pinfold Meadow and surrounding land as potato fields, which may explain the lack of reports in the early years or until 2005. The site consists of 1.1 hectares of acid grassland where sward is kept short by rabbits.

The loss of three small colonies around Formby and Freshfield have probably been caused by the growth of rush in the damper areas or being overgrown with rank grasses, so causing the loss of the potential larval foodplants. Though nationally declining, single sightings in unexpected areas (Freshfield Dune Heath and the fire break south of Ainsdale NNR office) plus the main site of Pinfold Meadow and the general abundance of the potential foodplants there would suggest that the well-being of the Ainsdale NNR will hold the key to the future of this moth on the Sefton Coast.

Surveys of the areas with scattered records should be encouraged, as should the search for larvae. Searches at the former White Moss area would also be useful.

Richard Walker

54.003 Cistus Forester *Adscita geryon*

■ **LANCS STATUS** Rare / very local resident   1885–2022

Given the status of Least Concern (Randle *et al.*, 2019), this Nationally Scarce day-flying moth is difficult to monitor and is easily overlooked. The first Lancashire record was in the Silverdale area in 1885 by J. C. Melvill (Ellis, 1890), although the exact location of this and further early 20th century Silverdale records are unknown. It was next seen at Warton Crag, the only current known location, in 1961, and a strong colony was reported from Yealand Hall Allotment in 1974. The latter area has been the subject of butterfly transect walks since c. 2000 by B. Hancock, but with no sign of this species.

Occasional sightings continued on a few parts of Warton Crag up to 1990, and then ceased completely. In 1995, a small colony was found on a rather remote, sheltered, western slope with short limestone grassland and abundant rock-rose. This was on an RSPB Butterfly Transect route and led to annual reports of the moth. The largest one-day counts were of 30 on 11 June 2011 and 16 on 11 June 2016. Since 2017, most records were of two or three individuals during the first two weeks of June. It is curious that it now seems to occur in just this one very small area, as the larval foodplant, Common Rock-rose, is widespread over the Crag.

All records are of day-flying moths, with no evidence of attraction to light. It has been seen flying low over the calcareous grassland on sunny days, frequently stopping to feed on Common Bird's-foot-trefoil and hawkbit. In 2005, egg laying was observed, a small batch being placed on the underside of rock-rose leaves. Despite this, and a few searches in 2022, the larvae have not been found in Lancashire, but are known to mine leaves, leaving visible blotches, whilst feeding externally on the underside in the late summer. It consumes the whole leaf when completing larval development in spring.

Over the last three years, the moth has proved increasingly difficult to find. In June 2022 despite many hours of searching, on three separate occasions, just one worn individual was found, on 8 June 2022 (S. Graham & B. Hancock).

Rock-rose plants, at the location of the remaining colony, have diminished despite scrub clearance designed to increase the limestone grassland flora. A succession of cold, dry springs has resulted in many desiccated plants by June, and mild, wet winters have encouraged competitive plants, especially bramble. These factors may well account for its decline and the moth's future here appears precarious.

The Morecambe Bay limestone outcrops provide plentiful, potential suitable habitat, yet the moth only occurs, otherwise, in low numbers on a few sites in adjacent Cumbria (C. Clay pers. comm.). In Yorkshire, the established colony in upper Wharfedale appears stable (C. Fletcher pers. comm.). Elsewhere, nationally, the species has a very scattered and local distribution from parts of southern England, the North Wales coast, Derbyshire and Durham.

Brian Hancock

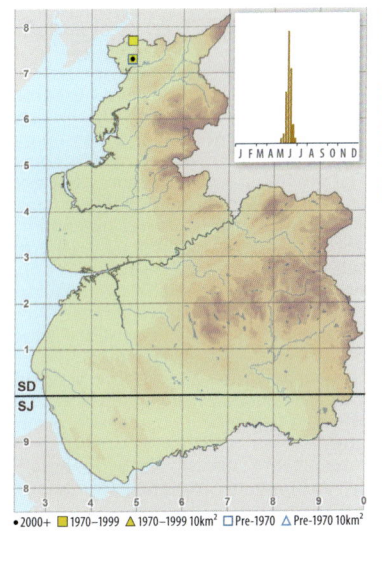

• 2000+   □ 1970–1999   ▲ 1970–1999 10km^2 □ Pre-1970   △ Pre-1970 10km^2

ZYGAENIDAE

54.008 Six-spot Burnet *Zygaena filipendulae*

■ **LANCS STATUS** Frequent / widespread resident   c.1857–2022

First noted in the Liverpool area (Gregson, 1857) and later, on the (former) dunes at Crosby, where it was described as excessively abundant by J.W. Ellis on 22 July 1881. At the same site, Mansbridge (1940) noted it to be present in 1,000s on 5 August 1916. Over the next century and a half, it has been reported from scattered sites across the county, but most frequently in coastal and lowland areas. By 1983, it was still, on rare occasions, noted as highly abundant, such as on the Seaforth Dock pools area by P. Smith. Counts into the low 1,000s continued to be occasionally reported up to 2012, but not since. Sites with long-term record sets include Ainsdale, Crosby and Flixton, with all now suffering a considerable decline in reported numbers. Habitat loss or scrub encroachment are the main contributary factors at two of these. More widely, in recent decades, counts of up to 100 have not been unusual, although one exceptional record was of an estimated 3,000 at Fleetwood Nature Park on 26 June 2011 (B. Dyson). This demonstrates the value of large, flower-rich, grassy, brownfield sites, many of which are under threat of redevelopment.

■ **HABITAT** Grassland on the coast, brownfield sites, parks, verges and limestone areas.

■ **FOODPLANTS** Common Bird's-foot-trefoil.

Single brooded. Flies during the day, mostly in sunny conditions.

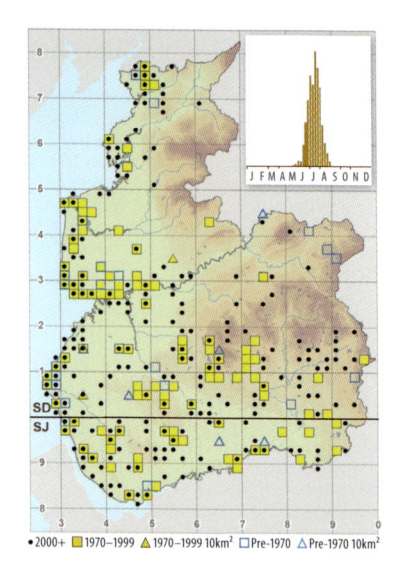

• 2000+   ■ 1970–1999   ▲ 1970–1999 10km²   □ Pre-1970   △ Pre-1970 10km²

54.009 Narrow-bordered Five-spot Burnet *Zygaena lonicerae*

■ **LANCS STATUS** Frequent / widespread resident   c.1857–2022

Confusion with *Z. trifolii* (Five-spot Burnet) seems to be a factor in early recording in the county with a record of this species having been seen 'in profusion on Hale Marsh in June' (Greyson, 1857); this is repeated in Ellis (1890) and Mansbridge (1940). As far as *Z. lonicerae* is concerned, both Ellis and Mansbridge (loc. cit.) report it to be confined to Cheshire, although they fail to mention a possible *lonicerae*, mentioned by Gregson (loc. cit.) near St Helens, on a dry hill-side on the road to Eccleston. This is accepted here as the first Lancashire record. A national assessment of *Zygaena* records appears in Heath & Emmet (1985), where there are no confirmed Lancashire records of *Z. trifolii* shown on the 10km dot map, but plenty of *Z. lonicerae*. The next reports were not until c.1939, at Lytham and Didsbury in 1940, with a few scattered sightings elsewhere up to 1970. Records then increased gradually from across much of the county, including from St Annes, Risley Moss, Speke, Thornton Cleveleys, and some from upland sites, e.g., Haslingden. Counts of a few hundred were occasional until c.2013, but have been rare since then.

■ **HABITAT** Rough grassland, on sand dunes, brownfield sites, parks, wet commons, etc.

■ **FOODPLANTS** Meadow Vetchling and Red Clover; once on Common Bird's-foot-trefoil.

Single brooded. Flies in sunshine. Still misidentified as Five-spot Burnet on occasions.

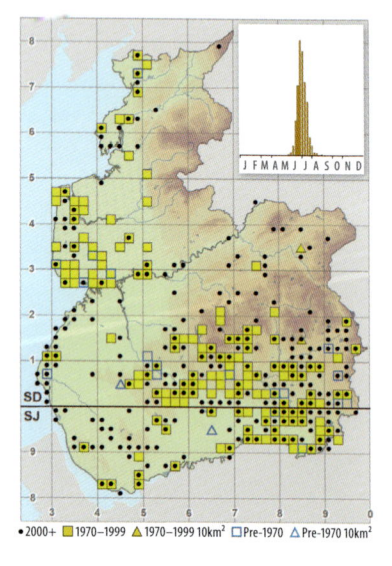

• 2000+   ■ 1970–1999   ▲ 1970–1999 10km²   □ Pre-1970   △ Pre-1970 10km²

The Moths of Lancashire

62.001 Bee Moth *Aphomia sociella*

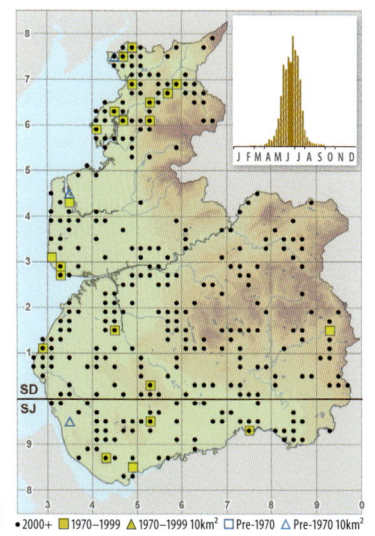

■ **LANCS STATUS** Frequent / widespread resident   1887–2022
Noted on the Banks of R Wyre in 1887 (Ellis, 1890) and in Silverdale in 1910. Next seen at Crosby in 1943 and St Annes in 1978, with a wider expansion of range into lowland areas commencing in the mid-1980s. Although it arrived in Littleborough and Claughton in 1998, it was 2006 before the main expansion into upland areas took place. Mainly seen in single figures, with occasional higher adult counts all post-2005.
■ **HABITAT** Woodland, hedgerows, gardens and houses.
■ **FOODPLANTS** On the comb in nests of bees and wasps.
Single brooded over an extended flight period. Comes to light. Larvae were found in a bird nest box used by *Bombus hortorum* with only a few moribund bees remaining.

• 2000+ □ 1970–1999 ▲ 1970–1999 10km^2 □ Pre-1970 △ Pre-1970 10km^2

62.003 Stored Nut Moth *Aphomia gularis*

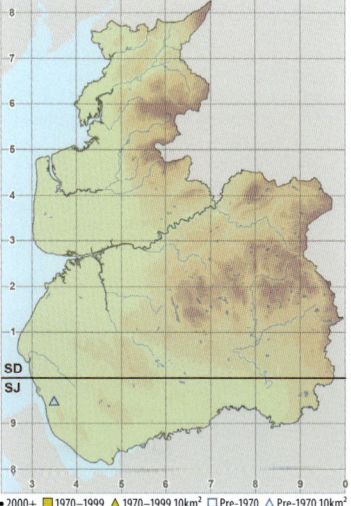

■ **LANCS STATUS** Adventive   1950
There is only one report of this species in the county, that being of a singleton from within the Liverpool Docks area in 1950 by C.M. Jones (Michaelis, 1955).
■ **HABITAT** Formerly, warehouses and food stores.
■ **FOODPLANTS** Not recorded in Lancashire. H.N. Michaelis (loc. cit.) noted that it is likely to occur among imports of nuts, wheat and dried fruit.
Its previous distribution nationally was mainly centred around the London Docks area where it was found resting on walls of warehouses. Elsewhere, attracted to an outside light close to an area where purchased bird seed was being stored, which seemed the most likely origin of the moth.

• 2000+ □ 1970–1999 ▲ 1970–1999 10km^2 □ Pre-1970 △ Pre-1970 10km^2

62.004 Rice Moth *Aphomia cephalonica*

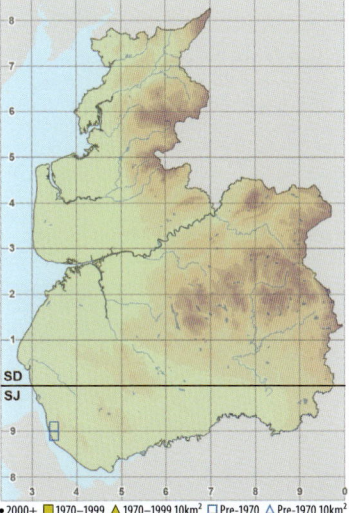

■ **LANCS STATUS** Adventive   1917–1921
Probably previously overlooked in docks in the region, the first reference to this species relates to its discovery at Liverpool Docks in 1917 (Mansbridge, 1940). At this site it was noted as 'abundant in warehouses and near docks'. It is only otherwise mentioned when it was found to be present, in great numbers, in a warehouse in Blackstock St., Liverpool, infesting groundnuts in 1921.
■ **HABITAT** Formerly, warehouses and food stores.
■ **FOODPLANTS** Groundnuts. Elsewhere it has been noted feeding on a wide variety of dried, stored foods, including cacao, dried fruit, nutmeg, rice, nuts, maize, chocolate and ships biscuits. Nationally it is reported as having two overlapping broods within warehouses, resting by day on walls and rafters, looking like a splinter of wood (Goater, 1986).

• 2000+ □ 1970–1999 ▲ 1970–1999 10km^2 □ Pre-1970 △ Pre-1970 10km^2

PYRALIDAE

62.005 Lesser Wax Moth *Achroia grisella*

■ **LANCS STATUS** Occasional / local resident   c.1857–2022

First noted in the Liverpool district (Gregson, 1857) among beehives. It was not reported again until 1995, as larvae in a beehive at Rufford, by I. McLean, and again in 2009. The 68 other recent reports have all been of single sightings at light in mainly lowland areas. Sites with frequent records include Billinge, Chorlton, Crosby, Flixton, Longton and Wigan. Reports from five sites in lowland VC59 and VC60 during 2022.

■ **HABITAT** Gardens and other localities where beehives are kept.
■ **FOODPLANTS** Within beehives, on the wax.

Probably single brooded. Comes to light, mainly as a singleton. Larvae have also been found, in August, amongst stored bee-keeping equipment.

• 2000+   ■ 1970–1999   ▲ 1970–1999 10km²   □ Pre-1970   △ Pre-1970 10km²

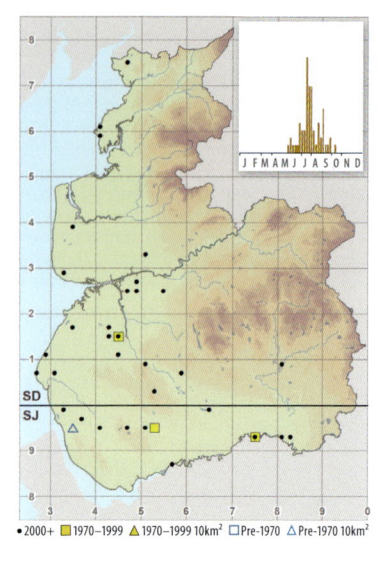

62.006 Wax Moth *Galleria mellonella*

■ **LANCS STATUS** Occasional / local resident   1911–2022

Found as larvae in Burnley in 1911 by A.E. Wright who bred them through; included a melanic form. Noted next in 1953 at Quernmore and Martin Mere in 1989. In 2007, larvae were found in a Rufford beehive for the first time and were also present in 2009 (I. McLean). All other records have been at light, but with only one upland sighting, at Briercliffe on 27 June 2015 (T. Lally). Sites with more than one report include Chorlton (3), Flixton (18), Fulwood (2), Preston (2), Roby (3) and Swinton (2).

■ **HABITAT** Gardens and other localities where bees are kept.
■ **FOODPLANTS** Combs within beehives.

Possibly double brooded. Comes to light, usually as a singleton.

• 2000+   ■ 1970–1999   ▲ 1970–1999 10km²   □ Pre-1970   △ Pre-1970 10km²

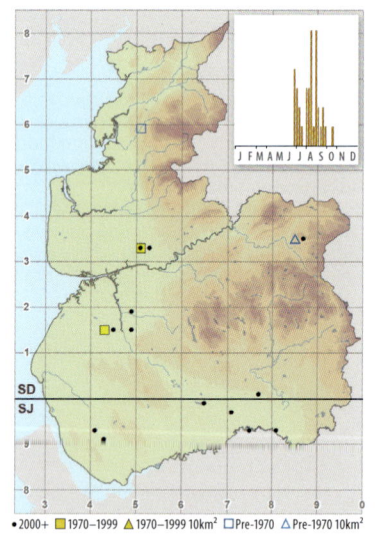

62.007 *Cryptoblabes bistriga*

■ **LANCS STATUS** Occasional / local resident   c.1857–2023

First noted at Woolton, Liverpool (Gregson, 1857), then Brockholes Wood, nr. Preston c.1886. It was seen in Silverdale c.1910, but not again until 1986, at Gait Barrows. Since then, it has been found mainly at single scattered sites. The exceptions were at Hale and Preston. At the former site, 45 moths came to light between 2008 and 2016, while at the latter, 17 were seen from 1996 to 2014. Since then, there have been only two records, at Briercliffe and Mill Houses in 2018 and Sunderland Point in 2023.

■ **HABITAT** Woodland, scrub and hedgerows with oak standards.
■ **FOODPLANTS** Oak. An adult has been beaten from a young birch (Gregson, 1857).

Single brooded. Comes to light, mainly as singletons, but with seven on one occasion.

• 2000+   ■ 1970–1999   ▲ 1970–1999 10km²   □ Pre-1970   △ Pre-1970 10km²

334 *The Moths of Lancashire*

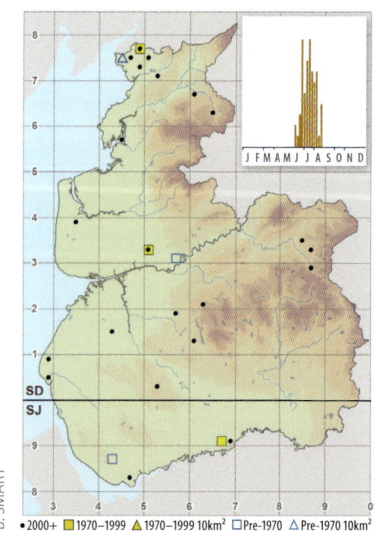

62.011 *Ortholepis betulae*

■ **LANCS STATUS** Scarce / very local resident   c.1859–2009

First noted in Great Woolden Wood and Botany Bay Wood by J. Chappell c.1859 and on Chat Moss by J.B. Hodgkinson c.1887 (Ellis, 1890). W. Mansbridge found it on Formby Moss in 1923. A larva, found at Stalmine by M. Evans, was bred emerging on 28 June 1984. Since then, it has been found at Holcroft Moss on 13 July 1991 (L.W. Hardwick), twice at Warton Crag in 1993, twice at Winmarleigh Moss, on 18 July 1996 and as a larva in 2002, and at Astley Moss on 2 July 2009 (K. McCabe and G. Riley).
■ **HABITAT** Mosses and, rarely, other areas with birch scrub.
■ **FOODPLANTS** Birch sp.
Single brooded. Comes very sparingly to light.

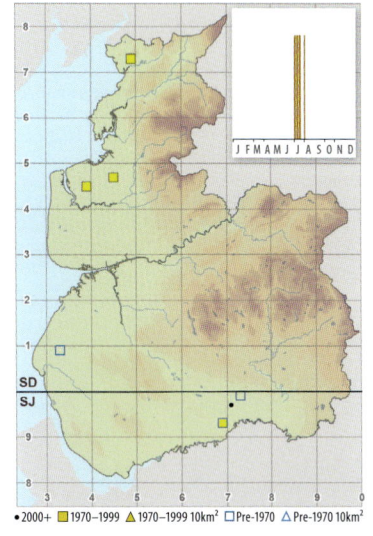

● 2000+   ■ 1970–1999   ▲ 1970–1999 10km²   □ Pre-1970   △ Pre-1970 10km²

62.012 *Pyla fusca*

■ **LANCS STATUS** Occasional / widespread resident   1844–2022

First reported from White Moss, Manchester on 31 May 1844 by R.S. Edleston. Next, Ellis (1890) noted it as common on heaths and mosses and c.1940 it was found on Formby Moss and at Ainsdale. Since then, it has been noted widely across the county, in small numbers, at sites close to moorland and mosses as well as elsewhere. In some years it disperses widely, this habit increasing in recent years, such as in 1996–97, 2004–08 and 2017–22. There may be a link to increasingly regular moorland fires.
■ **HABITAT** Moorland and mosses, but has a great tendency to wander.
■ **FOODPLANTS** Heather sp. An association with areas of burnt *Erica* is known nationally.
Probably single brooded with a possible second brood on occasions. Comes to light.

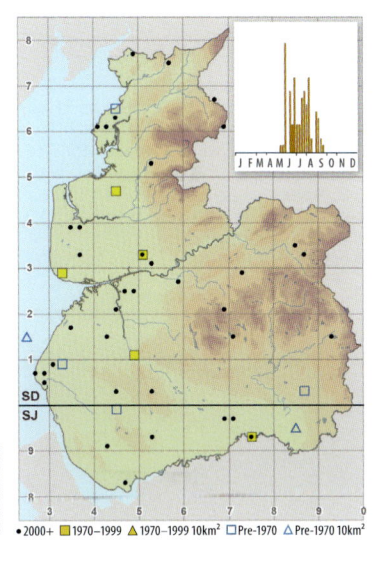

● 2000+   ■ 1970–1999   ▲ 1970–1999 10km²   □ Pre-1970   △ Pre-1970 10km²

62.023 *Pempelia palumbella*

■ **LANCS STATUS** Considered Extinct   c.1845–1957

First noted as scarce on Chat Moss in July (Edleston, 1846), but no year was listed. A Manchester area record (Stainton, 1859), probably refers to Edleston's record. It was then recorded from Silverdale by Mansbridge (1940) but with no further details. Mansbridge (loc. cit.) also reports it as a well-known immigrant, which is now considered unlikely. Later, in Leech & Michaelis (1957), W. Mansbridge is credited as finding it occasionally at Formby Moss and at Ainsdale, but with no specific dates. Considered a probable former resident, last seen sometime between 1940 and 1957.
■ **HABITAT** Lowland raised bogs.
■ **FOODPLANTS** Not reported in Lancashire.

Nationally, the species is mainly associated with heathers, from where it can be readily disturbed on sunny, summer days. It occurs very locally across England and Wales where suitable habitat occurs and has been noted in 2022 in low numbers in both Cheshire and Yorkshire. This suggests that daytime searches on the lowland mosses of Lancashire could well be productive.

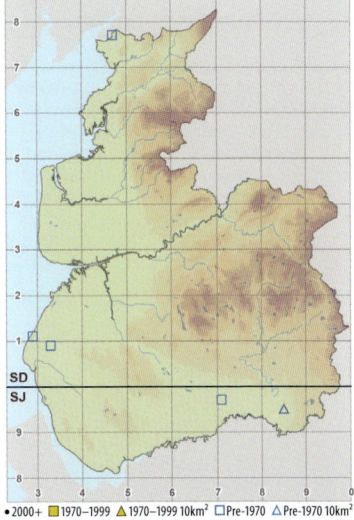

● 2000+   ■ 1970–1999   ▲ 1970–1999 10km²   □ Pre-1970   △ Pre-1970 10km²

The Moths of Lancashire

PYRALIDAE

62.024 *Rhodophaea formosa*

■ **LANCS STATUS** Rare / vagrant   2020

This species is considered a new arrival into the county from the south; it was new to Cheshire in 2010 (cheshire-moth-charts.co.uk, accessed September 2023). Lancashire's only record is of one that came to an MV light run alongside Stanning's Folly Wood on Ridley Lane, east of Bretherton, on 30 July 2020 (A. Barker).
■ **HABITAT** Woodland edge and hedgerows.
■ **FOODPLANTS** Not recorded in Lancashire.
Considered single brooded nationally. Comes to light. Nationally, the larva can be searched for in August to September on elm scrub or hedgerows, the larva spinning silk on a leaf upperside sometimes causing it to curl upwards.

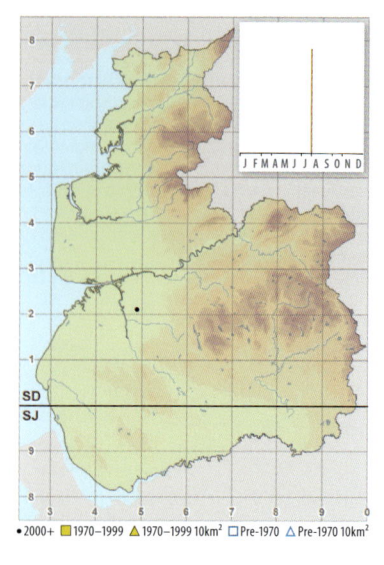

62.025 *Dioryctria sylvestrella*

■ **LANCS STATUS** Rare / vagrant   2022

Recorded as new to the county on 17 July 2022, when one came to a garden light trap in Longton, Preston (R. Boydell). It was new to Cheshire in 2012 and had been noted there on two further occasions, so is considered a new arrival as part of a national expansion of range northwards, rather than an overlooked resident. It is, however, also known to be migratory at times and this can't be ruled out considering the large number of migrant species seen in Lancashire during 2022.
■ **HABITAT** Unknown in the county.
■ **FOODPLANTS** Not recorded in Lancashire.
Comes to light. It is superficially similar to other *Dioryctria* species.

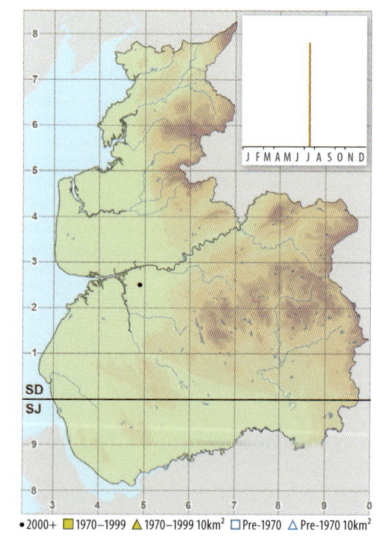

62.027 *Dioryctria simplicella*

■ **LANCS STATUS** Occasional / local resident   1986–2020

First recorded at Gait Barrows in 1986 by E. Emmett, followed by Billinge (C.A. Darbyshire) and Caton (M. Dempsey) in 1990. Since then, 22 of the 32 remaining county records have been from Sefton Coast pine woods, from Altcar to Ainsdale. The Silverdale area, in comparison, had six records. All reports involved low single numbers, with a few scattered, lowland reports of singletons elsewhere.
■ **HABITAT** Coniferous woodland and plantations.
■ **FOODPLANTS** Not recorded in Lancashire.
Single brooded. Comes to light. Easily confused with other *Dioryctria* species and early records of *D. abietella* from Silverdale and Freshfield might refer to this species.

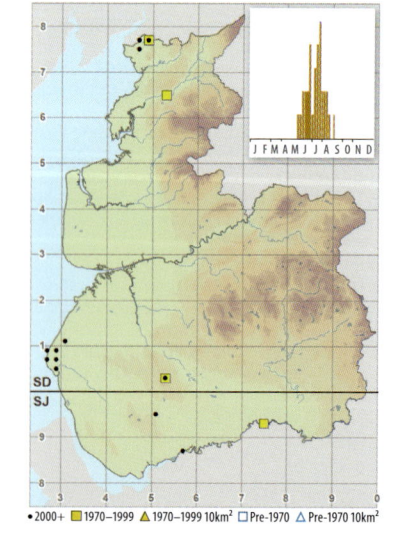

336   *The Moths of Lancashire*

62.028 *Dioryctria abietella*

■ **LANCS STATUS** Occasional / local resident; possible migrant　　　　**1986–2022**
First confirmed report was from Gait Barrows on 13 August 1986 and thereafter, scattered reports of mainly singletons received from across the county. It is found more frequently in some years than others, with only a few regular sites, mainly on the Sefton Coast from Formby and Ainsdale. Other locations with more than one or two records include Docker Moor, Gait Barrows and north Preston.
■ **HABITAT** Where pine species occur, but also on occasions away from such areas.
■ **FOODPLANTS** Not recorded in Lancashire.
Single brooded. Comes to light. Easily confused with other *Dioryctria* spp. and early records from Silverdale and Freshfield are omitted as they may refer to *D. simplicella*.

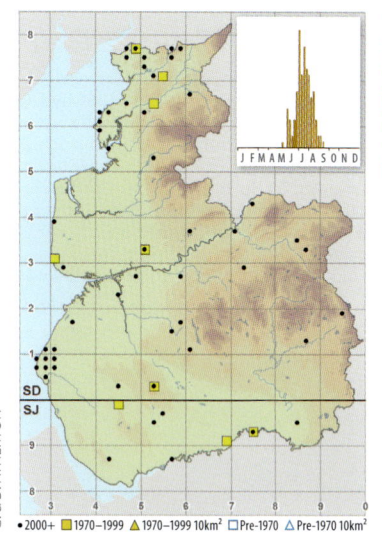

62.029 *Phycita roborella*

■ **LANCS STATUS** Occasional / widespread resident　　　　**1977–2022**
Probably overlooked historically. First noted at Yealand Storrs in 1977 and Risley Moss on 24 July 1982 (both M. Hull). Found at Gait Barrows in 1983 and, since 1989, noted regularly at Billinge. Elsewhere reports were more intermittent and scattered across lowland areas, with few upland sightings. Other sites with occasional, but nowhere near annual records include Flixton, Gait Barrows, Hale and Leighton Moss.
■ **HABITAT** Deciduous woodland and mature hedgerows with oak standards.
■ **FOODPLANTS** Not recorded in Lancashire
Single brooded, but may rarely have a small second brood in September. Comes to light, mainly in small single figures.

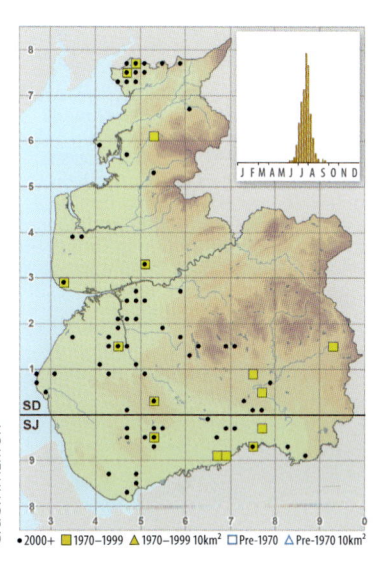

62.030 *Hypochalcia ahenella*

■ **LANCS STATUS** Rare / very local resident　　　　**2011**
A light trapping session held at the Lancashire Wildlife Trust's Cross Hill Quarry Local Nature Reserve on the 3 June 2011, by G. Jones, produced the only known record for this species in the county. Six were seen, confirming a breeding population was present, but two subsequent visits to this very sheltered site, during daylight hours, have failed to relocate the species. An exposed, limestone hill to the south has since been light trapped in the hope of rediscovering the moth, but to no avail.
■ **HABITAT** Disused section of inland limestone quarry.
■ **FOODPLANTS** Not recorded in Lancashire
Single brooded nationally. Comes to light.

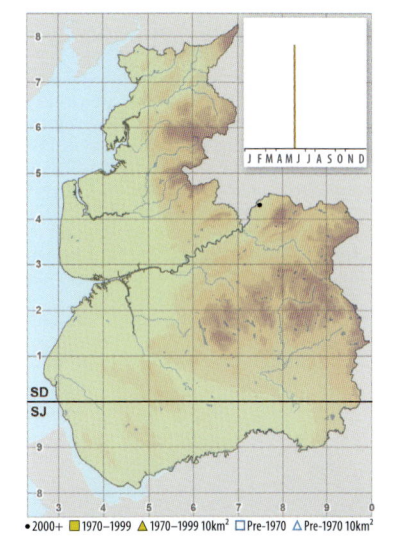

62.034 *Acrobasis repandana*

■ **LANCS STATUS** Occasional / local resident   c.1859–2022

First recorded on Chat Moss c.1859 by J. Chappell, under *A. tumidella* Zinck (Ellis, 1890). Mansbridge (1940), alters the name and authority, without explanation and presumably in error, to *A. tumidana* Schiff. Next reported in 1910 by A.E. Wright, also under *tumidana*, from Pendle Hill (Britten, c.1950). In 1992 it was seen in Billinge, followed by Hollingworth Lake and Widnes (2001), Gait Barrows (2002) and eight other mainly lowland sites, the last being Heaton Moor on 11 July 2022 (S. Roberts).
■ **HABITAT** Deciduous woodland.
■ **FOODPLANTS** Not recorded in Lancashire.
Single brooded. Comes to light, usually as singletons.

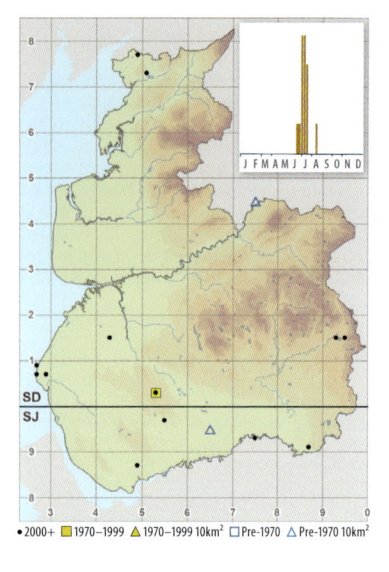

62.035 *Acrobasis advenella*

■ **LANCS STATUS** Frequent / widespread resident   c.1889–2022

First found alongside the R Wyre by J.B. Hodgkinson (Ellis, 1890) and then Silverdale in 1910, but was not reported again until 1961, at Leighton Moss. There were seven records in the 1980s and from 1990 a rapid expansion occurred across lowland parts, reaching the eastern uplands in 1998, at Littleborough. It is now established widely wherever Hawthorn occurs and is a regular at most garden light traps.
■ **HABITAT** Hedgerows, scrubland, brownfield sites and gardens.
■ **FOODPLANTS** Hawthorn, on the leaves and in the berries.
Single brooded. Comes to light, sometimes in numbers. Counts of over 20 were once regular at Flixton (1999 to 2011), the maximum being 67 on 4 August 2004.

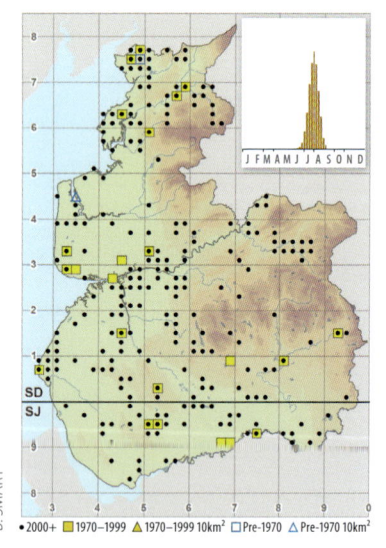

62.037 *Acrobasis marmorea*

■ **LANCS STATUS** Occasional / very local resident   1947–2022

First reported in the Silverdale area in 1947 (Britten, c.1950) and by H.N. Michaelis in 1959. Next noted at Yealand Hall Allotment (M. Hull *et al.*) in 1977, with the first of several sightings at Gait Barrows in 1993 (R. Petley-Jones). Other nearby sites include Jack Scout, Leighton Moss, Silverdale Moss, Silverdale village, Warton Crag, Warton village and Yealand Conyers. Single records were made on upland limestone at Dalton Crags on 19 July 2021 and 7 July 2022 (J. Patton *et al.*).
■ **HABITAT** Scrubland and hedgerows on the northern limestone.
■ **FOODPLANTS** Not recorded in Lancashire.
Single brooded. Comes to light, usually as singletons.

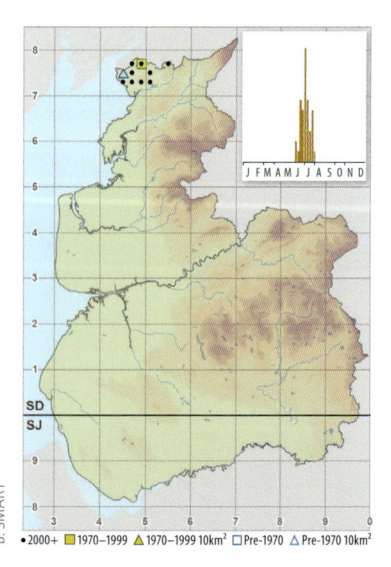

338   *The Moths of Lancashire*

62.038 *Acrobasis consociella*

■ **LANCS STATUS** Occasional / local resident   c.1887–2022
Noted originally in the Silverdale area, with four records between c.1887 and 1977, the first of these by J.B. Hodgkinson (Ellis, 1890). Next, seen at Risley Moss in 1995, Gait Barrows in 1999 and Silverdale in 2002. In 2006, the species began a slow range expansion, mainly confined to the Mersey Valley area, with several seen at sites such as Astley Moss, Billinge, Chorlton and Flixton. Since then, it has also been noted at Docker Moor and Freshfield in 2009, Preston in 2013 and Oldham Edge in 2021.
■ **HABITAT** Deciduous woodland and other locations with young oak growth.
■ **FOODPLANTS** Oak, in conspicuous, often communal spinnings.
Single brooded. Comes to light. Larvae found late August to June, often on saplings.

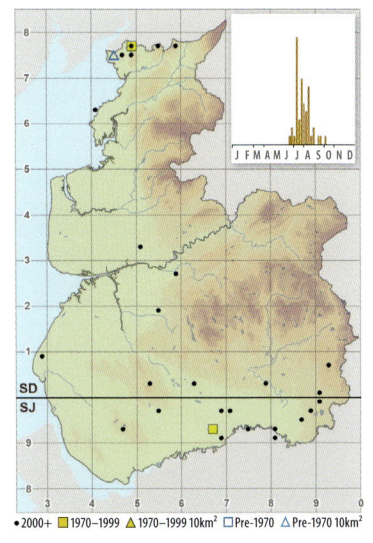

• 2000+ ▪ 1970–1999 ▲ 1970–1999 10km² □ Pre-1970 △ Pre-1970 10km²

62.039 *Apomyelois bistriatella*

■ **LANCS STATUS** Scarce / very local resident   1999–2016
Records have been few and scattered, the first being at Heysham N.R. on 6 July 1999 (P.J. Marsh), followed by Gait Barrows on 23 August 1999 (R. Petley-Jones). It next came to light in Flixton on 28 July 2001 and again on 30 May 2016 (K. McCabe). In between these two records, it was noted at Freshfield Dune Heath on 15 July 2009 (S. Palmer) and Heysham Head on 21 July 2014 (P.J. Marsh).
■ **HABITAT** Dune heathland, brownfield sites with gorse and birch, and scrubland.
■ **FOODPLANTS** Not recorded in Lancashire.
Comes to light. Nationally, it is associated with *Daldinia* sp. fungus growing on burnt gorse and birch, probably explaining its erratic occurrence in the county.

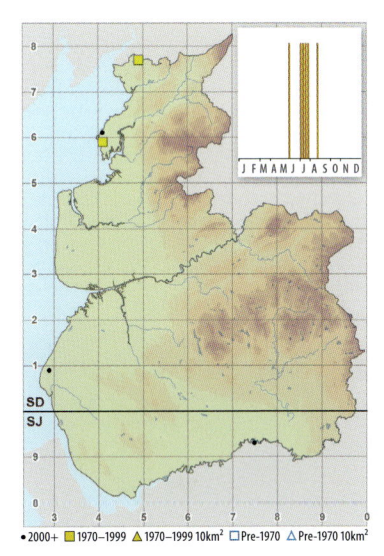

• 2000+ ▪ 1970–1999 ▲ 1970–1999 10km² □ Pre-1970 △ Pre-1970 10km²

62.040 Locust Bean Moth *Apomyelois ceratoniae*

■ **LANCS STATUS** Adventive   c.1905
The only documented record is the one mentioned in Mansbridge (1940) where the species is reported as 'Occasionally but regularly found in dried dates. Most of the specimens bred have been var. *pryerella* Vaughan' – presumably a misspelling of *pyralella* Vaughan. The location is given as Liverpool and the recorder W. Mansbridge. No indication was given in the account of the date range when this was found but it is believed to refer to around 1905. The wording above demonstrates that it was found on more than one occasion and possibly in other years prior to 1940.
■ **HABITAT** Probably established as a pest in a dock storehouse.
■ **FOODPLANTS** Dried dates.
This species was accidentally introduced in the docks of London as well as in Liverpool and could be found resting in dried fruit warehouses in the past. Unusually for an adventive, it was seen at light in the wild more often than the majority of other warehouse pests (Goater, 1986).

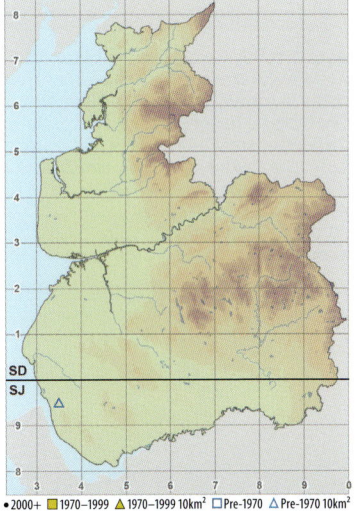

• 2000+ ▪ 1970–1999 ▲ 1970–1999 10km² □ Pre-1970 △ Pre-1970 10km²

PYRALIDAE

62.042 Thistle Ermine *Myelois circumvoluta*

- **LANCS STATUS** Frequent / widespread resident 1991–2022

First recorded at Worsley by G. Riley on 12 July 1991, it soon spread widely across lowland parts, entering VC60 in 1996 and Worsthorne, in the east, in 1998. It has since become an annual visitor to many light traps across the county, although it does remain more local in the east and north-east of the county. 16 adults were found on thistle flowerheads at King's Moss on 20 June 2005 (C.A. Darbyshire).
- **HABITAT** Rough ground, meadows, roadside verges, gardens and brownfield sites.
- **FOODPLANTS** Spear Thistle and a burdock sp.

Single brooded. Comes to light, usually in small numbers; nectars on thistle. The larva has been found in thistle stems in most months from August to March.

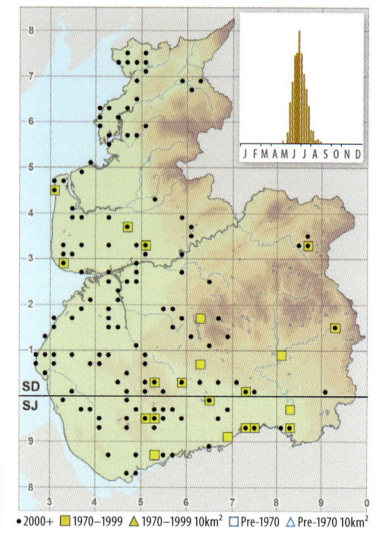

62.048 *Euzophera pinguis*

- **LANCS STATUS** Frequent / widespread resident 1938–2022

First recorded in 1938 in Didsbury by H.N. Michaelis and then in Formby in 1950 (K.C. Greenwood). It was not until 1997 that it started appearing annually and spreading across lowland parts of the county. In coastal areas it arrived in St Annes in 2001, Preston in 2003 and Bay Horse, and the Silverdale area, in 2007. To the east, it reached Rochdale in 2009 but remains quite local in upland areas. Now occurs annually in many lowland gardens, from the Mersey Valley to Silverdale.
- **HABITAT** Woodland, hedgerows and brownfield sites.
- **FOODPLANTS** Not recorded in Lancashire.

Single brooded. Comes to light, mainly in small numbers.

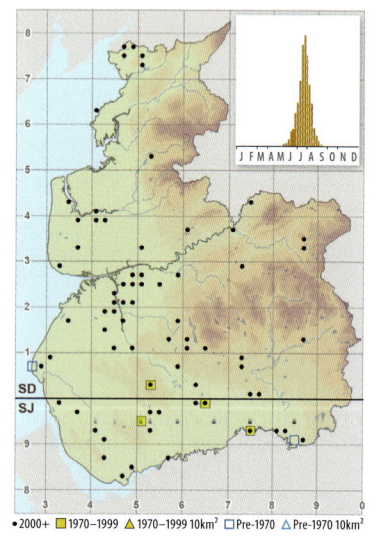

62.054 *Homoeosoma sinuella*

- **LANCS STATUS** Occasional / local resident 1984–2022

A relatively recent arrival in the county following a national expansion of range. First noted in Formby in 1984 by L.W. Hardwick, then St Annes in 1996 and Flixton in 1997. Since then, it has occupied many coastal and estuarine areas and moved inland along the Mersey and Ribble valleys and is probably still very gradually expanding its range. 17 seen at Fleetwood in 2022 and 20 at St Annes in 2014 during the day.
- **HABITAT** Sandy or limestone grassland banks and some dry brownfield sites.
- **FOODPLANTS** Not recorded in Lancashire.

Probably single brooded with an extended flight period. Comes to light in small numbers, but readily disturbed during the daytime in sunny conditions.

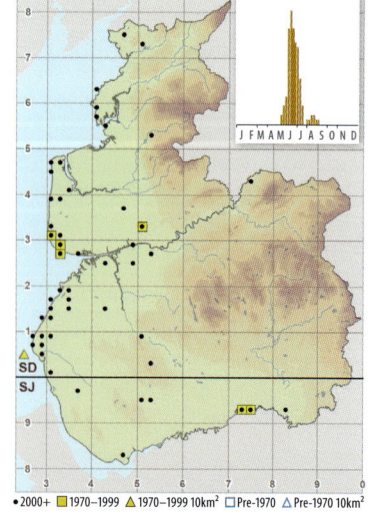

340 The Moths of Lancashire

62.057 *Phycitodes maritima*

■ **LANCS STATUS** Occasional / local resident   **1930–2022**
First recorded commonly as a larva at Formby from 1930–37 by W. Mansbridge (Leech & Michaelis, 1957) and next in 1984 at Gait Barrows (E. Emmett). Since 2006, searches of ragwort flower-heads for larvae proved this to be more widespread than first thought, the patchy distribution suggesting it is probably under-recorded in lowland areas. Other sites with regular records include Chorlton and Preston.
■ **HABITAT** Rough ground, dune grassland, meadows, gardens and brownfield sites.
■ **FOODPLANTS** Common Ragwort.
Single brooded. Comes to light. Larvae found in densely spun flowerheads from July to September. Separable as larvae; adults require dissection to exclude *P. saxicola*.

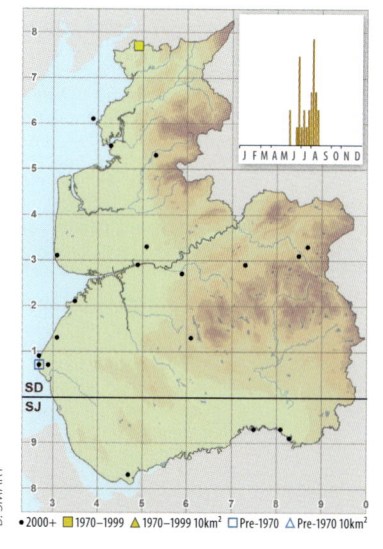

62.058 *Phycitodes binaevella*

■ **LANCS STATUS** Frequent / widespread resident   **1922–2022**
First noted in the county at Formby in 1922 by W. Tyerman, and again in 1930 (W. Mansbridge). It was found at Morecambe in 1959, Crosby in 1975, Yealand Storrs in 1977, Martin Mere in 1986 and Billinge from 1987. The range increase has been gradual, east and north in lowland VC59 and south from the northern limestone areas. It was found on upland northern limestone in 2005 and Rishton in 2018. Double figures are rare, but seen at Parr, with 19 on 19 August 2018 (R. Banks).
■ **HABITAT** Dune and limestone grassland, rough ground and brownfield sites.
■ **FOODPLANTS** Not recorded in Lancashire.
Single brooded. Comes to light, usually in small numbers.

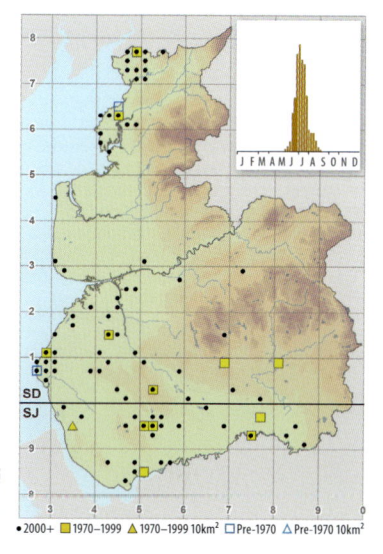

62.059 *Phycitodes saxicola*

■ **LANCS STATUS** Occasional / local resident   **1933–2022**
First noted in Formby in 1933 by W. Mansbridge, the genitalia checked by F.N. Pierce. Thereafter, not seen again until 1996 in St. Annes (D. & J. Steeden), followed by reports from Billinge, Parr, Rossall and Chorlton. Predominantly a lowland species, but with occasional upland records, in Rishton and Briercliffe. Sites with the most regular reports include Billinge, Flixton, Morecambe, Southport and St Annes.
■ **HABITAT** Dune grassland, rough ground, meadows, brownfield sites and gardens.
■ **FOODPLANTS** Common Ragwort and Yarrow.
Mainly single brooded. Comes to light in small numbers. Larvae found from late July to mid-August and at one site noted as only one larva per flowerhead.

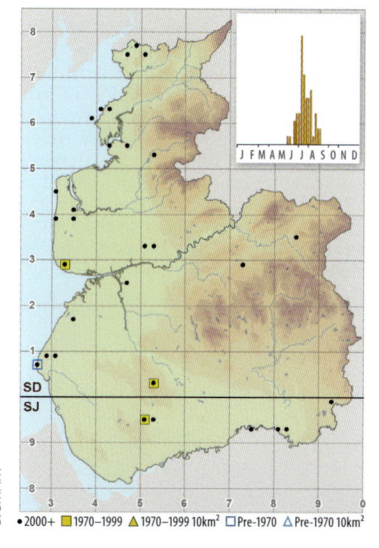

62.062 Indian Meal Moth *Plodia interpunctella*

■ **LANCS STATUS** Adventive c.1857–2022

All early records relate to its presence on warehouse walls and doors where grain and fruit were stored in the Liverpool and Manchester Dock areas, with the first in Chapel St., Liverpool (Gregson, 1857). Seen flying around in pet shops and garden centres that sell bird seed, in Liverpool (1990), Preston (1995), Clifton (2004) and Worsley in 2006. At the latter site larvae were abundant in grain bins.

■ **HABITAT** Indoors, where seed is stored.

■ **FOODPLANTS** Stored bird seed, peanuts and once on stored dates of Iraqi origin. Continuously brooded. Has been caught in garden light traps fairly frequently, but this is usually at sites where bird food has been stored nearby.

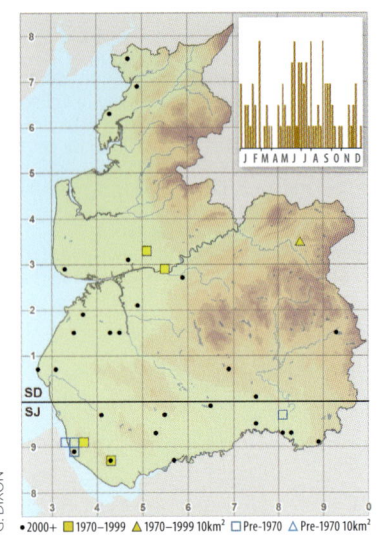

62.063 Mediterranean Flour Moth *Ephestia kuehniella*

■ **LANCS STATUS** Adventive 1917–2001

As with many adventive species, the first reports came from Liverpool Docks where it was abundant in warehouses and corn and rice mills in 1917 (Mansbridge, 1940). In the same year it was noted at Pendleton, in stored flour, by A.W. Imms. There were two reports from the Liverpool area in 1972 and 1975 and one was found indoors in Preston in 1995. It was next reported at Moor Park Mill, Preston with several seen in 1997 and a large infestation occurred in Heysham from 1999 to 2001 (P.J. Marsh).

■ **HABITAT** Indoors, where stored food products are kept.

■ **FOODPLANTS** Flour, corn and rice. Associated with stored bird food on one occasion. Continuously brooded indoors with around 100 adults seen on occasions.

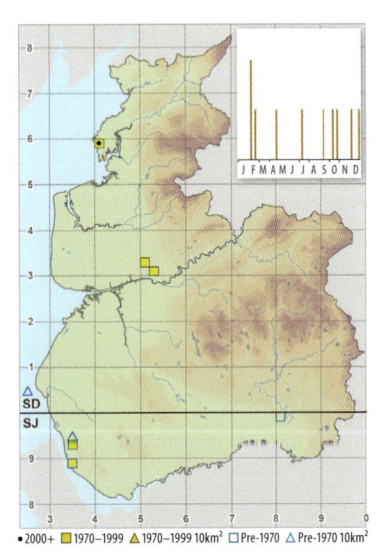

62.064 Cacao Moth *Ephestia elutella*

■ **LANCS STATUS** Adventive; occasional / local resident c.1853–2022

Noted by all early authors as plentiful in the streets near docks and in warehouses, the first such report in 1853 in Liverpool (Byerley, 1854). Found at Preston Docks pre-1890 and, as well as further Liverpool records, was seen in Burnley in 1920 and Formby pre-1957. All subsequent sightings have been outdoors, the first of these in the daytime at Winmarleigh Moss on 18 July 1996, c.1km from horse stables. The others were all singletons in garden light traps at ten scattered, lowland sites.

■ **HABITAT** Formerly indoors, since 1996 in gardens and probably stables.

■ **FOODPLANTS** Stored food products and animal feed.

Indoors continuously brooded, probably double brooded outdoors. Comes to light.

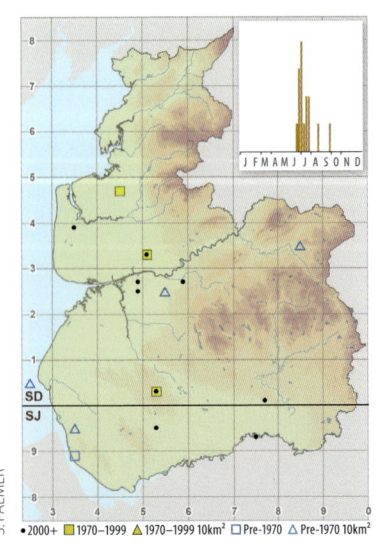

62.065 *Ephestia woodiella*

■ **LANCS STATUS** Scarce / very local resident   2010–2022

First recorded in a Leighton Moss MV light trap on 3 July 2010 (S. Palmer). It was not recorded again until one came to light at Great Sankey on 22 May 2017 (J. Mitchell-Lisle). Since then, found on four further occasions, at Formby in June 2020 and Chorlton, Preston and Parr in June 2022, all as singletons. This is suggestive of a range expansion occurring in the county and perhaps more widely.

■ **HABITAT** Gardens and adjacent to human habitation.

■ **FOODPLANTS** Not recorded in Lancashire.

Single brooded. Comes to light. Similar looking to *E. elutella*. This sp. has also, until quite recently, been known under the synonyms of *E. parasitella* and *E. unicolorella*.

● 2000+   ■ 1970–1999   ▲ 1970–1999 10km²   □ Pre-1970   △ Pre-1970 10km²

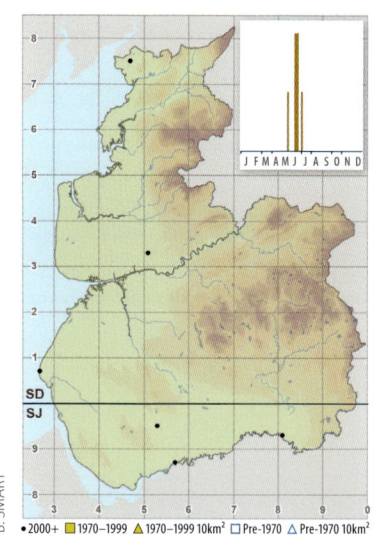

62.066 Raisin Moth *Cadra figulilella*

■ **LANCS STATUS** Adventive   c.1860–c.1940

Said to be the least common of the *Ephestia* species (Goater, 1986), there are only two reports of this moth in the county. C.S. Gregson found and named this species as new to science, based on specimens from Liverpool (Gregson, 1871). He noted 'That this fine and distinct species has now been plentiful upon warehouse walls, in Liverpool, for several years', so from at least the 1860s. The only other report came from Mansbridge (1940), with no date, where it was documented that 'I once bred this from raisins purchased in Liverpool'.

■ **HABITAT** Indoors, where dried fruits were stored.

■ **FOODPLANTS** Raisins. Also bred by Gregson, from ova laid by gravid females in warehouses, using figs and raisins. When bred, he noted it as double brooded and that the second brood produced small specimens (Gregson, loc. cit.).

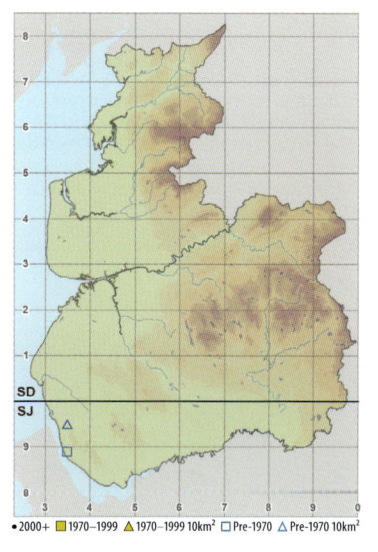

● 2000+   ■ 1970–1999   ▲ 1970–1999 10km²   □ Pre-1970   △ Pre-1970 10km²

62.067 Dried Currant Moth *Cadra cautella*

■ **LANCS STATUS** Adventive   c.1886–2022

First reported as abundant in Preston in an oil cake mill by J.B. Hodgkinson c.1886 (Ellis, 1890). Then noted as very abundant in Liverpool warehouses in 1900, and again in 1917, by W. Mansbridge. On 8 May 2022 one came to light at Parr, St Helens (R. Banks) and over the following two weeks a further seven individuals were found, apparently associated with stored bird food in a garden shed.

■ **HABITAT** Indoors, in warehouses.

■ **FOODPLANTS** Oil Cake.

Searches for the larvae of the Parr moths were made amongst stored bird food without success; no other likely pabulum seemed available.

● 2000+   ■ 1970–1999   ▲ 1970–1999 10km²   □ Pre-1970   △ Pre-1970 10km²

The Moths of Lancashire   **343**

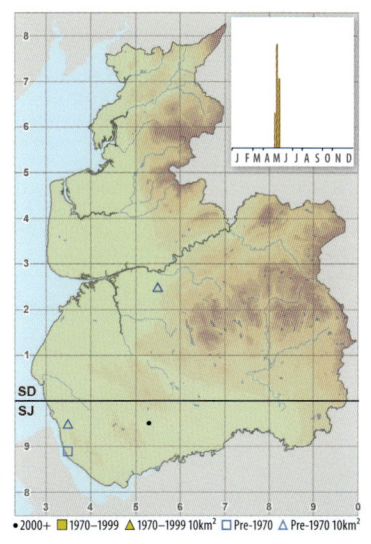

PYRALIDAE

62.068 Dried Fruit Moth *Cadra calidella*

■ **LANCS STATUS** Adventive   c.1887–1949

First reported from Prestwich by J.C. Melvill (Ellis, 1890) and later, at Liverpool Docks as an abundant warehouse pest in 1917 by W. Mansbridge and W.A. Tyerman, where it was found as larvae feeding on raisins. It was noted that many of the adventive species, including *C. calidella* (under the name *E. ficella*), had not been seen so commonly on the walls for about ten years (Mansbridge, 1918). The only other record came from the Didsbury area of Manchester, in 1949 by H.N. Michaelis, who found it in a grocer's shop (Mansbridge & Wright, 1950).

■ **HABITAT** Indoors, in warehouses.
■ **FOODPLANTS** Raisins.

More widely, it was noted as an uncommon species in warehouses, as far north as the Clyde (Goater, 1986). Since then, it has been reported that there were no recent records of the species in Britain (Langmaid, *et al.*, 2018). As with most of these very similar adventive, pyralid species, retention of a specimen will be essential to establish a correct identification.

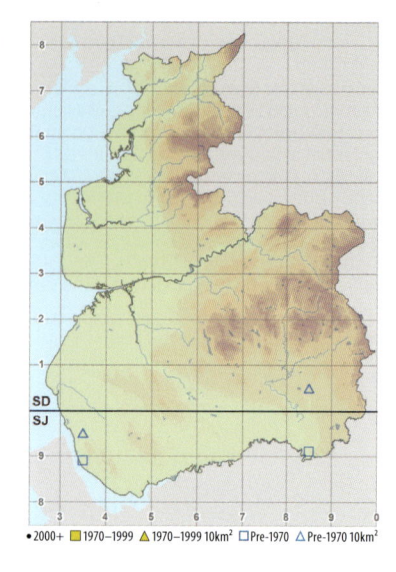

62.069 *Anerastia lotella*

■ **LANCS STATUS** Occasional / very local resident   c.1887–2022

First noted at St Annes by J.B. Hodgkinson (Ellis, 1890), then at Crosby in 1907, Formby in 1921, Ainsdale in 1957 and Freshfield in 1959. Most of these locations have remained the main source of records to the present day, excepting Crosby and Freshfield, with no reports respectively since 1907 and 1960. It was seen at Fairhaven in 1987 and 1996 but not since, and twice in the Southport area, in 2012 and 2014.

■ **HABITAT** Sand dunes.
■ **FOODPLANTS** Not recorded in Lancashire.

Single brooded. Comes to light, mostly in low single figures. The largest count, of 30, were attracted to light at Birkdale Green Beach on 25 July 2012 (G. Jones).

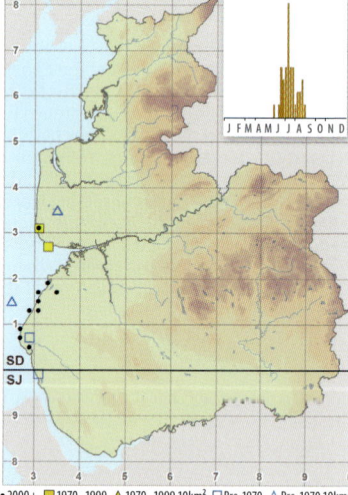

62.072 Meal Moth *Pyralis farinalis*

■ **LANCS STATUS** Occasional / widespread resident   1845–2022

First recorded 'on the coast' in 1845 (Hodgkinson, 1845) and later, abundantly in stables and granaries (Ellis, 1890). By 1940, it was noted as common at such sites, including grain stores. It was also noted as common at farm buildings in the Formby area in the 1950s (Leech & Michaelis, 1957). Since then, it has been found widely across lowland parts, but numbers have declined; it is rare in the uplands. Regular sites include Billinge, Flixton, Heysham, Parr, Preston, St Annes and St Helens.

■ **HABITAT** Farms, stables, grain stores and gardens.
■ **FOODPLANTS** Not recorded in Lancashire.

Single brooded. Comes to light, usually as singletons.

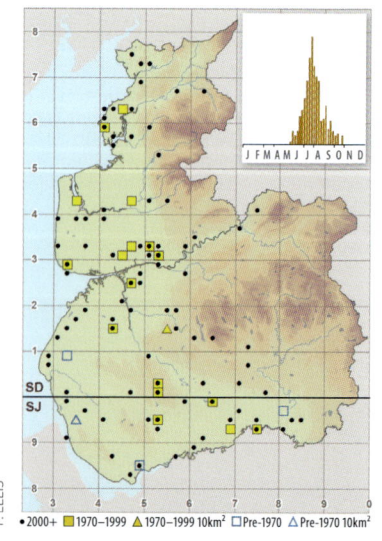

The Moths of Lancashire

62.073 Small Tabby *Aglossa caprealis*

■ **LANCS STATUS** Considered Extinct   c.1859–1922

There are only two historic records of this species in the county, both under *A. cuprealis* Hübner, and the demise of this moth in Lancashire mirrors the significant decline across much of England. At best, it was always considered scarce and local in the north-west of England (Mansbridge, 1940). The first record was from Barton, near Manchester, probably around the mid-19th century by J. Chappell, but the precise date is undocumented. The only other record in Lancashire was from Old Trafford, Manchester, in 1922 by G.W. Wynne (Mansbridge, loc. cit.), with an unconfirmed earlier report from the same district in 1913.

■ **HABITAT** Unknown in the county.
■ **FOODPLANTS** Not recorded in Lancashire.

Nationally, this species was associated with barns, outhouses and stables in low-lying farmland. The larva feed from a silken gallery amongst vegetable debris, preferring damp material (Goater, 1986).

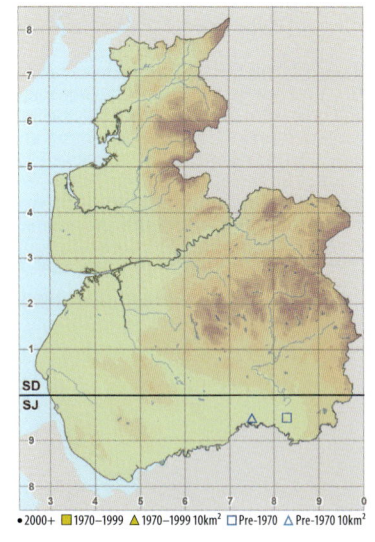

● 2000+   ■ 1970–1999   ▲ 1970–1999 10km²   □ Pre-1970   △ Pre-1970 10km²

62.074 Large Tabby *Aglossa pinguinalis*

■ **LANCS STATUS** Scarce / very local resident   c.1857–2022

First recorded in the Liverpool district and noted as plentiful in stables by Gregson (1857) and Ellis (1890). It was noted in Moss Side in 1915, Old Trafford in 1920 and, although noted as fairly common by Mansbridge (1940), no further records were documented until 1973 in St Annes and the Burnley area in 1979. Since then, it has been found occasionally as singletons in outbuildings or attracted to light, in Ainsdale, Formby, Southport (most frequently) and St Annes.

■ **HABITAT** Stables and other outbuildings.
■ **FOODPLANTS** Not recorded in Lancashire.

Single brooded, a very late one was found on Altcar Moss on 12 September 2015.

● 2000+   ■ 1970–1999   ▲ 1970–1999 10km²   □ Pre-1970   △ Pre-1970 10km²

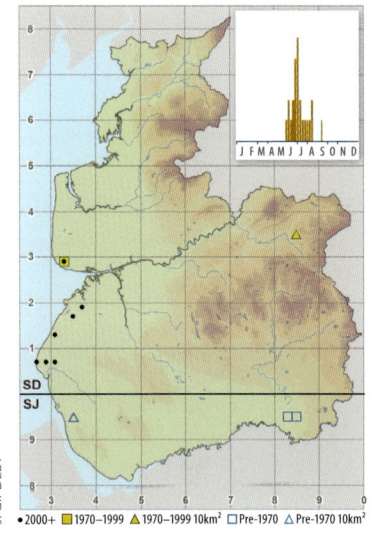

62.075 Gold Triangle *Hypsopygia costalis*

■ **LANCS STATUS** Frequent / widespread resident   1938–2022

First recorded in Formby in 1938 by G.C. Fraser, then from Blackburn, Ormskirk and Wigan c.1950, Silverdale in 1959 and Burnley c.1979. Annual records remained low until 1998, when 67 were received. This increase peaked in 2004 at 114 reports, with a steady decline thereafter. There were only seven reports in 2017, but still present in Worsthorne in 2022 alongside eleven other records from mostly lowland sites.

■ **HABITAT** Woodland, scrub, hedgerows, gardens and brownfield sites.
■ **FOODPLANTS** Not recorded in Lancashire.

Single brooded. A small second brood occurred during its peak period. Comes to light in low numbers, but 13 were found at Parr on 15 July 2003 (R. Banks).

● 2000+   ■ 1970–1999   ▲ 1970–1999 10km²   □ Pre-1970   △ Pre-1970 10km²

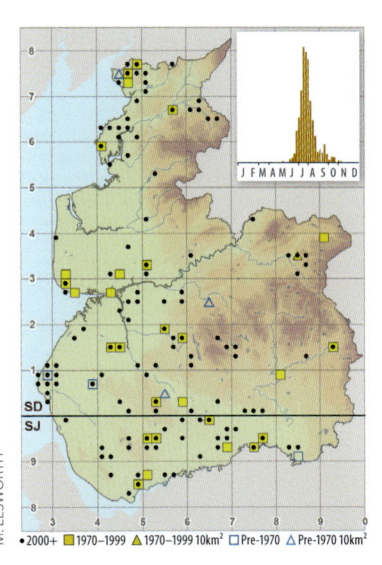

PYRALIDAE

62.076 *Hypsopygia glaucinalis*

■ **LANCS STATUS** Occasional / widespread resident   c.1857–2022

First noted at Hale c.1857 by C.S. Gregson, then Barton Moss c.1859, Didsbury in 1934 and other sites to 1950. Seen at Billinge in 1989 and gradually expanded its range within lowland VC59, appearing almost annually at Billinge, Flixton and Parr. There have been 16 records in VC60, all since 2002, and most of these from Leighton Moss. It is scarce in the uplands with only eight records throughout, all since 2004.

■ **HABITAT** Woodland, scrubland, hedgerows and gardens.
■ **FOODPLANTS** Not recorded in Lancashire.

Probably double brooded. Comes to light, mainly in small numbers; eleven seen in Flixton on 10 October 2018 (K. McCabe). A late May sighting at Worsley was unusual.

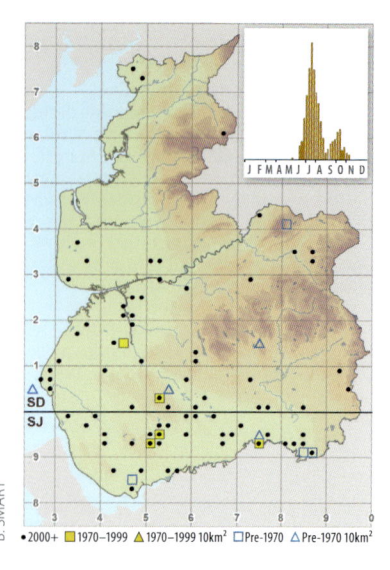

62.077 *Endotricha flammealis*

■ **LANCS STATUS** Rare / vagrant   2007–2023

Until very recently, the only record related to one attracted to a MV light trap in St Annes on 11 August 2007 (D. & J. Steeden). On 28 July 2023, a singleton was attracted to a garden light trap in Hale by C. Cockbain, followed by another, on 11 August 2023 to light in Eccleston, St Helens (T. Ferguson). Elsewhere, it has been seen a few times in Cheshire, is quite widely distributed in parts of south and east Yorkshire and it has been noted once in Cumbria.

■ **HABITAT** Unknown.
■ **FOODPLANTS** Not recorded in Lancashire.

Probably single brooded. Comes to light.

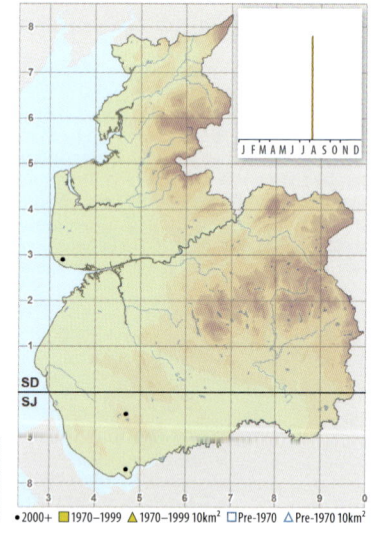

CRAMBIDAE

63.002 *Loxostege sticticalis*

■ **LANCS STATUS** Scarce / migrant   c.1859–2022

First noted at Ashton-under-Lyne by J. Chappell around 1859, with other 19th century reports from Lytham St Annes and Chat Moss. Thereafter, unrecorded for over 100 years, but has been noted on seven occasions in the last 30 years. Recorded again at Lytham St Annes in 1995 and 2000, and at Parr, St Helens in 2001. In 2019 the moth turned up at Sutton Leach, St Helens, with others at Freshfield and Worsthorne in 2020. 2022 saw two more reported; from Fallowfield, Manchester and Green Bank, Thrushgill. All post-2000 records are of moths to light.

■ **HABITAT** Unknown in the county.
■ **FOODPLANTS** Not recorded in Lancashire.

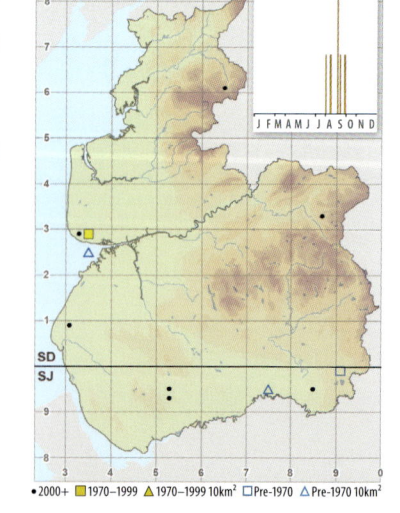

346   *The Moths of Lancashire*

63.003 *Pyrausta cingulata*

■ **LANCS STATUS** Occasional / very local resident   c.1887–2022
Recorded 'at Silverdale, among wild sage' by J.B. Hodgkinson (Ellis, 1890); status given as local. Virtually all the subsequent 129 records are from Morecambe Bay limestone with most from Gait Barrows, Warton Crag and Yealand Hall Allotment. Single records from Jack Scout 1998, Trowbarrow Quarry 2012, Dalton Crags 2021 and three at Cote Stones, Warton 2015. A bizarre, but confirmed, record, was of a singleton recorded at Flixton in 2016 by B. Hilton, the origin of which is unknown.
■ **HABITAT** Limestone grassland.
■ **FOODPLANTS** Unrecorded within the county.
Double brooded. Mostly recorded during day, flying in sunshine. Occasional at light.

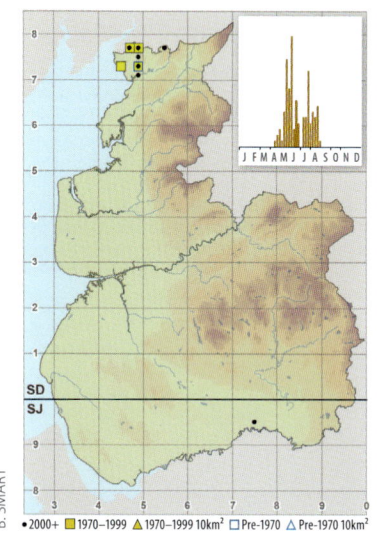

63.005 *Pyrausta despicata*

■ **LANCS STATUS** Occasional / local resident   c.1890–2022
The earliest report of this species in the county is from Ellis (1890), noting the moth to be common and generally distributed, but without listing specific sites. Recorded from Silverdale and Formby pre-1940. The only other 20th century records were from Ainsdale and Freshfield in the 1950s and Gait Barrows and Heysham N.R. in the 1990s. Since 2008, far more frequently recorded, with new sites at Fairhaven and Ashton Moss, and on higher ground at Worsaw Hill, Green Bank, Dalton Crags, etc.
■ **HABITAT** Grassland, sand dunes.
■ **FOODPLANTS** Unrecorded within the county.
Double brooded with overlapping generations. Noted during day; also at light.

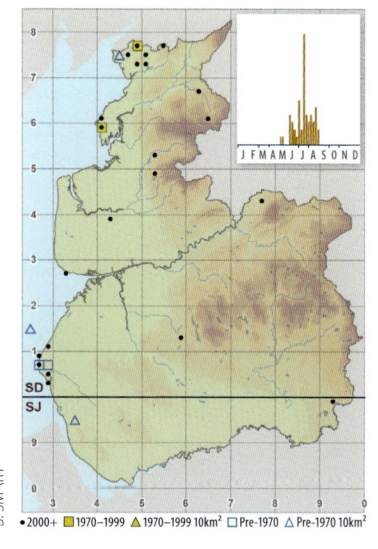

63.006 *Pyrausta aurata*

■ **LANCS STATUS** Frequent / widespread resident   c.1857–2022
A now familiar moth, having rapidly increased in range and numbers during the last 40 years. First noted at Lydiate, Liverpool, where described as scarce by C.S. Gregson (1857). Other than from Silverdale pre-1940, potentially a misidentification of the following species, not known again until at Ainsdale 1985 by S.H. Hind. Since, found to be common wherever mint and Marjoram grow, e.g., gardens at Billinge, Swinton, Briercliffe, Chorlton, etc. Often noted in gardens flying in sunshine, but also at light.
■ **HABITAT** Gardens, grassland.
■ **FOODPLANTS** Mint, Marjoram. Larvae typically noted between June and September. Double brooded. A late larval record from Chorlton, at light, on 24 November 2003.

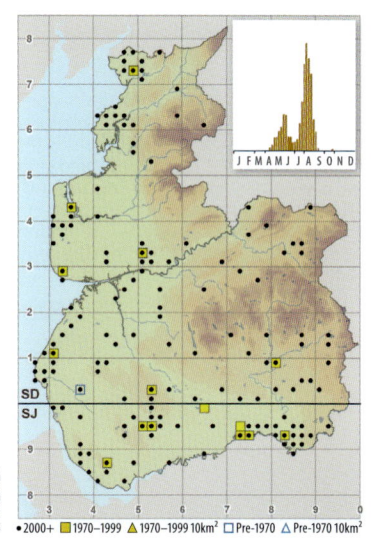

The Moths of Lancashire **347**

CRAMBIDAE

63.007 *Pyrausta purpuralis*

■ **LANCS STATUS** Occasional / local resident   c.1857–2022

First recorded at Ditton, nr. Widnes (Gregson, 1857), and on numerous occasions at Silverdale from 1880s onwards. Adjudged by Mansbridge (1940) to be common, with presence noted at Formby. Now, occasional on Sefton Coast; less frequent on Fylde coast. Most records are from Gait Barrows, Warton Crag, and other limestone sites such as Yealand Hall Allotment and Dalton Crags. Occasionally noted on moorland sites such as Docker Moor, and in lowland gardens, e.g., St Helens and Flixton.
■ **HABITAT** Limestone grassland, coastal grassland, gardens.
■ **FOODPLANTS** Unrecorded within the county.
Double brooded. Flies in sunshine; occasional at light.

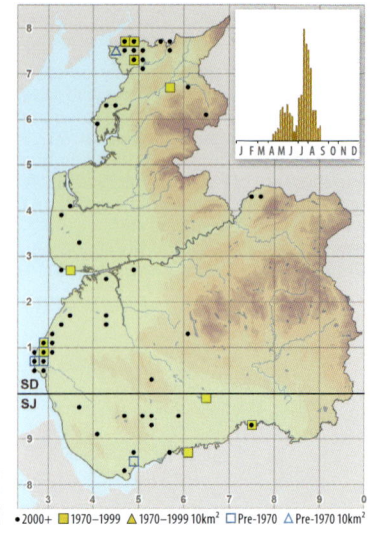

63.008 *Pyrausta ostrinalis*

■ **LANCS STATUS** Scarce / very local resident   c.1887–2022

A difficult moth to separate from the previous, with a good view of the underside needed, hence potentially under-recorded. All early records came from Silverdale, the first by J.C. Melvill (Ellis, 1890). Recorded at Gait Barrows in 1984, with most subsequent records from this site and, to a lesser extent, from Warton Crag. The few other records come from Myers Allotment, Jack Scout, Dalton Crags, Yealand Hall Allotments and Heald Brow. Unrecorded outside the limestone sites of north Lancs.
■ **HABITAT** Limestone grassland.
■ **FOODPLANTS** Unrecorded within the county.
Double brooded. Recorded flying in sunshine and occasionally at light.

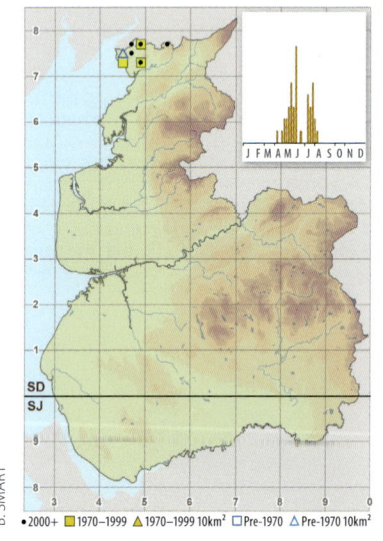

63.014 *Sitochroa palealis*

■ **LANCS STATUS** Rare / migrant   c.1859–2008

Noted at Barton Moss c.1859 by J. Chappell, with a further 19th century record from Seaforth on 12 August 1876 by W. Whitwick. The only other confirmed records were from Penketh, where J. Clarke recorded the moth twice. The first turned up at light on 14 July 2008; the second appeared eight days later. The moth is a migrant species, but also resident in southern England and capable of forming temporary colonies elsewhere. Possibly moving north, as there are also five post-2018 Cheshire records.
■ **HABITAT** Coastal grassland, waste ground.
■ **FOODPLANTS** Unrecorded within the county.
Nationally, single brooded.

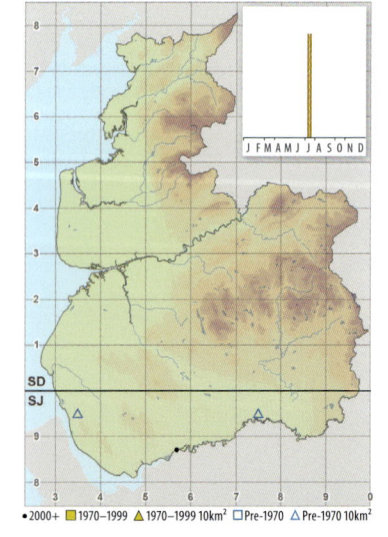

348 *The Moths of Lancashire*

63.016 *Anania fuscalis*

■ **LANCS STATUS** Occasional / local resident   c.1859–2022

First recorded c.1859 in Manchester by J. Chappell, and later by J.B. Hodgkinson, finding it common among Yellow-rattle in the Preston district (Ellis, 1890). Other pre-1940 records from Southport and Formby, and it is in the Sefton Coast dunes that the moth is most often recorded. Larvae found in foodplant seed-heads, Birkdale dunes, 1 July 2022. Unrecorded from the Fylde dunes since early 1980s. Few records elsewhere, e.g., Heysham N.R., Horwich, Longridge, Abram and Haskayne.
■ **HABITAT** Sand dunes, open woodland, meadows.
■ **FOODPLANTS** Yellow-rattle.
Single brooded. Recorded in day, easily disturbed from vegetation. Also, at light.

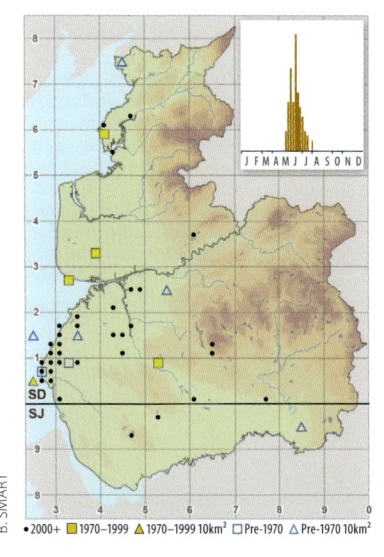

63.017 *Anania lancealis*

■ **LANCS STATUS** Rare / very local resident   2022

An unsurprising addition to the county fauna in 2022. This species is extending its range north and was first recorded in Yorkshire and Cheshire in 2005 and 2014. The moth was recorded at a MV trap in a garden at Parr, St Helens on 16 May 2022 by R. Banks, with photographic evidence sufficient for confirmation. The next few years are likely to see further Lancashire records, following the pattern established in neighbouring counties.
■ **HABITAT** Unknown in the county.
■ **FOODPLANTS** Not recorded in Lancashire.
Nationally, single brooded. The single record was of one at light.

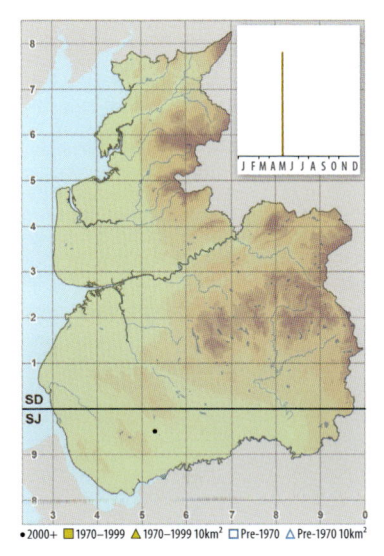

63.018 *Anania coronata*

■ **LANCS STATUS** Frequent / widespread resident   c.1857–2022

First noted from Liverpool district around Elder hedges (Gregson, 1857). Reported to be common at street lamps and waste ground in 1920s Manchester by G.W. Wynne. Early authors found the moth to be common and generally distributed. Records significantly increased from the 1970s onwards, initially from lowland areas, but recently from upland sites too. The moth has been recorded on 3,851 occasions, all but 263 from this century. High count of 22 at St Helens on 26 June 2011 (D. Owen).
■ **HABITAT** Woodland, hedgerows, scrub, gardens.
■ **FOODPLANTS** Elder; once on bindweed. The larva spins a web on the leaf's underside.
Single brooded. Comes to light. Larvae recorded August and September.

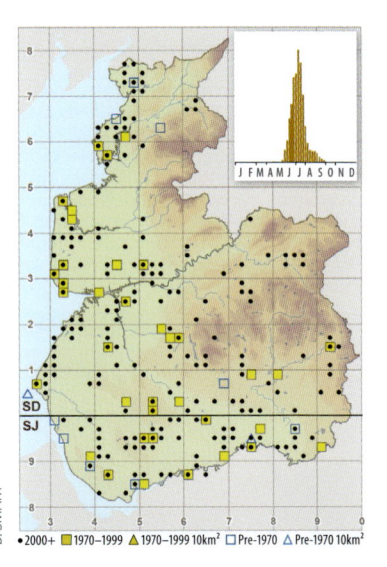

The Moths of Lancashire   **349**

CRAMBIDAE

63.020 *Anania perlucidalis*

■ **LANCS STATUS** Occasional / local resident   2002–2022

This recent arrival has been recorded 277 times since noted at Woolston Eyes on 13 July 2002 by D. Taylor. The moth has rapidly become established throughout the south of the county with regular garden records at Flixton, Great Sankey, Hale, Abram, etc., and on higher ground at Rishton, Worsthorne, Briercliffe and Clitheroe. A smaller number of records in the north; thus far limited to Lytham St Annes and the Fylde area, Preston, Longridge, Sunderland Point and Warton, nr. Carnforth.
■ **HABITAT** Damp grassland and marshes.
■ **FOODPLANTS** Unrecorded within the county.
Single brooded. Comes to light. Approx. 50 seen at dusk, Flixton 2017 (K. McCabe).

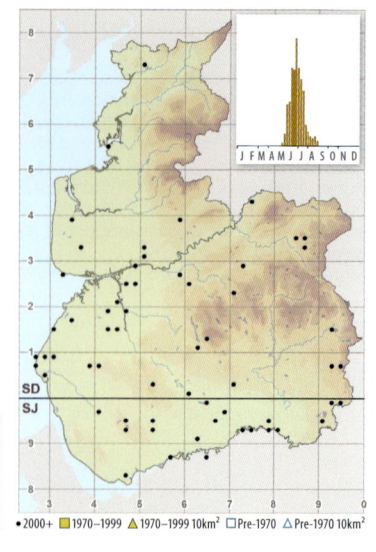

• 2000+   ☐ 1970–1999   △ 1970–1999 10km²   ☐ Pre-1970   △ Pre-1970 10km²

63.021 *Anania terrealis*

■ **LANCS STATUS** Scarce / very local resident   c.1940–2022

First noted from Preston (Mansbridge, 1940). Next, at Levenshulme, Manchester in 1968. Was regular at Gait Barrows from 1984 to 2009, but only noted twice since. Recorded at Flixton in 2001, and 70 times since (from a total of 96 Lancs records), including 13 from 2022. Other Manchester sightings at Chorlton and Swinton from 2002 to 2010. The apparent absence of the only known foodplant, Goldenrod, from these Manchester sites, indicates the use of an alternative, as yet unknown, foodplant.
■ **HABITAT** Limestone grassland, scrub, waste ground, gardens.
■ **FOODPLANTS** Unrecorded within the county.
Single brooded. Comes to light. High count of six at Flixton on July 2017 (K. McCabe).

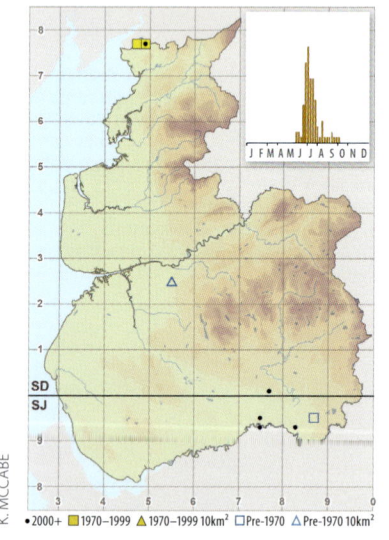

• 2000+   ☐ 1970–1999   △ 1970–1999 10km²   ☐ Pre-1970   △ Pre-1970 10km²

63.022 *Anania crocealis*

■ **LANCS STATUS** Occasional / local resident   c.1857–2022

First recorded in the woods at Kirkby, Lydiate and Eccleston (Gregson, 1857). Early authors noted the moth to be local, with other pre-1940 records from Manchester, Preston, Silverdale, Formby and Blackpool. Post-1990, numbers and distribution have increased, although primarily from western, lowland sites, mirroring foodplant distribution. Particularly populous around Rixton, Sefton Coast and Silverdale.
■ **HABITAT** Damp grassland, coastal scrub, limestone grassland, woodland clearings.
■ **FOODPLANTS** Common Fleabane, Ploughman's Spikenard.
Single brooded. Comes to light. Moth also recorded at dusk on foodplant. Larvae recorded in May and June; once in August as an early instar prior to overwintering.

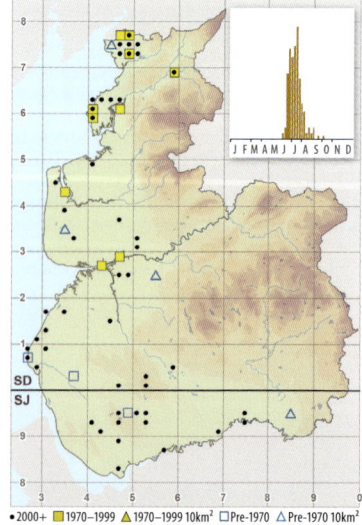

• 2000+   ☐ 1970–1999   △ 1970–1999 10km²   ☐ Pre-1970   △ Pre-1970 10km²

350   *The Moths of Lancashire*

63.024 *Anania funebris*

■ **LANCS STATUS** Occasional / very local resident   **c.1889–2022**

This Red Data Book species, proposed as Nationally Scarce B (Davis, 2012), was formerly a widespread and sometimes locally common species of open woodland where Goldenrod occurred in England, Wales and parts of Scotland. In recent years it has only been noted, very locally, in south-east England, Wales, Lancashire, Cumbria and western Scotland.

It was first reported in Lancashire in the late 1800s from Silverdale by S.J. Capper (Ellis, 1890), and later described as 'locally common on limestone' (Mansbridge, 1940). The population in the Morecambe Bay limestone area is therefore nationally important and is centred on the Silverdale and Arnside AONB.

Monitoring of this species has always been opportunistic rather than organised. Of 279 Lancashire records, 153 are from Gait Barrows and 98 from Warton Crag, both sites with records to the present day. The few other reports have come from Trowbarrow Quarry (1970–2016), Yealand Hall Allotment (2002–2019), Silverdale Golf Course and surrounding area (2003–2007) and a single record at Butterfly Conservation's reserve at Myers Allotment, in 2007. The highest count was 31 at Gait Barrows on 21 May 2008.

Anania funebris is a day-flying species and nearly all records are as a result of sightings on sunny days, only a small number being recorded at light. Its small size and low-flying habits can make it difficult to spot. The best time to search is on a warm, windless day in late May or early June. The limital dates are 23rd April and 25th July, although these dates should be considered exceptional.

A search at Gait Barrows on 25 July 2022 by B. Hancock, S. Palmer, J. Patton and B. Smart found three larvae feeding on the leaves of Goldenrod. The utilised plants were well developed and in partial shade close to limestone rocks and scrub cover. Only a single larva was found on each plant, grazing on the underside of lower leaves. Two larvae were kept but rearing was unsuccessful. A parasitic wasp emerged from one, later identified by Mark Shaw as a male *Apanteles obscurus* (Nees, 1834), quite a common species, known to parasitise a range of leaf-feeding Crambidae.

Generally, the Lancashire population appears stable on its two main sites, albeit without some of the higher counts from the first decade of this century. Potential risks include encroaching scrub and increasing shade from maturing trees, both of which could adversely affect foodplant numbers and microclimates. The potential effects of climate change, such as an increasing tendency for dry springs, on the moth and its foodplant require investigation.

Surveying of previously occupied sites is recommended, as is the establishment of regular monitoring at the two main sites. Habitat management should include the coppicing of woodland and scrub removal, creating wide, sheltered glades and encouraging the wider growth of Goldenrod. Ensuring physical connections exist between known, former and potential sites is considered crucial to ensure the future of this moth within the county.

John Girdley

● 2000+   ☐ 1970–1999   △ 1970–1999 10km²   ☐ Pre-1970   △ Pre-1970 10km²

CRAMBIDAE

63.025 Small Magpie *Anania hortulata*

■ **LANCS STATUS** Frequent / ubiquitous resident   c.1890–2022

A distinctive visitor to the moth trap, familiar even to those who take little notice of micro-moths. First noted by Ellis (1890), although no specific sites reported, with status described as 'common and generally distributed'. Other records to 1940 from Ditton, Barrowford and Formby. Mansbridge (1940) considered it common. Records post-2000 confirm a strong upland and lowland presence, with no obvious decline.
■ **HABITAT** Scrub, woodland edges, hedgerows, gardens.
■ **FOODPLANTS** Common Nettle, Hedge Bindweed; once on Mugwort.
Single brooded, although rare September and October records suggest possible small second brood in suitable years. Comes to light. Easily disturbed from vegetation.

● 2000+   ▪ 1970–1999   ▲ 1970–1999 10km²   □ Pre-1970   △ Pre-1970 10km²

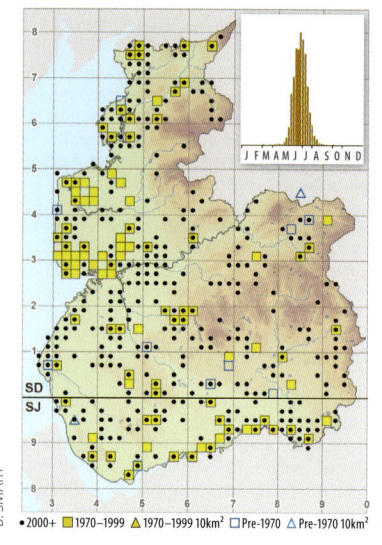

63.028 European Corn-borer *Ostrinia nubilalis*

■ **LANCS STATUS** Rare / migrant   2022

This species, now an established resident in parts of south-east England, occasionally reaches more northern counties as a migrant from European populations. First recorded in Lancashire on 5 September 2022, when a moth attended the light trap in J. Girdley's Longton garden. Another turned up in Wales on the same day, possibly part of the same wave of migration. The moth has been recorded on several occasions in Yorkshire and Cumbria, but not in Cheshire.
■ **HABITAT** Unknown in the county
■ **FOODPLANTS** Not recorded in Lancashire.
Single brooded. Comes to light.

● 2000+   ▪ 1970–1999   ▲ 1970–1999 10km²   □ Pre-1970   △ Pre-1970 10km²

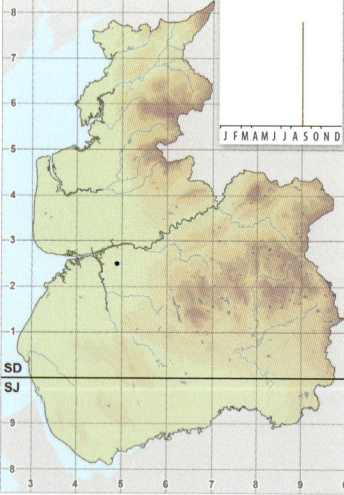

63.029 Bordered Pearl *Paratalanta pandalis*

■ **LANCS STATUS** Rare / very local resident   1906–2012

Noted by Stainton (1859) from the Manchester district. However, it is doubtful whether this was a Lancashire record. Lack of suitable habitat, plus an 1887 record taken near Altrincham in VC58 (possibly Stainton's site), suggest it wasn't. All other records are in Silverdale area, with first by A.E. Wright in 1906. Mansbridge (1940) noted it to be very rare. Recorded at Gait Barrows six times since 1984, first by E. Emmett. Has not been noted again since final Gait Barrows record on 24 May 2012.
■ **HABITAT** Limestone woodland clearings.
■ **FOODPLANTS** Not recorded in Lancashire.
Single brooded. All Gait Barrows records are of singletons to light.

● 2000+   ▪ 1970–1999   ▲ 1970–1999 10km²   □ Pre-1970   △ Pre-1970 10km²

The Moths of Lancashire

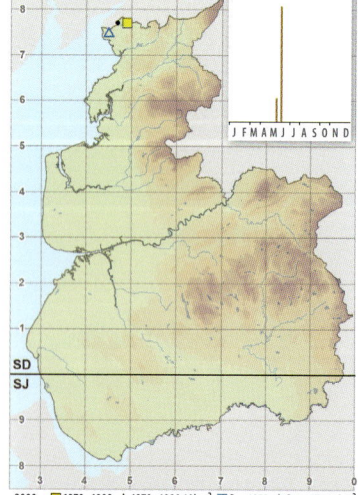

63.031 Rusty-dot Pearl *Udea ferrugalis*

■ **LANCS STATUS** Frequent / migrant   c.1853–2022

Annual numbers vary hugely, from one record in 2018, to 174 in 2006, and 79 in 2022. First reported by N. Cooke from mosses and gardens near Warrington (Byerley, 1854). Described by Ellis (1890) as local and not common, suggesting he may have been unaware of the moth's migrant status. Britten (c.1950) noted the moth to be often absent but occasionally widespread. Capable of showing up anywhere in the county, although distribution is rather more scattered away from the coast.
■ **HABITAT** Coastal locations, gardens, etc.
■ **FOODPLANTS** Not recorded in Lancashire.
Single brooded. Comes readily to light. High count of 19 at Sunderland Point in 2017.

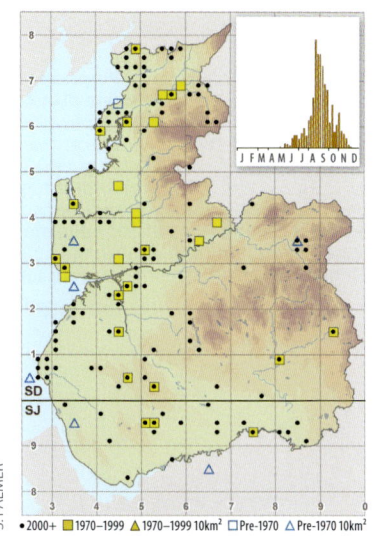

63.033 *Udea lutealis*

■ **LANCS STATUS** Abundant / ubiquitous resident   c.1857–2022

First noted from the Liverpool district, where Gregson (1857) reported breeding it from Bladder Campion, noting its abundance around ditches. Later, described as common and general by Mansbridge (1940). Recorded throughout in all habitats, with high counts of 100 during day at Dalton, nr. Parbold in 2014 (C. Darbyshire), and at Dunscar, Bolton in 2011 (D. Lumb). Larvae noted from late April to mid-June.
■ **HABITAT** Waste ground, woodland edges, rough grassland, gardens.
■ **FOODPLANTS** Colt's-foot, Black Knapweed, Mugwort, Ribwort Plantain, Creeping Thistle, Bladder Campion. Bred from pupa in spinning on Hedge Woundwort 2019.
Single brooded. Easily disturbed from vegetation during the day. Also, to light.

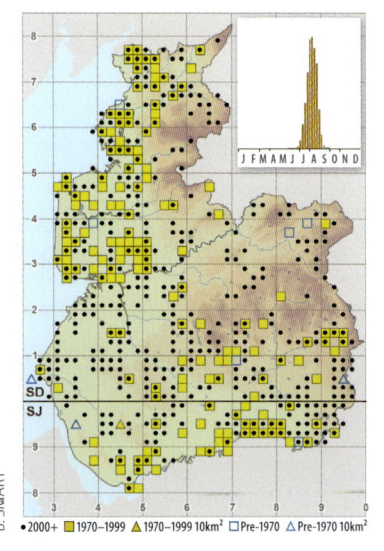

63.034 *Udea prunalis*

■ **LANCS STATUS** Frequent / widespread resident   c.1857–2022

Reported in abundance at Hale (Gregson, 1857). Whilst early authors described it as common and generally distributed, few sites outside the Liverpool district were recorded. Noted at Silverdale in 1909 by W. Mansbridge, and at Formby and Oldham during 1950s. Increased light trap use confirmed wide distribution, including upland sites such as Docker Moor. Population appears stable, with 100 records in 2022.
■ **HABITAT** Woodland, hedgerows, scrub, gardens.
■ **FOODPLANTS** Blackthorn.
Single brooded. Most records are at light. Disturbed from Blackthorn hedge by S.H. Hind in Tame Valley July 2021. Overwinters as a larva; recorded October and May.

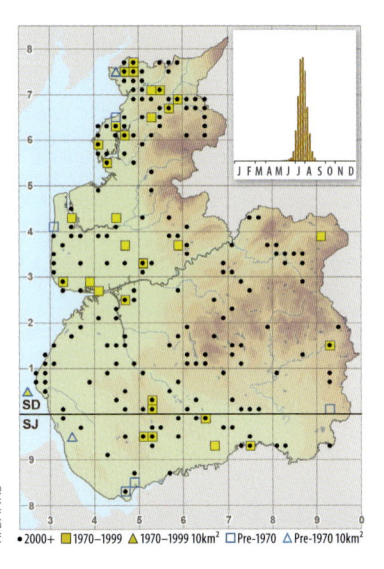

The Moths of Lancashire **353**

CRAMBIDAE

63.037 *Udea olivalis*

■ **LANCS STATUS** Occasional / widespread resident   c.1857–2022

First recorded in Liverpool, where plentiful in damp places, particularly in woods (Gregson, 1857). Britten (c.1950) found it fairly common in woods and 'wastes' where Dog's Mercury grows, scarcer near the coast. Other than the initial Liverpool records and one from Silverdale in 1910, further unrecorded until noted at Yealand by M. Hull in 1981. The 875 subsequent records show it to be most frequent on limestone, but also in deciduous wooded river valleys in uplands, e.g., Mill Houses.
■ **HABITAT** Limestone woodland, hedgerows, waste ground, gardens.
■ **FOODPLANTS** Unrecorded within the county.
Single brooded. Most modern records taken at light. Occasionally noted at dusk.

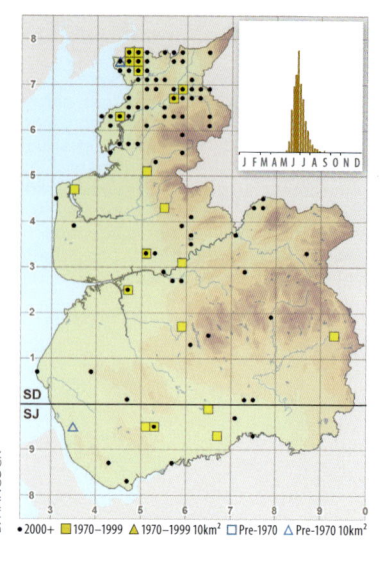

63.038 Mother of Pearl *Patania ruralis*

■ **LANCS STATUS** Abundant / ubiquitous resident   c.1887–2022

Early authors agreed the moth was common among nettles, although few sites stated. First noted from Liverpool district (Ellis, 1890), with the few pre-1970 sites including Didsbury, Formby and Morecambe. Numbers appear to have increased since 1970 with over 10,000 well-distributed records, showing the moth to be extremely common in a variety of habitats. High count of 305 from Flixton in 2005.
■ **HABITAT** Damp grassland, woodland edges, waste ground, gardens, etc.
■ **FOODPLANTS** Common Nettle.
Single brooded. Comes readily to light and easily disturbed from vegetation during day and at dusk. The overwintering larvae have been recorded from late July to June.

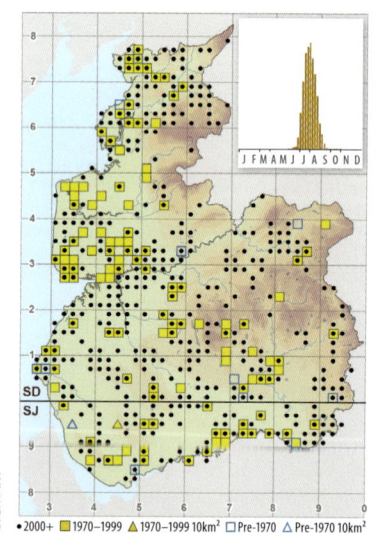

63.0381 *Patania aegrotalis*

■ **LANCS STATUS** Adventive   1889

This species has been recorded on one occasion in the county, and this remains the only British record. The find was documented by Mansbridge (1940) under the moth's former name *Botys mutualis*. He wrote, 'A specimen of this European species is recorded by C.S. Gregson as having been captured near Bolton in 1889.' The species, actually from southern Africa rather than Europe, is considered an accidental import but the circumstances surrounding this find are not known. In Goater (1986) it is mentioned that the specimen was exhibited by C.G. Barrett at the London Entomological Society in March, 1890, with the record originally published by Tutt (1890).
■ **HABITAT** Unknown in the county.
■ **FOODPLANTS** Not recorded in Lancashire.

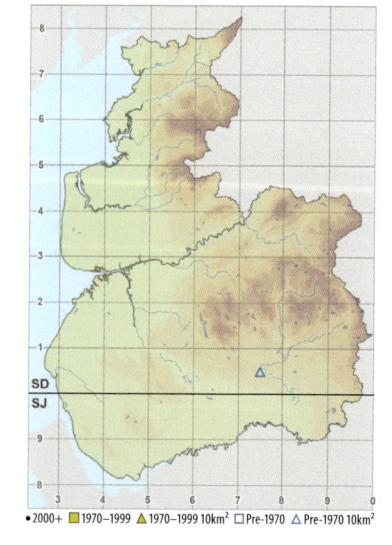

354 *The Moths of Lancashire*

63.042 *Diplopseustis perieresalis*

■ **LANCS STATUS** Adventive   2008–2021

This native of Australia and south-east Asia has been recorded from much of western Europe, likely with the introduction of a New Zealand sedge, *Carex secta* (Speidel *et al.*, 2007). The moth was first recorded in the county at Billinge on 23 July 2008, by C.A. Darbyshire from his garden light trap. The moth was subsequently dissected and confirmed as the third British record. Further recorded by K. Hughes at Catterall on 6 August 2021, and two months later at Flixton on 16 October 2021 by K. McCabe.

■ **HABITAT** Unknown in the county.
■ **FOODPLANTS** Not recorded in Lancashire.
All records are of moths to light.

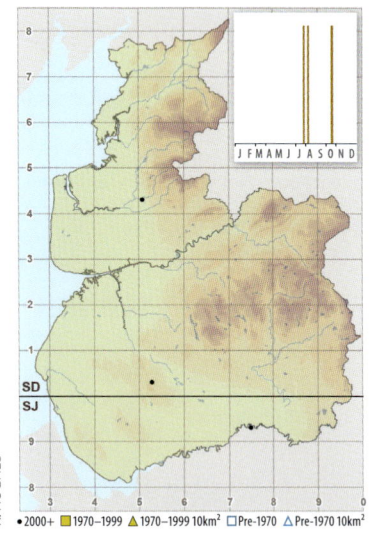

63.046 *Duponchelia fovealis*

■ **LANCS STATUS** Adventive   1999–2022

First recorded in Lancs on 6 May 1999 by P. Pugh at Pennington, three years after the first British record, with another at the same address in 2003. Both were indoors, and the finder wondered if their appearance was connected to the presence of cut flowers. Numbers have increased since, with 13 further records, including three at Warton, nr. Carnforth and twice at Longridge. The presence of these multiple records suggests possible local breeding, likely connected to horticultural trade.

■ **HABITAT** Unknown in the county.
■ **FOODPLANTS** Not recorded in Lancashire.
Eight of the 15 records have been found indoors; the others at light.

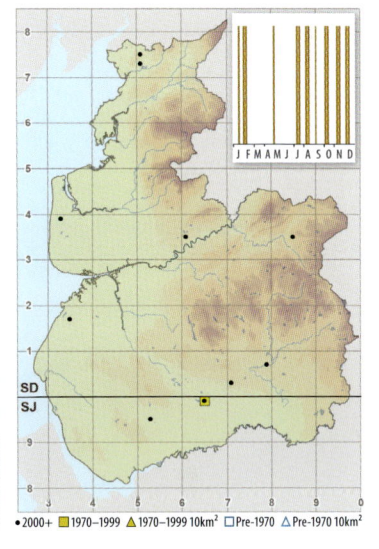

63.048 *Palpita vitrealis*

■ **LANCS STATUS** Occasional / migrant   2001–2022

First noted at the start of the current century with records almost annual since. Recorded at Flixton on 16 October 2001 by K. McCabe, with others at Bispham, Morecambe, Heysham, Lancaster and Leighton Moss in the following ten years. More recently, recorded at Catterall, Woolton, Manchester, Great Sankey and Normoss, nr. Blackpool. With only 22 records, it is perhaps surprising that five sites have multiple records, including six from Heysham.

■ **HABITAT** Most frequently from coastal locations, including gardens.
■ **FOODPLANTS** Not recorded in Lancashire.
Breeding unrecorded in Britain. Where noted, all records are of moths to light.

The Moths of Lancashire   **355**

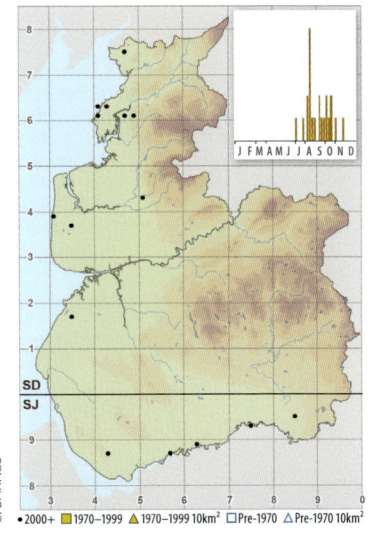

CRAMBIDAE

63.051 *Antigastra catalaunalis*

■ **LANCS STATUS** Rare / migrant   **2006**
Only recorded in the county on a single occasion. This was of a moth trapped at light at Fowley Common, nr. Culcheth on 22 September 2006 by J.D. Wilson. The only Cheshire record was four days later, presumably part of the same movement.
■ **HABITAT** Unknown in the county.
■ **FOODPLANTS** Not recorded in Lancashire.

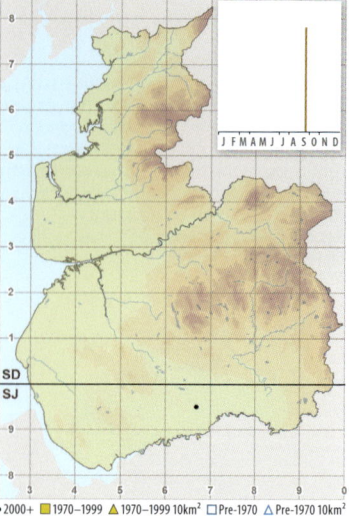

63.052 **Rush Veneer** *Nomophila noctuella*

■ **LANCS STATUS** Frequent / migrant   **c.1890–2022**
One of our commonest migrants, with 170 records from 2022. Numbers vary, with only four in 2020 and in 2021, but 186 in 2003, 288 in 2006 and 468 in 2000. First reported in the county by Ellis (1890), although no sites listed. Later, at Simonswood Moss 1905, Formby 1950 and Morecambe 1959. Subsequent light trapping has vastly increased number of records, with a count of 40 from Lightfoot Green 16 June 1996.
■ **HABITAT** Whilst most frequent in coastal locations, the moth has been found in all habitats. Commonly recorded in gardens throughout the county.
■ **FOODPLANTS** Not recorded in Lancashire.
Moths recorded in all months between 3 February (2004) and 29 November (1979).

63.054 **Box-tree Moth** *Cydalima perspectalis*

■ **LANCS STATUS** Frequent / widespread resident   **2014–2022**
First recorded on 2 July 2014 by G. Riley at Swinton, this initially adventive species is spreading north at a dramatic rate. Other early VC59 records came from Clitheroe and Hoghton. The moth reached VC60 by 2019 with records at Lightfoot Green and Bilsborrow (in 2020). Some idea of the rapid increase over the last few years can be seen by looking at annual totals of records from 2014 to 2022: 1, 0, 2, 2, 7, 11, 32, 94, 180. Larvae now easy to find on defoliated Box hedges in Manchester and elsewhere.
■ **HABITAT** Gardens, parks, wherever Box has been planted.
■ **FOODPLANTS** Box.
Double brooded. Comes to light. Larva recorded April, May, and August to October.

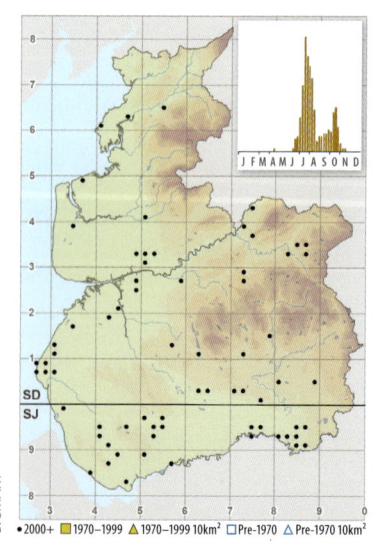

356   *The Moths of Lancashire*

63.057 Garden Pebble *Evergestis forficalis*

■ **LANCS STATUS** Frequent / widespread resident   c.1857–2022

First recorded in the Liverpool district by Gregson (1857), and from Oldham by J.T. Rodgers in 1883. Described by Ellis (1890) as 'frequent wherever there are kitchen-gardens', a reference to its foodplants. Mansbridge (1940) was more specific, stating it to be common where Horse-radish is grown. The larvae favour various Brassicaceae and have been noted late June and August to September. A widespread species, even noted in upland areas. Most frequent by the coast.

■ **HABITAT** Gardens, waste ground, coastal sites.
■ **FOODPLANTS** Larvae on various Brassicaceae including Sea Radish.

The brood situation is complex and may, on occasions, include a second generation.

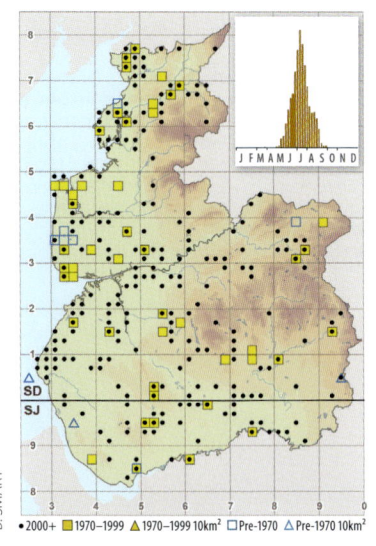

63.058 *Evergestis extimalis*

■ **LANCS STATUS** Rare / migrant   c.1900–2023

Recorded at Silverdale by A.E. Wright, sometime around the start of the 20th century, with the record documented by Mansbridge (1940). Considered to almost certainly be a migrant (Sterling & Parsons, 2012). Recent decades have seen increased numbers in Yorkshire, where it has on occasion become temporarily established; mainly on the coast, occasionally wandering inland. This may well have been the source of a surprising record on 9 August 2023, when a singleton was trapped at Worsthorne by G. Gavaghan.

■ **HABITAT** Unknown in the county.
■ **FOODPLANTS** Not recorded in Lancashire.

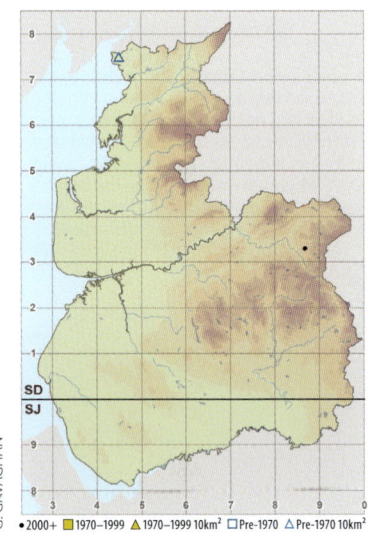

63.060 *Evergestis pallidata*

■ **LANCS STATUS** Scarce / local resident   c.1910–2022

First recorded in the Silverdale area by A.E. Wright, probably in early 1900s. Next from Formby in 1949 and 1951, the latter at light; both recorded by G. de C. Fraser. Other pre-2000 records from Morecambe, Claughton, Bolton-le-Sands, Cuerden and Hollins, Bury. This century, records from Lancaster, Leighton Moss, Yealand Conyers, Southport, Martin Mere, Oldham, Briercliffe, Worsthorne, Great Eccleston and Gorply Clough, nr. Todmorden, suggest the moth is becoming more widespread.

■ **HABITAT** Damp woodland scrub.
■ **FOODPLANTS** Unrecorded within the county.

Single brooded. All records since 1949 have been of moths at light.

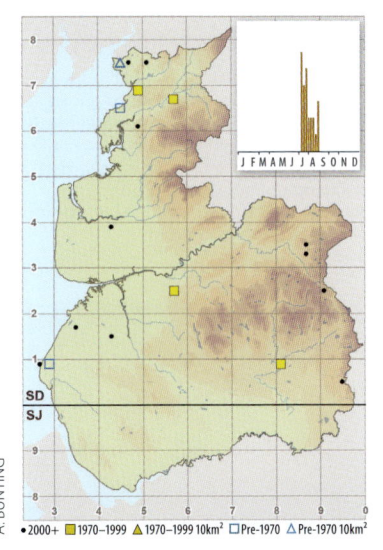

The Moths of Lancashire

CRAMBIDAE

63.061 Old World Webworm *Hellula undalis*

■ **LANCS STATUS** Rare / migrant   **2006**

Another species with just a solitary Lancashire record, from 29 September 2006, a productive year for many migratory moths. The moth was trapped at light in St Helens by D. Owen. This was one of 18 individuals recorded nationally during the year, the second highest annual total since 1990. Other than one Irish record, the remainder were confined to southern England (Clancy, 2008). The species is unrecorded from neighbouring counties.
■ **HABITAT** Unknown.
■ **FOODPLANTS** Unrecorded within the county.

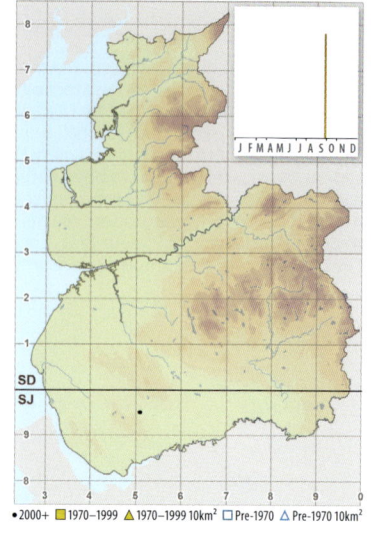

• 2000+   ☐ 1970–1999   △ 1970–1999 10km^2 ☐ Pre-1970   △ Pre-1970 10km^2

63.062 *Scoparia subfusca*

■ **LANCS STATUS** Occasional / widespread resident   **c.1859–2022**

First noted at Cheetham Hill and Pendleton by J. Chappell (Ellis, 1890) and described as local and not common. There are few pre-1970 records, with the moth noted from Formby, Burnley and Morecambe. More regular from mid-1990s onwards, particularly at former industrial sites at Middleton N.R. and St Helens. Also, recorded at upland sites, where care needed to avoid confusion with larger, poorly marked upland variants of *S. ambigualis*. Evidence of declining numbers since around 2010.
■ **HABITAT** Waste ground, scrub, woodland.
■ **FOODPLANTS** Not recorded in Lancashire.
Single brooded. Comes to light and easily disturbed during day.

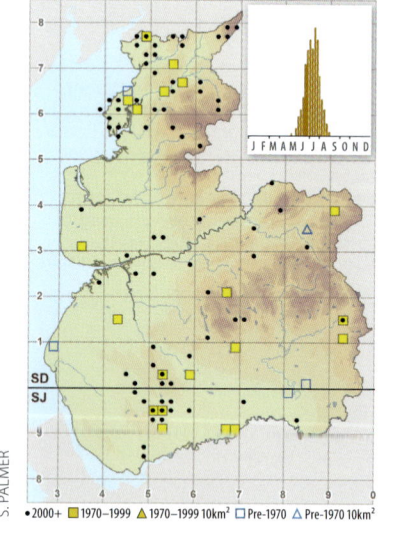

• 2000+   ☐ 1970–1999   △ 1970–1999 10km^2 ☐ Pre-1970   △ Pre-1970 10km^2

63.063 *Scoparia basistrigalis*

■ **LANCS STATUS** Rare / very local resident   **2016**

Recorded once by R. Walker at Formby, with a moth coming to light on 24 July 2016. The moth, a female, was subsequently dissected to confirm identity. This remains the only confirmed record of this mainly southern species. Few records from Cumbria and Yorkshire and unrecorded in Cheshire.
■ **HABITAT** Unknown in the county.
■ **FOODPLANTS** Not recorded in Lancashire.
Nationally, single brooded.

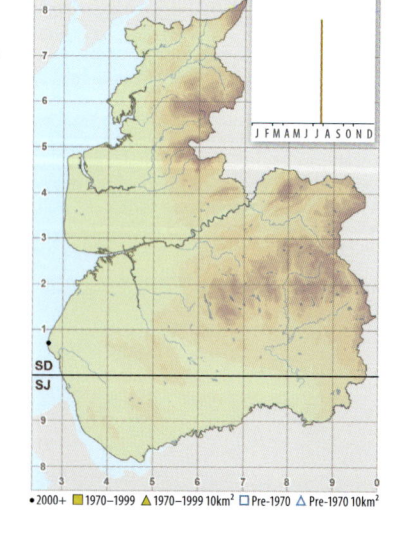

• 2000+   ☐ 1970–1999   △ 1970–1999 10km^2 ☐ Pre-1970   △ Pre-1970 10km^2

358 *The Moths of Lancashire*

63.064 *Scoparia ambigualis*

■ **LANCS STATUS** Abundant / ubiquitous resident   c.1857–2022
Noted by Gregson (1857) in the Liverpool district, with other records to 1940 from Longridge, Silverdale and Formby. Described by Mansbridge (1940) as abundant everywhere. Status appears little changed. Commonly recorded throughout, from sand dunes to moorlands, with apparent range extension in the latter habitat since 2000, and no evidence of decline elsewhere. High count of approximately 200 from Docker Moor on 18 July 2015, recorded by P.J. Marsh.
■ **HABITAT** Damp grassland, scrub, waste ground, woodland edges, gardens, etc.
■ **FOODPLANTS** Unrecorded within the county.
Single brooded. Easily disturbed from vegetation during day. Comes readily to light.

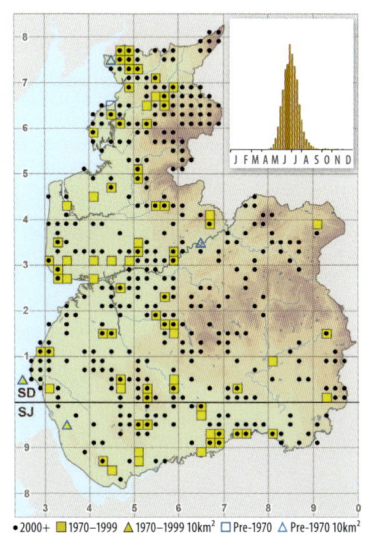

63.065 *Scoparia ancipitella*

■ **LANCS STATUS** Rare / very local resident   1990–2012
Initially noted at Birk Bank, then at Caton, in June and July 1990, by M. Dempsey. Recorded six times in a nine-day period in August 2001 at three adjacent sites in Hindburndale by S. Palmer and P.J. Marsh. The multiple records reflect increased effort following the first of the 2001 records. Only noted once since, in 2012, at Mill Houses, close to the 2001 locations. Whilst the lack of earlier or later records is surprising, it is likely that the moth is an overlooked, long-term resident.
■ **HABITAT** Old deciduous woodland in upland river valleys.
■ **FOODPLANTS** Unrecorded within the county.
Single brooded. Comes to light. High count of three in two traps, Hindburndale 2001.

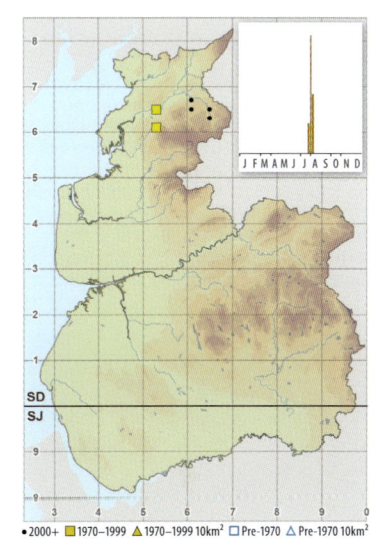

63.066 *Scoparia pyralella*

■ **LANCS STATUS** Occasional / local resident   c.1887–2022
Earliest records all from Silverdale, first by J.B. Hodgkinson (Ellis, 1890). Next, from Yealand Hall Allotment 1977, Formby 1984 and Martin Mere 1987. First noted at Flixton in 1995, with 160 records in total, but none since 2018, mirroring a sharp decline across the county. There were 148 Lancs records from 1998 to 2002, but only 40 from 2018 to 2022. Recent records suggest range also contracting, with the moth now largely restricted to Sefton Coast and limestone uplands around Dalton Crags.
■ **HABITAT** Grassland, sand dunes.
■ **FOODPLANTS** Unrecorded within the county.
Single brooded. Mostly at light but also disturbed from vegetation during day.

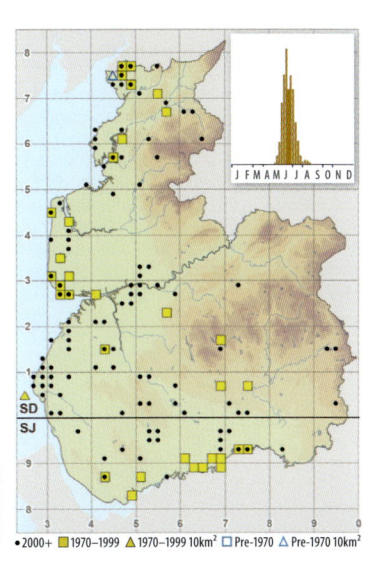

The Moths of Lancashire   **359**

CRAMBIDAE

63.067 *Eudonia lacustrata*

■ **LANCS STATUS** Frequent / widespread resident   c.1887–2022

First noted at Silverdale by J.B. Hodgkinson (Ellis, 1890), and later described as common in the limestone districts (Mansbridge, 1940). Pre-1980 records outside Morecambe Bay were limited to Formby in 1950. Regular at Claughton during 1990s. The first VC59 records were at Parr in 1997 and Flixton in 1998, with 60 records at latter site before end of 2000. Range and numbers have since increased massively, with over 1000 Lancs records in 2021–2022. Similar process has occurred in Cheshire.

■ **HABITAT** Woodland, grassland, gardens.
■ **FOODPLANTS** Not recorded in Lancashire.
Single brooded. Comes readily to light.

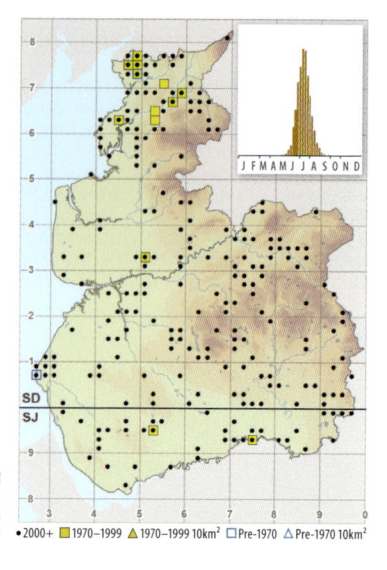

63.068 *Eudonia murana*

■ **LANCS STATUS** Rare / very local resident   2015–2016

All confirmed records are from a single upland limestone area, with early records from Manchester (Stainton, 1859), Longridge (Ellis, 1890) and Burnley (Mansbridge, 1940) considered unconfirmed. The lack of early *E. truncicolella* records seems to support this, as identity of the two may have been confused. Recorded at Leck Fell on 5 July 2015 by S. Palmer and P.J. Marsh, with three further in 2015 and two in 2016. Subsequent absence likely related to lack of further searches for the species.

■ **HABITAT** Limestone moorland.
■ **FOODPLANTS** Unrecorded within the county.
Single brooded. Recorded at light and at dusk.

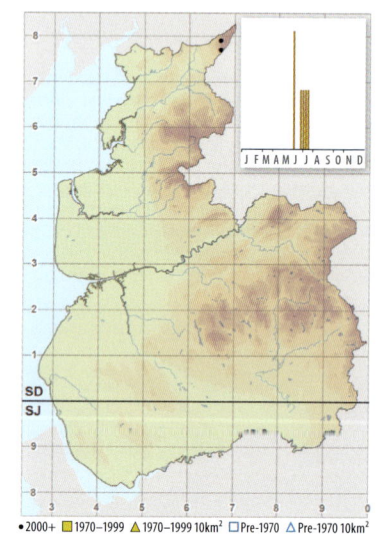

63.069 *Eudonia angustea*

■ **LANCS STATUS** Frequent / widespread resident   c.1857–2022

First documented as plentiful throughout the Liverpool district (Gregson, 1857), with other early authors describing it as common and generally distributed. Britten (c.1950) reported it to be common in coastal sandhills, less so inland. Nevertheless, there are few records from this period; from Crosby 1887, Silverdale 1909, Ainsdale and Formby in 1957, but nowhere else until Billinge 1989. Increasingly found inland in 1990s as numbers also increased, a process that appears to be continuing.

■ **HABITAT** Coastal grassland, sand dunes, waste ground, gardens.
■ **FOODPLANTS** Unrecorded within the county.
May be double brooded. High count of 13 at Sunderland on 14 September 2016.

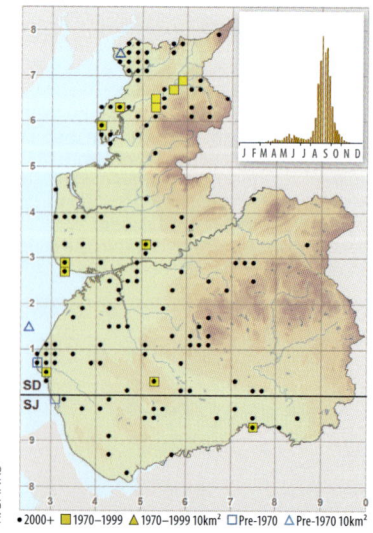

360   *The Moths of Lancashire*

63.072 *Eudonia delunella*

■ **LANCS STATUS** Rare / very local resident   **c.1857–1982**
The species was first recorded from Croxteth and Allerton Hall, Liverpool (Gregson, 1857). It has only been noted on one further occasion, at Risley Moss by M. Hull in 1982. The lack of any other recent records tallies with the experience of some adjacent counties; only once in Cheshire since 1947 (in 2006), and no Yorkshire records since 1904 (cheshire-chart-maps.co.uk; yorkshiremoths.co.uk, both accessed 1 September 2023). To the north, the moth is resident in Cumbria, with around 70 records since 2010 (S. Colgate, pers. comm.). However, these populations seem unlikely to be the source of the Risley Moss individual. It is possible that this species is extinct in the county.
■ **HABITAT** Woodland.
■ **FOODPLANTS** Not recorded in Lancashire.
Single brooded. The 1982 record was taken at a MV light trap.

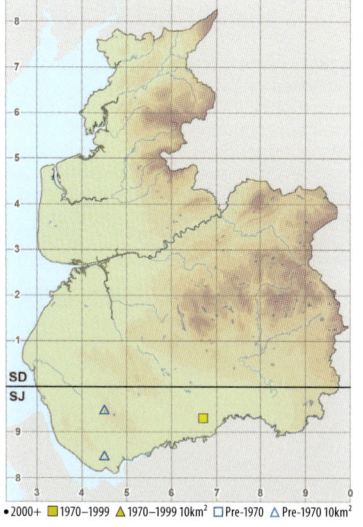

63.073 *Eudonia truncicolella*

■ **LANCS STATUS** Frequent / widespread resident   **c.1910–2022**
Many *Scoparia* and *Eudonia* species were little recorded prior to the last 40 years; whether scarce or under-recorded is unclear. A.E. Wright found the moth fairly common at Pendle HIll around 1910. Other early records from Formby 1920 and Warton Crag 1940. Mansbridge reported the moth to be fairly common throughout, with a liking for pine woods (1940). Numbers and range have increased from 1980s onwards, with over 1,400 records throughout lowland and upland areas.
■ **HABITAT** Woodland, scrub, moorland, gardens.
■ **FOODPLANTS** Unrecorded within the county.
Single brooded. High count of 27 to light at Mill Houses 21 August 2012 (P.J. Marsh).

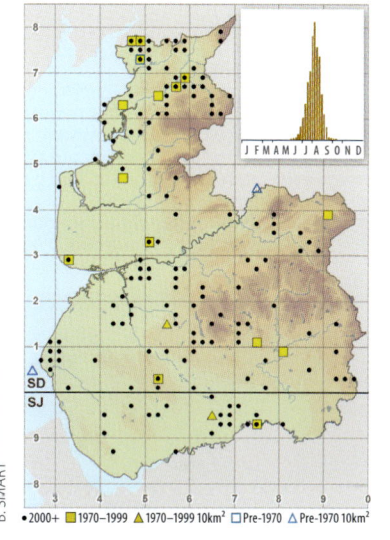

63.074 *Eudonia mercurella*

■ **LANCS STATUS** Abundant / widespread resident   **c.1857–2022**
Early records were mainly coastal, with first from Liverpool (Gregson, 1857), and later, from Silverdale and Morecambe. The only pre-1970 inland records were from Preston in 1880s and Oldham in 1955. Known distribution increased from 1980s onwards with many new sites including lowland gardens (Flixton, etc.), and upland sites such as Green Bank, where approx. 200 recorded on 19 July 2022 (P.J. Marsh, J. Roberts).
■ **HABITAT** Woodland, moorland, gardens, scrub, etc.
■ **FOODPLANTS** Mosses, including *Hypnum cupressiforme*. Also recorded in 2019 from an old, disused blackbird nest in a hawthorn hedge in Chorlton (B. Smart, J. Agar).
Single, extended generation. Comes to light. Larvae recorded from February to May.

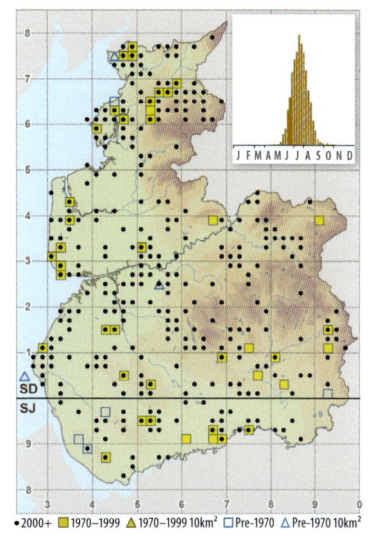

CRAMBIDAE

63.075 *Eudonia pallida*

■ **LANCS STATUS** Occasional / local resident   c.1857–2022

Recorded by Gregson (1857) at Simonswood Moss, with other 19th century records from Crosby, Chat Moss, Penwortham and Pilling Moss. Noted by Ellis (1890) to be local but common where it occurs. Later, reported to be local in marshy places (Mansbridge, 1940). Further recorded at Preston, Freshfield and Leigh in 1950s. Numbers have sharply increased from 2005 onwards and continue to do so. Range also increasing, although mainly limited to coastal and inland lowland sites.

■ **HABITAT** Coastal grassland, marshes.
■ **FOODPLANTS** Not recorded in Lancashire.

Single brooded. Comes to light. High count of eight at Middleton N.R. 6 August 2014.

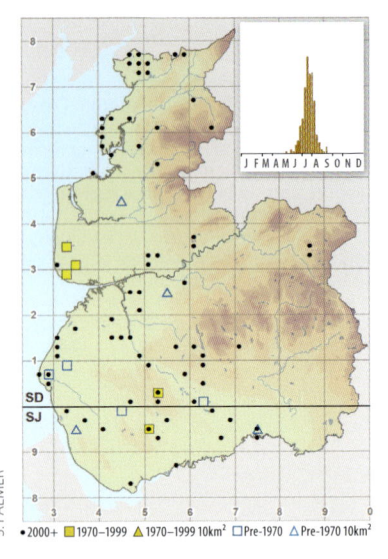

• 2000+ ▪ 1970–1999 ▲ 1970–1999 10km² ▫ Pre-1970 △ Pre-1970 10km²

63.076 *Euchromius ocellea*

■ **LANCS STATUS** Rare / migrant   1998–2006

This species mainly reaches southern counties as a migrant from European populations, but occasionally travels further north. First noted in Lancashire on 14 February 1998 when two moths were recorded at Lightfoot Green, Preston, by S. Palmer. The second, and final record, was from Fowley Common, nr. Culcheth, where a singleton was trapped at light by J.D. Wilson on 20 September 2006.

■ **HABITAT** Unknown.
■ **FOODPLANTS** Unrecorded within the county.

Both records were of moths trapped at light.

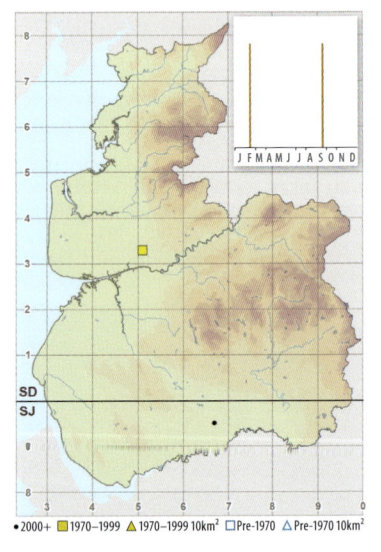

• 2000+ ▪ 1970–1999 ▲ 1970–1999 10km² ▫ Pre-1970 △ Pre-1970 10km²

63.077 *Chilo phragmitella*

■ **LANCS STATUS** Occasional / local resident   c.1887–2022

First noted by G.A. Harker at Crosby Marsh (Ellis, 1890). Next from Gathurst, nr. Wigan in 1935, and Leighton Moss, Silverdale and Morecambe in the 1950s. Not recorded again until the 1980s at Lytham St Annes and Gait Barrows. This century has seen range increase, although the moth remains largely confined to lowland sites. 16 records prior to 2000 and 420 from 2000 onwards demonstrate a significant increase in numbers. High count of 50 to light at Leighton Moss on 21 June 2003.

■ **HABITAT** Reedbeds, riverbanks and other wetland habitats.
■ **FOODPLANTS** Not recorded in Lancashire.

Single brooded. Virtually all post-1959 records are of moths to light.

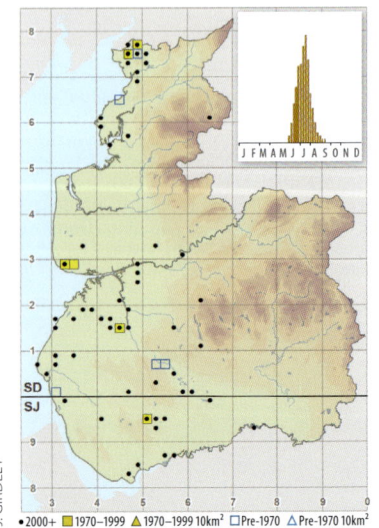

• 2000+ ▪ 1970–1999 ▲ 1970–1999 10km² ▫ Pre-1970 △ Pre-1970 10km²

The Moths of Lancashire

63.079 *Calamotropha paludella*

■ **LANCS STATUS** Occasional / local resident 2004–2022

Formerly restricted to southern England and East Anglia (Goater, 1986), this species has dramatically extended its range north and west in the last few decades. 2004 saw the first records in Lancs and Cheshire, following the first Yorkshire record in 1995. Trapped at Leigh by J.D. Wilson on 7 August 2004. Also noted at Martin Mere, Flixton, Hale and Birkdale Green Beach prior to 2010. Numbers have continued to increase, with over half the 252 records coming in last four years, primarily from lowland sites.

■ **HABITAT** Reedbeds and other wetland sites.

■ **FOODPLANTS** Unrecorded within the county.

Single brooded. Comes to light. High count of 8 at Flixton 23 July 2019 (K. McCabe).

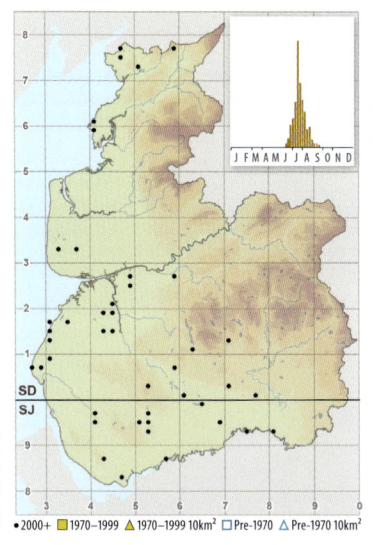

● 2000+   ■ 1970–1999   ▲ 1970–1999 10km²   □ Pre-1970   △ Pre-1970 10km²

63.080 *Chrysoteuchia culmella*

■ **LANCS STATUS** Abundant / ubiquitous resident c.1857–2022

Gregson (1857) reported its occurrence throughout Liverpool district, with other 19th century records from Chat Moss, Oldham, Prestwich, Southport. Noted as common but local (Ellis, 1890), and later, as common in meadows and grassy lanes (Mansbridge, 1940). 550 records to 2000, with over 10,000 since. Recorded in good numbers in gardens throughout, e.g., Tarleton, Hale, Worsthorne, Morecambe, Heysham, Flixton, Alkrington, Rishton, etc. High count of 600 at Pilling 1 July 2000.

■ **HABITAT** Rough grassland, waste ground.

■ **FOODPLANTS** Unrecorded. Larva (ex-ova) fed on various grasses at Chorlton 2013. Single brooded. Comes to light. Easily disturbed from grassland during the day.

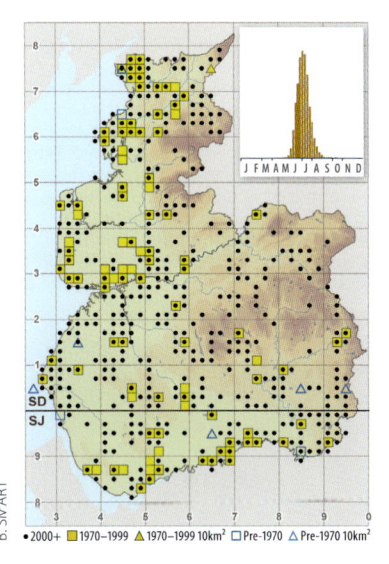

● 2000+   ■ 1970–1999   ▲ 1970–1999 10km²   □ Pre-1970   △ Pre-1970 10km²

63.081 *Crambus pascuella*

■ **LANCS STATUS** Frequent / widespread resident c.1890–2022

Ellis (1890) and Mansbridge (1940) both reported abundance on heaths and mosses without noting specific sites. Recorded from Liverpool in 1919 by F.N. Pierce. There were no further records until the 1980s, when recorded at Risley Moss (1982), and thereafter, at Scorton, Ince Moss, Formby, Winmarleigh Moss, Gait Barrows, Martin Mere, etc. Numbers of this formerly lowland moth have continued to increase, in gardens and elsewhere. Now also found on upland sites, e.g., Leck Fell, Docker Moor.

■ **HABITAT** Heaths, mosses, waste ground, damp grassland.

■ **FOODPLANTS** Unrecorded within the county.

Single brooded. Comes to light and easily disturbed from vegetation during day.

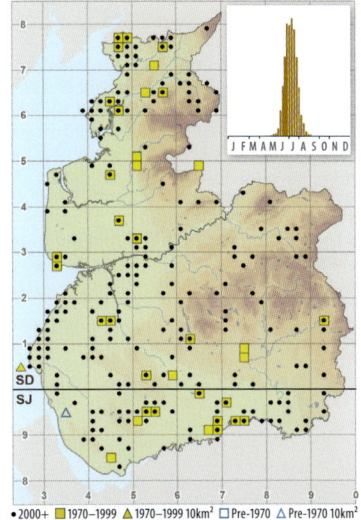

● 2000+   ■ 1970–1999   ▲ 1970–1999 10km²   □ Pre-1970   △ Pre-1970 10km²

CRAMBIDAE

The Moths of Lancashire **363**

63.083 *Crambus uliginosellus*

■ **LANCS STATUS** Rare / very local resident 2006–2023

Recorded on four occasions pre-2023, the first at Bay Horse on 30 June 2006 by N.A.J. Rogers. Further noted at the same site, by the same recorder, on 23 and 25 June 2010. Next, at Leighton Moss on 7 July 2012 (S. Palmer), and Silverdale in 2023 (J. Patton). Recorded at Ainsdale on 28 June 2023 by M. & S. Jackson; initially considered the first VC59 record. However, a specimen from Formby Dune Heath on 12 July 2015 by R. Walker, was dissected by S. Palmer in 2023 and also confirmed as this species.

■ **HABITAT** Damp grassland and dunes.
■ **FOODPLANTS** Unrecorded within the county.
Single brooded. Care is required to differentiate this moth from the previous species.

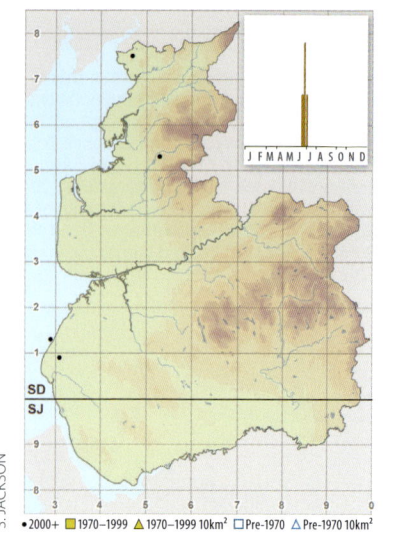

63.084 *Crambus ericella*

■ **LANCS STATUS** Rare / very local resident 2021–2022

First noted at Dalton Crags, in the Lancashire portion of Hutton Roof, by J. Patton on 17 June 2021, when ten moths were netted during the day after being flushed from vegetation. Further noted by J. Patton nearby at Crag House Allotment on 8 July 2021, when five moths were recorded, including four at light, and again on 14 June 2022. It is likely that this species is a previously overlooked resident, rather than a new arrival to the county.

■ **HABITAT** Limestone crags.
■ **FOODPLANTS** Not recorded in Lancashire.
Moths recorded at light and during day, disturbed from vegetation.

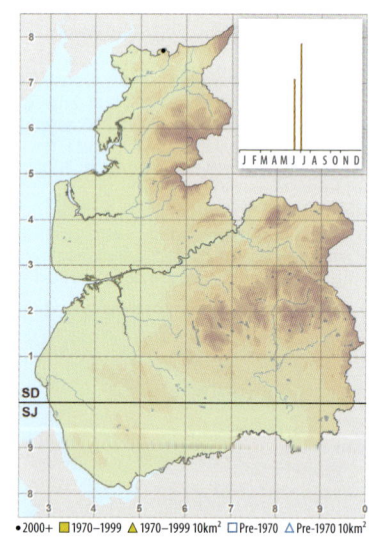

63.085 *Crambus pratella*

■ **LANCS STATUS** Scarce / very local resident 2009–2022

The moth was noted to be present on Cheshire sand-hills historically, and so may possibly have been overlooked in Lancashire, with care needed to separate it from the following species. First recorded on 4 July 2009 in the dunes at Formby Point by S. Tomlinson. All 48 subsequent records have been confined to the Sefton Coast with sites including Ainsdale Dunes, Birkdale Green Beach, Altcar and Freshfield. High count of 18 at light at Formby on 28 May 2017, noted by R. Walker and T. Davenport.

■ **HABITAT** Sand dunes.
■ **FOODPLANTS** Unrecorded within the county.
Single brooded. Netted during day and recorded at light.

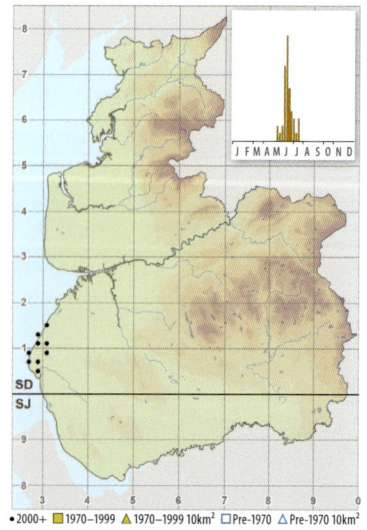

63.086 *Crambus lathoniellus*

■ **LANCS STATUS** Frequent / widespread resident   c.1857–2022
First noted in the Liverpool district, where abundant on waste land and damp pastures (Gregson, 1857). Further noted pre-1940 at Crosby, Oldham, Simonswood Moss, Silverdale, Warton Crag. Early authors reported status to be common or abundant. The few records from 1940–1980 probably reflect recorder effort as modern records continue to show abundance at lowland and upland grassland sites. Easily disturbed from grasses with 16 counts of 100+, mostly during the day.
■ **HABITAT** Grasslands, gardens.
■ **FOODPLANTS** Unrecorded. Larva (ex-ova) fed on unknown grasses in Chorlton 2013.
Single brooded. Noted during day; also comes to light.

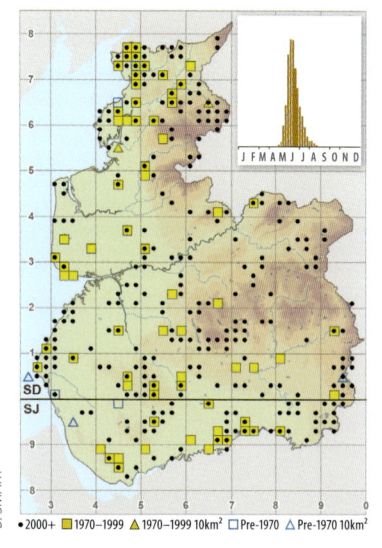

• 2000+   ■ 1970–1999   ▲ 1970–1999 10km²   □ Pre-1970   △ Pre-1970 10km²

63.087 *Crambus hamella*

■ **LANCS STATUS** Scarce / very local resident   c.1859–2022
First recorded by J. Chappell at Chat Moss (Ellis, 1890), with early authors noting it to be local and infrequent, and at Cadishead Moss in 1882. First modern record in 1995 at Risley Moss, by L.W. Hardwick and S. McWilliam. Nine records at Cadishead Moss from 2009–2012, with a similar number from K. McCabe's Flixton garden since 1996, approx. 2 miles from the Lancs. mosses, the likeliest source. Continued presence in Flixton, with three in 2022, suggests continued presence in their mossland habitat.
■ **HABITAT** Mosses.
■ **FOODPLANTS** Not recorded in Lancashire.
Single brooded. Minimum daytime counts of 100 at Cadishead Moss in 2011.

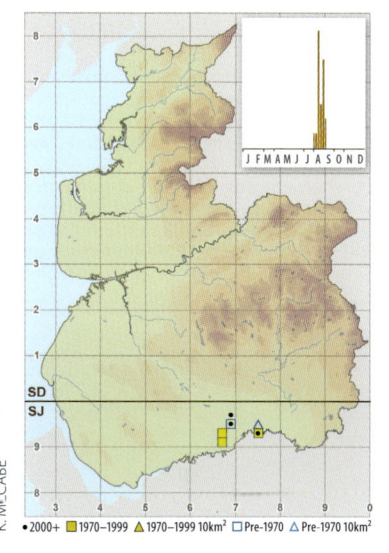

• 2000+   ■ 1970–1999   ▲ 1970–1999 10km²   □ Pre-1970   △ Pre-1970 10km²

63.088 *Crambus perlella*

■ **LANCS STATUS** Frequent / widespread resident   c.1847–2022
Forewings shining white; sometimes streaked. The latter form (f. *warringtonellus*), was noted by N. Cooke at Rixton Moss 1847 and published as a separate species (Stainton, 1848). Ellis (1890), noted both, finding *C. perlella* local on saltmarshes and mosses and *warringtonellus* common on mosses and heaths. Mansbridge, following amalgamation of the two, reported it common on mosses and marshes (1940). Most frequent in coastal and lowland areas; occasional in upland areas, e.g., Docker Moor.
■ **HABITAT** Mosses, marshes, grassland.
■ **FOODPLANTS** Unrecorded within the county.
Single brooded. Noted in day and at light. 123 to light at Heysham N.R. 25 July 2012.

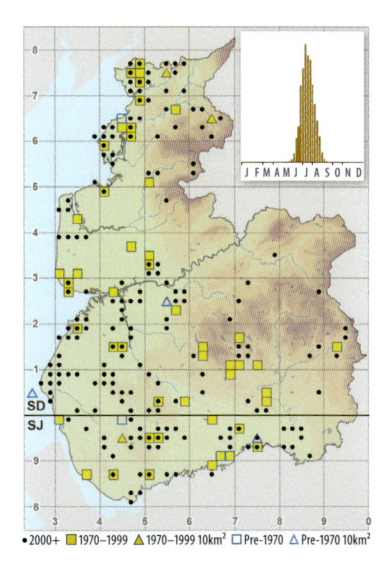

• 2000+   ■ 1970–1999   ▲ 1970–1999 10km²   □ Pre-1970   △ Pre-1970 10km²

CRAMBIDAE

63.089 *Agriphila tristella*

■ **LANCS STATUS** Frequent / widespread resident   c.1890–2022

Ellis (1890) noted its abundance everywhere, without mentioning specific sites. Other records to 1940 came from Sefton Park, Formby, Silverdale, Middle Hulton. Light trapping post-1955 resulted in many new sites in varied habitats including gardens. Most high counts are from limestone and moorlands, e.g., 300 at Yealand Conyers and 200 at Mill Houses, both in 2006. However, such counts seem a thing of the past, with 60 at Green Bank in 2019 the highest from the last decade.

■ **HABITAT** Grassland, moorland, waste ground, gardens.
■ **FOODPLANTS** Unrecorded within the county.

Single brooded. Comes readily to light. Often disturbed from vegetation during day.

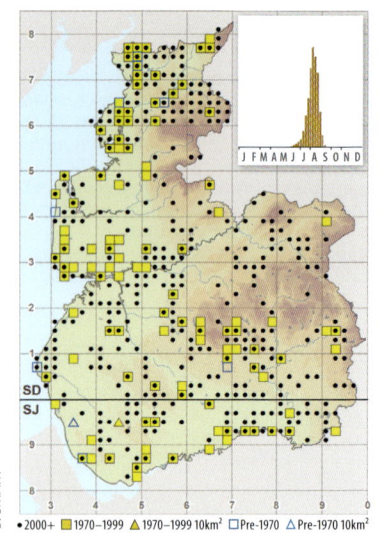

● 2000+   ■ 1970–1999   ▲ 1970–1999 10km²   □ Pre-1970   △ Pre-1970 10km²

63.090 *Agriphila inquinatella*

■ **LANCS STATUS** Occasional / local resident   c.1887–2022

Recorded at Silverdale by J.B. Hodgkinson (Ellis, 1890), with the author describing distribution as local. Next recorded at Formby in the 1930s by W. Mansbridge, then at Wigan 1950, Risley Moss 1982 and Claughton 1986. 17 records from 1990s, 47 to 2010, 80 to 2020, and 28 from 2021–2022 show a species gradually increasing in numbers and range. Apparent preference for coastal and other lowland sites, but with a few records in eastern upland areas, e.g., Littleborough, Burnley, Mill Houses.

■ **HABITAT** Dry grassland, sand dunes, waste ground.
■ **FOODPLANTS** Unrecorded within the county.

Single brooded. Comes to light with high count of 34 at Gait Barrows on 25 July 2014.

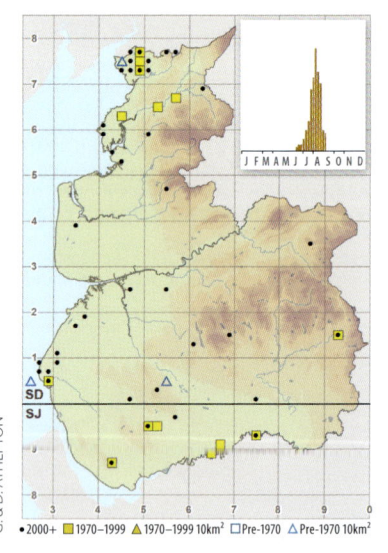

● 2000+   ■ 1970–1999   ▲ 1970–1999 10km²   □ Pre-1970   △ Pre-1970 10km²

63.091 *Agriphila latistria*

■ **LANCS STATUS** Occasional / local resident   c.1857–2022

Reported from the Crosby sand-hills by Gregson (1857), with all further pre-1990 records from Sefton Coast. Described as local by early authors. Became more widely recorded from 1990s onwards with records during the decade from Heysham N.R., and inland, lowland locations such as Risley Moss, Flixton, Mere Sands Wood, Bold Moss and Parr, St Helens. Numbers have declined since around 2015, with almost all subsequent records of singletons only, and no records at all in 2021.

■ **HABITAT** Dry grassland, sand dunes.
■ **FOODPLANTS** Unrecorded within the county.

Single brooded. High count of 10 at light at Heysham N.R. 30 July 2001 (P.J. Marsh).

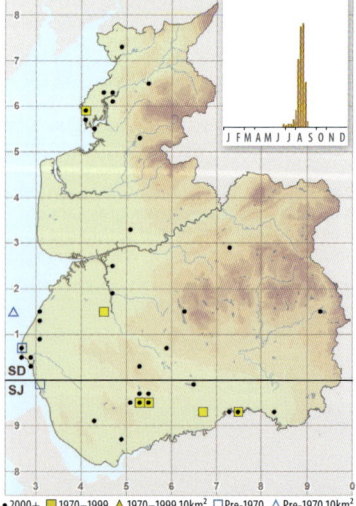

● 2000+   ■ 1970–1999   ▲ 1970–1999 10km²   □ Pre-1970   △ Pre-1970 10km²

63.092 *Agriphila selasella*

■ **LANCS STATUS** Occasional / local resident   c.1857–2022

First noted at Sefton meadows, on the banks of the River Alt (Gregson, 1857). Other 19th century records from Fleetwood and Preston saltmarshes by J.B. Hodgkinson and from Chorlton. Described as local but usually common by Mansbridge (1940). Unrecorded again until Morecambe 1993. Numbers significantly increased this century with 256 records compared to just five pre-2000, over a quarter of which are from Sunderland Point, associated with saltmarsh (J. Girdley, P.J. Marsh, J. Roberts).
■ **HABITAT** Damp grassland, saltmarshes.
■ **FOODPLANTS** Unrecorded within the county.
Single brooded. Noted at light and in day. 102 at light, Sunderland Point 23 July 2013.

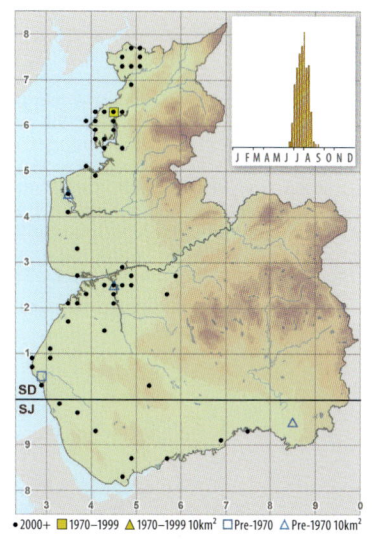

63.093 *Agriphila straminella*

■ **LANCS STATUS** Abundant / ubiquitous resident   c.1857–2022

Very familiar to anyone out walking in grassland sites from July onwards, as every step induces a few moths to fly out. First documented from the Liverpool district (Gregson, 1857). All early authors noted abundance, but specific sites not recorded until noted at Formby in 1957. Highest count of approx. 1,500 on Whit Moor, nr. Caton by S. Palmer 13 July 2009. Other 4-figure counts from Warton Crag and White Moss, Skelmersdale. Declining numbers in VC60 lowland sites noted by Marsh (2022).
■ **HABITAT** Grasslands, sand dunes, moorland, waste ground, gardens, etc.
■ **FOODPLANTS** Larva noted beneath moss at Rixton 8 May 2009; feeding not observed.
Single brooded. Easily disturbed by day. Readily to light, including garden traps.

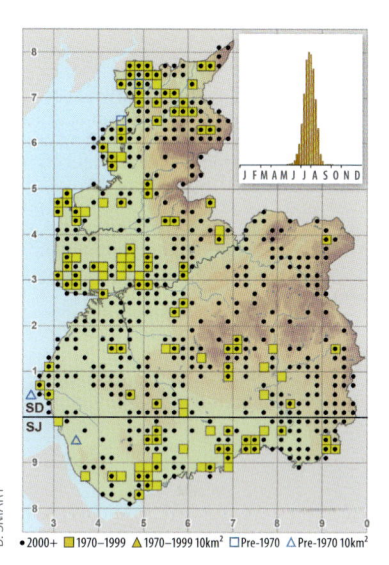

63.095 *Agriphila geniculea*

■ **LANCS STATUS** Frequent / local resident   c.1887–2022

First noted by Ellis (1890) as common on the sand-hills. Recorded from Crosby 1907 by W. Mansbridge and at Southport 1923 by H.L. Burrows. First noted outside Sefton Coast in 1987, from Lancaster and Fairhaven. Only recorded 13 times pre-2000, but over 1,220 records since, including 83 counts in double figures, with many records from Fylde, Formby, Heysham N.R., Silverdale, Hale, Woolton, and at inland sites too, such as Roby, Flixton, Billinge. Occasional in limestone uplands, e.g., Dalton Crags.
■ **HABITAT** Coastal grassland, limestone grassland, dry grasslands, gardens.
■ **FOODPLANTS** Not recorded in Lancashire.
Single brooded. Comes to light. High count of 51 from St Annes, 2013 (J. Steeden).

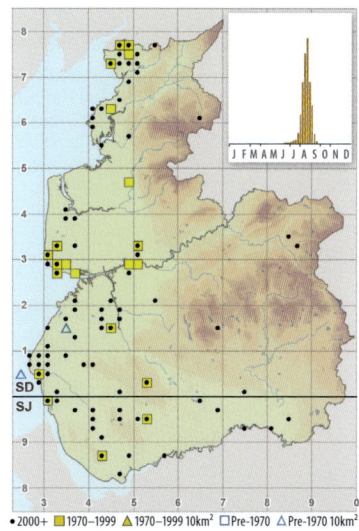

The Moths of Lancashire **367**

CRAMBIDAE

63.099 *Catoptria pinella*

■ **LANCS STATUS** Occasional / local resident   c.1859–2022

First records from Manchester; at Withington and Chorlton by J. Chappell, and from Didsbury by J.B. Hodgkinson, all in 19th century. No subsequent Manchester records. Whilst numbers increasing this century, the moth is now found almost exclusively at Morecambe Bay and northern upland sites. The only exceptions this century are singletons from Horwich, Flixton and Formby in VC59. Regularly recorded VC60 sites include Docker Moor, Yealand Conyers, Gait Barrows, Silverdale and Leighton Moss.
■ **HABITAT** Limestone grassland, woodland clearings.
■ **FOODPLANTS** Unrecorded within the county.
Single brooded. Mainly noted at light, with high count of 16 from Gait Barrows 2002.

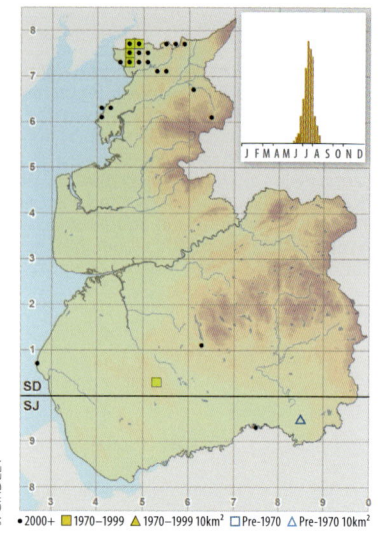

63.100 *Catoptria margaritella*

■ **LANCS STATUS** Frequent / local resident   1882–2022

First noted in 1882 by Gregson at Cadishead Moss, with Ellis reporting it common on south-west Lancashire mosses and heaths (1890). Recorded from Formby Moss c.1900, Simonswood Moss 1947, Holcroft Moss 1991 and Astley Moss in 2000. However, records have dried up from Chat Moss complex where last noted in 2015. Most recent records are from moorland areas following first such record, in 1996 at Longridge Fell, and often abundant. Occasional wanderer, reaching nearby gardens.
■ **HABITAT** Upland wet to boggy grassland sites, lowland mosses.
■ **FOODPLANTS** Unrecorded within the county.
Single brooded. 1,250 trapped at light at Green Bank 20 July 2022 by P.J. Marsh.

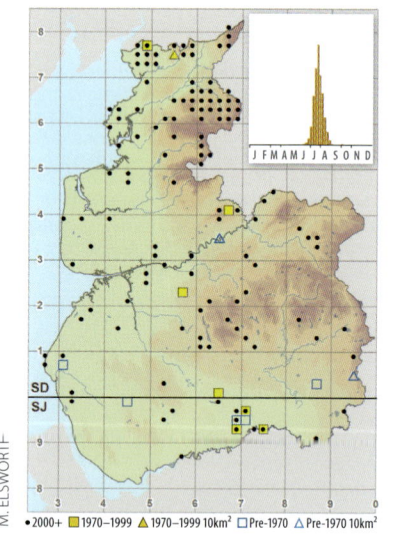

63.102 *Catoptria falsella*

■ **LANCS STATUS** Frequent / widespread resident   c.1859–2022

First noted on a wall at Hough End Hall, Chorlton by J. Chappell (Ellis, 1890). Mansbridge (1940) reported it to be local on moss-covered walls, although few early records. Noted at Silverdale 1940. Later, at Gait Barrows 1986 and Caton 1990. Over 1,650 subsequent records, with multiple records in many gardens, e.g., Parr, Flixton, Holmeswood and Warton, nr. Carnforth, etc. Range has increased sharply since 2000. Previously, only in extreme north and south, now the most widespread of the genus.
■ **HABITAT** Woodland, scrub, gardens, on moss-covered walls and roofs.
■ **FOODPLANTS** Not recorded in Lancashire.
Single brooded. Comes to light, with 37 at old sewage works, Ormskirk 17 July 2005.

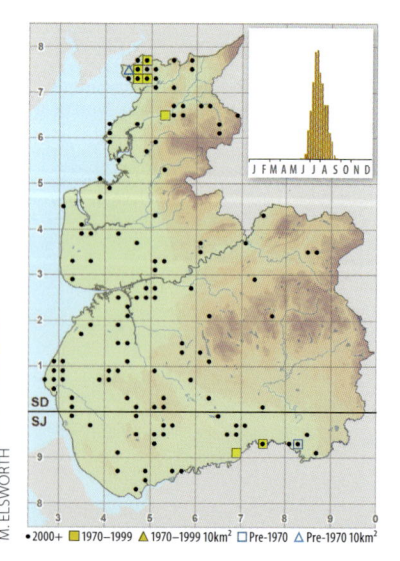

368 *The Moths of Lancashire*

63.110 *Pediasia aridella*

- **LANCS STATUS** Occasional / local resident   c.1889–2022

First found by J.B. Hodgkinson on saltmarshes near Preston (Ellis, 1890). Initially, thought to be *P. contaminella*, an error corrected by A.E. Wright in 1920. Noted on a few occasions during the 1950s at Formby and Birkdale, but not again until 2003 when noted at saltmarsh in Heaton, nr. Lancaster by P.J. Marsh. Subsequently, regular at Heysham N.R. and Sunderland Point, occasional on Sefton Coast, and a couple from Lytham St Annes in 2019. Recorded 69 times in total; 64 from 2003 onwards.
- **HABITAT** Saltmarshes and sand dunes.
- **FOODPLANTS** Not recorded in Lancashire.

Single brooded. Comes to light. 97 at Sunderland Point on 23 July 2013 by J. Girdley.

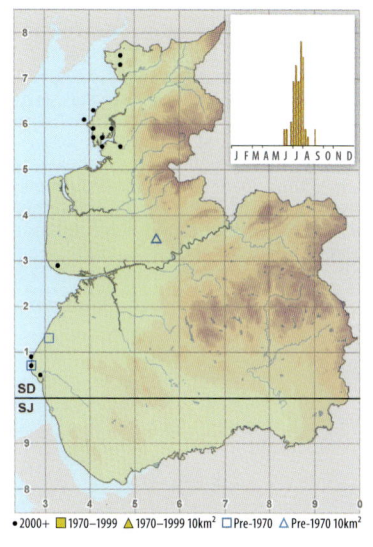

63.112 *Platytes alpinella*

- **LANCS STATUS** Rare / vagrant   2012

Nationally, this is a predominantly coastal moth. Perhaps surprising, therefore, that the only Lancashire record was inland, at Lightfoot Green, Preston. This was a moth trapped at light by S. & C. Palmer on 11 August 2012. However, this may have been a wanderer from the Yorkshire populations to the east. Unrecorded from Cumbria and Cheshire.
- **HABITAT** Unknown in the county.
- **FOODPLANTS** Not recorded in Lancashire.

Nationally, single brooded.

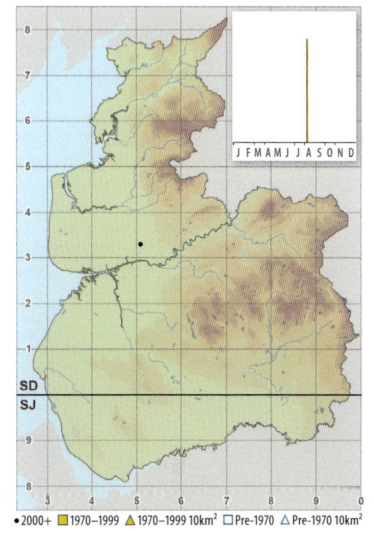

63.114 Brown China-mark *Elophila nymphaeata*

- **LANCS STATUS** Frequent / widespread resident   c.1857–2022

Abundant in the Liverpool district in mid-19th century (Gregson, 1857). Other early authors found it common everywhere. Reported at Middle Hulton by M. Morris in 1925, Ditton 1937 and Formby 1940. Other pre-1970 records from Leighton Moss and canal-side in Pendleton showed connection with waterbodies. Commonly recorded since the 1970s; even noted at upland sites such as Wray and Longridge Fell.
- **HABITAT** Edges of ponds, streams, canals and other freshwater sites.
- **FOODPLANTS** Water lilies and pondweed (*Potamogeton* sp.).

Single brooded. Easily disturbed from waterside vegetation. Comes to light. Larvae recorded in March, May, July and October, forming cases from leaves of foodplant.

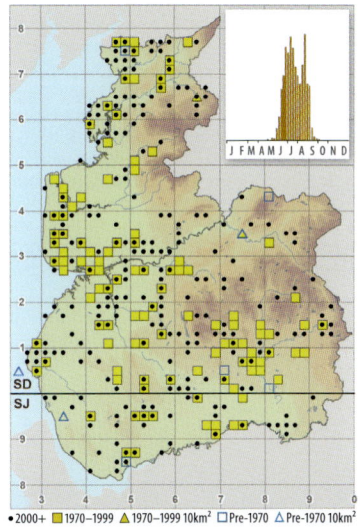

CRAMBIDAE

63.115 Water Veneer *Acentria ephemerella*

■ **LANCS STATUS** Frequent / widespread resident   1950–2022

Possibly over-looked in earlier years, as not reported until 1950, at Yealand Storrs, by an unknown recorder. H. Michaelis noted it at Didsbury in 1955 and at Bolton in 1959. Rarely recorded until 1990s, when noted from Lytham, Preston, Billinge, Gait Barrows, etc. Post-2000 records show range extending into eastern, more elevated areas. The aquatic larva and female remain unrecorded in Lancashire.

■ **HABITAT** Ponds, lakes, marshes and other freshwater sites. Disperses widely.

■ **FOODPLANTS** Unrecorded within the county.

Single brooded. Regularly found swarming in huge numbers at light, with highest estimated count of 500 at Billinge on 22 July 2013, recorded by C. Darbyshire.

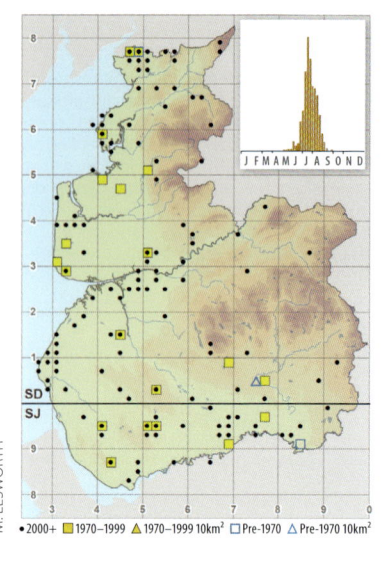

● 2000+   ▢ 1970–1999   ▲ 1970–1999 10km²   ▢ Pre-1970   △ Pre-1970 10km²

63.116 Small China-mark *Cataclysta lemnata*

■ **LANCS STATUS** Frequent / widespread resident   c.1857–2022

Reported from Liverpool district 'in every old pit' (Gregson, 1857). Ellis (1890) noted the moth be abundant everywhere, whereas Mansbridge (1940) considered it far more local. Britten (c.1950) noted it as widespread around ponds and canals. Early records from Middle Hulton and Burnage in 1925, Ditton 1929 and Simonswood Moss 1940. Recorded fairly regularly since, increasingly so since late 1990s; often attending garden traps, travelling some distance from the ponds where early stages are spent.

■ **HABITAT** Ponds, reedbeds, ditches.

■ **FOODPLANTS** Larvae feed on duckweed in cases made from leaves of the foodplant.

Single brooded. Comes to light and noted in day. Larvae from 31 March to 1 June.

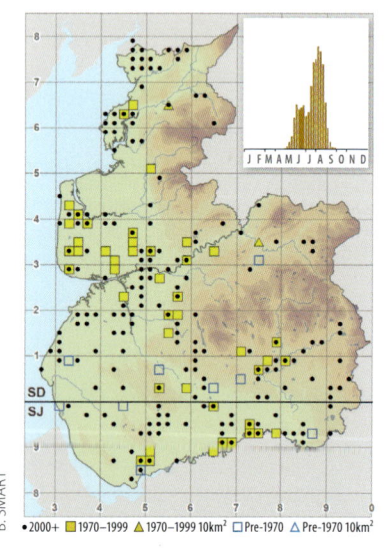

● 2000+   ▢ 1970–1999   ▲ 1970–1999 10km²   ▢ Pre-1970   △ Pre-1970 10km²

63.117 Ringed China-mark *Parapoynx stratiotata*

■ **LANCS STATUS** Occasional / local resident   c.1859–2022

Early records from Old Trafford by J. Chappell c.1859, Silverdale, Crosby and Preston Canal, nr. Salwick (Ellis, 1890), and considered local. Remained infrequent; only 15 records pre-2019. However, records have increased significantly with 55 from 2019–2022. Mainly from Leighton Moss, where the moth appears a new arrival, as first noted in 2009, then on 39 occasions from 2019 onwards. Other post-2018 records from Silverdale and Great Eccleston, and in VC59, Longton, Speke and Worsthorne.

■ **HABITAT** Edges of ponds, canals, lakes and other freshwater sites.

■ **FOODPLANTS** Unrecorded within the county.

Single brooded. Comes to light. High count of 13 at Silverdale Moss 2019 by C. Clay.

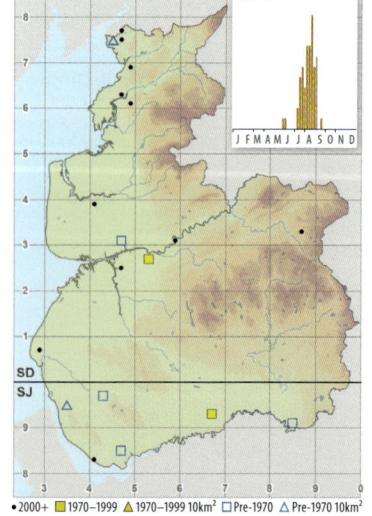

● 2000+   ▢ 1970–1999   ▲ 1970–1999 10km²   ▢ Pre-1970   △ Pre-1970 10km²

63.1175 *Parapoynx bilinealis*

■ **LANCS STATUS** Adventive　　　　　　　　　　　　　　　　**1983**

A male of this species was recorded flying in the quarantine section of Bolton Museum Aquarium by E.G. Hancock on 26 August 1983 (Hancock, 1984). It was thought likely to have been introduced with Water Wisteria *Synnema triflorum*, aquatic plants imported from Singapore. This was the second British record of this species, the first having been found in a strip-light insect trap in Hampshire in 1977. The current location of the Lancashire specimen remains unknown, following unsuccessful examination of the collections at Bolton, World and Manchester Museums.

■ **HABITAT** Unknown. However, as a native of Asia, it is likely that the species can only survive in the U.K. in warm greenhouses. Hancock further noted 'I could find no evidence of immature stages and under the circumstances it would appear unlikely that the species could become established in this area.'

■ **FOODPLANTS** Water Wisteria.

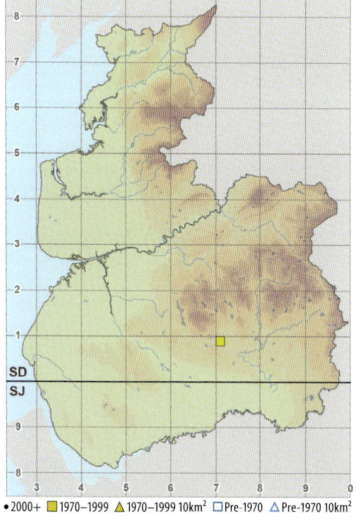

63.118 Beautiful China-mark *Nymphula nitidulata*

■ **LANCS STATUS** Occasional / widespread resident　　　　**c.1887–2022**

Described by Ellis (1890) as fairly common and generally distributed, although no sites mentioned. Recorded in 1920s at Carleton by W.G. Clutten and at Bolton by M. Morris, with a few 1950s records from Whalley and Gait Barrows. Only ten records pre-1980, but over 200 since, with Leighton Moss the location for 80 of these. Other sites include Hawes Water, Pennington Flash, Martin Mere, Mill Houses and gardens at Flixton, Hoghton, Worsthorne, Horwich, Briercliffe, Great Eccleston, etc.

■ **HABITAT** Edges of rivers, streams, lakes, etc.

■ **FOODPLANTS** Not recorded in Lancashire.

Single brooded. Easily disturbed from vegetation and comes to light.

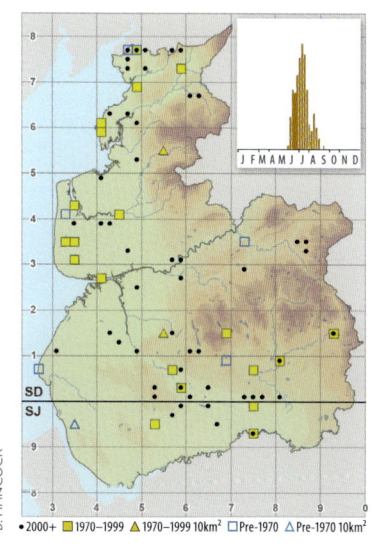

63.121 *Donacaula forficella*

■ **LANCS STATUS** Occasional / local resident　　　　　　　**c.1857–2022**

Whilst noted from Liverpool district in most swampy places where mint grows (Gregson, 1857), the only specific site mentioned is in VC58. Later, reported at Preston, Withington, Pendleton (Ellis, 1890). No further records until 1990 at Birk Bank (M. Dempsey), Lightfoot Green 1997 and Littleborough 1999. This century has seen an upturn in fortunes at extensive reedbed sites, particularly Leighton Moss and Martin Mere. Also, occasional at garden light traps, e.g., Rishton, Rufford, etc.

■ **HABITAT** Reedbeds.

■ **FOODPLANTS** Not recorded in Lancashire.

Single brooded. Comes to light. Occasionally disturbed from vegetation.

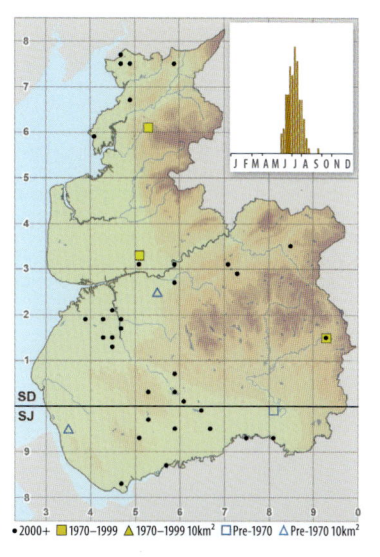

The Moths of Lancashire　**371**

CRAMBIDAE

63.122 *Donacaula mucronella*

■ **LANCS STATUS** Scarce / local resident   c.1887–2022

First noted by the canal in Preston by T. Cooper (Ellis, 1890). Similar habitat preferences as previous species, so somewhat surprisingly unrecorded from Martin Mere. The eight VC59 records include Holcroft Moss 1992 and Wigan Flashes 1999. Mainly recorded in the north of the county, with regular sites around Leighton Moss and Silverdale Moss. The few records from elsewhere include Thrushgill, Gait Barrows, Mere Sands Wood, and gardens at Hale, Flixton, Silverdale and Warton, nr. Carnforth, etc.

■ **HABITAT** Reedbeds.
■ **FOODPLANTS** Not recorded in Lancashire.

Single brooded. Most records at light; once at dusk during a search by torchlight.

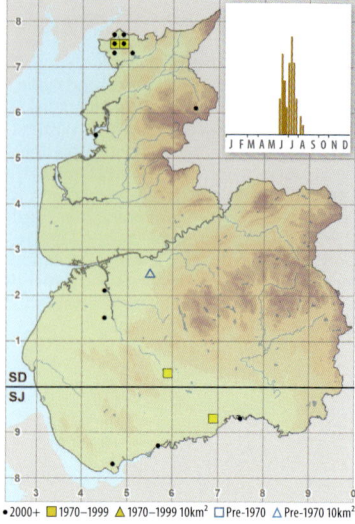

DREPANIDAE

65.001 Scalloped Hook-tip *Falcaria lacertinaria*

■ **LANCS STATUS** Occasional / local resident   1846–2022

The first records were from Chat Moss in July 1846 (Edleston, 1846), and of larvae at Heysham Moss (Gregson, 1883). Noted at Rixton Moss in 1894 and Silverdale in 1925. Primarily a lowland species, with most recent records from Chat Moss complex and Sefton Coast. Also recorded on 98 occasions (of a total of 351) at K. McCabe's Flixton garden, with a larva found nearby, on birch, at Ash Hill, Flixton on 11 October 2015.

■ **HABITAT** Mosses, sand dunes, birch scrub.
■ **FOODPLANTS** Birch, with a preference for seedlings.

Double brooded with second generation far more numerous. Comes to light. Larvae recorded on eleven occasions, between June and October.

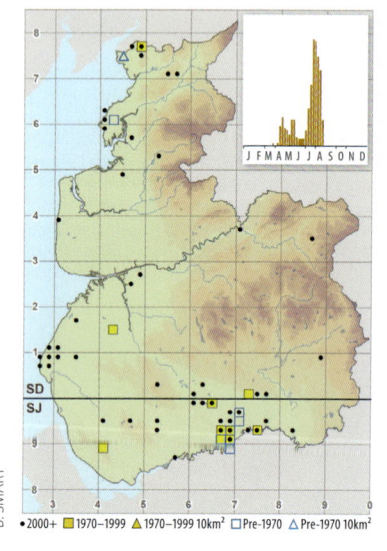

65.002 Oak Hook-tip *Watsonalla binaria*

■ **LANCS STATUS** Occasional / widespread resident   c.1940–2022

Recorded around 1940 in Didsbury by H.N. Michaelis, and in Prescot by the Rev. R. Freeman. The lack of older records suggests a possible new arrival. H. Britten noted considerable extension in range post-1945. Subsequently, recorded in good numbers throughout, primarily in lowland areas. It seems the moth is likely declining. The 44 records from 2022 are approximately 60% of the annual totals from 10–15 years ago, with most counts of singletons.

■ **HABITAT** Oak woodland, hedgerows.
■ **FOODPLANTS** Not recorded in Lancashire

Double brooded. Most records are of moths to light.

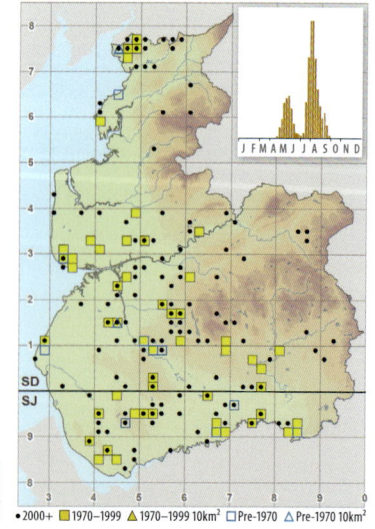

65.003 Barred Hook-tip *Watsonalla cultraria*

■ **LANCS STATUS** Occasional / local resident   1983–2022
A new arrival in the county, as nationally, range is expanding northwards; likely from Cumbria too, as most Lancs records are in VC60. Noted at Gait Barrows in 1983 and 1984 by E. Emmett, and next at Formby in 2000. VC60 records include Silverdale, Bay Horse, Yealand Conyers and Dalton, nr. Burton-in-Kendal. Only 20 of the 121 records are from VC59, e.g., Cuerden Valley Park, Hoghton, Briercliffe, Flixton and Chorlton. 28 records from 2022 suggests that range and frequency are still increasing.
■ **HABITAT** Woodland containing Beech.
■ **FOODPLANTS** Unrecorded within the county.
Double brooded. Where noted, all records have been at light.

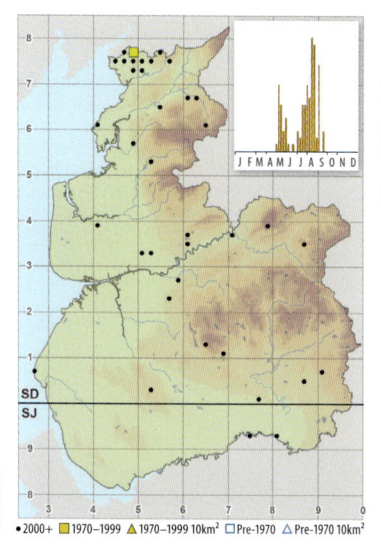

65.005 Pebble Hook-tip *Drepana falcataria*

■ **LANCS STATUS** Frequent / widespread resident   c.1843–2022
First records from Pilling (Hodgkinson, 1844), Chat Moss (Edleston, 1846) and Simonswood Moss, Kirkby (Gregson, 1857). Ellis (1890) noted the moth to be more common than Scalloped Hook-tip and found in the same localities. Since, noted to be frequent, particularly where natural birch growth occurs. Many garden records, including over 800 from Flixton by K. McCabe. Maximum count of 27 at Formby on 8 August 2013 by R. Walker and A. & S. Parsons. Larvae noted from June to October.
■ **HABITAT** Birch woodland, hedgerows, gardens.
■ **FOODPLANTS** Silver Birch, Downy Birch. Once on Alder.
Double brooded. Comes to light. Occasionally disturbed from vegetation.

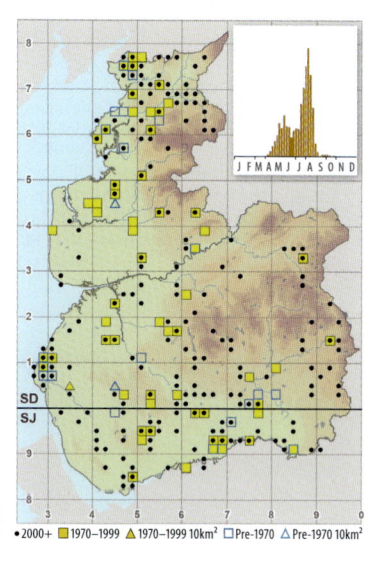

65.007 Chinese Character *Cilix glaucata*

■ **LANCS STATUS** Frequent / widespread resident   c.1857–2022
First reported from the Liverpool district by Gregson (1857), noting the moth to be 'plentiful in lanes.' Ellis (1890) noted it to be 'common…occasionally abundant.' Recorded on 138 occasions in 2022, although counts possibly declining with the highest in 2022 just five, compared to 25 at Higher Tatham by P.J. Marsh in August 2004. Well-represented in upland areas, indeed that is where the highest counts occur. Also present in lowland areas, and a frequent visitor to garden traps.
■ **HABITAT** Scrub, hedgerows, woodland.
■ **FOODPLANTS** Blackthorn.
Double brooded. Comes to light. Larva from Didsbury, 1929 (L. Nathan).

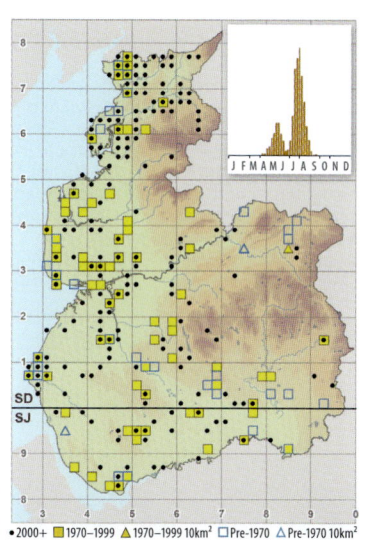

The Moths of Lancashire **373**

DREPANIDAE

65.008 Peach Blossom *Thyatira batis*

■ **LANCS STATUS** Frequent / widespread resident   c.1843–2022
First records from Preston (Hodgkinson, 1844) and Hale (Gregson, 1857). Noted to be generally distributed and fairly common (Ellis, 1890). Currently, distribution and population appear stable, with 83 records from 2022. Recorded across lowland and upland parts of the county, with latter sites including Docker Moor, Wray, Dalton Crags and Todmorden. Notable high counts of 15 from Heysham N.R. 2004 by P.J. Marsh, and 11 at Ainsdale in 2021 by R. Moyes and C. Daly.
■ **HABITAT** Woodland, scrub, hedgerows, gardens.
■ **FOODPLANTS** Bramble.
Single brooded. Comes to light. Larvae noted from July to September.

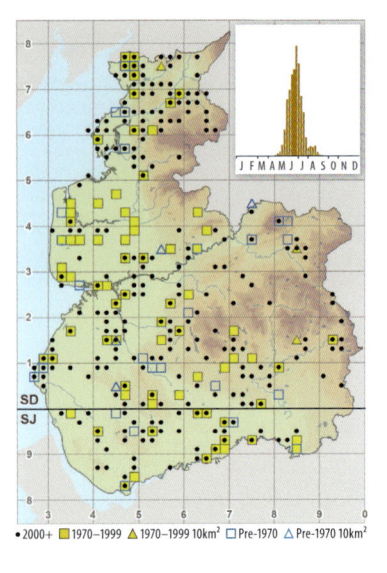

• 2000+   ▢ 1970–1999   △ 1970–1999 10km²   ▢ Pre-1970   △ Pre-1970 10km²

65.009 Buff Arches *Habrosyne pyritoides*

■ **LANCS STATUS** Frequent / widespread resident   c.1857–2022
First recorded from Hale and Dingle, Liverpool (Gregson, 1857). Noted to be generally distributed (Ellis, 1890), with Britten (c.1950) reporting the moth to be fairly common in most areas except high ground. Whilst most recent records are from lowland areas, there are others from more elevated sites too, such as at Docker Moor on 21 July 2022. Population appears stable in terms of range and numbers. Maximum counts of 51 from Flixton, 2009 (K. McCabe) and 33 from Ainsdale, 2022 (R. Moyes).
■ **HABITAT** Woodland, rough grassland, gardens.
■ **FOODPLANTS** Bramble. Larva noted at Huncoat 17 September 2022 (C. Jameson).
Single brooded. Comes regularly to light.

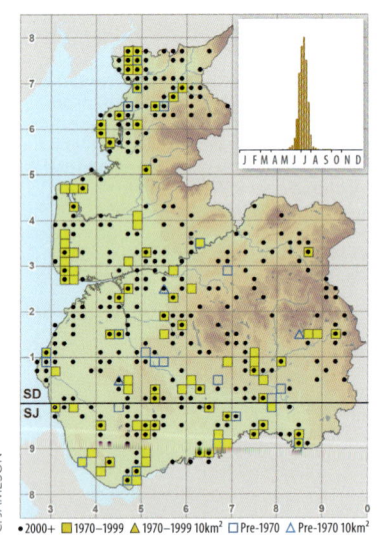

• 2000+   ▢ 1970–1999   △ 1970–1999 10km²   ▢ Pre-1970   △ Pre-1970 10km²

65.010 Figure of Eighty *Tethea ocularis*

■ **LANCS STATUS** Occasional / widespread resident   c.1890–2022
Earliest records are from Ribchester and Preston in the late 19th century, recorder probably J.B. Hodgkinson (Britten, c.1950), but not again until Formby, 4 June 1952 by G. de C. Fraser. Recorded at Parbold in 1961. Gradual spread elsewhere with late 1960s records from Manchester, Flixton, Liverpool, Lytham, Longridge and Leighton Moss, as part of the national range expansion (Randle *et al.*, 2019). Annual counts variable, e.g., 58 records from 2022 was half that of 2009 but double that of 2012.
■ **HABITAT** Woodland, hedgerows, scrub.
■ **FOODPLANTS** Poplars including White Poplar.
Single brooded. Comes to light. Larvae in July and August.

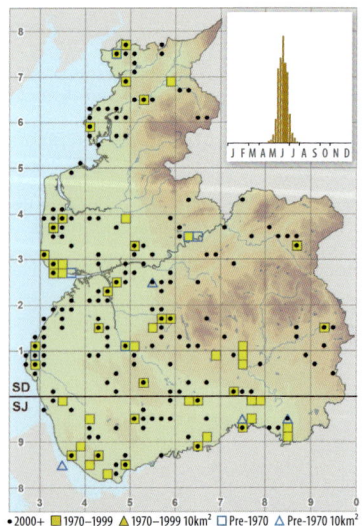

• 2000+   ▢ 1970–1999   △ 1970–1999 10km²   ▢ Pre-1970   △ Pre-1970 10km²

374   The Moths of Lancashire

65.011 Poplar Lutestring *Tethea or*

■ **LANCS STATUS** Considered Extinct   1879–1890

The first record we have of this species is from Dutton, nr. Ribchester in August 1879. The author reports beating two small Aspens over a sheet, and 'down came…a cloud of *Cymatophora or* (now *Tethea or*) larvae' (Hodgkinson, 1880). Based on Hodgkinson's observations, it was later noted that the moth 'occurs occasionally about Preston' (Ellis, 1890). In the absence of further records, the moth is likely extinct within the county. The moth remains present in a few scattered locations in north Cumbria and north-east Yorkshire. Unrecorded from Cheshire.
■ **HABITAT** Woodland.
■ **FOODPLANTS** Aspen.
Difficult to determine from the few Lancs records, but nationally, single brooded.

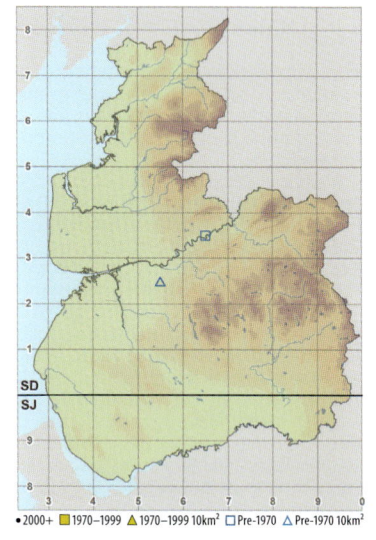

● 2000+   ■ 1970–1999   ▲ 1970–1999 10km²   □ Pre-1970   △ Pre-1970 10km²

65.012 Satin Lutestring *Tetheella fluctuosa*

■ **LANCS STATUS** Scarce / very local resident   1986–2022

First reported from Gait Barrows in 1986 by B. Hudson, with a further record from the same location on 10 May 1993 by R. Petley-Jones, and subsequently from 1994, 1995, 1998, 1999, 2001 and 2003. The next records were on 12 June 2014, when two Robinson MV traps were used at Gait Barrows from dusk until 1am, with 12 moths trapped by G. Jones. Recorded from a Silverdale garden on two occasions during July 2022 by J. Patton, only the second known site for this species.
■ **HABITAT** Limestone woodland.
■ **FOODPLANTS** Foodplant unrecorded within the county.
Single brooded.

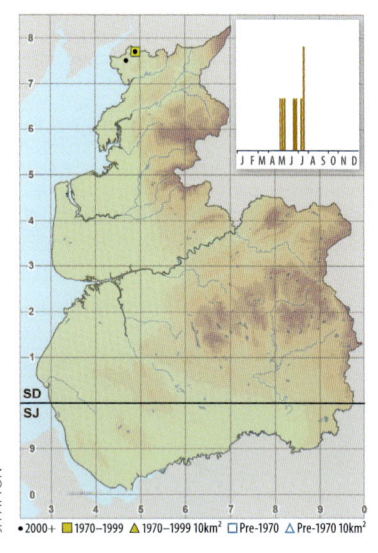

● 2000+   ■ 1970–1999   ▲ 1970–1999 10km²   □ Pre-1970   △ Pre-1970 10km²

65.013 Common Lutestring *Ochropacha duplaris*

■ **LANCS STATUS** Frequent / widespread resident   c.1843–2022

Recorded from Preston around 1843 (Hodgkinson, 1844), with further 19th century records from Roby, Huyton (Gregson, 1857), Worsley and Chat Moss (Ellis, 1890). Described as abundant on the mosses (Mansbridge, 1940), but with no further sites given. Recorded at Formby in 1942 and from Warton Crag in 1959. Recent records suggest it is widespread in birch woodland, with frequency appearing stable. The 89 records from 2022 include many garden sightings, e.g., Rishton, Flixton, Bacup, etc.
■ **HABITAT** Woodland, scrub, mosses.
■ **FOODPLANTS** Birch; once on Alder, at Chorlton Water Park, 14 August 2016.
Single brooded. Comes to light. Larvae recorded during August and September.

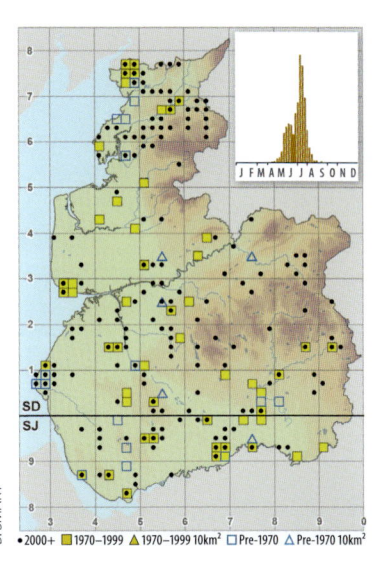

● 2000+   ■ 1970–1999   ▲ 1970–1999 10km²   □ Pre-1970   △ Pre-1970 10km²

DREPANIDAE

65.014 Oak Lutestring *Cymatophorina diluta*

■ **LANCS STATUS** Scarce / local resident　　　　　　　　　　c.1859–2023

A nationally declining species where regional extinctions have left VC60, and South Cumbria, as the only extant known populations in the northern United Kingdom. It was formerly known as Lesser Lutestring and the British form is of the subspecies *hartwiegi*.

Historic VC59 records comprised unknown numbers at Mere Clough, Bury, where described as uncommon by J. Chappell c.1859, Rochdale (1880), Knowsley (1899) and Bolton (1955). The first VC60 records were from a Rothamsted trap at Leighton Moss comprising 12 individuals from 1968–1978 (J. Briggs *et al.*). 1980 saw the first record from the Lune Valley catchment, followed by a scattering of single records from Burton Wood, Claughton and Hornby during the late 20th century.

At the start of the 21st century, an extensive programme of tetrad recording was commenced in north-east Lancashire by P.J. Marsh. This included a visit to native oak-dominated Cragg Wood, Littledale, on 26 Aug 2001, where 52 were recorded during the night. Subsequent double-figure light-trap sessions included the Furnessford area of Hindburndale, Littledale Valley bottom and Mallowdale, in upper Roeburndale, as well as further visits to Cragg Wood, the latter including 31 in an overnight actinic trap. Garden traps within the Lune catchment area, for example Wray and Mill Houses, also recorded occasional wandering individuals. Oddly, though, extensive recording in oak-woodland in upper Hindburndale failed to record this species.

With no pre-2000 recording effort, it has not been possible to relate the Lune catchment records to the noted national decrease since 1970 (Randle *et al.*, 2019). However, the status in the Arnside and Silverdale AONB, with just five records 1996–2019 and none at Leighton Moss since 1978, is surely indicative of decline there. The lack of very recent records, other than nine in the Birk Bank area on 23 August 2022 (J. Patton), reflects a lack of monitoring rather than any suggested decrease.

The early stages have not been searched for during the post-2000 recording effort and there are no previously documented records of larvae in the county. The literature suggests the larvae emerge from eggs overwintering on oak twigs in April, and thence hide in spun leaves during the day, feeding at night. They are reported to pupate around the end of June amongst the leaves on the tree (Henwood & Sterling, 2020). Limital dates for adults ranged from 27 July to 23 September. Like others of its genus, it is an early flyer. It has not been tried at the key Lancashire sites, but early-night sugaring has historically been found to be productive and may, of course, have been the methodology for the old VC59 records (South, 1907). A remarkable sighting occurred on 3 September 2023 as one came to light at Hale in the far south of VC59 (C. Cockbain).

Pete Marsh

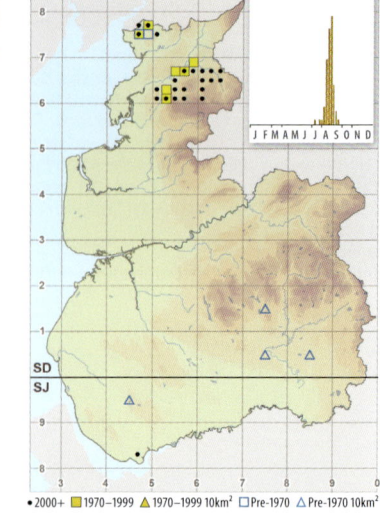

65.015 Frosted Green *Polyploca ridens*

■ **LANCS STATUS** Considered Extinct   c.1857

The only record we have of this species is of 'one specimen taken in the Boor's Wood, Hale, by Frederick Hitchmough' (Gregson, 1857). The actual date of capture is unknown. Mansbridge (1940) later described the moth's regional status as very rare. The likelihood is that this species is now extinct in Lancashire. Similarly, the moth has not been recorded in Yorkshire or Cheshire for over 100 years. It remains present, although very local, in Cumbria. Usually occurring in low numbers, an exception was a count of nine to light at Derwent Water in 2018.
■ **HABITAT** Unknown in the county.
■ **FOODPLANTS** Not recorded in Lancashire.
Nationally, single brooded. Randle *et al.* (2019) report that the national long-term abundance trend for this species shows a severe decline.

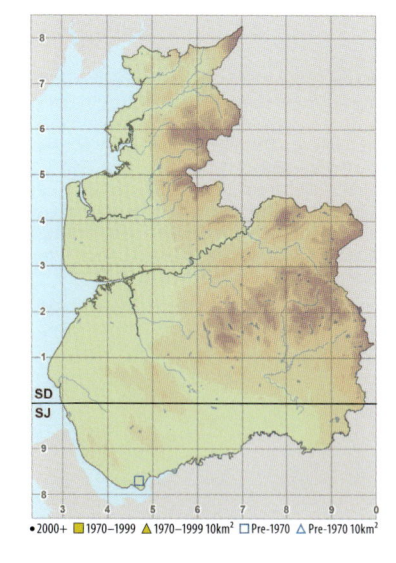

DREPANIDAE

● 2000+   ▪ 1970–1999   ▲ 1970–1999 10km²   □ Pre-1970   △ Pre-1970 10km²

65.016 Yellow Horned *Achlya flavicornis*

■ **LANCS STATUS** Occasional / local resident   c.1843–2022

Retrospectively, noted from Heysham Moss c.1843 by C.S. Gregson (1883), and from Liverpool (Gregson, 1857). Early authors noted the moth as common on the mosses, occasionally elsewhere. Britten (c.1950) reported it as plentiful among birch in mosses and woods. There has been no obvious decline in range or abundance since, with many upland records as well as those from lowland mosses. Absent from Fylde coast. High count of 68 to light at Freshfield Dune Heath 21 March 2010 by G. Jones.
■ **HABITAT** Woodland, moorland, mosses, dunes.
■ **FOODPLANTS** Birch.
Single brooded. Larvae recorded during May and June. Moth comes readily to light.

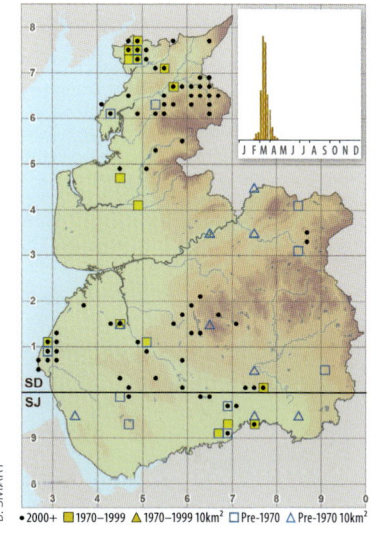

● 2000+   ▪ 1970–1999   ▲ 1970–1999 10km²   □ Pre-1970   △ Pre-1970 10km²

66.001 December Moth *Poecilocampa populi*

■ **LANCS STATUS** Frequent / widespread resident   1843–2022

First recorded in Preston (Hodgkinson, 1844) and, thereafter, at many other sites, such as West Derby, Manchester, Bolton, Lancaster, Silverdale, Nelson and Lytham. From the 1970s, records increased notably in lowland areas, but it wasn't until after 2000 that this also occurred in higher parts of the county, but to a lesser extent. Double figure counts, at light, have been regular since the 1960s and occasionally it can occur abundantly, such as 120 at Mill Houses on 27 November 2013 (P.J. Marsh).
■ **HABITAT** Woodland, hedgerows, gardens, parks and brownfield sites.
■ **FOODPLANTS** Not recorded in Lancashire. Ova from a caught female reared on oak.
Single brooded. Comes to light.

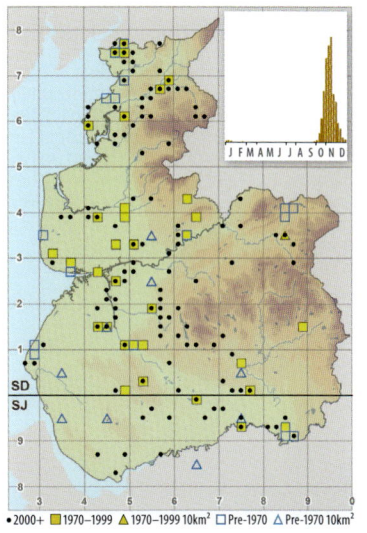

LASIOCAMPIDAE

● 2000+   ▪ 1970–1999   ▲ 1970–1999 10km²   □ Pre-1970   △ Pre-1970 10km²

The Moths of Lancashire   **377**

LASIOCAMPIDAE

66.002 Pale Eggar *Trichiura crataegi*

■ **LANCS STATUS** Scarce / very local resident; lowland pop. Extinct   c.1859–2015
Two distinct populations existed; in the lowlands and a single upland site, the latter with the dark-winged form. Lowland moths first noted nr. Preston (Stainton, 1859) and later, in the Silverdale area by J.C. Melvill. The final, isolated, lowland record was from Heysham N.R. on 28 August 1998 (P.J. Marsh). Subsequently, recorded annually at Leck Fell from 22 July 2001 to 12 September 2015, numbers peaking at 40 in Aug. 2005. Since then, searched for once at Leck Fell, in 2022, with a negative result.
■ **HABITAT** Heather moorland near limestone upland. Formerly lowland hedgerows.
■ **FOODPLANTS** Not recorded in Lancashire.
Single brooded. Comes to light. The life cycle of the upland moths is unknown.

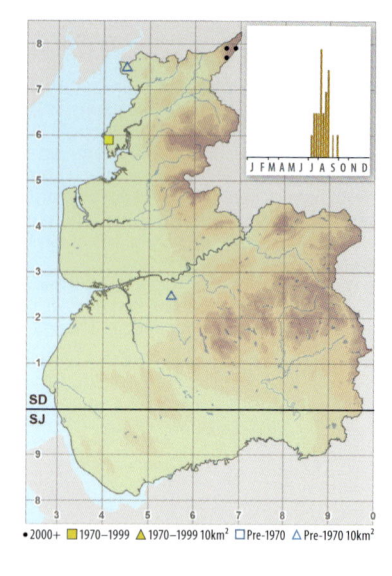

66.003 Lackey *Malacosoma neustria*

■ **LANCS STATUS** Considered Extinct   c.1890–1909
The only published details pertaining to this species were in Ellis (1890). This account covers Cheshire, Lancashire and parts of south Cumbria, where it is given a general status of 'local, but tolerably common where it does occur'. The Lancashire locations listed were Blackpool and Lytham, by J. Chappell, and the records were probably made in the second half of the 19th century. The NHM, London has a single specimen in the Rothschild, Cockayne and Kettlewell coll., labelled 'Lancs., 1909', and the recorder is noted as F. Womersley. No further details are known.
■ **HABITAT** Coastal scrub and hedgerows based on the historic locations.
■ **FOODPLANTS** Not recorded in Lancashire.

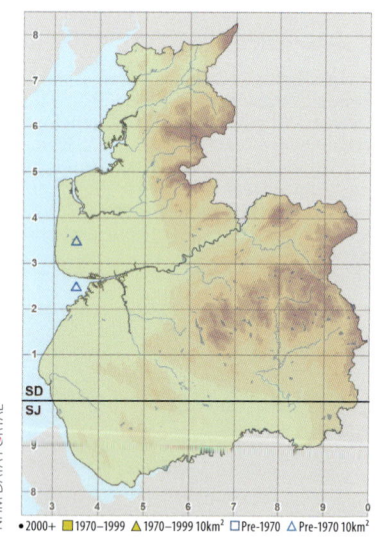

66.005 Small Eggar *Eriogaster lanestris*

■ **LANCS STATUS** Considered Extinct   c.1890–c.1970
The documented historic records summarise its geographical range in the county in the 19th century, but without any dates. Those listed in Ellis (1890) were from Blackpool and Morecambe (J. Chappell) and the Fylde district, where J.B. Hodgkinson found it common 'from Preston to Fleetwood.' Mansbridge (1940) added Silverdale (local) in 1909 and Warton (common), the latter by C.F. Forsythe. The final record is listed in Briggs (1985) from Leighton Moss in 1968, although in N.L. Birkett's card index, Tullie House Museum, Carlisle, it is given as 1970. Still found in south Cumbria.
■ **HABITAT** Scrub and hedgerows.
■ **FOODPLANTS** Not recorded in Lancashire.

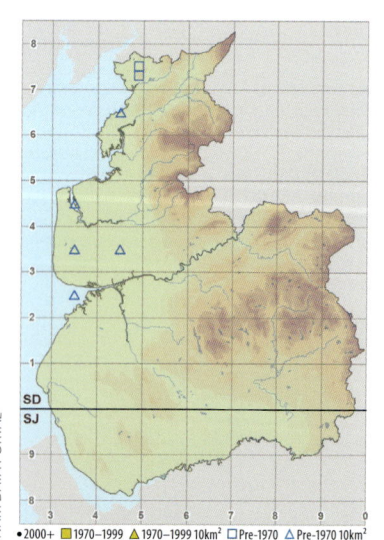

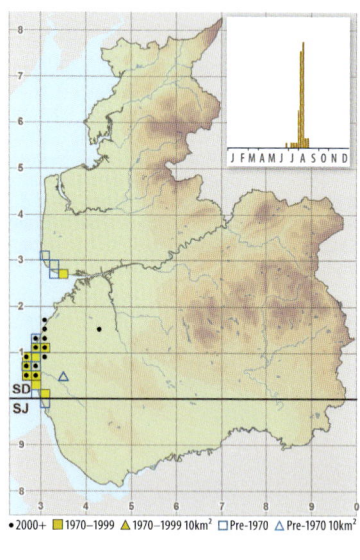

66.006 Grass Eggar *Lasiocampa trifolii*

■ **LANCS STATUS** Occasional / very local resident　　　　　　　　1856–2023

A Nationally Scarce moth, allocated a status of Least Concern in Randle *et al.*, 2019), the British population lies in pockets along the south coast of England and south Wales, with the Lancashire population at the northern limit of its national range.

The first record was from Crosby on 3 May 1856 by C.S. Gregson (Gregson, 1856b), and the expansive sand dunes at Formby and Ainsdale have remained its Lancashire stronghold to the present day. Ellis (1890) noted it as occurring on coastal sandhills from Blackpool and Lytham to Wallasey. The Fylde population declined considerably in the 19th Century, and Mansbridge (1940) states that the last record for St Annes was in 1888. Development, most likely the creation of both marine lakes, along with the 'cupidity of collectors' (Ellis, 1890) were cited as reasons for losses at St Annes and Crosby. However, it is likely that it was not lost completely from the Fylde and a record from the dunes at St Annes in 1985 appears to confirm this, although this was to be the last record from VC60.

The national separation of populations has seen differences in larval foodplant preferences emerge. In Lancashire, the main foodplant is thought to be Creeping Willow although there are no records of the larva being found on anything other than Broom. Elsewhere, it is noted that the species overwinters as an egg, the larva hatching in early spring but can do so as early as January, feeding at night and resting openly during the day (Henwood & Sterling, 2020). Mid- to late instar larvae can often be found basking in sunshine in late spring and early summer, unconcerned by predators due to protective, irritant hairs.

● 2000+　□ 1970–1999　△ 1970–1999 10km^2　□ Pre-1970　△ Pre-1970 10km^2

Pupation is undertaken in a tough cocoon, attached to vegetation and lasts around six weeks. Males emerge during the day and, like other Lasiocampids, will readily respond to 'calling' females from late afternoon onwards. Peak adult flight period is the first two weeks of August, although there is a record of one to light at Birkdale Green Beach on 26th June 2012 (G. Jones). Adults come readily to light and can reach double figures, with the highest count of 54 recorded at Formby on 12th August 2000 (Lancashire Moth Group). Noted at Martin Mere on 13 August 2023.

The population at Formby and Ainsdale appears relatively stable, although the larger catches of 20–50 plus in the early 2000s seem to be a thing of the past, but no standardised surveys have been undertaken. This stability is probably due to both the extent and protected status of this nationally important habitat. However, sand dunes are fragile habitats, vulnerable to a variety of pressures. Unchecked recreational disturbance causes considerable damage to these delicate dune ecosystems, and scrub growth, especially of non-native Sea Buckthorn, results in the loss of floristically rich dune grasslands that are critically important for the continued success of the species.

Graham Jones

LASIOCAMPIDAE

The Moths of Lancashire

LASIOCAMPIDAE

66.007 Oak Eggar *Lasiocampa quercus*

■ **LANCS STATUS** Occasional / widespread resident   1842–2022

First seen on White Moss (Eddleston, 1843), then Chat Moss and Freshfield, amongst many others; early authors report it as abundant. The development of golf courses on the dunes is blamed for declines (Mansbridge, 1940), and it is likely that the early loss of many lowland mosses in VC59 and recent large moorland fires, have done likewise. Still widely recorded on the moors and dunes, but rarely in double figures.
■ **HABITAT** Heather moors, mosses, sand dunes and sometimes, limestone pavement.
■ **FOODPLANTS** Heather, Bilberry, birch, Bramble, once Privet and reportedly, Soft Rush. Single brooded. Comes to light. The subspecific status of Northern and Oak Eggar is discussed in Emmet & Heath (1991), but the position in Lancashire is unclear.

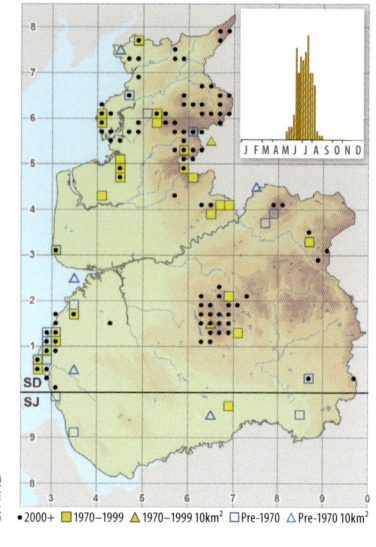

66.008 Fox Moth *Macrothylacia rubi*

■ **LANCS STATUS** Occasional / widespread resident   1842–2022

Noted by early authors to be abundant as larvae. First recorded in 1842, on White Moss (Edleston, 1843). Later, reported widely and equally on dunes and moors, with regular sightings at Ainsdale and Formby for example; has declined at St Annes in recent years. Remains widely distributed on the moors, particularly in the north. Numbers of records received have declined during the last decade, throughout.
■ **HABITAT** Moorland, mosses and sand dunes.
■ **FOODPLANTS** Creeping Willow, Bilberry and Heath Rush. Larva rarely observed feeding. Single brooded. Comes to light. Males fly on sunny days and larvae are present from mid-June to April, but most frequently found in September.

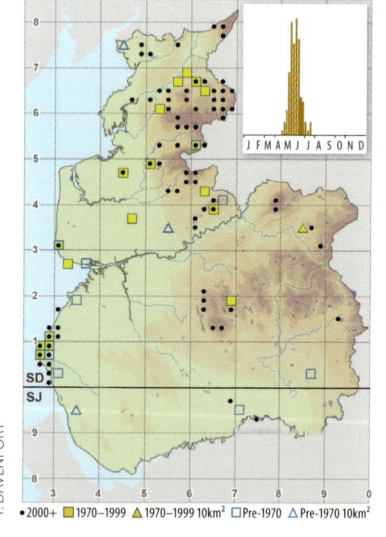

66.010 Drinker *Euthrix potatoria*

■ **LANCS STATUS** Occasional / widespread resident   c.1857–2022

All of the many early records were from lowland, particularly coastal, areas, the first in Liverpool (Gregson, 1857). Mansbridge (1940) noted it as abundant as a larva in these areas. It continued to be widely reported in lowland districts, but it was not until after 2000 that regular upland records appeared, such as at Botton Mill, Green Bank and Leck Fell. Double figure counts, always uncommon, are now increasingly infrequent. Last large count was of 35 adults at Goodber Common on 4 July 2009.
■ **HABITAT** Damp, rough grassland, dunes, brownfield sites, moss and moorland edges.
■ **FOODPLANTS** Grasses, including Reed Canary-grass, and once on Bramble. Single brooded. Comes to light. Larvae are most frequently found in the Spring.

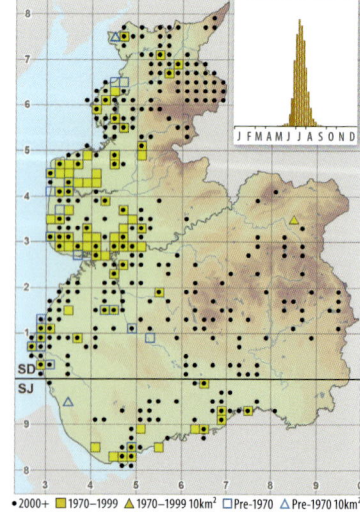

66.011 Small Lappet *Phyllodesma ilicifolia*

■ **LANCS STATUS** Nationally Extinct
Whilst visiting Peterborough Museum and Art Gallery a few years ago, Steve Garland spent some time examining the Lepidoptera specimens (S. Garland pers. comm.). Included in the collection were two Small Lappet specimens, a male and a female, each with undated data labels indicating they had been collected from Chat Moss. The recorder's name was difficult to decipher but appeared to be J. Muleburm. These specimens were unknown to the national Lepidoptera community and were also the only record of this moth having been found in Lancashire.
■ **HABITAT** Unknown, but nationally associated with mosses containing Bilberry.
■ **FOODPLANTS** Not recorded in Lancashire.

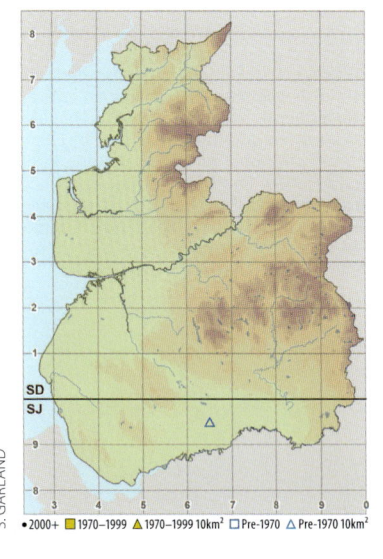

LASIOCAMPIDAE

68.001 Emperor Moth *Saturnia pavonia*

■ **LANCS STATUS** Occasional / widespread resident   1842–2022
Early authors successively noted it as abundant, common, then generally distributed on mosses and heaths, from mid-19th to mid-20th centuries. First reported April 1842 from Boggart Hole Clough and White Moss (Edleston, 1843a). Since then, it has been widely recorded in suitable habitat, but in gradually declining numbers. Records in 2021–22 include the results of pheromone use. This helped with distributional data but led to higher counts than usual based on previously used fieldwork techniques.
■ **HABITAT** Mosses, moors and sand dunes.
■ **FOODPLANTS** Heather, Bilberry, birch and Bramble.
Single brooded. Comes to light. Males fly by day.

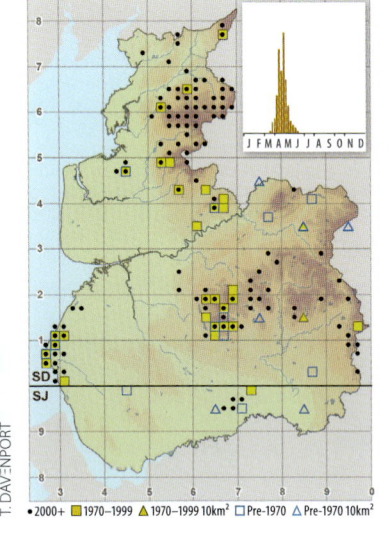

SATURNIIDAE

69.001 Lime Hawk-moth *Mimas tiliae*

■ **LANCS STATUS** Frequent / widespread resident   1893–2022
First noted in Prescot by R. Freeman in 1893 (Mansbridge, 1940), but it was not until 1941 that the next was seen, in Formby. The main arrival in the county commenced much later when it was found at Leighton Moss in 1969 (P.J. Marsh) and thereafter, from sites such as Didsbury, Bolton, Wavertree and Chorley. It reached some upland areas around 1996 but is still rather local at such sites. Mostly occurs in low single figures, with the largest count, of seven, in Formby on 30 May 2003 (G. Jones, *et al.*).
■ **HABITAT** Woodland, parks, amenity plantings and gardens.
■ **FOODPLANTS** Lime and Alder.
Single brooded. Comes to light. Larvae found from mid-July to early September.

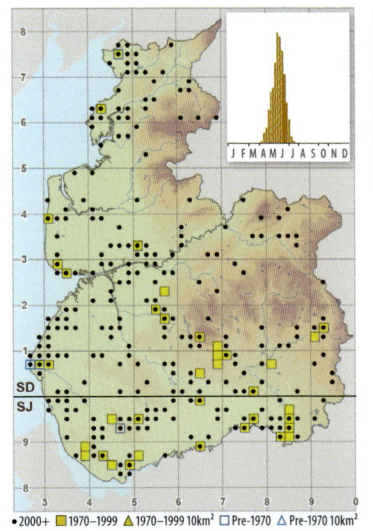

SPHINGIDAE

The Moths of Lancashire **381**

SPHINGIDAE

69.002 Eyed Hawk-moth *Smerinthus ocellata*

■ **LANCS STATUS** Frequent / widespread resident   c.1857–2022

Noted as common among willows in the Liverpool district (Gregson, 1857) and subsequently found widely across lowland areas, to the present day. Noted as common throughout the county by Ellis (1890), but the first documented upland report was not until 1912, in Burnley. It remains rather local on high ground. Mostly found in low single figures, but 22 attracted to light at Ainsdale in 2013 (R. Moyes).
■ **HABITAT** Woodland, scrubby dunes, parks, gardens and brownfield sites.
■ **FOODPLANTS** Sallows, including Creeping Willow, Apple and once on strawberry. Single brooded, with a few records up to late August. Comes to light. Larvae have been found in August and September.

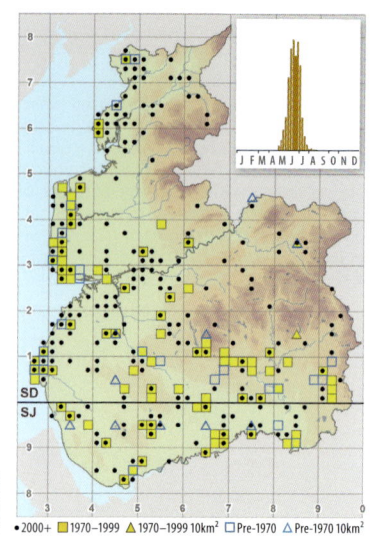

● 2000+   ▢ 1970–1999   △ 1970–1999 10km²   ▢ Pre-1970   △ Pre-1970 10km²

69.003 Poplar Hawk-moth *Laothoe populi*

■ **LANCS STATUS** Frequent / widespread resident   c.1857–2022

Noted as 'common in the Liverpool area' (Gregson, 1857), this remaining the case, across the county, to the present day. There has been no apparent change in the wider distribution, any new sites generated by increased light trapping, infilling rather than expanding the known range. Found mostly in single figures, but 33 came to light at Leighton Moss on 14 July 2013 (L. Bagley), a site with many sallows.
■ **HABITAT** Woodland, sand dunes, parks, gardens and brownfield sites with scrub.
■ **FOODPLANTS** Goat Willow, Grey Willow and poplars, including Black- and White. Considered single brooded, with occasional later records until the early 1990s. Thereafter, probably double brooded almost annually, with a few October records.

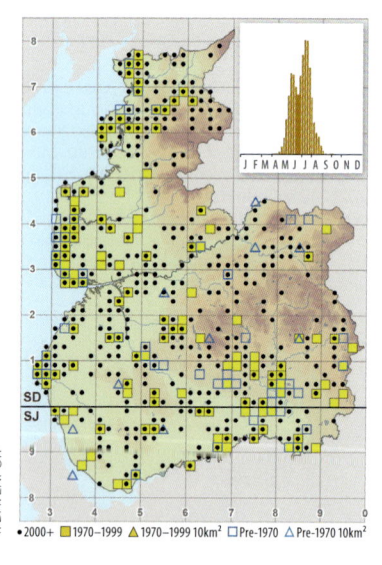

● 2000+   ▢ 1970–1999   △ 1970–1999 10km²   ▢ Pre-1970   △ Pre-1970 10km²

69.004 Convolvulus Hawk-moth *Agrius convolvuli*

■ **LANCS STATUS** Occasional / migrant   1841–2023

First reported on Chat Moss in 1841 (R.S. Edleston), with around 150 recorded across the county since then. All but nine relate to singletons, the most found on any one occasion being three at a few coastal sites in the 1950s. Years with five or more site reports were 1859, 1953, 2001, 2003, 2006 and 2022. In 2006, ten site reports were received, with two-thirds of these in one week in early October. In 2022, five adults were recorded between mid-September and early November.
■ **HABITAT** Not applicable. Readily attracted to some garden-planted *Nicotiana*.
■ **FOODPLANTS** Bindweed sp.

Comes to light. Larva found on five occasions, in 1900, 1967, 2022, and twice in 2023.

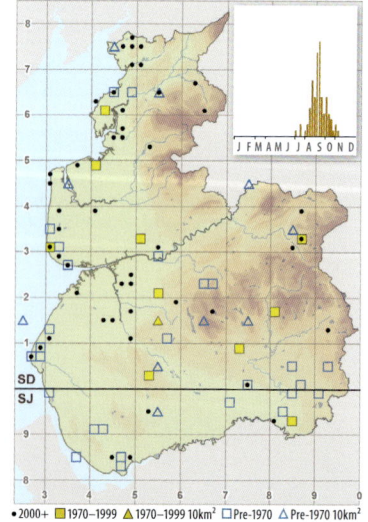

● 2000+   ▢ 1970–1999   △ 1970–1999 10km²   ▢ Pre-1970   △ Pre-1970 10km²

382 The Moths of Lancashire

69.005 Death's-head Hawk-moth *Acherontia atropos*

■ **LANCS STATUS** Occasional / migrant   1842–2023

First reported from Heaton Park, Middleton (VC59) on 22 September 1842 and since then, there have been 119 site reports. These came from 40 locations up to 1899, 66 site records between 1900 and 1999, and 14 from 2000 to 2023. It was most frequently recorded in 1908 with seven site reports, these widely spread across the county. All but one record of an adult involved singletons. 120 larvae and pupae were found in the Blackpool area over a few days of searching in 1868 (Thorpe, 1870). A recent report was of a pupa dug from a garden potato crop at Bolton-le-Sands on 18 September 2022 (S. & B. Garland). A larva was in Silverdale in 2023. Comes to light.

■ **FOODPLANTS** Duke of Argyll's Teaplant and potato crops.

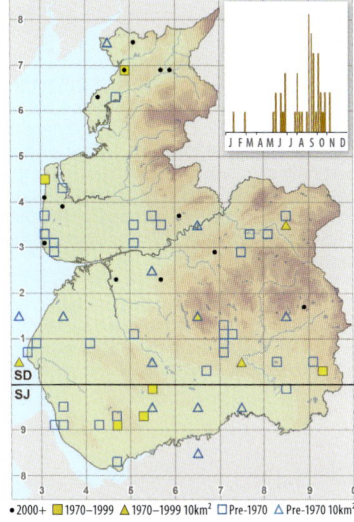

69.006 Privet Hawk-moth *Sphinx ligustri*

■ **LANCS STATUS** Rare / vagrant   1870–2006

This species was first reported at Whittle-le-Woods, near Chorley, in 1870 by J.B. Hodgkinson (Ellis, 1890) and was followed by one in Nelson Park in 1926 by W.G. Clutten (Mansbridge, 1940). In 1970, a singleton was reported by P. Summers in the Parbold area (Fairhurst, 1982) and the final report came from Feniscowles, near Blackburn, when one came to a 22w Actinic trap run by W.C. Aspin in his garden on 29 June 2006. The species has shown signs of a range expansion northward nationally and was noted in south Cumbria in recent years (Randle *et al*., 2019).

■ **HABITAT** Unknown in the county.

■ **FOODPLANTS** Not recorded in Lancashire.

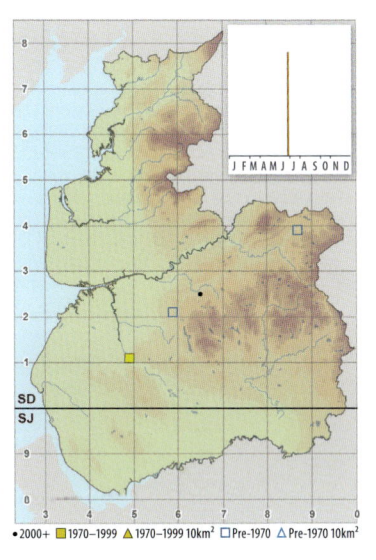

69.007 Pine Hawk-moth *Sphinx pinastri*

■ **LANCS STATUS** Scarce / very local resident   1907–2023

There were two pre-1985 sightings; the first at St Annes in 1907, the details of which were discovered by Pratt (2002) and originally published in Curtis (1930). The second was a female attracted to a lighted window in Chorley on 8 July 1984 (C. Smith) and identified by P. Kirk; the latter specimen is in World Museum's Lepidoptera collection. During 2023, eleven came to light at Ainsdale NNR (R. Moyes, *et al*.), one in Formby (R. Walker) and one in Silverdale (D. Talbot). The Sefton Coast records are indicative of the moth establishing a resident population in the county.

■ **FOODPLANTS** Not recorded in Lancashire.

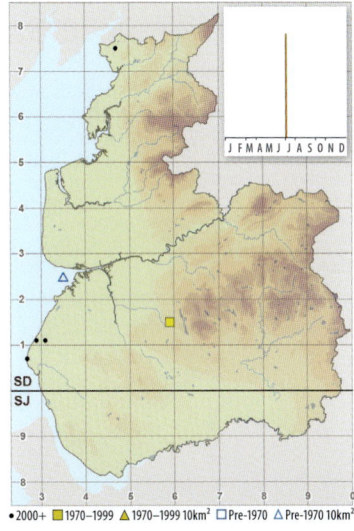

SPHINGIDAE

69.008 Narrow-bordered Bee Hawk-moth *Hemaris tityus*

■ **LANCS STATUS** Considered Extinct   **pre-1890**

The single record known to be within the Lancashire vice county boundaries relates to larvae found at Ribbleton Moor, Preston, 'an area now cultivated', by J.B. Hodgkinson (Ellis, 1890). The date is most likely to be in the 1840s or '50s, but the exact location to which this refers has not been located and is now almost certainly developed. Two other (unmapped) reports are documented from the Manchester district (probably outside VC59) and Hale Moss (just in VC69). These two sites are mentioned by H.T. Stainton in 1902, within J. W. Tutt's British Lepidoptera Volume 3. Unusually, no reference is made to them by Ellis (1890) or Mansbridge (1940).
■ **HABITAT** Unknown.
■ **FOODPLANTS** Unknown, but probably Devil's-bit Scabious.

Historically distributed in many scattered localities across the British Isles, with records in Cheshire up to 1919, Cumbria post-1999 and Yorkshire up to 2004 (cheshire-moth-charts.co.uk and yorkshiremoths.co.uk - accessed in September 2023).

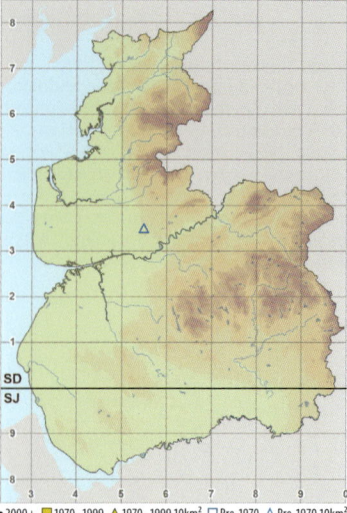

69.010 Humming-bird Hawk-moth *Macroglossum stellatarum*

■ **LANCS STATUS** Frequent / widespread migrant   **c.1857–2022**

First was on Crosby sandhills (Gregson, 1857). Since then, an irregular visitor, usually in single figures per-year up to the 1980s. Annual since 1998 and often in double figures. Large influxes occurred in 2006 (309 noted), 2020 and 2022 (189 each). Has bred on at least nine occasions between 1911 and 2022 suggesting that later season observations may, on occasions, relate to home bred individuals.
■ **HABITAT** Larvae associated with rough ground and coastal grassland.
■ **FOODPLANTS** Lady's Bedstraw, Cleavers, Hedge Bedstraw.

It may, rarely, overwinter as an adult. Often observed nectaring on various flowers including Buddleia, Red Valerian, Sweet Williams and Verbena.

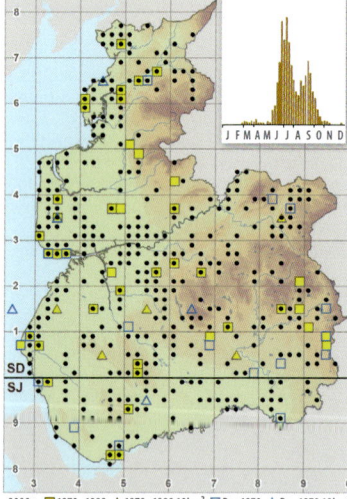

69.011 Oleander Hawk-moth *Daphnis nerii*

■ **LANCS STATUS** Scarce/ migrant   **1847–1954**

There have been 17 records in the county, with the first at Agecroft in 1847 by J.C. Melvill, the specimen now in NHM, London. Subsequent records include one in Bolton in 1885, Chorley Gardens c.1897, Ainsdale in 1913, Prestwich in 1917, Wigan in 1923, Blackpool in 1947, Lytham in 1949 and Leyland in 1954 (J.H. Vine-Hall).
■ **HABITAT** Not applicable.
■ **FOODPLANTS** Not recorded in Lancashire.

In 1903, one was captured on board SS Achilles within Liverpool Docks. The only larval record refers to those being found in a greenhouse in Manchester by T.G. Mason around 1896. Very few records had full dates.

384   The Moths of Lancashire

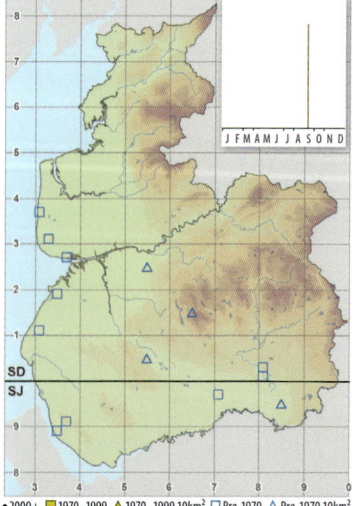

69.013 Spurge Hawk-moth *Hyles euphorbiae*

■ **LANCS STATUS** Rare / migrant   1850–1865

The county database contains two historic records of this species. The first was a larva found in August 1850 (Gregson, 1850) and, at the time, this was reported by Gregson to be near Formby. However, other authors elaborated on this, with the first of these probably obtaining the information direct from Gregson. Byerley (1854) and Britten (c.1950), who presumably quoted Byerley, mention it related to two larvae on, or near *Euphorbia paralias* (Sea Spurge) and near *E. portlandica* (Portland Spurge), between Little Brighton and Hightown, or in Hightown. The other was in Chorley Gardens in 1865 by A. Reade. The specimen, with data label, was thought to be in Bolton Museum, but an examination of the main Museum collection in August 2023 only included specimens of this moth bred from foreign localities.

■ **HABITAT** Larvae in sand dunes.

■ **FOODPLANTS** Sea Spurge and, possibly, Portland Spurge.

This is one of the rarer migrant species in Britain, but with recent indications that it may be established as a very local resident in coastal parts of south-east England.

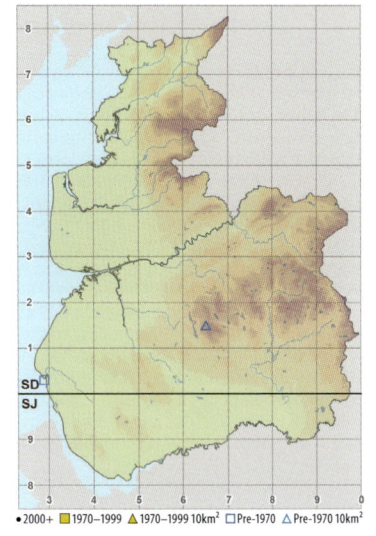

● 2000+   ☐ 1970–1999   ▲ 1970–1999 10km²   ☐ Pre-1970   △ Pre-1970 10km²

69.014 Bedstraw Hawk-moth *Hyles gallii*

■ **LANCS STATUS** Occasional / migrant   1842–2021

The first was a male at Whitefield, Bury on 15 September 1842 by R. S. Edleston. During the rest of the 19th century, it was noted on 16 occasions, 25 times in the 20th century and, more recently, from 16 sites, such as Flixton, Preston, Chipping, Hutton and Crossdale Grains. Ten larvae were found at Bispham on 1 September 2008. The most recent was an adult in Wilpshire on 17 July 2021 (T. & J. Davie).

■ **HABITAT** Not applicable.

■ **FOODPLANTS** Rosebay Willowherb and Fuchsia.

Larvae have been found in September and October on 21 occasions between 1888 and 2020, with most of these in 1973 at sites such as Crosby, Fleetwood and Nelson.

● 2000+   ☐ 1970–1999   ▲ 1970–1999 10km²   ☐ Pre-1970   △ Pre-1970 10km²

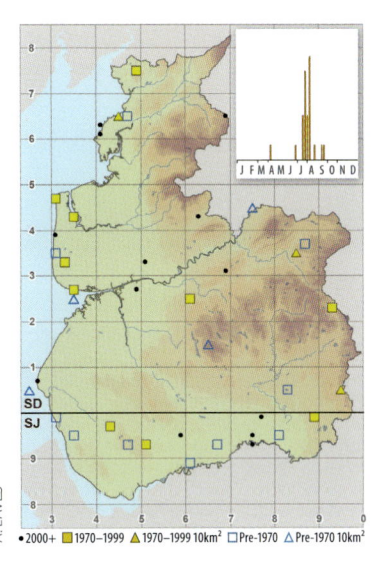

69.015 Striped Hawk-moth *Hyles livornica*

■ **LANCS STATUS** Occasional / migrant   1844–2022

First noted in Chorlton and Withington in June 1844 by R.S. Edleston (Ellis, 1890) and next in Preston in 1846. These were followed by six other 19th century records and 22 in the 20th century. The latter included four larval records, from Ainsdale, Middleton (Manchester), Oldham and Freshfield, none of which document the larval foodplant. Since then, singletons have been attracted to light at Martin Mere on 17 June 2006, Formby on 23 July 2020 and 31 March 2021, Silverdale on 14 May 2022, Ribchester on 5 June 2022 (G. Anderton) and Silverdale on 4 August 2022 (J. Stonehouse).

■ **HABITAT** Not applicable.

■ **FOODPLANTS** Not recorded in Lancashire.

● 2000+   ☐ 1970–1999   ▲ 1970–1999 10km²   ☐ Pre-1970   △ Pre-1970 10km²

The Moths of Lancashire **385**

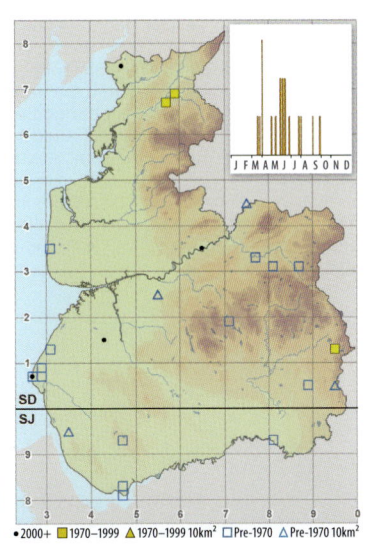

69.016 Elephant Hawk-moth *Deilephila elpenor*

■ **LANCS STATUS** Frequent / widespread resident   1853–2022

First noted at Adlington Hall, Chorley by E.C. Buxton (junior) in June 1853 (Buxton, 1853) then, in the Liverpool district (Gregson, 1857), at Withington, Rixton and Kenyon c.1859 (J. Chappell) and thereafter widely across the county to the present day. There are plenty of sites with long–term datasets, including Billinge (C.A. Darbyshire), Formby (R. Walker), Hale (C. Cockbain), Flixton (K. McCabe), Longton (R. Boydell), Morecambe (J. Holding), Parr (R. Banks) and Widnes (P. Hillyer). Mostly noted in single figures, at light traps, with no apparent change in distribution or numbers over the years. Larger counts of 20 or more come mostly from sites well inland and include 20 at Rivington on 21 June 1984 (P. Kirk), 24 at Rishton on 20 June 2020 (D. Bickerton) and 29 at Belmont on 13 June 2004 (S. Martin). The largest count was of 33 at Green Bank on 9 June 2016 (P.J. Marsh).

■ **HABITAT** A wide range, from coast to the uplands, including gardens.

■ **FOODPLANTS** Rosebay Willowherb, Bogbean, Fuchsia, Great Willowherb, Himalayan Balsam, Hoary Willowherb, Tobacco Plant and Water Lily.
Single brooded. Comes to light. Larvae, occasionally in a green form, found from mid-July to mid-October. One adult displaying intermediate features with Small Elephant Hawk-moth was found at Green Bank on 2 September 2019 (P.J. Marsh).

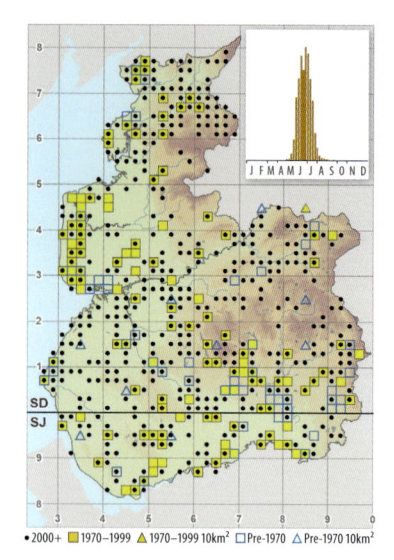

69.017 Small Elephant Hawk-moth *Deilephila porcellus*

■ **LANCS STATUS** Frequent / widespread resident   1843–2022

First noted in the Preston area in 1843 (Hodgkinson, 1844) and later, from Crosby and Lytham in 1850, Quernmore and Clougha in 1901 (C.H. Forsyth), Silverdale in 1907 (W.P. Stocks) and Ainsdale (W.A. Tyerman) and Southport (W. Mansbridge) in 1912. These locations are good examples of the two separated populations present in the county since recording began. Until the end of the 20th century, lowland sightings were mainly restricted to coastal sand dunes and lowland limestone, while further inland nearly all records were from moorland or nearby localities. This began to change in 2000, as reports arrived from sites such as Dalton (Wigan), Parr and Pennington up to 2002, followed by a range of sites including Hutton, Flixton, Hoghton, Preston and Longridge. The last site, trapped regularly since the early 2000s, produced its first records (11 in all) between 2019 and 2022. Records are still equally split between lowland and upland sites, but the smaller numbers noted from the formerly unoccupied zone in the middle are gradually increasing. Larval sightings are rare, the only recent one being from Drinkwater Park, Prestwich on 11 August 2019 (M. Murray).

■ **HABITAT** Coastal dunes, limestone grassland, heather moorland and upland grassland.

■ **FOODPLANTS** Bedstraw sp.; the species involved will vary with habitat and altitude. Single brooded. Comes to light, and sometimes wanders. Larvae July to September. The photo below, right, shows a comparison of this and the previous species.

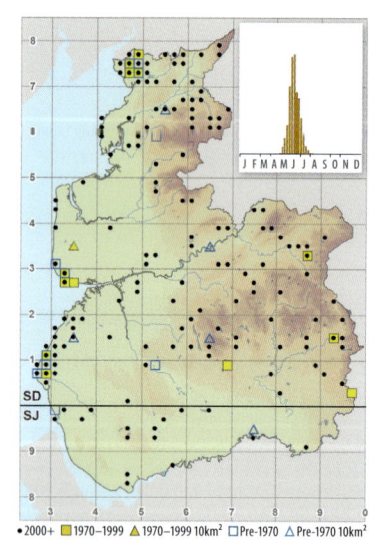

69.018 Silver-striped Hawk-moth *Hippotion celerio*

■ **LANCS STATUS** Scarce / migrant   1838–1963
First found in the Newton Heath area in 1838 by R.S. Edleston (Edleston, 1846a) and later, two larvae were noted in the same area by W. Jamieson in September 1844 (The Zoologist, page 1346). All subsequent records were of adults, from Preston in 1850 and 1885 (J.B. Hodgkinson), in Lytham on 27 September 1870 (W. Gregson), at Bolton in 1870 and 1873 (W. Johnson), in Southport on 17 July 1871 (recorder unknown, identified by E. Bell) and in September 1873 (H. Burton). Other records reported were one in the Oldham area in September 1880 (J. Taylor) and, in the same town, at Suthers St., by H. Horsfall in 1884. One was found in Market Square, Burnley in 1883 by A.E. Wright, at Crosby in September 1885 (G.A. Harker), in Manchester in 1885 and 1887, the latter by J. Chappell, and Lancaster in 1898 (C.H. Forsythe). Since the beginning of the 20th century, there have only been two records, one in 1940 from the Burnley area by J.C. Lavin and the last, found in an outhouse in Blackpool by D. Cowie on 3 November 1963.
■ **HABITAT** Not applicable.
■ **FOODPLANTS** The two larvae found in 1844 were not feeding at the time of discovery.

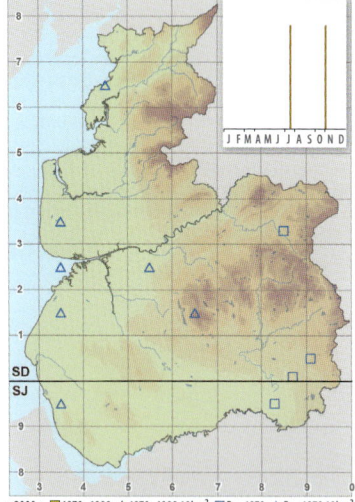

SPHINGIDAE

70.002 Purple-bordered Gold *Idaea muricata*

■ **LANCS STATUS** Rare / very local resident   c.1843–2018
First reported from Pilling Moss (Hodgkinson, 1844). Other 19th century records from Chat Moss, Simonswood Moss, Rixton Moss, Risley Moss, Heysham Moss. Decline of specialist mossland habitat in VC59 has led to loss of Chat Moss populations; a singleton in 1964 being the only post-1900 VC59 record. First noted at Winmarleigh and Cockerham Mosses, 1973 by P.J. Hodge. All subsequent records are from these sites. A single post-2010 record (in 2018) suggests the need for further survey work.
■ **HABITAT** Mosses. Photo of Cockerham Moss in August 2010 below.
■ **FOODPLANTS** Not recorded in Lancashire.
Single brooded. High count of 20 from Winmarleigh Moss, 27 June 1988 (M. Evans).

GEOMETRIDAE

70.004 Least Carpet *Idaea rusticata*

■ **LANCS STATUS** Rare / vagrant   2008
Recent years have seen this moth experience a significant and rapid increase in range and abundance nationally, spreading from its stronghold in south-east England (Randle *et al.*, 2019), with the first records from Cheshire in 1999 and Yorkshire in 2006. The sole Lancashire sighting was of two moths noted on walls near a garden light trap, run at Hale on 25 July 2008 by C. Cockbain. Unlike neighbouring counties, where multiple records followed the initial finds, this has remained our only sighting to date. Because of this, the Hale moths are considered likely to be vagrants, rather than residents.
■ **HABITAT** Unknown in the county.
■ **FOODPLANTS** Not recorded in Lancashire.
Nationally, single brooded with the possibility of an occasional, small second generation.

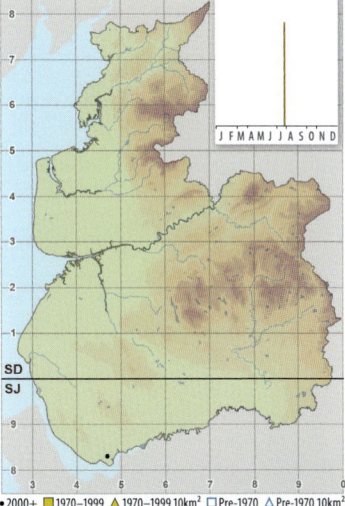

The Moths of Lancashire   **387**

GEOMETRIDAE

70.006 Dwarf Cream Wave *Idaea fuscovenosa*

■ **LANCS STATUS** Occasional / very local resident   c.1943–2022

Lancs populations are almost exclusively from the Sefton Coast, between Crosby and Southport. First noted at Formby around 1943 by G. de C. Fraser, Freshfield in 1960 and Crosby in 2007. Far more frequently recorded since 2007, with 91 subsequent records, compared to seven pre-2007. Sefton Coast counts appear stable, with five records of ten moths in 2022 from Formby, Freshfield and Birkdale Green Beach. May be spreading, as two VC60 records in last decade; Middleton 2015 and Lytham 2017.

■ **HABITAT** Grasslands and scrub adjacent to fixed dunes.
■ **FOODPLANTS** Not recorded in Lancashire.
Single brooded. Vast majority of records have been to light.

● 2000+   ▢ 1970–1999   △ 1970–1999 10km²   ▢ Pre-1970   △ Pre-1970 10km²

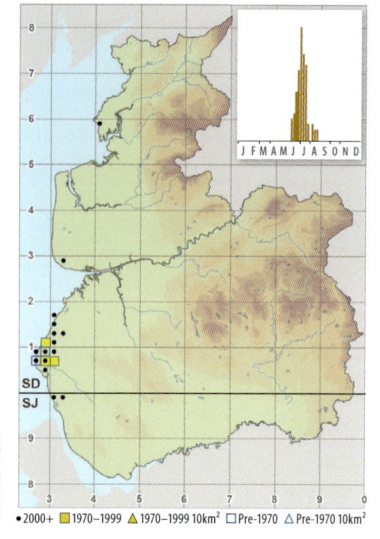

70.008 Small Dusty Wave *Idaea seriata*

■ **LANCS STATUS** Frequent / widespread resident   c.1857–2022

Noted from Liverpool by Gregson (1857) where 'not uncommon in hedgerows'. Further recorded at Lytham, Prestwich, Preston (Ellis, 1890). Bred from larva at Crosby, 1920 by W. Mansbridge, foodplant undocumented. Status described as fairly well distributed and locally common (Mansbridge, 1940). A well-recorded species with stable population. Frequent in gardens, usually in small numbers. Records less frequent from upland habitats in the east and north-east.

■ **HABITAT** Hedgerows, scrub, gardens.
■ **FOODPLANTS** Larvae (ex-ova) fed on Enchanter's Nightshade 2022 (B. Hancock).
Double brooded. Comes readily to light. Occasional at lighted windows, road signs, etc.

● 2000+   ▢ 1970–1999   △ 1970–1999 10km²   ▢ Pre-1970   △ Pre-1970 10km²

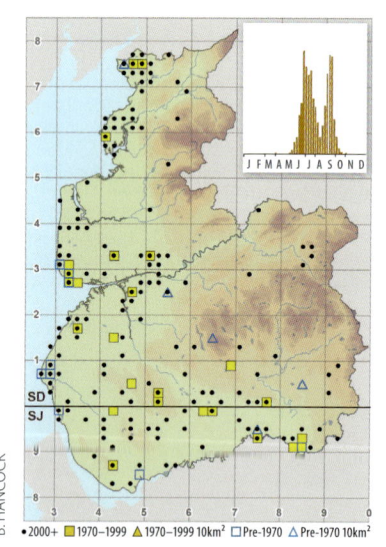

70.009 Satin Wave *Idaea subsericeata*

■ **LANCS STATUS** Occasional / local resident   c.1859–2022

First recorded from Withington by J. Chappell c.1859, and from Barlow Moor by J.B. Hodgkinson. These inland records from Manchester were accepted by Mansbridge and Ellis. Mansbridge (1940) blamed subsequent decline to habitat loss for golf. Subsequent records primarily coastal, mainly from Sefton Coast. Less frequent from Lytham St. Annes (last record 2015), and Morecambe Bay, where noted from Gait Barrows, Sunderland, Heysham N.R., Yealand Conyers, etc., but not since 2018.

■ **HABITAT** Coastal grassland, scrub.
■ **FOODPLANTS** Not recorded in Lancashire.
Single brooded. Comes to light. Occasionally disturbed from vegetation in day.

● 2000+   ▢ 1970–1999   △ 1970–1999 10km²   ▢ Pre-1970   △ Pre-1970 10km²

388 *The Moths of Lancashire*

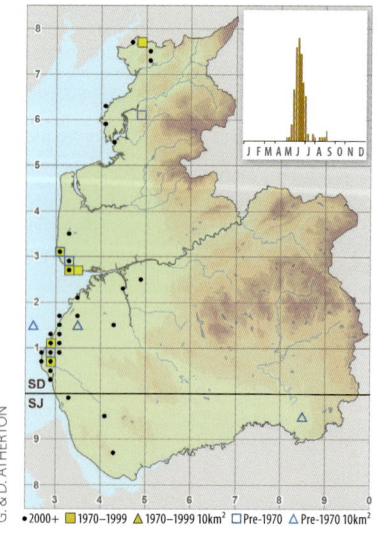

70.010 Dotted Border Wave *Idaea sylvestraria*

■ **LANCS STATUS** Considered Extinct c.1855–1905
J.W. Ellis (1890) noted mid-19th century records by J. Chappell and J.B. Hodgkinson from Chat Moss, and described the moth as 'rare, probably extinct, the wood where it occurred having been destroyed by fire.' The wood in question was Traverse's Wood, with the fire considered possibly a deliberate act by keepers draining the Moss for game (Gregson, 1883). R. Leverton (pers. comm.) reported unsuccessfully searching for it in the 1960s on a degraded and drained remnant of Chat Moss.
■ **HABITAT** Mosses. Skinner (2009) notes a deep brown form with distinct cross-lines from the Lancs & Cheshire mosses (ab. *circellata* Guenée). Four Lancs specimens, without any further information, are in the NHM; two of which are shown here.

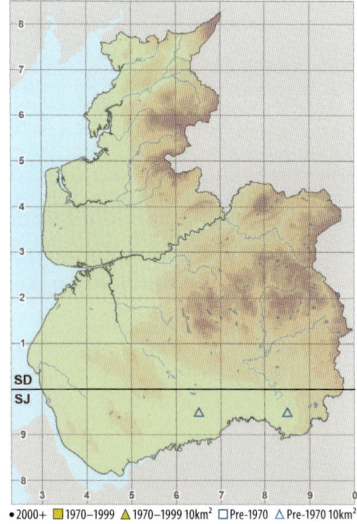

70.011 Single-dotted Wave *Idaea dimidiata*

■ **LANCS STATUS** Frequent / widespread resident c.1857–2022
Found in Liverpool district 'freely in mixed hedges where nettles grow' (Gregson, 1857). Early authors reported it as common and generally distributed. Status appears unchanged with over 7000 records. Generally, well distributed, although patchier in south-eastern districts. Noted from lowland and upland areas, with nine records from Docker Moor examples of the latter. High count of 195 from Heysham N.R., 2002.
■ **HABITAT** Rough grassland, scrub, gardens.
■ **FOODPLANTS** Bramble, Ivy, Cow Parsley.
Single brooded, with extended flight period. Comes readily to light. Larvae recorded from February to June, with moths reared from Bramble and Ivy (B. Smart).

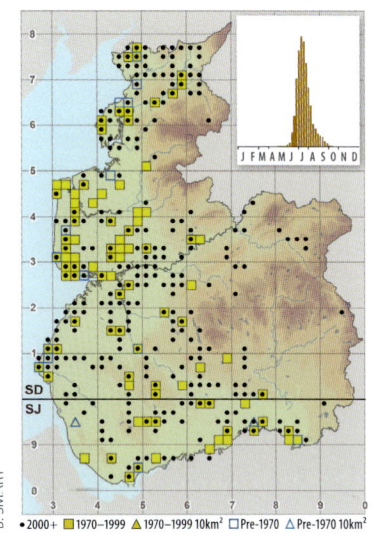

70.013 Small Fan-footed Wave *Idaea biselata*

■ **LANCS STATUS** Frequent / widespread resident c.1857–2022
Noted as plentiful in the Liverpool district by Gregson (1857) and, by other early authors, as common and generally distributed, although specific sites not given. Recorded at Didsbury by H.N. Michaelis in 1940 and at Formby 1946. Increasingly recorded since 1950s. Particularly common at coastal and other lowland sites. Present in smaller numbers at upland sites such as Longridge and Docker Moor. A very common garden visitor, e.g., over 1,000 records at Flixton.
■ **HABITAT** Scrub, woodland, gardens.
■ **FOODPLANTS** Larvae unrecorded other than ex-ova, Chorlton 2012; fed on Dandelion.
Single brooded. Comes to light. Netted in day and at dusk. Dark form shown below.

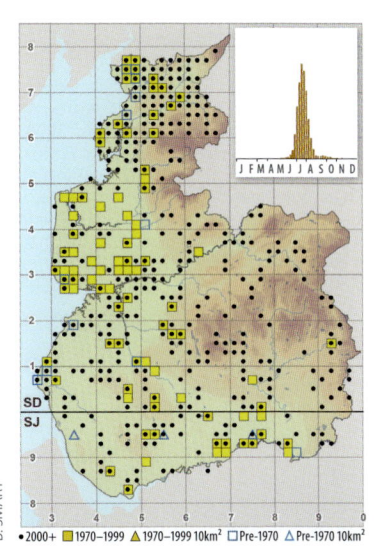

70.015 Small Scallop *Idaea emarginata*

■ **LANCS STATUS** Considered Extinct   c.1859–1950

This species was first reported from Irlam and Barlow Moor, Manchester around 1859 by J. Chappell (Ellis, 1890). The author described its distribution as very local. Further noted at Ainsdale in 1908 by W.A. Tyerman, with the moth later considered scarce and local (Mansbridge, 1940). The only subsequent record was on 29 July 1950 at Formby by M.G. Fraser (Leech and Michaelis, 1957). The likelihood, therefore, is that the moth is no longer present in the county.

Regarding neighbouring counties, the moth remains local in Yorkshire with recent records from the south and east of the county (yorkshiremoths.co.uk, accessed 1 September 2023). It has been recorded in Cheshire on 45 occasions, most recently in 2019 (cheshire-chart-maps.co.uk, accessed 1 September 2023). Unrecorded in Cumbria.

■ **HABITAT** Damp woodland, scrub.
■ **FOODPLANTS** Not recorded in Lancashire.

Difficult to determine from the few Lancashire records, but nationally, single brooded.

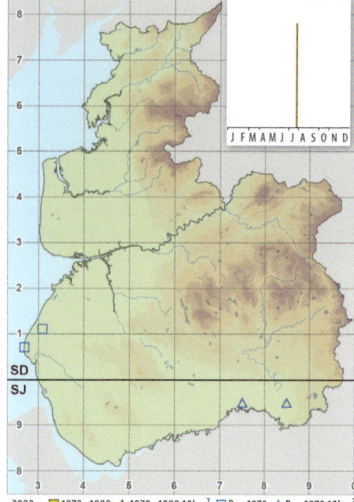

● 2000+   ▪ 1970–1999   ▲ 1970–1999 10km²   □ Pre-1970   △ Pre-1970 10km²

70.016 Riband Wave *Idaea aversata*

■ **LANCS STATUS** Abundant / ubiquitous resident   c.1857–2022

First noted in the Liverpool district (Gregson, 1857). Described by early authors as abundant everywhere. This remains the case, with over 20,000 records throughout the county and in all habitats, lowland and upland. Whilst its range remains stable, there appears a recent reduction in numbers. Of the 15 highest counts, only one has occurred since 2004, with the highest of 143 at Flixton in 2001 by K. McCabe.

■ **HABITAT** Scrub, woodland edges, hedgerows, gardens, etc.
■ **FOODPLANTS** Ivy. Larvae found late May to early June.

Has a small second generation, more numerous in certain years, e.g., 28 October records in 2018, but only seven since. Found in both banded and unbanded forms.

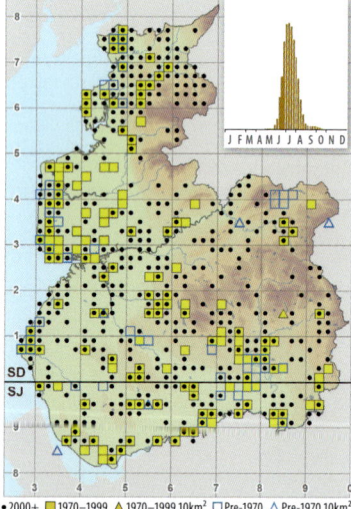

● 2000+   ▪ 1970–1999   ▲ 1970–1999 10km²   □ Pre-1970   △ Pre-1970 10km²

70.018 Plain Wave *Idaea straminata*

■ **LANCS STATUS** Occasional / local resident   1846–2022

Noted as rare from Chat Moss (Edleston, 1846). Later, reported from Preston and Bolton and described as scarce (Ellis, 1890). Further records to 1970 from Silverdale, Ainsdale, Bispham and Risley Moss. From Gait Barrows in 1983 and Botany Bay Wood in the 1990s. Regular at Astley Moss in early 21st century, but not noted since 2015. There is a single 2021 record from nearby Cadishead Moss. Other recent records from limestone sites, e.g., Silverdale, and upland sites, in particular Docker Moor.

■ **HABITAT** Scrub, rough grassland and mosses.
■ **FOODPLANTS** Not recorded in Lancashire.

Single brooded. Comes to light. Occasionally netted at dusk. Similar to Riband Wave.

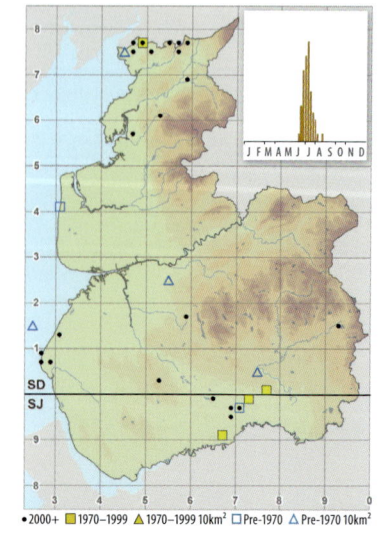

● 2000+   ▪ 1970–1999   ▲ 1970–1999 10km²   □ Pre-1970   △ Pre-1970 10km²

70.023 Mullein Wave *Scopula marginepunctata*

■ **LANCS STATUS** Occasional / local resident   c.1886–2022
This lowland species was first recorded at Silverdale by J.B. Hodgkinson (Ellis, 1890). Mansbridge (1940) noted 'very occasional occurrence' in N. Lancs without mentioning other sites. Recorded at Warton, nr. Carnforth 1961 and Leighton Moss 1970. Most subsequent records from Morecambe Bay, with other post-2000 sites including Formby, Hale, St Helens, Longton, Lytham St Annes. Possibly increasing in number. 70 records in last decade, compared to 32 in previous. Absent in east Lancs.
■ **HABITAT** Coastal habitats, woodland, grassland.
■ **FOODPLANTS** Not recorded in Lancashire.
Double brooded. Almost all records noted at light.

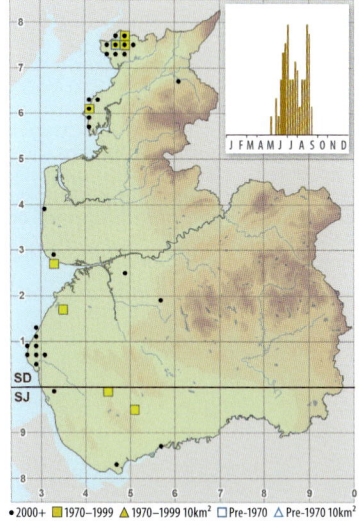

70.024 Small Blood-vein *Scopula imitaria*

■ **LANCS STATUS** Occasional / local resident   c.1857–2022
First noted as 'abundant in hedges about West Derby' in Liverpool (Gregson, 1857). Next, from Eaves Wood, Silverdale by A.E. Wright 1905, with all subsequent records from southern and western parts. Less frequent in north-west where known from Lytham and Morecambe Bay but apparently absent between these areas. The ten highest counts all from Sefton Coast, the stronghold for this species. Also, noted in gardens at Hale, West Didsbury, Billinge, etc. Increasing in number, if not range.
■ **HABITAT** Lowland habitats including scrub, gardens, etc.
■ **FOODPLANTS** Not recorded in Lancashire.
Double brooded; second generation increasingly strong since 2014. Comes to light.

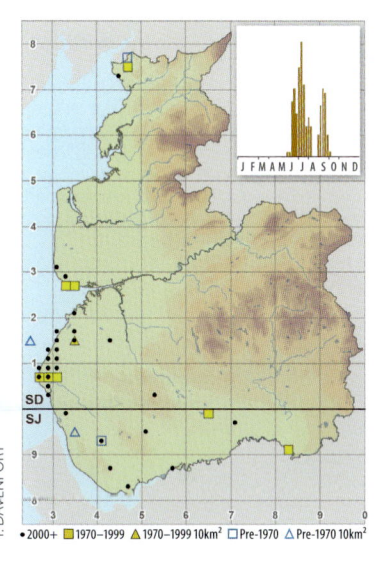

70.025 Lesser Cream Wave *Scopula immutata*

■ **LANCS STATUS** Occasional / very local resident   1906–2022
First recorded on 17 July 1906 at White Moss, Yealand by A.E. Wright (Britten, c.1950), with other pre-1950 records from Silverdale and Formby. The latter record, from May 1948 by M.J. Leech, remains the only record for this moth outside the Morecambe Bay limestone area. Regular at Leighton Moss, Yealand, Gait Barrows and Hawes Water. Records are almost all of singletons, with only two or three sightings most years. High count of 30 from Hawes Water 30 June 2010 (S. Palmer).
■ **HABITAT** Damp meadows.
■ **FOODPLANTS** Not recorded in Lancashire.
Single brooded. Comes to light. Recorded flying at dusk and in late afternoon.

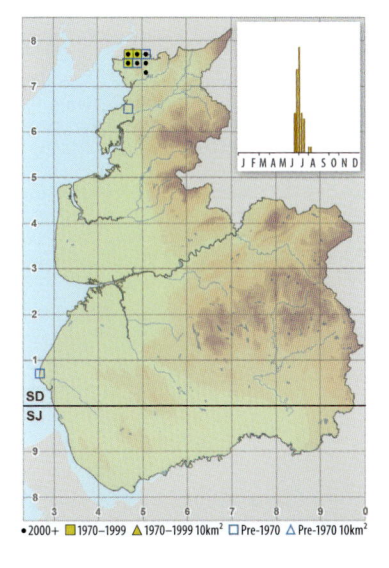

The Moths of Lancashire   **391**

GEOMETRIDAE

70.026 Smoky Wave *Scopula ternata*

■ **LANCS STATUS** Occasional / local resident   1846–2022

First recorded in June 1846 at Chat Moss (Edleston, 1846), and subsequently from Rixton Moss, Simonswood Moss and Heysham Moss. The earliest authors noted the moth to be plentiful on the mosses (e.g., Gregson, 1857). However, unrecorded from south Lancs mosses since late 19th century. Although still known from Winmarleigh and Cockerham Mosses, this is now predominantly a moorland species, with recent records from Birk Bank, Wray, Anglezarke Moor, Jeffrey Hill, Longridge Fell, etc.
■ **HABITAT** Moorland. Formerly common on mossland, now scarce.
■ **FOODPLANTS** Bilberry.
Single brooded. At light and netted in day. Larva at Longridge Fell, May 1996.

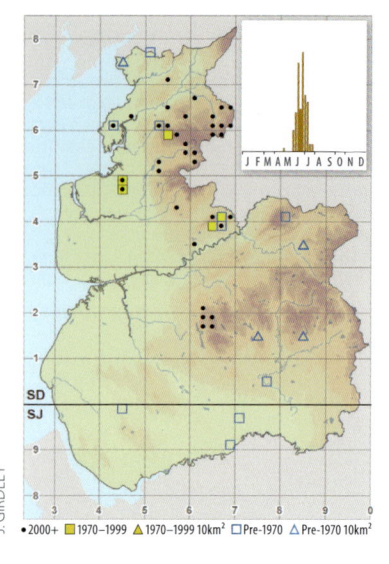

• 2000+   ▪ 1970–1999   ▲ 1970–1999 10km^2 ▫ Pre-1970   △ Pre-1970 10km^2

70.027 Cream Wave *Scopula floslactata*

■ **LANCS STATUS** Occasional / local resident   c.1857–2022

Reported in the Liverpool district by Gregson (1857). Ellis (1890) and Mansbridge (1940) noted the moth to be fairly common, with early 20th century records from Silverdale, Heysham, Knowsley and Sales Wood, nr. Prescot. Britten (c.1950) gave status as common in woods and woodland edges. Current distribution shows the moth to be regular around Silverdale and on Sefton Coast, with occasional records from Lytham St Annes, Longton, Mill Houses, Docker Moor, Birk Bank, etc.
■ **HABITAT** Woodland edges, scrub.
■ **FOODPLANTS** Not recorded in Lancashire.
Single brooded. Most records at light. Occasionally recorded by dusking.

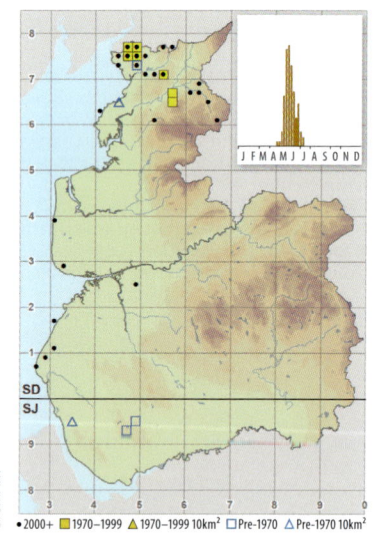

• 2000+   ▪ 1970–1999   ▲ 1970–1999 10km^2 ▫ Pre-1970   △ Pre-1970 10km^2

70.029 Blood-vein *Timandra comae*

■ **LANCS STATUS** Frequent / widespread resident   1946–2022

Although present in Cheshire from mid-19th century, not recorded in Lancs until 1946 at Preston by H.N. Michaelis. Further pre-1970 sightings at Formby, Didsbury, Leigh, Wrightington and Silverdale. Significantly increased in abundance and range since 2000 with just 21 Lancashire records prior and 2,179 since. Predominantly a lowland species. Occasional at more elevated sites such as Wray, Rossendale and Rishton.
■ **HABITAT** Damp grassland, scrub, hedgerows, gardens.
■ **FOODPLANTS** Dock, bistort. Bred from larva found on dock, Rixton 7 September 2009. Double brooded. Comes to light, with high count of 25 recorded at Glazebury on 9 August 2013 by R. Rhodes. Easily disturbed from vegetation.

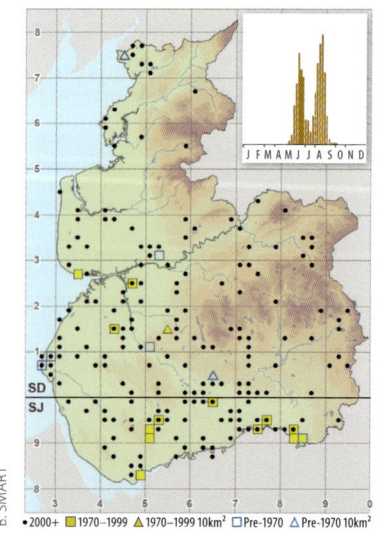

• 2000+   ▪ 1970–1999   ▲ 1970–1999 10km^2 ▫ Pre-1970   △ Pre-1970 10km^2

The Moths of Lancashire

70.032 Birch Mocha *Cyclophora albipunctata*

■ **LANCS STATUS** Rare / vagrant   **2020**

Recorded on a sole occasion. This was a moth trapped at MV light on the National Trust site at Formby by R. Walker on 7 June 2020. Being a single record, true status is hard to assess. Is this the first sighting of a small resident population, or was it a vagrant from the small Cheshire and Wirral populations to the south? Unless further individuals go on to be recorded on the Sefton Coast, then the latter seems more likely. The species is very local in Cheshire and Yorkshire.

■ **HABITAT** Unknown in the county.
■ **FOODPLANTS** Not recorded in Lancashire.

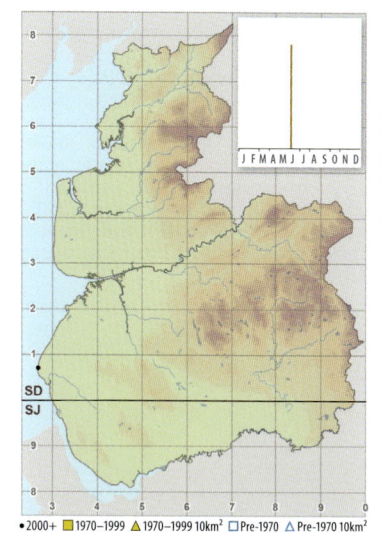

70.036 Maiden's Blush *Cyclophora punctaria*

■ **LANCS STATUS** Scarce / local resident   **c.1857–2022**

C.S. Gregson reported a few taken around Hale; 'first by Mr Nixon' (Gregson, 1857). Not noted again until 1 June 2016 when recorded from Kirkby by J. Dempsey, the first of a wave moving north. Next from Formby in 2017 and Briercliffe in 2018. 2021 saw records from Longton, Cuerden Valley Park, Worsthorne and Gressingham. Further range expansion in 2022 saw gardens in Silverdale and Flixton added to the list of sites, with four singletons at the latter between 13 August and 8 September.

■ **HABITAT** Woodland, hedgerows.
■ **FOODPLANTS** Not recorded in Lancashire.
Double brooded. Comes to light.

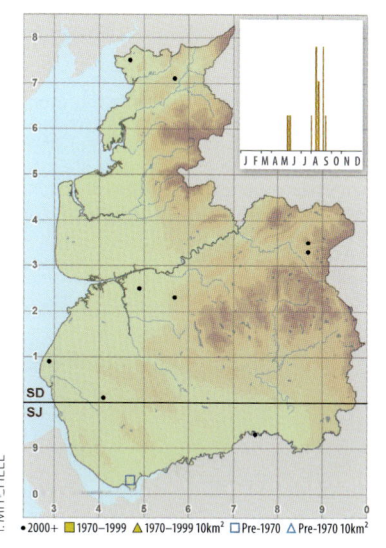

70.037 Clay Triple-lines *Cyclophora linearia*

■ **LANCS STATUS** Occasional / local resident   **2001–2022**

Another moth that has increased its range northwards, leading to rapid colonisation of scattered areas of the county during this century. First recorded on 2 July 2001 from Mere Sands Wood by I. Kippax, and from Pennington Flash the following year. In 2008, further noted at Dalton Crags, Swinton and Singleton. Numbers have increased since, with multiple records from Ainsdale and Docker Moor, and many others from the Silverdale area, including Yealand Conyers, Leighton Moss and Warton.

■ **HABITAT** Woodland, hedgerows and gardens.
■ **FOODPLANTS** Not recorded in Lancashire.
Single brooded. Occasional second generation. Virtually all records at light.

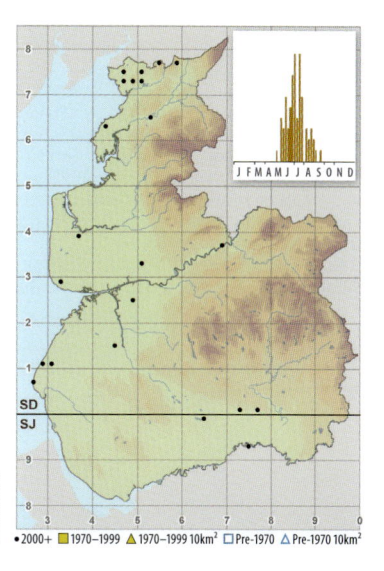

GEOMETRIDAE

70.038 Vestal *Rhodometra sacraria*

■ **LANCS STATUS** Occasional / migrant c.1857–2022
First noted in Liverpool district (Gregson, 1857). Reported from Huyton in August 1867, when 'returning about eleven o'clock at night… after sugaring… a moth flew to the light of my lantern' (Capper, 1867). There were no records from 1872 to 1969, then 28 records from the remainder of the 20th century, and 168 in this. Numbers vary considerably from year to year with highest annual counts of 52 in 2006, 24 in 2022 and 20 in 2000. Unrecorded in 2002, 2007, 2010, 2012, 2018, 2021.
■ **HABITAT** Moths recorded in wide variety of habitats throughout the county. Larvae could potentially be found on coastal grassland, but unrecorded as yet. Attracted to light.

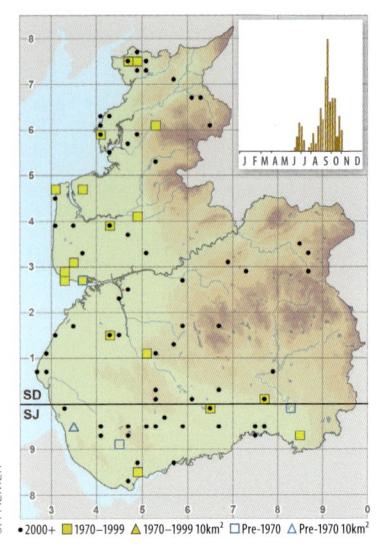

70.039 Oblique Striped *Phibalapteryx virgata*

■ **LANCS STATUS** Considered Extinct c.1857–1952
Noted as common in the dunes (Gregson, 1857). Ellis (1890) later described the moth as 'common on the Lancashire and Cheshire sand-hills wherever the yellow bedstraw (*Galium verum*) is plentiful.' Mansbridge (1940) reported the moth to be common, but also noted the lack of post-1887 records. A specimen from 1890 has since been noted in the NHM's collection, taken by an unknown recorder, possibly from the Crosby area. The final record, from Hall Road, Crosby, was taken by H.W. Wilson in 1952.
■ **HABITAT** Sand dunes.
■ **FOODPLANTS** Not recorded in Lancashire.

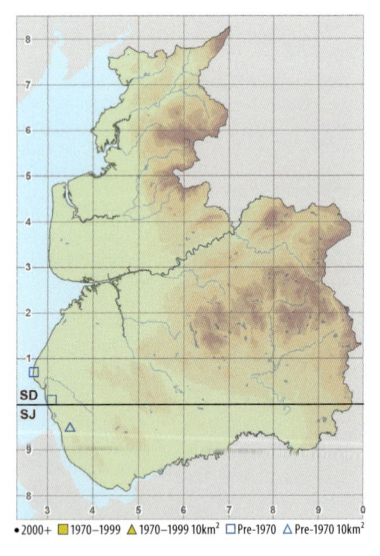

70.040 Lead Belle *Scotopteryx mucronata*

■ **LANCS STATUS** Rare / very local resident; possibly Extinct 1954–1973
First recorded on 22 June 1954 by J.H. Vine-Hall at Warton Crag. The specimen is in the NHM, London and has recently been verified by R. Leverton and J. Patton. Our other three records hail from Carr House Green Common, recorded by D. & J. Steeden. The first was on 18 July 1972, when a moth was attracted to car headlights at the roadside; identity confirmed by J. Heath following dissection. Two more individuals noted in 1973, on 25 June and 16 July, both disturbed from Gorse.
■ **HABITAT** Rough grassland, scrub.
■ **FOODPLANTS** Not recorded in Lancashire.
Single brooded. Attracted to light and disturbed from Gorse, the foodplant.

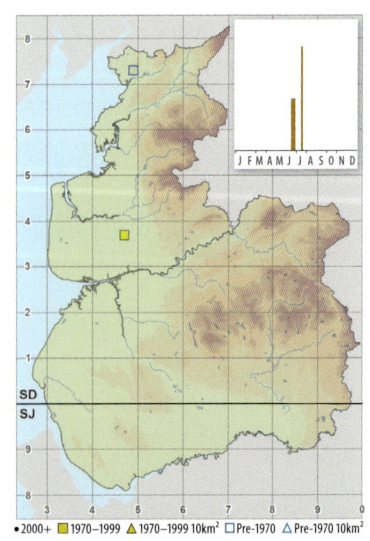

70.041 July Belle *Scotopteryx luridata*

■ **LANCS STATUS** Scarce / local resident 1948–2021

Recorded four times from Warton Crag between 1948 and 1969 by J.H. Vine-Hall. Not noted again until 2002, when trapped at light by P.J. Marsh at Smithy Wood, nr. Wennington, and 2004 from Bay Horse, nr. Dolphinholme. It was a surprise when, in 2006, a moth was recorded at Freshfield. This area then became regular with nine further noted from Freshfield Dune Heath, although none since 2017. The only post-2017 records are from Yealand Conyers, Wray, Docker Moor and Dalton Crags.

■ **HABITAT** Grassland, scrub, moorland.
■ **FOODPLANTS** Not recorded in Lancashire.

Single brooded. Comes to light.

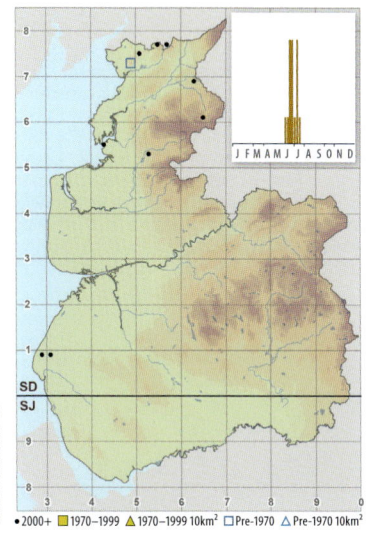

• 2000+ ■ 1970–1999 ▲ 1970–1999 10km² □ Pre-1970 △ Pre-1970 10km²

70.045 Shaded Broad-bar *Scotopteryx chenopodiata*

■ **LANCS STATUS** Abundant / widespread resident c.1857–2022

Recorded in the Liverpool district by Gregson (1857). Ellis (1890) and Mansbridge (1940) described the moth as common on waste lands and generally distributed. This remains an expected find in the appropriate habitat and has been recorded in every decade since 1900. There have been some high counts, including 100 at Upholland on 21 July 2004 by C.A. Darbyshire, but, perhaps significantly, less numerous in the last decade or so. Noted in lowland and upland areas, primarily the former.

■ **HABITAT** Rough grassland and scrub.
■ **FOODPLANTS** Larvae unrecorded other than ex-ova, Chorlton 2013; fed on vetch.

Single brooded. Occasional to light; easily disturbed from vegetation during the day.

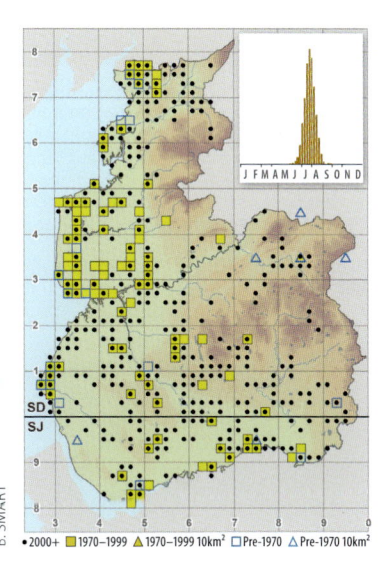

• 2000+ ■ 1970–1999 ▲ 1970–1999 10km² □ Pre-1970 △ Pre-1970 10km²

70.046 Oblique Carpet *Orthonama vittata*

■ **LANCS STATUS** Occasional / local resident c.1857–2022

'Plentiful around lamps' at Otterspool (Gregson, 1857). Other authors described it as scarce, noting many locations lost to golf courses and building (Mansbridge, 1940). Although fairly widely distributed, there appears to be a western and northern bias, with less than 10% of the 833 records from VC59, and absence from Mersey Valley. The VC60 records contain 498 from Leighton Moss. Other sites with over ten records include Martin Mere, High Tatham, Docker Moor, Heysham, and Hawes Water.

■ **HABITAT** Damp grassland, marshes and moorland.
■ **FOODPLANTS** Bred ex-larva from Ainsdale, 1913, although foodplant not noted.

Double brooded. Comes to light. High count of 27 at Silverdale Moss 11 August 2022.

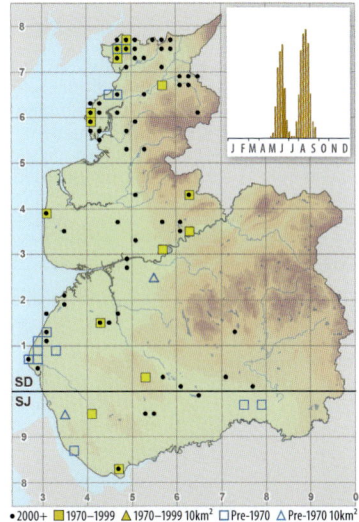

• 2000+ ■ 1970–1999 ▲ 1970–1999 10km² □ Pre-1970 △ Pre-1970 10km²

The Moths of Lancashire

GEOMETRIDAE

70.047 Gem *Nycterosea obstipata*

■ **LANCS STATUS** Occasional / migrant   c.1859–2022

First noted by J. Chappell in Manchester, with other late 19th century records from Preston, Hale, Warrington, and noted as scarce (Ellis, 1890). Also, recorded from Birkdale by J.C. Melvill in August 1878. Britten (c.1950) described the moth as rare, except in the years of 1922 and 1947, when considered widespread. There are twelve known records from 1900–1970, 20 from the remainder of the 20th century and 62 to date from this century. Peak years are 1998 with ten records and 2006 with 17. Scarce in recent years, with just one in 2018, followed by two singletons in 2022.

■ **HABITAT** May occur anywhere, although records are primarily from lowland areas.
■ **FOODPLANTS** Unrecorded within the county.

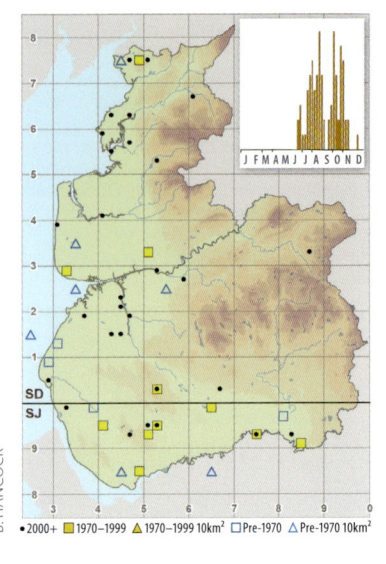

70.048 Red Carpet *Xanthorhoe decoloraria*

■ **LANCS STATUS** Scarce / very local resident   c.1891–2022

First recorded at Trawden in the 1890s by W.G. Clutten, Burnley in 1906 (A.E. Wright) and Pendle Hill in 1940 (B.D.S. Smith). All early records are from VC59, but these ended in 1940. Recorded from Claughton in 1990 by M. Broomfield, the first VC60 record. Current distribution is restricted to the northern uplands. Most recent records have been from high ground, for example Leck Fell House (420m), but lower former sites White Moss (245m) and Caton Moor (300m) not checked recently.

■ **HABITAT** Moorland.
■ **FOODPLANTS** Unrecorded within the county.
Single brooded. All post-1990 records have been to light.

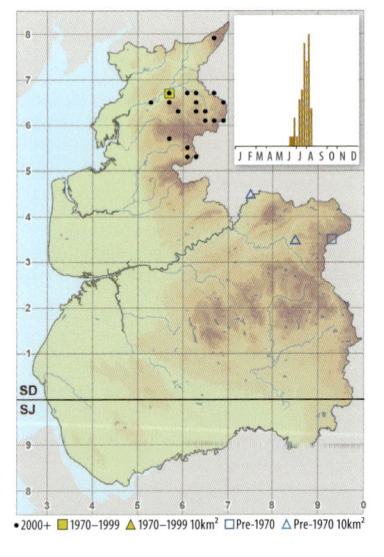

70.049 Garden Carpet *Xanthorhoe fluctuata*

■ **LANCS STATUS** Abundant / ubiquitous resident   c.1857–2022

First noted from the Liverpool district, with the author reporting it to be 'found everywhere' (Gregson, 1857). Mansbridge (1940) described it 'as one of our commonest moths, especially in gardens.' Whilst numbers of records remain extremely high (711 in 2022) counts do seem to be down, with the highest counts dating from 15–20 years ago. Whilst noted from many upland sites, e.g., Leck Fell (although unrecorded at the very top), the moth is more numerous at lower altitudes.

■ **HABITAT** Rough grassland, scrub, woodland edges, gardens.
■ **FOODPLANTS** Unrecorded. Bred from larvae 1913 and 1922; foodplant not noted.
Double brooded. Vast majority of records are at light; often to a lit window.

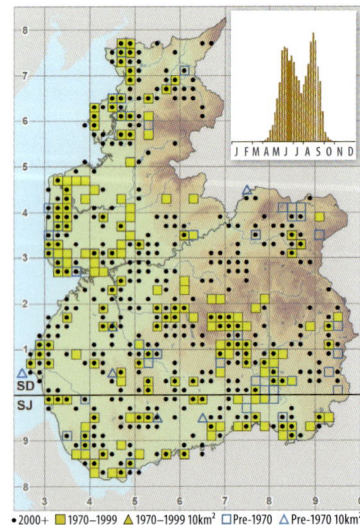

70.051 Red Twin-spot Carpet *Xanthorhoe spadicearia*

■ **LANCS STATUS** Occasional / local resident   1879–2022
First recorded at Green Thorn Fell (Hodgkinson, 1880). Described as 'generally distributed, but scarcely common' (Ellis, 1890). Mansbridge (1940) later noted it to be 'only common on the moors.' Unlike in southern England, where found in a wide variety of habitats, this is a moorland specialist in Lancs. The only exception is a single record from Freshfield Dune Heath in 2017, identity confirmed by S. Palmer.
■ **HABITAT** Moorland, rough grassland.
■ **FOODPLANTS** Bred by Mansbridge in 1912. Foodplant not documented.
Mainly single-brooded (late May-early July), with very limited second brood, likely due to its northern, upland status. Comes to light, with a few daytime sightings.

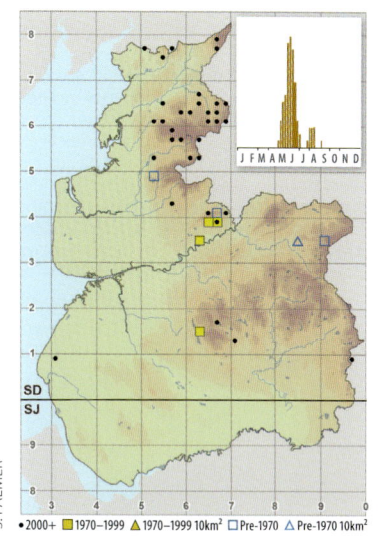

• 2000+   ■ 1970–1999   ▲ 1970–1999 10km²   □ Pre-1970   △ Pre-1970 10km²

70.052 Dark-barred Twin-spot Carpet *Xanthorhoe ferrugata*

■ **LANCS STATUS** Frequent / widespread resident   c.1857–2022
First record from Liverpool (Gregson, 1857). Later, described as very common and widespread (Mansbridge, 1940). Unlike the previous species, this is primarily a moth of lowland areas, but with some upland records also, e.g., Rishton, Longridge Fell, Docker Moor, Mill Houses, etc. Sites with over 100 records include Martin Mere, Mere Sands Wood, Leighton Moss, Heysham N.R., and gardens in Hale, Parr, Flixton, Lightfoot Green and Adlington. High count of 46 at latter site, 2010 (P. Krischkiw).
■ **HABITAT** Rough grassland, woodland, gardens.
■ **FOODPLANTS** Not recorded in Lancashire.
Double brooded. Comes readily to light.

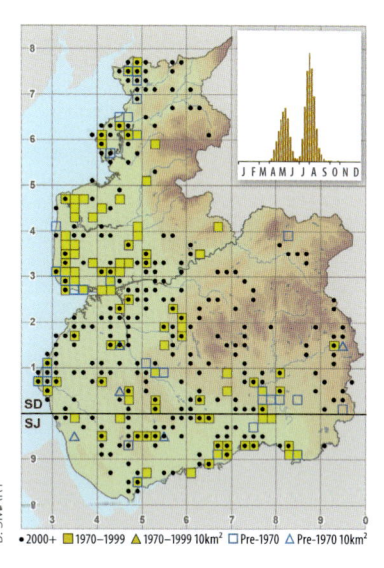

• 2000+   ■ 1970–1999   ▲ 1970–1999 10km²   □ Pre-1970   △ Pre-1970 10km²

70.053 Flame Carpet *Xanthorhoe designata*

■ **LANCS STATUS** Abundant / ubiquitous resident   c.1857–2022
Present in the Liverpool district, where plentiful in many thorn hedges (Gregson, 1857). Later, described as local by Ellis (1890). Recorded widely over the following decades at sites such as Stretford, Chat Moss, Burnley and Preston. The increase in light trapping since the 1960s has demonstrated the moth's presence across the whole county. There appears no obvious threat to the species, with the two highest counts, of 28 from Flixton and 27 from Horrocks Fold, Bolton, both from 2021.
■ **HABITAT** Woodlands, hedgerows, gardens, moorland, brownfield sites, etc.
■ **FOODPLANTS** Not recorded in Lancashire.
Double brooded. Comes to light. Early stages unrecorded other than ex-ova.

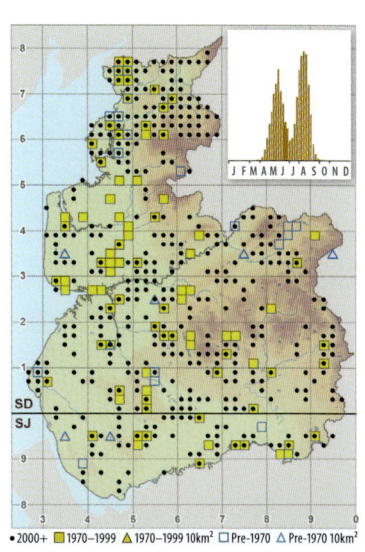

• 2000+   ■ 1970–1999   ▲ 1970–1999 10km²   □ Pre-1970   △ Pre-1970 10km²

The Moths of Lancashire

GEOMETRIDAE

70.054 Silver-ground Carpet *Xanthorhoe montanata*

■ **LANCS STATUS** Abundant / ubiquitous resident　　c.1857–2022

First noted from Liverpool district by Gregson (1857) and described as common and abundant by early authors. Remains so, with over 11,000 records distributed throughout lowland and upland sites. Following the initial record, it was next noted at Burnley, Oldham, Lytham and Ditton, and continued to be recorded regularly in every decade. As with many other species, the quantity of records holds up well, but potentially masks a small decline in numbers from high point of early 21st century.

■ **HABITAT** Damp grassland, woodland, scrub, gardens.

■ **FOODPLANTS** Yarrow, Honeysuckle. Emerged from the latter in 2002 (C.A. Darbyshire). Single brooded with extended generation. Easily disturbed from vegetation in day.

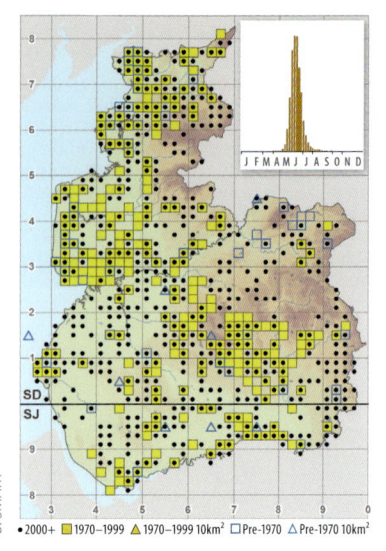

70.055 Large Twin-spot Carpet *Xanthorhoe quadrifasiata*

■ **LANCS STATUS** Occasional / local resident　　2007–2022

A recent arrival into the county as range has expanded northwards. First recorded on 19 July 2007 at Parr by R. Banks. Next noted at Flixton in 2011 by K. McCabe and recorded at the site on 51 occasions in total. Only noted on ten occasions elsewhere, with other VC59 sites at Chorlton, Freshfield, Cadishead Moss and Astley Moss. The sole VC60 record, presumably part of the ongoing expansion, arrived on 19 June 2022, when recorded at Yealand Conyers by B. Hancock.

■ **HABITAT** Damp woodland, scrub.

■ **FOODPLANTS** Not recorded in Lancashire.

Single brooded. Where noted, all records have been of moths attracted to light.

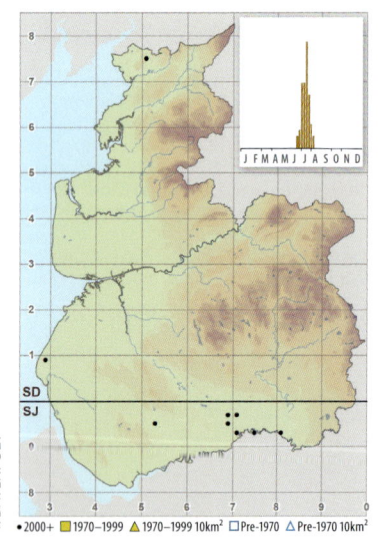

70.059 Yellow Shell *Camptogramma bilineata*

■ **LANCS STATUS** Frequent / widespread resident　　c.1857–2022

Very often noted flying in day or at dusk, and easily disturbed when walking through appropriate habitat. Considered as abundant by most early authors. Britten (c.1950) gave its status as common. First noted in the Liverpool district by Gregson (1857), with other pre-1940 sites including Oldham, Hopwood, Lytham, Hightown and Ditton, nr. Widnes. As the map suggests, the moth is relatively scarce in upland areas. Numbers show a reduction from 10–20 years ago.

■ **HABITAT** Rough grassland, scrub.

■ **FOODPLANTS** Unrecorded. Larva found in leaf-litter at Pinfold Meadow, Ainsdale 2024. Single extended generation. Recorded frequently at light and during the day.

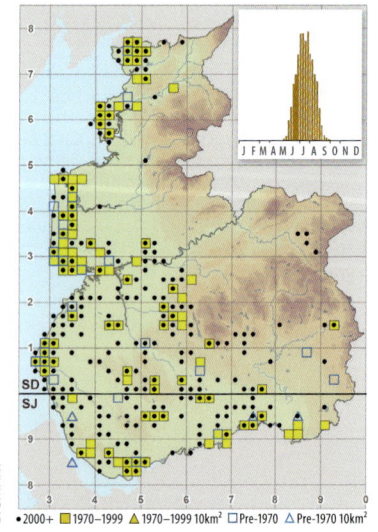

398　*The Moths of Lancashire*

70.060 Small Argent and Sable *Epirrhoe tristata*

■ **LANCS STATUS** Scarce / very local resident 1879–2021

First recorded at Green Thorn Fell on 28 June 1879, with the author noting presence of fir plantations (Hodgkinson, 1879). Reported from Chat Moss 1887, the only VC59 record. From Clougha 1920, Littledale 1981, Nottage Crag 1997 and Thrushgill 2014. Recorded on five occasions at the latter site, including high count of ten in 2014, but not since 2017. Noted from Burnslack by B. Dyson during a June 2021 daytime visit.
■ **HABITAT** Sheltered moorland, favouring mid-stage conifer plantation rides. Suitable sites rendered less so as conifers mature and shade the area (Marsh & White, 2019).
■ **FOODPLANTS** Not recorded in Lancashire.
Single brooded. Noted flying in mid- to late afternoon in weak sunshine.

• 2000+ ☐ 1970–1999 ▲ 1970–1999 10km² ☐ Pre-1970 △ Pre-1970 10km²

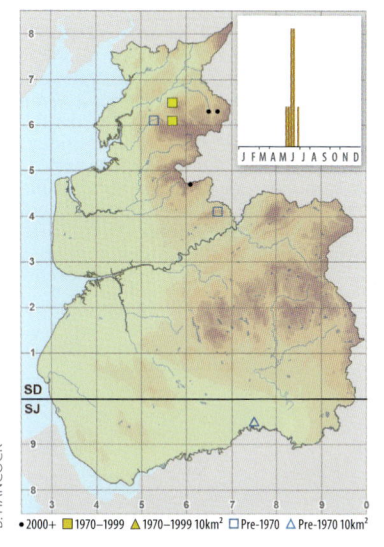

70.061 Common Carpet *Epirrhoe alternata*

■ **LANCS STATUS** Abundant / ubiquitous resident c.1887–2022

Reported by Ellis (1890) to be common and generally distributed, although no sites noted. Recorded at Ditton c.1929 by F.M.B. Carr and shortly after, from Lytham St Annes. Recorded regularly in every subsequent decade. Highest count is 93 by P.J. Marsh at Heysham N.R. on 1 June 2003. Numbers of records remain high, over 600 in 2022, although actual counts are declining. Recent records are mainly of singletons, with records of four or more rare in recent years.
■ **HABITAT** Rough grassland, scrub, woodland, moorland, gardens, etc.
■ **FOODPLANTS** Larvae (ex-ova), from Chorlton 2010, fed on Cleavers with moth reared.
Double brooded. Recorded at light. Easily disturbed from vegetation during day.

• 2000+ ☐ 1970–1999 ▲ 1970–1999 10km² ☐ Pre-1970 △ Pre-1970 10km²

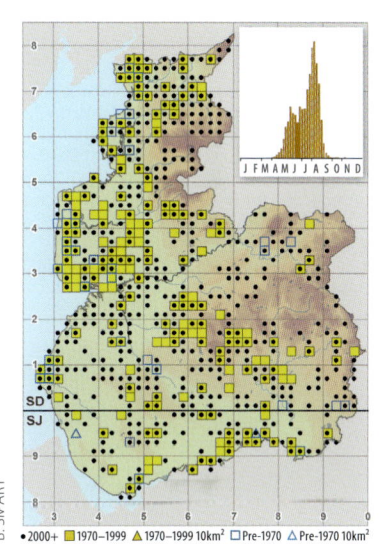

70.062 Wood Carpet *Epirrhoe rivata*

■ **LANCS STATUS** Rare / very local resident 2016

A moth, recorded at light on 10 June 2016 by D. Bickerton at Rishton, was confirmed as this species by R. Walker on the basis of forewing and hindwing markings. Compared to the previous species, the white cross-band beyond the central band on the forewing is broader and lacks a distinct grey line within. This line is also absent from the broad white cross-band on the hindwing. There are no other accepted records of this species, in larval or adult form, but it is potentially under-recorded due to the difficulty in identification.
■ **HABITAT** Unknown in the county.
■ **FOODPLANTS** Not recorded in Lancashire.

• 2000+ ☐ 1970–1999 ▲ 1970–1999 10km² ☐ Pre-1970 △ Pre-1970 10km²

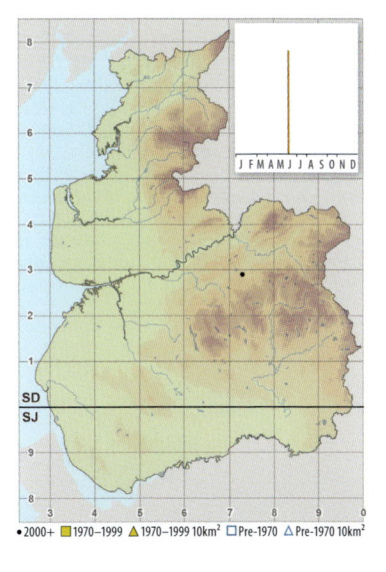

70.063 Galium Carpet *Epirrhoe galiata*

■ **LANCS STATUS** Occasional / local resident c.1857–2022

Reported as 'common on the coast' (Gregson, 1857), with later authors describing it as local, but common where it occurs. Recorded in the Burnley area by A.E. Wright in 1900. Bred by M.J. Leech from Formby area in 1950s (Leech & Michaelis, 1957); foodplant not noted. Larvae found on Lady's Bedstraw at Marton in 1940s by W.A. Watson, although 1970s searches at same site unsuccessful. Most recent records from uplands: Ease Gill, Hutton Roof, Leck Fell. Last Gait Barrows record from 2009.
■ **HABITAT** Moorland, grassland.
■ **FOODPLANTS** Lady's Bedstraw.
Twice in August, suggesting small, second brood may occur, as in southern England.

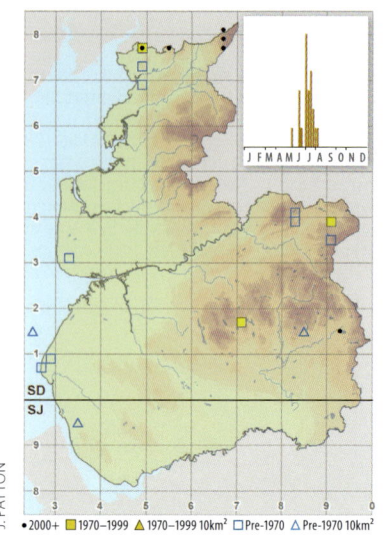

70.065 Sharp-angled Carpet *Euphyia unangulata*

■ **LANCS STATUS** Rare / very local resident; possible vagrant 1845–2021

First recorded on 21 July 1845 by W.E. Hodgkinson from the Preston area. His brother, J.B. Hodgkinson, noted the moth in 1872 from the Cleveleys area. Other 19th century records from Lytham St Annes and Bolton, with status described as local and rarely common (Ellis, 1890). Subsequently, appeared lost to the county, as unrecorded for over 130 years. The appearance of a single moth at Heysham N.R. on 8 July 2021, trapped by R. Neville, was therefore a very welcome surprise.
■ **HABITAT** Coastal woodland.
■ **FOODPLANTS** Not recorded in Lancashire.
Nationally, single brooded. The 2021 moth was trapped using black light.

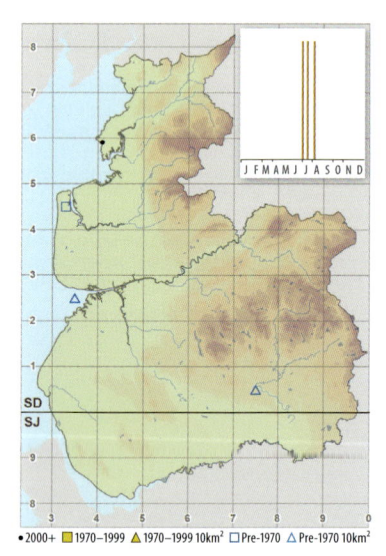

70.066 Shoulder Stripe *Earophila badiata*

■ **LANCS STATUS** Occasional / local resident 1846–2022

Noted in Prestwich and described as scarce (Edleston, 1846). Also, from Liverpool district, 'where brambles grow' (Gregson, 1857). Other early authors described it as generally distributed and fairly common. Pre-1970 records from Burnley, Lytham, Holden Clough, Silverdale, Carnforth, Longton and Formby. Most recent records from Silverdale and far north of the county including upland areas. Occasional on Sefton Coast and scarce in Mersey Valley, with last Flixton record in 2009.
■ **HABITAT** Hedgerows, woodland edges, scrub, gardens.
■ **FOODPLANTS** Burnet Rose, Dog Rose. Larvae recorded in May and June.
Single brooded. Attracted to light. Reared from larva on Dog Rose at Formby 2022.

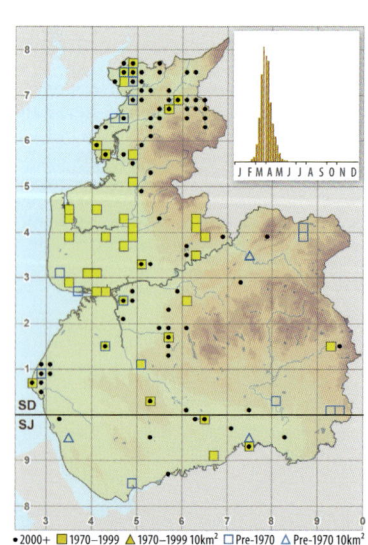

70.067 Streamer *Anticlea derivata*

■ **LANCS STATUS** Occasional / local resident   c.1859–2022
First noted from Hoghton, around 1859, by J. Chappell, and from Preston (Ellis, 1890). Status later noted as local and not common (Mansbridge, 1940). Recorded post-1970 at Claughton, Leighton Moss, Bolton-le-Sands, Longridge, with 21st century records showing similar distribution to previous species. Sparsely scattered in south-east Lancs. Comes to light, with a high count of 12 from Higher Tatham 2004 (P.J. Marsh).
■ **HABITAT** Hedgerows, woodland edges, scrub, gardens.
■ **FOODPLANTS** Dog Rose. Early instar was beaten from foodplant by B. Smart on 3 June 2022 at Formby, with larva of previous species also noted. Both moths were bred. Single brooded but with one very late record; 8 August 2021 at Lancaster (M. Sole).

• 2000+   □ 1970–1999   ▲ 1970–1999 10km²   □ Pre-1970   △ Pre-1970 10km²

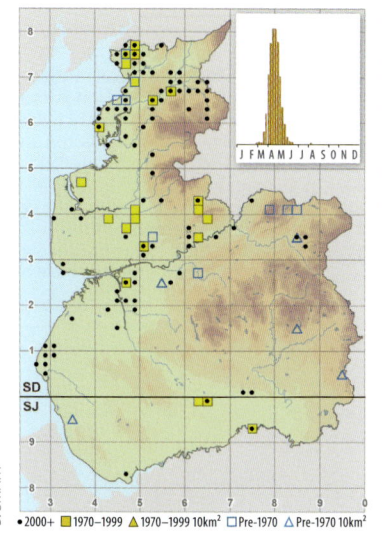

70.068 Beautiful Carpet *Mesoleuca albicillata*

■ **LANCS STATUS** Occasional / local resident   1843–2022
Early records from Chat Moss 1843 (Edleston, 1844) and from Preston, Silverdale, Hoghton Tower, Hale and Halewood (Ellis, 1890), with the latter author noting the moth to be 'rather local, and not usually common.' Mansbridge (1940), described it as widely distributed but not very common. Appears to have suffered a recent downturn in fortunes with marginally fewer 21st than 20th century records, a worrying statistic given increasing county-wide recording activity post-2000.
■ **HABITAT** Woodland clearings, scrub.
■ **FOODPLANTS** Bred from larva on a few occasions although foodplant not documented. Single brooded. Comes to light. Occasionally disturbed in daytime.

• 2000+   □ 1970–1999   ▲ 1970–1999 10km²   □ Pre-1970   △ Pre-1970 10km²

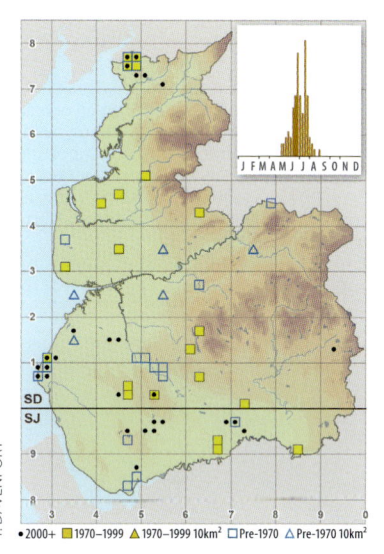

70.069 Dark Spinach *Pelurga comitata*

■ **LANCS STATUS** Occasional / local resident   c.1857–2022
This predominantly lowland species, found around the western and southern edges of the county, has shown signs of decline in recent years, e.g., no Mersey Valley records from 2022. The moth was first reported from Liverpool by Gregson (1857). Later, from Salford, Crosby, Preston. Britten (c.1950) noted a tendency to disappear from certain areas for some time before reappearing; likely linked to dependence on goosefoots and oraches, a group of plants requiring disturbed or coastal habitats.
■ **HABITAT** Coastal sites, waste ground.
■ **FOODPLANTS** Not recorded in Lancashire.
Single brooded. Comes to light.

• 2000+   □ 1970–1999   ▲ 1970–1999 10km²   □ Pre-1970   △ Pre-1970 10km²

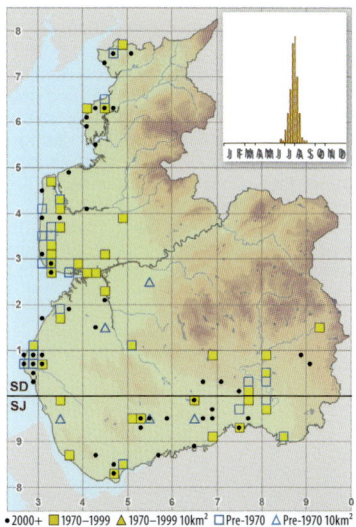

GEOMETRIDAE

70.070 Mallow *Larentia clavaria*

■ **LANCS STATUS** Scarce / very local resident   c.1843–2010

Recorded from Preston by J.B. Hodgkinson, firstly from around 1843 (Hodgkinson, 1844). Of 22 subsequent records, 19 are from the 20th century, including at Ainsdale, Lancaster, Silverdale, Marton and Lytham St Annes, and only three from this, all from the Sefton Coast. Noted at Southport in 2004 by A. Forrest, Altcar 2007 by G. Jones and Formby 2010 by M. Dean and R. Hill. The absence of any subsequent records marks this out as a very vulnerable species within the county.

■ **HABITAT** Rough coastal grassland.

■ **FOODPLANTS** Bred from larvae on Mallow at Marton in 1940s by W.A. Watson. Single brooded. All recent records have been of moths to light.

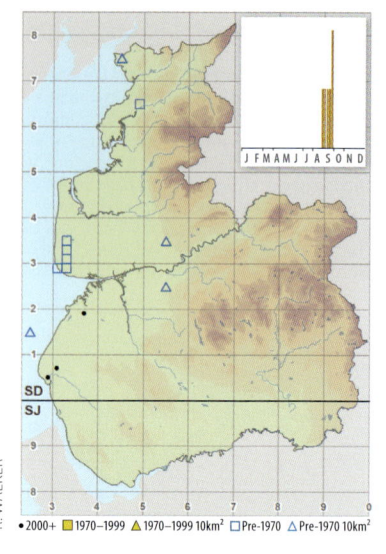

• 2000+ □ 1970–1999 △ 1970–1999 10km^2 □ Pre-1970 △ Pre-1970 10km^2

70.072 Grey Mountain Carpet *Entephria caesiata*

■ **LANCS STATUS** Scarce / very local resident   1876–2021

First reported by R. Kay (1876) when over 50 larvae beaten from heather in the Bury area, with moths noted nectaring on heather at Turton in 1880. Ellis (1890) described the species as common on Lancs moorlands. Other pre-1970 records came from Bolton, Burnley, Longridge and Quernmore. However, unrecorded from the West Pennine Moors since 2006. Possibly now restricted to the highest ground around Leck Fell and Green Bank. Once from Warton Crag, in 1974, presumably a wanderer.

■ **HABITAT** Moorland.

■ **FOODPLANTS** Heather. Will eat Bilberry in captivity (Kay, 1876). Single brooded. Comes to light. Larvae recorded in late May and early June 1876.

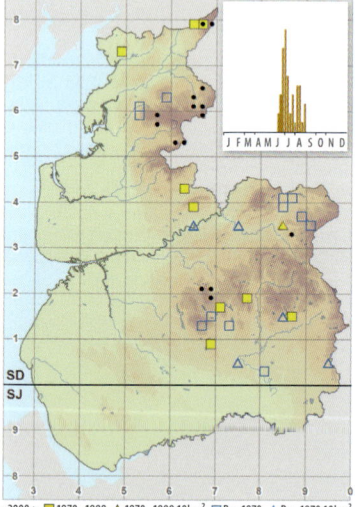

• 2000+ □ 1970–1999 △ 1970–1999 10km^2 □ Pre-1970 △ Pre-1970 10km^2

70.074 July Highflyer *Hydriomena furcata*

■ **LANCS STATUS** Abundant / widespread resident   c.1857–2022

Recorded throughout, in lowland and upland areas. First noted in Liverpool (Gregson, 1857) with other 19th century records at Longridge, Turton, Oldham and Blackpool. Later, described as abundant (Mansbridge, 1940). The spread of light trapping since 1960s has revealed this to be a common garden visitor, seemingly under no imminent threat. High count 130 at Docker Moor 2014. Larvae easy to find on birch at Astley Moss 2008, Bilberry at Healey Dell 2017, and Creeping Willow at Ainsdale 2021.

■ **HABITAT** Scrub, dunes, hedgerows, woodland, moorland.

■ **FOODPLANTS** Bilberry, Grey Willow, Creeping Willow, Heather, birch, Hazel. A single extended generation. Comes to light. Larvae recorded from April to June.

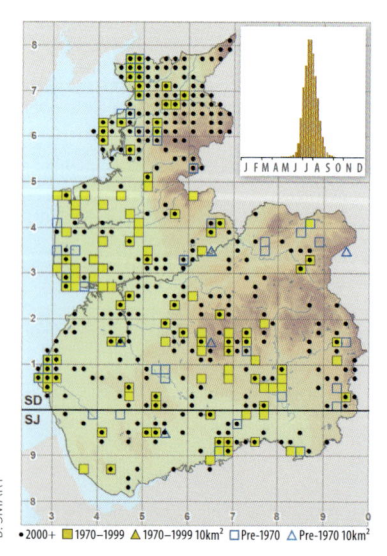

• 2000+ □ 1970–1999 △ 1970–1999 10km^2 □ Pre-1970 △ Pre-1970 10km^2

The Moths of Lancashire

70.075 May Highflyer *Hydriomena impluviata*

- **LANCS STATUS** Frequent / widespread resident c.1857–2022

First noted from Liverpool district 'where alders and willows grow' (Gregson, 1857). Other early records from Longridge, Burnley, Preston. Described as common throughout, with a predominant melanic form (Mansbridge, 1940). Has increased in frequency and range post-2000, with almost 90% of the 2,588 records coming from this century. A slight reduction in the last five years suggests this process may have stalled. Melanic forms still common, although the ratio to nominate form is variable.
- **HABITAT** Alder carr, wet woodlands.
- **FOODPLANTS** Alder.

Single brooded. Most records are of moths at light. Larvae recorded in late August.

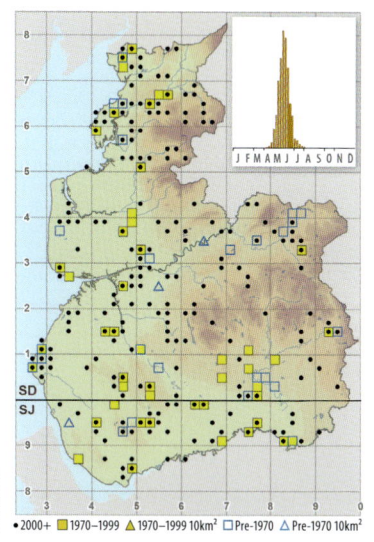

70.076 Ruddy Highflyer *Hydriomena ruberata*

- **LANCS STATUS** Scarce / local resident c.1857–2019

Far less common than previous two species, with only 18 records this century and 42 in total. First reported from Woolton, in 'damp places where willows grow' (Gregson, 1857). Records from last 20 years are largely from Heysham N.R. and, to a lesser extent, Sefton Coast. Other sites include Holcroft Moss in 2004, Mere Sands Wood in 2011, Heysham Moss in 2016 and Middleton N.R. in 2017. The lack of any records since 2019 is of concern, as is the apparent disappearance from Fylde district.
- **HABITAT** Mossland, damp woodland, scrub.
- **FOODPLANTS** Sallows and willows. Larvae noted in September.

Single brooded. Comes to light. Netted in car headlights, Fylde 1973 by D&J Steeden.

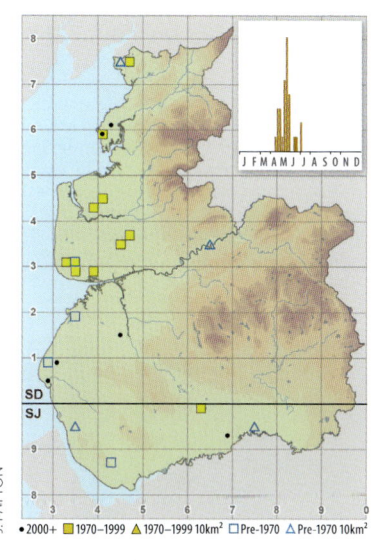

70.077 Pine Carpet *Pennithera firmata*

- **LANCS STATUS** Occasional / local resident c.1886–2022

Ellis noted this to be reasonably common in pine woods, without noting sites (1890). The map shows post-2000 range extension from the previously known upland and northern sites. Some of this may be due to the advent of light trapping, with almost all records obtained by this method. Prior to 1960 only known from Silverdale, Fylde and Sabden, with Leighton Moss, Gait Barrows and Bowland Fells added before 2000. Subsequent lowland records include Chorlton, Heaton Moor, Formby, Parr, Pilling.
- **HABITAT** Conifer plantations, moorland. Also, associated with isolated, mature pines.
- **FOODPLANTS** Not recorded in Lancashire.

Single extended generation with limital dates 12 May and 14 Dec.

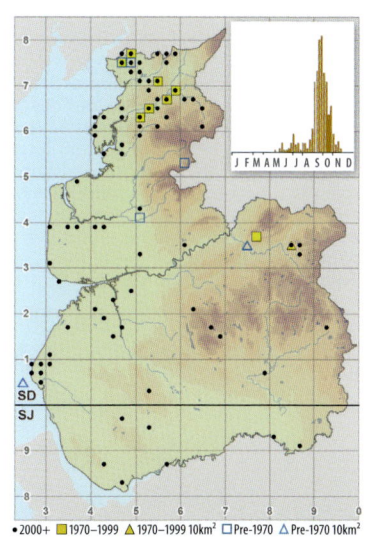

GEOMETRIDAE

The Moths of Lancashire

GEOMETRIDAE

70.078 Chestnut-coloured Carpet *Thera cognata*

■ **LANCS STATUS** Considered Extinct   1918–1926

First reported from Silverdale by A.E. Wright in 1918 (Mansbridge, 1920). Further records from the site in 1919, when bred from larvae by W. Mansbridge, and in 1922. Again bred from larvae in 1926 by H.B. Stocks, also at Silverdale; foodplant not recorded (Mansbridge, 1927). There are no subsequent county records. Nationally, it is a creature of moorland and has been recorded nearby at Hutton Roof and Farleton Knott in Cumbria, in one case just 20 metres from the VC60 boundary!

■ **HABITAT** Lowland limestone.

■ **FOODPLANTS** Bred from larva, presumably on Juniper, although not explicitly noted. Single brooded. All records from daytime fieldwork. Elsewhere, noted at light.

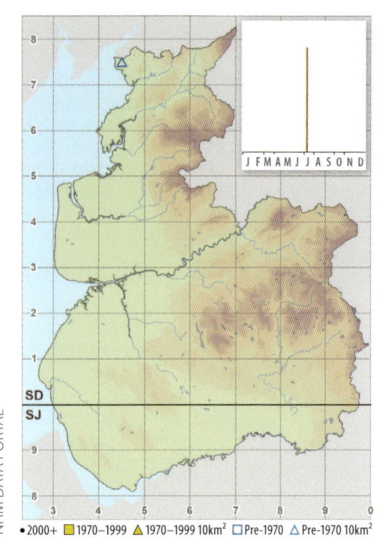

70.079 Spruce Carpet *Thera britannica*

LANCS STATUS Abundant / widespread resident   c.1947–2022

Early reports did not clearly differentiate between this and following species. First confirmed in 1940s at Simonswood Moss, recorder unknown. The only other pre-1980 records are from Slyne, Bolton-le-Sands and Ainsdale. Subsequently, range has expanded and the moth can now be expected almost anywhere, taking advantage of the widespread planting of conifers, whether in formal plantations or of single trees in parks and gardens, etc. Recorded on 115 occasions from 2022 alone.

HABITAT Conifer plantations, woodland, scrub, parks, gardens.

FOODPLANTS Larvae unrecorded other than ex-ova, Chorlton 2013; fed on pine. Double brooded. Virtually all records are of moths to light.

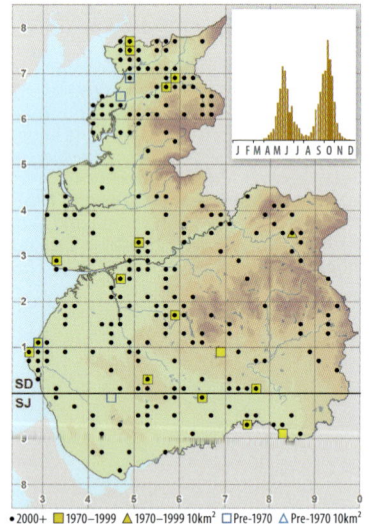

70.081 Grey Pine Carpet *Thera obeliscata*

■ **LANCS STATUS** Frequent / widespread resident   1896–2022

More widely recorded than the previous species pre-1980, with a more modest subsequent increase in range and frequency. 82 records from 2022. First recorded at Manchester in 1896 by T.G. Mason. Recorded in 1917 at Burnley and as larvae at Silverdale. Pre-1970 records from Didsbury, Prestwich and Formby demonstrated ability to survive outside the conifer plantations. Now a regular visitor to garden traps in lowland and upland areas. High count of 50 at Freshfield Dune Heath in 2009.

■ **HABITAT** Conifer plantations, woodland, parks, gardens.

■ **FOODPLANTS** Pine.

Double brooded. Comes readily to light. Larvae recorded April and July.

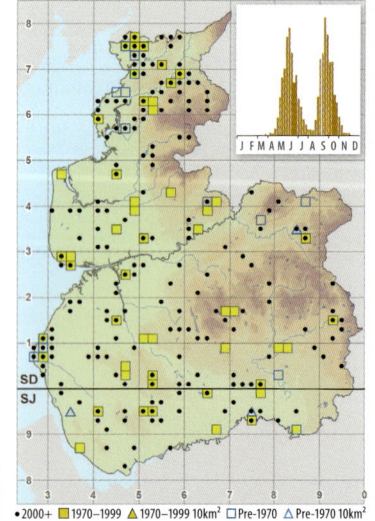

404 *The Moths of Lancashire*

70.082 Juniper Carpet *Thera juniperata*

- **LANCS STATUS** Frequent / widespread resident   c.1940–2022

Described by Mansbridge (1940) as scarce and local amongst Juniper in the Silverdale area, noting a record by G. Laxom from Warton, as well as one by A.E. Wright, later considered erroneous. Next recorded at Leighton Moss 1968 by J. Briggs, and during 1970s from Poulton-le-Fylde, Hoghton and Didsbury. Post-1990 records show the moth is now present throughout, with many garden records, suggesting larval use of cultivated Juniper. However, numbers of records have declined since 2015.
- **HABITAT** Moorland, parks and gardens.
- **FOODPLANTS** Not recorded in Lancashire.

Single brooded. Comes readily to light.

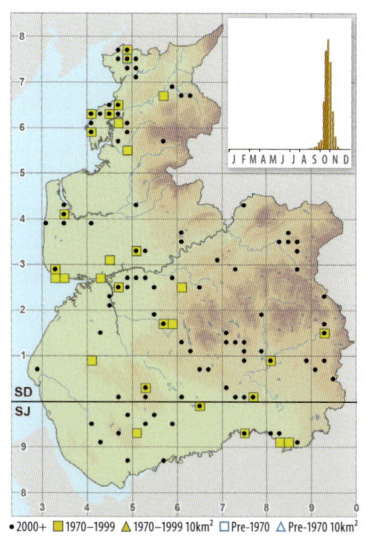

70.083 Cypress Carpet *Thera cupressata*

- **LANCS STATUS** Rare / very local resident. Possibly adventive   2016–2022

Recorded by R. Pyefinch on five occasions at Tarleton, showing likely existence of a local breeding population, presumably linked to nearby *Leylandii*. First recorded on 28 October 2016, and again on 15 November 2018, 19 June 2020 and 12 October 2021. The most recent record, from 11 November 2022, was of two moths at light. Unclear if records due to inadvertent introduction or genuine range expansion. Similar pattern in Cheshire and Yorkshire with just a handful of records since 2015.
- **HABITAT** Parks, gardens.
- **FOODPLANTS** Not recorded in Lancashire.

Double brooded. All records are of moths to light.

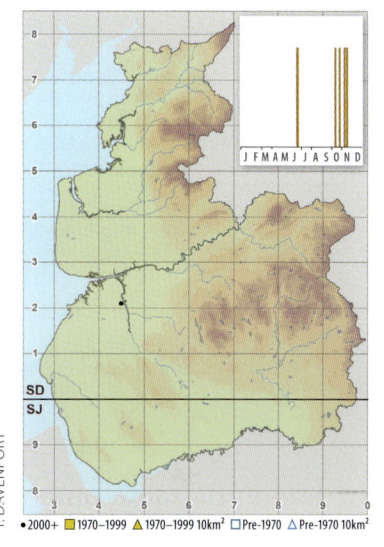

70.084 Blue-bordered Carpet *Plemyria rubiginata*

- **LANCS STATUS** Frequent / widespread resident   1843–2022

First noted in Preston (Edleston, 1843), and later from Liverpool, Ormskirk and Wyre estuary. Described as common and generally distributed (Mansbridge, 1940). Although rarely present in high numbers, populations appear stable, possibly with slightly increasing range. 2022 was the year with the highest number of records, 33. High counts of 6 from Flixton in 2022 and Warton Crag in 2014. Noted from lowland and upland areas, with the latter including Wray and Rishton.
- **HABITAT** Hedgerows, woodland, scrub, banks of streams and lakes, etc.
- **FOODPLANTS** Not recorded in Lancashire.

Single brooded. Almost all records are of moths to light.

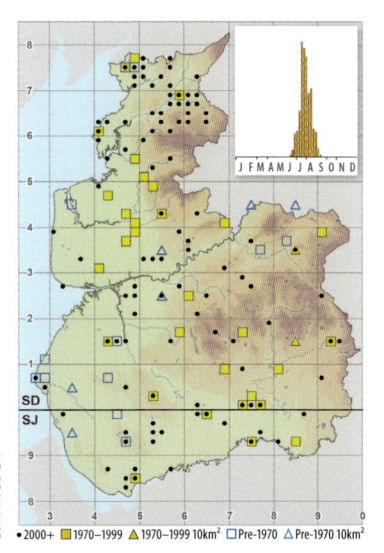

The Moths of Lancashire

GEOMETRIDAE

70.085 Barred Yellow *Cidaria fulvata*

■ **LANCS STATUS** Occasional / widespread resident   c.1857–2022

First reported from the Liverpool district (Gregson, 1857). Later, noted as generally distributed and sometimes common (Mansbridge, 1940). The 1186 records show a predominantly lowland species, with just a few records from the West Pennines. National decline mirrored by slight fall in total records over the last 15 years and by the experience of K. McCabe's Flixton garden, with 38 records in total, but none since 2009. Bred from larva beaten from rose at Middleton N.R. on 24 May 2022.

■ **HABITAT** Scrub, woodland, gardens.
■ **FOODPLANTS** Rose.

Single brooded. Most records are of moths at light.

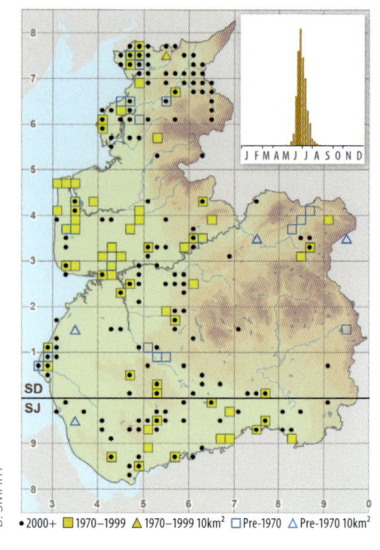

● 2000+   ▢ 1970–1999   △ 1970–1999 10km²   ▢ Pre-1970   △ Pre-1970 10km²

70.086 Broken-barred Carpet *Electrophaes corylata*

■ **LANCS STATUS** Occasional / widespread resident   c.1859–2022

Recorded from the mid-1800s at Chat Moss and Prestwich by J. Chappell, and at Hoghton by J.B. Hodgkinson. Britten (c.1950) described the moth as common in birch woods. The distribution map suggests an increase in upland records this century, with sites including Mill Houses, Green Bank, Docker Moor and Briercliffe, but an apparent decline in lowland records. The Flixton experience is again illustrative (as with Barred Yellow), with 72 records to 2006 and just a singleton in 2020 since.

■ **HABITAT** Woodland, moorland and scrub.
■ **FOODPLANTS** Not recorded in Lancashire.

Single brooded. Comes readily to light with a high count of 20 from Chorley in 1999.

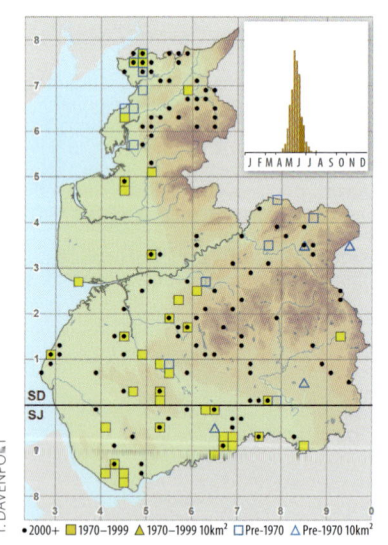

● 2000+   ▢ 1970–1999   △ 1970–1999 10km²   ▢ Pre-1970   △ Pre-1970 10km²

70.087 Purple Bar *Cosmorhoe ocellata*

■ **LANCS STATUS** Occasional / widespread resident   c.1857–2022

Noted in Liverpool district to be far from plentiful but widely distributed (Gregson, 1857). Ellis & Mansbridge noted the moth to be common throughout. Pre-1970 records from Burnley area, Nelson, Silverdale, Freshfield, Slyne, Lytham, Ditton and Wrightington. Subsequent numbers and range appear stable, with healthy populations in northern upland areas and on Sefton Coast reflected in the high counts for this species of 70 at Formby in 2005 and 19 at Docker Moor in 2021.

■ **HABITAT** Moorland, woodland, scrub, hedgerows, sand dunes.
■ **FOODPLANTS** Not recorded in Lancashire.

Double brooded. Comes to light.

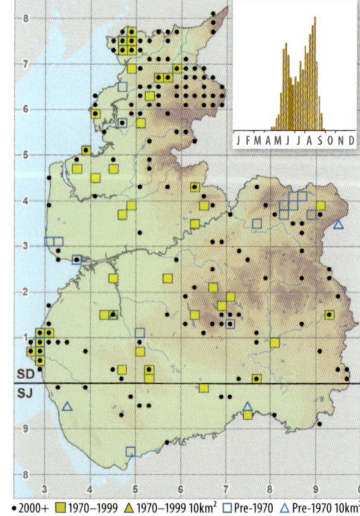

● 2000+   ▢ 1970–1999   △ 1970–1999 10km²   ▢ Pre-1970   △ Pre-1970 10km²

The Moths of Lancashire

70.088 Netted Carpet *Eustroma reticulata*

■ **LANCS STATUS** Scarce / very local resident   2009–2022

This species is noted as nationally Endangered (Fox *et al.*, 2019). It is a Red Data Book and UK BAP Priority Species.

The NHM, London, has a specimen of the moth labelled Carnforth, 1895 H. Murray, although its true origins are uncertain. Otherwise, the species is only known historically from north Wales, where it now appears extinct, and from Cumbria, where its foodplant flourishes in damp woodland glades and the population appears reasonably secure.

The moth remained unconfirmed from Lancashire until 4 September 2009, when J. and M. Elsworth found a larva on Touch-me-not Balsam, the larval foodplant, in woodland near Carnforth. A search of the surrounding area revealed more plants and larvae at a private site 1km away. The following year, adults came readily to actinic light at both sites and were easily netted as they flew slowly over the patches of Balsam in the second half of July.

Larvae initially feed on the leaves, later preferring the explosive seed pods. Though well camouflaged, they often rest between a leaf and the stem making a distinctive triangle. At the Lancashire sites an annual larval count is made during early September. Numbers fluctuate but the trend is downward.

As an annual, the Touch-me-not Balsam benefits from regular ground disturbance to reduce competition from more vigorous plants. Work parties are held in late autumn to disturb the ground by raking and pulling out brambles, ferns, grasses, nettles and tree seedlings and selective felling. This has not been enough to prevent the extinction of the plants in several areas. Mechanical scraping back to bare soil and reseeding failed at both sites.

In 2010 the Carnforth woodland had five small metapopulations of plants and the private site three, but by 2017 the private site had lost the moth and most of its plants, while the Carnforth site was reduced to three small patches of Balsam that in 2022 produced the lowest count of larvae of 18 compared to 240 in 2010.

An intriguing story has come to light that may explain the origin of our Lancashire moths and the Museum specimen. A notebook of Henry Murray, a saddler and taxidermist from Carnforth in the 1880's, has been transcribed (Hancock, 2018). He had a very profitable side-line selling Netted Carpet specimens to Victorian collectors avid for rare species. Here, we speculate that the nearest source of wild examples was probably near Windermere. As a result, although he had some success with captive breeding, it is possible he used local, naturally occurring Balsam or perhaps introduced the plant and presumably the moth, to a suitable woodland, the nearest of which could be our present site. If the wood was subject to regular coppicing and ground disturbance the plant and moth could have survived until now.

Lack of traditional woodland management and climate change probably accounts for the moth's current decline. Hard winters favour germination of the Balsam, whilst mild wet ones give competing vegetation an advantage. In the last 10 years, we are just about retaining the moth with some focused management, but for how long? Continued monitoring and small-scale management in late autumn provide the best chance of survival for this precarious colony.

Brian Hancock

GEOMETRIDAE

• 2000+   ▪ 1970–1999   ▲ 1970–1999 10km²   ☐ Pre-1970   △ Pre-1970 10km²

The Moths of Lancashire   **407**

GEOMETRIDAE

70.089 Phoenix *Eulithis prunata*

■ **LANCS STATUS** Occasional / widespread resident c.1857–2022

The earliest records are from Hale and Crosby (Gregson, 1857), and later from Bury and Prestwich (Ellis, 1890). Mansbridge (1940) considered the moth to be local and not common. Lancaster and Formby were added to the list of sites in 1910 and 1928. Increased light trapping, post-1950s, has shown the moth to be well distributed, although with a coastal bias. Records are more scattered inland but occur at all altitudes. Noted in an allotment greenhouse at Norcross by A.E. Smith, 11 July 2009.
■ **HABITAT** Woodland, allotments, gardens.
■ **FOODPLANTS** Bred ex-larva 1930 and 1964, although foodplant not documented. Single brooded. Comes to light.

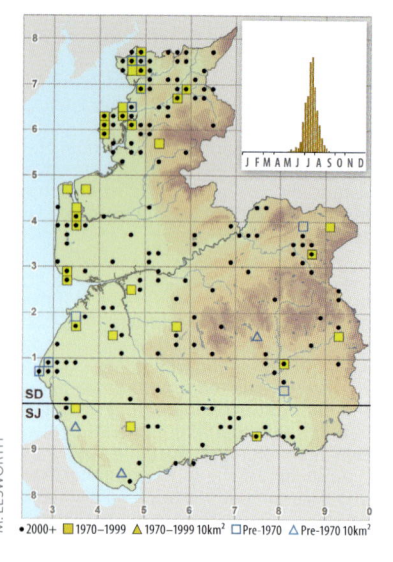

70.090 Chevron *Eulithis testata*

■ **LANCS STATUS** Occasional / local resident c.1857–2022

First recorded from Liverpool (Gregson, 1857), with early authors describing it as common on moors, mosses and coastal sand-hills. Declining nationally, the local picture is less clear. Numbers of records and of moths appear stable compared to 20 years ago. However, this may be deceptive due to an increased number of recorders. There is particular concern over the loss of lowland populations, as demonstrated by K. McCabe's Flixton garden, with 20 records from 1995 to 2000 and only three since.
■ **HABITAT** Moorland, rough grassland, scrub.
■ **FOODPLANTS** Creeping Willow.
Single brooded. Comes to light. Larvae recorded 17 May 1975 at Formby (M. Hull).

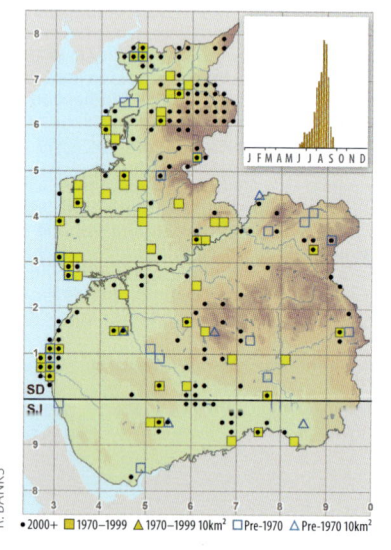

70.091 Northern Spinach *Eulithis populata*

■ **LANCS STATUS** Occasional / local resident c.1857–2022

Noted to be uncommon in Liverpool (Gregson, 1857), with other 19th century records from Irlam, Turton and Longridge Fell. Status variously described as local, but common where it occurs (Ellis, 1890), and abundant where good areas of Bilberry grow (Mansbridge, 1940). Has declined in distribution and abundance, with range now concentrated in higher ground in the east. Some hang on in a few lowland areas, such as Sefton Coast, Lancaster and Flixton, but with counts significantly down.
■ **HABITAT** Moorland, wooded areas where Bilberry grows.
■ **FOODPLANTS** Larvae noted on moorland near Nelson in 1930s; foodplant unrecorded. Single brooded. Comes to light. Larvae recorded May and early June (A. Brindle).

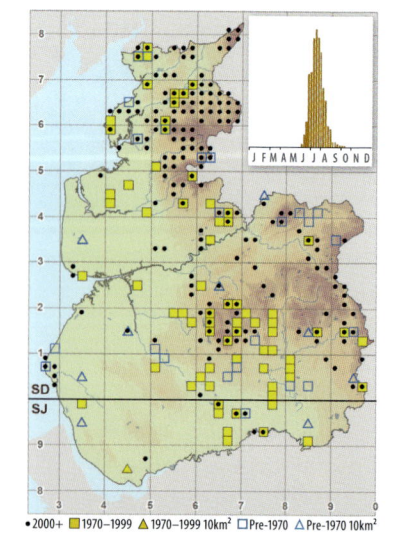

408 *The Moths of Lancashire*

70.092 Spinach *Eulithis mellinata*

LANCS STATUS Occasional / local resident   c.1859–2022
First recorded by J. Chappell around 1859 at Pendlebury and Withington (Ellis, 1890). Also, noted at Crosby and Preston in the late 1800s. Early authors described the moth as local. Since then, reported mainly from lowland sites, particularly in gardens and in coastal areas. Numbers have declined significantly in recent years. Since 2020, most have been in the Manchester area, although here too it has declined. The few recent records elsewhere include moths at Clitheroe, Hale, Formby and Leighton Moss, etc.
HABITAT Allotments, gardens, rough grassland.
FOODPLANTS Larva on Black Currant at Hutton, 19 April 2008 (A. Barker).
Single brooded. Comes to light. Noted on garden Gooseberry by R. Leverton 1965.

• 2000+   □ 1970–1999   △ 1970–1999 10km²   □ Pre-1970   △ Pre-1970 10km²

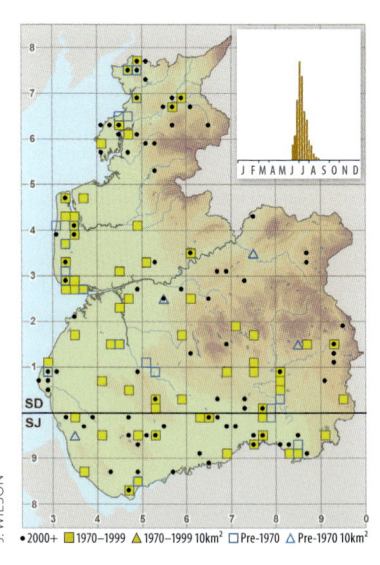

70.093 Barred Straw *Gandaritis pyraliata*

■ **LANCS STATUS** Frequent / widespread resident   c.1857–2022
Reported as plentiful in Liverpool woods (Gregson, 1857), and from Irlam, Pendleton and Preston (Ellis, 1890), with the latter describing distribution as somewhat local. Britten (c.1950) described status as widespread and common where found. Primarily a lowland species, more frequent in the north. Distribution post-2000 shows some infilling of previously unrecorded tetrads. Numbers appear stable, possibly increasing, with 920 records from last five years, compared to 690 from 2000–2005.
■ **HABITAT** Rough grassland, scrub, gardens.
■ **FOODPLANTS** Larvae (ex-ova), from Chorlton 2012, fed on Cleavers; moth reared.
Single brooded. Mainly recorded at light; occasionally disturbed from vegetation.

• 2000+   □ 1970–1999   △ 1970–1999 10km²   □ Pre-1970   △ Pre-1970 10km²

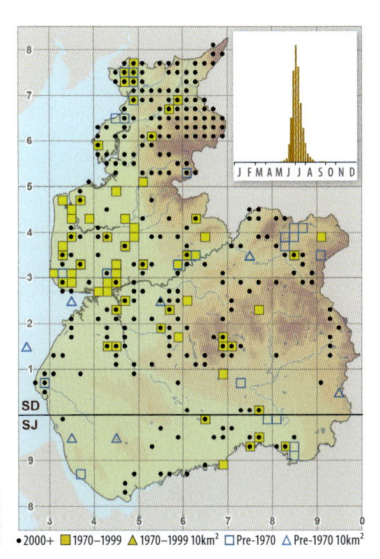

70.094 Small Phoenix *Ecliptopera silaceata*

■ **LANCS STATUS** Abundant / widespread resident   c.1843–2022
This species, described by Mansbridge (1940) as local and uncommon, increased significantly in the post-war (WW2) years. This increase was noted by Britten (c.1950) who linked it to the spread of Rosebay Willowherb. Earlier records from Preston (Hodgkinson, 1844), Rainhill, Hale, Hoghton and Bury. Comes readily to light, and records since the advent of trapping show the moth to be well distributed throughout. Slightly more frequently recorded in VC59. Numbers appear stable.
■ **HABITAT** Scrub, waste ground, woodland clearings, gardens.
■ **FOODPLANTS** Larvae on Himalayan Balsam at Ormskirk October 2020 (J. Kirk).
Double brooded. Maximum count of 62 at Glazebury in 2013 by R. Rhodes.

• 2000+   □ 1970–1999   △ 1970–1999 10km²   □ Pre-1970   △ Pre-1970 10km²

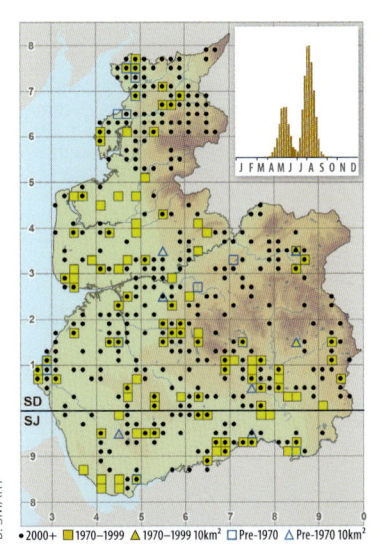

GEOMETRIDAE

70.095 Red-green Carpet *Chloroclysta siterata*

■ **LANCS STATUS** Frequent / widespread resident   c.1857–2022
First recorded at Knowsley and Speke (Gregson, 1857), with only two other records, from Manchester and Lytham St Annes, prior to 1964. Early authors described status as very local and uncommon. Records remained sparse until mid-1990s and mainly from lowland areas. Subsequently, there has been significantly increased abundance, with the moth now likely to be encountered almost anywhere in the county. High count of 66 to light at Gait Barrows on 22 October 2012, recorded by G. Jones.
■ **HABITAT** Woodland, hedgerows and gardens.
■ **FOODPLANTS** Oak, Sycamore and Blackthorn. Larvae noted June and July. Single brooded with the female overwintering. Comes to light throughout the year.

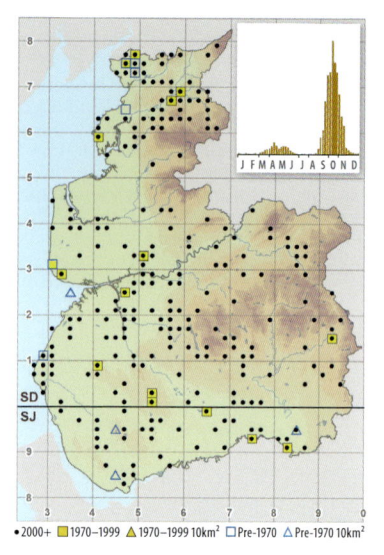

70.096 Autumn Green Carpet *Chloroclysta miata*

■ **LANCS STATUS** Scarce / local resident   c.1857–2022
Noted where willows grow (Gregson, 1857); location unclear, but possibly Crosby, the site of F.N. Pierce's 1887 record. Early authors found it local and scarce. Although records suggest coastal bias to 1960s, the last 20 years of targeting upland valley and fell-edge sites, especially Lune catchment, Bowland and, e.g., Worsthorne, Belmont, has produced regular small numbers. Infrequent on Sefton Coast; last in 2018 at Formby. The green form of Red-green Carpet may be confused with this species.
■ **HABITAT** Moorland, upland woodland, scrub.
■ **FOODPLANTS** Not recorded in Lancashire.
Single brooded. Most recent records are of moths to light. The female overwinters.

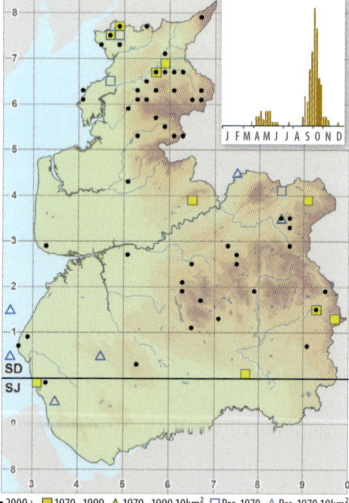

70.097 Common Marbled Carpet *Dysstroma truncata*

■ **LANCS STATUS** Abundant / ubiquitous resident   c.1857–2022
Early authors considered the moth common and generally distributed. It was first noted from Liverpool (Gregson, 1857), with other early records from Oldham, Ditton and St Annes. Range and abundance have significantly increased since the 1970s. Records from the last 40 years show the first generation appearing roughly a fortnight earlier, with the second generation flying well into November. A peculiarly early sighting was a moth at M. Sole's Lancaster kitchen window on 7 January 2021.
■ **HABITAT** Woodland, hedgerows, rough grassland, scrub, moorland, gardens, etc.
■ **FOODPLANTS** Geranium, Black Currant, Gorse, Heather.
Double brooded. Readily to light. High count 311 from Rishton 8 September 2016.

410   *The Moths of Lancashire*

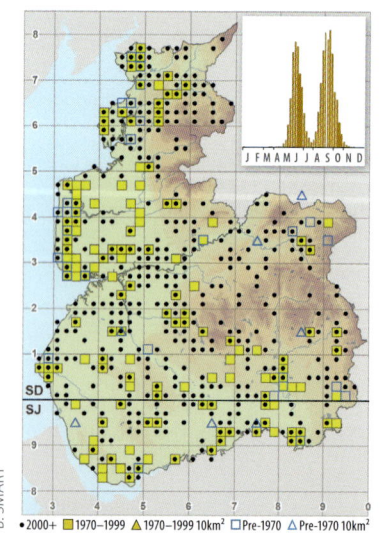

70.098 Dark Marbled Carpet *Dysstroma citrata*

- **LANCS STATUS** Occasional / local resident　　　　　　c.1843–2022

First noted in Preston area (Hodgkinson, 1843). Next recorded by J. Baldwin, nectaring on heather at Turton in 1880. Care needed to separate from the previous species, and it may be that some of the early records were misidentified. Confirmed records from the last few decades show range is largely limited to upland areas, with regularly recorded sites including Green Bank, Docker Moor, Briercliffe and Wray, and very infrequent lowland records. Numbers of records significantly declining.
- **HABITAT** Moorland, woodland, scrub.
- **FOODPLANTS** Unrecorded within the county.

Single brooded. Attracted to light with high count of 20 at Roeburndale in 2003.

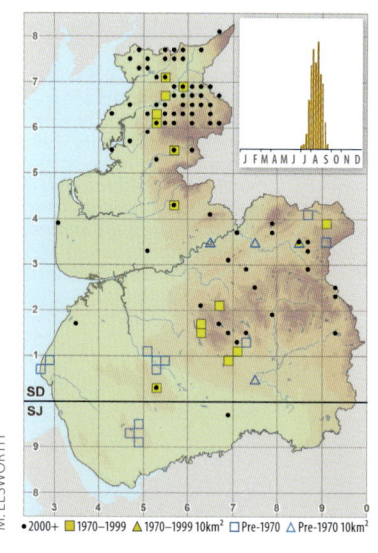

70.099 Beech-green Carpet *Colostygia olivata*

- **LANCS STATUS** Scarce / local resident　　　　　　c.1843–2022

Another moth first noted by J.B. Hodgkinson, and documented in his list from the Preston area (Hodgkinson, 1844). Other 19th century records from Bolton, Clougha, Withnell and Silverdale. Noted to be local in north Lancs and moorland districts (Mansbridge, 1940). The 28 records from this century demonstrate the moth to be most frequent in the uplands, with multiple records from Dalton Crags, Mill Houses and Leck Fell. Also noted at lowland limestone sites, particularly Gait Barrows.
- **HABITAT** Moorland, upland and lowland limestone grassland.
- **FOODPLANTS** Not recorded in Lancashire.

Single brooded. Readily to light. High count of 20 from Leck Fell 2002 by S. Hayhow.

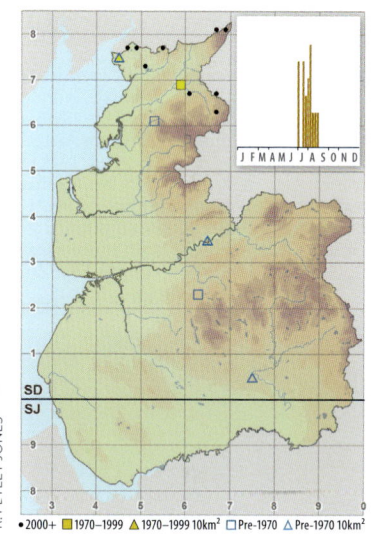

70.100 Green Carpet *Colostygia pectinataria*

- **LANCS STATUS** Frequent / widespread resident　　　　　　c.1857–2022

Noted by Gregson (1857) as plentiful in woods around Liverpool where ferns grow. Other early authors described the moth as common and general in mixed woods, and reported it from Whalley, Burnley, Nelson and Colne. There has been a large increase in abundance during this century with distribution nearing ubiquity. 1,685 records from 2018 to 2022, compared to 1,171 from 2008 to 2012.
- **HABITAT** Rough grassland, scrub, gardens.
- **FOODPLANTS** Larvae (ex-ova), from Chorlton 2013, fed on Cleavers; moth reared. Double brooded pattern much clearer than 25 years or so ago, when the moth appeared to be univoltine. Comes to light. Maximum count of 35 at Flixton 2021.

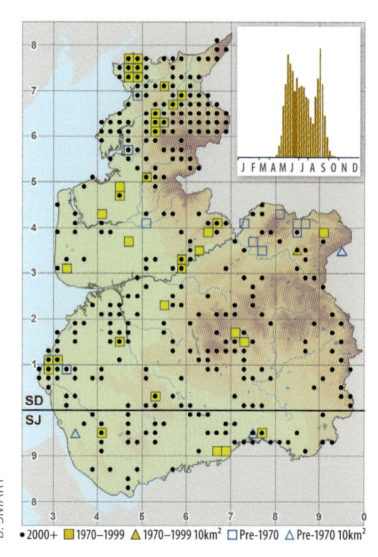

GEOMETRIDAE

70.101 Mottled Grey *Colostygia multistrigaria*

■ **LANCS STATUS** Occasional / local resident   c.1857–2022

Noted to be plentiful under tufts of grass in Liverpool during March (Gregson, 1857). Bred from larva in 1906 at Burnley; foodplant not documented. Described as common and general (Mansbridge, 1940), although Britten (c.1950) later reported the moth as uncommon. Subsequent records have shown the moth to be most frequent on moorland, Morecambe Bay limestone and at coastal sites, in that order. Records elsewhere are scattered and few.

■ **HABITAT** Moorland, calcareous grassland, dunes and heaths.
■ **FOODPLANTS** Not recorded in Lancashire.

Single brooded. Comes to light, sometimes abundantly, e.g., 112 at Green Bank 2021.

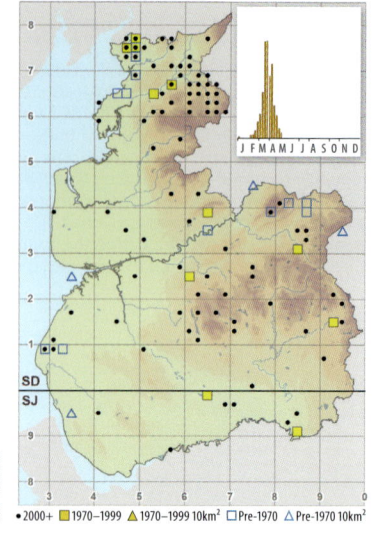

70.102 Striped Twin-spot Carpet *Coenotephria salicata*

■ **LANCS STATUS** Occasional / local resident   c.1886–2022

First records noted from Silverdale, Withnell, Fleetwood and on the moors at Bolton, with status described as local (Ellis, 1890). Mansbridge (1940) reported it to be local in north Lancs and moorland districts. Noted from Blacko in 1961, but not again in VC59. Regularly recorded further north at Leck Fell, Green Bank, Dalton Crags, Baines Crag, Warton Crag, Yealand Conyers, etc. Numbers appear on the increase but may at least partially reflect increased recording effort at the sites mentioned above.

■ **HABITAT** Moorland, limestone sites.
■ **FOODPLANTS** Not recorded in Lancashire.

Double brooded. Comes to light. High count 115 near Hutton Roof 2021 (J. Patton).

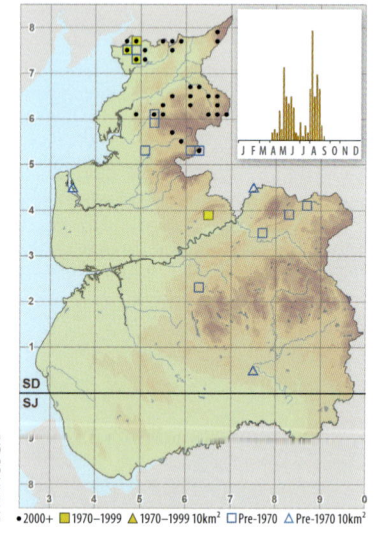

70.103 Water Carpet *Lampropteryx suffumata*

■ **LANCS STATUS** Occasional / local resident   c.1857–2022

First noted from woods in the Liverpool district (Gregson, 1857), with other 19th century records from Chat Moss, Worsley, Bolton, Huyton and Preston. Noted by Mansbridge (1940) to be local and seldom common. Records from this century appear concentrated in northern upland and lowland sites, and on the Sefton Coast, with scattered distribution elsewhere. Recent years have shown declining records and numbers. Highest count is 27 from Freshfield Dune Heath 2015 by C. Fletcher.

■ **HABITAT** Rough grassland, moorland, limestone pavement, woodland, heaths.
■ **FOODPLANTS** Not recorded in Lancashire.

Single brooded. Virtually all post-1965 records are of moths trapped at light.

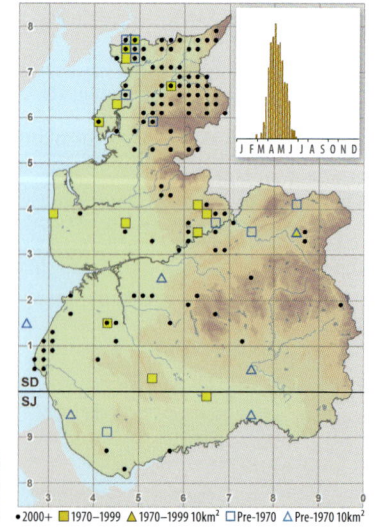

412   *The Moths of Lancashire*

70.104 Devon Carpet *Lampropteryx otregiata*

■ **LANCS STATUS** Occasional / local resident 2008–2022

A recent arrival in the county as part of significant range extension, far stronger in the north of the county. This suggests likely spread from Cumbria, where the moth was noted 17 years before the first Lancs record on 1 June 2008 of two moths at Lord's Lot (J. Girdley, P. Marsh, S. Palmer). Annual at the same site until 2015. Other frequent sites include Leighton Moss, Silverdale Moss, and on higher ground at Mill Houses, Green Bank, Docker Moor, etc. Twice noted during 2021 at Hoghton, VC59.
■ **HABITAT** Damp grassland, woodland, moorland.
■ **FOODPLANTS** Unrecorded within the county.
Double brooded. Comes to light. High count of 17 at White Moss, Yealand in 2010.

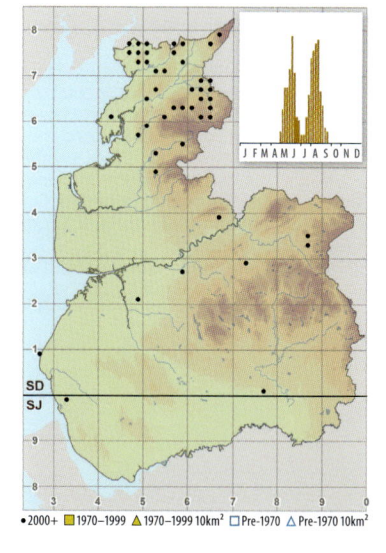

70.105 Northern Winter Moth *Operophtera fagata*

■ **LANCS STATUS** Occasional / local resident c.1859–2022

First reported from Chat Moss c.1859 by J. Chappell (Ellis, 1890). Early authors described the moth as local. Distribution is scattered, linked to mature birch, with presence in most areas other than the Fylde coast. Regular at Mill Houses and known from other upland sites, e.g., Docker Moor. Probably under-recorded as flying when many have put their traps away, and possibly with limited attraction to light. As likely to be encountered by searching for larvae or moths in natural birch woodland.
■ **HABITAT** Woodland, hedgerows, moorland.
■ **FOODPLANTS** Birch. Larvae recorded in May, with moth reared in Chorlton 2010.
Single brooded. Highest count of 75 at Brinscall, 2014 by G. Jones using torchlight.

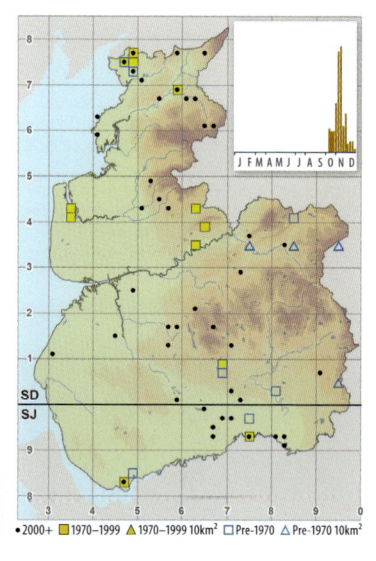

70.106 Winter Moth *Operophtera brumata*

■ **LANCS STATUS** Abundant / ubiquitous resident c.1857–2022

Far more common than the previous moth, the male is a familiar sight at lighted windows from November. The female is wingless. First recorded in Liverpool, with Gregson (1857) noting it to be present 'everywhere'. This remains the case, especially in lowland areas. No evident threat to status. Larvae recorded on over 140 occasions.
■ **HABITAT** Woodland, hedgerows, scrub, parks, gardens.
■ **FOODPLANTS** Extremely polyphagous. Recorded on birch, willows, Blackthorn, Hazel, Apple, Sycamore, Hawthorn, Field Maple, Bilberry, Heather, poplar, Aspen, Ash, oak, Rowan, elm, Privet, rose, Gorse, Honeysuckle, Bird Cherry. Larvae April to June. Single brooded. Comes to light. Searches by torchlight can be particularly successful.

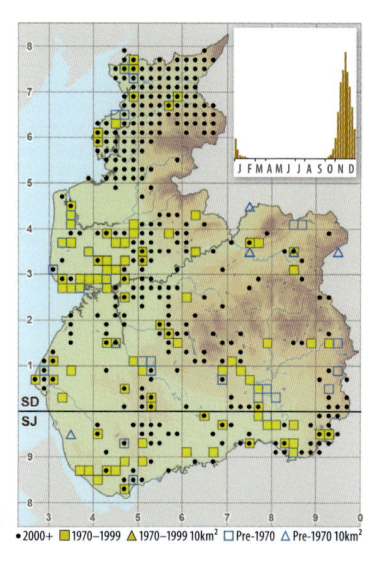

The Moths of Lancashire

GEOMETRIDAE

70.107 November Moth *Epirrita dilutata*

■ **LANCS STATUS** Frequent / widespread resident   1940–2022

Early authors classed the moth as common. However, the status of the *Epirrita* genus has been difficult to interpret due to taxonomic changes and the requirement for dissection. The first confirmed record, therefore, is from Didsbury in 1940, recorded by H.N. Michaelis, followed by Morecambe in 1959 (C.J. Goodall) and Slyne in 1965 (P.J. Marsh). Whilst regular at some upland locations such as Rishton, Wray and Higher Tatham, lowland records predominate. See also Appendix 6.

■ **HABITAT** Woodland, hedgerows, scrub, gardens.
■ **FOODPLANTS** Blackthorn.

Single brooded. Comes readily to light. Larvae found in April and May.

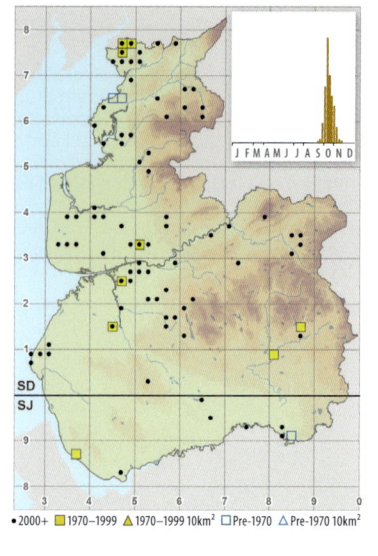

70.108 Pale November Moth *Epirrita christyi*

■ **LANCS STATUS** Occasional / local resident   2001–2022

Not separated from the previous species until 1906. Again, only dissected specimens have been included. First confirmed records were from Roeburndale, 28 September 2001 (S. Palmer & P.J. Marsh). Next, at Jack Scout, October 2001. The first confirmed VC59 records came from Longton 2005 and Billinge 2006. Generally, it appears more frequent in upland woodland than in lowland areas. Difficult to assess current status, due to the small number of dissected specimens. See also Appendix 6.

■ **HABITAT** Moorland, mature woodland.
■ **FOODPLANTS** Unrecorded in Lancashire.

Single brooded. All records are of moths to light.

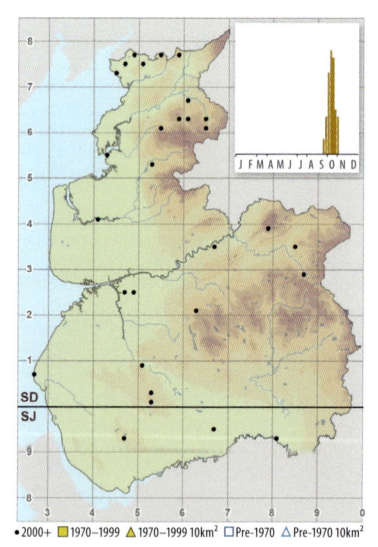

70.109 Autumnal Moth *Epirrita autumnata*

■ **LANCS STATUS** Occasional / local resident   1972–2022

Mid-20th century reports from Liverpool, Burnley and Formby, but no indication of dissection. The first confirmed record therefore, is from Aigburth 1972, by I.D. Wallace. From Lightfoot Green in 1997 and Mere Sands Wood in 1999. Occasional at upland sites, e.g., Wray and Higher Tatham, but absent from higher ground in VC59. Lowland sites with multiple records include Billinge, Formby and Flixton. Difficult to assess current status, as few dissected specimens. See also Appendix 6.

■ **HABITAT** Moorland, woodland, scrub, gardens.
■ **FOODPLANTS** Not recorded in Lancashire.

Single brooded. Almost all records are to light.

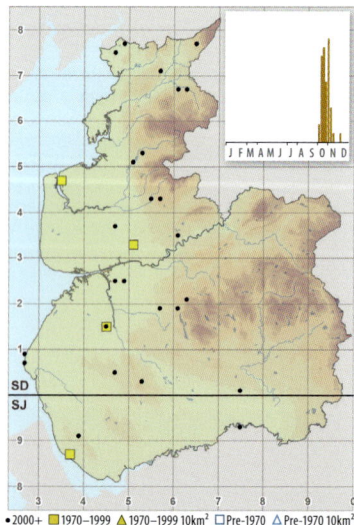

70.110 Small Autumnal Moth *Epirrita filigrammaria*

■ **LANCS STATUS** Occasional / local resident   1876–2022

First documented by R. Kay (1876), when over 50 larvae were beaten from heather in the Bury area. Moths also noted nectaring on heather by J.W. Baldwin in 1880, at Turton, nr. Bury. Later, from Oldham and Burnley areas in early 1900s. Recorded from Formby in 1936 by G. de C. Fraser, assumed to be a wanderer from the moors. The same presumably applies to a Flixton record from 1995. All 21st century records are from upland sites such as Briercliffe, Green Bank, Leck Fell and Docker Moor.
■ **HABITAT** Moorland.
■ **FOODPLANTS** Heather. Bred from larva at Burnley by A.L. Goodson in 1910.
Single brooded. Can be identified with care, taking into account date and habitat.

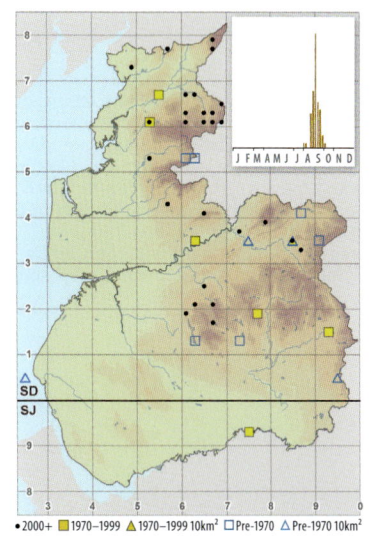

70.111 Small White Wave *Asthena albulata*

■ **LANCS STATUS** Scarce / very local resident   c.1859–2022

Always reported as local, early records nevertheless showed slightly wider range than currently known, with pre-1900 records from Chat Moss, by J. Chappell around 1859, Preston and Silverdale. Subsequent records have shown the Silverdale area, particularly Gait Barrows, to be the stronghold for this species, but with evidence of recent decline, as only recorded twice since 2018. There are also a few records from other areas, i.e., Mill Houses 2005, Pennington 1999 and Formby in 1982 and 2009.
■ **HABITAT** Limestone woodland, broadleaved woodland.
■ **FOODPLANTS** Not recorded in Lancashire.
Single brooded. Comes to light, and readily disturbed from vegetation in day.

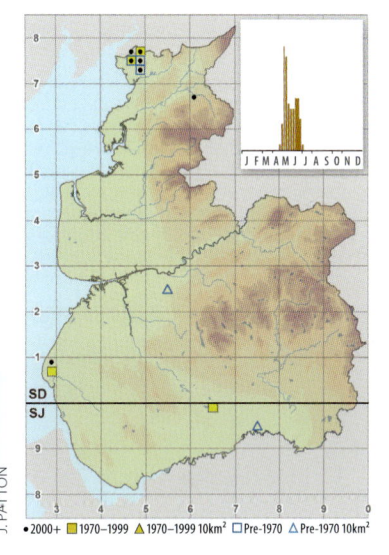

70.112 Dingy Shell *Euchoeca nebulata*

■ **LANCS STATUS** Occasional / widespread resident   1845–2022

First recorded at Chat Moss in July 1845 by R.S. Edleston, noting the species to be abundant (1846). Later, noted at Hoghton Woods by J.B. Hodgkinson in June 1865. Other early records from Prestwich, Preston, Yealand and Formby. Britten (c.1950) gave status as widespread but never common. Occasional in the uplands around Wray, but primarily known from southern lowland areas, particularly around Manchester. Absent from Fylde coast since 1999. Numbers appear stable.
■ **HABITAT** Damp woodland, edges of waterbodies.
■ **FOODPLANTS** Larva beaten from Alder by B. Smart at Chorlton 28 September 2014.
Single brooded. Comes to light. High count of 27 from Flixton from June 2003.

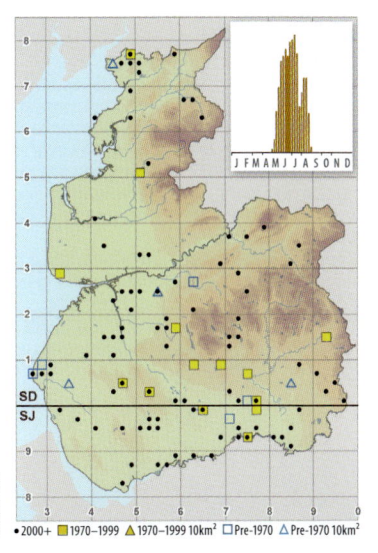

The Moths of Lancashire   **415**

GEOMETRIDAE

70.113 Waved Carpet *Hydrelia sylvata*

■ **LANCS STATUS** Rare / very local resident　　　　　　　　**1845–2018**

First noted at Chat Moss, where reported as scarce around birch (Edleston, 1846). Other 19th century records from Worsley, Silverdale, Preston and Burnley, and again in 1925 from the Silverdale area. Noted by Mansbridge as scarce and local (1940). There was a long period of absence before the moth reappeared in the county, with a garden record from Heysham by D.J. Holding on 14 June 2014. The only subsequent record was from Warton, nr. Carnforth by J. Patton on 8 June 2018.
■ **HABITAT** Unknown.
■ **FOODPLANTS** Not recorded in Lancashire.
Single brooded. Recent records may represent wanderers from south Cumbria.

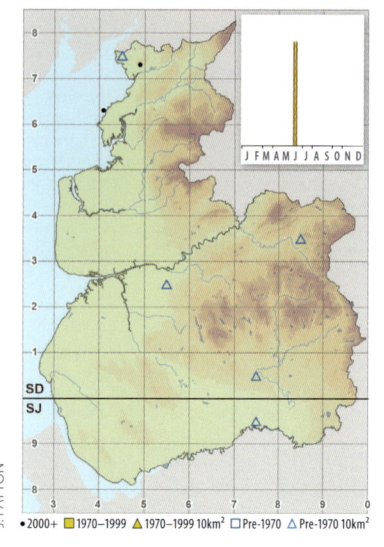

● 2000+　□ 1970–1999　▲ 1970–1999 10km²　□ Pre-1970　△ Pre-1970 10km²

70.114 Small Yellow Wave *Hydrelia flammeolaria*

■ **LANCS STATUS** Occasional / widespread resident　　　　　**c.1843–2022**

First noted within J.B. Hodgkinson's list from the Preston area (1844). Later reported as abundant at Chat Moss (Edleston, 1846). Mansbridge (1940) described the moth as local and never abundant. Noted at Formby 1953, where associated with Alder. Records increased significantly during 1990s, and early 2000s saw the moth recorded in upland areas for the first time, including 21 records at Rishton from 2020 to 2022. However, whilst range has expanded, numbers have declined in last 20 years.
■ **HABITAT** Woodland, Alder carr, edges of waterbodies, hedgerows, gardens, scrub.
■ **FOODPLANTS** Not recorded in Lancashire.
Single brooded. All recent records have been of moths to light.

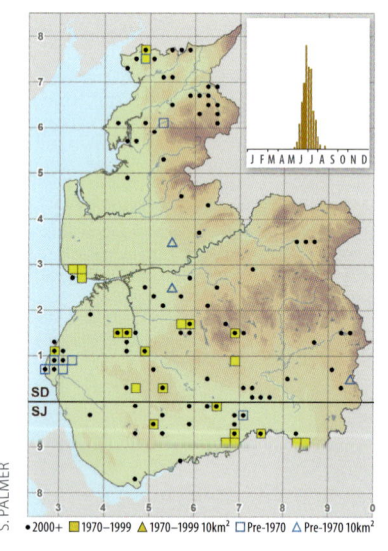

● 2000+　□ 1970–1999　▲ 1970–1999 10km²　□ Pre-1970　△ Pre-1970 10km²

70.115 Welsh Wave *Venusia cambrica*

■ **LANCS STATUS** Occasional / local resident　　　　　　　　**c.1859–2022**

Noted to be very local, with first records at Hoghton by J. Chappell and at Longridge, where decline from previous abundance noted by J.B. Hodgkinson (Ellis, 1890). Later reported as freely taken where Rowan occurs in woods (Mansbridge, 1940). Extensive recording 2000–2017 saw many records and high numbers in northern VC60 uplands, e.g., Docker Moor, Green Bank, with declining numbers of records since linked to reduced effort (P.J. Marsh, pers. comm.). Other regular sites include Oswaldtwistle.
■ **HABITAT** Moorland.
■ **FOODPLANTS** Not recorded in Lancashire.
Single brooded. Noted in the day and to light. High count 57 from Docker Moor 2017.

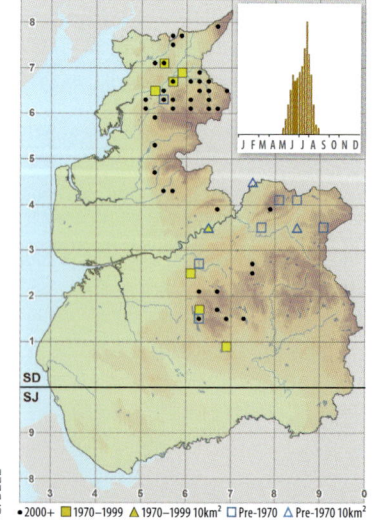

● 2000+　□ 1970–1999　▲ 1970–1999 10km²　□ Pre-1970　△ Pre-1970 10km²

416　*The Moths of Lancashire*

70.116 Blomer's Rivulet *Venusia blomeri*

■ **LANCS STATUS** Occasional / very local resident   c.1843–2022

First noted from Preston area (Edleston, 1844), with other 19th century records from Hoghton, Dutton and Brockholes Wood. Next recorded from Silverdale in 1928 and 1957, with most subsequent records from northern limestone areas. No evidence of further Preston records since 1975, but an isolated record from Lytham St Annes in 1988. Regular sites this century include Gait Barrows, Warton Crag and Hyning Scout Wood, Yealand Conyers. Whilst numbers of records stable, counts have declined.

■ **HABITAT** Woodland, hedgerows.

■ **FOODPLANTS** Foodplant unrecorded within the county.

Single brooded. Comes to light. High count of 14 from Hyning Scout Wood 2012.

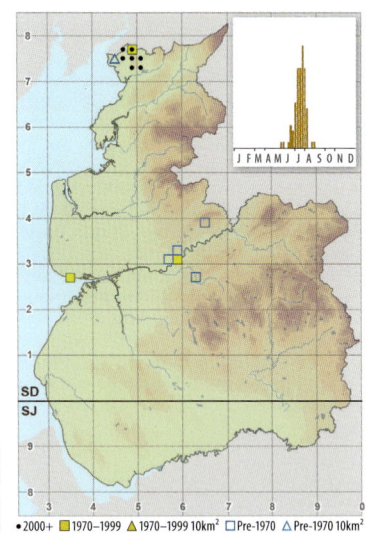

70.118 Brown Scallop *Philereme vetulata*

■ **LANCS STATUS** Scarce / very local resident   1972–2022

A relatively recent addition to the county's fauna, first noted in 1972 at Silverdale by C. Whitehead & C. Jones-Parry. Larvae found to be common on Buckthorn at Yealand Hall Allotment in 1975 by W.A. Watson (Creaser, 1976). Also from Leighton Moss 1975 and Gait Barrows 1984. A record from Dalton Crags in 2022 extends the known range, taking in an upland site to add to the lowland limestone sites where previously confined. One at Docker Moor in 2014 was likely a wanderer as foodplant absent.

■ **HABITAT** Limestone scrub and woodland.

■ **FOODPLANTS** Buckthorn.

Single brooded. Comes to light. High count of 32 from Myers Allotment in 2012.

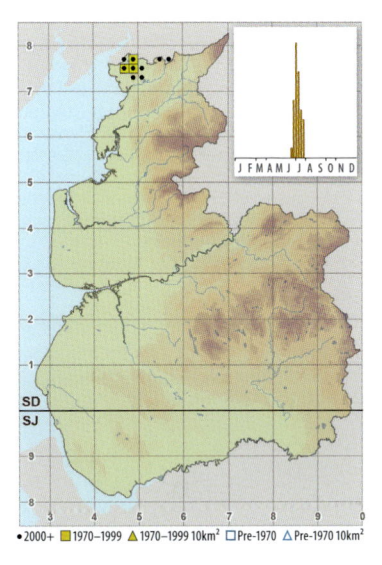

70.119 Dark Umber *Philereme transversata*

■ **LANCS STATUS** Scarce / very local resident   1905–2021

First noted at Silverdale in 1905 by A.E. Wright, later confirmed by Mansbridge (1940) to be the only Lancashire record at the time. Next noted at Yealand Hall Allotment, Leighton Moss, Warton Crag and Gait Barrows, including seven at the latter on 14 July 2008. During the last ten years, most likely encountered at Hyning Scout Wood, Yealand Conyers and Silverdale. Whilst never common, numbers appear to have decreased at its favoured limestone sites, with no records at all from 2022.

■ **HABITAT** Limestone woodland and scrub.

■ **FOODPLANTS** Larva on Buckthorn. Adult emerged 2003 at Warton Crag (S. Palmer).

Single brooded. Most records are at light. Larvae noted from early May to early June.

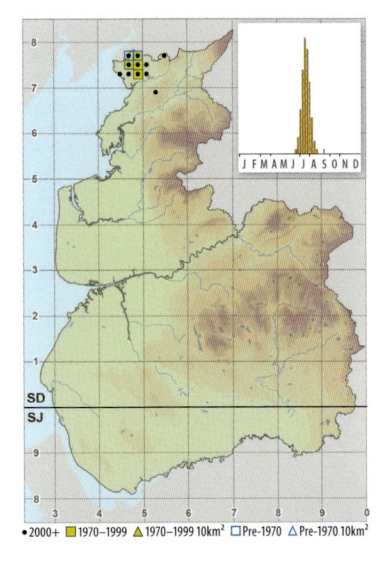

GEOMETRIDAE

70.120 Argent and Sable *Rheumaptera hastata*

■ **LANCS STATUS** Considered Extinct   c.1859–1968

First recorded at Chat Moss by J. Chappell around 1859 and by J.C. Melvill a few decades later (Ellis, 1890). It was also reported from the Manchester area in 1896 by T.G. Mason. Ellis and other early authors considered the species scarce and very local. Not noted again until 1955, at White Moss, Yealand, although the recorder is unknown. The moth was next recorded flying near birches at Warton Crag on 13 May 1961 by C.J. Goodall, with, sadly, the final record at Leighton Moss RSPB Reserve in 1968, noted by J. Briggs. Unsuccessful searches by S. Palmer for three consecutive years at Winmarleigh Moss in the early 2000s and the continued lack of records leaves little doubt that this distinctive moth is now extinct in the county. Scarce and declining in all the neighbouring counties, although all have seen a few records in the current century.

■ **HABITAT** Mosses.
■ **FOODPLANTS** Not recorded in Lancashire.
Difficult to determine from the few Lancashire records, but nationally, single brooded.

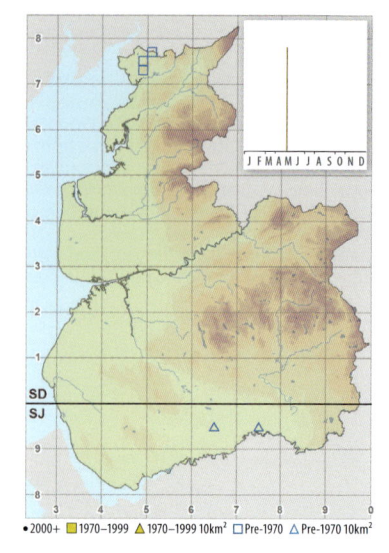

70.121 Scallop Shell *Rheumaptera undulata*

■ **LANCS STATUS** Occasional / local resident   c.1843–2022

Earliest records from Preston area (Hodgkinson, 1844) and from Botany Bay Wood, nr. Astley by J. Chappell (Ellis, 1890). Mansbridge (1940) noted the moth to be local and in some years fairly common. Mainly a lowland species, although range does appear to have extended into more upland areas since 2000, e.g., Brierfield, Belmont, Rishton, Docker Moor, etc. K. McCabe's Flixton records show some decline, from high point of 14 records during 1997, to just two in each of the last three years.

■ **HABITAT** Damp woodland.
■ **FOODPLANTS** Grey Willow. Moths bred from spinnings 2007 at Rixton (B. Smart). Single brooded. Comes to light. Larvae recorded during September.

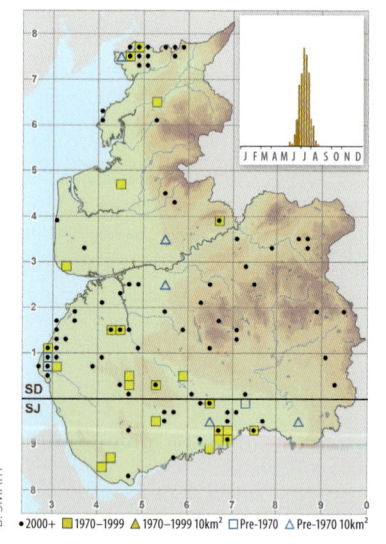

70.123 Tissue *Triphosa dubitata*

■ **LANCS STATUS** Occasional / local resident   c.1857–2022

First noted in Liverpool, nectaring on ragwort (Gregson, 1857). Later, considered to be uncommon but generally distributed (Mansbridge, 1940). An unusual geometrid in that it hibernates as an adult. Can be searched for at overwintering sites, e.g., found by G. Hedges at Ease Gill in 2016 in a cave above ground and in a pothole 15 m below ground. Also, found in bat hibernacula and abandoned mine workings at Warton Crag, with high count of 30 at the latter by R. Fellows in 2021. Occasional at light.

■ **HABITAT** Woodland, scrub.
■ **FOODPLANTS** Not recorded in Lancashire.
Single brooded. Moths noted nectaring on heather, Buddleia, ragwort, etc.

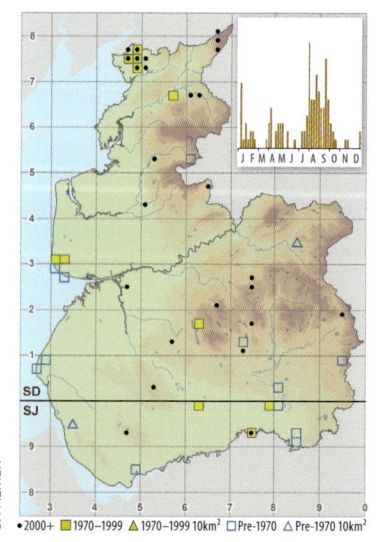

418 *The Moths of Lancashire*

70.128 Pretty Chalk Carpet *Melanthia procellata*

■ **LANCS STATUS** Occasional / very local resident   2001–2022

Whilst national distribution trends are in long-term decline (Randle *et al.*, 2019), the moth has recently extended into the Silverdale area, likely from populations in south Cumbria, and seems to be doing reasonably well. First recorded at Jack Scout on 8 July 2001 by C. Cockbain and L. Sivell. The next five years saw additional records from Silverdale, Leighton Moss and Yealand Redmayne. The only moth found outside this area was a singleton trapped at light at St Annes 2013 and photographed (A. Baines).
■ **HABITAT** Limestone woodland.
■ **FOODPLANTS** Traveller's-joy.
Single brooded. High count of 12 at dusk by Traveller's-joy at Jack Scout (B. Hancock).

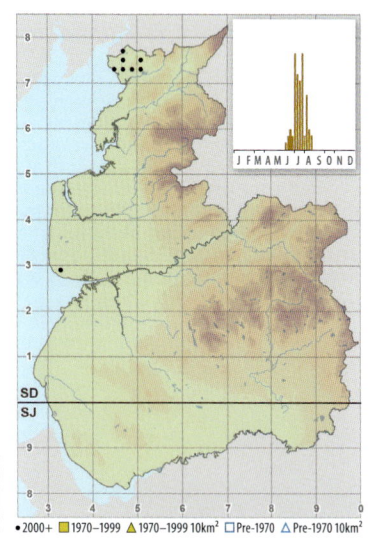

70.130 Chimney Sweeper *Odezia atrata*

■ **LANCS STATUS** Frequent / local resident   c.1859–2022

This day-flying moth was noted in Ellis (1890) to be local, but common where it does occur, with records listed from Pendleton (by J. Chappell), Prestwich, Bolton and Withnell. Other early authors commented similarly, noting preference for the eastern uplands. This seems increasingly the case as the moth has been lost to a number of lowland sites, with counts down from the high point of early 21st century. Maximum count of 350 noted at Yarrow Valley Park, Chorley 2010 by E. Langrish & S. Priestley.
■ **HABITAT** Moorland, damp grassland.
■ **FOODPLANTS** Not recorded in Lancashire.
Single brooded. Only a handful of records at light; predominantly noted during day.

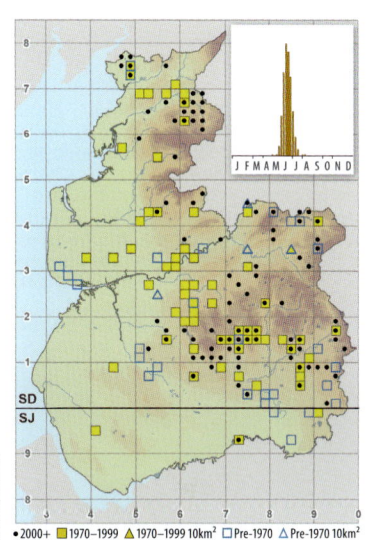

70.131 Twin-spot Carpet *Mesotype didymata*

■ **LANCS STATUS** Occasional / local resident   c.1857–2022

First reported in Liverpool by Gregson (1857), whilst R. Kay later obtained numerous larvae from heather in 1872 at Bury. Mansbridge (1940) noted it to be common everywhere but preferring rough ground. Britten (c.1950s) reported it as plentiful on the moorlands. Overall, range remains similar, with most recent records on moorland, e.g., Pendle Hill, Docker Moor, Grizedale Fell. However, it appears in sharp decline in Fylde, on the limestone sites of Morecambe Bay, and possibly elsewhere.
■ **HABITAT** Moorland, rough grassland.
■ **FOODPLANTS** Heather.
Single brooded. Recorded in day, disturbed from vegetation, and at light.

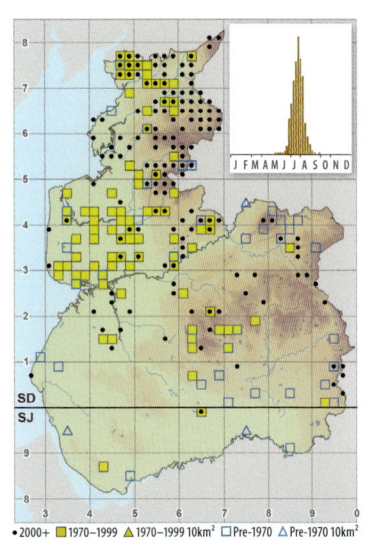

GEOMETRIDAE

70.132 Rivulet *Perizoma affinitata*

■ **LANCS STATUS** Frequent / widespread resident   c.1843–2022

19th century records from the Preston area (Hodgkinson, 1844), Hoghton Woods, Manchester, Lancaster and Broadgreen, Liverpool. Light-trapping since mid-1950s has shown the moth to be widespread across the county, initially in lowland areas, but more recently in upland areas such as Mill Houses and Green Bank. Sites regularly recorded in the last decade include Hale, Worsthorne, Crook O' Lune, Sunderland Point, Rishton and Hoghton, although counts showing signs of decline in this period.
■ **HABITAT** Open woodland, hedgerows, verges, scrub.
■ **FOODPLANTS** Bred from larva at Silverdale 1969 (R.E. Leonard); foodplant not noted.
Single brooded. Almost all records are of moths to light.

• 2000+   ▢ 1970–1999   △ 1970–1999 10km²   ▢ Pre-1970   △ Pre-1970 10km²

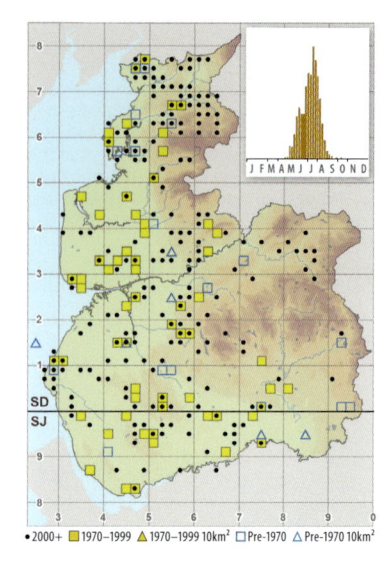

70.133 Small Rivulet *Perizoma alchemillata*

■ **LANCS STATUS** Frequent / widespread resident   c.1843–2022

Rather commoner than the previous species. First recorded in the Preston area (Hodgkinson, 1844). Noted by Gregson (1857) to be plentiful, and by Mansbridge (1940) as well distributed and occasionally locally common. Britten (c.1950) agreed, referring to carrs and woods as the main habitat. Primarily a lowland moth, with a few more records from upland areas in subsequent years. A common garden species, recorded regularly at Dolphinholme, Parr, Rishton, Flixton, Billinge and Belmont.
■ **HABITAT** Open woodland, carrs, hedgerows, scrub, gardens.
■ **FOODPLANTS** Unrecorded within the county.
Single brooded. Comes readily to light. High count of 43 at Flixton 6 July 1998.

• 2000+   ▢ 1970–1999   △ 1970–1999 10km²   ▢ Pre-1970   △ Pre-1970 10km²

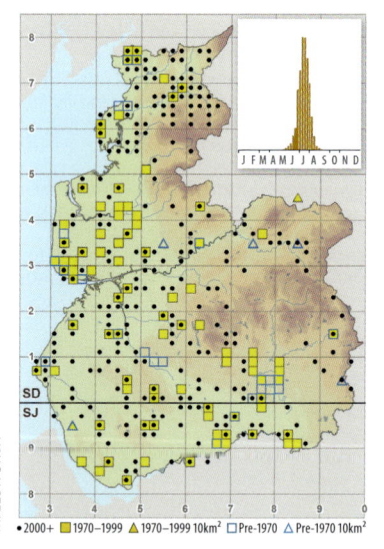

70.134 Barred Rivulet *Perizoma bifaciata*

■ **LANCS STATUS** Occasional / local resident   c.1857–2022

Reported sparingly from Liverpool district by C.S. Gregson (1857). Beaten from an old hedge in Preston by J.B. Hodgkinson in 1867, with the same recorder tapping larvae from the foodplant on the Wyre estuary 1872. Status noted by Mansbridge (1940) as local and scarce. Recent decline with just 27 records in last six years, compared to 34 in 2001 alone. Distribution shows a few more upland records; otherwise unchanged.
■ **HABITAT** Scrub, dry grassland, waste ground.
■ **FOODPLANTS** Larvae noted feeding in Red Bartsia flowers and seedheads in Chorlton on 5 September 2012, with adult emerging two years later (B. Smart).
Single brooded. Comes to light with high count of 22 from Hollins Green 1996.

• 2000+   ▢ 1970–1999   △ 1970–1999 10km²   ▢ Pre-1970   △ Pre-1970 10km²

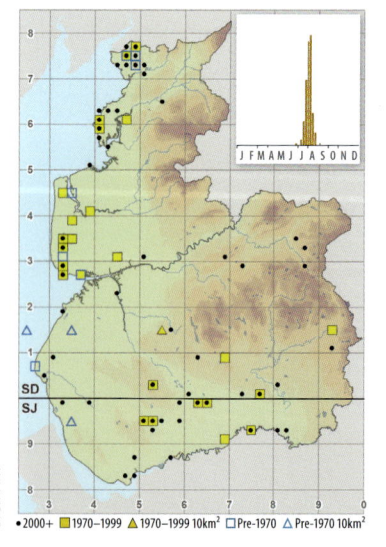

70.137 Grass Rivulet *Perizoma albulata*

■ **LANCS STATUS** Occasional / local resident c.1843–2022

First noted from Preston area (Hodgkinson, 1844); later, from Liverpool, Burnley, Oldham and Silverdale. Early authors reported it to be common wherever foodplant grows. The last few decades have seen range contraction as the moth has all but disappeared from sites on higher ground. Now largely a coastal species, with the highest concentrations at Sefton and Fylde dunes, e.g., approx. 100 recorded at Lytham in 2022 by the Nelson Naturalists. Frequently netted in daytime and at dusk.
■ **HABITAT** Sand dunes, calcareous grasslands.
■ **FOODPLANTS** Larvae in seeds of Yellow Rattle in June and early July.
Single brooded. A late record noted at Martin Mere, 26 September 2009 (A. Bunting).

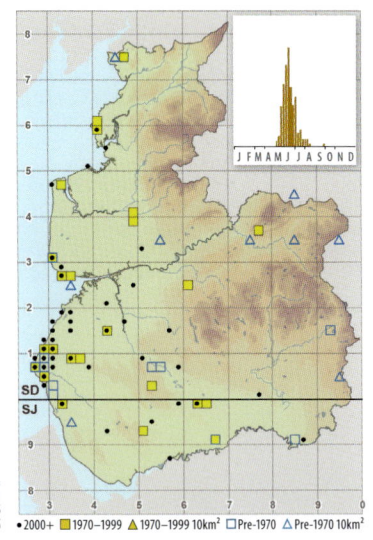

70.138 Sandy Carpet *Perizoma flavofasciata*

■ **LANCS STATUS** Frequent / widespread resident c.1857–2022

First recorded from Liverpool by Gregson (1857) and described as plentiful. Other early authors noted the moth to be common everywhere. Nevertheless, this appears primarily a lowland species, avoiding the fells, with distribution particularly sparse in eastern areas of VC59. 2022 records show presence in Chorlton, Flixton, St Helens, Hale, Formby, Mere Sands Wood, Chorley, Longridge, Pilling, Yealand Conyers, Nether Kellet, Lancaster, etc. Population appears stable.
■ **HABITAT** Woodland edges and rides, hedgerows, grassland.
■ **FOODPLANTS** Red Campion, feeding on seeds within the capsule.
Single brooded. Comes to light. Larvae recorded June to mid-July.

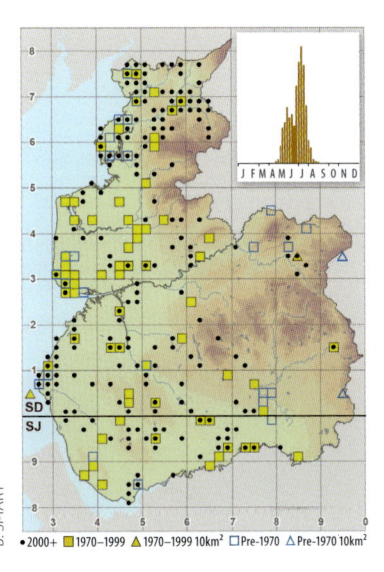

70.139 Barred Carpet *Martania taeniata*

■ **LANCS STATUS** Occasional / very local resident c.1874–2022

First recorded by J. Sidebotham at Silverdale in the mid-1870s. Reported as very local in the north of the county by Ellis (1890), and as local and scarce by Mansbridge (1940). Subsequently, noted at Warton Crag in 1967, Leighton Moss 1968 and Gait Barrows 1993. Although records have increased in the last decade, with 93 from 2013–2022 compared to 20 from 2003–2012, counts of individuals are showing signs of decline from the high point of 2015. Limited to Morecambe Bay limestone area.
■ **HABITAT** Limestone woodland.
■ **FOODPLANTS** Not recorded in Lancashire.
Single brooded. Occasionally noted in day and at dusk. Comes readily to light.

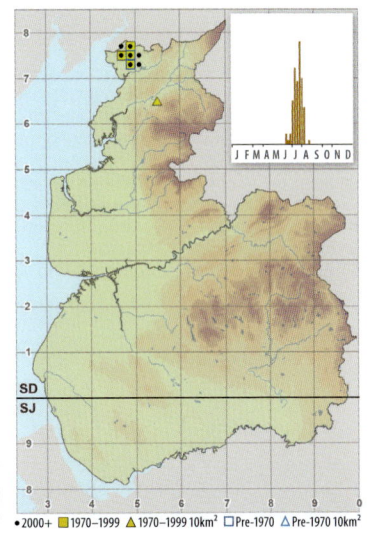

The Moths of Lancashire **421**

GEOMETRIDAE

70.141 Double-striped Pug *Gymnoscelis rufiasciata*

■ **LANCS STATUS** Frequent / ubiquitous resident c.1853–2022

First noted near Warrington by N. Cooke (Byerley, 1854), then the Preston district c.1887 and Silverdale in 1910. Mansbridge (1940) described it as confined to the north of the county where it was common. Since the 1970s it has expanded its range considerably and also increased in numbers, particularly in lowland areas. It is usually found in single figures, but since 1994 counts of over 20 have occurred at sites such as Flixton and Heysham N.R., the latter attracting 36 to light on 28 August 2008.

■ **HABITAT** Woodland, meadows, scrubland, gardens and brownfield sites.
■ **FOODPLANTS** Gorse, Red Clover, Ivy flowers and rose.
Double, with a small, occasional third brood in October/November. Comes to light.

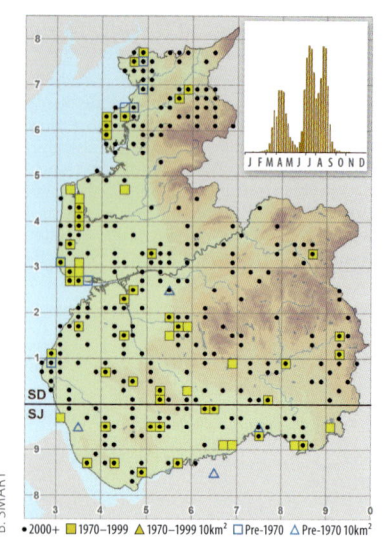

70.142 V-Pug *Chloroclystis v-ata*

■ **LANCS STATUS** Frequent / ubiquitous resident 1921–2022

Having spread into north-west England in the early 20th century, it was first noted in Lancashire in 1921 at Formby by W.A. Tyerman. The next record was not until 1940 at Ditton, followed by Silverdale in 1944 and St Annes in 1948. Thereafter, it spread steadily and widely across lowland parts and reached upland areas, such as Littleborough, in 1998. On 25 June 2020, 17 came to light in Flixton (K. McCabe).

■ **HABITAT** Scrub, woodland, hedgerows, gardens and brownfield sites.
■ **FOODPLANTS** Hawthorn and the flowers of Burnet Rose.
Double brooded, the first being much smaller than the second. A possible third brood occurred, from October to early November, between 2001 and 2011. Comes to light.

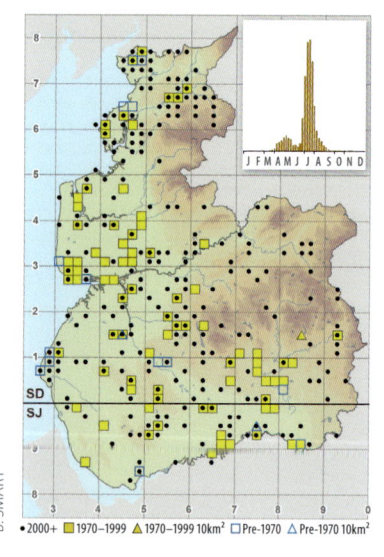

70.143 Sloe Pug *Pasiphila chloerata*

■ **LANCS STATUS** Occasional / local resident 1976–2022

Not recognised as British until 1971. First noted in Lancashire, with larvae beaten from Blackthorn blossom, on 4 May 1976 at White Moss, Yealand by J. Briggs. Reported in the following year at the same site by others. Increasing awareness led to records from Leighton Moss in 1980, while larval searches revealed its presence on Warton Bank (nr. Lytham), Out Rawcliffe and Preesall from 1986 to 1991. It appears to be a lowland species, but may be overlooked amongst Green Pug elsewhere.

■ **HABITAT** Scrubland, older hedgerows and scrubby brownfield sites.
■ **FOODPLANTS** Blackthorn blossom.
Single brooded. Flies at dusk; comes to light in small numbers.

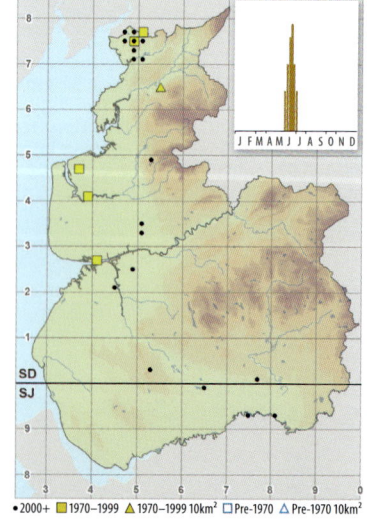

422 *The Moths of Lancashire*

70.144 Green Pug *Pasiphila rectangulata*

■ **LANCS STATUS** Frequent / ubiquitous resident　　　c.1857–2022
Noted at Hale and Linacre, as plentiful in old orchards, by Gregson (1857). Over the following century records were rather sparse, from sites such as Formby, Didsbury, Lytham, Wrightington and Warton (Carnforth). Since the 1960s, it has been noted across much of the county, being regular and annual at most light traps. Larger counts are rare, but 38 were seen at Flixton on 6 July 2001 (K. McCabe).
■ **HABITAT** Woodland, scrub, hedgerows, orchards, gardens and brownfield sites.
■ **FOODPLANTS** On the flowers of Apple and Blackthorn.
Single brooded, although there have been 18 late August to early September records between 1984 and 2021, with most in the last decade. Comes to light.

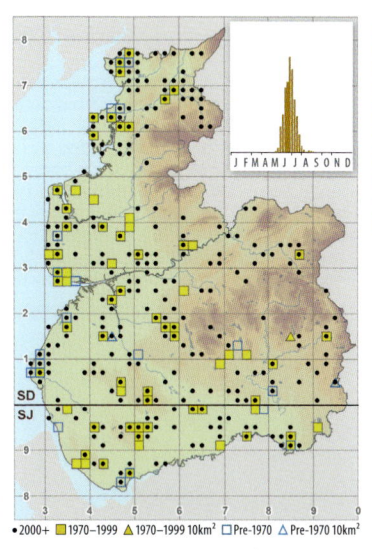

70.145 Bilberry Pug *Pasiphila debiliata*

■ **LANCS STATUS** Occasional / very local resident　　　c.1862–2022
First noted on Chat Moss by C. Campbell (Campbell, 1862). Not seen again until a search at Lord's Lot Wood on 23 June 2014 produced 20 adults flying at dusk (B. Hancock & P. Stevens), with a few larvae found there in May 2015. On 25 June 2018, 25 were seen flying at Birk Bank at dusk, and it has been found here every year since then. The only other recent sites were at Docker Moor, in 2014 and 2020, Longridge Fell in 2019 and one resting on sallow, near Bilberry, at Mossley in 2021 (S.H. Hind).
■ **HABITAT** Bilberry within or very close to woods and trees, in sheltered situations.
■ **FOODPLANTS** Bilberry.
Single brooded. Best searched for at dusk, when it flies low over Bilberry.

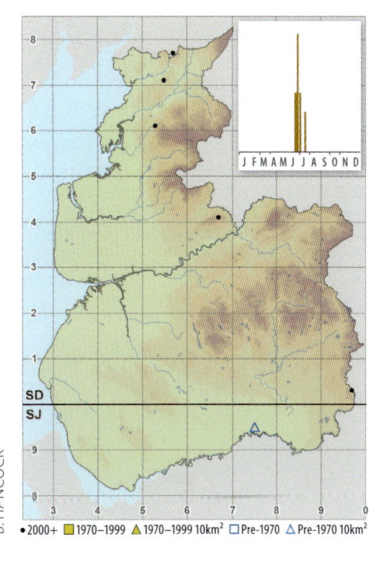

70.146 Haworth's Pug *Eupithecia haworthiata*

■ **LANCS STATUS** Occasional / very local resident　　　1904–2022
First noted in the Carnforth area in 1904 (H. Murray) and in 1906 at Lathom by W.A. Tyerman, but not reported again until 27 July 2004, at Yealand Conyers (B. Hancock). Subsequent searches have shown it to be well established in the Silverdale and Yealand areas, with up to 25 recorded at dusk around flowers of the larval foodplant. An isolated sighting was made in Blackrod in 2008 and it has since become regular in small numbers at Formby from 2013, with records from Flixton in 2015 and 2016.
■ **HABITAT** Hedgerows and scrub.
■ **FOODPLANTS** Traveller's-joy. It may be utilising garden *Clematis* spp. in VC59.
Single brooded. Comes to light and can be found flying around the foodplant at dusk.

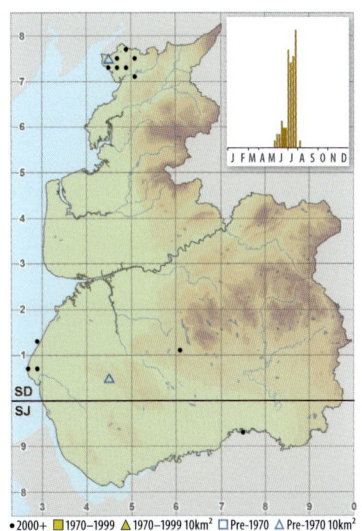

The Moths of Lancashire

GEOMETRIDAE

70.147 Slender Pug *Eupithecia tenuiata*

■ **LANCS STATUS** Frequent / widespread resident　　　　　c.1857–2022

First noted as plentiful where sallows occur in the Liverpool district (Gregson, 1857), and later found in Bury and Preston (Ellis, 1890). Into the 20th century, records were few and Mansbridge (1940) regarded it as scarce across the region. By the 1970s, it was regular at Leighton Moss, with later records from sites such as Swinton, Leigh and Billinge. Since 1990, it has been recorded widely and regularly at many sites, usually in small numbers. The largest count was 34 at light in Preston on 18 July 2022.

■ **HABITAT** Woodland, scrub, gardens, edges of wetland areas and brownfield sites.
■ **FOODPLANTS** In shoots and catkins of sallow spp., including Goat Willow.
Single brooded. Comes to light. Possibly under-recorded due to identification issues.

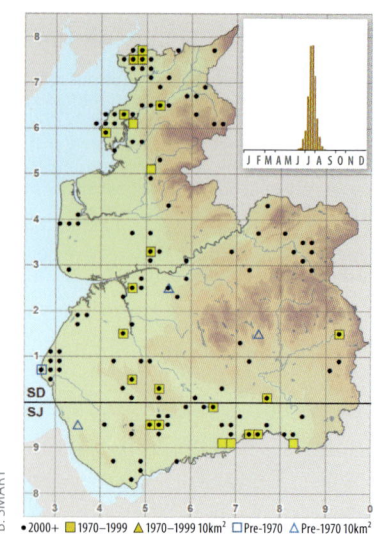

• 2000+　□ 1970–1999　▲ 1970–1999 10km²　□ Pre-1970　△ Pre-1970 10km²

70.148 Maple Pug *Eupithecia inturbata*

■ **LANCS STATUS** Scarce / very local resident　　　　　　1988–2022

Mirroring a recent increase in Cheshire, it was first noted at Lytham Hall on 7 September 1988 by D. & J. Steeden. The next was found on 11 August 2012 at Spring Wood, Whalley (A. Barker). Since then, it has been reported from Flixton on five occasions (K. McCabe), four of these in 2021, at Longridge in 2019 by D. Lambert and at Maple Farm Nursery, near Westby, on 19 July 2022 (J. Steeden).

■ **HABITAT** Woodland and parks.
■ **FOODPLANTS** Not recorded in Lancashire.
Single brooded. Comes to light. It is associated with mature Field Maple elsewhere, including just across the Cumbrian border, at Beetham.

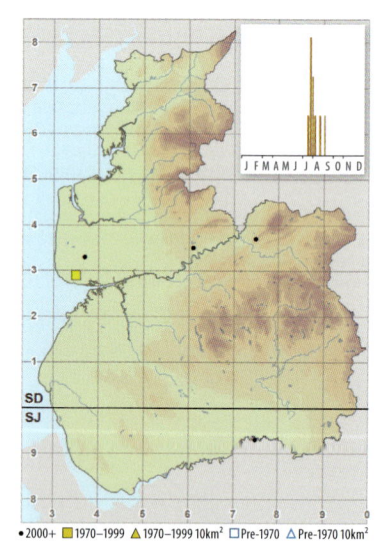

• 2000+　□ 1970–1999　▲ 1970–1999 10km²　□ Pre-1970　△ Pre-1970 10km²

70.149 Cloaked Pug *Eupithecia abietaria*

■ **LANCS STATUS** Scarce / very local resident and rare migrant　　1998–2020

First recorded, at light, in Pennington on 19 July 1998 by P. Pugh, followed by one in Briercliffe on 5 July 2001 (T. Lally). In 2008 two separate moths came to light in Silverdale on 4 July and 13 July 2008 (T. Riden), suggesting the possibility of local breeding (Hancock, 2018). Next reported from Formby on 21 June 2017, Preston on 1 June 2018, Warton Crag on 11 July 2018, and Woolton on 3 July 2020 (S. McWilliam).

■ **HABITAT** Unknown, all records being from garden light traps.
■ **FOODPLANTS** Not recorded in Lancashire.
Single brooded. Comes to light. A 1982 record from Didsbury in a Rothamsted trap is considered unconfirmed.

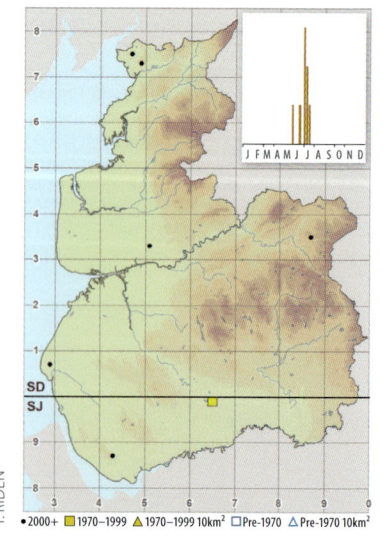

• 2000+　□ 1970–1999　▲ 1970–1999 10km²　□ Pre-1970　△ Pre-1970 10km²

The Moths of Lancashire

70.150 Toadflax Pug *Eupithecia linariata*

■ **LANCS STATUS** Occasional / local resident   **c.1857–2022**
First noted as plentiful amongst Toadflax in the Liverpool district (Gregson, 1857), but not again until 1935 at Formby and St Annes and Didsbury in 1940. Since then, mainly in near-coastal and Mersey Valley areas. Has declined markedly in the 21st century, the last sightings for areas with former populations being Billinge (1992), Blackpool (2004), Flixton (2005) and St Helens area (2012). Still regular in the Heysham area.
■ **HABITAT** Open grassy places on dry, often sandy, waste ground and brownfield sites.
■ **FOODPLANTS** Common Toadflax.
Single brooded. Comes to light. Larvae have been found in September. It is possible that some small, brightly marked Foxglove Pugs have been misidentified as Toadflax.

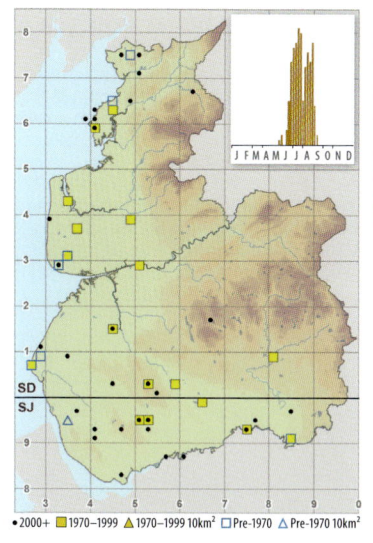

• 2000+   ☐ 1970–1999   △ 1970–1999 10km²   ☐ Pre-1970   △ Pre-1970 10km²

70.151 Foxglove Pug *Eupithecia pulchellata*

■ **LANCS STATUS** Frequent / widespread resident   **c.1857–2022**
Present in the Liverpool district (Gregson, 1857) and then recorded widely over the following decades at sites such as Prestwich, Dutton, Longridge, Burnley and Pendle Hill. It was considered as locally common by all early authors up to 1950. Since then, the use of light traps has demonstrated that it is present across much of the county. Since the 1980s, it has been noted regularly at many sites including Adlington, Bay Horse, Billinge, Claughton, Flixton, Green Bank, Hale, High Tatham and Wray.
■ **HABITAT** Woods, hedgerows, gardens, moorland, mosses, dunes and brownfield sites.
■ **FOODPLANTS** Foxglove flowers.
Single brooded with an extended flight period. Comes to light. Larva found in June.

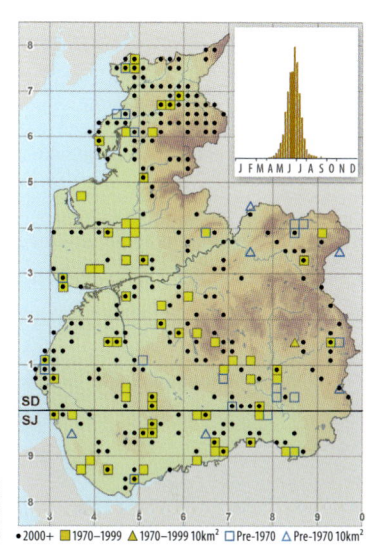

• 2000+   ☐ 1970–1999   △ 1970–1999 10km²   ☐ Pre-1970   △ Pre-1970 10km²

70.153 Lead-coloured Pug *Eupithecia plumbeolata*

■ **LANCS STATUS** Considered Extinct   **c.1845–1865**
First noted by Hodgkinson (1866), as follows: 'During the last week in June (1865), Mr. Gregson and I went to Pelling (Pilling) Moss for *plumbeolata*, which we soon met with in its old haunt, where 20 years ago I used to take it among *Melanpyrum* (cow-wheat)'. Gregson (1857) noted it as, 'not uncommon among *Myrica gale*' (Bog Myrtle) on mosses in the Liverpool area in July, perhaps suggesting some confusion over the identity. However, this record was widely accepted by early authors and the habitat would seem possible for cow-wheat, the nationally known larval foodplant. The only other records are listed by Ellis (1890), where he describes it as local, from Mereclough, Prestwich and Bury by J. Chappell (c.1859). Mansbridge (1940) reports no further records post-1890.
■ **HABITAT** Mosses.
■ **FOODPLANTS** Not recorded in the county. Moths noted among Common Cow-wheat.
Elsewhere in northern England, it has shown a major historical decline, as has been the case elsewhere (Randle *et al.* 2019). There have been single records in the 21st century in Cumbria and Yorkshire, cheshire-moth-charts.co.uk and yorkshiremoths.co.uk accessed in September 2023, so the species may be present in some areas where the larval foodplant occurs, such as at Birk Bank in VC60.

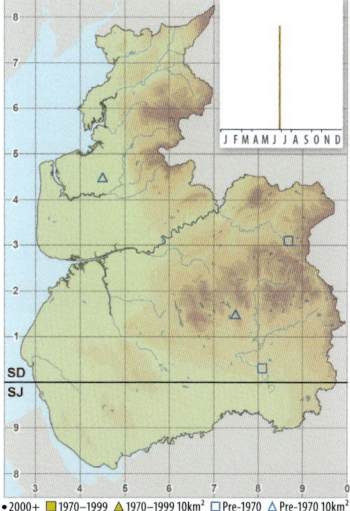

• 2000+   ☐ 1970–1999   △ 1970–1999 10km²   ☐ Pre-1970   △ Pre-1970 10km²

The Moths of Lancashire   **425**

GEOMETRIDAE

70.154 Marsh Pug *Eupithecia pygmaeata*

■ **LANCS STATUS** Occasional / local resident   c.1857–2022
First recorded at Crosby (Gregson, 1857) and later at several other sites, including Formby where it was common, up to 1940. Since then, most records have been from lowland areas, although it occasionally occurs well inland, such as recent sightings at Rowley and Medlock Valley. Its main colonies are at Flixton, where numbers have declined since 2013, and Fleetwood and Heysham, which have no regular monitoring.
■ **HABITAT** Brownfield sites, rough grassland in fields and meadows and dune grassland.
■ **FOODPLANTS** Common Mouse-ear (egg-laying observed on the flower).
Mainly single brooded, with an infrequent second brood in late summer. A day-flyer, settling on mouse-ear; 15 seen at Flixton in 1999 and 14 at Fleetwood in 2010.

● 2000+   ☐ 1970–1999   △ 1970–1999 10km²   ☐ Pre-1970   △ Pre-1970 10km²

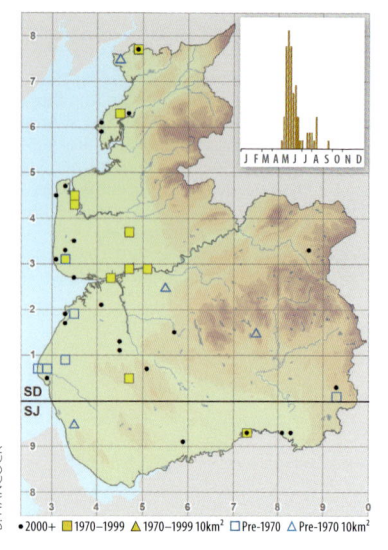

70.155 Netted Pug *Eupithecia venosata*

■ **LANCS STATUS** Occasional / local resident   c.1857–2022
First found at Hale by Mr. Nixon (Gregson, 1857). From then, to the present day, noted from Liverpool to the Silverdale area in scattered, coastal or near-coastal sites and in parts of central and southern VC59. Although regular at Leighton Moss in a Rothamsted trap in the 1970s, it is erratic in appearance at light but, where the foodplant occurs, it can be found flying at dusk (Hancock, 2018). Almost annual near Carnforth from 2013 to 2021 and recently at Yealand Conyers (both B. Hancock).
■ **HABITAT** Open grassy locations, brownfield sites and dry verges.
■ **FOODPLANTS** Bladder Campion.
Single brooded. Comes to light. Flies at dusk near the foodplant. Larvae seen in July.

● 2000+   ☐ 1970–1999   △ 1970–1999 10km²   ☐ Pre-1970   △ Pre-1970 10km²

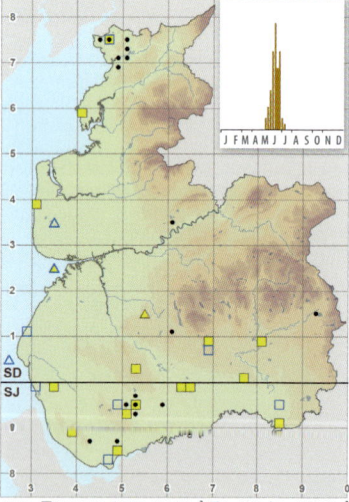

70.156 Brindled Pug *Eupithecia abbreviata*

■ **LANCS STATUS** Frequent / widespread resident   c.1887–2022
First noted in the Liverpool area (Ellis, 1890) then, as fairly plentiful in oak woods across the region (Mansbridge, 1940). Since then, noted widely in locations such as Ainsdale, Briercliffe, Claughton, Flixton, Hale, Leighton Moss, Mere Sands Wood, Longton, Preston, Royton and Silverdale. Recent removal of many hedgerow oaks during extensive suburban housing development have led to localised declines.
■ **HABITAT** Oak woodland and, to a lesser extent, mature oak standards in hedges.
■ **FOODPLANTS** Oak sp.
Single brooded. Comes to light, with high count of 166 recorded at Herring Head Wood on 2 April 2016 (P.J. Marsh). Larvae found from mid-May to June.

● 2000+   ☐ 1970–1999   △ 1970–1999 10km²   ☐ Pre-1970   △ Pre-1970 10km²

426   *The Moths of Lancashire*

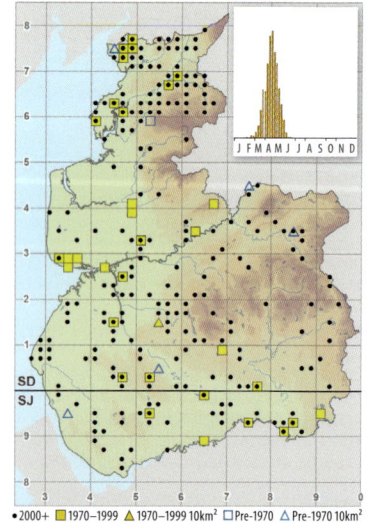

70.157 Oak-tree Pug *Eupithecia dodoneata*

■ **LANCS STATUS** Frequent / widespread resident **1985–2022**
Moving northwards nationally, it arrived in Cheshire in 1968 and was first noted in Lancashire, at Risley Moss, on 18 May 1985 (L.W. Hardwick). It was next seen in Billinge and St Annes in 1987 and Lancaster in 1990. Within a decade, it was present in many lowland areas and was moving into higher parts, arriving in Worsthorne in 2001. More recently, the range expansion has slowed in upland areas, but included reaching east to Walsden by 2005 and north to Roeburndale in 2011 (P. Marsh).
■ **HABITAT** Hedgerows, scrubland and brownfield sites.
■ **FOODPLANTS** Hawthorn.
Single brooded. Comes to light. Larva found in June.

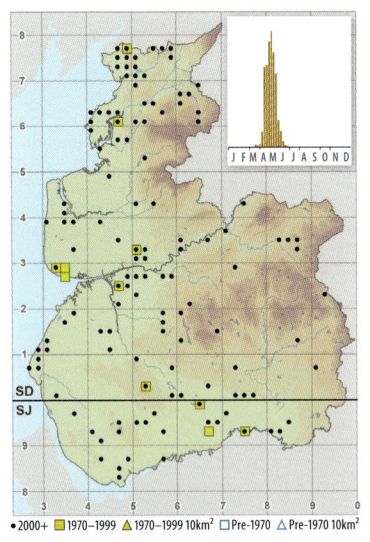

70.158 Juniper Pug *Eupithecia pusillata*

■ **LANCS STATUS** Frequent / local resident **c.1860–2022**
Found in Preston c.1860 (J.B. Hodgkinson) and then more widely in the Silverdale area, the latter to the present day. Elsewhere, seen at sites such as Didsbury in 1977, Myerscough in 1979 and Staples in 1981 and gradually becoming quite widespread. From around 2010, a steady decline set in, with only two records in the whole of VC59 during 2022. Its core area is once again the limestone area of VC60, with only a few records from other sites, such as in St Annes, Preston and Bolton-le-Sands.
■ **HABITAT** Gardens, parks and where Common Juniper grows in limestone areas.
■ **FOODPLANTS** Common Juniper. Adults found amongst cultivated garden Juniper spp.
Single brooded; one September record. Nectars on ragwort. Larva found in May.

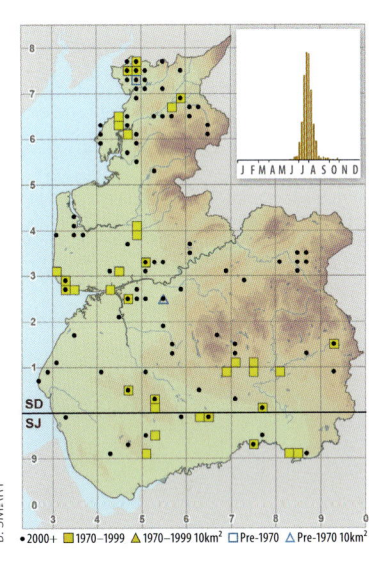

70.159 Cypress Pug *Eupithecia phoeniceata*

■ **LANCS STATUS** Rare / vagrant **2022**
A species that utilises garden cypresses, it was discovered as new to Britain in Cornwall in 1959 and has since spread gradually northwards (Randle *et al.*, 2019). It was first noted in Cheshire in 2019 and southern Yorkshire in 2022. The first, and so-far only record in Lancashire, relates to a singleton at MV light in a Briercliffe garden on 19 August 2022 (T. Lally). It is considered very likely that it will spread to other parts of the county in coming years either naturally or via the horticultural trade.
■ **HABITAT** Unknown at present.
■ **FOODPLANTS** Not recorded in Lancashire.
Single brooded nationally. Comes to light.

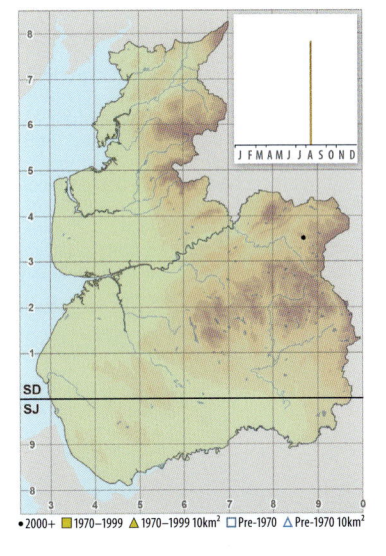

The Moths of Lancashire **427**

GEOMETRIDAE

70.160 White-spotted Pug *Eupithecia tripunctaria*

■ **LANCS STATUS** Frequent / widespread resident   c.1859–2022

First noted in Barton, Manchester by J. Chappell c.1859. Later, at Hulme in 1868, Colne in 1900, Silverdale in 1918 and Burnley and Formby pre-1940. From the 1950s to 1970s, noted at sites such as Astley Bridge, Euxton, Moss Bank, Overton and, on several occasions, at Slyne. Distribution and numbers have remained fairly constant since then, with regular sites including Flixton, Billinge, Briercliffe, Haydock, Leighton Moss, north Preston, Rishton, Royton, St Helens area and Yealand Conyers.
■ **HABITAT** Gardens, meadows, verges, scrubland and brownfield sites.
■ **FOODPLANTS** Traveller's-joy (1st brood); Angelica, Hogweed, ragwort (2nd brood).
Double brooded. Comes to light in single figures. Larvae found in July and September.

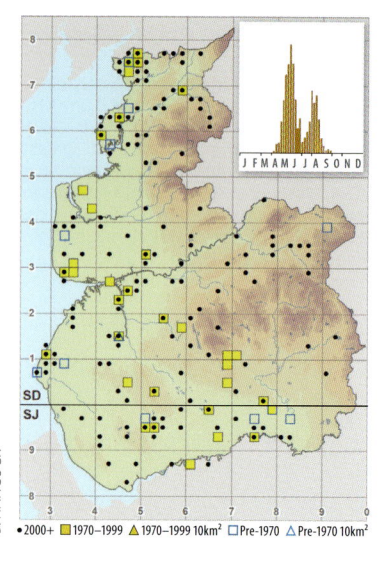

70.161 Golden-rod Pug *Eupithecia virgaureata*

■ **LANCS STATUS** Frequent / widespread resident   c.1887–2022

First noted at Crosby (W. Johnson) and Lytham, c.1887, then Liverpool in 1900 and the Silverdale area from 1918 onwards. Records in the 1970s and 1980s were mainly from Billinge, Leighton Moss and Silverdale. Indications were that it commenced a range expansion in the late 1980s, arriving at Claughton in 1990 and later becoming annual at many lowland sites. It was first noted in Wray in 2001 and Burnley in 2002.
■ **HABITAT** Rough ground, grassland, hedgerows, gardens, scrub and brownfield sites.
■ **FOODPLANTS** First brood, Hogweed. Second brood, ragwort and Goldenrod.
Double brooded; a possible small third brood on occasions in October. Comes to light in single figures and nectars on ragwort. Larvae in May, August and September.

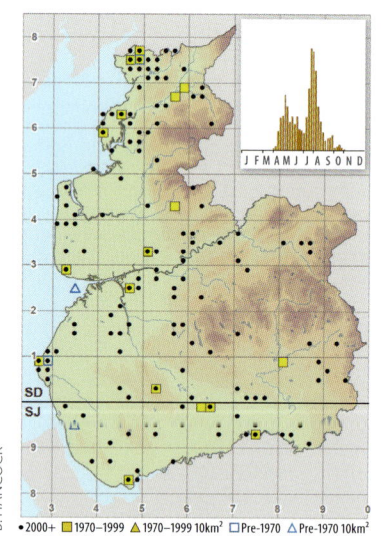

70.162 Dwarf Pug *Eupithecia tantillaria*

■ **LANCS STATUS** Occasional / local resident   1953–2019

Noted near Silverdale in 1953 by H.N. Michaelis, this as part of a gradual expansion in range nationally. Next seen at Leighton Moss in 1968, Billinge in 1987 and St Helens in 1999. Since then, sightings have been infrequent and mostly associated with conifer plantations. Docker Moor was the most regular site for a while, between 2014 and 2017, but there have only been a few records in the county since then.
■ **HABITAT** Conifer woods and plantations; occasionally well away from such areas.
■ **FOODPLANTS** Not recorded in Lancashire.
Single brooded. Comes to light, usually in low numbers, but over 40 were found at Crossgill Hairpin, Littledale on 20 May 2008 (J. Girdley and P. Marsh).

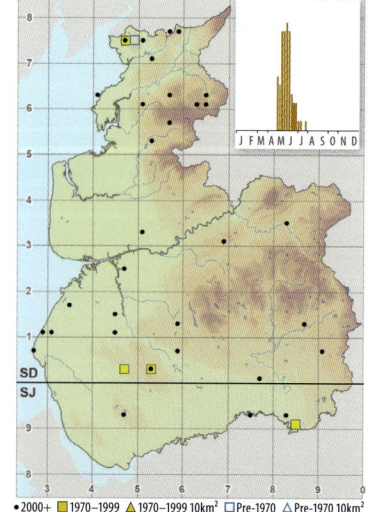

428 *The Moths of Lancashire*

70.163 Larch Pug *Eupithecia lariciata*

■ **LANCS STATUS** Occasional / local resident 1879–2022
First noted, among Larches, at Green Thorn Fell on 28 June 1879 (Hodgkinson, 1879c), then at sites such as Burnley (1910), Silverdale (1913), Pendle Hill (1920) and Sabden (1937). Not found again until 1973 in Leyland, at Gait Barrows in 1986 and Billinge in 1991; this erratic occurrence and scattered distribution has continued to the present day. Seen most regularly in Briercliffe, Docker Moor and Worsthorne.
■ **HABITAT** Areas where many Larches occur, in parks, woodland and plantations.
■ **FOODPLANTS** Larch.
Single brooded; seen once in late September. Comes to light, usually in low single figures. 35 came to light at Thrushgill, near a Larch plantation, on 25 May 2004.

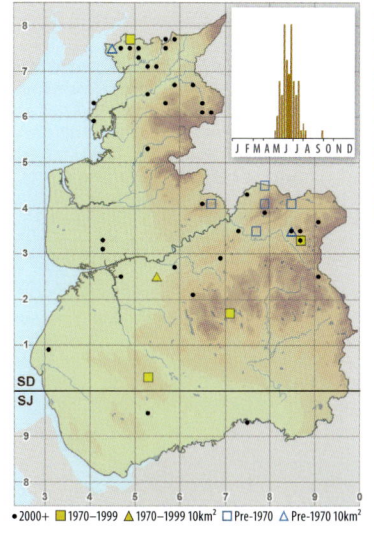

70.165 Pimpinel Pug *Eupithecia pimpinellata*

■ **LANCS STATUS** Rare / very local resident c.1889–2013
A record of it being 'not common' in the Manchester district, by J. Chappell, is listed in Ellis (1890) with no further details. The only other report was of one attracted to light at Heysham N.R. on 25 July 2013 (P.J. Marsh). This area has records of Burnet Saxifrage (Greenwood, 2012), the national larval foodplant, but the current status of the plant here is unknown (R. Neville pers. comm.).
■ **HABITAT** Open grassland on a single brownfield site.
■ **FOODPLANTS** Not recorded in Lancashire.
Single brooded. Seen once at light, a 160w Blended bulb. Possibly overlooked where the larval foodplant occurs, sparingly, on limestone grassland in north Lancashire.

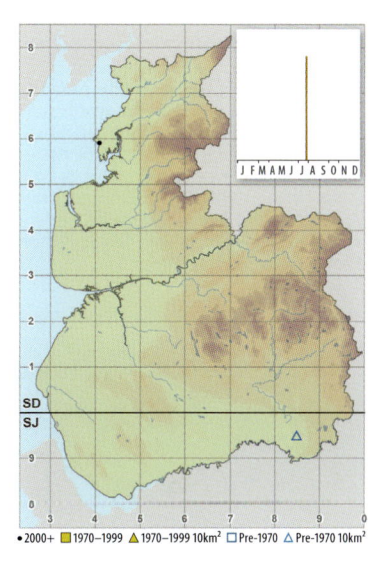

70.166 Plain Pug *Eupithecia simpliciata*

■ **LANCS STATUS** Occasional / very local resident c.1886–2022
Noted as present in the Preston and Longridge areas by J.B. Hodgkinson c.1886 (Ellis, 1890), and by W.P. Stocks at St Annes in 1913. Since, then there have been several scattered reports, including from Blackpool in 1952, Didsbury in 1958 and Warton Bank, nr. Lytham in 1982 and 1984. Noted occasionally at some sites during the early 2000s, such as Flixton, Parr and Pennington. Since 2016, when it was found at Bolton-le-Sands (S. Garland), it has only otherwise been seen in Flixton (K. McCabe).
■ **HABITAT** Saltmarsh edges, waste ground and arable field margins.
■ **FOODPLANTS** Orache sp.; Spear-leaved Orache noted at one site where an adult found.
Single brooded. Comes to light. Larva found at Fairhaven in 1987 (D. & J. Steeden).

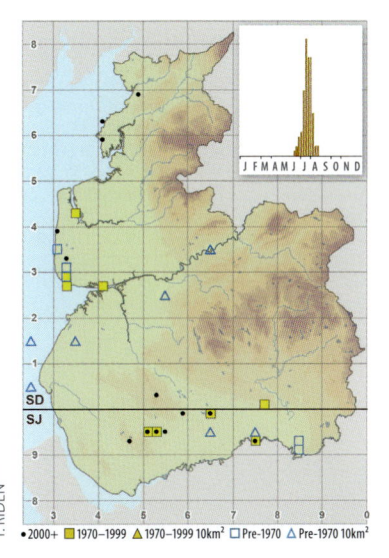

The Moths of Lancashire **429**

GEOMETRIDAE

70.168 Narrow-winged Pug *Eupithecia nanata*

■ **LANCS STATUS** Frequent / widespread resident   c.1843–2022

First noted in the Preston area (Hodgkinson, 1844), the Liverpool district in c.1854, Chat Moss c.1859 and Oldham 1900, all early authors reporting it as common on heaths and mosses throughout. This has remained the case to the present day, but since the 1970s it appears to have started utilising garden planted heathers, well away from the naturally occurring plants, such as in Billinge, Flixton and Fulwood.

■ **HABITAT** Mosses, moorland, lowland dune heath; sometimes gardens.
■ **FOODPLANTS** Heather.

Double brooded, probably overlapping. Comes to light and easily disturbed by day. 75 attracted to light at Leck Fell on 9 July 2005. Larvae found in June and September.

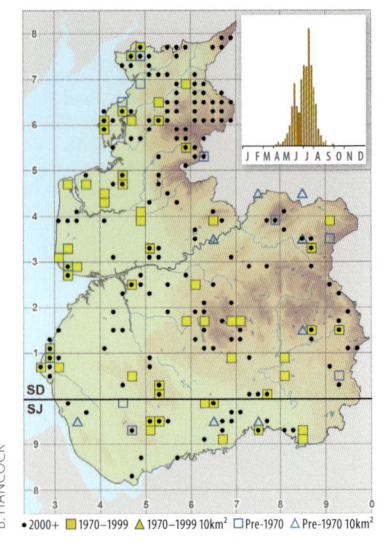

• 2000+   ▪ 1970–1999   ▲ 1970–1999 10km²   ☐ Pre-1970   △ Pre-1970 10km²

70.169 Angle-barrred Pug *Eupithecia innotata*

■ **LANCS STATUS** Occasional / widespread resident   c.1853–2022

Formerly known as Ash Pug, it was first noted near Warrington by N. Cooke (Byerley, 1854). Thereafter, to 1900, found in Prestwich, Rusholme, Stretford, Withington, Preston, Lytham and Burnley. Since then, reports to the present day have been widespread but erratic in occurrence. Sites with more frequent sightings include Salford in the 1960s (R. Leverton), St. Annes from 1991 to 2002 (D. & J. Steeden), Morecambe in 2014 (J. Patton) and Green Bank in 2020 and 2021 (P.J. Marsh).

■ **HABITAT** Coastal areas with Sea Buckthorn and wooded or scrubby areas with Ash.
■ **FOODPLANTS** Sea Buckthorn.

Single brooded, possibly a small second brood in August. Comes occasionally to light.

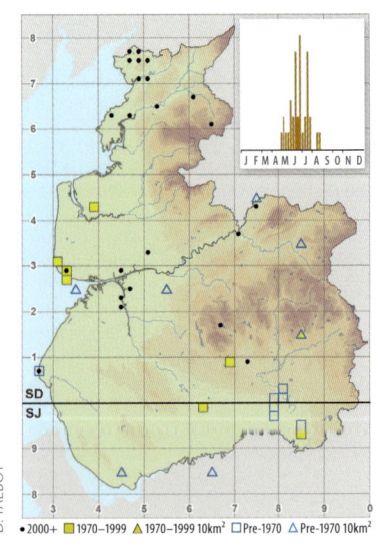

• 2000+   ▪ 1970–1999   ▲ 1970–1999 10km²   ☐ Pre-1970   △ Pre-1970 10km²

70.171 Ochreous Pug *Eupithecia indigata*

■ **LANCS STATUS** Occasional / widespread resident   1887–2022

First noted on Chat Moss in 1887 (unknown), on Rixton Moss in 1900 (W. Mansfield), Knowsley Park in 1910 (J. Cotton), Silverdale area in 1920 (W.P. Stocks) and Formby in 1923 (W. Mansbridge). Since then, found at scattered sites across the county, predominantly in lowland parts of VC59. Sites with more regular records include Ainsdale, Flixton, Formby, Hardy Grove, Longton, Mere Sands Wood and Southport.

■ **HABITAT** Areas with pines on sand dunes, mosses, moors, woods, gardens and parks.
■ **FOODPLANTS** Not recorded in Lancashire.

Single brooded. Comes to light in single figures, with the largest count of nine to MV light at Hesketh Bank on 29 June 2019 (R. Yates).

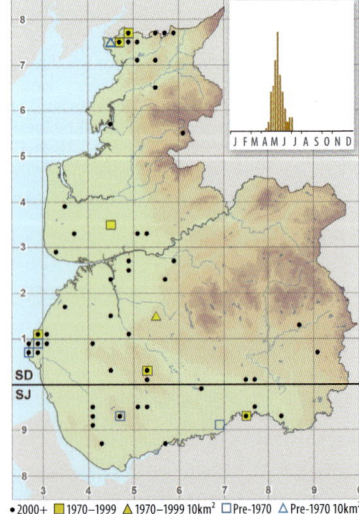

• 2000+   ▪ 1970–1999   ▲ 1970–1999 10km²   ☐ Pre-1970   △ Pre-1970 10km²

430   *The Moths of Lancashire*

70.172 Thyme Pug *Eupithecia distinctaria*

■ **LANCS STATUS** Scarce / very local resident   1887–2022

Reported from the Silverdale area by J.B. Hodgkinson in 1887 and near Carnforth on 11 June 1897 by H. Murray. It was first found at Gait Barrows in 1922, with further records from there to 2008. All subsequent records have been on or near the northern limestone, including on Warton Crag, Carnforth slag heap, Yealand Conyers, Cote Stones and Crag House Allotment. Found in a Silverdale garden on 11 July 2022.
■ **HABITAT** Limestone grassland; a coastal brownfield site with Wild Thyme present.
■ **FOODPLANTS** Not recorded in Lancashire. Adults seen amongst Wild Thyme.
Single brooded. Comes to light. Twelve were found resting on Wild Thyme plants at Cote Stones on 19 June 2017 during a dusk torchlight search (B. Hancock).

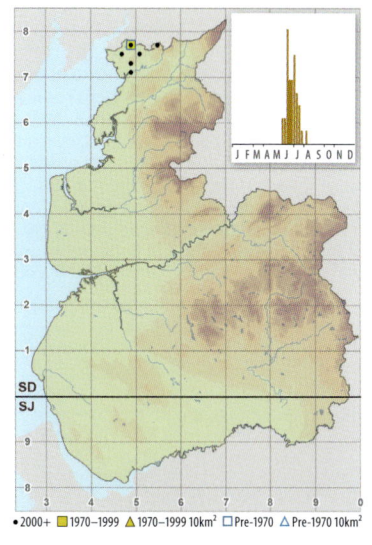

70.173 Lime-speck Pug *Eupithecia centaureata*

■ **LANCS STATUS** Frequent / widespread resident   c.1857–2022

Found in Liverpool (Gregson, 1857), St Annes in 1908, Southport in 1938, and later noted as fairly common on sandhills (Mansbridge, 1940). Became more frequent inland at sites such as Didsbury in 1940, Moss Bank in 1947, Wrightington in 1949, Salford in 1960, Bury in 1978 and Burnley in 1979. Since then, although regular in many coastal areas, and within lowland parts of VC59, it can be more variable in occurrence in inland VC60 and is infrequent in upland areas.
■ **HABITAT** Rough and waste ground, meadows, dune grassland and brownfield sites.
■ **FOODPLANTS** Ragwort, Sea Aster, Sea Holly and Sea Lavender, feeding on the flowers.
Single brooded. Comes to light. Larvae found late August to September.

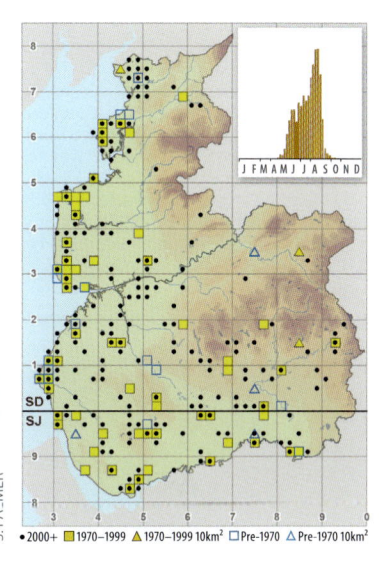

70.175 Triple-spotted Pug *Eupithecia trisignaria*

■ **LANCS STATUS** Scarce / very local resident   c.1887–2021

Reported near Preston c.1887 (J.B. Hodgkinson), with Mansbridge (1940) noting it to be scarce and local. Not seen again until 30 July 2008 in north Preston (S. Palmer), at a time when angelica was growing nearby. It was then noted twice in 2012, at Morecambe and Lord's Lot Wood, the latter site also producing records in 2014, at Hoghton in 2015, Longton (twice) in 2021 and Halebank on 3 July 2021 (D. Hunt).
■ **HABITAT** Damp verges, tracks and ditches; occasionally in gardens.
■ **FOODPLANTS** Wild Angelica.
Single brooded. Comes to light as singletons; nectars at dusk on Wild Angelica. Four larvae found on 4 September 2014 at Lord's Lot Wood (B. Hancock).

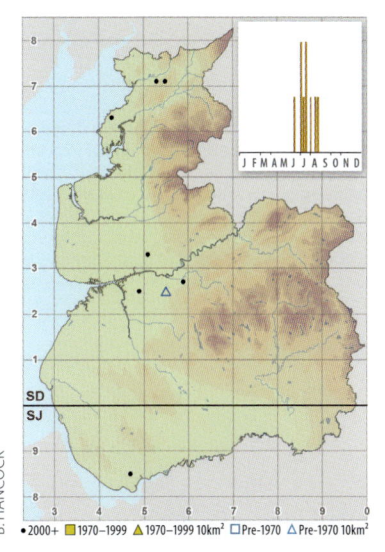

The Moths of Lancashire   **431**

GEOMETRIDAE

70.176 Freyer's Pug *Eupithecia intricata*

■ **LANCS STATUS** Frequent / widespread resident 1993–2022

The southern form of this moth has spread northwards over recent decades and was first noted in Lancashire in 1993 at Swinton (S. Christmas). It was then found in Longton on 12 July 1994, Flixton on 24 May 1995, arriving in the east, at Littleborough, in 1998. It was first noted in VC60 at St Annes on 18 June 2000 and since then, has occurred in most garden light traps and occasionally elsewhere.

■ **HABITAT** Gardens, parks and other amenity planted areas.

■ **FOODPLANTS** Not recorded in Lancashire.

Single brooded. Comes to light, occasionally in double figures. The largest count was of 26 to light in Royton on 7 June 2004 (R. Hart).

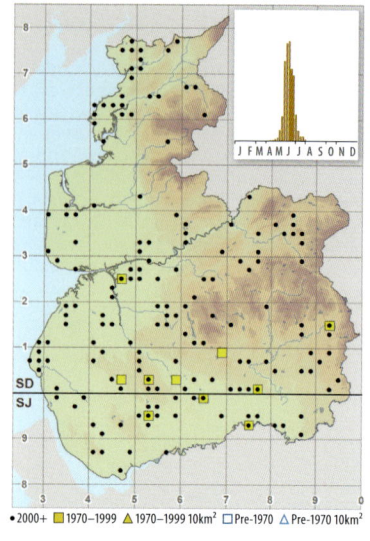

• 2000+   □ 1970–1999   △ 1970–1999 10km²   □ Pre-1970   △ Pre-1970 10km²

70.177 Satyr Pug *Eupithecia satyrata*

■ **LANCS STATUS** Scarce / very local resident c.1857–2022

First seen on Simonswood Moss (Gregson, 1857), then in the Manchester district c.1859 and Ainsdale in 1912. A record from the Silverdale area c.1887 was accepted by Ellis (1890) but was very unusual. It was next noted in the Formby area on 14 May 1950 (H.N. Michaelis) and at Freshfield Dune Heath on 4 June 2004 (G. Jones). Since then, only found on an area of dry upland heath adjacent to open heather moors, at Docker Moor (P.J. Marsh *et al.*) from 2020 to 2022, with eight seen on 29 June 2021.

■ **HABITAT** Dry Heather moorland, dune heath and formerly, mosses.

■ **FOODPLANTS** Not recorded in Lancashire; larvae have accepted Heather in captivity.

Single brooded. Comes to light. In Scotland, readily seen flying by day over Heather.

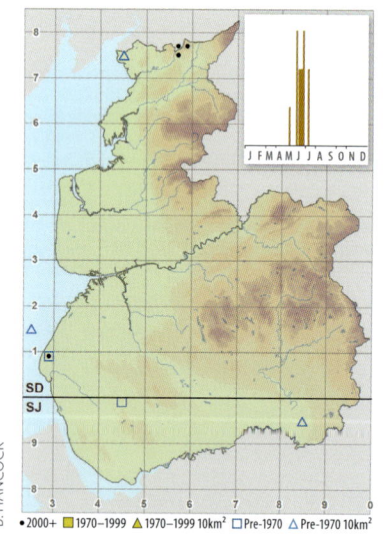

• 2000+   □ 1970–1999   △ 1970–1999 10km²   □ Pre-1970   △ Pre-1970 10km²

70.179 Wormwood/Ling Pug *Eupithecia absinthiata*

■ **LANCS STATUS** Frequent / widespread resident. c.1857–2022

Includes Ling Pug, noted historically as a separate species (e.g., Mansbridge, 1940), feeding commonly on Heather on the mosses; this is currently considered a form of Wormwood. The nominate form of Wormwood was first noted in Liverpool (Gregson, 1857), then at St Annes, Southport, Withington and Silverdale. Since the 1930s, it has been reported widely across lowland parts, although less frequently in upland areas.

■ **HABITAT** Rough ground, dunes, mosses, moors, gardens and brownfield sites.

■ **FOODPLANTS** Flowers of Common Ragwort, Yarrow, Mugwort, Oxeye Daisy, Heather.

Single brooded. Comes to light. May and early June specimens should be carefully checked to exclude Currant Pug. Larvae have been found from August to September.

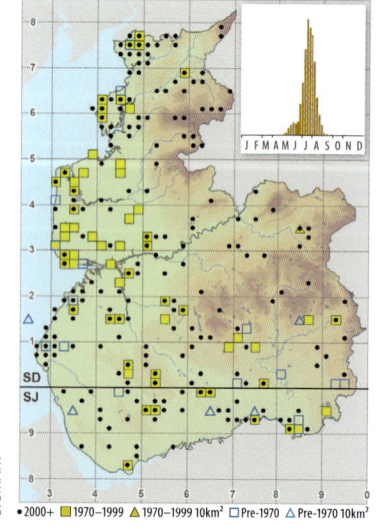

• 2000+   □ 1970–1999   △ 1970–1999 10km²   □ Pre-1970   △ Pre-1970 10km²

70.180 Bleached Pug *Eupithecia expallidata*

- **LANCS STATUS** Rare / very local resident   1896–2020

First noted in Carnforth area in August 1896 and 1898 (H. Murray). Next, recorded in 1905 at Silverdale, but not seen again until 1986, at Gait Barrows (I. Rutherford). Since then, Gait Barrows has remained its main site, although it has also been reported from The Row, Silverdale on 28 July 2005 (R. Griffiths) and at Kay's Nursery, Silverdale on 9 August 2009 (M. Elsworth). A specimen from Formby Point on 13 August 2013 (not mapped) has not proved possible to confirm prior to publishing.
- **HABITAT** Limestone grassland.
- **FOODPLANTS** Goldenrod.

Single brooded. Comes to light and can be netted at dusk. Larvae, Sept. to October.

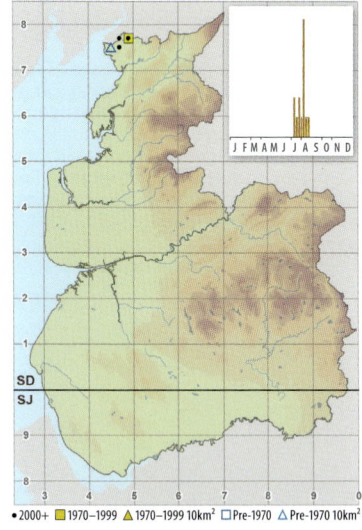

70.181 Valerian Pug *Eupithecia valerianata*

- **LANCS STATUS** Occasional / local resident   1865–2021

Reported from Pilling Moss on 25 June 1865 (Hodgkinson, 1865) and then at Bewsey, Warrington, in 1904 (J. Collins). Not found again until 1948 in Formby and in 1953 near Silverdale (B.B. Snell). It was noted at Martin Mere, at light, in 2004 and, in the following year, larvae were located there in July and bred (A. Bunting). Since then, seen at six scattered sites including Littleborough, Docker Moor and Cinderbarrow, the largest count being nine at the latter site at dusk, on 14 June 2018 (B. Hancock).
- **HABITAT** Wetland, damp hedgerows, ditches and canalside.
- **FOODPLANTS** Common Valerian.

Single brooded. Comes to light; can be netted amongst the foodplant at dusk.

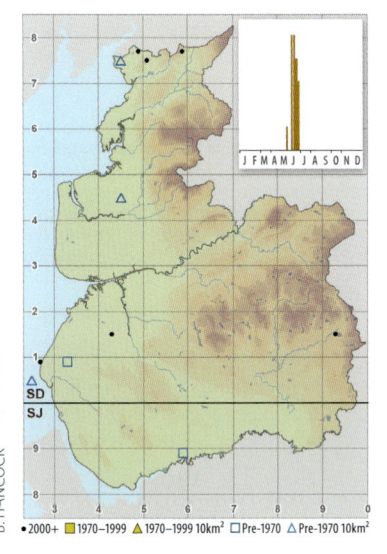

70.182 Currant Pug *Eupithecia assimilata*

- **LANCS STATUS** Frequent / widespread resident   c.1857–2022

Reported from the Liverpool district (Gregson, 1857) and then from sites such as Eccles in 1859, Dutton in 1879, Preston in 1887, St Annes in 1891, Pendle in 1940 and Nelson in 1961. Thereafter, noted widely across the county, particularly from garden sites, such as in Claughton, Flixton, Formby, Morecambe, Billinge, Parr, Briercliffe, Hesketh Bank, Swinton, Bay Horse and Yealand Conyers. The largest count was of 13 attracted to light in a Formby garden on 28 May 2011 (R. Walker).
- **HABITAT** Gardens, allotments and hedgerows.
- **FOODPLANTS** Black Currant. Elsewhere, frequently found on Hop.

Double brooded. Comes to light. Larvae found in September.

The Moths of Lancashire

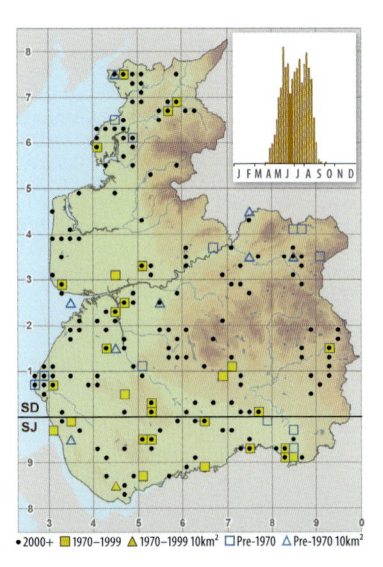

GEOMETRIDAE

70.183 Common Pug *Eupithecia vulgata*

■ **LANCS STATUS** Frequent / widespread resident   c.1857–2022
First noted in the Liverpool district (Gregson, 1857), followed by records from sites such as Oldham, Burnley, Chadderton, Lytham and Didsbury, up to 1940. All early authors regarded it as abundant everywhere, including Mansbridge (1940). This wide distribution has been maintained to the present day. Many sites report it annually, such as Billinge, Lancaster, Poulton-le-Fylde, Longton, St Helens, Wilpshire, Worsthorne, Wray and Yealand Conyers, but it is now rarely abundant.
■ **HABITAT** In a wide range of habitats in both lowland and upland areas.
■ **FOODPLANTS** Not recorded in Lancashire.
Single brooded. Comes to light. Largest recent count, 39 at Heysham on 1 July 2003.

● 2000+   ▢ 1970–1999   ▲ 1970–1999 10km²   ▢ Pre-1970   △ Pre-1970 10km²

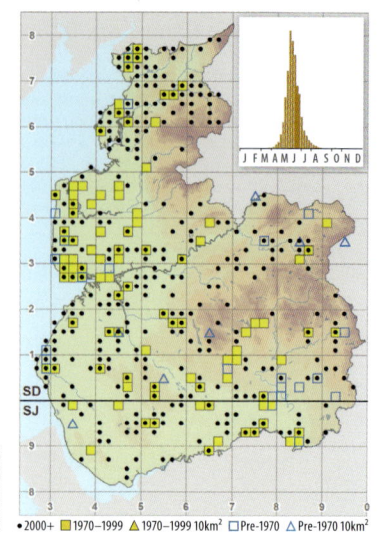

70.184 Mottled Pug *Eupithecia exiguata*

■ **LANCS STATUS** Frequent / widespread resident   c.1857–2022
First noted in the Liverpool district as abundant in woods (Gregson, 1857), although Mansbridge (1940) later commented on a lack of Lancashire records since Ellis (1890). A few were noted in the Silverdale area during the 1950s to '80s but it was otherwise only reported as singletons in that period from Formby, Didsbury, Hoghton, Bolton, Leigh, Pennington and Myerscough. Since 1995, however, its range has increased considerably, with reports from many lowland and a few upland sites.
■ **HABITAT** Woodland, gardens, hedgerows, scrub, parks and brownfield sites.
■ **FOODPLANTS** Hawthorn, Beech, Grey Willow, Blackthorn and Snowberry.
Single brooded. Comes to light, mostly in single figures. Larvae mid-Aug. to mid-Oct.

● 2000+   ▢ 1970–1999   ▲ 1970–1999 10km²   ▢ Pre-1970   △ Pre-1970 10km²

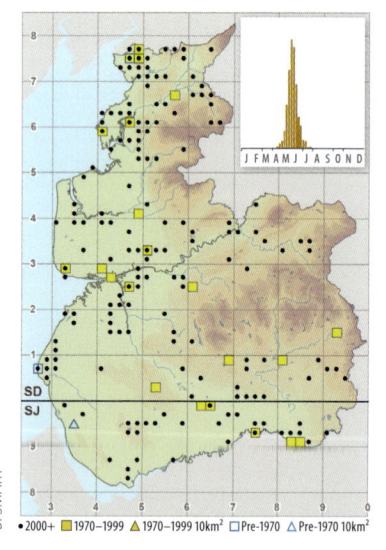

70.187 Tawny-speckled Pug *Eupithecia icterata*

■ **LANCS STATUS** Occasional / local resident   c.1859–2022
First noted in the Irlam area by J. Chappell c.1859 and later, at Cleveleys (Hodgkinson, 1872a), followed by a well scattered range of sites in mainly lowland areas, up to the 21st century. From 2000, its range started contracting, with several upland locations noted until c.2005. Since then, most records have been near-coastal, with a few sites further inland and, on higher ground, at Tarnbrook, Dalton Crag and Green Bank.
■ **HABITAT** Coasts, limestone grassland, brownfield sites and, occasionally, moorland.
■ **FOODPLANTS** Considered highly likely to feed on Yarrow; larvae found but not bred.
Single brooded. Comes to light, usually in single figures, but 25 were noted at Formby Point on 30 August 2005 (M. Hammond). Nectars on ragwort.

● 2000+   ▢ 1970–1999   ▲ 1970–1999 10km²   ▢ Pre-1970   △ Pre-1970 10km²

434   *The Moths of Lancashire*

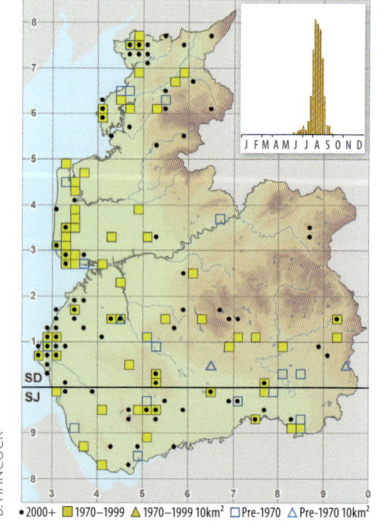

70.188 Bordered Pug *Eupithecia succenturiata*

- **LANCS STATUS** Occasional / local resident   c.1857–2022

Noted in Liverpool (Gregson, 1857), at Fleetwood in 1872, Lytham in 1887 and Southport and Formby in 1900. Predominantly a lowland species, which is quite widespread in VC59, mainly coastal in VC60, but generally rather erratic in occurrence. Seen mostly in single figures. Recorded almost annually in Billinge, Flixton and Parr, but much less frequently at sites such as Crosby, Formby, Hale, Heysham N.R., Leigh, Mere Sands Wood, Orrell, Royton, Southport and St Annes.
- **HABITAT** Rough, disturbed and arable land, coastal areas and brownfield sites.
- **FOODPLANTS** Mugwort.

Single brooded. Comes to light. Larvae, often parasitised, found Sept. and Oct.

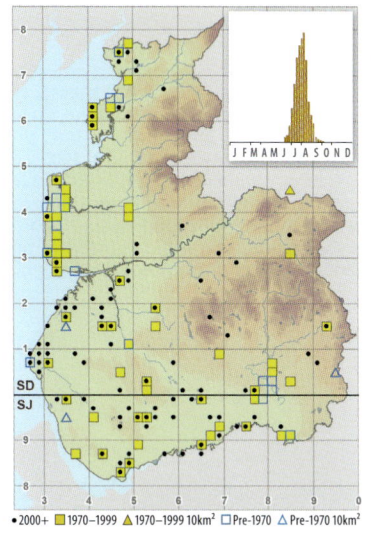

• 2000+ ☐ 1970–1999 △ 1970–1999 10km² ☐ Pre-1970 △ Pre-1970 10km²

70.189 Shaded Pug *Eupithecia subumbrata*

- **LANCS STATUS** Occasional / very local resident   c.1913–2022

First noted in the Silverdale area by W.P. Stocks c.1913, at St Annes in 1944, Formby in 1968 and, on 15 July 1969, at Bolton-le-Sands (C.J. Goodall). Since then, it has occurred fairly regularly at Ainsdale and almost annually at Heysham N.R. Other sites with far fewer, and sometimes single, records have included Formby, Middleton N.R., Silverdale, Southport, Sunderland Point and Yealand Conyers.
- **HABITAT** Coastal grassland and brownfield sites in near-coastal areas.
- **FOODPLANTS** Common Knapweed.

Single brooded. Comes to light, mostly in single figures, although 20 were seen on Green Beach, Birkdale on 23 June 2022 (J. Girdley). Larvae, late July and August.

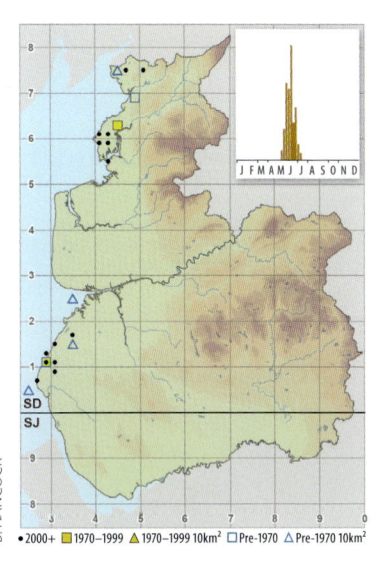

• 2000+ ☐ 1970–1999 △ 1970–1999 10km² ☐ Pre-1970 △ Pre-1970 10km²

70.190 Grey Pug *Eupithecia subfuscata*

- **LANCS STATUS** Frequent / widespread resident   c.1857–2022

Noted in the Liverpool area (Gregson, 1857), Simonswood in 1915 and Eccleston Mere in 1917. Mansbridge (1940) reported it common in woods, including melanics. Since then, it has been noted regularly from across the county, usually in single figures. Sites with long-term almost annual sightings include Billinge, Hardy Grove, Leigh, Orrell, Pennington, Preston, St Helens area, Worsthorne and Yealand Conyers.
- **HABITAT** Wide range of habitats throughout the county.
- **FOODPLANTS** Common Ragwort, Common Fleabane, Hemp-agrimony, Bramble, Ground-elder, melilot sp. and hawkweed sp.

Single brooded. Comes to light. Larvae found from mid-August to September.

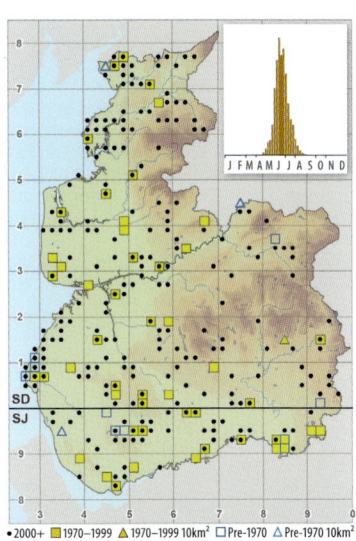

• 2000+ ☐ 1970–1999 △ 1970–1999 10km² ☐ Pre-1970 △ Pre-1970 10km²

The Moths of Lancashire

GEOMETRIDAE

70.191 Manchester Treble-bar *Carsia sororiata*

■ **LANCS STATUS** Occasional / local resident 1845–2021

First noted at Pilling on 21 July 1845 by W. Hodgkinson, followed by the lowland raised bogs at Rixton, Simonswood, Cadishead and Chat Mosses. Apart from the Pilling colony, where last seen in 1987, all other lowland colonies were lost by 1934. In the uplands, seen nr. Bolton in 1900, Longridge Fell in 1955, where still present, and other moorland mires in east VC60, including strong colonies at Holdron and Loftshaw Mosses. Seen at Docker Moor in 2021. Probably overlooked in some areas.
■ **HABITAT** Damp moorland mires and, formerly, lowland raised bogs.
■ **FOODPLANTS** Larva once swept from Cranberry.
Single brooded. Mostly recorded in flight by day; occasionally comes to light.

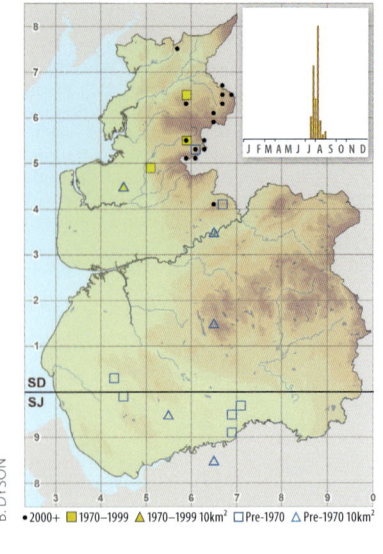

70.192 Treble-bar *Aplocera plagiata*

■ **LANCS STATUS** Occasional / widespread resident c.1857–2022

Noted in Childwall (Gregson, 1857) and several other sites, including Didsbury, up to 1940. Since then, reports have been mainly from lowland river-valley sites and northern coastal areas. Sites with records vary considerably from year to year, but Carnforth, Flixton and Heysham N.R. have been most regular in recent decades. Of these, Heysham is the most prolific, producing nightly counts in low double figures, with peak counts reaching 52 on 3 September 2015 (P.J. Marsh *et al.*).
■ **HABITAT** Disturbed or waste ground, rough grassland and brownfield sites.
■ **FOODPLANTS** St. John's-wort sp.
Double brooded. Comes to light. 10 larvae were found at Wesham on 19 July 1984.

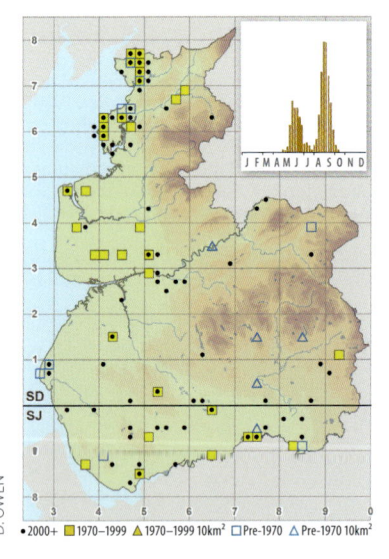

70.195 Streak *Chesias legatella*

■ **LANCS STATUS** Scarce / local resident c.1859–2022

Noted in Bury c.1859 (J. Chappell) and then, Preston, Manchester, Crosby, Carnforth and Gathurst. Mansbridge (1940) reported it local among Broom. Regular in Billinge from 1986 to 2008 (C. Darbyshire) and Flixton from 1994 to 2014 (K. McCabe). Elsewhere, occasional and erratic, with most records in VC59. Significant decline has occurred since 2014, with only seven noted from seven sites throughout the county.
■ **HABITAT** Rough grassland, sand dunes, brownfield sites and edges of open woodland.
■ **FOODPLANTS** Broom.
Single brooded. Comes to light in small numbers; it may wander. Larvae found in May. Eleven were attracted to light in Leigh on 18 October 1987 (J.D. Wilson).

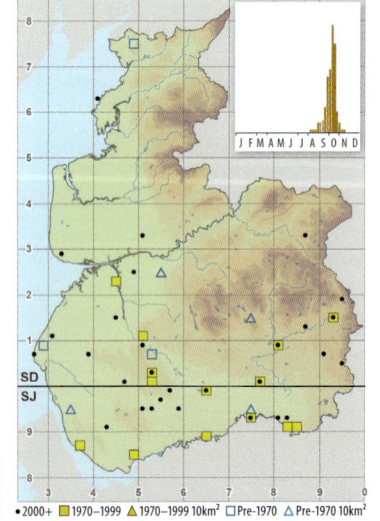

436 *The Moths of Lancashire*

70.198 Seraphim *Lobophora halterata*

■ **LANCS STATUS** Occasional / widespread resident 1972–2022
First noted in Fulwood Park, Liverpool in 1972 (I.D. Wallace), then at Speke and Didsbury in 1977. It appears to have spread into the county from the south although earlier records from further north are known (Randle *et al.*, 2019). During the 1980s, its spread was slow and mainly involved central VC59. By 1992 it arrived in St Annes, followed by Preston in 1995, Bolton in 1999, Formby in 2000, Lancaster in 2001, Silverdale and Littleborough in 2002 and, in the north-east, at Perry Moor in 2004.
■ **HABITAT** Sand dunes, gardens, parks, planted shelterbelts and woodland.
■ **FOODPLANTS** Not recorded in Lancashire, but adults are regular close to poplars. Single brooded. Comes to light, mostly in single figures.

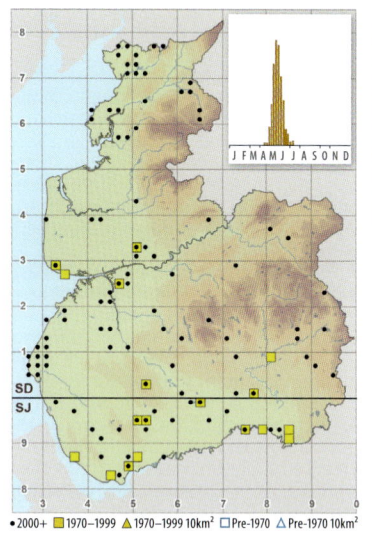

70.199 Small Seraphim *Pterapherapteryx sexalata*

■ **LANCS STATUS** Rare / very local resident c.1857–2023
Recorded once, historically, from Croxteth Woods, Liverpool 'in May' (Gregson, 1857). Neither the year nor the recorder (which could have been Gregson) were mentioned in the text. During the later stages of work on this book, details arrived of one attracted to light at Hesketh Bank, Southport on 14 June 2023 (R. Yates). This suggests it is probably still resident in very small numbers in the county.
■ **HABITAT** Unknown in the county.
■ **FOODPLANTS** Not recorded in Lancashire.
Searches of sallows in other damp locations would be worth considering. Also rarely noted in Cheshire and South Cumbria during the twentieth century.

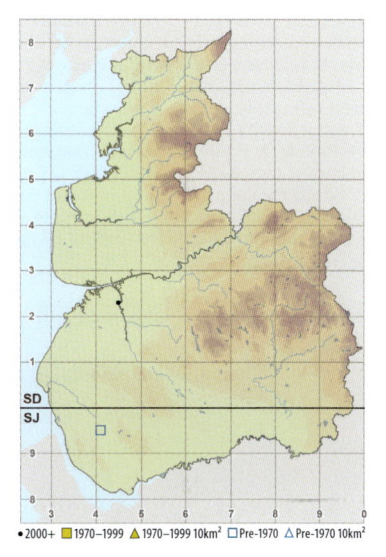

70.200 Yellow-barred Brindle *Acasis viretata*

■ **LANCS STATUS** Frequent / widespread resident c.1910–2022
First noted in the Silverdale area from 1910 to 1922 by A.E. Wright and W.P. Stocks, and then in Freshfield and Formby from 1931 to 1947. It was next reported from Morecambe in 1960 and Liverpool in 1972. Since then, the species has spread widely across the county, particularly in lowland areas. It has also increased in numbers being reported, becoming a regular and annual moth in many garden light-traps.
■ **HABITAT** Hedgerows, gardens, parks and woodland.
■ **FOODPLANTS** Ivy flowers, Holly flowers and immature berries; also Hawthorn. Double brooded. Comes to light. 21 noted at Torrisholme on 21 May 2018 (A. Draper). Larvae have been found in June and September.

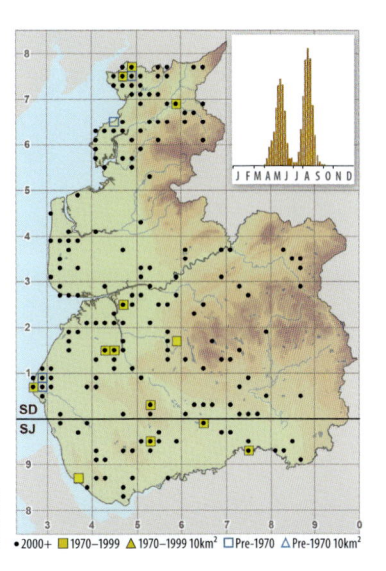

The Moths of Lancashire **437**

GEOMETRIDAE

70.201 Barred Tooth-striped *Trichopteryx polycommata*

■ **LANCS STATUS** Frequent / very local resident　　　　　　1882–2022

Although restricted to the habitats of the Morecambe Bay limestone area in the north-west, it also occurs in scattered locations across England, Wales and Scotland (Randle *et al.*, 2019). The Arnside and Silverdale AONB is the main stronghold for this species in Lancashire and it was first noted from the Carnforth area in 1882 (specimen in NHM, London). From 1900 to the early 1990s, adults were rarely encountered in any significant numbers, although it was noted at Warton Crag (1959 and 1965), Silverdale marshes (1961), Leighton Moss (1971–80) and Gait Barrows (1984).

As light trapping became more widespread in the 2000s, records increased, and it was found at Hutton Roof around the same time. In recent years, targeted light trapping has identified significant populations at Gait Barrows and Warton Crag, with up to 71 individuals recorded from a single trap in 2017 (J. Patton).

The adults are nocturnal and are readily attracted to light, sometimes in large numbers. For many years, it was thought that the larvae fed on Wild Privet in Lancashire as they do in the south of their range and so early searches were limited to this foodplant, although it is not an abundant plant in Lancashire. Scottish populations are known to utilise Ash. This, and the frequent observations of freshly emerged adults on the lower trunks of Ash in well-established, mixed deciduous woods between 2014 and 2017 (J. Patton), changed the direction of the larval searches.

An organised search at Warton Crag in June 2018 yielded the first larval record for Lancashire, with a single larva being found feeding on Ash regrowth on the limestone pavement (C. Palmer). Larvae can be located shortly after leaves unfurl from the buds in June, where the distinctive, semi-circular feeding cut-outs can be found, along with frass deposits on the surface of the leaves. Individuals have since been found at Thrang End and Gait Barrows.

A pheromone lure to attract males was developed at Canterbury University in 2017 and, after initial trials, was used in a mark-release-recapture study, predominantly in Lancashire (Clay & Wain, 2019). Population estimates were calculated from recapture data for Sharp's Lot, Silverdale, revealing a healthy population estimate of 240 males. Pheromone lures also proved an effective method for surveying the wider area, gaining records from nine new tetrads and impressive numbers were frequently attracted at known sites, with a maximum of 125 males at Warton Crag (18 Apr 2018).

Whilst the Lancashire population of Barred Tooth-striped appears strong, Ash Dieback *Hymenoscyphus fraxineus* was confirmed in the Arnside and Silverdale AONB in 2012 and it remains unclear as to how the moth will respond as extensive numbers of trees die. Monitoring of this population will be crucial in the coming years.

Justine Patton

70.202 Early Tooth-striped *Trichopteryx carpinata*

■ **LANCS STATUS** Frequent / widespread resident   c.1857–2022
Reported from Simonswood Moss (Gregson, 1857), Atherton in 1859 and Silverdale and Quernmore in 1910. By the 1960s it was regular at Leighton Moss, at Claughton during the late 1980s and at Flixton from the 1990s to the present day. Elsewhere, it has since been noted at many new sites, including Ainsdale, Bolton, Briercliffe, Herring Head Wood, Longridge, Mere Sands Wood, Mill Houses and Warton Crag. 136 came to light at Docker Moor on 18 April 2018 (P.J. Marsh).
■ **HABITAT** Woods, mosses, moorland edge, sand dune slacks and scrubland.
■ **FOODPLANTS** Creeping Willow and Goat Willow (in coastal areas).
Single brooded. Comes to light. Larvae found in June.

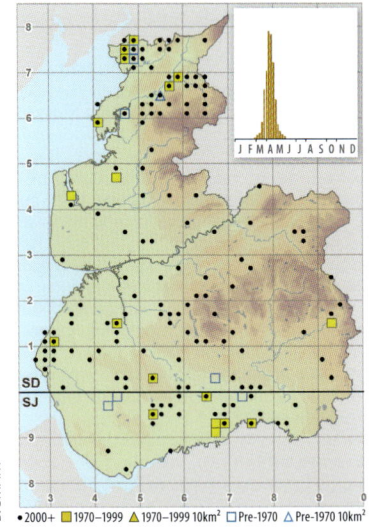

70.203 Orange Underwing *Archiearis parthenias*

■ **LANCS STATUS** Occasional / widespread resident   1849–2022
First noted at Woolton on 1 May 1849 (C.S. Gregson), with Mansbridge (1940) reporting this the only county record. Next seen at Warton (Carnforth) in 1961 (H.S. Robinson), White Moss (Yealand) in 1963, Warton Crag in 1965 and Leighton Moss in 1968. Since then, noted quite regularly at sites such as Ainsdale, Bold Moss, Gait Barrows, Heysham Moss, Mere Sands Wood, Rainford, Red Moss and Warton Crag.
■ **HABITAT** Mosses, moorland edge, birch woodland and scrub.
■ **FOODPLANTS** Birch sp.
Single brooded. Flies around birch trees in sunny conditions and attracted to sallow blossom. 50 seen at Knots Wood on 4 April 1996 (P.J. Marsh). Larva found mid-May.

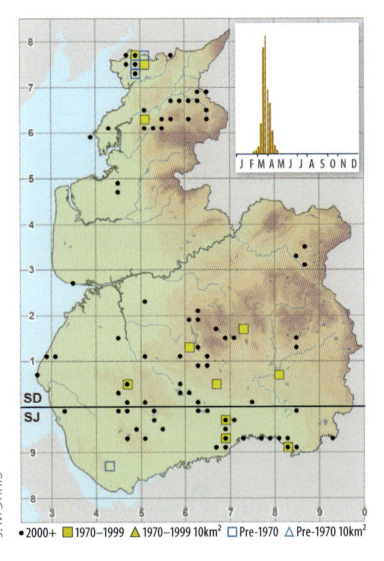

70.205 Magpie *Abraxas grossulariata*

■ **LANCS STATUS** Occasional / widespread resident   c.1857–2022
Found in Liverpool (Gregson, 1857) and later, many other parts of the county. Noted as common in old gardens (Mansbridge, 1940). In Salford, R. Leverton reported it as common in gardens in 1955, but absent from the nearby wider countryside. A decline in sightings from lowland areas took place during the 1980s and '90s. The situation has since stabilised, with a meagre increase occurring in the last two decades.
■ **HABITAT** Gardens, allotments and parks; rarely on heather moorland.
■ **FOODPLANTS** Blackcurrant, Flowering Currant, Quince, Gooseberry, Japanese Spindle.
Single brooded. Comes to light, usually in single figures. 40 were at Littleborough on 16 June 1999 (I. Kimber). Larvae found from September to early June.

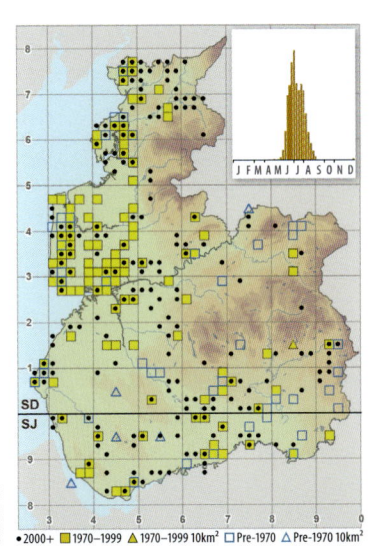

The Moths of Lancashire

GEOMETRIDAE

70.206 Clouded Magpie *Abraxas sylvata*

■ **LANCS STATUS** Scarce / local resident   1843–2023

First noted in the Preston area in 1843 (J.B. Hodgkinson) then, at Pendlebury Wood 16 June 1844 (R.S. Edleston), Chat Moss in 1846, Bolton-le-Moors area in 1850, Croxteth c.1857 and Hoghton Woods in 1865. Thereafter, over the next century, found at many scattered sites, including Blackpool, Burnley, Formby, Oldham, Longridge, Preston, Rochdale and Silverdale. The number of reports declined considerably during the 1980s and 1990s, and since 2000 only five were reported from sites in VC59, the last two being Formby on 5 August 2018 (M. Dean) and Rishton on 19 June 2023 (D. Bickerton). In VC60, it has remained a regular at its strongholds in the wooded northern valleys, but recording effort has eased in these area since 2018, so its current status there is uncertain. Elsewhere, it has only been noted in ones or twos at light from Green Bank on 9 July 2019 (P.J. Marsh), Leighton Moss on 3 August 2019 (P. Ashton), Mill Houses on 13 June 2021 (J. Roberts) and Warton village, Silverdale on 20 July 2022 (P. Stevens). Nearly all records relate to single figures coming to light with the largest reported count being 40 at Dalton (VC60) on 16 June 2003 (P.J. Marsh), a site where five larvae were found on elm on 3 September 2008.

■ **HABITAT** Deciduous woodland.
■ **FOODPLANTS** Elm sp., including Wych Elm.

Single brooded. Comes to light. Larvae found from late August to early October.

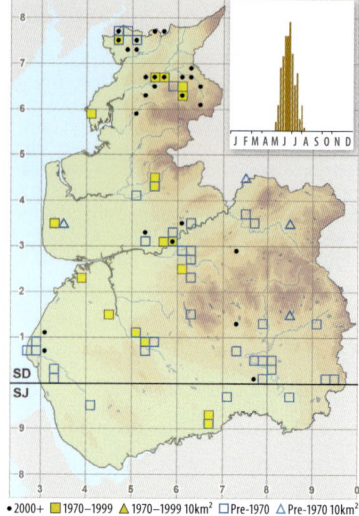

• 2000+   □ 1970–1999   ▲ 1970–1999 10km²   □ Pre-1970   △ Pre-1970 10km²

70.207 Clouded Border *Lomaspilis marginata*

■ **LANCS STATUS** Frequent / widespread resident   c.1857–2022

Noted in the Liverpool area (Gregson, 1857) with other early authors describing it as common throughout, but with few supporting records. From 1940 to the end of the 1980s, many more sightings with specific site and date details were submitted from scattered sites across the county. These included records from Burscough, Formby, Didsbury, St Annes, Salford, Rufford, Wrightington, Slyne, Lancaster, Blackpool, Leighton Moss and Billinge. Since then, it has been found almost annually, at many sites, including Flixton, Hale, Longton, Swinton, Abram, Rishton, Adlington, Preston, Mere Sands Wood, Ainsdale and Worsthorne. Most records are of single figure counts at light, but double figure counts are not uncommon and include 27 at Flixton on 22 July 1996 (K. McCabe), 28 at Horwich on 17 July 2022 (K. Haydock & J. Mills), 30 at Skelmersdale on 10 July 1989 (C.A. Darbyshire), 35 at Hic Bibi on 24 July 2009 (E. Langrish and A. Barker) and 38 to light at National Trust, Formby on 5 July 2014 (A. & S. Parsons).

■ **HABITAT** Damp woodland, scrub, gardens, parks and brownfield sites.
■ **FOODPLANTS** Sallow sp., including Grey Willow.

Single brooded over an extended period with a few sightings from mid-September early October. Comes to light. Larva found mid-August to September.

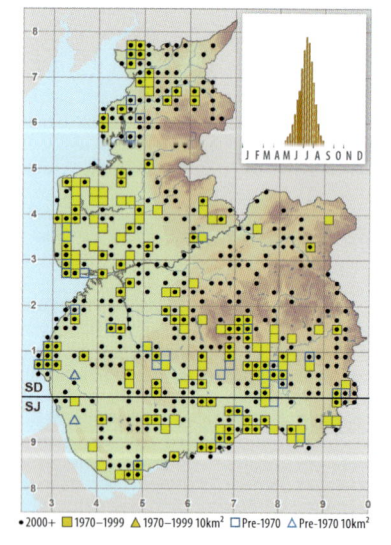

• 2000+   □ 1970–1999   ▲ 1970–1999 10km²   □ Pre-1970   △ Pre-1970 10km²

440   *The Moths of Lancashire*

70.208 Scorched Carpet *Ligdia adustata*

■ **LANCS STATUS** Occasional / local resident   c.1886–2022

First reported in the Silverdale area c.1886 (J.C. Melvill), with all but five of the 166 records from this general area to the present day. Other sites included Burnley in 1910, Sabden in 1975 and Middlewood, nr. Wray, on 29 April 2011 (G. Jones), these having no subsequent records. In the Silverdale area, it was found at Warton Crag in 1965, followed by Leighton Moss in 1968, both sites noting it regularly since then. Other nearby sites with regular records include Gait Barrows and Yealand Conyers.
■ **HABITAT** Woodland, scrub and hedgerows, mostly on limestone.
■ **FOODPLANTS** Not recorded in Lancashire.
Double brooded. Comes to light, mostly in low single figures.

●2000+ ▢1970–1999 △1970–1999 10km^2 ▫Pre-1970 △Pre-1970 10km^2

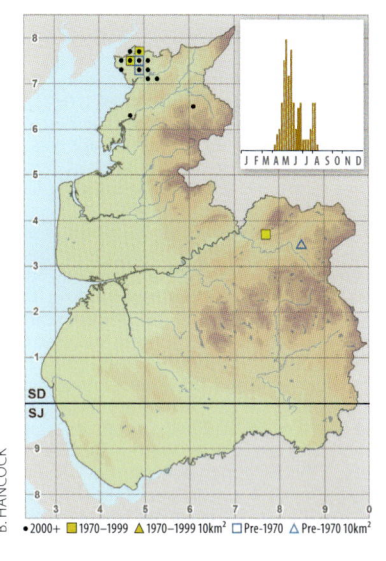

70.212 Sharp-angled Peacock *Macaria alternata*

■ **LANCS STATUS** Rare / former resident; vagrant   1904–2007

Easily confused with Peacock Moth. First confirmed from the Carnforth area in 1904 by H. Murray, the specimen being present, via a H.C. Huggins bequest, in the NHM, London. Later, one was confirmed by N. Birkett, from the Silverdale area on 7 July 1971; the recorder is unknown. The only other record, considered a vagrant, came to light at Hardy Grove, Swinton on 25 August 2007 (G. Riley). Apart from south Cumbria, the nearest populations are 100km east, in Lincolnshire and Yorkshire.
■ **HABITAT** Damp, deciduous woods and limestone woodland.
■ **FOODPLANTS** Not recorded in Lancashire. The Swinton moth was accidentally included under Peacock Moth in Randle *et al.* (2019).

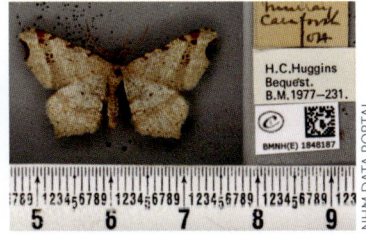

●2000+ ▢1970–1999 △1970–1999 10km^2 ▫Pre-1970 △Pre-1970 10km^2

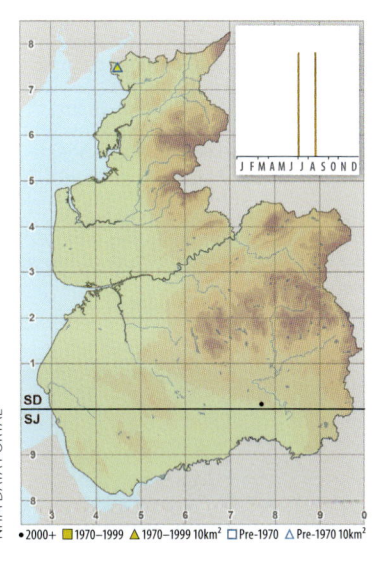

70.214 Tawny-barred Angle *Macaria liturata*

■ **LANCS STATUS** Occasional / widespread resident   c.1857–2022

Noted at Hale (Gregson, 1857) and, to the end of the 19th century, at Lydiate, Chat Moss, Longridge Fell and Silverdale. Since then, reported widely across the county, with particular concentrations in the Formby area pinewoods and around conifer plantations in north-east VC60. Other regular sites include Gait Barrows, Hardy Grove, Preston, Mere Sands Wood, Rainford, Billing and Yealand Redmayne.
■ **HABITAT** Coniferous woodland, gardens near isolated, mature pines.
■ **FOODPLANTS** Not recorded in Lancashire; often found near pine and spruce plantings.
Single brooded; possibly an occasional small second brood in September. Comes to light, mostly in single figures. 13 seen at Docker Moor on 30 June 2015 (P.J. Marsh).

●2000+ ▢1970–1999 △1970–1999 10km^2 ▫Pre-1970 △Pre-1970 10km^2

The Moths of Lancashire   **441**

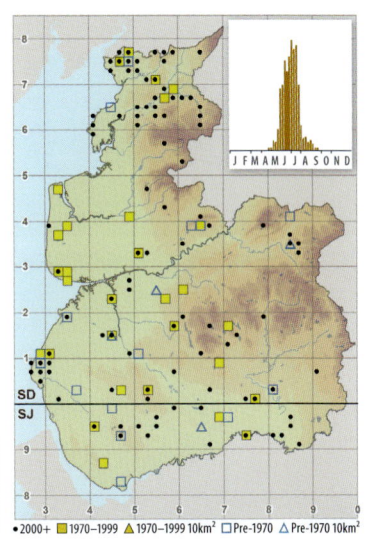

GEOMETRIDAE

70.215 V-Moth *Macaria wauaria*

■ **LANCS STATUS** Scarce/ local resident   c.1857–2023

Noted as Nationally Endangered (Fox *et al.*, 2019), with the decline most severe to the south of a line between the Bristol Channel and the Humber (Randle *et al.*, 2019). In Lancashire, however, it appears to have remained relatively stable since at least 1970, in an area to the north of Carnforth and west of the M6 motorway. This area incorporates the villages of Silverdale, Warton, Yealand Redmayne and Yealand Conyers.

This was, however, not always the case. The earliest reports (Gregson, 1857) describe it as 'plentiful in gardens' in the Liverpool district, as 'abundant wherever currants grow' (Ellis, 1890), 'common in old gardens' in the Burnley area c.1910 and 'fairly common' at Chadderton in 1915 (Mansbridge, 1940). From 1920, there were a series of reports from over 30 well-scattered sites, continuing into the second half of the 20th century. These included 63 at a garden light trap in Morecambe from 1959 to 1965 (C.J. Goodall) and 24 in the Leighton Moss Rothamsted light trap between 1970 and 1978. Subsequently, 17 were seen in Claughton between 1996 and 2003, 27 at Littleborough (1998–2000), 20 at Worsthorne (2000–2008) and 22 at Brierfield (2001–2004). Despite this range of records, with one exception, the moth was last seen away from the Silverdale area at Heysham N.R. in 2012 (P. Marsh *et al.*); this exception being one at Gressingham, in the Lune Valley, on 4 June 2023 by S. Piner.

In contrast, after a belated first record in 1961 at Warton (H.S. Robinson), the strength of the current population in the Silverdale area has been exemplified by 83 at Leighton Moss MV trap (1990–2022), 40 at Newcroft, Warton (2014–2022) and other 2022 records in Silverdale and Yealand garden light traps, as well as Leek Hill and Hyning Scout Woods.

• 2000+   ■ 1970–1999   ▲ 1970–1999 10km²   □ Pre-1970   △ Pre-1970 10km²

The early stages are mostly undocumented in Lancashire, other than a vague 'bred, ex-larvae' 1927 report and Anne Smith's discovery of three larvae on Red Currant at Normoss (Fylde) on 9 June 2008. All other reports originate from light traps, but the species can be skittish and may be disturbed by other moths.

Reasons for this rapid and extreme decline and presumed local extinctions are speculative but may include insecticides on commercial and garden currant crops and climate change (Randle *et al.*, 2019). The possibility that a reduced number of gardeners are keeping fruit bushes might also need evaluating. Relying exclusively on garden light trap data to make pronouncements in Lancashire is possibly premature. This is highlighted by the Norcross larvae, 44 years after the last adult record in the surrounding area, and suggests a low-level presence needs to be considered. As a result, it is recommended that widespread larval searches of currant species, especially 'wild' examples, would enable a better understanding of the status of this moth in the county. The recent (2023) Lune valley record, where it was previously last noted in 2006, illustrates the possibility that apparent absence may simply reflect changes in the amount of light trapping.

Pete Marsh

442   *The Moths of Lancashire*

70.218 Latticed Heath *Chiasmia clathrata*

■ **LANCS STATUS** Occasional / widespread resident 1843–2022
First noted in Preston in 1843 (Hodgkinson, 1844), then at several widespread sites up to 1910. Infrequently found to the 1950s, when noted in Morecambe and Formby. Since the early 1970s, reports have indicated a widely scattered distribution but few with regular sightings except, more recently, at Briercliffe, Oswaldtwistle, Royton, Heysham, Rishton and Flixton, the last with c.100 records between 2003 and 2022. Unable to assess any population trends due to the random nature of recording effort.
■ **HABITAT** Dune grassland, meadows, verges, waste ground and brownfield sites.
■ **FOODPLANTS** Vetch, probably Meadow Vetchling.
Single brooded. Flies by day, but will come to light. Larvae found in September.

● 2000+   ☐ 1970–1999   △ 1970–1999 10km²   ☐ Pre-1970   △ Pre-1970 10km²

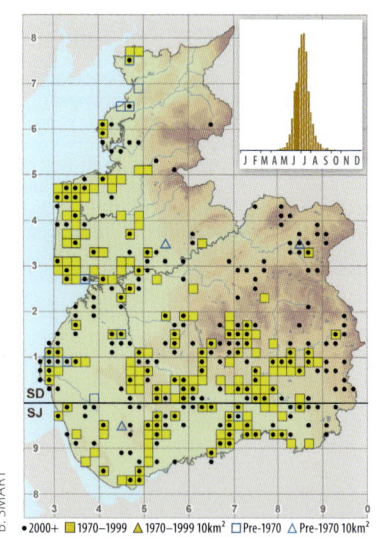

70.222 Brown Silver-line *Petrophora chlorosata*

■ **LANCS STATUS** Frequent / widespread resident c.1857–2022
First noted in Liverpool area (Gregson, 1857); later at Chat Moss, Silverdale, Holden Clough, Sabden and Thursden. Local in near-coastal, south and central lowland areas; more frequently encountered in the north and in upland areas. In recent years, regularly seen at sites such as Astley Moss, Billinge, Birk Bank, Botton Mill, Flixton, Gait Barrows, Green Bank, Heysham, Leighton Moss and Mere Sands Wood.
■ **HABITAT** Moors, woods, scrub, drier parts of mosses and some limestone areas.
■ **FOODPLANTS** Bracken.
Single brooded. Comes to light; readily disturbed during the day. 75 were seen on Yealand Hall Allotment on 5 June 2010 (B. Hancock, G. Jones). Larvae found in July.

● 2000+   ☐ 1970–1999   △ 1970–1999 10km²   ☐ Pre-1970   △ Pre-1970 10km²

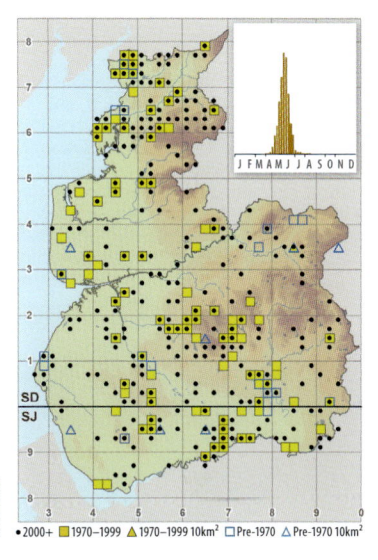

70.223 Barred Umber *Plagodis pulveraria*

■ **LANCS STATUS** Occasional / local resident c.1859–2022
Probably a more widespread species of old woodland in the past, with records from Hoghton Towers c.1859, by J. Chappell, and Brockholes Wood, Preston c.1886 (J.B. Hodgkinson). Since then, there have been isolated records in Longridge (1980) and Littleborough (2018), with all other records from the woods of northern VC60. Most of these were from Gait Barrows, Leighton Moss and Warton Crag, but it has also been noted, less frequently, from sites such as Mill Houses and Docker Moor.
■ **HABITAT** Limestone woodland and occasionally deciduous woods elsewhere.
■ **FOODPLANTS** Not recorded in Lancashire.
Single brooded. Comes to light. 25 at Gait Barrows on 24 May 2012 (G. Jones).

● 2000+   ☐ 1970–1999   △ 1970–1999 10km²   ☐ Pre-1970   △ Pre-1970 10km²

The Moths of Lancashire

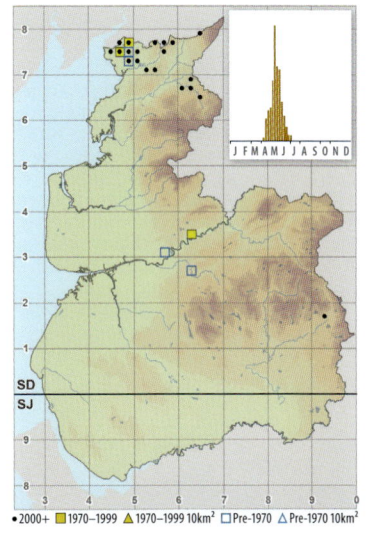

GEOMETRIDAE

70.224 Scorched Wing *Plagodis dolabraria*

■ **LANCS STATUS** Occasional / widespread resident   c.1910–2022

First noted at Yealand (A.E. Wright) and nr. Lancaster (C.H Forsyth), both c.1910, and then Nelson in 1930. The only other VC60 reports before 1993, were at Warton Crag in 1961 and Gait Barrows in 1984. It then moved slowly south and east, reaching Dolphinholme in 2006. In the south, it appeared in Liverpool in 1972 and Widnes in 1993, presumably via Cheshire. This population then spread gradually north and east, reaching Royton in 2003 (R. Hart), Preston in 2016 and Briercliffe in 2020 (T. Lally).
■ **HABITAT** Woodland and scrub; occasionally gardens.
■ **FOODPLANTS** Not recorded in Lancashire.
Single brooded. Comes to light. 13 found at Mill Houses on 21 May 2022 (P.J. Marsh).

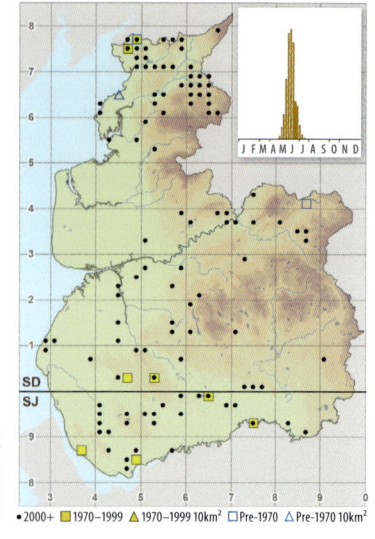

70.226 Brimstone Moth *Opisthograptis luteolata*

■ **LANCS STATUS** Abundant / ubiquitous resident   c.1857–2022

Found in Liverpool (Gregson, 1857) and later, noted as abundant everywhere by all early authors, up to 1940. Around the 1950s and '60s, it was reported regularly in Bispham, Liverpool, Morecambe, Salford, Slyne and St. Annes. Since the 1990s, it has been annual in occurrence at many sites such as Carnforth, Clitheroe, Hale, Heysham Hutton, Kirkham and Leigh, with no discernible change in distribution or abundance.
■ **HABITAT** Gardens, hedgerows, scrub, woodland and brownfield sites.
■ **FOODPLANTS** Blackthorn, Hawthorn, Rowan, Crab Apple, Buckthorn and Cherry Plum.
Double brooded. Comes to light. 62 found at Ainsdale on 7 Sept. 2021 (R. Moyes and C. Daly). Larvae found in August, September and October and once in early February.

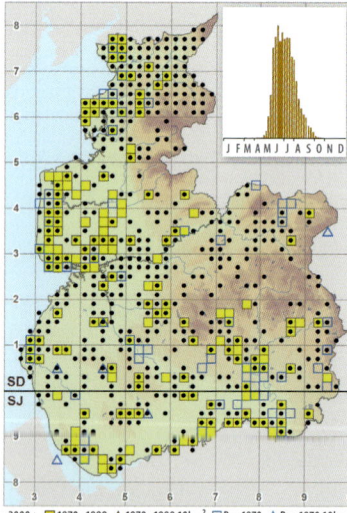

70.227 Bordered Beauty *Epione repandaria*

■ **LANCS STATUS** Occasional / local resident   c.1857–2022

Noted as not uncommon on the coast in the Liverpool district (Gregson, 1857) and then, amongst other sites, at Chat Moss c.1859, Preston area in 1879, Bootle in 1886, Southport in 1911 and Silverdale in 1913. Mansbridge (1940) noted it as generally distributed, but not common. Since the 1990s, it has been regular at a few sites such as Flixton, Great Sankey, Heysham N.R., Leighton Moss and Mere Sands Wood.
■ **HABITAT** Damp to wet woodland and scrub in lowland areas, including dune slacks.
■ **FOODPLANTS** Not recorded in Lancashire.
Single brooded. Comes to light, mostly in single figures. Ten were attracted to light at Heysham N.R. on 21 August 2002 (P.J. Marsh).

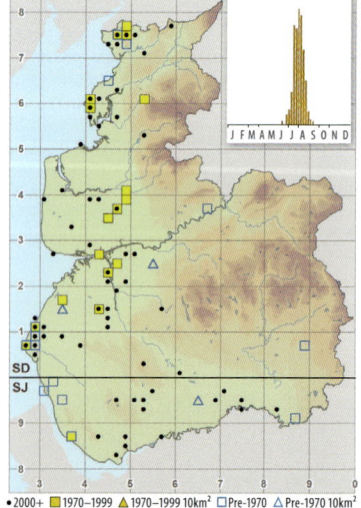

444 *The Moths of Lancashire*

70.229 Speckled Yellow *Pseudopanthera macularia*

■ **LANCS STATUS** Occasional / very local resident　　　　c.1886–2022
Noted in the Silverdale area c.1886 by J.C. Melvill and, in 1910, by C.H. Forsyth and W. Mansbridge. Since the 1960s it has been reported from Eaves Wood, Silverdale Golf Course, Yealand Hall Allotment, Dalton Crags and Challan Hall Allotments, but is most frequently encountered at Gait Barrows and on Warton Crag.
■ **HABITAT** Open woodland and scrub on limestone.
■ **FOODPLANTS** Not recorded in Lancashire.
Single brooded. Flies by day in sunny conditions. Largest count was from a Warton Crag butterfly transect, when 37 were seen on 25 May 2018 (RSPB team). A Mere Sands Wood record is considered unconfirmed.

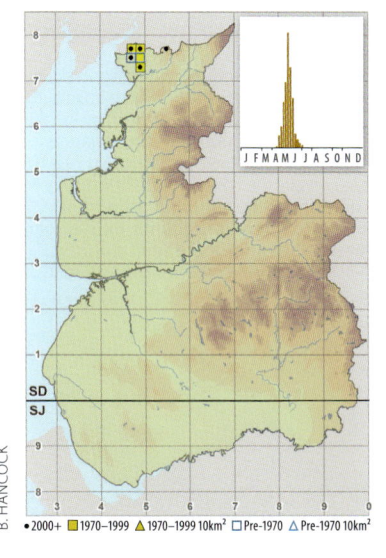

70.231 Lilac Beauty *Apeira syringaria*

■ **LANCS STATUS** Occasional / local resident　　　　c.1857–2020
First noted at Hale and Warbrick Moor (Gregson, 1857), then the Manchester district c.1859 by J. Chappell. Found at Ribchester in 1879, Silverdale in 1910 and near Irlam in 1940. Since then, very infrequent in VC59, with only six post-1999 records, the last at Flixton in 2019. In the 1990s, the small, but long-established population centred around Leighton Moss extended eastwards, reaching Claughton in 1994 and Mill Houses in 2005. Since 2015 there have only been eleven records from seven sites.
■ **HABITAT** Scrub, woodland and gardens, mainly on limestone.
■ **FOODPLANTS** Not recorded in Lancashire.
Single brooded; possible small second broods in Sept. 2000 and 2003. Comes to light.

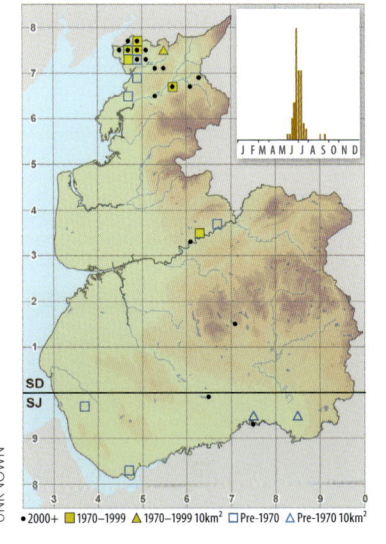

70.233 August Thorn *Ennomos quercinaria*

■ **LANCS STATUS** Scarce / local resident　　　　c.1859–2022
First noted in the Manchester district c.1859 by J. Chappell, it was later found in Burnley in 1910, Pendle Hill in 1940 and in the Nelson area around the same time. Prior to the 21st century, it was only otherwise seen in Morecambe (1959), Risley Moss (1961) and Allerton in 1982. Since then, there have been only 29 records from 18 widespread locations, the most recent at Great Sankey and Flixton in 2022.
■ **HABITAT** Woodland, hedgerows and gardens.
■ **FOODPLANTS** Not recorded in Lancashire.
Single brooded. Comes to light, nearly always as singletons. An occasionally misidentified species, rarely seen more than once a year at any site.

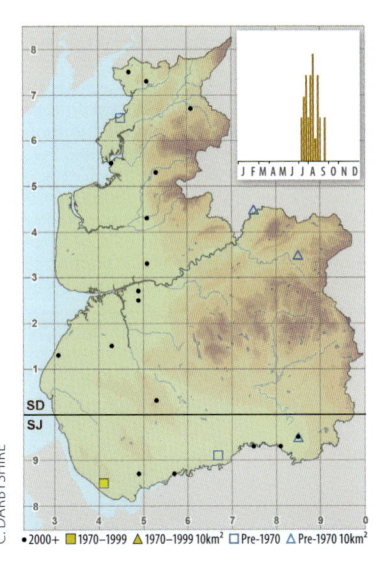

The Moths of Lancashire　**445**

GEOMETRIDAE

70.234 Canary-shouldered Thorn *Ennomos alniaria*

■ **LANCS STATUS** Frequent / widespread resident 1843–2022

Found in Preston in 1843 (Hodgkinson, 1844), at Old Swan, Liverpool c.1857, Risley Moss and Chat Moss c.1859, Crosby in 1886, Withington in 1913 and Lytham in 1939. It was not reported from upland areas until 1940, at Colne, suggesting it has expanded into these parts in more recent years. It is now found annually at many sites across the county, including Astley Moss, Briercliffe, Brockhall Village, Hoghton, Longton, Martin Mere, Morecambe, Orrell, Poulton-le-Fylde, Worsthorne and Wray.
■ **HABITAT** Woods, hedgerows, mosses, gardens, scrub and damp brownfield sites.
■ **FOODPLANTS** Alder; the adult is also found commonly amongst birch.
Single brooded. Comes to light. 29 seen at Heysham 18 Sept. 2002. Larva, June.

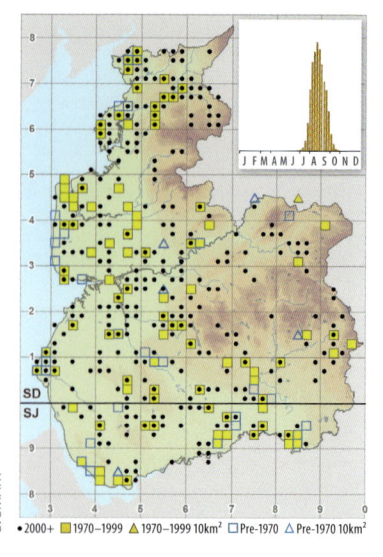

● 2000+   ▪ 1970–1999   ▲ 1970–1999 10km²   ▫ Pre-1970   △ Pre-1970 10km²

70.235 Dusky Thorn *Ennomos fuscantaria*

■ **LANCS STATUS** Frequent / widespread resident c.1857–2022

First noted at Walton, Liverpool (Gregson, 1857), then the Manchester district c.1859 and West Didsbury in 1915. As recently as 1940, considered a scarce species (Mansbridge, 1940). Slowly expanding its range from Freshfield in 1952, it reached Hoghton in 1970, Leighton Moss in 1977, Myerscough in 1981, St Annes in 1982 and then started moving inland and into higher parts. The first indication of this was a record from Longridge in 1979, then Milnrow in 1993 and Worsthorne by 2006.
■ **HABITAT** Woods, hedgerows, scrub, gardens and brownfield sites.
■ **FOODPLANTS** Ash; once on Privet.
Single brooded. Comes to light. 21 seen at Leighton Moss on 12 August 2018.

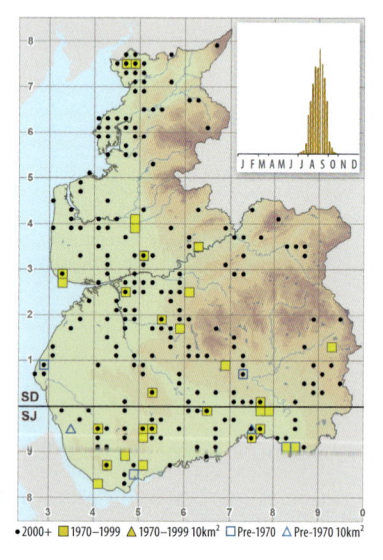

● 2000+   ▪ 1970–1999   ▲ 1970–1999 10km²   ▫ Pre-1970   △ Pre-1970 10km²

70.236 September Thorn *Ennomos erosaria*

■ **LANCS STATUS** Frequent / widespread resident c.1857–2022

First noted at Hale (Gregson, 1857) and then on Chat Moss c.1859, but then not until 1940 in Didsbury, Formby in 1950, Wrightington in 1954 and Rufford and Carnforth, in 1961. Since then, it has been found in scattered localities in lowland areas and mainly in wooded river valleys in higher parts. It is less frequent in coastal areas. Sites with the most regular reports include Bay Horse, Billinge, Hale, Higher Tatham, Hoghton, Longton, Mill Houses, north Preston and Wray.
■ **HABITAT** Woodland, hedgerows, parks and scrub.
■ **FOODPLANTS** Not recorded in Lancashire.
Single brooded. Comes to light, usually in low single figures.

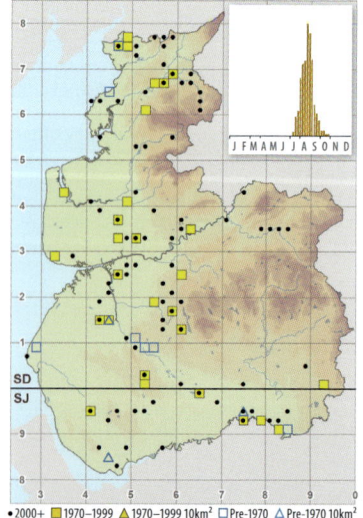

● 2000+   ▪ 1970–1999   ▲ 1970–1999 10km²   ▫ Pre-1970   △ Pre-1970 10km²

446 *The Moths of Lancashire*

70.237 Early Thorn *Selenia dentaria*

■ **LANCS STATUS** Frequent / widespread resident   c.1857–2022

Noted at Speke, Hale and Warbrick Moor prior to 1857 (Gregson, 1857), with further early sightings in Carnforth, Ditton, Blackpool, Burnley and Formby. Mansbridge (1940) reported it as common throughout, noting an occasional second brood and various forms. In recent decades, recorded regularly at most sites, including Abram, Bispham, Great Sankey, Hough Green, Longton, Rishton, Singleton and Urmston.
■ **HABITAT** Woodland, gardens, parks, hedgerows, scrub and brownfield sites.
■ **FOODPLANTS** Alder and a *Prunus* sp.
Double brooded. Comes to light. 32 were seen at Parr, St Helens on 5 September 2018 (R. Banks). Larvae found in June and August.

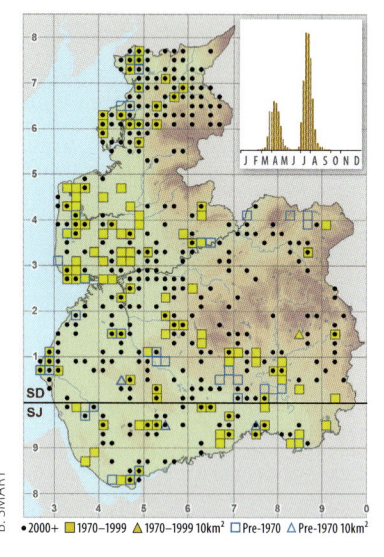

70.238 Lunar Thorn *Selenia lunularia*

■ **LANCS STATUS** Scarce / local resident   1879–2022

Found at Dutton in 1879 (Hodgkinson, 1880a), then Chat and Simonswood Mosses and Pendlebury, c.1886, followed by Burnley, Silverdale and Knowsley Park, c.1910. In VC59, last noted at Whalley in 1975 and Sabden in 1976. In VC60, apart from one at Galgate in 2018 and Clitheroe in 2020, its range has contracted northwards, with records from four Silverdale area sites, and at Dalton, Green Bank and Docker Moor.
■ **HABITAT** Woodland and scrub, particularly on limestone and moorland edge.
■ **FOODPLANTS** Not recorded in Lancashire.
Single brooded; two possible second brood individuals were noted at Leighton Moss on 18 July 2019 (RSPB) and Warton Crag on 28 July 2019 (J. Patton). Comes to light.

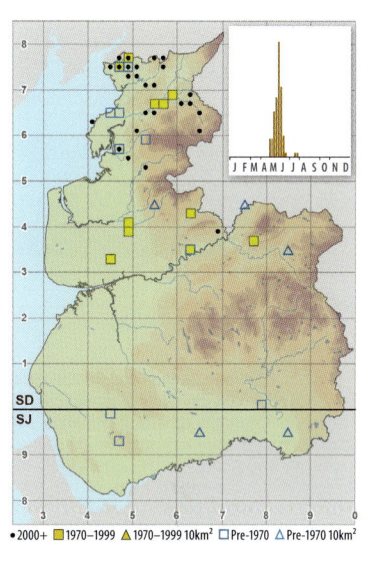

70.239 Purple Thorn *Selenia tetralunaria*

■ **LANCS STATUS** Frequent / widespread resident   1938–2022

First noted report in 1938, at Brock by S. Coxey (and again in 1949), then at Blackpool in 1954 and Risley Moss and Parbold in 1961. From 1970 onwards, it spread gradually north and east, reaching Gait Barrows in 1983, Bury in 1994, Briercliffe in 2001 and Oldham in 2004. It is now widespread, with increasingly regular reports from several sites such as Billinge, Leighton Moss, Mere Sands Wood, Mill Houses and Silverdale.
■ **HABITAT** Woodland, gardens, parks, scrub and brownfield sites.
■ **FOODPLANTS** Hawthorn and Alder.
Double brooded. Comes to light, usually in single figures. 20 were seen at Mill Houses on 14 August 2013 (P. Marsh). Larvae found in June.

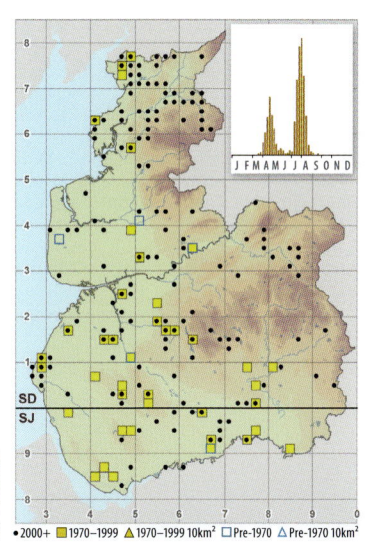

GEOMETRIDAE

The Moths of Lancashire   **447**

GEOMETRIDAE

70.240 Scalloped Hazel *Odontopera bidentata*

■ **LANCS STATUS** Frequent / widespread resident   c.1857–2022

First found in the Liverpool area by C.S. Gregson prior to 1857 who noted it as plentiful. Then, in Burnley in 1887 (A.E. Wright), Crumpsall in 1890, Simonswood Moss in 1910 and Urmston in 1914. Mansbridge (1940) reported it to be abundant and general throughout, with additional records from sites such as Rainhill Stoops, Ditton, Fallowfield, Moss Side, Huyton and Lytham. Since the 1950s, it has been widely recorded from scattered locations across the county, but is infrequent or absent from many lowland agricultural areas. Elsewhere, it can be annual in occurrence, including at locations such as Bispham, Briercliffe, Claughton, Flixton, Leighton Moss, Mere Sands Wood, Morecambe, Royton, Swinton, Worsthorne and Yealand Conyers. Although attracted readily to light, numbers are almost always in single figures. Larger counts include 17 at Docker Moor on 11 June 2015, 18 at Littleborough on 11 May 1999 and 20 at Lord's Lot Wood on 21 May 2012 (J. Girdley). The melanic form (ab. *nigra*) has been noted at several sites but is infrequent in occurrence, usually as singletons.

■ **HABITAT** Woodland, hedgerows, scrub, gardens and brownfield sites.
■ **FOODPLANTS** Clematis; recorded elsewhere on a wide variety of woody plants.
Single brooded, with a very small number of August records reported in recent decades. Comes to light. Larvae found from August to October.

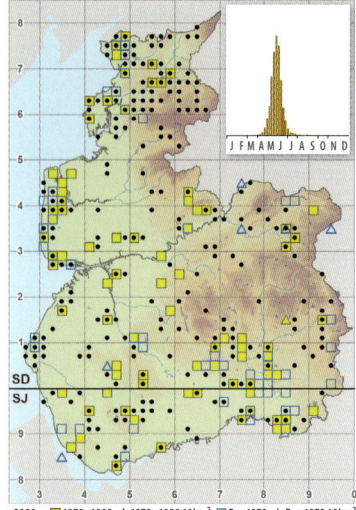

• 2000+   ▪ 1970–1999   ▲ 1970–1999 10km²   ▫ Pre-1970   △ Pre-1970 10km²

70.241 Scalloped Oak *Crocallis elinguaria*

■ **LANCS STATUS** Frequent / widespread resident   c.1857–2022

First noted in Liverpool (Gregson, 1857) and then, nectaring on heather at Turton Toppings in 1880 (J. Baldwin), Crumpsall in 1890 (G.W. & T.G. Mason) and Springhead in 1894 (H. Horsfall). Mansbridge (1940) listed several other additional locations and described it as generally distributed and often common. Since then, it has remained a widely recorded species, particularly in lowland parts and in deciduous woods within upland river valleys. It has been noted almost annually at many sites, including Bispham (B. Brigden), Briercliffe (T. Lally & G. Turner), Carnforth (L. Lyon & J. Rae), Claughton (M. Broomfield), Flixton (K. McCabe), Hale (C. Cockbain), High Tatham (P.J. Marsh), Lancaster, Longton (R. Boydell), St Helens (D. Owen & R. Banks), Worsthorne (G. Gavaghan) and Yealand Conyers (B. Hancock) and many others. Numbers attracted to light have always been mostly in single figures but double-digit counts have become very infrequent in the last decade. The last large counts being 17 at Hough Green, Widnes on 24 July 2006 and 19 at Aughton Old Hall on 11 July 2004 (M. Broomfield). Occasionally noted until mid-September.

■ **HABITAT** Gardens, woodland, scrub, hedgerows and brownfield sites.
■ **FOODPLANTS** Heather, Rowan and Cotoneaster.
Single brooded. Comes to light. Larvae found from late March to May.

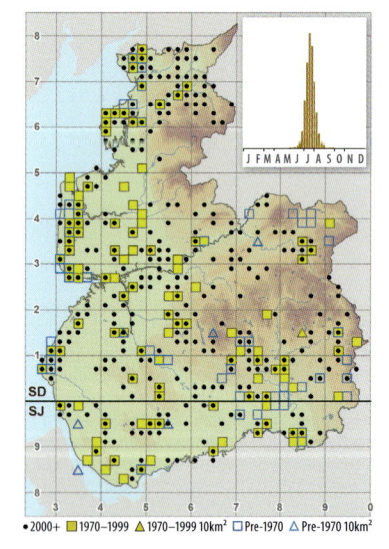

• 2000+   ▪ 1970–1999   ▲ 1970–1999 10km²   ▫ Pre-1970   △ Pre-1970 10km²

70.243 Swallow-tailed Moth *Ourapteryx sambucaria*

- **LANCS STATUS** Frequent / widespread resident c.1857–2022

Noted in Liverpool (Gregson, 1857), at St Annes in 1908, Sabden and Chadderton in 1910, Fallowfield in 1924 and Blackburn in 1930. Considered common by all early authors, except Brindle (1939), who reported it as rare but increasing in East Lancashire. Since then, noted widely across the county, but less so in higher parts.
- **HABITAT** Woodland, gardens, hedgerows, scrub and brownfield sites.
- **FOODPLANTS** Ivy.

Single brooded until 2002. Double brooded from 2003 to 2006 and then in most years from 2012 to the present day. Comes to light. The largest count was of 43 at Hough Green on 19 July 1990 (S. McWilliam). Larvae found September and May.

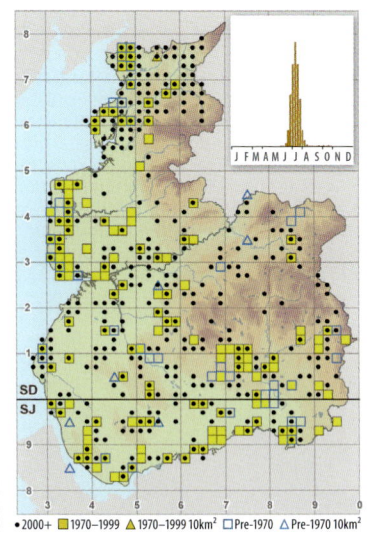

70.244 Feathered Thorn *Colotois pennaria*

- **LANCS STATUS** Frequent / widespread resident c.1857–2022

Reported as plentiful at gas lamps in West Derby and Aigburth (Gregson, 1857), and later noted widely, in Preston, Lancaster, Burnley, Withington, Silverdale, Formby and Lytham, up to 1939. Mansbridge (1940), reported it as common on the mosses around birch. Since then, it has been found widely across the county, with coverage in north Lancashire demonstrating it is likely to be found in all but the highest parts.
- **HABITAT** Woodland, hedgerows, scrub, gardens and brownfield sites.
- **FOODPLANTS** Sallow, birch, poplar, Alder and Heather.

Single brooded. Comes to light. The largest count was of 75 at Gait Barrows on 27 October 2014 (G. Jones). Larvae found April to July.

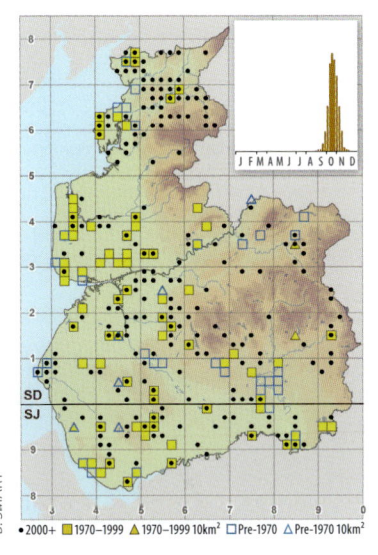

70.245 March Moth *Alsophila aescularia*

- **LANCS STATUS** Frequent / widespread resident c.1857–2022

Noted in the Liverpool (Gregson, 1857) and Manchester areas, c.1859, and as common throughout by early authors, including Mansbridge (1940). Since then, it has been found at many lowland sites and, less frequently, in upland areas where it is mostly associated with wooded river valleys. Recent poorly planned development schemes, resulting in extensive tree and hedgerow loss are causing localised declines.
- **HABITAT** Woodland, gardens, hedgerows, scrub and brownfield sites.
- **FOODPLANTS** Sallow, Sycamore, Hawthorn and Birch.

Single brooded. Comes to light. The largest count was of 65 at Aughton Wood on 25 February 2015 (J. Patton). Larvae found in May and June. The female is wingless.

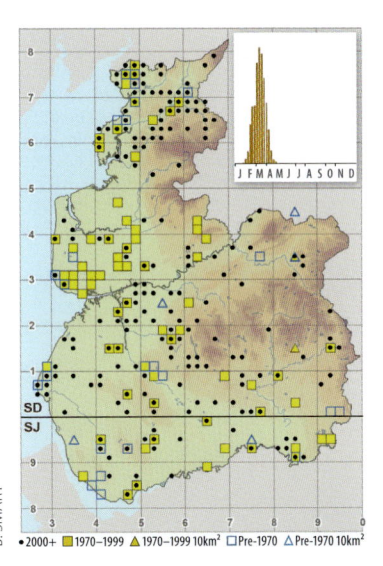

GEOMETRIDAE

The Moths of Lancashire

GEOMETRIDAE

70.246 Small Brindled Beauty *Apocheima hispidaria*

■ **LANCS STATUS** Occasional / very local resident   2004–2022
Considered an overlooked resident, first noted on 5 February 2004 at High Tatham (P.J. Marsh). Since then, targeted searches of deciduous woodland in VC60 have produced regular sightings in several old oak woods of the Hindburn and Roeburn catchments, with Mill Houses the most frequent. Noted at Warton Crag in 2018 and Silverdale in 2021 (both J. Patton), but not found in Aughton or Burton Woods.
■ **HABITAT** Oak woodland.
■ **FOODPLANTS** Not recorded in Lancashire.
Single brooded. Comes to light. Largest count was of 22 at Mill Houses on 22 February 2019 and 10 March 2022 (P.J. Marsh and J. Roberts).

● 2000+   ▫ 1970–1999   △ 1970–1999 10km^2 ☐ Pre-1970   △ Pre-1970 10km^2

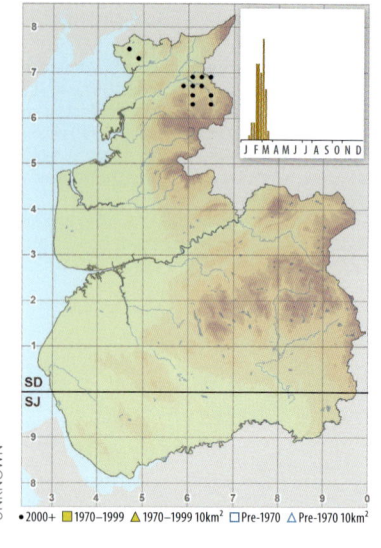

70.247 Pale Brindled Beauty *Phigalia pilosaria*

■ **LANCS STATUS** Frequent / widespread resident   c.1857–2022
Reported as plentiful in Liverpool (Gregson, 1857) and later, from Oldham, Burnley, Rochdale, Allerton, Sabden and Formby, to 1940. Since then, noted widely across the county, with a recent, occasional, tendency to emerge in small numbers from late November. The early flight period will have led to under-recording in more remote areas. Early light trapping in north VC60 demonstrates how widespread it can be.
■ **HABITAT** Woodland, hedgerows, scrub, gardens and brownfield sites.
■ **FOODPLANTS** Birch, oak, Hawthorn, Bird Cherry, sallow and Sycamore.
Single brooded. Comes to light. The largest count was of 113 at Mill Houses on 18 February 2014 (P.J. Marsh). Larvae have been found from April to mid-July.

● 2000+   ▫ 1970–1999   △ 1970–1999 10km^2 ☐ Pre-1970   △ Pre-1970 10km^2

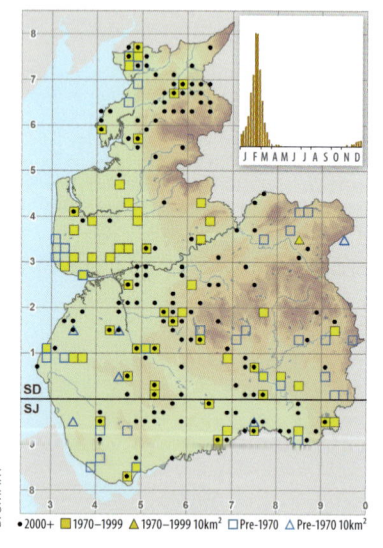

70.248 Brindled Beauty *Lycia hirtaria*

■ **LANCS STATUS** Scarce / local resident   1905–2022
Bred from larva on unspecified plant at Simonswood Moss in 1905 (J.J. Richardson). Not recorded again until 16 April 1993 at Swinton (S. Christmas) and at Billinge in 1995 (C. Darbyshire). Since 2004 it has been found on 16 occasions in well-scattered locations, with most of these since 2019. Sites include Flixton, Great Sankey, Norden, Longridge, St Helens and Silverdale. Appears to be slowly expanding range, but it is unusual for it to be seen at the same site on more than one or two occasions.
■ **HABITAT** Hedgerows and gardens.
■ **FOODPLANTS** Not recorded in Lancashire.
Single brooded. Comes to light and only ever seen as singletons.

● 2000+   ▫ 1970–1999   △ 1970–1999 10km^2 ☐ Pre-1970   △ Pre-1970 10km^2

450   *The Moths of Lancashire*

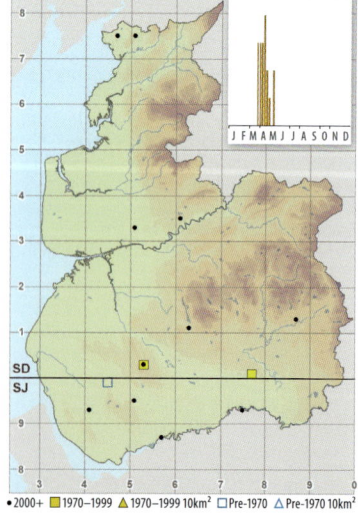

70.250 Belted Beauty *Lycia zonaria*

■ **LANCS STATUS** Occasional / very local resident   c.1857–2023

Present in Cheshire in 1832, this moth was not reported in Lancashire until Gregson (1857) found it 'plentiful on the sandhills of Crosby' where it remained, despite being heavily collected, until about 1915. Gregson later admitted to introducing it there and elsewhere (Formby, Hightown and Churchtown) with females collected from Lytham on unspecified dates (Gregson, 1879 & 1882). Between 1862 and 1921 it was reported from Formby, Ainsdale, Longton Marsh, St Annes and Anchorsholme (Blackpool). It was discovered at Anchorsholme, again, in 1963, but searches for it there in the 1970s were unsuccessful (J. Steeden pers. comm.). The declines and subsequent extinctions at all but one of the Lancashire coastal sites were all as a direct result of habitat loss during coastal development (e.g., Mansbridge, 1940) and until 1975 it was considered lost to the county.

On 6 July 1975, botanist Pat Livermore found a larva north of Sunderland Point and further singletons of both adult and larva were noted here, intermittently, until 1993. In 2002, an organised survey (Palmer, 2002) confirmed the presence of a colony along the 2km stretch of saltmarsh from Sunderland Point to Potts Corner (Howe et al., 2004). Although monitored annually since 2002, regular fixed transect counts were not introduced here until 2008. Searches at potentially suitable sites elsewhere in the county and region have all proved negative (Kimpton, 2000 and 2001).

The Belted Beauty was formerly considered a sand dune specialist in Britain, as is the case at its only other remaining colonies in western Ireland and NW Scotland. Near Sunderland Point however, it occurs atypically on an accreting saltmarsh consisting of Sea Rush dominated vegetation and extensive short sward areas, with Red Fescue dominating. Eggs are laid in batches of 30 or so in fallen rush seedheads, detritus or cracks in posts (Lawson, 2011). Larvae feed from May to July on a wide range of plants (see box A). The pupae overwinter underground and, like all other stages, experience tidal inundations to a greater or lesser extent. The flight period commences about two to three weeks later here than was historically the case at Crosby.

The males sometimes fly at night or during calm, mild, early mornings. Predation by birds occurs and a Common Lizard was observed catching and consuming a female. Males are occasionally caught in spiders' webs at rush-top height; on one occasion ants were seen 'bothering' a larva (Lawson, loc. cit.). In 2014, significant trench digging was proposed through the marsh for landfall of off-shore windfarm cables; this was resolved by use of underground Horizontal Direct Drilling following a national enquiry (Skelcher, 2021).

Analysis of records from the Sunderland Point area (Callaghan, 2020) concluded the population was stable, but experienced considerable peaks and trough in numbers. For example, a maximum count of 1691 moths on 18th April 2010 compares markedly with 2018 to 2020, where single day counts did not exceed 90. Covid restrictions limited recent counts.

Box A
Larval foodplants - Autumn Hawkbit, Sea Plantain, Common Sea-lavender, Sea Aster, White Clover, Meadow Vetchling, Common Bird's-foot-trefoil, Distant Sedge, Parsley Water-dropwort, Sea Arrowgrass and Saltmarsh Rush inflorescences.

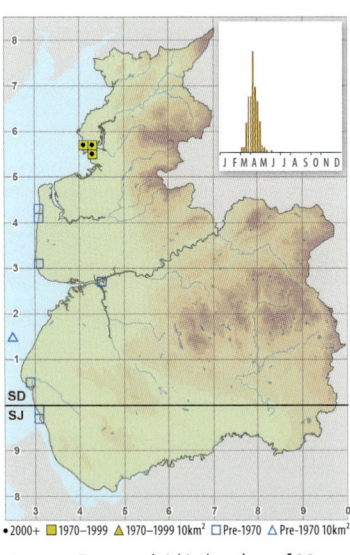

GEOMETRIDAE

70.251 Oak Beauty *Biston strataria*

■ **LANCS STATUS** Occasional / widespread resident c.1853–2022

Noted in Knowsley Park by N. Cooke c.1853 (Byerley, 1854), rare in Preston (Ellis, 1890) and Quernmore in 1910 (Mansbridge, 1940). During the 1950s and '60s, it was found mostly in the Silverdale area, to the east of Lancaster, and in the Formby area, it being very infrequent elsewhere. From the 1990s onwards, it has gradually increased in range and regularity of occurrence in much of lowland VC59 and north VC60, and elsewhere to a much lesser extent.

■ **HABITAT** Woodland, hedgerows, gardens, scrub and brownfield sites.
■ **FOODPLANTS** Oak.

Single brooded. Comes to light. 43 at Mill Houses on 29 March 2006. Larvae in May.

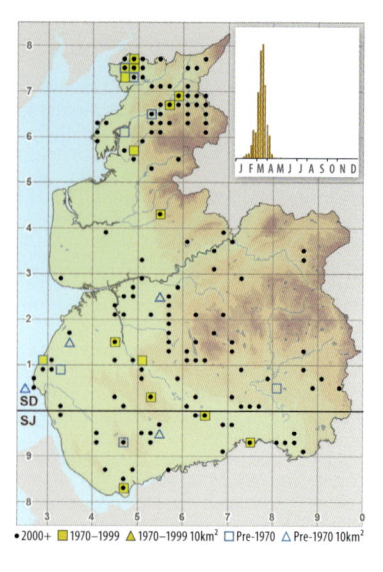

70.252 Peppered Moth *Biston betularia*

■ **LANCS STATUS** Frequent / widespread resident c.1848–2022

R.S. Edleston, c.1848, discovered a melanic form in central Manchester (Hooper, 2002), possibly the first British record of this form. Later, (Ellis, 1890) noted the species as 'fairly common but scarcer northwards', mostly melanics. Mansbridge (1940) stated it was abundant and widespread. Has remained widespread since, but less numerous, the largest count (38) from Docker Moor in July 2014. In Billinge, melanics predominated c.1985 (C. Darbyshire) but decreased, with the last in 2005.

■ **HABITAT** Woodland, hedgerows, gardens, mosses, scrub and brownfield sites.
■ **FOODPLANTS** Birch, Rowan, rose, Hawthorn and once on Mugwort.

Single brooded. Comes to light. Larvae July to September and once in mid-June.

70.253 Spring Usher *Agriopis leucophaearia*

■ **LANCS STATUS** Occasional / widespread resident 1861–2022

Found in Prestwich by J. Chappell in 1861 and later, in Preston (1886), Burnley (1910), Didsbury (1934) and Sabden (1940). Mansbridge (1940) noted it as common in oak woods. Since then, reported occasionally from many sites and, more regularly, at Billinge, Burton Wood, Cuerden, Euxton, Meal Bank (Wray) and Mill Houses.

■ **HABITAT** Oak woodland, hedgerows with oak standards and gardens.
■ **FOODPLANTS** Oak sp.

Single brooded. Comes to light. 41 seen at Euxton on 19 January 2008 (E. Langrish). As with other very early and late season sp., the distribution map depicts where recorder search effort occurred; probably very under-recorded. Larva found in May.

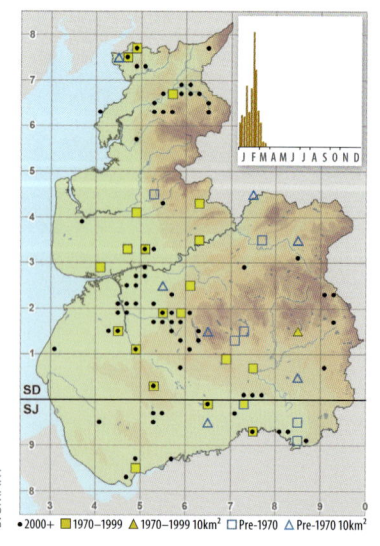

452 *The Moths of Lancashire*

70.254 Scarce Umber *Agriopis aurantiaria*

■ **LANCS STATUS** Occasional / widespread resident   c.1857–2022

Noted in Aigburth and West Derby (Gregson, 1857) and later, in Preston, Silverdale Burnley, Lancaster, Withington, Colne and Didsbury up to 1940. Mansbridge (1940) noted it as rather local. Since the 1980s, it has been found at many scattered sites across the county. Those with more regular sightings include Billinge, Claughton, Euxton, Flixton, Leighton Moss, Mere Sands Wood and Mill Houses.
■ **HABITAT** Woodland and well-established scrub.
■ **FOODPLANTS** Lime, Rowan, poplar and Sycamore.
Single brooded. Comes to light. Larvae found March to May. The largest count was of 131 attracted to actinic light at Aughton Woods on 19 November 2014 (J. Patton).

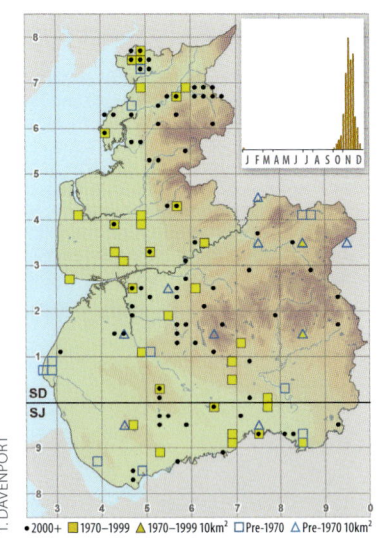

70.255 Dotted Border *Agriopis marginaria*

■ **LANCS STATUS** Frequent / widespread resident   1843–2022

First noted at Preston in 1843 (Hodgkinson, 1844) and later, from Childwall, Oldham and St Annes; all early authors to 1940 noting it as abundant and widely distributed. Since then, it has remained widely distributed in all but the higher parts of the county, but numbers have decreased in recent decades in some lowland areas. Sites with long-term datasets include Billinge, Flixton, Mere Sands Wood and Wray.
■ **HABITAT** Woodland, gardens, scrub and brownfield sites.
■ **FOODPLANTS** Hawthorn, birch, sallow and Bird Cherry.
Single brooded. Comes to light. 66 were found at Worsley on 21 February 2012 by I. Walker. Larvae found in May and June.

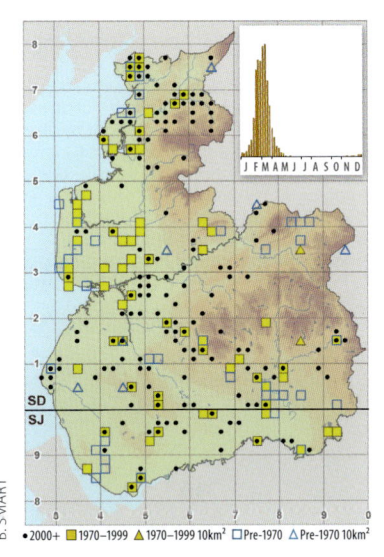

70.256 Mottled Umber *Erannis defoliaria*

■ **LANCS STATUS** Frequent / widespread resident   c.1857–2022

Reported to be plentiful at gas lamps in Liverpool (Gregson, 1857), with both Ellis (1890) and Mansbridge (1940) noting it as locally abundant to abundant in deciduous woods. Since the 1950s, its coastal distribution has reduced to some extent and since the early 2000s, records from some urban and suburban areas have also declined.
■ **HABITAT** Woodland, hedgerows, scrub, gardens, moorland and brownfield sites.
■ **FOODPLANTS** Oak sp., Turkey Oak, birch sp., Hazel, Aspen, Hawthorn, sallow, Apple, Field Maple, Sycamore, elm sp., lime, Beech, Rowan and Alder.
Single brooded. Comes to light. The largest count was of 1,164 to actinic light at Aughton Woods on 17 December 2014 (J. Patton). Larvae found mid-April to June.

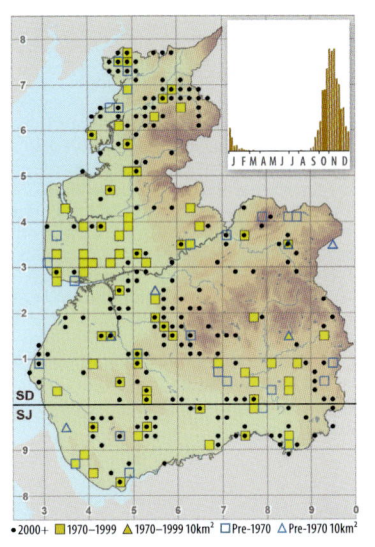

The Moths of Lancashire   **453**

GEOMETRIDAE

70.257 Waved Umber *Menophra abruptaria*

■ **LANCS STATUS** Occasional / local resident   c.1857–2022
First noted at Hale (Gregson, 1857) and then Burnley, Knowsley and Lancaster in 1890. Next seen in 1961 nr. Carnforth, Morecambe in 1964 and Ainsdale in 1965. In VC59, the distribution had always been very localised but, starting with Hale in 2007, there has been a slow increase in sites, including Rishton in 2014. In VC60, it occupied several northern sites in the 1960s, but is now spreading further south and east, as well as increasing in regularity, reaching Heysham in 1992 and Perry Moor in 2004.
■ **HABITAT** Scrub, woodland edge and much less frequently in gardens and parks.
■ **FOODPLANTS** Not recorded in Lancashire.
Single brooded; four also seen in July and August. Comes to light in low numbers.

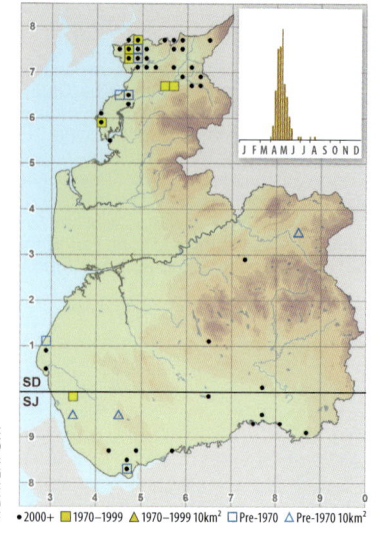

70.258 Willow Beauty *Peribatodes rhomboidaria*

■ **LANCS STATUS** Frequent / widespread resident   c.1857–2022
Noted as abundant in the Liverpool district (Gregson, 1857) and later reported as common to abundant throughout by all other early authors. Since the 1950s, it has been found across much of lowland Lancashire and in upland river valley areas, the increased distribution since the late 1990s probably indicative of increased light trapping. Maximum numbers reported have declined to some extent.
■ **HABITAT** Woods, hedgerows, gardens, parks and scrub.
■ **FOODPLANTS** Mugwort and Ivy.
Single brooded until c.1994, with almost annual double brooding thereafter. Comes to light; 76 at Heysham on 6 Sept. 2002 (P.J. Marsh). Larvae found October to April.

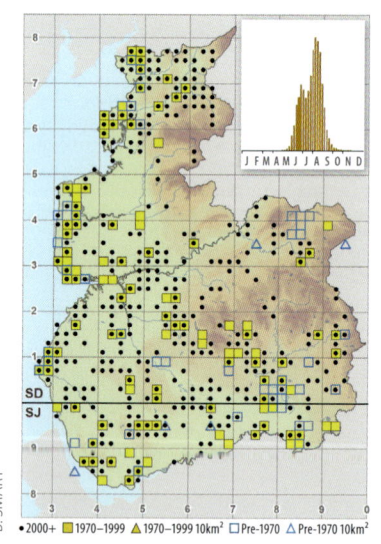

70.262 Bordered Grey *Selidosema brunnearia*

■ **LANCS STATUS** Considered Extinct   1843–c.1857
First reported from near Pilling, an extensive mossland area, in 1843 by J.B. Hodgkinson (Hodgkinson, 1844) and later at the same site, by his brother W. Hodgkinson. The latter find involved 30 moths on 21 July 1845 (Hodgkinson, 1845). Elsewhere, it was noted as present on the mosses near Warrington (none specifically named) by N. Cooke in 1853 and as local on Rainford Moss and upon a small moss near Eccleston by C.S. Gregson (1857). The mention of records from Pilling Moss in Ellis (1890) are considered as referring to the 1843 and 1845 records, but no mention is made to the dates. All subsequent references to this species occurring in N. Lancashire (Mansbridge, 1940), relate to two sites in what is now Cumbria (VC69) and it is still present at a few sites in this area.
■ **HABITAT** Lowland raised bogs.
■ **FOODPLANTS** Not recorded in Lancashire.
Considered to be single brooded, with all records relating to daytime sightings. Its loss in much of Lancashire is likely due to extensive draining of mossland sites, large areas being transformed for agricultural use and later, industrialised peat extraction.

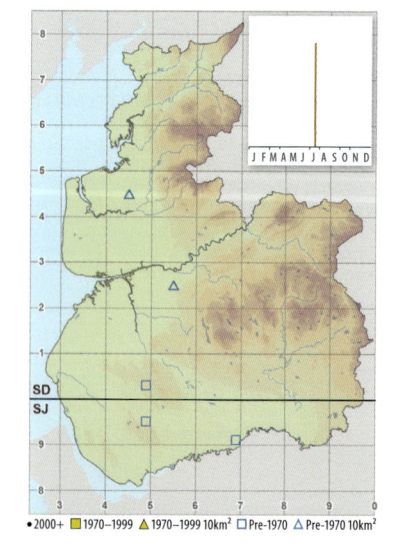

70.264 Satin Beauty *Deileptenia ribeata*

■ **LANCS STATUS** Occasional / local resident   1955–2022

Considered to be a relatively recent arrival and probably linked with the extensive planting of conifers. First noted in 1955 at Thrang End and then at Bolton-le-Sands on 5 August 1968 (C.J. Goodall). Thereafter, found at many well-scattered sites, but with the main concentrations in northern VC60. These sites, mostly with records to the present day, include Gait Barrows from 1986, Leighton Moss from 2001, Docker Moor from 2006, Yealand Conyers from 2008 and Warton Crag from 2014.
■ **HABITAT** Deciduous and coniferous woods.
■ **FOODPLANTS** Not recorded in Lancashire.
Single brooded. Comes to light. 35 were seen at Leek Hill Wood on 2 July 2014.

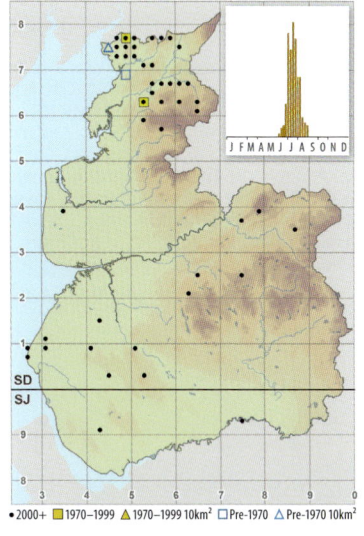

70.265 Mottled Beauty *Alcis repandata*

■ **LANCS STATUS** Frequent / widespread resident   c.1857–2022

Reported as abundant in the Liverpool district (Gregson, 1857) and subsequently noted as common everywhere (Mansbridge, 1940). Since then, has been found at most light trap sites, including in moorland areas. Over 30 sites have long-term datasets, including Bay Horse, Briercliffe, Burrow Heights, Fulwood, Hutton, Longton, Poulton-le-Fylde, Rishton, Royton, Wigan, Worsthorne and Wray.
■ **HABITAT** Woodland, gardens, scrub, parks, moorland, hedgerows and brownfield sites.
■ **FOODPLANTS** Birch sp., Bilberry, Blackthorn, Heather and Hawthorn.
Single brooded. Comes to light. 150 were attracted to actinic light at Warton Crag on 17 July 2015 (P. Stevens). Larvae found from September to mid-May.

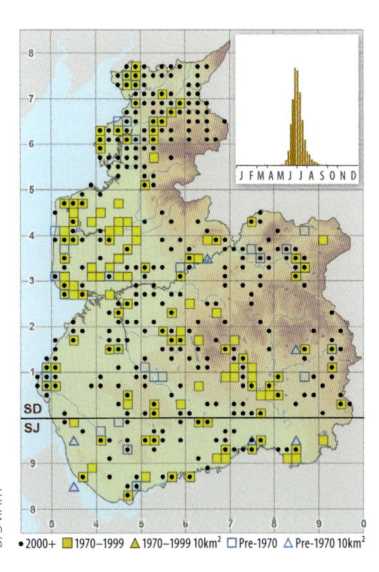

70.270 Engrailed *Ectropis crepuscularia*

■ **LANCS STATUS** Frequent / widespread resident   c.1857–2022

Early authors considered Small Engrailed a separate species, but this no longer seems to be the case (Agassiz *et al.*, 2022); the records are here combined. Noted from Hale (Gregson, 1857) then, Prestwich, Worsley, Longridge and Lancaster amongst others, to 1940. In more recent decades, noted at an increasing number of sites but overall numbers of records have generally stayed the same.
■ **HABITAT** Woodland, hedgerows, gardens, scrub and brownfield sites.
■ **FOODPLANTS** Alder, birch, oak, poplar and Alder Buckthorn.
Double brooded; rare small third brood. Comes to light, with 52 at actinic light on Warton Crag 7 July 2017 (P. Stevens). Larvae, April to May, late June & Aug.-Sept.

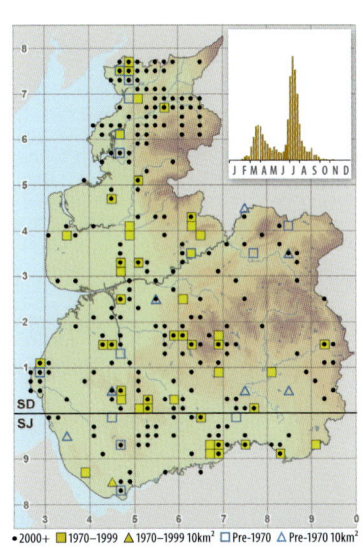

GEOMETRIDAE

70.272 Square Spot *Paradarisa consonaria*

■ **LANCS STATUS** Occasional / very local resident   1910–2019

First noted at Quernmore in 1910 (C.H. Forsyth), an area in which it has not been seen since. Not reported again until 1971 and 1972, when found in the Silverdale area by C. Jones-Parry, and then at Leighton Moss on 1 May 2003 (K. Briggs). In 2008, it was noted in Dalton, Gait Barrows, Lord's Lot Wood, Henridding and Silverdale (J. Girdley *et al.*). This overall distribution remains much the same to the present day, with a few more sites added, such as Eaves Wood, Warton Crag and Docker Moor.

■ **HABITAT** Deciduous woodland.
■ **FOODPLANTS** Not recorded in Lancashire.

Single brooded. Comes to light, in single figures.

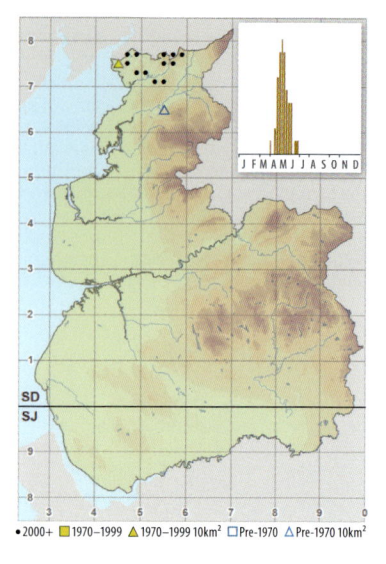

70.274 Grey Birch *Aethalura punctulata*

■ **LANCS STATUS** Occasional / widespread resident   c.1857–2022

Reported from Croxteth and Knowsley (Gregson, 1857) and from Chat Moss and Hoghton Tower c.1859. In 1886 it was found in Silverdale and Preston, with few more sites added prior to 1940, when Mansbridge listed it as local and nowhere common. Next seen at Leighton Moss in 1967 (P.J. Marsh) and thereafter, more widely from 1980 onwards. Only a few sites have recorded it regularly, usually in single figures, these being Astley Moss, Flixton, Gait Barrows, Mere Sands Wood and Mill Houses.

■ **HABITAT** Birch woodland, scrub and moorland and moss edges.
■ **FOODPLANTS** Not recorded in Lancashire.

Single brooded. Comes to light; rests on birch trunks. Seen once in January.

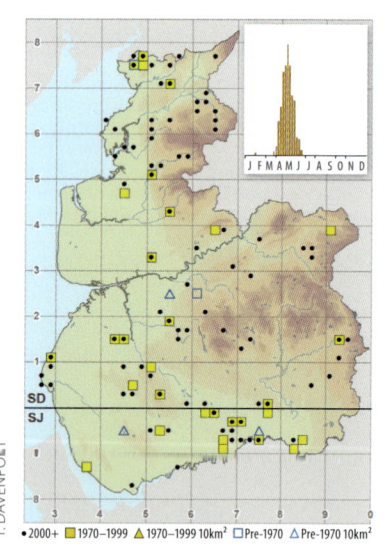

70.275 Common Heath *Ematurga atomaria*

■ **LANCS STATUS** Frequent / widespread resident   c.1857–2022

Found on Simonswood Moss (Gregson, 1857) and later, on southern lowland mosses and some eastern moorland localities, up to 1920. Noted in Silverdale in 1922 and from Hightown in 1930. Since then, found on most remaining lowland mosses, such as Winmarleigh and Heysham. Coastal records are mostly from the Formby area. In upland areas, it is regular on heather moors, but is considered under-recorded.

■ **HABITAT** Moorland, mosses, dune heath and, less frequently, flower rich grassland.
■ **FOODPLANTS** Heather.

Single brooded. Flies in sunny conditions; over 1,000 seen on Darwen Moor on 6 June 2004 (S. Martin). Occasionally comes to light. Larvae found August to September.

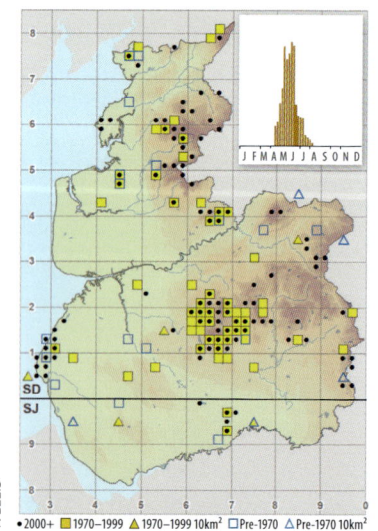

456 *The Moths of Lancashire*

70.276 Bordered White *Bupalus piniaria*

■ **LANCS STATUS** Occasional / widespread resident   c.1857–2022
First noted on Kirkby Moss (Gregson, 1857), in Silverdale on 26 June 1909, Burnley in 1910 and near Nelson in the 1930s. Mansbridge (1940) documented it as common in pine woods throughout the region. Since then, it has been recorded at scattered localities across the county, with concentrations of records from the Formby and Ainsdale area pine woods, Billinge, Briercliffe, Skelmersdale and Worsthorne.
■ **HABITAT** Coniferous woodland and where isolated mature pine trees occur.
■ **FOODPLANTS** Pine sp.
Single brooded. Comes to light and flies around trees by day. 81 were attracted to light at Ainsdale on 25 June 2011 (J. Clews *et al.*). Larva found in September.

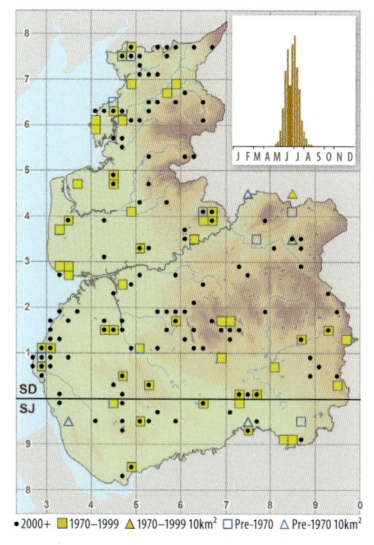

70.277 Common White Wave *Cabera pusaria*

■ **LANCS STATUS** Frequent / widespread resident   c.1857–2022
Found in Liverpool (Gregson, 1857) then, nr. Ribchester in 1879, Knowsley in 1905, Silverdale in 1909, Burnley in 1910, Longworth in 1925 and Sabden in 1936. All early authors considered it common and widespread and that has remained the case to the present day, with many sites recording it annually. Those with long-term datasets include Ainsdale, Bay Horse, Billinge, Flixton, Hale, Longton, Preston and Worsthorne.
■ **HABITAT** Wide range, including woodland, gardens, scrub, mosses and moorland edge.
■ **FOODPLANTS** Birch sp., Alder and Hazel.
Single brooded. Comes to light. 34 were seen at Todmorden on 23 July 2010 (B. Leecy). Larvae have been found from late July to mid-September.

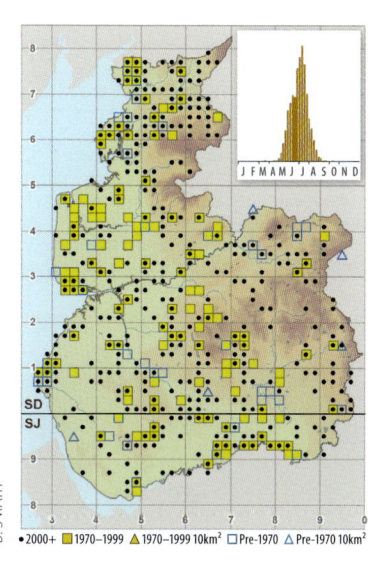

70.278 Common Wave *Cabera exanthemata*

■ **LANCS STATUS** Frequent / widespread resident   c.1857–2022
Noted in Liverpool (Gregson, 1857), then Burnley in 1910, Formby in 1919, Belmont in 1925 and Lytham in 1939. Early authors considered it common and general across the region. Since then, noted widely, particularly in damper parts of lowland areas and, less frequently, in similar upland localities. Sites with long-term, annual reports include Briercliffe, Flixton, Heysham, Leighton Moss, Mere Sands Wood and Swinton.
■ **HABITAT** Woodland, scrub, gardens and brownfield sites, particularly in wetter areas.
■ **FOODPLANTS** Willow sp., including Grey Willow.
Single brooded with, possibly, an occasional second brood. Comes to light. 50 seen on Winmarleigh Moss 17 June 2017 (B. Dyson). Larvae, late Aug. to mid-September.

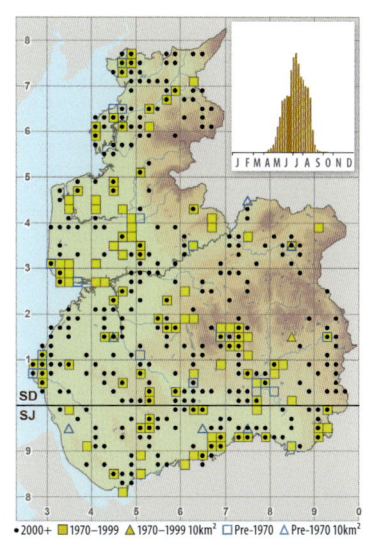

The Moths of Lancashire

GEOMETRIDAE

70.279 White-pinion Spotted *Lomographa bimaculata*

■ **LANCS STATUS** Occasional / widespread resident   1960–2022

Noted at Warton Crag on 22 May 1960 (C.J. Goodall), but present in south Cumbria in the late 19th century. Until 2000, only reported in the Silverdale area, from which it has since spread to the south and east. On 2 May 2000, found in Woolton, Liverpool (S. McWilliam), presumably arriving from the south, and spreading north and east, reaching Royton and Preston in 2003; the range expansion having since stalled.

■ **HABITAT** Hedgerows, woodland, scrub and gardens.

■ **FOODPLANTS** Not recorded in Lancashire.

Single brooded, with an occasional, small, second brood from late July to early September since 2005. Comes to light. 23 at Warton Crag on 29 April 2014 (J. Patton).

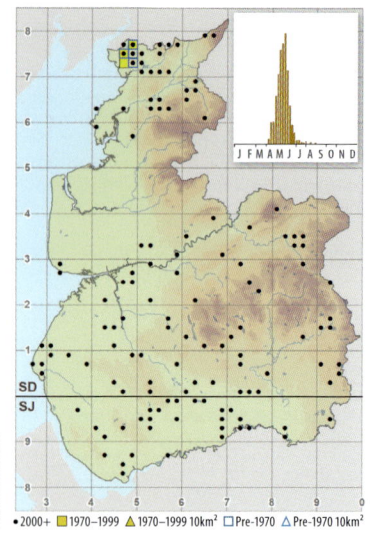

• 2000+   □ 1970–1999   △ 1970–1999 10km²   □ Pre-1970   △ Pre-1970 10km²

70.280 Clouded Silver *Lomographa temerata*

■ **LANCS STATUS** Frequent / widespread resident   1843–2022

Found in Preston in 1843 (Hodgkinson, 1844), then at Hale, Botany Bay Wood and Silverdale by 1910. Mansbridge (1940) noted it as local and not common. Reports increased during the 1960s and, with this, its range expanded, reaching Burnley by 1979. Since then, it has moved into many new areas and can occur in double figures at favoured sites. Despite this, it remains local in some parts of the county.

■ **HABITAT** Hedgerows, scrub, woodland and gardens.

■ **FOODPLANTS** Hawthorn and Blackthorn.

Single brooded. Comes to light. 52 seen at Royton on 5 June 2004 (R. Hart). Larvae have been found in August.

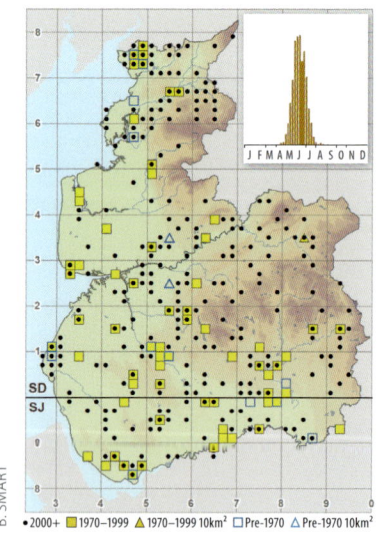

• 2000+   □ 1970–1999   △ 1970–1999 10km²   □ Pre-1970   △ Pre-1970 10km²

70.282 Early Moth *Theria primaria*

■ **LANCS STATUS** Occasional / local resident   c.1857–2022

Noted as plentiful in old lanes in the Liverpool area (Gregson, 1857), then in Stretford c.1859, Preston and Crosby c.1886 (Ellis, 1890), and Colne and Sabden c.1940. Since then, found when specifically targeted, near larger, old, rural hedgerows and occasionally in gardens. It has declined or been lost from areas where development has led to extensive hedgerow removal, but is also probably under-recorded.

■ **HABITAT** Hedgerows, scrub, rural gardens, woodland edge and brownfield sites.

■ **FOODPLANTS** Hawthorn and Blackthorn.

Single brooded. Comes to light. 27 males noted in car headlights at Tatham on 27 January 2006 (P.J. Marsh). Larvae found late April and May.

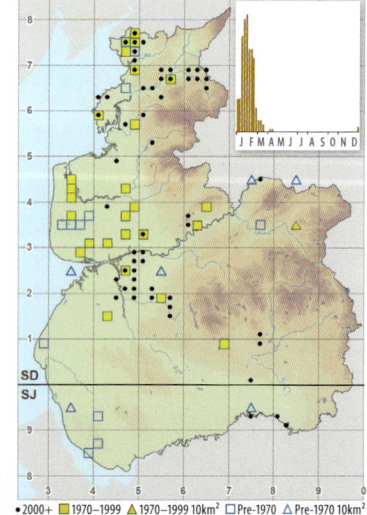

• 2000+   □ 1970–1999   △ 1970–1999 10km²   □ Pre-1970   △ Pre-1970 10km²

458   *The Moths of Lancashire*

70.283 Light Emerald *Campaea margaritaria*

- **LANCS STATUS** Frequent / widespread resident   c.1857–2022

Noted in Liverpool (Gregson, 1857), then reported as common and well distributed in the region by all early authors. A wider expansion in range in recent decades may have been partly linked to increased light trapping, but record numbers have also increased at several long-term trap sites, such as Flixton, Hale, Parr and Preston.
- **HABITAT** Woodland, gardens, scrub and brownfield sites.
- **FOODPLANTS** Elm sp., including Wych Elm, Ash and Sallow.

Single brooded until the early 1990s; double brooded in every year since 2006. The numbers of second brood individuals have increased considerably since 2020. Comes to light. 68 seen at Myers Allotment on 2 July 2011. Larvae, late August to mid-April.

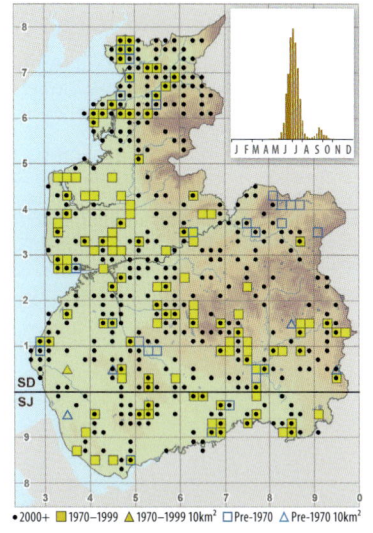

• 2000+   ■ 1970–1999   ▲ 1970–1999 10km²   □ Pre-1970   △ Pre-1970 10km²

70.284 Barred Red *Hylaea fasciaria*

- **LANCS STATUS** Occasional / widespread resident   1843–2022

Reported from the Preston area in 1843 (Hodgkinson, 1844), followed by Chat Moss, Ribchester, Silverdale and Formby. Up to 1940, all early authors considered it frequent in fir woods and widely distributed. Since the late 1980s, records increased in geographical range and quantity, initially from sites such as Billinge, Bilsborrow, Claughton and Skelmersdale. More recent, regular reports have come from Ainsdale, Briercliffe, Docker Moor, Formby and Leighton Moss, generally in small numbers.
- **HABITAT** Coniferous woodland.
- **FOODPLANTS** Not recorded in Lancashire.

Single brooded. Comes to light. 23 seen at Ainsdale on 14 July 2017 (R. Moyes *et al.*).

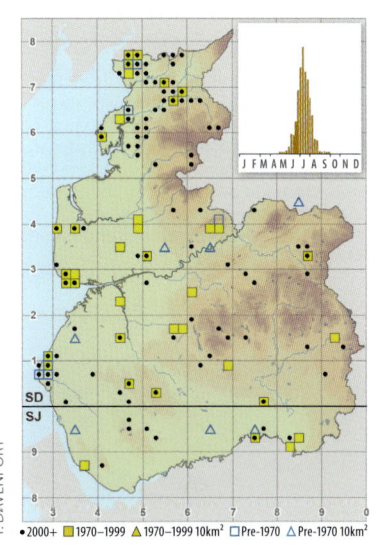

• 2000+   ■ 1970–1999   ▲ 1970–1999 10km²   □ Pre-1970   △ Pre-1970 10km²

70.287 Annulet *Charissa obscurata*

- **LANCS STATUS** Occasional / very local resident   c.1886–2022

Noted, once, on the banks of the R Wyre c.1886 by J.B. Hodgkinson. It was then seen in the Silverdale district in 1910, an area that remains its stronghold to the present day. During this time, it has been noted on 84 occasions with the majority of records from Gait Barrows and Warton Crag, and once as far east as Plain Quarry, Burton-in-Kendal. Elsewhere, it came to light in Hornby on one occasion in 1993 (J. Newton).
- **HABITAT** Herb rich grassland on the northern limestone.
- **FOODPLANTS** Not recorded in Lancashire.

Single brooded. Comes to light in single figures, the highest count being eight to actinic light on 21 July 2012 at Yealand Hall Allotments (B. Hancock).

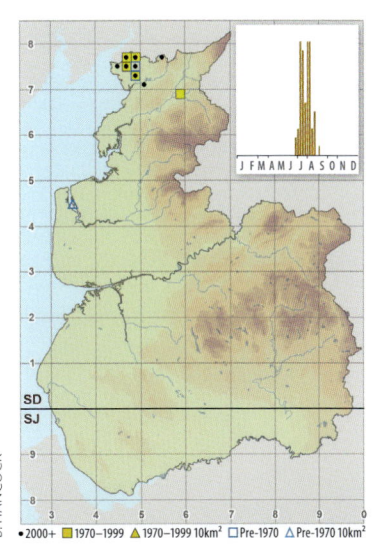

• 2000+   ■ 1970–1999   ▲ 1970–1999 10km²   □ Pre-1970   △ Pre-1970 10km²

GEOMETRIDAE

70.288 Brussels Lace *Cleorodes lichenaria*

■ **LANCS STATUS** Rare / very local resident c.1886–2017

The first was noted in the Preston district c.1886 by J.B. Hodgkinson, this being the only historic record, the precise details of which are unknown. On 25 July 2002, one came to light at Flixton (K. McCabe), followed by a singleton at Formby on 5 July 2010 (T. Davenport). It was then reported from Ainsdale on 3 July 2016, when two came to light (R. Moyes). The final sighting was of a singleton attracted to light at Sunderland Point on 16 June 2017 (P.J. Marsh).
■ **HABITAT** Gardens and scrub adjacent to the coast, and a lowland brownfield site.
■ **FOODPLANTS** Not recorded in Lancashire.
Single brooded. Comes to light.

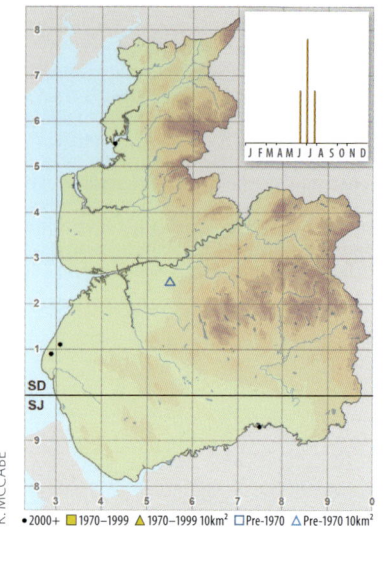

● 2000+ ▢ 1970–1999 ▲ 1970–1999 10km² ▢ Pre-1970 △ Pre-1970 10km²

70.292 Grey Scalloped Bar *Dyscia fagaria*

■ **LANCS STATUS** Occasional / local resident 1844–2022

First noted at White Moss, Manchester, on 31 May 1844 (Edleston, 1844), then from Chat Moss, Longridge Fell, Pilling Moss, Clougha and Pendle Hill, amongst others. Mansbridge (1940) noted it as fairly frequent on mosses but, by 1975, the loss or degradation of many lowland mosses led to its apparent demise at such sites. Since then, scattered and erratic sightings have been noted from several upland areas, although it does wander on occasions. Most recent reports are from Thrushgill.
■ **HABITAT** Lowland raised bogs and moorland.
■ **FOODPLANTS** Heather.
Single brooded. Comes to light in low single figures. Larva found in May.

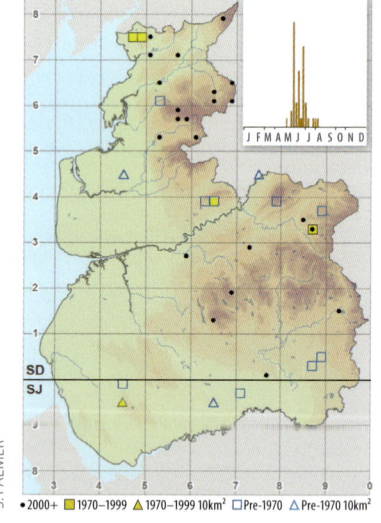

● 2000+ ▢ 1970–1999 ▲ 1970–1999 10km² ▢ Pre-1970 △ Pre-1970 10km²

70.295 Grass Wave *Perconia strigillaria*

■ **LANCS STATUS** Rare / very local resident 1843–2021

The first was near Pilling in 1843 (Hodgkinson, 1844), then noted as abundant at Chat Moss in 1846. Later seen on Risley Moss, Ecclestone Mere, Rainford Moss and near Bury. The last known records in VC59 were in 1900, although it may have hung on for a decade or two (Mansbridge, 1940). This leaves the adjacent mosses at Winmarleigh and Cockerham as its last known sites, and where three were seen on 19 June 2021.
■ **HABITAT** Lowland raised bog and once, historically, on moorland.
■ **FOODPLANTS** Not recorded in Lancashire.
Single brooded. Flies low over the mosses in sunny conditions with a peak of 35 seen on Cockerham Moss by day on 16 June 2010 (S. Palmer). Occasionally comes to light.

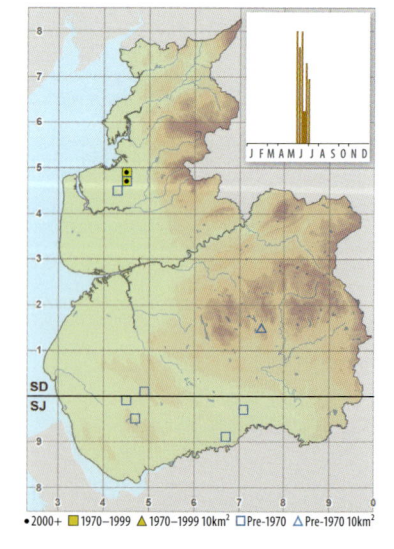

● 2000+ ▢ 1970–1999 ▲ 1970–1999 10km² ▢ Pre-1970 △ Pre-1970 10km²

460   *The Moths of Lancashire*

70.297 Grass Emerald *Pseudoterpna pruinata*

■ **LANCS STATUS** Scarce / very local resident   **c.1886–2022**

Noted in Silverdale and Morecambe c.1886 by J.B. Hodgkinson and later, in Bury and Blackburn in 1920. From the 1960s, it was regular at Leighton Moss and later, at Heysham, Carr House Common and Flixton. A rapid decline became apparent c.2000 and it was last seen in VC59 in 2016. The decline was slower in VC60, with it now found at only a few sites, the main ones being Warton Crag and Docker Moor.
■ **HABITAT** Scrub on coastal dunes, limestone grassland, moorland and brownfield sites.
■ **FOODPLANTS** Gorse and Broom.

Single brooded. Comes to light in single figures. In 1989, Heysham N.R. noted 101 between 21 June and 23 September (P.J. Marsh). Larvae found mid-Oct. to mid-June.

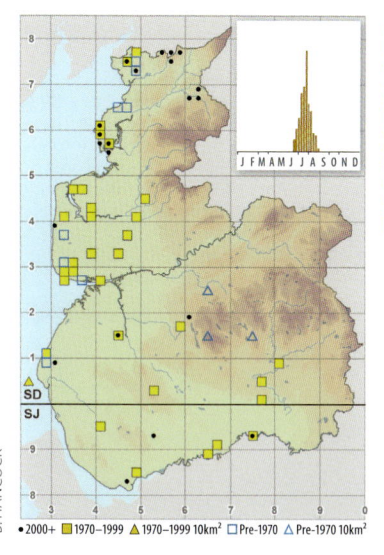

70.299 Large Emerald *Geometra papilionaria*

■ **LANCS STATUS** Occasional / widespread resident   **1843–2022**

Noted in Preston, 1843 (Hodgkinson, 1844), then Woolton, Chat Moss and Didsbury. From 1950–1979, records increased in number and range, possibly in line with the commencement of light trapping. Regular at several sites from the 1990s, such as Astley Moss, Gait Barrows, Leighton Moss, Flixton and Mere Sands Wood; irregularly noted at many others. Mating pair seen 9 September 2021 (K. Haydock & J. Mills)
■ **HABITAT** Woodland, particularly on limestone, scrub, hedgerows and brownfield sites.
■ **FOODPLANTS** Birch sp.

Single brooded. Comes to light, usually in single figures. 14 were found on Cadishead Moss on 27 June 2011 (K. McCabe). Larvae found in May and June.

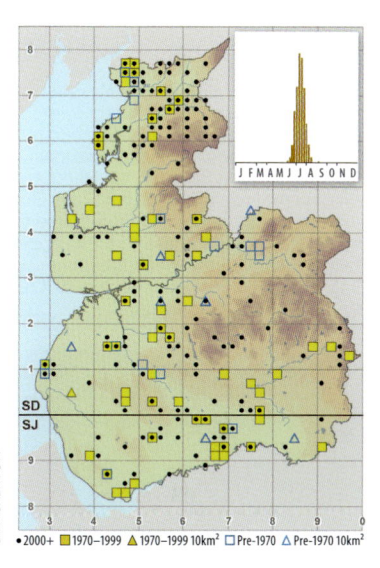

70.300 Blotched Emerald *Comibaena bajularia*

■ **LANCS STATUS** Scarce / local resident   **1954–2022**

First recorded at Didsbury on 1 July 1954 by H.N. Michaelis, this being suggestive of a recent arrival in the county or a vagrant, rather than an overlooked resident. It was not found again until 9 July 1979 at Mere Sands Wood (A. Fairhurst) and since then, there have been 17 records, all but one in VC59 between 1986 and 2022. The single VC60 record was from Leighton Moss on 5 July 1998 (G. Powell). Recorded on three occasions at Clincton Woods, Widnes (P. Hillyer *et al.*), in 2018, 2019 and 2022.
■ **HABITAT** Woodland and scrub.
■ **FOODPLANTS** Not recorded in Lancashire.

Single brooded. Comes to light. Three were seen at Widnes on 17 June 2022.

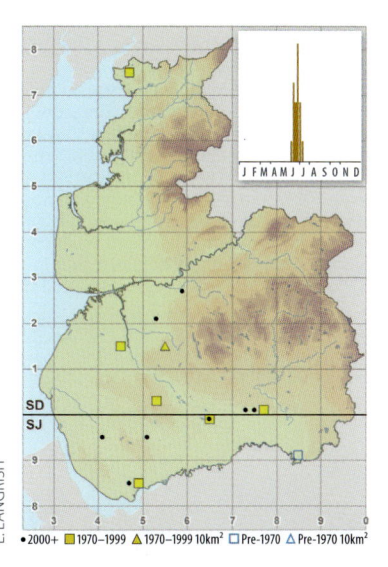

The Moths of Lancashire **461**

GEOMETRIDAE

70.302 Small Emerald *Hemistola chrysoprasaria*

■ **LANCS STATUS** Rare / vagrant   2020

A single record of one attracted to a MV light trap in a Clitheroe garden on 25 June 2020 (J. Morris). In adjacent counties, noted recently in south Cumbria, but has not been seen in Cheshire (cheshire-moth-charts.co.uk, accessed April 2023) and there were only two relatively recent records from sites in Yorkshire (yorkshiremoths.co.uk accessed April 2023). In Yorkshire it is suggested it may occur as a very localised resident utilising *Clematis* sp., a possibility than cannot be dismissed here.
■ **HABITAT** Unknown.
■ **FOODPLANTS** Not recorded in Lancashire.
Single brooded. Comes to light.

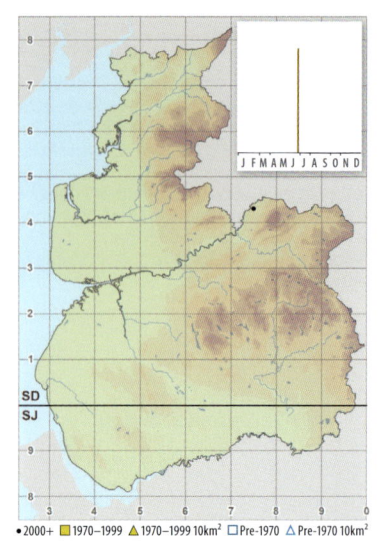

70.303 Little Emerald *Jodis lactearia*

■ **LANCS STATUS** Occasional / local resident   c.1857–2022

Noted as abundant in lanes near Liverpool (Gregson, 1857), near Silverdale in 1909, as common at Lytham Hall in 1910 and at Simonswood Moss in 1912. In more recent decades, seen regularly in only two main areas, the moorland-edge woods east of Lancaster (such as Bay Horse and Birk Bank) and (with most sightings), open, scrubby areas on the limestone, east to Dalton. Elsewhere, irregular reports received from 18 well scattered localities, too widespread to suggest vagrancy from other sites.
■ **HABITAT** Woodland rides and scrub; occasionally woody moorland edges.
■ **FOODPLANTS** Hawthorn and Sycamore.
Single brooded. Flies by day and comes to light. 16 seen at Birk Bank on 31 May 2004.

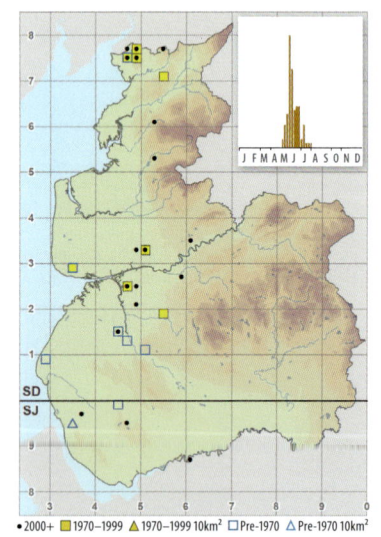

70.305 Common Emerald *Hemithea aestivaria*

■ **LANCS STATUS** Frequent / widespread resident   c.1857–2022

Noted in the Liverpool district (Gregson, 1857), with Ellis (1890) reporting it from Aigburth, Fleetwood and Lytham. Mansbridge (1940) referred to it as local and not common, except at Heysham where it was fairly common. Since then, it has spread into most lowland areas, but it remains local in higher parts, such as Littleborough from 2001, Overtown in 2002 and Todmorden in 2009.
■ **HABITAT** Hedgerows, scrub, parks, gardens and brownfield sites.
■ **FOODPLANTS** Oak sp., Hawthorn, Meadowsweet, Rowan and Raspberry.
Single brooded. Comes to light, mainly in single figures. 44 seen at Heysham N.R. on 27 June 2006, but numbers there have declined since then. Larvae found July to June.

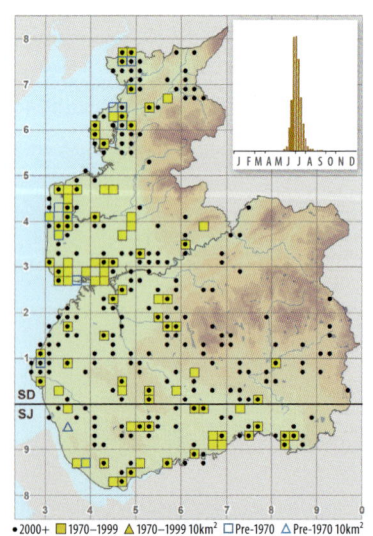

71.003 Puss Moth *Cerura vinula*

■ **LANCS STATUS** Occasional / widespread resident   c.1843–2022
First noted from Preston area (Hodgkinson, 1844). Noted as plentiful where sallows and willows grow (Gregson, 1857). Next, from Oldham area in the late 19th century. Recent records show frequent in coastal areas, such as Birkdale and Ainsdale dunes. Also, from upland sites, e.g., Thrushgill. An interesting record was of several larvae in a willow circle nr. Lune Aqueduct 2014 (B. Nixie). Overall, numbers appear stable.
■ **HABITAT** Woodland, hedgerows, sand dunes, brownfield sites.
■ **FOODPLANTS** Poplars including Balsam Poplar, Black-poplar hybrids and White Poplar, and willows including White Willow and Creeping Willow.
Single brooded. Comes to light. Many larval records from June to September.

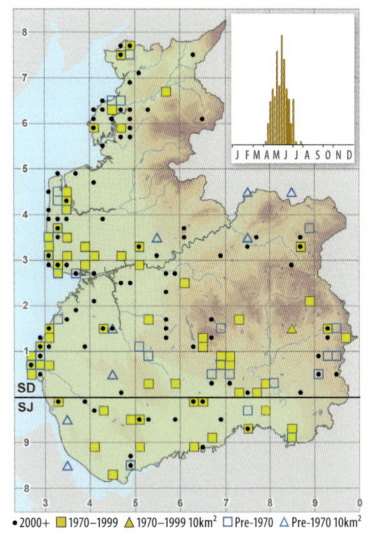

71.005 Sallow Kitten *Furcula furcula*

■ **LANCS STATUS** Frequent / widespread resident   c.1848–2022
First reported from Preston district by Hodgkinson (1849a) with other 19th century records from Warrington (Byerley, 1854), Chat Moss and Lytham St Annes. Later, noted at Ainsdale and Crosby (Mansbridge, 1940), demonstrating apparent early preference for coastal sites, with inland records less common. Mainly a lowland species, although the last few decades have seen first records from eastern, more elevated, parts of the county, such as Wray, Thrushgill, Briercliffe and Rishton.
■ **HABITAT** Woodland, hedgerows, scrub.
■ **FOODPLANTS** Willows including Grey Willow, Balsam Poplar.
Double brooded with larvae May-June and August-September. Moth comes to light.

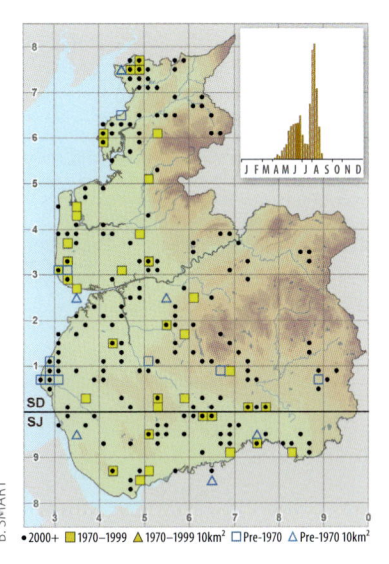

71.006 Alder Kitten *Furcula bicuspis*

■ **LANCS STATUS** Occasional / local resident   c.1847–2022
Found on an Alder trunk in the Preston area, the first British record, by J. Cooper (Doubleday, 1847). Vacated cocoons found nr. Hurst Green 1879 by J.B. Hodgkinson. Described as occasional around Preston, with 'as many as eight or nine being taken in a week by Mr Hodgkinson' (Ellis, 1890). Few records thereafter, until the 21st century, with 120 subsequent reports mainly in upland areas of north and east. Recorded during 2022 at Lancaster, Longridge, Cuerden Valley Park and Widnes.
■ **HABITAT** Woodland, moorland.
■ **FOODPLANTS** Empty cocoons on Alder. Larva recorded at Red Moss, nr. Horwich, 1999.
Single brooded; occasional August record suggests possible small second brood.

The Moths of Lancashire **463**

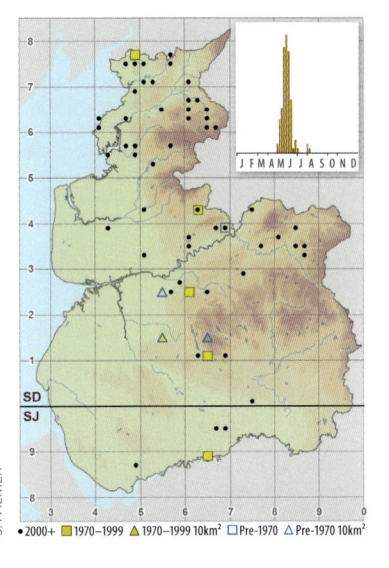

NOTODONTIDAE

71.007 Poplar Kitten *Furcula bifida*

■ **LANCS STATUS** Occasional / local resident   c.1848–2022

In a description of the previous species, Hodgkinson (1849a) also mentions previously finding Poplar and Sallow Kittens in the Preston area. Both Mansbridge and Ellis reported this to be a commoner species than the Sallow Kitten, which suggests quite a reversal in the fortunes of the two since. Nevertheless, appears to be slowly increasing in frequency this century. Almost all records are from VC59, with the only 21st century exceptions being two trapped at Carnforth in 2012 by E. Lyon.

■ **HABITAT** Woodland, sand dunes.
■ **FOODPLANTS** Unrecorded. Larvae noted at Formby sand dunes in 1930 and 1947. Double brooded. Comes to light. Larvae recorded during late August.

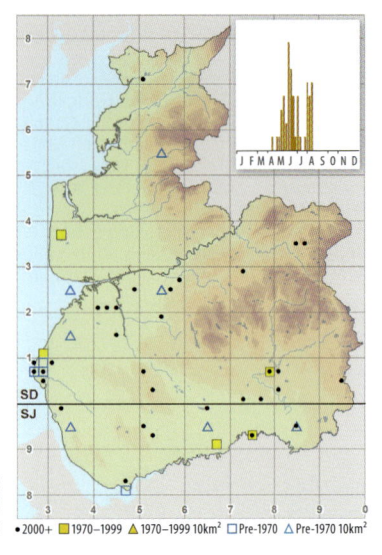

● 2000+ ☐ 1970–1999 ▲ 1970–1999 10km² ☐ Pre-1970 △ Pre-1970 10km²

71.010 Marbled Brown *Drymonia dodonaea*

■ **LANCS STATUS** Rare / vagrant.   1968

There is just a single Lancashire record for this species. This was an individual trapped at light by J. Briggs at Leighton Moss in 1968. The lack of any other records and the moth's presence in Cumbria, make it appear more likely that this was a vagrant or possibly a very temporary resident, rather than an isolated individual from an unknown Lancashire colony.

■ **HABITAT** Limestone woodland.
■ **FOODPLANTS** Not recorded in Lancashire.

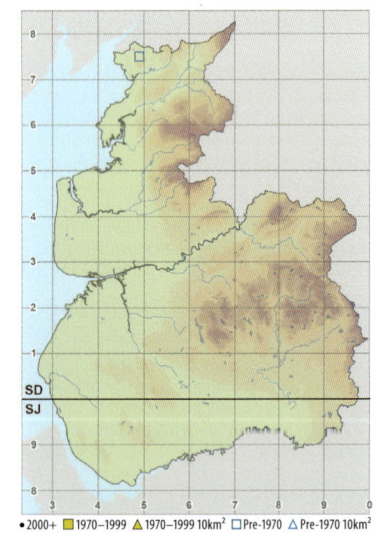

● 2000+ ☐ 1970–1999 ▲ 1970–1999 10km² ☐ Pre-1970 △ Pre-1970 10km²

71.011 Lunar Marbled Brown *Drymonia ruficornis*

■ **LANCS STATUS** Occasional / local resident   1947–2022

Distribution has expanded across the UK in recent years. First recorded in 1947 at the Formby sand dunes by R. Pritchard. Later, at Leighton Moss in 1968, Silverdale in 1970 and Warrington in 1985. The population explosion of the early 21st century has possibly slowed with only four 2021 records and 13 from 2022. Distribution suggests two separate populations; in the far north of VC60 and in the western half of VC59, with absence from much of Bowland, the Fylde, and the West Pennine Moors.

■ **HABITAT** Woodland.
■ **FOODPLANTS** Oak. Larvae noted at Rixton in 2010 and at Chorlton in 2020. Single brooded. Comes to light. Larvae recorded in mid-June.

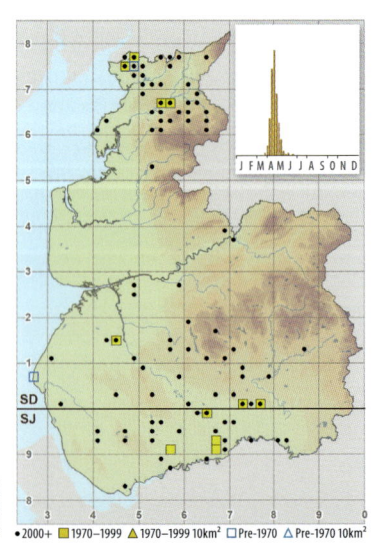

● 2000+ ☐ 1970–1999 ▲ 1970–1999 10km² ☐ Pre-1970 △ Pre-1970 10km²

The Moths of Lancashire

71.012 Iron Prominent *Notodonta dromedarius*

■ **LANCS STATUS** Frequent / widespread resident c.1843–2022
Recorded by C.S. Gregson at Heysham Moss c.1843. Also noted at Simonswood Moss (Gregson, 1857), with other early records from Chat Moss, Burnage, Burnley, Crosby, etc. Reported as very common on the mosses (Mansbridge, 1940). Subsequent records have shown wider habitat variety, including upland sites at Longridge Fell, Leck Fell, Docker Moor. Some sites, e.g., Flixton, show evidence of recent decline.
■ **HABITAT** Woodland, mosses, moorland, scrub, gardens.
■ **FOODPLANTS** Birch, Alder, Hazel, Grey Willow.
Double brooded with second generation more numerous. Comes to light. High count of 22 at Dolphinholme 2013. Larvae have been recorded from July to September.

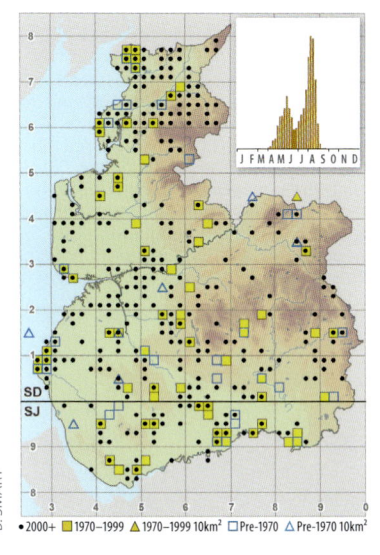

71.013 Pebble Prominent *Notodonta ziczac*

■ **LANCS STATUS** Frequent / widespread resident c.1843–2022
This species was first noted around 1843 at Preston by J.B. Hodgkinson (1844), and later, at Heysham Moss (Gregson, 1857). In the late 19th century, it was also noted at St Helens, Manchester, Liverpool, Simonswood Moss and Formby. Both Ellis (1890) and Mansbridge (1940) described the moth as generally distributed. This remains the same to the present day. However, counts have declined, dramatically so at Flixton, from a maximum annual total of 172 in 1998 to just ten in 2022 (K. McCabe).
■ **HABITAT** Woodland, hedgerows, gardens, moorland, sand dunes.
■ **FOODPLANTS** Willows including Grey Willow, Aspen. Larvae June and August-October.
Double brooded. Comes to light, with high count of 27 at Ainsdale, 6 May 2018.

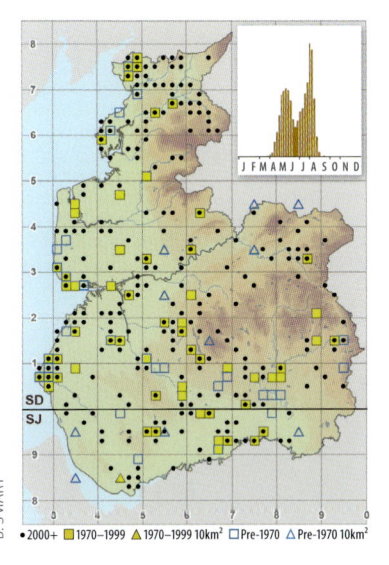

71.016 Great Prominent *Peridea anceps*

■ **LANCS STATUS** Scarce / very local resident 1977–2022
Only known from the Morecambe Bay limestone area, with earliest records from Leighton Moss in 1977 by J. Briggs and from Gait Barrows in 1983. Recorded seven times during the 1990s, the species has since become established, likely having spread from nearby Cumbrian colonies. Other frequently recorded sites are Yealand Conyers and Silverdale. Concerningly, numbers seem to be tailing off, with just four records from 2020 to 2022, compared to 25 in the three previous years.
■ **HABITAT** Limestone woodland.
■ **FOODPLANTS** Foodplant unrecorded within the county.
Single brooded. All records are of moths attracted to light.

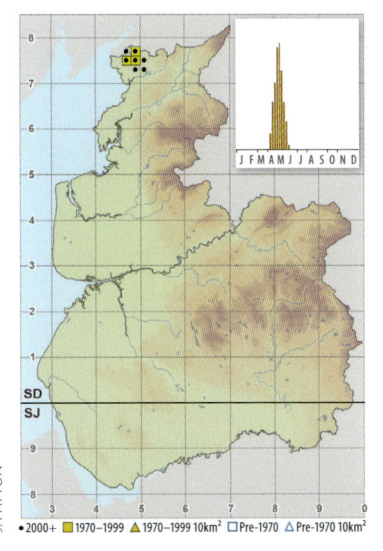

The Moths of Lancashire **465**

NOTODONTIDAE

71.017 Swallow Prominent *Pheosia tremula*

■ **LANCS STATUS** Frequent / widespread resident   c.1843–2022

The earliest record is from the Preston area (Hodgkinson, 1844), with Gregson (1857) noting larvae on sallows and poplars. Mansbridge (1940) reported a preference for coastal sites. Both he and Ellis agreed the moth was uncommon, whilst noting it from Manchester, Preston and the Sefton Coast. Light trapping revealed populations to be widespread, with most gardens recording the moth. Less common on moorland and mosses than the following species. Records show slight increase over last few years.
■ **HABITAT** Woodland, hedgerows, scrub, sand dunes, brownfield sites.
■ **FOODPLANTS** Willows and poplars including Black-poplar var., White Poplar and Aspen.
Double brooded. Comes to light. Larvae have been recorded from August-October.

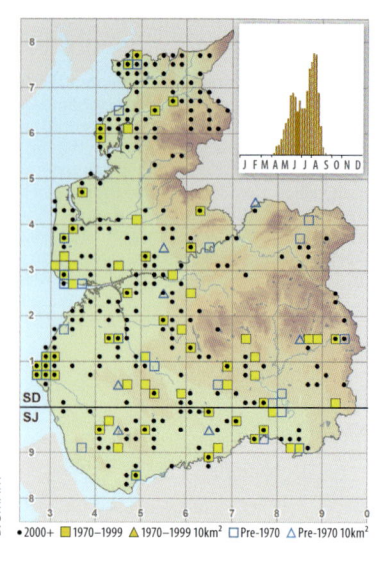

71.018 Lesser Swallow Prominent *Pheosia gnoma*

■ **LANCS STATUS** Frequent / widespread resident   c.1857-2022

Larvae noted on birch at Liverpool (Gregson, 1857); later from Chat Moss and Risley Moss by J. Chappell c.1859. Early authors noted its affinity for the mosslands, finding the moth less common elsewhere. The advent of light trapping produced many new records, often in gardens away from the mosses. Remains well distributed throughout, including on higher ground, but with a decline in total records from 2020-2022 to around 60% of a decade earlier, and an even greater reduction in abundance.
■ **HABITAT** Woodland, mosses, moorland, sand dunes, scrub, brownfield sites.
■ **FOODPLANTS** Birch.
Double brooded. Comes readily to light. Larvae found August to September.

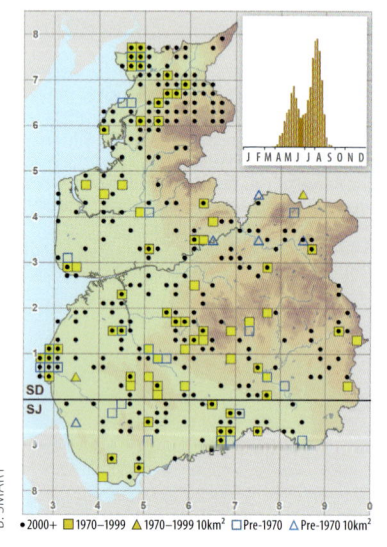

71.020 Pale Prominent *Pterostoma palpina*

■ **LANCS STATUS** Occasional / local resident   1888–2022

First recorded by T. Baxter at St Annes 1888 and considered very scarce (Mansbridge, 1940). Not further noted until 1956 at Silverdale by H.N. Michaelis, and at Ainsdale, Leighton Moss and Carnforth in the 1960s. Increasingly recorded this century, but the map demonstrates current absence from much of southern VC60 and eastern VC59. Primarily a lowland species, although highest count, 29 in 2017, was at Docker Moor.
■ **HABITAT** Woodland, hedgerows, scrub, gardens.
■ **FOODPLANTS** Grey Willow. Larva noted on 5 September 2022 at Chorlton (B. Smart).
Double brooded. Early records suggested the first generation was more numerous. However, since the mid-1990s, the second has been larger. Comes to light.

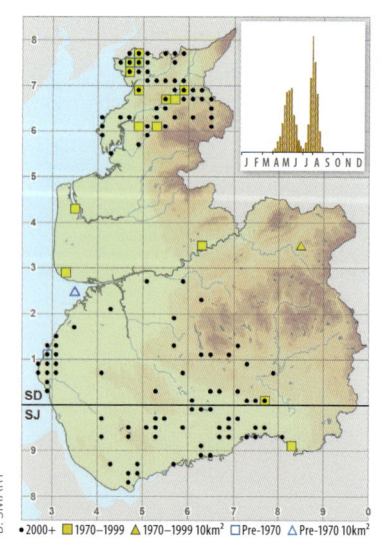

71.021 Coxcomb Prominent *Ptilodon capucina*

■ **LANCS STATUS** Frequent / widespread resident   c.1843–2022

Noted on Heysham Moss around 1843 by C.S. Gregson. Fairly common, especially as larvae on the mosses (Ellis, 1890). Mansbridge (1940) considered the species common and reported larvae found near Liverpool. Thereafter, regular to light with an increase in distribution likely linked to increased recording. Shows marked decline over the past two decades with 2022 numbers around 60% of those from 2002.
■ **HABITAT** Mosses, scrubby moorland edge, woodland, gardens, brownfield sites.
■ **FOODPLANTS** Birch, willows, Sycamore, Hazel. Larvae found August to October. Single brooded with extended flight period. Comes to light; occasionally in double figures in suitable habitat, e.g., Gait Barrows, Mere Sands Wood.

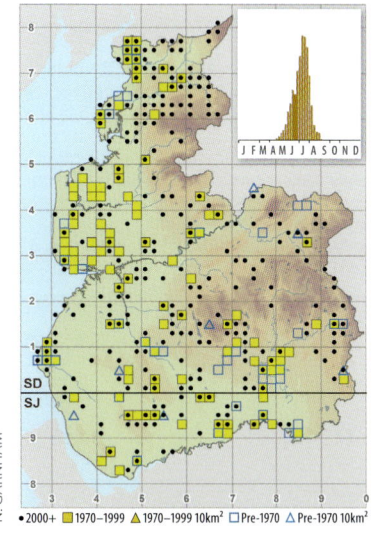

71.023 Scarce Prominent *Odontosia carmelita*

■ **LANCS STATUS** Scarce / very local resident   1891–2019

First recorded in Carnforth 1891 (recorder unknown), and at Silverdale in 1920 and 1940. Increased records following the advent of light trapping has shown Morecambe Bay limestone area to be its stronghold, with records from Warton Crag, Silverdale Moss, Gait Barrows and Leighton Moss. Noted 34 times this century, but not since 2019. There is a single VC59 record, from Simonswood Moss, nr. Kirkby in the mid-20th century. More recently, on forestry sites at Docker Moor and Lord's Lot Wood.
■ **HABITAT** Birch woodland in limestone areas. Occasionally, acidic woodland.
■ **FOODPLANTS** Not recorded in Lancashire.
Single brooded. Attracted to light in small numbers.

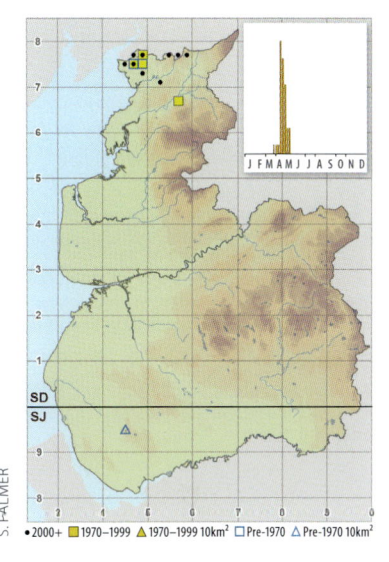

71.025 Buff-tip *Phalera bucephala*

■ **LANCS STATUS** Frequent / widespread resident   c.1857–2022

Larvae common on willow and oak in the mid-19th century (Gregson, 1857), with other historical authors concurring with this status. Frequently recorded during 20th century in southern and central Lancs. No Morecambe Bay records noted until 1961 at Carnforth. Current status appears stable, with moth well distributed throughout. High count of 32 at Mere Sands Wood 1998, with larger counts of the gregarious larvae, including up to 100 by a disused railway line nr. St Helens, 1999 (S. McWilliam).
■ **HABITAT** Woodland, hedgerows, parks, gardens, mosses and other damp habitats.
■ **FOODPLANTS** Willows (inc. Grey Willow, Osier), oak, birch, Hawthorn, Hazel, lime, rose. Single brooded. Occasionally noted in day and at dusk. Comes readily to light.

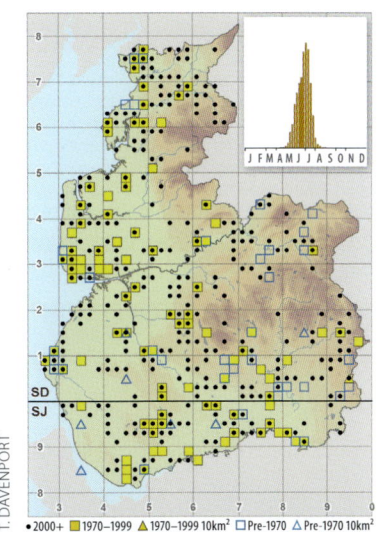

NOTODONTIDAE

71.027 Chocolate-tip *Clostera curtula*

■ **LANCS STATUS** Occasional / local resident   2010–2022

A recent arrival with the first record on 23 April 2010 at Astley Moss by G. Riley. Noted at Pennington Flash 2011, Hale and Formby in 2014, Parr and Ainsdale in 2015 and Flixton in 2016. The first VC60 record came in 2021 at Heysham, surprisingly far north in the Vice-county. The only other VC60 records are from Longridge in 2022. Early indications from 2023 are that range is continuing to expand. Reasons for this species' successful colonisation are unclear, but likely related to climate change.
■ **HABITAT** Damp woodland.
■ **FOODPLANTS** Aspen. Larva recorded by B. Smart at Chorlton Water Park 25 June 2016. Double brooded. Comes readily to light with maximum count of six at Formby 2022.

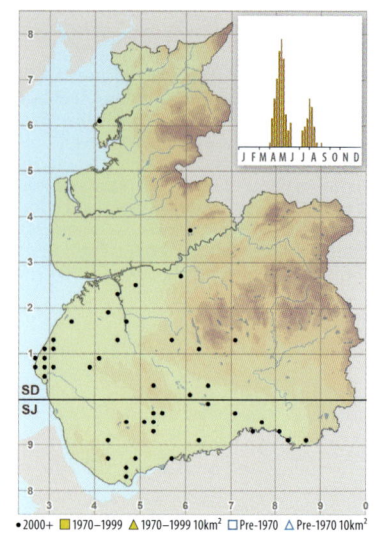

• 2000+   ☐ 1970–1999   △ 1970–1999 10km²   ☐ Pre-1970   △ Pre-1970 10km²

71.028 Small Chocolate-tip *Clostera pigra*

■ **LANCS STATUS** Rare / very local resident.   c.1857–1984

A single moth reported from Kirkby (Gregson, 1857), and later accepted by Ellis and Mansbridge, was our only record of this species, until noted from Southport sand dunes in 1968, recorder unknown. Next noted when trapped at Ainsdale dunes in 1975 by A.C. Aldridge and W.A. Watson. The 1975 moth was photographed, with the slide later exhibited by Watson at the Lancashire and Cheshire Entomological Society meeting of 12 December 1978 (LCES Annual Report 1978/79). The only subsequent report of the species in the county was a larval record, also from Ainsdale, in 1984 by M. Hadley. Unfortunately, we have no record of the foodplant, but given the site and its preference nationally, the likelihood is that Creeping Willow was utilised. It is possible that investigation of spun leaves on this plant may yet produce positive results. Recorded in Cumbria and Yorkshire with a few recent records at the latter. Unrecorded from Cheshire.
■ **HABITAT** Sand dunes.
■ **FOODPLANTS** Creeping Willow (probable).
Single brooded. As it is nearly 40 years since last recorded, extinction is a possibility.

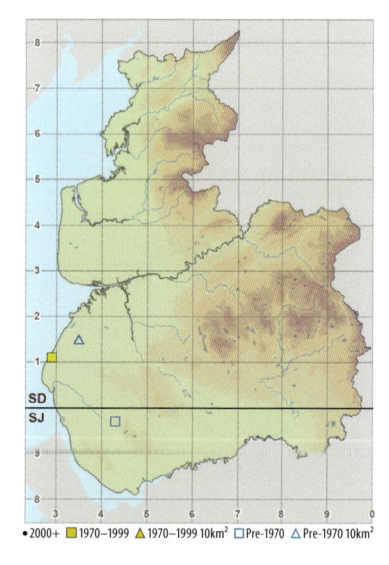

• 2000+   ☐ 1970–1999   △ 1970–1999 10km²   ☐ Pre-1970   △ Pre-1970 10km²

EREBIDAE

72.001 Herald *Scoliopteryx libatrix*

■ **LANCS STATUS** Frequent / widespread resident   c.1857–2022

Plentiful near old ditches in Liverpool (Gregson, 1857), with other authors finding it generally distributed but uncommon. Noted pre-1940 from Burnley, Pendle Hill and Ditton. Records increased significantly with advent of trapping, from lowland and upland areas. However, 2020–2022 records about 25% down from 20 years earlier. High count of 52 noted during bat surveys of tunnels at Whitefield 5 October 2012. Many other high counts from Jackhouse, Oswaldtwistle 2013 and 2014 (M. Memory).
■ **HABITAT** Woodland, hedgerows, brownfield sites, gardens, etc.
■ **FOODPLANTS** Willows, White Poplar, Aspen.
Single brooded. Comes to light. Recorded overwintering in caves, outhouses, etc.

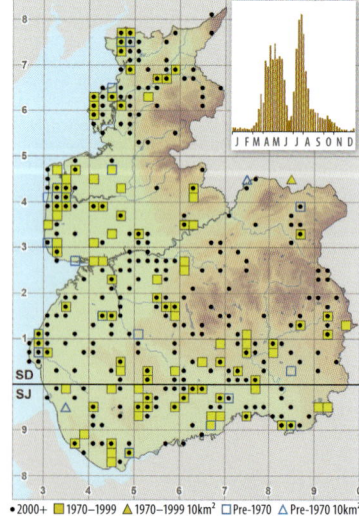

• 2000+   ☐ 1970–1999   △ 1970–1999 10km²   ☐ Pre-1970   △ Pre-1970 10km²

468   *The Moths of Lancashire*

72.002 Straw Dot *Rivula sericealis*

■ **LANCS STATUS** Frequent / widespread resident c.1857–2022
First records from Hale (Gregson, 1857), Manchester (Stainton, 1859), Silverdale 1910 and Formby 1941. Early authors considered it scarce and local. Increasingly well distributed this century, with range including Sefton Coast, Forest of Bowland and eastern, upland sites such as Green Bank, Pendle Hill, Worsaw Hill.
■ **HABITAT** Grassland.
■ **FOODPLANTS** Not recorded in Lancashire.
Formerly single brooded. However, the evidence of 40 October records in 2022, compared to none later than August in 2001, strongly suggests the moth is now double brooded. Comes to light. Easily disturbed from low vegetation during day.

• 2000+   ▪ 1970–1999   ▲ 1970–1999 10km²   ▫ Pre-1970   △ Pre-1970 10km²

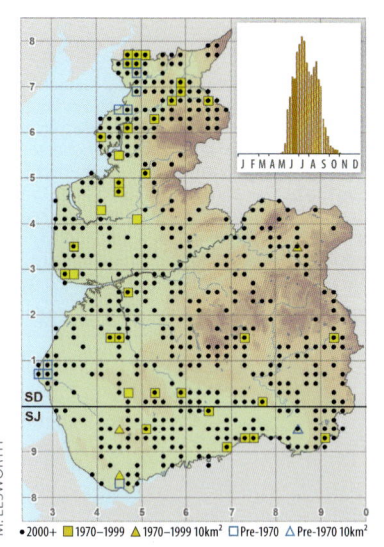

72.0021 *Orodesma apicina*

■ **LANCS STATUS** Adventive 1858
Noted on 15 August 1858, when a female moth was captured in the Chorley area, on the banks of the Leeds and Liverpool Canal by Thomas West. The recorder, unable to name the moth, showed it to a number of other entomologists, who were no more successful (Worthing, 1861). The moth's origins were a mystery. Hodgkinson (1861) wrote that as Chorley is 25 miles from a port, and the canal traffic was largely local coal barges, it was likely to be a native species. However, a couple of weeks later, he wrote that, Mr Doubleday had informed him that this was not a British species, but 'an exotic form, not at all likely to occur here unless imported' (Hodgkinson, 1861a). Doubleday also commented that he had a male of the same species which was taken near London and must have been imported from either Australia or India (Heath & Emmet, 1983c). Unsurprisingly, the moth has not been subsequently recorded in the county, or indeed elsewhere in the country.

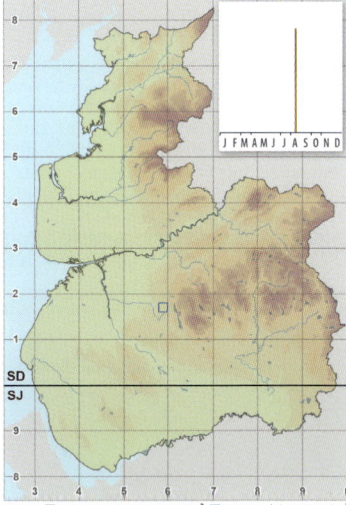

• 2000+   ▪ 1970–1999   ▲ 1970–1999 10km²   ▫ Pre-1970   △ Pre-1970 10km²

72.003 Snout *Hypena proboscidalis*

■ **LANCS STATUS** Abundant / widespread resident c.1857–2022
First reported in the Liverpool district (Gregson, 1857), with few documented records elsewhere until 1940. Despite this, Mansbridge (1940) reported it as common everywhere among nettles. In recent years, recorded far more widely, most frequently in lowland parts of the county, but with a presence in upland areas too. Overall, it seems that status has changed little since mid-19th century. High count of 66 at Bay Horse, nr. Dolphinholme on 3 July 2009 by N.A.J. Rogers.
■ **HABITAT** Woods, rough grassland, scrub, gardens, etc.
■ **FOODPLANTS** Nettle. Larva beaten from foodplant, Chorlton 17 August 2010 (B. Smart). Double brooded. Comes to light. Noted during day, disturbed from vegetation.

• 2000+   ▪ 1970–1999   ▲ 1970–1999 10km²   ▫ Pre-1970   △ Pre-1970 10km²

The Moths of Lancashire **469**

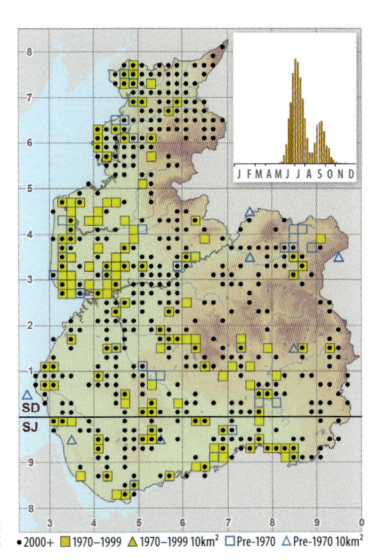

EREBIDAE

72.007 Beautiful Snout *Hypena crassalis*

■ **LANCS STATUS** Occasional / local resident 2001–2022
Recent expansion from the south, and possibly from Cumbria too, with the moth now fairly well distributed in upland areas. Occasional records in unlikely habitats, e.g., Mere Sands Wood and central Manchester, likely wanderers from higher ground. First noted at Trough Summit by S. Palmer on 2 July 2001, with other records to 2010 from Bay Horse, Gait Barrows and Lord's Lot. Counts of moths and records have increased post-2010 with regular sightings at Docker Moor and Lord's Lot Wood.
■ **HABITAT** Moorland, acidic woodland.
■ **FOODPLANTS** Not recorded in Lancashire.
Single brooded. Comes to light. Netted during day on Bilberry moorland.

● 2000+ ▪ 1970–1999 ▲ 1970–1999 10km² ☐ Pre-1970 △ Pre-1970 10km²

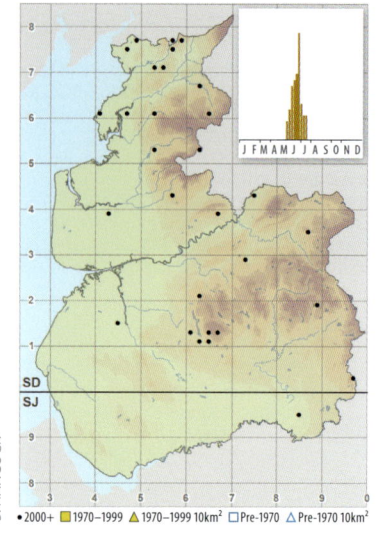

72.009 White Satin Moth *Leucoma salicis*

■ **LANCS STATUS** Frequent / widespread resident c.1857–2022
Noted among the dunes at Crosby by Gregson (1857). Ellis and Mansbridge noted its abundance on the coast, particularly as larvae, and a few inland records from Urmston and Burnley. The status appears little changed. Primarily a lowland species, with a coastal bias, and occasionally from Pennine towns, e.g., Rishton. Prone to population irruptions, with a spectacular count of 1,000 or so at Ainsdale Dunes 2011.
■ **HABITAT** Sand dunes, hedgerows, open woodland.
■ **FOODPLANTS** Creeping Willow, White Willow, Black-poplar hybrid, Ontario Poplar.
Single brooded, although a second brood of 'thousands' noted in the sand-hills by Gregson (1870). Comes to light. Larvae recorded from late April to June.

● 2000+ ▪ 1970–1999 ▲ 1970–1999 10km² ☐ Pre-1970 △ Pre-1970 10km²

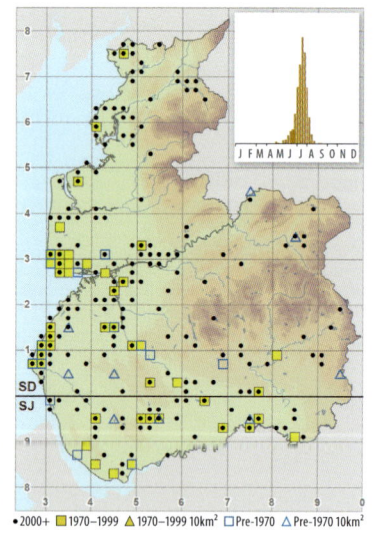

72.010 Black Arches *Lymantria monacha*

■ **LANCS STATUS** Scarce / local resident 1898–2022
Other than a possible adventive 1898 record from Carnforth by H. Murray, this appears to be a recent arrival in the county, spreading north. The first modern record came in 1998, when noted at Adlington, nr. Chorley by M. Maher and M. Lightowler. The 22 records from this century show presence in lowland areas at Silverdale, Sefton Coast, and Flixton, with two records from higher ground at Mill Houses. Other sites of single records include Martin Mere, Marton, Fazakerley and Warrington.
■ **HABITAT** Broadleaved and pine woodland.
■ **FOODPLANTS** Not recorded in Lancashire.
Single brooded. Comes to light, with all counts restricted to one or two.

● 2000+ ▪ 1970–1999 ▲ 1970–1999 10km² ☐ Pre-1970 △ Pre-1970 10km²

470 The Moths of Lancashire

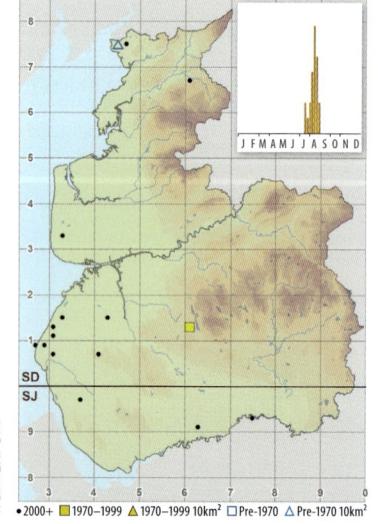

72.011 Gypsy Moth *Lymantria dispar*

■ **LANCS STATUS** Adventive c.1846–2017

There is a specimen of this moth, a male, in J.B. Hodgkinson's collection from Warrington in 1846. It is not known whether the moth was collected or bred (Ellis, 1890). The only other Lancashire record is of a larva found by A. Baines on 4 August 2017. It was noted among various plants in the garden department of a DIY superstore in Blackpool, so was presumably an accidental introduction with plant material.

■ **HABITAT** Unknown.
■ **FOODPLANTS** Not recorded in Lancashire.

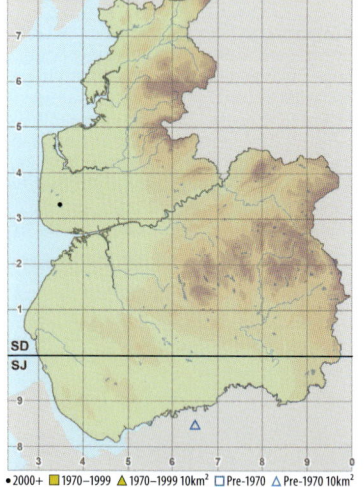

72.012 Brown-tail *Euproctis chrysorrhoea*

■ **LANCS STATUS** Rare / local resident and possible adventive 1984–2022

First recorded on 26 July 1984 at Heysham N.R. by P.J. Marsh, possibly of adventive origin. Recently, the national range has been expanding northwards, and a further specimen was noted at Hazelrigg Weather Station, nr. Lancaster on 18 July 2022 by J. Patton and S. Sharp. Both of the Lancashire moths were recorded at light.

■ **HABITAT** Coastal woodland, hedgerows.
■ **FOODPLANTS** Not recorded in Lancashire.

Single brooded. Occasionally, Yellow-tail females have been noted to have a brown, rather than a yellow tip to the 'tail', a feature previously documented by Mansbridge (1940). However, they lack the mainly brown abdomen typical of this species.

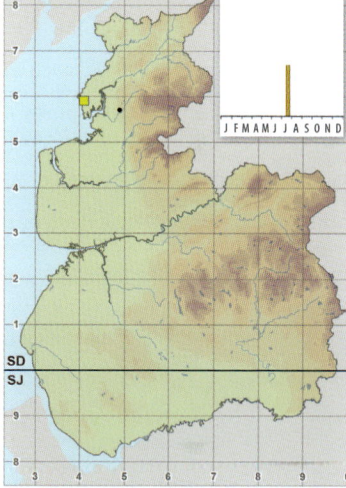

72.013 Yellow-tail *Euproctis similis*

■ **LANCS STATUS** Frequent / widespread resident c.1857–2022

First recorded in Liverpool (Gregson, 1857), then at Manchester, Burnley and Simonswood Moss. Early authors found the moth common, occasionally abundant. Found throughout, including on higher ground in VC60. Highest count of 52 to light at Wray in 2008. Females occasionally noted with dark tail-tufts, resembling Brown-tail.

■ **HABITAT** Woodland, hedgerows, scrub, gardens, sand dunes, etc.
■ **FOODPLANTS** Willows (inc. Creeping Willow), Hazel, Hawthorn, birch (inc. Silver Birch), oak (inc. Pedunculate Oak), Alder, Heather, Rowan, Blackthorn, Bramble, Sycamore, Field Maple, Meadowsweet, Wild Rose, Barberry, Bog Myrtle, Larch and Holly.

Single brooded. Comes to light. Larvae recorded April to September.

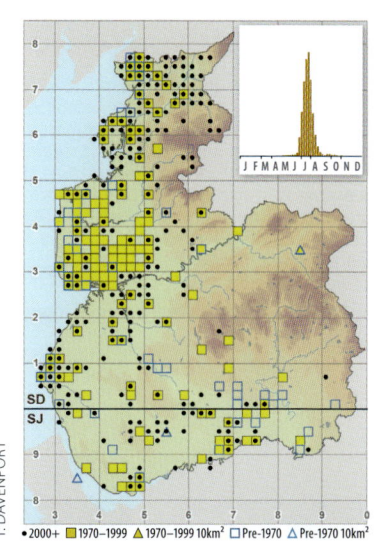

The Moths of Lancashire   471

EREBIDAE

72.015 Pale Tussock *Calliteara pudibunda*

■ **LANCS STATUS** Frequent / widespread resident   c.1857–2022
Early records from Liverpool (Gregson, 1857), Chat Moss, Preston and Lytham St Annes. Ellis and Mansbridge noted it to be generally distributed but uncommon. Post-2000 records show some expansion inland, although remains uncommon in eastern and central parts of the county. Highest count of 25 at Ainsdale 2017, with other high totals from Crosby, Dalton Crags, Yealand Hall, Docker Moor, etc.
■ **HABITAT** Woodland, hedgerows, gardens, parks, etc.
■ **FOODPLANTS** Goat Willow, Grey Willow, Beech, Hawthorn, Sycamore. Also, found fully-grown beneath lime trees, and once on Red Clover.
Single brooded. Comes to light. Larvae from July to October.

• 2000+   ☐ 1970–1999   △ 1970–1999 10km²   ☐ Pre-1970   △ Pre-1970 10km²

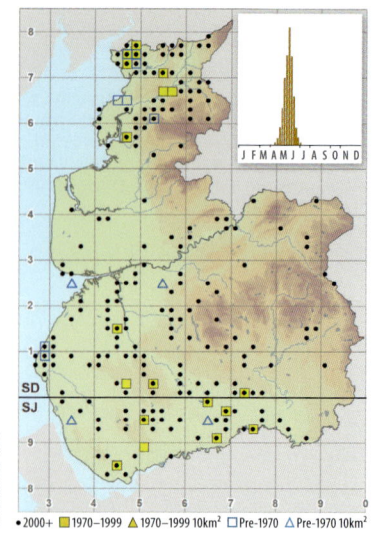

72.016 Dark Tussock *Dicallomera fascelina*

■ **LANCS STATUS** Scarce / very local resident   c.1845–2022
First reported from the coast, although specific location unclear (Hodgkinson, 1845a). Larvae noted in profusion on Creeping Willow (Gregson, 1857). Other 19th century records from Carnforth, Formby, Chat Moss, Blackpool etc. Ellis (1890) reported it common in the dunes; Mansbridge (1940) felt it was becoming less so. Recorded a little more often this century, likely related to recorder effort rather than increasing numbers, but now appears restricted to Sefton and Fylde coasts.
■ **HABITAT** Sand dunes.
■ **FOODPLANTS** Creeping Willow, Hawthorn, poplar.
Single brooded. Comes to light. Larvae recorded during May and June.

• 2000+   ☐ 1970–1999   △ 1970–1999 10km²   ☐ Pre-1970   △ Pre-1970 10km²

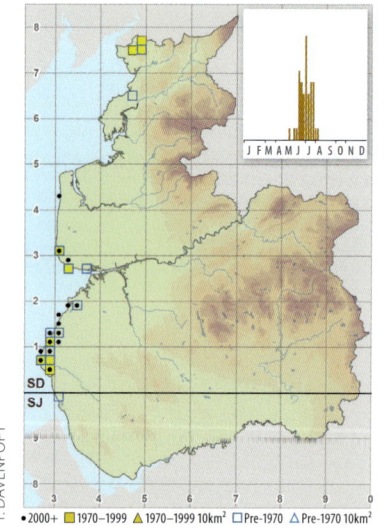

72.017 Vapourer *Orgyia antiqua*

■ **LANCS STATUS** Frequent / widespread resident   c.1857–2022
First recorded in the Liverpool district (Gregson, 1857). Mansbridge (1940) reported it to be widely distributed and common; more so in large towns. Most frequent in coastal and other lowland sites. Larvae may occur in numbers, defoliating all nearby plants, especially when in enclosed area, as at St Helens Hospital grounds in 2020.
■ **HABITAT** Woodland, hedgerows, field edges, scrub, parks, gardens, etc.
■ **FOODPLANTS** Rose, Bramble, Raspberry, Ash, dock, Dandelion, Sycamore, sallow, birch, Hawthorn, Pedunculate Oak, Alder, Wych Elm, Globe Buddleia, Primrose.
Single brooded. Male flies in sunshine; occasional at light. Flightless, apterous female, sometimes noted with eggs on window frames, etc. Larvae June to September.

• 2000+   ☐ 1970–1999   △ 1970–1999 10km²   ☐ Pre-1970   △ Pre-1970 10km²

The Moths of Lancashire

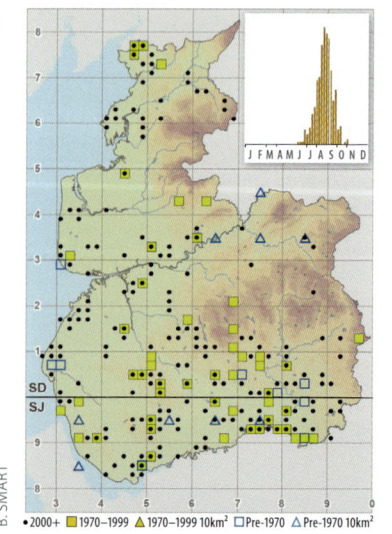

72.019 Buff Ermine *Spilosoma lutea*

■ **LANCS STATUS** Abundant / widespread resident c.1857–2022
First noted in Liverpool, with Gregson (1857) reporting it to be profuse in gardens. Other early authors gave the status as common to abundant. The many pre-1940 sites included Oldham, Crumpsall, Burnley, Nelson and Lytham. The map shows it to be well distributed throughout all the date ranges. Whilst numbers of records are increasing, 874 in 2022, actual totals of moths may possibly be in slight decline.
■ **HABITAT** Woodland, hedgerows, scrub, parks, gardens.
■ **FOODPLANTS** Sycamore, Norway Maple, sallows, thistles, Deadly Nightshade, Monk's-hood, Cucumber.
Single brooded but with occasional September record (12 since 2000). Comes to light.

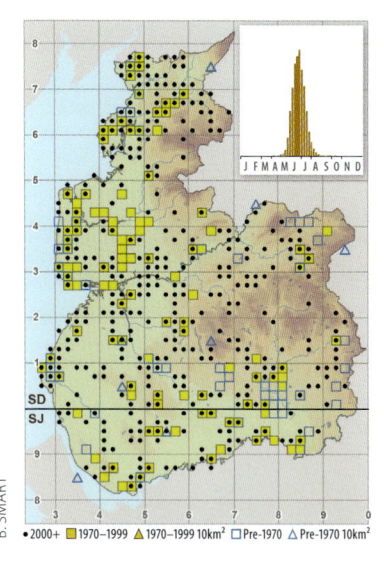

● 2000+   ■ 1970–1999   △ 1970–1999 10km^2 □ Pre-1970   △ Pre-1970 10km^2

72.020 White Ermine *Spilosoma lubricipeda*

■ **LANCS STATUS** Abundant / widespread resident c.1857–2022
First found in Liverpool (Gregson, 1857), then Oldham 1883, Crumpsall 1890, Burnley 1902 and Lytham St Annes 1914. Ellis (1890) considered it abundant and widespread and that has remained the case to the present day. Recorded annually at many sites in lowland and upland parts of the county, e.g., Yealand Conyers, Green Bank, Wray, Bay Horse, Heysham, Adlington, Worsthorne, Martin Mere, Formby, etc.
■ **HABITAT** Woodland, gardens, scrub, grassland, moorland.
■ **FOODPLANTS** Unknown. Fed on ragwort in captivity. Larvae noted late July to October. Mainly single brooded but with a probable small and occasional second brood. Comes to light. High count of 99, trapped at Longridge on 2 June 2004 (D. Lambert).

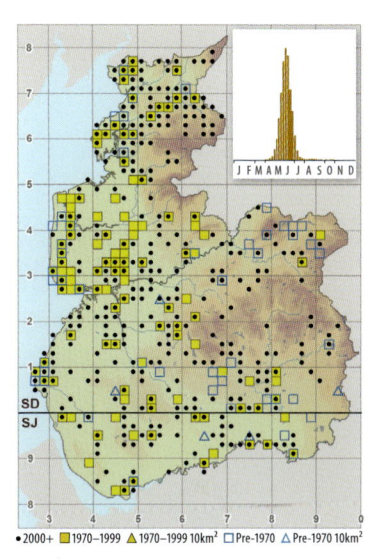

● 2000+   ■ 1970–1999   △ 1970–1999 10km^2 □ Pre-1970   △ Pre-1970 10km^2

72.022 Muslin Moth *Diaphora mendica*

■ **LANCS STATUS** Frequent / widespread resident c.1843–2022
First reported from the Preston area (Hodgkinson, 1844) with other 19th century records from Blackpool, Chorley, Bolton and Withington. Considered local by early authors. Now, widely distributed in lowland areas but virtually absent from higher ground. High counts of 25 in Flixton 1997 and 2011. Most records relate to males trapped at light. The females, primarily day-fliers, have been recorded far less at light, but occasionally noted flying in woodland or resting on tree trunks, fences, etc.
■ **HABITAT** Grasslands, open woodland, scrub, gardens, parks.
■ **FOODPLANTS** Sea Radish, Bramble. Larvae noted August and September.
Single brooded. Two late September records suggest small, occasional second brood.

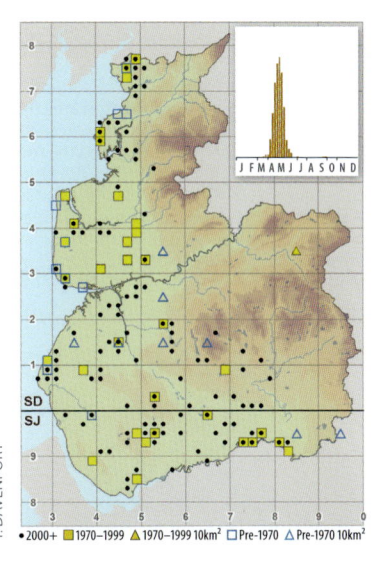

● 2000+   ■ 1970–1999   △ 1970–1999 10km^2 □ Pre-1970   △ Pre-1970 10km^2

EREBIDAE

72.023 Clouded Buff *Diacrisia sannio*

■ **LANCS STATUS** Scarce / very local resident   **1846–2023**

19th century records suggested moth confined to moss and moorland (Ellis, 1890). First recorded in 1846 from Chat Moss by R.S. Edleston. Also, at Hale (Gregson, 1857), Silverdale 1885 and Clougha Pike 1906. Only 18 records post-1900, with late 20th century records from Leighton Moss, Cockerham Moss, etc, and from Heysham Moss in 2009. Recorded in uplands at Thrushgill, Docker Moor and Longridge, in early 21st century, and at Ainsdale Dunes in 2021. Recorded at Green Bank 2023 (P.J. Marsh).
■ **HABITAT** Sand dunes, moorland, grassland, mosses.
■ **FOODPLANTS** Unrecorded, although larvae abundant at Barton Moss 1859 (J. Chappell). Single brooded. Recorded during day and at light, mainly of singletons only.

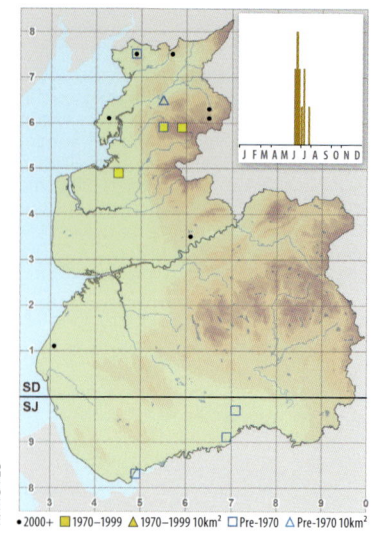

72.024 Ruby Tiger *Phragmatobia fuliginosa*

■ **LANCS STATUS** Frequent / widespread resident   **1844–2022**

First recorded in Manchester (Edleston, 1844). Early writers noted it to be common, with preferred habitat reported as waste lands (Gregson, 1857), heaths and moors (Ellis, 1890). Mansbridge (1940) noted larvae abundant on mosses and sand-hills, especially in autumn. Increased trapping showed it to be well distributed throughout. Population appears stable, with one-third of all records from 2018 to 2022.
■ **HABITAT** Open woodland, hedgerows, scrub, moorland, parks, gardens, etc.
■ **FOODPLANTS** Creeping Thistle, Common Ragwort, Betony, dock.
Double brooded with a small first generation. Comes readily to light. Larvae often encountered in the field and noted in all months other than December.

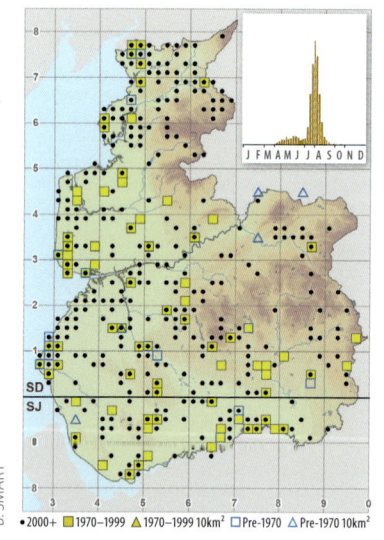

72.025 Wood Tiger *Parasemia plantaginis*

■ **LANCS STATUS** Occasional / local resident   **c.1843–2022**

19th century records from Preston area (Hodgkinson, 1844), Chat Moss, Risley Moss, Leighton Moss, with moorland records from Longridge Fell and Blackstone Edge, Littleborough. Early authors noted the moth on heaths and mosses. Never common, though possibly under-recorded. Recorded at Winmarleigh Moss, last in 2014. All subsequent records from uplands, e.g., Cant Clough, nr. Burnley and Jeffrey Hill.
■ **HABITAT** Moorland, rough grassland, open woodland, formerly on the mosses.
■ **FOODPLANTS** Unrecorded. Hundreds of larvae were noted by R. Foster on the fells at Woodyards, Over Wyresdale 5 July 2022, many being predated by gulls.
Single brooded. Occasional at light. Far more often encountered during day.

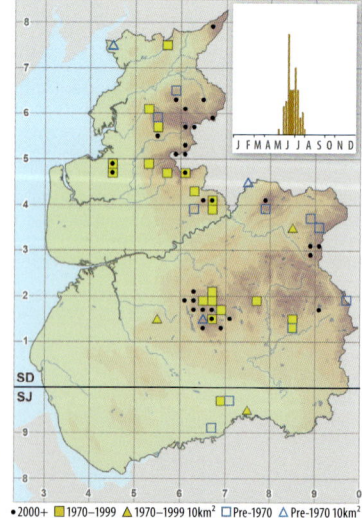

474 The Moths of Lancashire

72.026 Garden Tiger *Arctia caja*

■ **LANCS STATUS** Occasional / widespread resident   c.1857–2022

Has experienced significant decline nationally during this century, although locally, distribution and numbers of records are little changed compared to 20 years earlier. The data from K. McCabe's Flixton garden is more concerning, declining from 66 in 1997 to 3 in 2022. Early authors considered the moth abundant, with first records from Liverpool (Gregson, 1857), Lancaster, Crosby and Lathom. More recent records include Rishton, Ainsdale, Lytham St Annes, Leighton Moss and Dalton Crags.
■ **HABITAT** Open woodland, rough grassland.
■ **FOODPLANTS** Nettle, Comfrey, dock, Apple, sallow, ragwort, Colt's-foot, Cleavers.
Single brooded. Most records are at light. Larvae recorded August to June.

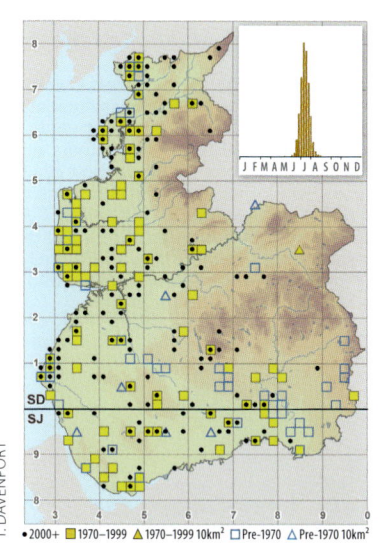

72.0272 Isabelline Tiger *Pyrrharctia isabella*

■ **LANCS STATUS** Adventive   c.1905–1906

Larvae were found near Carnforth by H. Murray, and recorded by Dr T.A. Chapman (LCES, 1906). It seems larvae were found for a number of years up to 1906, with an adult being reared by Chapman (Heath & Emmet, 1979). These were the first British records of this American species. It is thought likely that mode of entry related to importation of timber. There have been no further records of the moth in the county. The only further UK record was of larvae imported 'in recent years' on American oak by an Edinburgh cooper with three moths reared (Heath & Emmet, 1979).
■ **HABITAT** Unknown.
■ **FOODPLANTS** Not recorded in Lancashire.

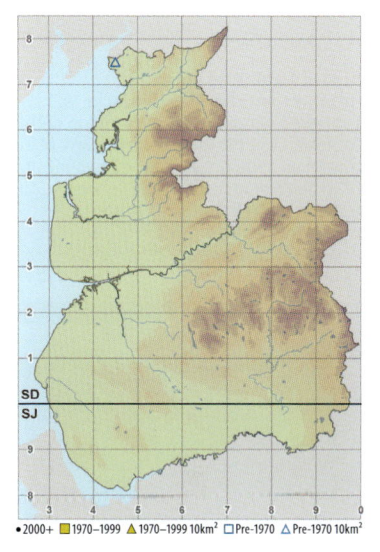

72.029 Scarlet Tiger *Callimorpha dominula*

■ **LANCS STATUS** Scarce / local resident   2016–2023

Spreading northwards from south-west England, the moth has been recorded in the county on ten occasions in the last ten years. First noted at Greenbank allotments, Wavertree by S. Collier on 6 Jul 2016. The following year, on 16 June, a further moth arrived in the light trap of J. Mitchell-Lisle at Great Sankey, Warrington. Recorded at Fishwick LNR on 25 June 2020 by T. Blackburn, the first for VC60. 2023 saw a significant increase in south-west VC59 with 7 records.
■ **HABITAT** Allotments, gardens.
■ **FOODPLANTS** Unrecorded within the county.
Single brooded. Comes to light but flies by day; often seen at rest on vegetation.

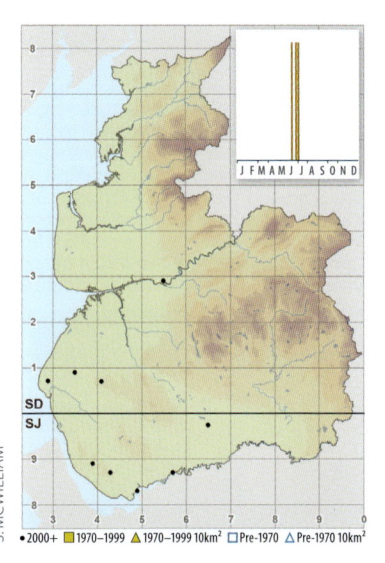

The Moths of Lancashire   **475**

EREBIDAE

72.031 Cinnabar *Tyria jacobaeae*

- **LANCS STATUS** Abundant / widespread resident   c.1857–2022

A very familiar species, with the gregarious yellow and black larvae noted as often in the field as the colourful day-flying moths. First reported from the Crosby sand-hills by Gregson (1857). Other early authors found it plentiful on the coast and upon the mosses, but uncommon inland. Now commonly found wherever Common Ragwort grows, less so in upland areas. Most high counts relate to larvae, e.g., approx. 500 at King Moss, nr. Rainford 12 July 2006 by C.A. Darbyshire.
- **HABITAT** Grasslands, sand dunes, waste ground, scrub, gardens, etc.
- **FOODPLANTS** Common Ragwort, Oxford Ragwort (latter at Preston 2022, D.&J. Earl). Single brooded. Flies in sunshine. Occasional at light.

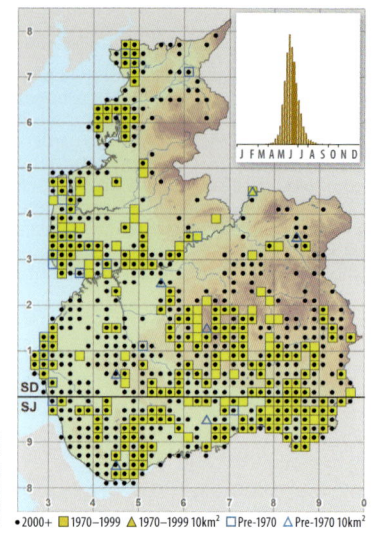

• 2000+   ▫ 1970–1999   ▲ 1970–1999 10km²   ☐ Pre-1970   △ Pre-1970 10km²

72.034 Crimson Speckled *Utetheisa pulchella*

- **LANCS STATUS** Rare / migrant   1871–2011

First recorded near Manchester in 1871 by J. Thorpe, noting, 'I have in my possession a very fine female in good condition of this very rare insect, captured on the 8th of September, in the railway canal-yard, Middleton Station, by one of the workmen employed, who boxed it for mere curiosity, being attracted by its beauty' (Thorpe, 1871). Further recorded near Fulwood College on 9 August 1961 by M.J. Ainscough and at Formby Point on 25 August 2011 by R. Hill, with the moth photographed.
- **HABITAT** Unknown.
- **FOODPLANTS** Not recorded in Lancashire.

All records are of moths noted during the day.

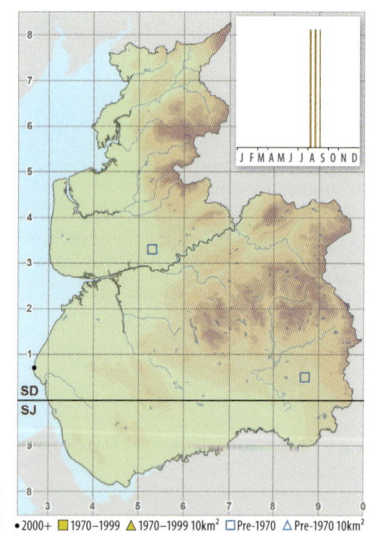

• 2000+   ▫ 1970–1999   ▲ 1970–1999 10km²   ☐ Pre-1970   △ Pre-1970 10km²

72.036 Muslin Footman *Nudaria mundana*

- **LANCS STATUS** Frequent / local resident   c.1885–2022

First reported at Silverdale by J.C. Melvill and at Preston by J.B. Hodgkinson (Ellis, 1890). Noted by Mansbridge (1940) to be common only in north Lancs. As with other lichen-feeding species the moth is doing well, with increasing records and slowly increasing distribution. Mainly found in the north, largely, though not exclusively, on higher ground. Occasional outlying records, e.g., Rochdale and Ormskirk, both 2013.
- **HABITAT** Stony habitats, woodland, moorland.
- **FOODPLANTS** Unrecorded. Larva swept from grassy bank with hawthorn nearby at Carnforth saltmarsh on 3 June 2014 by S. Palmer.

Single brooded. High count of 91 at light at Mill Houses, Wray by P.J. Marsh, 2008.

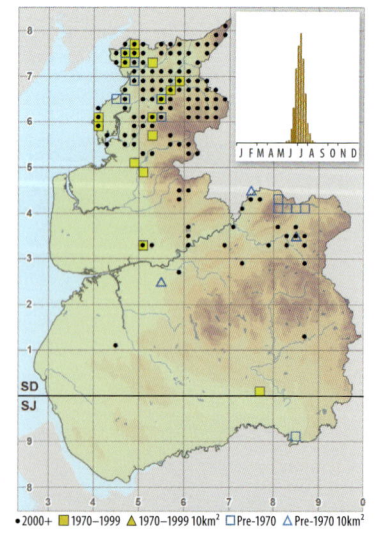

• 2000+   ▫ 1970–1999   ▲ 1970–1999 10km²   ☐ Pre-1970   △ Pre-1970 10km²

476   *The Moths of Lancashire*

72.037 Round-winged Muslin *Thumatha senex*

■ **LANCS STATUS** Occasional / local resident   1948–2022

Another lichen-feeder, doing extremely well since its arrival in the county from the south. First recorded by M.J. Leech at Formby in July 1948, and on three further occasions in the early 1950s, suggesting temporary colonisation. Not recorded again until 1 July 2001 when trapped at light at Risley Moss by P. Pugh. Only three more records before 2013, but 118 times since, with regular recordings from Great Sankey, Flixton and Leighton Moss. Absent from uplands and much of north Lancashire.

■ **HABITAT** Grassland, reedbeds, marshes.
■ **FOODPLANTS** Not recorded in Lancashire.

Single brooded. Comes to light. Mostly singletons; 6 at Great Sankey 23 June 2020.

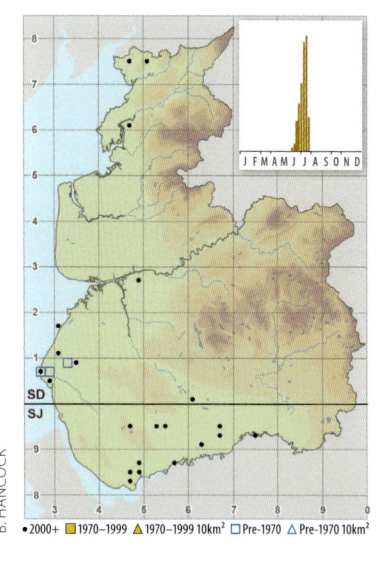

72.038 Four-dotted Footman *Cybosia mesomella*

■ **LANCS STATUS** Scarce / very local resident   1846–2022

Noted at Chat Moss in June 1846 (Edleston, 1846), with early records apparently confined to mosslands, e.g., Rixton Moss, Risley Moss and Simonswood Moss (Ellis, 1890). However, likely linked to deterioration of the mosses, no further reports from these sites since 1880s. Next recorded in 1920 at Scotforth, then at Gait Barrows in 1984 and 1998. Subsequently, noted eight times this century, all from northern limestone sites, e.g., Silverdale, Leighton Moss, Yealand Conyers, Dalton Crags.

■ **HABITAT** Damp woodland, limestone woodland. Formerly on mosses.
■ **FOODPLANTS** Not recorded in Lancashire.

Single brooded. All 21st century records are of moths taken at light.

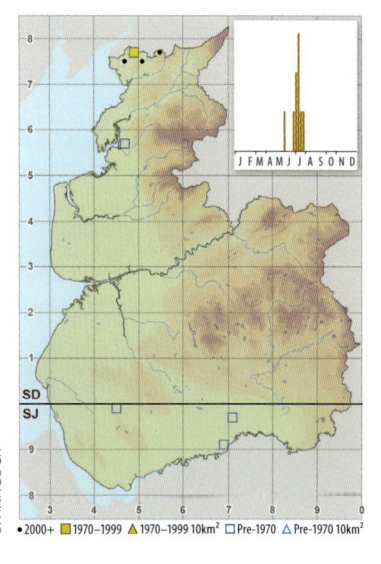

72.041 Four-spotted Footman *Lithosia quadra*

■ **LANCS STATUS** Increasingly frequent / widespread vagrant and migrant   1897–2022

Recorded at Middleton, nr. Rochdale in 1897 by 'J.T.' Seven 20th century records from Lancaster, Formby, Billinge, etc., all considered migrants. However, with 72 records since 2016, including 54 from 2022, it is unlikely that all have migrated from mainland Europe. In recent years, colonies have become established in southern England and Wales, and it is likely many of our arrivals have a UK origin. It is even possible that a colony or two may have become temporarily established in coastal areas of Lancs.

■ **HABITAT** Coastal locations, scrub.
■ **FOODPLANTS** Not recorded in Lancashire.

Single brooded. High count of eight to light, all males, at Ainsdale 26 July 2022.

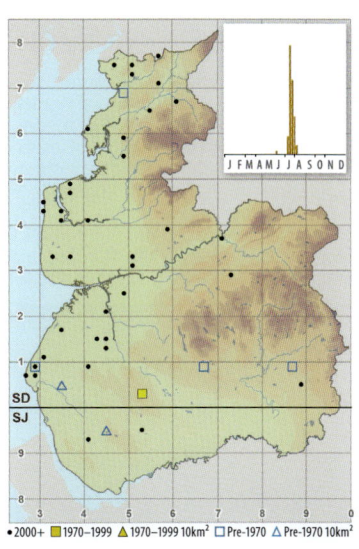

The Moths of Lancashire   **477**

EREBIDAE

72.042 Red-necked Footman *Atolmis rubricollis*

■ **LANCS STATUS** Occasional / local resident　　　　　1986–2022

A recent arrival into the county, first noted at Gait Barrows by P. Kirk in 1986, and at Ainsdale the following year. Recorded 68 times this century, including a few upland records at Longridge, Dalton Crags, Docker Moor, etc. An impressive 54 moths noted on a half-mile long transect at Thrushgill Plantation on 2 July 2011 by John Girdley. Also noted in West Pennines area with records from Rishton, Belmont and Jackhouse, Oswaldtwistle. Three records in 2022, from Green Bank, Formby and Gait Barrows.
■ **HABITAT** Coniferous plantations, broad-leaved woodlands.
■ **FOODPLANTS** Unrecorded. Larva noted at Formby 13 September 2022 by R. Walker. Single brooded. Flies in sunshine. May also be noted at rest on bushes, or at light.

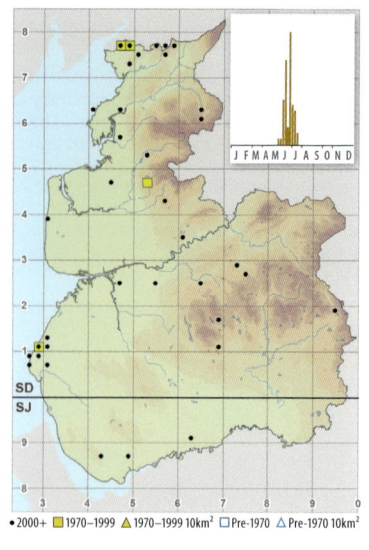

72.043 Buff Footman *Eilema depressa*

■ **LANCS STATUS** Frequent / widespread resident　　　　　2000–2022

Arrived in Lancs as part of rapid national range expansion at the onset of this century. Recorded at light in St Helens on 8 August 2000 by R. Banks. The following year, counts of eleven made at Leighton Moss and Silverdale Moss in the north of the county. Now, widely distributed throughout and recorded over 1500 times, including from Billinge, Flixton, Orrell, Crosby, Mill Houses, Docker Moor etc., with a count of over 2,500 from two traps at the latter site on 20 August 2015 by P.J. Marsh.
■ **HABITAT** Woodland, scrub, moorland.
■ **FOODPLANTS** Unrecorded. Bred from larva at Adlington 13 May 2011, P. Krischkiw. Single brooded. Virtually all records are of moths at light.

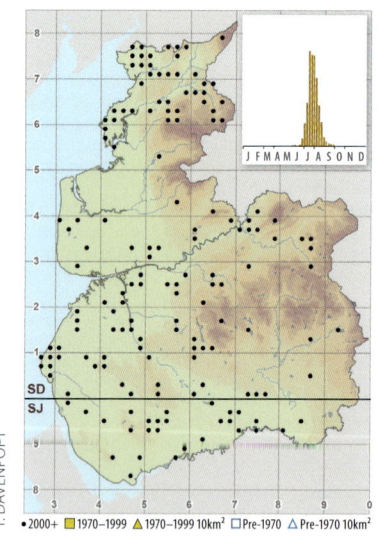

72.044 Dingy Footman *Eilema griseola*

■ **LANCS STATUS** Frequent / widespread resident　　　　　1889–2022

There is a specimen in the Natural History Museum recorded by 'Capper' at Liverpool in 1889. Otherwise, this is another species to have expanded north in recent decades, with the first modern record from Freshfield Dune Heath on 14 August 2004 by G. Jones. 2006 saw VC60 records at Leighton Moss and Heysham N.R., with increasingly rapid spread throughout the county. Recorded on over 3,700 occasions.
■ **HABITAT** Damp woodland.
■ **FOODPLANTS** Larva beaten from Witches Broom on birch, Winmarleigh Moss by J. Steeden on 19 June 2021. Larva noted under trap at Flixton 5 April 2020, K. McCabe. Single brooded. Comes to light. High count 220 at Leighton Moss on 20 July 2022.

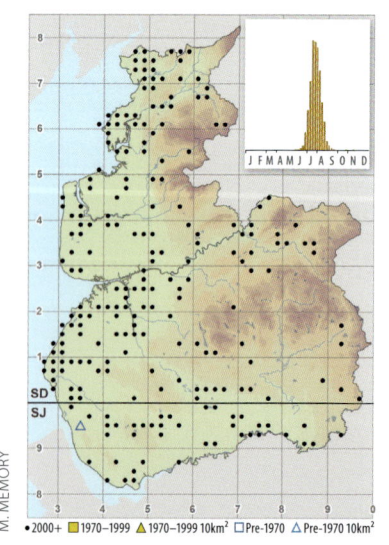

The Moths of Lancashire

72.045 Common Footman *Eilema lurideola*

■ **LANCS STATUS** Frequent / widespread resident   c.1857–2022
Gregson (1857) reported larvae plentiful around West Derby, Liverpool on lichen-coated Beech trunks. Ellis (1890) considered it generally distributed in south Lancashire but scarce in the north. Other pre-1970 locations were Prestwich, Formby, Freshfield, Morecambe, and Silverdale. Distribution map shows a significant increase from 1970s onwards into all parts of the county including upland regions. High counts of 200 plus at Heysham N.R. in 2002 and Rindle Wood at Astley Moss in 2013.
■ **HABITAT** Woodland, parks, gardens.
■ **FOODPLANTS** Larvae noted on oak, Hawthorn, and on mosses and lichens.
Single brooded. Frequent at light. Larvae recorded from February to June.

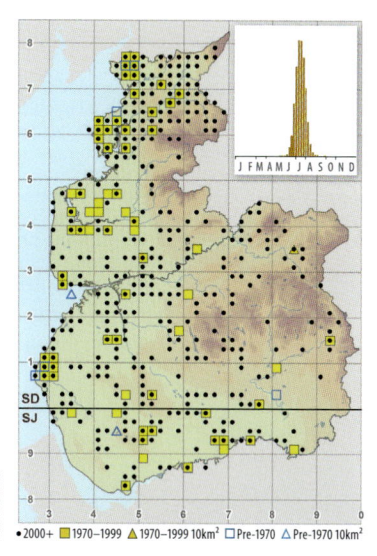

● 2000+   ■ 1970–1999   ▲ 1970–1999 10km²   □ Pre-1970   △ Pre-1970 10km²

72.046 Scarce Footman *Eilema complana*

■ **LANCS STATUS** Occasional / local resident   1846–2022
Recorded at Silverdale in 1921 and at Formby in 1938. Numbers and range have increased in the last decade with 32 records up to 2017 and 75 since. Remains limited to lowland sites; regular at Sefton Coast, Silverdale and in south Manchester gardens. 19th century records from the mosses, including at Chat Moss (Edleston, 1846), were considered by Gregson (1860) to represent a distinct species, the Northern Footman *Lithosia sericea* Gregs. This is now regarded as a form of Scarce Footman.
■ **HABITAT** Sand dunes, woodland.
■ **FOODPLANTS** Unrecorded. Moth bred from larva at Chat Moss in 1892.
Single brooded. Comes to light with high count of nine from Formby, 10 August 2021.

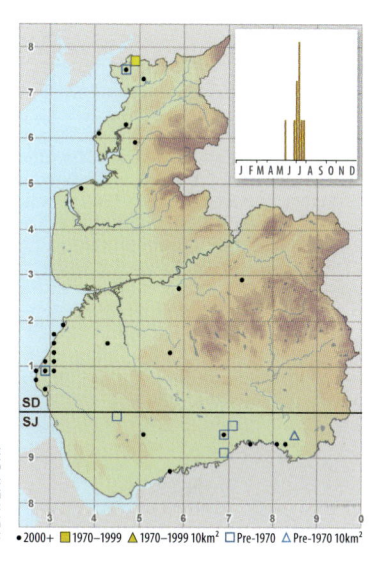

● 2000+   ■ 1970–1999   ▲ 1970–1999 10km²   □ Pre-1970   △ Pre-1970 10km²

72.049 Orange Footman *Eilema sororcula*

■ **LANCS STATUS** Occasional / local resident   2007–2022
First records from Billinge 23 May 2007 by C.A. Darbyshire and at Wigan 2012 by G. and B. Wynn, possibly of vagrants from the south. However, subsequent Sefton Coast records suggest the moth now resident in this area of the county. Noted at Ainsdale in 2017 and frequently since, with high count of 29 to light on 8 May 2022 by C. Daly & R. Moyes. Singletons at other sites: Lightfoot Green, Flixton, Briercliffe, Silverdale and Runcorn, may represent vagrants, or the first signs of further colonisation.
■ **HABITAT** Woodland, parks and gardens.
■ **FOODPLANTS** Not recorded in Lancashire.
Single brooded. All records are of moths at light.

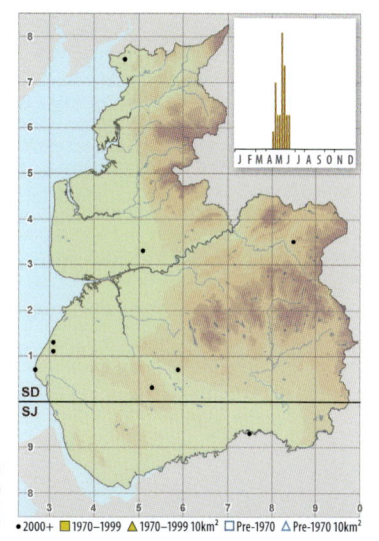

● 2000+   ■ 1970–1999   ▲ 1970–1999 10km²   □ Pre-1970   △ Pre-1970 10km²

The Moths of Lancashire   479

EREBIDAE

72.053 Fan-foot *Herminia tarsipennalis*

■ **LANCS STATUS** Frequent / widespread resident   c.1857–2022
Reported nr. Sefton Park, Liverpool by W. Skellon (Gregson, 1857). Also, from Aigburth (Ellis, 1890), with Mansbridge (1940) describing the moth as fairly well-distributed but less common than the following species. Pre-1970 records from Slyne, Lancaster, Silverdale, Blackpool, Rufford, Formby and Manchester. The post-1970 and post-2000 records show increased presence in upland areas, e.g., Docker Moor.
■ **HABITAT** Woodland, hedgerows, parks, gardens.
■ **FOODPLANTS** Beaten from Ivy, showing preference for old, brown leaves in captivity. Single brooded with possible small, second brood represented by 13 mid- to late September records post-2000. Overwintering larvae noted in September and May.

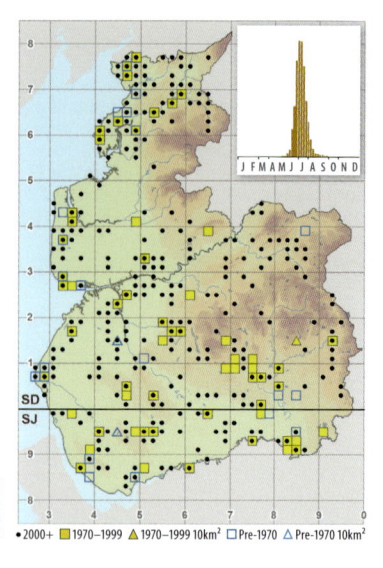

● 2000+   ☐ 1970–1999   △ 1970–1999 10km²   ☐ Pre-1970   △ Pre-1970 10km²

72.055 Small Fan-foot *Herminia grisealis*

■ **LANCS STATUS** Frequent / widespread resident   1846–2022
First recorded from Chat Moss (Edleston, 1846), with other 19th century records from Liverpool, Hale, Simonswood Moss and Preston. Early authors considered the species common and generally distributed. Remains so, although numbers of records this century are approx. 50% of the previous species, a turnaround in comparative frequency from Mansbridge's 1940 comments (see Fan-foot). Occurs in lowland and upland areas, with high counts of 15 at Rainford on 30 July 2005 and 1 July 2006.
■ **HABITAT** Woodland, scrub, gardens.
■ **FOODPLANTS** Not recorded in Lancashire.
Mainly single brooded with rare second-generation from September to October.

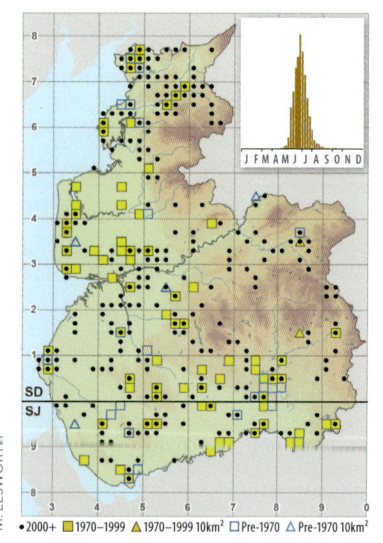

● 2000+   ☐ 1970–1999   △ 1970–1999 10km²   ☐ Pre-1970   △ Pre-1970 10km²

72.060 Marsh Oblique-barred *Hypenodes humidalis*

■ **LANCS STATUS** Occasional / local resident   1854–2022
Never common; only recorded 78 times. First noted 1854 at Warrington by N. Cooke and at Pilling Moss by J.B. Hodgkinson shortly after. Other than a couple at Billinge in 1999, 20th century records almost exclusively from the mosses. Post-2000, initially regular at Astley and Cadishead Mosses, but unrecorded there since 2011. 2022 records from Heysham Moss, Westby and Flixton. Recorded nine times at the latter, likely wandering from the mosses. Infrequent upland records, e.g., Mill Houses 2008.
■ **HABITAT** Mosses, marshes and boggy moorland.
■ **FOODPLANTS** Not recorded in Lancashire.
Single brooded. Comes to light with high count of 50 at Astley Moss 14 July 2008.

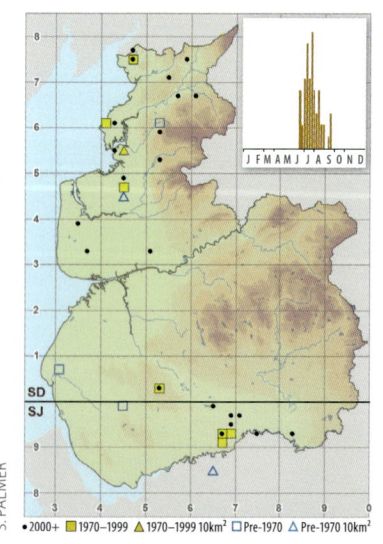

● 2000+   ☐ 1970–1999   △ 1970–1999 10km²   ☐ Pre-1970   △ Pre-1970 10km²

480   *The Moths of Lancashire*

72.061 Pinion-streaked Snout *Schrankia costaestrigalis*

■ **LANCS STATUS** Frequent / widespread resident 1980–2022
In keeping with national range expansion, the last few decades have seen this moth become far more widely distributed in the county, where previously very local. First reported by C.J. Cadbury at Leighton Moss in 1980, with the only other 1980s records from Gait Barrows and Leighton Moss. New sites in the 1990s included Ainsdale, Flixton, Orrell Water Park, Heysham N.R., Longshaw and Rixton Clay Pits. 158 records from 2022 including some upland sightings at Green Bank, Briercliffe and Rishton.
■ **HABITAT** Wet ground, marshes, boggy moorland.
■ **FOODPLANTS** Not recorded in Lancashire.
Double brooded. High count of 18 at light at Leighton Moss 2018 by B. Hancock.

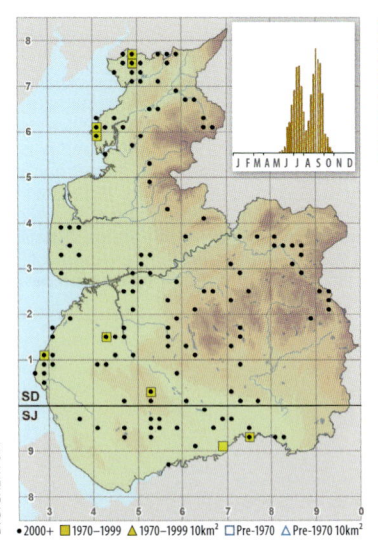

● 2000+ ■ 1970–1999 ▲ 1970–1999 10km² □ Pre-1970 △ Pre-1970 10km²

72.063 Blackneck *Lygephila pastinum*

■ **LANCS STATUS** Frequent / widespread resident 1988–2022
First at Swinton 10 July 1988 by S. Christmas, then Flixton 1997. All other records from this century, 1210 in total. Other early records from Bold Moss, Holcroft Moss, reclaimed tips at Stretford and Chorlton, and Yealand Conyers, the first VC60 record. Primarily a lowland species, although occasional at Rishton, Longridge, etc.
■ **HABITAT** Damp woodland, scrub, reclaimed tips, brownfield sites.
■ **FOODPLANTS** Feeding not observed. Larvae found resting on rushes, oak seedling and at base of Cock's-foot, in each case amongst Tufted Vetch, the usual foodplant. Single brooded, although a few late individuals to early September may represent small second brood. Comes to light. Larvae recorded from September to November.

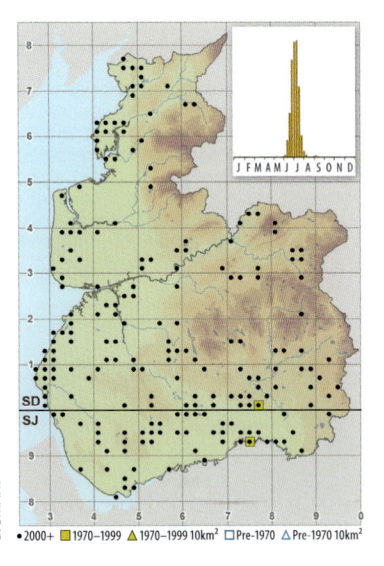

● 2000+ ■ 1970–1999 ▲ 1970–1999 10km² □ Pre-1970 △ Pre-1970 10km²

72.066 Waved Black *Parascotia fuliginaria*

■ **LANCS STATUS** Scarce / very local resident 1995–2016
First noted in 1995 with range expansion nationally from the south; first Cheshire record in 1993. Recorded 6 August 1995 by J. Whiteside at Hoghton, with others at Astley Green and Everton in 1997 and 1998. Next noted at Flixton by K. McCabe on 11 August 2007, with two further records at same location in 2012. Recorded by the lights at Carnforth Railway Station 23 August 2008 by T. Hutchison and annually from 2013 to 2016. Hopefully, presence retained in the county although unrecorded since.
■ **HABITAT** Damp woodland, gardens.
■ **FOODPLANTS** Not recorded in Lancashire.
Single brooded. Comes to light, including to lighting in underpass.

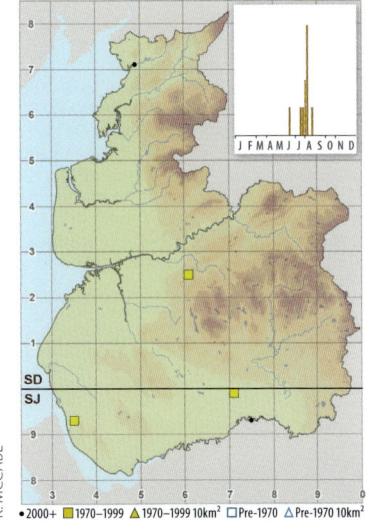

● 2000+ ■ 1970–1999 ▲ 1970–1999 10km² □ Pre-1970 △ Pre-1970 10km²

EREBIDAE

72.067 Small Purple-barred *Phytometra viridaria*

■ **LANCS STATUS** Scarce / very local resident   c.1857–2022

First reported from Simonswood Moss, nr. Kirkby by Gregson (1857) with other early records from Lytham, Southport, Silverdale and Clougha. Noted as common on the mosslands and some of the sand-hills (Ellis, 1890). All subsequent records from Sefton and Fylde coasts and from Morecambe Bay. Most from Sefton Coast with 45 of the 53 records this century. The moth's future at Morecambe Bay appears very uncertain with the only record this century at Yealand 9 August 2020 by A. Mather.
■ **HABITAT** Sand dunes, limestone grassland.
■ **FOODPLANTS** Not recorded in Lancashire.
Likely double brooded. Flies in sunshine and also comes to light.

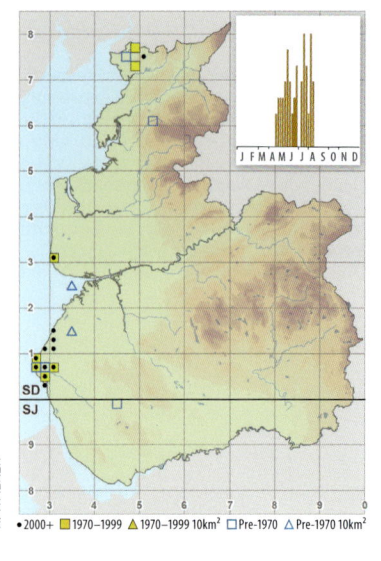

72.069 Beautiful Hook-tip *Laspeyria flexula*

■ **LANCS STATUS** Frequent / widespread resident   2001–2022

Another lichen-feeding species to have dramatically extended its range this century, with first Lancs record on 5 July 2001 at Pennington by P. Pugh. Next at Belmont, Crosby and Flixton in 2006. Now recorded on 828 occasions, including 175 records in 2022 alone. There are a few upland records from Wray, High Tatham, Longridge, Thrushgill, etc. High count of 15 at Longton on 4 July 2021 by J. Girdley.
■ **HABITAT** Woodland, hedgerows, gardens.
■ **FOODPLANTS** Larva recorded feeding on lichens on a branch at Lightshaw Meadows, Wigan by H. Nelson on 11 March 2022.
Mainly single brooded, with a much smaller second brood. Comes readily to light.

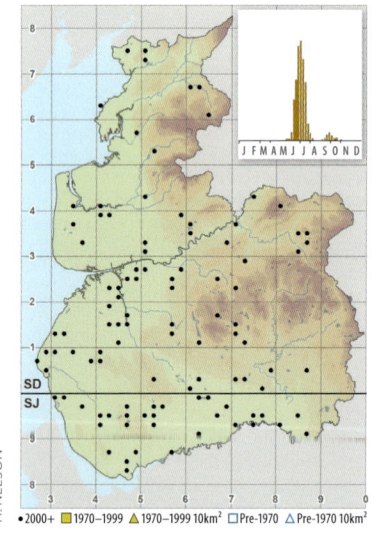

72.073 Small Marbled *Eublemma parva*

■ **LANCS STATUS** Scarce / migrant   1998–2019

This migrant has been noted on 16 occasions, all from the past three decades. The first was on 22 June 1998 at Longton by E.J. Roskell, with the next two records, a pair at Lightfoot Green and a singleton at Gait Barrows, occurring the following month. Next noted at Pennington in 2003 and 2005. Recorded from three different sites at Silverdale in 2019, raising the possibility of being locally bred. Other post-2010 records from Whitworth, Lytham St. Annes, Longton, Catterall, Crosby and Freshfield.
■ **HABITAT** Unknown in the county.
■ **FOODPLANTS** Not recorded in Lancashire.
Single brooded. All records are of moths at light.

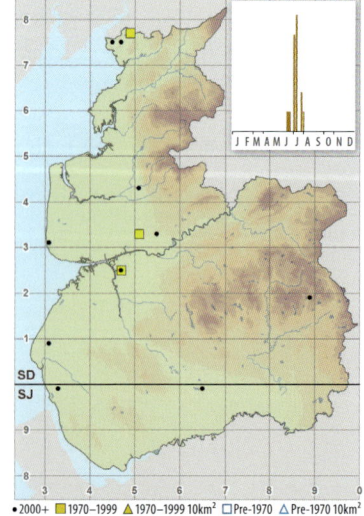

482 *The Moths of Lancashire*

72.076 Clifden Nonpareil *Catocala fraxini*

■ **LANCS STATUS** Scarce / migrant, possible vagrant   c.1840–2023
This spectacular migrant species was netted in flight at Newton Heath (Edleston, 1846d), with six other 19th century records from Manchester, Bolton and Chorley. Next, recorded at Lytham St Annes in 1948, but not again until 2022. Became extinct in UK in mid-20th century but has since recolonised southern English counties. The influx of 2022, with records from St Helens, Bamber Bridge, Ainsdale, Hale and Didsbury, suggests the possibility of resident colonies in or close to the county.
■ **HABITAT** Woodland, hedgerows, parks, gardens.
■ **FOODPLANTS** Not recorded in Lancashire.
Single brooded. 2023 saw 14 further records, including four in VC60.

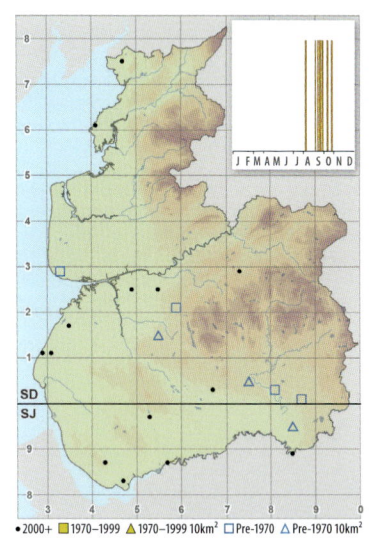

72.078 Red Underwing *Catocala nupta*

■ **LANCS STATUS** Frequent / widespread resident   1972–2022
First recorded in Maghull, Liverpool 1972 by R. Murphy at a time of national range expansion, and next at Burnley in 1979. The first VC60 records were from four sites on the Fylde in 1992. The start of the 21st century saw a rapid spread of distribution, mostly in lowland areas, and a sharp increase in records, with 212 records in the first four years of this century. However, there is evidence of more recent decline, as the four years from 2019 to 2022 generated just 103 records.
■ **HABITAT** Woodland, hedgerows, parks, gardens.
■ **FOODPLANTS** Not recorded in Lancashire.
Single brooded. Comes to light with high of seven at Flixton on 2 September 1996.

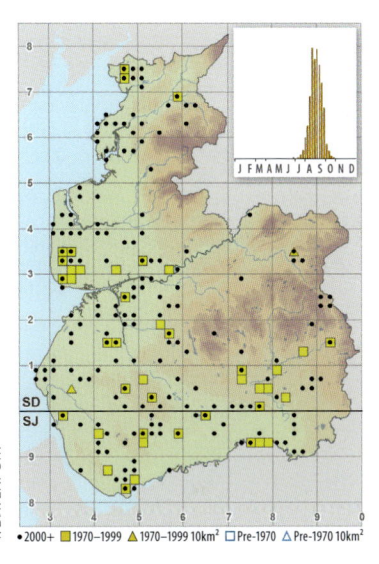

72.083 Burnet Companion *Euclidia glyphica*

■ **LANCS STATUS** Occasional / local resident   c.1857–2022
This day-flying moth was first noted at Speke (Gregson, 1857), and later considered scarce by Mansbridge (1940), with Carnforth the only additional location. Recorded at Warton Crag 1966 and regularly since. Generally restricted to coastal and some inland lowland locations. Recent examples of the latter are from the Mersey Valley at Flixton, Chorlton and Stretford tip, and six at Westhoughton in 2015. Has increased in recent years with over half the county records from 2017 onwards.
■ **HABITAT** Grassland, scrub, brownfield sites.
■ **FOODPLANTS** Not recorded in Lancashire.
Single brooded. Flies in sunshine and easily disturbed from vegetation.

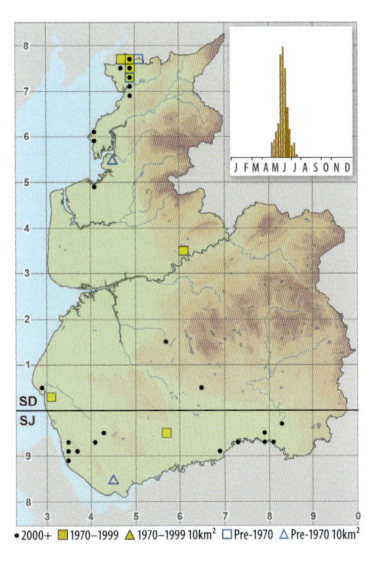

The Moths of Lancashire

EREBIDAE

72.084 Mother Shipton *Callistege mi*

■ **LANCS STATUS** Occasional / widespread resident   c.1853–2022

First noted by N. Cooke at Simonswood Moss, nr. Kirkby in the mid-19th century (Byerley, 1854), with other pre-1900 records from Lydiate, Hale, Ainsdale and Silverdale. Bred from larva at Chat Moss in 1936. Early authors considered the moth local, but common where it occurs. Mainly noted in lowland areas, with occasional upland records, the first at Pendle Hill 1940; more recently at Dalton Crags and Rivington Moor. 21st century larval records from Cadishead Moss and Astley Moss.
■ **HABITAT** Scrub, grassland, mosses.
■ **FOODPLANTS** Grasses. Larvae noted July and August, often swept from grassland. Single brooded. Day-flying moth with a small number of records at light.

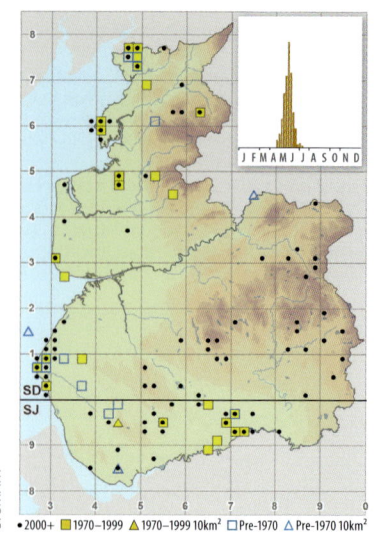

NOCTUIDAE

73.001 Spectacle *Abrostola tripartita*

■ **LANCS STATUS** Frequent / widespread resident   c.1886–2022

Noted in the Bolton and Preston areas (W. Johnson and J.B. Hodgkinson respectively) pre-1887, and infrequently elsewhere to 1940. At that time considered much less common than Dark Spectacle. Since then, has become more widespread and now occasionally reaches double figures, such as the 28 at Flixton on 15 August 2020.
■ **HABITAT** A wide range of habitats where nettles occur.
■ **FOODPLANTS** Common Nettle; bred once from Hawthorn.
Single brooded, with possible small second brood in September, rarely into October, in the last two decades. April records commenced and increased at the same time. Comes to light. See note about scientific name confusion under Dark Spectacle.

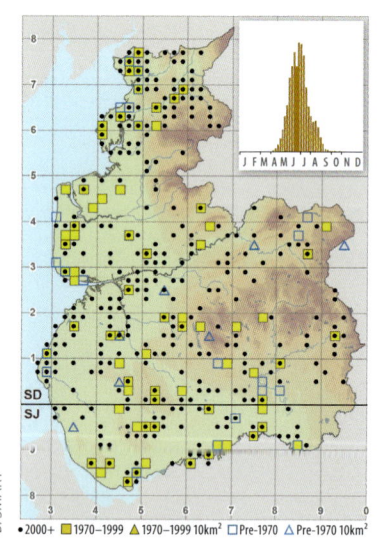

73.002 Dark Spectacle *Abrostola triplasia*

■ **LANCS STATUS** Frequent / widespread resident   1853–2022

First noted in 1853 at Chorley by E.C. Buxton and Liverpool by N. Cooke. Later, seen in Bolton, Bury, Rochdale, Preston, Crosby, Oldham, Burnley and Silverdale, all prior to 1940. Since then, it has been noted widely across the county. Recent regular sites include Bispham, Hale, St Helens area, Walmer Bridge, Wray and Yealand Conyers. 15 were noted in Worsthorne on 23 July 1999 and 14 at Rishton on 30 June 2022.
■ **HABITAT** A wide range of habitats. Larvae noted, historically, in weedy gardens.
■ **FOODPLANTS** Not recorded in Lancashire.
Single brooded. Comes to light, usually in single figures. Between 1972 and 1993 the name *triplasia* was incorrectly applied to *tripartita* (Agassiz et al., 2013).

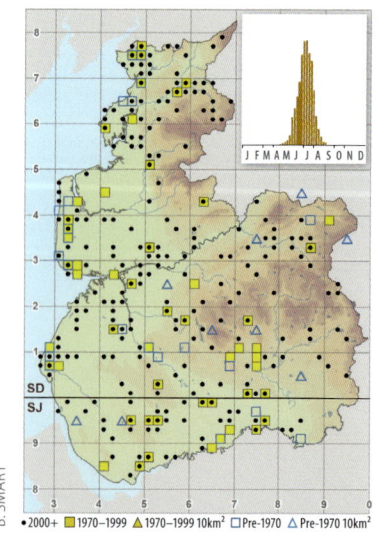

73.003 Ni Moth *Trichoplusia ni*

■ **LANCS STATUS** Scarce / migrant　　　　　　　　　　　　　c.1871–2017

Gregson (1889) noted finding this moth c.1871 on the wall of St George's Church, Liverpool, this being the first Lancashire record. It was next seen at Formby on 8 August 1953, in Flixton on 19 August 1996 and Orrell on 21 August 2003. 2011 was a good year for this migrant species, it being found in Preston on 28 September, Yealand Redmayne on 1 October and Yealand Conyers the following day. Only three other records are known, these being from Hale on 8 September 2015, at Heysham on 27 June 2017 and the final one at Formby on 4 July 2017 (T. Davenport).
■ **HABITAT** Unknown, all records being in garden light traps.
■ **FOODPLANTS** Not recorded in Lancashire.

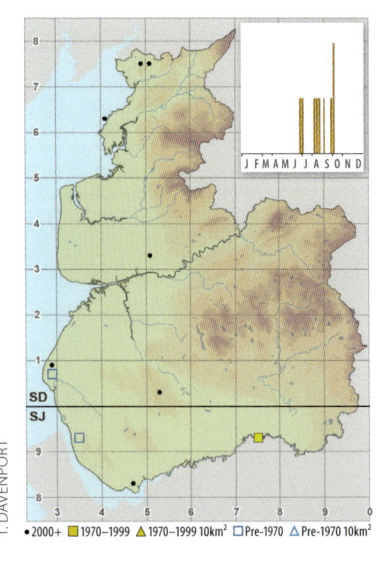

73.008 Golden Twin-spot *Chrysodeixis chalcites*

■ **LANCS STATUS** Rare / migrant or adventive　　　　　　　　2003

The only record of this species in Lancashire occurred on 5 July 2003 when one was attracted to a Robinson MV light trap run in a Heysham garden by D.J. & B. Holding.
■ **HABITAT** Unknown.
■ **FOODPLANTS** Not recorded in Lancashire.

The origin of this moth cannot be attributed with certainty and may possibly have involved an adventive origin. D. J. Holding investigated this possibility with his immediate neighbours, and neither he nor they had purchased any imported plants earlier that year.

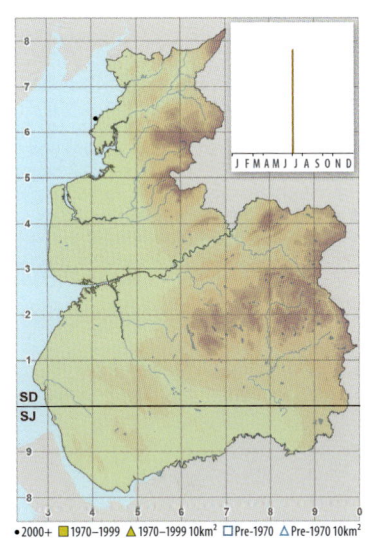

73.011 Scarce Burnished Brass *Diachrysia chryson*

■ **LANCS STATUS** Rare / vagrant　　　　　　　　　　　　　　c.1866

This scarce British resident has only, otherwise, been reported from parts of south-west Wales, central southern and eastern England. It was therefore unexpected to find a historic record for Lancashire listed and accepted in Ellis (1890). It is documented that it was beaten out of Honeysuckle near Preston by Mr. Hodgkinson c.1866. Assuming this to be J.B. Hodgkinson, who was a very well-respected and experienced Lepidopterist of this time, he would have been aware of the unlikely nature of such a record. Mansbridge (1940), also includes this record but adds 'that it must be regarded as an accidental visitor only'.
■ **HABITAT & FOODPLANTS** Unknown in the county.

This species is not known to be migratory and although it has become more numerous in south-west Wales in recent years, its 19th century distribution is still similar to that of today, with the exception that it has apparently been lost from East Anglia. The record is therefore accepted with reservation and on the same basis as that mentioned by Mansbridge (loc. cit.).

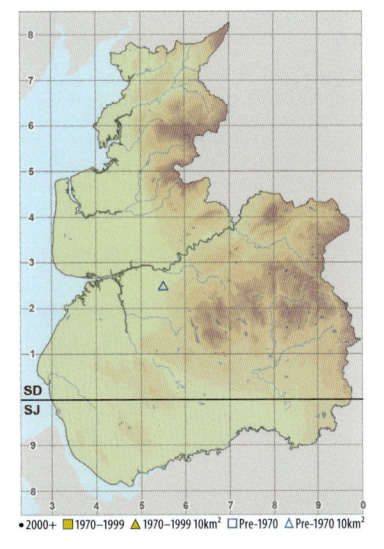

NOCTUIDAE

73.012 Burnished Brass *Diachrysia chrysitis*

■ **LANCS STATUS** Frequent / widespread resident   c.1857–2022

First reported in Liverpool (Gregson, 1857), with early authors noting it as generally distributed but not usually common. It was later considered common across the region (Mansbridge, 1940), this situation remaining relatively unchanged to the present day. Reported annually at most sites, but overall numbers have declined. Single-night counts in double figures are now restricted to a few upland areas.
■ **HABITAT** Wide ranging, at all altitudes in the county.
■ **FOODPLANTS** Not recorded in Lancashire.

Regularly double brooded, the first much larger. Comes to light. Suggestions that a separate species occurs within this taxon remain unresolved at time of publishing.

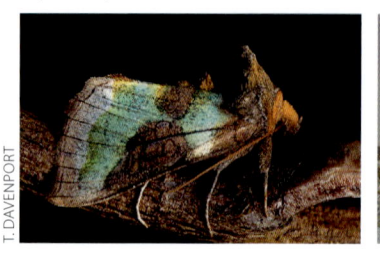

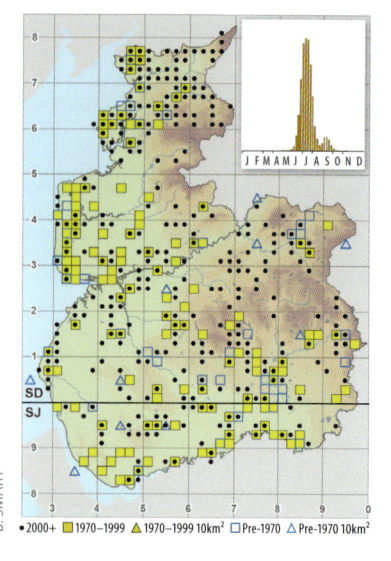

73.014 Golden Plusia *Polychrysia moneta*

■ **LANCS STATUS** Occasional / local resident   1917–2022

First noted in Britain in late 19th century, spreading north and reaching Manchester by 1917 (R. Tait). Subsequently noted in gardens across VC59, including Nelson in 1934 and Burnley in 1940. In VC60, it was reported from Lytham in 1939, Morecambe (1959) and Carnforth (1961). Records have declined a little since 2012 but include those from Flixton, Longton, Tarleton, Warton (nr. Silverdale) and Yealand Conyers.
■ **HABITAT** Gardens and parks.
■ **FOODPLANTS** Delphinium and Monk's-hood.

Single brooded with an occasional small second brood in September. Comes to light. Larvae found from May to mid-July including 30 in 1941 at Roughlee by C. Baldwin.

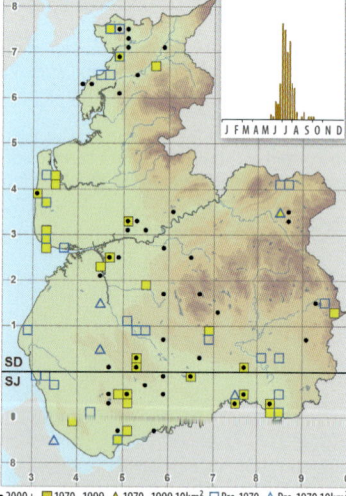

73.015 Silver Y *Autographa gamma*

■ **LANCS STATUS** Frequent / primary and secondary migrant   c.1857–2022

First noted in Liverpool (Gregson, 1857), and since then, reported from most of the county. These could relate to primary migrants and/or British (including Lancashire) bred individuals. Very large, single day, influxes were noted in 1994 (400 at Leighton Moss), 1996 (250 at Preston), 2002 and 2006 (100 plus at Heysham), 2008 (342 at Fleetwood), 2009 (150 on West Pennine Moors) and 2020 (102 at Worsthorne).
■ **HABITAT** As a primary migrant, it can appear anywhere and may subsequently breed.
■ **FOODPLANTS** Alder and Hogweed.

Potentially, at least double brooded. Comes to light. Nectars during the day and at dusk on plants such as Buddleia and Verbena. Larvae found from July to October.

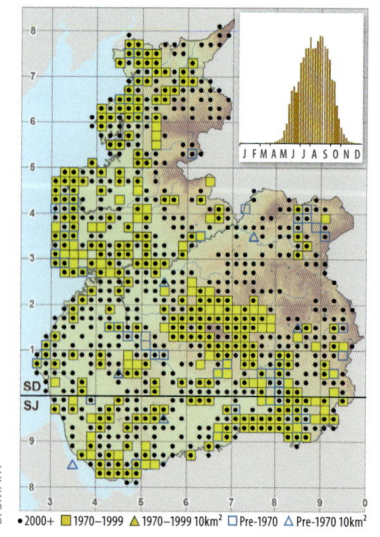

73.016 Beautiful Golden Y *Autographa pulchrina*

■ **LANCS STATUS** Frequent / widespread resident   c.1859–2022
First noted in Manchester by J. Chappell c.1859 and subsequently, from Prestwich, Rochdale, Preston, Sefton Park, Burnley and Silverdale amongst others, up to the early 20th century. Recorded sparingly until the late 1950s, since when, records were received from an increasingly widespread range of sites. From the 1990s onwards it was noted annually at many sites, including Flixton, Belmont, Claughton, Leighton Moss, Preston, Briercliffe, Rishton, High Tatham, Longridge and Mill Houses.
■ **HABITAT** Woodland, scrub, gardens, moorland, mosses and brownfield sites.
■ **FOODPLANTS** Not recorded in the wild in Lancashire.
Single brooded. Comes to light, usually in single figures. 68 seen at Royton in 2001.

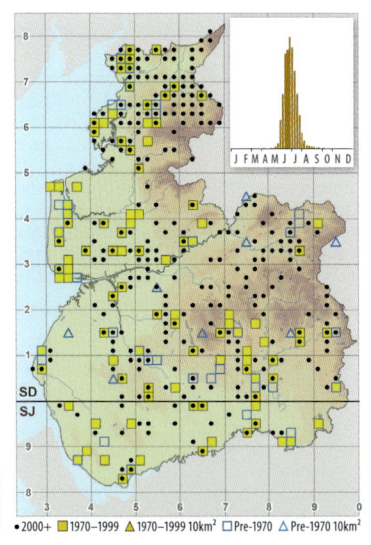

73.017 Plain Golden Y *Autographa jota*

■ **LANCS STATUS** Frequent / widespread resident   1853–2022
First noted in Chorley in 1853 (E.C. Buxton) and in Liverpool by Gregson (1857). In the late 19th century also noted in Manchester, Rochdale, Oldham and Preston. No specific records were detailed by Ellis (1890) or Mansbridge (1940), but said to be generally distributed. The distribution remains the same to the present day, although numbers have declined slightly since 2014. Noted over many years at sites such as Billinge, Wray, Worsthorne, Woolmer Bridge, Bay Horse and Orrell.
■ **HABITAT** Scrub, woodland, gardens, hedgerows, parks and brownfield sites.
■ **FOODPLANTS** Not recorded in Lancashire.
Single brooded. Comes to light, mostly in single figures; 39 on 18 July 2013 at Wray.

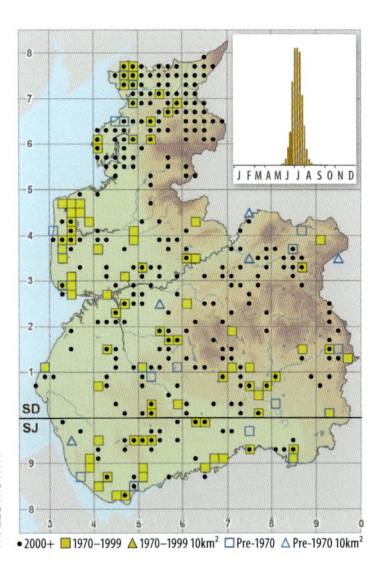

73.018 Gold Spangle *Autographa bractea*

■ **LANCS STATUS** Occasional / local resident   c.1889–2022
First noted at Longridge and Preston by J.B. Hodgkinson c.1890, then Daisy Nook, Oldham, Formby, Fairhaven, Nelson, St Annes and Silverdale, up to 1954. Records gradually increased over the next few decades, particularly in upland areas. It then suffered a county-wide collapse in 2007. Upland populations began their recovery in 2008, this not happening in lowland areas until 2011. 2013 was a good year all round, but since then, a slow decline has set in. The decline is particularly noticeable in lowland areas, where reports were only received from Pilling and Lancaster in 2022.
■ **HABITAT** Moorland, upland grassland, hedgerows
■ **FOODPLANTS** Not recorded in Lancashire. The species is single brooded. Comes to light.

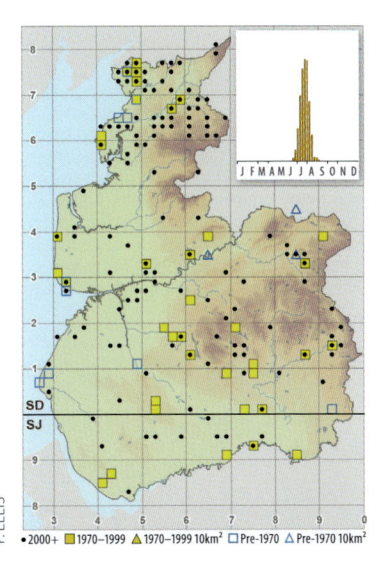

NOCTUIDAE

73.021 Scarce Silver Y *Syngrapha interrogationis*

■ **LANCS STATUS** Scarce / local resident 1857–2021

First noted as common in the Preston district (Pugh, 1857) and at Longridge Fell in 1885. Elsewhere, it was found on Chat Moss, Clougha and later, in 1979, at Burnley. Further reports from Longridge Fell came in 1919 and from 1953–56, but none since. Apart from isolated singletons at Crossdale Grains (2012) and Thrushgill (2019), all other reports have been from Worsthorne (2000–2021), Littleborough (2006–2013), Cutgate (2013), Rishton (2017), Briercliffe (2018–2021) and Crompton Moor (2021).

■ **HABITAT** Moorland areas and once, historically, on a lowland raised bog.
■ **FOODPLANTS** Not recorded in Lancashire.

Single brooded. Comes to light, usually as singletons; five at Worsthorne in 2006.

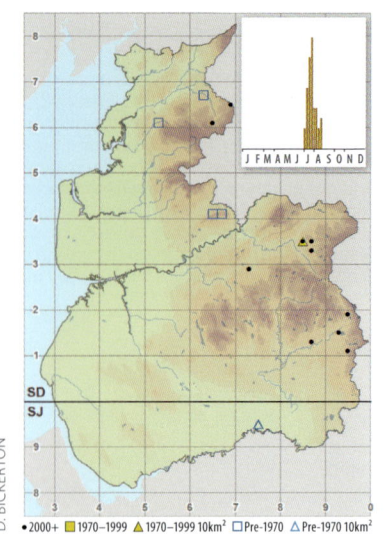

• 2000+   ☐ 1970–1999   △ 1970–1999 10km²   ☐ Pre-1970   △ Pre-1970 10km²

73.022 Gold Spot *Plusia festucae*

■ **LANCS STATUS** Frequent / widespread resident 1889–2022

Specimens in NHM, London provided the earliest records, from Warrington, Bolton and Thornton, between 1889 and 1921. Since then, it has been found to occur across the whole county, with long-term datasets from sites such as Hale, Bispham, Abram, Longton, Flixton, Heysham, Rishton, High Tatham, Adlington, Billinge, Parr and many more. The second brood is larger than the first and the population seems stable.

■ **HABITAT** A wide range of damp or wet habitats.
■ **FOODPLANTS** Cock's-foot Grass and Purple Moor-grass.

Double brooded. Comes to light, sometimes in high numbers. Larvae found in April and May. See note about identification and early records under Lempke's Gold Spot.

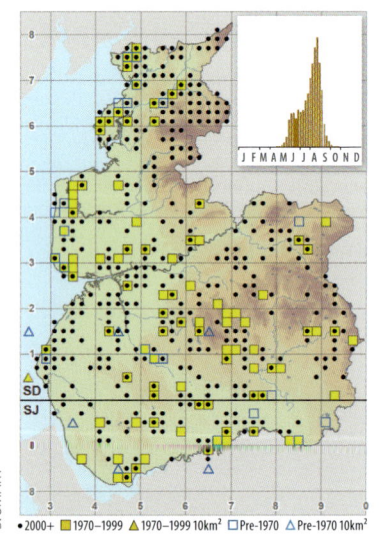

• 2000+   ☐ 1970–1999   △ 1970–1999 10km²   ☐ Pre-1970   △ Pre-1970 10km²

73.023 Lempke's Gold Spot *Plusia putnami*

■ **LANCS STATUS** Frequent / local resident 1921–2022

Not recognised as separate from Gold Spot until 1965. Any listed below from before then were confirmed from retained specimens. First noted in Thornton Cleveleys on 2 July 1921 by O.N. Hawkins (specimen in NHM, London), then St Annes in 1955 (I. Rutherford) and Rufford in 1961. Since then, noted throughout the county, most frequently in wet, upland areas. Found annually elsewhere in scattered localities and low numbers. Has declined or been lost from some lowland sites in recent years.

■ **HABITAT** Wetland habitats; mosses, boggy moorland and wet brownfield sites.
■ **FOODPLANTS** Not recorded in Lancashire.

Single brooded. Comes to light. Care required in identifying this species.

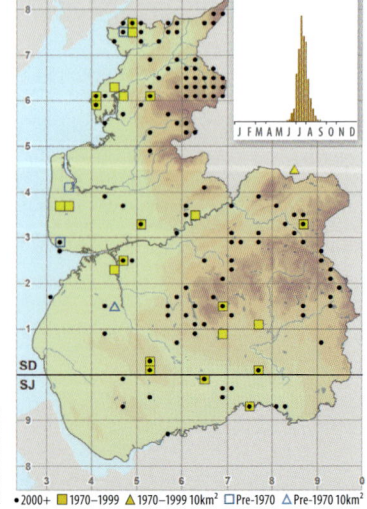

• 2000+   ☐ 1970–1999   △ 1970–1999 10km²   ☐ Pre-1970   △ Pre-1970 10km²

488   *The Moths of Lancashire*

73.024 Marbled White Spot *Protodeltote pygarga*

■ **LANCS STATUS** Occasional / widespread resident 2004–2022
This species has spread northwards in Britain, reaching Crosby on 14 June 2004 (J. Donnelly). It was next seen in Swinton in 2006, Crosby in 2007 and Astley Moss in 2008. Numbers of records increased steadily from 2009, with it reaching Formby and Yealand Conyers in 2013. It has since spread widely, often in leaps and bounds, being noted in Read and Docker Moor in 2015 and Briercliffe and Thrushgill in 2021.
■ **HABITAT** Mosses, moorland and limestone woodland; appears transient elsewhere.
■ **FOODPLANTS** Purple Moor-grass and Reed Canary-grass.
Single brooded; possible small second brood in 2017 and 2021. Comes to light. 14 at Cadishead Moss in 2010 (K. McCabe). Larvae found in August and September.

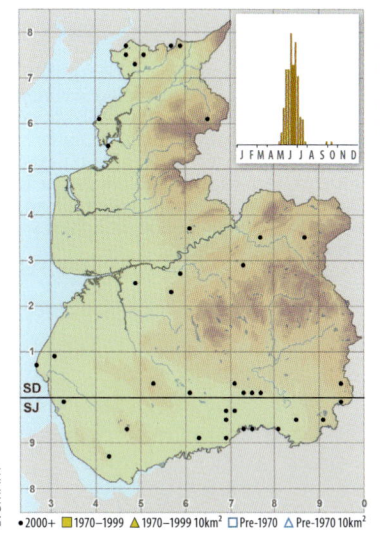

● 2000+ ■ 1970–1999 ▲ 1970–1999 10km² □ Pre-1970 △ Pre-1970 10km²

73.026 Silver Hook *Deltote uncula*

■ **LANCS STATUS** Occasional / widespread resident 1947–2022
First noted at Freshfield in 1947 then later, between Altcar and Birkdale. 30 came to light at Freshfield Dune Heath in 2004 (G. Jones). Away from this core area, it has been regular at several sites including Carr House Common, Leighton Moss, St Annes and Middleton N.R. Most other reports are post-2000 and consisting of one or two seen at over 35 locations. This suggests it is a mobile species, expanding its range.
■ **HABITAT** Marshy grassland, dune slacks, moorland, mosses and wet brownfield sites.
■ **FOODPLANTS** Not recorded in Lancashire.
Single brooded; a small, partial second brood on occasions from mid-August to early September. Comes to light; disturbed during the day.

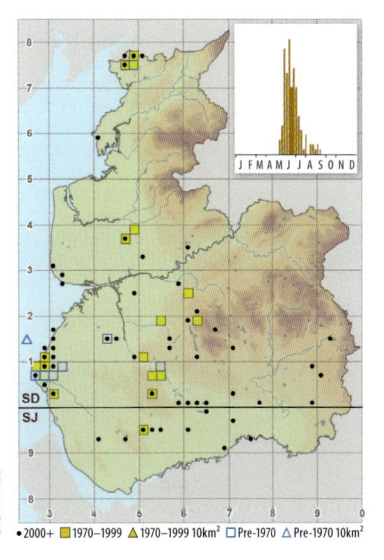

● 2000+ ■ 1970–1999 ▲ 1970–1999 10km² □ Pre-1970 △ Pre-1970 10km²

73.032 Nut-tree Tussock *Colocasia coryli*

■ **LANCS STATUS** Frequent / local resident c.1857–2022
Noted as larvae by C.S. Gregson (c.1857) in Croxteth Pk. Later, in Liverpool (1910), at Parbold (1960–61), Oldham (1994) and Lytham (2010). These reports not repeated. The current population is restricted to north Lancashire, where it is regular in the Silverdale area, east to Docker Moor, and from north Lancaster to Hornby. Outliers south to Dolphinholme and Beacon Fell indicate a possible recent expansion of range.
■ **HABITAT** Deciduous woodland, scrub and gardens.
■ **FOODPLANTS** Oak, birch sp. and Wych Elm.
Single brooded, with occasional, small, second brood in July. Comes to light, mostly in small numbers, but 28 seen at Docker Moor in 2017. Larvae found in July and August.

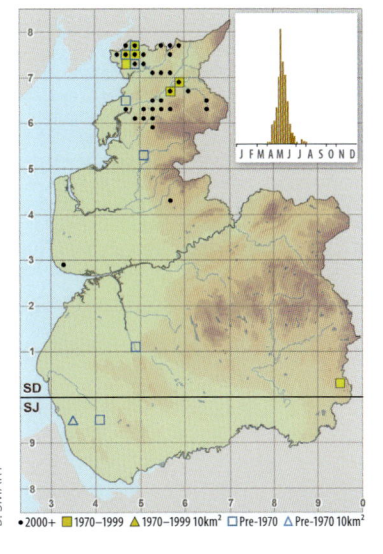

● 2000+ ■ 1970–1999 ▲ 1970–1999 10km² □ Pre-1970 △ Pre-1970 10km²

NOCTUIDAE

73.033 Figure of Eight *Diloba caeruleocephala*

■ **LANCS STATUS** Occasional / local resident　　　　c.1857–2022

Found in Liverpool, where noted as common at gas lamps (Gregson, 1857) and later, as 'generally distributed and locally abundant in old thorn hedges' (Mansbridge, 1940). By the 1950s, a decline was noted by Britten (c.1950), who reported it to be 'less common'. In recent decades, the rate of decline has increased in VC59, with few post-2017 records. Although the decline is also particularly evident in the Fylde area, it is still regular and relatively widespread, in small numbers, in other parts of VC60.

■ **HABITAT** Hedgerows, scrub, woodland and brownfield sites.
■ **FOODPLANTS** Hawthorn.

Single brooded. Comes to light. Habitat loss is a factor in the decline in some areas.

● 2000+　□ 1970–1999　▲ 1970–1999 10km^2　□ Pre-1970　△ Pre-1970 10km^2

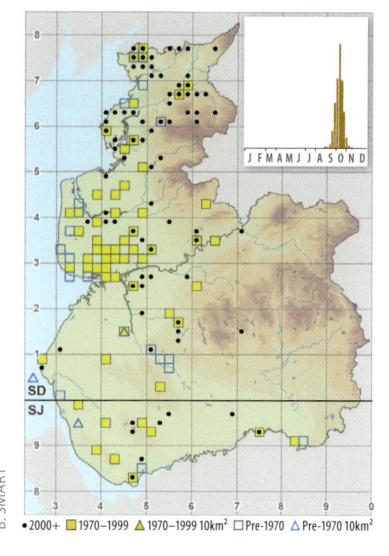

73.036 Alder Moth *Acronicta alni*

■ **LANCS STATUS** Frequent / widespread resident　　　　1850–2022

First noted at Speke in June 1850 by C.S. Gregson, from Warrington in 1853 and Holme in 1868. Mansbridge (1940) reported it to be occasional, but since then, it has commenced a gradual expansion of range, including into upland areas. This became particularly noticeable from 2003 at sites such as Gait Barrows, Leighton Moss, Rishton, Thrushgill and Yealand Conyers as well as other sites in the last decade.

■ **HABITAT** Woodland, scrub, brownfield sites, parks and occasionally gardens.
■ **FOODPLANTS** Alder, lime, Copper Beech, Blackthorn, birch sp. and elm sp.

Single brooded. Comes to light, mostly in low numbers, and to sugar. 16 seen at Thrushgill on 9 June 2016 (P.J. Marsh). Larvae have been noted in July and August.

● 2000+　□ 1970–1999　▲ 1970–1999 10km^2　□ Pre-1970　△ Pre-1970 10km^2

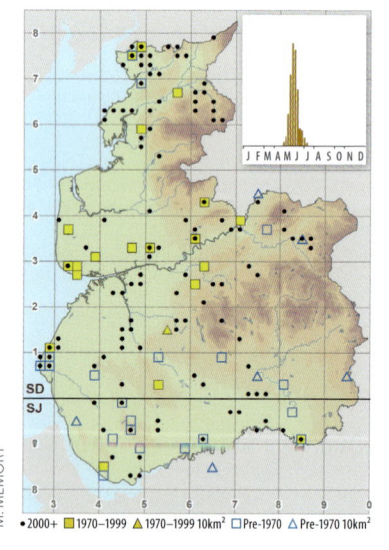

73.037 Dark Dagger *Acronicta tridens*

■ **LANCS STATUS** Occasional / widespread resident　　　　1965–2022

See note on identification under Grey Dagger. Records from Warrington (1853) and Formby (1910) are excluded as no mention is made of the stage, adult or larva. First confirmed from Bispham on 7 June 1965 by G. Bowden. Since then, records of larvae or dissected adults indicate a scattered distribution, in small numbers, across parts of lowland Lancashire, but with only three upland records, the latter between 2010 and 2021. It is considered an under-recorded species. See also Appendix 6.

■ **HABITAT** A wide range, including coastal areas, gardens, hedgerows and moorland.
■ **FOODPLANTS** Hawthorn, Rowan, Wild Plum and larva found next to garden rose.

Single brooded. Comes to light. Larvae found late July to mid-October.

● 2000+　□ 1970–1999　▲ 1970–1999 10km^2　□ Pre-1970　△ Pre-1970 10km^2

490　*The Moths of Lancashire*

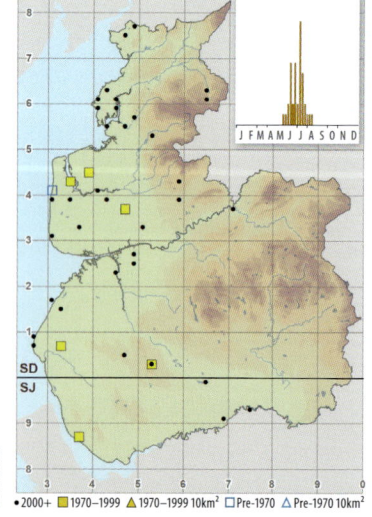

73.038 Grey Dagger *Acronicta psi*

■ **LANCS STATUS** Frequent / widespread resident   c.1857–2022

Grey and Dark Dagger have identical external markings, but are readily separable as larvae. First noted in the Liverpool area (Gregson, 1857) and since then, found across much of the county, in low numbers. It is under-recorded due to the identification problems, but dissections of a few specimens, and larval records, at many sites show it is more numerous and widespread than Dark Dagger. See also Appendix 6.

■ **HABITAT** Woodland, hedgerows, scrub, gardens and brownfield sites.

■ **FOODPLANTS** Rowan, birch, Plum, Cherry, oak, Hazel, lime, sallow, rose, Hawthorn, Alder, Blackthorn, elm, Pear, Apple, *Eucryphia*, Cotoneaster, Pyracantha, blueberry cultivar.

Single brooded. Comes to light. Larvae found from July to October.

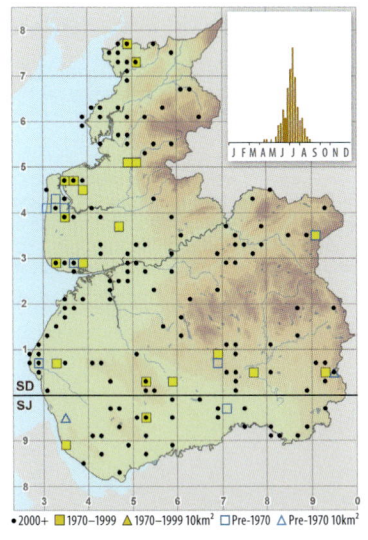

• 2000+   ■ 1970–1999   ▲ 1970–1999 10km²   □ Pre-1970   △ Pre-1970 10km²

73.039 Sycamore *Acronicta aceris*

■ **LANCS STATUS** Frequent / local resident   2017–2022

As part of a national northward expansion in range, this species was first recorded on 1 July 2017 at Aintree, Liverpool by K. Fairclough and a larva was found in Chorlton on 13 August 2017 by D. Riding. The distribution and number of records received has increased considerably each year since then, reaching 70 records in 2022. It has now populated most of southern VC59, reaching Hightown Dunes in 2018, Briercliffe in 2021 and Tarleton, where it was seen as a larva, on 7 August 2022 (B. Pyefinch).

■ **HABITAT** Gardens, parks and amenity planted areas.

■ **FOODPLANTS** Not recorded in Lancashire, despite 32 larval records.

Single brooded. Comes to light in low numbers. Larvae found August and September.

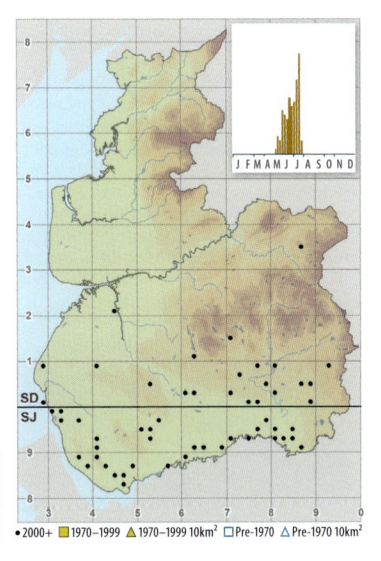

• 2000+   ■ 1970–1999   ▲ 1970–1999 10km²   □ Pre-1970   △ Pre-1970 10km²

73.040 Miller *Acronicta leporina*

■ **LANCS STATUS** Frequent / widespread resident   1843–2022

First noted in Preston in 1843 by J.B. Hodgkinson and then, up to 1940, found mostly on lowland mosses. During the 1950s, site numbers increased steadily and there were indications of a spread into higher parts, with a record from Nelson in 1954. From 1999, it was a regular at Worsthorne and has, since then, established itself across much of the county, being annual in small numbers at many sites.

■ **HABITAT** Scrub, woodland, gardens, moorland edges and brownfield sites.

■ **FOODPLANTS** Birch sp.

Single brooded. Comes to light. Larvae found in July to September. Of two larvae found on 16 June 2007, one emerged in May 2008 and the other on 29 May 2009.

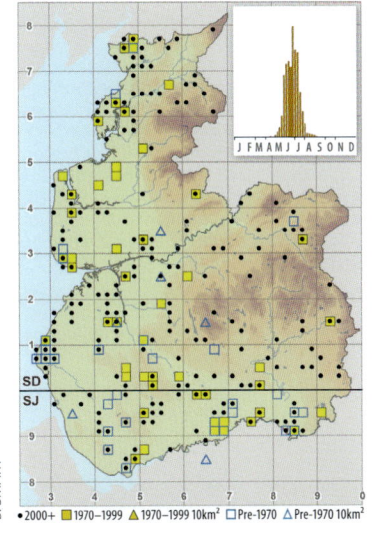

• 2000+   ■ 1970–1999   ▲ 1970–1999 10km²   □ Pre-1970   △ Pre-1970 10km²

The Moths of Lancashire   **491**

NOCTUIDAE

73.042 Light Knot Grass *Acronicta menyanthidis*

■ **LANCS STATUS** Occasional / local resident   1843–2022

Noted at Heysham Moss in 1843 by J.B. Hodgkinson and, later that century, widely on the Manchester and Liverpool area mosses. From 1879, found at upland sites such as Withnell Fold, Clougha, Burnley and Oldham. Few records were received up to 2000, but included Beacon Fell and Winmarleigh Moss. Since then, there have been five well-scattered lowland records, possibly wanderers, all others being from moorland sites, including Leck Fell, Cross of Greet, Thrushgill and Worsthorne.

■ **HABITAT** Moorland, mosses and upland marshy areas.

■ **FOODPLANTS** Heather, willow sp., birch sp., and once on Bogbean (Ellis, 1890). Single brooded. Comes to light in single figures. 18 seen at Grizedale Head in 2004.

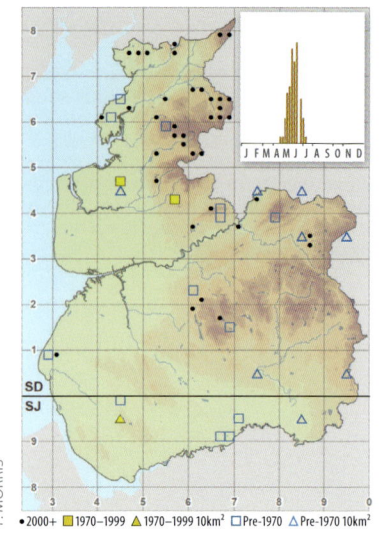

73.045 Knot Grass *Acronicta rumicis*

■ **LANCS STATUS** Frequent / widespread resident   1844–2022

First found in Manchester (R.S. Edleston) in 1844 and thereafter reported from many parts of the county. Noted as common as a larva in gardens in the Salford area in the 1950s (R. Leverton). In recent decades, it has remained regular at sites such as Ainsdale, Formby, Thrushgill and Worsthorne, but other sites, such as Flixton and Billinge, have experienced a marked decline into the early 21st century.

■ **HABITAT** Coastal dunes, moorland, grassland, wetlands and scrubland.

■ **FOODPLANTS** Bramble, Marsh Thistle, Borage, Water Mint, Common Nettle and birch. Double brooded, second brood being small. Comes to light, usually in single figures. Larvae late May to early September and late September to early November.

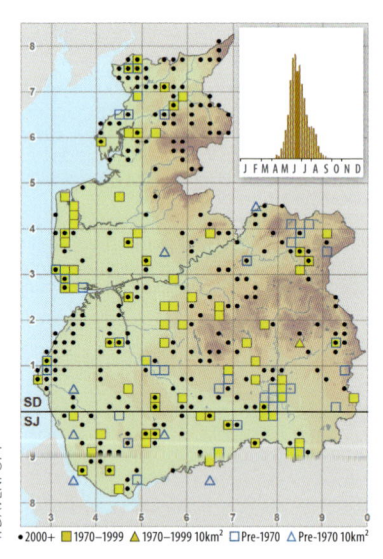

73.046 Poplar Grey *Subacronicta megacephala*

■ **LANCS STATUS** Frequent / widespread resident   c.1857–2022

Noted in Liverpool (Gregson, 1857) where White Poplars grow. Later, in the 19th century, at Oldham, Peel Green, Thursden, St Annes and Crumpsall, amongst others. Noted as often abundant in towns where poplars were planted (Mansbridge, 1940). Since then, reported widely to the present day. Sites with regular, long-term record sets include Martin Mere, St Helens area, Billinge, Swinton, Orrell and Morecambe.

■ **HABITAT** Woodland, wetland sites, amenity plantings, parks, scrub and gardens.

■ **FOODPLANTS** Poplars and willows, including White Poplar, Aspen and Crack-willow. Single brooded, with an extended flight period. Comes to light, almost always in single figures. Larvae have been found from late July to early September.

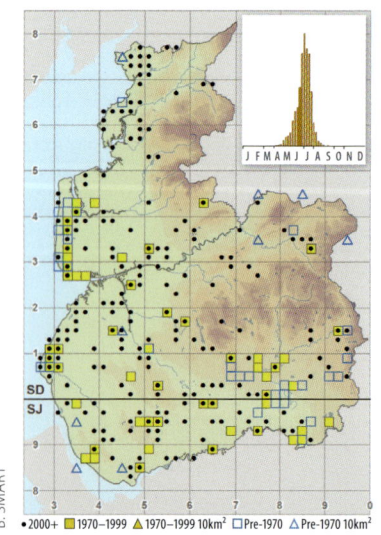

73.047 Coronet *Craniophora ligustri*

■ **LANCS STATUS** Occasional / local resident   c.1857–2022

Recorded from Speke, Hale and Rainhill (Gregson, 1857), these being the only Liverpool area sightings until one in 2021. Since the late 1990s, it has been noted widely and regularly in northern parts of the county such as Carnforth, Gait Barrows, Thrushgill, Leighton Moss, Silverdale, Warton Crag and Yealand Conyers. Away from its core northern area it is very infrequent, but perhaps increasing. In recent years, it has been found in eight scattered localities across the mid-section of the county.
■ **HABITAT** Limestone scrub, moorland edge and river valley woods mostly in the north.
■ **FOODPLANTS** Not recorded in Lancashire.
Single brooded. Comes to light usually in single figures.

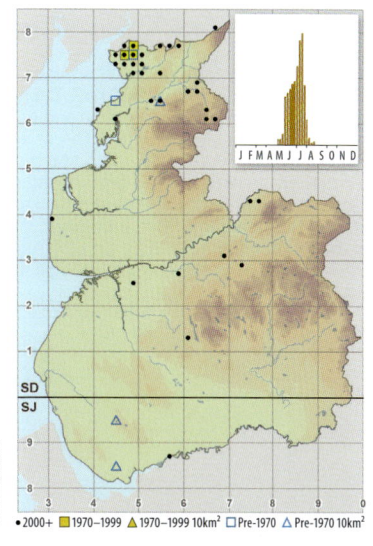

73.048 Small Yellow Underwing *Panemeria tenebrata*

■ **LANCS STATUS** Occasional / local resident   c.1857–2022

First noted in Orrell (Gregson, 1857) and later, to 1910, in Chorley, Hornby, Torrisholme, Oldham and Ainsdale, among others. Not noted again until 1951, at Normoss, and thereafter at many scattered sites across mainly lowland parts. As a daytime flyer it is probably under-recorded, but where suitable sites are visited regularly, such as in Flixton from 1999–2019 (K. McCabe), it is readily observed.
■ **HABITAT** Meadows, rough ground, dune grassland and brownfield sites.
■ **FOODPLANTS** Not recorded in Lancashire.
Single brooded. Visits flowers of mouse-ear (the nationally known foodplant), clovers and vetches by day. 25 were seen at Pickering's Pasture on 28 May 1994 (P. Hillyer).

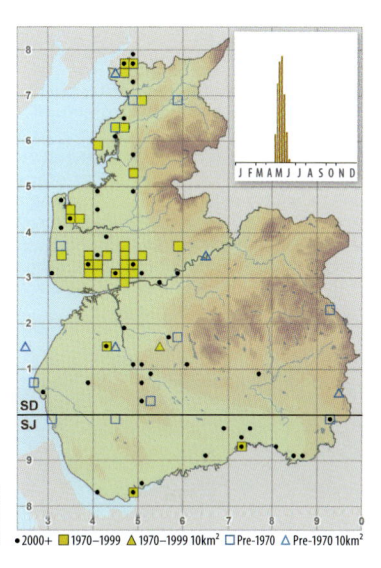

73.050 Wormwood *Cucullia absinthii*

■ **LANCS STATUS** Considered Extinct   1952–1999

This nationally scarce moth arrived in Lancashire following a northward expansion in range after the Second World War (Heath & Emmet, 1983b). First noted in Formby in 1952 (D. Dixon), followed by St. Annes in 1955 and then, more widely, from sites such as Salford, Bolton and Flixton in the 1960s. Since then, it was found at Woolston Eyes in 1984, Widnes in 1990, Worsley in 1994, Flixton from 1995 to 1999 and Pennington and Salford in 1999. The last record was from Flixton on 25 July 1999 (K. McCabe).
■ **HABITAT** Brownfield sites and other disturbed ground.
■ **FOODPLANTS** Not recorded in Lancashire.
Single brooded. Comes to light in low single figures.

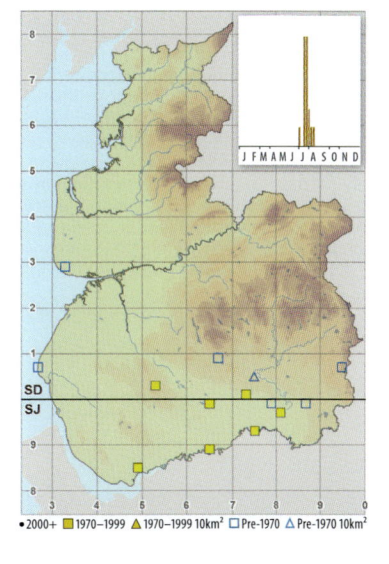

The Moths of Lancashire   **493**

NOCTUIDAE

73.052 Shark *Cucullia umbratica*

■ **LANCS STATUS** Occasional / local resident   1853–2022

Recorded from Chorley in 1853 by E. Buxton, in the Liverpool district (Gregson, 1857) and later, from many other sites, including Rochdale in 1887. Mansbridge (1940) noted it as generally distributed. Thereafter, it was reported widely across lowland parts, and fewer upland sites such as Littleborough in 1999 and Briercliffe in 2014. Since then, there has been a gradual reduction in records, which are now mostly from near-coastal sites. The most inland recent report was from Horwich in 2016.
■ **HABITAT** Gardens, marshes, brownfield sites and sandy grassland.
■ **FOODPLANTS** Not recorded in Lancashire.
Single brooded. Comes to light in single figures. Attracted to Honeysuckle flowers.

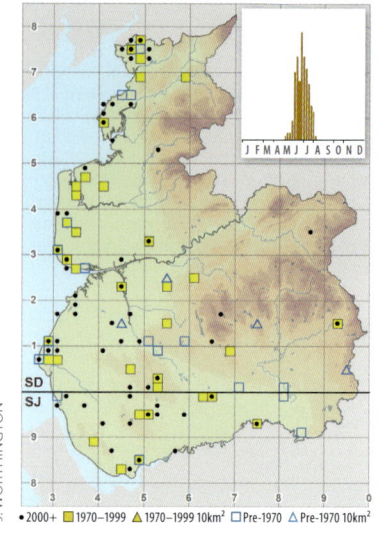

• 2000+   ☐ 1970–1999   △ 1970–1999 10km²   ☐ Pre-1970   △ Pre-1970 10km²

73.053 Chamomile Shark *Cucullia chamomillae*

■ **LANCS STATUS** Occasional / local resident   1853–2022

First noted in 1853 near Warrington by N. Cooke and then, from widely scattered localities in coastal and lowland areas to the present day. Reports increased in the early 2000s, boosted by larval searches in the Bickerstaffe and Skelmersdale areas in 2007 (C. Darbyshire). Records have declined notably since then, which may be partly due to an over-reliance on light trapping and few recent larval searches.
■ **HABITAT** Disturbed ground, coastal and estuarine areas, fields and brownfield sites.
■ **FOODPLANTS** Scentless, Scented and (probably) Sea Mayweed, and Corn Chamomile.
Single brooded. Comes to light in low single figures. Larvae found, sometimes commonly, in June and July, making up over one third of the 169 records.

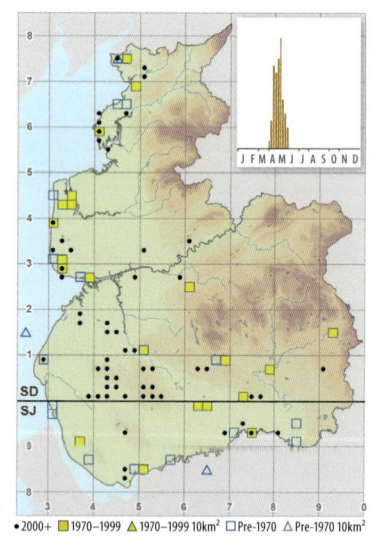

• 2000+   ☐ 1970–1999   △ 1970–1999 10km²   ☐ Pre-1970   △ Pre-1970 10km²

73.058 Mullein *Cucullia verbasci*

■ **LANCS STATUS** Frequent / local resident   1940–2022

First noted in 1940 by G. Pringle in 10km square SJ79, then at Nelson by C. Baldwin in 1943. Between then and 2002, it remained a scarce moth with nine records mostly from the Silverdale area. From 2004, its range gradually increased and by 2019 it had reached many lowland, and a few upland, sites with most records being of larvae. This demonstrates its ability to move large distances to reach plants in new areas.
■ **HABITAT** Rough ground, often on limestone, gardens and brownfield sites,
■ **FOODPLANTS** Great Mullein, Buddleia and *Verbascum chaixii* var. 'milkshake'.
Single brood. Comes to light. Often seen as larvae, from late May to early July, and can be common where found with 120 at Middleton N.R. on 7 June 2015 (J. Patton).

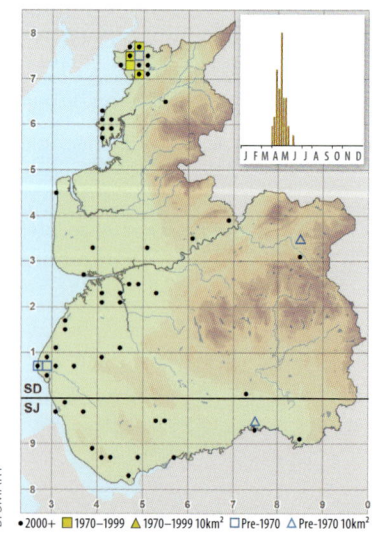

• 2000+   ☐ 1970–1999   △ 1970–1999 10km²   ☐ Pre-1970   △ Pre-1970 10km²

494   *The Moths of Lancashire*

73.059 Toadflax Brocade *Calophasia lunula*

■ **LANCS STATUS** Rare / very local resident 2018–2023

A new arrival in the county as part of a national northward expansion of range. The first came to a garden actinic light trap in Great Sankey on 8 August 2018 (D. Riley). Not recorded again until 2023, when a few records suggested colonisation was in progress in the county. One was found at Widnes on 2 June (D. Kelly), then it occurred on 6 and 8 August at Great Sankey (D. & A. Riley) and finally at Woolton, Liverpool on 10 August (S. McWilliam).
■ **HABITAT** Unknown.
■ **FOODPLANTS** Not recorded in Lancashire.
Comes to light, all present records of singletons.

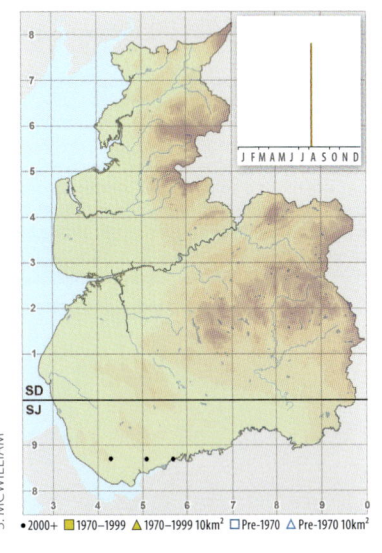

73.061 Anomalous *Stilbia anomala*

■ **LANCS STATUS** Occasional / local resident 1879–2022

First noted near Ribchester in 1879 (Hodgkinson, 1880a) and later, at Longridge and Silverdale. Later, found in Catlow in 1939 (A. Brindle) and Pendle, Formby, Southport, Morecambe, Leighton Moss, amongst others, up to the early 1990s. Most since then have been from moorland areas such as Docker Moor, Green Bank and Worsthorne, and less frequently, from limestone sites such as Gait Barrows and Dalton Crag.
■ **HABITAT** Moorland; less often on limestone and dune grassland.
■ **FOODPLANTS** Tufted and Wavy Hair-grass. Reported use of mouse-ear is very unlikely.
Single brooded. Comes to light. During 2009 and 2010, up to 27 larvae (this on 21 February 2009) were found at various sites south of Todmorden, by B. Leecy.

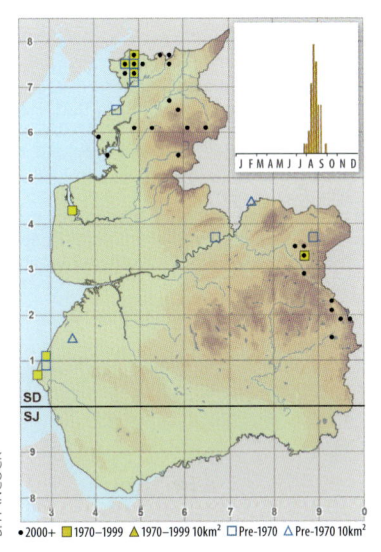

73.062 Copper Underwing *Amphipyra pyramidea*

■ **LANCS STATUS** Frequent / widespread resident 1984–2022

Earliest record was in 1940, before *A. berbera* was split as a separate species, in 1968. First confirmed record from Lytham in 1984 (E. Stuart, genitalia det. M. Hull) then, one in Bolton in 1987. Although noted on occasions before 2009, it was not until after then that extensive checks of critical features commenced. This proved it was widespread, but it is probably under-recorded. See also Appendix 6.
■ **HABITAT** Woodland, hedgerows, gardens, parks, scrubland and brownfield sites.
■ **FOODPLANTS** Ash, Goat Willow, Hawthorn and elm sp.
Single brooded. Comes to light. Larvae noted April to June. Can be found at the same sites as *A. berbera* and considerable care is needed to ensure accurate identification.

The Moths of Lancashire **495**

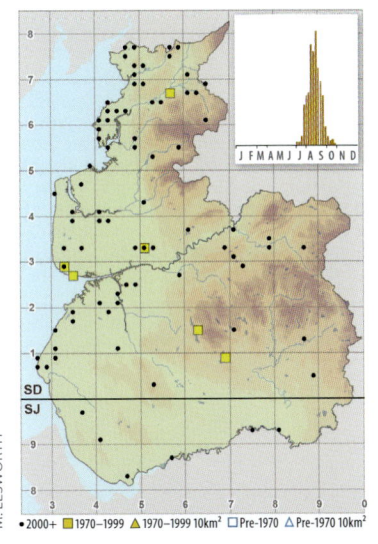

NOCTUIDAE

73.063 Svensson's Copper Underwing *Amphipyra berbera*

■ **LANCS STATUS** Frequent / widespread resident   1983–2022

First separated from Copper Underwing in 1968, but not identified in Lancashire until 16 June 1983 when a larva was found and bred in Bolton (Hancock and Wallace, 1986). Next, in Ansdell (1988), Bolton (1992), St Annes (1995) and Brock Bottom in 1999. From 2009, critical examination of specimens by recorders increased in the county, but the moth is still probably under-recorded. See also Appendix 6.

■ **HABITAT** Woodland, hedgerows, gardens, parks, scrubland and brownfield sites.
■ **FOODPLANTS** Blackthorn, elm sp., White Lilac (flowers), oak sp., Goat Willow and Ivy.
Single brooded. Comes to light. This and Copper Underwing can be found at the same sites and great care is needed to ensure accurate recording. Larvae, April to June.

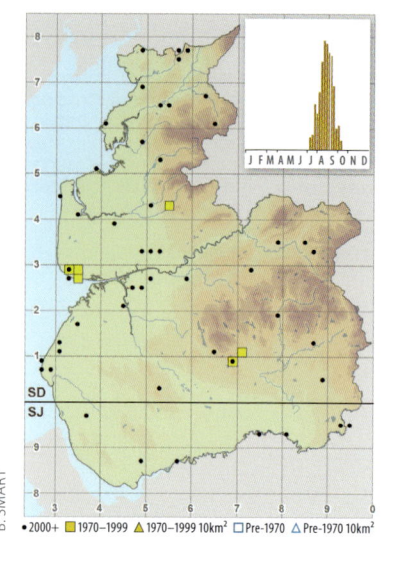

73.064 Mouse Moth *Amphipyra tragopoginis*

■ **LANCS STATUS** Frequent / widespread resident   1843–2022

First found in the Preston area in 1843 (Hodgkinson, 1844), then Withington in 1846, Liverpool in 1853 and Oldham in 1892, with Ellis (1890) noting it as abundant and Mansbridge (1940) common throughout. It has continued to be widely reported from across the county, being locally common at a few sites such as Swinton into the early 1990s. Overall numbers have been lower since then, with a slight upturn since 2020.

■ **HABITAT** Woodland, scrubland, moorland, sand dunes and gardens.
■ **FOODPLANTS** Willow sp. and sowthistle sp.
Single brooded. Comes to light and sugar. 42 came to a Freckleton light trap on 27 August 1985 M. Evans. Larvae have been found from late May to early July.

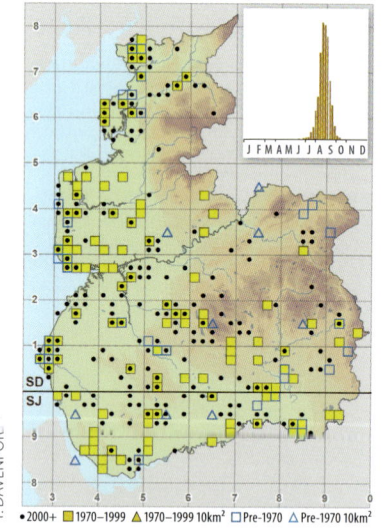

73.065 Sprawler *Asteroscopus sphinx*

■ **LANCS STATUS** Scarce / local resident   c.1853–2022

First noted near Warrington (Byerley, 1854) and Preston in 1885. It was later found in Longridge in 1976 and 1979 and at Burrow Heights in 2001. However, since 1952, the core sites have been in the Silverdale area and wooded valleys of the Rivers Roeburn and Hindburn. Noted most frequently at Mill Houses and Leighton Moss, but at the latter not since 2015. There has been an overall decline in reports since 2017.

■ **HABITAT** Mature deciduous woodland, on limestone and in northern river valleys.
■ **FOODPLANTS** Not recorded in Lancashire.
Single brooded. Comes to light, usually in low single figures. The largest count was of 42 to MV light at Mill Houses, near Wray on 30 October 2016 (P.J. Marsh).

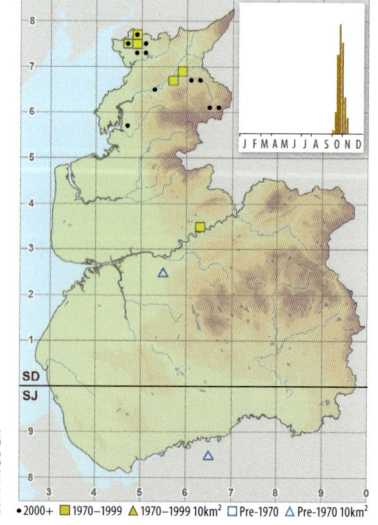

73.068 Green-brindled Crescent *Allophyes oxyacanthae*

■ **LANCS STATUS** Frequent / widespread resident 1843–2022
First noted in the Preston area in 1843 (Hodgkinson, 1844), near Liverpool in 1853, Rochdale in 1880, Manchester and St Annes in 1892 and Oldham in 1894. All authors between 1890 and 1950, considered it common and generally distributed and it has remained such to the present day. Numbers have declined in recent years in some suburban gardens. 30 came to actinic light on Warton Crag on 12 October 2018.
■ **HABITAT** Hedgerows, scrub, gardens, woodland and brownfield sites.
■ **FOODPLANTS** Blackthorn and Hawthorn.
Single brooded. Comes to light, usually in single figures. Nectars at Ivy blossom and is attracted to sugar. Larvae have been found in April and May.

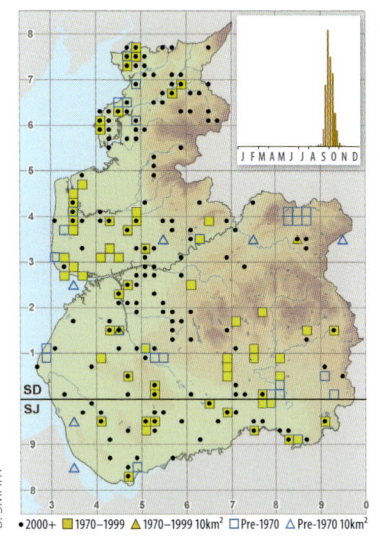

• 2000+ ☐ 1970–1999 ▲ 1970–1999 10km² ☐ Pre-1970 △ Pre-1970 10km²

73.069 Early Grey *Xylocampa areola*

■ **LANCS STATUS** Frequent / widespread resident c.1857–2022
First in the Liverpool area by Gregson (1857) then, not until 1922, in Silverdale and 1935 in St Annes. Ellis (1890) noted it as tolerably common while Mansbridge (1940) noted it as not common except in the north into Cumbria. During the 1950s, reports from lowland areas were increasing steadily, and included a few more inland sites, such as Bolton in 1952. By the 1980s, it was regular in Swinton and since then, has spread to much of the county, wherever the foodplants occur.
■ **HABITAT** Woodland, gardens, parks, larger hedgerows, scrubland and brownfield sites.
■ **FOODPLANTS** Honeysuckle and Snowberry.
Single brooded. Comes to light, often in good numbers. Larvae found in June.

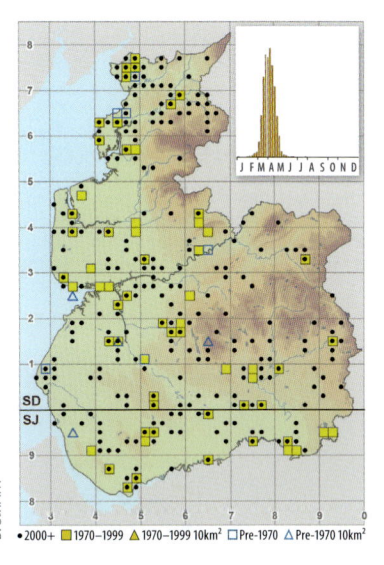

• 2000+ ☐ 1970–1999 ▲ 1970–1999 10km² ☐ Pre-1970 △ Pre-1970 10km²

73.070 Bordered Sallow *Pyrrhia umbra*

■ **LANCS STATUS** Occasional / local resident c.1857–2022
First noted in Crosby (Gregson, 1857), then at other coastal areas including Ainsdale, Silverdale and Formby. From the 1950s to c.2012, reported frequently from Lytham St Annes and, since 2000, has been regular in small numbers at Ainsdale, Bispham, Formby, Heysham, Sunderland Point and in the Silverdale area. Occasional inland records suggest wanderers, or that it may be utilising additional larval foodplants.
■ **HABITAT** Coastal embankments, sandy and limestone grassland; possibly river shingle.
■ **FOODPLANTS** Larva found close to a restharrow plant.
Single brooded. Comes to light and to sugar. Usually in low single figures, but 20 came to light at Birkdale Green Beach in June 2011 and 18 at the same site in 2022.

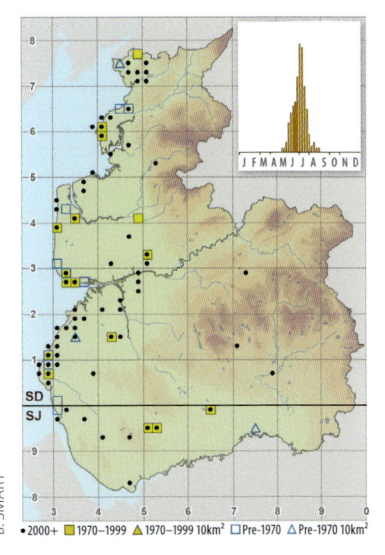

• 2000+ ☐ 1970–1999 ▲ 1970–1999 10km² ☐ Pre-1970 △ Pre-1970 10km²

NOCTUIDAE

73.074 Bordered Straw *Heliothis peltigera*

■ **LANCS STATUS** Occasional / migrant   1876–2022

First noted at Blackpool in June 1876 by J.W. Aspinwall (Ellis, 1890), followed by records from Farington, Lytham, Platt Fields Park and St Annes, up to the late 1950s. Between then and 1999 it was noted on 16 occasions. 2000 was a good '*peltigera*' year, with seven records, as was 2015 with 26 records and 2022 with nine records. All of these were dwarfed by the 52 reports received in 2006, from the 19 May to 30 September. The largest overnight count was of three at Hough Green on 3 July 2006.
■ **HABITAT** Gardens, where larvae, on rare occasions, utilise planted annuals.
■ **FOODPLANTS** Marigold and Tobacco Plant.
Comes to light, nearly always as singletons.

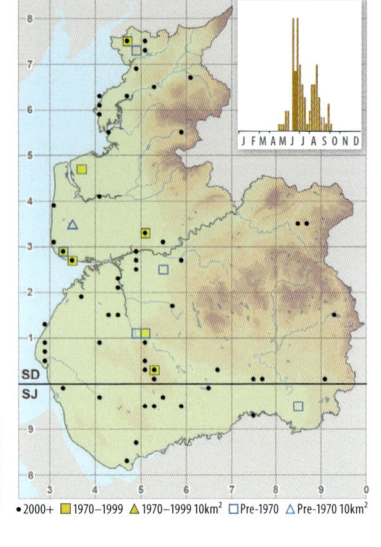

73.076 Scarce Bordered Straw *Helicoverpa armigera*

■ **LANCS STATUS** Occasional / migrant; rare adventive.   1840–2022

The first Lancashire record was also the first in Britain, when found on the door of an outhouse in Oldfield Lane, Salford in September 1840 by J. Thomas (Edleston, 1843b). Between then and 1999 there were only nine documented records, three of these being in 1950. Since 1999, it has been regular, but not annual, with reports usually in low single figures. The exceptions were in 2022, when twelve records were received and in 2006, from late June to early November, when 189 records were received.
■ **HABITAT** Unknown.
■ **FOODPLANTS** In box of imported Tomatoes.
Comes readily to light in small numbers. Largest count, eight at Hale on 20 Aug. 2006.

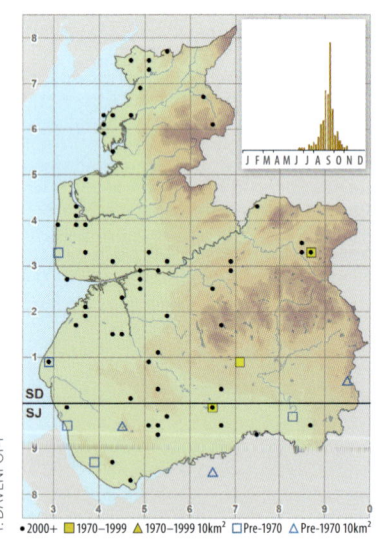

73.084 Marbled Beauty *Bryophila domestica*

■ **LANCS STATUS** Frequent / widespread resident   c.1857–2022

First noted in the Liverpool district (Gregson, 1857) and reported by early authors, up to 1940, as common throughout, except in East Lancashire; here it was noted as scarce. Within a few decades, it was found throughout the county, including the east, as a widespread and regular species, for example being very frequent in Swinton during the 1980s and '90s (S. Christmas). Double figure counts of 25 or more were not uncommon until 2011 or so, but are now much less frequently encountered.
■ **HABITAT** Woodland and areas with man-made stone or brick structures.
■ **FOODPLANTS** On lichens on walls (including a garage wall) and tree trunks.
Single brooded. Comes to light. Several records in May and one in October.

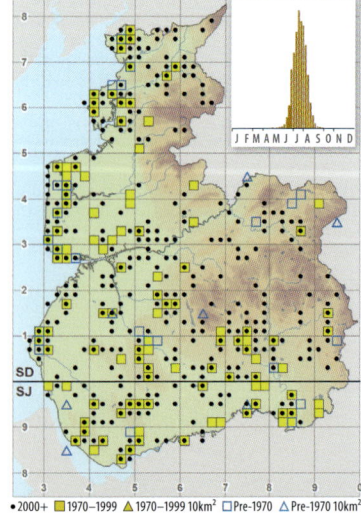

498 *The Moths of Lancashire*

73.085 Marbled Green *Bryopsis muralis*

■ **LANCS STATUS** Rare / very local resident **2018–2022**
Has been spreading rapidly northwards in Britain and was first confirmed in the county at Fazakerley on 2 August 2018, with one at a garden light trap (L. Ward). One was then noted in Formby on 24 July 2019 by T. Davenport and, most recently, one was seen resting on Heysham South Harbour wall on 26 July 2022 (K. Eaves).
■ **HABITAT** Unknown; nationally associated with rocky places and dry-stone walls.
■ **FOODPLANTS** Not recorded in Lancashire.
Single brooded. Comes to light. A record from 2009 was not possible to verify as the moth was not retained or photographed to ensure the rather similar, widespread and sometimes green-tinged, Marbled Beauty had been fully excluded.

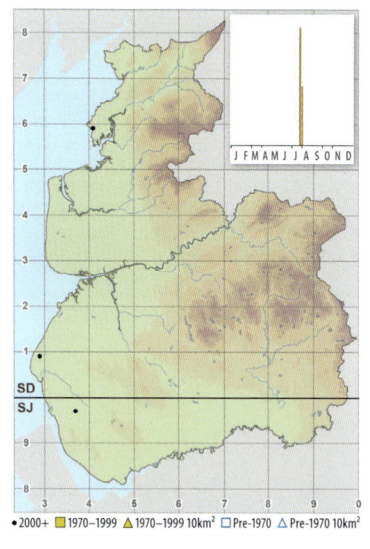

73.087 Small Mottled Willow *Spodoptera exigua*

■ **LANCS STATUS** Occasional / migrant **1884–2022**
Prior to 1996, there were only twelve reports of this migratory moth, mostly from coastal localities. The first of these was on 16 September 1884 at Crosby (G.A. Harker). Since 1996, records have increased considerably, with fewer blank years and the moths more widely spread across the county. It was also notable that six years within this period had records reaching double figures, including 42 in 2006 and 25 during 2015. Most were attracted to light and many involved singletons, but 5 were attracted to MV light at Sunderland Point on 12 July 2015 (P.J. Marsh & J. Roberts).
■ **HABITAT** Unknown.
■ **FOODPLANTS** Not recorded in Lancashire.

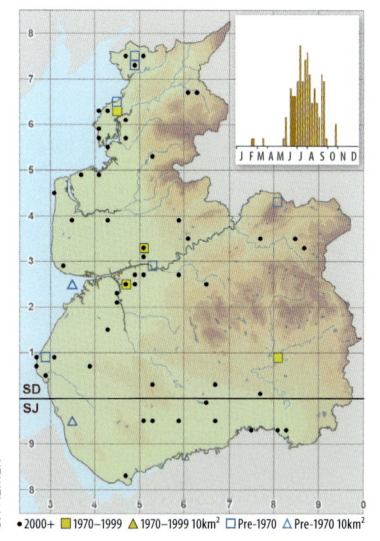

73.089 Mediterranean Brocade *Spodoptera littoralis*

■ **LANCS STATUS** Adventive; rare migrant **1871–2022**
Recorded twice in the county. An imported larva was found in a fruit warehouse in Liverpool in 1871 (Heath & Emmet, 1983). It was subsequently bred by C.S. Gregson and the specimen is in the NHM, London. The only other record was of a slightly worn individual that came to a MV light trap in Flixton, Manchester on 2 September 2022 (K. McCabe). The specimen was retained and confirmed by critical examination to exclude *Spodoptera litura*. The origin of the Flixton moth is unknown, but occurred during a period when nine common migrant species were noted in the county.
■ **HABITAT** Unknown.
■ **FOODPLANTS** Not recorded in Lancashire, but once associated with imported fruit.

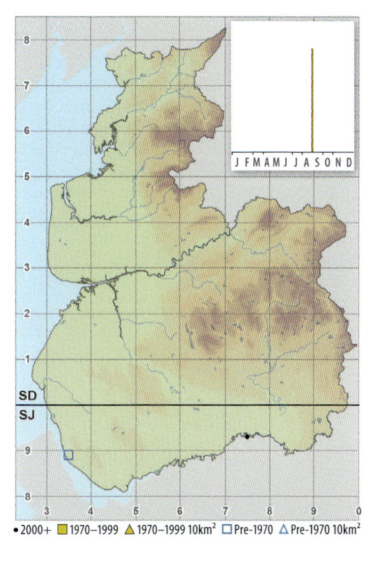

The Moths of Lancashire

NOCTUIDAE

73.0892 Asian Cotton Leafworm *Spodoptera litura*

■ **LANCS STATUS** Adventive 1978

The following was published in Heath & Emmet (1983a). 'The species was first correctly identified in Britain in 1978; previous records probably refer to *Spodoptera littoralis* (Mediterranean Brocade) with which it was confused until 1963. One larva of *S. litura* was found in September and three in November 1978 at Manchester on unnamed aquatic plants, imported from Singapore. Both the imago and the larva closely resemble those of *S. littoralis*'.
■ **HABITAT** Not applicable.
■ **FOODPLANTS** On unknown imported aquatic plants.

The original report of this was published in Seymour, P.R. & Kilby, L.J., 1978. *Insects and other invertebrates intercepted in check inspections of imported plant material in England and Wales 1978*. Rep. MAFF. Pl. Path. Lab. 12: 1–33. This document has not been examined.

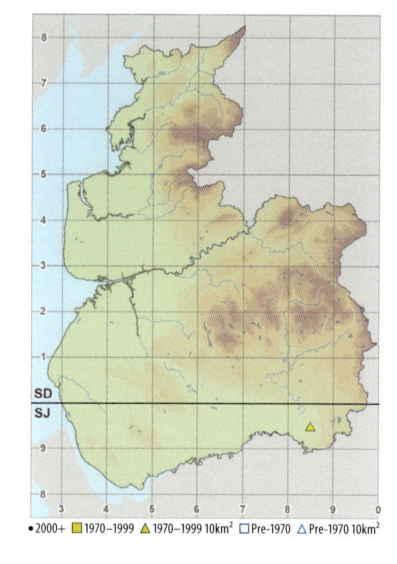

● 2000+   ■ 1970–1999   ▲ 1970–1999 10km²   □ Pre-1970   △ Pre-1970 10km²

73.092 Mottled Rustic *Caradrina morpheus*

■ **LANCS STATUS** Frequent / widespread resident c.1857–2022

First noted in the Liverpool district (Gregson, 1857) and then, from Withington, Crosby, Pendleton, Preston, Prestwich and St Annes up to 1908. Increased light trapping from the 1950s confirmed the early indications of a widespread distribution in both lower and higher parts of the county. There has been a decline in numbers recorded at light from peaks in the 1970s to early 2000s. Counts of 20 or so are now uncommon although 22 were found at Gorse Hill N.R. on 2 July 2019 by S. Haselton.
■ **HABITAT** Wide range of habitats, but particularly favouring lowland grassland.
■ **FOODPLANTS** Common Nettle.

Single brooded. Comes to light. Adult observed 'nectaring' on honeydew.

● 2000+   ■ 1970–1999   ▲ 1970–1999 10km²   □ Pre-1970   △ Pre-1970 10km²

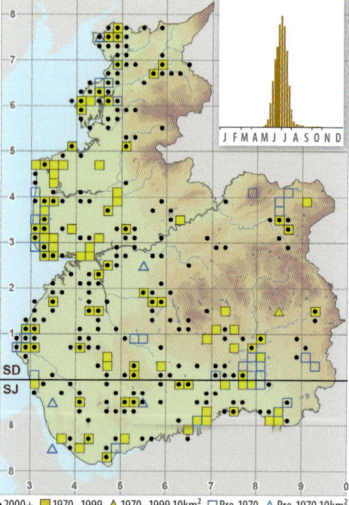

73.095 Pale Mottled Willow *Caradrina clavipalpis*

■ **LANCS STATUS** Frequent / widespread resident 1846–2022

First seen in 1846 nectaring on ragwort at Cheetham Hill by R.S. Edleston. Later, noted as abundant everywhere by Ellis (1890) and Mansbridge (1940). Since the 1950s, found to be widely distributed, but mostly in single figures. Sites with regular reports include Billinge, Briercliffe, Catterall, Hale, Flixton, Hutton, Longton, Urmston and Pennington. Double digit counts rare in recent decades, although 17 nr. Chorley in 2008. Randle *et al.* (2019) suggest it could also, probably, be a migrant species.
■ **HABITAT** Gardens and grassland.
■ **FOODPLANTS** Plantain sp.

Double brooded. Comes to light. Larvae found August, emerging September.

● 2000+   ■ 1970–1999   ▲ 1970–1999 10km²   □ Pre-1970   △ Pre-1970 10km²

The Moths of Lancashire

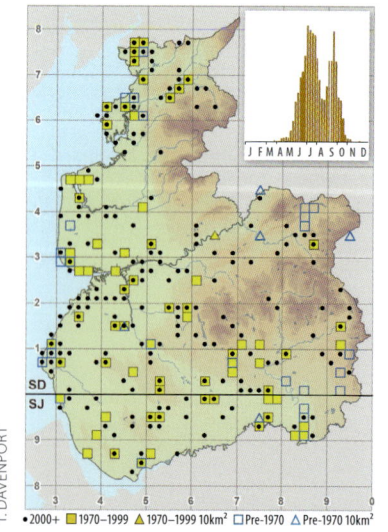

73.096 Uncertain *Hoplodrina octogenaria*

■ **LANCS STATUS** Frequent / widespread resident   c.1857–2022

First found in the Liverpool district (Gregson, 1857) and noted as 'not common' by both Ellis (1890) and Mansbridge (1940). Other early sites were limited to the Preston area, Eccles and Ainsdale. Since the late 1970s it has been reported widely and regularly from sites such as Bispham, Carnforth, Ellenbrook, Fazakerley, Flixton, Formby, Hale, Longton, Morecambe, Swinton and Worsley with no obvious changes in numbers or distribution in that time. Can be difficult to separate from worn Rustic.
■ **HABITAT** A wide range of habitats.
■ **FOODPLANTS** Dandelion.
Single brooded. Comes to light, mostly in single figures.

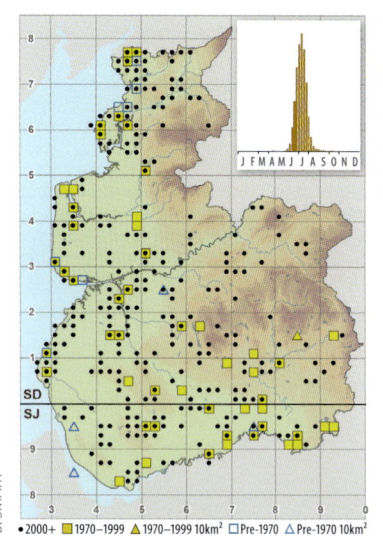

73.097 Rustic *Hoplodrina blanda*

■ **LANCS STATUS** Frequent / widespread resident   c.1857–2022

First noted in the Liverpool district (Gregson, 1857) and subsequently, up to the end of the 19th century, at Withington, Pendleton, Preston and Salford. Since then, it has been noted widely across the county in both lowland and upland districts, with many sites recording it regularly, including Hesketh Bank, St Helens, Sunderland Point, Walmer Bridge and Worsthorne. 49 were noted at Mere Sands Wood on 9 July 2001 (I. Kippax) but, since 2015, single night catches have rarely exceeded single figures.
■ **HABITAT** A wide range of habitats.
■ **FOODPLANTS** Not recorded in Lancashire.
Single brooded. Comes to light. This sp. and Uncertain can be difficult to separate.

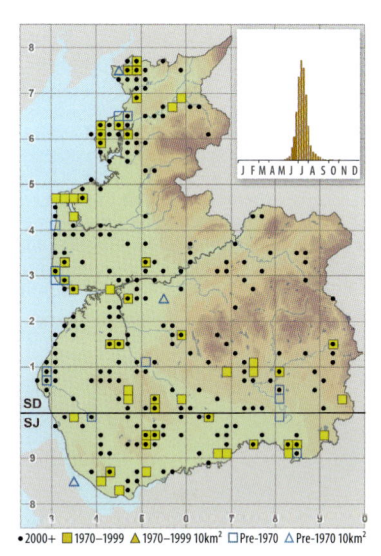

73.099 Vine's Rustic *Hoplodrina ambigua*

■ **LANCS STATUS** Frequent / local resident   1996–2022

First arrived in Britain in the 1940s, reaching Lancashire, at Heysham, on 27 August 1996 (P.J. Marsh). Next, at Freshfield (G. & D. Atherton) in 2007, followed by a few in the Formby area up to 2017. Since 2018, numbers and distribution have increased rapidly and it is now regular coastally, from Hale to Southport; less often, north to Silverdale. It moved inland around 2020, reaching Green Bank (north VC60) in 2021 and Victoria Park, Manchester (P. Pemberton) and Briercliffe (G. Turner) in 2022.
■ **HABITAT** A wide range of dry, grassy, open habitats.
■ **FOODPLANTS** Not recorded in Lancashire.
Double brooded, the first being much smaller than the second. Comes to light.

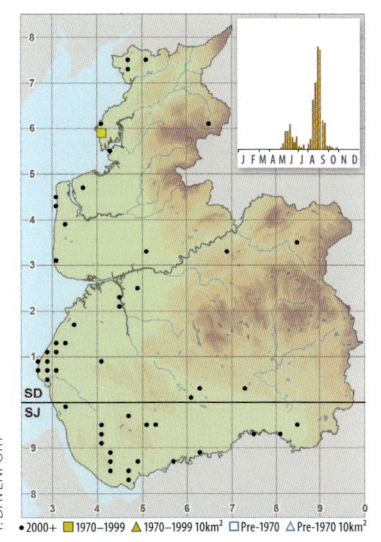

NOCTUIDAE

73.100 Silky Wainscot *Chilodes maritima*

■ **LANCS STATUS** Occasional / local resident   1982–2022
First noted at Leighton Moss in 1982 (Briggs, 1985), this site later producing over half the 680 county records. Seen at Gait Barrows in 1983 and 1994, but nowhere else other than Leighton Moss, until Flixton in 1997, Pennington in 1998 and Wigan Flashes and Parr in 1999. It has since spread slowly across lowland VC59 and reached Bispham (VC60) in 2004. Now also resident at Brockholes N.R. and Middleton N.R.
■ **HABITAT** Reedbeds; regularly wanders, turning up well away from such habitat.
■ **FOODPLANTS** Not recorded in Lancashire.
Single brooded, the few September records indicate a possible second brood. Comes to light. All three forms, ab. *nigristriata*, ab. *bipunctata* and ab. *wismariensis* occur.

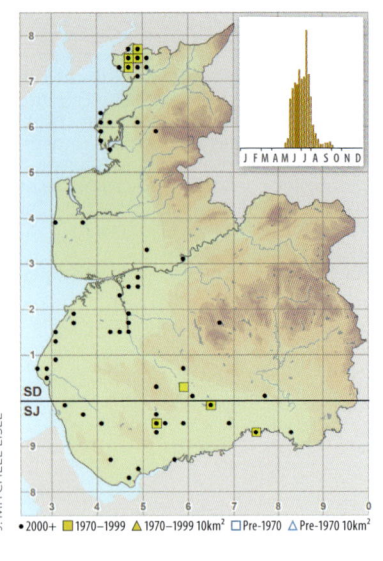

73.101 Treble Lines *Charanyca trigrammica*

■ **LANCS STATUS** Occasional / local resident   c.1857–2022
Noted as plentiful on the coast in the Liverpool area (Gregson, 1857) and later, from Prestwich, Preston, Huyton and Silverdale. Ellis (1890) notes it as very local overall. Since then, mainly associated with sites within 10km or so of the coast, with the largest counts on dune grassland and in limestone areas. Locations with many records in recent years include Crosby, Formby, Leighton Moss, Silverdale, Southport, St Annes and Yealand Conyers. Has recently shown signs of moving further inland.
■ **HABITAT** Sand dunes, limestone grassland and dry grassland.
■ **FOODPLANTS** Not recorded in Lancashire.
Single brooded. Comes to light. 40 to light at Ainsdale 25 May 2012 (J. Clews *et al.*).

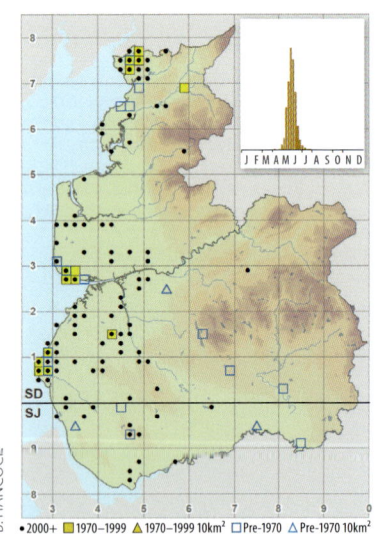

73.102 Brown Rustic *Rusina ferruginea*

■ **LANCS STATUS** Frequent / local resident   c.1857–2022
First reported in Woolton (Gregson, 1857) and later, from many scattered sites prior to 1940, including Carnforth, Longridge, Crosby, St Annes and Burnley. Since then, widely reported from mainly rural, lowland areas, particularly at sites such as Leighton Moss, Ainsdale and Formby. Since 2000, found regularly at some northern moorland edge and wooded sites, such as at Green Bank and Docker Moor, but remains very infrequent in East Lancashire and in most urban and suburban areas.
■ **HABITAT** Sand dunes, woodland, northern moors and limestone areas.
■ **FOODPLANTS** Not recorded in Lancashire.
Single brooded. Comes to light. 44 were seen at Green Bank on 29 June 2022.

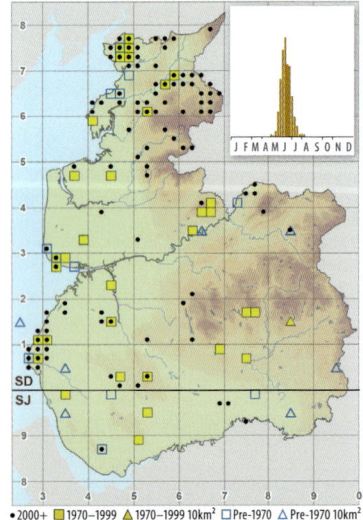

73.105 Bird's Wing *Dypterygia scabriuscula*

■ **LANCS STATUS** Scarce / local resident   c.1859–2022
First noted in Agecroft around 1859 by J. Chappell, then the Manchester area in 1885, Carnforth in 1929 (the only one in VC60) and several in the Formby area in the 1940s. All early authors considered it as very occasional or scarce. Recorded twice in the 1970s, from Didsbury and Wavertree. From 1986, it has been noted on occasions at several lowland VC59 sites, centred around the Mersey Valley. Most frequently, but not annually, reported in Hale, Flixton and Eccleston.
■ **HABITAT** Scrubland near gardens, brownfield sites and on the coast.
■ **FOODPLANTS** Not recorded in Lancashire.
Single brooded. Comes to light, with a maximum of two on three occasions.

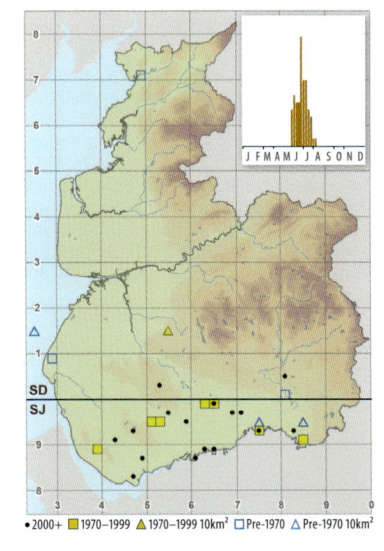

73.106 Orache Moth *Trachea atriplicis*

■ **LANCS STATUS** Rare / migrant   2009
A former resident in England, this species is now only known as a migratory species, recorded mostly along the south coast of England. The single Lancashire record relates to a specimen attracted to a Skinner MV light trap in Broadgreen, Liverpool on 4 July 2009 by D. Hardy. Butterfly Conservation provided data on others reported in Britain during 2009 (L. Evans-Hill pers. comm.). One was seen over three nights in South-east Yorkshire in late June, and four others were noted in Dorset, the Isle of Wight and Kent between 16 July and 1 August 2009.
■ **HABITAT** Unknown.
■ **FOODPLANTS** Not recorded in Lancashire.

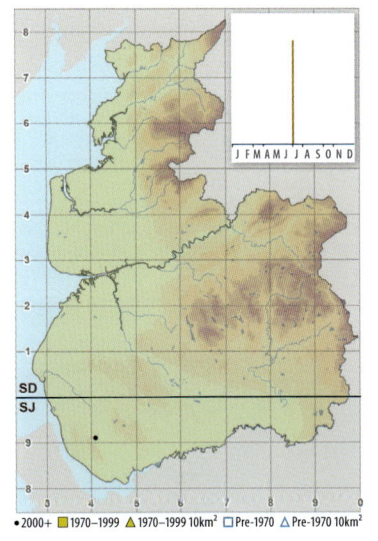

73.107 Old Lady *Mormo maura*

■ **LANCS STATUS** Occasional / widespread resident   c.1857–2022
First noted in Liverpool (Gregson, 1857), with widespread reports over the following decades. Mansbridge (1940) noted a decline in numbers. Since then, increased light trapping and sugaring has added many new sites, but the overall, general distribution has changed little. Recent regular sites include Hale, Longton, Great Crosby and St Helens. The long-term trap site at Flixton has shown an increase in annual numbers since about 2017, possibly linked to a local increase in Ivy (K. McCabe pers. comm.).
■ **HABITAT** A wide range of habitats including woodland, scrub, gardens and parks.
■ **FOODPLANTS** Ivy.
Single brooded. Comes to light and sugar. 35 seen at Swinton on 18 August 2009.

The Moths of Lancashire   **503**

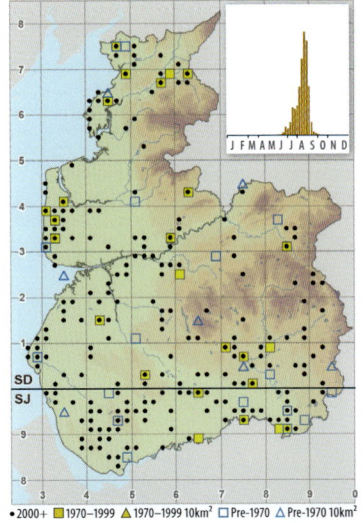

73.109 Straw Underwing *Thalpophila matura*

■ **LANCS STATUS** Frequent / widespread resident c.1857–2022
Noted as common on Crosby sandhills (Gregson, 1857), with most early records being from lowland and coastal areas. Mansbridge (1940) noted there were 'few recent records and certainly not common'. Since then, confirmed as predominantly a species of coastal areas and much of VC59, where it extends on occasions, onto higher ground. Sites include Hale, Formby, Flixton, Heysham, the St Helens area and St Annes. Peak numbers have declined at its main site, Heysham, in the last decade.
■ **HABITAT** Grassland in coastal areas, lowland river valleys and brownfield sites.
■ **FOODPLANTS** Not recorded in Lancashire.
Single brooded. Comes readily to light and sugar; observed nectaring on ragwort.

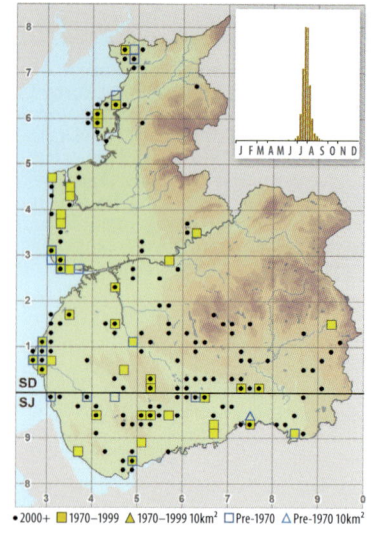

73.110 Saxon *Hyppa rectilinea*

■ **LANCS STATUS** Rare / vagrant c.1889–2008
A single historic record is documented from the Old Trafford area by J. Chappell (Ellis, 1890). This, together with a few south-west Yorkshire records (pre-1886), are the most southerly reports in the British Isles. The only other Lancashire sighting was at light in a rural garden on the southern edge of Hawes Water Moss, Silverdale (P. & A. Palmer) on 5 June 2008. The species has been noted in adjacent parts of Cumbria, post-2000, suggesting it may have a very localised population in Lancashire.
■ **HABITAT** Unknown in Lancashire, but it is associated with mosses, open woodland and moorland in Cumbria and Scotland (Randle, et al., 2019).
■ **FOODPLANTS** Not recorded in Lancashire.

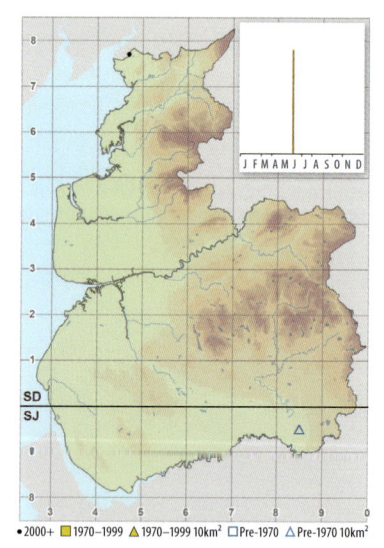

73.113 Angle Shades *Phlogophora meticulosa*

■ **LANCS STATUS** Frequent / ubiquitous resident c.1857–2022
First noted in Liverpool (Gregson, 1857) and as common everywhere by all authors up to 1940. Since the 1980s, annual at most light trapped sites to the present day, a few of these with datasets of around 30 years. One such, in Flixton, recorded a large decline in annual numbers from 2000-18 and a small, but steady increase since then.
■ **HABITAT** A wide range of habitats from the coast to moorland, including gardens.
■ **FOODPLANTS** Dock sp., willow sp., pine sp. (sapling), Ivy, St John's-wort sp., Hogweed, Yarrow, Polyanthus, goosefoot, rose, forget-me-not and Hollyhock.
Double brooded. Comes to light, the largest single count being 54 at Middle Gill Bridge on 14 October 2017. Attracted to sugar; seen nectaring on Ivy and Buddleia.

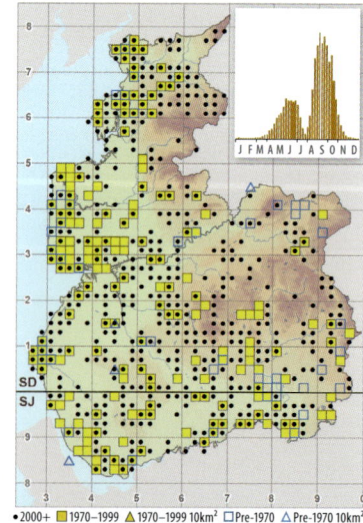

The Moths of Lancashire

73.114 Small Angle Shades *Euplexia lucipara*

■ **LANCS STATUS** Frequent / widespread resident　　　　c.1857–2022
First noted in the Liverpool district as common at sugar (Gregson, 1857) and subsequently as abundant or common throughout by all other early authors. Sites with regular reports over the years include Bispham, Clitheroe, Longton, Flixton, Morecambe, Great Crosby, Littleborough, Rishton, Belmont, Wilpshire, Yealand Conyers and many more. 31 came to light at Briercliffe on 9 July 2005 (T. Lally).
■ **HABITAT** A wide range including woodlands, gardens, moorland and brownfield sites.
■ **FOODPLANTS** Only found on Bracken; probably also on a wide range of other plants. Single brooded, with some indications of a small, occasional, second brood in autumn since c.1972. Comes to light; attracted to sugar; seen nectaring on Red Valerian.

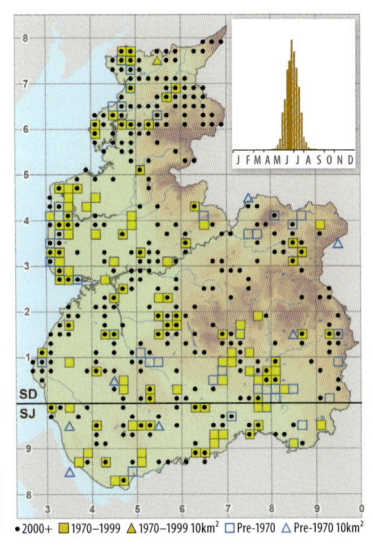

73.118 Haworth's Minor *Celaena haworthii*

■ **LANCS STATUS** Frequent / local resident　　　　1843–2022
First recorded from Heysham Moss in 1843 (Gregson, 1883a) and, in the same year, in the Preston area (J.B. Hodgkinson). It was later reported from Simonswood Moss (where it 'swarmed to acetylene light'), Bold Moss, Chat Moss, Bolton, Burnley area and Turton Toppings. Since 1940, it has been found at many moorland localities in single figures. Larger counts were 35 at Cross of Greet in 2001, 22 at Belmont in 2005 and 26 at Leck Fell in 2015. Occasionally wanders well away from its usual haunts.
■ **HABITAT** Moorland and lowland raised bogs; most of the latter have since been lost.
■ **FOODPLANTS** Not recorded in Lancashire.
Single brooded. Comes to light and nectars by day on thistles, ragwort and Heather.

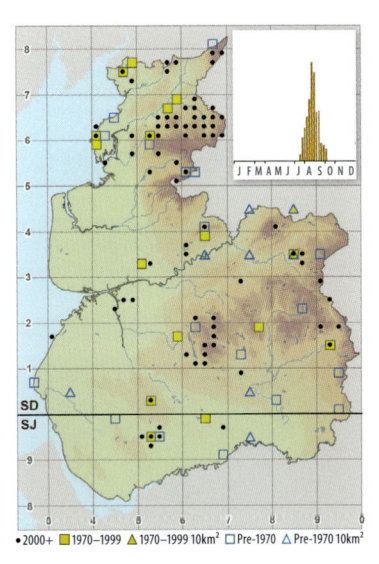

73.119 Crescent *Helotropha leucostigma*

■ **LANCS STATUS** Occasional / widespread resident　　　　1885–2022
First noted in 1885 at Morecambe (J.B. Hodgkinson) and Huyton (S.J. Capper), and in Warrington (c.1889), with Ellis (1890) reporting it as local in the region. It was then found at Eccleston Mere, Levenshulme, Prescot and Freshfield, but with Mansbridge (1940) reporting it as scarce. Since then, it has been noted at scattered sites across the county, most frequently in Flixton, Hale, Leigh, Leighton Moss, Lower Burgh Meadows and Orrell. 49 came to light at Little Woolden Moss on 26 July 2021.
■ **HABITAT** Wet areas on moorland, commons, marshes, mosses and brownfield sites.
■ **FOODPLANTS** Not recorded in Lancashire.
Single brooded. Comes to light, mostly in single figures.

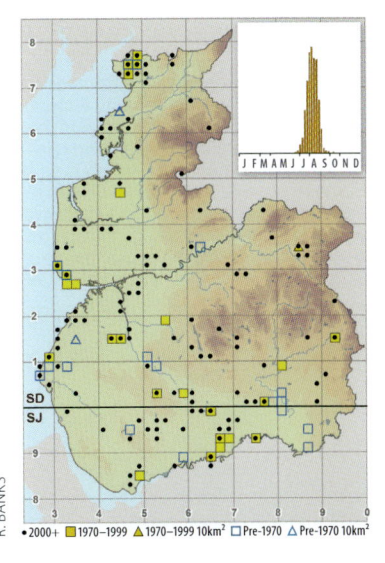

The Moths of Lancashire **505**

NOCTUIDAE

73.120 Dusky Sallow *Eremobia ochroleuca*

■ **LANCS STATUS** Scarce / local resident   1978–2020

A relatively recent addition to the county, probably associated with a strong expansion of range in eastern England (Randle *et al.* 2019). First reported at Risley Moss in 1978 and 24 July 1982 by M. Diamond and M. Hull respectively. Infrequent since then, at Swinton and Marton Mere in 1996, St Helens area, Heysham and Flixton from 2000 to 2003, Worsthorne in 2005, Fowley Common in 2006, Flixton in 2019, Prescot on 4 August 2020 and Whit Moor on 9 August 2020 (N. Garnham).
■ **HABITAT** Rough grassland, roadside verges and brownfield sites.
■ **FOODPLANTS** Not recorded in Lancashire.
Single brooded. Comes to light; observed nectaring on Yarrow and Black Knapweed.

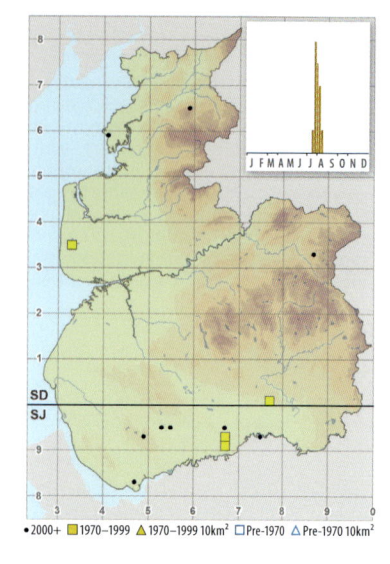

73.121 Frosted Orange *Gortyna flavago*

■ **LANCS STATUS** Frequent / widespread resident   c.1857–2022

First noted in Liverpool (Gregson, 1857) and, over the next 170 years, reported widely across the county. During this period, the distribution has appeared relatively stable, taking the introduction of light trapping into account. In the 1960s in Morecambe, the early 1970s in Longridge and late 1980s at Heysham, annual counts were often well into double figures. Since around 2010 annual counts at many lowland sites have declined noticeably, although upland areas appear more stable.
■ **HABITAT** Rough ground, damp meadows, gardens and brownfield sites.
■ **FOODPLANTS** Pupae found in thistle and ragwort stems; larvae in Mugwort roots.
Single brooded. Comes to light. Britten (c.1950) advises common in the larval stage.

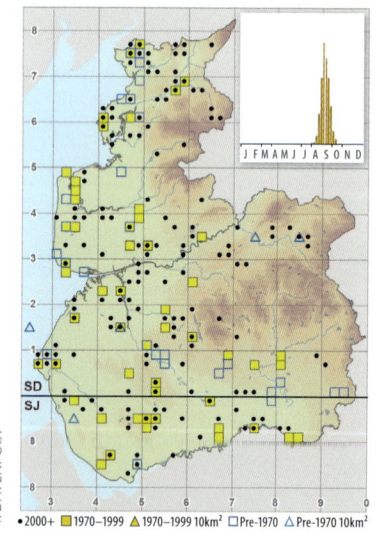

73.123 Rosy Rustic *Hydraecia micacea*

■ **LANCS STATUS** Frequent / widespread resident   1844–2022

Reported from Manchester on 30 August 1844 and Cheetham Hill, where it was abundant on ragwort flowers, in 1846 (both R.S. Edleston). Early authors noted it as common throughout, including at gas lamps. Since the 1950s, it has been reported regularly across the county, mainly in garden light traps, in lowland and upland areas. Numbers observed have declined, with reports of 20 plus rare in recent decades. The most notable in this period was 36 at Silverdale Moss on 18 August 2018 (J. Patton).
■ **HABITAT** Damp meadows, gardens, mosses, moorland edge and brownfield sites.
■ **FOODPLANTS** Meadowsweet, in the stem.
Single brooded. Comes readily to light but mostly in single figures.

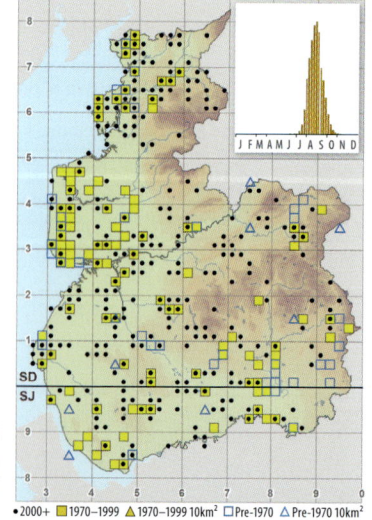

73.124 Butterbur *Hydraecia petasitis*

■ **LANCS STATUS** Occasional / local resident 1844–2022
Found nr. Failsworth in August 1844, 'where Butterbur plants grow' (Eddleston, 1844) and, in 1854, bred from larvae found nr. Manchester (C.S. Gregson). Then, noted from scattered localities across the region up to 1990, such as Bolton, Leighton Moss, Lancaster, Preston, Oldham and Burnley, with seven at Carr House Common in 1984 (M. Evans). Since 1991, found at 15 sites, most frequently in Flixton, Abram, Leigh and Worsthorne. Reports have declined since 2014 but it may be overlooked.
■ **HABITAT** Banks of streams and rivers.
■ **FOODPLANTS** Butterbur.
Single brooded. Comes to light. Torch searches amongst Butterbur can be effective.

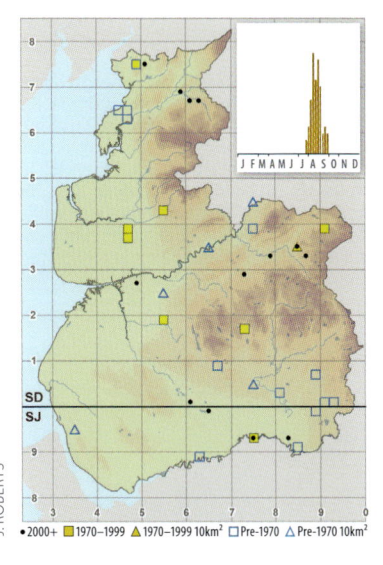

73.126 Saltern Ear *Amphipoea fucosa*

■ **LANCS STATUS** Occasional / local resident 1899–2022
Due to potential hybridisation between Saltern and Large Ear there is plenty of scope for confusion with these two species. Saltern Ear appears to be mainly coastal, from Crosby to Silverdale and was noted as very common and variable in the Formby area (Leech & Michaelis, 1957). The earliest known record, following dissection by K. Harrison in 1977, is of one from Chat Moss in 1899 by F. Warmesley. This specimen is in World Museum. See also Appendix 6.
■ **HABITAT** Generally saltmarshes and sand dunes.
■ **FOODPLANTS** Not recorded in Lancashire.
Single brooded. Comes to light. Dissection required but is not always definitive.

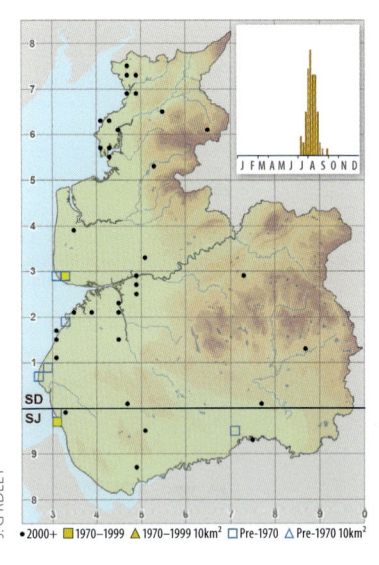

73.127 Large Ear *Amphipoea lucens*

■ **LANCS STATUS** Frequent / local resident 1895–2022
Noted at Rixton Moss in 1895 by H. Massey. This specimen is in World Museum and was dissected by K. Harrison. Since then, noted widely on moors and mosses, with 48 at light on Cockerham Moss on 12 August 2010 and 25 at Crossdale Grains on 27 August 2013. Also reported coastally, at Formby and Seaforth for example, where Saltern Ear also occurs. Hybridisation with Saltern Ear has been noted elsewhere, making identification difficult. See also Appendix 6.
■ **HABITAT** Heather moorland and lowland raised bogs.
■ **FOODPLANTS** Not recorded in Lancashire.
Single brooded. Comes to light. Dissection required but is not always definitive.

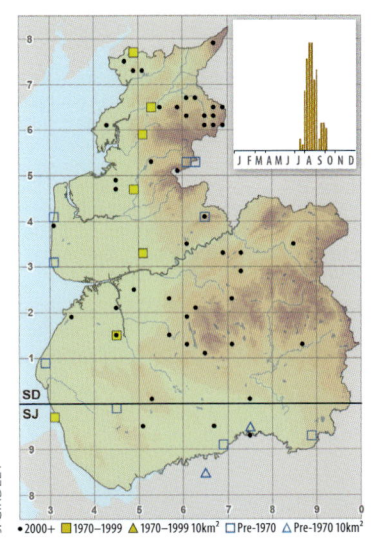

NOCTUIDAE

73.128 Ear Moth *Amphipoea oculea*

■ **LANCS STATUS** Occasional / widespread resident   1917–2022

The earliest confirmed report was by J.E.R. Allen at Bolton in 1917, dissected by K.C. Greenwood. Since then, found in Didsbury in 1940, Formby in 1946, Southport in 1951 and Morecambe in 1959. Until 2006, all records were from lowland, often near-coastal sites or upland limestone. Since then, it has been found at Churn Clough in 2006, Mill Houses in 2011 and at Briercliffe in 2019. The Sefton and Silverdale coast areas remain its stronghold. Will usually require dissection for confirmation. See also Appendix 6.

■ **HABITAT** Sandy and limestone grassland, meadows and brownfield sites.

■ **FOODPLANTS** Not recorded in Lancashire.

Single brooded. Comes to light.

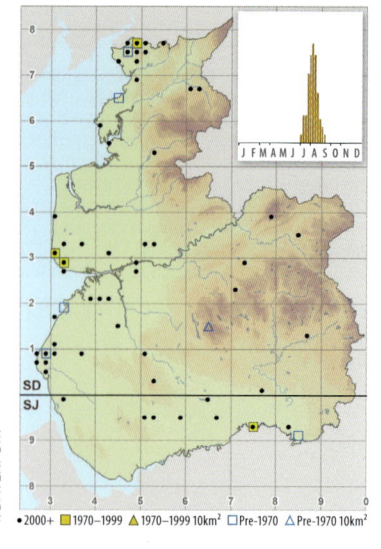

● 2000+   □ 1970–1999   ▲ 1970–1999 10km²   □ Pre-1970   △ Pre-1970 10km²

73.129 Crinan Ear *Amphipoea crinanensis*

■ **LANCS STATUS** Scarce / very local resident   1897–2013

The least frequently encountered of the ear moths, with the first confirmed record from Bolton in 1897 by J.E.R. Allen (dissected, F.N. Pierce). In 1917 several were found on Pendle Hill by W.G. Clutten and singles at Bacup on 28 August 1950 and Great Harwood 1 September 1951 by D. Bryce. Since then, only noted from Sunderland Point in 2008, Leck Beck in 2011, Morecambe on 27 August 2013 and Crossdale Grains on 11 September 2013 (T. Whitaker and P.J. Marsh). Comes to light. See also Appendix 6.

■ **HABITAT** Moorland and less frequently near saltmarshes.

■ **FOODPLANTS** Not recorded in Lancashire.

Single brooded. Require dissection to confirm identity.

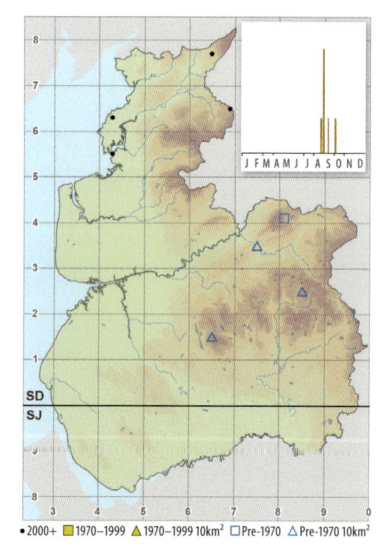

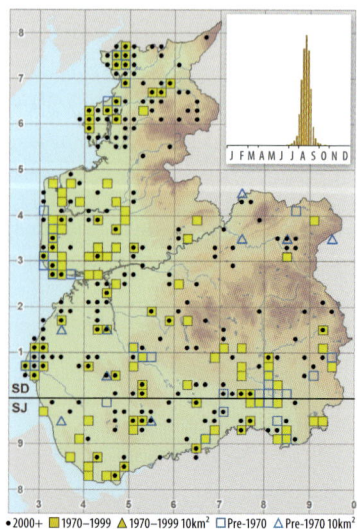

● 2000+   □ 1970–1999   ▲ 1970–1999 10km²   □ Pre-1970   △ Pre-1970 10km²

73.131 Flounced Rustic *Luperina testacea*

■ **LANCS STATUS** Frequent / widespread resident   1844–2022

First reported in Manchester on 30 August 1844 (Edleston, 1844). All other early authors considered it to be at least common and widespread, with only Britten (c. 1950) adding 'scarce in the north-east'. Since then, it has been recorded widely in small numbers across the county, being particularly prevalent in near-coastal areas and in the Mersey valley. A marked decline in annual records commenced at some regularly recorded sites during the early 2000s, stabilising at lower levels by 2015.

■ **HABITAT** Meadows, rough grassland, gardens, sand dunes and brownfield sites.

■ **FOODPLANTS** Not recorded in Lancashire.

Single brooded. Comes to light.

● 2000+   □ 1970–1999   ▲ 1970–1999 10km²   □ Pre-1970   △ Pre-1970 10km²

73.132 Sandhill Rustic *Luperina nickerlii*

■ **LANCS STATUS** Occasional / very local resident 1889–2022

Four sub-species of Sandhill Rustic occur in the British Isles, *gueneei* being the one found on North Wales and Lancashire coasts. *Gueneei* was also the first ssp. to be discovered, at Rhyl around 1860, and was subsequently found at several locations from Anglesey to Fleetwood. The first documented record in Lancashire was on 1st August 1889 at St Annes by T. Baxter (South, 1889).

In the early 20th century, Charles Rothschild listed two Lancashire sites (Fleetwood and St Annes) as deserving of special protection, specifically because they supported populations of *gueneei* (Rothschild & Marren, 1997). Both colonies were subsequently lost to development, the Fleetwood area c.1910 and St Annes area c.1921, but *gueneei* survived in Lancashire.

In recent years, a substantial colony has been well recorded on the Sefton Coast where its favoured habitat of low protodunes, regularly inundated at hight tide, is actively forming in areas such as Birkdale Green Beach. Much of the information that follows comes from an intensive study conducted on this colony in 2008 (Burkmar, 2009).

The larva of *gueneei* feed below the surface of the sand on the roots of Sand Couch and, to a lesser extent, Common Saltmarsh-grass. Adults can be found by searching frontal dunes by torchlight after 10 pm through to dawn, when they make their way to the base of grasses where they remain, virtually indetectable, during the day. At night they are usually found resting on their foodplants and generally sit tight when approached. Both sexes fly and peak activity is between midnight and 2 am. Once flying, the moths are very restless and harder to approach. Most mating pairs are found between 2 and 3 am. Light trapping is effective but should be used sparingly because of the potentially disruptive effects on a rare moth that relies on a restricted linear habitat.

Mass emergences have occasionally been witnessed with many freshly emerged moths found over the course of a single night. For a rare and sparsely distributed animal, synchronised emergence has an evolutionary advantage. For *gueneei*, it seems likely that the tidal cycle could play a role in this synchronicity, but the precise mechanism is unknown. Observations of strongly flying females alighting and immediately ovipositing indicate that dispersal by flight is significant. There is also the possibility that eggs or small larvae could be passively distributed to suitable new sites when autumn storms dramatically reconfigure dune landscapes. It is possible that the Sefton Coast population could contribute to the return of an established colony of *gueneei* at St Annes. Further reading – see Spalding (2023).

Richard Burkmar

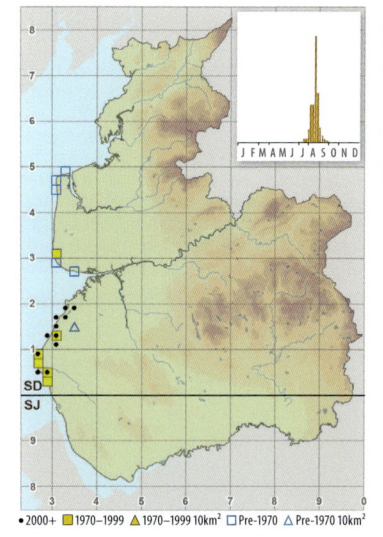

• 2000+ ▢ 1970–1999 ▲ 1970–1999 10km² ▢ Pre-1970 △ Pre-1970 10km²

NOCTUIDAE

73.134 Large Wainscot *Rhizedra lutosa*

■ **LANCS STATUS** Frequent / widespread resident   1843–2022

Noted in Preston in 1843 by R.S. Edleston and thereafter, recorded widely across lowland parts. During the early 20th century, it was noted as somewhat local, but abundant where it occurs (Mansbridge, 1940). Since then, light trapping has shown it to be present in many lowland areas, but it is very local in the uplands. Numbers in traps infrequently reach double figures, except at a few favoured sites such as Leighton Moss and Sunderland Point, the latter with 28 on 29 September 2011.
■ **HABITAT** Reedbeds and other wet locations with reed growth.
■ **FOODPLANTS** Not recorded in Lancashire.
Single brooded. Comes to light, usually in single figures.

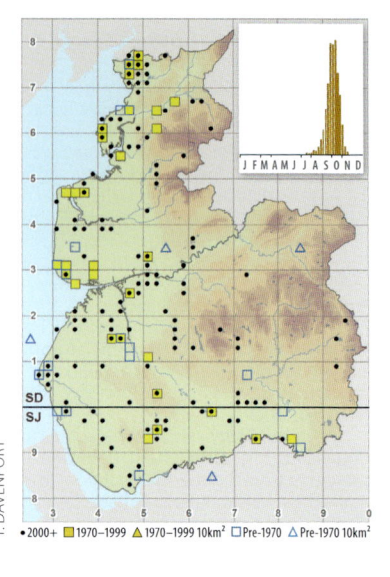

● 2000+   ▪ 1970–1999   ▲ 1970–1999 10km²   ▫ Pre-1970   △ Pre-1970 10km²

73.136 Bulrush Wainscot *Nonagria typhae*

■ **LANCS STATUS** Frequent / widespread resident   c.1857–2022

First recorded in the Liverpool district (Gregson, 1857). Since then, it has been noted regularly at many sites across lowland parts of the county and, to a lesser extent, upland sites such as Belmont, Briercliffe, Littleborough and Worsthorne. Most regular and usually annual at sites such as Billinge, Flixton, Hale, Heysham N.R., Leighton Moss, Longridge, Longton, Mere Sands Wood and Middleton N.R. amongst others. Vacated pupal cases can readily be found in Bulrush stems.
■ **HABITAT** Wetlands, including watery ditches and ponds.
■ **FOODPLANTS** Bulrush.
Single brooded. Comes to light in single figures.

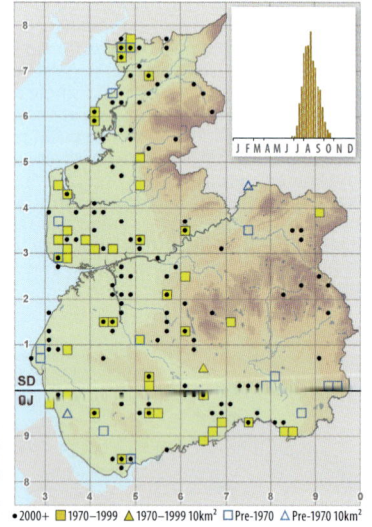

● 2000+   ▪ 1970–1999   ▲ 1970–1999 10km²   ▫ Pre-1970   △ Pre-1970 10km²

73.137 Fen Wainscot *Arenostola phragmitidis*

■ **LANCS STATUS** Occasional / local resident   1895–2022

Noted in Warrington, in 1895 by T. Acton and by J. Collins, as common, in 1900. Since then, Leighton Moss has been its most regular site, from 1959 to the present day. Smaller numbers have been seen at St Annes (1954–1995), Morecambe (1959–66), Martin Mere (1980–2014), Heysham (1989 to 2014), Great Sankey (2014 to 2022) and many other scattered coastal or lowland sites. Rarely wanders further inland, but has reached sites such as Adlington in 2013 (P. Krischkiw) and Wray in 2006 (P.J. Marsh).
■ **HABITAT** Wetlands.
■ **FOODPLANTS** Not recorded in Lancashire.
Single brooded. Comes to light. 26 seen at Morecambe on 7 August 2013 (J. Patton).

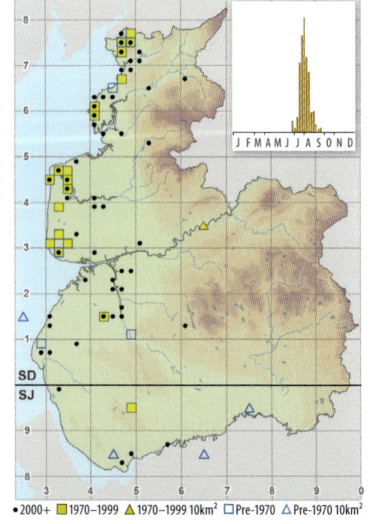

● 2000+   ▪ 1970–1999   ▲ 1970–1999 10km²   ▫ Pre-1970   △ Pre-1970 10km²

510   *The Moths of Lancashire*

73.138 Lyme Grass *Longalatedes elymi*

■ **LANCS STATUS** Occasional / very local resident 1911–2022

During the research for this book, no historical mention was found of the presence of this primarily east coast moth in Lancashire. It was only during a check of photographs of specimens displayed on the Natural History Museum, London website (see Rothschild *et al.*), accessed in March 2022, that photographs of three specimens, labelled 'Lancashire coast' and dated July 1911 (S.A. Wallis), were discovered. Despite extensive fieldwork in the intervening years on the sand-dunes of the Sefton Coast, and in the St Annes area, its presence remained unknown in the county for almost another century. This changed on 20 June 2009 (and subsequently), when it was found in St Annes by P. Davies. It was next noted at Crosby on 23 June 2010 by C. Fletcher and has since been regularly recorded, sometimes in double figures, on the dunes at Formby, Ainsdale, Birkdale and Southport. A maximum count of 38 were recorded at Ainsdale on 17 June 2017 (R. Walker *et al.*). These records are indicative of a range expansion and population increase, but its Lancashire status between 1912 and 2008 remains a mystery.

■ **HABITAT** On the coastal side of sand dunes.
■ **FOODPLANTS** Not recorded in Lancashire.

Single brooded. Comes to light. The early records have been mapped as a centralised 10km blue triangle near Formby, but could equally have been caught in VC60.

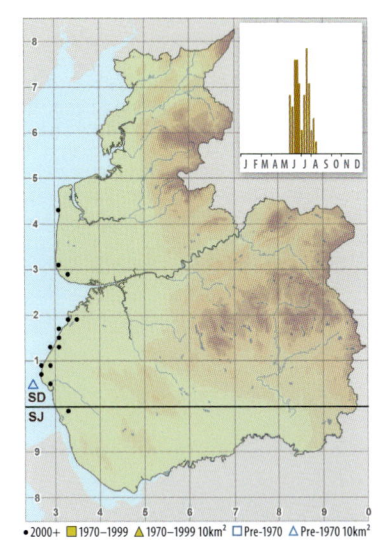

• 2000+ ☐ 1970–1999 △ 1970–1999 10km² ☐ Pre-1970 △ Pre-1970 10km²

73.139 Twin-spotted Wainscot *Lenisa geminipuncta*

■ **LANCS STATUS** Scarce / local resident 1928–2022

There are three early records, accepted by Mansbridge (1940), from Crosby in 1928, 1931 and 1933 by H.W. Wilson. Supporting specimens have not been traced but it is noted that confirmed records exist from Cheshire in 1919 and 1932, with the specimens having been displayed at meetings of the Lancashire and Cheshire Entomological Society. As also happened in Cheshire, it was not seen again until the 21st century, in Lancashire this being on 25 August 2017 at Stadt Moers Country Park (R. Moyes *et al.*). These records are in line with a national range expansion (Randle *et al.*, 2019). Since then, there have been nine reports, all in VC59 and all but one involving singletons. The first of these was at Flixton on 13 August 2000 (K. McCabe), a site with a second record in 2021. The next report was of two noted at Great Sankey on 20 August 2021 (J. Mitchell-Lisle), with two further reports, from a nearby site, in August 2022 (A. & D. Riley). The most northerly of the records came from Southport on 21 and 22 August 2021 (A. Pryce) and on 11 August 2022 (R. Moyes). A day later, one came to light at Abram, Wigan on 12 August 2022 (A.J. Smith).

■ **HABITAT** Nationally, associated with reedbeds.
■ **FOODPLANTS** Not recorded in Lancashire.

Single brooded. Comes sparingly to light in low numbers.

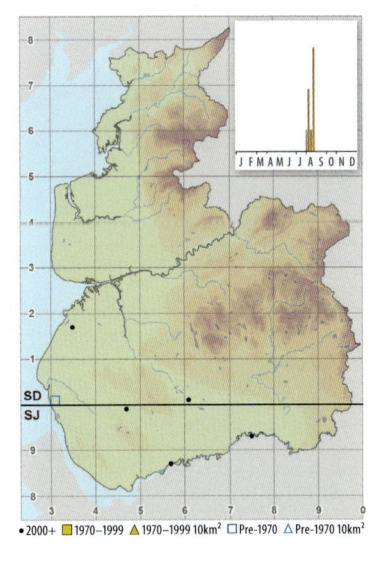

• 2000+ ☐ 1970–1999 △ 1970–1999 10km² ☐ Pre-1970 △ Pre-1970 10km²

NOCTUIDAE

73.141 Brown-veined Wainscot *Archanara dissoluta*

■ **LANCS STATUS** Occasional / local resident   c.1933–2021

Although reported as showing a nationally significant decline since 1990 (Randle *et al*., 2019), the picture in Lancashire has been one of a moderate increase in range. First noted at Crosby c.1933 by H.W. Wilson, it was then found in the Formby, Ainsdale and Freshfield areas between 1935 and 1949, at Leigh in 1953 and Chorley in 1983. In 1996 it was seen in Flixton and, thereafter, appeared at several scattered sites in south-west Lancashire, then at Leighton Moss in 2019 and Cleveleys in 2020.
■ **HABITAT** Wetlands.
■ **FOODPLANTS** Not recorded in Lancashire.
Single brooded. Comes sparingly to light, almost always as singletons.

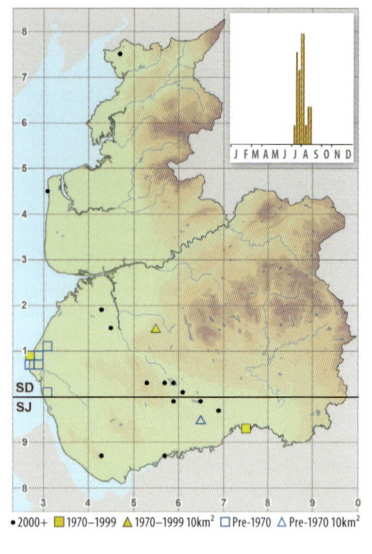

73.142 Small Rufous *Coenobia rufa*

■ **LANCS STATUS** Occasional / widespread resident   1928–2022

First found on 21 July 1928 at Nelson by W.G. Clutten and later, at Burnley and Pendle (c.1940), Formby Moss (1956) and Kirkby (1968). More recently, it came to light at Billinge in 1991 and became a regular in Flixton from 1995. Thereafter, it continued to spread across mainly lowland parts, establishing strong colonies at sites such as Martin Mere, Mere Sands Wood and Leighton Moss, but it remains very local in upland areas. Large counts are rare, but include 18 at Sunderland Point in 2019.
■ **HABITAT** Wetlands and damp meadows.
■ **FOODPLANTS** Not recorded in Lancashire.
Single brooded. Comes to light, almost always in low single figures.

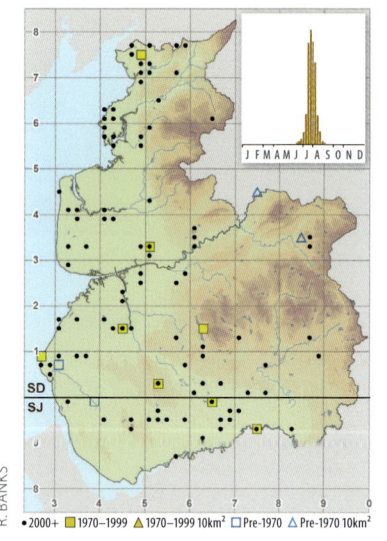

73.144 Small Wainscot *Denticucullus pygmina*

■ **LANCS STATUS** Frequent / widespread resident   c.1857–2022

First noted at Simonswood Moss (Gregson, 1857) and later, from many upland sites, including Burnley, Oldham and Rochdale. Possibly initially overlooked in lowland areas, as it was later found to be common at sites such as Leighton Moss, Astley Moss, Heysham N.R. and Mere Sands Wood. It can appear almost anywhere on rare occasions, but is regular and frequent on lowland wetlands and in upland areas. Counts of up to 32 have been regular in recent years at Green Bank (P.J. Marsh).
■ **HABITAT** Damp moorland, mosses and other wet or marshy habitats.
■ **FOODPLANTS** Not recorded in Lancashire.
Single brooded. Comes to light, usually in single figures.

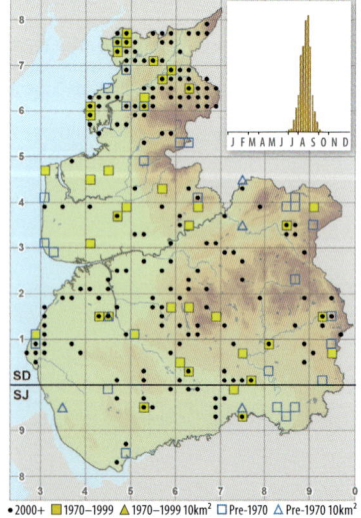

512   *The Moths of Lancashire*

73.146 Least Minor *Photedes captiuncula*

1909–2022

■ **LANCS STATUS** Occasional / very local resident

Formerly a Red Data Book species, but recently updated to Nationally Rare with a threat status of Endangered due to the limited number and fragmented nature of populations and observed decline (Fox et al., 2019; Randle et al., 2019). Occurs as two sub-species; ssp. *tincta* Kane is restricted to the Burren area, western Ireland, with ssp. *expolita* Stainton being found in the UK (Waring & Townsend, 2017).

In Lancashire, it is predominantly restricted to the limestone grasslands of the Arnside and Silverdale AONB and Hutton Roof. First reported in the Silverdale area on 26 June 1909 (W. Mansbridge), it was not observed again until the 1950s at Warton Crag (Rev. Vine-Hall). Thereafter, observations were sporadic, becoming more regular in the 1990s and 2000s, but difficulties in locating this species arise due to its small size and rapid, erratic flight when disturbed.

Adults are mainly diurnal, but also fly at dusk and have twice been known to come to light, well away from the nearest suitable habitat, with singletons recorded from near Preston and Middleton N.R. in 2013. Historically, larvae were thought to feed on Glaucous Sedge (Waring & Townsend, 2003) but had not been found in the wild until 2014, where final instar larvae were found feeding in the stems of Glaucous Sedge and Blue Moor-grass at the Burren, Ireland (Henwood & Sterling, 2020).

In light of the recent status re-assessment and new information on larval feeding, funding was provided by the Tanyptera Trust and Butterfly Conservation in 2022 for the first targeted larval and adult searches, focussing on the Morecambe Bay limestone area (Patton, 2023). Larvae were found feeding within the stems of Blue Moor-grass at Hutton Roof, Warton Crag and Gait Barrows. These were the first to be found in the wild and successfully bred in the UK, but further work is required to establish if Glaucous Sedge is also used as a foodplant. Underdeveloped and dead flowerheads indicate larval presence but is not indicative of Least Minor as other species of Lepidoptera also feed on Blue Moor-grass. Larvae move from stem-to-stem, feeding head down at the base.

Adults were only observed at Warton Crag during the 2022 survey, despite several repeat visits to the key sites, Gait Barrows and Thrang End. It has been identified from the surveys that a variety of environmental conditions are utilised, from small, discrete patches of Blue Moor-grass on limestone escarpments, to extensive, south-facing limestone grasslands, but also areas that are moderately scrubbed over. Habitat management will be key to the persistence of Least Minor in the county as too much scrub will out-compete the larval foodplant.

Justine Patton

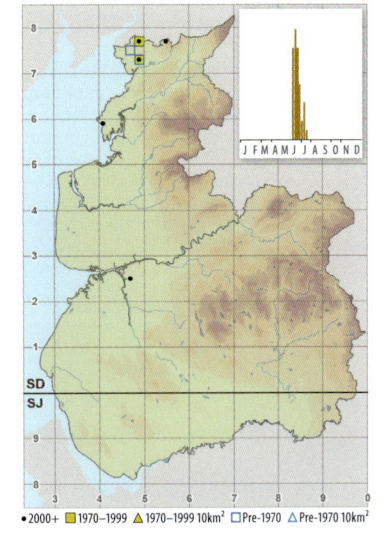

• 2000+ ■ 1970–1999 ▲ 1970–1999 10km² □ Pre-1970 △ Pre-1970 10km²

The Moths of Lancashire 513

NOCTUIDAE

73.147 Small Dotted Buff *Photedes minima*

■ **LANCS STATUS** Frequent / widespread resident   c.1857–2022

First noted at Halewood and Altcar dunes (Gregson, 1857) and later in the 19th century, at Pendleton, Oldham, Bolton and Preston. Subsequently recorded widely and regularly in the higher parts of the county, particularly in recent decades, at sites such as Adlington, Bay Horse, Briercliffe, Green Bank, Mill Houses and Worsthorne. More local in lowland areas such as at Flixton, Leigh, Leighton Moss and Middleton N.R. Usually in single figures, but 19 came to light at Docker Moor on 18 July 2015.

■ **HABITAT** Damp grassland in meadows, marshes and brownfield sites.

■ **FOODPLANTS** Not recorded in Lancashire.

Single brooded. Comes to light.

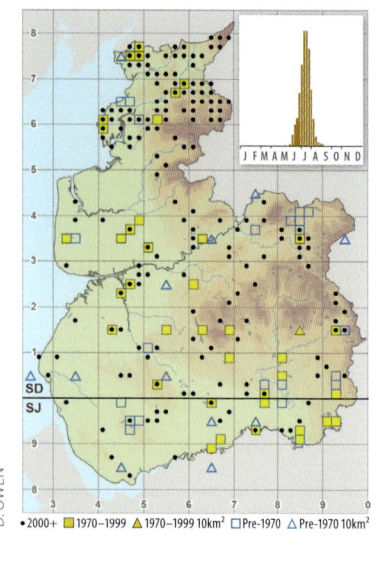

73.154 Dusky Brocade *Apamea remissa*

■ **LANCS STATUS** Frequent / widespread resident   1843–2022

Recorded in Preston (Hodgkinson, 1844) and from scattered sites such as Chorley, Rochdale, Oldham, Formby and Burnley through to the 1940s. Since then, noted widely across the county, usually in single figures. Larger counts of 30 plus have been infrequent but only one is known post 2014, in Worsthorne on 26 June 2020. A few regularly trapped lowland garden sites, such as Flixton and Parr, have shown an overall decrease in annual records over the last decade.

■ **HABITAT** A range of grassy habitats across the county.

■ **FOODPLANTS** Not recorded in Lancashire.

Single brooded. Comes to light. Quite variable in the strength of the darker markings.

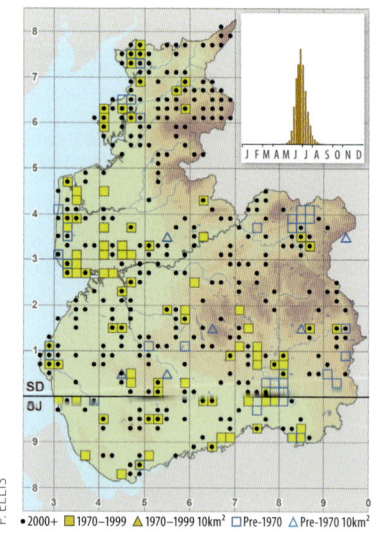

73.155 Clouded Brindle *Apamea epomidion*

■ **LANCS STATUS** Occasional / local resident   1843–2022

Noted near Preston in 1843 by J.B. Hodgkinson and at a few other scattered sites including Silverdale, Longworth Clough and Oldham, prior to 1940. Since then, three-quarters of the 238 county records have come from the Silverdale limestone area or within northern river valleys of the Roeburn and Hindburn. Regular sites include Leighton Moss, Gait Barrows and Mill Houses. Elsewhere, it is infrequent in occurrence, such as in a few parts of the Ribble valley. Occurs in low single figures.

■ **HABITAT** Open woodland, meadows, moorland edge and scrub.

■ **FOODPLANTS** Not recorded in Lancashire.

Single brooded. Comes to light. Eight were seen at Gait Barrows on 18 June 2014.

514 *The Moths of Lancashire*

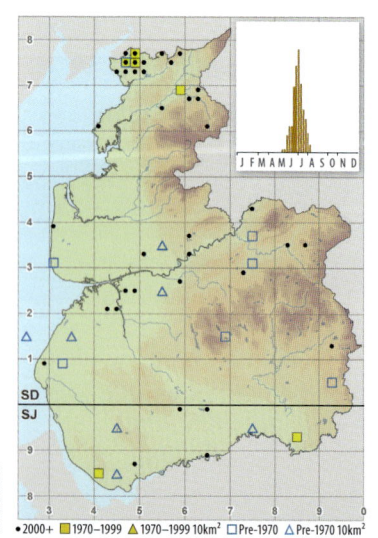

73.156 Clouded-bordered Brindle *Apamea crenata*

■ **LANCS STATUS** Frequent / ubiquitous resident 1843–2022
All early authors report this to be common to abundant everywhere, with the first documented record in Preston (Hodgkinson, 1844). Since then, it has found in most parts of the county with many regular sites in both upland and lowland areas. Numbers are generally in single figures, but on occasions larger counts have occurred with Leck Fell, producing counts of over 100 at light, twice in 2004 and once in 2015.
■ **HABITAT** Occurs in most habitats in lowland and upland areas.
■ **FOODPLANTS** Grasses, including Wavy Hair-grass.
Single brooded, often flying into late August in upland areas. Comes to light. It occurs commonly in two forms in the county, the darker form being ab. *combusta*.

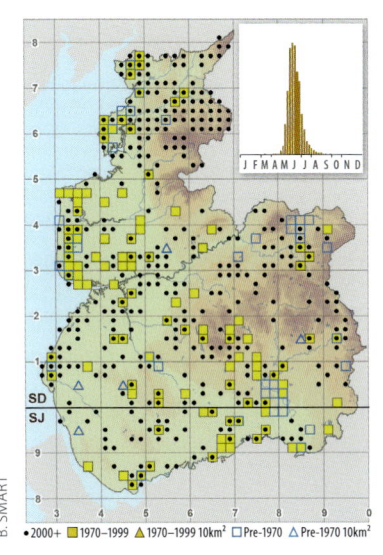

73.157 Large Nutmeg *Apamea anceps*

■ **LANCS STATUS** Rare / very local resident c.1885–2009
Described by Ellis (1890) as local, with one undated record from the Preston area (J.B. Hodgkinson). Later Mansbridge (1940) noted it as scarce, adding one record from Simonswood in 1900 by J. Cotton. Later, found in the St Annes area by T. Baxter in 1905 and Didsbury by H.N. Michaelis in 1940. Only recent records, both singletons to light, were from Heysham on 10 June 2005 (D.J. Holding) and Swinton on 30 May 2009 (G. Riley). Overall, abundance decreased nationally (Randle *et al.*, 2019).
■ **HABITAT** Unknown.
■ **FOODPLANTS** Not recorded in Lancashire.
Single brooded. Comes to light.

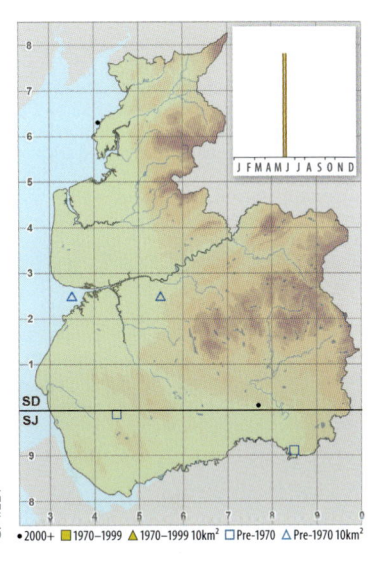

73.158 Rustic Shoulder-knot *Apamea sordens*

■ **LANCS STATUS** Frequent / widespread resident c.1857–2022
First noted in the Liverpool area as abundant (Gregson, 1857), other early authors noting it as very common to common in the county. Later, reported as fairly common in East Lancs (Brindle, 1948). From the 1960s, it has been recorded widely, at sites such as Nelson, Skelmersdale, Ainsdale, St Annes, Lancaster and Silverdale, but is less frequently encountered in upland areas. A overall decline in numbers has occurred in the last decade or so with double figures catches now an unusual occurrence.
■ **HABITAT** Grassy habitats across the county, including gardens.
■ **FOODPLANTS** Not recorded in Lancashire.
Single brooded. Comes to light, usually in single figures.

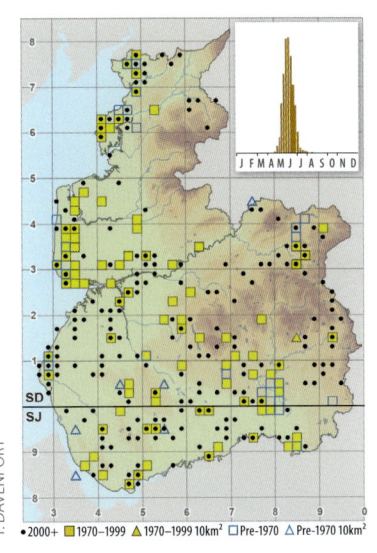

NOCTUIDAE

73.159 Small Clouded Brindle *Apamea unanimis*

■ **LANCS STATUS** Frequent / widespread resident   c.1857–2022
Recorded from Croxteth and Hale (Gregson, 1857), followed by Chat Moss, Lostock, Withington, Rochdale and St Helens, amongst others, during the late 19th century. Since then, it has been widely reported throughout the county, but to a lesser extent in upland areas, a notable exception being Briercliffe. Elsewhere, it has occurred very regularly at sites such as Flixton, Leighton Moss, Mere Sands Wood, Parr and Preston. Although always infrequent, double figure counts have not been noted since 2014.
■ **HABITAT** Damp grassland.
■ **FOODPLANTS** Not recorded in Lancashire.
Single brooded. Comes to light.

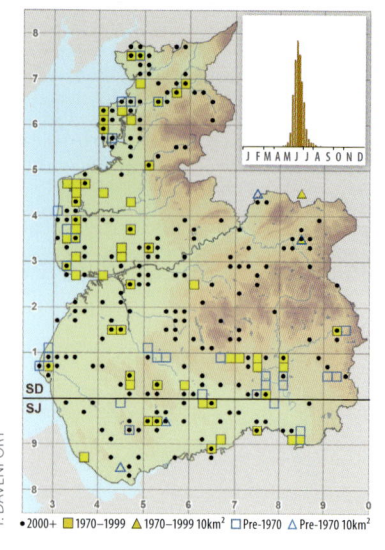

73.160 Slender Brindle *Apamea scolopacina*

■ **LANCS STATUS** Frequent / widespread resident   c.1859–2022
First reported from Agecroft c.1859 by J. Chappell, with a few other records into the 1940s, from Ditton, Heywood and Preston. Records increased from the 1950s onwards, but it remained scattered and rather irregular in occurrence until the mid-1990s. Since 1994, it has spread across many eastern lowland areas, with a similar increase in upland sites post-2000. It remains infrequently encountered in lowland areas lacking mature deciduous woods, such as the Fylde and western parts of VC59.
■ **HABITAT** Wooded areas with grassy edges or glades and scrubland.
■ **FOODPLANTS** Millet sp., Great Wood-rush.
Single brooded. Comes to light.

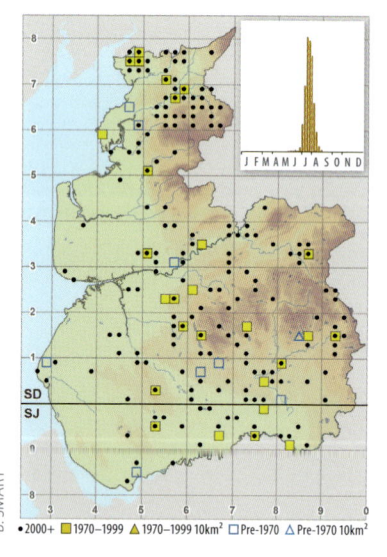

73.161 Crescent Striped *Apamea oblonga*

■ **LANCS STATUS** Occasional / local resident   c.1885–2021
First noted in the Preston area pre-1885 by J.B. Hodgkinson (Ellis, 1890). Since then, recorded mostly on coastal and estuarine saltmarshes. Few sites have regular sightings, but include Sunderland Point, Little Singleton and Birkdale Green Beach. Seven larvae were found at the latter site from 19-25 April 2009 (G. Jones); one was reared. 15 adults were seen at dusk and at actinic light on 21 July 1985 at Freckleton Marsh (M. Evans), otherwise all adult records are of low single figures to light.
■ **HABITAT** Saltmarsh and sandy, sparsely vegetated, coastline.
■ **FOODPLANTS** Saltmarsh-grass sp.
Single brooded. Comes to light. Rare inland, as an accidental introduction or vagrant.

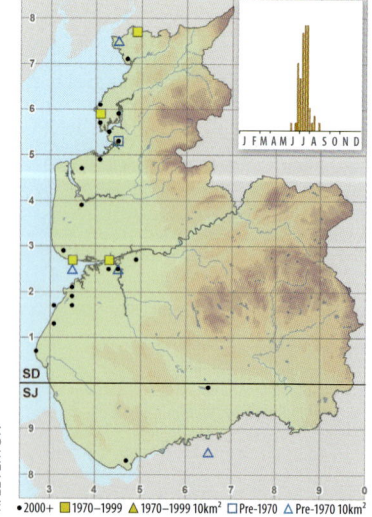

73.162 Dark Arches *Apamea monoglypha*

■ **LANCS STATUS** Abundant / ubiquitous resident   c.1857–2022
All early authors report this to be abundant everywhere, the first record being from the Liverpool district (Gregson, 1857). Since then, it has been noted regularly, widely and annually at many sites across the county. Numbers vary from year to year, but the regular, large counts of the late 1990s are, mostly, a thing of the past.
■ **HABITAT** Many different habitats across the county.
■ **FOODPLANTS** Found amongst grass roots on one occasion, but species not identified.
Single brooded, with a small, fairly regular second brood from 2001 onwards. Comes to light, often in double figures. Attracted to sugar and flowers, such as Buddleia and Heather. A dark form, ab. *aethiops*, does occur but is rarely noted.

● 2000+   ■ 1970–1999   ▲ 1970–1999 10km^2 □ Pre-1970   △ Pre-1970 10km^2

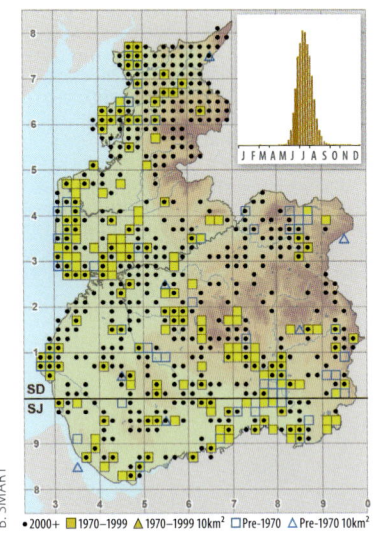

73.163 Light Arches *Apamea lithoxylaea*

■ **LANCS STATUS** Frequent / widespread resident   c.1857–2022
First recorded in Liverpool (Gregson, 1857) and, up to 1940, reported as common across the county by all early authors. From 1950 to the present day, the distribution has remained similar, with the species noted regularly at sites such as Billinge, Hale, Heysham, Longton, Parr and Preston. At the turn of the 20th century, Flixton recorded 65 on 20 July 1996. Since c.2015, double digit counts have been infrequent and numbers recorded at most sites have declined to lower single figures.
■ **HABITAT** Wide range of grassy habitats, favouring coastal and dry, lowland areas.
■ **FOODPLANTS** Not recorded in Lancashire.
Single brooded. Comes to light. Attracted to sugar and flowers, such as privet.

● 2000+   ■ 1970–1999   ▲ 1970–1999 10km^2 □ Pre-1970   △ Pre-1970 10km^2

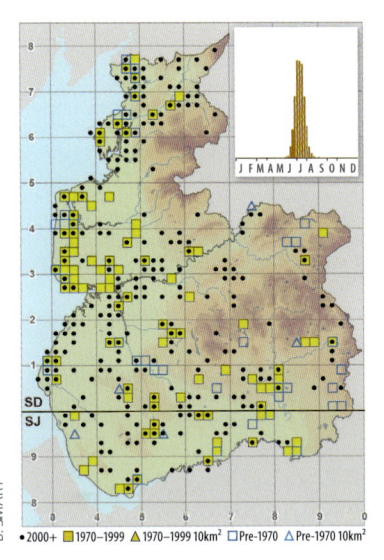

73.164 Reddish Light Arches *Apamea sublustris*

■ **LANCS STATUS** Occasional / very local resident   1843–2019
First noted nr. Preston in 1843 (Hodgkinson, 1844), then Boars Wood, Hale (Gregson, 1857). These, and a few others away from northern VC60, are unusual and not repeated. Found in the Silverdale area from 1889, with nine records at Leighton Moss from 1968–94 and Heysham in 2000. Since then, from Jack Scout in 2003, Yealand Conyers 2004-14 and on Warton Crag from 2014-19, including 24 in two actinic light traps on 20 June 2018 (J. Patton). Noted in Silverdale on 15 July 2019 (J. & J. Webb).
■ **HABITAT** Limestone grassland.
■ **FOODPLANTS** Not recorded in Lancashire.
Single brooded. Comes to light.

● 2000+   ■ 1970–1999   ▲ 1970–1999 10km^2 □ Pre-1970   △ Pre-1970 10km^2

The Moths of Lancashire   **517**

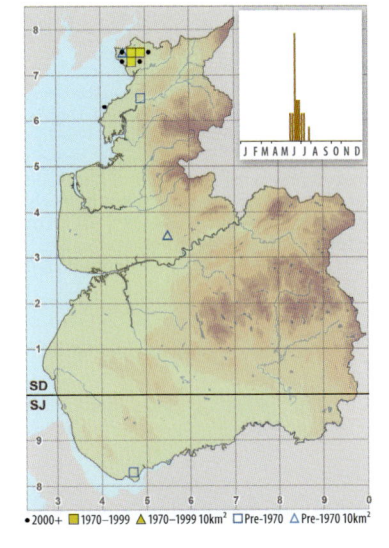

73.165 Confused *Apamea furva*

■ **LANCS STATUS** Occasional / local resident c.1843–2022

Noted near Preston c.1843 (Hodgkinson, 1844), then Jenny Brown's Point (1908) and Clougha in 1910. Sporadic sightings have continued since then, with the majority of the 66 records from lowland and upland limestone sites, moorland edge and a few from coastal dunes. Upland sites include Leck Fell, Docker Moor, Green Bank and Worsaw Hill, near Clitheroe, while lowland limestone reports are mainly from Gait Barrows. Can be confused with the plain form of Dusky Brocade.
■ **HABITAT** Moorland and limestone grassland; occasionally coastal sandy areas.
■ **FOODPLANTS** Not recorded in Lancashire.
Single brooded. Comes to light, mostly as singletons.

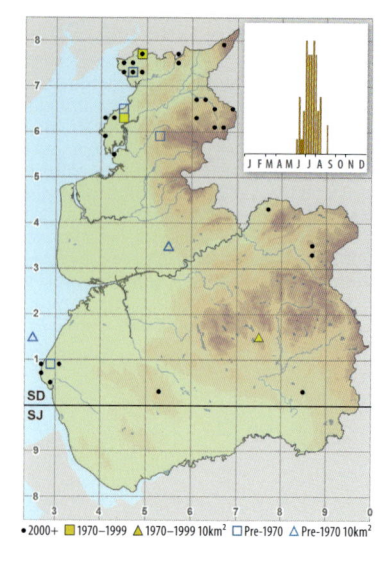

73.168 Double Lobed *Lateroligia ophiogramma*

■ **LANCS STATUS** Frequent / widespread resident 1940–2022

First noted in Didsbury in 1940 by H.N. Michaelis, the same year as the first Cheshire record. Not reported again until 1949 in Formby, at Chorlton in 1950 and at Clifton, nr. Kearsley in 1960. It continued to spread more widely from the early 1960s and by the late 1990s records increased in upland areas. It is now regularly encountered across much of the county, mostly in single figures. Larger counts, from Abram in 2013 and 2020 (J. Smith) and Worsthorne in 2021 (G. Gavaghan), are exceptional.
■ **HABITAT** Woodland, marshes, flashes, canals, gardens and damp meadows.
■ **FOODPLANTS** Not recorded in Lancashire.
Single brooded. Comes to light.

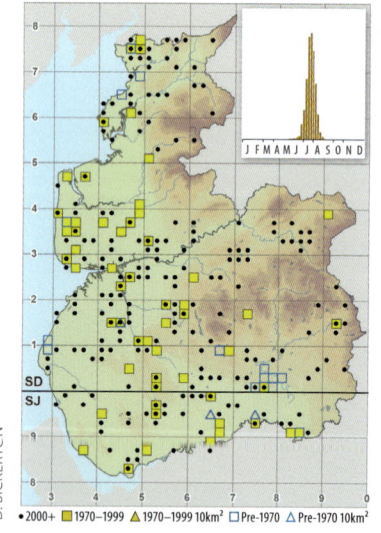

73.169 Common Rustic *Mesapamea secalis*

■ **LANCS STATUS** Considered frequent / widespread resident 1958–2022

Identification difficulties between this species and *M. didyma* meant that only confirmed records were available for inclusion. These consisted of 114 specimens (0.66% of total reports). First confirmed on 18 July 1958 by G. Bowden at Bispham and the most recent on 22 August 2022 at Ainsdale by R. Moyes and C. Daly. The small number of confirmed records means the assessment of status and relative abundance of this sp. and *M. didyma* are provisional and require further research. See also Appendix 6.
■ **HABITAT** Grassland of various types across the county.
■ **FOODPLANTS** Not recorded in Lancashire.
Considered single brooded. Comes to light and nectars on ragwort.

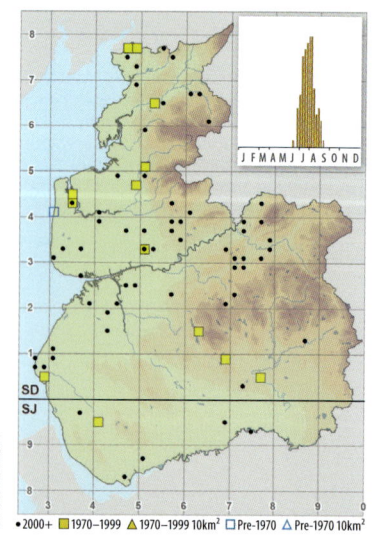

73.170 Lesser Common Rustic *Mesapamea didyma*

■ **LANCS STATUS** Considered occasional / widespread resident   **1958–2022**
Identification difficulties between this species and *M. secalis*, mean that only confirmed records were available for inclusion. These consisted of 79 specimens (0.45% of the total reports). At localities where regular critical examination took place, this species was found to be much less frequently encountered than *M. secalis*. The first confirmed record was from 8 August 1958 at Bispham by G. Bowden and the most recent from 19 August 2022 at Banks by R. Walker. See also Appendix 6.
■ **HABITAT** Grassland of various types across the county.
■ **FOODPLANTS** Not recorded in Lancashire.
Single brooded. Comes to light.

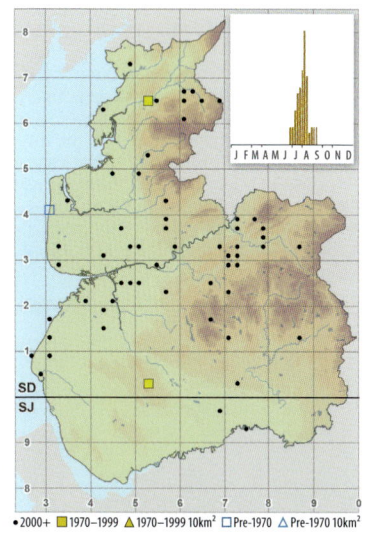

73.171 Rosy Minor *Litoligia literosa*

■ **LANCS STATUS** Frequent / widespread resident   **1844–2022**
First found in the Manchester area (Edleston, 1844), then Rochdale in 1870 and Liverpool in 1885. Mansbridge (1940) noted it as plentiful on sandhills, but scarce inland. More recently, it has been recorded widely, but is patchy in appearance. Regular sites include Bay Horse, Bispham, Briercliffe, Flixton, Pilling and Worsthorne.
■ **HABITAT** Grassy habitats in coastal areas, moorland, scrubland and brownfield sites.
■ **FOODPLANTS** Not recorded in Lancashire.
Single brooded. Comes to light and nectars on ragwort. Larger counts, such as 42 at Ainsdale on 1 August 2021 (R. Moyes), have all been coastal, but it does reach double figures in some upland areas, such as Belmont (S. Martin) and Briercliffe (T. Lally).

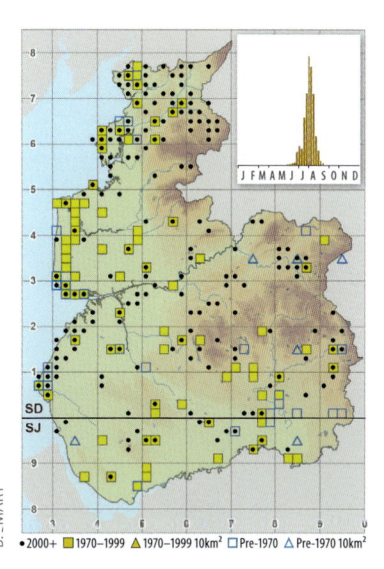

73.172 Cloaked Minor *Mesoligia furuncula*

■ **LANCS STATUS** Frequent / widespread resident   **c.1857–2022**
Reported from the Liverpool district (Gregson, 1857) and later, in Oldham in 1883 and St Annes in 1913. Since then, has been found to occur across the county, in both lowland and upland areas. Found annually at many sites, such as Billinge, Flixton, Hale, Heysham, Parr, Preston, St Annes, Sunderland Point and Worsthorne.
■ **HABITAT** Grassland across the county, including gardens and brownfield sites.
■ **FOODPLANTS** Not recorded in Lancashire.
Single brooded. Comes to light and sugar; nectars on ragwort and Creeping Thistle. Larger counts, which include 30 at Brockholes N.R. in 2009 and 40 at Flixton on 23 July 2010 (K. McCabe), have become very infrequent since 2011.

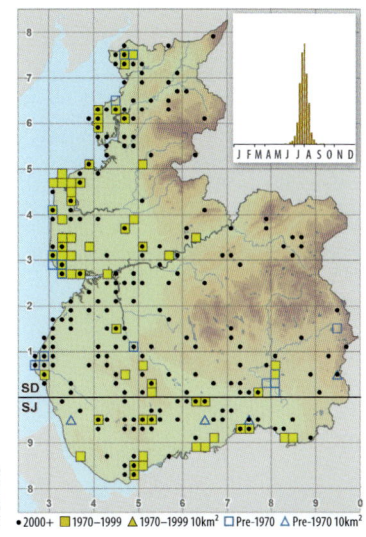

The Moths of Lancashire   **519**

NOCTUIDAE

73.173 Marbled Minor *Oligia strigilis*

■ **LANCS STATUS** Considered frequent / widespread resident   1913–2022
Identification difficulties between this sp., Tawny Marbled Minor and Rufous Minor, have meant only 186 specimens were available for assessment, 1.45% of the total dataset of aggregated records of the three species. The first confirmed record was from Silverdale in 1913 by W.P. Stocks and the most recent from Formby NT on 7 August 2022 by R. Walker. This appears to be the most frequently encountered of the three species but more data is required to be certain. See also Appendix 6.
■ **HABITAT** Grasslands across the county in lowland and upland areas.
■ **FOODPLANTS** Not recorded in Lancashire.
Single brooded. Comes to light.

• 2000+   ☐ 1970–1999   ▲ 1970–1999 10km²   ☐ Pre-1970   △ Pre-1970 10km²

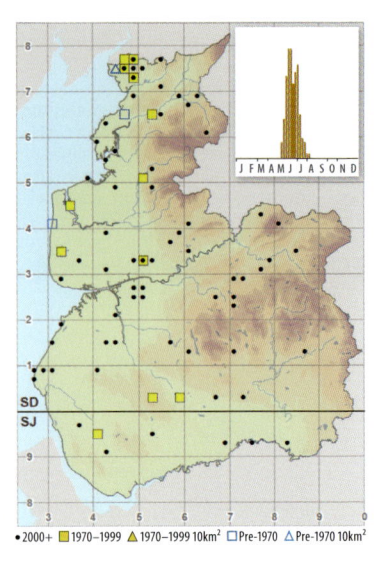

73.174 Tawny Marbled Minor *Oligia latruncula*

■ **LANCS STATUS** Considered local / widespread resident   1913–2022
Identification difficulties between this sp., Marbled Minor and Rufous Minor, have meant only 120 specimens were available for assessment in the text, 0.93% of the total dataset of aggregated records of the three species. The first confirmed records were from Withington and Silverdale in 1913 (W.P. Stocks) and the most recent from Formby on 7 August 2022 by R. Walker. It seems likely that this species will be found to occur across much of the county. See also Appendix 6.
■ **HABITAT** Grasslands across the county in lowland and upland areas.
■ **FOODPLANTS** Not recorded in Lancashire.
Single brooded. Comes to light.

• 2000+   ☐ 1970–1999   ▲ 1970–1999 10km²   ☐ Pre-1970   △ Pre-1970 10km²

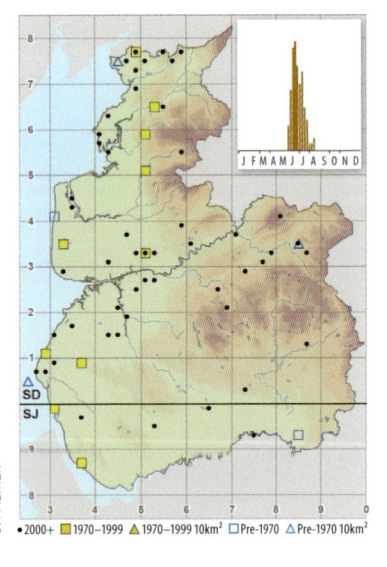

73.175 Rufous Minor *Oligia versicolor*

■ **LANCS STATUS** Considered local / widespread resident   1968–2022
Identification difficulties between this sp., Marbled Minor and Tawny Marbled Minor, have meant only 146 specimens were available for inclusion in this text, 1.14% of the total dataset of aggregated records of the three species. The first confirmed record was from Burscough in 1968 by R.M. Palmer and the most recent in Briercliffe on 19 July 2022, by G. Turner. The limited data suggests this moth may occur across much of the county. See also Appendix 6.
■ **HABITAT** Grasslands across the county in lowland and upland areas.
■ **FOODPLANTS** Not recorded in Lancashire.
Single brooded. Comes to light.

• 2000+   ☐ 1970–1999   ▲ 1970–1999 10km²   ☐ Pre-1970   △ Pre-1970 10km²

The Moths of Lancashire

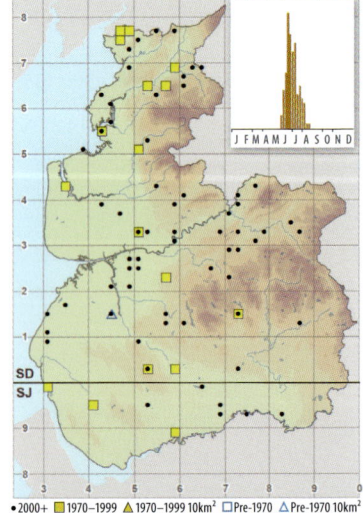

73.176 Middle-barred Minor *Oligia fasciuncula*

■ **LANCS STATUS** Frequent / widespread resident   c.1857–2022
Found in the Liverpool district where it was present in profusion on the coast (Gregson, 1857). Since then, it has been recorded across the whole county from grassy coastal sites to moorland edge. It is found on both acidic and calcareous soils and is regularly recorded in most garden light traps each year. There is perhaps some indication of an overall drop in numbers, as the large catches of 100 to 500 reported on a few occasions up to 2015 are no longer repeated.
■ **HABITAT** Most grassland types, including coastal, gardens, parks and brownfield sites.
■ **FOODPLANTS** Not recorded in Lancashire.
Single brooded. Comes strongly to light, often in double figures.

● 2000+   ▫ 1970–1999   △ 1970–1999 10km²   ▫ Pre-1970   △ Pre-1970 10km²

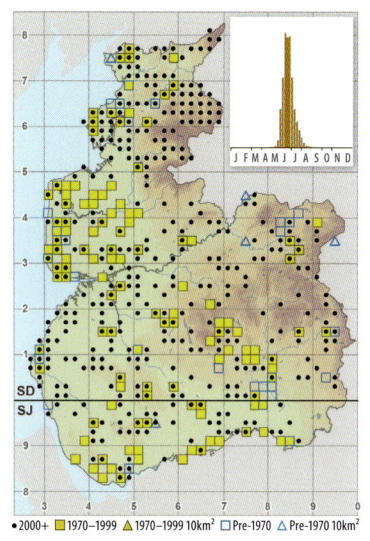

73.179 Orange Sallow *Tiliacea citrago*

■ **LANCS STATUS** Occasional / widespread resident   c.1857–2022
Found in Aigburth, Liverpool (Gregson, 1857), Withington in 1859, Myerscough in 1885 and Burnley in 1910. Since then, noted irregularly but widely across the county. Locations with more than a few records in recent years include Flixton, Fulwood, Gait Barrows, Hardy Grove, Leighton Moss, Longton, Victoria Park (Manchester), Preston, Pennington, Swinton and Yealand Conyers. More frequent in some years than others.
■ **HABITAT** Woodland, gardens and parks, where lime trees grow.
■ **FOODPLANTS** Not recorded in Lancashire.
Single brooded. Comes to light. Nearly always as singletons, although 14 came to light at Bretherton on 24 September 2008 (E. Langrish).

● 2000+   ▫ 1970–1999   △ 1970–1999 10km²   ▫ Pre-1970   △ Pre-1970 10km²

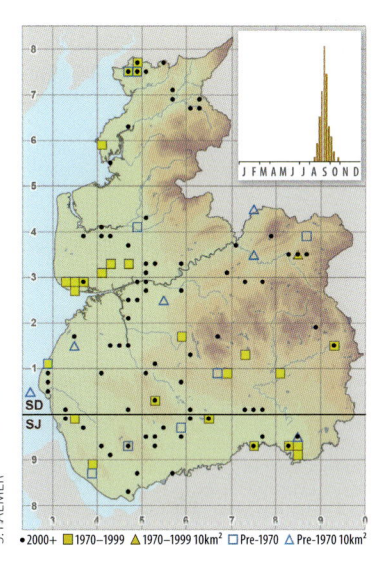

73.180 Barred Sallow *Tiliacea aurago*

■ **LANCS STATUS** Occasional / widespread resident   c.1857–2022
The first report was of two at Lydiate, Liverpool (Gregson, 1857). Not seen again until 1983, at Gait Barrows, with the only other report from the Silverdale area being at Leighton Moss in 2005. In VC59, it was next noted in Netherton in 1986, Flixton in 1995 and St Helens in 2002, these few heralding a much wider expansion of range. It arrived at Hale in 2006 (C. Cockbain) and has since spread east and north, reaching Longridge in 2011 and Victoria Park, Manchester in 2021 (P. Pemberton).
■ **HABITAT** Woodland, hedgerows, scrub and brownfield sites.
■ **FOODPLANTS** Not recorded in Lancashire.
Single brooded and comes to light.

● 2000+   ▫ 1970–1999   △ 1970–1999 10km²   ▫ Pre-1970   △ Pre-1970 10km²

The Moths of Lancashire

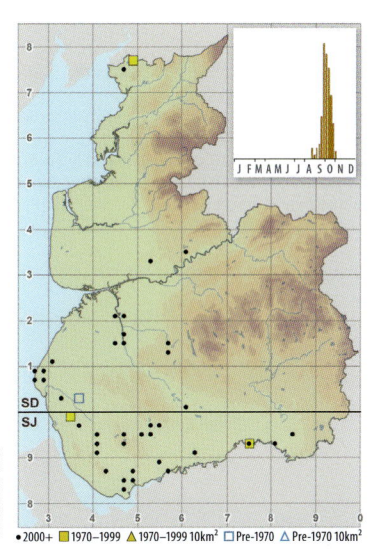

73.181 Pink-barred Sallow *Xanthia togata*

■ **LANCS STATUS** Frequent / widespread resident   c.1857–2022

Noted as plentiful between Hightown and Crosby (Gregson, 1857) and later, from Warrington in 1889 (J. Collins), Eccleston Mere in 1904 (J. Cotton), Oldham in 1915 (J. Cottam) and Ditton in 1929 (F.M.B. Carr). Mansbridge (1940) reported it to be generally distributed but not as common as Sallow (*Cirrhia icteritia*). The numbers of records received increased consistently throughout the rest of the 20th century and beyond, as has the recorded distribution. There are indications that the species has expanded its range into some higher parts of the county, but increased recording effort may also be a factor. Although annual at many locations that run light traps in the autumn, it is usually only found in single figures. Double-digit counts were not unusual at some sites, such as Flixton, Heysham N.R., Leighton Moss and Mere Sands Wood, but these have been less frequent in recent years.

■ **HABITAT** Damp woodland, scrub, parks, gardens and brownfield sites.
■ **FOODPLANTS** Sallow catkins and Sycamore.

Single brooded, flying through into mid-November at times. Comes to light and nectars on honeydew, Ivy blossom, ragwort, *Spartina* and Umbellifers. Larvae have been found in April and early May.

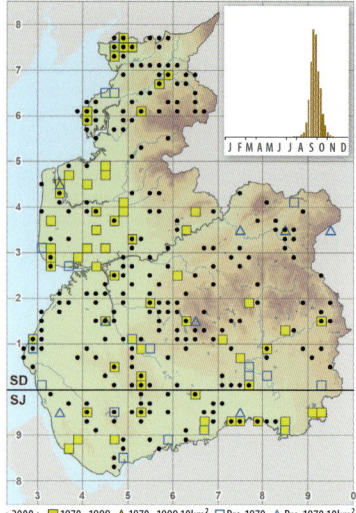

73.182 Sallow *Cirrhia icteritia*

■ **LANCS STATUS** Frequent / widespread resident   c.1853–2022

First noted in Hale (Byerley, 1854) and elsewhere, at Crosby in 1857 (C.S. Gregson) Warrington in 1889 (J. Collins), Rochdale in 1890 (J. Taylor), Alexandra Park, Oldham in 1892 (W.F. Windle) and Eccleston Mere in 1904 (J. Cotton). Widely recorded during the mid-20th century across the county, from sites such as Formby, Colne, Irlam, Crosby, Didsbury, Burnley, Wrightington, St. Annes and Warton (Silverdale). Since then, regular at many sites, often annually, such as Billinge, Flixton, Leigh, Leighton Moss, Mere Sands Wood, Morecambe, Orrell, Preston, Rishton, St Helens area and Worsthorne. Usually seen in low single figures, particularly since 2012, but higher counts prior to then were not unusual, such as 44 at Leigh on 19 August 1987 (J.D. Wilson) and 71 at Glazebury on 22 August 2011 (R. Rhodes and J.D. Wilson). A form devoid of most markings, ab. *flavescens*, has been noted as singletons from seven scattered locations across the county, including most of the records from Bay Horse (N. Rogers).

■ **HABITAT** Woodland, parks, rough ground with scrub, gardens and brownfield sites.
■ **FOODPLANTS** Willow sp.

Single brooded. Comes readily to light, sugar and nectars on flowers and fruit, such as Apple Mint, blackberries, *Spartina*, willowherbs, ragwort and Creeping Thistle. Larvae have been found in April and May.

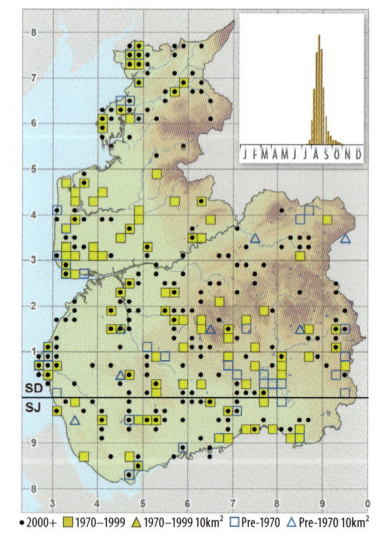

522   *The Moths of Lancashire*

73.183 Dusky-lemon Sallow *Cirrhia gilvago*

■ **LANCS STATUS** Scarce / local resident   1936–2021

Present in Cheshire during the late 19th century, the first in Lancashire was not until 1936, in Formby (G. de C. Fraser), with further records from this site. Next seen in Lytham (1939), Pendle Hill (1940), Didsbury (1953) and Parbold (1961). Since 2003, there have been only 21 records, with over a third of these from Hale (C. Cockbain). Of the remaining sites, only Great Eccleston (M. Wilby) and Sutton Leach (S. Briers) have more than single records and the last was seen at Hoghton in 2021 (G. Dixon).
■ **HABITAT** Coastal woodland, scrub, woodland and gardens.
■ **FOODPLANTS** Not recorded in Lancashire.
Single brooded. Comes to light in small numbers.

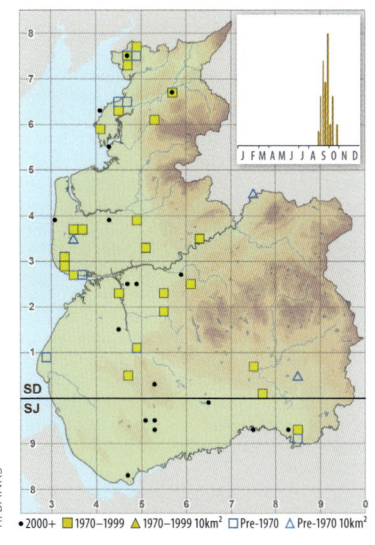

73.186 Beaded Chestnut *Agrochola lychnidis*

■ **LANCS STATUS** Occasional / local resident   c.1857–2022

Noted as abundant in the Liverpool area (Gregson, 1857), later reports suggested it was locally common elsewhere (Mansbridge, 1940). Found regularly at St Annes, Morecambe and Salford in the 1950s and '60s. From the 1990s, it proved to be rather local in VC60, but more widespread in lowland VC59 including regular sightings at sites such as Billinge, Flixton, Hale, near Longton and Mere Sands Wood. However, in recent years, away from those favoured sites, records and its range have decreased.
■ **HABITAT** Coastal grassland, scrub, disturbed ground, gardens and brownfield sites.
■ **FOODPLANTS** Not recorded in Lancashire.
Single brooded. Comes to light and sugar.

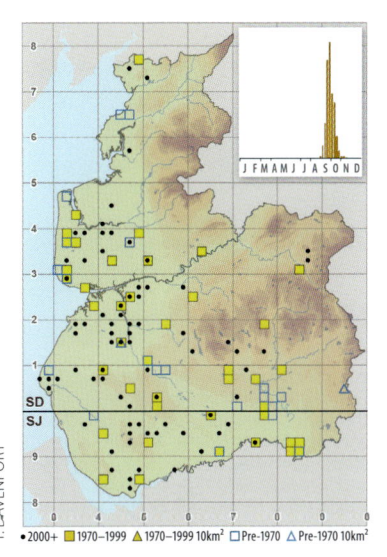

73.187 Brown-spot Pinion *Anchoscelis litura*

■ **LANCS STATUS** Scarce / very local resident   c.1857–2022

First found in Croxteth (Gregson, 1857) and considered common to abundant by early authors, including Mansbridge (1940). Remained widely recorded from Formby to Rochdale and Liverpool to Silverdale into the late 20th century, with a study in St Annes (by sugaring) in September 1970 recording up to 51 on some nights (D. & J. Steeden). From 2000, site reports declined widely. In Hale it has remained annual since 2005 (C. Cockbain), but only noted from Hale and Westby (once each) in 2022.
■ **HABITAT** Scrub and gardens.
■ **FOODPLANTS** Not recorded in Lancashire.
Single brooded. Comes to light and to sugar.

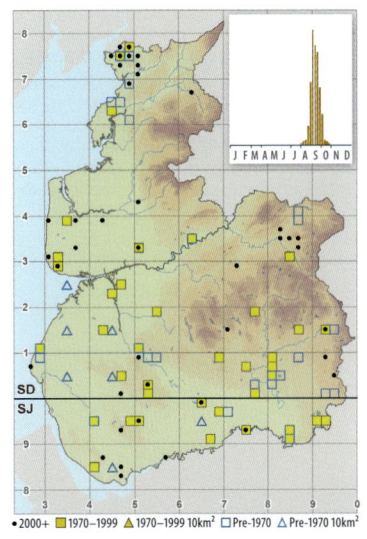

73.188 Flounced Chestnut *Anchoscelis helvola*

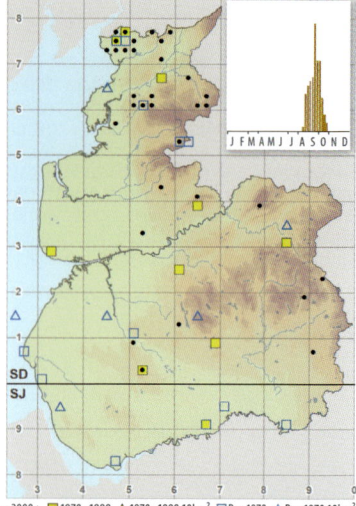

■ **LANCS STATUS** Occasional / local resident   c.1857–2022

The first record was from Hale (Gregson, 1857) and subsequently noted from Crosby in 1886 (F.N. Pierce), Burnley in 1887 (A.E. Wright) and Lancaster and Silverdale (also 1887) by C.H. Forsyth and W.P. Stocks respectively. It was described as 'not generally common' by Ellis (1890) and by Mansbridge (1940). Between 1940 and the early 1990s, records were scattered and infrequent, from sites such as Bolton in 1950 (S. Coxey), Chat Moss and Didsbury between 1950 and 1954 (H.N. Michaelis), Clougha in 1962 (C.J. Goodall), Hoghton in 1970 (J. Whiteside), Leighton Moss in 1972 (J. Briggs), St Annes in 1974 (W.A. Watson), Gait Barrows in 1983 (E. Emmett) and Longridge Fell in 1985 (G. Carefoot). Reports of this species have declined considerably in the last two decades or so, but it is still seen on occasions at locations such as Baines Crag, Gait Barrows, Green Bank, Leighton Moss, Warton Crag and Yealand Conyers. Usually occurs in single figures, but 12 came to light at Baines Crag on 31 August 2010 (J. Girdley) and 18 were seen at a very wet sheet and light session at Gait Barrows on 27 September 2019 (J. Patton).
■ **HABITAT** Woodland and scrub on limestone, moorland and formerly, on mosses.
■ **FOODPLANTS** Not recorded in Lancashire.
Single brooded. Comes to light and sugar. Wanders from prime habitat on occasions.

• 2000+   ☐ 1970–1999   △ 1970–1999 10km²   ☐ Pre-1970   △ Pre-1970 10km²

73.189 Red-line Quaker *Leptologia lota*

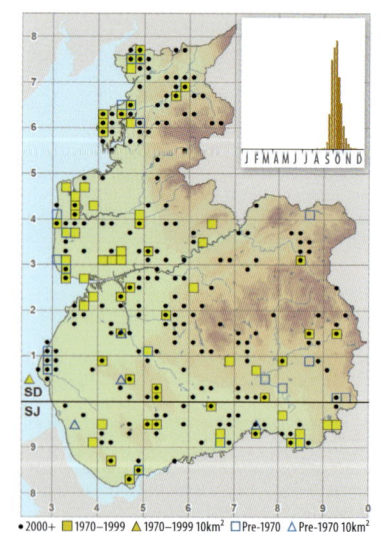

■ **LANCS STATUS** Frequent / widespread resident   c.1857–2022

First reports were of it being plentiful in the Liverpool district (Gregson, 1857), and thereafter it was referred to as common and generally distributed by Ellis (1890) and Mansbridge (1940). 19th and early 20th century records were few but included Middleton, nr. Rochdale, in 1879 and 1880 (J. Taylor), Ditton in 1929 (F.M.B. Carr) and larvae found at Freshfield on 20 May 1933 (L. Nathan). As with many common species, few records are documented in these early books. Since 1939, it has been reported from many lowland and fewer upland areas, with light trapping responsible for most of these. Early confirmation that it was widespread came from sites such as Didsbury in 1940 (H.N. Michaelis), Formby in 1946 (G. de C. Fraser), St Annes in 1949, Warton, nr. Silverdale in 1961, Slyne in 1964 and Wrightington in 1969. Since the late 20th century, many sites have accumulated large, long-term datasets, including Adlington, Billinge, Flixton, Green Bank, Hale, High Tatham, Leighton Moss and Preston. Most reports are of low single figures. Double figure counts occurred prior to 2015, but not since.
■ **HABITAT** Woodland, scrub, gardens, wet areas with sallows and brownfield sites.
■ **FOODPLANTS** Creeping, Goat & Grey Willow.
Single brooded. Comes to light and attracted to Ivy blossom, sugar and wine ropes. Noted as infrequent to scarce at honeydew by R. Leverton in Manchester in 1961.

• 2000+   ☐ 1970–1999   △ 1970–1999 10km²   ☐ Pre-1970   △ Pre-1970 10km²

73.190 Yellow-line Quaker *Leptologia macilenta*

■ **LANCS STATUS** Frequent / widespread resident   c.1859–2022
First found in Agecroft by J. Chappell c.1859 (Ellis, 1890) and thereafter, over the next century, in lowland and upland sites including Woolton, Bolton, Burnley, Chorley, Lancaster and Nelson. With the arrival of increased light trapping across the county in the early 1990s, records likewise increased, but usually involved single figure counts. Surprisingly, larger counts have tended to be in the woods adjoining moorland, with 52 in one trap at Botton Head Farm on 6 October 2015 (P.J. Marsh).
■ **HABITAT** Deciduous woodland, hedgerows, gardens, mosses and brownfield sites.
■ **FOODPLANTS** Crack-willow.
Single brooded. Comes to light and nectars on Ivy blossom.

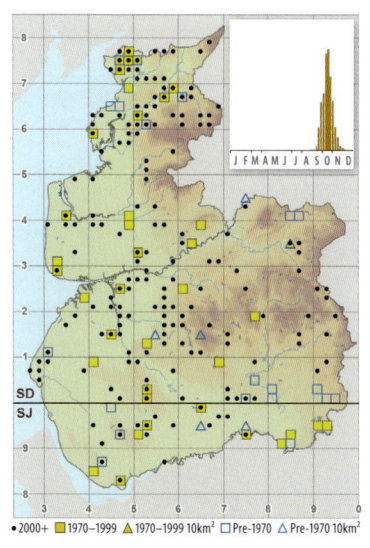

●2000+ ■1970–1999 ▲1970–1999 10km² □Pre-1970 △Pre-1970 10km²

73.192 Brick *Sunira circellaris*

■ **LANCS STATUS** Frequent / widespread resident   c.1857–2022
First recorded in the Liverpool district (Gregson, 1857) and since then has been noted widely across lowland and upland areas. Many sites have recorded it regularly over the years, including Belmont, Billinge, Flixton, Fulwood, Hale, Leigh, Longton, Parr, Mill Houses, Preston, Royton, Swinton and Worsthorne.
■ **HABITAT** Damp woodland, scrub, brownfield sites and gardens.
■ **FOODPLANTS** Sallow catkins.
Single brooded. Comes to light and Ivy blossom. Larvae found from April to early May. Usually occurs in single figures, but counts of over ten have been rare since 2014. On 3 November 2004, 37 were attracted to a light trap in Leigh (J.D. Wilson).

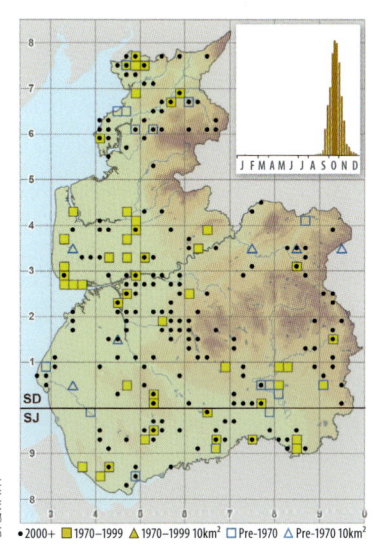

●2000+ ■1970–1999 ▲1970–1999 10km² □Pre-1970 △Pre-1970 10km²

73.193 Lunar Underwing *Anchoscelis lunosa*

■ **LANCS STATUS** Frequent / local resident   1900–2022
First noted in Oldham in 1900 (J. Taylor), followed by Knowsley and Ainsdale (1910), with Mansbridge (1940) reporting it as scarce. However, records increased in the 1930s, from Formby in 1934, Ditton in 1935 and Lytham in 1939. Since then, found regularly, mainly in coastal areas and lowland VC59, occasionally in numbers, such as in Billinge, Flixton, Hale, Heysham and Parr. It is very local in upland areas. Numbers have declined since 2006; double-figure counts are now almost unknown.
■ **HABITAT** Rough grassland, on the coast, brownfield sites and gardens.
■ **FOODPLANTS** Not recorded in Lancashire.
Single brooded. Comes to light. 60 noted at light in Heysham on 13 September 2003.

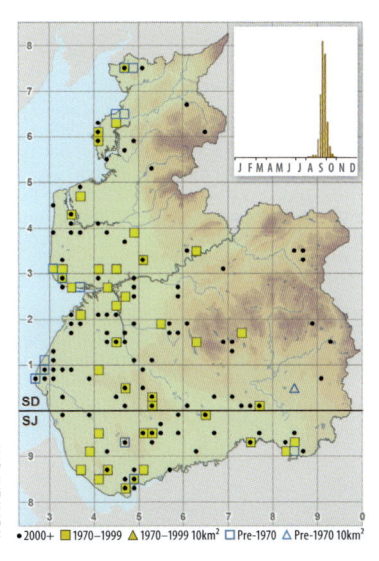

●2000+ ■1970–1999 ▲1970–1999 10km² □Pre-1970 △Pre-1970 10km²

The Moths of Lancashire   **525**

73.194 Chestnut *Conistra vaccinii*

■ **LANCS STATUS** Frequent / widespread resident   c.1857–2022

Reported as common to abundant across the region by all early authors up to 1940, it was first noted in Liverpool (Gregson, 1857). Since then, it has continued to be widely recorded, most frequently in deciduous woodland. Sites with good numbers of records include Adlington, Ellenbrook, Fulwood, Mere Sands Wood, Mill Houses, Silverdale, Swinton and Worsley. The largest single catch was of 49 attracted to actinic light in Euxton on 3 April 2010 (E. Langrish).
■ **HABITAT** Woodland, scrub, gardens and brownfield sites.
■ **FOODPLANTS** Sycamore.
Single brooded; overwinters as an adult. Comes to light and Ivy and sallow blossom.

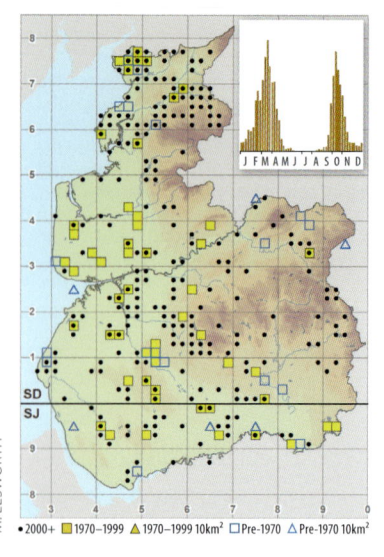

73.195 Dark Chestnut *Conistra ligula*

■ **LANCS STATUS** Occasional / widespread resident   c.1859–2022

First noted in Agecroft by J. Chappell c.1859 and next, at Preston in 1885 and Silverdale in 1913. Reported quite widely during the 1960s and 1970s at sites such as Hoghton, Leighton Moss and Slyne. Seen in Hale from 1987 onwards and from the mid-1990s increasingly recorded at light in places like Great Eccleston, Higher Tatham and Longton. 14 were seen on Ivy blossom at Whittle-le-Woods in 2008 (E. Langrish).
■ **HABITAT** Woodland edge, scrub, hedgerows, brownfield sites and gardens.
■ **FOODPLANTS** Blackthorn.
Single brooded, overwintering as an adult. Comes to light. More frequently encountered in the autumn than in the following spring. Larvae found in April.

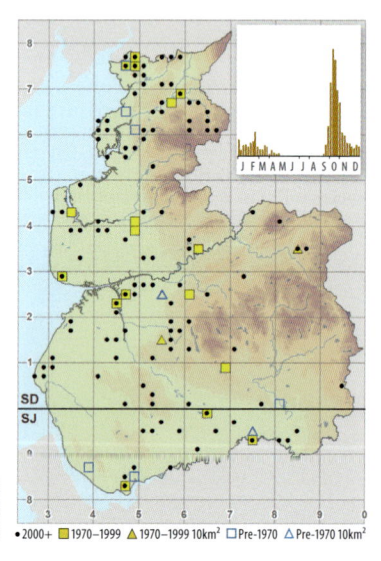

73.200 Tawny Pinion *Lithophane semibrunnea*

■ **LANCS STATUS** Scarce / local resident   1970–2022

First noted in St Annes on 12 October 1970 (D. & J. Steeden), followed by Leighton Moss (1978) and Billinge (1991). In 1994, noted in Flixton, this being the sole site with records (21 in all) until September 1998. A rapid, but rather limited, increase in distribution then commenced, peaking in 2010 at ten sites, a few in upland areas. These reduced to six sites in 2010, followed by a decline in range and records, with only single records in 2019, '20 and '21 and two records in 2022, all in lowland areas.
■ **HABITAT** Woodland, copses, mature hedgerows, parks and gardens.
■ **FOODPLANTS** Not recorded in Lancashire.
Single brooded, overwintering as an adult. Comes to light and Ivy blossom.

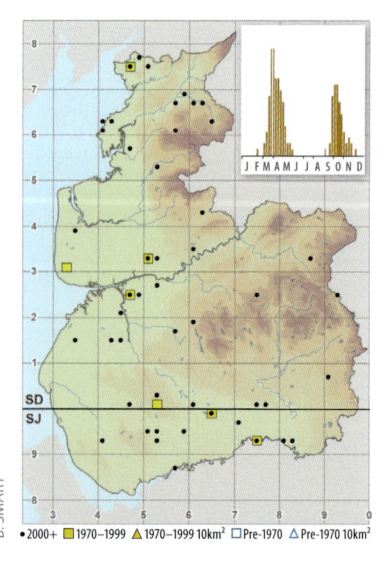

73.201 Pale Pinion *Lithophane socia*

■ **LANCS STATUS** Occasional / widespread resident   c.1885–2022

First noted c.1885 in Liverpool by W. Johnson (Ellis, 1890), then Silverdale in 1900 (W.P. Stocks). Next, in St Annes (1945), but then not until 1993-94 at Leighton Moss, Gait Barrows and Claughton. These heralded a significant range expansion, mirrored nationally. The main arrival was possibly from the south, with records from Woolton in 1996 and Flixton in 1999. Since then, it has spread across the county, including upland wooded areas. A few moths appear to survive, on occasions, to mid-summer.
■ **HABITAT** Woodland, parks, mature hedgerows, shelterbelts and gardens.
■ **FOODPLANTS** Not recorded in Lancashire.
Single brooded, overwintering as an adult. Comes to light and Ivy blossom.

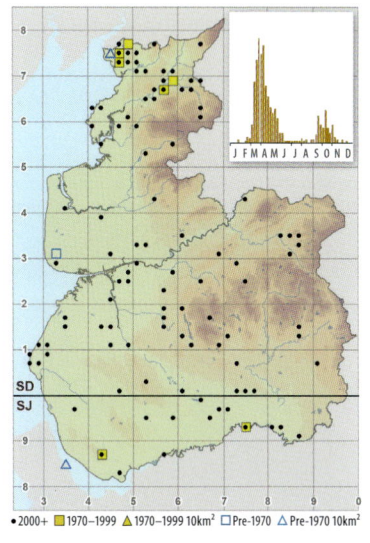

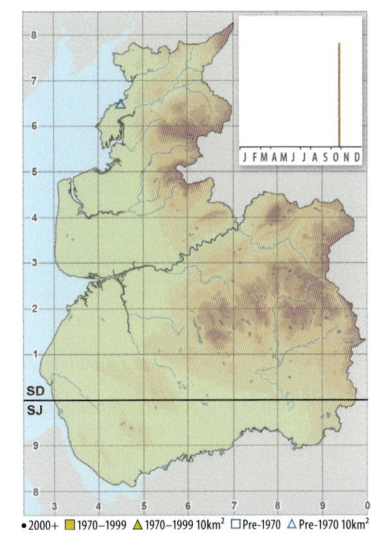

• 2000+ ☐ 1970–1999 △ 1970–1999 10km² ☐ Pre-1970 △ Pre-1970 10km²

73.202 Grey Shoulder-knot *Lithophane ornitopus*

■ **LANCS STATUS** Occasional / widespread resident   1885–2022

Found in the Manchester district by J.B. Hodgkinson in 1885 (Ellis, 1890), in Silverdale in 1916 and 1926, and in the Burnley area in 1979. Until 2005 it remained a very local species, with fewer than six records per year. Although reports were variable in number up to 2016, it has, since then, significantly increased its range and between 2017 and 2022 records also increased noticeably. Locations with regular sightings include Ainsdale, Flixton, Hale, Leighton Moss, Parr, Longton and Yealand Conyers.
■ **HABITAT** Deciduous woodland, copses and mature hedgerows with oak standards.
■ **FOODPLANTS** Oak.
Single brooded; overwinters as adult. Comes to light. Rests on walls and tree trunks.

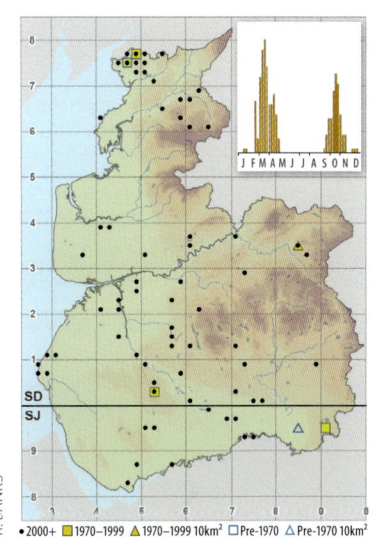

• 2000+ ☐ 1970–1999 △ 1970–1999 10km² ☐ Pre-1970 △ Pre-1970 10km²

73.203 Conformist *Lithophane furcifera*

■ **LANCS STATUS** Rare / migrant   1901

A single report of two specimens, on 22 October 1901, was noted by R. South (South, 1902) and is the only known record of this species in the county. South noted that 'Mr. C.H. Forsythe, of Lancaster, recently sent me a very nice photograph of a moth which he was unable to identify, and which I recognised as *Xylina conformis* (a synonym of *L. furcifera*). He states that he took two specimens when beating ivy blossom late at night, nearly 12 o'clock, on Oct. 22 last'. The wording of this report does leave some doubt about the locality for this sighting. From information published in Mansbridge (1940), it appears that Claude H. Forsythe lived and worked in the Lancaster area from at least the 1890s to 1910 and probably well before and after those dates. There are many records of his mentioned in Mansbridge (loc. cit.), only a few of which are considered unconfirmed. This record is accepted on the basis that the photo was presented to and the species identified by R. South, the species is migratory and the reasonable assumption they were found in the Lancaster area.
■ **HABITAT** Unknown.
■ **FOODPLANTS** Not recorded in Lancashire.

• 2000+ ☐ 1970–1999 △ 1970–1999 10km² ☐ Pre-1970 △ Pre-1970 10km²

The Moths of Lancashire   **527**

73.206 Blair's Shoulder-knot *Lithophane leautieri*

■ **LANCS STATUS** Frequent / widespread resident 1980–2022
First noted at Formby in 1980 (B. Holloway), with the next not until 2 October 1992 at Swinton (S. Christmas). This was followed by a record in Billinge the next day and on 8 October 1992 it came to light in Hornby, north VC60 (J. Newton). Despite this, it wasn't until 1995 that the main arrival commenced in VC59, with VC60 following the same trend from 2000. Double figures were regular in some traps between 1996 and 2018 but since then, no sites have experienced counts of greater than single figures.
■ **HABITAT** Gardens and parks.
■ **FOODPLANTS** Not recorded in Lancashire.
Single brooded and comes readily to light. First recorded in Britain in 1951.

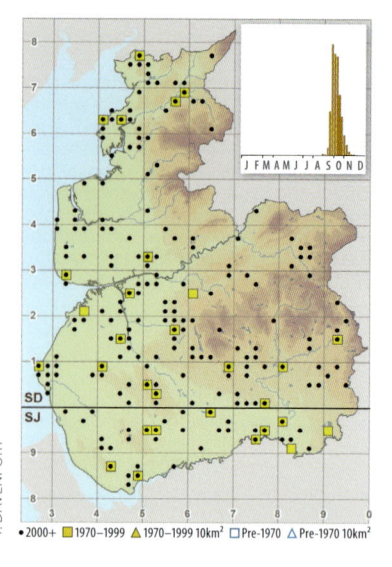

73.207 Golden-rod Brindle *Xylena solidaginis*

■ **LANCS STATUS** Scarce / local resident c.1859–2020
Formerly regular in moorland areas from the late 19th century. The first was in Bolton (c.1859) by J. Chappell, followed by Bury, Nelson, Oldham and Pendle. A few noted in the Trough of Bowland in 1962 and 1963. Sightings declined after the mid-1980s with only a few records from Birk Bank (1993), Worsthorne and Baines Crag, both in 2001, Mallowdale in 2004, Longridge Fell and Crossdale Grains in 2012 and Green Bank in 2019. Eight lowland records were noted from 1861 to 2020, all assumed wanderers.
■ **HABITAT** Heather moors.
■ **FOODPLANTS** Not recorded in Lancashire.
Single brooded. Comes to light. The early July record from Formby was exceptional.

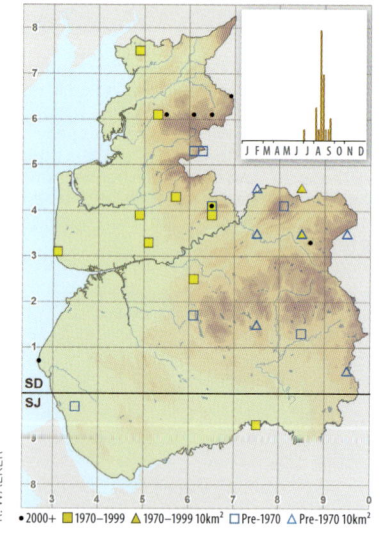

73.208 Sword-grass *Xylena exsoleta*

■ **LANCS STATUS** Considered Extinct c.1859–c.1968
Given as abundant in Liverpool area Byerley (1854), which may relate to the Cheshire part of this region. First confirmed in Lancashire when reported from Agecroft and a few other sites by J. Chappell c.1859, followed by Goosnargh c.1879 (R. Standen). More widely, given as 'generally distributed but not common' by Ellis (1890). Even in 1913 it was being reported as more common than Red Sword-grass and generally distributed (Mansbridge, 1940). As a result, the decline of this species appears to have caught early recorders by surprise, and the earlier distribution of the moth remained undocumented. A female was noted on sallow blossom in Oldham (1915), followed by single reports from Nelson (1927), Didsbury (1934) and the Blackpool area in 1933 and 1935. It was noted as rare in Formby (Leech & Michaelis, 1940), but with no specific dates given. The final record seems likely to have been at Leighton Moss in 1968 (Briggs, 1985), but even this was poorly documented, as the 1985 list could be interpreted to suggest it was still present in that year, which now seems unlikely. Nationally, it is now considered extinct in England, with the very few recent English records considered as migrants.
■ **HABITAT** Unknown in Lancashire.
■ **FOODPLANTS** Not recorded in Lancashire.

528 *The Moths of Lancashire*

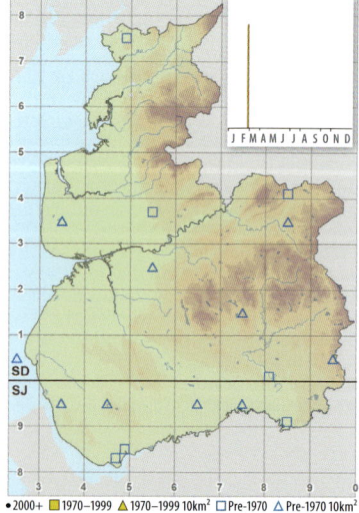

73.209 Red Sword-grass *Xylena vetusta*

■ **LANCS STATUS** Occasional / widespread resident   c.1856–2022

Occasional, mostly at sugar, in the Crosby area (Gregson, 1857) and has remained 'of intermittent occurrence' since then (Mansbridge, 1940). Possibly, due to more widespread light trapping, the frequency and distribution of records increased to some extent from the late 1990s. However, in more recent years it has decreased in numbers at light, with fewer also seen on Ivy blossom. Since 2019 it has been regular at only two sites, Green Bank and Great Eccleston, and occasional at a few others.
■ **HABITAT** Mosses, moorland edge, dune slacks, damp woodland, ditches and gardens.
■ **FOODPLANTS** Not recorded in Lancashire.
Comes readily to Ivy blossom and sugar, but more sparingly attracted to light.

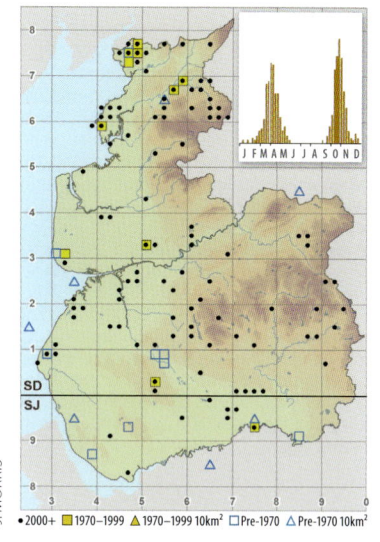

● 2000+   ☐ 1970–1999   △ 1970–1999 10km²   ☐ Pre-1970   △ Pre-1970 10km²

73.210 Satellite *Eupsilia transversa*

■ **LANCS STATUS** Frequent / widespread resident.   c.1857–2022

Reported as common, in Liverpool (Gregson, 1857), and elsewhere, by Mansbridge (1940). It continued to be widely recorded from the 1980s onwards, at sites such as Billinge, Claughton, Euxton, Flixton, Leighton Moss, Royton and Swinton. Since then, the distribution has remained unchanged, but numbers have declined at some sites in recent years. White- and orange-spotted forms are equally common; once seen with pink spots at Mill Houses. Adults once sieved from leaf litter in winter.
■ **HABITAT** Deciduous woods, parks, scrub, moorland edges, brownfield sites & gardens.
■ **FOODPLANTS** Ash, Beech, Blackthorn, Bramble, Hawthorn, lime, Norway Maple, oak.
Comes strongly to Ivy blossom, sugar and tree sap; less strongly to light.

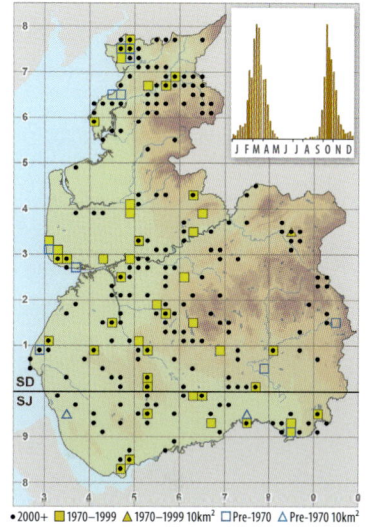

● 2000+   ☐ 1970–1999   △ 1970–1999 10km²   ☐ Pre-1970   △ Pre-1970 10km²

73.211 Angle-striped Sallow *Enargia paleacea*

■ **LANCS STATUS** Unknown   c.1895

Two specimens, labelled as caught in 'Lancashire', were exhibited at the South London Entomological and Natural History Society meeting on 24 October 1895 (Entomologist, Vol. 28: 346). The exhibitor, Mr Oldham, also displayed specimens from other locations across England, but no further details were published as to the date or precise locality in which these moths were found. This does raise the slim possibility that the location may not have been within the Vice County boundaries of VC59 or VC60 due to the historic realignment of county boundaries.
■ **HABITAT** Not known.
■ **FOODPLANTS** Not recorded in Lancashire.
Nationally, its distribution extends, mainly, from the Midlands to north-east Yorkshire and, separately, across the Highlands of Scotland. It is associated with woods and heaths containing mature birch trees and is also considered an occasional migrant. Strangely, no reference was made in any of the early books covering the moths of Lancashire, despite this fully verified and published record. For mapping purposes only, the record has been placed centrally in the county.

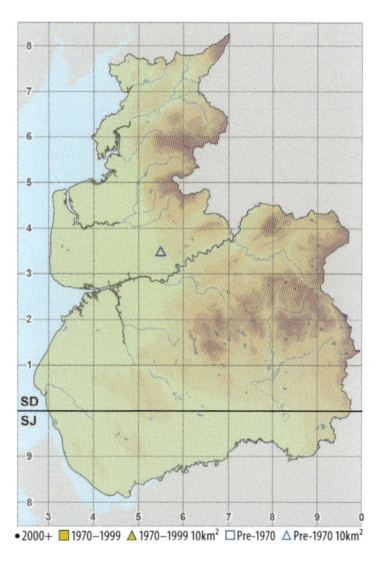

● 2000+   ☐ 1970–1999   △ 1970–1999 10km²   ☐ Pre-1970   △ Pre-1970 10km²

73.212 Double Kidney *Ipimorpha retusa*

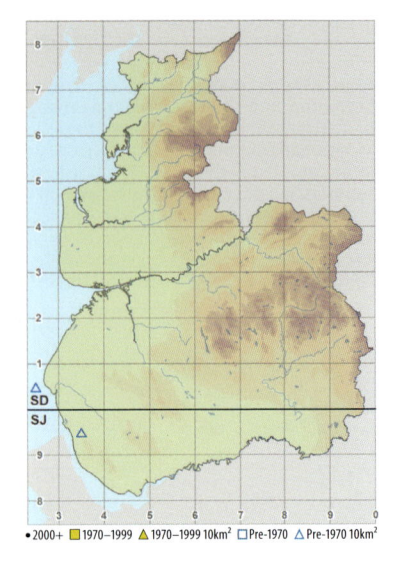

■ **LANCS STATUS** Considered Extinct c.1860–c.1886

Reported by J.B. Hodgkinson as found by C.S. Gregson in the Formby area (Ellis, 1890) without any date being given. 1860 is an approximated year, based on Gregson's active period and that it was not published in Gregson (1857). The only other records were also without a date but have been estimated to be around 1886 where Ellis (loc. cit.) refers to it being found occasionally in the Liverpool district, by William Johnson.

■ **HABITAT** Not known.
■ **FOODPLANTS** Not recorded in Lancashire.

There are slight reservations about the accuracy of these records, but they are included on the basis that they were accepted by Hodgkinson and Ellis, both very experienced lepidopterists of this era. The lack of more recent records is in line with a national contraction in range southwards, away from Cheshire, south-east in a band to the Midlands, which is evident on the distribution maps in Randle *et al.* (2019) and mentioned in the text. This moth is associated with various willow species nationally, occurring in damper habitats. With the amount of modern-day light trapping in south-west Lancashire it is unlikely to have been overlooked in recent years.

73.213 Olive *Ipimorpha subtusa*

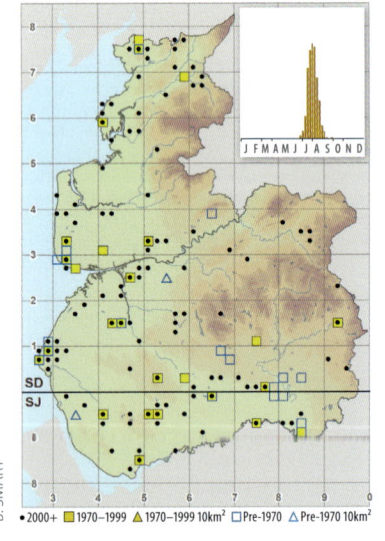

■ **LANCS STATUS** Occasional / widespread resident c.1859–2022

First noted at Chat Moss by J. Chappell (c.1859) with larvae reported as plentiful at some sites, such as Dutton (nr. Longridge) in 1879 (Hodgkinson, 1880) and Formby in 1931 by H.W. Wilson. More recently, light trapping has shown it to be scattered, in low numbers, over much of lowland Lancashire, including near annual reports from Flixton and Preston, but absent or very localised in upland areas.

■ **HABITAT** Coastal dunes with poplar trees, woodland, amenity plantings and gardens.
■ **FOODPLANTS** Aspen, Balsam Poplar and Black-poplar varieties.

Single brooded, with one very late record on 25 September 2001 in Orrell. The larvae have been found in May and early June.

73.214 White-spotted Pinion *Cosmia diffinis*

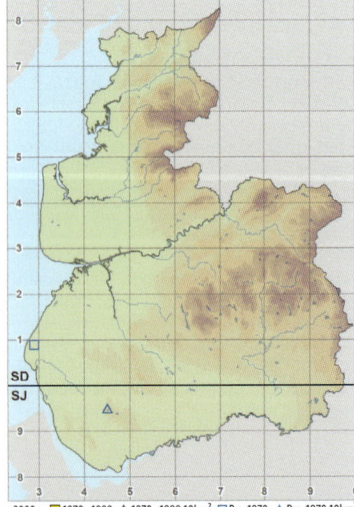

■ **LANCS STATUS** Considered Extinct 1852–1952

There are only two county records, exactly one hundred years apart. The first relates to one bred from a larva found near Rainhill, St Helens by C.S. Gregson, the moth emerging in August 1852 (Gregson, 1857). The second record reports the capture of a single imago at Freshfield by G. de C. Fraser in August 1952 (Leech & Michaelis, 1957).

■ **HABITAT** Unknown.
■ **FOODPLANTS** Not documented, but the 1852 larva was presumably bred from an elm. The extensive loss of elm, following the second arrival of Dutch elm disease in the 1960s (a more virulent form than the first), is likely to have led to the local extinction of this already scarce moth. Nationally, the population declined to the extent it was only found in two southern counties by 1990. Since then, there have been signs of a small recovery, but it is felt extremely unlikely that it will make it back into Lancashire again, particularly as the previous Lancashire records were right on the north-western periphery of the moth's former national range.

530 *The Moths of Lancashire*

73.215 Lesser-spotted Pinion *Cosmia affinis*

■ **LANCS STATUS** Considered Extinct   c.1855

The single county record was of one 'On sugar in a young Elm plantation between Hightown and Sephton' (Sefton), prior to 1856 by C.S. Gregson (Gregson, 1857). He adds that eggs from this moth were taken and placed in the plantations at Wallasey (Cheshire).
■ **HABITAT** Formerly amongst elms in coastal dunes.
■ **FOODPLANTS** Not recorded in Lancashire.
The loss of most mature elms due to the two bouts of Dutch elm disease may well have affected this species, although it could have already been lost before then. Nationally, the species was known as far north as southern and central Cumbria but as with other closely related species, the moth declined and was lost over a wide part of its more northerly range. More recently, it has been reported, post 2000, in Yorkshire, https://yorkshiremoths.co.uk/ (accessed 21 Dec. 2022) where it is suggested that 'other elm-rich areas .... should be targeted, as it is quite elusive and only ever appears in small numbers'. There are two recent 2022 records not too far away in Cheshire – see Cheshire-moth-charts.co.uk (accessed 3 September 2023). This species may yet be re-found in Lancashire.

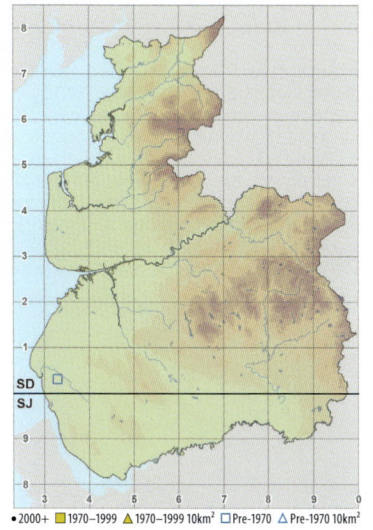

• 2000+   ■ 1970–1999   ▲ 1970–1999 10km²   □ Pre-1970   △ Pre-1970 10km²

73.216 Dun-bar *Cosmia trapezina*

■ **LANCS STATUS** Frequent / widespread resident   c.1857–2022

Found abundantly in the Liverpool area (Gregson, 1857), this also reported by Ellis (1890) and Mansbridge (1940), with no sites or dates documented. Subsequently, recorded regularly at light traps throughout the county and occasionally seen nectaring on Buddleia. Can be locally abundant at times, with over 100 near Thimble Hall on 2 August 2013 and Dalton (Wigan area) on 9 August 2013 (C. Darbyshire).
■ **HABITAT** Woodland, gardens and other habitats with trees and shrubs.
■ **FOODPLANTS** Apple, Beech, Bird Cherry, Blackthorn, elm, Field Maple, Goat Willow, Grey Willow, Hawthorn, Hazel, oak, Silver Birch and Sycamore.
Single brooded. A few have been recorded in early June, but this is exceptional.

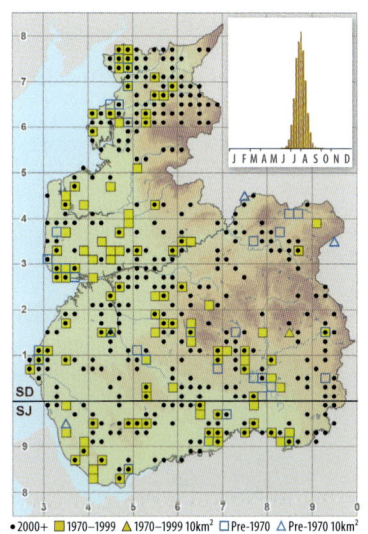

• 2000+   ■ 1970–1999   ▲ 1970–1999 10km²   □ Pre-1970   △ Pre-1970 10km²

73.217 Lunar-spotted Pinion *Cosmia pyralina*

■ **LANCS STATUS** Considered Extinct   1952

The only report of this species that is considered confirmed was noted in Michaelis and Leech (1957). It refers to one found in Freshfield by G. de. C. Fraser in 1952. No further details are known and the subsequent loss of elm, following Dutch elm disease, make it unlikely that this species has survived in the area. Regular light trapping has been taking place in Ainsdale in recent years, an area with elm trees present. If anywhere, this would be a likely location for the moth to reappear.
■ **HABITAT** Woodland or hedgerow adjacent to coastal dunes.
■ **FOODPLANTS** Not recorded in Lancashire.
A historic record from the Silverdale area was assessed by N. Birkett (Card Index, unpublished) and considered likely to be a misidentification. Further details can be found in the section of this book covering unverified records. Of the three elm-feeding pinions, this is the one that appears to have suffered least from Dutch elm disease in terms of distribution, although data still suggests a long-term decline. Despite this, its current national distribution shows an extensive recent set of records from the Midlands southwards, as well as occasional records further north, including from an area in Cheshire post 1999 (Randle *et al.*, 2019).

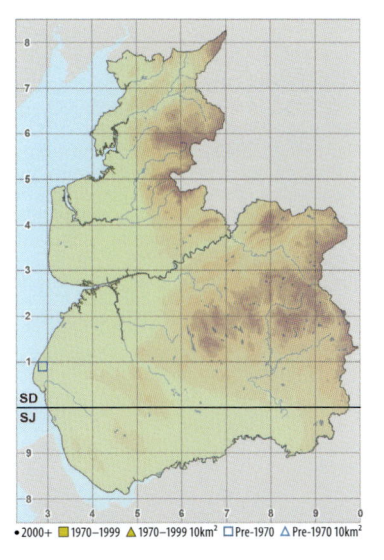

• 2000+   ■ 1970–1999   ▲ 1970–1999 10km²   □ Pre-1970   △ Pre-1970 10km²

NOCTUIDAE

73.219 Centre-barred Sallow *Atethmia centrago*

■ **LANCS STATUS** Frequent / widespread resident   1870–2022
Noted in Hulme, Manchester in 1870 by C. Campbell. Later, considered scarce and mostly recorded as a larva (Mansbridge, 1940). From 1960, a gradual increase in range and the number of reports became apparent, such that by the 1990s it could be found in many lowland areas and more locally elsewhere. Usually recorded in low single figures, but over 20 were attracted to a garden light trap adjacent to mature deciduous woodland in Dalton, north Lancashire, on 15 August 2003 (S. Palmer).
■ **HABITAT** Woodland, gardens and mature hedgerows, wherever Ash grows.
■ **FOODPLANTS** Early larval records do not specify the foodplant, but Ash is assumed.
Single brooded. Very late records were noted in Hale at the end of October 2016.

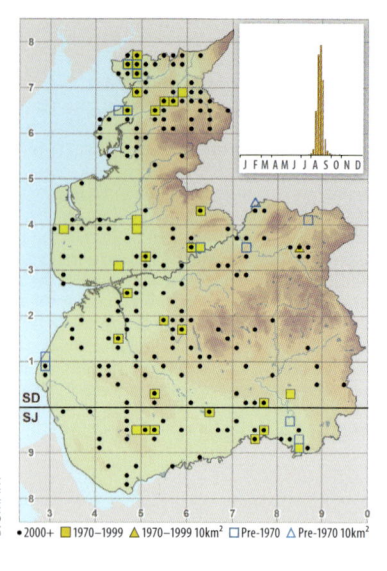

● 2000+   ■ 1970–1999   ▲ 1970–1999 10km²   □ Pre-1970   △ Pre-1970 10km²

73.220 Minor Shoulder-knot *Brachylomia viminalis*

■ **LANCS STATUS** Occasional / local resident   c.1857–2022
First noted as larva in Liverpool where sallows abundant (Gregson, 1857), but considered very local by Ellis (1890). Later, reported as more widely distributed by Mansbridge (1940). Found regularly at Leighton Moss since the 1960s, producing about one-third of the 208 county records. Seen much less frequently elsewhere at sites such as Chorley, Adlington, Gait Barrows, Ainsdale, Docker Moor, Mill Houses, Roeburndale and Mere Sands Wood. 19 came to light at Birkdale on 12 August 2012.
■ **HABITAT** Lowland wetlands, limestone scrub, wooded river valleys, moorland edge.
■ **FOODPLANTS** Willows.
Single brooded. Attracted to light in small numbers; observed nectaring on heather.

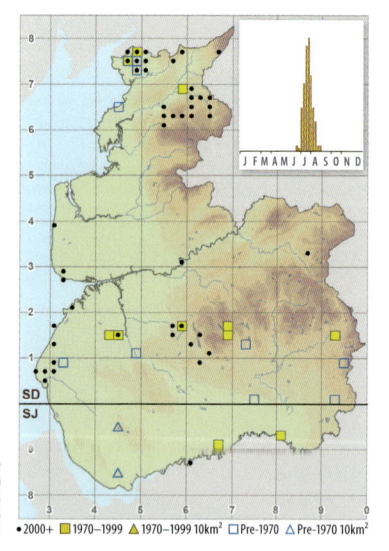

● 2000+   ■ 1970–1999   ▲ 1970–1999 10km²   □ Pre-1970   △ Pre-1970 10km²

73.221 Suspected *Parastichtis suspecta*

■ **LANCS STATUS** Occasional / widespread resident   c.1859–2022
First noted on Barton Moss c.1859 (J. Chappell) and considered scarce (Ellis, 1890). By the early 20th century, reported as occasionally common on the mosses by Mansbridge (1940). Has been noted more widely since the early 1990s, usually only in small numbers, and is regular at sites such as Docker Moor, Freshfield Dune Heath, Ellenbrook and Astley Moss. Particularly frequent at Flixton and Mere Sands Wood.
■ **HABITAT** Scattered birch or willow scrub on the edges of mosses, dunes, wetlands, moorland and, less frequently, on brownfield sites.
■ **FOODPLANTS** Birch.
Single brooded. Comes to light. Nectars on ragwort. Single late record 27 Sept. 2019.

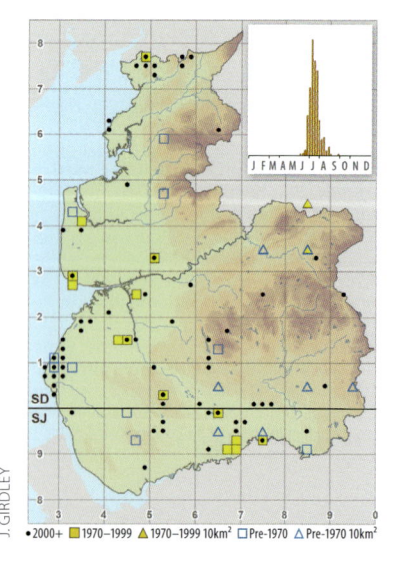

● 2000+   ■ 1970–1999   ▲ 1970–1999 10km²   □ Pre-1970   △ Pre-1970 10km²

The Moths of Lancashire

73.222 Dingy Shears *Apterogenum ypsillon*

■ **LANCS STATUS** Frequent / widespread resident   c.1859–2022

First recorded in Withington by J. Chappell. Also noted from Preston and Prestwich, with status considered uncommon (Ellis, 1890). Mansbridge (1940) gave additional locations as Lostock, Warrington and Crosby, noting the moth to be not often recorded but probably common among willows. Mainly a lowland species with few records from higher ground. Records mainly of singletons. Population appears healthy with the last five years providing increased total records of between 49 and 73 annually.

■ **HABITAT** Damp woodland, riverbanks.

■ **FOODPLANTS** Not recorded in Lancashire.

Single brooded. Comes to light. High count of eight, from Morecambe 2018 (C. Clay).

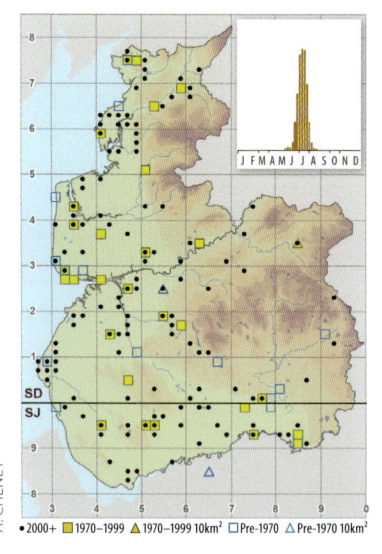

• 2000+   ☐ 1970–1999   △ 1970–1999 10km²   ☐ Pre-1970   △ Pre-1970 10km²

73.224 Merveille du Jour *Griposia aprilina*

■ **LANCS STATUS** Occasional / widespread resident   1844–2022

First recorded at Preston (Hodgkinson, 1844) and later at Liverpool, where noted as plentiful at sugar (Gregson, 1857). Thereafter, at many other sites, such as Burnley, Manchester, Parbold, Lytham, Lancaster, Silverdale, Slyne. The highest nightly counts, including 11 at Mill Houses and 8 at Lower Salter, are from upland sites. Occurrence in the south, and at more urban sites, is patchier and more sporadic, with approximately three-quarters of Lancashire records coming from VC60.

■ **HABITAT** Woodland, hedgerows, parks.

■ **FOODPLANTS** Unknown. Larva found in crack of unknown tree trunk at Hopwood, 2002. Single brooded. Comes to light. Larvae recorded from late May and June.

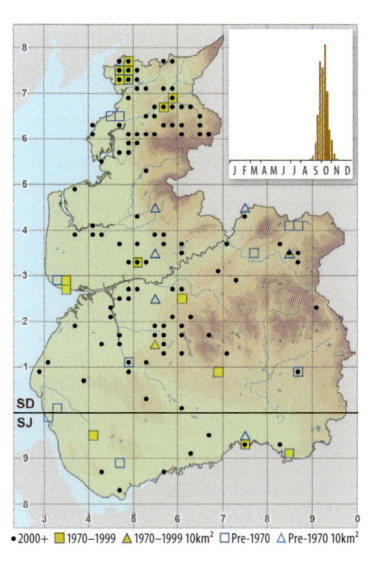

• 2000+   ☐ 1970–1999   △ 1970–1999 10km²   ☐ Pre-1970   △ Pre-1970 10km²

73.225 Brindled Green *Dryobotodes eremita*

■ **LANCS STATUS** Occasional / widespread resident   c.1857–2022

Reported in Liverpool by Gregson (1857), noting the attraction to sugar. Not recorded again until 1910 in Burnley, and 1913 in Silverdale. Noted at Formby and Lytham St Annes in the 1930s. Records remained scarce until the 1990s, when it started to appear in multiple locations annually. Since when, it has been regular at Claughton, Billinge, Mere Sands Wood, Green Bank and Lightfoot Green. Appears to be absent from the West Pennine Moors and much of the surrounding areas.

■ **HABITAT** Woodland, parks, gardens.

■ **FOODPLANTS** Not recorded in Lancashire.

Single brooded. Comes to light, with seven trapped at Dalton, 2006 (C.A. Darbyshire).

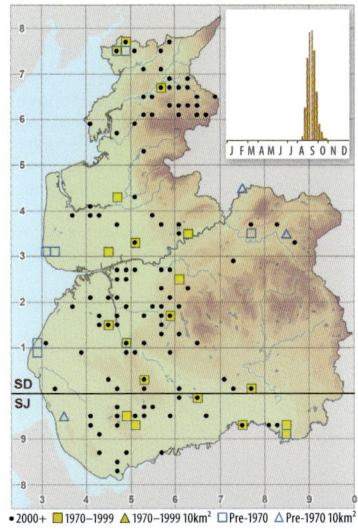

• 2000+   ☐ 1970–1999   △ 1970–1999 10km²   ☐ Pre-1970   △ Pre-1970 10km²

The Moths of Lancashire

NOCTUIDAE

73.228 Grey Chi *Antitype chi*

■ **LANCS STATUS** Occasional / widespread resident   1843–2022

First noted in Preston area by J.B. Hodgkinson in 1843 (Hodgkinson, 1854), with later 1800s records from Liverpool, Turton, Oldham and Middleton, nr. Rochdale. Regular throughout 20th century, most frequently from eastern and upland areas. Absent from Fylde and scarce on Sefton Coast. The species appears to be declining, particularly in VC59, with 2018-2022 records down from a decade earlier. Bred from early instar larva on Common Rock-rose at Warton Crag, 11 April 2022 (J. Patton).

■ **HABITAT** Moorland, limestone grassland.

■ **FOODPLANTS** Dyers Greenweed (Hodgkinson, 1880a), Common Rock-rose.

Single brooded. Comes readily to light. Larvae recorded from April to June.

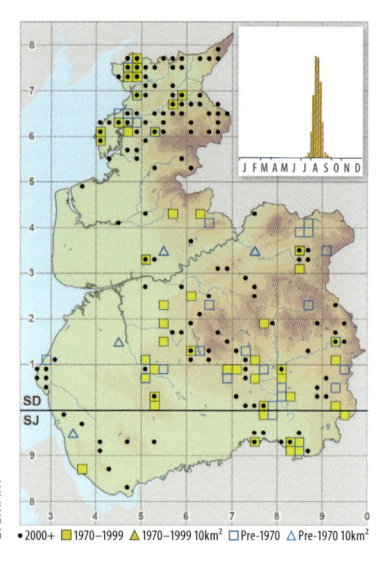

• 2000+   □ 1970–1999   △ 1970–1999 10km^2 □ Pre-1970   △ Pre-1970 10km^2

73.231 Deep-brown Dart *Aporophyla lutulenta** /
73.232 Northern Deep-brown Dart *Aporophyla lueneburgensis**

■ **LANCS STATUS** Scarce / local resident   c.1857–2021

Historically, all records were ascribed to Deep-brown Dart, with the Northern Deep-brown Dart considered a variety (var. *lueneburgensis*). This was the approach taken by Heath and Emmet (1983). However, Skinner (1984) 'followed the generally accepted opinion of most Continental authors' and treated the two as separate species.

At that time, it appeared that both species occurred in the county, and both in reasonable numbers, although with Deep-brown Dart largely absent from northern, upland sites. However, Randle *et al.* (2019), following discussion with the relevant County Moth Recorders, considered distribution of the pair, particularly in northern England, to be uncertain. Hence, Lancashire (and Yorkshire) records were omitted from distribution maps in the NMRS Atlas.

More recently, Boyes *et al.* (2021) suggested that whilst the two are separate species, the only British species is *A. lueneburgensis*, with initial DNA sequencing suggesting that *A. lutulenta* is confined to south-eastern Europe.

At present, both species remain on the UK list (Agassiz *et al.*, 2013), and so both are listed here. It may be that this status will need to be reassessed in the future.*

The Deep-brown Dart was first reported from the Liverpool district on gas lamps and at sugar (Gregson, 1857) and was considered by Ellis (1890) to be uncommon. A moth recorded as var. *lueneburgensis* by J.B. Hodgkinson at Ditton (Hodgkinson, 1881), would later be considered as Northern Deep-brown Dart. Mansbridge (1940) commented that this variety was the only form known within the county. Records of the species complex remained scarce until the mid-1990s when they started to increase, before declining again in the last 15 years. Of the 116 records we have of Northern Deep-brown Dart, all but 19 are from Flixton by K. McCabe between 1995 and 2004. Moths typical of Deep-brown Dart were then recorded at Flixton in 2005 and 2006, with none of either type at this location since. The last county record for Northern Deep-brown Dart is from Green Bank Farm, nr. Thrushgill in 2015, with the last Deep-brown Dart recorded at Worsthorne in 2021.

Arguably of more concern than ongoing taxonomic debate, is the increasing scarcity of the species complex, with only two records of each from the last decade, mirroring a national decline.

■ **HABITAT** Rough grassland, moorland, gardens.

■ **FOODPLANTS** Early stages and foodplant unrecorded in Lancashire.

Single brooded. Almost all records are of moths to light.

• 2000+   □ 1970–1999   △ 1970–1999 10km^2 □ Pre-1970   △ Pre-1970 10km^2

*In January 2024, *Aporophyla lutulenta* was removed from the British list (Agassiz *et al.*, 2024). All previous UK records are now considered to refer to *A. lueneburgensis* which has been allocated the vernacular name Deep-brown Dart.

The Moths of Lancashire

73.233 Black Rustic *Aporophyla nigra*

■ **LANCS STATUS** Frequent / widespread resident   c.1857–2022

Early records from Crosby, Lydiate and Speke (Gregson, 1857). Further 19th century records from Botany Bay, Worsley and Chat Moss. Increased numbers and extended range, evident with the advent of trapping, continued this century, e.g., in Preston, where first noted in 2000. However, some sign of declining numbers in last decade as 25 highest counts all preceded this period. Annual total in C. Cockbain's Hale garden fell from 114 moths in 2009 to 22 in 2022. A few upland records, e.g., at Mill Houses.

■ **HABITAT** Rough grassland, moorland, sand dunes, gardens and brownfield sites.

■ **FOODPLANTS** Sweet Vernal Grass. Larva recorded June 2009 by B. Leecy, Todmorden. Single brooded. Readily to light with high count of 37 at Orrell in 2004 by P.J. Alker.

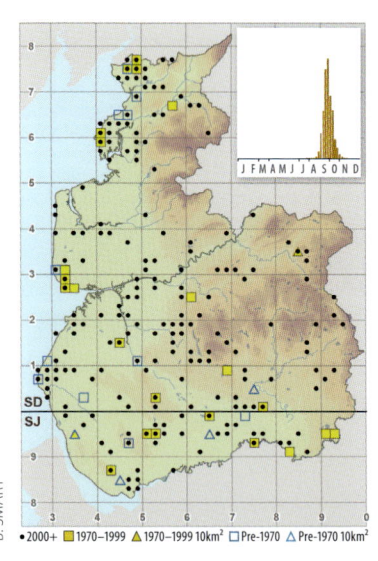

73.234 Brindled Ochre *Dasypolia templi*

■ **LANCS STATUS** Scarce / local resident   c.1857–2022

First recorded at Liverpool (Gregson, 1857), with other early records from Burnley, Manchester, Preston and Nelson. The first 21st century record was at Queens Park, St Helens by C. Davies in 2006, followed by others from Formby in 2013 and 2016. More recent records, from 2019 onwards, have come from a regular MV trap on moorland at Green Bank, upper Hindburndale (five autumn males, one spring female), with singletons at Tarnbrook (moorland) and Sunderland (coastal) in 2021.

■ **HABITAT** Moorland, sand dunes, rough grassland.

■ **FOODPLANTS** Not recorded in Lancashire.

Single brooded. Recorded at light. Early records to gas lamps.

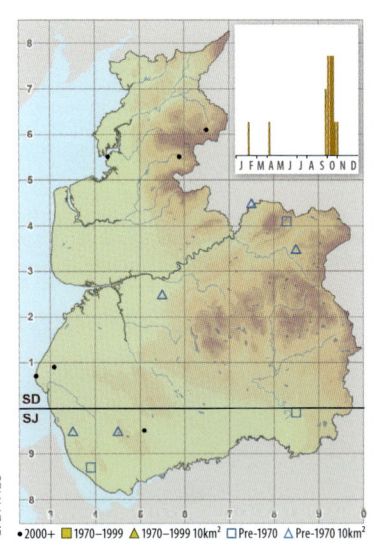

73.235 Feathered Ranunculus *Polymixis lichenea*

■ **LANCS STATUS** Occasional / local resident   1848–2022

Primarily a coastal species with inland records at Dolphinholme, Mere Sands Wood and Billinge presumably representing wanderers. First record was of a moth beaten from gorse nr. Lytham in September 1848 by Dr Nelson (Hodgkinson, 1849). 'May be taken on the coast where stonecrop grows' (Gregson, 1857). Ellis, Mansbridge and Britten described it as local, found only near the coast. Regular at Southport, Formby, Heysham, Carnforth, etc., although reduced counts suggest likely recent decline.

■ **HABITAT** Sand dunes, coastal grassland.

■ **FOODPLANTS** Although larva has been recorded, foodplant not noted.

Single brooded. Comes to light. Larva noted from March to May.

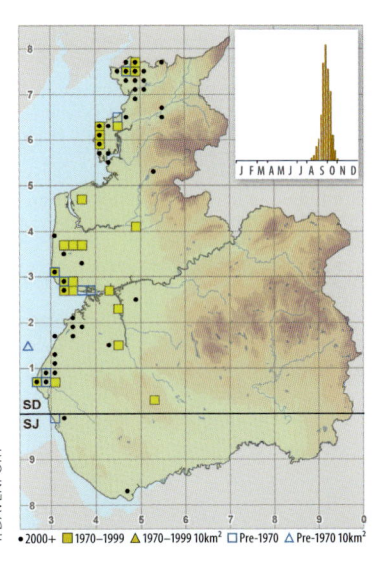

NOCTUIDAE

73.237 Large Ranunculus *Polymixis flavicincta*

■ **LANCS STATUS** Rare / vagrant   c.1857–2006

First reported from the Liverpool district, attracted to sugar at Allerton Hall, Woolton, with two more recorded between Ditton and Hale (Gregson, 1857). These records were noted by Ellis (1890) and Mansbridge (1940), without the addition of any further sightings. It was therefore quite a surprise when the moth was trapped at light on 26 September 2006 by G. & D. Atherton at Haydock. The lack of any further records suggests this was likely a vagrant, rather than a resident from a previously undetected local colony.

■ **HABITAT** Unknown.
■ **FOODPLANTS** Not recorded in Lancashire.

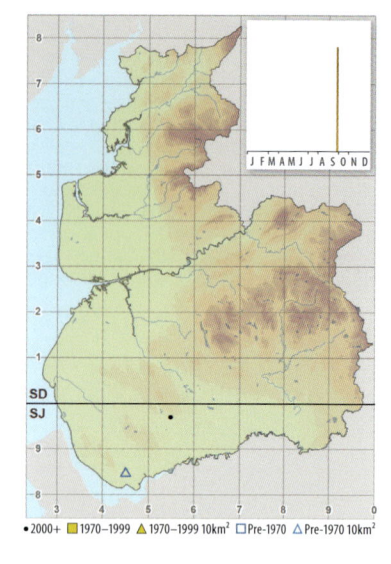

• 2000+   ▫ 1970–1999   ▵ 1970–1999 10km²   ▫ Pre-1970   △ Pre-1970 10km²

73.238 Dark Brocade *Mniotype adusta*

■ **LANCS STATUS** Occasional / local resident   c.1843–2022

Early records from Preston (Hodgkinson, 1844) and White Moss, Manchester (Edleston, 1844). Noted as abundant at Cheetham Hill 1846, and in Liverpool woods on sugar (Gregson, 1857). Mansbridge (1940) reported it to be 'common on mosses, occasional elsewhere.' Post-2000, considered ubiquitous, sometimes abundant during extensive northern moorland recording, e.g., upper Hindburndale (P.J. Marsh, pers. comm.). Recent VC59 records from Rishton, Horwich, Oldham and Formby.

■ **HABITAT** Moorland, upland grassland, sand dunes.
■ **FOODPLANTS** Not recorded in Lancashire.

Single brooded, with occasional August record. Comes to light.

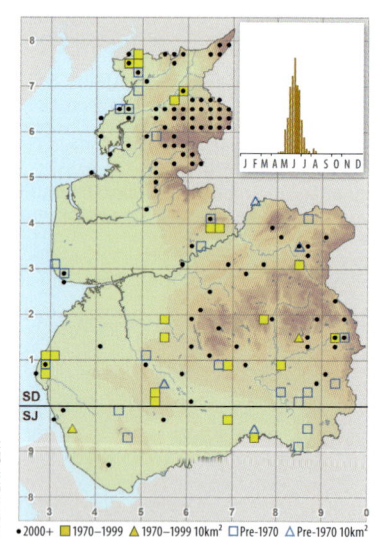

• 2000+   ▫ 1970–1999   ▵ 1970–1999 10km²   ▫ Pre-1970   △ Pre-1970 10km²

73.241 Pine Beauty *Panolis flammea*

■ **LANCS STATUS** Frequent / widespread resident   c.1857–2022

First noted from Liverpool district, 'where Scotch firs grow' (Gregson, 1857). Other early authors reported it as common among firs. Recorded from 1900-1950 at Nelson, Pendle Hill, Formby, Simonswood Moss, Appley Bridge and Longridge. Next, noted at Slyne 1964 by P.J. Marsh. Subsequent records at light have shown scattered distribution, although with concentration in the pine woods of the Sefton Coast; less so in the uplands around conifer plantations. A frequent visitor to some garden traps.

■ **HABITAT** Pine woods, isolated pines in gardens, parks, scrub, etc.
■ **FOODPLANTS** Pine. Feeding noted on cones, rather than needles, 2022 S.H. Hind.

Single brooded. Comes to light. Larvae from late April to July.

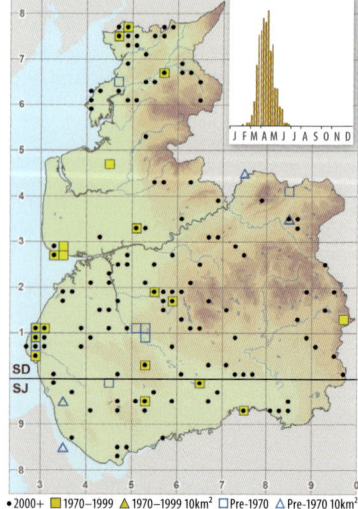

• 2000+   ▫ 1970–1999   ▵ 1970–1999 10km²   ▫ Pre-1970   △ Pre-1970 10km²

73.242 Clouded Drab *Orthosia incerta*

■ **LANCS STATUS** Abundant / ubiquitous resident   c.1857–2022
First reported from the Liverpool district, on sallows in bloom (Gregson, 1857). Noted by Ellis (1890) to be generally distributed. Mansbridge (1940) documented its status as abundant and general, noting many variations, and records from Simonswood Moss, Burnley and Warrington. The moth has subsequently been recorded in high numbers throughout the county in a variety of habitats; from the upland counts of 72 at Higher Tatham on 8 April 2004 and 24 May 2005 (P.J. Marsh), to the 70 at Mere Sands Wood (Kippax & Boydell) and at Ainsdale dunes (G. Jones), both in 2011. Whilst remaining abundant and extremely widely distributed, numbers appear to have declined over the last decade. The 1,822 records from 2018 to 2022 compare unfavourably to the 3,109 from 2008 to 2012.
■ **HABITAT** Woodland, hedgerows, scrub, parks, gardens, etc.
■ **FOODPLANTS** Blackthorn, Hawthorn, Alder, Plum, willow, birch, lime.
Single brooded. Comes readily to light and also to sallow blossom. Larvae have been recorded from May to early July.

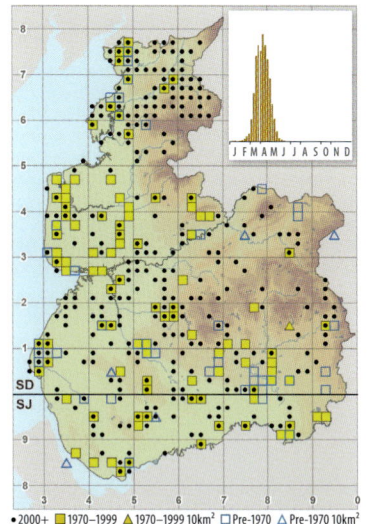

73.243 Blossom Underwing *Orthosia miniosa*

■ **LANCS STATUS** Scarce / local resident   c.1890–2022
First noted from Simonstone, nr. Burnley in the late 1800s by A.E. Wright, and from Silverdale in 1914. The status was given by Mansbridge (1940) as rare, with Britten (c.1950) reporting the moth to be local in Silverdale woods. It was further recorded at Formby in 1942, at Leighton Moss in 1968 and at Gait Barrows in 1983. Subsequent records have been mainly on the lowland limestone woodland of Morecambe Bay and occasionally in nearby gardens. Twice recorded from higher ground, at Higher Tatham in 2006 and at Thrushgill in 2021 (P.J. Marsh).
A presumed wanderer was trapped at light in Lancaster by I. Mitchell on 14 April 2022, the only record in the county for that year.
■ **HABITAT** Limestone woodland, upland woods.
■ **FOODPLANTS** Oak. Gregarious feeding of about 50 larvae on oak was noted at Gait Barrows on 9 May 2020 by J. Patton.
Single brooded. Comes to light. The highest adult count of seven were trapped by S. Willis at Leighton Moss on 20 April 1994.

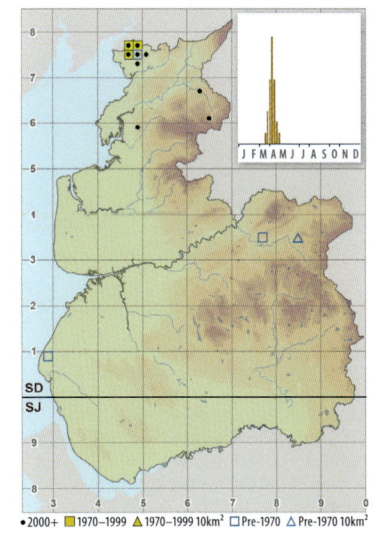

The Moths of Lancashire **537**

NOCTUIDAE

73.244 Common Quaker *Orthosia cerasi*

■ LANCS STATUS Abundant / ubiquitous resident   c.1857–2022

An extremely frequently recorded, early spring species. Early recorders noted the same, with Ellis (1890) describing it as the most common of the genus. Reported to be plentiful in Liverpool oak woods (Gregson, 1857). Other 19th century records from Oldham and Middleton, nr. Rochdale. Current distribution demonstrates a strong presence throughout upland and lowland areas, with highest count of 257 at Brinscall in 2011 (E. Langrish *et al.*). Since 2006, recorded on six occasions between Oct. - Dec.
■ HABITAT Woodland, hedgerows, scrub, parks, gardens, etc.
■ FOODPLANTS Oak, birch, willow, Alder, Bird Cherry, Hawthorn, Sycamore.
Single brooded. Comes readily to light. Larvae noted from May to July.

● 2000+   ▪ 1970–1999   ▲ 1970–1999 10km²   □ Pre-1970   △ Pre-1970 10km²

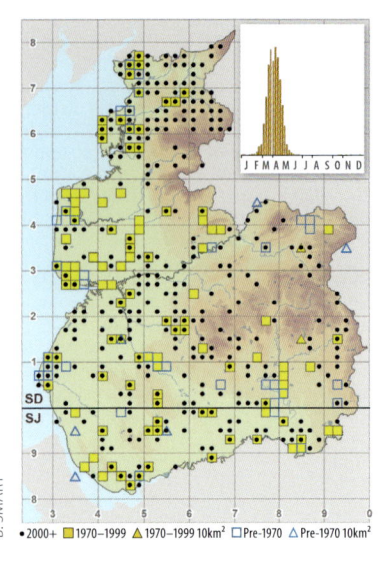

73.245 Small Quaker *Orthosia cruda*

■ LANCS STATUS Frequent / widespread resident   c.1857–2022

Noted in profusion in Liverpool woodlands (Gregson, 1857). Recorded at Middleton, nr. Rochdale by J. Taylor in 1870; later, from Silverdale, Ditton, Hollingworth Lake and Pendle Hill. Described by Ellis (1890) as generally distributed and by Mansbridge (1940) as abundant throughout. Remains particularly abundant in uplands, with 18 records of 200+ moths to light at Mill Houses; maximum 406 on 24 March 2014, by P.J. Marsh. Regular in gardens, with long series from Flixton, Heysham, Longton, etc.
■ HABITAT Woodland, hedgerows, scrub.
■ FOODPLANTS Oak, Silver Birch.
Single brooded. Comes to light. Larvae recorded in late April and May.

● 2000+   ▪ 1970–1999   ▲ 1970–1999 10km²   □ Pre-1970   △ Pre-1970 10km²

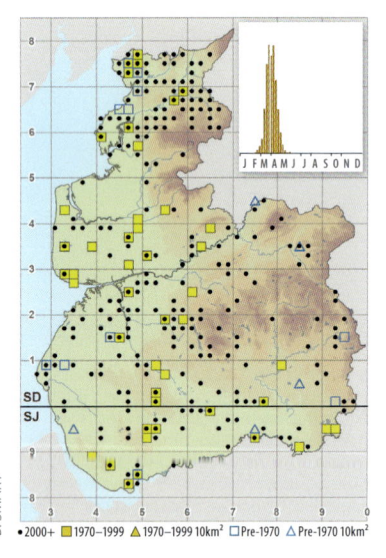

73.246 Lead-coloured Drab *Orthosia populeti*

■ LANCS STATUS Occasional / very local resident   c.1857–2022

One of the less commonly recorded members of the genus, with the moth first noted at Hale and Lydiate (Gregson, 1857) and other records to 1900 from Withington, Chat Moss, Bolton, Longridge, Prestwich, Preston and Lancaster. Not recorded again until 2002, when trapped at Chorlton (B. Smart). Subsequently, regular at Flixton and Hale. In the north, found among old Aspens at Herring Head Wood in 2014; thereafter, annual at the same site with a single-trap maximum of 24 in 2015 (P.J. Marsh).
■ HABITAT Woodlands rich in Aspen.
■ FOODPLANTS Not recorded in Lancashire.
Single brooded. Moths recorded at light.

● 2000+   ▪ 1970–1999   ▲ 1970–1999 10km²   □ Pre-1970   △ Pre-1970 10km²

538   *The Moths of Lancashire*

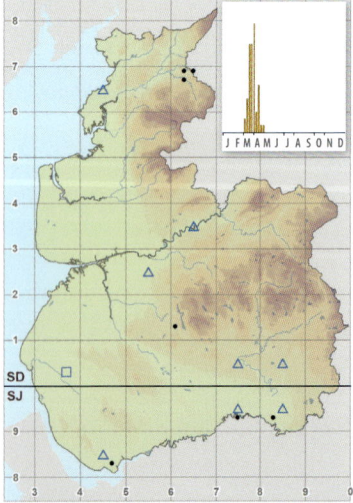

73.247 Powdered Quaker *Orthosia gracilis*

■ **LANCS STATUS** Frequent / widespread resident   c.1843–2022
First recorded from the Preston area (Hodgkinson, 1844), then at Crosby (Gregson, 1857). Noted as common on sand-hills and also generally distributed (Mansbridge, 1940). Other early records from Chat Moss, Huyton, Lytham St Annes and Silverdale. Whilst range has extended into eastern and upland areas of the county since 2000, abundance appears to have declined, with fewer annual records than 15 years ago and mostly singletons. Highest count of 41, from Heysham in 2001 (P.J. Marsh).
■ **HABITAT** Scrub, grassland, damp woodland, sand dunes.
■ **FOODPLANTS** Bramble, loosestrife, Creeping Willow, Mugwort, Common Fleabane. Single brooded. Comes to light. Larvae have been recorded during May and June.

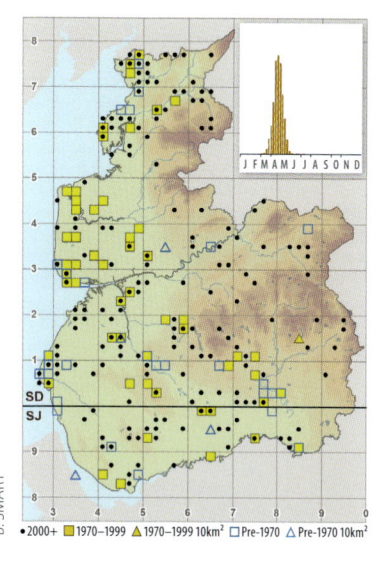

73.248 Northern Drab *Orthosia opima*

■ **LANCS STATUS** Scarce / very local resident   c.1857–2011
Reported on sallow bloom on the coast, with Crosby noted as the best site for the moth (Gregson, 1857). Noted in early 20th century at Lancaster and Lytham St Annes. Later, recorded at Pendle Hill in 1940, Formby 1950 and Great Harwood 1950. Although never common, there were 13 records during 1970s and 1980s, mainly from Sefton and Fylde coasts, with an outlier from Bolton in 1977. Only recorded twice since, from Belmont, nr. Bolton in 2000 and Ravenmeols, Formby in 2011.
■ **HABITAT** Open grassland, sand dunes.
■ **FOODPLANTS** Bred from larva at Formby 1951, although foodplant unrecorded. Single brooded. Moth recorded on Creeping Willow catkins at Fairhaven in 1976.

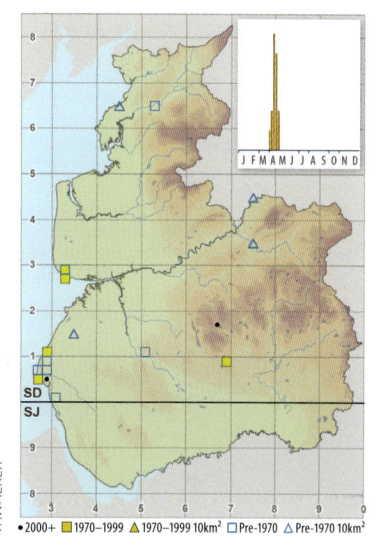

73.249 Hebrew Character *Orthosia gothica*

■ **LANCS STATUS** Abundant / ubiquitous resident   c.1857–2022
First reported from Liverpool by Gregson (1857). Early authors noted the moth to be common and generally distributed. Widely and frequently recorded since the 1960s, including in upland areas, e.g., 104 at Mill Houses on 18 March 2014. Comes readily to light, with a high count of 129 at Brinscall, 25 March 2011 by E. Langrish, A. Barker and S. Priestley. Numbers of records remain high, although with reduced counts.
■ **HABITAT** Woodland, hedgerows, moorland, parks, gardens, etc.
■ **FOODPLANTS** Hawthorn, sallow, Honeysuckle, Viper's-bugloss, Rowan, Ivy, Dandelion, Creeping Thistle. Larval feeding noted late April to July; once in August. Single brooded. The few autumn records likely represent early emergence.

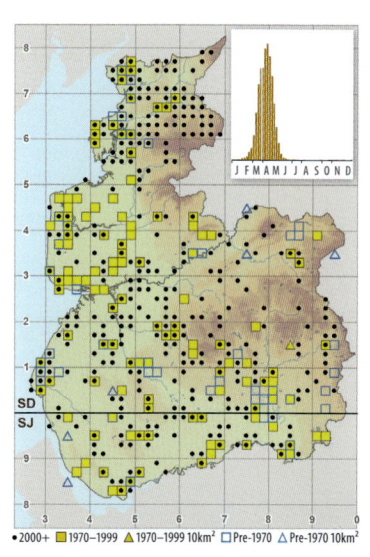

The Moths of Lancashire   **539**

73.250 Twin-spotted Quaker *Anorthoa munda*

■ **LANCS STATUS** Frequent / widespread resident   c.1843–2022

First reported from Preston area (Hodgkinson, 1844). Ellis and Mansbridge described the moth as local and not common. Britten (c.1950) reported it as common in lowland areas, but only noted twice in East Lancashire. As the map shows, its range has increased in last 25 years, including in eastern parts. Upland records include Green Bank and Docker Moor. However, remains more frequent in lowland areas, with many records from Mere Sands Wood, Heysham N.R., Flixton, Hale, Preston, etc.

■ **HABITAT** Woodland, hedgerows, parks.

■ **FOODPLANTS** Birch, Bramble, Grey Willow, Ash, poplar, Alder Buckthorn.

Single brooded. Comes to light. High count 140 at Ainsdale 2011. Larvae May to June.

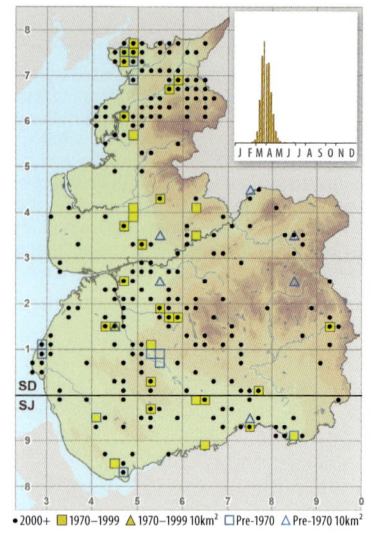

• 2000+   □ 1970–1999   ▲ 1970–1999 10km²   □ Pre-1970   △ Pre-1970 10km²

73.252 Hedge Rustic *Tholera cespitis*

■ **LANCS STATUS** Scarce / local resident   c.1857–2022

Recorded at Crosby in mid-1850s (Gregson, 1857), with other 19th century records from Withington, Preston, Carnforth and Knowsley. Although declining nationally, it appears at least stable here, with 35 records in last five years, compared to eleven in the same period a decade earlier. Nevertheless, reliant on upland sites, in particular Green Bank and Docker Moor, with high count of 20 at the latter in 2020. Noted at a few lowland sites in 2022: Chorlton, St Helens, Great Sankey, Formby, Lytham.

■ **HABITAT** Rough grassland, moorland.

■ **FOODPLANTS** Not recorded in Lancashire.

Single brooded. Almost all records at light.

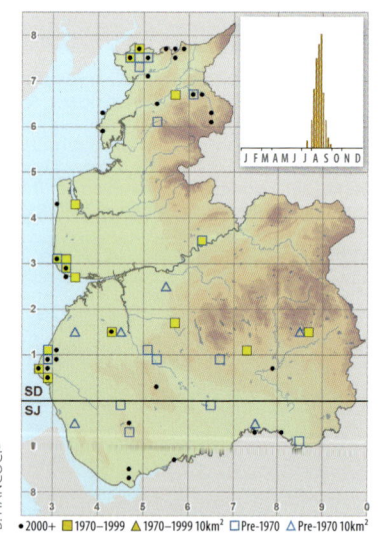

• 2000+   □ 1970–1999   ▲ 1970–1999 10km²   □ Pre-1970   △ Pre-1970 10km²

73.253 Feathered Gothic *Tholera decimalis*

■ **LANCS STATUS** Occasional / local resident   c.1843–2022

First reported from the Preston district (Hodgkinson, 1844); later, from old grasslands in Liverpool area, and at light (Gregson, 1857). Reported as common and generally distributed (Ellis, 1890), with Mansbridge (1940) stating fairly common. Other pre-1970 records at Formby, Didsbury, Slyne, etc. Numbers of records currently stable, but mainly of singletons. Possibly reduced abundance in the last decade as all double figure counts occurred prior to this. Appears increasingly local in VC59.

■ **HABITAT** Rough grassland.

■ **FOODPLANTS** Not recorded in Lancashire.

Single brooded. Comes to light. Highest count of 40 at Formby 2005 (M. Hammond).

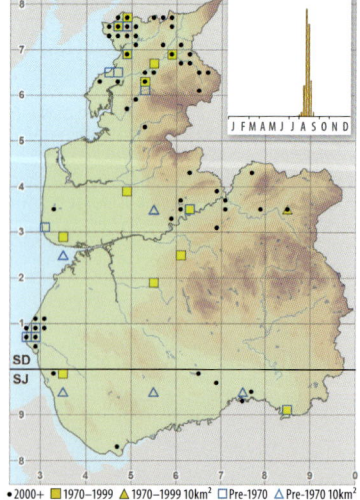

• 2000+   □ 1970–1999   ▲ 1970–1999 10km²   □ Pre-1970   △ Pre-1970 10km²

540 *The Moths of Lancashire*

73.254 Antler Moth *Crapteryx graminis*

■ **LANCS STATUS** Frequent / widespread resident   c.1857–2022
First reported from Liverpool district (Gregson, 1857), with abundance reported by other early authors. A.E. Wright noted that larvae could be swept in their 1000s from the moors above Rossendale in 1917 and Burnley in 1918. Very widely recorded throughout from 1960s onwards with increasing use of light trapping. The first record to MV light was 1949 at Wrightington by A. Fairhurst. Whilst widespread, the highest counts are on moorland, with approx. 200 at Leck Fell on 6 August 2006 (P.J. Marsh).
■ **HABITAT** Acidic grassland on moorland and open countryside.
■ **FOODPLANTS** Not noted in the county. Larvae recorded during May.
Single brooded. Comes to light. Occasionally noted during day, nectaring on ragwort.

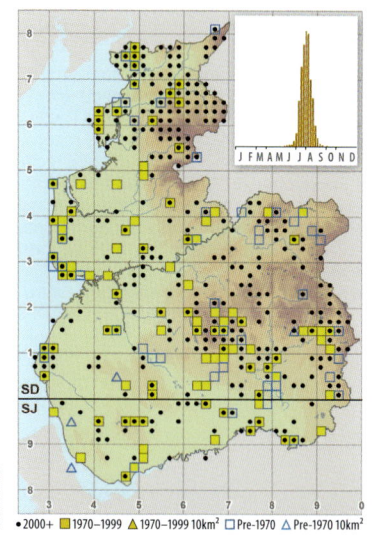

73.255 Nutmeg *Anarta trifolii*

■ **LANCS STATUS** Occasional / widespread resident   1940–2022
A relatively recent addition to the county's fauna, added in 1940, when recorded at Lytham St Annes by A. Watson. Further recorded at Wrightington in 1949, Formby 1951 and Crosby 1951. Evidence of further range expansion came in the next few decades with records from Silverdale, Hoghton, Prestwich, Croxteth and Warton, nr. Carnforth. 132 records from 2022 confirm moth to be well established, primarily at lowland sites. Frequent at light with maximum of 16 at Singleton 2010 by B. Dyson.
■ **HABITAT** Coastal grassland, brownfield sites, recently disturbed ground.
■ **FOODPLANTS** Orache.
Single brooded. Larva recorded in September 2017 at Sunderland Point by J. Patton.

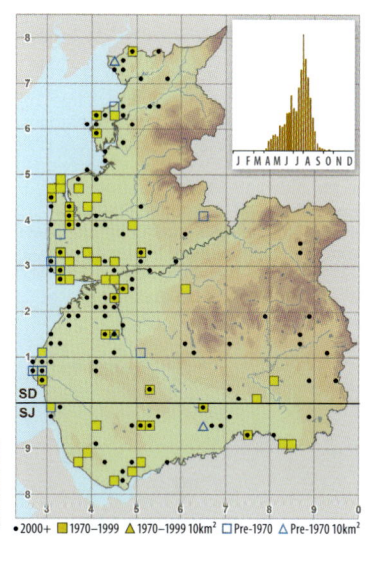

73.257 Beautiful Yellow Underwing *Anarta myrtilli*

■ **LANCS STATUS** Occasional / local resident   c.1843–2022
First reported from Preston area (Hodgkinson, 1844). R.S. Edleston noted abundance on Chat Moss in 1846, with Gregson (1857) similarly reporting abundance on heaths and mosses. Early 20th century records from Burnley, Pendle Hill and Freshfield. The species appears to have declined in the last decade. Most recently noted at Gull Moss, Briercliffe, Ainsdale and Winmarleigh Moss in 2020, and at Astley Moss in 2022.
■ **HABITAT** Moorland, mosses, heaths.
■ **FOODPLANTS** Heather.
Probably two overlapping generations. Uncommon at light, flying during the day. Larvae recorded April and June to October; often swept from Heather.

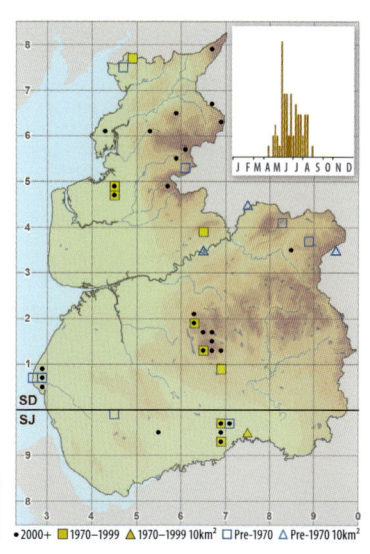

NOCTUIDAE

73.259 Pale Shining Brown *Polia bombycina*

■ LANCS STATUS Considered Extinct   c.1857

C.S. Gregson (1857) documented two records of this species from the mid-19th century, both on valerian flowers. The first was at Hale, the other from Lydiate. These remain the only Lancashire records of this species. Unrecorded in Cheshire and Cumbria with the last Yorkshire record in 1971 (yorkshiremoths.co.uk, accessed 1 September 2023). The chances of finding the moth again in Lancashire are extremely slim, as it has undergone massive decline and range retraction; formerly known from Scotland and Northumberland in the early 19th century, but now known only from a very few sites in southern England.
■ HABITAT Unknown.
■ FOODPLANTS Not recorded in Lancashire.

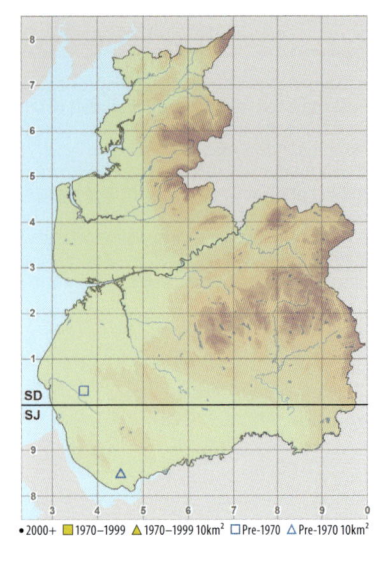

73.260 Silvery Arches *Polia hepatica*

■ LANCS STATUS Considered Extinct   1928–c.1950

Recorded from Silverdale in 1928 by C.E. Stott and in the 1950s by S. Coxey, and noted in the card indexes maintained by H. Britten and N.L. Birkett. There is an earlier report of a larva found by C.S. Gregson, J.B. Hodgkinson and W. Ashworth at Withnall Birch Clough, nr. Chorley on 20 April 1856 (Gregson, 1856). However, as the only potential record from VC59, and with no indication given of what it was feeding on or whether it was bred, this is considered unverified. Whilst still present on adjacent mosses in Cumbria, the moth appears to be extinct within Lancashire, and indeed has undergone a severe decline nationally (Randle *et al.*, 2019). The last Yorkshire and Cheshire records were from 2011 and 1975 respectively (yorkshiremoths.co.uk; cheshire-chart-maps.co.uk, both accessed 1 September 2023).
■ HABITAT Woodland.
■ FOODPLANTS Not recorded in Lancashire.
Difficult to determine from the few Lancashire records, but nationally, single brooded.

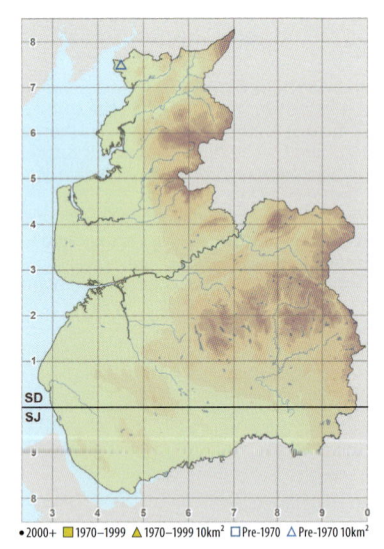

73.261 Grey Arches *Polia nebulosa*

■ LANCS STATUS Occasional / local resident   c.1843–2022

First reported from the Preston area (Hodgkinson, 1844). Noted by Gregson (1857) to be plentiful in woods at sugar. Status given by Ellis (1890) as fairly common and generally distributed, whereas Mansbridge (1940) found it to be much more local. Far more widely reported since the mid-1990s. Remains primarily a lowland species with multiple records from Astley Moss, Rainford, Flixton, Ainsdale, Leighton Moss, Gait Barrows, Yealand Conyers, etc. Absent from Fylde and much of east Lancs.
■ HABITAT Woodland.
■ FOODPLANTS Larva noted at sugar at Ainsdale on 4 March 2022, foodplant unknown. Single brooded. Comes readily to light. High count of ten from Warton Crag 2013.

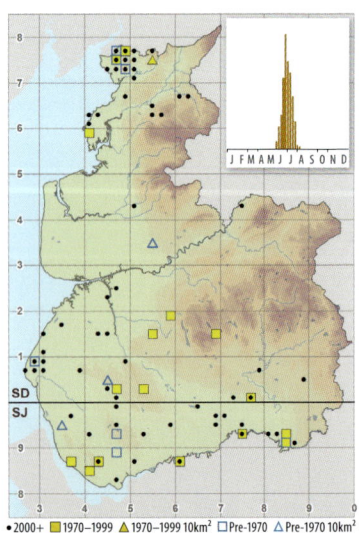

73.264 Pale-shouldered Brocade *Lacanobia thalassina*

■ **LANCS STATUS** Frequent / widespread resident   1844–2022
First recorded on 25 May 1844 at White Moss, Manchester (Edleston, 1844). Later, noted as abundant in Liverpool (Gregson, 1857). Other early authors reported it to be common and generally distributed. Early light trapping noted the moth at Barrowford 1951, Morecambe 1959, Slyne 1965. Has declined over the last 20 years, particularly in the south. Formerly common at Flixton, but unrecorded since 2020. Appears more stable at upland sites, with a high count of eleven at Tarnbrook in 2020 (R. Foster).
■ **HABITAT** Moorland, woodland, rough grassland.
■ **FOODPLANTS** Not recorded in Lancashire.
Single brooded. Almost all records are of moths to light.

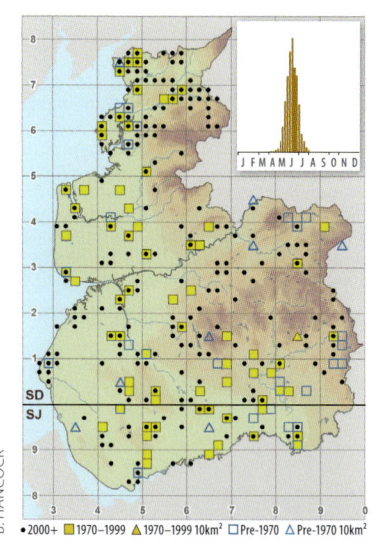

73.265 Beautiful Brocade *Lacanobia contigua*

■ **LANCS STATUS** Considered Extinct   c.1889
There is a single accepted Lancashire record, by J. Chappell from Barlow Moor in south Manchester in the late 1800s (Ellis, 1890), with the author confirming the moth's status as rare. The Britten card index (c.1950) includes a further record from Silverdale, but without date or recorder, and is considered unconfirmed. The species is unrecorded in Cheshire and scarce in Yorkshire and Cumbria, with the last confirmed record in the latter county at Allithwaite Quarry, nr. Grange-over-Sands in 2017 (S. Colgate, pers. comm.). Nationally, this is a declining species (Randle *et al.*, 2019).
■ **HABITAT** Unknown.
■ **FOODPLANTS** Not recorded in Lancashire.
Misidentifications have been a problem with this species, both of historical specimens and in more recent years.

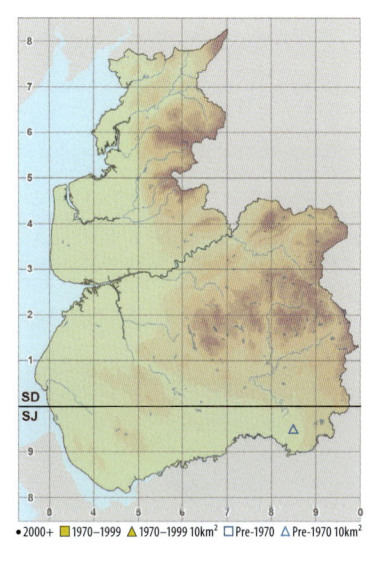

73.266 Dog's Tooth *Lacanobia suasa*

■ **LANCS STATUS** Frequent / widespread resident   1844–2022
First recorded at Chorlton, Manchester (Edleston, 1844). Later, reported as being taken in the Liverpool district at sugar (Gregson, 1857), and at Lancaster and Stretford (Ellis, 1890). Into the 20th century, noted from Silverdale and Lytham St Annes, and regarded as local and uncommon (Mansbridge, 1940). Seems to be increasing in number with approximately half the 2,300 plus records coming since 2015. Primarily recorded in lowland areas; particularly in coastal and estuarine regions.
■ **HABITAT** Coastal grassland, scrub, salt marshes.
■ **FOODPLANTS** Large larva at base of Sea Rush tussock July 2002; feeding not observed.
Double brooded. Comes to light. Larvae recorded in July and September.

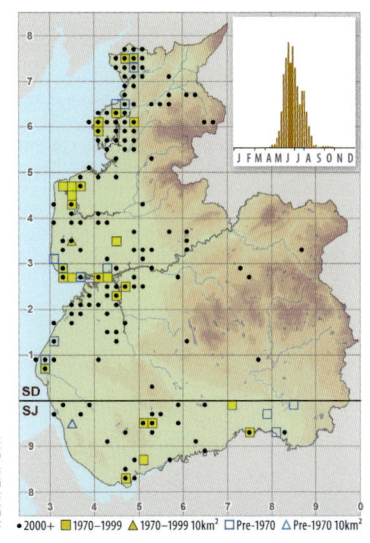

The Moths of Lancashire   543

NOCTUIDAE

73.267 Bright-line Brown-eye *Lacanobia oleracea*

■ **LANCS STATUS** Abundant / ubiquitous resident   c.1857–2022

Recorded throughout the county in numbers, in lowland and upland areas, using a wide variety of herbaceous plants. First noted in the Liverpool district by Gregson (1857), later noting it to be found in profusion in weedy gardens. Other early authors noted frequency to be common or abundant. Whilst numbers of records have increased this century, abundance seems to have declined in the last 15 years, with the highest counts prior to this, e.g., 58 at Mere Sands Wood, 6 July 2000 (D. Rigby).
■ **HABITAT** Scrub, brownfield sites, cultivated land, saltmarshes, gardens, etc.
■ **FOODPLANTS** Bindweed, skullcap, orache, tomato plants, Ash.
Single brooded. Comes readily to light. Larvae found from July to early October.

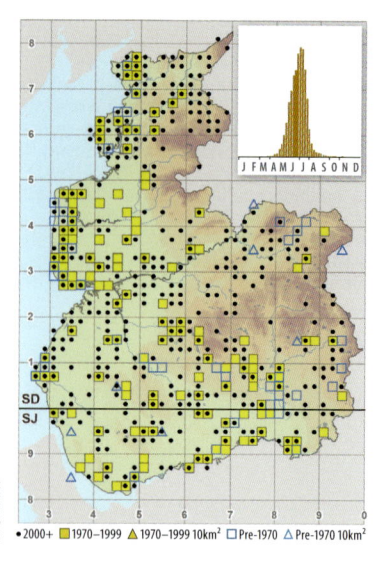

• 2000+   ▪ 1970–1999   ▲ 1970–1999 10km²   ▫ Pre-1970   △ Pre-1970 10km²

73.270 Dot Moth *Melanchra persicariae*

■ **LANCS STATUS** Frequent / widespread resident   c.1853–2022

First recorded by N. Cooke from Warrington (Byerley, 1854), with other early records from Liverpool, Ditton, Manchester and Salford. Recorded at light in Morecambe 1959, one of 20 similar records from the same site in two years, showing the revolutionary impact light trapping had on moth recording. Distribution revealed to be primarily in lowland areas, although with a few records from Pennine foothills.
■ **HABITAT** Rough grassland, scrub, woodland, gardens.
■ **FOODPLANTS** Clematis, birch, Alder Buckthorn, rock-rose, bistort, Hollyhock, willow, Hawthorn, Blackthorn, dock and Pelargonium.
Single brooded. Comes to light. Larva from late July to October; once in early January.

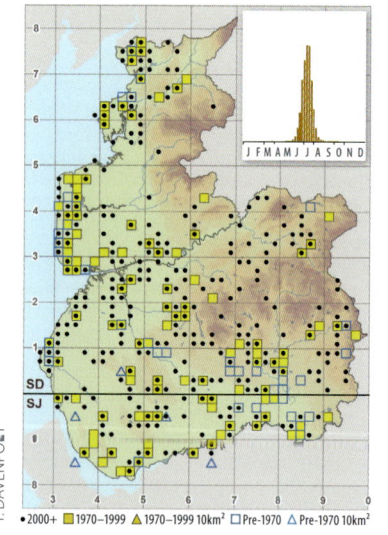

• 2000+   ▪ 1970–1999   ▲ 1970–1999 10km²   ▫ Pre-1970   △ Pre-1970 10km²

73.271 Broom Moth *Ceramica pisi*

■ **LANCS STATUS** Frequent / widespread resident   1846–2022

First recorded from Chat Moss in July 1846 (Edleston, 1846), and thereafter from Liverpool, Oldham, Rixton, etc. Noted by Gregson (1857) as plentiful on the mosses, with other early authors reporting common and well distributed. Most early records were from lowland areas, with the first upland records from Burnley in 1910, Pendle Hill 1940 and Nelson 1961. In recent years, these moorland sites have become the most reliable locations, with the highest daily counts, e.g., 305 at Thrushgill in 2016.
■ **HABITAT** Mosses, heaths, moorland, woodland.
■ **FOODPLANTS** Bilberry, Heather, Soft Rush, Yellow Flag, Common Ragwort.
Single brooded. Comes to light. Larvae recorded July to October.

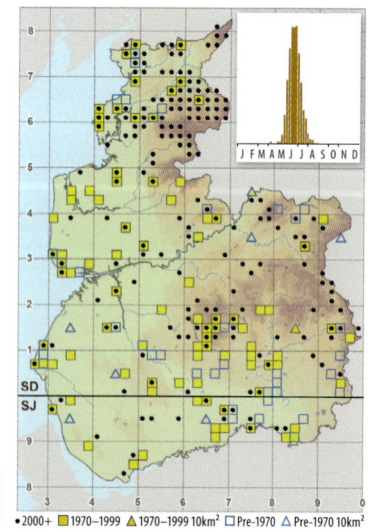

• 2000+   ▪ 1970–1999   ▲ 1970–1999 10km²   ▫ Pre-1970   △ Pre-1970 10km²

544   The Moths of Lancashire

73.272 Glaucous Shears *Papestra biren*

■ **LANCS STATUS** Occasional / local resident   1846–2022

First noted at Cheetham Hill (Edleston, 1846), with other early records from lowland mosses at Chat Moss and Simonswood Moss. Also, from Longridge, Withnell, Chorley, with status reported as rather local (Ellis, 1890). Described as 'common on the moors' (Britten c.1950). Post-2000 recording confirms status to be ubiquitous on moorland. Also, liable to drift to lowland traps in easterlies, e.g., Heysham, Preston, Silverdale, Freshfield, Longton. No records from VC59 mosses for 80 years.
■ **HABITAT** Moorland, heathland.
■ **FOODPLANTS** Not recorded in Lancashire.
Single brooded. Comes to light. Highest count of 31 from Thrushgill, 24 May 2019.

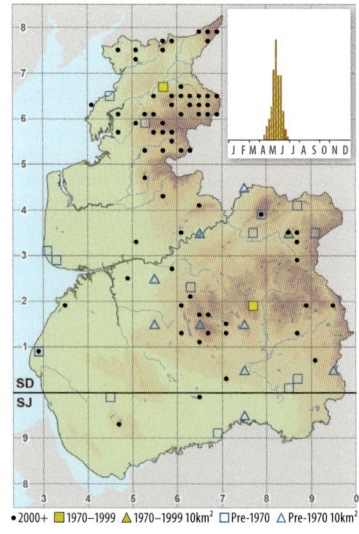

73.273 Shears *Hada plebeja*

■ **LANCS STATUS** Occasional / local resident   c.1857–2022

Noted to be plentiful on tree trunks in Liverpool district (Gregson, 1857), with other pre-1920 records from Oldham, Burnley, Ainsdale, Southport and Silverdale. Early light trapping saw records from Lytham St Annes, Slyne and Leighton Moss. Other than a few Wigan records, the moth now appears absent from Greater Manchester. The main concentrations of records are from Sefton Coast and lowland limestone of Morecambe Bay, with smaller numbers in the West Pennines and Bowland Fells.
■ **HABITAT** Open grassland, sand dunes, limestone grassland, moorland.
■ **FOODPLANTS** Although bred from larva in 1949 by A. Fairhurst, foodplant unrecorded.
Single brooded. Comes readily to light. Larvae in June and early July.

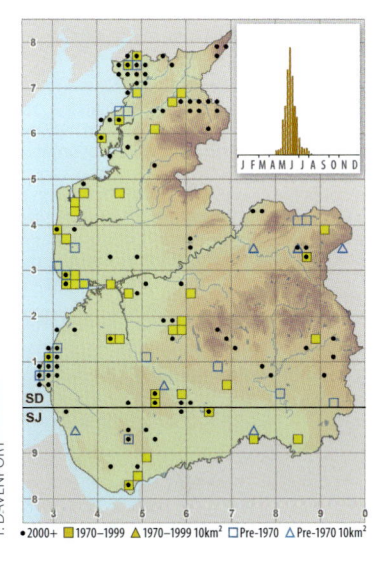

73.274 Cabbage Moth *Mamestra brassicae*

■ **LANCS STATUS** Frequent / widespread resident   c.1857–2022

First noted in Liverpool, where abundant (Gregson, 1857). Other early authors reported abundance everywhere. Found nectaring on heather at Turton 1880 by J.W. Baldwin and on the roof of World Museum in 1920, per E.G. Hancock. Favours lowland habitats. Largely absent from the fells, although it occurs in some of the districts beneath the Pennines, e.g., Brierfield, Rishton, and once at Docker Moor. Numbers remain high but have significantly declined this century. Comes to light.
■ **HABITAT** Grassland, scrub, woodland edges, gardens, etc.
■ **FOODPLANTS** Dock, unknown crucifer (probably Hairy Bittercress).
At least two overlapping broods. Larvae June to July and September to October.

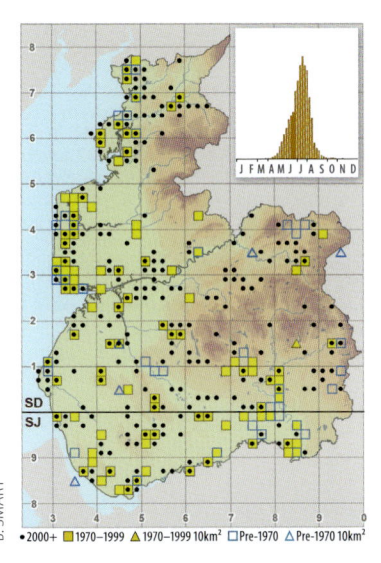

The Moths of Lancashire **545**

NOCTUIDAE

73.275 White Colon *Sideridis turbida*

■ **LANCS STATUS** Occasional / very local resident   c.1857–2022

First reported from Liverpool district, 'freely under banks on the sand hills' (Gregson, 1857). Larvae found at Fairhaven dunes in August 1914, common on Sea Plantain. Further records from Lytham St Annes 1939, Formby 1950, Morecambe 1960 and Heysham 1991. Restricted to coastal locations. Whilst a fifth of all records have been from 2020 onwards, these are now largely confined to the Sefton Coast. The only recent exception was from Lytham in 2020, the first Fylde record since 2009.

■ **HABITAT** Coastal grassland, sand dunes, salt marshes.
■ **FOODPLANTS** Sea Plantain.

Single brooded, possibly with small second generation. Comes to light.

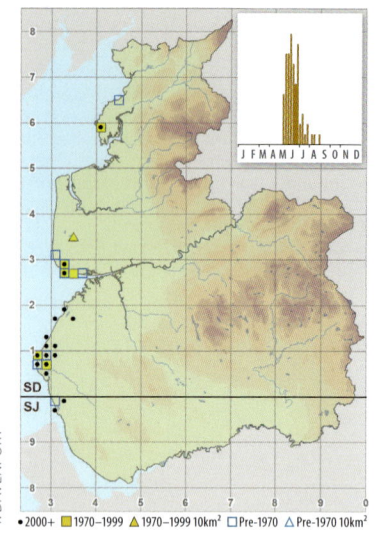

73.276 Campion *Sideridis rivularis*

■ **LANCS STATUS** Frequent / widespread resident   c.1843–2022

First recorded at Preston (Hodgkinson, 1844), with other 19th century records from Oldham, Liverpool and Middleton, nr. Rochdale. Noted by Mansbridge (1940) to be generally distributed, but less common than the Lychnis. Current distribution confirms presence in most lowland and upland areas, including six Docker Moor records between 2015 and 2021. As with many other species, numbers of records have held up but high counts have declined from a peak at the start of this century.

■ **HABITAT** Coastal grassland, waste ground, woodland rides, gardens.
■ **FOODPLANTS** White Campion. Feeds in the seed-heads.

Single brooded. Comes readily to light. Larvae recorded in July.

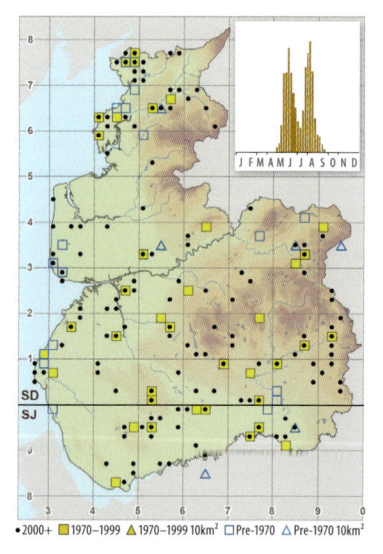

73.277 Bordered Gothic *Sideridis reticulata*

■ **LANCS STATUS** Considered Extinct   c.1857–1912

The first report of this species was by C.S. Gregson, of 'a single specimen taken by myself between Rainhill and Ditton, in July' (Gregson, 1857). It was next recorded by C. Stott from Fleetwood in 1892, and then in 1907 from Formby. The recorder of the latter is unclear, with initials given as B.S., possibly referring to B.B. Snell. The final record, from 1912, was of two moths attracted to light at Ainsdale sand dunes, recorded by W.A. Tyerman. The species has been recorded from Cheshire and Yorkshire, with the most recent records being from 1912 and 1979 respectively (cheshire-chart-maps.co.uk; yorkshiremoths.co.uk, both accessed 2 September 2023). The moth is now extinct as a resident in the UK, only occurring as an occasional migrant to southern England (Randle *et al.*, 2019).

■ **HABITAT** Coastal grasslands.
■ **FOODPLANTS** Not recorded in Lancashire.

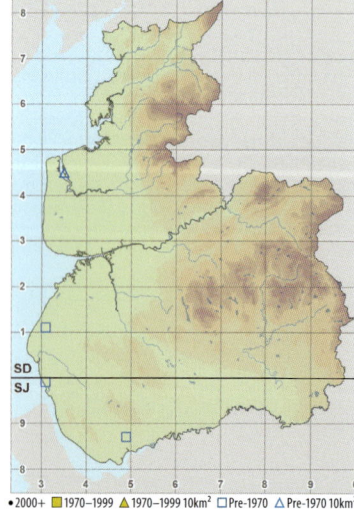

73.279 Broad-barred White *Hecatera bicolorata*

■ **LANCS STATUS** Occasional / local resident   c.1885–2022
First noted by F.N. Pierce and J.W. Ellis in the Crosby sand-hills (Ellis, 1890) and then from Carnforth, Chorlton, Formby, Ditton and Lytham St Annes by 1940. Increased light trapping from mid-1950s onwards confirmed wide distribution in lowland areas, particularly coastal, but also largely absent from upland areas. Almost all counts are of singletons with highest count of six from Heysham N.R. 29 June 2002 (P.J. Marsh). Numbers of records have declined from peaks in the 1990s to early 2000s.
■ **HABITAT** Disturbed rough grassland, fixed sand dunes, sandy soils.
■ **FOODPLANTS** Not recorded in Lancashire.
Single brooded. Comes readily to light.

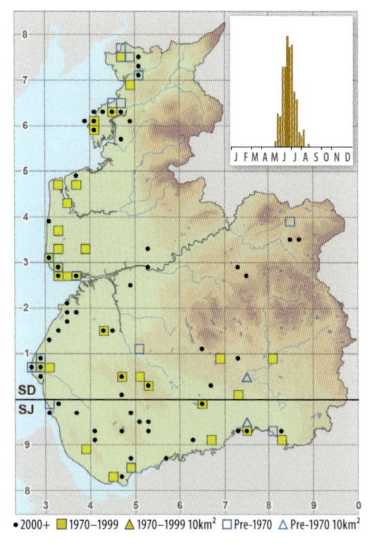

● 2000+   ■ 1970–1999   ▲ 1970–1999 10km²   □ Pre-1970   △ Pre-1970 10km²

73.280 Small Ranunculus *Hecatera dysodea*

■ **LANCS STATUS** Occasional / local resident   2005–2023
A very recent arrival in the county, and one that is continuing to extend its range north and east since the first record at Waterloo, nr. Crosby on 25 June 2005 by R. Ayres. Further recorded at St Helens and Crosby in July 2005. Of the 63 records to date, the only VC60 report was at Carleton Crematorium in 2012. Recorded in south Manchester since 2021; from Fallowfield, Victoria Park, Chorlton and Flixton. A 2022 Briercliffe record, the first from upland parts, suggests widening distribution.
■ **HABITAT** Waste ground, gardens, allotments.
■ **FOODPLANTS** Prickly Lettuce. Larvae noted in Chorlton and Salford, August 2023.
Single brooded. Most records are of moths to light.

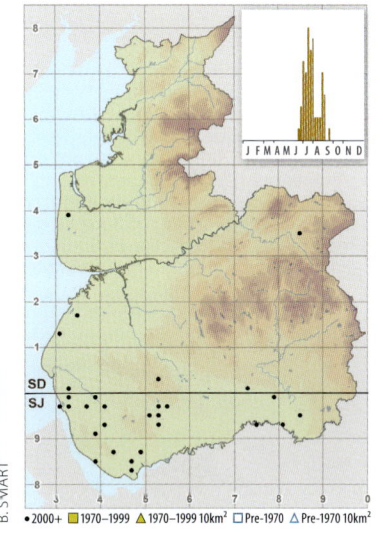

● 2000+   ■ 1970–1999   ▲ 1970–1999 10km²   □ Pre-1970   △ Pre-1970 10km²

73.281 Lychnis *Hadena bicruris*

■ **LANCS STATUS** Frequent / widespread resident   c.1843–2022
First recorded in the Preston area by J.B. Hodgkinson (1844). It was later noted to be plentiful around Liverpool, where 'catchfly' (Bladder Campion) occurs (Gregson, 1857). Other early authors described it as common and general, with Britten (c.1950) noting it plentiful on the coast. Remains most numerous in lowland areas, including gardens where larvae can commonly be recorded. Counts appear stable, possibly increasing. Occasional on higher ground, e.g., Thrushgill, Belmont, etc.
■ **HABITAT** Disturbed ground, scrub, woodland rides, gardens.
■ **FOODPLANTS** Red Campion. Feeds in seed-heads.
Single brooded. Comes to light. Larvae have been recorded in June and July.

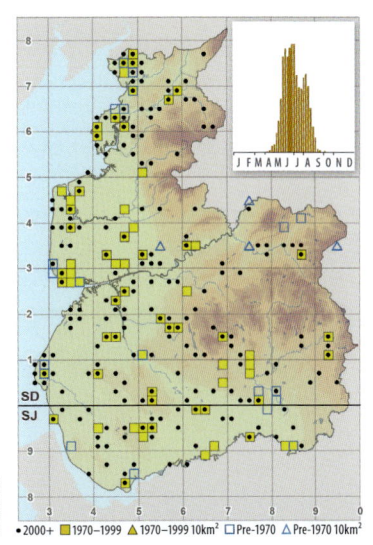

● 2000+   ■ 1970–1999   ▲ 1970–1999 10km²   □ Pre-1970   △ Pre-1970 10km²

The Moths of Lancashire   **547**

73.282 Varied Coronet *Hadena compta*

■ **LANCS STATUS** Rare / very local resident, possibly adventive　　2015–2020

Another recent arrival, with only three records to date, each from a different area of the county. The first was from the Fylde coast, at Bispham, on 19 June 2015 by D. & I. Smith. The next record was from the Mersey Valley, at Flixton, on 25 June 2020 by K. McCabe, and then a few weeks later, in Morecambe Bay, at Silverdale, on 6 July 2020 by J. Stonehouse. With the larvae feeding on Sweet Williams, it is possible the moth has arrived with plants, which could account for the unusual distribution.
■ **HABITAT** Unknown.
■ **FOODPLANTS** Not recorded in Lancashire.
Single brooded. All records are of singletons, trapped at light.

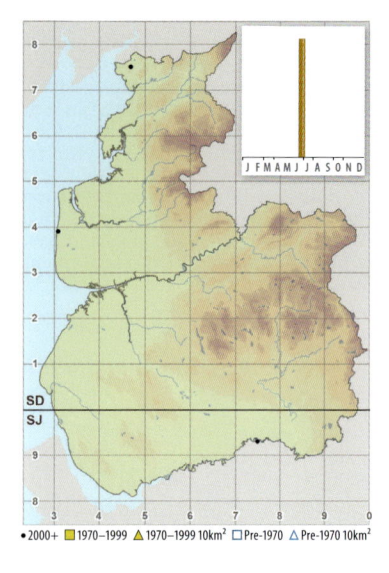

73.283 Marbled Coronet *Hadena confusa*

■ **LANCS STATUS** Occasional / local resident　　c.1853–2022

Earliest records by N. Cooke from Warrington (Byerley, 1854) and at Chorley in June 1853 (E.C. Buxton, 1853), with other 19th century records from Bickerstaffe, Leigh, Preston, Lostock. Described by Ellis (1890) as scarce. Recorded at Silverdale and Ainsdale in early 20th century, but not again until at MV light in Lytham St Annes by C.I. Rutherford 1953. Subsequent records have confirmed this as a lowland species, with post-2000 records concentrated in coastal regions in the north of the county.
■ **HABITAT** Open ground at coastal sites, gardens.
■ **FOODPLANTS** Not recorded in Lancashire.
Single brooded. Comes to light. Also noted at dusk around Bladder Campion.

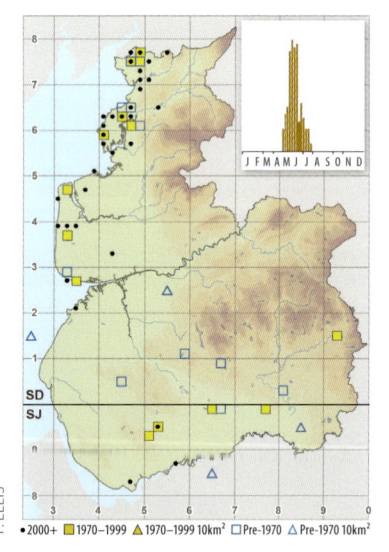

73.286 Tawny Shears/Pod Lover *Hadena perplexa*

■ **LANCS STATUS** Occasional / local resident　　c.1857–2022

First noted at Formby and Hale (Gregson, 1857), with other early records from Crosby, Prescot, Morecambe and Middleton, nr. Rochdale. Now, most frequent at coastal sites, e.g., Sunderland Point, Bispham and Formby. Occasional inland, e.g., Billinge 1986, Parr 2001, Ormskirk 2011, with a colony noted on Bladder Campion on a slagheap at Agecroft Colliery in 1963 by R. Leverton. A few darker individuals from the north Lancashire coast have been recorded as the sub-species, Pod Lover.
■ **HABITAT** Open grassland, sand dunes, waste ground.
■ **FOODPLANTS** Bladder Campion. Reared from larva at Carnforth 2016 (B. Hancock).
Double brooded. Most records have been at light. Larva recorded in July.

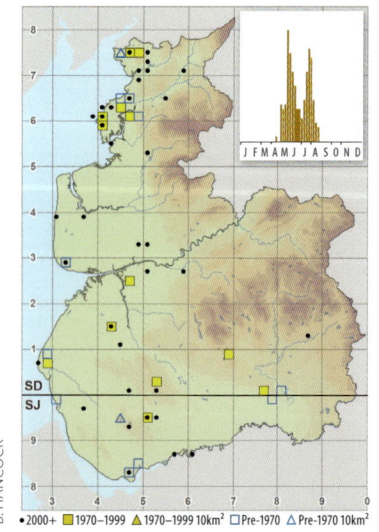

548　*The Moths of Lancashire*

73.288 Double Line *Mythimna turca*

■ **LANCS STATUS** Rare / vagrant   1961–2023

The first two records were both of singletons from Parbold, indicating the possible presence of a colony, now likely historical. The moth was first noted in 1961 by A.R. Fairhurst. The second, on 2 July 1973 by P. Summers, was subsequently exhibited at the Lancashire and Cheshire Entomological Society meeting at the Adelphi Hotel in October 1973 (*LCES Journal*, 1973–74). The lack of any subsequent records cast doubt on its continued presence in the county. However, a further moth was trapped at St Helens on 29 June 2023 by R. Banks, likely a wanderer from colonies to the south.
■ **HABITAT** Unknown.
■ **FOODPLANTS** Not recorded in Lancashire.

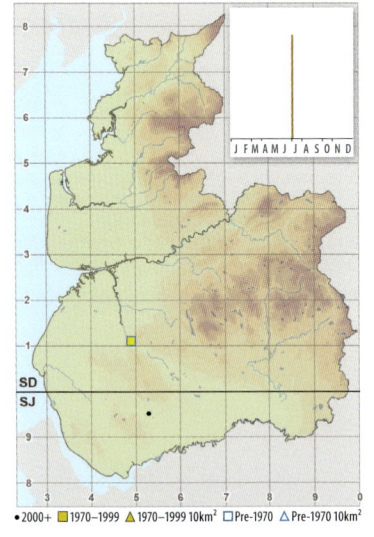

73.289 Striped Wainscot *Mythimna pudorina*

■ **LANCS STATUS** Scarce / local resident   2009–2022

A recent addition to the county fauna, the first confirmed recording was on 1 July 2009 at Lightfoot Green by S. Palmer. Further noted in 2011 at Cadishead Moss, Heysham and White Moss, nr. Yealand. Next, from Bay Horse, nr. Dolphinholme in 2012. In 2016, two moths were recorded from the MOD land at Altcar on the Sefton coast. High count of eight from Docker Moor on 29 June 2021 by P.J. Marsh. A previous record on a Martin Mere list from 1989 is considered unconfirmed.
■ **HABITAT** Damp grassland, marshes, moorland.
■ **FOODPLANTS** Not recorded in Lancashire.
Single brooded. All records are of moths to light.

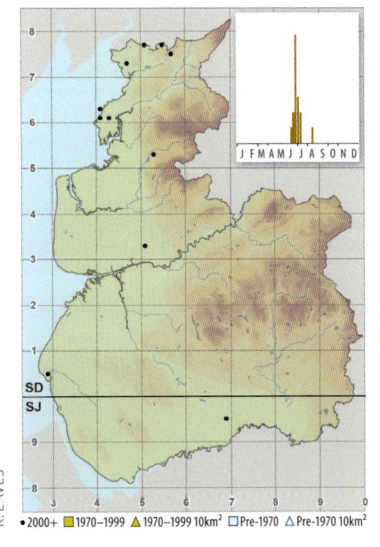

73.290 Brown-line Bright-eye *Mythimna conigera*

■ **LANCS STATUS** Rare / very local resident   c.1886–2007

First noted at Crosby on ragwort flowers, and described as local and uncommon (Ellis, 1890). Also, from Preston by J.B. Hodgkinson around the same time. Further recorded at Knowsley and Eccles, with building works and golf courses blamed for declining frequency (Mansbridge, 1940). Unfortunately, the species is still declining, with the only 21st century records coming from Torrisholme, Arkholme, Leighton Moss, Flixton and Little Singleton, and none at all since 2007.
■ **HABITAT** Grassland.
■ **FOODPLANTS** Bred ex-larva 1907 at Hale Bank by W. Mansbridge. Foodplant not noted. Single brooded. Comes to light.

The Moths of Lancashire   **549**

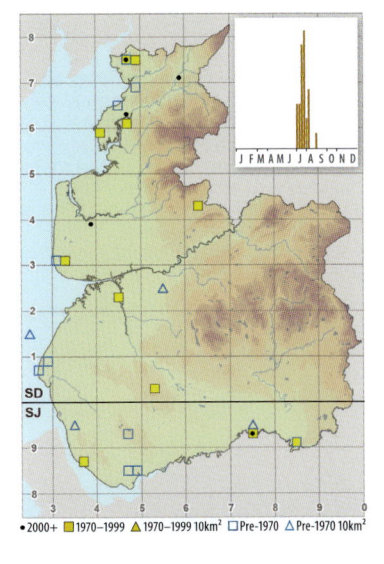

NOCTUIDAE

73.291 Common Wainscot *Mythimna pallens*

■ **LANCS STATUS** Frequent / widespread resident   c.1857–2022

First recorded in the Liverpool district by Gregson (1857). Next, from Burnley, Oldham, Warrington and Middleton, nr. Rochdale. Early authors considered it common to abundant everywhere. Whilst less common than the following species, still averages around 400 records a year. A few counts from 2022 exceeded 30; in one case, at Longton, 55. During this century the second generation has increased in number, almost reaching the peaks of the first. Most frequent in lowland areas.

■ **HABITAT** Grassland, woodland.
■ **FOODPLANTS** Unrecorded within the county.
Double brooded. Readily attracted to light.

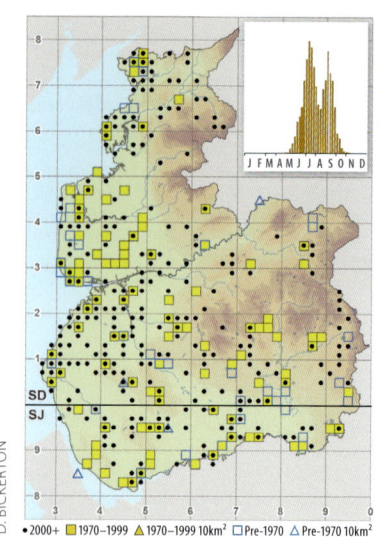

73.293 Smoky Wainscot *Mythimna impura*

■ **LANCS STATUS** Abundant / ubiquitous resident   c.1857–2022

Reported to be plentiful in the Liverpool district on thistle and rush flowers at dusk (Gregson, 1857). Other early authors described the moth as common to abundant. Light trapping since the mid-20th century has shown it to be extremely widespread. Found in all habitats, with high totals including 239 at Belmont on 4 August 2004 (S. Martin) and 230 at Worsthorne on 10 August 2020 (G. Gavaghan). High totals have also been noted at Heysham N.R. and Mere Sands Wood, but not for over a decade.

■ **HABITAT** Grassland, sand dunes, gardens.
■ **FOODPLANTS** Cock's-foot and other unidentified grasses.
Single brooded, but with extended generation. Larvae found from March to May.

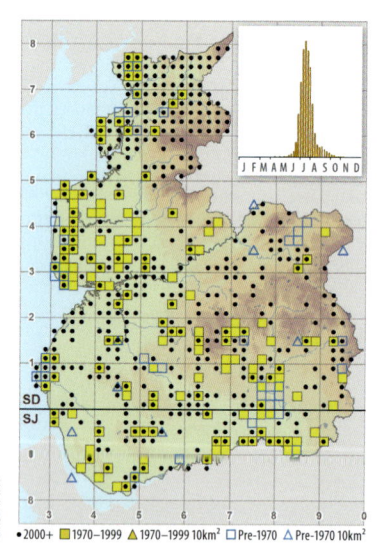

73.294 Southern Wainscot *Mythimna straminea*

■ **LANCS STATUS** Occasional / local resident   1958–2022

Recorded by C.G.M. de Worms on 23 August 1958 at Leighton Moss. All subsequent records came from Leighton Moss area, bar a singleton in Morecambe 1965, until 1980s when moth began to appear elsewhere e.g., Gait Barrows, Caton, Leigh, Hale, Woolston Eyes, Heysham N.R. Most frequent in coastal areas, although with recent inland records from St Helens, Flixton, Abram, Longridge, Brockholes N.R., etc. Increased numbers this century appear part of a national expansion northwards.

■ **HABITAT** Reedbeds, marshes.
■ **FOODPLANTS** Not recorded in Lancashire.
Single brooded. Almost all records at light.

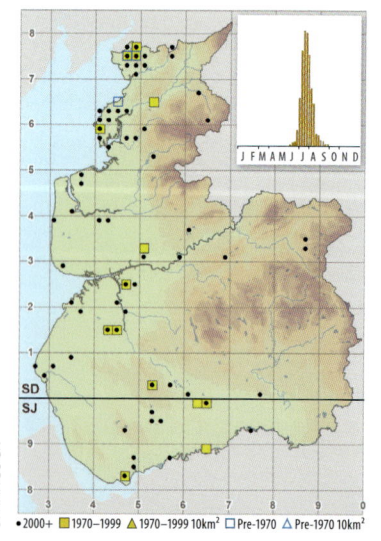

The Moths of Lancashire

73.295 Delicate *Mythimna vitellina*

■ **LANCS STATUS** Occasional / migrant 2001–2022

Recorded in the county on 24 occasions, the first on 1 November 2001, when a moth was trapped at light by B. Cockburn at Burrow Heights, nr. Bailrigg. Eleven of the records came from 2006, a year of a huge national influx. Records are mainly coastal, but with a few inland records too, occasionally on higher ground, such as at High Tatham. Perhaps surprisingly for a moth migrating from southern Europe, two-thirds of the records are from the north of the county.

■ **HABITAT** Coastal grassland.
■ **FOODPLANTS** Not recorded in Lancashire.

Migration noted in late summer and autumn, with a peak in late September.

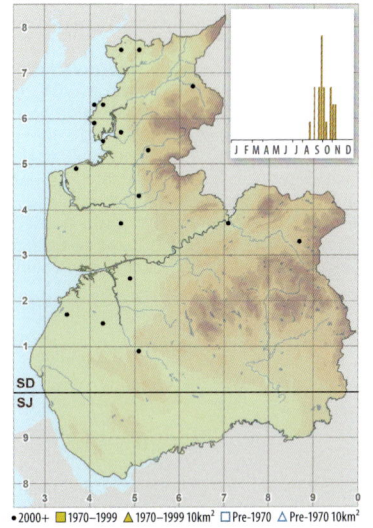

73.296 White-speck *Mythimna unipuncta*

■ **LANCS STATUS** Occasional / migrant 1966–2016

First recorded at Slyne by P.J. Marsh on 17 September 1966, with the second the following night by J.G. Huddleston at Caton. Next, at Woolston Eyes in 1983, but not again until 2000, which was the best year for this moth, with 18 of the 37 records. As with the previous species, most records are from north of the county; exceptions being at Flixton, Haydock, Dalton, etc. Recorded from upland areas at Briercliffe, Todmorden, Mill Houses, as well as on lowland sites.

■ **HABITAT** Grassland.
■ **FOODPLANTS** Not recorded in Lancashire.

Comes to light. Also noted on wine ropes and at Ivy blossom.

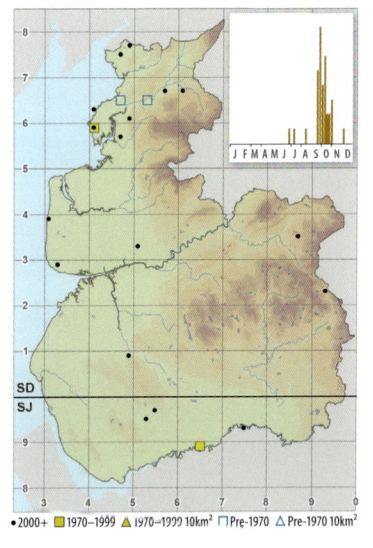

73.297 White-point *Mythimna albipuncta*

■ **LANCS STATUS** Rare / migrant 2015–2022

Recorded in the county on just three occasions, all from the southern half. The first was a moth trapped at light in Hoghton by G. Dixon on 23 August 2015. The second was at Great Sankey, Warrington on 26 June 2021 by J. Mitchell-Lisle. Further recorded at light on 22 October 2022 at Hale by C. Cockbain. Possibly under-recorded due to the difficulty in separating from the next species.

■ **HABITAT** Unknown.
■ **FOODPLANTS** Unrecorded within the county.

All records are of moths to light in garden traps.

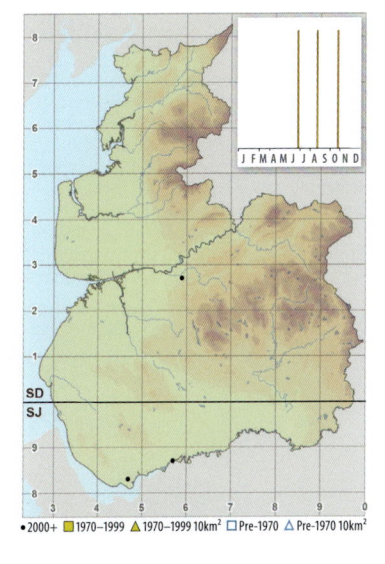

NOCTUIDAE

73.298 Clay *Mythimna ferrago*

■ **LANCS STATUS** Frequent / widespread resident c.1843–2022

First recorded from the Preston area (Hodgkinson, 1844). Other 19th century records from Liverpool, Warrington and Middleton, nr. Rochdale. Reported as abundant near the coast (Gregson, 1857) with other early authors describing it as common and widely distributed. H. Britten (c.1950) noted it to be less common in the east. Pre-2000 records suggested the moth was largely absent from Pennine areas. However, it has since moved into these upland areas and is now widely recorded throughout.

■ **HABITAT** Coastal grassland, scrub, gardens.

■ **FOODPLANTS** Unrecorded. Larva swept from Heather at Swinton in 2004 (B. Leecy). Single brooded. Readily to light. High count of 44 at Peel Hall 7 July 2018 (G. Dunbar).

• 2000+ ☐ 1970–1999 △ 1970–1999 10km² ☐ Pre-1970 △ Pre-1970 10km²

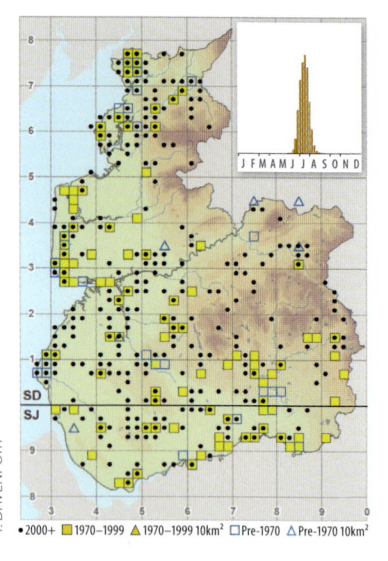

73.299 Shore Wainscot *Mythimna litoralis*

■ **LANCS STATUS** Scarce / local resident c.1857–2022

An exclusively coastal species, first noted from the Crosby sand-hills by Gregson (1857). Larvae noted from Lytham to the Mersey by Ellis (1890), with Mansbridge (1940) later reporting abundant larvae feeding on Marram Grass in the dunes. The 28 records from 2020 onwards are all from the Sefton Coast, with the last records from St Annes and Heysham in 2019. Almost all recent records are of moths to light.

■ **HABITAT** Coastal grassland.

■ **FOODPLANTS** On roots of Marram Grass on 22 May 1976 at Formby Point (LCES report). Single brooded with occasional second brood including six moths in 2016 from September and October. Also noted during September in 2021 and 2022.

• 2000+ ☐ 1970–1999 △ 1970–1999 10km² ☐ Pre-1970 △ Pre-1970 10km²

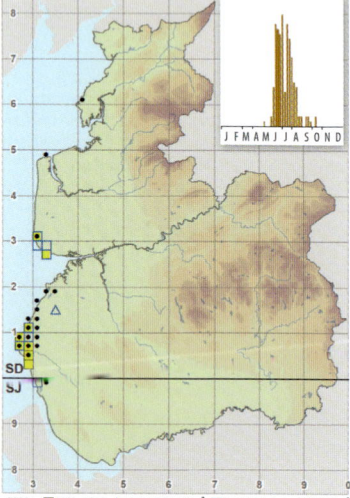

73.301 Shoulder-striped Wainscot *Leucania comma*

■ **LANCS STATUS** Frequent / widespread resident c.1857–2022

Noted as abundant in Liverpool district (Gregson, 1857), with other early authors describing it as common on the coast and scarcer inland. Further early records from Knowsley, Simonswood Moss and Lancaster. Recorded in Didsbury 1940 and Salford 1950, confirming existence of populations in the Manchester area. Remains frequent in Mersey Valley. Upland sites include Longridge Fell, Tarnbrook and Briercliffe.

■ **HABITAT** Damp woodland, marshes.

■ **FOODPLANTS** Not recorded in Lancashire.

Appears single brooded, although with the odd autumn straggler. Comes readily to light with high count of 247 from Heysham N.R. on 16 July 2022 by P.J. Marsh.

• 2000+ ☐ 1970–1999 △ 1970–1999 10km² ☐ Pre-1970 △ Pre-1970 10km²

552 *The Moths of Lancashire*

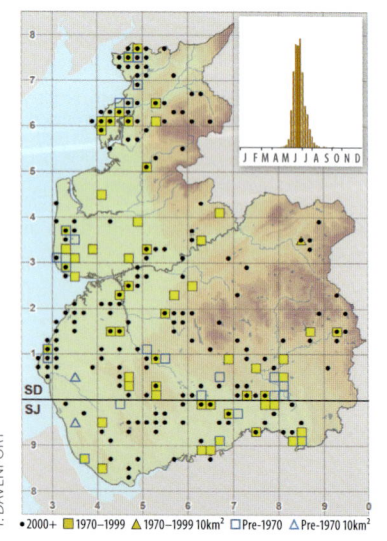

73.302 Obscure Wainscot *Leucania obsoleta*

■ **LANCS STATUS** Occasional / local resident   1984–2022

A recent arrival to the county, first recorded at Gait Barrows in 1984 by E. Emmett. Next noted at Leighton Moss in 1993 and frequently at that site thereafter. The first VC59 records came from the Wigan Flashes in 1999, Bold Moss, St Helens in 2000 and Mere Sands Wood 2003. Records remain restricted to lowland areas other than a singleton trapped by G. Gavaghan in 2021 from higher ground at Worsthorne in the east of the county, suggesting possible range expansion as numbers increase.
■ **HABITAT** Reedbeds, marshes and other wetland sites.
■ **FOODPLANTS** Not recorded in Lancashire.
Single brooded. Comes to light, with high count of 19 at Leighton Moss, June 2009.

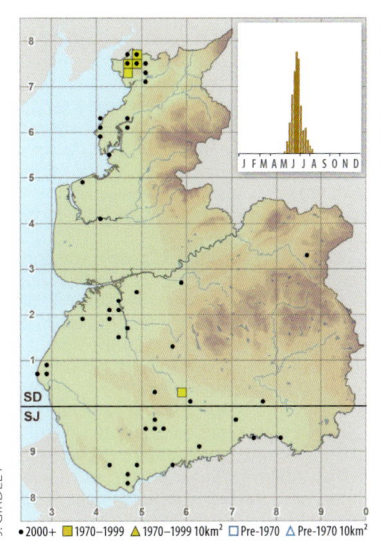

73.304 Cosmopolitan *Leucania loreyi*

■ **LANCS STATUS** Rare / migrant   2008–2020

There are only two records of this migrant within the county. Although primarily a visitor to southern England, both records come from northern parts of the county. The moth was first noted on 19 August 2008 at Heysham N.R. by A. Draper. The only other record was from 9 November 2020 at Wray by Ga. Jones. Both moths were photographed.
■ **HABITAT** Unknown.
■ **FOODPLANTS** Not recorded in Lancashire.
Both moths were trapped at light.

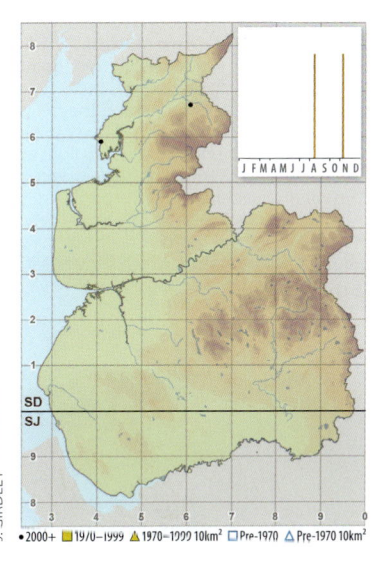

73.307 Pearly Underwing *Peridroma saucia*

■ **LANCS STATUS** Occasional / migrant   c.1857–2022

Gregson (1857) documented the capture of this moth by Mr Robson and Mr Almond in the Liverpool area, with others from Pendleton and Chorlton by J. Chappell shortly after (Ellis, 1890). Ellis gave status as generally distributed but nowhere common. The status of migrant not documented in any of the early works. Recorded 50 times in 20th century and 220 in this, with 2000 and 2006 being particularly good years. 19 records from 2022. Bred from ova laid by female trapped in Chorlton 2004.
■ **HABITAT** Unknown due to migrant status.
■ **FOODPLANTS** Not recorded in Lancashire.
Appear to be two waves of migration; in early summer and mid-autumn.

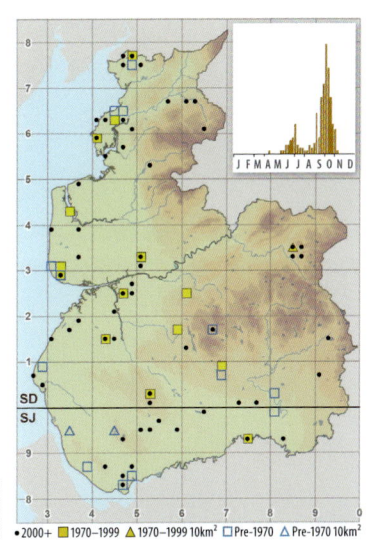

NOCTUIDAE

The Moths of Lancashire   **553**

73.308 Portland Moth *Actebia praecox*

■ **LANCS STATUS** Occasional / very local resident   c.1857–2022

First documented on the Crosby sandhills (Gregson, 1857) and in 1885 at Ainsdale, Birkdale and Freshfield/Formby (Ellis, 1890), it has since been recorded from much of the Sefton Coast, including Cabin Hill in 1954, Altcar from 1987 and Green Beach (Birkdale) in 2009. The Formby area has received most recording effort with regular counts of low single figures to the present day. Larger single-night counts produced 31 at light in 2003, 16 in 2004, 12 in 2005 and 2010 and two counts of 14 in 2013.

Away from the Sefton Coast, it was formerly found with some regularity on or near the dunes of Lytham St Anne's, but always in small numbers. First noted here in 1906 by T. Baxter and in 1908 by A.E. Wright, it was regular until the early 1970s but with few subsequent records, the last report being by J. Wardle in 2005. The moth occasionally wanders, with Morecambe seeing examples in 1940, 1959 and 1965 and Mere Sands Wood in 1969, 1985 and 1989. Elsewhere, singles have been found at Freckleton, Gait Barrows, Leighton Moss, Slyne and Warton (nr. Carnforth); the last vagrant noted was at Haskayne in 2004.

Habitat loss reduced the moth's distribution over the years with significant areas of dunes lost to development at Crosby in the 1920s and 1960s. The same occurred in the Lytham St Anne's area where it was last seen in 2005. Recently, at Crosby, the dunes have started to rebuild in places and the moth was recorded nearby in 2004 and 2006.

This is a coastal species in Lancashire, occurring on sand dune systems near slacks with established vegetation. Ellis took about two dozen larvae at Crosby in the sand around the foodplant, sallow. Small burrows lead from the foodplant to a small mound where the larvae lay hidden during the day. He observed they ate through the leaf, separating it from the plant and devouring the rest by holding it between their pro-legs (Ellis, 1881). Creeping Willow was observed as the larval foodplant in 2012 (G. Jones) and the reference to 'sallow', above, is considered to refer to this plant as well. The larva can be found between late April and late June.

The moth is readily attracted to light and quickly settles. It varies little in colour and size but an example of a pale green form is in the F.N. Pierce collection (World Museum), collected 'near Liverpool'.

There may be less than a dozen widely scattered populations in Britain (Randle, *et al.* 2019), but the Sefton Coast still has extensive sand dunes and slacks with established vegetation which suggests it presently looks secure in this area.

Richard Walker

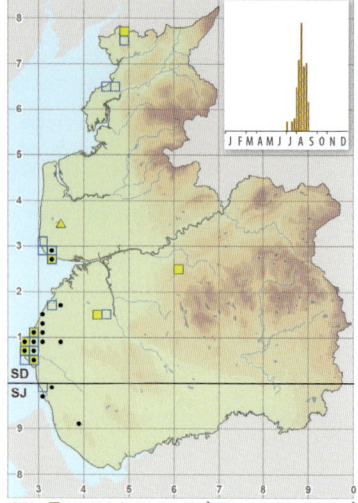

• 2000+   ☐ 1970–1999   △ 1970–1999 10km^2 ☐ Pre-1970   △ Pre-1970 10km^2

73.311 Coast Dart *Euxoa cursoria*

■ **LANCS STATUS** Considered Extinct　　　　　　　　　1845–1975

Appears a significant loss for the county. This species has been recorded on 41 occasions, but not since 1975. First reported from the coast, location uncertain, by J.B. Hodgkinson (1845). Ellis & Mansbridge reported it to be common on the sand-hills, with early records from the Sefton and Fylde coasts. The moth was last recorded in 1964 at Formby, with a larval record from May 1975, also from Formby, by J. Thompson. This is a species also known to have declined nationally.
■ **HABITAT** Fixed sand dunes.
■ **FOODPLANTS** Larvae recorded 1975; foodplant not documented. Previously recorded at light and at Common Ragwort flowers.

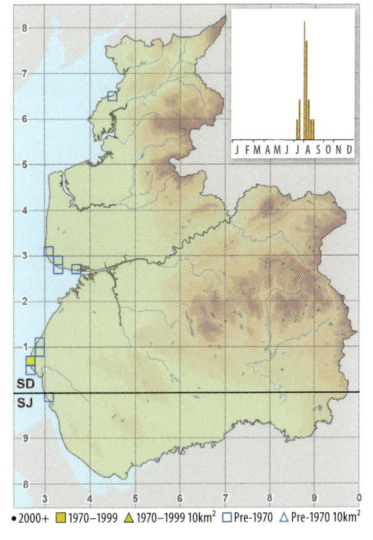

73.312 Square-spot Dart *Euxoa obelisca*

■ **LANCS STATUS** Rare / very local resident　　　　　　c.1857–2014

Only recorded on five occasions, scattered along the coast. First reported from Crosby (Gregson, 1857), and from Lytham St Annes in 1903 by A.E. Wright. Not noted again until August 1998 when twice trapped at light by P.J. Marsh at Heysham N.R. The final record, the only one from the 21st century, was trapped at Southport by R. Moyes on 15 August 2014 and subsequently dissected. The scattered distribution suggests a few small colonies may yet persist, possibly poorly attracted to light.
■ **HABITAT** Sand dunes, coastal grassland.
■ **FOODPLANTS** Not recorded in Lancashire.
Single brooded. All records are of singletons.

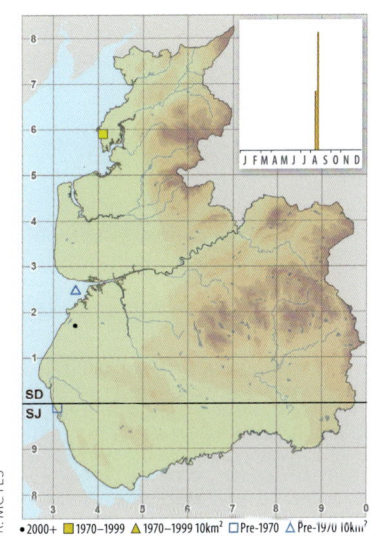

73.313 White-line Dart *Euxoa tritici*

■ **LANCS STATUS** Occasional / local resident　　　　　c.1857–2022

Mainly from coastal locations, although occasionally inland. Plentiful on the Crosby sand-hills (Gregson, 1857), and at Lytham St Annes 1908. First recorded inland by R. Leverton 1958 at Irlams o' th' Height. Also, at Wrightington, Lightfoot Green, Flixton, Chorlton, Fulwood, St Helens; most, if not all, are likely to be wanderers from the coast. Regular coastal sites include Lytham, Heysham, Sunderland Point and Sefton Coast. 55 records from 2020–2022, less than half that of the 144 from 2010–2012.
■ **HABITAT** Coastal grassland, sand dunes.
■ **FOODPLANTS** Bred ex-larva 1923; foodplant not documented.
Single brooded. Comes readily to light. Recorded by 'dusking' at Southport 2020.

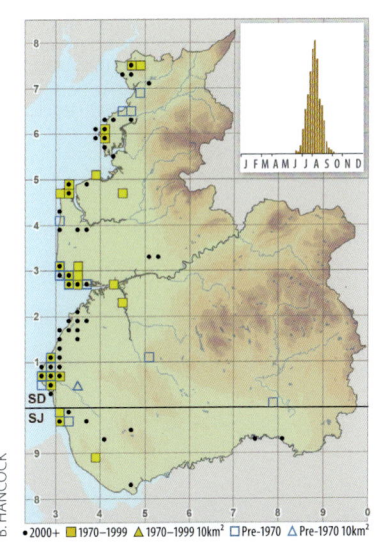

NOCTUIDAE

73.314 Garden Dart *Euxoa nigricans*

■ **LANCS STATUS** Scarce / local resident   c.1843–2022

Declining nationally and locally. Only noted seven times from 2013 to 2022 compared to 70 from 1983 to 1992. Further evidence of decline can be seen in C.A. Darbyshire's records from Billinge where it was recorded 47 times since 1985 but not at all since 1996. This species was first noted by J.B. Hodgkinson (1844) at Preston, and later on ragwort flowers at Cheetham Hill (Edleston, 1846). Ellis (1890) and Mansbridge (1940) reported it to be fairly common on sandhills and mosses. Predominantly, a lowland species, although with a few records from the Burnley area, last in 2002. It seems to be just about hanging on in the Mersey Valley, with recent records from Flixton (24 and 29 August 2021, K. McCabe) and Chorlton (3 August 2021, 3 August 2022 and 24 August 2022, S. Marsden). The only other post-2010 records are from Rainford, St Helens in 2019 and at Poulton-le-Fylde in 2021. Overall, its prospects look bleak.

■ **HABITAT** Rough open ground, damp grassland, arable fields, gardens.
■ **FOODPLANTS** Not recorded in Lancashire.

Single brooded. Recorded at light and nectaring on Common Ragwort.

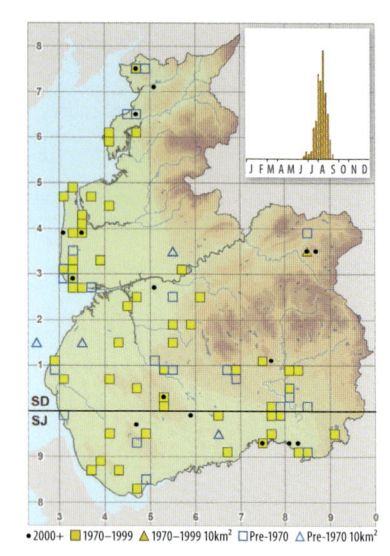

73.317 Heart and Dart *Agrotis exclamationis*

■ **LANCS STATUS** Abundant / ubiquitous resident   c.1857–2022

The earliest records are from the Liverpool district (Gregson, 1857), with other 19th century records coming from Preston (Ellis, 1890) and Middleton, nr. Rochdale (J. Taylor). Extensively recorded from 1910 onwards with over 30,000 records in total. Early authors noted the moth to be abundant throughout the county (Ellis, 1890; Mansbridge, 1940). This appears to still be the case, although there is some evidence of a recent decline in this abundance. The 80 highest nightly totals are all pre-2010, with the highest, 298, from Flixton on 24 July 1996 (K. McCabe). The highest record post-2010 was 90 moths to light at Heysham in 2015 (D.J. Holding). The highest nightly total from 2022 was 51 at Maple Farm Nursery, Westby on 23 June (J. Steeden). Nevertheless, this remains an extremely widely distributed moth, with the highest counts in lowland habitats including gardens. Those from upland sites are typically in single figures.

■ **HABITAT** Rough grassland, arable fields, gardens, etc.
■ **FOODPLANTS** Larvae ex-ova, fed on dock and Dandelion leaves 2021 (J. Worthington).

This species has become increasingly double brooded with 40 September and October records from 2021 to 2022 alone. Comes readily to light.

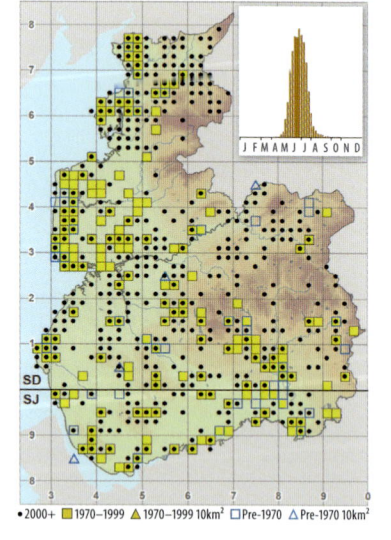

73.319 Turnip Moth *Agrotis segetum*

■ **LANCS STATUS** Frequent / widespread resident; occasional migrant   c.1857–2022
First noted by Gregson (1857) in the Liverpool area, in profusion on the coast. Later, described as common and generally distributed, occasionally abundant (Ellis, 1890). Britten (c.1950) considered it common in agricultural areas. Increased light trapping has helped to monitor the increased range of the moth, extending into the uplands during the last 50 years. High count of 24 at Parr on 30 August 2018. Comes to light.
■ **HABITAT** Open woodland, parks, farmland, gardens, etc.
■ **FOODPLANTS** Not recorded in Lancashire.
Double brooded. Second brood individuals scarce up to 1990s. Now form significant percentage of annual total, with 25% of 2022 records from September onwards.

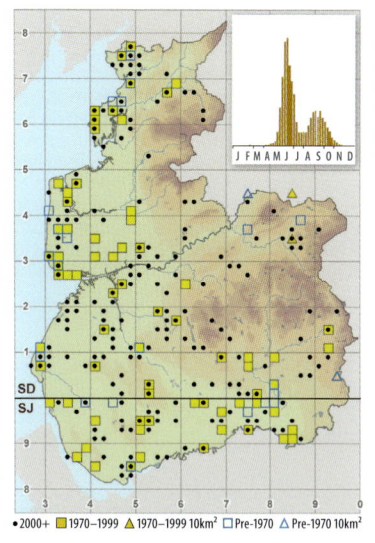

73.320 Heart and Club *Agrotis clavis*

■ **LANCS STATUS** Frequent / local resident   c.1857–2022
Gregson (1857) noted the moth to be abundant at sugar in Crosby, with other early records from Preston, Knowsley, Heysham, etc. Described as local and not common (Mansbridge, 1940). Only recorded 18 times pre-2000 but almost 600 times since, corresponding to nationally increasing range of this mainly southern species. Local distribution is mainly coastal with a few inland records, including at Cadishead Moss in 2022, and on higher ground at Worsaw Hill and Dalton Crags in 2021.
■ **HABITAT** Dry grassland, sand dunes, gardens.
■ **FOODPLANTS** Not recorded in Lancashire.
Single brooded. Comes to light with high count of 36 at Formby on 3 July 2014.

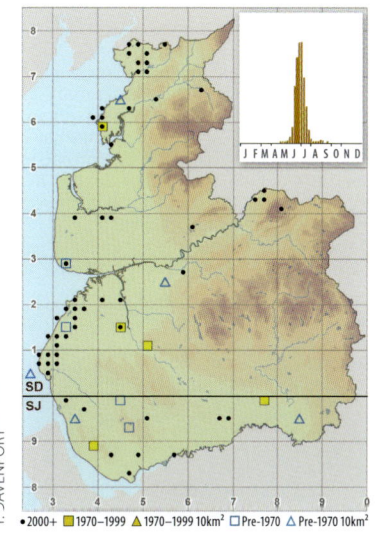

73.322 Archer's Dart *Agrotis vestigialis*

■ **LANCS STATUS** Occasional / local resident   1885–2022
Noted by Gregson in the Crosby sand-hills on 3 August 1885, and by Ellis (1890) as abundant on sugar and ragwort blossom on the coast, from the Mersey to Lytham. Recorded by H.N. Michaelis in Formby, nectaring on ragwort in 1950. The moth remains regular on the Sefton and Fylde coasts, with smaller numbers from Heysham. Occasional inland records, from Preston, Belmont, Hoghton, etc., are presumably wanderers from the coast. Appears to have declined in number in last few decades.
■ **HABITAT** Sand dunes.
■ **FOODPLANTS** Not recorded in Lancashire.
Single brooded. Comes to light with high count of 120 from Formby 30 August 2005.

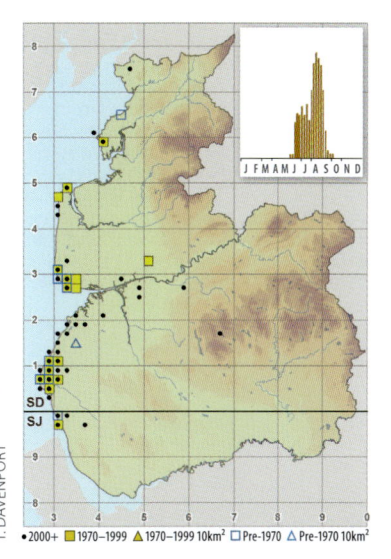

The Moths of Lancashire 557

73.323 Sand Dart *Agrotis ripae*

■ **LANCS STATUS** Scarce / very local resident   c.1900–2022

Noted by T. Baxter c.1900 at Lytham St Annes, where the moth was often recorded, although not since 1996. First noted on Sefton Coast at Ainsdale and Formby in 1952, then in 1982 at Formby, and more frequently from 2000 onwards. Larvae and pupae recorded in the sand of Birkdale Green Beach between 2007 and 2010, with moths reared (G. Jones). Whilst the moth has been noted at Formby, Ainsdale and Birkdale in the last five years, its continued absence from the Fylde coast is of some concern.
■ **HABITAT** Sand dunes, just beyond the strandline.
■ **FOODPLANTS** Common at Fairhaven dunes on Sea Plantain, 1914.
Single brooded. Comes to light. Larvae found in August; pupae in April.

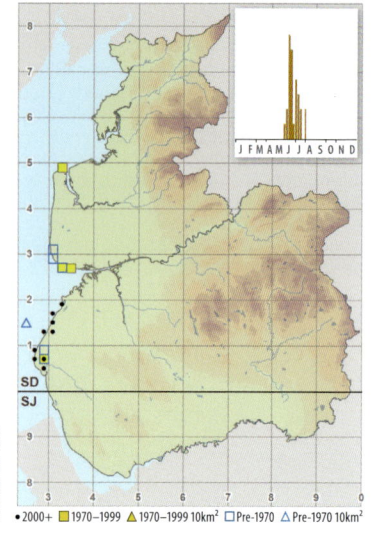

73.324 Crescent Dart *Agrotis trux*

■ **LANCS STATUS** Rare / very local resident   1887–2021

A rarely recorded coastal species that has twice occurred inland. First noted on the Crosby sand-hills in 1887, nectaring on ragwort, by A.W. Hughes, but not again until 24 July 2004 at Lytham St Annes by D. & J. Steeden. Next, at Heysham by D.J. Holding on 9 July 2014. More surprising locations were inland sites at Swinton on 29 July 2014 by G. Riley and Chorlton on 26 July 2021 by S. Marsden. Presumably, these moths represent the occasional inland wanderers reported by Randle *et al.* (2019).
■ **HABITAT** Coastal sites, gardens.
■ **FOODPLANTS** Not recorded in Lancashire.
Single brooded. Recent records to light; all of singletons.

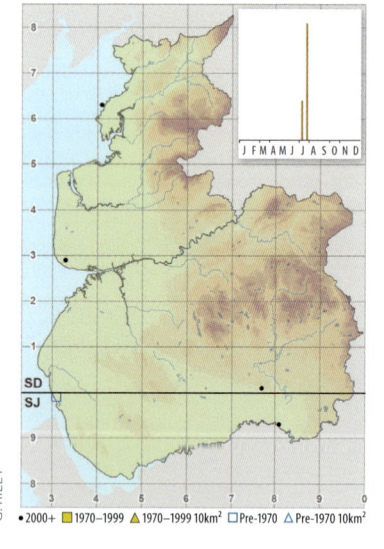

73.325 Shuttle-shaped Dart *Agrotis puta*

■ **LANCS STATUS** Frequent / widespread resident   1901–2022

The earliest record is from Knowsley Park by F.C. Thompson in 1901. Other records to 1950 from Heysham, Ditton, Formby, Wrightington and Lytham St Annes. Described by Mansbridge (1940) as infrequent. Records increased as trapping became more widespread, initially in lowland areas. High count of 79 from Southport on 2 August 2011 by R. Moyes. The last few decades have seen range extending as the moth has started to colonise eastern upland areas with records from Wray, Worsaw Hill, etc.
■ **HABITAT** Grassland, open woodland, gardens.
■ **FOODPLANTS** Not recorded in Lancashire.
Double brooded. Comes readily to light.

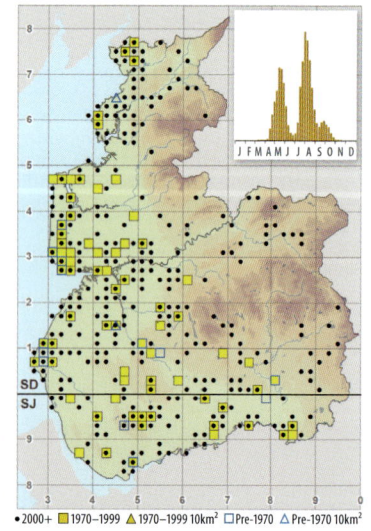

558   *The Moths of Lancashire*

73.327 Dark Sword-grass *Agrotis ipsilon*

■ **LANCS STATUS** Frequent / migrant   **c.1857–2022**

Recorded in Liverpool as plentiful on sugar in September and on sallows in spring (Gregson, 1857), with other early records from Oldham and Middleton (J. Taylor). Mansbridge (1940) found it widely distributed, some years common, particularly on the coast. Numbers vary from year to year, with the largest number of records, 174, from 2011. Difficult to assess preferred habitat due to migrant status, lack of larval records and capacity to turn up anywhere throughout county, including uplands.
■ **HABITAT** Unknown. Frequently in garden traps.
■ **FOODPLANTS** Not recorded in Lancashire.
Comes to light. Recorded every month with a peak in autumn. 50 records from 2022.

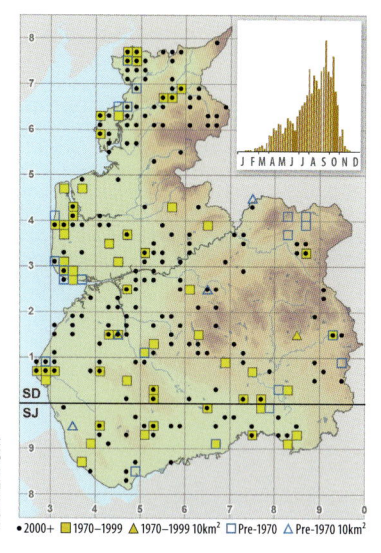

73.328 Flame *Axylia putris*

■ **LANCS STATUS** Frequent / ubiquitous resident   **c.1857–2022**

First noted in the Liverpool district (Gregson, 1857), then by J. Taylor at Middleton, nr. Rochdale in 1885. Status reported as generally distributed and fairly common (Ellis, 1890), although by 1940 reportedly not common (Mansbridge, 1940). Since the 1960s, distribution and abundance have increased, extending into upland areas, and emerging earlier in the year with the suggestion of a small second generation.
■ **HABITAT** Grassland, hedgerows, gardens.
■ **FOODPLANTS** Not recorded in Lancashire.
Probably double brooded. Highest count is of 86 to light at Green Bank, 23 June 2022 (P.J. Marsh). Bred from pupa found in soil at Billinge, 8 April 2002 (C.A. Darbyshire).

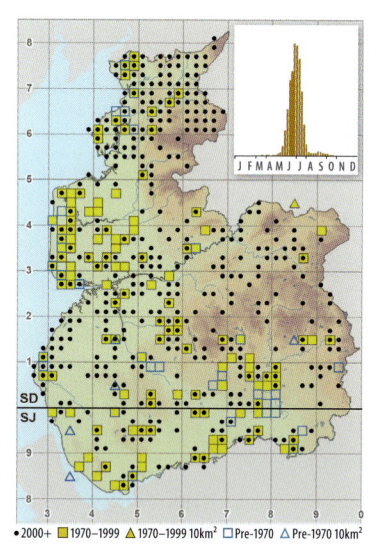

73.329 Flame Shoulder *Ochropleura plecta*

■ **LANCS STATUS** Abundant / ubiquitous resident   **c.1857–2022**

First reported from Liverpool by Gregson (1857), noting its abundance among blackberry bushes. Also, at Oldham, attracted to sugar in 1883 by J.T. Rogers. Early authors agreed the moth was common and widely distributed. However, all pre-1950 records were from VC59, other than a couple from Lytham St Annes. From 1960s onwards, light trapping revealed the moth to be also common in north of the county; at Lancaster, Morecambe, Slyne, Green Bank, Longridge Fell, Winmarleigh Moss, etc.
■ **HABITAT** Woodland, moorland, grassland, hedgerows, gardens.
■ **FOODPLANTS** Foodplant unrecorded within the county.
Double brooded, with increasing second generation since late 1990s. Comes to light.

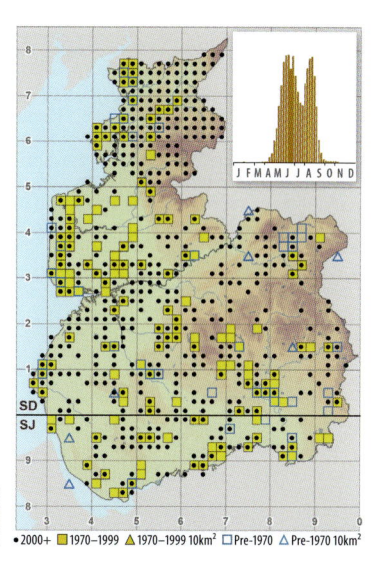

The Moths of Lancashire   **559**

NOCTUIDAE

73.331 Barred Chestnut *Diarsia dahlii*

■ **LANCS STATUS** Occasional / local resident   c.1857–2022

Recorded at 'Birch-wood', Woolton (Gregson, 1857) and noted as very local (Ellis, 1890). Also, from Lostock 1888 and Silverdale 1928. Not noted again until 1968, when trapped at Clougha and Bolton-le-Sands by C.J. Goodall. First recorded in VC59 at Helmshore 1971 and most recently from Freshfield Dune Heath 2004 and Hoghton 2018. Vast majority of records are from northern upland areas e.g., Mill Houses, Green Bank, and on limestone at Warton Crag, where populations appear healthy.
■ **HABITAT** Woodland, moorland, limestone pavement.
■ **FOODPLANTS** Not recorded in Lancashire.
Single brooded. Comes to light with high count of 41 at Warton Crag 2017 (J. Patton).

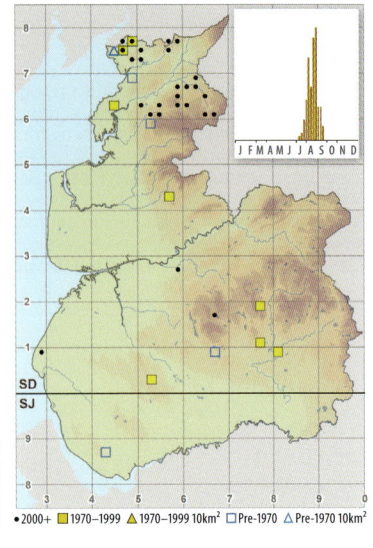

73.332 Purple Clay *Diarsia brunnea*

■ **LANCS STATUS** Frequent / widespread resident   c.1843–2022

First recorded from the Preston area by J.B. Hodgkinson (1844). Ellis (1890) described the moth as common and generally distributed, without mentioning further sites. Recorded during the early 20th century at Burnley, Oldham, Ditton, Roughlee and Didsbury. Increasingly recorded from the 1990s onwards, confirming the moth's presence at moorland, lowland and coastal sites, although with few records from the Fylde area. High count of 35 from Warton Crag by J. Patton on 25 June 2014.
■ **HABITAT** Woodland, moorland, gardens.
■ **FOODPLANTS** Ivy, Black Currant.
Single brooded. Comes to light. Larvae recorded in March and April.

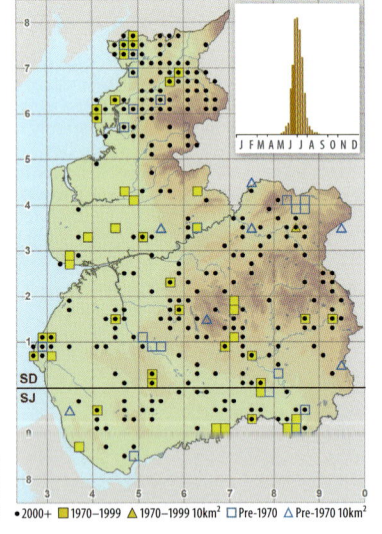

73.333 Ingrailed Clay *Diarsia mendica*

■ **LANCS STATUS** Frequent / widespread resident   c.1857–2022

Recorded from the Liverpool district (Gregson, 1857); noted as common, especially in 'heathy places.' Other records to 1910 from Preston, Burnley and Simonswood Moss. Early authors described it as common and generally distributed. Since 2000, noted to be particularly frequent in upland areas such as Leck Fell and Mill Houses. Also, from lowland areas such as Heysham N.R. and Mere Sands Woods. Highest count of 160 at Warton Crag 2016. Distribution patchy in the Mersey Valley with declining frequency.
■ **HABITAT** Rough grassland, moorland, gardens.
■ **FOODPLANTS** Heather.
Single brooded. Comes readily to light. Larvae found from mid-March to mid-May.

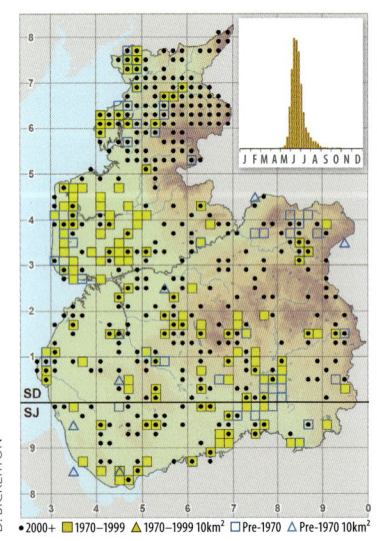

73.334 Small Square-spot *Diarsia rubi*

■ **LANCS STATUS** Abundant / ubiquitous resident   c.1843–2022

First recorded from Preston (Hodgkinson, 1844) and Liverpool (Gregson, 1857). Noted by Mansbridge (1940) to be generally distributed and common, particularly on the coast. Records since light trapping began have confirmed its presence throughout, with the highest counts at coastal sites e.g., Sunderland, Heysham N.R.

A further species on the British list is Fen Square-spot (*Diarsia florida*), first recorded locally at Docker Moor in 2015, and considered a single brooded, midsummer species of blanket bogs and marshes, typically larger and paler than Small Square-spot. However, the two do not significantly differ in genitalia or DNA barcoding.

Potential specimens of Fen Square-spot from Lancashire were analysed in 2021, courtesy of the Tanyptera Trust, and DNA matches were made to each of the two species. However, as the DNA profile of the two did not differ significantly in the first place this did not take us much further forward.

Nevertheless, it seems that those specimens formerly recorded as Fen Square-spot do differ in colouration and phenology. The decision was made by the Lancashire County Moth Recorders to amalgamate all records as Small Square-spot, whilst detailing potentially distinctive characteristics such as size, colour and flight period, to aid identification of any potential Fen Square-spot records should definitive evidence of its existence be found in the future (Hedges and Patton, 2022).

■ **HABITAT** Coastal grassland, damp meadows, marshes, woodland, gardens.
■ **FOODPLANTS** Unrecorded. Larva beneath moss on poplar log, Chorlton 15 April 2009. Double brooded. Occasional in November, possibly as a small third generation.

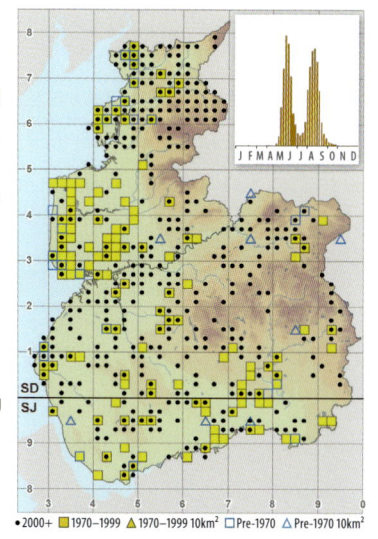

• 2000+ ☐ 1970–1999 ▲ 1970–1999 10km² ☐ Pre-1970 △ Pre-1970 10km²

73.336 Red Chestnut *Cerastis rubricosa*

■ **LANCS STATUS** Frequent / widespread resident   c. 1843–2022

Having been impressed by the first volume of *The Zoologist* in 1843, J.B. Hodgkinson requested of the editor in the second; 'I beg leave to hand you … a list of the rarer Lepidoptera that have been taken by my brother and myself in the neighbourhood of Preston' (Hodgkinson, 1844). Within this list was the first county record of this species. The moth was next noted a few years later in Prestwich Woods, where it was reported to be abundant around sallows during April (Edleston, 1846). Gregson (1857) also found it plentiful at sallow blossom.

The species now appears far less common at lowland sites, where generally only singletons are recorded. The only lowland double figure counts are from K. McCabe's Flixton garden in the 1990s. More recently, from 2020 to 2022, the latter site has only seen twelve individuals in total, with no count above two.

In upland areas, however, it is the commonest early spring noctuid with a high of 104 at Longridge Fell on 16 April 2014, and a peak of 65 (on 17 April 2022) in over 20 double-figure trap nights between 2019 and 2022 at Green Bank (P.J. Marsh, J. Roberts).

■ **HABITAT** Mainly moorland, but also woodland, hedgerows, gardens.
■ **FOODPLANTS** Not recorded in Lancashire.
Single brooded. One very late, but confirmed, record from 8 November 2020 at Yealand Conyers (B. Hancock).

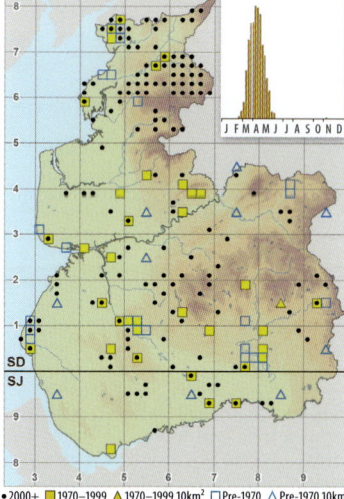

• 2000+ ☐ 1970–1999 ▲ 1970–1999 10km² ☐ Pre-1970 △ Pre-1970 10km²

The Moths of Lancashire

73.337 White-marked *Cerastis leucographa*

■ **LANCS STATUS** Rare / very local resident   1886–2022

This species was recorded in 1886 by J.B. Hodgkinson from the Preston area, but that seemed to have been a one-off. Whether it represented a small resident population, a migrant, or a vagrant from the Cumbrian or Yorkshire populations was unclear. It was therefore quite a surprise when a second, a male with its distinctively bipectinate antennae, turned up on 23 April 2022 at Green Bank, Upper Hindburndale, recorded by P.J. Marsh and J. Roberts.
■ **HABITAT** Woodland.
■ **FOODPLANTS** Unknown in the county.
Nationally, single brooded. The 2022 specimen was recorded at light.

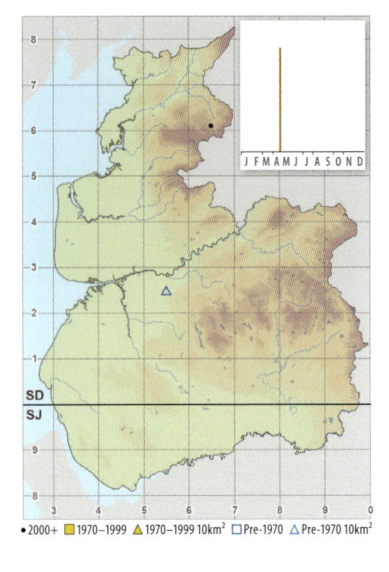

● 2000+   ■ 1970–1999   ▲ 1970–1999 10km²   □ Pre-1970   △ Pre-1970 10km²

73.338 True Lover's Knot *Lycophotia porphyrea*

■ **LANCS STATUS** Abundant / widespread resident   c.1857–2022

First reported from the Liverpool district as plentiful where heather grows (Gregson, 1857). Early records from Chat Moss, Longridge, Nelson and Withington. Coastal records from Lytham St Annes 1939 and Formby 1946. Most recent records are from moorland sites, e.g., Docker Moor, Green Bank, Mill Houses, etc, where it can be abundant, with highest count of 500 at Leck Fell on 9 July 2005 (P.J. Marsh) and several 100 plus sightings. Many garden records, presumably as a wanderer.
■ **HABITAT** Moorland. Occasional on mosses, e.g., Cadishead Moss in 2022.
■ **FOODPLANTS** Larvae swept from Heather at Astley Moss 2001 and Longridge Fell 2003.
Single brooded. Comes to light. Larvae noted from September to April.

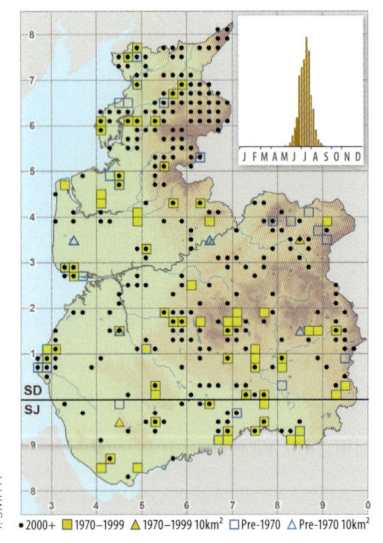

● 2000+   ■ 1970–1999   ▲ 1970–1999 10km²   □ Pre-1970   △ Pre-1970 10km²

73.339 Dotted Rustic *Rhyacia simulans*

■ **LANCS STATUS** Rare / very local resident   1847–2002

Reported by C.S. Gregson in his garden at Old Swan, Liverpool 1847 (Gregson, 1848). Noted at Lytham St Annes; described as scarce (Ellis, 1890). Next recorded in 1959, at Formby. The moth was regular at Billinge and Leigh in 1980s, but not seen since 1995. Recorded at Preesall, Fylde from 1985 to 1987, with all subsequent records in VC59, most recently at Littleborough and Hollins. However, records came to an abrupt stop in 2002, leading to concerns that the moth may possibly be extinct in the county.
■ **HABITAT** Open grassland, woodland.
■ **FOODPLANTS** Not recorded in Lancashire.
Single brooded. Comes to light.

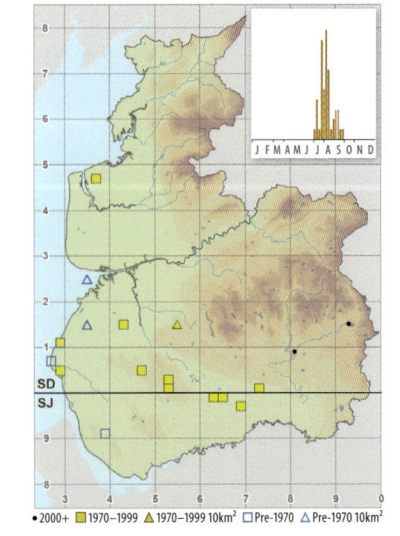

● 2000+   ■ 1970–1999   ▲ 1970–1999 10km²   □ Pre-1970   △ Pre-1970 10km²

73.341 Northern Rustic *Standfussiana lucernea*

■ **LANCS STATUS** Scarce / local resident   c.1880–2022

Only recorded on 30 occasions, the first at Turton, nr. Bolton in the late 1800s, nectaring on heather (J.W. Baldwin, 1880). Later, noted from Burnley and Oldham at start of the 20th century. Mansbridge (1940) documented a record at Catlow, nr. Nelson by A. Brindle, where the moth was noted to be 'not uncommon'. Post-2000 records from West Pennine moors (Belmont, Edenfield, Whitworth), Bowland Fells (Green Bank), the Lancashire portion of Hutton Roof and, in 2022, Yealand Conyers.
■ **HABITAT** Old quarries, moorland.
■ **FOODPLANTS** Not recorded in Lancashire.
Single brooded. All records from last 40 years were taken at light, all of singletons.

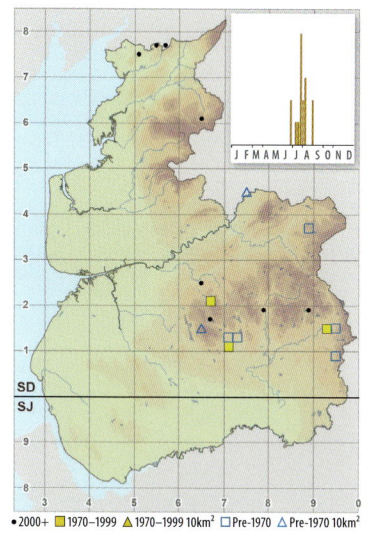

73.342 Large Yellow Underwing *Noctua pronuba*

■ **LANCS STATUS** Abundant / ubiquitous resident   c.1857–2022

Familiar to all moth trappers as an extremely common visitor to the light trap, often in large numbers, and usually disturbing much of the catch. First noted in Liverpool (Gregson, 1857), with early authors finding it common to abundant everywhere. More recently, recorded in all habitats, usually in good numbers, e.g., 156 in the dunes at Ainsdale in 2019, 720 from Green Bank on the fells in 2004, 866 at Pilling and 875 at Rishton, these latter two garden totals both recorded on 3 August 2019.
■ **HABITAT** Can be found in all habitats.
■ **FOODPLANTS** Yellow Melilot, Foxglove, forget-me-not, goosefoot.
Single brooded with extended generation. Larvae recorded from October to March.

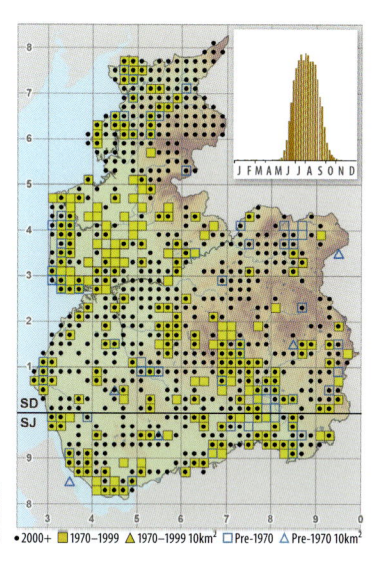

73.343 Broad-bordered Yellow Underwing *Noctua fimbriata*

■ **LANCS STATUS** Frequent / widespread resident   c.1857–2022

First recorded in the Liverpool district (Gregson, 1857). Other late 1800's records from Withington, Chat Moss, Crosby, Lytham St Annes, Preston. Early authors noted it to be generally distributed but not common. 106 records in 2002, 90 in 2012 and 242 in 2022 suggest increasing frequency, although does not reach anything like the numbers of the previous species. Widely distributed through lowland and upland areas. Highest count of 28 at light in Horwich on 23 August 2013 by R. Burkmar.
■ **HABITAT** Woodland, scrub, gardens.
■ **FOODPLANTS** Dog's Mercury, Sycamore seedlings.
Single brooded. Comes to light. Larvae recorded from March to May.

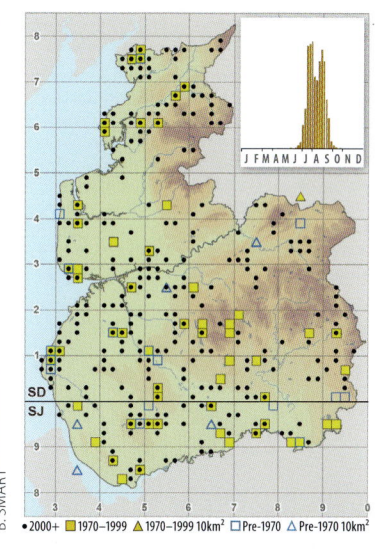

The Moths of Lancashire   **563**

73.345 Lesser Yellow Underwing *Noctua comes*

■ **LANCS STATUS** Abundant / ubiquitous resident   c.1857–2022
First recorded from Liverpool district where present 'in profusion' (Gregson, 1857). Later, noted as 'an abundant and generally distributed species, especially common in the larval stage' (Mansbridge, 1940). This remains true, as an early spring torch walk will reveal numerous larvae on grasses and low-growing plants in most habitats. Possibly declining in number; most high counts are from around 20 years ago.
■ **HABITAT** Gardens, meadows, scrub, woodland, moorland, etc.
■ **FOODPLANTS** As with other yellow underwing larvae, very polyphagous. Recorded on Dandelion, Creeping Willow, Ivy, Chard, Tulip, violet, clover and grasses.
Single brooded with extended generation. Larvae noted from January to May.

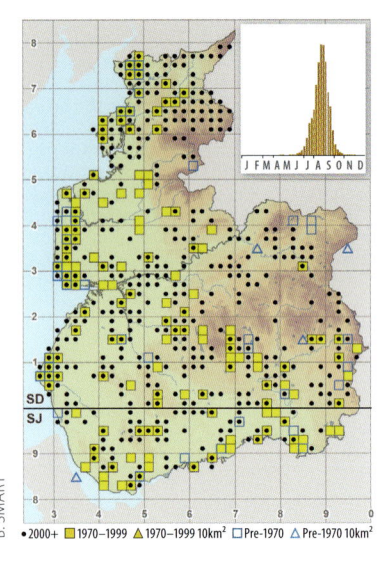

73.346 Least Yellow Underwing *Noctua interjecta*

■ **LANCS STATUS** Frequent / widespread resident   c.1885–2022
Earliest records from the Fleetwood area by J.B. Hodgkinson (Ellis, 1890), and at Lancaster, Hest Bank and Heysham by C.H. Forsyth (Mansbridge, 1940). The latter author described the moth as local and infrequent. Historically, less frequent in VC59, with first records from Oldham 1913, as larvae, and Formby dunes 1962. Light trapping has since shown the moth to be well distributed throughout, but with some evidence of recent decline. High count of 62 at Pennington, 5 August 1999 (P. Pugh).
■ **HABITAT** Hedgerows, scrub, gardens.
■ **FOODPLANTS** Not recorded in Lancashire.
Single brooded. Comes readily to light.

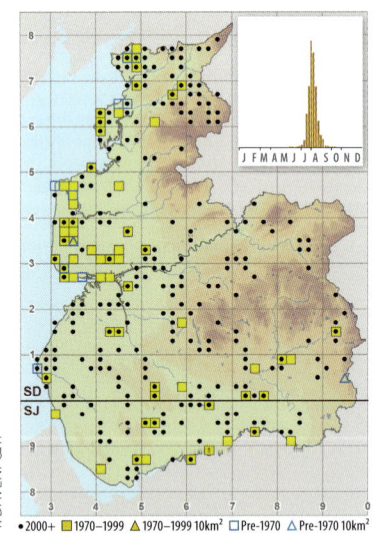

73.348 Lesser Broad-bordered Yellow Underwing *Noctua janthe*

■ **LANCS STATUS** Abundant / ubiquitous resident   c.1857–2022
Reported to be plentiful around Garston, Liverpool (Gregson, 1857). Next, recorded at Preston by Ellis (1890), giving status as not generally common. By 1940, known from many other sites, e.g., Ditton, Silverdale and Holden Clough, nr. Oldham, and noted to be widely distributed (Mansbridge, 1940). Remains plentiful throughout, although most of the high counts, such as 140 to light at Flixton 1997 by K. McCabe, are from around the turn of this century, suggesting possible declining abundance.
■ **HABITAT** Grasslands, hedgerows, gardens.
■ **FOODPLANTS** Larvae found in flower beds and in leaf litter; foodplants not recorded.
Single brooded. Comes to light. Larvae recorded from December to early May.

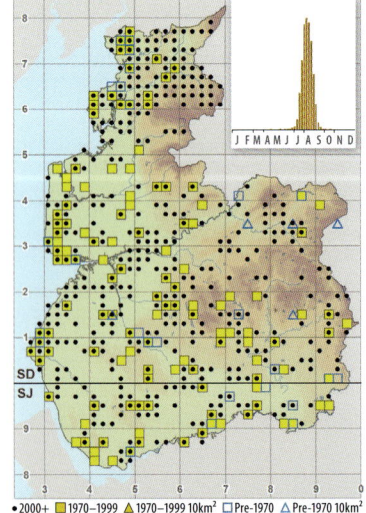

73.349 Stout Dart *Spaelotis ravida*

■ **LANCS STATUS** Considered Extinct   c.1885–1997

Only recorded in the county on six occasions, the first of which was in the late 19th century at Rusholme, Manchester by G.W. Adams (Ellis, 1890), with no further records documented by Mansbridge (1940). The moth was also recorded from Ditton around 100 years ago, date uncertain, by F.M.B. Carr. The next record, in 1972, was a moth noted at sugar in Thornton by J. Thompson. It was later recorded at Woolston Eyes in 1986 (recorder unknown) and to light at Chorley on 3 July 1986 by F. Smith. The final Lancashire record for this species was at Flixton on 16 August 1997, recorded at light by K. McCabe. Nationally, 'on the verge of extinction, if not already extinct' (Randle *et al.*, 2019).

■ **HABITAT** Marshes, grassland.
■ **FOODPLANTS** Not recorded in Lancashire.
Single brooded. Twice noted at light, once at sugar.

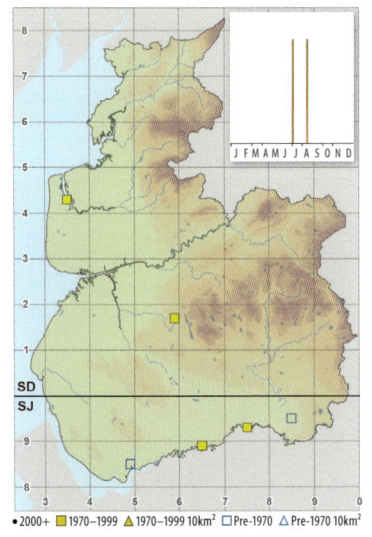

73.350 Great Brocade *Eurois occulta*

■ **LANCS STATUS** Occasional / migrant   c.1859–2019

First recorded at Chorlton by J. Chappell, with other early records from Stretford and Crosby (Ellis, 1890). Ellis noted the moth to be generally distributed but not common, suggesting migrant status not understood. The moth is resident in parts of Scotland but there is no evidence of the same here. All records are of adults, mainly at light, and almost all in August. Records particularly numerous in certain years, e.g., 11 in 2006 and 7 in 2007, but only once since, at Rishton on 3 August 2019 (D. Bickerton).

■ **HABITAT** Unknown.
■ **FOODPLANTS** Not recorded in Lancashire.
Single brooded. Mostly singletons, although three at Dolphinholme 2007.

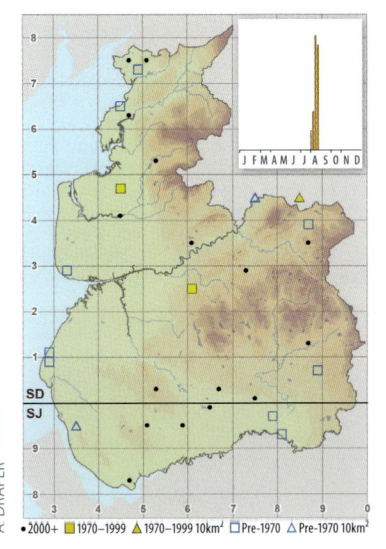

73.351 Double Dart *Graphiphora augur*

■ **LANCS STATUS** Occasional / local resident   c.1857–2022

First recorded in the Liverpool district, attracted by sugar (Gregson, 1857), with Ellis (1890) describing the moth as common and generally distributed. Widely recorded in 20th century, with reports from most lowland and a few upland sites. However, post-2000, it has become more local and concentrated in the west, e.g., Heysham, Sefton Coast, Middleton N.R. Occasional inland records at Mere Sands Wood, Docker Moor, Flixton and Abram, nr. Wigan, etc. Numbers of records have declined in last decade.

■ **HABITAT** Woodland, gardens, scrub, marshes, etc.
■ **FOODPLANTS** Unrecorded. Bred ex-larva at Withington 1925 by L. Nathan.
Single brooded. Comes to light. High count of 17 at Middleton N.R. 2018 (J. Mason).

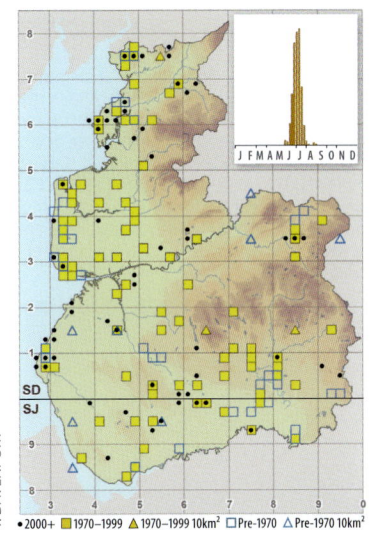

NOCTUIDAE

73.352 Green Arches *Anaplectoides prasina*

■ **LANCS STATUS** Frequent / widespread resident   c.1843–2022

First reported in Preston area (Hodgkinson, 1844), and at Chat Moss, where noted as scarce (Edleston, 1846). Other 19th century records from Woolton, Hale, Lytham St Annes and Botany Bay Wood, Astley. Early authors described it as local and nowhere common. The post-1970 period saw distribution increasing, particularly in eastern upland areas, with high counts from Rishton, Wray, Worsthorne, Belmont. Counts show some localised decline in last decade; but increases elsewhere, e.g., Preston.

■ **HABITAT** Woodland.

■ **FOODPLANTS** Dog's Mercury, at Eaves Wood 2 April 2004 (S. Palmer); moth reared. Single brooded. Comes readily to light.

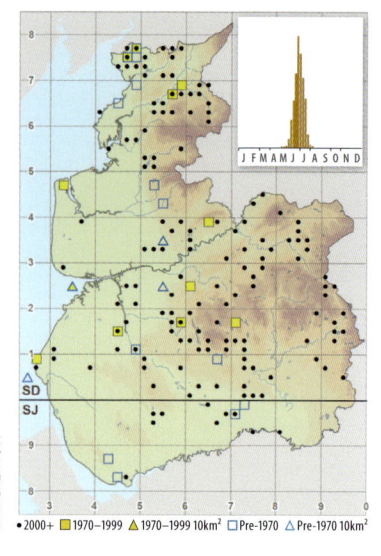

• 2000+   ☐ 1970–1999   △ 1970–1999 10km²   ☐ Pre-1970   △ Pre-1970 10km²

73.353 Dotted Clay *Xestia baja*

■ **LANCS STATUS** Frequent / widespread resident   c.1843–2022

First recorded by J.B. Hodgkinson (1844) in the Preston area, and from Liverpool, with Gregson (1857) noting its abundance everywhere. Britten (c.1950) reported the moth to be common in woods, mosses and sheltered parts of moorland. It remains widely recorded at coastal sites such as Heysham and Ainsdale, and in upland areas such as Docker Moor. However, numbers of records from 2021–2022 (402) were barely more than half those of 2011–2012 (724), with decline mainly apparent in lowland areas.

■ **HABITAT** Damp grassland, woodland and scrub.

■ **FOODPLANTS** Bog Myrtle, at Cockerham Moss 1 June 2011 (G. Jones); moth reared. Single brooded. Comes to light, including 130 at Worsthorne 2021 (G. Gavaghan).

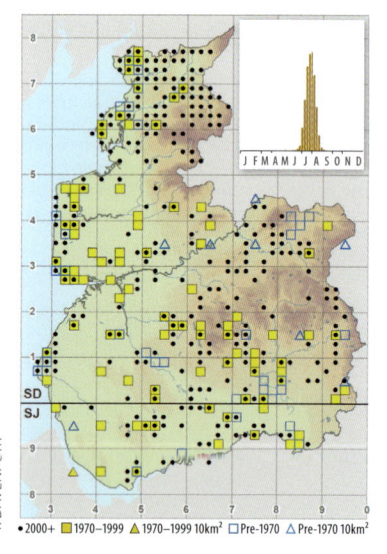

• 2000+   ☐ 1970–1999   △ 1970–1999 10km²   ☐ Pre-1970   △ Pre-1970 10km²

73.355 Neglected Rustic *Xestia castanea*

■ **LANCS STATUS** Occasional / local resident   c.1857–2022

First from Hale (Gregson, 1857); other early records from Pendleton, Chat Moss, Rixton Moss and Dutton, nr. Ribchester. Noted by Ellis (1890) to be scarce. Only recorded 24 times prior to 2000, then an increase due to extensive moorland recording in VC60 (P.J. Marsh, Ga. Jones *et al.*). Largest single night counts of around 50 on Longridge Fell (2012), 29 at Docker Moor (2021) and 22 on Baines Crag (2017). Smaller numbers on lowland heaths, dunes and mosses and sporadic in garden traps.

■ **HABITAT** Moorland, lowland mosses, heaths.

■ **FOODPLANTS** Larvae swept from Heather during March and April. Single brooded. Comes readily to light.

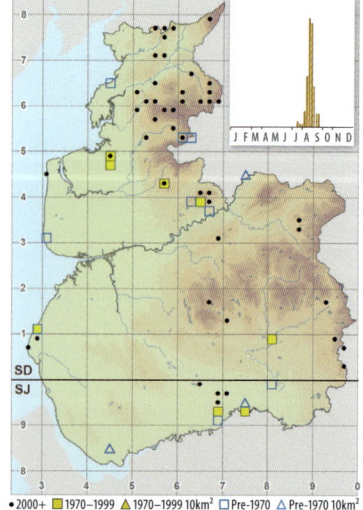

• 2000+   ☐ 1970–1999   △ 1970–1999 10km²   ☐ Pre-1970   △ Pre-1970 10km²

The Moths of Lancashire

73.356 Heath Rustic *Xestia agathina*

- **LANCS STATUS** Occasional / local resident   c.1857–2022

First noted at Simonswood Moss, nr. Kirkby (Gregson, 1857). Reported by Ellis (1890) to be scarce, with Mansbridge (1940) noting that it was commonly found as a larva. Remains so, with 25 larval records this century. P.J. Marsh (pers. comm.) reports a preference for tall, unmanaged Heather, with counts of 20+ from Docker Moor bothy in 2018 and 2021 contrasting with lower numbers on managed moorland. Lowland sites include Astley and Cadishead Mosses, Freshfield Dune Heath and Hoghton.
- **HABITAT** Moorland, lowland mosses and heaths.
- **FOODPLANTS** Larvae swept from Heather and Cross-leaved Heath. Adult reared 2011.

Single brooded. Comes readily to light. Larvae recorded during March and April.

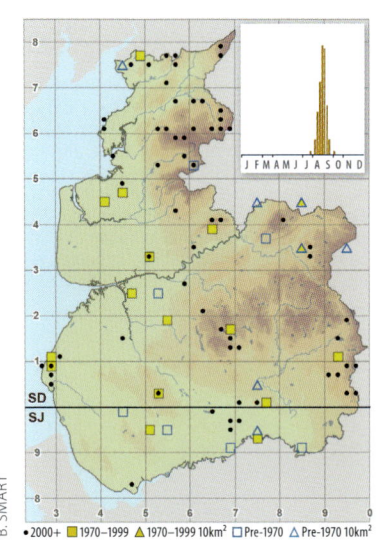

73.357 Square-spot Rustic *Xestia xanthographa*

- **LANCS STATUS** Abundant / ubiquitous resident   c.1844–2022

First records from Manchester (Edleston, 1844) and Hale (Gregson, 1857), with the latter describing this as 'the most abundant noctuid we have'. Noted nectaring on heather at Turton in 1880. Ellis (1890) and Mansbridge (1940) both assessed it to be abundant throughout. Has remained so, with highest count of 342 at Fowley Common, nr. Culcheth 25 August 2006 by J.D. Wilson. Other high totals from Flixton, Heysham and Green Bank, the latter confirming significant upland presence.
- **HABITAT** A wide range, including grassland and gardens.
- **FOODPLANTS** Unrecorded. Bred ex-larva, West Didsbury 1925 (L. Nathan).

Single brooded. Comes to light in numbers. Larvae recorded from January to April.

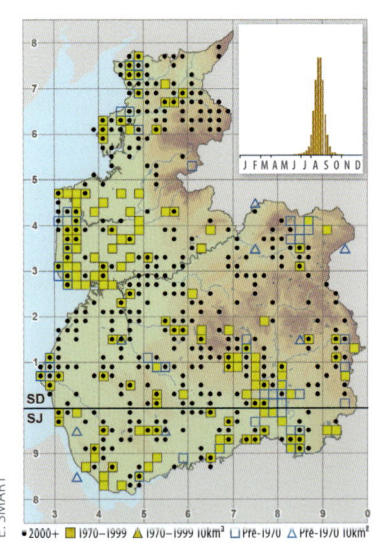

73.358 Six-striped Rustic *Xestia sexstrigata*

- **LANCS STATUS** Frequent / widespread resident   c.1857–2022

The earliest records are from the Liverpool district (Gregson, 1857), and at Preston, Burnley, Eccles, Chat Moss and Lytham St Annes. Mansbridge (1940), considered it widespread, and particularly common on the coast. Records since the middle of last century confirm its presence in all habitats, as with the previous species, although generally with much lower counts. Numbers have declined of late with the highest counts from around 20 years ago, including 49 at Worsley 14 August 2005 (I. Walker).
- **HABITAT** Woodland, grassland, gardens.
- **FOODPLANTS** Not recorded in Lancashire.

Single brooded. Comes readily to light.

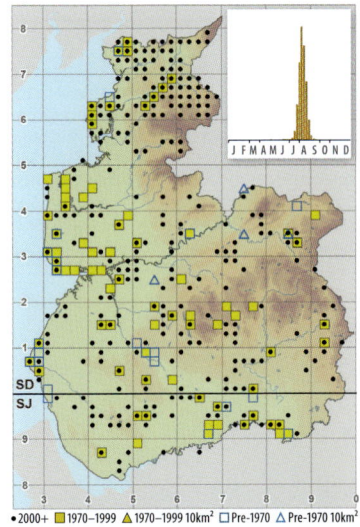

NOCTUIDAE

73.359 Setaceous Hebrew Character *Xestia c-nigrum*

■ **LANCS STATUS** Abundant / widespread resident   c.1857–2022

Reported from Liverpool (Gregson, 1857), with most early authors noting abundance. Brindle (1950) disagreed, reporting it scarce in East Lancs. It remains true that range is patchier in eastern, upland areas, although it is not scarce. There have been high counts of around 500 in lowland areas, such as Great Eccleston and Halebank, during the last few years. Possibly increasing in abundance, although the picture is variable.
■ **HABITAT** A wide range, including grassland and gardens.
■ **FOODPLANTS** Larva on Tulip leaves, at Flixton 11 April 2004 (K. McCabe); moth reared. Double brooded, with second brood now more numerous. Occasionally noted in late autumn, e.g., at Leighton Moss 11 November 2022 by B. Hancock. Comes to light.

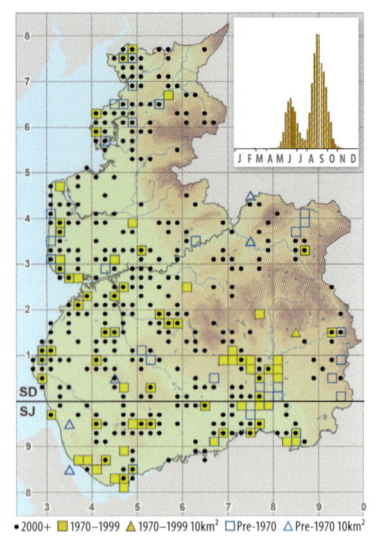

• 2000+   ■ 1970–1999   ▲ 1970–1999 10km²   □ Pre-1970   △ Pre-1970 10km²

73.360 Triple-spotted Clay *Xestia ditrapezium*

■ **LANCS STATUS** Occasional / local resident   1921–2022

First noted in 1921 at Hightown by H.B. Prince. All records were from Sefton Coast until late 1950s, although not seen there again until 2010 at Formby. Noted at Gait Barrows, Warton Crag and Leighton Moss since 1959, all on limestone at Morecambe Bay, and more recently, at Dalton Crags and Docker Moor. Also, recorded from Longridge 2020 and Flixton 2012– 2013, whether as vagrants or representatives of isolated populations is unclear. Care needed to distinguish from Double Square-spot.
■ **HABITAT** Limestone woodland and grassland.
■ **FOODPLANTS** Not recorded in Lancashire.
Single brooded. Comes to light. Larva at Freshfield 29 April 1933 (W. Mansbridge).

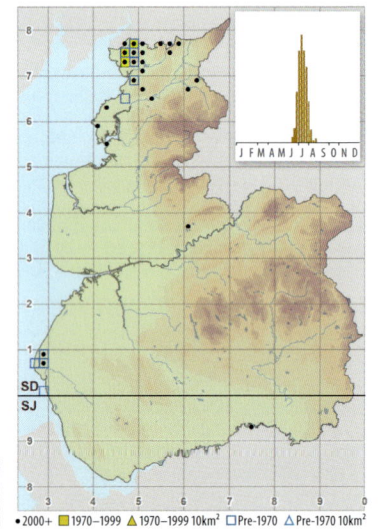

• 2000+   ■ 1970–1999   ▲ 1970–1999 10km²   □ Pre-1970   △ Pre-1970 10km²

73.361 Double Square-spot *Xestia triangulum*

■ **LANCS STATUS** Abundant / widespread resident   c.1843–2022

First recorded by Hodgkinson (1844) in the Preston area. Early authors reported the species to be uncommon, although numerous sites listed, including Liverpool, Longridge, Lytham St Annes and Prestwich. Mansbridge (1940) noted particular scarcity in the east of the county. The map shows range extending to eastern, more upland, areas in recent decades, with records increasing rapidly from 2000; possibly stalling in last five years. High counts of 46 at Ainsdale in 2011 and Hutton in 2012.
■ **HABITAT** Woodland, hedgerows, gardens.
■ **FOODPLANTS** Hawthorn (Gregson, 1857), although plant named as 'whitethorn'.
Single brooded. Most records are of moths to light. Larvae recorded March and April.

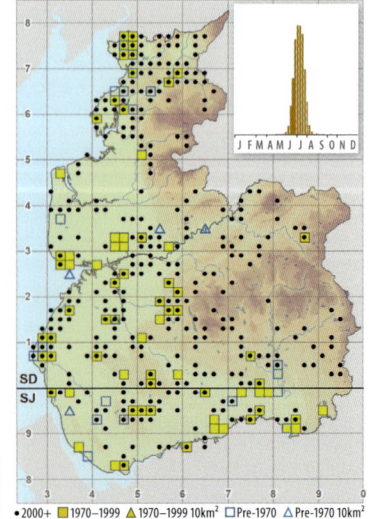

• 2000+   ■ 1970–1999   ▲ 1970–1999 10km²   □ Pre-1970   △ Pre-1970 10km²

568   *The Moths of Lancashire*

73.365 Autumnal Rustic *Eugnorisma glareosa*

■ **LANCS STATUS** Frequent / widespread resident c.1857–2022
First recorded from Woolton (Gregson, 1857), with other 19th century records from Lostock, Longridge and Turton. Noted by Ellis (1890) to be local and not common, with Mansbridge (1940) giving status as occasional. Widely distributed, other than a near absence from the Fylde, where only once noted, in 2003 at Lytham. Counts of records and moths have declined post-2000, with 2022 records primarily from upland sites, e.g., Green Bank, Leck Fell, with others at Ainsdale, gardens, etc.
■ **HABITAT** Rough grassland, moorland.
■ **FOODPLANTS** Not recorded in Lancashire.
Single brooded. Comes to light. High count of 90 at Docker Moor 1 September 2021.

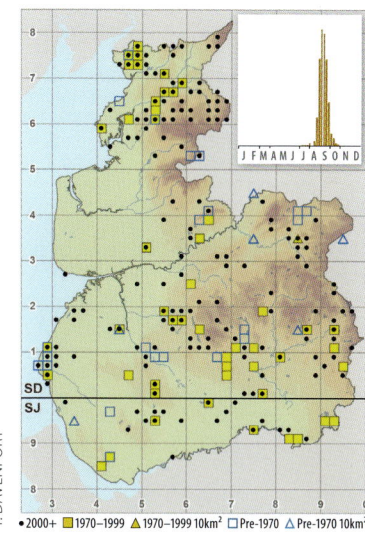

73.368 Gothic *Naenia typica*

■ **LANCS STATUS** Frequent / widespread resident c.1857–2022
First recorded from the Liverpool district (Gregson, 1857), with abundance around brambles noted. Mansbridge (1940) reported the moth to be common everywhere, although all pre-1950 records were in southern and central areas. Light trapping demonstrated a more extensive range; including at Morecambe in 1959. 21st century records have confirmed its ongoing presence in most lowland and upland habitats throughout, although with declining counts (none above four since 2014).
■ **HABITAT** Woodland, hedgerows, scrub, waste ground, marshes, gardens.
■ **FOODPLANTS** Great Willowherb, Butterbur, Heather, primrose.
Single brooded. Comes readily to light. Larvae found in March and April.

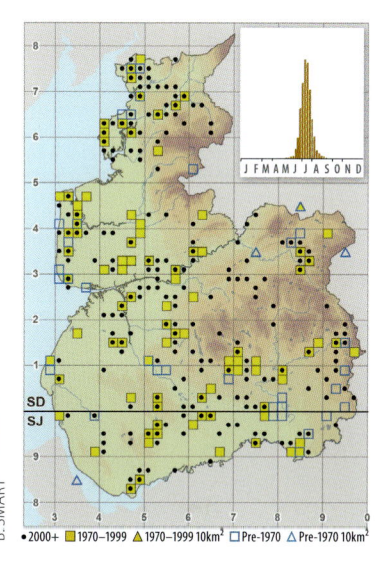

74.002 Kent Black Arches *Meganola albula*

■ **LANCS STATUS** Rare / vagrant 2022–2023
First recorded in Preston on 22 July 2022 by S. & C. Palmer to MV light trap during a period of very warm weather when at least one other non-resident species, Four-spotted Footman, was widely reported in the county. A second came to light at Abram, Wigan on 14 July 2023 (J. Smith). Randle *et al.* (2019), describe the moth as rapidly expanding its range northwards from a southern English base. Its arrival in Lancashire is considered as a likely continuation of this northward expansion, accelerated by the prevailing climatic conditions in those summers.
■ **HABITAT** Unknown.
■ **FOODPLANTS** Not recorded in Lancashire.

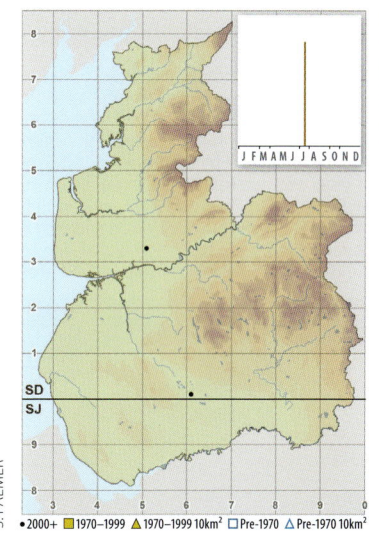

The Moths of Lancashire 569

NOLIDAE

74.003 Short-cloaked Moth *Nola cucullatella*

■ **LANCS STATUS** Occasional / widespread resident   c.1859–2022

First found in the Manchester area (Stainton, 1859) and since recorded regularly throughout the county. Early authors, including Leech & Michaelis (1957), list it as fairly common. Britten (c.1950) agreed, but added it was scarce in the eastern hills. This status has remained unchanged in recent years except that there has been a slight increase in records from upland areas. Usually found in low single figures.
■ **HABITAT** Hedgerows, woodland edge, scrub and brownfield sites.
■ **FOODPLANTS** Hawthorn. The few larval records have all been in May.
Single brooded, with an occasional, small second brood in 1972, 1998 and four other years post 2000. Comes to light. One very early record on 19 May 2012, in Formby.

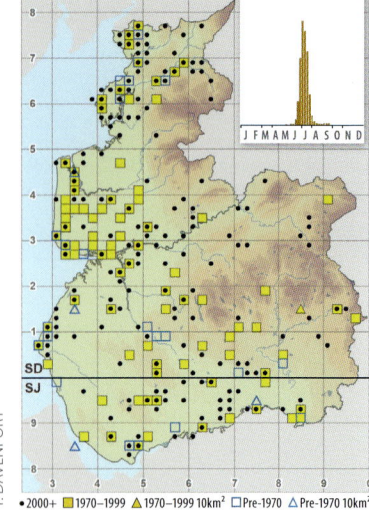

74.004 Least Black Arches *Nola confusalis*

■ **LANCS STATUS** Occasional / widespread resident   1909–2022

First found in Ainsdale by W.A. Tyerman (1909) and Silverdale by W.P. Stocks (1911). It was not until after 1948 that a more general expansion in range occurred. Initially this was restricted to coastal and lowland areas but reached Claughton, for example, on the edge of higher ground, in 1987. It was widespread in lowland areas by the 1990s and reached many parts of the uplands by the early 2000s. Usually found in small numbers, but in prime habitat can reach double figures, e.g., Silverdale Moss.
■ **HABITAT** Open woodland, scrub, suburban and rural gardens.
■ **FOODPLANTS** Not recorded in Lancashire.
Single brooded; a small second brood has appeared in recent years in coastal areas.

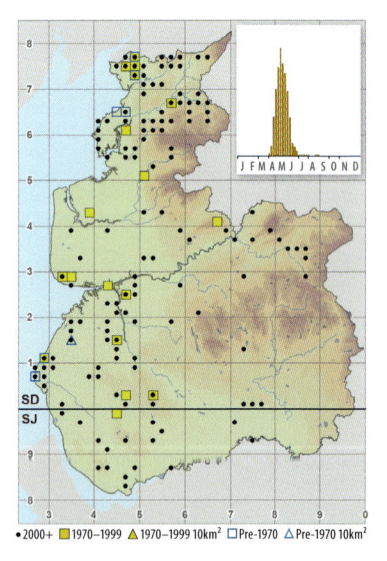

74.007 Scarce Silver-lines *Bena bicolorana*

■ **LANCS STATUS** Occasional / widespread resident   1961–2022

First recorded, in Formby in 1961 by M. Pons, followed by one in Didsbury in 1977. By the late 1990s it had spread to several parts of the Manchester and Liverpool districts but was very infrequent in occurrence. The range expansion continued, reaching Preston by 2000 and Leighton Moss in 2013. The species is usually a 'once-a-year' moth at most sites. It remains scarce in the upland areas, with singles noted in Rochdale in 2006 and 2013 and in the Burnley area in 2020 and 2022.
■ **HABITAT** Deciduous woodland and older hedgerows or gardens with mature oaks.
■ **FOODPLANTS** Oak sp. Larvae have been found in September and May.
Single brooded, with possible second brood individuals in September 2003 and 2014.

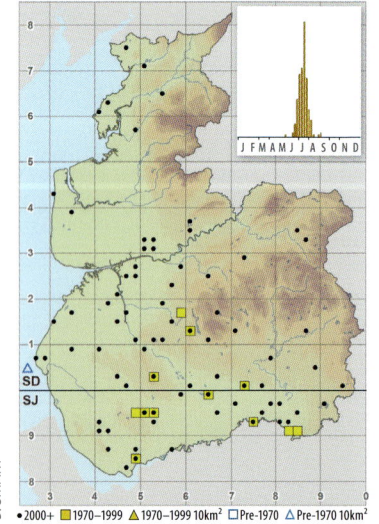

570 *The Moths of Lancashire*

74.008 Green Silver-lines *Pseudoips prasinana*

■ **LANCS STATUS** Occasional / widespread resident   c.1857–2022

First in Hale (Gregson, 1857), followed by singles in Bolton and Silverdale (1886), and Sabden (1939). Light trapping has helped to determine its distribution, but it also appears to have expanded its range in recent years, avoiding treeless upland areas. A second brood was first suspected in 1982 at Salwick and on eleven occasions since then. Generally found in low single numbers, with eight at Docker Moor the largest single count. An early individual was noted in Flixton on 24 March 2020 (K. McCabe).
■ **HABITAT** Deciduous woods, parks, gardens and tree-lined edges of mosses and moors.
■ **FOODPLANTS** Beech, birch, oak. Larva found in August and September.
Single brooded, with an occasional second brood. Readily attracted to light.

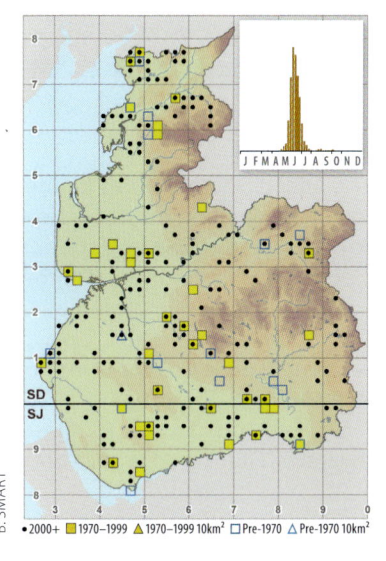

74.009 Oak Nycteoline *Nycteola revayana*

■ **LANCS STATUS** Occasional / widespread resident   1938–2022

With a few 19th century records from all adjacent counties, it was perhaps surprising this moth was not noted in Lancashire until 1938, in Silverdale per. H.N. Michaelis. The next was not found until 1980 in Formby. It was only during the early 1990s, in Billinge, and mid-1990s, in Flixton, that it started appearing with any regularity, perhaps significantly at MV light. Since then, it has been reported widely from lowland regions in small numbers, but remains very local in upland areas.
■ **HABITAT** Deciduous woodland and areas with scattered mature Oak trees.
■ **FOODPLANTS** Oak. Larvae infrequently found, in June.
Single brooded, with an extended flight period, overwintering as an adult.

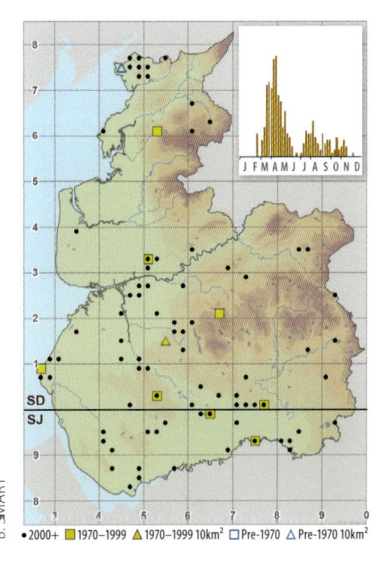

74.011 Cream-bordered Green Pea *Earias clorana*

■ **LANCS STATUS** Rare / vagrant   2002

Only one county record, of a singleton attracted to MV light on 17 August 2002 in Flixton (K. McCabe). At the time, the nearest known sites were over 50km to the east, in Yorkshire and 80km to the south, in Shropshire. Perhaps significantly, it was reported to be expanding its range in Yorkshire during this decade (Randle *et al.*, 2019). The presence of wetland habitat in the Mersey Valley, close to this site, may in the future prove suitable for this species, if this hasn't already occurred.
■ **HABITAT** Not known, having come to a garden light trap on the edge of suburban south-west Manchester.
■ **FOODPLANTS** Not recorded in Lancashire.

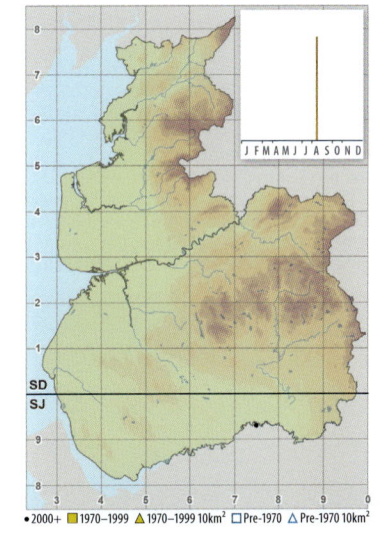

The Moths of Lancashire   **571**

Addendum: New Lancashire records for 2023

The original intention was for the book to include records up to and including 2022 only, with 2023 devoted to writing the book. However, as the year progressed, it became obvious that we also needed to include the eleven new species that were recorded in Lancashire during the year, although we were unable to slot these into the main section according to their ABH number order. These eleven species are therefore presented here in this section. Whilst over half are migrants, perhaps encouraged to venture further north in an era of global warming, at least four appear to be hitherto unrecorded residents, suggested there may be other unrecorded species already present in the county, that will be detected in future years.

We have added a map for these new species to show the very few, generally single, recorded locations. Flight graphs have not been included as all dates recorded are listed in the text.

PLUTELLIDAE

18.006 *Rhigognostis incarnatella*

■ **LANCS STATUS** Rare / very local resident   2013–2023

Added to the Lancashire list when a female was caught at light at Ainsdale NNR on 15 November 2023 by R. Moyes. Dissection (by S. Palmer) was required to separate the moth from the rare resident, *R. annulatella*.
Subsequently, a moth trapped at the National Trust site in Formby on 8 August 2013 by R. Walker, previously thought to be *R. annulatella*, was dissected and found to be the first county record of *R. incarnatella*.
Searches for larvae on Dame's-violet and other Brassicaceae may help to reveal the moth's true status in the county.

● 2000+ ■ 1970–1999 ▲ 1970–1999 10km² □ Pre-1970 △ Pre-1970 10km²

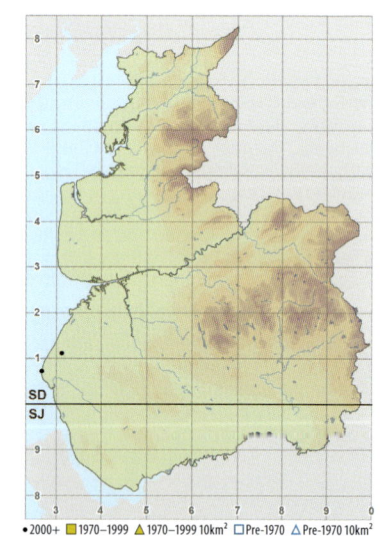

CHOREUTIDAE

48.003 *Prochoreutis sehestediana*

■ **LANCS STATUS** Rare / very local resident   2023

Extensive larval spinnings were noted on 3 July 2023 on Skullcap, growing on the southern bank of Chorlton Water Park, Manchester (B. Smart). Two larvae were retained and quickly completed feeding and spun their white pupal cocoons. Both moths emerged on 17 July 2023. One was dissected by S. Palmer and confirmed as the first Lancashire record for this species.
P. myllerana, a very similar moth, had previously been recorded at the same site, on the northern bank of the lake, although not since 2007. The long-term future of both species at this site will rely on the continued presence of the foodplant, Skullcap, growing in partially shaded, often disturbed ground, at the water's edge.

● 2000+ ■ 1970–1999 ▲ 1970–1999 10km² □ Pre-1970 △ Pre-1970 10km²

572 *The Moths of Lancashire*

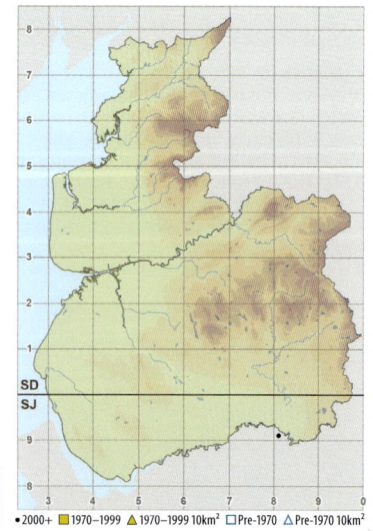

49.356 *Grapholita lobarzewskii*

■ **LANCS STATUS** Scarce / very local resident   2023

This moth is one of 14 that entered a pheromone trap during the night of 13 June 2023 in Chorlton, Manchester (B. Smart). It is likely to be significant that the gardens either side both contain apple trees, the national foodplant of this species.
There have been similar recent finds in neighbouring counties, with the first Yorkshire record in 2022 and the first Cheshire record in 2018. It is unclear whether the moth is actually spreading, or if the records are just a result of increasing use of the newly developed pheromone lure for this moth.

• 2000+   □ 1970–1999   ▲ 1970–1999 10km²   □ Pre-1970   △ Pre-1970 10km²

TORTRICIDAE

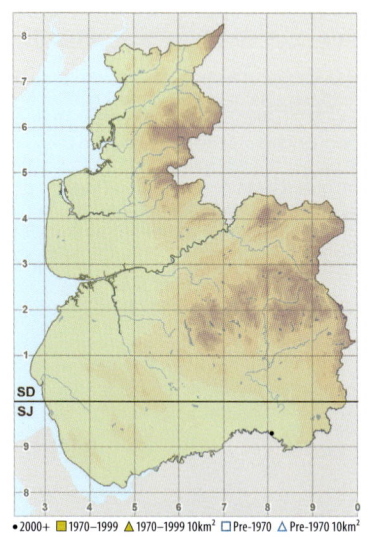

52.011 Red-belted Clearwing *Synanthedon myopaeformis*

■ **LANCS STATUS** Rare / very local resident   2023

A male was recorded in B. Smart's Chorlton garden on 16 June 2023. Although the lure used was for the Large Red-belted Clearwing, there is no doubt about the identity of the moth. It had white palps, a red band across the abdomen and lacked any red on the upperside of the forewings.
Although previous occurrence in the county was reported by Skinner (1983) and Waring and Townsend (2003), this may relate to records from Grange, Cumbria. The first VC60 record, a female (right), was also noted this year, not at a lure, but spotted at a garden centre in Brock, nr. Garstang on 25 June 2023 (A. Baines).

• 2000+   □ 1970–1999   ▲ 1970–1999 10km²   □ Pre-1970   △ Pre-1970 10km²

SESIIDAE

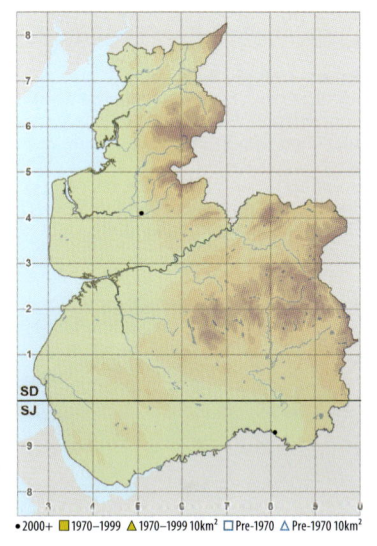

63.047 *Spoladea recurvalis*

■ **LANCS STATUS** Rare / migrant   2023

This species was first trapped by B. Hancock at a Silverdale garden on 8 October 2023. The moth had an amazingly successful few days as further individuals turned up the following night at Hesketh Bank, near Southport, and on 10 October, at Leighton Moss (R. Fellows). The Hesketh Bank moth, recorded by R. Yates, was the first in VC59. All three individuals were trapped at light.
The species can occur as an adventive (Parsons & Clancy, 2023) but there seems no doubt these records were part of a wider migratory movement as the moth was also recorded in Scotland, Cumbria, Durham, Leicestershire, Norfolk and elsewhere during the same week.

• 2000+   □ 1970–1999   ▲ 1970–1999 10km²   □ Pre-1970   △ Pre-1970 10km²

CRAMBIDAE

The Moths of Lancashire

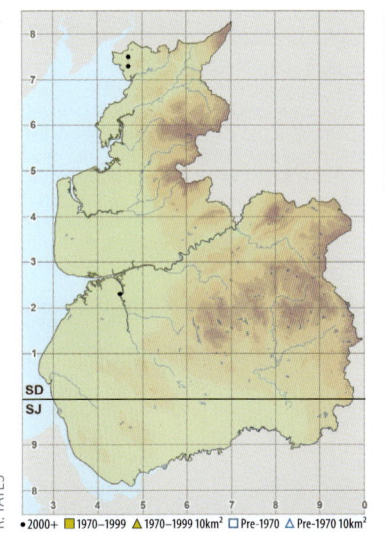

GEOMETRIDAE

70.064 Cloaked Carpet *Euphyia biangulata*

■ **LANCS STATUS** Rare / very local resident, possible vagrant   2023

Trapped at light in a Southport garden on 28 July 2023 by J. Worthington. Perhaps, less of a surprise than some of the other new arrivals from this year, as the nearest populations are relatively close, in the Wirral portion of Cheshire (VC58).
Whether this is a wanderer from south of the Mersey, where the moth has been recorded twice this century, although not since 2011 (cheshire-chart-maps.co.uk, accessed 19 March 2024), or part of a hitherto undetected local population is uncertain. Larval searches on stitchworts in the Southport area and elsewhere could be helpful.

● 2000+   □ 1970–1999   ▲ 1970–1999 10km²   □ Pre-1970   △ Pre-1970 10km²

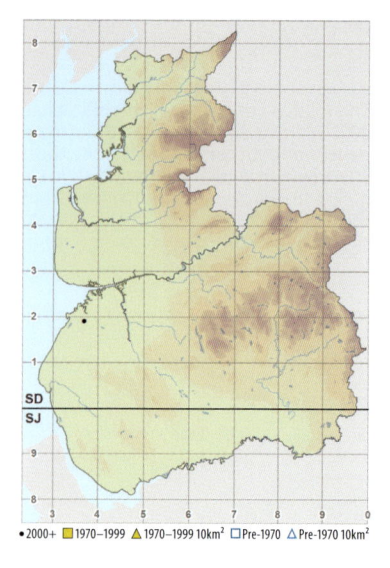

70.217 Rannoch Looper *Macaria brunneata*

■ **LANCS STATUS** Rare / migrant   2023

Recorded by J. Bowen at his light trap in Horwich on the night of 16 June 2023. Photographs of the moth, a female, were compared by R. Walker with a Scottish set specimen. Upperside and underside views confirmed the identification.
Whilst populations do occur in central Scotland, this was part of a mass migration event from mainland Europe, resulting in many national records of the species around the same date, including reports from Cumbria, Yorkshire, Cheshire, Derbyshire and Nottinghamshire.

● 2000+   □ 1970–1999   ▲ 1970–1999 10km²   □ Pre-1970   △ Pre-1970 10km²

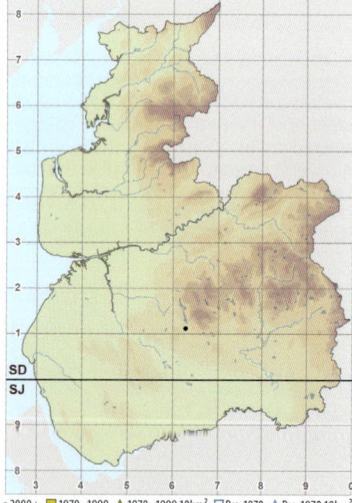

EREBIDAE

72.074 Beautiful Marbled *Eublemma purpurina*

■ **LANCS STATUS** Rare / migrant   2023

Only recorded in the UK for the first time in 2001, at Portland, Dorset (Randle *et al.*, 2019), with subsequent records restricted to the south of England. It was therefore an unexpected addition to the county list when one appeared in P. Ellis's moth trap at Preesall on 6 September 2023, the most northerly UK record to date.
This was a time of high migrant activity, with airstreams moving up from Iberia, leaving widespread deposits of Saharan dust and a number of interesting migrants locally, of which this was the undoubted highlight.
Single brooded. Occasional at light. Far more often encountered during day.

● 2000+   □ 1970–1999   ▲ 1970–1999 10km²   □ Pre-1970   △ Pre-1970 10km²

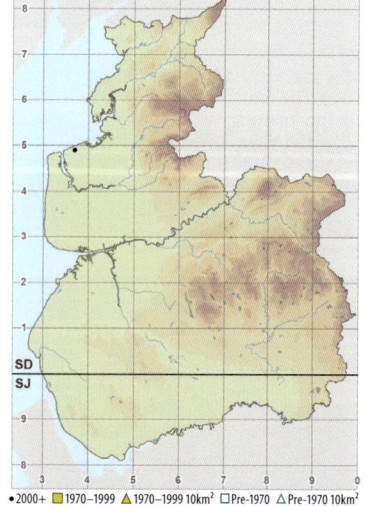

73.004 Slender Burnished Brass *Thysanoplusia orichalcea*

■ **LANCS STATUS** Rare / migrant   **2023**

Recorded at MV light by D. Lambert at Longridge on the night of 19 August 2023, this was the first report of this distinctive migrant in northern England, with the 100 or so previous British records mainly from the Isles of Scilly, Cornwall and elsewhere on the English south coast (Randle *et al.*, 2019).
The few specimens previously recorded in the north of the UK included one from Gwynedd in 2017 and another from County Down in 1989.

S. PALMER

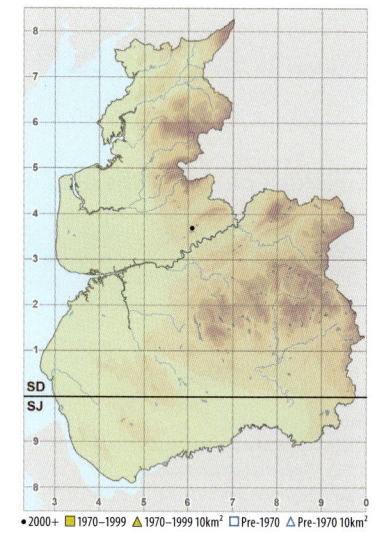

• 2000+   ▪ 1970–1999   ▲ 1970–1999 10km^2 ▫ Pre-1970   △ Pre-1970 10km^2

73.009 Tunbridge Wells Gem *Chrysodeixis acuta*

■ **LANCS STATUS** Rare / migrant   **2023**

The first county specimen of this moth, a migrant species native to the Canary Islands, Africa and further east, was recorded at Hesketh Bank on 9 October by R. Yates. This was a very successful night for the recorder as the MV trap also held Vestal, Scarce Bordered Straw, and the first VC59 record of *Spoladea recurvalis*. Nationally, the moth is considered a scarce immigrant (Randle *et al.*, 2019), with previous records from Cheshire in 2015 and from Wakefield, Yorkshire in 2022 (cheshire-chart-maps.co.uk; yorkshiremoths.co.uk, both accessed 19 March 2024).

R. YATES

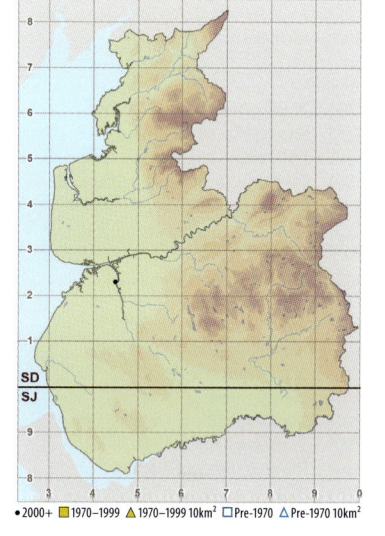

• 2000+   ▪ 1970–1999   ▲ 1970–1999 10km^2 ▫ Pre-1970   △ Pre-1970 10km^2

73.082 Tree-lichen Beauty *Cryphia algae*

■ **LANCS STATUS** Rare / probable migrant   **2023**

Came to light at Flixton on 19 July 2023, recorded by K. McCabe. It is unclear whether this represents a vagrant from colonies in south-east England or a migrant.
The English populations became established during the early part of this century, and are expanding north and west. As with a number of other lichen-feeders, this moth is increasing in numbers and range. Since 2018, there have been a number of records from southern parts of Yorkshire, but not in our other neighbouring counties. The sole Cheshire record is from 1859 (cheshire-chart-maps.co.uk; yorkshiremoths.co.uk, both accessed 19 March 2024).

B. SMART   B. SMART

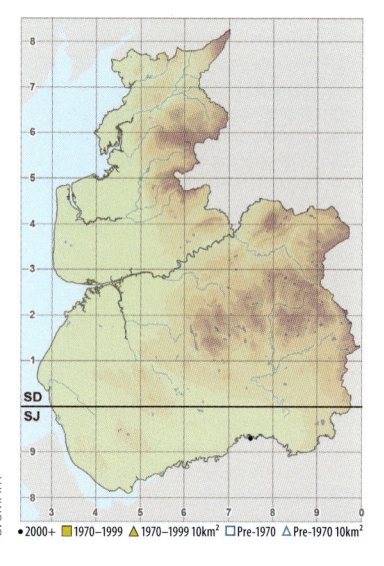

• 2000+   ▪ 1970–1999   ▲ 1970–1999 10km^2 ▫ Pre-1970   △ Pre-1970 10km^2

NOCTUIDAE

Appendix 1

Checklist of the Moths of Lancashire

MICROPTERIGIDAE
1.001 *Micropterix tunbergella* (Fabricius, 1787)
1.002 *Micropterix mansuetella* (Zeller, 1844)
1.003 *Micropterix aureatella* (Scopoli, 1763)
1.004 *Micropterix aruncella* (Scopoli, 1763)
1.005 *Micropterix calthella* (Linnaeus, 1761)

ERIOCRANIIDAE
2.001 *Dyseriocrania subpurpurella* (Haworth, 1828)
2.002 *Paracrania chrysolepidella* (Zeller, 1851)
2.003 *Eriocrania unimaculella* (Zetterstedt, 1839)
2.004 *Eriocrania sparrmannella* (Bosc, 1791)
2.005 *Eriocrania salopiella* (Stainton, 1854)
2.006 *Eriocrania cicatricella* (Zetterstedt, 1839)
2.007 *Eriocrania semipurpurella* (Stephens, 1835)
2.008 *Eriocrania sangii* (Wood, 1891)

HEPIALIDAE
3.001 *Triodia sylvina* (Linnaeus, 1761) Orange Swift
3.002 *Korscheltellus lupulina* (Linnaeus, 1758) Common Swift
3.003 *Korscheltellus fusconebulosa* (De Geer, 1778) Map-winged Swift
3.004 *Phymatopus hecta* (Linnaeus, 1758) Gold Swift
3.005 *Hepialus humuli* (Linnaeus, 1758) Ghost Moth

NEPTICULIDAE
4.002 *Stigmella lapponica* (Wocke, 1862)
4.003 *Stigmella confusella* (Wood, 1894)
4.004 *Stigmella tiliae* (Frey, 1856)
4.005 *Stigmella betulicola* (Stainton, 1856)
4.006 *Stigmella sakhalinella* Puplesis, 1984
4.007 *Stigmella luteella* (Stainton, 1857)
4.008 *Stigmella glutinosae* (Stainton, 1858)
4.009 *Stigmella alnetella* (Stainton, 1856)
4.010 *Stigmella microtheriella* (Stainton, 1854)
4.011 *Stigmella prunetorum* (Stainton, 1855)
4.013 *Stigmella malella* (Stainton, 1854) Apple Pygmy
4.014 *Stigmella catharticella* (Stainton, 1853)
4.015 *Stigmella anomalella* (Goeze, 1783) Rose Leaf Miner
4.017 *Stigmella centifoliella* (Zeller, 1848)
4.020 *Stigmella paradoxa* (Frey, 1858)
4.023 *Stigmella crataegella* (Klimesch, 1936)
4.024 *Stigmella magdalenae* (Klimesch, 1950)
4.025 *Stigmella nylandriella* (Tengström, 1848)
4.026 *Stigmella oxyacanthella* (Stainton, 1854)
4.028 *Stigmella minusculella* (Herrich-Schäffer, 1855)
4.030 *Stigmella hybnerella* (Hübner, 1796)
4.032 *Stigmella floslactella* (Haworth, 1828)
4.034 *Stigmella tityrella* (Stainton, 1854)
4.035 *Stigmella salicis* (& cluster) (Stainton, 1854)
4.036 *Stigmella myrtillella* (Stainton, 1857)
4.038 *Stigmella obliquella* (Heinemann, 1862)
4.039 *Stigmella trimaculella* (Haworth, 1828)
4.040 *Stigmella assimilella* (Zeller, 1848)
4.041 *Stigmella sorbi* (Stainton, 1861)
4.042 *Stigmella plagicolella* (Stainton, 1854)
4.043 *Stigmella lemniscella* (Zeller, 1839)
4.044 *Stigmella continuella* (Stainton, 1856)
4.045 *Stigmella aurella* (Fabricius, 1775)
4.049 *Stigmella aeneofasciella* (Herrich-Schäffer, 1855)
4.051 *Stigmella poterii* (Stainton, 1857)
4.053 *Stigmella incognitella* (Herrich-Schäffer, 1855)
4.054 *Stigmella perpygmaeella* (Doubleday, 1859)
4.055 *Stigmella hemargyrella* (Kollar, 1832)
4.056 *Stigmella speciosa* (Frey, 1858)
4.059 *Stigmella svenssoni* (Johansson, 1971)
4.060 *Stigmella ruficapitella* (Haworth, 1828)
4.061 *Stigmella atricapitella* (Haworth, 1828)
4.062 *Stigmella samiatella* (Zeller, 1839)
4.063 *Stigmella roborella* (Johansson, 1971)
4.065 *Trifurcula cryptella* (Stainton, 1856)
4.066 *Trifurcula eurema* (Tutt, 1899)
4.068 *Trifurcula immundella* (Zeller, 1839)
4.069 *Trifurcula beirnei* Puplesis, 1984
4.071 *Bohemannia pulverosella* (Stainton, 1849)
4.072 *Bohemannia quadrimaculella* (Boheman, 1853)
4.074 *Etainia sericopeza* (Zeller, 1839)
4.075 *Etainia louisella* (Sircom, 1849)
4.076 *Etainia decentella* (Herrich-Schäffer, 1855)
4.077 *Fomoria weaveri* (Stainton, 1855)
4.078 *Fomoria septembrella* (Stainton, 1849)
4.082 *Ectoedemia intimella* (Zeller, 1848)
4.085 *Ectoedemia argyropeza* (Zeller, 1839)
4.089 *Ectoedemia albifasciella* (Heinemann, 1871)
4.090 *Ectoedemia subbimaculella* (Haworth, 1828)
4.091 *Ectoedemia heringi* (Toll, 1934)
4.094 *Ectoedemia angulifasciella* (Stainton, 1849)
4.095 *Ectoedemia atricollis* (Stainton, 1857)
4.096 *Ectoedemia arcuatella* (Herrich-Schäffer, 1855)
4.099 *Ectoedemia occultella* (Linnaeus, 1767)
4.100 *Ectoedemia minimella* (Zetterstedt, 1839)

OPOSTEGIDAE
5.001 *Opostega salaciella* (Treitschke, 1833)
5.004 *Pseudopostega crepusculella* (Zeller, 1839)

HELIOZELIDAE
6.001 *Antispila metallella* ([Denis & Schiffermüller], 1775)
6.003 *Heliozela sericiella* (Haworth, 1828)
6.004 *Heliozela resplendella* (Stainton, 1851)
6.005 *Heliozela hammoniella* Sorhagen, 1885

ADELIDAE
7.001 *Nemophora degeerella* (Linnaeus, 1758)
7.003 *Nemophora cupriacella* (Hübner, [1819])
7.005 *Nemophora minimella* ([Denis & Schiffermüller], 1775)
7.006 *Adela reaumurella* (Linnaeus, 1758)
7.007 *Adela cuprella* ([Denis & Schiffermüller], 1775)
7.008 *Adela croesella* (Scopoli, 1763)
7.009 *Cauchas fibulella* ([Denis & Schiffermüller], 1775)
7.010 *Cauchas rufimitrella* (Scopoli, 1763)
7.011 *Nematopogon pilella* ([Denis & Schiffermüller], 1775)
7.012 *Nematopogon schwarziellus* Zeller, 1839
7.014 *Nematopogon metaxella* (Hübner, [1813])
7.015 *Nematopogon swammerdamella* (Linnaeus, 1758)

INCURVARIIDAE
8.001 *Incurvaria pectinea* Haworth, 1828
8.002 *Incurvaria masculella* ([Denis & Schiffermüller], 1775)

8.003   *Incurvaria oehlmanniella* (Hübner, 1796)
8.004   *Incurvaria praelatella* ([Denis & Schiffermüller], 1775)
8.005   *Phylloporia bistrigella* (Haworth, 1828)

PRODOXIDAE
9.001   *Lampronia capitella* (Clerck, 1759) Currant Shoot Borer
9.002   *Lampronia luzella* (Hübner, [1817])
9.003   *Lampronia corticella* (Linnaeus, 1758) Raspberry Moth
9.004   *Lampronia morosa* Zeller, 1852
9.006   *Lampronia fuscatella* (Tengström, 1848)

TISCHERIIDAE
10.001 *Tischeria ekebladella* (Bjerkander, 1795)
10.002 *Tischeria dodonaea* Stainton, 1858
10.003 *Coptotriche marginea* (Haworth, 1828)
10.006 *Coptotriche angusticollella* (Duponchel, [1843])

PSYCHIDAE
11.001 *Diplodoma laichartingella* (Goeze, 1783)
11.002 *Narycia duplicella* (Goeze, 1783)
11.004 *Dahlica inconspicuella* (Stainton, 1849) Lesser Lichen Case-bearer
11.005 *Dahlica lichenella* (Linnaeus, 1761) Lichen Case-bearer
11.006 *Taleporia tubulosa* (Retzius, 1783)
11.007 *Bankesia conspurcatella* (Zeller, 1850)
11.009 *Luffia lapidella* (Goeze, 1783)
11.012 *Psyche casta* (Pallas, 1767)
11.016 *Acanthopsyche atra* (Linnaeus, 1767)

TINEIDAE
12.006 *Infurcitinea argentimaculella* (Stainton, 1849)
12.010 *Morophaga choragella* ([Denis & Schiffermüller], 1775)
12.011 *Triaxomera fulvimitrella* (Sodoffsky, 1830)
12.012 *Triaxomera parasitella* (Hübner, 1796)
12.014 *Nemaxera betulinella* (Paykull, 1785)
12.015 *Nemapogon granella* (Linnaeus, 1758) Corn Moth
12.016 *Nemapogon cloacella* (Haworth, 1828) Cork Moth
12.017 *Nemapogon koenigi* Capuşe, 1967
12.021 *Nemapogon clematella* (Fabricius, 1781)
12.022 *Nemapogon picarella* (Clerck, 1759)
12.025 *Trichophaga tapetzella* (Linnaeus, 1758) Tapestry Moth
12.026 *Tineola bisselliella* (Hummel, 1823) Common Clothes Moth
12.027 *Tinea pellionella* Linnaeus, 1758 Case-bearing Clothes Moth
12.0272 *Tinea translucens* Meyrick, 1917
12.028 *Tinea dubiella* Stainton, 1859
12.0291 *Tinea lanella* Pierce & Metcalfe, 1934
12.030 *Tinea pallescentella* Stainton, 1851 Large Pale Clothes Moth
12.032 *Tinea semifulvella* Haworth, 1828
12.033 *Tinea trinotella* Thunberg, 1794
12.034 *Niditinea fuscella* (Linnaeus, 1758) Brown-dotted Clothes Moth
12.036 *Monopis laevigella* ([Denis & Schiffermüller], 1775) Skin Moth
12.038 *Monopis obviella* ([Denis & Schiffermüller], 1775)
12.039 *Monopis crocicapitella* (Clemens, 1859)
12.040 *Monopis imella* (Hübner, [1813])
12.044 *Haplotinea insectella* (Fabricius, 1794)
12.0442 *Lindera tessellatella* Blanchard, 1852
12.046 *Oinophila v-flava* (Haworth, 1828) Yellow V Moth
12.047 *Psychoides verhuella* Bruand, 1853
12.048 *Psychoides filicivora* (Meyrick, 1937)

ROESLERSTAMMIIDAE
13.002 *Roeslerstammia erxlebella* (Fabricius, 1787)

BUCCULATRICIDAE
14.001 *Bucculatrix cristatella* (Zeller, 1839)
14.002 *Bucculatrix nigricomella* (Zeller, 1839)
14.003 *Bucculatrix maritima* Stainton, 1851
14.007 *Bucculatrix albedinella* (Zeller, 1839)
14.008 *Bucculatrix cidarella* (Zeller, 1839)
14.009 *Bucculatrix thoracella* (Thunberg, 1794)
14.010 *Bucculatrix ulmella* Zeller, 1848
14.012 *Bucculatrix bechsteinella* (Bechstein & Scharfenberg, 1805)
14.013 *Bucculatrix demaryella* (Duponchel, 1840)

GRACILLARIIDAE
15.002 *Caloptilia cuculipennella* (Hübner, 1796)
15.003 *Caloptilia populetorum* (Zeller, 1839)
15.004 *Caloptilia elongella* (Linnaeus, 1761)
15.005 *Caloptilia betulicola* (Hering, 1928)
15.006 *Caloptilia rufipennella* (Hübner, 1796)
15.007 *Caloptilia azaleella* (Brants, 1913) Azalea Leaf Miner
15.008 *Caloptilia alchimiella* (Scopoli, 1763)
15.009 *Caloptilia robustella* Jäckh, 1972
15.010 *Caloptilia stigmatella* (Fabricius, 1781)
15.012 *Caloptilia semifascia* (Haworth, 1828)
15.014 *Gracillaria syringella* (Fabricius, 1794)
15.015 *Aspilapteryx tringipennella* (Zeller, 1839)
15.016 *Euspilapteryx auroguttella* Stephens, 1835
15.017 *Calybites phasianipennella* (Hübner, [1813])
15.018 *Povolnya leucapennella* (Stephens, 1835)
15.019 *Acrocercops brongniardella* (Fabricius, 1798)
15.022 *Callisto denticulella* (Thunberg, 1794)
15.025 *Parornix betulae* (Stainton, 1854)
15.028 *Parornix anglicella* (Stainton, 1850)
15.029 *Parornix devoniella* (Stainton, 1850)
15.030 *Parornix scoticella* (Stainton, 1850)
15.032 *Parornix finitimella* (Zeller, 1850)
15.033 *Parornix torquillella* (Zeller, 1850)
15.034 *Phyllonorycter harrisella* (Linnaeus, 1761)
15.036 *Phyllonorycter heegeriella* (Zeller, 1846)
15.039 *Phyllonorycter quercifoliella* (Zeller, 1839)
15.040 *Phyllonorycter messaniella* (Zeller, 1846)
15.041 *Phyllonorycter platani* (Staudinger, 1870)
15.043 *Phyllonorycter oxyacanthae* (Frey, 1856)
15.044 *Phyllonorycter sorbi* (Frey, 1855)
15.046 *Phyllonorycter blancardella* (Fabricius, 1781)
15.048 *Phyllonorycter junoniella* (Zeller, 1846)
15.049 *Phyllonorycter spinicolella* (Zeller, 1846)
15.050 *Phyllonorycter cerasicolella* (Herrich-Schäffer, 1855)
15.051 *Phyllonorycter lantanella* (Schrank, 1802)
15.052 *Phyllonorycter corylifoliella* (Hübner, 1796)
15.053 *Phyllonorycter leucographella* (Zeller, 1850) Firethorn Leaf Miner
15.054 *Phyllonorycter viminiella* (Sircom, 1848)
15.055 *Phyllonorycter viminetorum* (Stainton, 1854)
15.056 *Phyllonorycter salicicolella* (Sircom, 1848)
15.057 *Phyllonorycter dubitella* (Herrich-Schäffer, 1855)
15.058 *Phyllonorycter hilarella* (Zetterstedt, 1839)
15.060 *Phyllonorycter ulicicolella* (Stainton, 1851)
15.061 *Phyllonorycter scopariella* (Zeller, 1846)
15.063 *Phyllonorycter maestingella* (Müller, 1764)
15.064 *Phyllonorycter coryli* (Nicelli, 1851) Nut Leaf Blister Moth
15.065 *Phyllonorycter esperella* (Goeze, 1783)
15.066 *Phyllonorycter strigulatella* (Lienig & Zeller, 1846)
15.067 *Phyllonorycter rajella* (Linnaeus, 1758)
15.069 *Phyllonorycter anderidae* (Fletcher, 1885)
15.070 *Phyllonorycter quinqueguttella* (Stainton, 1851)
15.073 *Phyllonorycter lautella* (Zeller, 1846)
15.074 *Phyllonorycter schreberella* (Fabricius, 1781)
15.075 *Phyllonorycter ulmifoliella* (Hübner, [1817])
15.076 *Phyllonorycter emberizaepenella* (Bouché, 1834)
15.078 *Phyllonorycter tristrigella* (Haworth, 1828)
15.079 *Phyllonorycter stettinensis* (Nicelli, 1852)
15.080 *Phyllonorycter froelichiella* (Zeller, 1839)
15.081 *Phyllonorycter nicellii* (Stainton, 1851)
15.082 *Phyllonorycter klemannella* (Fabricius, 1781)
15.083 *Phyllonorycter trifasciella* (Haworth, 1828)
15.084 *Phyllonorycter acerifoliella* (Zeller, 1839)
15.085 *Phyllonorycter joannisi* (Le Marchand, 1936)
15.086 *Phyllonorycter geniculella* (Ragonot, 1874)

15.089 *Cameraria ohridella* Deschka & Dimić, 1986 Horse-chestnut Leaf-miner
15.090 *Phyllocnistis saligna* (Zeller, 1839)
15.092 *Phyllocnistis unipunctella* (Stephens, 1834)

YPONOMEUTIDAE
16.001 *Yponomeuta evonymella* (Linnaeus, 1758) Bird-cherry Ermine
16.002 *Yponomeuta padella* (Linnaeus, 1758) Orchard Ermine
16.003 *Yponomeuta malinellus* Zeller, 1838 Apple Ermine
16.004 *Yponomeuta cagnagella* (Hübner, [1813]) Spindle Ermine
16.005 *Yponomeuta rorrella* (Hübner, 1796) Willow Ermine
16.007 *Yponomeuta plumbella* ([Denis & Schiffermüller], 1775)
16.008 *Yponomeuta sedella* Treitschke, 1832
16.010 *Zelleria hepariella* Stainton, 1849
16.011 *Zelleria oleastrella* (Millière, 1864)
16.014 *Pseudoswammerdamia combinella* (Hübner, 1786)
16.015 *Swammerdamia caesiella* (Hübner, 1796)
16.017 *Swammerdamia pyrella* (Villers, 1789)
16.019 *Paraswammerdamia albicapitella* (Scharfenberg, 1805)
16.020 *Paraswammerdamia nebulella* (Goeze, 1783)
16.021 *Cedestis gysseleniella* Zeller, 1839
16.022 *Cedestis subfasciella* (Stephens, 1834)
16.023 *Ocnerostoma piniariella* Zeller, 1847
16.024 *Ocnerostoma friesei* Svensson, 1966

YPSOLOPHIDAE
17.002 *Ypsolopha nemorella* (Linnaeus, 1758)
17.003 *Ypsolopha dentella* (Fabricius, 1775) Honeysuckle Moth
17.005 *Ypsolopha scabrella* (Linnaeus, 1761)
17.006 *Ypsolopha horridella* (Treitschke, 1835)
17.007 *Ypsolopha lucella* (Fabricius, 1775)
17.008 *Ypsolopha alpella* ([Denis & Schiffermüller], 1775)
17.009 *Ypsolopha sylvella* (Linnaeus, 1767)
17.010 *Ypsolopha parenthesella* (Linnaeus, 1761)
17.011 *Ypsolopha ustella* (Clerck, 1759)
17.012 *Ypsolopha sequella* (Clerck, 1759)
17.013 *Ypsolopha vittella* (Linnaeus, 1758)
17.014 *Ochsenheimeria taurella* ([Denis & Schiffermüller], 1775)
17.015 *Ochsenheimeria urella* Fischer von Röslerstamm, 1842
17.016 *Ochsenheimeria vacculella* Fischer von Röslerstamm, 1842 Cereal Stem Moth

PLUTELLIDAE
18.001 *Plutella xylostella* (Linnaeus, 1758) Diamond-back Moth
18.003 *Plutella porrectella* (Linnaeus, 1758)
18.005 *Rhigognostis annulatella* (Curtis, 1832)
18.006 *Rhigognostis incarnatella* (Steudel, 1873)
18.007 *Eidophasia messingiella* (Fischer von Röslerstamm, 1840)

GLYPHIPTERIGIDAE
19.001 *Orthotelia sparganella* (Thunberg, 1788)
19.002 *Glyphipterix thrasonella* (Scopoli, 1763)
19.003 *Glyphipterix fuscoviridella* (Haworth, 1828)
19.004 *Glyphipterix equitella* (Scopoli, 1763)
19.005 *Glyphipterix haworthana* (Stephens, 1834)
19.007 *Glyphipterix simpliciella* (Stephens, 1834) Cocksfoot Moth
19.008 *Glyphipterix schoenicolella* Boyd, 1858
19.010 *Digitivalva pulicariae* (Klimesch, 1956)
19.011 *Acrolepiopsis assectella* (Zeller, 1839) Leek Moth
19.014 *Acrolepia autumnitella* Curtis, 1838

ARGYRESTHIIDAE
20.001 *Argyresthia laevigatella* Herrich-Schäffer, 1855
20.002 *Argyresthia glabratella* (Zeller, 1847)
20.004 *Argyresthia arceuthina* Zeller, 1839
20.005 *Argyresthia trifasciata* Staudinger, 1871
20.006 *Argyresthia dilectella* Zeller, 1847
20.007 *Argyresthia cupressella* Walsingham, 1890 Cypress Tip Moth
20.010 *Argyresthia ivella* (Haworth, 1828)
20.011 *Argyresthia brockeella* (Hübner, [1813])
20.012 *Argyresthia goedartella* (Linnaeus, 1758)
20.013 *Argyresthia pygmaeella* ([Denis & Schiffermüller], 1775)
20.014 *Argyresthia sorbiella* (Treitschke, 1833)
20.015 *Argyresthia curvella* (Linnaeus, 1761)
20.016 *Argyresthia retinella* Zeller, 1839
20.017 *Argyresthia glaucinella* Zeller, 1839
20.018 *Argyresthia spinosella* Stainton, 1849
20.019 *Argyresthia conjugella* Zeller, 1839 Apple Fruit Moth
20.020 *Argyresthia semifusca* (Haworth, 1828)
20.021 *Argyresthia pruniella* (Clerck, 1759) Cherry Fruit Moth
20.022 *Argyresthia bonnetella* (Linnaeus, 1758)
20.023 *Argyresthia albistria* (Haworth, 1828)
20.024 *Argyresthia semitestacella* (Curtis, 1833)

LYONETIIDAE
21.001 *Lyonetia clerkella* (Linnaeus, 1758) Apple Leaf Miner
21.004 *Leucoptera laburnella* (Stainton, 1851) Laburnum Leaf Miner
21.005 *Leucoptera spartifoliella* (Hübner, [1813])
21.008 *Leucoptera malifoliella* (Costa, [1836]) Pear Leaf Blister Moth

PRAYDIDAE
22.001 *Atemelia torquatella* (Lienig & Zeller, 1846)
22.002 *Prays fraxinella* (Bjerkander, 1784) Ash Bud Moth
22.003 *Prays ruficeps* (Heinemann, 1854)

BEDELLIIDAE
24.001 *Bedellia somnulentella* (Zeller, 1847)

SCYTHROPIIDAE
25.001 *Scythropia crataegella* (Linnaeus, 1767) Hawthorn Moth

AUTOSTICHIDAE
27.001 *Oegoconia quadripuncta* (Haworth, 1828)

OECOPHORIDAE
28.004 *Denisia similella* (Hübner, 1796)
28.005 *Denisia albimaculea* (Haworth, 1828)
28.007 *Denisia subaquilea* (Stainton, 1849)
28.008 *Metalampra italica* Baldizzone, 1977
28.009 *Endrosis sarcitrella* (Linnaeus, 1758) White-shouldered House-moth
28.010 *Hofmannophila pseudospretella* (Stainton, 1849) Brown House-moth
28.011 *Borkhausenia minutella* (Linnaeus, 1758)
28.012 *Borkhausenia fuscescens* (Haworth, 1828)
28.014 *Crassa unitella* (Hübner, 1796)
28.015 *Batia lunaris* (Haworth, 1828)
28.019 *Esperia sulphurella* (Fabricius, 1775)
28.022 *Alabonia geoffrella* (Linnaeus, 1767)
28.024 *Tachystola acroxantha* (Meyrick, 1885)
28.025 *Pleurota bicostella* (Clerck, 1759)

CHIMABACHIDAE
29.001 *Diurnea fagella* ([Denis & Schiffermüller], 1775)
29.002 *Diurnea lipsiella* ([Denis & Schiffermüller], 1775)
29.003 *Dasystoma salicella* (Hübner, 1796)

LYPUSIDAE
30.003 *Agnoea josephinae* (Toll, 1956)
30.004 *Amphisbatis incongruella* (Stainton, 1849)

PELEOPODIDAE
31.001 *Carcina quercana* (Fabricius, 1775)

DEPRESSARIIDAE
32.001 *Semioscopis avellanella* (Hübner, 1793)
32.002 *Semioscopis steinkellneriana* ([Denis & Schiffermüller], 1775)
32.006 *Exaeretia allisella* Stainton, 1849
32.007 *Agonopterix ocellana* (Fabricius, 1775)

32.008 *Agonopterix liturosa* (Haworth, 1811)
32.009 *Agonopterix purpurea* (Haworth, 1811)
32.010 *Agonopterix conterminella* (Zeller, 1839)
32.011 *Agonopterix scopariella* (Heinemann, 1870)
32.012 *Agonopterix atomella* ([Denis & Schiffermüller], 1775)
32.013 *Agonopterix carduella* (Hübner, [1817])
32.015 *Agonopterix subpropinquella* (Stainton, 1849)
32.016 *Agonopterix propinquella* (Treitschke, 1835)
32.017 *Agonopterix arenella* ([Denis & Schiffermüller], 1775)
32.018 *Agonopterix heracliana* (Linnaeus, 1758)
32.019 *Agonopterix ciliella* (Stainton, 1849)
32.024 *Agonopterix assimilella* (Treitschke, 1832)
32.025 *Agonopterix nanatella* (Stainton, 1849)
32.026 *Agonopterix kaekeritziana* (Linnaeus, 1767)
32.029 *Agonopterix umbellana* (Fabricius, 1794)
32.030 *Agonopterix nervosa* (Haworth, 1811)
32.031 *Agonopterix alstromeriana* (Clerck, 1759)
32.032 *Agonopterix angelicella* (Hübner, [1813])
32.035 *Agonopterix yeatiana* (Fabricius, 1781)
32.036 *Depressaria radiella* (Goeze, 1783) Parsnip Moth
32.038 *Depressaria badiella* (Hübner, 1796)
32.039 *Depressaria daucella* ([Denis & Schiffermüller], 1775)
32.040 *Depressaria ultimella* Stainton, 1849
32.042 *Depressaria pulcherrimella* Stainton, 1849
32.043 *Depressaria sordidatella* Tengström, 1848
32.044 *Depressaria douglasella* Stainton, 1849
32.045 *Depressaria albipunctella* ([Denis & Schiffermüller], 1775)
32.047 *Depressaria chaerophylli* Zeller, 1839
32.050 *Telechrysis tripuncta* (Haworth, 1828)

COSMOPTERIGIDAE
34.001 *Pancalia leuwenhoekella* (Linnaeus, 1761)
34.003 *Euclemensia woodiella* (Curtis, 1830)
34.004 *Limnaecia phragmitella* Stainton, 1851
34.0111 *Anatrachyntis badia* (Hodges, 1962)
34.014 *Sorhagenia janiszewskae* Riedl, 1962

GELECHIIDAE
35.001 *Aproaerema sangiella* (Stainton, 1863)
35.002 *Aproaerema cinctella* (Clerck, 1759)
35.003 *Aproaerema larseniella* (Gozmány, 1957)
35.004 *Aproaerema taeniolella* (Zeller, 1839)
35.010 *Aproaerema anthyllidella* (Hübner, [1813])
35.011 *Anacampsis populella* (Clerck, 1759)
35.012 *Anacampsis blattariella* (Hübner, 1796)
35.013 *Anacampsis temerella* (Lienig & Zeller, 1846)
35.017 *Neofaculta ericetella* (Geyer, [1832])
35.018 *Hypatima rhomboidella* (Linnaeus, 1758)
35.020 *Anarsia spartiella* (Schrank, 1802)
35.022 *Dichomeris marginella* (Fabricius, 1781) Juniper Webber
35.026 *Acompsia cinerella* (Clerck, 1759)
35.028 *Brachmia blandella* (Fabricius, 1798)
35.031 *Helcystogramma rufescens* (Haworth, 1828)
35.032 *Pexicopia malvella* (Hübner, [1805]) Hollyhock Seed Moth
35.034 *Sitotroga cerealella* (Olivier, 1789) Angoumois Grain Moth
35.035 *Chrysoesthia drurella* (Fabricius, 1775)
35.036 *Chrysoesthia sexguttella* (Thunberg, 1794)
35.038 *Bryotropha domestica* (Haworth, 1828)
35.039 *Bryotropha politella* (Stainton, 1851)
35.040 *Bryotropha terrella* ([Denis & Schiffermüller], 1775)
35.041 *Bryotropha desertella* (Douglas, 1850)
35.042 *Bryotropha boreella* (Douglas, 1851)
35.045 *Bryotropha basaltinella* (Zeller, 1839)
35.046 *Bryotropha senectella* (Zeller, 1839)
35.047 *Bryotropha affinis* (Haworth, 1828)
35.048 *Bryotropha umbrosella* (Zeller, 1839)
35.049 *Bryotropha similis* (Stainton, 1854)
35.050 *Aristotelia ericinella* (Zeller, 1839)
35.052 *Aristotelia brizella* (Treitschke, 1833)

35.053 *Isophrictis striatella* ([Denis & Schiffermüller], 1775)
35.056 *Metzneria lappella* (Linnaeus, 1758)
35.058 *Metzneria metzneriella* (Stainton, 1851)
35.060 *Apodia martinii* Petry, 1911
35.061 *Ptocheuusa paupella* (Zeller, 1847)
35.065 *Monochroa cytisella* (Curtis, 1837)
35.066 *Monochroa tenebrella* (Hübner, [1817])
35.068 *Monochroa tetragonella* (Stainton, 1885)
35.071 *Monochroa lucidella* (Stephens, 1834)
35.076 *Monochroa suffusella* (Douglas, 1850)
35.077 *Monochroa hornigi* (Staudinger, 1883)
35.079 *Oxypteryx wilkella* (Linnaeus, 1758)
35.080 *Oxypteryx unicolorella* (Duponchel, [1843])
35.081 *Oxypteryx atrella* ([Denis & Schiffermüller], 1775)
35.085 *Athrips mouffetella* (Linnaeus, 1758)
35.089 *Prolita sexpunctella* (Fabricius, 1794)
35.091 *Sophronia semicostella* (Hübner, [1813])
35.092 *Mirificarma lentiginosella* (Zeller, 1839)
35.093 *Mirificarma mulinella* (Zeller, 1839)
35.094 *Aroga velocella* (Zeller, 1839)
35.095 *Chionodes distinctella* (Zeller, 1839)
35.096 *Chionodes fumatella* (Douglas, 1850)
35.097 *Gelechia rhombella* ([Denis & Schiffermüller], 1775)
35.101 *Gelechia sororculella* (Hübner, [1817])
35.103 *Gelechia cuneatella* Douglas, 1852
35.105 *Gelechia nigra* (Haworth, 1828)
35.107 *Psoricoptera gibbosella* (Zeller, 1839)
35.109 *Scrobipalpa acuminatella* (Sircom, 1850)
35.113 *Scrobipalpa salicorniae* (Hering, 1889)
35.114 *Scrobipalpa instabilella* (Douglas, 1846)
35.115 *Scrobipalpa nitentella* (Fuchs, 1902)
35.116 *Scrobipalpa obsoletella* (Fischer von Röslerstamm, 1841)
35.117 *Scrobipalpa atriplicella* (Fischer von Röslerstamm, 1841)
35.118 *Scrobipalpa ocellatella* (Boyd, 1858)
35.119 *Scrobipalpa samadensis* (Pfaffenzeller, 1870)
35.120 *Scrobipalpa artemisiella* (Treitschke, 1833) Thyme Moth
35.123 *Scrobipalpa costella* (Humphreys & Westwood, 1845)
35.126 *Phthorimaea operculella* (Zeller, 1873) Potato Tuber Moth
35.127 *Tuta absoluta* (Meyrick, 1917)
35.128 *Caryocolum alsinella* (Zeller, 1868)
35.129 *Caryocolum viscariella* (Stainton, 1855)
35.131 *Caryocolum marmorea* (Haworth, 1828)
35.132 *Caryocolum fraternella* (Douglas, 1851)
35.133 *Caryocolum blandella* (Douglas, 1852)
35.137 *Caryocolum tricolorella* (Haworth, 1812)
35.141 *Teleiodes vulgella* ([Denis & Schiffermüller], 1775)
35.143 *Teleiodes luculella* (Hübner, [1813])
35.145 *Neotelphusa sequax* (Haworth, 1828)
35.146 *Teleiopsis diffinis* (Haworth, 1828)
35.147 *Carpatolechia decorella* (Haworth, 1812)
35.148 *Carpatolechia fugitivella* (Zeller, 1839)
35.149 *Carpatolechia alburnella* (Zeller, 1839)
35.150 *Carpatolechia notatella* (Hübner, [1813])
35.151 *Carpatolechia proximella* (Hübner, 1796)
35.152 *Pseudotelphusa scalella* (Scopoli, 1763)
35.153 *Pseudotelphusa paripunctella* (Thunberg, 1794)
35.154 *Xenolechia aethiops* (Humphreys & Westwood, 1845)
35.157 *Recurvaria leucatella* (Clerck, 1759)
35.159 *Exoteleia dodecella* (Linnaeus, 1758)
35.160 *Stenolechia gemmella* (Linnaeus, 1758)

BATRACHEDRIDAE
36.001 *Batrachedra praeangusta* (Haworth, 1828)
36.002 *Batrachedra confusella* Berggren, Aarvik, Huemer, Lee & Mutanen, 2022

COLEOPHORIDAE
37.005 *Coleophora lutipennella* (Zeller, 1838)
37.006 *Coleophora gryphipennella* (Hübner, 1796)
37.007 *Coleophora flavipennella* (Duponchel, [1843])

The Moths of Lancashire 579

37.009 *Coleophora milvipennis* Zeller, 1839
37.012 *Coleophora limosipennella* (Duponchel, [1843])
37.013 *Coleophora siccifolia* Stainton, 1856
37.014 *Coleophora coracipennella* (Hübner, 1796)
37.015 *Coleophora serratella* (Linnaeus, 1761)
37.016 *Coleophora spinella* (Schrank, 1802) **Apple & Plum Case-bearer**
37.017 *Coleophora prunifoliae* Doets, 1944
37.020 *Coleophora fuscocuprella* Herrich-Schäffer, 1855
37.022 *Coleophora lusciniaepennella* (Treitschke, 1833)
37.027 *Coleophora potentillae* Elisha, 1885
37.028 *Coleophora juncicolella* Stainton, 1851
37.029 *Coleophora orbitella* Zeller, 1849
37.030 *Coleophora binderella* (Kollar, 1832)
37.033 *Coleophora trifolii* (Curtis, 1832)
37.034 *Coleophora frischella* (Linnaeus, 1758)
37.035 *Coleophora alcyonipennella* (Kollar, 1832) **Clover Case-bearer**
37.038 *Coleophora lineolea* (Haworth, 1828)
37.044 *Coleophora discordella* Zeller, 1849
37.046 *Coleophora deauratella* Lienig & Zeller, 1846
37.048 *Coleophora mayrella* (Hübner, [1813])
37.049 *Coleophora anatipennella* (Hübner, 1796) **Pistol Case-bearer**
37.050 *Coleophora albidella* ([Denis & Schiffermüller], 1775)
37.051 *Coleophora kuehnella* (Goeze, 1783)
37.052 *Coleophora ibipennella* Zeller, 1849
37.053 *Coleophora betulella* Heinemann, [1876]
37.054 *Coleophora currucipennella* Zeller, 1839
37.055 *Coleophora pyrrhulipennella* Zeller, 1839
37.063 *Coleophora albicosta* (Haworth, 1828)
37.066 *Coleophora laricella* (Hübner, [1817]) **Larch Case-bearer**
37.068 *Coleophora adjunctella* Hodgkinson, 1882
37.069 *Coleophora caespititiella* Zeller, 1839
37.070 *Coleophora tamesis* Waters, 1929
37.071 *Coleophora glaucicolella* Wood, 1892
37.072 *Coleophora otidipennella* (Hübner, [1817])
37.073 *Coleophora alticolella* Zeller, 1849
37.074 *Coleophora taeniipennella* Herrich-Schäffer, 1855
37.078 *Coleophora maritimella* Newman, 1873
37.080 *Coleophora virgaureae* Stainton, 1857
37.082 *Coleophora asteris* Mühlig, 1864
37.083 *Coleophora saxicolella* (Duponchel, [1843])
37.086 *Coleophora versurella* Zeller, 1849
37.087 *Coleophora vestianella* (Linnaeus, 1758)
37.088 *Coleophora atriplicis* Meyrick, [1928]
37.090 *Coleophora artemisicolella* Bruand, [1855]
37.093 *Coleophora peribenanderi* Toll, 1943
37.098 *Coleophora inulae* Wocke, [1876]
37.099 *Coleophora striatipennella* Nylander [1848]
37.102 *Coleophora argentula* (Stephens, 1834)
37.104 *Coleophora adspersella* Benander, 1939
37.106 *Coleophora paripennella* Zeller, 1839
37.108 *Coleophora salicorniae* Heinemann & Wocke, [1876]

ELACHISTIDAE
38.001 *Perittia obscurepunctella* (Stainton, 1848)
38.004 *Elachista argentella* (Clerck, 1759)
38.005 *Elachista triatomea* (Haworth, 1828)
38.007 *Elachista subocellea* (Stephens, 1834)
38.008 *Elachista triseriatella* Stainton, 1854
38.013 *Elachista cingillella* (Herrich-Schäffer, 1855)
38.015 *Elachista gangabella* Zeller, 1850
38.016 *Elachista subalbidella* Schläger, 1847
38.017 *Elachista adscitella* Stainton, 1851
38.018 *Elachista bisulcella* (Duponchel, [1843])
38.022 *Elachista gleichenella* (Fabricius, 1781)
38.023 *Elachista biatomella* (Stainton, 1848)
38.024 *Elachista poae* Stainton, 1855
38.025 *Elachista atricomella* Stainton, 1849
38.026 *Elachista kilmunella* Stainton, 1849
38.028 *Elachista alpinella* Stainton, 1854

38.029 *Elachista luticomella* Zeller, 1839
38.030 *Elachista albifrontella* (Hübner, [1817])
38.032 *Elachista apicipunctella* Stainton, 1849
38.033 *Elachista subnigrella* Douglas, 1853
38.036 *Elachista humilis* Zeller, 1850
38.037 *Elachista canapennella* (Hübner, [1813])
38.038 *Elachista rufocinerea* (Haworth, 1828)
38.039 *Elachista maculicerusella* (Bruand, 1859)
38.040 *Elachista trapeziella* Stainton, 1849
38.041 *Elachista cinereopunctella* (Haworth, 1828)
38.042 *Elachista serricornis* Stainton, 1854
38.043 *Elachista scirpi* Stainton, 1887
38.044 *Elachista eleochariella* Stainton, 1851
38.045 *Elachista utonella* Frey, 1856
38.046 *Elachista albidella* Nylander, [1848]
38.047 *Elachista freyerella* (Hübner, [1825])

PARAMETRIOTIDAE
39.001 *Blastodacna hellerella* (Duponchel, 1838)
39.002 *Blastodacna atra* (Haworth, 1828) **Apple Pith Moth**
39.003 *Spuleria flavicaput* (Haworth, 1828)
39.005 *Chrysoclista linneella* (Clerck, 1759)
39.006 *Chrysoclista lathamella* Fletcher, 1936

MOMPHIDAE
40.001 *Mompha conturbatella* (Hübner, [1819])
40.002 *Mompha ochraceella* (Curtis, 1839)
40.003 *Mompha lacteella* (Stephens, 1834)
40.004 *Mompha propinquella* (Stainton, 1851)
40.006 *Mompha jurassicella* (Frey, 1881)
40.007 *Mompha bradleyi* Riedl, 1965
40.008 *Mompha subbistrigella* (Haworth, 1828)
40.009 *Mompha sturnipennella* (Treitschke, 1833)
40.010 *Mompha epilobiella* ([Denis & Schiffermüller], 1775)
40.011 *Mompha langiella* (Hübner, 1796)
40.012 *Mompha miscella* ([Denis & Schiffermüller], 1775)
40.013 *Mompha locupletella* ([Denis & Schiffermüller], 1775)
40.014 *Mompha terminella* (Humphreys & Westwood, 1845)
40.015 *Mompha raschkiella* (Zeller, 1839)

BLASTOBASIDAE
41.002 *Blastobasis adustella* Walsingham, 1894
41.003 *Blastobasis lacticolella* Rebel, 1940
41.005 *Blastobasis rebeli* Karsholt & Sinev, 2004

STATHMOPODIDAE
42.002 *Stathmopoda pedella* (Linnaeus, 1761)

SCYTHRIDIDAE
43.001 *Scythris fallacella* (Schläger, 1847)
43.002 *Scythris grandipennis* (Haworth, 1828)
43.004 *Scythris picaepennis* (Haworth, 1828)

ALUCITIDAE
44.001 *Alucita hexadactyla* Linnaeus, 1758 **Twenty-plume Moth**

PTEROPHORIDAE
45.004 *Platyptilia gonodactyla* ([Denis & Schiffermüller], 1775) **Triangle Plume**
45.008 *Gillmeria pallidactyla* (Haworth, 1811) **Yarrow Plume**
45.009 *Gillmeria ochrodactyla* ([Denis & Schiffermüller], 1775) **Tansy Plume**
45.010 *Amblyptilia acanthadactyla* (Hübner, [1813]) **Beautiful Plume**
45.011 *Amblyptilia punctidactyla* (Haworth, 1811) **Brindled Plume**
45.012 *Stenoptilia pterodactyla* (Linnaeus, 1761) **Brown Plume**
45.013 *Stenoptilia bipunctidactyla* (Scopoli, 1763) **Twin-spot Plume**
45.019 *Stenoptilia millieridactyla* (Bruand, 1861) **Saxifrage Plume**
45.021 *Stenoptilia zophodactylus* (Duponchel, 1840) **Dowdy Plume**
45.023 *Marasmarcha lunaedactyla* (Haworth, 1811) **Crescent Plume**
45.027 *Oxyptilus laetus* (Zeller, 1847) **Scarce Light Plume**
45.028 *Capperia britanniodactyla* (Gregson, 1869) **Wood Sage Plume**

45.030 *Pterophorus pentadactyla* (Linnaeus, 1758) White Plume
45.033 *Merrifieldia leucodactyla* ([Denis & Schiffermüller], 1775) Thyme Plume
45.034 *Merrifieldia baliodactylus* (Zeller, 1841) Dingy White Plume
45.037 *Oidaematophorus lithodactyla* (Treitschke, 1833) Dusky Plume
45.038 *Hellinsia tephradactyla* (Hübner, [1813]) Plain Plume
45.040 *Hellinsia lienigianus* (Zeller, 1852) Mugwort Plume
45.041 *Hellinsia osteodactylus* (Zeller, 1841) Small Goldenrod Plume
45.043 *Adaina microdactyla* (Hübner, [1813]) Hemp Agrimony Plume
45.044 *Emmelina monodactyla* (Linnaeus, 1758) Common Plume

SCHRECKENSTEINIIDAE
46.001 *Schreckensteinia festaliella* (Hübner, [1819])

EPERMENIIDAE
47.001 *Phaulernis fulviguttella* (Zeller, 1839)
47.004 *Epermenia aequidentellus* (Hofmann, 1867)
47.005 *Epermenia chaerophyllella* (Goeze, 1783)
47.006 *Epermenia falciformis* (Haworth, 1828)

CHOREUTIDAE
48.001 *Anthophila fabriciana* (Linnaeus, 1767)
48.002 *Prochoreutis myllerana* (Fabricius, 1794)
48.003 *Prochoreutis sehestediana* (Fabricius, [1777])
48.005 *Tebenna micalis* (Mann, 1857)
48.007 *Choreutis pariana* (Clerck, 1759) Apple Leaf Skeletonizer

TORTRICIDAE
49.001 *Olindia schumacherana* (Fabricius, 1787)
49.002 *Isotrias rectifasciana* (Haworth, 1811)
49.004 *Ditula angustiorana* (Haworth, 1811) Red-barred Tortrix
49.005 *Epagoge grotiana* (Fabricius, 1781)
49.008 *Philedone gerningana* ([Denis & Schiffermüller], 1775)
49.009 *Capua vulgana* (Frölich, 1828)
49.010 *Philedonides lunana* (Thunberg, 1784)
49.013 *Archips podana* (Scopoli, 1763) Large Fruit-tree Tortrix
49.015 *Archips xylosteana* (Linnaeus, 1758) Variegated Golden Tortrix
49.016 *Archips rosana* (Linnaeus, 1758) Rose Tortrix
49.018 *Choristoneura hebenstreitella* (Müller, 1764)
49.020 *Argyrotaenia ljungiana* (Thunberg, 1797)
49.021 *Ptycholomoides aeriferana* (Herrich-Schäffer, 1851)
49.022 *Ptycholoma lecheana* (Linnaeus, 1758)
49.023 *Pandemis cinnamomeana* (Treitschke, 1830)
49.024 *Pandemis corylana* (Fabricius, 1794) Chequered Fruit-tree Tortrix
49.025 *Pandemis cerasana* (Hübner, 1786) Barred Fruit-tree Tortrix
49.026 *Pandemis heparana* ([Denis & Schiffermüller], 1775) Dark Fruit-tree Tortrix
49.028 *Syndemis musculana* (Hübner, [1799])
49.029 *Lozotaenia forsterana* (Fabricius, 1781)
49.030 *Cacoecimorpha pronubana* (Hübner, [1799]) Carnation Tortrix
49.031 *Zelotherses paleana* (Hübner, 1793) Timothy Tortrix
49.033 *Aphelia viburnana* ([Denis & Schiffermüller], 1775) Bilberry Tortrix
49.035 *Clepsis senecionana* (Hübner, [1819])
49.037 *Clepsis spectrana* (Treitschke, 1830) Cyclamen Tortrix
49.038 *Clepsis consimilana* (Hübner, [1817])
49.039 *Epiphyas postvittana* (Walker, 1863) Light Brown Apple Moth
49.040 *Lozotaeniodes formosana* (Frölich, 1830)
49.042 *Neosphaleroptera nubilana* (Hübner, [1799])
49.043 *Exapate congelatella* (Clerck, 1759)
49.044 *Tortricodes alternella* ([Denis & Schiffermüller], 1775)
49.045 *Eana osseana* (Scopoli, 1763)
49.047 *Eana incanana* (Stephens, 1852)
49.048 *Eana penziana* (Thunberg & Becklin, 1791)
49.049 *Cnephasia incertana* (Treitschke, 1835) Light Grey Tortrix
49.050 *Cnephasia stephensiana* (Doubleday, [1849]) Grey Tortrix
49.051 *Cnephasia asseclana* ([Denis & Schiffermüller], 1775) Flax Tortrix
49.052 *Cnephasia pasiuana* (Hübner, [1799])
49.054 *Cnephasia genitalana* Pierce & Metcalfe, 1922
49.056 *Cnephasia conspersana* Douglas, 1846
49.057 *Cnephasia longana* (Haworth, 1811)
49.059 *Tortrix viridana* Linnaeus, 1758 Green Oak Tortrix

49.060 *Aleimma loeflingiana* (Linnaeus, 1758)
49.061 *Acleris holmiana* (Linnaeus, 1758)
49.062 *Acleris forsskaleana* (Linnaeus, 1758)
49.063 *Acleris bergmanniana* (Linnaeus, 1758)
49.064 *Acleris caledoniana* (Stephens, 1852)
49.065 *Acleris comariana* (Lienig & Zeller, 1846) Strawberry Tortrix
49.066 *Acleris laterana* (Fabricius, 1794)
49.067 *Acleris abietana* (Hübner, [1822])
49.069 *Acleris sparsana* ([Denis & Schiffermüller], 1775)
49.070 *Acleris rhombana* ([Denis & Schiffermüller], 1775) Rhomboid Tortrix
49.071 *Acleris emargana* (Fabricius, 1775)
49.072 *Acleris effractana* (Hübner, [1799])
49.073 *Acleris schalleriana* (Linnaeus, 1761)
49.075 *Acleris umbrana* (Hübner, [1799])
49.077 *Acleris variegana* ([Denis & Schiffermüller], 1775) Garden Rose Tortrix
49.078 *Acleris aspersana* (Hübner, [1817])
49.079 *Acleris shepherdana* (Stephens, 1852)
49.080 *Acleris hastiana* (Linnaeus, 1758)
49.082 *Acleris hyemana* (Haworth, 1811)
49.083 *Acleris ferrugana* ([Denis & Schiffermüller], 1775)
49.084 *Acleris notana* (Donovan, 1806)
49.086 *Acleris logiana* (Clerck, 1759)
49.087 *Acleris literana* (Linnaeus, 1758)
49.090 *Eulia ministrana* (Linnaeus, 1758)
49.091 *Pseudargyrotoza conwagana* (Fabricius, 1775)
49.092 *Phtheochroa inopiana* (Haworth, 1811)
49.094 *Phtheochroa sodaliana* (Haworth, 1811)
49.095 *Phtheochroa rugosana* (Hübner, [1799])
49.096 *Hysterophora maculosana* (Haworth, 1811)
49.097 *Cochylimorpha straminea* (Haworth, 1811)
49.100 *Phalonidia curvistrigana* (Stainton, 1859)
49.101 *Phalonidia manniana* (Fischer von Röslerstamm, 1839)
49.103 *Phalonidia affinitana* (Douglas, 1846)
49.105 *Gynnidomorpha vectisana* (Humphreys & Westwood, 1845)
49.108 *Gynnidomorpha alismana* (Ragonot, 1883)
49.109 *Agapeta hamana* (Linnaeus, 1758)
49.110 *Agapeta zoegana* (Linnaeus, 1767)
49.111 *Eupoecilia angustana* (Hübner, [1799])
49.114 *Aethes hartmanniana* (Clerck, 1759)
49.120 *Aethes smeathmanniana* (Fabricius, 1781)
49.121 *Aethes tesserana* ([Denis & Schiffermüller], 1775)
49.122 *Aethes dilucidana* (Stephens, 1852)
49.123 *Aethes beatricella* (Walsingham, 1898)
49.124 *Aethes francillana* (Fabricius, 1794)
49.127 *Aethes cnicana* (Westwood, 1854)
49.128 *Aethes rubigana* (Treitschke, 1830)
49.129 *Cochylidia rupicola* (Curtis, 1834)
49.132 *Cochylidia implicitana* (Wocke, 1856)
49.133 *Thyraylia nana* (Haworth, 1811)
49.134 *Cochylis roseana* (Haworth, 1811)
49.137 *Neocochylis dubitana* (Hübner, [1799])
49.139 *Cochylichroa atricapitana* (Stephens, 1852)
49.142 *Falseuncaria ruficiliana* (Haworth, 1811)
49.144 *Eudemis profundana* ([Denis & Schiffermüller], 1775)
49.146 *Apotomis semifasciana* (Haworth, 1811)
49.149 *Apotomis turbidana* Hübner, [1825]
49.150 *Apotomis betuletana* (Haworth, 1811)
49.151 *Apotomis capreana* (Hübner, [1817])
49.152 *Apotomis sororculana* (Zetterstedt, 1839)
49.153 *Apotomis sauciana* (Frölich, 1828)
49.154 *Orthotaenia undulana* ([Denis & Schiffermüller], 1775)
49.155 *Hedya salicella* (Linnaeus, 1758)
49.156 *Hedya nubiferana* (Haworth, 1811) Marbled Orchard Tortrix
49.157 *Hedya pruniana* (Hübner, [1799]) Plum Tortrix
49.158 *Hedya ochroleucana* (Frölich, 1828)
49.159 *Hedya atropunctana* (Zetterstedt, 1839)
49.160 *Celypha rufana* (Scopoli, 1763)
49.161 *Celypha striana* ([Denis & Schiffermüller], 1775)
49.164 *Celypha cespitana* (Hübner, [1817])

The Moths of Lancashire **581**

49.166 *Celypha lacunana* ([Denis & Schiffermüller], 1775)
49.167 *Celypha rivulana* (Scopoli, 1763)
49.172 *Phiaris schulziana* (Fabricius, [1777])
49.173 *Phiaris micana* ([Denis & Schiffermüller], 1775)
49.174 *Phiaris palustrana* (Lienig & Zeller, 1846)
49.178 *Stictea mygindiana* ([Denis & Schiffermüller], 1775)
49.179 *Olethreutes arcuella* (Clerck, 1759)
49.180 *Piniphila bifasciana* (Haworth, 1811)
49.183 *Lobesia abscisana* (Doubleday, [1849])
49.184 *Lobesia reliquana* (Hübner, [1825])
49.185 *Lobesia littoralis* (Westwood & Humphreys, 1845)
49.186 *Endothenia gentianaeana* (Hübner, [1799])
49.187 *Endothenia oblongana* (Haworth, 1811)
49.188 *Endothenia marginana* (Haworth, 1811)
49.190 *Endothenia ustulana* (Haworth, 1811)
49.191 *Endothenia nigricostana* (Haworth, 1811)
49.192 *Endothenia ericetana* (Humphreys & Westwood, 1845)
49.193 *Endothenia quadrimaculana* (Haworth, 1811)
49.194 *Bactra lancealana* (Hübner, [1799])
49.195 *Bactra furfurana* (Haworth, 1811)
49.196 *Bactra lacteana* Caradja, 1916
49.197 *Bactra robustana* (Christoph, 1872)
49.199 *Eucosmomorpha albersana* (Hübner, [1813])
49.200 *Enarmonia formosana* (Scopoli, 1763) Cherry Bark Tortrix
49.201 *Ancylis unguicella* (Linnaeus, 1758)
49.202 *Ancylis uncella* ([Denis & Schiffermüller], 1775)
49.204 *Ancylis obtusana* (Haworth, 1811)
49.207 *Ancylis geminana* (Donovan, 1806)
49.209 *Ancylis diminutana* (Haworth, 1811)
49.211 *Ancylis myrtillana* (Treitschke, 1830)
49.212 *Ancylis apicella* ([Denis & Schiffermüller], 1775)
49.214 *Ancylis badiana* ([Denis & Schiffermüller], 1775)
49.215 *Ancylis achatana* ([Denis & Schiffermüller], 1775)
49.216 *Ancylis mitterbacheriana* ([Denis & Schiffermüller], 1775)
49.218 *Eriopsela quadrana* (Hübner, [1813])
49.219 *Thiodia citrana* (Hübner, [1799])
49.220 *Rhopobota myrtillana* (Humphreys & Westwood, 1845)
49.221 *Rhopobota stagnana* ([Denis & Schiffermüller], 1775)
49.222 *Rhopobota ustomaculana* (Curtis, 1831)
49.223 *Rhopobota naevana* (Hübner, [1817]) Holly Tortrix
49.224 *Spilonota ocellana* ([Denis & Schiffermüller], 1775) Bud Moth
49.225 *Spilonota laricana* (Heinemann, 1863)
49.228 *Epinotia sordidana* (Hübner, [1824])
49.229 *Epinotia caprana* (Fabricius, 1798)
49.230 *Epinotia trigonella* (Linnaeus, 1758)
49.231 *Epinotia brunnichana* (Linnaeus, 1767)
49.232 *Epinotia maculana* (Fabricius, 1775)
49.233 *Epinotia solandriana* (Linnaeus, 1758)
49.234 *Epinotia abbreviana* (Fabricius, 1794)
49.237 *Epinotia signatana* (Douglas, 1845)
49.238 *Epinotia cruciana* (Linnaeus, 1761) Willow Tortrix
49.239 *Epinotia mercuriana* (Frölich, 1828)
49.240 *Epinotia immundana* (Fischer von Röslerstamm, 1839)
49.242 *Epinotia nanana* (Treitschke, 1835)
49.243 *Epinotia demarniana* (Fischer von Röslerstamm, 1840)
49.244 *Epinotia subocellana* (Donovan, 1806)
49.245 *Epinotia tetraquetrana* (Haworth, 1811)
49.246 *Epinotia pygmaeana* (Hübner, [1799])
49.247 *Epinotia subsequana* (Haworth, 1811)
49.248 *Epinotia tenerana* ([Denis & Schiffermüller], 1775) Nut Bud Moth
49.249 *Epinotia ramella* (Linnaeus, 1758)
49.251 *Epinotia rubiginosana* (Herrich-Schäffer, 1851)
49.252 *Epinotia tedella* (Clerck, 1759)
49.254 *Epinotia bilunana* (Haworth, 1811)
49.255 *Epinotia nisella* (Clerck, 1759)
49.257 *Zeiraphera griseana* (Hübner, [1799]) Larch Tortrix
49.259 *Zeiraphera ratzeburgiana* (Saxesen, 1840) Spruce Bud Moth
49.260 *Zeiraphera isertana* (Fabricius, 1794)
49.261 *Crocidosema plebejana* Zeller, 1847

49.264 *Eucosma obumbratana* (Lienig & Zeller, 1846)
49.265 *Eucosma cana* (Haworth, 1811)
49.266 *Eucosma hohenwartiana* ([Denis & Schiffermüller], 1775)
49.269 *Eucosma campoliliana* ([Denis & Schiffermüller], 1775)
49.272 *Eucosma tripoliana* (Barrett, 1880)
49.276 *Eucosma aspidiscana* (Hübner, [1817])
49.279 *Gypsonoma dealbana* (Frölich, 1828)
49.280 *Gypsonoma oppressana* (Treitschke, 1835)
49.281 *Gypsonoma sociana* (Haworth, 1811)
49.283 *Gypsonoma aceriana* (Duponchel, [1843])
49.284 *Epiblema sticticana* (Fabricius, 1794)
49.285 *Epiblema scutulana* ([Denis & Schiffermüller], 1775)
49.286 *Epiblema cirsiana* (Zeller, 1843)
49.288 *Epiblema foenella* (Linnaeus, 1758)
49.289 *Epiblema costipunctana* (Haworth, 1811)
49.290 *Epiblema turbidana* (Treitschke, 1835)
49.292 *Notocelia cynosbatella* (Linnaeus, 1758)
49.293 *Notocelia tetragonana* (Stephens, 1834)
49.294 *Notocelia uddmanniana* (Linnaeus, 1758) Bramble Shoot Moth
49.295 *Notocelia roborana* ([Denis & Schiffermüller], 1775)
49.296 *Notocelia incarnatana* (Hübner, [1800])
49.297 *Notocelia rosaecolana* (Doubleday, 1850)
49.298 *Notocelia trimaculana* (Haworth, 1811)
49.299 *Pseudococcyx posticana* (Zetterstedt, 1839)
49.300 *Pseudococcyx turionella* (Linnaeus, 1758) Pine Bud Moth
49.301 *Retinia resinella* (Linnaeus, 1758) Pine Resin-gall Moth
49.304 *Clavigesta purdeyi* (Durrant, 1911) Pine Leaf-mining Moth
49.305 *Rhyacionia buoliana* ([Denis & Schiffermüller], 1775) Pine Shoot Moth
49.306 *Rhyacionia pinicolana* (Doubleday, 1850)
49.307 *Rhyacionia pinivorana* (Lienig & Zeller, 1846) Spotted Shoot Moth
49.309 *Dichrorampha plumbana* (Scopoli, 1763)
49.310 *Dichrorampha sedatana* Busck, 1906
49.311 *Dichrorampha aeratana* (Pierce & Metcalfe, 1915)
49.313 *Dichrorampha acuminatana* (Lienig & Zeller, 1846)
49.315 *Dichrorampha simpliciana* (Haworth, 1811)
49.318 *Dichrorampha vancouverana* McDunnough, 1935
49.319 *Dichrorampha flavidorsana* Knaggs, 1867
49.320 *Dichrorampha alpinana* (Treitschke, 1830)
49.321 *Dichrorampha petiverella* (Linnaeus, 1758)
49.322 *Dichrorampha plumbagana* (Treitschke, 1830)
49.323 *Dichrorampha alpestrana* (Herrich-Schäffer, 1851)
49.324 *Cydia nigricana* (Fabricius, 1794) Pea Moth
49.325 *Cydia ulicetana* (Haworth, 1811)
49.332 *Cydia coniferana* (Saxesen, 1840)
49.334 *Cydia cosmophorana* (Treitschke, 1835)
49.335 *Cydia strobilella* (Linnaeus, 1758) Spruce Seed Moth
49.338 *Cydia pomonella* (Linnaeus, 1758) Codling Moth
49.341 *Cydia splendana* (Hübner, [1799])
49.342 *Cydia fagiglandana* (Zeller, 1841)
49.345 *Lathronympha strigana* (Fabricius, 1775)
49.347 *Grapholita compositella* (Fabricius, 1775)
49.349 *Grapholita internana* (Guenée, 1845)
49.351 *Grapholita lunulana* ([Denis & Schiffermüller], 1775)
49.354 *Grapholita jungiella* (Clerck, 1759)
49.3541 *Grapholita molesta* (Busck, 1916) Oriental Fruit Moth
49.355 *Grapholita lathyrana* (Hübner, [1813])
49.356 *Grapholita lobarzewskii* (Nowicki, 1860)
49.357 *Grapholita funebrana* Treitschke, 1835 Plum Fruit Moth
49.358 *Grapholita tenebrosana* Duponchel, [1843]
49.359 *Grapholita janthinana* (Duponchel, 1835)
49.360 *Pammene splendidulana* (Guenée, 1845)
49.362 *Pammene giganteana* (Peyerimhoff, 1863)
49.363 *Pammene argyrana* (Hübner, [1799])
49.364 *Pammene suspectana* (Lienig & Zeller, 1846)
49.366 *Pammene obscurana* (Stephens, 1834)
49.367 *Pammene fasciana* (Linnaeus, 1761)
49.368 *Pammene herrichiana* (Heinemann, 1854)
49.371 *Pammene rhediella* (Clerck, 1759) Fruitlet Mining Tortrix
49.372 *Pammene populana* (Fabricius, 1787)

49.373 *Pammene spiniana* (Duponchel, [1843])
49.375 *Pammene regiana* (Zeller, 1849)
49.376 *Pammene aurita* Razowski, 1991
49.378 *Pammene ochsenheimeriana* (Lienig & Zeller, 1846)
49.379 *Pammene aurana* (Fabricius, 1775)
49.380 *Pammene gallicana* (Guenée, 1845)
49.381 *Strophedra weirana* (Douglas, 1850)
49.382 *Strophedra nitidana* (Fabricius, 1794)

COSSIDAE
50.001 *Cossus cossus* (Linnaeus, 1758) Goat Moth
50.002 *Zeuzera pyrina* (Linnaeus, 1761) Leopard Moth

SESIIDAE
52.002 *Sesia apiformis* (Clerck, 1759) Hornet Moth
52.003 *Sesia bembeciformis* (Hübner, [1806]) Lunar Hornet Moth
52.006 *Synanthedon spheciformis* ([Denis & Schiffermüller], 1775) White-barred Clearwing
52.007 *Synanthedon culiciformis* (Linnaeus, 1758) Large Red-belted Clearwing
52.008 *Synanthedon formicaeformis* (Esper, 1782) Red-tipped Clearwing
52.011 *Synanthedon myopaeformis* (Borkhausen, 1789) Red-belted Clearwing
52.012 *Synanthedon vespiformis* (Linnaeus, 1761) Yellow-legged Clearwing
52.013 *Synanthedon tipuliformis* (Clerck, 1759) Currant Clearwing
52.014 *Bembecia ichneumoniformis* ([Denis & Schiffermüller], 1775) Six-belted Clearwing

ZYGAENIDAE
54.002 *Adscita statices* (Linnaeus, 1758) Forester
54.003 *Adscita geryon* (Hübner, [1813]) Cistus Forester
54.008 *Zygaena filipendulae* (Linnaeus, 1758) Six-spot Burnet
54.009 *Zygaena lonicerae* (Scheven, 1777) Narrow-bordered Five-spot Burnet

PYRALIDAE
62.001 *Aphomia sociella* (Linnaeus, 1758) Bee Moth
62.003 *Aphomia gularis* (Zeller, 1877) Stored Nut Moth
62.004 *Aphomia cephalonica* (Stainton, 1866) Rice Moth
62.005 *Achroia grisella* (Fabricius, 1794) Lesser Wax Moth
62.006 *Galleria mellonella* (Linnaeus, 1758) Wax Moth
62.007 *Cryptoblabes bistriga* (Haworth, 1811)
62.011 *Ortholepis betulae* (Goeze, 1778)
62.012 *Pyla fusca* (Haworth, 1811)
62.023 *Pempelia palumbella* ([Denis & Schiffermüller], 1775)
62.024 *Rhodophaea formosa* (Howarth, 1811)
62.025 *Dioryctria sylvestrella* (Ratzeburg, 1840)
62.027 *Dioryctria simplicella* Heinemann, 1863
62.028 *Dioryctria abietella* ([Denis & Schiffermüller], 1775)
62.029 *Phycita roborella* ([Denis & Schiffermüller], 1775)
62.030 *Hypochalcia ahenella* ([Denis & Schiffermüller], 1775)
62.034 *Acrobasis repandana* (Fabricius, 1798)
62.035 *Acrobasis advenella* (Zincken, 1818)
62.037 *Acrobasis marmorea* (Haworth, 1811)
62.038 *Acrobasis consociella* (Hübner, [1813])
62.039 *Apomyelois bistriatella* (Hulst, 1887)
62.040 *Apomyelois ceratoniae* (Zeller, 1839) Locust Bean Moth
62.042 *Myelois circumvoluta* (Fourcroy, 1785) Thistle Ermine
62.048 *Euzophera pinguis* (Haworth, 1811)
62.054 *Homoeosoma sinuella* (Fabricius, 1794)
62.057 *Phycitodes maritima* (Tengström, 1848)
62.058 *Phycitodes binaevella* (Hübner, [1813])
62.059 *Phycitodes saxicola* (Vaughan, 1870)
62.062 *Plodia interpunctella* (Hübner, [1813]) Indian Meal Moth
62.063 *Ephestia kuehniella* Zeller, 1879 Mediterranean Flour Moth
62.064 *Ephestia elutella* (Hübner, 1796) Cacao Moth
62.065 *Ephestia woodiella* Richards & Thomson, 1932
62.066 *Cadra figulilella* (Gregson, 1871) Raisin Moth
62.067 *Cadra cautella* (Walker, 1863) Dried Currant Moth
62.068 *Cadra calidella* (Guenée, 1845) Dried Fruit Moth
62.069 *Anerastia lotella* (Hübner, [1813])
62.072 *Pyralis farinalis* (Linnaeus, 1758) Meal Moth
62.073 *Aglossa caprealis* (Hübner, [1809]) Small Tabby
62.074 *Aglossa pinguinalis* (Linnaeus, 1758) Large Tabby
62.075 *Hypsopygia costalis* (Fabricius, 1775) Gold Triangle
62.076 *Hypsopygia glaucinalis* (Linnaeus, 1758)
62.077 *Endotricha flammealis* ([Denis & Schiffermüller], 1775)

CRAMBIDAE
63.002 *Loxostege sticticalis* (Linnaeus, 1761)
63.003 *Pyrausta cingulata* (Linnaeus, 1758)
63.005 *Pyrausta despicata* (Scopoli, 1763)
63.006 *Pyrausta aurata* (Scopoli, 1763)
63.007 *Pyrausta purpuralis* (Linnaeus, 1758)
63.008 *Pyrausta ostrinalis* (Hübner, 1796)
63.014 *Sitochroa palealis* ([Denis & Schiffermüller], 1775)
63.016 *Anania fuscalis* ([Denis & Schiffermüller], 1775)
63.017 *Anania lancealis* ([Denis & Schiffermüller], 1775)
63.018 *Anania coronata* (Hufnagel, 1767)
63.020 *Anania perlucidalis* (Hübner, [1809])
63.021 *Anania terrealis* (Treitschke, 1829)
63.022 *Anania crocealis* (Hübner, 1796)
63.024 *Anania funebris* (Ström, 1768)
63.025 *Anania hortulata* (Linnaeus, 1758) Small Magpie
63.028 *Ostrinia nubilalis* (Hübner, 1796) European Corn-borer
63.029 *Paratalanta pandalis* (Hübner, [1825]) Bordered Pearl
63.031 *Udea ferrugalis* (Hübner, 1796) Rusty-dot Pearl
63.033 *Udea lutealis* (Hübner, [1809])
63.034 *Udea prunalis* ([Denis & Schiffermüller], 1775)
63.037 *Udea olivalis* ([Denis & Schiffermüller], 1775)
63.038 *Patania ruralis* (Scopoli, 1763) Mother of Pearl
63.0381 *Patania aegrotalis* (Zeller, 1852)
63.042 *Diplopseustis perieresalis* (Walker, 1859)
63.046 *Duponchelia fovealis* Zeller, 1847
63.047 *Spoladea recurvalis* (Fabricius, 1775)
63.048 *Palpita vitrealis* (Rossi, 1794)
63.051 *Antigastra catalaunalis* (Duponchel, 1833)
63.052 *Nomophila noctuella* ([Denis & Schiffermüller], 1775) Rush Veneer
63.054 *Cydalima perspectalis* (Walker, 1859)
63.057 *Evergestis forficalis* (Linnaeus, 1758) Garden Pebble
63.058 *Evergestis extimalis* (Scopoli, 1763) Box-tree Moth
63.060 *Evergestis pallidata* (Hufnagel, 1767)
63.061 *Hellula undalis* (Fabricius, 1781) Old World Webworm
63.062 *Scoparia subfusca* Haworth, 1811
63.063 *Scoparia basistrigalis* Knaggs, 1866
63.064 *Scoparia ambigualis* (Treitschke, 1829)
63.065 *Scoparia ancipitella* (La Harpe, 1855)
63.066 *Scoparia pyralella* ([Denis & Schiffermüller], 1775)
63.067 *Eudonia lacustrata* (Panzer, 1804)
63.068 *Eudonia murana* (Curtis, 1827)
63.069 *Eudonia angustea* (Curtis, 1827)
63.072 *Eudonia delunella* (Stainton, 1849)
63.073 *Eudonia truncicolella* (Stainton, 1849)
63.074 *Eudonia mercurella* (Linnaeus, 1758)
63.075 *Eudonia pallida* (Curtis, 1827)
63.076 *Euchromius ocellea* (Haworth, 1811)
63.077 *Chilo phragmitella* (Hübner, [1810])
63.079 *Calamotropha paludellam* (Hübner, [1824])
63.080 *Chrysoteuchia culmella* (Linnaeus, 1758)
63.081 *Crambus pascuella* (Linnaeus, 1758)
63.083 *Crambus uliginosellus* Zeller, 1850
63.084 *Crambus ericella* (Hübner, [1813])
63.085 *Crambus pratella* (Linnaeus, 1758)
63.086 *Crambus lathoniellus* (Zincken, 1817)
63.087 *Crambus hamella* (Thunberg, 1788)
63.088 *Crambus perlella* (Scopoli, 1763)
63.089 *Agriphila tristella* ([Denis & Schiffermüller], 1775)
63.090 *Agriphila inquinatella* ([Denis & Schiffermüller], 1775)
63.091 *Agriphila latistria* (Haworth, 1811)
63.092 *Agriphila selasella* (Hübner, [1813])
63.093 *Agriphila straminella* ([Denis & Schiffermüller], 1775)

63.095 *Agriphila geniculea* (Haworth, 1811)
63.099 *Catoptria pinella* (Linnaeus, 1758)
63.100 *Catoptria margaritella* ([Denis & Schiffermüller], 1775)
63.102 *Catoptria falsella* ([Denis & Schiffermüller], 1775)
63.110 *Pediasia aridella* (Thunberg, 1788)
63.112 *Platytes alpinella* (Hübner, [1813])
63.114 *Elophila nymphaeata* (Linnaeus, 1758) Brown China-mark
63.115 *Acentria ephemerella* ([Denis & Schiffermüller], 1775) Water Veneer
63.116 *Cataclysta lemnata* (Linnaeus, 1758) Small China-mark
63.117 *Parapoynx stratiotata* (Linnaeus, 1758) Ringed China-mark
63.1175 *Parapoynx bilinealis* Snellen, 1876
63.118 *Nymphula nitidulata* (Hufnagel, 1767) Beautiful China-mark
63.121 *Donacaula forficella* (Thunberg, 1794)
63.122 *Donacaula mucronella* ([Denis & Schiffermüller], 1775)

DREPANIDAE
65.001 *Falcaria lacertinaria* (Linnaeus, 1758) Scalloped Hook-tip
65.002 *Watsonalla binaria* (Hufnagel, 1767) Oak Hook-tip
65.003 *Watsonalla cultraria* (Fabricius, 1775) Barred Hook-tip
65.005 *Drepana falcataria* (Linnaeus, 1758) Pebble Hook-tip
65.007 *Cilix glaucata* (Scopoli, 1763) Chinese Character
65.008 *Thyatira batis* (Linnaeus, 1758) Peach Blossom
65.009 *Habrosyne pyritoides* (Hufnagel, 1766) Buff Arches
65.010 *Tethea ocularis* (Linnaeus, 1767) Figure of Eighty
65.011 *Tethea or* ([Denis & Schiffermüller], 1775) Poplar Lutestring
65.012 *Tetheella fluctuosa* (Hübner, [1803]) Satin Lutestring
65.013 *Ochropacha duplaris* (Linnaeus, 1761) Common Lutestring
65.014 *Cymatophorina diluta* ([Denis & Schiffermüller], 1775) Oak Lutestring
65.015 *Polyploca ridens* (Fabricius, 1787) Frosted Green
65.016 *Achlya flavicornis* (Linnaeus, 1758) Yellow Horned

LASIOCAMPIDAE
66.001 *Poecilocampa populi* (Linnaeus, 1758) December Moth
66.002 *Trichiura crataegi* (Linnaeus, 1758) Pale Eggar
66.003 *Malacosoma neustria* (Linnaeus, 1758) Lackey
66.005 *Eriogaster lanestris* (Linnaeus, 1758) Small Eggar
66.006 *Lasiocampa trifolii* ([Denis & Schiffermüller], 1775) Grass Eggar
66.007 *Lasiocampa quercus* (Linnaeus, 1758) Oak Eggar/Northern Eggar
66.008 *Macrothylacia rubi* (Linnaeus, 1758) Fox Moth
66.010 *Euthrix potatoria* (Linnaeus, 1758) Drinker
66.011 *Phyllodesma ilicifolia* (Linnaeus, 1758) Small Lappet

SATURNIIDAE
68.001 *Saturnia pavonia* (Linnaeus, 1758) Emperor Moth

SPHINGIDAE
69.001 *Mimas tiliae* (Linnaeus, 1758) Lime Hawk-moth
69.002 *Smerinthus ocellata* (Linnaeus, 1758) Eyed Hawk-moth
69.003 *Laothoe populi* (Linnaeus, 1758) Poplar Hawk-moth
69.004 *Agrius convolvuli* (Linnaeus, 1758) Convolvulus Hawk-moth
69.005 *Acherontia atropos* (Linnaeus, 1758) Death's-head Hawk-moth
69.006 *Sphinx ligustri* Linnaeus, 1758 Privet Hawk-moth
69.007 *Sphinx pinastri* Linnaeus, 1758 Pine Hawk-moth
69.008 *Hemaris tityus* (Linnaeus, 1758) Narrow-bordered Bee Hawk-moth
69.010 *Macroglossum stellatarum* (Linnaeus, 1758) Humming-bird Hawk-moth
69.011 *Daphnis nerii* (Linnaeus, 1758) Oleander Hawk-moth
69.013 *Hyles euphorbiae* (Linnaeus, 1758) Spurge Hawk-moth
69.014 *Hyles gallii* (Rottemburg, 1775) Bedstraw Hawk-moth
69.015 *Hyles livornica* (Esper, [1804]) Striped Hawk-moth
69.016 *Deilephila elpenor* (Linnaeus, 1758) Elephant Hawk-moth
69.017 *Deilephila porcellus* (Linnaeus, 1758) Small Elephant Hawk-moth
69.018 *Hippotion celerio* (Linnaeus, 1758) Silver-striped Hawk-moth

GEOMETRIDAE
70.002 *Idaea muricata* (Hufnagel, 1767) Purple-bordered Gold
70.004 *Idaea rusticata* ([Denis & Schiffermüller], 1775) Least Carpet
70.006 *Idaea fuscovenosa* (Goeze, 1781) Dwarf Cream Wave
70.008 *Idaea seriata* (Schrank, 1802) Small Dusty Wave

70.009 *Idaea subsericeata* (Haworth, 1809) Satin Wave
70.010 *Idaea sylvestraria* (Hübner, [1799]) Dotted Border Wave
70.011 *Idaea dimidiata* (Hufnagel, 1767) Single-dotted Wave
70.013 *Idaea biselata* (Hufnagel, 1767) Small Fan-footed Wave
70.015 *Idaea emarginata* (Linnaeus, 1758) Small Scallop
70.016 *Idaea aversata* (Linnaeus, 1758) Riband Wave
70.018 *Idaea straminata* (Borkhausen, 1794) Plain Wave
70.023 *Scopula marginepunctata* (Goeze, 1781) Mullein Wave
70.024 *Scopula imitaria* (Hübner, [1799]) Small Blood-vein
70.025 *Scopula immutata* (Linnaeus, 1758) Lesser Cream Wave
70.026 *Scopula ternata* Schrank, 1802 Smoky Wave
70.027 *Scopula floslactata* (Haworth, 1809) Cream Wave
70.029 *Timandra comae* Schmidt, 1931 Blood-vein
70.032 *Cyclophora albipunctata* (Hufnagel, 1767) Birch Mocha
70.036 *Cyclophora punctaria* (Linnaeus, 1758) Maiden's Blush
70.037 *Cyclophora linearia* (Hübner, [1799]) Clay Triple-lines
70.038 *Rhodometra sacraria* (Linnaeus, 1767) Vestal
70.039 *Phibalapteryx virgata* (Hufnagel, 1767) Oblique Striped
70.040 *Scotopteryx mucronata* (Scopoli, 1763) Lead Belle
70.041 *Scotopteryx luridata* (Hufnagel, 1767) July Belle
70.045 *Scotopteryx chenopodiata* (Linnaeus, 1758) Shaded Broad-bar
70.046 *Orthonama vittata* (Borkhausen, 1794) Oblique Carpet
70.047 *Nycterosea obstipata* (Fabricius, 1794) Gem
70.048 *Xanthorhoe decoloraria* (Esper, [1806]) Red Carpet
70.049 *Xanthorhoe fluctuata* (Linnaeus, 1758) Garden Carpet
70.051 *Xanthorhoe spadicearia* ([Denis & Schiffermüller], 1775) Red Twin-spot Carpet
70.052 *Xanthorhoe ferrugata* (Clerck, 1759) Dark-barred Twin-spot Carpet
70.053 *Xanthorhoe designata* (Hufnagel, 1767) Flame Carpet
70.054 *Xanthorhoe montanata* ([Denis & Schiffermüller], 1775) Silver-ground Carpet
70.055 *Xanthorhoe quadrifasiata* (Clerck, 1759) Large Twin-spot Carpet
70.059 *Camptogramma bilineata* (Linnaeus, 1758) Yellow Shell
70.060 *Epirrhoe tristata* (Linnaeus, 1758) Small Argent and Sable
70.061 *Epirrhoe alternata* (Müller, 1764) Common Carpet
70.062 *Epirrhoe rivata* (Hübner, [1813]) Wood Carpet
70.063 *Epirrhoe galiata* ([Denis & Schiffermüller], 1775) Galium Carpet
70.064 *Euphyia biangulata* (Haworth, 1809) Cloaked Carpet
70.065 *Euphyia unangulata* (Haworth, 1809) Sharp-angled Carpet
70.066 *Earophila badiata* ([Denis & Schiffermüller], 1775) Shoulder Stripe
70.067 *Anticlea derivata* ([Denis & Schiffermüller], 1775) Streamer
70.068 *Mesoleuca albicillata* (Linnaeus, 1758) Beautiful Carpet
70.069 *Pelurga comitata* (Linnaeus, 1758) Dark Spinach
70.072 *Larentia clavaria* (Haworth, 1809) Mallow
70.072 *Entephria caesiata* ([Denis & Schiffermüller], 1775) Grey Mountain Carpet
70.074 *Hydriomena furcata* (Thunberg, 1784) July Highflyer
70.075 *Hydriomena impluviata* ([Denis & Schiffermüller], 1775) May Highflyer
70.076 *Hydriomena ruberata* (Freyer, [1831]) Ruddy Highflyer
70.077 *Pennithera firmata* (Hübner, [1822]) Pine Carpet
70.078 *Thera cognata* (Thunberg, 1792) Chestnut-coloured Carpet
70.079 *Thera britannica* (Turner, 1925) Spruce Carpet
70.081 *Thera obeliscata* (Hübner, [1787]) Grey Pine Carpet
70.082 *Thera juniperata* (Linnaeus, 1758) Juniper Carpet
70.083 *Thera cupressata* (Geyer, [1831]) Cypress Carpet
70.084 *Plemyria rubiginata* ([Denis & Schiffermüller], 1775) Blue-bordered Carpet
70.085 *Cidaria fulvata* (Forster, 1771) Barred Yellow
70.086 *Electrophaes corylata* (Thunberg, 1792) Broken-barred Carpet
70.087 *Cosmorhoe ocellata* (Linnaeus, 1758) Purple Bar
70.088 *Eustroma reticulata* ([Denis & Schiffermüller], 1775) Netted Carpet
70.089 *Eulithis prunata* (Linnaeus, 1758) Phoenix
70.090 *Eulithis testata* (Linnaeus, 1761) Chevron
70.091 *Eulithis populata* (Linnaeus, 1758) Northern Spinach
70.092 *Eulithis mellinata* (Fabricius, 1787) Spinach
70.093 *Gandaritis pyraliata* ([Denis & Schiffermüller], 1775) Barred Straw
70.094 *Ecliptopera silaceata* ([Denis & Schiffermüller], 1775) Small Phoenix
70.095 *Chloroclysta siterata* (Hufnagel, 1767) Red-green Carpet
70.096 *Chloroclysta miata* (Linnaeus, 1758) Autumn Green Carpet
70.097 *Dysstroma truncata* (Hufnagel, 1767) Common Marbled Carpet

70.098 *Dysstroma citrata* (Linnaeus, 1761) Dark Marbled Carpet
70.099 *Colostygia olivata* ([Denis & Schiffermüller], 1775) Beech-green Carpet
70.100 *Colostygia pectinataria* (Knoch, 1781) Green Carpet
70.101 *Colostygia multistrigaria* (Haworth, 1809) Mottled Grey
70.102 *Coenotephria salicata* (Hübner, [1799]) Striped Twin-spot Carpet
70.103 *Lampropteryx suffumata* ([Denis & Schiffermüller], 1775) Water Carpet
70.104 *Lampropteryx otregiata* (Metcalfe, 1917) Devon Carpet
70.105 *Operophtera fagata* (Scharfenberg, 1805) Northern Winter Moth
70.106 *Operophtera brumata* (Linnaeus, 1758) Winter Moth
70.107 *Epirrita dilutata* ([Denis & Schiffermüller], 1775) November Moth
70.108 *Epirrita christyi* (Allen, 1906) Pale November Moth
70.109 *Epirrita autumnata* (Borkhausen, 1794) Autumnal Moth
70.110 *Epirrita filigrammaria* (Herrich-Schäffer, 1846) Small Autumnal Moth
70.111 *Asthena albulata* (Hufnagel, 1767) Small White Wave
70.112 *Euchoeca nebulata* (Scopoli, 1763) Dingy Shell
70.113 *Hydrelia sylvata* ([Denis & Schiffermüller], 1775) Waved Carpet
70.114 *Hydrelia flammeolaria* (Hufnagel, 1767) Small Yellow Wave
70.115 *Venusia cambrica* Curtis, 1839 Welsh Wave
70.116 *Venusia blomeri* (Curtis, 1832) Blomer's Rivulet
70.118 *Philereme vetulata* ([Denis & Schiffermüller], 1775) Brown Scallop
70.119 *Philereme transversata* (Hufnagel, 1767) Dark Umber
70.120 *Rheumaptera hastata* (Linnaeus, 1758) Argent and Sable
70.121 *Rheumaptera undulata* (Linnaeus, 1758) Scallop Shell
70.123 *Triphosa dubitata* (Linnaeus, 1758) Tissue
70.128 *Melanthia procellata* ([Denis & Schiffermüller], 1775) Pretty Chalk Carpet
70.130 *Odezia atrata* (Linnaeus, 1758) Chimney Sweeper
70.131 *Mesotype didymata* (Linnaeus, 1758) Twin-spot Carpet
70.132 *Perizoma affinitata* (Stephens, 1831) Rivulet
70.133 *Perizoma alchemillata* (Linnaeus, 1758) Small Rivulet
70.134 *Perizoma bifaciata* (Haworth, 1809) Barred Rivulet
70.137 *Perizoma albulata* ([Denis & Schiffermüller], 1775) Grass Rivulet
70.138 *Perizoma flavofasciata* (Thunberg, 1792) Sandy Carpet
70.139 *Martania taeniata* (Stephens, 1831) Barred Carpet
70.141 *Gymnoscelis rufifasciata* (Haworth, 1809) Double-striped Pug
70.142 *Chloroclystis v-ata* (Haworth, 1809) V-Pug
70.143 *Pasiphila chloerata* (Mabille, 1870) Sloe Pug
70.144 *Pasiphila rectangulata* (Linnaeus, 1758) Green Pug
70.145 *Pasiphila debiliata* (Hübner, [1817]) Bilberry Pug
70.146 *Eupithecia haworthiata* Doubleday, 1856 Haworth's Pug
70.147 *Eupithecia tenuiata* (Hübner, [1813]) Slender Pug
70.148 *Eupithecia inturbata* (Hübner, [1817]) Maple Pug
70.149 *Eupithecia abietaria* (Goeze, 1781) Cloaked Pug
70.150 *Eupithecia linariata* ([Denis & Schiffermüller], 1775) Toadflax Pug
70.151 *Eupithecia pulchellata* Stephens, 1831 Foxglove Pug
70.153 *Eupithecia plumbeolata* (Haworth, 1809) Lead-coloured Pug
70.154 *Eupithecia pygmaeata* (Hübner, [1799]) Marsh Pug
70.155 *Eupithecia venosata* (Fabricius, 1787) Netted Pug
70.156 *Eupithecia abbreviata* Stephens, 1831 Brindled Pug
70.157 *Eupithecia dodoneata* Guenée, [1858] Oak-tree Pug
70.158 *Eupithecia pusillata* ([Denis & Schiffermüller], 1775) Juniper Pug
70.159 *Eupithecia phoeniceata* (Rambur, 1834) Cypress Pug
70.160 *Eupithecia tripunctaria* Herrich-Schäffer, 1852 White-spotted Pug
70.161 *Eupithecia virgaureata* Doubleday, 1861 Golden-rod Pug
70.162 *Eupithecia tantillaria* Boisduval, 1840 Dwarf Pug
70.163 *Eupithecia lariciata* (Freyer, 1841) Larch Pug
70.165 *Eupithecia pimpinellata* (Hübner, [1813]) Pimpinel Pug
70.166 *Eupithecia simpliciata* (Haworth, 1809) Plain Pug
70.168 *Eupithecia nanata* (Hübner, [1813]) Narrow-winged Pug
70.169 *Eupithecia innotata* (Hufnagel, 1767) Ash Pug/Angle-barred Pug
70.171 *Eupithecia indigata* (Hübner, [1813]) Ochreous Pug
70.172 *Eupithecia distinctaria* Herrich-Schäffer, 1848 Thyme Pug
70.173 *Eupithecia centaureata* ([Denis & Schiffermüller], 1775) Lime-speck Pug
70.175 *Eupithecia trisignaria* Herrich-Schäffer, 1848 Triple-spotted Pug
70.176 *Eupithecia intricata* (Zetterstedt, 1839) Freyer's Pug
70.177 *Eupithecia satyrata* (Hübner, [1813]) Satyr Pug
70.179 *Eupithecia absinthiata* (Clerck, 1759) Wormwood/Ling Pug
70.180 *Eupithecia expallidata* Doubleday, 1856 Bleached Pug
70.181 *Eupithecia valerianata* (Hübner, [1813]) Valerian Pug
70.182 *Eupithecia assimilata* Doubleday, 1856 Currant Pug

70.183 *Eupithecia vulgata* (Haworth, 1809) Common Pug
70.184 *Eupithecia exiguata* (Hübner, [1813]) Mottled Pug
70.187 *Eupithecia icterata* (Villers, 1789) Tawny-speckled Pug
70.188 *Eupithecia succenturiata* (Linnaeus, 1758) Bordered Pug
70.189 *Eupithecia subumbrata* ([Denis & Schiffermüller], 1775) Shaded Pug
70.190 *Eupithecia subfuscata* (Haworth, 1809) Grey Pug
70.191 *Carsia sororiata* (Hübner, [1813]) Manchester Treble-bar
70.192 *Aplocera plagiata* (Linnaeus, 1758) Treble-bar
70.195 *Chesias legatella* (Fabricius, 1775) Streak
70.198 *Lobophora halterata* (Hufnagel, 1767) Seraphim
70.199 *Pterapherapteryx sexalata* (Retzius, 1783) Small Seraphim
70.200 *Acasis viretata* (Hübner, [1799]) Yellow-barred Brindle
70.201 *Trichopteryx polycommata* ([Denis & Schiffermüller], 1775) Barred Tooth-striped
70.202 *Trichopteryx carpinata* (Borkhausen, 1794) Early Tooth-striped
70.203 *Archiearis parthenias* (Linnaeus, 1761) Orange Underwing
70.205 *Abraxas grossulariata* (Linnaeus, 1758) Magpie Moth
70.206 *Abraxas sylvata* (Scopoli, 1763) Clouded Magpie
70.207 *Lomaspilis marginata* (Linnaeus, 1758) Clouded Border
70.208 *Ligdia adustata* ([Denis & Schiffermüller], 1775) Scorched Carpet
70.212 *Macaria alternata* ([Denis & Schiffermüller], 1775) Sharp-angled Peacock
70.214 *Macaria liturata* (Clerck, 1759) Tawny-barred Angle
70.215 *Macaria wauaria* (Linnaeus, 1758) V-Moth
70.217 *Macaria brunneata* (Thunberg, 1784) Rannoch Looper
70.218 *Chiasmia clathrata* (Linnaeus, 1758) Latticed Heath
70.222 *Petrophora chlorosata* (Scopoli, 1763) Brown Silver-line
70.223 *Plagodis pulveraria* (Linnaeus, 1758) Barred Umber
70.224 *Plagodis dolabraria* (Linnaeus, 1767) Scorched Wing
70.226 *Opisthograptis luteolata* (Linnaeus, 1758) Brimstone Moth
70.227 *Epione repandaria* (Hufnagel, 1767) Bordered Beauty
70.229 *Pseudopanthera macularia* (Linnaeus, 1758) Speckled Yellow
70.231 *Apeira syringaria* (Linnaeus, 1758) Lilac Beauty
70.233 *Ennomos quercinaria* (Hufnagel, 1767) August Thorn
70.234 *Ennomos alniaria* (Linnaeus, 1758) Canary-shouldered Thorn
70.235 *Ennomos fuscantaria* (Haworth, 1809) Dusky Thorn
70.236 *Ennomos erosaria* ([Denis & Schiffermüller], 1775) September Thorn
70.237 *Selenia dentaria* (Fabricius, 1775) Early Thorn
70.238 *Selenia lunularia* (Hübner, [1788]) Lunar Thorn
70.239 *Selenia tetralunaria* (Hufnagel, 1767) Purple Thorn
70.240 *Odontopera bidentata* (Clerck, 1759) Scalloped Hazel
70.241 *Crocallis elinguaria* (Linnaeus, 1758) Scalloped Oak
70.243 *Ourapteryx sambucaria* (Linnaeus, 1758) Swallow-tailed Moth
70.244 *Colotois pennaria* (Linnaeus, 1761) Feathered Thorn
70.245 *Alsophila aescularia* ([Denis & Schiffermüller], 1775) March Moth
70.246 *Apocheima hispidaria* ([Denis & Schiffermüller], 1775) Small Brindled Beauty
70.247 *Phigalia pilosaria* ([Denis & Schiffermüller], 1775) Pale Brindled Beauty
70.248 *Lycia hirtaria* (Clerck, 1759) Brindled Beauty
70.250 *Lycia zonaria* ([Denis & Schiffermüller], 1775) Belted Beauty
70.251 *Biston strataria* (Hufnagel, 1767) Oak Beauty
70.252 *Biston betularia* (Linnaeus, 1758) Peppered Moth
70.253 *Agriopis leucophaearia* ([Denis & Schiffermüller], 1775) Spring Usher
70.254 *Agriopis aurantiaria* (Hübner, [1799]) Scarce Umber
70.255 *Agriopis marginaria* (Fabricius, [1777]) Dotted Border
70.256 *Erannis defoliaria* (Clerck, 1759) Mottled Umber
70.257 *Menophra abruptaria* (Thunberg, 1792) Waved Umber
70.258 *Peribatodes rhomboidaria* ([Denis & Schiffermüller], 1775) Willow Beauty
70.262 *Selidosema brunnearia* (Villers, 1789) Bordered Grey
70.264 *Deileptenia ribeata* (Clerck, 1759) Satin Beauty
70.265 *Alcis repandata* (Linnaeus, 1758) Mottled Beauty
70.270 *Ectropis crepuscularia* ([Denis & Schiffermüller], 1775) Engrailed
70.272 *Paradarisa consonaria* (Hübner, [1799]) Square Spot
70.274 *Aethalura punctulata* ([Denis & Schiffermüller], 1775) Grey Birch
70.275 *Ematurga atomaria* (Linnaeus, 1758) Common Heath
70.276 *Bupalus piniaria* (Linnaeus, 1758) Bordered White
70.277 *Cabera pusaria* (Linnaeus, 1758) Common White Wave
70.278 *Cabera exanthemata* (Scopoli, 1763) Common Wave
70.279 *Lomographa bimaculata* (Fabricius, 1775) White-pinion Spotted
70.280 *Lomographa temerata* ([Denis & Schiffermüller], 1775) Clouded Silver

70.282 *Theria primaria* (Haworth, 1809) Early Moth
70.283 *Campaea margaritaria* (Linnaeus, 1761) Light Emerald
70.284 *Hylaea fasciaria* (Linnaeus, 1758) Barred Red
70.287 *Charissa obscurata* ([Denis & Schiffermüller], 1775) Annulet
70.288 *Cleorodes lichenaria* (Hufnagel, 1767) Brussels Lace
70.292 *Dyscia fagaria* (Thunberg, 1784) Grey Scalloped Bar
70.295 *Perconia strigillaria* (Hübner, [1787]) Grass Wave
70.297 *Pseudoterpna pruinata* (Hufnagel, 1767) Grass Emerald
70.299 *Geometra papilionaria* (Linnaeus, 1758) Large Emerald
70.300 *Comibaena bajularia* ([Denis & Schiffermüller], 1775) Blotched Emerald
70.302 *Hemistola chrysoprasaria* (Esper, 1795) Small Emerald
70.303 *Jodis lactearia* (Linnaeus, 1758) Little Emerald
70.305 *Hemithea aestivaria* (Hübner, 1789) Common Emerald

NOTODONTIDAE

71.003 *Cerura vinula* (Linnaeus, 1758) Puss Moth
71.005 *Furcula furcula* (Clerck, 1759) Sallow Kitten
71.006 *Furcula bicuspis* (Borkhausen, 1790) Alder Kitten
71.007 *Furcula bifida* (Brahm, 1787) Poplar Kitten
71.010 *Drymonia dodonaea* ([Denis & Schiffermüller], 1775) Marbled Brown
71.011 *Drymonia ruficornis* (Hufnagel, 1766) Lunar Marbled Brown
71.012 *Notodonta dromedarius* (Linnaeus, 1767) Iron Prominent
71.013 *Notodonta ziczac* (Linnaeus, 1758) Pebble Prominent
71.016 *Peridea anceps* (Goeze, 1781) Great Prominent
71.017 *Pheosia tremula* (Clerck, 1759) Swallow Prominent
71.018 *Pheosia gnoma* (Fabricius, [1777]) Lesser Swallow Prominent
71.020 *Pterostoma palpina* (Clerck, 1759) Pale Prominent
71.021 *Ptilodon capucina* (Linnaeus, 1758) Coxcomb Prominent
71.023 *Odontosia carmelita* (Esper, [1798] Scarce Prominent
71.025 *Phalera bucephala* (Linnaeus, 1758) Buff-tip
71.027 *Clostera curtula* (Linnaeus, 1758) Chocolate-tip
71.028 *Clostera pigra* (Hufnagel, 1766) Small Chocolate-tip

EREBIDAE

72.001 *Scoliopteryx libatrix* (Linnaeus, 1758) Herald
72.002 *Rivula sericealis* (Scopoli, 1763) Straw Dot
72.0021 *Orodesma apicina* Herrich-Schäffer, 1868
72.003 *Hypena proboscidalis* (Linnaeus, 1758) Snout
72.007 *Hypena crassalis* (Fabricius, 1787) Beautiful Snout
72.009 *Leucoma salicis* (Linnaeus, 1758) White Satin Moth
72.010 *Lymantria monacha* (Linnaeus, 1758) Black Arches
72.011 *Lymantria dispar* (Linnaeus, 1758) Gypsy Moth
72.012 *Euproctis chrysorrhoea* (Linnaeus, 1758) Brown-tail
72.013 *Euproctis similis* (Fuessly, 1775) Yellow-tail
72.015 *Calliteara pudibunda* (Linnaeus, 1758) Pale Tussock
72.016 *Dicallomera fascelina* (Linnaeus, 1758) Dark Tussock
72.017 *Orgyia antiqua* (Linnaeus, 1758) Vapourer
72.019 *Spilosoma lutea* (Hufnagel, 1766) Buff Ermine
72.020 *Spilosoma lubricipeda* (Linnaeus, 1758) White Ermine
72.022 *Diaphora mendica* (Clerck, 1759) Muslin Moth
72.023 *Diacrisia sannio* (Linnaeus, 1758) Clouded Buff
72.024 *Phragmatobia fuliginosa* (Linnaeus, 1758) Ruby Tiger
72.025 *Parasemia plantaginis* (Linnaeus, 1758) Wood Tiger
72.026 *Arctia caja* (Linnaeus, 1758) Garden Tiger
72.0272 *Pyrrharctia isabella* (Smith, 1797) Isabelline Tiger
72.029 *Callimorpha dominula* (Linnaeus, 1758) Scarlet Tiger
72.031 *Tyria jacobaeae* (Linnaeus, 1758) Cinnabar
72.034 *Utetheisa pulchella* (Linnaeus, 1758) Crimson Speckled
72.036 *Nudaria mundana* (Linnaeus, 1761) Muslin Footman
72.037 *Thumatha senex* (Hübner, [1808]) Round-winged Muslin
72.038 *Cybosia mesomella* (Linnaeus, 1758) Four-dotted Footman
72.041 *Lithosia quadra* (Linnaeus, 1758) Four-spotted Footman
72.042 *Atolmis rubricollis* (Linnaeus, 1758) Red-necked Footman
72.043 *Eilema depressa* (Esper, 1787) Buff Footman
72.044 *Eilema griseola* (Hübner, [1803]) Dingy Footman
72.045 *Eilema lurideola* (Zincken, 1817) Common Footman
72.046 *Eilema complana* (Linnaeus, 1758) Scarce Footman
72.049 *Eilema sororcula* (Hufnagel, 1766) Orange Footman
72.053 *Herminia tarsipennalis* Treitschke, 1835 Fan-foot

72.055 *Herminia grisealis* ([Denis & Schiffermüller], 1775) Small Fan-foot
72.060 *Hypenodes humidalis* Doubleday, 1850 Marsh Oblique-barred
72.061 *Schrankia costaestrigalis* (Stephens, 1834) Pinion-streaked Snout
72.063 *Lygephila pastinum* (Treitschke, 1826) Blackneck
72.066 *Parascotia fuliginaria* (Linnaeus, 1761) Waved Black
72.067 *Phytometra viridaria* (Clerck, 1759) Small Purple-barred
72.069 *Laspeyria flexula* ([Denis & Schiffermüller], 1775) Beautiful Hook-tip
72.073 *Eublemma parva* (Hübner, [1808]) Small Marbled
72.074 *Eublemma purpurina* ([Denis & Schiffermüller], 1775) Beautiful Marbled
72.078 *Catocala fraxini* (Linnaeus, 1758) Clifden Nonpareil
72.078 *Catocala nupta* (Linnaeus, 1767) Red Underwing
72.083 *Euclidia glyphica* (Linnaeus, 1758) Burnet Companion
72.084 *Callistege mi* (Clerck, 1759) Mother Shipton

NOCTUIDAE

73.001 *Abrostola tripartita* (Hufnagel, 1766) Spectacle
73.002 *Abrostola triplasia* (Linnaeus, 1758) Dark Spectacle
73.003 *Trichoplusia ni* (Hübner, [1803]) Ni Moth
73.004 *Thysanoplusia orichalcea* (Fabricius, 1775) Slender Burnished Brass
73.008 *Chrysodeixis chalcites* (Esper, [1803]) Golden Twin-spot
73.009 *Chrysodeixis acuta* (Walker, 1857) Tunbridge Wells Gem
73.011 *Diachrysia chryson* (Esper, 1798) Scarce Burnished Brass
73.012 *Diachrysia chrysitis* (Linnaeus, 1758) Burnished Brass
73.014 *Polychrysia moneta* (Fabricius, 1787) Golden Plusia
73.015 *Autographa gamma* (Linnaeus, 1758) Silver Y
73.016 *Autographa pulchrina* (Haworth, 1809) Beautiful Golden Y
73.017 *Autographa jota* (Linnaeus, 1758) Plain Golden Y
73.018 *Autographa bractea* ([Denis & Schiffermüller], 1775) Gold Spangle
73.021 *Syngrapha interrogationis* (Linnaeus, 1758) Scarce Silver Y
73.022 *Plusia festucae* (Linnaeus, 1758) Gold Spot
73.023 *Plusia putnami* Grote, 1873 Lempke's Gold Spot
73.024 *Protodeltote pygarga* (Hufnagel, 1766) Marbled White Spot
73.026 *Deltote uncula* (Clerck, 1759) Silver Hook
73.032 *Colocasia coryli* (Linnaeus, 1758) Nut-tree Tussock
73.033 *Diloba caeruleocephala* (Linnaeus, 1758) Figure of Eight
73.036 *Acronicta alni* (Linnaeus, 1767) Alder Moth
73.037 *Acronicta tridens* ([Denis & Schiffermüller], 1775) Dark Dagger
73.038 *Acronicta psi* (Linnaeus, 1758) Grey Dagger
73.039 *Acronicta aceris* (Linnaeus, 1758) Sycamore
73.040 *Acronicta leporina* (Linnaeus, 1758) Miller
73.042 *Acronicta menyanthidis* (Esper, 1798) Light Knot Grass
73.045 *Acronicta rumicis* (Linnaeus, 1758) Knot Grass
73.046 *Subacronicta megacephala* ([Denis & Schiffermüller], 1775) Poplar Grey
73.047 *Craniophora ligustri* ([Denis & Schiffermüller], 1775) Coronet
73.048 *Panemeria tenebrata* (Scopoli, 1763) Small Yellow Underwing
73.050 *Cucullia absinthii* (Linnaeus, 1761) Wormwood
73.052 *Cucullia umbratica* (Linnaeus, 1758) Shark
73.053 *Cucullia chamomillae* ([Denis & Schiffermüller], 1775) Chamomile Shark
73.058 *Cucullia verbasci* (Linnaeus, 1758) Mullein
73.059 *Calophasia lunula* (Hufnagel, 1766) Toadflax Brocade
73.061 *Stilbia anomala* (Haworth, 1812) Anomalous
73.062 *Amphipyra pyramidea* (Linnaeus, 1758) Copper Underwing
73.063 *Amphipyra berbera* Rungs, 1949 Svensson's Copper Underwing
73.064 *Amphipyra tragopoginis* (Clerck, 1759) Mouse Moth
73.065 *Asteroscopus sphinx* (Hufnagel, 1766) Sprawler
73.068 *Allophyes oxyacanthae* (Linnaeus, 1758) Green-brindled Crescent
73.069 *Xylocampa areola* (Esper, 1789) Early Grey
73.070 *Pyrrhia umbra* (Hufnagel, 1766) Bordered Sallow
73.074 *Heliothis peltigera* ([Denis & Schiffermüller], 1775) Bordered Straw
73.076 *Helicoverpa armigera* (Hübner, [1808]) Scarce Bordered Straw
73.082 *Cryphia algae* (Fabricius, 1775) Tree-lichen Beauty
73.084 *Bryophila domestica* (Hufnagel, 1766) Marbled Beauty
73.085 *Bryopsis muralis* (Forster, 1771) Marbled Green
73.087 *Spodoptera exigua* (Hübner, [1808]) Small Mottled Willow
73.089 *Spodoptera littoralis* (Boisduval, 1833) Mediterranean Brocade
73.0892 *Spodoptera litura* (Fabricius, 1775) Asian Cotton Leafworm
73.092 *Caradrina morpheus* (Hufnagel, 1766) Mottled Rustic
73.095 *Caradrina clavipalpis* (Scopoli, 1763) Pale Mottled Willow
73.096 *Hoplodrina octogenaria* (Goeze, 1781) Uncertain

73.097 *Hoplodrina blanda* ([Denis & Schiffermüller], 1775) Rustic
73.099 *Hoplodrina ambigua* ([Denis & Schiffermüller], 1775) Vine's Rustic
73.100 *Chilodes maritima* (Tauscher, 1806) Silky Wainscot
73.101 *Charanyca trigrammica* (Hufnagel, 1766) Treble Lines
73.102 *Rusina ferruginea* (Esper, 1785) Brown Rustic
73.105 *Dypterygia scabriuscula* (Linnaeus, 1758) Bird's Wing
73.106 *Trachea atriplicis* (Linnaeus, 1758) Orache Moth
73.107 *Mormo maura* (Linnaeus, 1758) Old Lady
73.109 *Thalpophila matura* (Hufnagel, 1766) Straw Underwing
73.110 *Hyppa rectilinea* (Esper, 1796) Saxon
73.113 *Phlogophora meticulosa* (Linnaeus, 1758) Angle Shades
73.114 *Euplexia lucipara* (Linnaeus, 1758) Small Angle Shades
73.118 *Celaena haworthii* (Curtis, 1829) Haworth's Minor
73.119 *Helotropha leucostigma* (Hübner, [1808]) Crescent
73.120 *Eremobia ochroleuca* ([Denis & Schiffermüller], 1775) Dusky Sallow
73.121 *Gortyna flavago* ([Denis & Schiffermüller], 1775) Frosted Orange
73.123 *Hydraecia micacea* (Esper, 1789) Rosy Rustic
73.124 *Hydraecia petasitis* Doubleday, 1847 Butterbur
73.126 *Amphipoea fucosa* (Freyer, 1830) Saltern Ear
73.127 *Amphipoea lucens* (Freyer, 1845) Large Ear
73.128 *Amphipoea oculea* (Linnaeus, 1761) Ear Moth
73.129 *Amphipoea crinanensis* (Burrows, 1908) Crinan Ear
73.131 *Luperina testacea* ([Denis & Schiffermüller], 1775) Flounced Rustic
73.132 *Luperina nickerlii* (Freyer, 1845) Sandhill Rustic
73.134 *Rhizedra lutosa* (Hübner, [1803]) Large Wainscot
73.136 *Nonagria typhae* (Thunberg, 1784) Bulrush Wainscot
73.137 *Arenostola phragmitidis* (Hübner, [1803]) Fen Wainscot
73.138 *Longalatedes elymi* (Treitschke, 1825) Lyme Grass
73.139 *Lenisa geminipuncta* (Haworth, 1809) Twin-spotted Wainscot
73.141 *Archanara dissoluta* (Treitschke, 1825) Brown-veined Wainscot
73.142 *Coenobia rufa* (Haworth, 1809) Small Rufous
73.144 *Denticucullus pygmina* (Haworth, 1809) Small Wainscot
73.146 *Photedes captiuncula* (Treitschke, 1825) Least Minor
73.147 *Photedes minima* (Haworth, 1809) Small Dotted Buff
73.154 *Apamea remissa* (Hübner, [1809]) Dusky Brocade
73.155 *Apamea epomidion* (Haworth, 1809) Clouded Brindle
73.156 *Apamea crenata* (Hufnagel, 1766) Clouded-bordered Brindle
73.157 *Apamea anceps* ([Denis & Schiffermüller], 1775) Large Nutmeg
73.158 *Apamea sordens* (Hufnagel, 1766) Rustic Shoulder-knot
73.159 *Apamea unanimis* (Hübner, [1813]) Small Clouded Brindle
73.160 *Apamea scolopacina* (Esper, 1788) Slender Brindle
73.161 *Apamea oblonga* (Haworth, 1809) Crescent Striped
73.162 *Apamea monoglypha* (Hufnagel, 1766) Dark Arches
73.163 *Apamea lithoxylaea* ([Denis & Schiffermüller], 1775) Light Arches
73.164 *Apamea sublustris* (Esper, 1788) Reddish Light Arches
73.165 *Apamea furva* ([Denis & Schiffermüller], 1775) Confused
73.168 *Lateroligia ophiogramma* (Esper, 1794) Double Lobed
73.169 *Mesapamea secalis* (Linnaeus, 1758) Common Rustic
73.170 *Mesapamea didyma* (Esper, 1788) Lesser Common Rustic
73.171 *Litoligia literosa* (Haworth, 1809) Rosy Minor
73.172 *Mesoligia furuncula* ([Denis & Schiffermüller], 1775) Cloaked Minor
73.173 *Oligia strigilis* (Linnaeus, 1758) Marbled Minor
73.174 *Oligia latruncula* ([Denis & Schiffermüller], 1775) Tawny Marbled Minor
73.175 *Oligia versicolor* (Borkhausen, 1792) Rufous Minor
73.176 *Oligia fasciuncula* (Haworth, 1809) Middle-barred Minor
73.179 *Tiliacea citrago* (Linnaeus, 1758) Orange Sallow
73.180 *Tiliacea aurago* ([Denis & Schiffermüller], 1775) Barred Sallow
73.181 *Xanthia togata* (Esper, 1788) Pink-barred Sallow
73.182 *Cirrhia icteritia* (Hufnagel, 1766) Sallow
73.183 *Cirrhia gilvago* ([Denis & Schiffermüller], 1775) Dusky-lemon Sallow
73.186 *Agrochola lychnidis* ([Denis & Schiffermüller], 1775) Beaded Chestnut
73.187 *Anchoscelis litura* (Linnaeus, 1761) Brown-spot Pinion
73.188 *Anchoscelis helvola* (Linnaeus, 1758) Flounced Chestnut
73.189 *Leptologia lota* (Clerck, 1759) Red-line Quaker
73.190 *Leptologia macilenta* (Hübner, [1809]) Yellow-line Quaker
73.192 *Sunira circellaris* (Hufnagel, 1766) Brick
73.193 *Anchoscelis lunosa* (Haworth, 1809) Lunar Underwing
73.194 *Conistra vaccinii* (Linnaeus, 1761) Chestnut
73.195 *Conistra ligula* (Esper, 1791) Dark Chestnut

73.200 *Lithophane semibrunnea* (Haworth, 1809) Tawny Pinion
73.201 *Lithophane socia* (Hufnagel, 1766) Pale Pinion
73.202 *Lithophane ornitopus* (Hufnagel, 1766) Grey Shoulder-knot
73.203 *Lithophane furcifera* (Hufnagel, 1766) Conformist
73.206 *Lithophane leautieri* (Boisduval, 1829) Blair's Shoulder-knot
73.207 *Xylena solidaginis* (Hübner, [1803]) Golden-rod Brindle
73.208 *Xylena exsoleta* (Linnaeus, 1758) Sword-grass
73.209 *Xylena vetusta* (Hübner, [1813]) Red Sword-grass
73.210 *Eupsilia transversa* (Hufnagel, 1766) Satellite
73.211 *Enargia paleacea* (Esper, 1791) Angle-striped Sallow
73.212 *Ipimorpha retusa* (Linnaeus, 1761) Double Kidney
73.213 *Ipimorpha subtusa* ([Denis & Schiffermüller], 1775) Olive
73.214 *Cosmia diffinis* (Linnaeus, 1767) White-spotted Pinion
73.215 *Cosmia affinis* (Linnaeus, 1767) Lesser-spotted Pinion
73.216 *Cosmia trapezina* (Linnaeus, 1758) Dun-bar
73.217 *Cosmia pyralina* ([Denis & Schiffermüller], 1775) Lunar-spotted Pinion
73.219 *Atethmia centrago* (Haworth, 1809) Centre-barred Sallow
73.220 *Brachylomia viminalis* (Fabricius, [1777]) Minor Shoulder-knot
73.221 *Parastichtis suspecta* (Hübner, [1817]) Suspected
73.222 *Apterogenum ypsillon* ([Denis & Schiffermüller], 1775) Dingy Shears
73.224 *Griposia aprilina* (Linnaeus, 1758) Merveille du Jour
73.225 *Dryobotodes eremita* (Fabricius, 1775) Brindled Green
73.228 *Antitype chi* (Linnaeus, 1758) Grey Chi
73.231 *Aporophyla lutulenta* ([Denis & Schiffermüller], 1775)
73.232 *Aporophyla lueneburgensis* (Freyer, 1848) Deep-brown Dart
73.233 *Aporophyla nigra* (Haworth, 1809) Black Rustic
73.234 *Dasypolia templi* (Thunberg, 1792) Brindled Ochre
73.235 *Polymixis lichenea* (Hübner, [1813]) Feathered Ranunculus
73.237 *Polymixis flavicincta* ([Denis & Schiffermüller], 1775) Large Ranunculus
73.238 *Mniotype adusta* (Esper, 1790) Dark Brocade
73.241 *Panolis flammea* ([Denis & Schiffermüller], 1775) Pine Beauty
73.242 *Orthosia incerta* (Hufnagel, 1766) Clouded Drab
73.243 *Orthosia miniosa* ([Denis & Schiffermüller], 1775) Blossom Underwing
73.244 *Orthosia cerasi* (Fabricius, 1775) Common Quaker
73.245 *Orthosia cruda* ([Denis & Schiffermüller], 1775) Small Quaker
73.246 *Orthosia populeti* (Fabricius, 1781) Lead-coloured Drab
73.247 *Orthosia gracilis* ([Denis & Schiffermüller], 1775) Powdered Quaker
73.248 *Orthosia opima* (Hübner, [1809]) Northern Drab
73.249 *Orthosia gothica* (Linnaeus, 1758) Hebrew Character
73.250 *Anorthoa munda* ([Denis & Schiffermüller], 1775) Twin-spotted Quaker
73.252 *Tholera cespitis* ([Denis & Schiffermüller], 1775) Hedge Rustic
73.253 *Tholera decimalis* (Poda, 1761) Feathered Gothic
73.254 *Cerapteryx graminis* (Linnaeus, 1758) Antler Moth
73.255 *Anarta trifolii* (Hufnagel, 1766) Nutmeg
73.257 *Anarta myrtilli* (Linnaeus, 1761) Beautiful Yellow Underwing
73.259 *Polia bombycina* (Hufnagel, 1766) Pale Shining Brown
73.260 *Polia hepatica* (Clerck, 1759) Silvery Arches
73.261 *Polia nebulosa* (Hufnagel, 1766) Grey Arches
73.264 *Lacanobia thalassina* (Hufnagel, 1766) Pale-shouldered Brocade
73.265 *Lacanobia contigua* ([Denis & Schiffermüller], 1775) Beautiful Brocade
73.266 *Lacanobia suasa* ([Denis & Schiffermüller], 1775) Dog's Tooth
73.267 *Lacanobia oleracea* (Linnaeus, 1758) Bright-line Brown-eye
73.270 *Melanchra persicariae* (Linnaeus, 1761) Dot Moth
73.271 *Ceramica pisi* (Linnaeus, 1758) Broom Moth
73.272 *Papestra biren* (Goeze, 1781) Glaucous Shears
73.273 *Hada plebeja* (Linnaeus, 1761) Shears
73.274 *Mamestra brassicae* (Linnaeus, 1758) Cabbage Moth
73.275 *Sideridis turbida* (Esper, 1790) White Colon
73.276 *Sideridis rivularis* (Fabricius, 1775) Campion
73.277 *Sideridis reticulata* (Goeze, 1781) Bordered Gothic
73.279 *Hecatera bicolorata* (Hufnagel, 1766) Broad-barred White
73.280 *Hecatera dysodea* ([Denis & Schiffermüller], 1775) Small Ranunculus
73.281 *Hadena bicruris* (Hufnagel, 1766) Lychnis
73.282 *Hadena compta* ([Denis & Schiffermüller], 1775) Varied Coronet
73.283 *Hadena confusa* (Hufnagel, 1766) Marbled Coronet
73.286 *Hadena perplexa* ([Denis & Schiffermüller], 1775) Tawny Shears/Pod Lover
73.288 *Mythimna turca* (Linnaeus, 1761) Double Line
73.289 *Mythimna pudorina* ([Denis & Schiffermüller], 1775) Striped Wainscot
73.290 *Mythimna conigera* ([Denis & Schiffermüller], 1775) Brown-line Bright-eye

73.291	*Mythimna pallens* (Linnaeus, 1758)	Common Wainscot
73.293	*Mythimna impura* (Hübner, [1808])	Smoky Wainscot
73.294	*Mythimna straminea* (Treitschke, 1825)	Southern Wainscot
73.295	*Mythimna vitellina* (Hübner, [1808])	Delicate
73.296	*Mythimna unipuncta* (Haworth, 1809)	White-speck
73.297	*Mythimna albipuncta* ([Denis & Schiffermüller], 1775)	White-point
73.298	*Mythimna ferrago* (Fabricius, 1787)	Clay
73.299	*Mythimna litoralis* (Curtis, 1827)	Shore Wainscot
73.301	*Leucania comma* (Linnaeus, 1761)	Shoulder-striped Wainscot
73.302	*Leucania obsoleta* (Hübner, [1803])	Obscure Wainscot
73.304	*Leucania loreyi* (Duponchel, 1827)	Cosmopolitan
73.307	*Peridroma saucia* (Hübner, [1808])	Pearly Underwing
73.308	*Actebia praecox* (Linnaeus, 1758)	Portland Moth
73.311	*Euxoa cursoria* (Hufnagel, 1766)	Coast Dart
73.312	*Euxoa obelisca* ([Denis & Schiffermüller], 1775)	Square-spot Dart
73.313	*Euxoa tritici* (Linnaeus, 1761)	White-line Dart
73.314	*Euxoa nigricans* (Linnaeus, 1761)	Garden Dart
73.317	*Agrotis exclamationis* (Linnaeus, 1758)	Heart and Dart
73.319	*Agrotis segetum* ([Denis & Schiffermüller], 1775)	Turnip Moth
73.320	*Agrotis clavis* (Hufnagel, 1766)	Heart and Club
73.322	*Agrotis vestigialis* (Hufnagel, 1766)	Archer's Dart
73.323	*Agrotis ripae* (Hübner, [1823])	Sand Dart
73.324	*Agrotis trux* (Hübner, [1824])	Crescent Dart
73.325	*Agrotis puta* (Hübner, [1803])	Shuttle-shaped Dart
73.327	*Agrotis ipsilon* (Hufnagel, 1766)	Dark Sword-grass
73.328	*Axylia putris* (Linnaeus, 1761)	Flame
73.329	*Ochropleura plecta* (Linnaeus, 1761)	Flame Shoulder
73.331	*Diarsia dahlii* (Hübner, [1813])	Barred Chestnut
73.332	*Diarsia brunnea* ([Denis & Schiffermüller], 1775)	Purple Clay
73.333	*Diarsia mendica* (Fabricius, 1775)	Ingrailed Clay
73.334	*Diarsia rubi* (Vieweg, 1790)	Small Square-spot
73.336	*Cerastis rubricosa* ([Denis & Schiffermüller], 1775)	Red Chestnut
73.337	*Cerastis leucographa* ([Denis & Schiffermüller], 1775)	White-marked
73.338	*Lycophotia porphyrea* ([Denis & Schiffermüller], 1775)	True Lover's Knot
73.339	*Rhyacia simulans* (Hufnagel, 1766)	Dotted Rustic
73.341	*Standfussiana lucernea* (Linnaeus, 1758)	Northern Rustic
73.342	*Noctua pronuba* (Linnaeus, 1758)	Large Yellow Underwing
73.343	*Noctua fimbriata* (Schreber, 1759)	Broad-bordered Yellow Underwing
73.345	*Noctua comes* Hübner, [1813]	Lesser Yellow Underwing
73.346	*Noctua interjecta* Hübner, [1803]	Least Yellow Underwing
73.348	*Noctua janthe* (Borkhausen, 1792)	Lesser Broad-bordered Yellow Underwing
73.349	*Spaelotis ravida* ([Denis & Schiffermüller], 1775)	Stout Dart
73.350	*Eurois occulta* (Linnaeus, 1758)	Great Brocade
73.351	*Graphiphora augur* (Fabricius, 1775)	Double Dart
73.352	*Anaplectoides prasina* ([Denis & Schiffermüller], 1775)	Green Arches
73.353	*Xestia baja* ([Denis & Schiffermüller], 1775)	Dotted Clay
73.355	*Xestia castanea* (Esper, 1798)	Neglected Rustic
73.356	*Xestia agathina* (Duponchel, 1827)	Heath Rustic
73.357	*Xestia xanthographa* ([Denis & Schiffermüller], 1775)	Square-spot Rustic
73.358	*Xestia sexstrigata* (Haworth, 1809)	Six-striped Rustic
73.359	*Xestia c-nigrum* (Linnaeus, 1758)	Setaceous Hebrew Character
73.360	*Xestia ditrapezium* ([Denis & Schiffermüller], 1775)	Triple-spotted Clay
73.361	*Xestia triangulum* (Hufnagel, 1766)	Double Square-spot
73.365	*Eugnorisma glareosa* (Esper, 1788)	Autumnal Rustic
73.368	*Naenia typica* (Linnaeus, 1758)	Gothic

NOLIDAE

74.002	*Meganola albula* ([Denis & Schiffermüller], 1775)	Kent Black Arches
74.003	*Nola cucullatella* (Linnaeus, 1758)	Short-cloaked Moth
74.004	*Nola confusalis* (Herrich-Schäffer, 1847)	Least Black Arches
74.007	*Bena bicolorana* (Fuessly, 1775)	Scarce Silver-lines
74.008	*Pseudoips prasinana* (Linnaeus, 1758)	Green Silver-lines
74.009	*Nycteola revayana* (Scopoli, 1772)	Oak Nycteoline
74.011	*Earias clorana* (Linnaeus, 1761)	Cream-bordered Green Pea

Appendix 2

Moth record contributors
Following the list of recorders, those local groups, societies and museums who have also contributed records are listed alphabetically.

For those recorders where a surname only is documented, the first year of their records is provided. Duplicates indicate records received from more than one individual with the same surname and initial/s.

Every effort has been made to include the names of all recorders who have submitted moth records to the relevant County Moth Recorder. We apologise if any names have been accidentally omitted.

L. Aaron
G.W. Adams
C. Adamson
S. Adamson
R. Adderley
C. Adshead
V. Aeyn
G. Agnew
M.J. Ainscough
I. Ainsworth
A. Airey
P. Aitchison
S. Aitken
A.C. Aldridge
D. Alford
J.J. Alfrey
J.J. Alker
P.J. Alker
D. Allan
D. Allen
J.E.R. Allen
R. Allen
R.H. Allen
S. Allen
T. Allen
J. Allman
M. Allmark
Allnut (1984)
N. Almond
C. Ambrose
T. Ambury
D. Anderson
P. Anderson
S. Anderson
A. Anderton
G. Anderton
J. Anderton
K. Anderton
P. Anderton
S. Andrew
T. Andrew
T. Angelis

S. Appley
P.A. Arak
J. Armitage
W. Armitt
S. Armo
T. Armo
G. Armstrong
K. Armstrong
A. Arthington
H. Ash
S. Ashcroft
K. Ashton
P. Ashton
W. Ashton
N. Ashurst
D. Ashworth
G.F. Ashworth
M. Ashworth
M.C. Asker
W.C. Aspin
J.W. Aspinwall
C. Atherton
D. Atherton
G. Atherton
J. Atherton
A. Atkins
S.V. Atkins
G. Atkinson
M. Atkinson
J. Attfield
G.R. Avery
P. Axford
R. Ayres
D. Bagley
L. Bagley
S. Bagshaw
E. Bailes
C. E. Bailey
A. Baines
C. Baines
N. Baines
L. Baker

M. Baker
W. Bal
D. Balding
C. Baldwin (Mr.)
C. Baldwin (Mrs)
D. Baldwin
J.W. Baldwin
I. Ball
J. Ball
L. Ball
T. Ball
A. & N. Bamforth
C. Bancroft
G.H. Band
S. Banister
P. Banks
R. Banks
T. Banks
J. Bannister
K. Bannister
M. Bannister
S. Bannister
M. Bannon
A. Barber
J. Barber
N. Barber
A. Barker
B. Barker
D. Barker
P. Barker
C. Barlow
H. Barlow
M. Barlow
A. Barnes
C. Barnes
D. Barnes
P.J. Barnes
P. Barnett
P. Baron
Z. Barrett
D. Barrington
P. Barrington

N. Barron
H. Barton
J.W. Bateman
A. Bates
D. Bates
A. Bateson
P. Bateson
C. Batty
A. Baxter
R. Baxter
T. Baxter
D. Bayliss
E. Bayton
C. Beamish
M. Beard
D. Beattie
J. Beattie
N. Beaumont-
 Swindlehurst
S. Beavan
A. Bedford
H. Bedford
S. Bedford
J. Beere
D. Beevers
E. Bell
G. Bell
M. Bell
S. Bell
W.J.P. Bell
M. Bellingham
S. Benner
C. Bennett
D. Bennett
M.E. Bennett
T. Bennett
L. Bennett-
 Margrave
D. Bennion
Benson (1947)
D. Benson
J. Benson

A. Bentley
D.P. Bentley
J. Bentley
I. Berry
K. Berry
P. Berry
D. Best
R. Betterton
K. Bevan
A. Beyga
D.A. Bickerton
H. Bickley
S. Biddolph
K. Bidgood
A. Billington
L. Bimson
A. Binns
E.R. Birch
J. Birch
J. Birchall
A. Birchwood
J.L.M. Bird
N.L. Birkett
A. Birks
E. Birnie
S. Birtwistle
A.J. Bissitt
G. Blackburn
T. Blackburn
J.M. Blackie
J. Blackledge
E. Blackman
L. Blackmore
N. Blackstock
L. Blacow
L. Blades
I. Blagden
K. Blamire
K.P. Bland
L. Bland
L. Blazejewski
R. Blinnby

The Moths of Lancashire **589**

A. Blomfield
M. Bloomfield
P.J. Bloomfield
A. Blundell
J.W. Blundell
J. Boardman
P. Boardman
S. Bolton
A.J. Bond
D. Bond
K.G.M. Bond
S. Bond
J. Booker
I. Boote
A. Booth
M. Boothman
R. Borrow
A. Boulton
C.H. Bowden
G. Bowden
J. Bowen
M. Bowers
Bowman (1904)
E. Bowman
T. Boyce
A.W. Boyd
T. Boyd
R. Boydell
S. Boyle
R. Bradbury
A. Bradley
C. Bradley
D.A. Bradley
S. Bradley
J. Bradshaw
W. Brady
A. Brandreth
N. Brannagan
K. Brannan
P.R. Brash
C. Brathmere
M. Breaks
J. Breen
T. Brereton
R.F. Bretherton
P. Brewster
P.D. Brian
K. Brides
C. Briers
H. Briers
R. Briers
S. Briers
B. Brigden
J. Briggs
K.B. Briggs
E. Brighton
P. Brighton

A. Brindle
P. Brindle
S. Brindle
J. Britch
A. Britt
N. Brittain
H. Britten
M.R. Britton
J. Broadhurst
P. Broadman
J.F. Brockholes
A. Brocklebank
P. Bromilow
A. Bromley
C. Brook
S. Brooker
C. Brookes
C. Brooks
K. Brooks
D.P. Broome
M. Broomfield
L. Brotherstone
A. Brown
C. Brown
D. Brown
E.R. Brown
H. Brown
I. Brown
S. Brown
C. Browne
A. Bruce
R. Brunt
D. Bryant
P. Bryant
D. Bryce
P. Bryers
J. Buchanan
J. Buckley
W. Buckley
S. Bullen
K. Bullimore
K. Bullock
D.S. Bunn
A. Bunting
J. Burgoine
F.D. Burk
S. Burke
T. Burke
R. Burkmar
E. Burney-Cumming
R. Burnley
M. Burns
H.L. Burrows
M. Burrows
D. Burt
H. Burton

P. Burton
P. Busby
C. Bushell
E. Butcher
J. Butcher
P. Butterworth
E.C. Buxton
C.J. Cadbury
S. Cadwallader
J. Calder
T. Calderbank
D. Callender
A.E. Cameron
B. Cameron
C. Campbell
G. Campbell
L. Campbell
M. Campbell
S. Campbell
L. Campion
A. Campuzano
A. Cannell
B. Capper
S.J. Capper
G. Carefoot
A. Cargill
B. Carlyle
D. Carman
D. Caron
F. Carr
J. Carr
S. Carr
A. Carter
H. Carter
J. Carter
K. Carter
L. Carter
N. Carter
R. Carter
S. Cartwright
D. Case
T. Casey
J.R. Cass
P. Cass
C. Caulton
D.M. Caunce
D. Causey
A. Cawthray
C. Cesar
J. Chadwick
M. Chadwick
L. & R. Chalmers
D. Chambers
K. Chambers
R. Chambers
T. Chambers
M. Champion

I. Chandler
D. Chapman
F.I. Chapman
D. Chappell
J. Chappell
R. Charles
S. Charlson
H. Charlton
T. Charnley
A. Cheney
R. Chick
J. Chivers
M. Chorley
J. Christian
E. Christie
S.E. Christmas
S. Clancy
T. Claret
A. Clark
A.R. Clark
D.J. Clark
J. Clark
P. Clark
R. Clark
D. Clarke
J. Clarke
J.H. Clarke
J.J. Clarke
P. Clarke
R.A. Clarke
S. Clarke
J. Clarke-Mackintosh
G. Clarkson
D. Claxton
C. Clay
M. Clay
K. Clayworth
C. Clee
B. Clegg
P. Clegg
T. Clegg
A. Cleme
J. Clerk
M. Clerk
G. Clewley
L. Clewley
C. Clews
J. Clews
J. Clifton
W. Close
W.G. Clutten
J. Coates
J. Cobham
C. Cockbain
R. Cockbain
B. Cockburn

S. Cockburn
A. Cocker
A. Cockroft
C. Coe
A. Cogan Barber
L. Coldrick
H.R.P. Collett
M. Colley
S. Collier
J. Collins
S. Collins
I. Colquhoun
R. Comont
W.A. Comstive
E. Connolly
V. Connor
A.J. Conway
R. Conway
T. Conway
B. Cook
C.S. Cook
L.M. Cook
P. Cook
G. Cooke
J. Cooke
N. Cooke
R.H. Cooke
S. Cooksey
B. Cooper
D. Cooper
J. Cooper
M. Cooper
T. Cooper
M.J. Copland
H.H. Corbett
A. Corder
P. Corkhill
M. Corley
P. Corner
J. Corser
R. Corser
J. Cosmo Melvill
R. Cottam
R. Cottingham
J. Cotton
M. Cotton
T. Coult
T.A. Coward
D.R. Cowden
D. Cowie
K. Cowley
M. Cox
S. Coxey
R. Coyle
B.H. Crabtree
D. Craddock
J.Crane

590 *The Moths of Lancashire*

A. Creaser
M. Creaser
R.C.R. Crewdson
R. Cribb
R. Cripps
L. Croft
R. & P. Crofts
L.W. Cromarty
F. Crompton
N. Crompton
I. Crook
B. Crookes
B. Crooks
A. Cross
B. Cross
J. Crosse
L. Cross
M. Cross
S. Cross
S. Crossley
B. Crowder
G. Crowder
Crozier (1917)
G. Crumpton
R. Cryer
M. Cubitt
M. Culkin
P. Culkin
A. Cull
H. Cull
R. Cullen
A. Culshaw
G. Cummings
J. Cummings
A. Cunnelly
M. Cunningham
A.J. Currie
S. Curson
C. Cutts
A. Dale
M. Dale
C. Daly
M.C. Dalley
T. Dallimore
A.A. Dallman
C. Daly
J. Daniels
B. Danson
C.A. Darbyshire
R. Darcie
P. Darnell
K. Darwin
P. Daunter
C. Davenport
T. Davenport
A.R. Davidson
L. Davidson

N. Davidson
J. Davie
T. Davie
A. Davies
C. Davies
H. Davies
J. Davies
K. Davies
M. Davies
N. Davies
P. Davies
V. Davies
A. Davis
E.G. Davis
G. Davis
H. Davison
K. Davison
P. Daw
B. Dawson
K. Dawson
R. Dawson
G.O. Day
B. Dean
M. Dean
W.F. Dean
M. Deans
B. Deed
Delamere (2000)
J. Dempsey
L. Dempsey
M. Dempsey
C. Denny
B. Derbyshire
C. Derri
C. Dewhurst
J. Devaney
G.M. Deville
M. Diamond
A. Dickinson
B. Dickinson
D. Dickinson
E. Dickinson
K. Dickinson
B.R. Dickson
B. Dineley
D. Dixon
D.B. Dixon
G. Dixon
J.M. Dixon
L. Dixon
J. Docherty
M.A. Dockery
A.M. Dodds
D. Dodsworth
J. Dombroskie
H. Done
A. Donegan

J. Donnelly
P. Donnelly
R. Donnelly
S. Doran
A. Dore
J. Dore
L. Doswell
S.P. Doudney
J.M. Douglas
Dowd (Mr. & Mrs.)
J. Dower
M. Downham
D. Downing
A. Dore
G. Doyle
A.J. Draper
G. Draper
D. Drinkwater
J. Drinkwater
L. Drinkwater
C. Driver
J. Driver
R. Du Feu
R.M. Duck
R. Duffy
C. Duhig
P. Dullaghan
G. Dunbar
P. Dunbar
G. Dunlop
J. Dunlop
J. Dunning
W. Dunstall
D. Dunstan
S. Dunstan
D. Dutton
I. Dykes
Dyson
Dyson
C. Dyson
M. Dyson
T. Eagan
D. & J. Earl
M. Earlam
D. Earley
H. Earnshaw
P. Eastham
D. Eaton
A. Eaves
J. Eaves
K. Eaves
S. Eaves
T. Eccles
A. Eckersley
P. Eddleston
R.S. Eddleston
S. Edgar

B. & R. Edge
C. Edmondson
M. Edmondson
S. Edmondson
M. Edmunds
P. Edmunds
Edwards (1908)
C.J. Edwards
D. Edwards
G. Edwards
J. Edwards
P. Edwards
W.F. Edwards
M. Elford
B. Elliott
J.W. Ellis
P. Ellis
S. Ellis
M. Ellison
R. Elsby
R. Else
J. Elsworth
M. Elsworth
W. Ely
A. Emmerson
A.M. Emmet
E. Emmett
B. English
K. Eric
A. Esslinger
M. Evans
R. Evans
B. Evens
D. Evm
J. Fairclough
K. Fairclough
R. Fairclough
L. Fairfax
K. Fairfield
A.R. Fairhurst
D.R. Farnworth
J. Farraday
C. Farrell
L. Farrell
S. Farrell
G. Farrer
E.A.W. Faulder
C. Fawcett
O. Fawcett
A. Fearnley
M. Fell
R. Fellows
C. Felton
D. Felton
S. Felton
S. Felton
D.W.H. Fennell

J. Fennell
J. Fenton
A. Ferguson
J. Ferguson
R. Ferguson
T. Ferguson
G. Fernell
E.H. Fielding
J. Fielding
E. Figg
Finch (Miss)
J. Firth
P. Firth
E. Fish
C. Fisher
D. Fisher
M. Fisher
J.R. Fishwick
M. Fishwick
M.G. Fitton
P. Flatt
D. Flenley
P. Flenley
C. Fletcher
P. Fletcher
S. Fletcher
R. Flight
S. Flint
C. Flynn
D. Flynn
G. Flynn
S.G. Flynn
M. Fogan
D. Foot
J. Foran
N. Forbes
H. Ford
J. Ford
L.T. Ford
P. Ford
W.K. Ford
B. Foreman
A. Forrest
K. Forrest
I. Forrester
J.R. Forshaw
B. Forsythe
C.F. Forsythe
A. Forti
M. Forty
A. Foster
C. Foster
K. Foster
P. Foster
R. Foster
S. Foster
Z. Foster

K. Foulkes	B. Gilbert	T. Green	J. Hallsworth	Hastings
K. Fowler	E. Gilchrist	P. Greenall	S. Halsey	P. Hatcher
A. Fowles	P. Gilchrist	P. Greenhalgh	A. Hamer	D. Hatton
B. Fox	V. Gilchris	A. Greening	S. Hamer	J. Hatton
E. Fox	R. Giles	L. Greening	M. Hammond	P. Hatton
D. Foy	S. Gilleard	N. Greening	C. Hampson	E. Haughton
D. Foy	D. Gillibrand	A. Greenwood	C. Hampton	R.J. Hawker
P. France	J. Gillmore	E.F. Greenwood	J. Hancell	A. Hawkes
P.M.A. Francis	V. Gilson	H. Greenwood	B. Hancock	G. Hawkins
P. Frank	J. Girdley	K.C. Greenwood	E.G. Hancock	K. Haydock
G. de C. Fraser	K. Gittens	R. Greenwood	G. Hancock	G. Hayers
M.G. Fraser	M. Gittens	R.S. Greenwood	P. Hancock	T. Hayes
S. Fraser	M.F. Gittos	E. Gregory	R. Hand	S. Hayhow
F. Frazer	J.C. Gladman	L. Gregory	A. Hankinson	A. Hayward
J.F.D. Frazer	D. Glasson	M. Gregory	S. Hannah	M. Hayward
M. Frazer	A. Gleave	N. Gregory	A. Hannan	D. Haywood
R. Freeman	W.G. Glutton	C.S. Gregson	K. Harbinson	G. Heath
C. H. Frost	N. Godden	W. Gregson	R. Hardcastle	J. Heath
J. Frost	A. Godfrey	M.H. Grice	R.W. Harding	W.H. Heathcote
A. Fryer	H. Godfrey	P. Grice	S. Harding	R.J. Heckford
H.F. & J.C. Fryer	J. Godfrey	K. Grieve	J. Hardman	G. Hedges
M. Fuller	N. Golding	G. Griffin	P. Hardman	B. Hedley
L. Fulton	C.J. Goodall	J.W. Griffin	L.W. Hardwick	R. Hedley
M. Furmidge	G. Goodwin	C.F. Griffith	D. Hardy	S. Heginbottom
P. Gahan	L. Gore	R. Griffiths	D.E. Hardy	D. Helm
K. Gallie	P.D. Goriup	R. Griffiths	E. Hardy	A. Hembrow
C. Gallimore	S. Gorman	J. Grime	L. Hardy	C. Hemingway
M. Garbutt	D. Gornal	R. Grime	P.B. Hardy	M. Hemingway
M. Gardener	G. Gorse	K. Grimshaw	B. Hargreaves	R. Hemming
K. Gardner	J. Gorse	R. Groomridge	H. Hargreaves	S. Henderson
M. Gardner	E. Gorton	S.J. Grove	I. Hargreaves	S. Hendry
B. Garland	M. Gosling	A. Grubb	G.A. Harker	T. Henshaw
E. Garland jnr.	A. Goth	A. Grundy	M.E. Harling	J. Heritage
S.P. Garland	F. Gould	D. Grundy	M.W. Harper	N. Heron
J.B. Garner	J. Goulding	P. Grundy	P. Harper	M. Hesketh
Richards	C. Gower	P. Guarnaccio	P.A. Harper	A. Heslop
N. Garnham	S. Grace	J.A. Gudgeon	D. Harris	J. Hewitt
C. Garside	A. Graham	J.P. Guest	H. Harris	R. Hewitt
E. Garston	C. Graham	P.G. Gutteridge	P. Harris	C. Heyes
A. Gartside	D. Graham	F. Hackett	A. Harrison	T. Heyes
K. Gartside	S. Graham	J. Hackland	B. Harrison	R. Heywood
P. Gateley	T. Graham	J. Haddon	J. Harrison	M. Hickling
G. Gavaghan	J. Grant	C. Hadfield	K. Harrison	S.J. Hickson
J. Gavaghan	V. Grantham	L. Hadfield	W. Harrison	J. Hide
G.F. Gee	G. Graves	M. Hadley	M. Harrop	W. Hide
L.K. Gentle	P. Gravett	L. Haines	C. Hart	S. Higginbottom
C.A.M. George	A. Greatrex	B. Haizelden	R. Hart	A. Higgins
S. Ghilks	P. Greaves	A. Hall	I. Hartley	J. Higgins
R. Ghorbal	A. Green	J.G. Hall	R. Hartley	R. Higgins
V. Giavanni	A.P. Green	N. Hall	S. Hartnett	I. Higginson
B. Gibbons	C. Green	P. Hall	E. Harvey	D. Higginson-Tranter
J. Gibbs	D.G. Green	P.M. Hall	M.C. Harvey	C. Highfield
A. Gibson	G.N. Green	R. Hall	A. Harwood	C. Higson
J. Gibson	J. Green	R. Hall	T. Harwood	I.C. Higson
M. Gibson	K. Green	R.A. Hall	S. Haselton	T. Higson
S. Gibson	M. Green	S. Hall	J.R. Haslam	D. Hill
Giddings (2005)	P. Green	S. Halliwell	R. Hassanzadeh	

M. Hill
N. Hill
P. Hill
P. Hill
R. Hill
K. Hiller
K. Hillier
S. Hilling
P.F. Hillyer
B. Hilton
I. Hilton
P. Hilton
R. Hilton
S. Hilton
Z. Hinchcliffe
L. Hinchey
W.D. Hincks
S.H. Hind
J. Hindle
G. Hiscocks
C. Hitchen
S. Hitchen
F. Hitchmough
R. Hobbs
P.J. Hodge
S. Hodges
J.B. Hodgkinson
P. Hodkinson
F. Hodson
N. Hodson
P. Hodson
G. Hogan
K.M. Holcroft
B. Holden
C. Holden
P. Holden
A. Holder
A.E. Holding
D.J. Holding
P. Holdridge
M. Holker
J. Holland
S. Hollinrake
B. Holloway
P. Holme
A. Holmes
P. Holmes
S. Holmes
J.H. Holness
G.A. Holt
J. Holt
M. Holt
D.T. Holyoak
R. Homan
B. Honeywell
C.H.E. Hopkins
P. Hopkins

E. Hopley
P. Hornby
G. Horne
K. Horne
C. Horner
C. Horsfall
D. Horsfall
H. Horsfall
G. Horsley
M.G. Horwood
G. Hosie
R. Hoult
T. Howard
J. Howarth
S. Howarth
A. Howell
F. Hoyle
J. Huddleston
A. Hudson
B. Hudson
L. Hudson
H.C. Huggins
A.W. Hughes
B. Hughes
C. Hughes
K. Hughes
L.E. Hughes
O. Hughes
T. Hughes
B. Hugo
P. Hugo
M. Hull
G. Hulme
T. Humphrey
D. Humphreys
J. Humphreys
A. Hunt
D. Hunt
H. Hunt
J. Hunt
N. Hunt
A. Hunter
A.R. Hunter
B. Hunter
L. Hunter
T. Hunter
V. Huntriss
P. Hurd
J. Hurst
P. Hurst
D. Hurter
A. Hutchinson
J. Hutchinson
L. Hutchinson
T. Hutchison
A. Huyton
P. Iddon

A.W. Imms
K. Imms
B. Ingleby
E. Ingleby
E. Irwin
S. Irwin
K. Isherwood
B. Ives
M. Jackson
P. Jackson
S. Jackson
S.H. Jackson
G. Jackson-Pitt
B. James
C. Jameson
D. Jameson
W. Jamieson
R. Jardine
D. Jarvis
J. Jarvis
R. Jefferson
T. Jeffree
I. Jeffrey
S. Jellett
M. Jenkinson
S.R. Jennings
F.L. Jepson
D. Jewell
A. Jewels
C. Jinks
L. Jinks
B. Johnson
C. Johnson
C.F. Johnson
C.M. Johnson
K. Johnson
N. Johnson
S. Johnson
T. Johnson
W. Johnson
D. Johnston
R. Johnstone
D. Jolly
S. Jolly
B. Jones
C. Jones
Co. Jones
C.M. Jones
D.K. Jones
E.E. Jones
Ga. Jones
G. Jones
Ja. Jones
Je. Jones
M. Jones
Matt. Jones
N. Jones

R. Jones
S. Jones
T. Jones
W.L. Jones
C. Jones-Parry
Jordan
K. Jordan
S. Judd
R. Jude
A. Jukes
C. Kaighin
D. Kay
R. Kay
C. Keane
E. Kearns
L. Keedy
J. Keery
A. Keightley
C. Kellett
D. Kelly
H.M. Kelly
J. Ken
A. Kennedy
G. Kennedy
R. Kennerley
R. Kenworthy
A. Keogh
P. Key
J. Kidd
L.N. Kidd
J. Kilgour
J. Killerby
C. Kilner
M. Kilroy
I. Kimber
P.M. Kinder
A. King
F. King
R. King
S. King
E. Kinniburgh
P. Kinsella
R. Kinsella
W.W. Kinsey
I. Kippax
J. Kirk
P. Kirk
S. Kirkby
A. Kirkham
E. Kirkham
J. Kirkham
N. Kirkham
P. Kirkham
R. Kirkham
A. Kirkman
B. Kitch
P. Kitching

L. Kitz
E. Knight
G. Knight
P. Knight
S. Knight
C. Knott
D. Knower
J. Knowler
L. Knowles
P. Knowles
P. Krischkiw
V. Krivtsov
B. Kydd
D.W. Kydd
A. Lacey
N. Laing
Laithwaite (1950)
T. Lally
A. Lamb
A. Lambert
B. Lambert
D. Lambert
A. Landwer
D. Lane
J.S. Lane
J.R. Langmaid
B. Langridge
E. Langrish
G. Lavender
J.C. Lavin
S. Lawrenson
J. Lawson
K. Lawson
T. Lawson
J.H. Lawton
A.S. Lazenby
M. Lazenby
F. Lea
K. Lea
A. Leach
S. Leadsom
G. Leah
R. Leatham
G. Leather
J. Leather
N. Leather
R. Lee
M.J. Leech
B. Leecy
J. Leedall
W.B. Lees
H. Lehnhart-Barnett
B. Leiderman
H.S. Leigh
R.E. Leonard
C. Lester

The Moths of Lancashire 593

R. Leverton	A. Maclennan	N. Mazza	A. Mercer	A.M. Moss (Rev.)
A. Lewis	P.S.R. MacQueen	B. McAllister	D. Mercer	P. Mostyn
C. Lewis	P. Macro	C. McCabe	D. Messenger	W. Mountfield
J. Lewis	N. Madden	J. McCabe	H. Metcalfe	J.O. Mountford
K. Lewis	P. & R. Maddock	K. McCabe	H.N. Michaelis	T. Moverley
R.J. Lewis	K. Maguire	L. McCabe	P. Michell	I. Mower
J. Leyland	M. Maher	A. McCafferty	J. Micklethwaite	R. Moyes
C. Liggett	K. Maidment	J. McCallum	F.A. Middlehurst	S. Moyle
J. Lightfoot	P. Major	K. McCartney	S. Middlehurst	B. Mudway
M. Lightowler	J. Makepeace	W. McCauley	Midgley (1896)	J. Muggleton
G. Lilley	D. Mallon	C. McCleary	K. Milburn	A. Muirhead
K. Limb	J. Malpass	C. McDonagh	J. Miles	A. Mulley
A. Lindsay	M. Malpass	S. McDonald	A. Millar	C. Mumford
P. Linnell	G. Manger	R. McDonnoll	A. Millard	E. Mumford
E. Linney	C. Manley	K. Mcelroy	P. Millard	J. Munden
G. Lintott	W.M. Mansbridge	J. McEvoy	A. Miller	S. Mundle
P. Liptrot	P. Mansfield	H. McGhie	C. Miller	I. Murat
K. List	R. Mansfield	J. McGill	D. Miller	C. Murphy
K. Lister	W. Mansfield	A. McGlynn	K. Miller	D. Murphy
P. Livermore	S. Manwaring	M. McGough	R. Miller	M.D. Murphy
K. Livesey	L.A. Marley	P. McGough	Millett (Mr.)	R. Murphy
M.G. Livingstone	R. Marley	S. McGowan	A. Mills	S. Murphy
S. Llewellyn	S.E. Marley	D.J. McGrath	J. Mills	Murray (c.1900)
J. Lloyd	A. Marriott	F. McGregor	W. Milne	H. Murray
N. Lloyd	S. Marsden	R. McHale	C. Mitchell	M. Murray
H. Lloyd-Cox	D.G. Marsh	G.Y. McInnes	I. Mitchell	A. Musgrove
L. Lochead	G. Marsh	C. McKay	J. Mitchell-Lisle	G. Musker
P. Lockwood	H. Marsh	L. McKee	S. Mitchell	P. Musker
D.V. Logunov	P.J. Marsh	P. McKeon	L. Moat	M. Myerscough
J. Lomas	A. Marshall	S.G. McLardy	E. Mole	Y. Mynett
F. C. Long	R. Marshall	I. McLean	R. Monaghan	W. Nadin
M. Longden	S. Marshall	C. McLennan	G.M. Monteiro	L. Nathan
H. Longmore	T. Marshell	J. McLeod	S. Moorcroft	G. Naylor
G.A. Longworth	R. Marten	S. McLoughlin	A.J. Moore	E. Neale
J. Longworth	B. Martin	C.E. McManus	D. Moore	J. Neill
A. Lord	J. Martin	S.E. McManus	J. Moore	D. Nelson
J. Lord	N. Martin	N. McMillan	S.M. Moore	J. Nelson
M. Loughlin	P. Martin	G. McMullan	M. Moores	M. Nelson
A. Lovelady	S.J. Martin	F. D. McMullen	S. Moores	R. Neville
J. Lowe	C. Martindale	S. McNair	A.J. Moors	B. Newing
J. Lowen	R. Martinez	C. McShane	H. Morbey-Ganley	L.W. Newman
R. Lowry	G.W. Mason	C. McWilliam	T. Morbey-Ganley	R.C. Newns
E. Loxham	J. Mason	L. McWilliam	J. Moreton	B. Newton
G. Loxham	J.C. Mason	S. J. McWilliam	K. Morley	J. Newton
K. Ludley	P. Mason	Meadowcroft	J. Morris	J.M. Newton
D. Lumb	T.G. Mason	(1963)	J.B. Morris	M. Newton
A. Lupton	H. Massey	O. Meakin	M. Morris	L. Nicholas
D.M. Luscombe	S. Massey	W. Meakin	N. Morris	Nicholson
B. Lynch	T. Mather	J. Meldrum	P. Morris	L. Nicholson
P. Lynch	S. Mathers	C. Melling	R.K.A. Morris	D. Nickeas
L. Lyon	G.F. Mathew	T. Melling	M. Morrison	M. Nightingale
G. Macdonald	Matthews (1946)	F. Mellor	S. Morrison	B. Nixie
M. MacDonald	C. Mavros	S. Mellor	V. Mortimer	J.W. Nixon
A. Machin	P. Mawby	E. Melmoth	G. Mortimore	S. Nixon
C. MacKay	T.H. Mawdsley	J.C. Melvill	D. Morton	T. Nixon
A. Mackin	M. Maynard	M. Memory	M. Moseley	P. Noble
I. MacKennon	A. Mayor	D. Menzies	A. Moss	P. Noblet

594 *The Moths of Lancashire*

F. Nolan
P. Norman
T. North
D. Norton
P. Nower
L Nunnerley
R. Nuttall
P. Oakes
J. O'Boyle
M. O'Brien
S. Oddy
S. O'Hara
M. Oldfield
O. O'Malley
E. Ormand
J. Ormerod
S. Orridge
A. Osborne
L. Otten
C. Ovens
D. Owen
J. Packham
A. Pacula
J. Padget
M. Page
R. Page
J. Palframan
A. Palmer
C.A. Palmer
C.L. Palmer
D.J. Palmer
E. Palmer
J. Palmer
M.J. Palmer
P. Palmer
R.M. Palmer
S.H. Palmer
S.M. Palmer
C. Paminter
J. Park
A. Parker
A.R. Parker
I. Parker
M. Parker
R. Parker
S. Parker
S. Parker
L. Parnell
H. Parr
G. Parry
T. Parry
A. Parsons
M.H. Parsons
M.S. Parsons
P. Parsons
S. Parsons
R. Partington

R. Pate
D. Patel
N. Patel
I. Paton
G. Patterson
J. Patton
P. Paul
C. Payne
K. Payne
C. Peacock
A. Pearson
K. Pedler
Peers (Mr. - 1864)
E.C. Pelham-
    Clinton
P. Pemberton
W. Pentelow
M. Pentland
J. Percy
K. Perkins
J. Perry
W. Perry
J. Pescod
J. Petley-Jones
R. Petley-Jones
E. Petrie
C. Pettipher
M. Pettipher
H. Philips
D. Phillips
H. Phillips
J. Phillips
D. Pickis
F.N. Pierce
A. Piggot
D. Pilling
S. Piner
S. Pinnington
M. Pitchard
C. Place
C.W. Plant
C. Platt
D. Player
J. Player
S. Plummer
J. Plumtree
L. Pocock
D. Pollard
G. Pollard
M.A.S. Pons
E. Ponting
R. Poole
G.T. Porritt
A. Porter
J.A. Porter
R. Porter
A. Powell

G. Powell
I. Powell
W. Powell
P.J. Pownall
L. Poxon
D.J. Poynton
B. Pratt
S. Preistley
J. Premium
M. Prescott
R. Prescott
J. Price
S. Price
R. Prichard
A. Priddey
A. Priest
S. Priestley
H.B. Prince
M. Prince
S. Prince
G. Pringle
A. Prior
S. Prior
M. Pritchard
R. Pritchard
D.A. Procter
T. Procter
L.B. Prout
A. Pryce
P. Pugh
T.R. Pugh
M. Pullan
A. Purnell
P. Pye
R. Pyefinch
G. Quartly-Bishop
L. Quarton
D. Quinlan
A. Quinn
T. Quinn
S. Quinton
C. Raby
E. Radford
A.J. Radlett
J. Rae
E. Rahulan
J. Rainford
M. Ralph
J. Randle
C. Rankin
K.I. Ransome
J. Ranson
A. Ratcliffe
A. Ravey
M. Rayner
A. Reade
A. Reading

P. Reddell
L. Redgrave
L. Reed
P. Rees
J. Reid
W. Reid
L. Renshaw
R.W. Rhodes
K. Rice
M. Richards
S. Richards
Richardson (2006)
C. Richardson
J.J. Richardson
P. Richardson
S. Richardson
D.M. Richmond
D. Rickards
T. Riden
D. Riding
D. Rigby
L. Rigby
M. Rigby
A. Rigg
A. Riley
D. Riley
E. Riley
G. Riley
G. Riley
J. Riley
J. Riley
K. Riley
O. Riley
S. Riley
M. Rimmer
P. Rimmer
J. Ripley
P. N. Rispin
C. Roach
B. Roberts
D. Roberts
H. Roberts
J. Roberts
S. Roberts
G. Robertson
A. Robinson
D. Robinson
G. Robinson
H.S. Robinson
J. Robinson
M. Robinson
S. Robinson
T. Robinson
D. Robson
K. Robson
S. Roebuck
J.T. Rodgers

A. Rogers
J. Rogers
K. Rogers
N.A.J. Rogers
R. Rogers
T. Rogers
K. Ronan
P. Rooney
N. Root
P. Round
L. Rose
E.J. Roskell
P. Ross
S. Roth
G. Rotheray
C. Rowan
L. Rowley
O. Rowson
G. Roxburgh
D. Roy
F. Royle
S. Royle
P. Rudd
E. Rudge
D. Rugman
D. Rumley
S. Runciman
A. Rush
M. Rush
S. Rush
K. Russell
M. Russell
T. Rustage
C.I. Rutherford
K. Ryan
L. Ryan
P. Rycroft
S. Ryder
I. Ryding
A. Sailing
S. Sale
F. Salt
F. Salvesen
L. Sammy
R.A. Sanderson
K. Sandham
A. Sarney
V. Sarre
R. Saundby
D. Savage
J. Sawyer
R. Saxon
A. & P. Schofield
E. Scholes
R. Scholes
A. Scott
K. Scott

P. Scowcroft
J. Scragg
S. Seal
J. Searle
A. Seaton
Seddon (1974)
E.A. Seddon
H.T. Seddon
C. Sefton
H.S. Sellon
M. Selway
T. Serjeant
J. Shackman
Shadbolt (1946)
B. Shannon
J. Sharp
S. Sharp
J. Sharpe
G. Sharples
C. Sharratt
M. Sharrock
A. Shaw
A.L. Shaw
B.T. Shaw
C. Shaw
J. Shaw
M.R. Shaw
S. Shaw
S. Shawcross
D. Shearer
R. Shenton
T.H. Sheppard
K. Sheriff
A. Sherwin
P. Sherwood
D. Shevelan
B. Shields
J. Shields
Simms (2010)
S.J. Simms
H. Simons
J. Simons
N. Simons
A. Simpson
D. Simpson
I. Simpson
S. Simpson
K. Singleton
Y. Sivaram
L. Sivell
G. Skelcher
E. Skidmore
P. Skidmore
R. Skilling
D. Skingsley
B. Skinner
A. Slade

D. Slade
D.J. Slade
H.S. Slade
P. Slade
C. Slater
P. Slater
R. Slatter
A. Small
M. Small
T. Small
Smallpiece (Mr.)
T. Smallshaw
B. Smart
H. Smart
A. Smith
A.J. Smith
B. Smith
B.D.S. Smith
C. Smith
D. Smith
D.A. Smith
E. Smith
F. Smith
I. Smith
I.F. Smith
J. Smith
K. Smith
M. Smith
P. Smith
P.H. Smith
R. Smith
S. Smith
S.G. Smith
L. Smithers
K. Smycki
B.B. Snell
B.R. Snell
J.E. Sneyd
M. Sneyd
M. Sole
G. Sorenson
R. South
T. Southward
N. Southworth
A. Spalding
L. Speight
G. Speirs
D. Spencer
K. Spencer
R. Spencer
J. Spickett
C. Spooner
A. Spottiswood
K. Sroczynska
J. Stagg
H.T. Stainton
A. Stamford

R. Standen
M. Standish
C. Stanford
A. Stanger
P. Stanton
B. Starkie
D. Starr
D. (C.F.) Steeden
J. (N.J.) Steeden
D. Steel
S. Steeles-Yates
D. Stenhouse
M. Stephens
P. Stephens
A. Stephenson
C. Stephenson
E.M. Stephenson
H. Stephenson
P.H. Sterling
B. Steve
I. Steve
P. Stevens
J. Stinger
G. Stirrup
R. Stock
W.P. Stocks
S. Stokes
P. Stone
S. Stone
J. Stonehouse
L. Stooke
A. Storey
C. Storey
G. Storey
F. Stoter
A. Stott
C.E. Stott
T. Stott
K. Street
B. Stringer
E. Stuart
F.B. Stubbs
F.H. Stubbs
J. Stubbs
B. Styan
K. Suchecka
P. Summers
D. Sumner
R. Sumner
F. Sunners
A. Surtees
C. Sutcliffe
E. Sutcliffe
M.J. Sutcliffe
S. Suttill
J. Sutton
J. Swales

J. Swan
H.R. Sweeting
A. Swift
J. Swift
J. Swindlehurst
I. Sykes
A. Symonds
J. Symons
A. Tabor
E. Tacy
K. Tacy
R. Tait
D. Talbot
P. Talbot
K. Talby
D. Tanglewest
Taylor (Miss)
A. Taylor
D. Taylor
F. Taylor
G. Taylor
J. Taylor
P. Taylor
R. Taylor
R. Templeton
S. Thackeray
A. Thomas
B. Thomas
G. Thomas
J. Thomas
S. Thomas
P. Thomason
F.C. Thompson
H. Thompson
J. Thompson
P. Thompson
R.J. Thompson
S. Thomson
J. Thorpe
S. Thorpe
L. Threadgold
J.H. Threlfall
J.C. Thurgarland
J. Tilt
I. Tim
P.F. Tinne
P. Tipping
M. Toal
S. Tobin
A. Toft
W. Tofts
C. Tomlinson
P.R. Tomlinson
S. Tomlinson
T. Tomlinson
A.E. Tonge
G.M. Tonks

S. Toole
L. Tooth
K. Topping
G. Tordoff
M. Tordoff
H. Torrance
J. Towers
D. Townley
B. Townson
J. Trampnow
D. Tranter
S. Tranter
J. Travis
S. Travis
C. Tresadern
R. Tretheway
G. Trevis
R. Tribbick
K. Trout
N. Trust
D. Tucker
T. Tunstall
C. Turner
G. Turner
G. Turner
H.B. Turner
H.J. Turner
J.R.G. Turner
K. Turner
L. Turner
P. Turner
P.A. Turner
R. Turner
J.W. Tutt
B. Tweedale
C. Twist
W.A. Tyerman
J. Tymon
C. Tynan
R. Tyrer
A. Tyrox
R. Underwood
M. Unsworth
D. Unwin
T. Unwin
W.F. Urwick
J.R. Valentine
S. van Mesdag
C. Varela
D.E. Vickers
J. Vine Hall
N. Virtue
N. Voaden
T. Vosterman
P. Waddington
M. Wain
A. Wainwright

D. Wainwright
C. Waldram
M.N. Walford
D.C. Walker
G. Walker
I. Walker
K.V. Walker
R. Walker
V. Walker
B. Wallace
D. Wallace
I.D. Wallace
T. Wallace
E. Wallwork
I. Walmsley
V. Walmsley
C. Walsh
D. Walsh
F.W. Walsh
I. Walsh
J. Walsh
M. Walsh
W. Walton
A. Wander
C. Warburton
R. Warburton
L. Warburton-Marsh
A. Ward
C. Ward
J. Ward
L. Ward
S. Ward
T. Ward
J. Wardle
S. Warford
P. Waring
F. Warmesley
J. Warne
B. Warren
R.G. Warren

S. Warren
S. Warrington
R.J. Wash
C. Washington
C. Watkin
G. Watkinson
B. Watson
H. Watson
N.I. Watson
P. Watson
R.W. Watson
W.A. Watson
J. Watt
K. Watts
S. Waymont
J. Webb
J. Webb
R. Webster
E. Weedy
P. Weetman
A. Weir
D.G. Weir
D. Wells
L. Wes
D. Wesley
N. West
P. West
T. West
W.H. Western
V. Weston
S. Westwood
A. Whaley
D.J. Wheeler
D. Whiley
T.M. Whitaker
A.D. White
C. White
I.M. White
J. S. White
L. White
S.J. White

C. Whitehead
J.E. Whitehead
S. Whitehouse
J. Whiteside
F. Whitfield
A. Whitlock
K. Whittacker
M. Whittaker
A. Whittle
C. Whittle
F.G. Whittle
I. Whittle
J. Whittle
C. Whitwell
W. Whitwick
J. Wigglesworth
M. Wilby
J. Wilcock
M. Wilcox
T. Wilcox
R. Wilding
M.A. Wilkins
C. Wilkinson
G. Wilkinson
K. Wilkinson
R. Wilkinson
S. Wilkinson
A. Willetts
A.H. Williams
B.S. Williams
C.B. Williams
S.B. Williams
E. Williams
J. Williams
H. Williams
K. Williams
L. Williams
P. Williams
S. Williams
G. Williamson
S. Willis

T. Wilmer
A. Wilson
G. Wilson
H.W. Wilson
J.D. Wilson
J.R. Wilson
K. Wilson
D. Wilton
E. Wilton
D. Windle
W.F. Windle
R. Winnall
D. Winnard
C. Winnick
C. Winterbottom
P. Wisniewski
P.N. Witter
K. Wollen
J. Wolstencroft
J. Wolstenholme
F. Womersley
N. Wong
D. Wood
G. Wood
J. Wood
J.C. Wood
J.P. Wood
K. Wood
L. Wood
M. Wood
P. Wood
W. Woodall
M. Woodhead
P. Woodruff
E. Woods
G. Woods
K. Woods
N. Woods
P. Woods
R. Woods
D.J. Woodward

J. Woodward
I.B. Wooldridge
R. Woolnough
I. Woosey
C. Workman
C.G.M de Worms
C. Wormwell
J. Worthington
G. Wotherspoon
D. Wrapson
A. Wright
A.E. Wright
D. Wright
E. Wright
E. Wright
J. Wright
M. Wright
N. Wright
A.L. Wrightson
D. Wrigley
B. Wynn
G. Wynn
J.F.G. Wynne
R. Wynne
G. Wyse
R. Yates
W. Yates
B. Yorke
G.H. Youden
D. Young
J. Young
M.R. Young
P. Young
T. Young
W. Youngs
D. Zachmollie
D. Zakrzewski
C. Zanmato

Local Groups, Societies, Museums etc. that have assisted with acquiring records

Bolton Museum
Chorley & District Natural History Society
Cliffe Castle Museum, Keighley
Croxteth Country Park Volunteer Group
Cuerden Valley Park (Friends of)
Dinton Pastures (BENHS collections)
Fleetwood Museum
Heysham Nature Reserve team
Lancashire and Cheshire Entomological Society
Lancashire and Cheshire Fauna Society
Manchester Museum
Merseyside Naturalist's Association
Natural History Museum, London
World Museum
Nelson Naturalists Society
North Lancashire Wildlife Group
Oldham Museum
Redmaids High Juniors School
Rochdale Field Naturalists' Society
Rossall School Natural History Society

Appendix 3

Photographic Contributors

The following list details all those who have very kindly submitted photographs for consideration of use in *The Moths of Lancashire* project. The authors are very grateful to everyone who has offered help in this respect and those whose photographs have been used are credited alongside the photographs in the book.

P. Alker
W.C. Aspin
S. Atkins
C. Atherton
D. Atherton
G. Atherton
A. Baines
N. Baines
R. Banks
C. Barnes
H. Barton
S. D. Beavan
A. Bedford
C. Bell
D. Bennett
D.A. Bickerton
S. Bond
J. Bowen
R. Boydell
P. Brash
S. Briers
A. Bunting
R. Burkmar
J. Chadwick
M. Champion
A. Cheney
J. Clarke
J. Clews
C. Cockbain
C.A. Darbyshire
T. Davenport
C. Davies
M. Dean
G. Dixon
A. Draper
B. Dyson

M. Dyson
A. Eaves
K. Eaves
P. Ellis
M. Elsworth
T. Ferguson
C. Fletcher
A. Forrest
R. Foster
S. Garland
N. Garnham
G. Gavaghan
J. Girdley
G. Griffin
R. Griffiths
B. Hancock
D. Hardy
C. Hart
S. Hartnett
R.J. Heckford
G. Hedges
R. Hill
L. Hinchey
S.H. Hind
D.J. Holding
A. Holmes
P. Hopkins
K. Hughes
T. Hutchison
M. Jackson*
S. Jackson*
C. Jameson
R. Johnstone*
G. Jones
J. Jones
D. Kelly

B. Kidd
I. Kimber
P. Krischkiw
T. Lally
E. Langrish
K. Lawson
R. Leverton
D. Logunov
J. Mackness
S. Marsden
P. J. Marsh
S. Martin
J. McAndrew
K. McCabe
C. McLennan
S.J. McWilliam
M. Memory
I. Mitchell
J. Mitchell-Lisle
J. Morris
P. Morris
R. Moyes
M. Murray
D. Nelson*
H. Nelson
R. Neville
D. Owen
A. Palmer
C.A. Palmer
P. Palmer
S.M. Palmer
J. Patton
P. Pemberton
R. Petley-Jones
S. Piner
S. Priestley

A. Pryce
P. Pugh
R. Pyefinch
T. Riden
A. Riley
D. Riley
G. Riley
J. Roberts
N.A.J. Rogers
J. Scragg
L. Sivell
B. Smart
B. Smith
I. Smith
I.F. Smith
J. Smith
P.H. Smith
P. Stevens
D. Talbot
D. Taylor
S. Tomlinson
G. Turner
P. Turner
N. Voaden
I. Walker
K. Walker
R.B. Walker
L. Ward
T. Whitaker
M. Wilby
J.D. Wilson
J. Worthington
G. Wynn
R. Yates

* Nelson Naturalists

Royal Collection Trust / © His Majesty King Charles III 2023.
Natural History Museum data portal (https://data.nhm.ac.uk).
The Trustees of the Natural History Museum (http://creativecommons.org/licenses/by/4.0/).

Photograph details (Species Accounts section)
Numbers relate to the ABH list, as used in the species accounts section from p.34 onwards.
A – adult, **O** – ova, **L** – larva, **M** – mine, **P** - pupa, **FS** – feeding signs, **SS** – set specimen, **H** – habitat, **ex** – reared from (e.g., ex-o, ex-l, ex-m), with date of emergence.
Foodplant is noted on photos of reared moths only where it differs from the following early-stage photo.

All dates are post-1950, unless stated otherwise.
Recorders' names are only listed where they differ from the photographer.
Red text indicates specimens in museum collections:

- BM – Bolton Museum
- NHM – Natural History Museum
- LWM – World Museum
- MM – Manchester Museum
- PM – Peterborough Museum

1.001 **A**, Warton Crag 17.5.15.
1.003 **A**, Astley Moss 11.5.05; **A**, Oswaldtwistle 29.5.23.
1.004 **A**♂, Chorlton 25.5.19; **A**♀, Chorlton 3.6.15.
1.005 **A**, Brockholes 11.5.22.

2.001 **A**, Chorlton 18.04.10; **M** (oak), Chorlton 29.5.21.
2.002 **A ex-m**, Warton Crag 4.4.10; **M** (Hazel), Gait Barrows 14.05.16.
2.003 **A ex-m**, Chorlton 24.3.11; **M** (birch), Chorlton 8.5.10.
2.004 **A ex-m**, Chorlton 24.4.06; **M** (birch), Chorlton 12.6.20.
2.005 **A ex-m**, Chorlton 13.4.07; **M** (birch), Ainsdale 4.6.21.
2.006 **A ex-m**, Chorlton 11.4.22; **M** (birch), Little Woolden Moss 12.5.16.
2.007 **A ex-m**, Chorlton 31.3.23; **M** (birch), Chorlton 3.5.22.
2.008 **A ex-m**, Chorlton 25.3.11; **M** (birch), Rochdale 2.5.20.

3.001 **A**♂, Chorlton 11.8.20; **A**♀, St Helens 13.8.22.
3.002 **A**♂, Sefton 16.5.17; **L**, (Mugwort roots), Rixton 5.2.08.
3.003 **A**, Sefton 4.7.15; **A**♂, Sefton 27.6.15; **A**♀, Little Woolden Moss 20.6.15.
3.004 **A**♂, Clitheroe 18.7.20; **A**♂, Gait Barrows 14.7.16; **A**♂, Oswaldtwistle 10.6.19.
3.005 **A**♂, Sefton 30.6.15; **A**♀, Warton, nr. Carnforth 4.8.12.

4.002 **M** (birch), Holcroft Moss 3.6.23; **M** (birch), Rixton 23.6.09.
4.003 **M** (birch), St Helens 10.9.20.
4.004 **A ex-m**, Eaves Wood 1.6.23; **M** (Small-leaved Lime), Eaves Wood 15.9.22.
4.005 **A ex-m**, Cadishead Moss 24.4.23; **M** (Downy Birch), Cadishead Moss 10.10.22.
4.006 **A ex-m**, Chorlton 5.5.23; **M** (birch), Chorlton 16.9.17.
4.007 **A ex-m**, Chorlton 27.5.08; **M** (birch), Gait Barrows 2.8.21.
4.008 **A ex-m**, Middleton NR 17.6.19; **M** (Alder), Chorlton 4.10.15.
4.009 **M** (Alder), Chorlton 6.10.21; **M**, as previous.
4.010 **A ex-m**, Chorlton 17.7.23; **M** (Hazel), Chorlton 1.12.18.
4.011 **A ex-m**, Gait Barrows 21.5.23; **M** (Blackthorn), Gait Barrows 5.8.22.
4.014 **A ex-m**, Gait Barrows 6.6.19; **M** (Buckthorn), Gait Barrows 30.9.18.
4.015 **A ex-m**, Dalton Crags 3.5.23, **M** (rose), Chorlton 6.11.15.
4.023 **A ex-m**, Flixton 27.5.19; **M** (Hawthorn), Chorlton 19.7.08.
4.024 **M** (Rowan), Cragg Wood 4.9.20.
4.025 **A ex-m**, Chorlton 12.4.10; **M** (Rowan), Rochdale 1.8.21.
4.026 **A ex-m**, Chorlton 19.5.22; **M** (Hawthorn), Chorlton 6.10.21.
4.030 **A**♀ **ex-m**, Chorlton 14.6.14; **M** (Hawthorn), Chorlton 28.9.15.
4.032 **A ex-m**, Chorlton 8.4.22; **M** (Hazel), Chorlton 6.10.21.
4.034 **A ex-m**, Chorlton 14.7.09; **M** (Beech), Chorlton 6.10.21.
4.035 **A ex-m**, Pennington 3.5.14; **A ex-m**, Manchester 29.5.22; **M** (Grey Willow), Irlam 2.11.15; **M** (Grey Willow), Manchester 6.11.21.
4.036 **A ex-m**, Cragg Wood 25.6.21; **M** (Bilberry), Rochdale 1.8.21.
4.038 **A ex-m**, Chorlton 23.4.09; **M** (Crack-willow), Chorlton 13.10.07.
4.039 **A ex-m**, Chorlton 24.5.22; **M** (poplar), Chorlton 6.10.21.
4.040 **A ex-m**, Chorlton 27.6.19; **M** (Aspen), Chorlton 8.9.18.

4.041 M (Rowan), Littleborough June 2001; **M** (Rowan), Littleborough 9.6.22.
4.042 A ex-m, Chorlton 8.5.18; **M** (Blackthorn), Chorlton 3.10.21.
4.043 A♀ ex-m, Chorlton 21.7.11; **M** (elm), Old Trafford 30.10.21.
4.044 A ex-m, Little Woolden Moss 27.6.18; **M** (birch), Little Woolden Moss 10.9.17.
4.045 A ex-m, Chorlton 24.2.19; **M** (Bramble), Chorlton 1.1.20.
4.049 A ex-m, Warton Crag 29.3.21; **M** (Agrimony), Warton Crag 2.10.20.
4.053 M (Apple), Appley Bridge 10.10.22; **M**, as previous.
4.054 A ex-m, St Helens 12.5.23; **M** (Hawthorn), St Helens 3.10.22.
4.055 M (Beech), Salford 24.9.16.
4.056 A ex-m, Ainsdale 9.5.23; **M** (Sycamore), Chorlton 8.7.12.
4.059 A ex-m, (Oak), Littleborough May 2002.
4.060 A ex-m, Chorlton 25.4.10; **M** (oak), Rixton 17.10.13.
4.061 A♀ ex-m, Chorlton 30.6.07; **M** (oak), Chorlton 1.11.13.
4.062 M (Sweet Chestnut), Chorlton 10.9.22; **M** (Sweet Chestnut), Chorlton 12.11.21.
4.063 A♀ ex-m, Chorlton 27.3.14; **M** (oak), Chorlton 10.11.13.
4.065 A♂ ex-m, Rixton 8.7.21; **M** (Greater Bird's-foot-trefoil), Rixton 16.7.20.
4.066 M (Common Bird's-foot-trefoil), Middleton NR 14.6.18; **M**, as previous.
4.068 A, Littleborough 19.8.01; **M with egg** (Broom), Formby 2.2.20.
4.071 A ex-m, Chorlton 20.4.09; **M** (Apple), Chorlton 28.6.20.
4.072 A, Chorlton 17.7.11.
4.074 A ex-m, Chorlton 26.7.22; **M with cocoon** (Norway Maple), Chorlton 14.7.15.
4.075 A ex-m, Chorlton 7.8.15; **M** (Field Maple), Chorlton 31.7.17.
4.076 A, Chorlton 12.6.20; **A**, St Helens 22.6.05.
4.077 A ex-m, Littledale 4.7.22; **M** (Cowberry), Littledale 28.3.22.
4.078 A ex-m, Rochdale 30.4.21; **M** (St John's-wort), Chorlton 28.8.21.
4.082 A ex-m, St Helens 18.6.14; **M** (Goat Willow), Chorlton 10.11.18.
4.085 A ex-m, Flixton 1.5.09; **M** (Aspen), St Helens 4.11.15.
4.089 A ex-m, Chorlton 26.5.06; **M** (oak), Littleborough 29.8.21.
4.090 A ex-m, Chorlton 14.5.16; **M** (oak), Rixton 30.10.13.
4.091 A ex-m, Billinge 6.5.21; **M** (oak), Chorlton 3.11.15.
4.094 M (rose), Gait Barrows 28.10.20; **M**, as previous.
4.095 A ex-m (Hawthorn), Chorlton 19.6.05; **M** (Pear), Chorlton 19.9.15.
4.096 A ex-m, Dalton Crags 21.6.23; **M** (Wild Strawberry), Warton Crag 25.10.17.
4.099 A♂ ex-m, Chorlton 23.4.09; **M** (birch), Chorlton 30.7.17.
4.100 A ex-m, Little Woolden Moss 28.5.19; **M** (birch), Chorlton 6.10.21.

5.001 A, Chorlton 1.7.14.
5.004 SS, Leighton Moss 22.7.06; **SS**, Formby, 10.7.1929 (W. Mansbridge) MM.

6.003 SS♂, Parbold, 19.5.73 (P. Summers) BM; **M** (oak), Cragg Wood 4.9.20.
6.004 A ex-m, Chorlton 8.7.07; **M** (Alder), Chorlton 16.10.17.
6.005 M (birch), Little Woolden Moss 10.8.16; **M** (birch), Rixton 2.9.20; **M** (birch), Chorlton 15.8.16.

7.001 A♀ ex-l, Chorlton 28.5.18; **A♂ ex-l**, Chorlton 27.5.18; **L** (oak leaf litter), Chorlton 27.3.17.
7.003 A, Clitheroe 29.07.19; **SS**, Gait Barrows 27.7.00.
7.005 Pair, Barley, nr. Pendle 19.7.23; **A♀**, as previous.
7.006 A♀, Chorlton 2.5.19; **A**, Rishton 18.5.21.
7.007 A, Middleton Wood 19.4.09.
7.008 SS, Gait Barrows 23.6.20.
7.009 A, Warton Crag 4.6.15; **A**, Hawes Water 29.5.22.
7.010 A, Didsbury 2.6.15; **L** (Cuckooflower), Chorlton 19.6.18.
7.011 A, Grains Bar 30.05.21.
7.012 SS, Longridge Fell 9.5.00; **A**, Hoghton 22.5.23.
7.014 A, White Moss 12.6.10; **A**, Leighton Moss 12.6.19.
7.015 A, Chorlton 24.5.21.

8.001 L (birch), Newhey 29.5.23; **M** (birch), Newhey 9.6.04.
8.002 A, Hoghton 29.4.14; **A**, Warton, nr. Carnforth 9.5.10; **A**, Flixton 17.5.19.
8.003 A, St Helens 26.5.05; **A**, Little Woolden Moss 19.6.20; **M**, Cragg Wood 4.9.20; **L** (Bilberry), Cragg Wood 27.10.20.
8.004 A, Silverdale 21.6.09; **L** (Wild Strawberry), Warton Crag 25.10.17.
8.005 SS, Formby 4.6.53 (H.N. Michaelis) MM; **M** (birch), Rochdale 22.8.15.

9.003 SS ex-l, Southport ,9.6.1932 (W. Mansbridge) MM; **SS ex-l**, as previous.

9.004 A, Sunderland Point 25.5.11; **L** (rose) Sunderland Point 28.4.19.
9.006 H, Astley Moss 24.3.08; **L** (birch), Astley Moss 31.3.07; **A**, Astley Moss 17.5.05 (K. McCabe); **Gall** (birch), Astley Moss 4.4.15; **L**, Astley Moss 31.3.07.

10.001 A ex-m, Chorlton 14.6.19; **M** (oak), Rixton 7.8.18.
10.002 M (oak), Chorlton 30.10.22; **M**, as previous.
10.003 A, Chorlton 18.5.20; **M** (Bramble), Freshfield 15.2.16.

11.001 SS, Preston 23.6.05; **L** (birch), Astley Moss 22.5.04, (K. McCabe).
11.002 A ex-l, Chorlton 6.5.15; **L** (*Lepraria*), Chorlton 27.2.20.
11.005 Case with exuvia, Bare 21.1.23; **Case**, as previous.
11.007 A ex-l, VC60 9.1.19; **L**, VC60 30.9.18.
11.009 A♀ ex-l, Manchester 5.7.16; **L** (*Lepraria*), Manchester 28.3.16.
11.012 A♂, Warton, nr. Carnforth 20.6.19; **Case**, Warton Crag 24.5.22.

12.006 A, Lytham 2001; **FS** (*Lepraria*), Stretford 17.4.23.
12.010 A ex-l, Astley Moss 8.5.08; **L** (bracket fungus on birch), Astley Moss 24.3.08.
12.011 A, St Helens 9.6.05; **A**, Holmeswood 3.6.18.
12.012 A, Hoghton 12.6.15; **SS**, Fulwood 18.6.13 (A. Powell).
12.014 A, Chorlton 20.7.22; **A**, as previous.
12.015 SS ex-l, Liverpool July 1922 (W. Mansbridge) MM.
12.016 A♂, Chorlton 16.6.22; **A♂**, Chorlton 29.5.22.
12.017 SS, Gait Barrows 22.5.00; **A**, Warton, nr. Carnforth 12.8.20.
12.021 A ex-l, Hawes Water 6.8.16; **FS** (Hazel Woodwart on Hazel) Hawes Water 2.4.17.
12.025 SS, Wavertree, Liverpool 20.8.1922 (W. Mansbridge) MM.
12.026 A ex-l, Rochdale 16.8.19; **L** (carpet), Rochdale 6.4.19.
12.027 A, Chorlton 14.6.20; **Case with exuvia**, Rochdale 12.6.22.
12.030 A, Chorlton 14.2.04; **A**, Hoddlesden 18.4.21.
12.032 A, Chorlton 26.6.20; **A**, St Helens 26.6.05.
12.033 A, Chorlton 12.5.19; **A**, Lancaster 29.5.08.
12.034 SS, Wavertree, Liverpool 5.8.1919 (recorder unknown) MM.
12.036 A, Heysham 8.6.07; **L** (from nestbox), Chorlton 8.2.20.
12.038 SS, Bold Moss 9.7.05 (R. Banks).
12.039 SS, Fazakerley 21.7.21 (L. Ward).
12.040 A, Sunderland Point 6.8.08; **SS ex-l**, Formby 10.9.1921 (W. Mansbridge) MM.
12.046 SS, Liverpool 1917 (F.N. Pierce) LWM.
12.047 A ex-l, Martin Mere 9.5.16; **L** (Hart's-tongue Fern), Martin Mere 13.3.16.
12.048 A, Preston 1.5.22; **L** (Hart's-tongue Fern), Rochdale 18.1.21.

13.002 A, St Helens 11.5.20; **M** (birch), St Helens 10.9.20; **Early instar L**, St Helens 16.9.20; **Final instar L**, St Helens 24.9.20.

14.001 A ex-p, Flixton 8.5.19; **Cocoon** (Yarrow), St Helens 15.5.19.
14.002 A ex-m, St Helens 17.4.15; **L** (Oxeye Daisy), Chorlton 1.3.21.
14.003 A ex-m, Fluke Hall 26.5.19; **M** (Sea Aster), Fluke Hall 28.4.19.
14.007 A ex-m, Formby 12.3.24; **M** (elm), Formby 26.8.22.
14.008 A ex-l, Chorlton 23.5.15; **L** (Alder), Chorlton 28.9.14.
14.009 A, Chorlton 4.5.23; **L** (lime), Chorlton 21.9.16.
14.010 A ex-l, St Helens 21.5.22; **M** (oak), St Helens 1.10.20.
14.012 A♂, Formby 3.6.22; **L** (Hawthorn), Formby 4.7.23.
14.013 A ex-l, Gait Barrows 15.5.22; **L** (birch), Gait Barrows 2.8.21.

15.002 A ex-l, Hawes Water 26.8.16; **FS** (Ash), Hawes Water 8.8.16.
15.003 A, Chorlton 31.10.16; **FS** (birch), Chorlton 20.8.21.
15.004 A ex-l, Little Woolden Moss 2.9.16; **FS** (Alder), Gait Barrows 21.7.19.
15.005 A ex-l, Gait Barrows 21.9.15; **FS** (birch), Gait Barrows 27.8.15.
15.006 A ex-l, Chorlton 16.7.16; **FS** (Sycamore), Gait Barrows 21.7.19.
15.007 A ex-l, Chorlton 16.7.23; **FS** (Azalea), Chorlton 2.7.23.
15.008 A, Gait Barrows 3.6.08.
15.009 A ex-l, Chorlton 6.5.23; **FS** (oak), Chorlton 13.10.22.
15.010 A, Chorlton 26.6.20; **FS** (Grey Willow), Cadishead Moss 13.9.15.
15.012 A, St Helens 13.8.22 (R. Banks); **A**, Woolton, Liverpool 19.7.22.
15.014 A, St Helens 20.5.20; **FS** (privet), Chorlton 3.12.17.
15.015 Pair, Chorlton 23.5.15; **M** (Ribwort Plantain), St Helens 13.4.21.
15.016 A ex-l, Rixton 7.5.18; **FS** (St John's-wort), Rixton 28.9.17.

15.017 A *(f. quadruplella)* **ex-l**, Ainsdale 9.9.21; **FS** (Redshank), St Helens 10.9.20.
15.019 A ex-m, Chorlton 21.6.22; **M** (oak), Chorlton 19.5.19.
15.022 A ex-l, Chorlton 5.6.18; **M** (Apple), Flixton 20.7.18.
15.025 A ex-l, Cadishead Moss 11.4.23; **L** (Downy Birch), Cadishead Moss 10.10.22.
15.028 A ex-l, Chorlton 21.7.23; **FS** (Hawthorn), Chorlton 26.8.21.
15.029 A ex-l, Chorlton 18.7.23; **FS** (Hazel), Chorlton 17.12.20.
15.030 A ex-l, Chorlton 4.5.23; **FS** (Apple), Chorlton 13.10.22.
15.032 A ex-l, Chorlton 26.7.20; **L** (Blackthorn), Chorlton 24.6.06.
15.033 A ex-l, Middleton NR 9.6.19; **L** (Blackthorn), Chorlton 21.8.17.
15.034 A ex-m, Rochdale 6.8.21; **M** (oak), Rochdale 1.8.21.
15.039 A ex-m, Formby 26.3.16; **M** (oak), Rixton 19.1.20.
15.040 A ex-m, Chorlton 12.10.22; **M** (Sweet Chestnut), Chorlton 16.9.22.
15.041 A ex-m, Manchester 24.10.22; **M** (London Plane), Chorlton 7.11.19.
15.043 A ex-m, Chorlton 28.8.21; **M** (Hawthorn), St Helens 19.9.19.
15.044 A ex-m, Chorlton 12.8.15; **M** (Rowan), Chorlton 24.8.20.
15.046 A♂ ex-m, Chorlton 9.4.23; **M** (Apple), Rixton 20.6.19.
15.049 A ex-m, (plum), Chorlton 9.4.14; **M** (Blackthorn), Gait Barrows 21.7.19.
15.050 A ex-m, Chorlton 31.8.21; **M** (cherry), Chorlton 20.10.20.
15.051 A ex-m, Salford (19.4.21); **M** (Laurustinus), Chorlton 19.3.20.
15.052 A ex-m, Rixton (6.8.16); **M** (Hawthorn), St Helens 1.10.20.
15.053 A ex-m, (Hawthorn), Chorlton 10.9.21; **M** (Firethorn), Didsbury 15.3.21.
15.054 A ex-m, Little Woolden Moss 17.7.16; **M** (Grey Willow), Rixton 16.7.20.
15.056 A ex-m, Docker Moor 16.3.19; **M** (Grey Willow), Rochdale 26.12.18.
15.057 A ex-m, Chorlton 23.7.16; **M** (Goat Willow), Chorlton 17.7.16.
15.058 A ex-m, Rochdale 2.3.12; **M** (sallow spp.), Docker Moor 30.9.18.
15.060 A♀, Ainsdale 10.6.22; **M** (Gorse), Ainsdale 18.3.22.
15.061 A♀ ex-m, Freshfield 9.5.22; **M** (Broom), Freshfield 14.4.22.
15.063 A ex-m, Dalton Crags 9.5.23; **M** (Beech), Gait Barrows 2.8.21.
15.064 A ex-m, Chorlton 8.7.23; **M** (Hazel), Chorlton 24.8.20.
15.065 A ex-m, Skelmersdale 18.4.23; **M** (Hornbeam), Chorlton 29.8.20.
15.066 A ex-m, Chorlton 15.3.14; **M** (Grey Alder), Chorlton 9.11.19.
15.067 A ex-m, Chorlton 28.4.14; **M** (Alder), Flixton 20.7.18.
15.069 A ex-m, Cadishead Moss 17.4.23; **M** (Downy Birch), Cadishead Moss 10.10.22.
15.070 A ex-m, Ainsdale 11.5.23; **M** (Creeping Willow), Ainsdale 20.9.22.
15.073 A ex-m, Rochdale 9.8.21; **M** (oak), Rixton 16.7.20.
15.074 A ex-m, Stretford 24.3.21; **M** (elm), Stretford 29.10.20.
15.075 A ex-m, Chorlton 10.1.22; **M** (birch), Chorlton 6.11.23.
15.076 A ex-m, Chorlton 20.2.23; **M** (Snowberry), Chorlton 12.11.17.
15.078 A ex-m, Skelmersdale 16.4.23; **M** (elm), Chorlton 15.10.16.
15.079 A ex-m, Manchester 19.5.22; **M** (Alder), Chorlton 9.11.20.
15.080 A ex-m, Chorlton 13.5.22; **M** (Alder), Birk Bank 4.9.20.
15.081 A ex-m, Chorlton 11.5.14; **M** (Hazel), Chorlton 6.11.20.
15.082 A ex-m, Chorlton 21.5.14; **M** (Alder), Chorlton 1.11.13.
15.083 A ex-m, Chorlton 12.9.19; **M** (Honeysuckle), Chorlton 2.9.17.
15.084 A ex-m, Chorlton 30.3.23; **M** (Field Maple), Chorlton 7.11.16.
15.085 A ex-m, Chorlton 5.9.21; **M** (Norway Maple), Chorlton 7.11.16.
15.086 A ex-m, Chorlton 22.7.16; **M** (Sycamore), Middleton NR 17.10.20.
15.089 A, Chorlton 11.5.21; **M** (Horse-chestnut), Leyland 29.9.20.
15.090 A ex-m, 25.8.22; **M** (Crack-willow), Chorlton 11.9.22.
15.092 A, Chorlton 20.4.21; **M** (poplar), Stretford 19.6.17.

16.001 A, Chorlton 24.6.14; **L** (Bird Cherry), Chorlton 17.6.09; **Web** (Bird Cherry), Chorlton 14.5.21.
16.002 A ex-p, Chorlton 26.6.14; **L** (Hawthorn), Ainsdale 2.6.21; **Web** (Hawthorn), Ainsdale 2.6.21.
16.003 A ex-l, Chorlton 22.6.22; **Web** (Apple), Chorlton 6.6.22.
16.004 A ex-l, Gait Barrows 26.6.16; **Web** (Spindle), Formby 30.7.21.
16.005 A ex-l, Chorlton 5.7.10; **Web** (White Willow), Chorlton 20.6.10.

16.007 A ex-l, Gait Barrows 6.7.16; **L** (Spindle), Gait Barrows 21.5.16.
16.008 A, Great Sankey 10.5.18.
16.010 A, Chorlton 3.8.04.
16.011 A, Preston 8.9.21; **A** (as previous).
16.014 A, Warton Crag 16.6.13.
16.015 A ex-l, Little Woolden Moss 2.7.16; **L** (birch), Little Woolden Moss 19.8.19.
16.017 A ex-l (cherry), Chorlton 9.4.17; **L** (Hawthorn), Hightown 22.6.21.
16.019 A ex-l, Middleton NR 26.6.22; **L** (Blackthorn), Middleton NR 24.5.22.
16.020 A ex-l, Chorlton 18.6.19; **L** (Hawthorn), Rochdale 6.6.20.
16.021 A, Chorlton 9.7.03; **A**, St Helens 22.6.07.
16.022 A, Formby 26.8.22; **M** (pine), Freshfield 2.2.20.
16.024 Pair, Clougha Pike 28.3.22; **P** (pine), Freshfield 17.2.19.

17.002 A, nr. Quernmore 14.8.12.
17.003 A, Chorlton 17.7.20; **L** (Snowberry), Chorlton 18.5.20.
17.005 A, St Helens 13.7.18; **L** (Hawthorn), Chorlton 17.5.19.
17.006 A, Tarleton 9.8.21.
17.008 A, St Helens 4.9.22; **A**, Haydock 21.8.05.
17.009 A ex-l, Chorlton 7.7.11; **L** (oak), Chorlton 12.6.11.
17.010 A ex-l, Warton Crag 22.6.08; **L** (oak), Warton Crag 23.5.08.
17.011 A ex-l, Chorlton 4.7.11; **L** (oak) 12.6.11.
17.012 A, Chorlton 10.8.14; **Cocoon** (beneath Sycamore), Leyland 11.11.22.
17.013 A, Morecambe 11.8.08; **A**, Preston 16.8.22.
17.014 SS, Warton, nr. Lytham 2.8.96.
17.015 A♂, Grains Bar 25.7.21.

18.001 A, Chorlton 16.7.16; **L** (Wild Mustard) 3.7.16.
18.003 A ex-l, Rixton 15.5.06; **L** (Dame's-violet) Rixton 2.5.06.
18.007 A, Flixton 17.6.18; **A**, Longton 27.6.17.

19.001 A, Leighton Moss 31.7.19; **A**, Yealand Conyers 24.7.16.
19.002 A, Chorlton 31.5.14; **A**, Lancaster 5.6.08.
19.003 Pair, Chorlton 31.5.15; **L** (Field Wood-rush), Chorlton 27.2.16.
19.004 SS, Heaton Norris 2.7.53 (W. Hincks) MM.
19.005 A ex-l, Little Woolden Moss 9.4.20; **FS** (cottongrass), Little Woolden Moss 15.3.20.
19.007 A ex-p, Chorlton 16.5.21; **FS** (Cock's-foot), Chorlton 29.4.19.
19.008 SS, Hawes Water 2.6.09.
19.011 A, Chorlton 9.8.20; **A**, St Helens 14.8.22.
19.014 A ex-m, Rixton 7.9.17; **M** (Bittersweet), Rixton 15.8.17.

20.001 A, Chorlton 7.6.03.
20.002♂ **A**, Swinton 25.5.11.
20.005 A, Chorlton 16.5.20; **M** (cypress), Chorlton 26.11.17.
20.006 A ex-l, Hawes Water 9.6.16; **L** (juniper), Hawes Water 10.4.16.
20.007 A ex-l, Little Woolden Moss 22.5.16; **L** (cypress), Little Woolden Moss 15.3.20.
20.011 A ex-l, Holcroft Moss 3.6.14; **L** (birch), Rochdale 11.4.11.
20.012 A ex-l, Astley Moss 4.6.08; **L** (birch), Chorlton 27.3.15.
20.013 A ex-l, Little Woolden Moss 3.6.16; **L** (Grey Willow), Little Woolden Moss 15.5.16.
20.014 A, Gait Barrows 7.7.07.
20.015 A ex-l, Chorlton 14.5.22; **L** (Apple), Chorlton 17.4.22.
20.016 A ex-l, Chorlton 28.5.08; **L** (birch), Chorlton 5.5.07.
20.017 SS, Eaves Wood 25.6.04.
20.018 A, St Helens 18.5.20; **A**, Preston 5.6.22.
20.019 A, St Helens 12.7.05; **A**, Littleborough 9.6.22.
20.020 A, Tarleton Moss 23.7.20; **A**, as previous.
20.021 A, St Helens 16.7.05; **L** (cherry), Chorlton 4.5.08.
20.022 A ex-l, Rochdale 11.6.16; **L** (Hawthorn), Rochdale 7.5.16.
20.023 A, Chorlton 16.8.21; **L** (Blackthorn), Chorlton 22.4.22.
20.024 A, Chorlton 4.9.02.

21.001 A ex-m (Hawthorn), Chorlton 12.7.23; **M** (Buckthorn), Eaves Wood 15.9.22.
21.004 A ex-m, Chorlton 26.4.10; **M** (Laburnum), Didsbury 23.6.14.
21.005 A ex-p, Littleborough May 2002; **Cocoon** (Broom), Littleborough May 2002.
21.008 A ex-m, St Helens 2.6.16; **M** (Hawthorn), Flixton 20.7.18.

22.002 A, Lancaster 17.7.08; **L** (Ash), Chorlton 2.5.10.
22.003 A, nr. Longridge 16.6.22; **A**, Chorlton 22.7.22.

24.001 A ex-m, Chorlton 9.9.04; **M** (bindweed), Chorlton 1.8.17.

25.001 A, Chorlton 3.7.13; **L** (Hawthorn), Chorlton 21.6.15.

27.001 A, Chorlton 26.6.20.

28.004 A, Green Bank 29.5.22.
28.005 A, Chorlton 28.5.02.

28.007 SS, Sefton Coast 2019 (C. Fletcher).
28.008 A, Brockhall 16.7.21.
28.009 A, St Helens 21.4.05; **L** (bird-food), Chorlton 23.2.19.
28.010 A, Chorlton 10.5.14; **L**, Chorlton 27.3.05.
28.012 A, Preston 23.6.22; **A**, Chorlton 11.7.15.
28.014 A, St Helens 18.7.16; **A**, Chorlton 22.6.23.
28.015 A, Chorlton 29.6.18; **A**, Freshfield 26.6.07.
28.019 A, Great Sankey 4.5.15; **L** (in rotten wood), Chorlton 22.1.06.
28.024 A ex-l, Chorlton 28.4.21; **L** (from nestbox), Chorlton 13.3.21.
28.025 A, Egerton 11.6.23; **A**, Astley Moss 17.5.08.

29.001 A, Astley Moss 4.4.15; **L** (oak), Rixton 20.6.19.
29.002 A, Warton Crag 4.2.21; **A**, Warton Crag 30.10.14.

30.003 A, Trowbarrow Quarry 30.6.23; **A**, Birk Bank 8.7.21.
30.004 A, Astley Moss 31.3.07.

31.001 A, Chorlton 26.6.20; **L** (Aspen), Chorlton 7.6.08.

32.001 A, EW, 20.3.11.
32.002 A, Gait Barrows 3.5.08; **A**, Herring Head Wood 28.3.23.
32.006 A, Cadishead Moss 26.7.10.
32.007 A ex-l, Ainsdale 2.7.21; **L** (Creeping Willow), Ainsdale 4.6.21.
32.008 H, Heysham Plateau July 2019.
32.010 A ex-l (Creeping Willow), Lytham 17.6.21; **L** (sallow) Chorlton 19.5.19.
32.011 A ex-l, Chorlton 19.7.11; **L** (Broom), Chorlton 12.6.11.
32.013 A ex-l, Warton Crag 15.7.10; **L** (Black Knapweed), Warton Crag 24.6.10.
32.015 SS, Flixton 16.3.04 (K. McCabe).
32.016 A, St Helens 14.9.22; **L** (Creeping Thistle), Chorlton 5.8.15.
32.017 A ex-l, Warton Crag 22.7.10; **L** (Black Knapweed), Warton Crag 24.6.10.
32.018 A ex-l, Preston 26.6.22; **L** (Cow Parsley), Chorlton 24.5.21.
32.019 A ex-l, Yealand 21.8.22; **L** (Hogweed), Yealand 25.7.22.
32.024 SS, Didsbury 24.08.55 (H. Michaelis) MM; **L** (Broom), Freshfield 18.3.22.
32.025 SS ex-l (Carline Thistle), Lytham 20.6.09.
32.026 SS ex-l, Preston 6.7.01; **L** (Black Knapweed), Warton Crag 4.6.15.
32.029 A ex-l, Freshfield 19.7.22; **L** (Gorse), Freshfield 10.6.22.
32.030 A, Preston 29.7.22; **L** (Gorse), Chorlton 9.5.22.

32.031 A ex-l, Westby 2022; **L** (Hemlock), Esprick 19.6.22.
32.032 A, Flixton 31.7.15.
32.035 A ex-l, Lytham 24.6.22; **L** (Hemlock Water Dropwort), Lytham 5.6.22.
32.036 A ex-l, Chorlton 12.7.08; **L** (Hogweed), Chorlton 17.7.15.
32.038 A, Worsaw Hill 20.7.21.
32.039 A, Preston 27.9.22; **L** (Hemlock Water Dropwort), Ainsdale 11.6.21.
32.040 A, Heysham 9.5.08.
32.042 A, Green Bank 10.7.21.
32.043 SS♂, Flixton 20.7.11 (K. McCabe).

34.001 A, Dalton Crags 26.4.22; **A**, Gait Barrows 21.5.08.
34.003 H, Kersal Moor by William Wyld 1852, Royal Collection Trust; **SS**, Kersal Moor, June 1829 (R. Cribb) Victoria Museum, Aus; **SS**, Victoria Museum; **SS**, MM.
34.004 A, St Helens 18.6.05; **L** (Bulrush), Little Woolden Moss 1.4.19.
34.0111 SS, Bootle 28.10.12 (C. Fletcher).
34.014 A ex-l, Gait Barrows 29.6.22; **L** (Buckthorn), Gait Barrows 31.5.22.

35.001 SS, Dalton Crags 19.7.21 (J. Patton); **L** (Common Bird's-foot-trefoil), Dalton Crags 26.4.22.
35.002 SS, Flixton 28.6.04 (K. McCabe).
35.003 SS, Freshfield 12.7.20 (R. Walker).
35.004 A ex-l, Middleton NR 24.6.21; **L** (Common Bird's-foot-trefoil), Middleton NR 17.5.21.
35.010 A, Chorlton 7.8.04; **L** (Kidney Vetch), Lytham 25.5.21.
35.011 A ex-l, Lytham 19.6.21; **L** (Creeping Willow), Birkdale 11.6.21; **FS** (willow), Chorlton 2.6.22.
35.012 A ex-l, Chorlton 27.6.15; **L** (birch), Chorlton 16.5.15; **FS**, Chorlton 29.4.19.
35.013 A ex-l, Ainsdale 15.6.21; **L** (Creeping Willow), Ainsdale 11.6.21; **H**, Ainsdale 4.7.23; **FS**, Ainsdale 7.6.21.
35.017 A, Freshfield 4.6.21; **L** (heather), Holcroft Moss 30.11.04.
35.018 A ex-l, Chorlton 4.7.10; **L** (Hazel), Chorlton 29.5.10.
35.020 A, Billinge 18.7.13; **A**, as previous.
35.022 A, Gait Barrows 29.5.08; **FS** (Juniper), Gait Barrows 31.5.22.
35.026 SS, Warton Crag 20.8.18 (J. Patton).
35.028 A, Rainford 2.7.05; **A**, Chorlton 29.9.14.
35.031 A, Chorlton 24.6.14; **L** (grasses), Chorlton 10.5.21.
35.035 A ex-m, Cadishead Moss 16.9.19; **M** (goosefoot), Cadishead Moss 31.8.19.

35.036 A ex-m, St Helens 3.6.18; **M** (orache), St Helens 6.9.17.
35.038 A, Chorlton 13.7.22; **A**, Chorlton 16.8.13.
35.039 A♂, Leck Fell 2015; **L** (*Rhytidiadelphus squarrosus*), Gannow Fell 17.4.15.
35.040 A, Salford 24.6.15; **L**, (*Rhytidiadelphus squarrosus*), Preston 2.4.23.
35.041 SS, Lytham 11.7.95 (S. Palmer).
35.042 A ex-l; Jeffrey Hill 28.6.15; **L** (*Rhytidiadelphus squarrosus*), Jeffrey Hill 2015.
35.046 A, Chorlton 22.7.03; **SS**, Flixton 15.7.15 (K. McCabe).
35.047 A, Chorlton 18.5.22; **A**, Lancaster 1.6.08.
35.048 SS, Formby 30.5.03 (S. Palmer).
35.049 A, Green Bank 30.6.21.
35.050 A, Freshfield 2007; **SS**, Winmarleigh Moss 18.7.96 (S. Palmer).
35.053 A, Flixton 31.7.12; **SS**, St Helens 13.7.22 (R. Banks).
35.056 A ex-l, Chorlton 9.6.15; **L** (burdock), Cragg Wood 4.9.21.
35.058 A ex-l, St Helens 12.6.04; **L** (Black Knapweed), Chorlton 25.1.20.
35.060 A ex-l, Rixton 5.7.15; **L** (fleabane), Rixton 20.1.15.
35.065 A, Dalton Crags 7.7.22; **A**, as previous.
35.066 A♂, Chorlton 25.6.20; **A**♀, Formby 7.6.14.
35.068 SS, Heysham 15.7.14 (S. Palmer); **L** (Sea-milkwort), Fluke Hall 28.4.19; **H**; **FS**, as previous.
35.071 A, Freshfield 26.6.07; **A**, Chorlton 27.7.13.
35.076 A, Astley Moss 17.5.07; **L** (cottongrass), Little Woolden Moss 11.3.22.
35.077 SS, Westby 16.6.22 (J. Steeden).
35.079 SS, Altcar 21.6.16 (R. Walker).
35.080 SS, Gait Barrows 25.6.15 (S. Palmer).
35.081 SS, Leighton Moss 25.7.05.
35.085 A, Lancaster 8.7.08; **L** (Honeysuckle), Chorlton 15.5.08.
35.089 A, Winmarleigh Moss 23.5.05; **A**, Cockerham Moss 11.5.10.
35.093 A ex-l, Chorlton 24.7.08; **L** (Broom), Chorlton 7.5.12.
35.094 A, Freshfield 5.7.10; **A**, Dolphinholme Aug 2011.
35.095 SS, Birkdale 27.6.05 (S. Palmer).
35.096 SS, Woodvale 2014 (S. Palmer).
35.097 A, Haydock 17.7.05.
35.101 A ex-l (Creeping Willow), Lytham 16.6.21; **L** (Grey Willow), Rixton 22.5.17.
35.103 A, Chorlton 4.9.14.
35.105 A, Swinton 31.7.11.
35.107 SS, Tarleton 29.7.22 (A. Barker); **L** (oak), Chorlton 19.5.19.
35.109 A ex-m, Stretford 2.6.16; **L** (Creeping Thistle), Chorlton 17.10.16.
35.113 A, Sunderland Point 2.7.10.
35.114 SS, Fluke Hall 23.6.16 (S. Palmer); **L** (Sea-purslane), Burrows Marsh 30.3.16.
35.115 A ex-l, Southport 22.6.21; **L** (orache), Southport 17.9.20.
35.116 A ex-l, Middleton NR 2018.
35.117 SS♀, Ainsdale 2.9.22 (R. Walker).
35.118 A, St Helens 2.9.22; **A**, Preston 7.9.22.
35.119 SS, Fluke Hall 23.6.16; **L** (Sea Plantain), Fluke Hall 28.4.19 (S. Palmer).
35.120 A, Clitheroe 2021; **A**, Dalton Crags 8.7.22.
35.123 A ex-l, Manchester 22.12.16; **L** (Bittersweet), Manchester 1.10.16.
35.126 SS ex-l (potato), Preston 1996 (S. Palmer).
35.128 SS ex-l (mouse-ear), Lytham 2016; **SS ex-l**, as previous.
35.129 A ex-l, St Helens 14.6.15; **L** (Red Campion), St Helens 23.4.15.
35.131 SS, Lytham 27.7.95 (S. Palmer).
35.132 SS, Potts Corner 2004 (S. Palmer).
35.137 A ex-l, Scorton 19.6.16; **L** (Greater Stitchwort), Scorton 30.3.16; **FS**, as previous.
35.141 A ex-l, Chorlton 15.10.18; **A ex-l**, Chorlton 11.6.22; **L** (Hawthorn), Chorlton 5.8.18.
35.143 A, Rainford 2.6.07; **L** (oak), Rixton 19.10.15.
35.145 A ex-l, Warton Crag 23.6.15; **FS** (Common Rock-rose), Warton Crag 24.5.22.
35.146 A, Chorlton 12.8.03; **A**, Longridge 16.6.22.
35.147 SS, Gait Barrows 25.7.14 (S. Palmer); **A**, Gait Barrows 22.4.12.
35.148 A, Chorlton 22.7.03; **A**, Preston 10.6.23.
35.149 A ex-l, Chorlton 19.6.21; **L** (birch), Chorlton 24.5.21.
35.150 A ex-l, Chorlton 14.6.21; **L** (Grey Willow), 27.9.20.
35.151 A ex-l, Rochdale 1.6.16; **L** (birch), Little Woolden Moss 15.9.21.
35.153 A, Astley Moss 7.6.08; **A**, Astley Moss 23.5.07.
35.157 A, Hale 16.7.06.
35.159 A ex-l, Freshfield 20.5.15; **M** (Scots Pine), Freshfield 31.3.15.
35.160 A, Chorlton 18.9.03; **A**, Preston 14.8.20.

36.001 A, Rixton 16.7.20; **L**, Chorlton 27.5.04.
36.0019 SS, St Helens 26.6.18 (R. Banks); **A** (agg.), Chorlton 24.6.20.

37.005 A, Chorlton 20.7.04; **A**, Chorlton 26.6.03.
37.006 FS (rose), Gait Barrows 18.10.15; **L** (rose), Warton Crag 14.5.16.
37.007 A ex-l, Chorlton 24.6.05; **L** (oak), Chorlton 21.5.05.
37.009 A ex-l, Little Woolden Moss 2.7.18; **L** (birch), Little Woolden Moss 5.9.16.
37.012 L (elm), Freshfield 28.5.21; **L**, as previous.
37.013 L (Apple), Chorlton 5.6.20; **L** (Apple), Catterall 14.7.06.

37.014 A ex-l, Chorlton 30.6.04; **L** (Blackthorn), Chorlton 2.6.04.
37.015 A ex-l (Alder), Chorlton 9.6.13; **L** (birch), Chorlton 14.5.21.
37.016 A♀ ex-l, Chorlton 17.6.11; **L** (cotoneaster), Chorlton 22.5.11.
37.017 L (Blackthorn), Gait Barrows 21.5.08.
37.020 A ex-l, Gait Barrows 10.6.19; **L** (Hazel), Gait Barrows 30.9.18.
37.022 A ex-l, Chorlton 23.6.19; **L** (Grey Willow), Chorlton 14.5.21.
37.027 L (Bramble), Davyhulme, 20.10.06.
37.028 SS, Trough Summit, 2.7.01; **L** (Heather), Birk Bank 28.3.22.
37.029 SS ex-l, Winmarleigh Moss 2004; **L** (birch), CW, 4.9.20.
37.030 A ex-l, St Helens 21.6.19; **L** (Alder), St Helens 15.5.19.
37.033 A, Chorlton 22.6.23; **L** (melilot), Chorlton 19.8.05.
37.034 SS, Hawes Water 25.5.12 (G. Jones).
37.035 A♂, Chorlton 17.6.23.
37.038 A ex-l, Middleton NR 11.7.21; **L** (Hedge Woundwort), Flixton 14.5.11.
37.044 A ex-l, Rixton 13.7.19; **L** (Bird's-foot trefoil), Lytham 25.5.21.
37.046 A♀ ex-l, Chorlton 26.6.10; **L** (Red Clover), Chorlton 19.8.09.
37.048 A, Rishton 17.6.21; **L** (White Clover), Chorlton 28.7.05.
37.050 A ex-l, Freshfield 27.6.21; **L** (Creeping Willow), Freshfield 28.5.21.
37.052 L (oak), Highfield Moss 13.10.07.
37.053 A♂, Chorlton 7.7.13, **L** (birch), Chorlton 15.5.10.
37.054 SS, Flixton 12.6.18 (K. McCabe).
37.055 SS, Winmarleigh Moss 7.6.98.
37.063 A, Chorlton 9.5.22; **L** (Gorse), Warton Crag 30.7.17.
37.066 A ex-l, Chorlton 23.4.14; **L** (Larch), Chorlton 21.3.15.
37.068 A ex-l, Fluke Hall 16.5.19; **L** (Saltmarsh Rush), Fluke Hall 28.4.19.
37.069 A ex-l, Rixton 8.5.09; **L** (Jointed Rush), Chorlton 28.8.15.
37.070 SS♂, Preston 13.5.15; **L** (Jointed Rush), Chorlton 5.8.15.
37.071 A ex-l, Rixton 25.6.06; **L** (rush), Chorlton 2.1.16.
37.072 A ex-l, Chorlton 7.5.11; **L** (Field Wood-rush), Chorlton 10.7.10.
37.073 A♂, Chorlton 13.6.22; **L** (rush), Chorlton 30.1.05.
37.074 SS♂, Preston 28.6.19; **L** (nr. rushes), Rixton 11.5.15.
37.078 SS ex-l (Sea Rush), Potts Corner 2006.
37.080 A ex-l (Sea Aster), Southport 12.8.21; **L** (Goldenrod), Gait Barrows 28.10.20.
37.083 A♂, Chorlton 22.7.13.
37.087 SS♀, Formby 14.8.20 (R. Walker).
37.088 SS, Stalmine 1.7.98.
37.090 A ex-l, Cadishead Moss 2.8.16; **L** (Mugwort), Cadishead Moss 14.9.13.
37.093 A ex-l, Chorlton 21.6.10; **L** (Creeping Thistle), Chorlton 13.10.20.
37.098 SS, Mere Sands Wood 9.7.21.
37.099 A ex-l, Chorlton 3.6.10; **L** (mouse-ear), Chorlton 25.7.10.
37.102 A ex-l, Flixton 8.7.09; **L** (Yarrow), Jack Scout 7.9.22.
37.104 A♂, Preston 27.7.23.
37.106 A ex-l, Rixton 12.6.15; **L** (Black Knapweed), Lytham 25.5.21.
37.108 A, Dolphinholme 5.8.06.

38.001 SS, Warton Crag 23.4.14; **M** (Honeysuckle), Warton Crag 24.6.18.
38.004 A, Chorlton 9.5.20; **M** (Red Fescue), St Helens 3.5.18.
38.005 A ex-m, Chorlton 1.6.18; **M** (Red Fescue), Chorlton 7.5.18.
38.008 A, Dalton Crags 7.7.22.
38.015 A ex-m, Dalton Crags 4.6.23; **M** (False Brome), Gait Barrows 28.10.20.
38.016 A ex-m, Little Woolden Moss 22.10.17; **M** (Purple Moor-grass), Little Woolden Moss 18.9.17.
38.017 A ex-m, Gait Barrows 16.5.18; **M** (Blue Moor-grass), Gait Barrows 5.4.18.
38.018 SS♂, Docker Moor 11.8.17.
38.022 A ex-m, Gait Barrows 16.5.21 (S. Palmer); **M** (sedge), Gait Barrows 10.4.21.
38.023 A ex-m, 24.4.22; **L ex-m**, 5.4.22; **M** (Glaucous Sedge), 13.9.21; **A ex-m**, 20.5.22. All from Worsaw Hill.
38.024 A ex-m, Stretford 21.4.23; **Cocoon**, Rochdale 10.4.21; **M** (Reed Sweet-grass), Stretford 17.4.23; **A ex-m**, Rochdale 27.4.21.
38.025 A ex-m, Chorlton 27.3.19; **M** (Cock's-foot), Chorlton 19.5.18.
38.026 SS, nr. Burnley 14.6.1913 (W. Mansbridge) MM.
38.029 A, Waddacar 21.6.22; **SS**, Didsbury 13.7.58 (H. Michaelis) MM.
38.030 A ex-m, Chorlton 30.5.17; **M** (Cock's-foot), Chorlton 1.5.17.
38.032 A♀, Chorlton 20.4.14; **A♂**, Chorlton 30.4.14.
38.033 SS, Sales Wood, Prescot 31.5.1919 (W. Mansbridge) MM.
38.036 A ex-m, Chorlton 10.5.18; **M** (Tufted Hair-grass), Chorlton 9.4.18.
38.037 A♂, Chorlton 6.9.21; **M** (Sheep's Fescue), Dalton Crags 26.4.22.

606 *The Moths of Lancashire*

38.038 A, Chorlton 22.5.13; **A**, Chorlton 18.5.19.
38.039 A ex-m, Chorlton 18.6.21; **L** (Reed Canary-grass), Chorlton 9.6.21.
38.040 SS ex-m, Brock Bottom 2012; **M** (Hairy Wood-rush), Brock Bottom 27.4.12.
38.041 A ex-m, Hawes Water 17.5.16; **M** (Glaucous Sedge), Warton Crag 2.4.17.
38.042 SS, Cockerham Moss 17.6.10.
38.043 A ex-m, Fluke Hall 17.5.19; **M** (Saltmarsh Rush), Fluke Hall, 28.4.19.
38.044 SS♂, Lytham 25.6.13.
38.045 SS♂, Flixton 16.8.17 (K. McCabe).
38.046 A, Chorlton 18.7.14; **M** (cottongrass), Little Woolden Moss 1.4.19.
38.047 A, Chorlton 27.5.05.

39.001 A, Lancaster 17.6.08; **L** (Hawthorn), Chorlton 18.9.05.
39.002 SS, Didsbury 4.8.56 (H. Michaelis) **MM**; **SS**, Didsbury 22.8.54 (H. Michaelis) **MM**.
39.003 A, Preston 14.5.22; **FS** (Hawthorn), Chorlton 28.2.15.
39.005 A, Manchester 14.8.22; **FS** (lime), Chorlton 4.5.23.
39.006 A, Cadishead Moss 20.6.11; **H**, St Helens 2005.

40.001 A ex-l, Ainsdale 12.6.21; **L** (Rosebay Willowherb), Ainsdale 28.5.21.
40.002 A, Chorlton 10.7.13.
40.003 SS, Gait Barrows 7.6.03.
40.004 A ex-m, St Helens 11.6.17; **M** (Great Willowherb), St Helens 5.4.17.
40.006 A ex-l, Chorlton 1.9.17; **L** (Great Willowherb), Chorlton 5.6.20.
40.007 A ex-l, Chorlton 27.8.17; **Gall** (Great Willowherb), Risley Moss 10.9.20.
40.008 A, Chorlton 20.5.20; **L** (Rosebay Willowherb), Chorlton 29.6.06.
40.009 A ex-l, Ainsdale 14.6.21; **Gall** (Rosebay Willowherb), Ainsdale 28.5.21.
40.010 A ex-p, Chorlton 18.8.15; **FS** (Great Willowherb), Chorlton 2.7.21.
40.011 A ex-m (Great Willowherb), Chorlton 28.8.15; **M** (Enchanter's Nightshade), Silverdale 12.6.22.
40.012 A, Warton Crag 24.5.22; **M** (Common Rock-rose), Warton Crag 25.10.17.
40.013 A ex-m, Chorlton 21.4.19; **M** (willowherb), Rixton 10.4.17.
40.014 A ex-m, Warton Crag 23.5.18; **M** (Enchanter's Nightshade), Warton Crag 8.8.16.
40.015 A ex-m, Chorlton 25.7.08; **M** (Rosebay Willowherb), Rochdale 25.9.20.

41.002 A, Chorlton 30.8.21; **L** (Yew), Didsbury 20.4.06.

41.003 A, Chorlton 27.10.22; **L** (acorn), Risley Moss 10.9.20.
41.005 A, Wigan 1.8.13; **A**, St Helens 31.7.19.

42.002 A, Chorlton 26.7.13; **A**, Hoghton 8.7.18.

43.001 H, Warton Crag 2022; **Foodplant**, Warton Crag 2022; **A**, Warton Crag 9.6.07; **Pair**, Warton Crag 23.4.21; **L** (Common Rock-rose), Warton Crag, coll. 17.5.12.
43.002 SS, Heysham Moss 24.6.15 (J. Patton).
43.004 SS, Gait Barrows 16.7.98.

44.001 A, Warton, nr. Carnforth 12.5.07; **L** (Honeysuckle), Chorlton 13.6.21.

45.004 A ex-l, Chorlton 29.4.20; **L** (Colt's-foot), St Helens 13.4.21.
45.008 A, Longton 4.7.21; **A**, Chorlton 25.7.02.
45.009 A, Flixton 13.8.12; **L** (Tansy), Middleton NR 24.5.22.
45.010 A, Chorlton 26.6.15; **L** (Herb-Robert), Chorlton 14.7.20.
45.011 A, Sunderland Point 3.7.09; **SS**, Gait Barrows 6.7.02 (K. McCabe).
45.012 A, Warton Crag 25.7.15.
45.013 A, Littleborough date unknown.
45.019 A, Leighton Moss 3.7.21; **L** (Mossy Saxifrage), Blacko 29.5.06.
45.021 A ex-l, Freshfield 11.8.14; **L** (centaury), Freshfield 28.7.14.
45.023 A, Chorlton 24.6.20.
45.027 SS, Silverdale 5.7.19 (J. Patton).
45.028 SS, Gait Barrows 30.6.06.
45.030 A, Chorlton 15.6.20; **A**, Chorlton 5.7.17.
45.033 A, Dalton Crags 17.6.21; **A**, Dalton Crags 7.7.22 (J. Patton).
45.037 A ex-l, Rixton 1.7.15; **L** (Common Fleabane), Rixton 20.6.19.
45.038 A, Gait Barrows 17.6.14.
45.040 A, St Helens 12.6.03.
45.041 A, Gait Barrows 25.7.22; **L** (Common Ragwort), Gait Barrows 5.8.22.
45.043 A ex-l, Rixton 5.6.15; **L** (Hemp Agrimony), Rixton 2.8.07.
45.044 A ex-l, Irlam Moss 13.9.15; **L** (bindweed), Chorlton 29.5.16.

46.001 A, Middleton NR 15.7.14 (J. Patton); **A**, Flixton 24.4.13.

47.001 A, Gait Barrows 6.8.16.
47.004 A, Formby 29.6.19; **SS**, Formby 29.6.19 (J. Girdley).
47.005 A ex-l (Wild Angelica), Rixton 1.10.09; **L** (Hogweed), Chorlton 26.8.21.

47.006 A, Flixton 20.8.21; **A**, St Helens 11.8.22.

48.001 A, Rishton 20.8.21; **L** (Nettle), Chorlton 23.5.15.
48.002 A ex-l, Chorlton 21.7.07; **L** (Skullcap), Chorlton 27.8.04.
48.005 A, Heysham Moss 28.8.22 (J. Patton); **A**, as previous.
48.007 A ex-l, St Helens 10.7.08; **L** (Hawthorn), St Helens 16.6.08.

49.001 A, Leighton Moss 6.7.09; **A**, Botton Mill, nr. Wray 21.6.09.
49.002 A, Gait Barrows 31.5.22; **A**, Chorlton 18.5.14.
49.004 A ex-l (Ash), Gait Barrows 14.9.16; **L** (birch), St Helens 10.9.20.
49.005 A, Mere Sands Wood 27.6.21.
49.008 A, Astley Moss 8.7.09; **A**, Loftshaw Moss 8.8.09.
49.009 A, Gait Barrows 23.5.07; **A**, Oswaldtwistle 30.5.19.
49.010 A, Grizedale 13.5.21; **A**, Grizedale 24.3.22.
49.013 A, Warton, nr. Carnforth 2.7.17; **L** (rose), Heysham 24.5.22.
49.015 A ex-l, Chorlton 10.6.21; **L** (birch), Chorlton 10.5.21.
49.016 A ex-l, Lytham 17.6.21; **L** (Creeping Willow), Lytham 14.6.21.
49.020 A, Longridge Fell 28.5.14.
49.021 A, Chorlton 7.7.03.
49.022 A ex-l (birch), Rixton 11.5.09; **L** (willow), Chorlton 16.5.15.
49.023 A, Flixton 22.9.20; **A**, Dalton Crags 27.6.08.
49.024 A ex-l (Alder Buckthorn), Chorlton 12.7.15; **L** (Alder), Chorlton 25.6.16.
49.025 A, Chorlton 31.5.20; **L** (sallow), Rixton 15.4.05.
49.026 A ex-l (rose), Southport 4.7.15; **L** (Blackthorn), Middleton NR 24.5.22.
49.028 A ex-l (birch), Chorlton 16.5.05; **L** (willow), Chorlton 21.9.19.
49.029 A, Lancaster 23.6.08; **L** (Heather), Birk Bank 28.3.22.
49.030 A, Preston 17.9.21; **A**, St Helens 29.4.05.
49.031 A, St Helens 11.6.05; **L** (grasses), Lytham 14.6.21.
49.033 A, Astley Moss 9.7.05.
49.035 SS, Langho 31.5.05.
49.037 A, Chorlton 9.6.03; **L** (dock), Chorlton 30.4.09.
49.038 A, Chorlton 24.6.14; **A**, St Helens 18.6.05.
49.039 A ex-l (Rosebay Willowherb), Chorlton 23.5.16; **L** (Bramble), Chorlton 5.1.20.
49.040 A, Lancaster 6.7.20; **A**, Preston July 2022.
49.042 A, St Helens 19.6.05.
49.043 A, St Helens 19.10.22; **A**, Chorlton 9.11.20.

49.044 A, Chorlton 19.2.14; **L** (oak), Chorlton 11.5.19.
49.045 A, Chorlton 11.8.04.
49.047 A, Mill Houses 6.7.14; **SS**, Gait Barrows 28.6.97.
49.048 A, Leck Fell 9.7.15.
49.049 A♀ ex-l (Creeping Thistle), Clitheroe 22.5.22; **L** (sallow), Chorlton 29.4.19.
49.050 A♂, Chorlton 16.7.13; **L** (Hogweed), Chorlton 30.5.15.
49.051 A ex-l (Mugwort), Chorlton 22.6.19; **L** (Ribwort Plantain), Chorlton 20.5.15.
49.052 gen. det. slide, Rishton 8.7.22.
49.054 SS♂, Tarleton 29.7.22 (A. Barker); **gen. det. slide**, as previous.
49.056 A♂ ex-l, Warton Crag 3.7.21; **L** (Oxeye Daisy), Warton Crag 24.6.18.
49.057 A♂, St Helens 12.7.04; **A♂**, Haydock 17.7.05.
49.059 A, Rishton 29.6.19; **A**, Chorlton 18.6.14.
49.060 A ex-l, Chorlton 11.6.19; **L** (oak), Chorlton 19.5.19.
49.061 A ex-l, Warton Crag 24.6.23; **L** (Blackthorn), Warton Crag 2.6.23.
49.062 A, Lancaster 7.7.07; **L** (Field Maple), Chorlton 13.5.22.
49.063 A, Chorlton 14.6.03.
49.064 SS ex-l (Bilberry), Jeffrey Hill 2012; **H**, Jeffrey Hill 2023.
49.065 A♂, Chorlton 9.7.13; **A**, Chorlton 14.9.03.
49.066 A♂, Preston 29.7.22; **L** (Meadowsweet), Chorlton 23.5.06.
49.067 A, Docker Moor 18.4.18.
49.069 A ex-l (Field Maple), Chorlton 1.9.15; **L** (Beech), Chorlton 28.6.09.
49.070 A ex-l, Chorlton 18.8.15; **L** (cherry), Chorlton 16.5.15.
49.071 A ex-l (sallow), Chorlton 4.7.16; **L** (Crack-willow), Chorlton 4.5.11.
49.073 A ex-l, Hawes Water 18.7.16; **FS** (Guelder Rose), Hawes Water 21.7.19.
49.075 A, Gait Barrows 22.4.09.
49.077 A, Chorlton 12.8.20; **L** (Blackthorn), Middleton NR 24.5.22.
49.078 A ex-l (Meadowsweet), Worsaw Hill 2022; **A**, Ainsdale 17.7.23.
49.080 A ex-l (Creeping Willow), Ainsdale 9.7.21; **L** (sallow), Chorlton 29.4.19.
49.082 A, Yealand Conyers 9.10.21; **A**, Heysham 13.10.11.
49.083 A ex-l, Chorlton 11.10.16; **L** (oak), Chorlton 28.8.21.
49.084 A, Little Woolden Moss 30.6.16; **L** (birch), Little Woolden Moss 17.9.16.
49.086 A, Flixton 25.10.21 (K. McCabe); **P** (birch), Chorlton 13.6.22.
49.087 A, Warton, nr. Carnforth 23.2.19; **A**, Catterall 17.4.23.

49.090 A, Astley Moss 8.5.11; **L** (birch), Little Woolden Moss 13.9.15.
49.091 Pair, Chorlton 13.6.22; **L** (Ash), Chorlton 17.10.15.
49.092 A, Heysham 3.7.16.
49.094 SS, Gait Barrows 20.6.00.
49.095 A, Hale 3.6.10.
49.096 A, Warton Crag 14.5.16; **A**, as previous.
49.097 A, Chorlton 12.8.20; **A**, Rossall 6.6.22.
49.100 A, Gait Barrows 2.7.08.
49.101 A, Bold Moss 23.5.07; **A**♀, Yealand Conyers 11.6.03.
49.103 SS, Fluke Hall 14.6.16.
49.105 SS, Burrows Marsh 9.5.12.
49.109 A, Chorlton 22.6.15; **A**, Gait Barrows 11.5.08.
49.110 A, Leighton Moss 7.7.19; **A**, Flixton 17.7.07.
49.111 A ex-l, Chorlton 3.6.21; **L** (Ribwort Plantain), Chorlton 11.9.20.
49.120 A, Fazakerley 3.8.21; **L** (Yarrow), St Helens 15.2.19.
49.123 A, Hale 27.6.09.
49.124 A, Formby 26.6.18.
49.127 A, Chorlton 7.6.19; **A**, Chorlton 7.6.19.
49.128 A, Leighton Moss 9.7.10; **A**, St Helens 9.7.05.
49.129 A, Gait Barrows 12.7.13.
49.132 A, Great Sankey 20.08.21.
49.133 A, Chorlton 24.5.10; **A**, Chorlton 31.5.15.
49.134 A ex-l, St Helens 30.7.19; **L** (Wild Teasel), Irlam 12.3.05.
49.137 A, Great Sankey 5.6.21; **A**, Chorlton 7.6.19.
49.139 A, Chorlton 16.8.21; **L** (ragwort), Chorlton 3.11.20.
49.144 A, St Helens 11.7.09; **A**, Chorlton 19.5.19.
49.146 A, Middleton NR 8.8.09.
49.149 A ex-l, Little Woolden Moss 2.6.16; **L** (birch), Little Woolden Moss 12.5.16.
49.150 A ex-l, Chorlton 13.6.19; **L** (birch), Chorlton 19.4.15.
49.151 A, Trowbarrow 14.7.12; **A**, Leighton Moss 26.6.10.
49.153 A, Trough summit 31.7.11; **A**, Parlick 11.7.22.
49.154 A, Rixton 23.5.19; **L** (birch), Little Woolden Moss 22.4.19.
49.155 A, Chorlton 24.6.03; **A**, Trafford Park 16.6.08.
49.156 A ex-l, St Helens 17.5.21; **L** (Rowan), St Helens 13.4.21.
49.157 A♀, Middleton NR 24.5.22; **L** (Blackthorn), Chorlton 2.5.10.
49.158 A ex-l (rose), Longridge 5.5.22; **A**, as previous.
49.159 A ex-l, Little Woolden Moss 5.7.16; **L** (birch), Little Woolden Moss 13.9.15.
49.160 A, Gait Barrows 18.6.13; **L ex-o**, final instar, Gait Barrows 17.8.11.
49.161 A, Chorlton 26.6.20; **A**, Staining 19.6.16.
49.164 A, Ainsdale 1.7.22; **A**, as previous.
49.166 A ex-l (bird's-foot-trefoil), Chorlton 14.6.15; **L** (Rosebay Willowherb), Ainsdale 2.6.21.
49.167 SS, Lytham 11.7.95.
49.172 A, Bowland 4.7.09; **A**, Cockerham Moss 11.5.10.
49.174 A, Clitheroe 15.7.19; **L** (*Dicranum scoparium*), Gannow Fell 2015.
49.179 A, Gait Barrows 8.6.15; **A**, Gait Barrows 24.5.07.
49.180 A, Little Crosby 29.6.09; **A**, Flixton 21.6.09.
49.183 A, Lancaster 13.7.20; **L** (Creeping Thistle), Chorlton 12.6.20.
49.184 A, Gait Barrows 21.5.08; **A**, as previous.
49.185 A, Chorlton 29.5.03; **A**, St Helens 30.8.22.
49.186 A ex-l, St Helens 1.8.19; **L** (Wild Teasel), Chorlton 11.1.20.
49.188 A ex-l, St Helens 9.5.21; **L** (Wild Teasel), St Helens 13.4.21.
49.190 SS♀, Gait Barrows 25.7.14.
49.191 A, Martin Mere 7.6.10; **L** (Hedge Woundwort), Rixton 28.1.19.
49.192 SS, Preston 7.7.95.
49.193 A, Chorlton 9.9.03; **A**, Preston 7.7.23.
49.194 A, Lancaster 1.7.09; **A**, Gait Barrows 22.5.20.
49.195 A, Middleton NR 30.6.10; **A**, Carleton 8.7.13.
49.196 SS, White Moss 12.6.10 (J. Girdley).
49.197 SS, Fluke Hall 14.6.16.
49.199 A, Gait Barrows 18.6.13.
49.200 A, Great Sankey 28.6.18; **L** (cherry), Rochdale 9.4.05.
49.201 A, Lee Fell 14.5.21.
49.202 A, Winmarleigh Moss 27.5.05.
49.204 A, Gait Barrows 9.6.09; **A**, Gait Barrows 1.6.11.
49.209 A, Leighton Moss 31.5.05; **A**, Gait Barrows 29.5.08.
49.211 SS, Longridge Fell 29.5.95; **L** (Bilberry), Cragg Wood 4.9.20.
49.212 A, Gait Barrows 19.7.19; **L** (Buckthorn), Gait Barrows 18.8.18.
49.214 A, Lancaster 16.7.20; **L** (Bush Vetch), Chorlton 4.11.21.
49.215 A, Chorlton 5.7.13; **A**, Chorlton 17.6.20.
49.216 A, Gait Barrows 27.5.18; **L** (oak), Gait Barrows 21.7.19.
49.218 A, Gait Barrows 5.5.10; **A**, Gait Barrows 6.5.08.
49.220 A, Grains Bar, Oldham 30.5.21; **A**, as previous.
49.222 A ex-l, Clougha 6.5.22; **L** (Cowberry), Clougha 28.3.22.
49.223 A ex-l (Hawthorn), Chorlton 18.6.23; **L** (Bilberry), Newhey 29.5.23.

49.224 A ex-l (cherry), Chorlton 6.6.19; **L** (oak), Rixton 14.5.22.
49.228 A ex-l, Chorlton 16.9.08; **L** (Alder), Chorlton 11.5.13.
49.229 A, Billinge 21.8.06; **L** (Creeping Willow), Lytham 25.5.21.
49.230 A, Gait Barrows 7.8.08.
49.231 A ex-l, Chorlton 14.6.09; **FS** (birch), Chorlton 23.5.15.
49.233 A♀ ex-l, Rochdale 9.7.15; **L** (birch), Rochdale 24.5.15.
49.234 A ex-l, Rochdale 24.5.22; **FS** (Wych Elm), Rochdale 2.5.22.
49.237 A ex-l, Chorlton 19.5.22; **L** (Blackthorn), Chorlton 22.4.22.
49.238 A ex-l, Ainsdale 27.6.21; **L** (Creeping Willow), Ainsdale 11.6.21.
49.240 A, Rishton 26.4.20; **L** (Alder), Salford 24.6.15.
49.242 A, Swinton 3.6.11.
49.243 A, Flixton 18.5.19; **A**, Medlock Vale 10.6.21.
49.244 A ex-l (willow), Oldham 3.4.22; **A**, Oswaldtwistle 9.7.20.
49.245 A ex-l, Little Woolden Moss 18.5.17; **L** (birch), Little Woolden Moss 5.9.16.
49.248 A ex-l, Chorlton 27.5.18; **L** (Hazel), Chorlton 6.4.15.
49.249 A, Gait Barrows 11.9.07; **L** (birch), Chorlton 19.4.15.
49.251 A, Chorlton 26.7.13.
49.252 A, Chorlton 17.6.22; **A**, Tame Valley 1.6.21.
49.254 A, Chorlton 31.5.20; **L** (birch), Chorlton 15.4.09.
49.255 A ex-l, Chorlton 5.7.05; **L** (sallow) 13.5.19.
49.257 A, Lord's Lot 9.7.11; **SS**, Cuerden 11.7.98.
49.259 A, Great Sankey 10.8.13; **A**, as previous.
49.260 A, Lancaster 25.6.20; **L** (oak), Rixton 14.5.22.
49.261 A, Heysham 2.9.05; **A**, Carnforth 25.10.07.
49.264 A, Chorlton 10.7.03.
49.265 A ex-l, Chorlton 4.6.11; **L** (Spear Thistle), Cadishead Moss 12.8.19.
49.266 A, Rixton 29.7.05; **A**, St Helens 15.7.09.
49.269 A, St Helens 26.7.05; **L** (ragwort), Cadishead Moss 12.8.19.
49.272 A ex-l, Southport 11.6.21; **L** (Sea Aster), Southport 17.9.20.
49.276 A, Gait Barrows 21.5.08; **L** (Goldenrod), Gait Barrows (J. Patton) 25.7.22.
49.279 A ex-l (oak), Chorlton 10.6.19; **L** (willow), Chorlton 9.5.22; **FS** (Hazel), Gait Barrows 18.10.15.
49.280 A ex-l, Chorlton 24.5.22; **FS** (poplar), Chorlton 24.9.20; **FS** (poplar), Chorlton 3.4.21;
49.281 A, Littleborough 2000.
49.283 A ex-l, Chorlton 1.7.15; **FS** (poplar), Chorlton 30.5.15.
49.284 A♀, Chorlton 27.6.04.

49.285 A, Lancaster 21.5.08; **L** (Creeping Thistle), Chorlton 11.3.19.
49.286 A ex-l, Rixton 10.5.06; **L** (Black Knapweed), Chorlton 25.1.20.
49.288 A, St Helens 10.7.05; **L** (Mugwort), Chorlton 7.4.19.
49.289 A ex-l, Chorlton 11.7.22; **L** (ragwort), Chorlton 26.10.20.
49.290 A, Chorlton 21.6.15; **A**, Longridge 16.6.22.
49.292 A, Lancaster 27.5.08; **L** (rose), Chorlton 1.5.06.
49.293 A, Gait Barrows 2.9.14.
49.294 A, Warton, nr. Carnforth 8.7.12; **FS** (Bramble), St Helens 13.4.21.
49.295 A ex-l, Sunderland Point 3.6.19; **L** (rose), Sunderland Point 28.4.19.
49.297 A ex-l, Chorlton 31.5.06; **L** (rose), Chorlton 25.4.22.
49.298 A ex-l, Chorlton 11.5.08; **L** (Hawthorn), Chorlton 4.4.08.
49.301 Host tree, Formby 2022; **Gall** (pine), Formby 22.4.22.
49.304 A ex-l, Ainsdale 21.7.21; **L** (pine), Ainsdale 22.6.21.
49.305 A ex-l, Freshfield 20.7.22; **L** (Scots Pine), Freshfield 10.6.22.
49.306 A, Formby 24.7.16; **A**, Dalton 2.8.08.
49.307 A, Formby 20.6.15; **A**, Chorlton 12.6.15.
49.309 A, Chorlton 1.6.11.
49.310 SS, Flixton 24.5.00 (K. McCabe).
49.311 A♂, Rossall 6.6.22.
49.313 A, Chorlton 30.5.20; **A**, as previous.
49.315 A, St Helens 18.8.21; **A**, Chorlton 16.8.03.
49.318 A, St Helens 3.7.06.
49.319 A, Flixton 2.8.09; **A♂**, Chorlton 23.7.22.
49.320 A♂, Chorlton 24.6.20.
49.321 A, Speke 24.7.20.
49.323 A, Longridge 16.6.22.
49.324 A ex-l, Chorlton 26.6.11; L (vetch), Chorlton 14.7.20.
49.325 A, Freshfield 28.5.21; **L** (Gorse), Ainsdale 4.7.23.
49.332 A, Formby 14.7.18.
49.334 SS, Beacon Fell 11.6.21.
49.335 A, Wigan 9.5.16.
49.338 A, Chorlton 5.7.13; **L** (Apple), Chorlton 30.8.16.
49.341 A, Leighton Moss 8.8.21; **L** (acorn), Chorlton 8.10.11.
49.342 A, Formby 5.8.16; **A**, Warton, nr. Carnforth 12.7.09.
49.345 A ex-l, Rixton 10.6.16; **L** (St John's-wort), Rixton 13.5.20.
49.347 A, Heysham 3.6.20; **L** (clover), Chorlton 26.9.06.
49.349 A, Warton Crag 15.5.18; **A**, Warton Crag 15.5.18.

49.351 A, Chorlton 14.5.21; **L** (vetch), 14.7.20.
49.354 A, Chorlton 27.3.17; **A**, Flixton 20.4.07.
49.3541 A, Chorlton 14.7.05.
49.357 A, Chorlton 12.6.22; **L** (Blackthorn), Chorlton 29.8.20.
49.358 A, Chorlton 2.6.22; **L** (rose), Formby 1.9.22.
49.359 A, Lancaster 23.6.20; **L** (Hawthorn), Chorlton 5.10.16.
49.360 SS♀, Thrushgill 4.6.15 (J. Patton).
49.362 A♂, Chorlton 25.3.22; **A**♂, Rivington 19.4.21.
49.363 A♂, Chorlton 21.5.23; **A**, Haydock 30.4.07.
49.364 A, Chorlton 30.4.22; **A**, Chorlton 6.5.23.
49.366 A♂, Holcroft Moss 13.5.05 (K. McCabe); **A**♂, as previous.
49.367 A, Silverdale 15.7.09; **L** (acorn), Rixton 1.9.06.
49.368 SS, Preston 10.5.11 (A. Powell).
49.371 A, Woodvale 10.5.06.
49.372 A ex-l, Ainsdale 11.7.21; **L** (Creeping Willow), Ainsdale 1.6.21.
49.373 SS♂, Gait Barrows 8.8.17 (J. Patton).
49.375 A ex-l, Rochdale 3.4.19; **L** (Sycamore), Chorlton 22.11.11.
49.376 A, Chorlton 13.7.20; **L** (in cocoon nr. Sycamore), Chorlton 14.1.06.
49.378 A, Swinton 31.5.11.
49.379 A, Little Woolden Moss 9.7.16; **L** (Hogweed), Worsley 28.7.23.
49.380 A ex-l, Middleton NR 20.7.21; **L** (Wild Carrot), Middleton NR 2.10.20.
49.381 A, Warton Crag 6.7.22 (J. Patton); **A**, as previous.

50.001 A, Sefton 20.7.14; **L**, Ainsdale 20.8.20.
50.002 A, St Helens 1.7.09; **A**, Woodvale 7.7.16.

52.003 A♂, Rishton 17.7.21; **A**♂, Haydock 10.7.13.
52.007 SS, Rixton 1.6.1902 (G.O. Day) LWM.
52.008 A, Ainsdale 21.6.14; **A**, St Helens 13.6.18.
52.012 A♂, Chorlton 4.6.22; **A**♂, Chorlton 21.6.23.
52.013 A♂, Sefton 17.6.21; **Pair**, Lancaster 11.7.07.
52.014 A, Liverpool 1.7.20; **A**, Liverpool 8.7.19.

54.002 H, Ainsdale 23.5.23; **A** (to pheromone), Ainsdale 7.7.16; **A**, Formby 23.7.17; **A**, Formby 2.7.11; **A**, Ainsdale 4.7.23.
54.003 H, Warton Crag 2021; **Ovipositing**, Warton Crag 10.6.06; **A**, Warton Crag 17.6.15; **A**, Warton Crag 4.6.15; **Ova** (rock-rose), Warton Crag 10.6.06.
54.008 A, Rixton 15.7.19; **L**, Ainsdale 21.6.14; **Cocoon**, Ainsdale 28.6.23.
54.009 A, Huncoat 1.6.21; **A**, Liverpool 1.8.12; **L** (Meadow Vetchling), Chorlton 24.5.03.

62.001 A♂, Chorlton 9.4.15; **L**, Chorlton 23.8.12.

62.005 A, Chorlton 5.8.03.
62.006 A, Chorlton 2.8.15.
62.007 A, Gait Barrows 18.6.13; **L** (oak), Rixton 27.8.14.
62.011 SS ex-l (birch), Winmarleigh Moss 7.6.03.
62.012 A♀, Hoghton 5.9.21 (G. Dixon); **A**, Darwen 7.7.21.
62.024 SS, Croston 30.7.20 (A. Barker).
62.025 A, Longton 17.7.22.
62.027 A, Formby 15.7.17.
62.028 A, Preston 6.6.21; **A**, Haydock 23.6.07.
62.029 A, Chorlton 13.7.04; **A**, Haydock 28.6.07.
62.030 SS, Clitheroe 3.6.11 (G. Jones).
62.034 A, Flixton 25.6.20.
62.035 A, Chorlton 10.7.15; **L** (Hawthorn), Chorlton 23.5.15.
62.037 A, Silverdale 2.7.08; **A**, Dalton Crags 7.7.22.
62.038 A ex-l, Chorlton 5.6.07; **L** (oak), Chorlton 20.5.22.
62.039 A, Flixton 30.5.16.
62.042 A, Chorlton 7.6.19; **L** (Spear Thistle), Chorlton 30.1.20.
62.048 A, St Helens 17.7.05; **A**, Chorlton 21.7.14.
62.054 A, Chorlton 7.6.04; **A**, Lytham 14.6.21.
62.057 A ex-l, Chorlton 14.6.10; **L** (ragwort), Chorlton 31.7.17.
62.058 A, Chorlton 16.7.03; **A**, Lancaster 19.7.07.
62.059 A ex-l, Chorlton 24.7.19; **L** (ragwort), Chorlton 23.7.18.
62.062 A, Chorlton 8.4.19; **A**, Hoghton 14.8.18.
62.063 SS, Heysham 24.2.00 (P. Marsh).
62.064 A♂, Hoghton 8.7.13; **SS**♂, Orrell 2.7.06 (P. Alker).
62.065 A♀, Preston 17.6.22; **A**♀, Chorlton 16.6.22.
62.067 A♂, St Helens 16.5.22 (R. Banks); **A**, St Helens 14.5.22 (R. Banks).
62.069 A, Formby 24.7.16; **A**, Ainsdale 16.6.17.
62.072 A, Chorlton 27.5.20; **A**, Preesall 10.9.19.
62.074 A, Ainsdale 4.7.15; **A**, Formby 30.6.21 (R. Walker).
62.075 A, St Helens 1.7.05; **A**, Warton, nr. Carnforth 9.7.12.
62.076 A, Great Sankey 16.10.16; **A**, Chorlton 19.9.03.
62.077 A, Hale 29.7.23; **A**, Eccleston 11.8.23.

63.002 A, Worsthorne 12.8.20.
63.003 A, Gait Barrows 4.6.15; **A**, Gait Barrows 15.7.22.
63.005 A, Formby 2018; **A**, Ainsdale 17.7.23.
63.006 A, Chorlton 19.8.10; **L** (mint), Preston 10.10.23.
63.007 A, Warton Crag 16.7.17; **A** underside, Warton Crag 16.7.17.
63.008 A, Gait Barrows 25.7.22; **A** underside, Gait Barrows 25.7.22.
63.014 A, Penketh 14.7.08.

63.016 A, Ainsdale 7.6.21; **L** (Yellow-rattle), Ainsdale 1.7.22.
63.017 A, St Helens 16.5.22.
63.018 A, Preesall 20.7.17; **L** (bindweed), Flixton 20.7.18.
63.020 A, Chorlton 25.6.13; **A**, Great Sankey 17.6.13.
63.021 A, Chorlton 25.6.10; **A**, Flixton 28.9.13.
63.022 A ex-l, Flixton 10.6.16; **L** (fleabane), Flixton 24.5.06.
63.024 H, Gait Barrows 25.7.22; **FS**, Gait Barrows 25.7.22; **A**, Gait Barrows 17.5.08; **L** (Goldenrod), Gait Barrows 25.7.22; **FS**, Gait Barrows 25.7.22.
63.025 St Helens 30.5.05; **L** (Nettle), Chorlton 15.9.18.
63.028 A, Longton 5.9.22.
63.029 A, Gait Barrows 24.5.12.
63.031 A, Turton Moor 8.9.23; **A**, Preston 7.9.22.
63.033 A, Lancaster 11.7.20; **L** (Raspberry), Chorlton 1.6.14.
63.034 A, Warton, nr. Carnforth 11.7.09; **A**, St Helens 25.7.05.
63.037 A, Warton, nr. Carnforth 31.5.08; **A**, Leighton Moss 6.6.06.
63.038 A, Lancaster 6.7.20; **L** (Nettle), Chorlton 27.5.04.
63.042 A, Flixton 16.10.21; **A**, Catterall 6.8.21.
63.046 A, Great Sankey 18.8.21; **A**, Warton, nr. Carnforth 10.2.20.
63.048 A, Morecambe 7.10.11; **A**, Lancaster 13.8.20.
63.051 A, Fowley Common 22.9.06 (D. Wilson); **A**, as previous.
63.052 A, Chorlton 10.9.21; **A**, Bilsborrow 15.9.06.
63.054 A, Chorlton 24.7.21; **L** (Box), Preston 13.9.21 (S. Palmer).
63.057 A, Warton, nr. Carnforth 18.6.11; **L** (*Brassica* spp.), Southport 17.9.20.
63.058 A, Worsthorne 9.8.23; **A**, as previous.
63.060 A, Leighton Moss 4.8.06; **A**, Martin Mere 14.7.09.
63.061 A, St Helens 29.9.06; **SS**, St Helens 29.9.06 (D. Owen).
63.062 A, Gait Barrows 11.7.21; **SS**, Gait Barrows 30.6.06.
63.063 SS♀, Formby 20.6.16 (R. Walker).
63.064 A, Chorlton 19.5.19; **A**, St Helens 14.6.05.
63.065 SS, Roeburndale 2.8.01.
63.066 A, St Helens 10.6.05; **A**, Haskayne 9.6.09.
63.067 A, Chorlton 10.7.13; **A**, Lancaster 2.7.20.
63.068 SS, Leck Fell 9.7.05.
63.069 A, Chorlton 12.8.20; **A**, St Helens 21.10.22.
63.073 A, Chorlton 7.8.15; **A**, Ashton-u-Lyne 18.7.14.
63.074 A, Chorlton 29.6.13; **L**, (moss spp.) Chorlton 15.5.20.
63.075 A, Warton, nr. Carnforth 17.7.19; **SS**, Formby 19.6.00.

63.076 A, Fowley Common 20.9.06; **SS**, Preston 14.2.98.
63.077 A, Silverdale 18.6.08; **A**, Heysham 1.7.09.
63.079 A, Chorlton 1.8.13; **A♂**, Pennington Flash 10.7.10.
63.080 A, Rixton 10.6.15; **L ex-o**, Rixton 15.9.13.
63.081 A, Lancaster 29.5.08; **SS♂**, Gait Barrows 12.6.98.
63.083 A, Dolphinholme 23.6.10; **A**, Ainsdale 28.6.23.
63.084 A, Dalton Crags 17.6.21.
63.085 A, Formby 2018; **A**, Formby 2018.
63.086 A, Gait Barrows 24.5.07; **L ex-o**, Chorlton 28.8.13.
63.087 A, Cadishead Moss 18.8.11; **A**, Flixton 13.8.20.
63.088 A (f.*warringtonellus*), Chorlton 5.7.03.
63.089 A, Yealand Conyers 20.7.04; **A**, Chorlton 26.7.13.
63.090 A, Yealand 31.7.10; **A**, Haydock 9.7.05.
63.091 A, Chorlton 11.8.04.
63.092 A, Rixton 7.8.18; **A**, Woolton, Liverpool 1.8.23.
63.093 A, Chorlton 18.7.21, **L** (beneath moss), Rixton 8.5.09.
63.095 A, Warton, nr. Carnforth 9.8.14; **A**, St Helens 11.8.21.
63.099 A, Yealand Conyers 27.6.23; **A**, Docker Moor 18.7.08.
63.100 A, Astley Moss 9.7.05; **A**, Warton, nr. Carnforth 8.8.10.
63.102 A, Chorlton 7.7.14; **A**, Warton, nr. Carnforth 29.7.12.
63.110 A, Sunderland Point 5.7.08; **SS**, Lancaster 9.7.03 (P.J. Marsh).
63.112 A, Preston 11.8.12; **SS**, as previous.
63.114 A, Warton, nr. Carnforth 4.7.21; **L** (*Potamogeton*), Rixton 30.5.10.
63.115 A, Chorlton 11.8.20; **A**, Warton, nr. Carnforth 27.7.08.
63.116 A ex-l, Rixton 4.7.09; **L** (duckweed), Chorlton 22.4.11.
63.117 A, Lancaster 11.8.07; **A**, Leighton Moss 28.8.19.
63.118 A, Gait Barrows 20.6.22; **A**, Yealand 21.6.20.
63.121 A, Rixton 15.7.21; **A**, Chorlton 19.6.14.
63.122 A, Leighton Moss 14.6.21; **A**, Leighton Moss 5.6.10.

65.001 A, Formby 9.5.09; **L** (birch), Astley Moss 27.6.08.
65.002 A, Chorlton 16.8.21; **A**, Chorlton 21.8.03.
65.003 A, Chorlton 4.9.22; **A**, Preston 13.8.22.
65.005 A, Formby 29.6.09; **L** (birch), Chorlton 13.8.18.
65.007 A, Preesall 7.8.20; **A**, Sefton 19.7.17.

65.008 A, Formby 4.7.17; **L** (Bramble), Chorlton 9.7.05.
65.009 A, Chorlton 21.6.14; **L** (Bramble), Huncoat 17.9.22.
65.010 A, Formby 6.6.20; **L** (White Poplar), Chorlton 3.8.15.
65.012 A, Silverdale 17.7.22; **A**, as previous.
65.013 A, Formby 20.5.20; **L** (birch), Chorlton 4.8.10.
65.014 A, Baines Crag 20.8.08; **A**, as previous; **H**, Birk Bank 16.7.23; **A**, Birk Bank 28.8.17; **A**, Hale 3.9.23.
65.016 A, Gait Barrows 31.3.08; **L** (birch), Astley Moss 6.5.08.

66.001 A, Formby 9.11.20; **L ex-o**, Clitheroe 9.5.18.
66.002 A, Leck Fell 13.8.08.
66.003 SS, Lancs 1909 (F. Womersley) NHM; as previous.
66.005 SS, Lytham (date unknown) NHM; as previous.
66.006 H, Birkdale Green Beach 22.5.23; **A**♀, Sefton 8.8.13; **A**♂, Ainsdale 21.7.22; **L**, Formby 6.5.11; **A**, Ainsdale 11.8.18.
66.007 A ex-l, Sefton 6.7.11; **L** (Heather), Mossley 30.3.21.
66.008 A♂, Sefton 23.5.17; **L**, Formby 14.10.12.
66.010 A♀, Formby 24.7.20; **L**, Formby 14.5.21.
66.011 SS, Chat Moss (date unknown) PM; as previous.

68.001 A♀, Formby 9.5.10; **L**, Formby 7.6.14.

69.001 A, Formby 20.5.20; **L** (Alder), Chorlton 24.7.16.
69.002 A, Formby 11.6.18; **L ex-o**, Clitheroe 12.8.19.
69.003 A, Rishton 29.5.21; **L**, Formby 8.9.20.
69.004 A, Belmont 12.9.05; **L**, Croston 31.08.23.
69.005 A ex-p (potato), Bolton-le-Sands 19.10.22; **L**, Eskrigge 25.8.06.
69.006 A, Ferniscowles 29.6.06; as previous.
69.007 A, Ainsdale 4.6.23; **A**, Silverdale 10.7.23.
69.010 A, Preesall 15.6.22; **L**, Preesall 1.8.17.
69.011 SS, Southport 14.9.53 (J.W. Blundell) LWM.
69.014 A, Wilpshire 17.7.21; **L** (fuschia), Bispham 1.9.08.
69.015 A, Silverdale 19.5.22; **A**, Formby 31.3.21.
69.016 A, Chorlton 12.6.15; **L**, Formby 4.8.20; **L** (Rosebay Willowherb), Whitworth 29.8.22.
69.017 A, Formby 1.6.17; **L**, Manchester 11.8.19; **A**, Rishton 31.5.19.

70.002 A, Cockerham Moss 8.7.10; **H**, Cockerham Moss 16.8.10.
70.006 A, Formby 30.6.18; **A**, Formby 16.6.09.
70.008 A, Warton, nr. Carnforth 5.6.11; **L ex-o**, Yealand Conyers 22.8.22.
70.009 A, Ainsdale 7.6.21; **A**, Freshfield 10.6.07.
70.010 SS, Lancs (no other details); **SS**, as previous.
70.011 A, Formby 20.7.09; **L** (Ivy), Chorlton 6.6.10.
70.013 A, Preston July 2022; **L ex-o**, Chorlton 8.9.12.
70.016 A, Chorlton 25.6.20; **L** (Ivy), Chorlton 31.5.10.
70.018 A, Gait Barrows 24.6.08.
70.023 A, Yealand Conyers 31.8.21; **A**, Warton, nr. Carnforth 2.9.18.
70.024 A, Great Sankey 24.7.15; **A**, Formby 20.9.20.
70.025 A, Leighton Moss 28.6.16.
70.026 A, Birk Bank 8.7.12; **A**, Mill Houses 26.6.09.
70.027 A, Gait Barrows 27.5.18; **A**, Gait Barrows 14.5.09.
70.029 A, Sefton 22.6.09; **L** (dock), Rixton 7.9.09.
70.032 A, Formby 7.6.20.
70.036 A, Haydock 13.7.23; **A**, Lancaster 12.6.23.
70.037 A, Great Sankey, 8.9.16; **A**, Dalton Crags 7.6.08.
70.038 A, Hurst Green 23.8.19; **A**, Preston 8.9.22.
70.039 SS, Liverpool July 1890 NHM.
70.040 SS, Warton Crag 22.6.54 (J.H. Vine-Hall) NHM.
70.041 A♀, Yealand Conyers 15.6.20; **A**, Freshfield 10.6.07.
70.045 A, Huncoat 8.7.21; **L ex-o**, Chorlton 18.4.14.
70.046 A, Leighton Moss 17.6.08; **A**, Sefton 2.9.16.
70.047 A, Orrell 17.8.06; **A**, Yealand Conyers 21.10.10.
70.048 A, Leck Fell 19.7.14; **A**, Leck Fell 9.7.05.
70.049 A, Formby 9.5.10; **L ex-o**, Chorlton 10.11.12.
70.051 A, Leck Fell 16.5.09; **A**, Grizedale 2.6.08.
70.052 A, Ainsdale 7.6.21; **L ex-o**, Chorlton 7.9.13.
70.053 A, Rishton 6.5.18; **L ex-o**, 9.7.13.
70.054 A, Sefton 21.5.17; **L ex-o**, Chorlton 15.2.15.
70.055 A, Chorlton 18.7.21; **A**, Formby 7.7.21.
70.059 A, Formby 25.8.21; **A**, Yealand Conyers 6.7.21.
70.060 A, Thrushgill 17.6.17; **A**, Thrushgill 29.6.19.
70.061 A, Warton, nr. Carnforth 16.8.08; **L ex-o**, Chorlton 19.6.10.
70.062 A, Rishton 10.6.16.
70.063 A, Leck Fell 5.7.05; **A**, Hutton Roof 9.8.21.
70.065 A, Heysham 8.7.21.
70.066 A, Yealand Conyers 31.3.21; **L** (rose), Formby 3.6.22.
70.067 A ex-l, Formby 3.4.23; **L** (rose), Formby 3.6.22.
70.068 A, Rindle Wood 27.6.10; **A**, Formby 28.4.10.
70.069 A, Formby 17.7.20; **A**, Great Sankey 4.8.13.
70.070 A, Southport 28.9.04; **SS**, Lytham 22.9.55 (C.I. Rutherford) LWM.
70.072 A, Thrushgill 26.6.09.
70.074 A ex-l (sallow), Holcroft Moss 2.6.14; **L** (Creeping Willow), Ainsdale 1.6.21.

70.075 A, Yealand Conyers 31.5.20; **L** (Alder), Chorlton 28.8.06.
70.076 A♂, Heysham 30.5.10; **A**, Heysham 22.5.16.
70.077 A, Yealand Conyers 25.9.17; **A**, Formby 19.9.21.
70.078 SS ex-l, Silverdale 20.7.1919 (W. Mansbridge) NHM.
70.079 A, Preston 8.9.21; **L ex-o**, Chorlton 25.3.13.
70.081 A ex-l, St Helens 30.4.15; **L** (pine), St Helens 24.4.14.
70.082 A, Chorlton 13.11.12; **A**, Yealand Conyers 13.10.02.
70.083 A, Tarleton 19.6.20; **A**, Tarleton 28.10.16.
70.084 A, Chorlton 30.7.20; **A**, Leighton Moss 31.7.05.
70.085 A ex-l, Middleton NR 18.6.22; **L** (rose), Middleton NR 24.5.22.
70.086 A, Rishton 1.6.21; **A**, Formby 28.5.17.
70.087 A, Formby 29.5.09; **A**, Silverdale 19.8.02.
70.088 A, 2.8.19; **Touch-me-not Balsam**, 31.7.19; **A**, 30.7.19; **L**, 1.9.14; **H**, 30.7.19. All photos from VC60.
70.089 A, Chorlton 17.7.20; **A**, Warton, nr. Carnforth 3.7.16.
70.090 A, Yealand 2.9.14; **A**, St Helens 19.8.05.
70.091 A, Birk Bank 10.6.07.
70.092 A, Chorlton 5.7.13; **A**, Pennington 26.6.09.
70.093 A, Preesall 1.7.21; **L ex-o**, Chorlton 13.4.12.
70.094 A, Rishton 28.5.21; **L ex-o**, Chorlton 3.9.12.
70.095 A, Leighton Moss 11.10.21; **L** (Sycamore), Formby 20.7.08.
70.096 A, Billinge 28.9.08; **A**, Formby 28.9.14.
70.097 A, Chorlton 10.9.21; **L** (Black Currant), 28.7.08.
70.098 A, Oswaldtwistle 6.8.12; **A**, Warton, nr. Carnforth 26.7.14.
70.099 A, Gait Barrows 9.8.13; **A**, Gait Barrows 27.7.07.
70.100 A, Chorlton 14.5.20; **L ex-o**, Chorlton 20.7.13.
70.101 A, Gait Barrows 31.3.08; **A**, nr. Wray 3.4.08.
70.102 A, Yealand 20.4.09; **A**, Yealand Conyers 15.7.21.
70.103 A, Yealand Conyers 17.4.21; **A**, Yealand Conyers 20.4.09.
70.104 A, Yealand Redmayne 3.6.10.
70.105 A♂, Chorlton 4.11.21; **L** (birch), Chorlton 9.5.10.
70.106 A♂, Chorlton 21.11.13; **L** (privet), Chorlton 16.5.03.
70.107 A♂, Preston 16.10.21; **L** (Blackthorn), Chorlton 13.4.07.
70.108 A♂, Chorlton 9.10.21.
70.109 A♂, Formby 24.10.22; **SS**, Liverpool 2000.
70.110 A, Green Bank 2.9.12.
70.111 A, Gait Barrows 17.5.08; **A**, Gait Barrows 11.5.22.

70.112 A, Chorlton 16.5.10; **L** (Alder), Chorlton 28.9.14.
70.113 A, Heysham 14.6.14; **A**, Warton Crag 8.6.18.
70.114 A, Rishton 17.6.22; **A**, Waddacar 21.6.22.
70.115 A, Lord's Lot 3.7.14; **A**, Crawshawbooth 11.8.23.
70.116 A, Yealand Conyers 16.7.09; **A**, Yealand Conyers 10.8.21.
70.118 A, Gait Barrows 8.7.09.
70.119 A, Silverdale 6.7.09.
70.121 A ex-l, Rixton 15.5.08; **L** (Grey Willow), 12.9.07.
70.123 A, Oswaldtwistle 30.7.20; **A**, Leighton Moss 19.8.08.
70.128 A, Silverdale 21.7.20; **L** (Traveller's-joy), Silverdale 12.6.21.
70.130 A, Warton Crag 24.6.10; **A**, Warton Crag 16.6.13.
70.131 A, Lord's Lot 16.7.12; **A**, Longridge 16.7.22.
70.132 A, Yealand Conyers 25.5.10.
70.133 A, Chorlton 28.6.03; **A**, Warton, nr. Carnforth 13.7.14.
70.134 A ex-l, Chorlton 29.7.14; **L** (Red Bartsia), Chorlton 3.10.15.
70.137 A, Ainsdale 7.6.21; **L** (Yellow-rattle), Ainsdale 1.7.22.
70.138 A, Formby 29.7.21; **L** (Red Campion), Astley Moss 9.7.05.
70.139 A, Hyning Scout Wood 27.7.13; **A**, Warton, nr. Carnforth 24.7.07.
70.141 A, Formby 9.4.20; **L** (Gorse), Middleton NR 24.5.22.
70.142 A, Yealand Conyers 14.4.20; **L** (ragwort), Rixton 3.9.12.
70.143 A, Warton Crag 2.5.08; **L** (Blackthorn), Silverdale 29.5.21.
70.144 A, Leighton Moss 29.6.16; **L** (Blackthorn), Chorlton 13.4.07.
70.145 A, Docker Moor 24.6.20; **L** (Bilberry), Lord's Lot 26.5.15.
70.146 A, Yealand Conyers 18.6.18; **L** (Traveller's-joy), Silverdale 28.7.22.
70.147 A ex-l, Chorlton 23.6.16; **L** (sallow), St Helens 13.4.21.
70.148 A, Flixton 23.7.21.
70.149 A, Woolton 3.7.20; **A**, Silverdale 13.7.08.
70.150 A, Yealand Conyers 30.8.05; **L**, (Common Toadflax) Heysham 1.9.14.
70.151 A, Warton, nr. Carnforth 5.6.11; **L** (Foxglove), Rixton 12.6.08.
70.154 A, Flixton 9.5.04 (K. McCabe); **A**, Flixton 4.6.10.
70.155 A, Yealand Conyers 12.6.16; **L** (Bladder Campion), Carnforth 17.7.15.
70.156 A, Yealand Conyers 24.3.21; **L** (oak), Chorlton 13.6.09.

70.157 A, Yealand Conyers 10.4.20; **L** (Hawthorn), Chorlton 2.6.09.
70.158 A ex-l, Yealand Conyers 20.7.20; **L** (Juniper), Hawes Water 21.5.16.
70.159 A, Briercliffe 19.8.22.
70.160 A ex-l, Warton Crag 21.4.14; **L** (angelica), Lord's Lot 10.9.20.
70.161 A ex-l, Warton Crag 10.5.19; **L** (Goldenrod), Warton Crag 24.8.20.
70.162 A, Chorlton 8.6.04; **A**, Docker Moor 12.6.15.
70.163 A, Burnley 10.6.21; **A**♀, Yealand Conyers 30.9.09.
70.165 A♀, Heysham 25.7.13.
70.166 A, Bolton-le-Sands 22.6.16; **A**, Silverdale 30.7.09.
70.168 A ex-o, Chorlton 12.6.14; **L ex-o**, Warton, nr. Carnforth 19.8.20.
70.169 A, Silverdale 29.6.09; **A**, Silverdale 26.6.16.
70.171 A, Chorlton 5.5.03; **A**, Yealand Redmayne 15.5.06.
70.172 A, Carnforth 8.6.20; **L ex-ova**, Carnforth 15.7.20.
70.173 A, Formby 5.8.09; **L** (Sea Aster), Fluke Hall 5.9.22.
70.175 A Lord's Lot 18.7.14; **L** (angelica), Lord's Lot 4.9.14.
70.176 A, Burnley 23.6.13; **A**, Warton, nr. Carnforth 27.5.20.
70.177 A ex-o, Docker Moor 3.5.22; **L ex-o**, Docker Moor 13.7.21.
70.179 A ex-l (Goldenrod), Warton Crag 30.6.16; **L** (Yarrow), Chorlton 14.8.10.
70.180 A ex-l, Gait Barrows 31.7.21; **L** (Goldenrod), Gait Barrows 23.9.10.
70.181 A, Yealand Conyers 22.6.21; **A**, as previous.
70.182 A ex-l (Black Currant), Chorlton 12.5.09; **L ex-o**, Yealand Conyers 27.7.22.
70.183 A, Yealand Conyers 14.5.20; **L**, Yealand Conyers 6.6.20.
70.184 A ex-l (Beech), Rixton 27.5.16; **L** (Blackthorn), Chorlton 20.9.21.
70.187 A, Yealand Conyers 29.7.11; **L** (Yarrow), Yealand Conyers 26.9.11.
70.188 A, Flixton 18.8.18; **L** (Mugwort), Chorlton 13.9.04.
70.189 Pair, Middleton NR 31.5.18; **L**, Middleton NR 6.8.17.
70.190 A, Yealand Conyers 10.6.20; **L ex-o**, Yealand Conyers 13.7.20.
70.191 A, nr. Tatham 21.7.19; **A**, Blaze Moss 12.8.09.
70.192 A, Formby 6.9.16; **A**, St Helens 26.5.17.
70.195 A, Billinge 17.10.08; **L** (Broom), Chorlton 12.5.04.
70.198 A, Chorlton 19.5.14; **A**, Warton, nr. Carnforth 14.6.15.
70.199 A, Hesketh Bank 13.6.23.
70.200 A, Chorlton 31.7.14; **L** (Hawthorn), Warton Crag 2.6.23.
70.201 A, Warton Crag 10.4.19; **L** (Ash), Thrangbrow 21.5.19; **A**, Yealand Conyers 18.4.20; **A**, Gait Barrows 20.4.21; **H**, Warton Crag 17.6.22.
70.202 A, Formby 1.5.09; **L** (Creeping Willow), Ainsdale 7.6.21.
70.203 A, Heysham 5.4.16; **A**, as previous.
70.205 A, Ainsdale 11.8.21; **L** (Japanese Spindle), Rochdale 3.4.10.
70.206 A, Dalton Crags 3.6.08; **A**, Rishton 19.6.23; **L** (elm), Dalton Crags 3.9.08.
70.207 A, St Helens 7.7.05; **A ex-l**, Little Woolden Moss 26.6.16; **L** (sallow), Little Woolden Moss 13.9.15.
70.208 A, Yealand Conyers 27.4.10; **A**, Yealand Conyers 10.5.21.
70.212 A, Swinton 25.8.07; **SS**, Carnforth 1904 (H. Murray) NHM.
70.214 A, Chorlton 19.6.14; **A**, Formby 10.7.10.
70.215 A, Silverdale 16.7.21; **H**, Yealand Conyers 7.10.23; **A**, Yealand Conyers 11.7.19; **A**, Gressingham 4.7.23; **A**, Leighton Moss 17.7.10.
70.218 A, Huncoat 30.6.22; **L** (vetch), Chorlton 5.9.12.
70.222 A, Rixton 23.6.09; **L** (Bracken), Rixton 11.7.11.
70.223 A, Gait Barrows 15.5.10; **A**, Gait Barrows 3.5.08.
70.224 A, Great Sankey 4.6.16; **A**, Formby 28.5.09.
70.226 A, Formby 4.9.21; **L** (Cherry Plum), Chorlton 27.9.22.
70.227 A, Ainsdale 17.7.23; **A**, Chorlton 24.8.12.
70.229 A, Gait Barrows 4.6.15; **A**, Warton Crag 27.5.21.
70.231 A, Yealand Conyers 17.6.20; **A**, VC60 June 2007.
70.233 A, Flixton 29.7.06; **A**, Billinge 27.8.08.
70.234 A, Sefton 3.9.21; **L** (Alder), Chorlton 23.6.14.
70.235 A, Formby 14.9.21; **A**, Preesall 6.8.21.
70.236 A, Chorlton 9.8.20; **A**, Chorlton 21.7.13.
70.237 A, Chorlton 14.7.15; **L** (*Prunus* spp.), Whitworth 29.8.22.
70.238 A, Clitheroe 22.5.20; **A**, Thrushgill 29.5.22.
70.239 A, Formby 6.5.16; **A**, Yealand Conyers 1.5.07.
70.240 A, Formby 20.5.20; **A**, Rishton 25.5.18; **L**, Chorlton 28.10.03.
70.241 A, Chorlton 10.7.20; **A**, Formby 15.7.09; **L ex-o**, Chorlton 19.4.14.
70.243 A, Reedley 7.7.21; **L** (Ivy), Chorlton 14.11.08.
70.244 A, Chorlton 18.10.14; **L** (sallow), Chorlton 24.4.10.
70.245 A, Formby 20.3.11; **A**, Chorlton 8.3.14.

70.246 A, Wray 29.2.12; **A**, Mill Houses 4.3.14.
70.247 A, Rishton 22.2.19; **L** (Hazel), Gait Barrows 14.5.16.
70.248 A, Preston 18.4.11; **A**, Yealand Conyers 15.4.21.
70.250 A♂, 18.3.11; **L**, 30.6.10; **H**, 24.3.09; **A**♀, 24.4.10; **A**♀, 28.3.11: Potts Corner (all).
70.251 A, Rishton 25.3.20; **L** (oak), Rixton 14.5.08.
70.252 A, Formby 22.5.09; **L** (elm), Formby 4.7.23.
70.253 A, Rishton 23.1.20; **L** (oak), Chorlton 22.5.19.
70.254 A, Chorlton 8.11.14; **A**, Formby 5.12.09.
70.255 A♂, Chorlton 31.3.14, **L** (sallow), Chorlton 13.6.09.
70.256 A, Chorlton 15.11.14; **L** (birch), Chorlton 2.5.10.
70.257 A, Warton, nr. Carnforth 24.5.10; **A**, Sefton 3.5.18 (R. Walker).
70.258 A, Formby 19.6.10; **L** (Ivy), Chorlton 31.3.16.
70.264 A, Lord's Lot 6.7.08; **A**, Hyning Scout Wood 3.7.14.
70.265 A, Rishton 22.6.22; **L** (Blackthorn), Chorlton 3.9.21.
70.270 A, Formby 17.7.16; **L** (Alder Buckthorn), Chorlton 7.8.10.
70.272 A, Gait Barrows 1.5.07; **A**, Eaves Wood 6.4.11.
70.274 A, Chorlton 13.5.04; **A**, Sefton 3.5.17.
70.275 A, Astley Moss 27.6.15; **A**, Winmarleigh Moss 6.5.18.
70.276 A, Formby 2.6.09; **L** (pine), Ainsdale 20.9.22.
70.277 A, Chorlton 7.6.14; **L** (birch), Chorlton 6.9.09.
70.278 A, Formby 16.5.17; **L** (sallow), Chorlton 7.8.10.
70.279 A, Warton Crag 5.5.04; **A**, Rishton 8.5.20.
70.280 A, Formby 5.6.17; **L** (Blackthorn), Chorlton 7.8.10.
70.282 A♂ & **A**♀ **ex-l**, Chorlton 28.2.10; **L** (Hawthorn), Chorlton 26.5.09.
70.283 A ex-l (sallow), Chorlton 11.4.11; **L** (Ash), Leighton Moss 18.10.14.
70.284 A, Warton, nr. Carnforth 22.6.14; **A**, Sefton 28.7.11.
70.287 A, Gait Barrows 27.7.07; **A**, Gait Barrows 8.7.09.
70.288 A, Ainsdale 3.7.16; **A**, Flixton 25.7.02.
70.292 A, Trough Road 9.6.09; **A**, Thrushgill 29.5.22.
70.295 A, Cockerham Moss 16.6.10.
70.297 A, Heysham 16.7.07; **A**, Heysham 8.8.10.
70.299 A, Reedley 27.7.21; **A**, Rishton 15.7.17.
70.300 A, Hoghton 23.6.09; **A**, Leyland 27.6.08.
70.302 A, Clitheroe 25.6.20.
70.303 A, Gait Barrows 19.6.06.

70.305 A, Formby 25.9.10; **L** (oak), Chorlton 4.4.15.
71.003 A♂, Formby 25.6.17; **L** (poplar), 7.6.21.
71.005 A, Chorlton 16.8.13; **L** (Grey Willow), Little Woolden Moss 26.9.15.
71.006 A, Clitheroe 18.5.16; **A**, Preston 26.5.23.
71.007 A, Swinton 14.8.10; **A**, St Helens 23.8.13.
71.011 A, Chorlton 21.4.14; **L** (oak), Chorlton 10.6.20.
71.012 A, Preesall 22.8.22; **L** (birch), Chorlton 4.8.10.
71.013 A, Rishton 17.6.22; **L** (sallow), Chorlton 12.6.11.
71.016 A, Gait Barrows 6.5.08; **A**, Silverdale 7.5.22.
71.017 A, Formby 1.5.09; **L** (Aspen), 7.8.10.
71.018 A, Chorlton 9.5.20; **L** (birch), Chorlton 20.9.08.
71.020 A, Formby 6.5.20; **L** (Grey Willow), Chorlton 5.9.22.
71.021 A, Chorlton 23.6.14; **L** (Sycamore), Lord's Lot 23.6.14.
71.023 A, Challen Hall Wood 4.5.16; **A**, Gait Barrows 3.5.08.
71.025 A, Formby 1.6.17; **L** (sallow), Ainsdale 28.8.13.
71.027 A, Ainsdale 11.8.15; **L** (Aspen), Chorlton 25.6.16.

72.001 A, Formby 12.5.20; **L** (White Poplar), Ainsdale 1.7.22.
72.002 A, Chorlton 10.6.04; **A**, Warton, nr. Carnforth 2.7.08.
72.003 A, Rishton 22.6.22; **L** (Nettle), Chorlton 17.8.10.
72.007 A, Silverdale 16.6.22; **A**, Docker Moor 10.6.21.
72.009 A, Ainsdale 1.5.13; **L** (willow), Ainsdale 8.5.11.
72.010 A, Little Eccleston 22.7.23; **A**, Formby 8.5.21.
72.011 L, Blackpool 4.8.17.
72.013 A, Yealand Conyers 31.7.20; **L**, Formby 1.5.14.
72.015 A, Holcroft Moss 22.5.04; **L**, Ainsdale 20.9.19.
72.016 A, Ainsdale 18.6.12; **L**, Ainsdale 23.6.17.
72.017 A♂, Great Sankey 12.9.15; **A**♀ (ovipositing), Formby 15.8.20; **L** (Birch), Ainsdale 4.7.23.
72.019 A, Formby 20.6.10; **L** (sallow), Rixton 10.8.10.
72.020 A, Formby 15.6.09; **L**, Chorlton 14.8.10.
72.022 A♀, Chorlton 29.4.23; **A**♂, Formby 20.5.09.
72.023 A, Ainsdale 22.6.18; **A**, Birkdale 14.6.21.
72.024 A, Chorlton 23.5.10 (K. McCabe); **L**, Chorlton 9.4.09.
72.025 A, Winmarleigh Moss 13.6.14; **L**, Woodyards, nr. Lancaster 7.7.22.
72.026 A, Ainsdale 8.8.17; **L**, Ainsdale 7.6.13.

72.029 A, Great Sankey 17.6.17; **A**, Liverpool 12.7.23.
72.031 A, Chorlton 16.5.14; **L** (ragwort), Ainsdale 24.6.11.
72.034 A, Formby 25.8.11; **A**, as previous.
72.036 A, Warton, nr. Carnforth 27.6.07; **A**, Yealand Conyers 3.7.04.
72.037 A, Great Sankey 20.7.13; **A**, Leighton Moss 18.7.21.
72.038 A, Silverdale 7.7.22; **A**, Leighton Moss 25.7.15.
72.041 A♂, Lancaster 17.7.22; **A**♀, Yealand Conyers 25.7.06.
72.042 A, Oswaldtwistle 6.7.21; **L**, Formby 13.9.22.
72.043 A, Chorlton 28.7.13; **A**, Formby 20.7.17.
72.044 A, Chorlton 31.7.17; **L**, Oswaldtwistle 12.5.23.
72.045 A, Formby 28.7.10; **L**, Preston 22.4.22.
72.046 A, Rishton 25.7.20; **A**, Formby 24.7.10.
72.049 A, Billinge 23.5.07; **A**, Formby 16.5.22.
72.053 A, Chorlton 7.6.14; **L** (Ivy), Chorlton 4.9.10.
72.055 A, St Helens 4.6.05; **A**, Warton, nr. Carnforth 27.6.09.
72.060 A, Chorlton 26.7.04; **A**, Preston 23.7.21.
72.061 A, Preston 8.9.22; **A**, Rishton 30.6.22.
72.063 A, Ainsdale 25.6.11; **L**, Chorlton 6.11.10.
72.066 A, Carnforth 23.8.08; **A**, Flixton 5.8.22.
72.067 A, Woodvale 18.5.10; **A**, Altcar 7.8.15.
72.069 A, Chorlton 5.7.13; **L**, Wigan 11.3.22.
72.073 A, Pennington, 20.6.05; **A**, Silverdale 24.7.19.
72.076 A, Silverdale 16.8.23; **A**, Ainsdale 19.9.22.
72.078 A, Great Sankey 20.8.10; **A**, Formby 3.8.22.
72.083 A, Chorlton 23.6.19; **A**, Chorlton 9.6.21.
72.083 A, Ainsdale 2.6.10; **L** (grasses), Astley Moss 9.7.05.

73.001 A, Chorlton 1.5.20; **L** (Hawthorn), Chorlton 8.8.10.
73.002 A, Rishton 20.5.18; **A**, Chorlton 23.5.20.
73.003 A, St Helens 4.9.17; **A**, Formby 4.7.17.
73.008 A, Heysham 5.7.03.
73.012 A, Sefton 1.7.11; **L**, Chorlton 17.5.14.
73.014 A, Great Sankey 9.7.16; **L** (Delphinium), Preston 17.5.07.
73.015 A, Birkdale 2.9.22; **L** (Alder), Chorlton 9.9.06.
73.016 A, Rishton 21.6.16; **L ex-o**, Chorlton 9.4.14.
73.017 A, Chorlton 29.6.18; **A**, Warton, nr. Carnforth 2.7.08.
73.018 A, Yealand Conyers 24.7.20; **A**, Preesall 2.8.19.
73.021 A, Worsthorne 19.7.06; **A**, Rishton 23.7.17.
73.022 A, Rishton 17.6.22; **L** (Cock's-foot), Chorlton 23.4.18.
73.023 A, Chorlton 1.7.14; **A**, Worsthorne 13.7.23;
73.024 A ex-l (Purple Moor-grass), Little Woolden Moss 14.5.18; **L** (Reed Canary-grass), Chorlton 21.8.22.

73.026 A, Sefton 25.6.20; **A**, Yealand 17.6.08.
73.032 A ex-l, Gait Barrows 18.5.16; **L** (oak), Gait Barrows 27.8.15.
73.033 A, Sefton 11.10.17; **A**, Mill Houses 20.9.14 (J. Roberts).
73.036 A, Rishton 31.5.19; **L**, Oswaldtwistle 12.8.17.
73.037 A ex-l, Rixton 4.6.08; **L** (Hawthorn), Rixton 25.7.07.
73.038 A♂, Preston 17.6.22; **L**, Formby 4.9.20.
73.039 A, Flixton 28.5.18; **L**, Heath Charnock 17.9.23.
73.040 A, Formby 10.7.21; **L** (birch), Chorlton 16.6.07.
73.042 A, Wardstone 12.6.09; **A**, Clitheroe 6.5.07.
73.045 A, Sefton 1.6.17; **L**, Formby 3.9.13.
73.046 A, Formby 2.7.16; **L** (Aspen), St Helens 21.8.14.
73.047 A, Silverdale 22.5.10; **A**, Yealand Conyers 16.6.04.
73.048 A, Gait Barrows 22.5.07; **A**, Chorlton 3.6.15.
73.050 A, Flixton 20.7.95.
73.052 A, Southport 8.6.11; **L**, Southport 3.7.23.
73.053 A, Formby 26.4.22; **L** (mayweed), Chorlton 21.6.14.
73.058 A, Formby 10.2.18; **L** (mullein), Preston 24.6.21 (S. Palmer).
73.059 A, Great Sankey 8.8.18; **A**, Woolton 10.8.23.
73.061 A, Gait Barrows 10.9.07; **A**, Yealand Conyers 24.8.21.
73.062 A, Preston 13.8.22; **L**, Warton, nr. Carnforth 13.5.11.
73.063 A ex-l (privet), Chorlton 2.7.05; **L** (Blackthorn), Chorlton 12.5.13.
73.064 A, Chorlton 16.8.13; **A**, Formby 7.9.21.
73.065 A, Gait Barrows 28.10.09; **A**, Yealand Conyers 22.10.10.
73.068 A, Chorlton 28.9.14; **L** (Blackthorn), Chorlton 2.5.10.
73.069 A, Preston 2021; **L** (Snowberry), Chorlton 20.5.08.
73.070 A ex-l, Birkdale 26.6.23; **L ex-o**, St Helens 20.7.14.
73.074 A, Yealand Conyers 10.5.15; **A**, Formby 6.9.17.
73.076 A, Warton, nr. Carnforth 24.9.16; **A**, Formby 11.10.18.
73.084 A, Chorlton 29.6.18; **L**, Oswaldtwistle 6.5.23.
73.085 A, Formby 24.7.19; **A**, Formby 24.7.19.
73.087 A, Chorlton 21.9.03; **A**, Longridge 16.6.22.
73.089 A, Flixton 2.9.22 (K. McCabe).
73.092 A, Lancaster 5.6.08; **L** (Nettle), Chorlton 27.9.22.
73.095 A, Formby 25.8.22; **A**, Formby 26.9.17.
73.096 A, Preston 17.6.22; **L** (Dandelion), Chorlton 20.4.14.

73.097 A, Chorlton 10.7.03.
73.099 A, St Helens 3.9.17; **A**, Formby 5.6.17.
73.100 A, Great Sankey 3.8.15; **A**, Great Sankey 27.7.17.
73.101 A, Formby 9.5.09; **A**, Yealand Conyers 31.5.04.
73.102 A, Gait Barrows 17.6.08; **A**, Yealand Conyers 29.5.04.
73.105 A, Chorlton 20.6.03.
73.106 A, Liverpool 4.7.09.
73.107 A ex-l, Chorlton 15.7.19; **L** (Ivy), Chorlton 13.4.08.
73.109 A, Chorlton 1.8.13; **A**, Formby 31.7.21.
73.110 A, Hawes Water 5.6.08.
73.113 A, Middleton NR 2.10.20; **L**, Sefton 3.9.13.
73.114 A, Rishton 18.6.21; **L** (Bracken), Rixton 11.7.11.
73.118 A, Brinscall 29.8.08; **A**, Littleborough Sept 2000.
73.119 A, Leighton Moss 8.8.21; **A**, St Helens 18.7.22.
73.120 A, Roeburndale 9.8.20.
73.121 A, Rishton 19.8.22; **A**, Formby 25.8.22.
73.123 A, Chorlton 6.8.13; **A**, Rishton 1.8.21.
73.124 A, Chorlton 14.9.03; **A**, Wray 23.9.06.
73.126 A♂, Sunderland Point 29.8.08; **A**, Longton 25.7.20.
73.127 A, Botton Mill 27.8.08; **A**, Leck Fell 29.8.08.
73.128 A♂, Chorlton 18.8.12; **A♂**, Formby 10.8.22.
73.129 A♂, Sunderland Point 22.8.08.
73.131 A, Chorlton 13.8.21; **A**, Chorlton 20.8.15.
73.132 A ex-l, Birkdale 2010; **Pair**, Southport 20.8.09; **A**, Southport 20.8.09; **L** (saltmarsh-grass), Birkdale 2009 (R. Burkmar, G. Jones); **H**, Birkdale Green Beach 2008.
73.134 A, Chorlton 26.10.14; **A**, Formby 14.10.09.
73.136 A, Chorlton 21.8.16; **A**, Sefton 21.8.10.
73.137 A, Great Sankey 8.8.13; **A**, Preesall 20.7.21.
73.138 A, Ainsdale 12.7.18; **A**, Birkdale 17.7.18; **H**, Ainsdale 23.5.23.
73.139 A, Southport 21.8.21; **A**, Southport 21.8.21; **A**, Great Sankey 20.8.21.
73.141 A, Great Sankey 1.8.13; **A**, Great Sankey 5.8.16.
73.142 A, Chorlton 2.8.13; **A**, Bold Moss 9.7.05.
73.144 A, Flixton 15.9.11; **A**, Sefton 5.8.13.
73.146 A♂ ex-l, Crag House Allotment (CHA) 1.7.22; **A**, Warton Crag June 2011; **L** (Blue Moor-grass), CHA 26.4.22; **L**, Gait Barrows 11.5.22; **H**, CHA 2022.
73.147 A, Chorlton 1.7.14; **A**, St Helens 15.7.09.
73.154 A, Warton, nr. Carnforth 10.6.18; **A**, Preesall 17.6.20.
73.155 A, Leighton Moss 21.6.03; **A**, Preston 18.6.06.
73.156 A, Preesall 5.5.17; **L**, Chorlton 7.12.09.

73.157 A, Heysham 12.6.05; **A**, Swinton 30.5.09.
73.158 A, Chorlton 18.5.14; **A**, Sefton 25.5.09.
73.159 A, Rishton 23.6.22; **A**, Formby 8.6.18.
73.160 A, Chorlton 29.6.18; **L**, Chorlton 4.6.15.
73.161 A♂ ex-l, Birkdale 2010; **L** (saltmarsh-grass), Birkdale 2009 (G Jones, R. Burkmar).
73.162 A, St Helens 7.7.05; **L** (grasses), Chorlton 15.4.09.
73.163 A, Chorlton 6.7.14; **L**, Chorlton 2.4.10.
73.164 A, Yealand Conyers 15.7.13.
73.165 A, Clitheroe 3.7.21.
73.168 A, Chorlton 26.7.13; **A**, Rishton 8.7.22.
73.169 A♂, Rishton 10.7.22; **A♂**, Rishton 22.7.22.
73.170 A♂, Preston 27.7.23; **A♂**, Preston 27.7.23.
73.171 A, Preesall 23.8.17; **A**, Chorlton 4.8.03.
73.172 A, Chorlton 11.7.03; **A**, St Helens 21.7.05.
73.173 A, Rishton 20.6.22; **A**, Preston 9.6.22.
73.174 A, Rishton 20.6.22; **A♂**, Preston 17.6.22.
73.175 A♂, Rishton 30.6.22; **L**, Chorlton 19.4.04.
73.176 A, Warton, nr. Carnforth 12.6.21; **A**, Chorlton 11.6.03.
73.179 A, Silverdale 21.9.22; **A**, Preston 6.9.21.
73.180 A, Chorlton 7.10.21; **A**, Formby 29.9.19.
73.181 A ex-l (Sycamore), Stretford 1.9.15; **A**, Sefton 18.9.20; **L** (sallow), Chorlton 11.4.11.
73.182 A, Chorlton 13.8.13; **A**, Sefton 21.8.17; **L** (sallow), Chorlton 17.4.10.
73.183 A, Chorlton 19.9.03; **A**, St Helens 25.9.18.
73.186 A, Flixton 27.9.10; **A**, Formby 29.9.11.
73.187 A, Yealand Conyers 25.9.06; **A**, Rishton 26.9.21.
73.188 A, Birk Bank 24.8.22; **A**, Leighton Moss 15.10.19; **A**, Crawshawbooth 17.9.23.
73.189 A, Formby 13.10.09; **A ex-l**, Little Woolden Moss 22.9.16; **L** (sallow), Little Woolden Moss 12.5.16.
73.190 A, Rixton 17.10.13; **A**, St Helens 21.10.22.
73.192 A, Sefton 16.10.15; **L** (sallow), Chorlton 2.5.10.
73.193 A, Chorlton 13.9.02; **A**, Formby 26.9.10.
73.194 A, Chorlton 1.3.14; **A**, Warton, nr. Carnforth 16.10.11.
73.195 A ex-l, Chorlton 12.11.11; **L** (Blackthorn), Chorlton 22.4.11.
73.200 A, Sefton 31.3.17; **A**, Chorlton 31.3.14.
73.201 A, Chorlton 30.3.14; **A**, Warton, nr. Carnforth 17.5.08.
73.202 A, Rishton 20.3.21; **A**, St Helens 30.4.17.
73.206 A, St Helens 28.9.04; **A**, Formby 28.9.16.
73.207 A, Crossdale Grains 18.8.12; **SS**, Formby 2.7.20.
73.209 A, Formby 22.10.09; **L**, nr. Tatham 21.7.19.
73.210 A, St Helens 22.3.18; **L** (Norway Maple), Chorlton 11.5.21.
73.213 A, Formby 7.8.13; **L** (poplar), Chorlton 30.5.15.

73.216 A, Formby 10.7.10; **L** (poplar), Chorlton 2.5.19.
73.219 A, Lancaster 20.8.20; **A**, Chorlton 22.8.13.
73.220 A, Mill Houses 23.7.08; **A**, Warton, nr. Carnforth 20.7.19.
73.221 A, St Helens 29.6.19; **A**, Woodvale 7.7.16.
73.222 A, Chorlton 10.7.15; **A**, Rainford 26.7.08.
73.224 A ex-o, Clitheroe 25.9.18; **L ex-o**, Clitheroe 9.5.18.
73.225 A, Rishton 3.9.22; **A**, Yealand Conyers 19.9.14.
73.228 A, Chorlton 1.8.13; **L** (rock-rose), Warton Crag 11.4.22 (J. Patton).
73.231 A, Worsthorne 24.9.11; **A**, Freshfield 18.9.09; **A**, Haydock 18.9.07.
73.233 A, Formby 19.9.09; **A**, Chorlton 26.9.13.
73.234 A, Green Bank 14.4.22; **A**, St Helens 11.10.06.
73.235 A, Yealand Conyers 12.10.20; **A**, Formby 26.9.09.
73.237 A, Haydock 26.9.06.
73.238 A, Thrushgill 29.5.22; **A**, Formby 22.6.11.
73.241 A, Formby 2.5.13; **L** (pine), Ainsdale 1.7.22.
73.242 A, Chorlton 14.3.20; **A**, Chorlton 23.2.14; **L** (Blackthorn), Chorlton 29.5.10.
73.243 A, Gait Barrows 31.3.08; **A**, Silverdale 10.4.10; **L** (oak), Gait Barrows 9.5.20.
73.244 A, Formby 30.3.21; **L** (Aspen), Chorlton 26.5.09.
73.245 A, Formby 21.3.12; **L** (birch), Chorlton 16.5.10.
73.246 A, Chorlton 17.3.04.
73.247 A, Chorlton 26.4.20; **L** (Garden Loosestrife), Chorlton 24.6.06.
73.248 A, Ravenmeols 10.4.11; **SS ex-l**, Formby 30.3.52 (B.B. Snell).
73.249 A, Formby 8.4.09; **L**, Warton, nr. Carnforth 15.5.09.
73.250 A, Rishton 20.3.21; **L** (birch), Chorlton 24.5.21.
73.252 A, Dalton Crags 19.8.08; **A**, Yealand Conyers 26.8.17.
73.253 A, Warton, nr. Carnforth 5.9.11; **A**, Formby 3.9.21.
73.254 A, Formby 10.8.21; **L**, Rochdale 17.5.20.
73.255 A, St Helens 16.8.17; **A**, Formby 30.7.21.
73.257 A, Bold Moss 30.5.05; **L** (Heather), Longridge Fell 26.7.19.
73.261 A, Formby 4.7.15; **L**, Ainsdale 4.3.22.
73.264 A, Formby 26.5.18; **A**, Yealand Conyers 20.5.20.
73.266 A, Chorlton 9.7.03; **A**, Southport 8.6.11.
73.267 A, Formby 22.5.22; **L** (orache), Chorlton 12.8.17.
73.270 A, Chorlton 1.7.14; **L**, Formby 25.9.22.
73.271 A ex-l, Astley Moss 25.6.06; **L**, Astley Moss 9.7.05.

73.272 A, Formby 11.5.22; **A**, Yealand Conyers 25.5.12.
73.273 A, Lancaster 28.5.08; **A**, Formby 19.5.22.
73.274 A, Formby 3.7.21; **L** (bitter-cress), Chorlton 2.8.09.
73.275 A, Formby 10.8.21; **A**, Formby 1.7.22.
73.276 A, Rishton 1.6.21; **A**, Warton, nr. Carnforth 31.8.14.
73.279 A, Yealand Conyers 31.5.20; **A**, Formby 29.6.13.
73.280 A, Chorlton 24.6.21 (S. Marsden); **L** (Prickly Lettuce), Chorlton 14.8.23.
73.281 A, Formby 28.7.22; **L** (Red Campion), Chorlton 25.6.22.
73.282 A, Flixton 25.6.20.
73.283 A, Yealand Conyers 31.5.04; **A**, Preesall 19.6.19.
73.286 A, Warton, nr. Carnforth 30.5.08; **A ex-l**, Carnforth 1.6.16.
73.288 A, St Helens 29.6.23.
73.289 A, Docker Moor 29.6.21; **A**, Heysham 29.6.22.
73.290 A, Flixton 24.7.07.
73.291 A, Chorlton 26.7.13; **A**, Rishton 26.9.21.
73.293 A, Preston 17.6.22; **L** (Cock's-foot), Chorlton 14.4.18.
73.294 A, Leighton Moss 1.7.09; **A**, Yealand Conyers 15.7.09.
73.295 A, Silverdale 26.6.23; **A**, Longton 23.9.17.
73.296 A, Flixton 22.9.00.
73.297 A, Hoghton 23.8.15.
73.298 A, Rishton 30.6.22; **A**, Formby 26.7.21.
73.299 A, Birkdale 18.6.12; **A**, Formby 6.6.16.
73.301 A, Chorlton 4.7.13; **A**, Formby 29.6.09.
73.302 A, Bold Moss 23.5.07; **A**, Mere Sands Wood 20.6.21.
73.304 A, Heysham 19.8.08 (A. Draper); **A**, as previous.
73.307 A, St Helens 20.8.22; **L ex-o**, Chorlton 5.12.04.
73.308 A, Freshfield 14.8.04; **A**, Formby 3.8.22; **L**, Birkdale 20.6.09; **H**, Birkdale Green Beach 2022; **H**, as previous.
73.311 SS ex-l (ragwort), Lytham 3.8.55 (C.I. Rutherford) LWM.
73.312 A, Southport 15.8.14; **A**, as previous.
73.313 A, Formby 18.8.20; **A**, Silverdale 30.7.11.
73.314 A, Chorlton 3.8.21; **A**, Flixton 16.8.12; **SS**, Lytham 30.7.55 (C.I. Rutherford) LWM.
73.317 A, Chorlton 10.5.22; **A**, Rishton 16.6.22; **L ex-o**, Southport 11.7.21.
73.319 A, Chorlton 20.6.13; **A**, Rishton 20.6.22.
73.320 A, Silverdale 19.6.22; **A**, Formby 22.6.11.
73.322 A, Formby 21.6.11; **A**, Formby 10.8.21.
73.323 A ex-p, Birkdale 18.5.09; **A**, Ainsdale 20.6.17.
73.324 A, Chorlton 26.7.21; **A**, Swinton 29.7.14.

73.325 A, Preston 17.10.21; **A**, Formby 9.5.09.
73.327 A, St Helens 28.9.17; **A**, Formby 18.8.09.
73.328 A, Chorlton 18.5.14; **A**, Warton, nr. Carnforth 14.6.09.
73.329 A, Chorlton 17.8.13; **L ex-o**, Chorlton 8.10.13.
73.331 A, Warton, nr. Carnforth 7.8.22; **A**, Yealand Conyers 19.8.17.
73.332 A ex-l, Chorlton 31.5.05; **L** (Ivy), Chorlton 8.3.20.
73.333 A, Rishton 17.6.22; **A**, Heysham 3.6.09.
73.334 A, Formby 10.8.10; **L**, Chorlton 15.4.09; **SS** (*D. florida*-type), Docker Moor 29.6.21 P. Marsh).
73.336 A, Chorlton 12.4.13; **A**, Sefton 18.4.18; **A**, Chorlton 1.4.04.
73.337 A, Green Bank 23.4.22 (P. Marsh).
73.338 A, Lancaster 24.7.08; **L** (Heather), Astley Moss 9.9.01.
73.339 A, Billinge 23.7.95.
73.341 A, Docker Moor 28.8.14; **A**, Yealand Conyers 27.7.22.
73.342 A, Chorlton 4.6.20; **L** (goosefoot), Chorlton 27.9.14.
73.343 A♀, Formby 28.7.22; **L**, Chorlton 2.4.08.
73.345 A ex-l, Chorlton 6.6.08; **L** (grasses), Chorlton 1.1.08.
73.346 A, Chorlton 2.8.15; **A**, Formby 22.7.09.
73.348 A, Formby 3.7.09; **L**, Chorlton 22.4.11.
73.350 A, Yealand Conyers 24.8.04; **A**, Lancaster 13.8.06.
73.351 A, Leighton Moss 21.6.03 (K. McCabe, S.H. Hind); **A**, Formby 19.6.10.
73.352 A, Rishton 15.6.16; **A**, Rishton 21.6.16.
73.353 A, Rishton 22.7.22; **A**, Formby 30.7.21.
73.355 A, Docker Moor 19.8.08; **L**, Besom Hill, Oldham 24.4.21.
73.356 A, Cadishead Moss 2.9.11; **L** (Heather), Birk Bank 28.3.22.
73.357 A, Formby 14.8.22; **L** (geranium), Chorlton 17.2.11.
73.358 A, Rishton 26.7.22; **A**, Chorlton 16.8.21.
73.359 A, Chorlton 26.8.13; **A**, Formby 7.9.21.
73.360 A, Formby 10.7.10; **A**, Yealand 28.6.09.
73.361 A, Formby 6.7.16; **L**, Chorlton 6.4.05.
73.365 A, Rishton 4.9.22; **A**, Formby 18.9.09.
73.368 A, Warton, nr. Carnforth 5.7.09; **L** (Great Willowherb), Chorlton 19.7.11.

74.002 A, Preston 22.7.22; **A**, as previous.
74.003 A, Chorlton 10.7.15; **A**, Formby 23.6.20.
74.004 A, Lancaster 22.4.08; **A**, Formby 23.5.17.
74.007 A, Great Sankey 29.6.21; **L** (oak), Chorlton 28.9.14.
74.008 A ex-l, Chorlton 2.5.11; **L** (oak), Chorlton 4.8.10.
74.009 A, Sefton 31.3.14 (A. Parsons); **L** (oak), Rixton 14.5.22.
74.011 A, Flixton 17.8.02.

Addendum
18.006 A♀, Ainsdale 15.11.23 (R. Moyes).
48.003 A ex-l, Chorlton 17.7.23; **L** (Skullcap), Chorlton 3.7.23.
49.356 A♂, Chorlton 13.6.23; **A**♂, as previous.
52.011 A♂, Chorlton 16.6.23; **A**♀, Brock 25.6.23.
63.047 A, Silverdale 8.10.23; **A**, Hesketh Bank 9.10.23.
70.064 A, Southport 28.7.23.
70.217 A, Horwich 16.6.23; **A**, as previous.
72.074 A, Preesall 6.9.23; **A**, as previous.
73.004 A, Longridge 19.8.23 (D. Lambert).
73.009 A, Hesketh Bank 9.10.23.
73.082 A, Flixton 19.7.23 (K. McCabe); **A**, as previous.

Appendix 4

Scientific names of foodplants mentioned in the text

Vernacular name	Scientific name
Agrimony	*Agrimonia eupatoria*
Alder	*Alnus glutinosa*
Alder, Grey	*Alnus incana*
Alder, Italian	*Alnus cordata*
Alder Buckthorn	*Frangula alnus*
Angelica, Wild	*Angelica sylvestris*
Apple	*Malus* spp.
Apple, Crab	*Malus sylvestris*
Arrowgrass, Sea	*Triglochin maritima*
Ash	*Fraxinus excelsior*
Aspen	*Populus tremula*
Aster, Sea	*Tripolium pannonicum*
Avens, Water	*Geum rivale*
Avens, Wood	*Geum urbanum*
Azalea	*Rhododendron* spp.
Balsam, Himalayan	*Impatiens glandulifera*
Balsam, Touch-me-not	*Impatiens noli-tangere*
Bartsia, Red	*Odontites vernus*
Basil, Wild	*Clinopodium vulgare*
Bay	*Laurus nobilis*
Bedstraw	*Galium* spp.
Bedstraw, Hedge	*Galium album*
Bedstraw, Lady's	*Galium verum*
Beech	*Fagus sylvatica*
Berberis	*Berberis* spp.
Betony	*Betonica officinalis*
Bilberry	*Vaccinium myrtillus*
Bindweed	*Calystegia* spp.
Bindweed, Field	*Convolvulus arvensis*
Bindweed, Hedge	*Calystegia sepium*
Bindweed, Large	*Calystegia silvatica*
Birch	*Betula* spp.
Birch, Downy	*Betula pubescens*
Birch, Silver	*Betula pendula*
Birch Polypore	*Fomitopsis betulina*
Bird Cherry	*Prunus padus*
Bird's-foot-trefoil, Common	*Lotus corniculatus*
Bird's-foot-trefoil, Greater	*Lotus pedunculatus*
Bistort, Common	*Bistorta officinalis*
Bitter-cress	*Cardamine* spp.
Bitter-cress, Large	*Cardamine amara*
Bittersweet	*Solanum dulcamara*
Black-poplar	*Populus nigra*
Black-poplar hybrid	*Populus* x *canadensis*
Blackthorn	*Prunus spinosa*
Blue Moor-grass	*Sesleria caerulea*
Bogbean	*Menyanthes trifoliata*
Bog-moss	*Sphagnum* spp.
Bog-myrtle	*Myrica gale*
Bog-rosemary	*Andromeda polifolia*
Borage	*Borago officinalis*
Boston-ivy	*Parthenocissus tricuspidata*
Box	*Buxus* spp.
Bracken	*Pteridium aquilinum*
Brambles	*Rubus fruticosus* agg.
Bridewort	*Spiraea* spp.
Brome, False	*Brachypodium sylvaticum*
Broom	*Cytisus scoparius*
Broom Fork-moss	*Dicranum scoparium*
Buckthorn	*Rhamnus cathartica*
Buddleja	*Buddleja davidii*
Bullace	*Prunus domestica*
Bulrush	*Typha latifolia*
Bulrush, Lesser	*Typha angustifolia*
Burdock	*Arctium* spp.
Burdock, Lesser	*Arctium minus*
Burnet-saxifrage	*Pimpinella saxifraga*
Butterbur	*Petasites hybridus*
Buttercup, Creeping	*Ranunculus repens*
Cabbage	*Brassica* spp.
Campion, Bladder	*Silene vulgaris*
Campion, Red	*Silene dioica*
Campion, White	*Silene latifolia*
Canary-grass, Reed	*Phalaris arundinacea*
Carrot, Wild	*Daucus carota*
Cedar	*Cedrus* spp.
Centaury	*Centaurium* spp.
Centaury, Common	*Centaurium erythraea*
Chamomile, Corn	*Anthemis arvensis*
Chard	*Beta* spp.
Cherry	*Prunus* spp.
Cherry Laurel	*Prunus laurocerasus*
Cherry Plum	*Prunus cerasifera*
Chervils	*Chaerophyllum* / *Anthriscus* spp.
Chestnut, Sweet	*Castanea sativa*
Cicely, Sweet	*Myrrhis odorata*
Cinquefoil	*Potentilla* spp.
Cinquefoil, Creeping	*Potentilla reptans*
Cinquefoil, Marsh	*Comarum palustre*
Cleavers	*Galium aparine*
Clematis	*Clematis* spp.
Clover	*Trifolium* spp.

Vernacular name	Scientific name	Vernacular name	Scientific name
Clover, Red	*Trifolium pratense*	Fuchsia	*Fuchsia*
Clover, White	*Trifolium repens*	Garlic Mustard	*Alliaria petiolata*
Club-rush, Sea	*Bolboschoenus maritimus*	Geranium	*Pelargonium* spp.
Cock's-foot	*Dactylis glomerata*	Glasswort	*Salicornia* spp.
Colt's-foot	*Tussilago farfara*	Goldenrod	*Solidago virgaurea*
Comfrey, Common	*Symphytum officinale*	Goldenrod, Canadian	*Solidago canadensis*
Cotoneaster	*Cotoneaster* spp.	Gooseberry	*Ribes uva-crispa*
Cottongrass	*Eriophorum* spp.	Goosefoot	*Chenopodium* spp.
Couch, Sand	*Elymus junceiformis*	Gorse	*Ulex europaeus*
Cow Parsley	*Anthriscus sylvestris*	Ground-elder	*Aegopodium podagraria*
Cowberry	*Vaccinium vitis-idaea*	Guelder-rose	*Viburnum opulus*
Crack-willow	*Salix euxina / x fragilis*	Hair-grass, Tufted	*Deschampsia cespitosa*
Cranberry	*Vaccinium oxycoccos*	Hair-grass, Wavy	*Avenella flexuosa*
Crosswort	*Cruciata laevipes*	Hart's-tongue	*Asplenium scolopendrium*
Cuckooflower	*Cardamine pratensis*	Hawkbit, Autumn	*Scorzoneroides autumnalis*
Cucumber	*Cucumis sativus*	Hawkweed	*Hieracium* spp.
Currant	*Ribes* spp.	Hawthorn	*Crataegus monogyna*
Currant, Black	*Ribes nigrum*	Hazel	*Corylus avellana*
Currant, Red	*Ribes rubrum*	Hazel Woodwart	*Hypoxylon fuscum*
Cypress	*Cupressus* spp.	Heath Plait-moss	*Hypnum jutlandicum*
Cypress, Leyland	*Cupressus x leylandii*	Heather	*Calluna vulgaris*
Cypress-leaved Plait-moss	*Hypnum cupressiforme*	Heath, Cross-leaved	*Erica tetralix*
Daisy, Oxeye	*Leucanthemum vulgare*	Heath-rush	*Juncus squarrosus*
Dame's-violet	*Hesperis matronalis*	Heath	*Erica* spp.
Dandelion	*Taraxacum* spp.	Hedge-parsley, Upright	*Torilis japonica*
Deadly Nightshade	*Atropa belladonna*	Hemlock	*Conium maculatum*
Delphinium	*Consolida* spp.	Hemp-agrimony	*Eupatorium cannabinum*
Dewberry	*Rubus caesius*	Herb-Robert	*Geranium robertianum*
Dock	*Rumex* spp.	Hogweed	*Heracleum sphondylium*
Dog-rose	*Rosa canina*	Holly	*Ilex aquifolium*
Dog's Mercury	*Mercurialis perennis*	Hollyhock	*Alcea rosea*
Dogwood	*Cornus* spp.	Honeysuckle	*Lonicera periclymenum*
Dropwort	*Filipendula vulgaris*	Honeysuckle, Himalayan	*Leycesteria formosa*
Duckweed	*Lemna* spp.	Hop	*Humulus lupulus*
Duke of Argyll's Teaplant	*Lycium barbarum*	Hornbeam	*Carpinus betulus*
Dyer's Greenweed	*Genista tinctoria*	Horse-chestnut	*Aesculus hippocastanum*
Elder	*Sambucus nigra*	Iris, Yellow	*Iris pseudacorus*
Elm	*Ulmus* spp.	Ivy, Common	*Hedera helix*
Elm, Small-leaved	*Ulmus minor*	Juniper	*Juniperus communis*
Elm, Wych	*Ulmus glabra*	Juniper (cultivar)	*Juniperus* spp.
Enchanter's-nightshade	*Circaea lutetiana*	Kidney Vetch	*Anthyllis vulneraria*
Eucryphia	*Eucryphia* spp.	Knapweed, Common or Black	*Centaurea nigra*
Fat-hen	*Chenopodium album*	Knapweed, Greater	*Centaurea scabiosa*
Fennel	*Foeniculum vulgare*	Laburnum	*Laburnum anagyroides*
Fescue	*Festuca* spp.	Larch, European	*Larix decidua*
Fescue, Red	*Festuca rubra*	Laurustinus	*Viburnum tinus*
Field-rose	*Rosa arvensis*	Lavender, Garden	*Lavandula angustifolia*
Firethorn	*Pyracantha coccinea*	Lavender, Sea	*Limonium* spp.
Fleabane, Common	*Pulicaria dysenterica*	Lettuce, Prickly	*Lactuca serriola*
Forget-me-not	*Myosotis* spp.	Lichen	*Lepraria* spp.
Forget-me-not, Wood	*Myosotis sylvatica*	Lilac	*Syringa vulgaris*
Foxglove	*Digitalis purpurea*	Lily, Water	*Nymphaeaceae*
Foxtail, Meadow	*Alopecurus pratensis*	Lime	*Tilia x europaea*

Vernacular name	Scientific name	Vernacular name	Scientific name
Lime, Small-leaved	*Tilia cordata*	Pondweed	*Potamogeton* spp.
Loosestrife	*Lysimachia* spp.	Poplar	*Populus* spp.
Lyme-grass	*Leymus arenarius*	Poplar, Eastern Balsam	*Populus balsamifera*
Male-fern	*Dryopteris filix-mas*	Poplar, Grey	*Populus* x *canescens*
Mallow, Common	*Malva sylvestris*	Poplar, Lombardy	*Populus nigra* 'Italica'
Maple, Field	*Acer campestre*	Poplar, White	*Populus alba*
Maple, Montpelier	*Acer monspessulanum*	Potato	*Solanum tuberosum*
Maple, Norway	*Acer platanoides*	Potato Tree	*Solanum crispum*
Marigold	*Calendula* spp.	Primrose	*Primula vulgaris*
Marjoram	*Origanum vulgare*	Primula	*Primula* spp.
Marram	*Ammophila arenaria*	Privet, Garden	*Ligustrum ovalifolium*
Mat-grass	*Nardus stricta*	Privet, Wild	*Ligustrum vulgare*
Mayweed, Scented	*Matricaria chamomilla*	Quince	*Cydonia oblonga*
Mayweed, Scentless	*Tripleurospermum inodorum*	Radish, Sea	*Raphanus raphanistrum* sspp. *maritimus*
Meadowsweet	*Filipendula ulmaria*		
Melilot	*Melilotus* spp.	Ragwort, Common	*Jacobaea vulgaris*
Melilot, Tall	*Melilotus altissimus*	Ragwort, Oxford	*Senecio squalidus*
Mexican Orange	*Choisya ternata*	Raspberry	*Rubus idaeus*
Millet spp.	*Milium* spp.	Redshank	*Persicaria maculosa*
Mint	*Mentha* spp.	Reed, Common	*Phragmites australis*
Mint, Water	*Mentha aquatica*	Restharrow	*Ononis* spp.
Monk's-hood	*Aconitum napellus*	Rocket, Sea	*Cakile maritima*
Moor-grass, Purple	*Molinia caerulea*	Rock-rose, Common	*Helianthemum nummularium*
Mouse-ear	*Cerastium* spp.	Rose	*Rosa* spp.
Mouse-ear, Common	*Cerastium fontanum*	Rose, Burnet	*Rosa spinosissima*
Mouse-ear, Little	*Cerastium semidecandrum*	Rose, Japanese	*Rosa rugosa*
Mouse-ear, Sea	*Cerastium diffusum*	Rose-of-Sharon	*Hypericum calycinum*
Mouse-ear-hawkweed	*Pilosella officinarum*	Rowan	*Sorbus aucuparia*
Mugwort	*Artemisia vulgaris*	Rush	*Juncus* spp.
Mullein, Great	*Verbascum thapsus*	Rush, Jointed	*Juncus articulatus*
Mullein, var. 'milkshake'	*Verbascum chaixii*	Rush, Saltmarsh	*Juncus gerardii*
Nettle, Common	*Urtica dioica*	Rush, Sharp-flowered	*Juncus acutiflorus*
Oak	*Quercus* spp.	Rye-grass	*Lolium* spp.
Oak, Holm	*Quercus ilex*	Sage, Wood	*Teucrium scorodonia*
Oak, Pedunculate	*Quercus robur*	Sallows	*Salix* spp.
Oak, Turkey	*Quercus cerris*	Saltmarsh-grass, Common	*Puccinellia maritima*
Orache	*Atriplex* spp.	Saltwort, Prickly	*Salsola kali*
Orache, Babington's	*Atriplex glabriuscula*	Saxifrage, Mossy	*Saxifraga hypnoides*
Orache, Common	*Atriplex patula*	Scurvy-grass	*Cochlearia* spp.
Orache, Grass-leaved	*Atriplex littoralis*	Sea-buckthorn	*Hippophae rhamnoides*
Osier	*Salix viminalis*	Sea-milkwort	*Lysimachia maritima*
Pea, Garden	*Lathyrus oleraceus*	Sea-purslane	*Atriplex portulacoides*
Pear	*Pyrus communis*	Sea-rush	*Juncus maritimus*
Pignut	*Conopodium majus*	Sedge	*Carex* spp.
Pine, Corsican	*Pinus nigra* sspp. *laricio*	Sedge, Distant	*Carex distans*
Pine, Scots	*Pinus sylvestris*	Sedge, Glaucous	*Carex flacca*
Plane, London	*Platanus* x *hispanica*	Sheep's-fescue	*Festuca ovina*
Plantain, Ribwort	*Plantago lanceolata*	Shield-fern, Soft	*Polystichum setiferum*
Plantain, Sea	*Plantago maritima*	Silverweed	*Potentilla anserina*
Ploughman's Spikenard	*Inula conyzae*	Skimmia, Japanese	*Skimmia japonica*
Plum	*Prunus* spp.	Skullcap spp.	*Scutellaria* spp.
Plum, Wild	*Prunus domestica*	Skullcap	*Scutellaria galericulata*
Polypody	*Polypodium* spp.	Snowberry	*Symphoricarpos albus*

Vernacular name	Scientific name	Vernacular name	Scientific name
Soft-grass, Creeping	*Holcus mollis*	Tulip	*Tulipa* spp.
Soft-rush	*Juncus effusus*	Turf-moss, Springy	*Rhytidiadelphus squarrosus*
Sorrel	*Rumex* spp.	Tutsan	*Hypericum androsaemum*
Sorrel, Sheep's	*Rumex acetosella*	Valerian, Common	*Valeriana officinalis*
Sowthistle	*Sonchus* spp.	Vernal-grass, Sweet	*Anthoxanthum odoratum*
Spindle	*Euonymus europaeus*	Vetch	*Vicia* spp.
Spindle, Evergreen	*Euonymus japonicus*	Vetch, Bush	*Vicia sepium*
Spruce, Sitka	*Picea sitchensis*	Vetch, Common	*Vicia sativa*
Spurge	*Euphorbia* spp.	Vetch, Tufted	*Vicia cracca*
Spurge, Portland	*Euphorbia portlandica*	Vetchling, Meadow	*Lathyrus pratensis*
Spurge, Sea	*Euphorbia paralias*	Violet	*Viola* spp.
St John's-wort	*Hypericum* spp.	Viper's-bugloss	*Echium vulgare*
St John's-wort, Perforate	*Hypericum perforatum*	Water-dropwort, Hemlock	*Oenanthe crocata*
Stitchwort	*Stellaria* spp.	Water-dropwort, Parsley	*Oenanthe lachenalii*
Stitchwort, Greater	*Stellaria holostea*	Whitebeam	*Sorbus aria* agg.
Strawberry, Barren	*Potentilla sterilis*	Whitebeam, Swedish	*Sorbus intermedia*
Strawberry, Garden	*Fragaria ananassa*	Wild Service-tree	*Sorbus torminalis*
Strawberry, Wild	*Fragaria vesca*	Willow	*Salix* spp.
Sundew, Round-leaved	*Drosera rotundifolia*	Willow, Bay	*Salix pentandra*
Sweet-grass, Reed	*Glyceria maxima*	Willow, Creeping	*Salix repens*
Sycamore	*Acer pseudoplatanus*	Willow, Goat	*Salix caprea*
Tansy	*Tanacetum vulgare*	Willow, Grey	*Salix cinerea*
Tare, Hairy	*Ervilia hirsuta*	Willow, Weeping	*Salix babylonica*
Teasel	*Dipsacus fullonum*	Willow, White	*Salix alba*
Thistle	*Carduus* spp. or *Cirsium* spp.	Willowherb, Broad-leaved	*Epilobium montanum*
Thistle, Carline	*Carlina vulgaris*	Willowherb, Great	*Epilobium hirsutum*
Thistle, Creeping	*Cirsium arvense*	Willowherb, Hoary	*Epilobium parviflorum*
Thistle, Marsh	*Cirsium palustre*	Willowherb, Rosebay	*Chamaenerion angustifolium*
Thistle, Spear	*Cirsium vulgare*	Wood-rush, Field	*Luzula campestris*
Thrift	*Armeria maritima*	Wood-rush, Great	*Luzula sylvatica*
Thyme	*Thymus* spp.	Wood-rush, Hairy	*Luzula pilosa*
Thyme, Wild	*Thymus drucei*	Woundwort, Field	*Stachys arvensis*
Toadflax, Common	*Linaria vulgaris*	Woundwort, Hedge	*Stachys sylvatica*
Tobacco	*Nicotiana* spp.	Yarrow	*Achillea millefolium*
Tomato	*Solanum lycopersicum*	Yellow-rattle	*Rhinanthus minor*
Tormentil	*Potentilla erecta*	Yellow-wort	*Blackstonia perfoliata*
Traveller's-joy	*Clematis vitalba*	Yew	*Taxus baccata*

Appendix 5

Unaccepted published data
Research into previously published, mostly historic, records for inclusion in *The Moths of Lancashire* produced a list of records requiring reassessment. Any that failed to reach an acceptable level of accuracy are listed below. It was considered important to list such records to highlight that the authors were aware of them and that they had not been overlooked. The list is also considered beneficial to guide future research or further reassessment if felt advisable.

The specific reasons for non-inclusion in the county database are as follows:
1. Records that have subsequently been split into two or more separate species following taxonomic changes and where no specimens were located.
2. Dots on published maps of difficult to identify species which have no supporting data and where no other reference or records to the species are known in the county. In most cases these relate to dots on maps in *The Moths and Butterflies of Great Britain and Ireland* series. It has also been found that several of those listed for VC60 referred to localities in VC69 (formerly Westmorland, now part of Cumbria). Additionally, keys to some leaf mines used criteria that are no longer considered reliable without breeding and sometimes also dissection.
3. Records considered unreliable without a specimen available for examination.
4. Authoritative statements published in subsequent peer-reviewed journals or books which considered the records unreliable.

Should further information come to light with respect to any of these records, the authors and relevant Vice County Moth Recorders would welcome receiving the relevant information to enable, if needed, publication of such amendments.

Nepiculidae
4.018 *Stigmella ulmivora* (Fologne, 1860) 1879, Dutton, *Entomologist* 13: 105, J.B. Hodgkinson. Pair in cop on Aspen tree.
4.022 *S. regiella* (Herrich-Schäffer, 1855), Preston, J.B. Hodgkinson (Ellis, 1890).
4.029 *S. desperatella* (Frey, 1856). VC60 dot (Heath, 1976).
4.047 *S. splendidissimella* (Herrich-Schäffer, 1855), J.H. Threlfall, Lytham (Ellis, 1890).
4.079 *Zimmermannia atrifrontella* (Stainton, 1851) 1879, *Entomologist* 13: 66, J.H. Threlfall, det. H. Stainton, 'bred from jar containing only hawthorn leaves'.

Prodoxidae
9.007 *Lampronia pubicornis* (Haworth, 1828). Data supporting the VC59 and VC60 map dots in Heath (1976) are unknown. K. Bland (pers. comm., 2023) provided details of all data from these areas in the *Incurvariid provisional Atlas* (Bland, 1986) and all refer to a location a few hundred metres north of the VC60 border, in Cumbria. He added that discussion at the time of production of the 1976 maps suggested sightings from Hest Bank and Formby dunes (none of which have been located). The larval foodplant occurs on the coast and searches would be worthwhile.

Tineidae
12.005 *Infurcitinea albicomella* (Stainton, 1851) (under *T. confusella*), Morecambe J.H. Threlfall. See *Entomologist's Gazette* 73: 74–90.
12.029 *Tinea flavescentella* Haworth, 1828. Female from Liverpool, no date and identified by F.N. Pierce. This relates to *T. translucens* Meyrick 1917 (Robinson, 1979).
12.031 *T. columbariella* Wocke, 1877. VC59 dot and text (Heath & Emmet, 1985).
12.037 *Monopis weaverella* (Scott, 1858). VC59 dot (Heath & Emmet, 1985).
12.042 *M. fenestratella* (Heyden, 1863). House in Liverpool c.1904 (*Lancs & Cheshire Entomological Society* Vol. 29-1905) and Burnley 1909 (Mansbridge, 1940).

Gracillariidae
15.024 *Parornix loganella* (Stainton, 1848). Simonswood, 'beaten off Rowan' (Gregson, 1857), Ellis (1890) and Mansbridge (1940). VC60 dot (Heath & Emmet, 1985). All recent attempts to locate this species have drawn a blank.

15.026 *P. fagivora* (Frey, 1861). VC60 dot (Heath & Emmet, 1985).

15.035 *Phyllonorycter roboris* (Zeller, 1839). VC59 and VC60 dot (Heath & Emmet, 1985). VC60 record is from a location in VC69.

15.038 *P. kuhlweiniella* (Zeller, 1839). VC59 and VC60 dot (Heath & Emmet, 1985). VC60 record is from a location in VC69.

15.042 *P. muelleriella* (Zeller, 1839). VC60 dot (Heath & Emmet, 1985).

15.045 *P. mespilella* (Hübner, [1805]). 1846 R.S. Edleston, *Zoologist* 1846: 1220 and Stainton (1859) in VC59. Also, Mansbridge (1940); this site is in VC69.

15.047 *P. hostis* Triberti, 2007. VC59 and VC60 dot (Heath & Emmet, 1985).

15.055 *P. viminetorum* (Stainton, 1854). VC60 dot (Heath & Emmet, 1985).

15.059 *P. cavella* (Zeller, 1846). VC59 and VC60 dot (Heath & Emmet, 1985).

15.071 *P. nigrescentella* (Logan, 1851). VC60 dot (Heath & Emmet, 1985).

15.087 *P. comparella* (Duponchel, [1843]). 1930, Formby, Mansbridge (1940). VC59 and VC60 dot (Heath & Emmet, 1985). Specimen (no abdomen) in Manchester Museum dated 26 August 1930, Formby (W. Mansbridge) was not *P. comparella*.

Argyresthiidae
20.009 *Argyresthia aurulentella* Stainton, 1849. VC60 dot in Emmet (1996) may have been mistakenly entered for a VC69 record.

Oecophoridae
28.013 *Crassa tinctella* (Hübner, 1796). 1844, under *Borkhausenia tinctella* in Mansbridge (1940). Possible confusion with *C. unitella* – no specimen located.

Depressariidae
32.034 *Agonopterix cnicella* (Treitschke, 1832). VC59 dot, Emmet & Langmaid (2002). Formby W.A. Tyerman, bred from sallow in 1919, *Lancs & Cheshire Naturalist* xii: 293 and Mansbridge (1940). The larva of this species feeds only on Sea Holly.

32.037 *Depressaria pimpinellae* Zeller, 1839. Pilling Moss, late June 1865, J.B. Hodgkinson, larvae on *Anthriscus sylvestris*, the wrong foodplant for this species.

Cosmopterigidae
34.012 *Sorhagenia rhamniella* (Zeller, 1839). Under *Chrysoclista rhamniella* Zell., c.1889, Silverdale, J.H. Threlfall (Ellis, 1890). Could refer to one of three species.

Gelechiidae
35.002/3 *Syncopacma vorticella* Lytham 1865, Preston c.1889 (both J.B. Hodgkinson) and Crosby (W. Mansbridge). Not possible to get to species level without specimen.

35.007 *S. albipalpella* (Herrich-Schäffer, 1854). Two in 1918, Formby, W.A. Tyerman, verified F.N. Pierce (Mansbridge, 1920). See Emmet & Langmaid (2002).

35.080 *Aristotelia unicolorella* Dup. var. *immaculatella* Doug. Three records c.1889 to c.1940 (Mansbridge 1940). Also, see Emmet & Langmaid (2002): 78–79.

35.086? *Gelechia nigricostella* FvR, may refer to *Athrips tetrapunctella* (Thunberg, 1794). Hale Marsh (Gregson 1857). Not included in Ellis (1890) or Mansbridge (1940).

35.087/88 *Neofriseria peliella* (Treitschke, 1835). In Stainton (1859) vol 2: 332 Manchester, well before *N. singula* was described. Also, beyond the known range of either *Neofriseria* spp. See also under *N. suppeliella* Wals., Mansbridge (1940): 261, referring to record from Burnley in 1912.

35.135 *Caryocolum proxima* (Haworth, 1828) under *maculiferella*. Given as Local; records from Preston district, Lytham July 1865, Swinton (Stott); Lytham, J.H. Threlfall, bred 1888 from *Cerastium semidecandrum*.

35.138 *C. junctella* (Douglas, 1851). In Stainton (1859) Manchester area, Mansbridge (1940) not uncommon on the sand hills. A 1922 specimen from Ainsdale by W. Mansbridge was dissected in 2022 and was found to be *C. fraternella*.

35.139/40 *C. hubneri* (Haworth, 1828). In Stainton (1859) Manchester district, under *Gelechia hubneri*, which is a synonym of *C. kroesmanniella*. No specimen known.

Batrachedridae
36.002 *Batrachedra pinicolella* (Zeller, 1839). See main Species section, **36.0019 *B. confusella*** (Berggren *et al.* 2022).

Coleophoridae
37.004 *Coleophora albella* (Thunberg, 1788). Under *Porrectaria leucapenella*, R.S. Edleston, Cheetham Hill June 1846, swampy fields, abundant. *Zoologist* 1846: 1220.

37.024 *C. vitisella* Gregson 1856. In Stainton (1859), on moors near Manchester. Almost certainly refers to Staley Brushes in VC58 (Gregson, 1857).

37.043 *C. niveicostella* Zeller 1839. In Stainton (1859) Manchester district.

37.075 *C. salinella* Stainton, 1859. Fleetwood J.B. Hodgkinson (Ellis, 1890).

37.081 *C. therinella* Tengström 1848. In Ellis (1890) and Mansbridge (1940). A synonym of *C. peribenanderi* Toll, 1943.

Elachistidae
38.003 *Stephensia brunnichella* (Linnaeus, 1767). VC60 dot in Emmet (1996).
38.010 *Elachista bedellella* (Sircom, 1848). In Byerley (1854) where given as common in the Liverpool district, but not repeated elsewhere. A VC60 dot in Emmet (1996).
38.012 *E. obliquella* Stainton, 1854. All early records considered unreliable.

Momphidae
40.005 *Mompha divisella* Herrich-Schäffer, 1854. VC60 dot in Emmet & Langmaid (2002), believed to relate to a site in VC69.

Pterophoridae
45.005 Goldenrod Plume *Platyptilia calodactyla* ([Denis & Schiffermüller], 1775). W. Mansbridge, 1922, Formby; accepted by Michaelis (1957). No specimen in World Museum and considered unconfirmed.
45.025 Small Plume *Oxyptilus parvidactyla* (Haworth, 1811). In Byerley (1854) as 'taken but seldom amongst gorse'. Not mentioned by subsequent authors.

Tortricidae
49.014 Brown Oak Tortrix *Archips crataegana* (Hübner, [1799]). By W. Mansbridge c.1908 in *Lancs & Cheshire Ent. Soc. Journal* 33: 29. Later, Mansbridge (1940) makes no reference to this record.
49.027 *Pandemis dumetana* (Treitschke, 1835). Kirkby Moss c.1904, W. Mansbridge, in *Entomologist* 38: 115. Mentioned in Bradley *et al.* (1973), but had been previously withdrawn by Mansbridge in the *Lancs & Cheshire Ent. Soc. Journal* (1907) 30th Annual Report and Proceedings, page 14.
49.068 *Acleris maccana* (Treitschke, 1835). In Stainton (1859) Lancashire (Manchester). Bradley *et al.* (1973) note 'unconfirmed and very dubious'.
49.148 *Apotomis lineana* ([Denis & Schiffermüller], 1775). The 'Lancashire' record, under *hartmanniana* L. (*scriptana* Hub.) in Bradley *et al.* (1979) is from Windermere (VC69). *Argyroploce corticana* Hübner is listed in Ellis (1890) as 'local, from a few sites', and in Mansbridge (1940) as 'generally common amongst birch'. This is a synonym of *A. lineana*, which is a willow feeder and of *Zeiraphera isertana* (an oak feeder). It is uncertain as to which species this name refers, although it is listed separately from *A. betuletana* and *A. turbidana*.
49.177 *Argyroploce arbutella* (Linnaeus, 1758). Mentioned as from Lancashire northwards in Meyrick (1928). Bradley *et al.* (1979) note that it is not mentioned in Mansbridge (1940) and that the origin of the record is unknown.
49.205 *Ancylis comptana* (Frölich, 1828). In Byerley (1854) attributed to N. Cooke, on moors and mosses. No further reference is made to this by later authors.
49.250 *Epinotia nigricana* (Herrich-Schäffer, 1851). Mentioned as occurring in Lancashire in Bradley *et al.* (1979). No supporting data has been found.
49.256 *E. cinereana* (Haworth, 1811). Under *Grapholita cinerana* Haw., Preston J.B. Hodgkinson.
49.262 *Phaneta pauperana* (Duponchel, [1843]). Records considered unconfirmed (Bradley *et al.*, 1979).
49.267 *Eucosma fulvana* Stephens, 1834. Lancaster area c.1940 C.H. Forsythe.
49.270 *E. aemulana* (Schläger, 1849). In Mansbridge (1940), A.E. Wright, Silverdale.
49.312 *Dichrorampha consortana* Stephens, 1852. Manchester (Stainton, 1859), Preston, J.B. Hodgkinson, *Entomologist's Monthly Magazine* 9:67. No mention in Bradley *et al.* (1979).
49.365 *Pammene albuginana* (Guenée, 1845). Caton 7 August 1986 E. Emmett. Genitalia slide in World Museum examined in 2022 - not this species.

Sesiidae
52.011 Red-belted Clearwing *Synanthedon myopaeformis* (Borkhausen, 1789). Noted in Lancashire (Skinner, 1983) and Waring & Townsend (2003). This is considered to refer to a record from Grange, now modern Cumbria, in 1880 (Mansbridge, 1940).

Zygaenidae
54.007 New Forest Burnet *Zygaena viciae* ([Denis & Schiffermüller], 1775). Under *Z. meliloti* Esper. Manchester, and on railway bank near Scorton, by J.B. Hodgkinson (Ellis, 1890).
54.010 Five-spot Burnet *Zygaena trifolii* (Esper, 1783). Hale Marsh (Gregson, 1857) considered unconfirmed at a county and national level.

Pyralidae
62.015 *Delplanqueia dilutella* ([Denis & Schiffermüller], 1775). 'In an old lane at Prescot' (Gregson, 1857). Species split in 2015 (*D. dilutella* and *D. inscriptella*).

62.019 *Pima boisduvaliella* (Guenée, 1845). W. Mansbridge, Formby June 1921. Britten (c.1950) noted it as a misidentification of *Anerastia lotella* (Hübner, [1813]).

62.033 *Acrobasis tumidana* ([Denis & Schiffermüller], 1775). Chat Moss by J. Chappell and Britten adds Pendle Hill area, A.E. Wright. No specimen located.

62.056 *Homoeosoma nimbella* (Duponchel, 1837). Ellis (1890), Mansbridge (1940). Morecambe (J.B. Hodgkinson) and Silverdale. No specimen located.

Crambidae

63.004 *Pyrausta sanguinalis* (Linnaeus, 1767). Britten (c.1950), unpublished card index, notes it as 'extinct on Crosby-Formby sandhills'. Neither Ellis (1890) or Mansbridge (1940) mention it in as occurring in Lancashire. Goater, 1986 lists Lancashire (no details), possibly repeated from Beirne (1952). All references to Lancashire are believed to relate to the colony on the Wallasey sandhills (VC58).

63.109 *Pediasia contaminella* (Hübner, 1796). Saltmarshes nr. Preston (Ellis, 1890); considered to relate to *P. aridella* (Thunberg, 1788).

63.1173 *Parapoynx crisonalis* (Walker, 1859). Middleton, to light 5.8.1951 M. Morris.

Endromidae

67.001 Kentish Glory *Endromis versicolora* (Linnaeus, 1758). *Entomologist's Weekly Intelligencer* 1: 28, Withnell Birch Clough, 20 April 1856, C.S. Gregson. Report of one, caught under unlikely circumstances. Not listed in Ellis (1890) or Mansbridge (1940).

Sphingidae

69.0151 White-lined Hawk-moth *Hyles lineata* (Fabricius, 1775). Once, Liverpool 1854. Considered referable to *H. livornica* (Esper, [1804]), previously known as *H. lineata livornica*.

Geometridae

70.005 Silky Wave *Idaea dilutaria* (Hübner, [1799]). Assumed error in Ellis (1890) and Mansbridge (1940) when translating synonyms.

70.007 Isle of Wight Wave *Idaea humiliata* (Hufnagel, 1767). Assumed error in Ellis (1890) and Mansbridge (1940) when translating synonyms.

70.012 Treble Brown Spot *Idaea trigeminata* (Haworth, 1809). In Ellis (1890), by J. Chappell, Barlow Moor and Irlam. All claimed since c.1950 have proved to be the dark-edged form of Single-dotted Wave *I. dimidiata* (Hufnagel, 1767).

70.021 Lace Border *Scopula ornata* (Scopoli, 1763). Ellis (1890) 'rare on Chat Moss', by J. Chappell.

70.043 Chalk Carpet *Scotopteryx bipunctaria* ([Denis & Schiffermüller], 1775). In *Lancashire & Cheshire Ent. Soc. Journal* 37: 18. Specimen displayed at a LCES exhibition on 15 December 1913 by R. Wilding, West Derby 1880. Also, Yealand, Silverdale 4 August 1926 by H.L. Burrows (N.L. Birkett Card Index). Neither listed in Mansbridge (1940).

70.057 Ruddy Carpet *Catarhoe rubidata* ([Denis & Schiffermüller], 1775). Gregson, 1857. Considered a typo for Ruddy Highflyer *Hydriomena ruberata* (Freyer, [1831]).

70.071 Yellow-ringed Carpet *Entephria flavicinctata* (Hübner, [1813]). Mansbridge (1940) 'scarce in July, C.H. Forsythe'. Considered misidentification.

70.122 Scarce Tissue *Rheumaptera cervinalis* (Scopoli, 1763). Formby, September 1948, M.J. Leach and G. de C. Fraser. Date suggests misidentification.

70.126 Small Waved Umber *Horisme vitalbata* ([Denis & Schiffermüller], 1775). J.B. Hodgkinson, Preston area, *Zoologist* 3: 1141. Considered very unlikely.

70.135 Heath Rivulet *Perizoma minorata* (Treitschke, 1828). Gait Barrows 18 August 1984 by E. E. Emmett (Emmett, 1991). Considered very unlikely.

70.136 Pretty Pinion *Perizoma blandiata* ([Denis & Schiffermüller], 1775). Ellis (1890) per J.B. Hodgkinson; from Accrington but no date or recorder detailed.

70.211 Peacock Moth *Macaria notata* (Linnaeus, 1758). VC59 dot in Randle *et al.* (2019) submitted in error and relates to Sharp-angled Peacock (*M. alternata*). Britten (c.1950) notes 'old record from Lancaster district', thought to refer to a VC69 record.

70.228 Dark Bordered Beauty *Epione vespertaria* (Linnaeus, 1767). Photo of specimen in Rothschild, Cockayne and Kettlewell coll. website, NHM London. Data label 'Burnley /12' (assumed 1912) by W.G. Clutten. Unlikely to be of Lancashire origin.

70.230 Orange Moth *Angerona prunaria* (Linnaeus, 1758). One old record from Prescot, in Britten (c.1950). No date, recorder or reference. Also, Carnforth, Murray, 1893 (specimen in Rothschild, Cockayne and Kettlewell coll. website, NHM London). First considered unconfirmed. Second very unlikely to be of Lancashire origin.

70.232 Large Thorn *Ennomos autumnaria* (Werneburg, 1859). In Rothschild, Cockayne and Kettlewell coll. website, NHM London and also in T. Baxter coll. (J. Steeden pers. comm.) indicated bred, St Annes. Both unlikely to be of Lancashire origin.

70.267 Great Oak Beauty *Hypomecis roboraria* ([Denis & Schiffermüller], 1775). In Mansbridge (1940), Corporation Wood and Quernmore Woods, nr. Lancaster C.H. Forsythe. Considered misidentifications.

70.268 Pale Oak Beauty *Hypomecis punctinalis* (Scopoli, 1763). Britten (c.1950), 'old records from Lancaster district' but no more details. Considered misidentifications.

70.293 Straw Belle *Aspitates gilvaria* ([Denis & Schiffermüller], 1775). Britten (c.1950), 'old record from Hale' but no more details. Considered misidentification.

70.301 Essex Emerald *Thetidia smaragdaria* (Fabricius, 1787). E.C. Buxton, wood on edge of Chat Moss 13 July 1850, *Zoologist* 8: 2882. A most unlikely record.

70.306 Small Grass Emerald *Chlorissa viridata* (Linnaeus, 1758). c.1983, Gait Barrows site list, E.E. Emmett (1991). No recorder or date known. Occurs on VC69 mosses.

Erebidae

72.014 Reed Tussock *Laelia coenosa* (Hübner, [1808]). Altcar, T. Townley (Gregson, 1857). Not listed in Mansbridge (1940) or elsewhere.

72.021 Water Ermine *Spilosoma urticae* (Esper, 1789). Given as 'very local at Oakcliffe Hall', G. Loxham (Mansbridge, 1940) and (South, 1948). Considered misidentification.

72.027 Cream-spot Tiger *Arctia villica* (Linnaeus, 1758). G. Loxham, Lancaster district, Ellis (1890). Mansbridge (1940) and South (1948). Unlikely; requires further research.

72.047 Hoary Footman *Eilema caniola* (Hübner, [1808]). Rothschild, Cockayne & Kettlewell coll. NHM London, female labelled Rixton Moss July 1892 J. Collins. Dissected by A. Zilli (NHM) on 23 September 2022 and reidentified as *E. complana*.

72.052 Dotted Fan-foot *Macrochilo cribrumalis* (Hübner, 1793) under *Herminia cribralis*, at or near Hale by Mr. Nixon c.1857. Considered misidentification.

72.056 Common Fan-foot *Pechipogo strigilata* (Linnaeus, 1758). Gait Barrows 1984 week 27 - in site list by E.E. Emmett, 1991. Considered misidentification.

72.081 Dark Crimson Underwing *Catocala sponsa* (Linnaeus, 1767). *Lancs & Cheshire Fauna Committee*, Annual Lep. Report of 1949 noted '1948, found by schoolboy and identified by Baron De Worms'. Leech and Michaelis (1957), with different details, state 'not indigenous to the area and possibly imported by rail or car'.

Noctuidae

73.263 Light Brocade *Lacanobia w-latinum* (Hufnagel, 1766). Silverdale 1922 W.P. Stocks & Morecambe 1959 C.J. Goodall (not in coll.), considered misidentifications.

73.344 Lunar Yellow Underwing *Noctua orbona* (Hufnagel, 1766). Historic records under (synonym) *N. orbona* refer to *N. comes*. Two 1950s records are unconfirmed.

Questionable Records

Appendix B 'Questionable Records' (Agassiz *et al.*, 2013) lists unconfirmed or erroneous British records. Those with a Lancashire origin are:

B16 *Depressaria emeritella* Stainton, 1849. 'Warrington area' (Gregson, 1857). These are considered very doubtful (Emmet & Langmaid, 2002).

B26 *Auximobasis normalis* (Meyrick, 1918). A singleton taken on the wall of the Liverpool Docks in July 1921 is the only British record (noted as September within *Lancashire and Cheshire Fauna Society report*, W. Mansbridge, 1924). Specimen not traced and the record cannot be confirmed (Emmet & Langmaid, 2002).

B103 Tawny Shoulder *Feltia subterranea* (Fabricius, 1794). c.1850, under *Agrotis annexa* Tr., from a well-sinker's window (on sandhills), in Mr Edleston's collection. Considered small variety of *A. ipsilon* (Hufnagel, 1766), in Heath & Emmet (1979).

Appendix 6

Aggregated macro-moth species

The following maps depict those macro-moth records submitted as aggregates, where species cannot be determined by external characters. The species mapped here are:

70.107x November Moth agg. (3 x *Epirrita* spp.) – Autumnal Moth, November Moth and Pale November Moth.

73.038x Dagger agg. (2 x *Acronicta* spp.) – Dark Dagger and Grey Dagger.

73.062x Copper Underwing agg. (2 x *Amphipyra* spp.) – Copper Underwing and Svensson's Copper Underwing.

73.127x Ear Moth agg. (4 x *Amphipoea* spp.) – Crinan Ear, Ear Moth, Large Ear and Saltern Ear.

73.169x Common Rustic agg. (2 x *Mesapamea* spp.) – Common Rustic and Lesser Common Rustic.

73.173x Marbled Minor agg. (3 x *Oligia* spp.) – Marbled Minor, Rufous Minor and Tawny Marbled Minor.

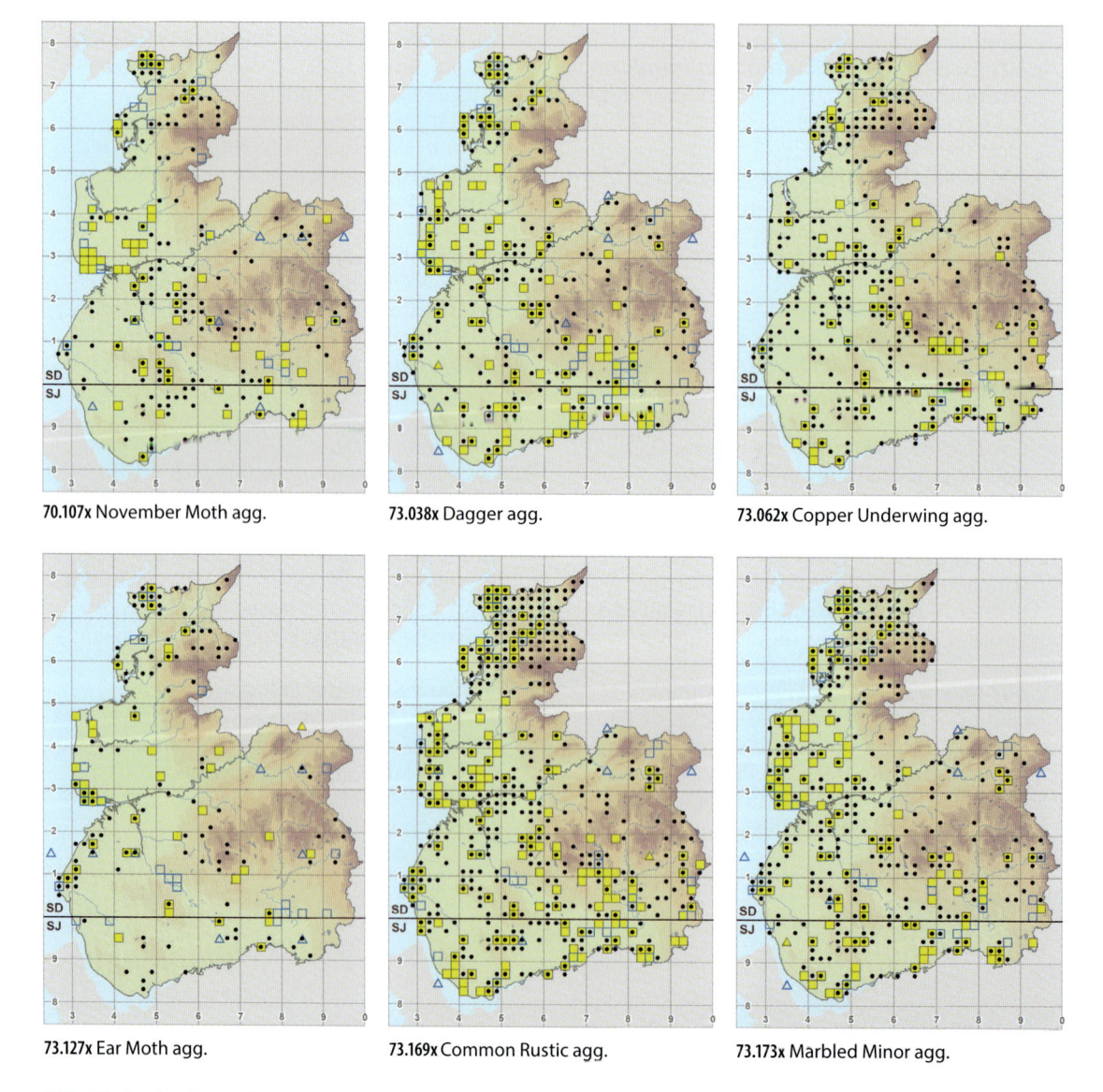

70.107x November Moth agg.

73.038x Dagger agg.

73.062x Copper Underwing agg.

73.127x Ear Moth agg.

73.169x Common Rustic agg.

73.173x Marbled Minor agg.

Appendix 7

Recording Moths

Much has changed in the field of moth recording since the first known Lancashire moth record, that of *Euclemensia woodiella* in 1829. Robert Cribb netted and pinned all that he could find, without consideration as to how this may affect the survival of the species. In this case, our subsequent knowledge that this was an adventive American species, probably introduced with timber, meant the colony's days were numbered anyway. Other techniques frequently used by early collectors of Lepidoptera included smoking – sending smoke into thatch, vegetation, etc, to induce the moths to fly out, digging for pupae – around tree roots and elsewhere, checking around gas lamps, 'sugaring' – applying a rich paste (often full of treacle, and alcohol too), 'sallowing' – netting the moths attracted to the spring blooms of willow trees, 'assembling' – involving using a freshly emerged female moth to attract males of the same species, and beating and sweeping vegetation, primarily in search of the larval stages.

Widespread access to electric powered light traps, from the 1950s onwards, has revolutionised moth recording, but has also tempted us to become less mobile in doing so. The instant success that can be gained by plugging in a device behind the back door, can at times be so spectacular that the rich rewards to be gained from searching further afield in unfamiliar habitats and by using some of these old techniques, are often neglected.

Light trapping

The majority of moth species are attracted to light, although there are exceptions, primarily although not exclusively, among the micro-lepidoptera. This attraction was noted by the Victorians recording at candle-light or with Tilley lamps, and by those early 20th century lepidopterists who used the headlights of the first motor-cars as primitive moth traps.

Heath trap, Wigan

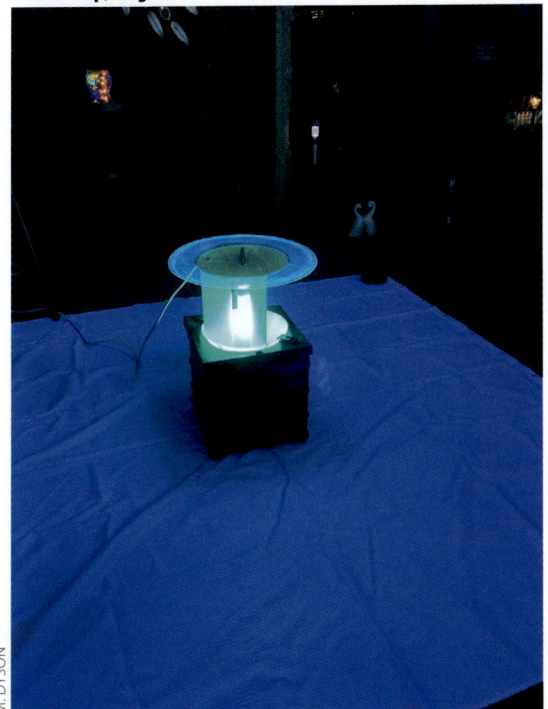

Skinner trap, Edenfield

Robinson trap, Rishton

Almost any light source will attract moths, but those with a high proportion of ultra-violet (UV) light will be most effective. To this end, mercury vapour (MV), actinic lights, and recently, various combinations of LED lights, have been used with traps of various types. There are Heath traps, Robinson traps, Skinner traps, and others, all using different constructions in an attempt to retain the moths attracted to the light, usually underneath the eggboxes placed within.

MV lights are extremely bright, to human and to moth eyes. Although very effective, their days are numbered as they are being gradually phased out in line with the Restriction of Hazardous Substances Directive. Actinic lights have now largely replaced MV lights, and are the standard bulb available to a purchaser of a new moth trap. Actinic lights give off UV light, meaning they are still effective in attracting many moths, but get less hot than MV lights, so less prone to exploding when wet, and are far less bright to the human eye. They are therefore more suitable for public displays, and for recorders wishing to minimise disturbance to their neighbours.

Positioning the trap is important too; keeping it out of overly sheltered spots where it may be less obvious to moths, but also avoiding the early morning sunlight that will encourage moths to disperse before they have been recorded, and later in the day, may even cause the demise of some of the catch. Bird predation is also a potential hazard if the trap is not checked at first light, particularly if regularly trapping at the same site. Placing the trap against a wall and on a white sheet or other similar surface, helps ensure that those moths resting outside the trap are not missed. One unusual location is that of Stuart Marsden's in Chorlton, Manchester. The Skinner trap is placed on top of a bay window overlooking a main road. Its elevated, south-facing position likely aiding the successful catches, which in recent years have included Crescent Dart, Garden Dart, Convolvulus Hawk-moth and Large Twin-spot Carpet.

The catch needs to be processed. All species should be documented, preferably with the number of individuals. If identification is uncertain, it is strongly advised that the moth should be retained until the recorder has had chance to discuss with an expert, as further investigation may be needed to confirm a significant county record. There are examples where potential county firsts have had to go unrecorded, even with a photograph, due to premature release of the moth. The records thus obtained should be sent to the CMR to form a permanent part of the Lancashire moth data. Once

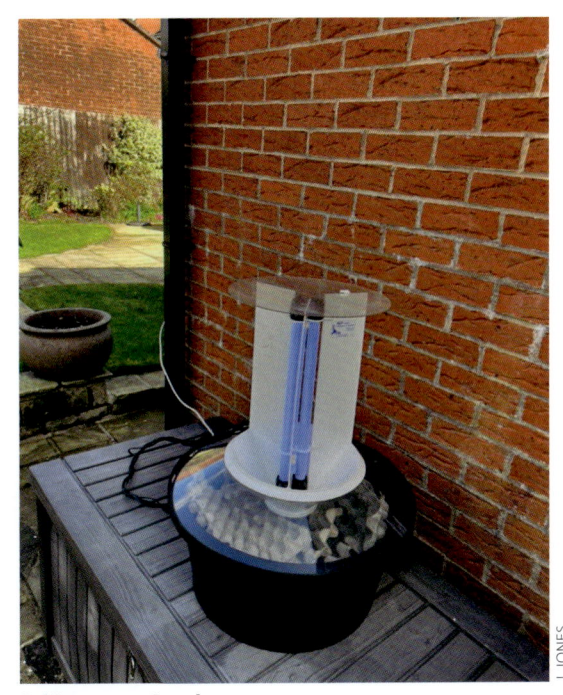

Skinner trap, Accrington

Robinson trap, Langho

identity is confirmed, moths should be released as soon as possible in suitable habitat close to the trap site.

It is possible on a muggy night in July, for a moth trap to record close on a hundred species. Kevin McCabe had a very special night on 18 July 2017, when he trapped 167 different species in his garden trap in Flixton.

Wonderful as the results of garden trapping may be, it has limitations. It doesn't often tell us much about the moth's preferred habitat, as individuals can be attracted from some distance away, and in some cases are also wind-borne, meaning species found in specialised habitats can surprisingly turn up in a trap a few miles away, for instance coastal and saltmarsh species recorded ten miles or so inland. 'The Manchester mosses are no more than 3–5 kilometres away, to the west of my garden and in the early years, moths were coming to the trap which I considered moss-land species. These included Neglected Rustic, Broom, Heath Rustic, Golden-rod Brindle and Marsh Oblique-barred. I recently caught my first Map-winged Swift in 2017, probably a wanderer from the moss-land' (McCabe, 2020). There is also an issue of diminishing returns, where after a few years of trapping at home, new species can become quite a rarity.

Pheromone lure trap, Chorlton

The use of light and sheet, often at remote car lay-bys (Marsh, 2022) and battery-operated traps, all allow recorders to leave their gardens and venture into new locations and different habitats, such as oak woodland, moorland and some of the specialist locations mentioned in the habitats chapter, always ensuring the approval of the landowner is first obtained.

The Moths of Lancashire **633**

Species recorded at pheromone lures in Lancashire

Nemaxera betulinella	TIP	*Pammene aurita*	AND, CUL, HOR, TIP
Nemapogon cloacella	FUN	*Pammene aurana*	SKI
Psychoides filicivora	TIP	Goat Moth	COS
Phyllonorycter blancardella	STR	Forester	Forester lure
Phyllonorycter trifasciella	ARG	Lunar Hornet Moth	LUN
Anthophila fabriciana	LUN	Large Red-belted Clearwing	CUL
Dichrorampha acuminatana	FUN	Red-tipped Clearwing	FOR
Cydia pomonella	STR	Red-belted Clearwing	CUL
Grapholita lobarzewskii	SKI	Yellow-legged Clearwing	AND, HOR, VES
Grapholita funebrana	ARG, FUN	Currant Clearwing	TIP
Grapholita tenebrosana	ARG, FUN	Six-belted Clearwing	API
Grapholita janthinana	MYO	Emperor Moth	EMP
Pammene giganteana	FUN, MOL	Barred Tooth-striped	Barred Tooth-striped lure
Pammene argyrana	ARG	Early Tooth-striped	Barred Tooth-striped lure
Pammene suspectana	FUN	Silver Y	NI
Pammene regiana	AND	Uncertain	AND

Pheromone lures

At present, still quite a limited part of moth recording, but as more and more artificial pheromones are devised, there is the potential for this technique to expand. Based upon the pheromones produced by female moths in nature, the use of these lures can be extremely successful, both as a method for searching for specific species, such as the Six-belted Clearwing, first located at Liverpool Festival Gardens in 2019 by P. Hopkins, or by studying the 'by-catch', i.e., those species other than the target for which the pheromone was developed. Examples include the two *Pammene* species, *giganteana* and *suspectana*, both attracted to the Plum Fruit Moth *Grapholita funebrana* lure (FUN), neither of which has ever been recorded in the county without the aid of lures.

The pheromone is impregnated into a 'bung', which can be bought from entomological (or pest-control!) suppliers. The bung may be placed in a netted bag and hung from foliage in a likely habitat. Alternatively, it may be placed in a pheromone trap, where, on a similar principle to a light-trap, the moth enters through a funnel-like structure, attracted by the pheromone lure in a cage at the top of the trap, and is retained for examination. Individual lures should only be used for a short period; up to 30 minutes should be sufficient in suitable conditions for most lures. Pheromone traps should certainly not be left unattended in direct sunlight.

Some of the species caught in Lancashire using this method are listed in the table above, together with the lures to which they have been attracted. Not all of these lures are commercially available.

Butterfly Conservation have produced guidance on the use of pheromones in moth recording. These guidelines can be found on

Herald, attracted to 'sugar' at Southport, 20 February 2023

Least Minor larval search, Crag House Allotment 2022

the Lancashire Moths Website/Guidance for Recorders page (https://www.lancashiremoths.co.uk/guidelines-for-recorders).

It should be remembered that on some specific sites, particularly SSSIs and National Nature Reserves, the use of pheromone lures requires prior permission from the landowner or manager.

Sugaring

This was a popular method amongst Victorian entomologists. The technique involved preparing a sweet and often intoxicating syrup, that could be pasted on to fence posts, tree trunks etc, to attract a variety of moths, mainly of the Noctuidae family. Common ingredients included molasses, rotten bananas, red wine and spirits. Gregson (1857) reports capturing a Lesser-spotted Pinion 'on sugar in a young Elm plantation between Hightown and Sephton'. This remains the sole Lancashire record for this species.

Excellent accounts of this technique were outlined by Steve Priestley (2018) and Andrew Barker (2019) in the Lancashire Moth Group Newsletters 35 and 36. Priestley, along with Eddie Langrish at Euxton, reported great success in VC59 during the 2010s, using a mixture of Guinness, sugar and black treacle, 'also very good to eat on a cold night but we always left some for the moths!' Those attracted at Euxton included '101 Satellite and 94 Chestnut on the same night in October 2011', plus hundreds more Copper Underwing (agg.).

Alternatively, Andrew Barker recommended an alcohol-free mixture of 1 jar of bramble jam, ½ a 250g box of molasses and 1 cup of Vimto, to make an irresistible compound; a more sober preparation for the moths, not to mention their recorders. He found this brew attracted large numbers of Copper Underwing (agg.), plus Large Yellow Underwing, Common Rustic (agg.), Red Underwing (often coming in later), Old Lady, Setaceous Hebrew Character, Dog's Tooth, Angle Shades, Dark Arches, Flame Shoulder, Flame Carpet, Mouse Moth, Dun-bar and Light Brown Apple Moth.

Fieldwork

This is somewhat of a catch-all term for searching for moths out in their habitat, rather than using attractants to bring them to you, and it can be done at any time of year. Fieldwork often involves searches for the early stages of moths, and can lead to interesting and sometimes novel findings. The photograph above shows Justine Patton, with the authors, engaging in a successful larval search for Least Minor in stems of Blue-moor Grass at Crag House Allotment, the first to be found in the UK.

Larvae of a number of other species can also be found by opening plant stems at the appropriate time of year (different for each species). Examples include Thistle Ermine in thistle and burdock stems, *Cochylichroa atricapitana* and *Epiblema costipunctana* in ragwort stems, *Endothenia*

Stigmella speciosa leaf-mine, Ainsdale 2022

Grapholita tenebrosana feeding in rose hips, Formby 2022

Coleophora juncicolella case, Birk Bank 2022

nigricostana in Hedge Woundwort stems, Tansy Plume in Tansy stems and Bulrush Wainscot in Bulrush stems. Others can be found within conspicuous galls in the plant stem, such as Hemp Agrimony Plume on Hemp Agrimony and *Mompha sturnipennella* on Rosebay Willowherb.

Numerous larvae can be found by searching through foliage and investigating leaves folded and twisted into a spinning secured with silk. Many of these will contain micro-moth larvae (unless

Some suggested targets for fieldwork

Species	Details	When to look
Eriocraniidae species on birch	Six species mine birch leaves. With care, all can be identified from a combination of leaf-mine and larva.	April-June
Heliozela sericiella	Oval cut-out with short mine at base of oak leaf.	June-October
Phyllonorycter lantanella	Leaf-mines on Laurustinus (*Viburnum tinus*).	All year round
Limnaecia phragmitella	Larvae feed gregariously in Reedmace (bulrush) heads. Presence indicated by spun white down hanging conspicuously from seed-head.	September-May. Easiest to find in spring.
Metzneria lappella	Larvae, and subsequent pupae, in burdock heads.	September-May
Grapholita tenebrosana	Collect rose hips and await larval emergence.	August-October
Phycitodes maritima	Larvae feed gregariously amongst spun ragwort flowers.	July-September
Barred Rivulet	Place Red Bartsia stems on paper and wait for frass to appear a few days later, indicating larvae present.	August-September
Triple-spotted Pug	Larvae on flowers and seeds of Wild Angelica.	August-early October
Sallow, Brick, Red-line Quaker, etc.	Collect willow catkins to look for larvae.	April-May
Lychnis, Sandy Carpet, etc.	Larvae in seed-heads of Red Campion.	June-August

Belted Beauty search, Potts Corner 2022

already vacated), such as many members of the Tortricidae. Others will house macro-moth larvae, such as Yellow Horned on birch, July Highflyer and Red-line Quaker on various willows, May Highflyer on Alder, and Oak Nycteoline on oak. Other larvae require an understanding of their feeding habits to detect their presence. For instance, Grass Rivulet and *Anania fuscalis* larvae can be found in the seeds of Yellow-rattle, Foxglove Pug and Toadflax Pug larvae can be found in the flowers that bear their name, Barred Rivulet larvae feed within the fruits of Red Bartsia and later on the seeds, and various *Cydia* and *Grapholita* larvae can be found in seed-pods and within fruit, such as the Plum Fruit Moth, whose larvae can be found in sloes and plums in late summer and early autumn.

Some other species, almost all micro-moths, feed within the leaf itself. Their presence is displayed by a leaf-mine, caused by their feeding techniques and the deposition of frass within the leaf, or in some cases, expelled from the leaf. The majority of these species will start and complete their feeding within a single leaf. The appearance of the mine, combined with host plant, timing and larval appearance, is often the best way to identify this species. Additionally, the adults of these species are often highly elusive with little evidence of attraction to light. Searching for leaf-mines can be done all year round, but is particularly successful in autumn. A significant number of species, primarily amongst the Nepticulidae and the Gracillariidae, have only been recorded in the county as leaf-mines. Examples include *Stigmella speciosa* on Sycamore and *Phyllonorycter schreberella* on elm.

Another old technique that can still pay dividends is beating for larvae. This involves placing a tray or upside-down umbrella beneath a branch and giving it a sharp tap, doing as little damage to the plant as possible. This method resulted in finds of Shoulder Stripe, Streamer and Barred Yellow larvae on rose in 2022. At times, beating can be very productive, although the numbers obtained by some of the Victorian entomologists have been very rarely reproduced in recent years. Perhaps the only record we have for Poplar Lutestring was of 'clouds of larvae' dislodged by beating Aspen at Ribchester in 1879 (Hodgkinson, 1880). Similarly, Kay (1876) reported beating 50 Small Autumnal Moth larvae from heather at Turton.

Other species can also be obtained via sweeping. The method involves passing a sweep net through low-growing vegetation, such as grasses and heather. Species obtained by this manner on heather have included larval cases of *Coleophora juncicolella* and *C. pyrrhulipennella*, larvae of *Neofaculta ericetella*, Neglected Rustic and Heath Rustic, as well as adults of *Amphisbatis incongruella* and *Prolita sexpunctella*.

Some interesting targets for fieldwork, that may potentially produce some new tetrad records, are listed in the table on p. 636.

Not all fieldwork involves searching for early stages. Some species do not travel far from their preferred habitat and are rarely trapped at light. One such is the macro-moth Belted Beauty. As described on p.451, this is a very special species in the county, and every year a targeted search is held for the adults in their saltmarsh habitat. These searches were initiated by Steve Palmer in 2002 and are now led by Ryan Clark. Details of the annual search, generally in early April, can be found on the Lancashire Moth Group website. The 2023 search, held on 10 April, recorded 57 males and 39 (flightless) females, the highest count since 2017.

'Dusking' is another aspect of fieldwork, and refers to walking around the likely habitat in the hour or so before night, perhaps with a head torch, and disturbing the vegetation. Many micro-moths and geometers will take to the wing and can be netted for further examination. Brian Hancock (2018) described dusking around specific foodplants as a particularly effective technique for locating rarely recorded and very local Pug moths. Within ten miles of his home in the north of the county, he has been able to catch Netted, Shaded, Thyme, Haworth's, Bleached, Pimpinel, Sloe, Valerian, Triple-spotted, Maple and Bilberry Pug in recent years, 'all without any light-traps.' Warm, still evenings are preferential for a successful dusking session.

All the techniques used above, including targeted searches for scarce species in their specialised habitats, have been used to record the 1,570 moths listed in this book. No doubt in the forthcoming years, these methods will produce a few more additions to the county list.

Bibliography and References

Bibliography and references (combined)

Agassiz, D. J. L., Beavan, S. D. & Heckford, R. J. (2013). *A Checklist of the Lepidoptera of the British Isles*. Royal Entomological Society, St Albans.

Agassiz, D. J. L., Beavan, S. D. & Heckford, R. J. (2022) Fifth update to *A Checklist of the Lepidoptera of the British Isles*, 2013 on account of subsequently published data. *Entomologist's Record and Journal of Variation*. Vol. 134. pp. 1–5.

Agassiz, D. J. L., Beavan, S. D. & Heckford, R. J. (2024) Seventh update to *A Checklist of the Lepidoptera of the British Isles*, 2013 on account of subsequently published and received data. *Entomologist's Record and Journal of Variation*. Vol. 136. pp. 1–6.

Anonymous. M.A.F.F. *IDENT Log books* (36). (1950–1981) Entomology Department Archives. World Museum, William Brown Street, Liverpool. MAFF Intercept Logs, handwritten & unpublished (a few example extracts provided by R. Walker in 2022, courtesy of World Museum).

Atkinson, H. W. (1901) *Rossall Fauna and Flora 1901* (copy supplied to SMP courtesy of Rossall School, Fleetwood). Privately published.

Auld, H. A. (1879) *Acronicta alni* (bred from Lime). *Entomologist*. Vol. 12. p. 251.

Averis, A., Averis, B., Birks, J., Horsfield, D., Thompson, D. & Yeo, M. (2004, reprint) *An Illustrated Guide to British Upland Vegetation*. Joint Nature Conservation Committee, Exeter.

Barker, A. (2019) Sugar line revisited. Lancashire Moth Group Newsletter. No. 36 p. 12

Baldwin, J. W. (1880) Lepidoptera at the Flowers of the Heath. *Young Naturalist*. Vol. 1. p. 387.

Beaumont, H. E. (2002) *Butterflies and Moths of Yorkshire, A Millennium Review*. Yorkshire Naturalists' Union, Dorset.

Beirne, B. P. (1952) *British Pyralid and Plume Moths*. Warne & Co., London.

Bell, E. (1871) *Chaerocampa celerio* at Southport. *Newman's Entomologist*. Vol. 5. p. 411.

Berggren, K., Aarvik, L., Huemer, P., Lee, K. M. & Mutanen, M. (2022) Integrative taxonomy reveals overlooked cryptic diversity in the conifer feeding *Batrachedra pinicolella* (Zeller, 1839) (Lepidoptera, Batrachedridae). *ZooKeys*. Vol. 1085. pp. 165–182.

Bevan, K. (1954) Notes on the Macrolepidoptera of the Fylde Coastal Area of West Lancashire. *Entomologist's Record and Journal of Variation*. Vol. 66. pp. 74–77.

Birkett, N. L. (undated) *Macro-lepidoptera Card Index*. Held at Tullie House, Museum, Carlisle, as scanned by G. Hedges, Tanyptera Trust.

Bland, K. (1986) *Preliminary Atlas of the Lepidoptera: Incurvarioidea of the British Isles*. Biological Records Centre, ITE Monks Wood, Huntingdonshire.

Bland K. P. (Ed.) (2015) *The Moths and Butterflies of Great Britain and Ireland Vol. 5 (Part 1)*. Brill. Leiden/Boston.

Bland K. P. (Ed.) (2015a) *The Moths and Butterflies of Great Britain and Ireland Vol. 5 (Part 2)*. Brill. Leiden/Boston.

Boyes, D., Crowley, L. & Mulhair, P. (2021) Brave New World: How DNA Barcoding is advancing our understanding of Insects. *Atropos*. Vol. 69. pp. 17–30.

Bradley, J. D., Tremewan, W. G. & Smith, A. (1973) *British Tortricoid Moths. Cochylidae and Tortricidae: Tortricinae*. Ray Society, London.

Bradley, J. D., Tremewan, W. G. & Smith, A. (1979) *British Tortricoid Moths. Tortricidae: Olethreutinae*. Ray Society, London.

Briggs, J. (1985) *Macrolepidoptera (Moths) of Leighton Moss Reserve*. (RSPB, unpublished).

Brindle, A. (1939) *The Lepidoptera of the Pendle Hill area 1939*. The same author also created an extensive Card Index, which is now held at World Museum.

Brindle, A. (1948) The Lepidoptera of the Pendle Hill area – Addenda and Corrigenda. *North Western Naturalist* (1948).

Britten, H. (c.1950) *Card Index of Lepidoptera*. Manchester Museum.

Burkmar, R. (2009) *Ecology and Habitat Requirements of the Sandhill Rustic Moth (Luperina nickerlii gueneei Doubleday, 1864) at Birkdale Green Beach*. MSc thesis, University of Birmingham, 2009.

Burrows, H. L. (1964) Some North Western Psychidae (Lepidoptera). *Manchester Entomological Society*. Vol. 59–61. pp. 25–27.

Buxton, E. C. (1853) Captures of Lepidoptera at Chorley. *Zoologist*. Vol. 11. pp. 4037.

Byerley, I. (1854) The Fauna of Liverpool. *Issue 7 of Proceedings of the Literary & Philosophical Society of Liverpool. Deighton and Laughton, 1854.*

Calderbank, T. (1870) *Deilephila gallii* near Bolton. *Entomologist* Vol. 5. pp. 181.

Callaghan, L. (2020) *The ecology of the belted beauty moth Lycia zonaria, and its sensitivity to the environment at the last remaining English colony.* Lancaster University, Lancaster.

Campbell, C. (1862) *Eupithecia debiliata*. *Zoologist*. Vol. 20 (1862). p. 8209.

Capper, S. J. (1867) *Sterrha sacraria* in Lancashire. *Entomologist*. Vol. 3. p. 347.

Capper, S. J. (1871) *Acronycta alni, Sphinx convolvuli* and *Deilephila gallii*. *Entomologist*. Vol. 5. (96) p. 417.

Capper, S. J. (1881) Abstract of the Inaugural Address: The History of the Liverpool Entomological Society. *5th Annual Report and Proceedings of the Lancashire and Cheshire Entomological Society*, p. 5–6.

Cash, J. (1873) *Where there's a will there's a way.* London.

Chadwick, D. (1860) On the rate of wages in Manchester and Salford, and the manufacturing districts of Lancashire 1839–1869. *Journal of the Statistical Society of London.* Vol. 23 (1). pp. 1–36.

Chappell, J. (1861) Captures near Manchester. *Entomologist's Weekly Intelligencer.* Vol. 9. p. 203.

Clancy, S. P. (2008) The Immigration of Lepidoptera to the British Isles in 2006. *Entomologist's Record and Journal of Variation.* Vol. 120. pp. 209–276.

Clay, C. & Wain, M. (2019) *Morecambe Bay Barred Tooth-striped Moth Survey 2019.* Tanyptera Trust Report, Liverpool.

Clutten, W. G. (1918) *The Moths and Butterflies of the Burnley District.* Burnley Natural History Society.

Clutten, W. G. & Wright, A. E. (1901) *The Moths and Butterflies of the Burnley district.* Burnley Natural History Society.

Cook, L. (2018) Lord Walsingham and the Manchester moth. *Entomologist's Gazette.* Vol. 69. pp. 47–63.

Cook, L. M. & Logunov, D. V. (2017) The Manchester Entomological Society (1902–1991), its story and historical context. *Russian Entomological Journal* 26(4): 365–388.

Cooke, N. (1856) Importation of Lepidoptera. *Entomologist's Weekly Intelligencer.* Vol. 1. p. 109.

Cooke, N. (1856a) Capture of *Tinea merdella*. *Entomologist's Weekly Intelligencer.* Vol. 1. p. 125.

Cottam, R., Horsfall, H., Taylor, J. & Windle, W. F. (1914) The Macro-lepidoptera of Oldham district. *Manchester Entomological Society Annual Reports 1904–1921.* pp. 41–50.

Creaser, A. (1976) Macrolepidoptera Report 1975. *Lancashire and Cheshire Fauna Society. Annual Report for 1975.* Vol. 69. pp. 11–14.

Curtis, W. P. (1930) Pairing of *Hyloicus pinastri* L., and further Records of its Occurrence in Britain. *Entomologist*. Vol. 63. pp. 184–185.

Dandy, J. E. (1969) *Watsonian Vice-Counties of Great Britain.* Ray Society, London.

Daniels, J. (1859) *Acherontia atropos*. *Entomologist's Weekly Intelligencer 1859–1860.* p. 27.

Davis, A. M. (2012) *A Review of the Status of Microlepidoptera in Britain.* Butterfly Conservation, Wareham. (Butterfly Conservation Report No. S12–02).

Doubleday, H. (1847) Occurrence of the true *Cerura bicuspis* in Britain. *Zoologist*. Vol. 5. p. 1863.

Edleston, R. S. (1843) Note on capture of Emperor Moth, by means of a captive female. *Zoologist*. Vol. 1. pp. 199–200.

Edleston, R. S. (1843a) Note on *Saturnia pavonia-minor* and *Lasiocampa rubi*. *Zoologist.* Vol. 1. p. 260.

Edleston, R. S. (1843b) Note on the capture of *Heliothis armigera* near Salford, (figured in Boisduval). *Zoologist.* Vol. 1. p. 260.

Edleston, R. S. (1844) Note on captures of Lepidopterous Insects near Manchester, in 1844. *Zoologist.* Vol. 2. pp. 734–735.

Edleston, R. S. (1846) Captures of Lepidopterous Insects near Manchester. *Zoologist.* Vol.4. pp. 1220–1223.

Edleston, R. S. (1846a) Capture of *Deilephila celerio* (Linn.) at Manchester. *Zoologist.* Vol. 4. p. 1346.

Edleston, R. S. (1846b) Remarks on 'the occurrence of *Lasiocampa trifolii* near Manchester. *Zoologist.* Vol. 4. p. 1226.

Edleston, R. S. (1846c). Habitat of *Gortyna petasidis*. *Zoologist.* Vol. 4. p. 1347.

Edleston, R. S. (1846d) Capture of *Catocala fraxini* at Manchester. *Zoologist.* Vol. 4. p. 1515.

Edleston, R. S. (1859) Observations on the Solenobiae of Lancashire, etc., *Zoologist.* Vol. 17. pp. 6462–6463.

Edleston, R. S. (1859a) *Depressaria rhodochrella*. *Entomologist's Weekly Intelligencer.* 1859–1860. p. 27.

Edleston, R. S. (1859b) *Phygas birdella. Entomologist's Weekly Intelligencer.* 1859–1860. p. 27.

Edleston, R. S. (1864) *Amphydasis betularia. Entomologist.* Vol. 2. p. 150.

Edmunds, M., Mitcham, T. & Morries, G. (2004) *Wildlife of Lancashire – Exploring the natural history of Lancashire, Manchester and North Merseyside.* Carnegie Publishing Ltd, Lancaster.

Ellis, J. W. (1881) *Agrotis praecox*, etc., *Young Naturalist.* Vol. 2. p. 244.

Ellis, J. W. (1890) The Lepidopterous Fauna of Lancashire and Cheshire. Liverpool.

Emmet, A. M. (1991) *The Moths and Butterflies of Great Britain and Ireland.* Vol. 7, Part 2. pp. 316–318. (*Lasiocampa quercus* Linn.) Harley Books, Essex.

Emmet, A. M. (1996) *The Moths and Butterflies of Great Britain and Ireland.* Vol. 3. pp. 146–151. (Key to species of the genus *Coleophora*). Harley Books, Essex.

Emmet, A. M. & Langmaid, J. R. (2002) *The Moths and Butterflies of Great Britain and Ireland.* Vol. 4 (Part 1). pp. 134–135 (*Depressaria emeritella*). Harley Books, Essex.

Emmet, A. M. & Langmaid, J. R. (2002a) *The Moths and Butterflies of Great Britain and Ireland* Vol. 4 (Part 2). Harley Books, Essex.

Emmett, E. E. (1984) The Silky Wainscot *Chilodes maritimus* (Tauscher) in North Lancashire, SD47. *Entomologist's Record and Journal of Variation.* Vol. 96. p. 221.

Emmett, E. E. (1991) *List of records of notable species in database of E. E. Emmett.* (unpublished).

Fairhurst, A. (1982) *Moths of the West Wigan area 1949 to 1982.* Unpublished.

Ford, L. T. (1941) In 'Records and full descriptions of Varieties and Aberrations', an Exhibit intended for the Annual Exhibition of the South London Entomological Society. *Entomologist's Record and Journal of Variation.* Vol. 53. pp. 6–8.

Fox, R., Parsons, M.S., & Harrower, C.A. (2019) *A Review of the Status of the Macro-moths of Great Britain.* Butterfly Conservation Report to Natural England. S19–17.

Fraser, G. de C. (1946) *The Moths and Butterflies of the Formby area.* [It has not been possible to locate where, or even if, this was published].

Goater, B. (1986). *British Pyralid Moths.* (*Oligostoma bilinealis* Snellen) p. 61. Harley Book, Essex.

Greening, N. (1870) *Deilephila gallii* at Warrington. *Entomologist.* Vol. 5 (82). p. 168.

Greenwood, E. (2012) *Flora of North Lancashire.* Carnegie Publishing Ltd, Lancaster.

Gregson, C. S. (1850) Captures of rare Lepidoptera near Liverpool. *Zoologist.* Vol. 8. p. 2898.

Gregson, C. S. (1850a) Capture of *Hydraecia petasitis* near Manchester. *Zoologist.* Vol. 8. pp. 2931–2932.

Gregson, C. S. (1856) Economy of *Micropterix* and *Argyresthia brockeella. Entomologist's Weekly Intelligencer.* Vol. 1. pp. 28–29.

Gregson, C. S. (1856a) *Lithocolletis vacciniella* and *Nepticula weaveri. Entomologist's Weekly Intelligencer.* Vol. 1. p. 29.

Gregson, C. S. (1856b) Captures of Lepidoptera. *Entomologist's Weekly Intelligencer.* Vol. 1. pp. 44–45.

Gregson, C. S. (1857) Comprising four elements.
1. On the lepidopterous insects of the district around Liverpool, with some of the causes of the abundance or scarcity of insects. *The Historic Society of Lancashire and Cheshire.* Vol. 7 (1854–1855). pp. 238–254.
2. On the lepidopterous insects of the district around Liverpool. *The Historic Society of Lancashire and Cheshire.* Vol. 8 (1855–1856). pp. 153–164.
3. On the lepidopterous insects of the district around Liverpool. *The Historic Society of Lancashire and Cheshire.* Vol. 9 (1856–1857). pp. 165–180.
4. On the lepidopterous insects of the district around Liverpool. *The Historic Society of Lancashire and Cheshire.* Vol. 10 (1857–1858). pp. 113–130.

Gregson, C. S. (1860) On a new *Lithosia* intermediate between *L. complana* and *complanula. Entomologist's Weekly Intelligencer.* Vol. 9. pp. 30–31.

Gregson, C. S. (1870) Notes on occasional second-broods in single-brooded lepidoptera. *Entomologist's Monthly Magazine.* Vol. 7–8. pp. 18–19.

Gregson, C. S. (1871) Description of an *Ephestia* new to Science. *Newman's Entomologist* (*Entomologist*). Vol. 94 and 95. p. 385.

Gregson, C. S. (1879) Entomological Notes for Beginners (March). *Young Naturalist.* Vol. 1. pp. 158–159.

Gregson, C. S. (1882) Note on the abundance of *Nyssia zonaria* on the Lancashire and Cheshire coast this season. *Young Naturalist.* Vol. 3. pp. 172–173.

Gregson, S. C. [*sic.* **C. S.**] (1883) Entomological localities No. 5, Chat Moss. *Young Naturalist.* Vol. 4. pp. 151–153.

Gregson, C. S. (1883a) Old Entomological localities, Heysham Moss. *Young Naturalist.* Vol. 5. pp. 44–45.

Gregson, C. S. (1887) *Semaria woeberana. Young Naturalist.* Vol. 8. pp. 203–204. (Generic and specific names misspelt and refer to *Semasia woberiana* (*Enarmonia formosana*)).

Gregson, C. S. (1889) *Plusia ni. Young Naturalist.* Vol. 10. pp. 44.

Gregson, W. (1870) *Chaerocampa celerio* at Lytham. *Entomologist.* Vol. 5 (84). p. 204.

Hancock, B. (2018) *Pug Moths of North-west England.* Lancashire and Cheshire Fauna Society, publication 124. Rishton.

Hancock, B. (2020) *Notable Pug records in North-west England 2018 – 2019.* PDF available for download on 'Lancashire Moths' website, accessed April 2023.

Hancock, E. G. (1984) *Oligostigma bilinealis* Snellen (Lepidoptera: Pyralidae), a second British record. *Entomologist's Gazette.* Vol. 35. p. 18.

Hancock, E. G. & Wallace, I. D. (1986) Unusual foodplants of Svensson's Copper Underwing (*Amphipyra berbera svenssoni* Fletcher) and The Mouse (*Amphipyra tragopoginis* [Clerck]), Lepidoptera: Noctuidae. *Entomologist's Record and Journal of Variation.* Vol. 98. pp. 7–8.

Hancock, E. G. (2018) Netted Carpet *Eustroma reticulatum* (D, & S., 1775), (Lep.: Geometridae) as a commodity. *Entomologist's Record and Journal of Variation.* Vol. 130. pp. 123–138.

Hardwick, L. W. (1990) *Lancashire and Cheshire Microlepidoptera Source Data* (unpublished, photocopied species notes).

Hart, C. (2011) *British Plume Moths.* British Entomological and Natural History Society, Reading.

Heath, J. (1976) *The Moths and Butterflies of Great Britain and Ireland Volume 1.* Curwen Press, London.

Heath, J., & Emmet, A. M. (1979) *The Moths and Butterflies of Great Britain and Ireland Volume 9.* p. 107 (*Pyrrharctia isabella*). Curwen Books, England.

Heath, J., & Emmet, A. M. (1983) *The Moths and Butterflies of Great Britain and Ireland Volume 10.* p. 277 (*Spodoptera littoralis*). Harley Books, Essex.

Heath, J., & Emmet, A. M. (1983a) *The Moths and Butterflies of Great Britain and Ireland Volume 10.* p. 278 (*Spodoptera litura*). Harley Books, Essex.

Heath, J., & Emmet, A. M. (1983b) *The Moths and Butterflies of Great Britain and Ireland Volume 10.* pp. 40–41. Harley Books, Essex.

Heath, J., & Emmet, A. M. (1983c) *The Moths and Butterflies of Great Britain and Ireland Volume 10.* p. 383 (*Orodesma apicina*). Harley Books, Essex.

Heath, A. M. & Emmet, A. M. (1985) *The Moths and Butterflies of Great Britain and Ireland Volume 2.* Harley Books, Essex.

Heath, A. M. & Emmet, A. M. (1985a) E. C. Pelham-Clinton in: *The Moths and Butterflies of Great Britain and Ireland Volume 2.* p. 168. Harley Books, Essex.

Heckford, R. J. (2010) *Endothenia oblongana* (Haworth, 1811) (Lepidoptera: Tortricidae) reared from a rootstock of *Plantago lanceolata* L. *Entomologist's Gazette.* Vol. 61. pp. 222–224.

Heckford, R. J & Beavan, S. (2012) *Celypha rufana* (Scopoli, 1763) (Lep.: Tortricidae): an account of the early stages in captivity. *Entomologist's Gazette.* Vol. 63. pp. 217–221.

Heckford, R. J & Beavan, S. (2013) *Celypha rufana* (Scopoli, 1763) (Lep.: Tortricidae): an account of the early stages in the wild. *Entomologist's Gazette.* Vol. 64. pp. 27–32.

Heckford, R. J & Beavan, S. (2013) *Scythris fallacella* (Schläger, 1847) (Lep.: Scythrididae): an account of the early stages. *Entomologist's Record and Journal of Variation.* Vol. 125. pp. 1–3.

Heckford, R. J., Beavan, S. & Palmer, S. M. (2015) *Bryotropha boreella* (Douglas, 1851) (Lep.: Gelechiidae): discovery of the larva. *Entomologist's Gazette.* Vol. 66. pp. 237–243.

Hedges, G. & Patton, J. (2022) Fen Square-spot *Diarsia florida* DNA barcoding and recording in Lancashire. *Lancashire Moth Group.* Newsletter 41. p. 16 (see link above under Lancashire Moth Group).

Henwood, B., & Sterling, P. (2020) *Field Guide to the Caterpillars of Great Britain and Ireland.* Bloomsbury, London.

Hodgkinson, J. B. (1844) Note on Captures of Lepidopterous Insects at Preston, in Lancashire. *Zoologist.* Vol. 2. pp. 685–686.

Hodgkinson, J. B. (1845) Capture of *Fidonia ericetaria* etc. near Preston. *Zoologist.* Vol. 3. pp. 1085–1086.

Hodgkinson, J. B. (1845a) Occurrence of *Lasiocampa trifolii* near Manchester. *Zoologist.* Vol. 3. pp. 1140–1141.

Hodgkinson, J. B. (1845b) Occurrence of *Acronycta salicis* near Preston. *Zoologist.* Vol. 3. pp. 1141.

Hodgkinson, J. B. (1849) Capture of *Polia lichenea*. *Zoologist*. Vol. 7. p. 2404.

Hodgkinson, J. B. (1849a) Capture of *Cerura bicuspis* at Preston. *Zoologist*. Vol. 7. p. 2500.

Hodgkinson, J. B. (1858) Lepidoptera near Preston. *Entomologist's Weekly Intelligencer*. Vol. 4. pp. 146–147.

Hodgkinson, J. B. (1861) The New Noctua taken near Manchester. *Entomologist's Weekly Intelligencer*. Vol. 9. pp. 146–147.

Hodgkinson, J. B. (1861a) The New Noctua. *Entomologist's Weekly Intelligencer*. Vol. 9. p. 163.

Hodgkinson, J. B. (1865) Captures of Lepidoptera in the North. *Entomologist's Monthly Magazine*. Vol.2. pp. 159–160 (part), concluded 186–188.

Hodgkinson, J. B. (1866) Captures at Lytham. *Entomologist's Monthly Magazine*. Vol. 3. p. 37.

Hodgkinson, J. B. (1866a) *Euchromia rufana* in Lancashire. *Entomologist's Monthly Magazine*. Vol. 3. p. 139.

Hodgkinson, J. B. (1872) *Depressaria douglasella* near Blackpool. *Entomologist's Monthly Magazine*. Vol. 9. p. 113.

Hodgkinson, J. B. (1872a) Captures of Lepidoptera near Fleetwood. *Entomologist's Monthly Magazine*. Vol. 9. p. 162.

Hodgkinson, J. B. (1875) A Coleophora new to Britain: *Coleophora tripoliella* (Hodgkinson). *Entomologist*. Vol. 12. pp. 55–57.

Hodgkinson, J. B. (1878) *Depressaria atomella*, a new species to Britain. *Entomologist's Monthly Magazine* Vol. 15. pp. 208–209.

Hodgkinson, J. B. (1879a) Two new Microlepidoptera. *Entomologist*. Vol. 12. pp. 55–57.

Hodgkinson, J. B. (1879b) Entomological rambles 1878. *Entomologist*. Vol. 12. pp. 126–128 (part), concluded 151–153.

Hodgkinson, J. B. (1879c) Northern notes on the season. *Entomologist*. Vol. 12. pp. 203–204.

Hodgkinson, J. B. (1879d) *Dicrorampha* [sic] *consortana* and *Retinia buoliana* at Preston. *Entomologist's Monthly Magazine*. Vol. 9. p. 67.

Hodgkinson, J. B. (1880a) Captures in North Lancashire in 1879. *Entomologist*. Vol. 13. pp. 105–109.

Hodgkinson, J. B. (1880b) *Nemophora pilella*. *Entomologist*. Vol. 13. p. 164.

Hodgkinson, J. B. (1880c) *Stigmonota scopariana* H.-S.: a tortrix new to the British fauna. *Entomologist*. Vol. 13. p. 162.

Hodgkinson, J. B. (1880d) *Stigmonota scopariana*. *Entomologist's Monthly Magazine*. Vol. 17. p. 38.

Hodgkinson, J. B. (1880e) *Stigmonota scopariana*. *Entomologist's Monthly Magazine*. Vol. 17. p. 70.

Hodgkinson, J. B. (1881) *Epunda lutulenta*, var. *luneburgensis*, in Lancashire. *Entomologist*. Vol. 14. p. 68.

Hodgkinson, J. B. (1884) *Laphygma exigua* in Lancashire. *Entomologist*. Vol. 17. P8. 274.

Hooper, J. (2002). *Of Moths and Men*. Fourth Estate, London.

Horsfall, H. & Windle, W. F. (1914) *Some notes on the Macro-lepidoptera of the Oldham District*. Manchester Entomological Society Ann. Reports 1904–1921. pp. 31–40.

Howe, M. A., Hinde, D., Bennett, D. & Palmer S. (2004) The conservation of the belted beauty *Lycia zonaria* (Lepidoptera: Geometridae) in the United Kingdom. *Journal of Insect Conservation*. Vol. 8. pp. 159–166.

Kay, R. (1876) *Oporabia filigrammaria* and *Larentia caesiata* near Bury, Lancashire. *Entomologist*. Vol. 9. pp. 158–159.

Kefford, W. R. (1871) *Melanthia rubiginata* var. *plumbata* etc., *Entomologist*. Vol 5 (86). pp. 227. (Notes of captures by a recorder who lived in Ormskirk – data not entered onto county database as location of capture is not stated).

Kimpton, A. (2000) A report on possible sites for the Belted Beauty moth (*Lycia zonaria britannica* Harrison) in the north west of England. Butterfly Conservation, Wareham. (Butterfly Conservation Report No. S00-18).

Kimpton, A. (2002) Survey for the Belted Beauty moth *(Lycia zonaria britannica Harrison)* in the north west of England: April to June 2001. Butterfly Conservation, Wareham. (Butterfly Conservation Report No. S02–04).

Lamb, J. (2018). *Lancashire, a journey into the wild*. Palatine Books, Lancaster.

Lancashire & Cheshire Ent. Soc. (1933). Meetings. *55th to 57th Annual Report and Proceedings of the Lancashire and Cheshire Entomological Society*. pp. 9–20.

Lancashire & Cheshire Ent. Soc., (1952). Field Meetings, Saturday 12th August 1950 at Formby, Lancs. *74th Annual Report and Proceedings of the Lancashire and Cheshire Entomological Society*. Session 1950/51. p. 9.

Lancashire & Cheshire Ent. Soc. (1955). William Mansbridge, M.Sc., F.R.E.S. A Tribute. *77th and 78th Annual Report and Proceedings of the Lancashire and Cheshire Entomological Society*. pp. 9–10.

Lancashire & Cheshire Ent. Soc. (1981). Indoor Meetings, Session 1978–79. *101st to 103rd Annual Report and Proceedings of the Lancashire and Cheshire Entomological Society.* pp. 33–42.

Lancashire Moth Group Newsletters 1–42 (1996–2023). Accessed on the Lancashire Moths website on 19 April 2023. https://www.lancashiremoths.co.uk/newsletters-and-other-Documents.

Langmaid, J. R., Palmer, S. M. & Young, M. R. (2018) *A Field Guide to the Smaller Moths of Great Britain and Ireland.* British Entomological and Natural History Society, Berkshire.

Lawson, K. (2011) *A study of the habitat requirements of the Belted Beauty Moth, Lycia zonaria britannica at Potts Corner, Middleton, Morecambe Bay 2010–2011.* BSc (Hons) dissertation. Blackpool and the Fylde College.

Lee, D. (2021) *Acleris sparsana* ([Denis & Schiffermüller], 1775): Flight times and voltinism re-examined. *Entomologist's Record and Journal of Variation.* Vol. 133. pp. 142–145.

Leech, M. J. & Michaelis, H. N. (1957) *The Lepidoptera of Formby.* The Raven Entomological and Natural History Society. Arbroath.

Mansbridge, W. (1905) Tortrices in the Liverpool district. *Entomologist.* Vol. 38. pp. 115–116.

Mansbridge, W. (1909) *Micro-lepidoptera in the Liverpool District. 33rd Annual Report and Proceedings of the Lancashire and Cheshire Entomological Society.* pp. 29–31.

Mansbridge, W. (1914) Local Records Lepidoptera. *Manchester Entomological Society. Ann. Reports 1904–1921.* p. 16.

Mansbridge, W. (1915) Report on the Lepidoptera for 1914. *Lancashire and Cheshire Fauna Committee. 1st Annual Report and Reports of the Recorders for 1914.* pp. 12–14.

Mansbridge, W. (1916) Report of the Recorder for Lepidoptera 1915. *Lancashire and Cheshire Fauna Committee. 3rd Annual Report.* pp. 89, 96–97 (report covering 1915).

Mansbridge, W. (1916a) Report of the Recorder for Lepidoptera. *Lancashire and Cheshire Fauna Committee. 3rd Annual Report.* pp. 9–11 (report covering 1916).

Mansbridge, W. (1916b) Revision of the Ellis list (part). *38th and 39th Annual Reports and Proceedings of the Lancashire and Cheshire Entomological Society.* pp. 1–11.

Mansbridge, W. (1917) Revision of the Ellis list (part). *40th and 41st Annual Reports and Proceedings of the Lancashire and Cheshire Entomological Society.* pp. 45–68.

Mansbridge, W. (1918) Report of the Recorder for Lepidoptera for 1917. Lancashire and Cheshire Fauna Committee. *4th Annual Report and Reports of the Recorders for 1917.* pp. 13–16.

Mansbridge, W. (1920) Lepidoptera, Report of the Recorder for 1918. *Lancashire and Cheshire Fauna Committee. 6th Annual Report and Report of the Recorders for 1919.* pp. 12–15.

Mansbridge, W. (1921) Lepidoptera, Report of the Recorder for 1919. *Lancashire and Cheshire Fauna Committee. 7th Annual Report and Reports of the Recorders for 1920.* pp. 46–50.

Mansbridge, W. (1921a) Revision of the Ellis list (Noctuidae, part). *42nd, 43rd and 44th Annual Reports and Proceedings of the Lancashire and Cheshire Entomological Society.* pp. 69–84.

Mansbridge, W. (1923) Revision of the Ellis list (Noctuidae, part). *45th and 46th Annual Reports and Proceedings of the Lancashire and Cheshire Entomological Society.* pp. 85–100.

Mansbridge, W. (1924) Notes on Lepidoptera for 1922–23. *Lancashire and Cheshire Fauna Committee. 10th Annual Report and Reports of the Recorders for 1923.* pp. 10–13.

Mansbridge, W. (1925) Report of the Recorder for Lepidoptera for 1924. *Lancashire and Cheshire Fauna Committee. 11th Annual Report and Reports of the Recorders for 1924.* p. 50.

Mansbridge, W. (1926) Report on Lepidoptera for 1925. *Lancashire and Cheshire Fauna Committee. 12th Annual Report and Reports of the Recorders for 1925.* pp. 1–2.

Mansbridge, W. (1927) Report of the Recorder for Lepidoptera. *Lancashire and Cheshire Fauna Committee. 13th Annual Report and Report of the Recorders for 1926.* pp. 2–3.

Mansbridge, W. (1928) Revision of the Ellis list (part). *47th to 51st Annual Reports and Proceedings of the Lancashire and Cheshire Entomological Society.* pp. 101–116.

Mansbridge, W. (1930) Report on Lepidoptera 1927–29. *16th Annual Report and Report of the Recorders for 1929.* pp. 10–11.

Mansbridge, W. (1931) Lepidoptera. *Lancashire and Cheshire Fauna Committee. 17th Annual Report and Report of the Recorders for 1930.* p. 14.

Mansbridge, W. (1932) Lepidoptera. *Lancashire and Cheshire Fauna Committee. 18th Annual Report and Report of the Recorders for 1931.* p. 28.

Mansbridge, W. (1933) Lepidoptera. *Lancashire and Cheshire Fauna Committee. 19th Annual Report and Report of the Recorders for 1932.* p. 37.

Mansbridge, W. (1934) Lepidoptera in 1932–33. *Lancashire and Cheshire Fauna Committee. 20th Annual Report and Report of the Recorders for 1933.* pp. 37–38.

Mansbridge, W. (1934a) Recent records of Lepidoptera new to Lancashire and Cheshire. *55th, 56th and 57th Annual Reports and Proceedings of the Lancashire and Cheshire Entomological Society.* pp. 30–32.

Mansbridge, W. (1934b) Revision of the Ellis list (Pyralidae to Momphidae). *55th, 56th and 57th Annual Reports and Proceedings of the Lancashire and Cheshire Entomological Society.* pp. 157–208.

Mansbridge, W. (1935) Lepidoptera in 1934. *Lancashire and Cheshire Fauna Committee.* pp. 23–24.

Mansbridge, W. (1937) Lepidoptera in 1935–36. *Lancashire and Cheshire Fauna Committee. 23rd Annual Report and Report of the Recorders for 1936.* pp. 37–40.

Mansbridge, W. (1937a) Continuation of the Revised (Ellis) Lepidoptera list: Oecophoridae to Nepticulidae. *58th, 59th and 60th Annual Reports and Proceedings of the Lancashire and Cheshire Entomological Society.* pp. 209–244.

Mansbridge, W. (1939) Lepidoptera in 1937. *Lancashire and Cheshire Fauna Committee. 24th Annual Report and Report of the Recorders for 1937.* pp. 29–30.

Mansbridge, W. (1939a) Lepidoptera in 1938. *Lancashire and Cheshire Fauna Committee. 25th Annual Report and Report of the Recorders for 1938.* pp. 38–40.

Mansbridge, W. (1940) Revision of Ellis, J. W., *The Lepidopterous Fauna of Lancashire and Cheshire.* Lancashire and Cheshire Entomological Society, Liverpool.

Mansbridge, W. (1940a) *The Lepidopterous Fauna of Lancashire and Cheshire: Porthesia similis Fues..* Variation. p. 22.

Mansbridge, W. (1944) The Lepidoptera of 1940–42. *Lancashire and Cheshire Fauna Committee, 26th Report.* pp. 12–15.

Mansbridge, W. (1947) The Lepidoptera of 1943–46. *Lancashire and Cheshire Fauna Committee, 27th Report.* pp. 15–19.

Mansbridge, W. & Wright, A. E. (1949) The Lepidoptera of 1947. *Lancashire and Cheshire Fauna Committee, 28th Report.* pp. 10–13.

Mansbridge, W. & Wright, A. E. (1949) The Lepidoptera of 1948. *Lancashire and Cheshire Fauna Committee, 28th Report.* pp. 14–21.

Mansbridge, W. & Wright, A. E. (1950) The Lepidoptera of 1949. *Lancashire and Cheshire Fauna Committee, 29th Report.* pp. 81–92.

Marsh, P. J. (2015) Fen Square-spot *Diarsia florida* and Small Square Spot *Diarsia rubi*. *Lancashire Moth Group Newsletter 30.* p. 7 (see link above under Lancashire Moth Group).

Marsh, P.J. & White, S.J. (Eds.) (2019) *The Butterflies and Day-flying Moths of Lancashire and North Merseyside.* Lancashire and Cheshire Fauna Society, Rishton.

Marsh, P. J. (2022) Observations from upland trapping in northern VC60. *Lancashire Moth Group Newsletter.* Vol. 41. p. 9.

McCabe, K. (2020) My garden in Flixton, Lancashire, VC59. Lancashire Moth Group Newsletter. No. 38. p. 14–15.

Meyrick, E. [1928]. *A revised handbook of British Lepidoptera.* London.

Michaelis, H. N. (1953) Microlepidoptera in Cheshire and South Lancashire in 1952. *Entomologist's Record and Journal of Variation.* Vol. 65. pp. 74–76.

Michaelis, H. N. (1953a) Lepidoptera Report, 1951. *30th Report of Lancashire and Cheshire Fauna Committee.* pp. 48–52.

Michaelis, H. N. (1954) Lepidoptera, 1952/53. *30th Report of Lancashire and Cheshire Fauna Committee.* pp. 53–56.

Michaelis, H. N. (1955) Records of Microlepidoptera from Lancashire and Cheshire. *Reports and Proceedings of the Lancashire and Cheshire Entomological Society 1953/1955.* pp. 55–69.

Michaelis, H. N. (1958) Microlepidoptera in Lancashire and Cheshire 1955–57. *Entomologist's Record and Journal of Variation.* Vol. 70. pp. 122–124.

Michaelis, H. N. (1962) Notes on species of Microlepidoptera found in Lancashire and Cheshire. *The Entomologist July 1962.* pp. 183–86 (originally published in *Transactions of the Manchester Entomological Society*, 1961).

Michaelis, H. N. (1966) Records of Lepidoptera from Lancashire, Cheshire and Wales. *Proceedings of the Lancashire and Cheshire Entomological Society 1966.* pp. 106–111.

Michaelis, H. N. (1977) The History and Present Status of Luperina nickerlii gueneei Doubleday in Britain. *Entomologist's Record and Journal of Variation.* Vol. 89. pp. 183–185.

Morgan, M. J. (1995) Hugh N. Michaelis 1904–1995. *116th to 120th Annual Report and Proceedings of the Lancashire and Cheshire Entomological Society.* pp. 105–6.

Natural England (2014) *Corporate report - National Character Area profiles.* Available at: https://www.gov.uk/government/publications/national-character-area-profiles-data-for-local-decision-making/national-character-area-profiles (Accessed: July 2023).

Natural England (2023) *Natural Character Area Profiles.* Available at: https://nationalcharacterareas.co.uk/ (Site in development, Accessed: July 2023).

NERC, 2006. Habitats and species of principal importance in England. [online] Section_41_of_the_Natural_Environment_and_Rural_Communities__NERC__Act_2006_habitats_and_species_of_principal_importance_in_England.ods (live.com).

Nieukerken, E.J. van, Mutanen, M. & Doorenweerd, C. *(2012)* DNA barcoding resolves species complexes in *Stigmella salicis* and *S. aurella* species groups and shows additional cryptic speciation in *S. salicis* (Lepidoptera: Nepticulidae). *Entomologisk Tidskrift.* Vol. 132 (4). pp. 235–255. Uppsala, Sweden 2012.

Nieukerken, E.J. van & Hartman, T. *(2019)* The *Salix* feeding species of *Stigmella* in Europe; almost as difficult to separate as their hosts (Lepidoptera: Nepticulidae). *Societas Europaea Lepidopterologica.* PDF, no note of having been published.

Palmer, S. M., Palmer, R. M. & Langmaid J. R. (2001) *Sorhagenia janiszewskae* Riedl. (Lep.: Cosmopterigidae) feeding on *Rhamnus cathartica* and new to Lancashire. *Entomologist's Record and Journal of Variation.* Vol. 113. pp. 81–82.

Palmer, S. M. (2002) *The Belted Beauty in Lancashire. Lancashire Moth Group Newsletter.* Vol. 6. pp. 4–5. Lancashire Moths website – accessed April 2023.

Palmer, S. M. (2002a) *A Provisional Checklist of the Lepidoptera of Lancashire.* Privately published.

Parsons, M. S. (1987) *Invertebrate Site Register, Report no. 96, part 1. Review of Invertebrate sites in England, Merseyside and Greater Manchester.* Nature Conservancy Council, Peterborough.

Parsons, M. S. (1995) *A review of the scarce and threatened ethmiine, stathmopodine and gelechiid moths of Great Britain.* UK Nature Conservation no. 16. Joint Nature Conservation Committee, Peterborough.

Parsons, M. S. & Clancy, S. (2023) *A Guide to the Pyralid and Crambid Moths of Britain and Ireland.* Atropos Publishing, Lizard.

Patton, J. (2023) *The Status and Distribution of the Endangered Macro-moth Least Minor Photedes captiuncula (Treitschke, 1825) (Lepidoptera: Noctuidae) in the Morecambe Bay Limestone Area 2022.* A report for the Tanyptera Project, World Museum, Liverpool.

Pierce, F. N. (1909) The Luperinas. *33rd Annual Report and Proceedings of the Lancashire and Cheshire Entomological Society.* pp.37–39 and plate.

Pierce, F. N. & Metcalfe, W. (1934) *Tinea merdella* Zell., and its allies. *Entomologist.* Vol. 67 (859). pp. 265–267.

Pierce, F. N., Metcalfe, W. & Diakonoff, A. (1938) *Corrections and Additions.* In Pierce, F. N. & Metcalfe, W., *The genitalia of British Pyrales with the Deltoids and Plumes.* pp. 65–67. E. W. Classey reprint, 1984.

Plant, C. W. (2008) *The Moths of Hertfordshire.* The Hertfordshire Natural History Society. Welwyn Garden City.

Plant, C. W. (2010) An examination of the status of *Diachrysia stenochrysis* (Warren, 1913) and its possible occurrence in Britain as a sibling of *D. chrysitis* (L.) (Lep.: Noctuidae). *Entomologist's Record and Journal of Variation.* Vol. 122. pp. 128–136.

Porter, J. (1997) *The Colour Identification Guide to Caterpillars of the British Isles.* Viking. Middlesex.

Pratt, C. R. (2002) A modern review of the history of the Pine Hawk-moth *Sphinx pinastri* L. (Lep.: Sphingidae) in Britain, with a European perspective. *Entomologist's Record and Journal of Variation.* Vol. 114. pp. 235–268.

Priestley, S. (2018) The mysteries of sugar. Lancashire Moth Group Newsletter. No. 35. p. 10–11.

Pugh, T. R. (1857) *Plusia interrogationis. Entomologist's Weekly Intelligencer.* Vol. ii. p. 115.

Randle, Z., Evans-Hill, L. J., Parsons, M. S., Tyner, A., Bourn, N. A. D., Davis, A. M., Dennis, E. B., O'Donnell, M., Prescott, T., Tordoff, G. M. & Fox, R. (2019) *Atlas of Britain and Ireland's Larger Moths.* Pisces Publications, Newbury.

Ridout, B.V. (2016) The 'Manchester Tinea', *Euclemensia woodiella* (Curtis, 1830) (Lepidoptera: Cosmopterigidae), an entomological mystery unravelled. *Entomologist's Gazette.* Vol. 67. pp. 257–265.

Roberts, H. & Wrightson, A. L. (unpublished manuscript) (1986) *A list of the Lepidoptera of the Burnley District* (includes a 1985 list of 75 species from a Towneley garden).

Robinson, G. (1979) Clothes-moths of the *Tinea pellionella* complex: a revision of the world's species (Lepidoptera: Tineidae). *Bulletin of the British Museum (Natural History), Entomology.* Vol. 38. pp. 57–127.

Robson, J. E. (1893) Naturalists of the Day XVI: J.B. Hodgkinson, F.E.S. *British Naturalist.* Vol 3. pp. 101–103.

Rose, M. E., Falconer, K. & Holder, J. (2011) *Ancoats, Cradle of industrialisation.* English Heritage, Swindon.

Rothschild, Cockayne & Kettlewell coll., Natural History Museum (2014). *Collection specimens* [Data set]. Natural History Museum. https://doi.org/10.5519/0002965

Rothschild, M. & Marren, P. (1997) *Rothschild's Reserves. Time and Fragile Nature.* pp. 137–138. Harley Books, Colchester, Essex.

Rutherford, C. I. (1994) *The Macro-moths in Cheshire 1961 to 1993.* Lancashire & Cheshire Entomological Society, Brentwood.

Salmon, M. A. (2000) *The Aurelian Legacy: British Butterflies and their Collectors.* Harley Books, Colchester.

Seymour, P. R. & Kilby, L. J. (1978) Insects and other invertebrates intercepted in check inspections of imported plant material in England and Wales during 1978. *Report MAFF. Plant Pathology Laboratory.* Vol. 12. 1978. pp. 1–33.

Sheldon, W. G. (1925) *Peronea comariana* Zeller, and its variation. *Entomologist.* Vol. 58. pp. 281–285.

Sidebotham, J. (1884) The story of *Oecophora woodiella. Entomologist.* Vol. 17. pp. 52–54.

Skelcher, G. (2021) *Monitoring of the Belted Beauty Moth population at Potts Corner salt marsh, Middleton, with respect to horizontal directional drilling works carried out beneath the marsh in June 2016.* Final report: review 2015–2021, a report for Orsted.

Skinner, B. (2009) *Colour Identification Guide to the Moths of the British Isles, 3rd revised and updated edition.* Apollo Books, Stenstrup, Denmark.

Smart, B. (2018) *Micro-moth Field Tips.* Lancashire and Cheshire Fauna Society, Rishton.

Smart, B. (2021) *Micro-moth Field Tips, volume 2.* Lancashire and Cheshire Fauna Society, Rishton.

Smart, B. (2022) The status and distribution of the nationally rare micro-moth *Anacampsis temerella* (Lienig and Zeller, 1846) (Lepidoptera: Gelechiidae) in Lancashire in 2021, with observations on other *Salix repens* L. feeding Lepidoptera. *Entomologist's Gazette.* Vol. 73. pp. 97–116.

Smith, P. S. & Lockford, P. (2021) Fifteen years of habitat, floristic and vegetation change on a pioneer sand-dune and slack system at Ainsdale, north Merseyside, UK. *British & Irish Botany.* Vol. 3(2). pp. 232–262.

South, R. (1889) *Luperina testacea* Var. *nickerlii* Freyer. *Entomologist.* Vol. 22. pp. 271–272.

South, R. (1902) *Xylina furcifera* (*conformis*) in Lancashire. *Entomologist.* Vol. 35. p. 25.

South, R. (1948) *The Moths of the British Isles, First Series, New Edition 1948.* F. Warne & Co., London.

Spalding, A. (2023) The Sandhill Rustic moth in Britain and Ireland. *British Wildlife.* Vol. 34. pp. 549–557.

Speed, J. & Hondius, J. (1610) *The Countie Palatine of Lancaster described and divided into Hundreds.* Published in Speed J., 1632. *Theatre of the Empire of Great Britaine.* London.

Speidel, W., van Nieukerken, E., Honey, M. R., Koster, S. (2007) *The exotic pyraloid moth Diplopseustis perieresalis* (Walker) expanding in the West Palaearctic Region (Crambidae: Spilomelinae). *Nota Lepidopterologica.* Vol. 29 (3/4). pp. 185–192.

Stainton, H. T. (1854) *Insecta Brittanica. Lepidoptera Tineina.* London

Stainton, H. T. (1855) New British species since 1835. *Cerura bicuspis* Borkhausen. *Entomologist's Annual.* 1855. p. 28.

Stainton, H. T. (1855a) New British species since 1835. *Psyche marginenigrella* Bruand. *Entomologist's Annual.* 1855. p. 30.

Stainton, H. T. (1855b) New British species since 1835. *Acidalia obsoletaria* Rambur and *Schrankia turfosalis* Wocke. *Entomologist's Annual.* 1855. p. 44.

Stainton, H. T. (1855c) New British species since 1835. *Elachista gregsoni* Stainton, n. spp. *Entomologist's Annual.* 1855. p. 70.

Stainton, H. T. (1859) *A Manual of British Butterflies and Moths* Vol. II. J. van Voorst, London.

Stainton, H. T. (1859a) A new *Tinea. Entomologist's Weekly Intelligencer.* Vol. 6. p. 183.

Stainton, H. T. (1872) Larva of *Depressaria yeatiana. Entomologist's Monthly Magazine.* Vol. 9. p. 113.

Stainton, H. T. (1882) On two of the species of *Gelechia* which frequent our salt-marshes. *Entomologist's Monthly Magazine.* Vol. 19. pp. 251–253.

Stainton, H.T. (1884) Note on a new *Nepticula* bred from Rose in Lancashire by Mr Hodgkinson. *Entomologist's Monthly Magazine.* Vol. 21. p. 103.

Stephenson, H. (1859) *Catocala fraxini. Entomologist's Weekly Intelligencer.* Vol. 1859–1860. p. 27.

Sterling, P. & Parsons, M. (2012) *Field Guide to the Micromoths of Great Britain and Ireland.* British Wildlife Publishing, Dorset.

Stringer, H. (1943) Observations on species of Lepidoptera infesting stored products. X. *Lindera tessellatella* Blanchard, a tineid new to Britain. *Entomologist.* Vol. 76. pp. 177–181.

Sumner, D. P. (1996) The Watsonian Vice-Counties of the North West. *Lancashire and Cheshire Fauna Society: Publication no. 95.* pp. 6–15.

Sutton, S. L. & Beaumont, H. E. (1989) *Butterflies and Moths of Yorkshire.* Yorkshire Naturalists' Union, Doncaster.

Thorpe, J. (1870) On breeding *Acherontia atropos. Newman's Entomologist.* Vol. 5 (80/81). p. 143.

Thorpe, J. (1871) *Deiopeia pulchella* at Middleton. *Newman's Entomologist.* Vol. 5. p. 412.

Threlfall, J. H. (1877) Tineina bred during 1876. *Entomologist.* Vol. 10. pp. 75–76.

Threlfall, J. H. (1878) Notes on the Tineina bred in 1877 and 1878. *Entomologist's Monthly Magazine.* Vol. 15. pp. 89–90.

Tutt, J. W. (1890) Current notes [re exhibition of *Botys mutualis*, Zell.]. *Entomologist's Record and Journal of Variation.* Vol. 1. p. 33.

Uffen, R.W.J. (1960) An alternative food-plant for *Isophrictis tanacetella* Schrank (Lep., Gelechiidae) in Britain. *Entomologist's Gazette.* Vol. 11. p. 120.

Walker, J. (1904) Some notes on the Lepidoptera of the Curtis collection of British insects. *Entomologist's Monthly Magazine.* Vol. 15 (series 2). pp. 187–194.

Waring, P. (1992) *Moth Conservation Project, News Bulletin 4, April 1991 – June 1992.* Joint Nature Conservation Committee, Peterborough.

Waring, P. & Townsend, M. (2003) (2nd Edition) *Field Guide to the Moths of Great Britain and Ireland.* British Wildlife Publishing, Dorset.

Waring, P. & Townsend, M. (2017) (3rd Edition) *Field Guide to the Moths of Great Britain and Ireland.* Bloomsbury Publishing Plc. London.

Watson, H. C. (1852) *Cybele Britannica* (Vol. 3) pp. 524–528. Longman & Co., London.

Watson, W. A. (1954) Macrolepidoptera of the Fylde Coastal Area. *Entomologist's Record and Journal of Variation.* Vol. 66. pp. 154–155.

White, J. S. (1871) *Sphinx convolvuli* near Droylsden. *Newman's Entomologist.* Vol. 5. p. 411.

Windle, W. F. (1912) Lepidoptera of Oldham Park. [Publication details unknown].

Worthing, A. (1861) A New Noctua. *Entomologist's Weekly Intelligencer.* Vol. 9. p. 131.

Wright, A. E. (1911) The Macrolepidoptera of North East Lancashire. *Report and Transactions of the Manchester Entomological Society 1911.* pp. 25–44 (based on a radius of six miles of the centre of Burnley).

Wright A. E. & Clutten, W. G. (1901) *The Moths and Butterflies of the Burnley District.* Burnley Literary and Philosophical Society Journal.

Journals and Proceedings

Information not specifically referenced in the species text, was also obtained from:
British Journal of Entomology and Natural History
Entomologist's Gazette
Entomologist's Record and Journal of Variation
Lancashire and Cheshire Entomological Society (journals)
Lancashire and Cheshire Fauna Committee / Society (journals)
Newman's Entomologist
The Entomologist's Annual
The Entomologist (including Newman's Entomologist)
The Entomologist's Monthly Magazine
The Entomologist's Weekly Intelligencer
The Naturalist
The Young Naturalist
The Zoologist

Index of species

Bold numbers refer to the main account for each species.

Abraxas grossulariata **439**
   sylvata **440**
Abrostola tripartita **484**
   triplasia **484**
Acanthopsyche atra **77**
Acasis viretata **437**
Acentria ephemerella **370**
Acherontia atropos **383**
Achlya flavicornis **377**
Achroia grisella **334**
Acleris abietana 28, **257**
   aspersana **260**
   bergmanniana **256**
   caledoniana **256**
   comariana **256**, 257
   effractana **258**
   emargana **258**
   ferrugana **261**
   forsskaleana **255**
   hastiana 258, **260**
   holmiana **255**
   hyemana 24, **261**
   laterana **257**
   literana **262**
   logiana **262**
   maccana 627
   notana **261**
   rhombana **258**
   schalleriana **259**
   shepherdana **260**
   sparsana **257**
   umbrana **259**
   variegana **259**
Acompsia cinerella **164**
Acrobasis advenella **338**
   consociella **339**
   marmorea **338**
   repandana **338**
   tumidana 628
Acrocercops brongniardella **95**
Acrolepia autumnitella 19, **128**
Acrolepiopsis assectella **128**
Acronicta aceris **491**
   alni **490**
   leporina **491**
   menyanthidis **492**
   psi **491**
   rumicis **492**
   tridens **490**
Actebia praecox **554**
Adaina microdactyla **237**
Adela croesella **66**
   cuprella **66**

   reaumurella **65**
ADELIDAE **64–68**
Adscita geryon **331**
   statices **330**
Aethalura punctulata **456**
Aethes beatricella **269**
   cnicana **269**
   dilucidana **268**
   francillana **269**
   hartmanniana **267**
   rubigana **270**
   smeathmanniana **268**
   tesserana **268**
Agapeta hamana **266**
   zoegana **267**
Aglossa caprealis **345**
   pinguinalis **345**
Agnoea josephinae **144**
Agonopterix alstromeriana **152**
   angelicella **152**
   arenella **149**
   assimilella **150**
   atomella 56, **148**
   carduella **148**
   ciliella **150**
   cnicella 626
   conterminella **147**
   heracliana **150**
   kaekeritziana **151**
   liturosa **147**
   nanatella **151**
   nervosa **152**
   ocellana **146**
   propinquella **149**
   purpurea **147**
   scopariella **148**
   subpropinquella **149**
   umbellana **151**
   yeatiana **153**
Agriopis aurantiaria **453**
   leucophaearia **452**
   marginaria **453**
Agriphila geniculea **367**
   inquinatella **366**
   latistria **366**
   selasella **367**
   straminella **367**
   tristella **366**
Agrius convolvuli **382**
Agrochola lychnidis **523**
Agrotis clavis **557**
   exclamationis **556**
   ipsilon **559**

   puta **558**
   ripae **558**
   segetum **557**
   trux **558**
   vestigialis **557**
Alabonia geoffrella **142**
Alcis repandata **455**
Alder Kitten **463**
Alder Moth **490**
Aleimma loeflingiana **255**
Allophyes oxyacanthae **497**
Alsophila aescularia **449**
Alucita hexadactyla **230**
ALUCITIDAE **230**
Amblyptilia acanthadactyla **232**
   punctidactyla **232**
Amphipoea crinanensis **508**
   fucosa **507**
   lucens **507**
   oculea **508**
Amphipyra berbera **496**
   pyramidea **495**
   tragopoginis **496**
Amphisbatis incongruella 18, 26, **145**
Anacampsis blattariella **161**
   populella **161**, 162
   temerella 7, 21, **162**
Anania coronata **349**
   crocealis **350**
   funebris 22, **351**
   fuscalis **349**
   hortulata **352**
   lancealis **349**
   perlucidalis **350**
   terrealis **350**
Anaplectoides prasina **566**
Anarsia spartiella **163**
Anarta myrtilli **541**
   trifolii **541**
Anatrachyntis badia **158**
Anchoscelis helvola **524**
   litura **523**
   lunosa **525**
Ancylis achatana **288**
   apicella **288**
   badiana **288**
   comptana 627
   diminutana **287**
   geminana **287**
   mitterbacheriana **289**
   myrtillana 24, **287**
   obtusana **286**
   uncella **286**

unguicella **286**
Anerastia lotella 21, **344**
Angerona prunaria 628
Angle Shades 29, **504**
Angle-barred Pug **430**
Angle-striped Sallow **529**
Angoumois Grain Moth **165**
Annulet **459**
Anomalous **495**
Anorthoa munda **540**
Anthophila fabriciana **239**
Anticlea derivata **401**
Antigastra catalaunalis **356**
Antispila metallella 3, **63**
Antitype chi **534**
Antler Moth **541**
Apamea anceps **515**
   crenata **515**
   epomidion **514**
   furva **518**
   lithoxylaea **517**
   monoglypha **517**
   oblonga **516**
   remissa **514**
   scolopacina **516**
   sordens **515**
   sublustris **517**
   unanimis **516**
Apeira syringaria **445**
Aphelia viburnana **248**
Aphomia cephalonica **333**
   gularis **333**
   sociella **333**
Aplocera plagiata **436**
Apocheima hispidaria **450**
Apodia martinii **171**
Apomyelois bistriatella **339**
   ceratoniae **339**
Aporophyla luneburgensis **534**
   lutulenta **534**
   nigra **535**
Apotomis betuletana **273**
   capreana **274**
   lineana 627
   sauciana **274**
   semifasciana **273**
   sororculana **274**
   turbidana **273**
Apple & Plum Case-bearer **195**
Apple Ermine **114**
Apple Fruit Moth **133**
Apple Leaf Miner **135**
Apple Leaf Skeletonizer **240**
Apple Pith Moth **221**
Apple Pygmy **43**
Aproaerema anthyllidella **160**
   cinctella **159**
   larseniella **160**
   sangiella **159**
   taeniolella **160**
Apterogenum ypsillon **533**

Archanara dissoluta **512**
Archer's Dart 21, **557**
Archiearis parthenias **439**
Archips crataegana 627
   podana **243**
   rosana **244**
   xylosteana **243**
Arctia caja **475**
   villica 629
Arenostola phragmitidis **510**
Argent and Sable **418**
Argyresthia albistria **135**
   arceuthina **129**
   aurulentella 626
   bonnetella **134**
   brockeella **131**
   conjugella **133**
   cupressella **130**
   curvella **132**
   dilectella **130**
   glabratella **128**, **129**
   glaucinella **133**
   goedartella **131**
   ivella **130**
   laevigatella **128**, 129
   pruniella **134**
   pygmaeella **131**
   retinella **132**
   semifusca **134**
   semitestacella **135**
   sorbiella **132**
   spinosella **133**
   trifasciata 28, 29, **129**
ARGYRESTHIIDAE **128–135**, 626
Argyroploce arbutella 627
   corticana 627
Argyrotaenia ljungiana **244**
Aristotelia brizella **170**
   ericinella **170**
   unicolorella 626
Aroga velocella **177**
Ash Bud Moth **137**
Ash Pug **430**
Asian Cotton Leafworm **500**
Aspilapteryx tringipennella **94**
Aspitates gilvaria 629
Asteroscopus sphinx **496**
Asthena albulata **415**
Atemelia torquatella **137**
Atethmia centrago **532**
Athrips mouffetella **176**
   tetrapunctella 626
Atolmis rubricollis **478**
August Thorn **445**
Autographa bractea **487**
   gamma **486**
   jota **487**
   pulchrina **487**
AUTOSTICHIDAE **138**
Autumn Green Carpet **410**
Autumnal Moth **414**

Autumnal Rustic **569**
Auximobasis normalis 629
Axylia putris **559**
Azalea Leaf Miner 28, **92**
Bactra furfurana **284**
   lacteana **284**
   lancealana **284**
   robustana 19, **285**
Bankesia conspurcatella **76**
Barred Carpet 22, **421**
Barred Chestnut **560**
Barred Fruit-tree Tortrix **246**
Barred Hook-tip **373**
Barred Red 21, **459**
Barred Rivulet 17, **420**
Barred Sallow **521**
Barred Straw **409**
Barred Tooth-striped 22, **438**
Barred Umber **443**
Barred Yellow **406**
Batia lunaris **142**
Batrachedra confusella **192**
   pinicolella 626
   praeangusta **191**
BATRACHEDRIDAE **191–192**, 626
Beaded Chestnut **523**
Beautiful Brocade **543**
Beautiful Carpet **401**
Beautiful China-mark 18, **371**
Beautiful Golden Y **487**
Beautiful Hook-tip **482**
Beautiful Marbled **574**
Beautiful Plume **232**
Beautiful Snout **470**
Beautiful Yellow Underwing 19, 21, 26, **541**
Bedellia somnulentella 17, **138**
BEDELLIIDAE **138**
Bedstraw Hawk-moth **385**
Bee Moth **333**
Beech-green Carpet **411**
Belted Beauty 19, **451**
Bembecia ichneumoniformis **329**
Bena bicolorana **570**
Bilberry Pug **423**
Bilberry Tortrix 24, **248**
Birch Mocha **393**
Bird's Wing **503**
Bird-cherry Ermine **113**
Biston betularia 3, **452**
   strataria **452**
Black Arches **470**
Black Rustic **535**
Blackneck 28, **481**
Blair's Shoulder-knot 28, 29, **528**
BLASTOBASIDAE **227–228**
Blastobasis adustella **227**
   lacticolella **228**
   rebeli **228**
Blastodacna atra **221**
   hellerella **221**

Bleached Pug 22, **433**
Blomer's Rivulet **417**
Blood-vein **392**
Blossom Underwing 22, **537**
Blotched Emerald **461**
Blue-bordered Carpet **405**
Bohemannia pulverosella **56**
    quadrimaculella **57**
Bordered Beauty **444**
Bordered Gothic **546**
Bordered Grey **454**
Bordered Pearl **352**
Bordered Pug **435**
Bordered Sallow **497**
Bordered Straw **498**
Bordered White 21, **457**
Borkhausenia fuscescens **141**
    minutella **141**
Box-tree Moth 28, **356**
Brachmia blandella **164**
Brachylomia viminalis **532**
Bramble Shoot Moth **307**
Brick **525**
Bright-line Brown-eye 29, **544**
Brimstone Moth 17, **444**
Brindled Beauty **450**
Brindled Green **533**
Brindled Ochre **535**
Brindled Plume **232**
Brindled Pug **426**
Broad-barred White **547**
Broad-bordered Yellow Underwing **563**
Broken-barred Carpet **406**
Broom Moth **544**
Brown China-mark **369**
Brown House-moth **140**
Brown Oak Tortrix **627**
Brown Plume **232**
Brown Rustic **502**
Brown Scallop **417**
Brown Silver-line **443**
Brown-dotted Clothes Moth **83**
Brown-line Bright-eye **549**
Brown-spot Pinion **523**
Brown-tail **471**
Brown-veined Wainscot **512**
Brussels Lace **469**
Bryophila domestica **498**
Bryopsis muralis **499**
Bryotropha affinis **169**
    basaltinella **168**
    boreella **168**
    desertella **167**
    domestica **166**
    politella 24, 164, **167**
    senectella **168**
    similis **169**
    terrella **167**
    umbrosella **169**
BUCCULATRICIDAE **87–90**

Bucculatrix albedinella **88**
    bechsteinella **90**
    cidarella **89**
    cristatella **87**
    demaryella **90**
    maritima 19, **88**
    nigricomella **88**
    thoracella **89**
    ulmella **89**
Bud Moth **291**
Buff Arches **374**
Buff Ermine **473**
Buff Footman **478**
Buff-tip **467**
Bulrush Wainscot 18, **510**
Bupalus piniaria **457**
Burnet Companion **483**
Burnished Brass **486**
Butterbur **507**
Cabbage Moth 29, **545**
Cabera exanthemata **457**
    pusaria **457**
Cacao Moth **342**
Cacoecimorpha pronubana **247**
Cadra calidella **344**
    cautella **343**
    figulilella **343**
Calamotropha paludella **363**
Callimorpha dominula **475**
Callistege mi **484**
Callisto denticulella **96**
Calliteara pudibunda **472**
Calophasia lunula **495**
Caloptilia alchimiella **92**
    azaleella **92**
    betulicola **91**
    cuculipennella **90**
    elongella **91**
    populetorum **91**
    robustella 92, **93**
    rufipennella **92**
    semifascia **93**
    stigmatella **93**
Calybites phasianipennella 17, **95**
Cameraria ohridella 28, **112**
Campaea margaritaria **459**
Campion **546**
Camptogramma bilineata **398**
Canary-shouldered Thorn **446**
Capperia britanniodactylus **234**
Capua vulgana **242**
Caradrina clavipalpis **500**
    morpheus **500**
Carcina quercana **145**
Carnation Tortrix **247**
Carpatolechia alburnella **188**
    decorella **188**
    fugitivella **188**
    notatella **189**
    proximella **189**
Carsia sororiata **436**

Caryocolum alsinella **184**
    blandella **185**
    fraternella **185**
    hubneri 626
    junctella 626
    kroesmanniella 626
    marmorea 21, **185**
    proxima 626
    tricolorella **186**
    viscariella **184**
Case-bearing Clothes Moth **81**
Cataclysta lemnata **370**
Catocala fraxini **483**
    nupta **483**
    sponsa 629
Catoptria falsella **368**
    margaritella **368**
    pinella **368**
Catorhoe rubidata 628
Cauchas fibulella **66**
    rufimitrella **67**
Cedestis gysseleniella **118**
    subfasciella **118**
Celaena haworthii **505**
Celypha cespitana **277**
    lacunana **278**
    rivulana **278**
    rufana **277**
    striana **277**
Centre-barred Sallow **532**
Ceramica pisi **544**
Cerapteryx graminis **541**
Cerastis leucographa **562**
    rubricosa **561**
Cereal Stem Moth **123**
Cerura vinula **463**
Chalk Carpet 628
Chamomile Shark **494**
Charanyca trigrammica **502**
Charissa obscurata **459**
Chequered Fruit-tree Tortrix **246**
Cherry Bark Tortrix **285**
Cherry Fruit Moth **134**
Chesias legatella **436**
Chestnut **526**
Chestnut-coloured Carpet **404**
Chevron **408**
Chiasmia clathrata **443**
Chilo phragmitella 18, **362**
Chilodes maritima **502**
CHIMABACHIDAE **143–144**
Chimney Sweeper **419**
Chinese Character **373**
Chionodes distinctella **178**
    fumatella **178**
Chlorissa viridata 629
Chloroclysta miata **410**
    siterata **410**
Chloroclystis v-ata **422**
Chocolate-tip **468**
CHOREUTIDAE **239–240, 572**

Choreutis pariana **240**
Choristoneura hebenstreitella **244**
Chrysoclista lathamella **222**
  linneella **222**
Chrysodeixis acuta **575**
  chalcites **485**
Chrysoesthia drurella 17, **166**
  sexguttella **166**
Chrysoteuchia culmella **363**
Cidaria fulvata **406**
Cilix glaucata **373**
Cinnabar **476**
Cirrhia gilvago **523**
  icteritia **522**
Cistus Forester 22, **331**
Clavigesta purdeyi 309, **310**
Clay **552**
Clay Triple-lines **393**
Cleorodes lichenaria **460**
Clepsis consimilana **249**
  senecionana **248**
  spectrana **249**
Clifden Nonpareil **483**
Cloaked Carpet **574**
Cloaked Minor **519**
Cloaked Pug **424**
Clostera curtula **468**
  pigra **468**
Clouded Border **440**
Clouded Brindle **514**
Clouded Buff 19, 26, **474**
Clouded Drab **537**
Clouded Magpie 23, **440**
Clouded Silver **458**
Clouded-bordered Brindle **515**
Clover Case-bearer **198**
Cnephasia asseclana **253**
  conspersana **254**
  genitalana **253**
  incertana **252**
  longana **254**
  pasiuana **253**
  stephensiana **252**
Coast Dart **555**
Cochylichroa atricapitana **272**
Cochylidia implicitana **270**
  rupicola **270**
Cochylimorpha straminea **264**
Cochylis roseana **271**
Cocksfoot Moth **127**
Codling Moth 7, 29, **316**
Coenobia rufa **512**
Coenotephria salicata **412**
Coleophora adjunctella 19, 173, **203**
  adspersella **209**
  albella **626**
  albicosta **202**
  albidella **200**
  alcyonipennella **198**
  alticolella **204**
  anatipennella **200**

argentula **209**
artemisicolella 26, 28, **207**
asteris **206**
atriplicis **207**
betulella **201**
binderella **197**
caespititiella **203**
coracipennella **194**, 195
currucipennella **201**
deauratella **199**
discordella **199**
flavipennella **193**
frischella **198**
fuscocuprella **195**
glaucicolella **204**
gryphipennella **192**
ibipennella **201**
inulae **208**
juncicolella **196**
kuehnella **200**
laricella **202**
limosipennella **193**
lineolea **198**
lusciniaepennella **196**
lutipennella **192**, 193
maritimella **205**
mayrella **199**
milvipennis **193**
niveicostella **626**
orbitella 5, **197**
otidipennella **204**
paripennella **209**
peribenanderi **208**
potentillae **196**
prunifoliae **194**, **195**
pyrrhulipennella **202**
salicorniae **210**
salinella **626**
saxicolella **206**
serratella **194**
siccifolia 17, **194**
spinella 194, **195**
striatipennella **208**
taeniipennella **205**
tamesis **203**
therinella **627**
trifolii **197**
versurella **206**
vestianella **207**
virgaureae **205**
vitisella **626**
COLEOPHORIDAE **192–210**, 626–627
Colocasia coryli **489**
Colostygia multistrigaria **412**
  olivata **411**
  pectinataria **411**
Colotois pennaria **449**
Comibaena bajularia **461**
Common Carpet **399**
Common Clothes Moth **81**
Common Emerald **462**

Common Fan-foot **629**
Common Footman **479**
Common Heath 26, **456**
Common Lutestring **375**
Common Marbled Carpet **410**
Common Plume **237**
Common Pug **434**
Common Quaker **538**
Common Rustic **518**
Common Swift 3, **38**
Common Wainscot **550**
Common Wave **457**
Common White Wave **457**
Conformist **527**
Confused **518**
Conistra ligula **526**
  vaccinii **526**
Convolvulus Hawk-moth **382**
Copper Underwing **495**
Coptotriche angusticollella **74**
  marginea **73**
Cork Moth **79**
Corn Moth **79**
Coronet **493**
Cosmia affinis **531**
  diffinis **530**
  pyralina **531**
  trapezina **531**
Cosmopolitan **553**
COSMOPTERIGIDAE **156–159**, 626
Cosmorhoe ocellata **406**
COSSIDAE **326–327**
Cossus cossus **326**
Coxcomb Prominent **467**
CRAMBIDAE **346–372**, **573**, 628
Crambus ericella **364**
  hamella 26, 29, **365**
  lathoniellus **365**
  pascuella **363**
  perlella **365**
  pratella **364**
  uliginosellus **364**
Craniophora ligustri **493**
Crassa tinctella **626**
  unitella **141**
Cream Wave **392**
Cream-bordered Green Pea 29, **571**
Cream-spot Tiger **629**
Crescent 18, **505**
Crescent Dart **558**
Crescent Plume **234**
Crescent Striped 19, **516**
Crimson Speckled **476**
Crinan Ear **508**
Crocallis elinguaria **448**
Crocidosema plebejana **300**
Cryphia algae **575**
Cryptoblabes bistriga **334**
Cucullia absinthii **493**
  chamomillae **494**
  umbratica **494**

verbasci **494**
Currant Clearwing 29, **329**
Currant Pug **433**
Currant Shoot Borer **70**
Cybosia mesomella **477**
Cyclamen Tortrix **249**
Cyclophora albipunctata **393**
    linearia **393**
    punctaria **393**
Cydalima perspectalis **356**
Cydia coniferana 3, **315**
    cosmophorana **316**
    fagiglandana **317**
    nigricana **315**
    pomonella 7, **316**
    splendana **317**
    trobilella **316**
    ulicetana **315**
Cymatophorina diluta **376**
Cypress Carpet **405**
Cypress Pug **427**
Cypress Tip Moth 28, **130**
Dahlica inconspicuella 3, **75**
    lichenella **75**
Daphnis nerii **384**
Dark Arches **517**
Dark Bordered Beauty **628**
Dark Brocade **536**
Dark Chestnut **526**
Dark Crimson Underwing **629**
Dark Dagger **490**
Dark Fruit-tree Tortrix **246**
Dark Marbled Carpet **411**
Dark Spectacle **484**
Dark Spinach **401**
Dark Sword-grass **559**
Dark Tussock **472**
Dark Umber **417**
Dark-barred Twin-spot Carpet **397**
Dasypolia templi **535**
Dasystoma salicella **144**
Death's-head Hawk-moth **383**
December Moth **377**
Deep-brown Dart **534**
Deilephila elpenor **386**
    porcellus **386**
Deileptenia ribeata **455**
Delicate **551**
Delplanqueia dilutella **627**
Deltote uncula **489**
Denisia albimaculea **139**
    similella **139**
    subaquilea **139**
Denticucullus pygmina **512**
Depressaria albipunctella **155**
    badiella **153**
    chaerophylli **156**
    daucella **154**
    douglasella **155**
    emeritella **629**
    pimpinellae **626**

    pulcherrimella **154**
    radiella **153**
    sordidatella **155**
    ultimella **154**
DEPRESSARIIDAE **145–156**, 626
Devon Carpet **413**
Diachrysia chrysitis **486**
    chryson **485**
Diacrisia sannio **474**
Diamond-back Moth 17, 29, **124**
Diaphora mendica **473**
Diarsia brunnea **560**
    dahlii **560**
    florida **561**
    mendica **560**
    rubi **561**
Dicallomera fascelina **472**
Dichomeris marginella **164**
Dichrorampha acuminatana **312**
    aeratana **312**
    alpestrana **314**
    alpinana **313**
    consortana **627**
    flavidorsana **313**
    petiverella **314**
    plumbagana **314**
    plumbana **311**
    sedatana **311**
    simpliciana 28, **312**
    vancouverana **313**
Digitivalva pulicariae **127**
Diloba caeruleocephala **490**
Dingy Footman **478**
Dingy Shears **533**
Dingy Shell **415**
Dingy White Plume **235**
Dioryctria abietella **337**
    simplicella **336**
    sylvestrella **336**
Diplodoma laichartingella **74**
Diplopseustis perieresalis **355**
Ditula angustiorana **241**
Diurnea fagella **143**
    lipsiella **144**
Dog's Tooth **543**
Donacaula forficella 18, **371**
    mucronella **372**
Dot Moth **544**
Dotted Border **453**
Dotted Border Wave **389**
Dotted Clay **566**
Dotted Fan-foot **629**
Dotted Rustic **562**
Double Dart **565**
Double Kidney **530**
Double Line **549**
Double Lobed **518**
Double Square-spot **568**
Double-striped Pug **422**
Dowdy Plume **233**
Drepana falcataria **373**

DREPANIDAE **372–377**
Dried Currant Moth **343**
Dried Fruit Moth **344**
Drinker **380**
Drymonia dodonaea **464**
    ruficornis **464**
Dryobotodes eremita **533**
Dun-bar **531**
Duponchelia fovealis **355**
Dusky Brocade **514**
Dusky Plume **236**
Dusky Sallow **506**
Dusky Thorn **446**
Dusky-lemon Sallow **523**
Dwarf Cream Wave **388**
Dwarf Pug **428**
Dypterygia scabriuscula **503**
Dyscia fagaria **460**
Dyseriocrania subpurpurella **35**
Dysstroma citrata **411**
    truncata **410**
Eana incanana **251**
    osseana **251**
    penziana **252**
    penziana f. *bellana* **252**
    penziana f. *colquhounana* **252**
Ear Moth **508**
Earias clorana **571**
Early Grey **497**
Early Moth **458**
Early Thorn **447**
Early Tooth-striped **439**
Earophila badiata **400**
Ecliptopera silaceata **409**
Ectoedemia albifasciella **59**
    angulifasciella **60**
    arcuatella 22, **61**
    argyropeza **59**
    atricollis **61**
    heringi **60**
    intimella **59**
    minimella **62**
    occultella **61**
    subbimaculella **60**
Ectropis crepuscularia **455**
Eidophasia messingiella **125**
Eilema caniola **629**
    complana **479**
    depressa **478**
    griseola **478**
    lurideola **479**
    sororcula **479**
Elachista adscitella **213**
    albidella 26, **220**
    albifrontella **216**
    alpinella 3, **215**
    apicipunctella **216**
    argentella **210**
    atricomella **215**
    bedellella **627**
    biatomella **214**

bisulcella **213**
canapennella **217**
cinereopunctella **219**
cingillella 22, **212**
eleochariella **220**
freyerella **221**
gangabella 22, **212**
gleichenella **213**
humilis **217**
kilmunella **215**
luticomella **216**
maculicerusella 18, **218**
obliquella 627
poae **214**
rufocinerea **218**
scirpi 14, 19, 173, **219**
serricornis **219**
subalbidella **212**
subnigrella **217**
subocellea **211**
trapeziella **218**
triatomea **211**
triseriatella **211**
utonella **220**
ELACHISTIDAE **210–221**, 627
Electrophaes corylata **406**
Elephant Hawk-moth **386**
Elophila nymphaeata **369**
Ematurga atomaria **456**
Emmelina monodactyla **237**
Emperor Moth 19, 21, 24, **381**
Enargia paleacea **529**
Enarmonia formosana **285**
Endothenia ericetana **283**
   gentianaeana **281**
   marginana **282**
   nigricostana **283**
   oblongana **282**
   quadrimaculana **283**
   ustulana 22, **282**
Endotricha flammealis **346**
ENDROMIDAE 628
Endromis versicolora 628
Endrosis sarcitrella **140**
Engrailed **455**
Ennomos alniaria **446**
   autumnaria 629
   erosaria **446**
   fuscantaria **446**
   quercinaria **445**
Entephria caesiata **402**
   flavicinctata 628
Epagoge grotiana **242**
Epermenia aequidentellus **238**
   chaerophyllella **239**
   falciformis **239**
EPERMENIIDAE **238–239**
Ephestia elutella **342**
   kuehniella **342**
   woodiella **343**
Epiblema cirsiana **305**

costipunctana **306**
foenella **305**
scutulana **305**
sticticana **304**
turbidana **306**
Epinotia abbreviana **294**
   bilunana **299**
   brunnichana **293**
   caprana **292**
   cinereana 627
   cruciana **294**
   demarniana **296**
   immundana **295**
   maculana **293**
   mercuriana **295**
   nanana **295**
   nigricana 627
   nisella **299**
   pygmaeana **297**
   ramella **298**
   rubiginosana **298**
   signatana **294**
   solandriana **293**
   sordidana **292**
   subocellana **296**
   subsequana **297**
   tedella **298**
   tenerana **297**
   tetraquetrana **296**
   trigonella **292**
Epione repandaria **444**
   vespertaria 628
Epiphyas postvittana **249**
Epirrhoe alternata **399**
   galiata **400**
   rivata **399**
   tristata **399**
Epirrita autumnata **414**
   christyi **414**
   dilutata **414**
   filigrammaria **415**
Erannis defoliaria **453**
EREBIDAE **468–484, 574**, 629
Eremobia ochroleuca **506**
Eriocrania cicatricella **37**
   salopiella **37**
   sangii **38**
   semipurpurella **37**
   sparrmannella **36**
   unimaculella **36**
ERIOCRANIIDAE **35–38**
Eriogaster lanestris **378**
Eriopsela quadrana **289**
Esperia sulphurella **142**
Essex Emerald 629
Etainia decentella 5, **58**
   louisella **57**
   sericopeza **57**
Eublemma parva **482**
   purpurina **574**
Euchoeca nebulata **415**

Euchromius ocellea **362**
Euclemensia woodiella **157–158**
Euclidia glyphica **483**
Eucosma aemulana 627
   aspidiscana **302**
   campoliliana **302**
   cana **301**
   fulvana 627
   hohenwartiana **301**
   obumbratana **301**
   tripoliana 19, **302**
Eucosmomorpha albersana **285**
Eudemis profundana **272**
Eudonia angustea **360**
   delunella **361**
   lacustrata **360**
   mercurella **361**
   murana 23, **360**
   pallida **362**
   truncicolella **361**
Eugnorisma glareosa **569**
Eulia ministrana **262**
Eulithis mellinata **409**
   populata **408**
   prunata **408**
   testata **408**
Euphyia biangulata **574**
   unangulata **400**
Eupithecia abbreviata **426**
   abietaria **424**
   absinthiata **432**
   assimilata **433**
   centaureata **431**
   distinctaria **431**
   dodoneata **427**
   exiguata **434**
   expallidata **433**
   haworthiata **423**
   icterata **434**
   indigata **430**
   innotata **430**
   intricata **432**
   inturbata **424**
   lariciata **429**
   linariata **425**
   nanata **430**
   phoeniceata **427**
   pimpinellata **429**
   plumbeolata **425**
   pulchellata **425**
   pusillata **427**
   pygmaeata **426**
   satyrata **432**
   simpliciata **429**
   subfuscata **435**
   subumbrata **435**
   succenturiata **435**
   tantillaria **428**
   tenuiata **424**
   tripunctaria **428**
   trisignaria **431**

valerianata **433**
venosata **426**
virgaureata **428**
vulgata **434**
Euplexia lucipara **505**
Eupoecilia angustana **267**
Euproctis chrysorrhoea **471**
   similis **471**
Eupsilia transversa **529**
Eurois occulta **565**
European Corn-borer **352**
Euspilapteryx auroguttella **94**
Eustroma reticulata **407**
Euthrix potatoria **380**
Euxoa cursoria **555**
   nigricans **556**
   obelisca **555**
   tritici **555**
Euzophera pinguis **340**
Evergestis extimalis **357**
   forficalis **357**
   pallidata **357**
Exaeretia allisella **146**
Exapate congelatella **250**
Exoteleia dodecella **191**
Eyed Hawk-moth **382**
Falcaria lacertinaria **372**
Falseuncaria ruficiliana **272**
Fan-foot **480**
Feathered Gothic **540**
Feathered Ranunculus **535**
Feathered Thorn **449**
Feltia subterranea **629**
Fen Square-spot **561**
Fen Wainscot **510**
Figure of Eight **490**
Figure of Eighty **374**
Firethorn Leaf Miner 28, **102**
Five-spot Burnet **627**
Flame **559**
Flame Carpet **397**
Flame Shoulder **559**
Flax Tortrix **253**
Flounced Chestnut **524**
Flounced Rustic 4, **508**
Fomoria septembrella **58**
   weaveri **58**, 101
Forester 21, **330**
Four-dotted Footman 18, 22, **477**
Four-spotted Footman **477**
Fox Moth 19, 21, **380**
Foxglove Pug **425**
Freyer's Pug 29, **432**
Frosted Green **377**
Frosted Orange 17, **506**
Fruitlet Mining Tortrix **323**
Furcula bicuspis **463**
   bifida **464**
   furcula **463**
Galium Carpet **400**
Galleria mellonella **334**

Gandaritis pyraliata **409**
Garden Carpet **396**
Garden Dart **556**
Garden Pebble 29, **357**
Garden Rose Tortrix **259**
Garden Tiger **475**
Gelechia cuneatella 29, **179**
   nigra **179**
   nigricostella **626**
   rhombella **178**
   sororculella **179**
GELECHIIDAE **159–191**, 626
Gem **396**
Geometra papilionaria **461**
GEOMETRIDAE **387–462**, **574**, 628–629
Ghost Moth **40**
Gillmeria ochrodactyla **231**
   pallidactyla **231**
Glaucous Shears 24, **545**
GLYPHIPTERIGIDAE **125–128**
Glyphipterix equitella **126**
   fuscoviridella **126**
   haworthana 26, **126**
   schoenicolella **127**
   simpliciella **127**
   thrasonella **125**
Goat Moth **326**
Gold Spangle **487**
Gold Spot **488**
Gold Swift **39**
Gold Triangle **345**
Golden Plusia 28, **486**
Golden Twin-spot **485**
Golden-rod Brindle **528**
Golden-rod Pug **428**
Goldenrod Plume **627**
Gortyna flavago **506**
Gothic **569**
Gracillaria syringella **94**
GRACILLARIIDAE **90–112**, 625–626
Graphiphora augur **565**
Grapholita compositella **318**
   funebrana 7, **320**
   internana **318**
   janthinana **320**
   jungiella **319**
   lathyrana 56, **319**
   lobarzewskii **573**
   lunulana **318**
   molesta **319**
   tenebrosana **320**
Grass Eggar 21, **379**
Grass Emerald **461**
Grass Rivulet **421**
Grass Wave 18, **460**
Great Brocade **565**
Great Oak Beauty **629**
Great Prominent 23, **465**
Green Arches **566**
Green Carpet **411**
Green Oak Tortrix 23, **254**

Green Pug **423**
Green Silver-lines **571**
Green-brindled Crescent **497**
Grey Arches **542**
Grey Birch **456**
Grey Chi **534**
Grey Dagger **491**
Grey Mountain Carpet **402**
Grey Pine Carpet 21, **404**
Grey Pug **435**
Grey Scalloped Bar **460**
Grey Shoulder-knot **527**
Grey Tortrix **252**
Griposia aprilina **533**
Gymnoscelis rufifasciata **422**
Gynnidomorpha alismana **266**
   vectisana 19, **266**
Gypsonoma aceriana **304**
   dealbana **303**
   oppressana **303**
   sociana **304**
Gypsy Moth **471**
Habrosyne pyritoides **374**
Hada plebeja **545**
Hadena bicruris **547**
   compta **548**
   confusa **548**
   perplexa **548**
Haplotinea insectella **85**
Haworth's Minor 24, **505**
Haworth's Pug **423**
Hawthorn Moth **138**
Heart and Club **557**
Heart and Dart 17, 29, **556**
Heath Rivulet **628**
Heath Rustic **567**
Hebrew Character 29, **539**
Hecatera bicolorata **547**
Hecatera dysodea **547**
Hedge Rustic **540**
Hedya atropunctana **276**
   nubiferana **275**
   ochroleucana **276**
   pruniana **276**
   salicella **275**
Helcystogramma rufescens **165**
Helicoverpa armigera **498**
Heliothis peltigera **498**
Heliozela hammoniella **64**
   resplendella **63**
   sericiella 23, **63**
HELIOZELIDAE **63–64**
Hellinsia lienigianus **236**
   osteodactylus **237**
   tephradactyla **236**
Hellula undalis **358**
Helotropha leucostigma **505**
Hemaris tityus **384**
Hemistola chrysoprasaria **462**
Hemithea aestivaria **462**
Hemp Agrimony Plume **237**

HEPIALIDAE **38–40**
Hepialus humuli **40**
Herald **468**
Herminia grisealis **480**
  tarsipennalis **480**
Hippotion celerio **387**
Hoary Footman **629**
Hofmannophila pseudospretella **140**
Holly Tortrix **291**
Hollyhock Seed Moth **165**
Homoeosoma nimbella **628**
  sinuella **340**
Honeysuckle Moth **119**
Hoplodrina ambigua **501**
  blanda **501**
  octogenaria **501**
Horisme vitalbata **628**
Hornet Moth **327**
Horse-chestnut Leaf-miner 28, **112**
Humming-bird Hawk-moth 3, **384**
Hydraecia micacea **506**
  petasitis **507**
Hydrelia flammeolaria **416**
Hydrelia sylvata **416**
Hydriomena furcata **402**
  impluviata **403**
  ruberata **403**
Hylaea fasciaria **459**
Hyles euphorbiae **385**
  gallii **385**
  lineata **628**
  livornica **385**
Hypatima rhomboidella **163**
Hypena crassalis **470**
  proboscidalis **469**
Hypenodes humidalis **480**
Hypochalcia ahenella **337**
Hypomecis punctinalis **628**
  roboraria **629**
Hyppa rectilinea **504**
Hypsopygia costalis **345**
  glaucinalis **346**
Hysterophora maculosana **264**
Idaea aversata **390**
  biselata **389**
  dilutaria **628**
  dimidiata **389**
  emarginata **390**
  fuscovenosa **388**
  humiliata **628**
  muricata **387**
  rusticata **387**
  seriata **388**
  straminata **390**
  subsericeata **388**
  sylvestraria **389**
  trigeminata **628**
Incurvaria masculella **69**
  oehlmanniella 48, **69**
  pectinea **68**
  praelatella **70**

INCURVARIIDAE **68–70**
Indian Meal Moth **342**
Infurcitinea albicomella **625**
  argentimaculella **77**
Ingrailed Clay **560**
Ipimorpha retusa **530**
  subtusa **530**
Iron Prominent **465**
Isabelline Tiger **475**
Isle of Wight Wave **628**
Isophrictis striatella **170**
Isotrias rectifasciana **241**
Jodis lactearia **462**
July Belle **395**
July Highflyer **402**
Juniper Carpet 29, **405**
Juniper Pug 29, **427**
Juniper Webber **164**
Kent Black Arches **569**
Kentish Glory **628**
Knot Grass **492**
Korscheltellus fusconebulosa **39**
  lupulina **38**
Laburnum Leaf Miner **136**
Lacanobia contigua **543**
  oleracea **544**
  suasa **543**
  thalassina **543**
  w-latinum **629**
Lace Border **628**
Lackey **378**
Laelia coenosa **629**
Lampronia capitella **70**
  corticella **71**
  fuscatella 26, 28, **72**
  luzella **71**
  morosa **71**
  pubicornis **625**
Lampropteryx otregiata **413**
  uffumata **412**
Laothoe populi **382**
Larch Case-bearer **202**
Larch Pug **429**
Larch Tortrix **299**
Larentia clavaria **402**
Large Ear **507**
Large Emerald **461**
Large Fruit-tree Tortrix **243**
Large Nutmeg **515**
Large Pale Clothes Moth **82**
Large Ranunculus **536**
Large Red-belted Clearwing **328**
Large Tabby **345**
Large Thorn **629**
Large Twin-spot Carpet **398**
Large Wainscot **510**
Large Yellow Underwing 29, **563**
Lasiocampa quercus **380**
  trifolii **379**
LASIOCAMPIDAE **377–381**
Laspeyria flexula **482**

Lateroligia ophiogramma **518**
Lathronympha strigana **317**
Latticed Heath **443**
Lead Belle **394**
Lead-coloured Drab **538**
Lead-coloured Pug **425**
Least Black Arches **570**
Least Carpet **387**
Least Minor 7, 22, **513**
Least Yellow Underwing **564**
Leek Moth 29, **128**
Lempke's Gold Spot **488**
Lenisa geminipuncta **511**
Leopard Moth **327**
Leptologia lota **524**
  macilenta **525**
Lesser Broad-bordered Yellow
  Underwing **564**
Lesser Common Rustic **519**
Lesser Cream Wave 22, **391**
Lesser Lichen Case-bearer **75**
Lesser Swallow Prominent **466**
Lesser Wax Moth **334**
Lesser Yellow Underwing 29, **564**
Lesser-spotted Pinion **531**
Leucania comma **552**
  loreyi **553**
  obsoleta **553**
Leucoma salicis **470**
Leucoptera laburnella **136**
  malifoliella **136**
  spartifoliella **136**
Lichen Case-bearer **75**
Ligdia adustata **441**
Light Arches **517**
Light Brocade **629**
Light Brown Apple Moth 28, 29, **249**
Light Emerald **459**
Light Grey Tortrix **252**
Light Knot Grass 24, **492**
Lilac Beauty 29, **445**
Lime Hawk-moth **381**
Lime-speck Pug **431**
Limnaecia phragmitella 18, **158**
Lindera tessellatella **85**
Ling Pug **432**
Lithophane furcifera **527**
  leautieri **528**
  ornitopus **527**
  semibrunnea **526**
  socia **527**
Lithosia quadra **477**
Litoligia literosa **519**
Little Emerald **462**
Lobesia abscisana **280**
  littoralis 29, **281**
  reliquana **281**
Lobophora halterata **437**
Locust Bean Moth **339**
Lomaspilis marginata **440**
Lomographa bimaculata **458**

temerata **458**
Longalatedes elymi **511**
Loxostege sticticalis **346**
Lozotaenia forsterana **247**
Lozotaeniodes formosana **250**
Luffia lapidella **76**
Lunar Hornet Moth 23, **327**
Lunar Marbled Brown **464**
Lunar Thorn **447**
Lunar Underwing **525**
Lunar Yellow Underwing **629**
Lunar-spotted Pinion **531**
Luperina nickerlii 4, **509**
    testacea 4, **508**
Lychnis **547**
Lycia hirtaria **450**
    zonaria **451**
Lycophotia porphyrea **562**
Lygephila pastinum **481**
Lymantria dispar **471**
    monacha **470**
Lyme Grass 21, **511**
Lyonetia clerkella **135**
LYONETIIDAE **135–136**
LYPUSIDAE **144–145**
Macaria alternata **441**
    brunneata **574**
    liturata **441**
    notata **628**
    wauaria **442**
Macrochilo cribrumalis **629**
Macroglossum stellatarum **384**
Macrothylacia rubi **380**
Magpie Moth **439**
Maiden's Blush **393**
Malacosoma neustria **378**
Mallow **402**
Mamestra brassicae **545**
Manchester Treble-bar 24, 25, **436**
Maple Pug **424**
Map-winged Swift **39**
Marasmarcha lunaedactyla **234**
Marbled Beauty **498**
Marbled Brown **464**
Marbled Coronet **548**
Marbled Green **499**
Marbled Minor **520**
Marbled Orchard Tortrix **275**
Marbled White Spot **489**
March Moth **449**
Marsh Oblique-barred **480**
Marsh Pug 28, **426**
Martania taeniata **421**
May Highflyer **403**
Meal Moth **344**
Mediterranean Brocade **499**
Mediterranean Flour Moth **342**
Meganola albula **569**
Melanchra persicariae **544**
Melanthia procellata **419**
Menophra abruptaria **454**

Merrifieldia baliodactylus **235**
    leucodactyla **235**
Merveille du Jour 23, **533**
Mesapamea didyma **519**
    secalis **518**
Mesoleuca albicillata **401**
Mesoligia furuncula **519**
Mesotype didymata **419**
Metalampra italica **140**
Metzneria lappella **171**
    metzneriella **171**
MICROPTERIGIDAE **34–35**
Micropterix aruncella **35**
    aureatella **34**
    calthella **35**
    mansuetella **34**
    tunbergella **34**
Middle-barred Minor **521**
Miller **491**
Mimas tiliae **381**
Minor Shoulder-knot **532**
Mirificarma lentiginosella 56, **177**
    mulinella **177**
Mniotype adusta **536**
Mompha bradleyi **224**
    conturbatella **223**
    divisella **627**
    epilobiella **225**
    jurassicella **224**
    lacteella **223**
    langiella **226**
    locupletella **226**
    miscella 22, **226**
    ochraceella **223**
    propinquella **224**
    raschkiella **227**
    sturnipennella **225**
    subbistrigella **225**
    terminella **227**
MOMPHIDAE **223–227**, 627
Monochroa cytisella **172**
    hornigi **174**
    lucidella **174**
    suffusella 26, **174**
    tenebrella **172**
    tetragonella 19, **173**
Monopis crocicapitella **84**
    fenestratella 625
    imella **85**
    laevigella **84**
    obviella **84**
    weaverella 625
Mormo maura **503**
Morophaga choragella **77**
Mother of Pearl **354**
Mother Shipton 28, **484**
Mottled Beauty **455**
Mottled Grey **412**
Mottled Pug **434**
Mottled Rustic **500**
Mottled Umber **453**

Mouse Moth **496**
Mugwort Plume **236**
Mullein **494**
Mullein Wave **391**
Muslin Footman **476**
Muslin Moth **473**
Myelois circumvoluta **340**
Mythimna albipuncta **551**
    conigera **549**
    ferrago **552**
    impura **550**
    litoralis **552**
    pallens **550**
    pudorina **549**
    straminea **550**
    turca **549**
    unipuncta **551**
    vitellina **551**
Naenia typica **569**
Narrow-bordered Bee Hawk-moth **384**
Narrow-bordered Five-spot Burnet 28, **332**
Narrow-winged Pug **430**
Narycia duplicella **74**
Neglected Rustic **566**
Nemapogon clematella **80**
    cloacella **79**
    granella **79**
    koenigi **79**
    picarella **80**
Nematopogon metaxella **68**
    pilella **67**
    schwarziellus **67**
    swammerdamella **68**
Nemaxera betulinella **78**
Nemophora cupriacella **65**
    degeerella **64**
    minimella **65**
Neocochylis dubitana **271**
Neofaculta ericetella **163**
Neofriseria peliella 626
    suppeliella 626
Neosphaleroptera nubilana **250**
Neotelphusa sequax 22, **187**
NEPTICULIDAE **40–62**, 625
Netted Carpet **407**
Netted Pug **426**
New Forest Burnet 627
Ni Moth **485**
Niditinea fuscella **83**
Noctua comes **564**
    fimbriata **563**
    interjecta **564**
    janthe **564**
    orbona 629
    pronuba **563**
NOCTUIDAE **484–569, 575**, 629
Nola confusalis **570**
    cucullatella **570**
NOLIDAE **569–571**
Nomophila noctuella **356**

Nonagria typhae **510**
Northern Deep-brown Dart **534**
Northern Drab **539**
Northern Eggar **380**
Northern Rustic **563**
Northern Spinach **408**
Northern Winter Moth 23, **413**
Notocelia cynosbatella **306**
    incarnatana **308**
    roborana **307**
    rosaecolana **308**
    tetragonana **307**
    trimaculana **308**
    uddmanniana **307**
Notodonta dromedarius **465**
    ziczac **465**
NOTODONTIDAE **463–468**
November Moth **414**
Nudaria mundana **476**
Nut Bud Moth **297**
Nut Leaf Blister Moth **105**
Nutmeg **541**
Nut-tree Tussock **489**
Nycteola revayana **571**
Nycterosea obstipata **396**
Nymphula nitidulata **371**
Oak Beauty 23, **452**
Oak Eggar 19, 21, 24, **380**
Oak Hook-tip **372**
Oak Lutestring 23, **376**
Oak Nycteoline **571**
Oak-tree Pug **427**
Oblique Carpet **395**
Oblique Striped **394**
Obscure Wainscot 18, **553**
Ochreous Pug 21, **430**
Ochropacha duplaris **375**
Ochropleura plecta **559**
Ochsenheimeria taurella 3, **123**
    urella **123**
    vacculella **123**
Ocnerostoma friesei 118, **119**
    piniariella 21, **118**
Odezia atrata **419**
Odontopera bidentata **448**
Odontosia carmelita **467**
OECOPHORIDAE **139–143**, 626
Oegoconia quadripuncta **138**
Oidaematophorus lithodactyla **236**
Oinophila v-flava **86**
Old Lady **503**
Old World Webworm **358**
Oleander Hawk-moth **384**
Olethreutes arcuella **280**
Oligia fasciuncula **521**
    latruncula **520**
    strigilis **520**
    versicolor **520**
Olindia schumacherana **241**
Olive **530**
Operophtera brumata **413**

fagata **413**
Opisthograptis luteolata **444**
Opostega salaciella **62**
OPOSTEGIDAE **62**
Orache Moth **503**
Orange Footman **479**
Orange Moth **628**
Orange Sallow **521**
Orange Swift **38**
Orange Underwing **439**
Orchard Ermine **113**
Orgyia antiqua **472**
Oriental Fruit Moth **319**
Orodesma apicina **469**
Ortholepis betulae **335**
Orthonama vittata **395**
Orthosia cerasi **538**
    cruda **538**
    gothica **539**
    gracilis **539**
    incerta **537**
    miniosa **537**
    opima **539**
    populeti **538**
Orthotaenia undulana **275**
Orthotelia sparganella 18, **125**
Ostrinia nubilalis **352**
Ourapteryx sambucaria **449**
Oxypteryx atrella **175**
    unicolorella 172, **175**
    wilkella **175**
Oxyptilus laetus **234**
    parvidactyla 627
Pale Brindled Beauty **450**
Pale Eggar 23, **378**
Pale Mottled Willow **500**
Pale November Moth **414**
Pale Oak Beauty 629
Pale Pinion **527**
Pale Prominent **466**
Pale Shining Brown **542**
Pale Tussock **472**
Pale-shouldered Brocade **543**
Palpita vitrealis **355**
Pammene albuginana 627
    argyrana **321**
    aurana **325**
    aurita **324**
    fasciana 23, **322**
    gallicana **325**
    giganteana 7, **321**
    herrichiana **323**
    obscurana **322**
    ochsenheimeriana **325**
    populana **323**
    regiana **324**
    rhediella **323**
    spiniana **324**
    splendidulana **321**
    suspectana 7, **322**
Pancalia leuwenhoekella **156**

Pandemis cerasana **246**
    cinnamomeana **245**
    corylana **246**
    dumetana 627
    heparana **246**
Panemeria tenebrata **493**
Panolis flammea **536**
Papestra biren **545**
Paracrania chrysolepidella **36**
Paradarisa consonaria **456**
PARAMETRIOTIDAE **221–222**
Parapoynx bilinealis **371**
    crisonalis 628
    stratiotata **370**
Parascotia fuliginaria **481**
Parasemia plantaginis **474**
Parastichtis suspecta **532**
Paraswammerdamia albicapitella **117**
    nebulella **117**
Paratalanta pandalis **352**
Parornix anglicella **96**
    betulae **96**
    devoniella **97**
    fagivora 626
    finitimella **97**, 98
    loganella 625
    scoticella **97**
    torquillella **98**
Parsnip Moth **153**
Pasiphila chloerata **422**
    debiliata **423**
    rectangulata **423**
Patania aegrotalis **354**
    ruralis **354**
Pea Moth 17, **315**
Peach Blossom **374**
Peacock Moth 628
Pear Leaf Blister Moth **136**
Pearly Underwing **553**
Pebble Hook-tip **373**
Pebble Prominent **465**
Pechipogo strigilata 629
Pediasia aridella 19, **369**
    contaminella 628
PELEOPODIDAE **145**
Pelurga comitata **401**
Pempelia palumbella **335**
Pennithera firmata **403**
Peppered Moth **452**
Perconia strigillaria **460**
Peribatodes rhomboidaria **454**
Peridea anceps **465**
Peridroma saucia **553**
Perittia obscurepunctella **210**
Perizoma affinitata **420**
    albulata **421**
    alchemillata **420**
    bifaciata **420**
    blandiata 628
    flavofasciata **421**
    minorata 628

Petrophora chlorosata **443**
Pexicopia malvella **165**
Phalera bucephala **467**
Phalonidia affinitana **265**
    curvistrigana **265**
    manniana **265**
Phaneta pauperana 627
Phaulernis fulviguttella **238**
Pheosia gnoma **466**
    tremula **466**
Phiaris micana **279**
    palustrana **279**
    schulziana 19, **278**
Phibalapteryx virgata **394**
Phigalia pilosaria **450**
Philedone gerningana **242**
Philedonides lunana 24, **243**
Philereme transversata **417**
    vetulata **417**
Phlogophora meticulosa **504**
Phoenix **408**
Photedes captiuncula **513**
    minima **514**
Phragmatobia fuliginosa **474**
Phtheochroa inopiana **263**
    rugosana **264**
    sodaliana **263**
Phthorimaea operculella **183**
Phycita roborella **337**
Phycitodes binaevella **341**
    maritima **341**
    saxicola **341**
Phyllocnistis saligna **112**
    unipunctella **112**
Phyllodesma ilicifolia **381**
Phyllonorycter acerifoliella **111**
    anderidae **26**, **107**
    blancardella **100**
    cavella 626
    cerasicolella **101**
    comparella 626
    coryli **105**
    corylifoliella **102**
    dubitella **104**
    emberizaepenella **108**
    tristrigella **109**
    esperella **106**
    froelichiella **109**
    geniculella **111**
    harrisella **98**
    heegeriella **98**
    hilarella **104**
    hostis 626
    joannisi **111**
    junoniella **101**
    klemannella **110**
    kuhlweiniella 626
    lantanella **102**
    lautella **107**
    leucographella **102**
    maestingella **105**

mespilella 626
messaniella **99**
muelleriella 626
nicellii **110**
nigrescentella 626
oxyacanthae **100**
platani **99**
quercifoliella **99**
quinqueguttella 21, **107**
rajella **106**
roboris 626
salicicolella **103**
schreberella **108**
scopariella 56, **105**
sorbi **100**
spinicolella **101**
stettinensis **109**
strigulatella **106**
trifasciella **110**
tristigella **109**
ulicicolella **104**
ulmifoliella **108**
viminetorum **103**, 626
viminiella **103**
Phylloporia bistrigella **70**
Phymatopus hecta **39**
Phytometra viridaria **482**
Pima boisduvaliella 628
Pimpinel Pug **429**
Pine Beauty 21, **536**
Pine Bud Moth **309**
Pine Carpet **403**
Pine Hawk-moth 21, **383**
Pine Leaf-mining Moth 21, **310**
Pine Resin-gall Moth **309**
Pine Shoot Moth 21, **310**
Pinion-streaked Snout **481**
Piniphila bifasciana 21, **280**
Pink-barred Sallow **522**
Pistol Case-bearer **200**
Plagodis dolabraria **444**
    pulveraria **443**
Plain Golden Y **487**
Plain Plume **236**
Plain Pug **429**
Plain Wave **390**
Platyptilia calodactyla 627
    gonodactyla **231**
Platytes alpinella **369**
Plemyria rubiginata **405**
Pleurota bicostella **143**
Plodia interpunctella **342**
Plum Fruit Moth 7, 17, **320**
Plum Tortrix **276**
Plusia festucae **488**
    putnami **488**
Plutella porrectella **124**
    xylostella **124**
PLUTELLIDAE **124–125**, **572**
Pod Lover **548**
Poecilocampa populi **377**

Polia bombycina **542**
    hepatica **542**
    nebulosa **542**
Polychrysia moneta **486**
Polymixis flavicincta **536**
    lichenea **535**
Polyploca ridens **377**
Poplar Grey **492**
Poplar Hawk-moth **382**
Poplar Kitten **464**
Poplar Lutestring **375**
Portland Moth 21, **554**
Potato Tuber Moth **183**
Povolnya leucapennella 3, **95**
Powdered Quaker **539**
PRAYDIDAE **137**
Prays fraxinella **137**
    ruficeps **137**
Pretty Chalk Carpet **419**
Pretty Pinion 628
Privet Hawk-moth **383**
Prochoreutis myllerana **240**, 572
    sehestediana 240, **572**
PRODOXIDAE **70–72**, 625
Prolita sexpunctella **176**
Protodeltote pygarga **489**
Pseudargyrotoza conwagana **263**
Pseudococcyx posticana **309**, 310
    turionella 5, **309**
Pseudoips prasinana **571**
Pseudopanthera macularia **445**
Pseudopostega crepusculella **62**
Pseudoswammerdamia combinella **116**
Pseudotelphusa paripunctella **190**
    scalella **189**
Pseudoterpna pruinata **461**
Psoricoptera gibbosella **180**
Psyche casta **76**
PSYCHIDAE **74–77**
Psychoides filicivora **86**
    verhuella **86**
Pterapherapteryx sexalata **437**
PTEROPHORIDAE **231–237**, 627
Pterophorus pentadactyla **235**
Pterostoma palpina **466**
Ptilodon capucina **467**
Ptocheuusa paupella **172**
Ptycholoma lecheana **245**
Ptycholomoides aeriferana **245**
Purple Bar **406**
Purple Clay **560**
Purple Thorn **447**
Purple-bordered Gold 18, 26, **387**
Puss Moth **463**
Pyla fusca **335**
PYRALIDAE **333–346**, 627–628
Pyralis farinalis **344**
Pyrausta aurata 29, **347**
    cingulata **347**
    despicata **347**
    ostrinalis 22, **348**

purpuralis **348**
sanguinalis 628
Pyrrharctia isabella **475**
Pyrrhia umbra **497**
Raisin Moth **343**
Rannoch Looper **574**
Raspberry Moth **71**
Recurvaria leucatella **190**
Red Carpet 23, **396**
Red Chestnut **561**
Red Sword-grass **529**
Red Twin-spot Carpet **397**
Red Underwing **483**
Red-barred Tortrix **241**
Red-belted Clearwing 7, **573**, 627
Reddish Light Arches 22, **517**
Red-green Carpet **410**
Red-line Quaker **524**
Red-necked Footman **478**
Red-tipped Clearwing 21, **328**
Reed Tussock 629
Retinia resinella **309**
Rheumaptera cervinalis 628
  hastata **418**
  undulata **418**
Rhigognostis annulatella **124**
  incarnatella **572**
Rhizedra lutosa **510**
Rhodometra sacraria **394**
Rhodophaea formosa **336**
Rhomboid Tortrix **258**
Rhopobota myrtillana **290**
  naevana **291**
  stagnana **290**
  ustomaculana **290**
Rhyacia simulans **562**
Rhyacionia buoliana **310**
  pinicolana **310**
  pinivorana **311**
Riband Wave **390**
Rice Moth **333**
Ringed China-mark **370**
Rivula sericealis **469**
Rivulet **420**
Roeslerstammia erxlebella **87**
ROESLERSTAMMIIDAE **87**
Rose Leaf Miner **44**
Rose Tortrix **244**
Rosy Minor **519**
Rosy Rustic **506**
Round-winged Muslin **477**
Ruby Tiger **474**
Ruddy Carpet 628
Ruddy Highflyer **403**
Rufous Minor **520**
Rush Veneer **356**
Rusina ferruginea **502**
Rustic **501**
Rustic Shoulder-knot **515**
Rusty-dot Pearl **353**
Sallow **522**

Sallow Kitten **463**
Saltern Ear **507**
Sand Dart 20, 21, **558**
Sandhill Rustic 4, 21, **509**
Sandy Carpet **421**
Satellite **529**
Satin Beauty 21, **455**
Satin Lutestring **375**
Satin Wave **388**
Saturnia pavonia **381**
SATURNIIDAE **381**
Satyr Pug 23, **432**
Saxifrage Plume **233**
Saxon **504**
Scallop Shell **418**
Scalloped Hazel **448**
Scalloped Hook-tip **372**
Scalloped Oak **448**
Scarce Bordered Straw **498**
Scarce Burnished Brass **485**
Scarce Footman **479**
Scarce Light Plume **234**
Scarce Prominent **467**
Scarce Silver Y 24, **488**
Scarce Silver-lines **570**
Scarce Tissue 628
Scarce Umber **453**
Scarlet Tiger **475**
Schrankia costaestrigalis **481**
Schreckensteinia festaliella **238**
SCHRECKENSTEINIIDAE **238**
Scoliopteryx libatrix **468**
Scoparia ambigualis **359**
  ancipitella 23, **359**
  basistrigalis **358**
  pyralella **359**
  subfusca **358**
Scopula floslactata **392**
  imitaria **391**
  immutata **391**
  marginepunctata **391**
  ornata 628
  ternata **392**
Scorched Carpet **441**
Scorched Wing **444**
Scotopteryx bipunctaria 628
  chenopodiata **395**
  luridata **395**
  mucronata **394**
Scrobipalpa acuminatella **180**
  artemisiella **183**
  atriplicella **182**
  costella **183**
  instabilella 19, **181**
  nitentella 19, **181**
  obsoletella **181**
  ocellatella **182**
  salicorniae **180**
  samadensis **182**
SCYTHRIDIDAE **229**–**230**
Scythris fallacella 22, **229**

  grandipennis **230**
  picaepennis **230**
Scythropia crataegella **138**
SCYTHROPIIDAE **138**
Selenia dentaria **447**
  lunularia **447**
  tetralunaria **447**
Selidosema brunnearia **454**
Semioscopis avellanella 22, **145**
  steinkellneriana **146**
September Thorn **446**
Seraphim **437**
Sesia apiformis **327**
  bembeciformis **327**
SESIIDAE **327**–**329**, **573**, 627
Setaceous Hebrew Character **568**
Shaded Broad-bar **395**
Shaded Pug **435**
Shark **494**
Sharp-angled Carpet **400**
Sharp-angled Peacock **441**
Shears **545**
Shore Wainscot 21, **552**
Short-cloaked Moth **570**
Shoulder Stripe **400**
Shoulder-striped Wainscot **552**
Shuttle-shaped Dart **558**
Sideridis reticulata **546**
  rivularis **546**
  turbida **546**
Silky Wainscot 18, **502**
Silky Wave 628
Silver Hook **489**
Silver Y **486**
Silver-ground Carpet **398**
Silver-striped Hawk-moth 3, **387**
Silvery Arches **542**
Single-dotted Wave **389**
Sitochroa palealis **348**
Sitotroga cerealella **165**
Six-belted Clearwing 7, **329**
Six-spot Burnet **332**
Six-striped Rustic **567**
Skin Moth **84**
Slender Brindle **516**
Slender Burnished Brass **575**
Slender Pug 28, **424**
Sloe Pug 17, **422**
Small Angle Shade **505**
Small Argent and Sable 11, 24, **399**
Small Autumnal Moth 24, **415**
Small Blood-vein **391**
Small Brindled Beauty 23, **450**
Small China-mark 3, **370**
Small Chocolate-tip 3, **468**
Small Clouded Brindle **516**
Small Dotted Buff **514**
Small Dusty Wave **388**
Small Eggar **378**
Small Elephant Hawk-moth **386**
Small Emerald **462**

660 *The Moths of Lancashire*

Small Fan-foot **480**
Small Fan-footed Wave **389**
Small Goldenrod Plume **237**
Small Grass Emerald 629
Small Lappet **381**
Small Magpie **352**
Small Marbled **482**
Small Mottled Willow **499**
Small Phoenix **409**
Small Plume 627
Small Purple-barred **482**
Small Quaker **538**
Small Ranunculus 28, **547**
Small Rivulet **420**
Small Rufous **512**
Small Scallop **390**
Small Seraphim **437**
Small Square-spot **561**
Small Tabby **345**
Small Wainscot **512**
Small Waved Umber 628
Small White Wave **415**
Small Yellow Underwing **493**
Small Yellow Wave **416**
Smerinthus ocellata **382**
Smoky Wainscot **550**
Smoky Wave 19, **392**
Snout **469**
Sophronia semicostella **176**
Sorhagenia janiszewskae 22, **159**
    rhamniella 626
Southern Wainscot 18, **550**
Spaelotis ravida **565**
Speckled Yellow 22, **445**
Spectacle **484**
SPHINGIDAE **381–387**, 628
Sphinx ligustri **383**
    pinastri **383**
Spilonota laricana **291**
    ocellana **291**
Spilosoma lubricipeda **473**
    lutea **473**
    urticae 629
Spinach **409**
Spindle Ermine **114**
Spodoptera exigua **499**
    littoralis **499**
    litura **500**
Spoladea recurvalis **573**
Spotted Shoot Moth **311**
Sprawler **496**
Spring Usher **452**
Spruce Bud Moth **300**
Spruce Carpet **404**
Spruce Seed Moth **316**
Spuleria flavicaput **222**
Spurge Hawk-moth **385**
Square Spot **456**
Square-spot Dart **555**
Square-spot Rustic **567**
Standfussiana lucernea **563**

Stathmopoda pedella **228**
STATHMOPODIDAE **228**
Stenolechia gemmella **191**
Stenoptilia bipunctidactyla **233**
    millieridactyla **233**
    pterodactyla **232**
    zophodactylus 4, **233**
Stephensia brunnichella 627
Stictea mygindiana **279**
Stigmella aeneofasciella **51**
    alnetella **42**
    anomalella **44**
    assimilella **49**
    atricapitella **54**
    aurella **51**
    betulicola **41**
    catharticella **44**
    centifoliella 4, **44**
    confusella **40**
    continuella **51**
    crataegella **45**
    desperatella 625
    floslactella **47**
    glutinosae **42**
    hemargyrella 47, **53**
    hybnerella **47**
    incognitella **52**
    lapponica **40**
    lemniscella **50**
    luteella **42**
    magdalenae **45**
    malella **43**
    microtheriella **43**
    minusculella **46**
    myrtillella **48**
    nylandriella **46**
    obliquella **49**
    oxyacanthella **46**
    paradoxa **45**
    perpygmaeella **52**
    plagicolella **50**
    poterii **52**
    prunetorum **43**, 50
    regiella 625
    roborella **55**
    ruficapitella **54**
    sakhalinella **41**
    salicis (& clusters) **48**
    samiatella **54**
    sorbi **50**
    speciosa **53**
    splendidissimella 625
    svenssoni **53**
    tiliae **41**
    tityrella **47**
    trimaculella **49**
    ulmivora 625
Stilbia anomala **495**
Stored Nut Moth **333**
Stout Dart 29, **565**
Straw Belle 629

Straw Dot **469**
Straw Underwing **504**
Strawberry Tortrix **256**
Streak **436**
Streamer **401**
Striped Hawk-moth **385**
Striped Twin-spot Carpet 24, **412**
Striped Wainscot 23, **549**
Strophedra nitidana **326**
    weirana **326**
Subacronicta megacephala **492**
Sunira circellaris **525**
Suspected **532**
Svensson's Copper Underwing **496**
Swallow Prominent **466**
Swallow-tailed Moth **449**
Swammerdamia caesiella **116**
    pyrella **117**
Sword-grass **528**
Sycamore **491**
Synanthedon culiciformis **328**
    formicaeformis **328**
    myopaeformis **573**, 627
    spheciformis **328**
    tipuliformis **329**
    vespiformis **329**
Syncopacma albipalpella 626
    vorticella 626
Syndemis musculana **247**
Syngrapha interrogationis **488**
Tachystola acroxantha **143**
Taleporia tubulosa **75**
Tansy Plume 28, **231**
Tapestry Moth **80**
Tawny Marbled Minor **520**
Tawny Pinion **526**
Tawny Shears **548**
Tawny Shoulder 629
Tawny-barred Angle 21, **441**
Tawny-speckled Pug **434**
Tebenna micalis **240**
Telechrysis tripuncta **156**
Teleiodes luculella **187**
    vulgella **186**
Teleiopsis diffinis **187**
Tethea ocularis **374**
    or **375**
Tetheella fluctuosa **375**
Thalpophila matura **504**
Thera britannica **404**
    cognata **404**
    cupressata **405**
    juniperata **405**
    obeliscata **404**
    primaria **458**
Thetidia smaragdaria 629
Thiodia citrana **289**
Thistle Ermine **340**
Tholera cespitis **540**
    decimalis **540**
Thumatha senex **477**

Thyatira batis **374**
Thyme Moth **183**
Thyme Plume **235**
Thyme Pug **431**
Thyraylia nana **271**
Thysanoplusia orichalcea **575**
Tiliacea aurago **521**
    citrago **521**
Timandra comae **392**
Timothy Tortrix **248**
Tinea columbariella **625**
    dubiella **82**
    flavescentella **625**
    lanella **82**
    pallescentella **82**
    pellionella **81**
    semifulvella **83**
    translucens **81**
    trinotella **83**
TINEIDAE **77–86**, **625**
Tineola bisselliella **81**
Tischeria dodonaea **73**
    ekebladella **73**
TISCHERIIDAE **73–74**
Tissue **418**
Toadflax Brocade **495**
Toadflax Pug **425**
TORTRICIDAE **241–326, 573**, **627**
Tortricodes alternella **251**
Tortrix viridana **254**
Trachea atriplicis **503**
Treble Brown Spot **628**
Treble Lines **502**
Treble-bar 17, **436**
Tree-lichen Beauty **575**
Triangle Plume **231**
Triaxomera fulvimitrella **78**
    parasitella **78**
Trichiura crataegi **378**
Trichophaga tapetzella **80**
Trichoplusia ni **485**
Trichopteryx carpinata **439**
    polycommata **438**
Trifurcula beirnei **56**
    cryptella 28, **55**
    eurema **55**
    immundella **56**
Triodia sylvina **38**
Triphosa dubitata **418**
Triple-spotted Clay **568**
Triple-spotted Pug **431**
True Lover's Knot **562**
Tunbridge Wells Gem **575**
Turnip Moth **557**
Tuta absoluta **184**
Twenty-plume Moth **230**
Twin-spot Carpet **419**
Twin-spot Plume **233**

Twin-spotted Quaker **540**
Twin-spotted Wainscot **511**
Tyria jacobaeae **476**
Udea ferrugalis **353**
    lutealis **353**
    olivalis **354**
    prunalis **353**
Uncertain **501**
Utetheisa pulchella **476**
Valerian Pug **433**
Vapourer **472**
Varied Coronet **548**
Variegated Golden Tortrix **243**
Venusia blomeri **417**
    cambrica **416**
Vestal **394**
Vine's Rustic **501**
V-Moth **442**
V-Pug **422**
Water Carpet **412**
Water Ermine **629**
Water Veneer **370**
Watsonalla binaria **372**
    cultraria **373**
Waved Black 29, **481**
Waved Carpet **416**
Waved Umber **454**
Wax Moth **334**
Welsh Wave **416**
White Colon **546**
White Ermine **473**
White Plume **235**
White Satin Moth **470**
White-barred Clearwing **328**
White-line Dart **555**
White-lined Hawk-moth **628**
White-marked **562**
White-pinion Spotted **458**
White-point **551**
White-shouldered House-moth **140**
White-speck **551**
White-spotted Pinion **530**
White-spotted Pug **428**
Willow Beauty **454**
Willow Ermine **114**
Willow Tortrix 21, **294**
Winter Moth **413**
Wood Carpet **399**
Wood Sage Plume **234**
Wood Tiger **474**
Wormwood 28, **493**
Wormwood Pug **432**
Xanthia togata **522**
Xanthorhoe decoloraria **396**
    designata **397**
    ferrugata **397**
    fluctuata **396**
    montanata **398**

    quadrifasiata **398**
    spadicearia **397**
Xenolechia aethiops **190**
Xestia agathina **567**
    baja **566**
    castanea **566**
    c-nigrum **568**
    ditrapezium **568**
    sexstrigata **567**
    triangulum **568**
    xanthographa **567**
Xylena exsoleta **528**
    solidaginis **528**
    vetusta **529**
Xylocampa areola **497**
Yarrow Plume **231**
Yellow Horned **377**
Yellow Shell **398**
Yellow V Moth **86**
Yellow-barred Brindle **437**
Yellow-legged Clearwing **329**
Yellow-line Quaker **525**
Yellow-ringed Carpet **628**
Yellow-tail **471**
Yponomeuta cagnagella **114**
    evonymella **113**
    malinellus **114**
    padella **113**
    plumbella **115**
    rorrella **114**
    sedella **115**
YPONOMEUTIDAE **113–119**
Ypsolopha alpella **121**
    dentella **119**
    horridella **120**
    lucella **120**
    nemorella **119**
    parenthesella **121**
    scabrella **120**
    sequella **122**
    sylvella **121**
    ustella **122**
    vittella **122**
YPSOLOPHIDAE **119–123**
Zeiraphera griseana **299**
    isertana **300**
    ratzeburgiana **300**
Zelleria hepariella **115**
    oleastrella **116**
Zelotherses paleana **248**
Zeuzera pyrina **327**
Zimmermannia atrifrontella **625**
Zygaena filipendulae **332**
    lonicerae **332**
    meliloti **627**
    trifolii **627**
    viciae **627**
ZYGAENIDAE **330–332**, **627**